DISABLED PERSONS AND THE LAW

State Legislative Issues

LAW, SOCIETY, AND POLICY

Volume 1 DISABLED PERSONS AND THE LAW: State Legislative Issues
Bruce Dennis Sales, D. Matthew Powell, Richard Van Duizend, and Associates

DISABLED PERSONS AND THE LAW

State Legislative Issues

Bruce Dennis Sales,
D. Matthew Powell,
Richard Van Duizend,
and Associates

Developmental Disabilities State Legislative Project
of the American Bar Association's
Commission on the Mentally Disabled

SPRINGER SCIENCE+BUSINESS MEDIA, LLC

Library of Congress Cataloging in Publication Data

Sales, Bruce Dennis.
Disabled persons and the law.

(Law, Society, and Policy; v. 1)
"Developmental Disabilities State Legislative Project of the American Bar Association's Commission on the Mentally Disabled."
Bibliography: p.
Includes index.
1. Handicapped—Legal status, laws, etc.—United States—States. I. Powell, D. Matthew. II. Van Duizend, Richard. III. American Bar Association. Commission on the Mentally Disabled. IV. Title. V. Series.
KF480.Z95S34 1982 346.7301'3 82-9089

ISBN 978-1-4757-0796-0 ISBN 978-1-4757-0794-6 (eBook)
DOI 10.1007/978-1-4757-0794-6

Originally published by Plenum Press, New York in 1982
Softcover reprint of the hardcover 1st edition 1982

ACKNOWLEDGMENT

Development of this manuscript was funded in part by Grant #54-P-71162/3 from the U.S. Department of Health and Human Services, Bureau of Developmental Disabilities. Special thanks are due to Alfred Rose, our project monitor from the Bureau, and to Daniel L. Skoler, past director of the Public Service Activities Division of the American Bar Association, for their constant help and concern. The content herein does not necessarily represent the policy of the Department or the American Bar Association or any of the other national organizations represented on the Project's Advisory Board unless specifically adopted as such.

The American Bar Association's Commission on the Mentally Disabled has been concerned since its inception with advancing the rights of mentally ill and developmentally disabled persons. Toward this end, the Commission created the Developmental Disabilities State Legislative Project to define these rights in the context of state legislation. The Commission members believe that this product will be of substantial help to state legislators, advocates, service providers, professors, and all others concerned with the role of the state in providing services for and protecting the rights of disabled citizens. Thus, we commend this book to your attention.

McNeill Smith, Chairman
American Bar Association
Commission on the Mentally Disabled

AUTHORS' NOTE

The research for this volume was conducted between 1977 and 1980. It is likely that there have been changes in pertinent state code provisions and federal statutes since this work has been completed. For example, the functions of the Secretary of the United States Department of Health, Education, and Welfare relating to education that are referred to in Chapter 5 have been transferred to the Secretary and Department of Education, and the state of Alaska enacted comprehensive guardianship legislation in 1981 (see Chapter 7). Although changes such as these will require readers to check specific citations before relying on them, they do not alter the import of this book nor of the scope of the law-related problems facing disabled persons.

PREFACE

The Developmentally Disabled Assistance and Bill of Rights Act of 1975 (P.L. 94-103) establishes an ambitious national priority for furtherance of services to, and the rights of, developmentally disabled persons. Its provisions for comprehensive planning, federal financial aid, fostering of training and improved service techniques, program coordination and evaluation and, perhaps most significantly, protection of vital rights of developmentally disabled persons requires a massive adjustment of state legislation and implementing regulations if the ultimate goals of this pioneering effort are to be realized.

At the same time national leadership in this field, as exemplified by the 1976 "Century of Decision" Report of the President's Committee on Mental Retardation,[1] and the evolving positions of national "consumer" and professional associations,[2] has shown an increased appreciation of the need for careful and constructive law reform as a precondition for achievement of the goals of P.L. 94-103. In addition, an explosion in court litigation on behalf of developmentally disabled and mentally ill persons has produced a number of constitutionally mandated principles (*e.g.*, right to habilitation, right to public education, right to "due process" and legal representation in decisions to instiutionalize) that literally beg for statutory articulation and "fine tuning" if they are to be assured full and uniform implementation, and thereby achieve a level of comprehensiveness and clarity often not attainable through a succession of litigated cases focused on specific factual situations or settings.

Unfortunately, very little national guidance exists for the enormous mission of legislative improvement that lies ahead. In fact, there are almost no up-to-date model laws or legislative guidelines[3] to which states can refer for guidance or, indeed, for information on the types of progressive bills currently being enacted or being considered in other states. Yet, it is clear that the next five years will see a considerable amount of legislative activity concerned with rights, services, and needs of developmentally disabled persons.

With financial support from the U.S. Department of Health and Human Services, Bureau of Developmental Disabilities, the American Bar Association's Commission on the Mentally Disabled initiated the Developmental Disabilities State Legislative Project to fill that void by providing detailed analyses of existing law, comprehensive discussions of more appropriate provisions and, where appropriate, actual model state legislation, that will safeguard the rights of developmentally disabled citizens and assure them equal access to quality services, consistent with the philosophy and requirements of P.L. 94-103 and other pertinent federal enactments.[4] This volume is the result of that effort. We hope that it will play an instrumental support role in several ways in the enactment of a variety of statutes by the states.

First, by reviewing all past legislative developments, it should help assure that any legislation advanced is well conceived and can draw on the

[1] PRESIDENT'S COMMITTEE ON MENTAL RETARDATION, MENTAL RETARDATION: CENTURY OF DECISION, 58-67, 133-46 (1976); *see also* THE MENTALLY RETARDED CITIZEN AND THE LAW (Kindred ed., 1976); A. STONE, MENTAL HEALTH AND LAW: A SYSTEM IN TRANSITION, 119-44 (1975); D. WEXLER, MENTAL HEALTH LAW: MAJOR ISSUES (1981).

[2] These groups include the National Association of Retarded Citizens, Council for Exceptional Children, United Cerebral Palsy Associations, American Association on Mental Deficiency, American Psychological Association, National Association of Social Workers, Epilepsy Foundation of America, National Society for Autistic Children and National Association of Coordinators of State Programs for the Mentally Retarded.

[3] A possible exception to this may be found in the area of special education. The Council for Exceptional Children developed a "Model State Law on the Education of Handicapped Children" (first published in 1971) and now available in CED's PUBLIC POLICY IN THE EDUCATION OF EXCEPTIONAL CHILDREN with numerous references to existing state statutes and litigation. For valuable policy, principles and standard "building blocks" for legislation in other areas, *see also*, AMERICAN ASSOCIATION ON MENTAL DEFICIENCY, POSITION PAPERS ON RIGHTS OF MENTALLY RETARDED PERSONS (1975); UNITED CEREBRAL PALSY ASSOCIATIONS, ORIENTATION-RESOURCE GUIDE FOR GOVERNMENTAL ACTIVITIES (1976); and NATIONAL ASSOCIATION OF COORDINATORS OF STATE PROGRAMS FOR THE MENTALLY RETARDED, MENTAL RETARDATION TRENDS IN STATE SERVICES (1976).

[4] *See, e.g.*, Education of All Handicapped Children Act of 1975, Pub. L. 94-142, 20 U.S.C. § 1401 (1976) (major federal support of public education); Rehabilitation Act of 1973, §§ 503, 504, 29 U.S.C. §§ 794-992 (1976) (prohibiting discrimination against handicapped persons in federally supported programs).

most advanced concepts, and outstanding work products from other states. Second, the manuscript identifies both the benefits and deficits of existing legislation, analyzes alternative approaches, and presents what we believe are the most appropriate approaches and, in most cases, models for states to adopt legislatively or at least use as the basis for revising and improving their current codes. Third, this manuscript should save considerable time and money for individual states who would otherwise have to duplicate these efforts in order to assure that their own formulations were sound. Finally, by analyzing alternative approaches, it should help states resolve policy questions that may well arise on which states can have legitimate differences in their legislative formulations.

In a very real sense, a test of our efforts and measure of its results and benefits will be the number of "have not" states that become "have" states in important legislative areas. No enactment, of course, will be or can hope to be solely (or even primarily) attributable to our writing. Legislative success is a matter of local responsibility and commitment and depends on a great deal more than the availability of sound models and guidance in the drafting of legislation.

Toward this end, our writing is directed to all of those individuals and groups who are concerned with legislative reform on behalf of developmentally disabled persons — program officials, service providers, citizen advocacy groups, professional associations. national, state and local lay groups, civic organizations, bar committees, legislative committees, law reform and public interest law offices, law professors, and state and local "decision-makers." Of these "decision-makers," a particular target is the "State Planning Councils" established under P.L. 94-103 in every state and charged both with (i) supervising the development of comprehensive plans under the Act and setting the direction and pace of program growth and (ii) serving as advocates for persons with a developmental disability.

Methodology

In approaching our task, we adopted the following plan:

(i) Cataloging Issues and Establishing Priorities.

In consultation with the research and development staff of the Bureau of Developmental Disabilities of the U.S. Department of Health and Human Services, State Planning Council program officials, and representatives of concerned professional and citizen advocacy groups, we developed a compilation of issues to be addressed in each major subject matter area to produce a comprehensive set of state laws serving developmentally disabled persons. Consultation was conducted through mail survey, individual contacts, attendance at workshops and meetings, and establishment of a working advisory board which met regularly throughout the project's duration (see listing of board members, *infra*). This board consisted of a carefully selected group of experts representing eleven national organizations. The major subject matter areas were, of course, well known but decisions such as grouping of topics and issues, and setting the priorities for dealing with them required careful analysis and recourse to a broad spectrum of perspectives.

(ii) Research on Current Status of Laws. Laws of the 50 states, the District of Columbia, and the possessions and territories were researched to ascertain how these issues are handled under current codes and statutes and which jurisdictions are basically "haves" and "have nots" in terms of adequate legislation in the various subject matter areas. While we accepted full responsibility for this cataloging and classification function, we hoped to minimize our labors in this area by tapping past work and enlisting the continuing collaboration of a number of organizations which we assumed had amassed such reviews of relevant legislative areas. Much to our surprise, however, with rare exception the compilation and reviews had not been undertaken by other groups. Thus, we had to perform this research and analysis without outside aid.

(iii) Development of Model/Guideline Units. Based on the initial consultation and priority-setting effort, analysis of the existing state laws, and the views of consumers, experts and organizations in the field, we undertook the drafting of introductions to the issues raised by each topic, detailed discussions of appropriate legislative provisions for all topics, and model state legislation for most of them. These models included, where appropirate, alternate sections and 'short' and 'long' forms for states with differing needs. Many of the models address only developmentally disabled persons. Some include other groups as well where the issues and problems are parallel and enactment of a broader statute appears more feasible and appropriate. Each chapter contains analytic and comparative commentary, highlights the most critical questions and choices behind statutory provisions, and offers practical information on provisions most likely to engender sensitivity and debate.

In order to insure accuracy and comprehensibility, each chapter typically went through four or five revisions. A member of the project staff, a special consultant, or a staff member with the advice of a special consultant (see listing of these consultants, *infra*) was assigned primary responsibility for preparing a draft based upon guidelines developed by the advisory board and the staff. This draft was then critically reviewed by staff and typically led to at least one and often two more revisions. This newest version was then presented to the advisory board for their comments and criticisms. Based upon their remarks, chapters underwent at least one more revision and often more. In a very real sense, then, this manuscript is the product of a group effort.

Topics Chosen for Inclusion

The topics chosen for inclusion can be classified broadly into two catagories. First were those topics (chapters 1-4) which needed to be addressed in order to insure that the principles of normalization and integration of developmentally disabled persons in the community would be maximized. Chapter 1 presents detailed discussions of various issues included within the topic of the personal and civil rights of disabled persons. Because of the breadth of these rights, the chapter is divided into eight subsections, each focusing on a separate personal and civil right's issue: marriage, annulment and divorce; relinquishment of parental rights; testamentary capacity; sterilization; abortion; emergency medical care and treatment for developmentally disabled infants; voting; and driver licencing.

As a reading of Chapter 1 will indicate, many of the personal and civil rights that most citizens enjoy have been traditionally denied to disabled persons because of invidious and discriminatory classifications that have no relevance to the issues with which a particular piece of legislation was concerned. Chapter 2 directly addresses the problem of discrimination against disabled persons in several major areas including employment, real estate, public accomodations, educational institutions, credit, insurance, and public services and assistance. The model statute presented in that chapter will provide states with statutory language that will effectively prohibit such invidious discrimination in these areas in the future.

Unfortunately, eliminating intentional discrimination by other individuals may still not be sufficient to protect the rights of those developmentally disabled persons who have a physical handicap. For them the structural environment may present barriers that will prohibit their access to various places in the community and thereby prohibit their enjoyment of their full rights as well as access to needed services. Thus, Chapter 3 discusses the issue of eliminating environmental barriers and presents a model act to accomplish this task. Relatedly, Chapter 4 discusses zoning for community homes for developmentally disabled persons and provides a model act to insure that such group homes can be located in appropriate residential environments to increase the probability that the residents of these dwellings will be appropriately integrated into the community.

The second category of topics chosen for inclusion within this manuscript (chapters 5-9) deals with the provision of generic and specialized services to developmentally disabled persons in order to insure that the intent of P.L. 94-103 is carried out and that developmentally disabled persons will have the greatest opportunity to realize their abilities and thereby enjoy the full benefits of citizenship. Chapter 5 discusses the education of exceptional persons, perhaps the most important service that can be provided to developmentally disabled minors; it is only through appropriate education that developmentally disabled persons, like all other individuals, will be able to achieve those skills necessary to fully assume their place in society. Because gifted and talented persons are discriminated against in this area, in a fashion similar to developmentally disabled persons, the model statute and the discussion in this chapter refer to both developmentally disabled as well as gifted and talented persons.

In addition to the right to education, developmentally disabled persons must have the right to receive other appropriate services in appropriate settings if they are to realize their maximum potential. Chapter 6 presents a discussion of the need for such a right to and admission to services as well as presenting a model statute on this topic. This will perhaps be the most controversial chapter in the volume. For example, in speaking of a right to services some argue that the services available for disabled persons should not include those provided in large residential facilities. Others argue that such facilities should be maintained for certain severely disabled individuals who, it is contended, cannot survive outside of the institution. This chapter tries to draw a balance. While permitting the continuation of large residential facilities, it places severe procedural restrictions and requirements on admissions to them while creating other service options. Although the staff and most members of the advisory board support this model statute, it should be noted that one member felt very strongly that the chapter would inhibit the provision of needed services.[5]

Chapter 7 presents a discussion of the issues relating to guardianship and conservatorship of developmentally disabled adults and minors, and presents a detailed model statute. The model act seeks to insure that individuals in need of a substitute decision-maker are provided with such a service while at the same time insuring against un-

---

[5] Specifically, this advisary board member felt that the chapter, if enacted into law, "would: (1) reduce parental authority and control over their children with the consequent likelihood of poorer services for such children; (2) increase the cost to taxpayers and/or reduce direct services to the handicapped because of the need for costly 'ancillary services' (e.g. payment of hearing officers); (3) increase adversarial procedures at a time when court dockets are already overcrowded, and innovative procedures are being suggested and tried in other legal fields in an attempt to reduce litigation; (4) increase the number of formal guardianships for handicapped adults, thereby serving to diminish individual rights and responsibilities; (5) eliminate services even though in some cases no better services will be available; and (6) set an arbitrary and meaningless maximum size for residential services, with no regard to the resident's individual needs. This member believes that the portions of Chapter 6 which should be enacted are Section 4, paragraph 1 and 2; and Section 16, paragraph 10."

warranted imposition. Chapter 8 presents a discussion of the protections that should be afforded to developmentally disabled persons in two very specialized areas — the criminal and juvenile justice systems. Some developmentally disabled individuals, as well as some individuals in the population-at-large, will come in contact with one or both of these systems. But because of special vulnerability, disabled persons are often less equipped than other individuals to enjoy the full rights that the law provides to alleged offenders. This chapter addresses that concern and presents model statutes to provide appropriate protections. The final chapter contains a discussion of advocacy for developmentally disabled persons, an essential service if these individuals are to enjoy, to the fullest, their rights and the appropriate provision of all other services.

It should be noted that the materials contained within this volume both reflect our judgments as to the most important topics to be considered as well as the resources available for this task. Some topics were excluded because they were not state legislative issues. Adequate transportation is a good example. This issue has been largely pursued within the context of federally subsidized mass transit programs and, as such, is primarily a federal issue which is not amenable to state legislative action. The issue is discussed, however, in Chapter 3 on eliminating environmental barriers, to the extent it impacts upon accessibility. Estate planning and will drafting techniques are other examples of topics outside of the scope of the project. While these issues involve state law, and are important to developmentally disabled persons and their parents, they are not problems which can be addressed by legislation and, thus, they were excluded.

The project was also guided in its selection of topics by a need to avoid duplication of the previous efforts of others. Some topics such as confidentiality of records[6], and experimentation on human subjects,[7] have already been addressed by other studies. Thus we declined to deal with these issues. This is not to argue that the work performed by others to date on these topics is definitive. Rather, we unfortunately had to conclude that further studies of these topics would not be feasible given the level of financial resources that we had to support our efforts. Finally, at least one area of legislation, the rights of disabled persons in residential facilities, revealed adequate legislative models among existing state statutes. Because of this finding, the decision was made not to develop a detailed chapter including model legislation. The Appendix does present, however, a review of existing state statutes on this topic which will allow the reader to easily identify those states which provide the most comprehensive and appropriate protections.

Principles Guiding the Drafting Process

Because of the diversity of the topics that needed to be addressed, and because of the concern the staff, the advisory board and the consultants shared for safeguarding the rights of developmentally disabled persons and for assuring them equal access to quality services, consistent with the philosophy and program of P.L. 94-103 and other pertinent law, we identified certain principles that guided all our drafting efforts. These were:

(1) Each Individual Should Be Given The Opportunity To Maximize His or Her Potential And Fully Develop His or Her Abilities. Every person has some potential which can be developed with proper education and training. Because of this fact, services should be provided whenever needed so that this potential can be fully realized. Control over the individual should never be substituted for adequate service delivery. As applied to developmentally disabled persons, this principle requires that these individuals should have access to comprehensive services designed to maximize potential and minimize handicaps. In addition, the services should be designed to develop the basic intellectual and personal self-help skills necessary to function in the least restrictive setting possible and to function within the larger community with the least possible supervision. These self-help skills should include personal care, housekeeping skills, money handling, use of transportation, vocational skills, recreational skills, adaptive interpersonal behavior and knowledge about access to other services. The absence of such services will often force a developmentally disabled person into an unproductive and unnecessarily dependent role. In fact, past experience shows that, in their absence, society will resort to segregative incarceration, a denial of the most fundamental rights of the developmentally disabled individual.

(2) Developmentally Disabled Persons Should Be Provided With As Normal An Environmental And Experiences As Possible. There should be the fullest opportunity for normalizing experiences in which programs and services assist developmentally disabled persons to become fully integrated into the general society. Since self-help and acqui-

---

[6] *See* CONTRACT RESEARCH CORPORATION, THE CONCEPTUAL FRAMEWORK: A REPORT ON THE SOCIOLOGICAL ISSUES RELATED TO THE PROTECTION OF CONFIDENTIALITY AND THE RIGHT TO PRIVACY OF DEVELOPMENTALLY DISABLED PERSONS (1977); CONTRACT RESEARCH CORPORATION, FEDERAL REGULATIONS: CONFIDENTIALITY AND THE RIGHT TO PRIVACY OF PERSONS WITH DEVELOPMENTAL DISABILITIES (1977) and CONTRACT RESEARCH CORPORATION, MODEL STATE CODE: PRIVACY ACT FOR THE DEVELOPMENTALLY DISABLED (1977). These studies were prepared for the U.S. Department of Health and Human Services, Bureau of Developmental Disabilities.

[7] The National Commission for the Protection of Human Subjects of Biomedical and Behavioral Research, created by Pub. L. 93-348. After considerable study, the Commission's recommendations published in 43 Fed. Reg. 11328 (March 17, 1978) have resulted in proposed regulations on the subject. *See* 43 Fed. Reg. 53950 (November 17, 1979).

sition of interpersonal skills, education, training and social adjustment can most effectively occur within a normalized environment, programs and services should also take place, as much as possible, in normal settings. Such settings will increase the interaction between developmentally disabled and non-developmentally disabled persons, thereby providing appropriate role models in social, self-help, and employment skills, while fostering the maximization of individual potential. This interaction also will help non-developmentally disabled individuals develop appropriate attitudes and acceptance of disabled individuals.

Services also should be delivered, insofar as possible, within the generic systems of service delivery in the community, rather than within specialized systems which may tend to isolate developmentally disabled persons and exaggerate differences. Generic service delivery systems (e.g., education, medical care, vocational training, recreation) already serve a highly diverse range of persons and needs. But, because of the tradition of segregating developmentally disabled persons from the mainstream of society, many such services have not accomodated these individuals' needs. Thus, adaption of generic service systems may be required to overcome past exclusionary practices and assumptions. The primary thrust must be to ensure that all generic services are flexible enough to accommodate the needs of developmentally disabled persons so that special services for these individuals will be exceedingly rare. For instance, education and training should be provided within the regular classroom or work situation unless the condition of the person makes it inappropriate to meet the person's developmental needs in this manner. Employment and economic productivity within the regular work force also should be maximized. For those who cannot enter the competitive labor market, sheltered workshops or daytime activities should be provided. Activity and productivity are a basic part of our lifestyle. Appropriate recreation and leisure activities should be available in the community. Finally, architectural and transportation barriers should be removed to maximize mobility within the community.

This principle further requires that developmentally disabled children be kept within their families when the developmentally disabled persons' needs can best be met in the home environment. This should aid in appropriate emotional development. Where alternative arrangements are required, they should be as homelike as possible with continued familial interaction encouraged. For adult developmentally disabled individuals, housing should reflect normal residential patterns to the extent possible and should be located within residential areas. Zoning restrictions should be removed when they limit this possibility.

(3) When Restrictions Must Be Placed On A Developmentally Disabled Person, The Alternative Which Is Least Restrictive Or Intrusive, Consistent With The Individuals Needs, Should be Imposed. People should be free to live as they please and pursue their own interests as they see fit. One individual's interest may, however, conflict with the communal interest or the interest of other individuals. Thus, some restrictions on individual freedom may be justified. There is a tendency, however, for those in power to impose broader restrictions than are necessary to meet communal interests. Where there are communal interests which may justify the regulation of individual conduct, these interests should be clearly identified so that impermissable interests can be rejected as the basis for differentiation.

Thus, where the individual cannot effectively participate in the systems of generic service delivery, he or she should receive specialized services or special adaptations of generic services based on an individual determination of the person's abilities and needs, so that the service alternative chosen is that which is least restrictive to the individual's rights. Placement in specialized services, generally, should be seen as temporary and should be accompanied by prescriptions to generic systems concerning adaptations needed in such systems to permit appropriate services to developmentally disabled persons. In addition, the delivery alternatives chosen should be periodically examined to assess the appropriateness of the choice given current needs and the availability of other alternatives. Restrictions should be maintained only where more informal, less coercive alternatives cannot achieve the identified needs.

(4) The Individual Shall Have Maximum Freedom of Choice Within His or Her Capacity And Within The Limitations Placed Upon All Persons By The Law. Although all persons are born dependent upon others, through the normal process of socialization, they become more responsible for themselves and make their own decisions about their lives. Unfortunately, this approach has not characterized the habilitation of persons with developmental disabilities. The traditional treatment of such persons has been characterized by paternalism and continued dependency.

Developmentally disabled persons should be encouraged to be independent and make their own decisions about their lives. Thus, they should participate as fully as possible in all decisions which may affect them. As noted above, this independence and decision-making ability can be fostered by normalization and maximizing individual potential.

Not all developmentally disabled persons will be able to function on their own, however. Some will need some degree of control and substituted decision-making. Yet "third-party consent" in lieu of personal decision-making allows the substitution of hidden agendas by other persons and should be carefully scrutinized. Where informed decision-making and consent are not possible by a developmentally disabled person, no person with a potential conflict of interest should be allowed to make these substituted decisions.

(5) The Law Should Be Used To Recognize The Equality of Developmentally Disabled Persons With Other Citizens. Developmentally disabled persons, like all other citizens, should be entitled to enjoy their personal and civil rights. For example, certain rights are basic to our functioning and are taken for granted by most persons including the right to procreate, marry, contract, make a will, vote, travel, and enjoy one's privacy. Yet, in many instances, states have imposed restrictions on these basic rights where a person is identified as being mentally ill or developmentally disabled. Yet, placing categorical disqualifications and restrictions on groups of individuals because of a label is unconscionable since most often we are unable to predict ability and performance based upon that label. Rather, there should be clearly specified behavioral standards which relate to the determination being made. For example, what standards of competency should be required to marry, or to make a testamentary disposition? These serve different purposes and, thus, should have different standards that are functionally related to the objectives being sought. Relatedly, discrimination against developmentally disabled persons in other areas (*e.g.*, employment, housing, insurance) must be stopped.

(6) Developmentally Disabled Persons Should Be Provided Advocacy Assistance When Needed To Protect And Effect The Exercise And Full Enjoyment of Their Rights. Many developmentally disabled persons will not be able to protect their rights or acquire the services they need without the assistance of an advocate including an attorney where appropriate. These rights and services mean little if they remain unavailable to those who do not have the ability to exercise or utilize them. Thus, advocacy assistance is essential. Advocates should represent the interests of the developmentally disabled person and should be independent of all service providers and other potential conflicts of interest. In addition, they should have the broadest access to information that is essential for pleading their client's interests.

Although these principles do not cover all of the issues that needed to be addressed in our work, they did provide a necessary framework from which we approached our drafting efforts. Other principles that guided the drafting of specific chapters are presented within them.

AMERICAN BAR ASSOCIATION COMMISSION ON THE MENTALLY DISABLED

McNeill Smith, Chairman

Bertram S. Brown	Paul Friedman
Thomas Bryant	Linda Glenn
June Jackson Christmas	Thomas A. Plaut
Judge Thomas Clark	Robert L. Sadoff
James D. Clements	Bruce D. Sales
Janet L. Coye	Russell E. Webb
Bruce Ennis	William L. Webb

Individuals Not Currently On The Commission But Who Served On It During Preparation Of This Manuscript Included:

Christopher Crowley	Mildred Mitchell-Bateman
Gregory R. Dallaire	Jonas Robitscher
Hayden H. Donahue	Judge Joseph Schneider
Gunnar Dybwad	Saleem A. Shah
Charles R. Halpern	Jerome J. Shestack
Nicholas Hobbs	Barry R. Vickrey
Karen C. Kincannon	Harold Visotsky
John H. Lashly	Helen Wright

The ABA Commission on the Mentally Disabled is a 15-member interdisciplinary body established in 1973 to initiate, on behalf of the Association, a public service program to foster improvements in the nation's mental disability system. The Commission has operated several action projects, all supported by private foundation or government funds, including the **Mental Disability Legal Resource Center** which publishes the **Mental Disability Law Reporter** and provides technical support to those involved in mental disability law reform, the **Developmental Disabilities State Legislative Project** which has culminated in this volume, the **Bar Activation/Bar Funding Program** which encouraged state and local bar associations to work for mental disability system reform and awarded modest project grants for legal services projects in the field, the **Guardianship and Protective Services Study** which examined implementation of statutes establishing limited guardianship, public guardianship and adult protective services, the **Pennsylvania Mental Health Advocacy Project** which provided institutional legal services in three Pennsylvania state hospitals and planned a statewide system of such services, and the **Developmental Disabilities Protection and Advocacy Plan Review Project** which examined the initial character of nine federally mandated state advocacy systems for developmentally disabled persons.

Commission activities have also included the submission of *amicus* briefs in leading Supreme Court cases on mental disability issues, the drafting of ABA policy positions on restrictive zoning laws regarding community group homes, limited guardianship, advocacy services for mentally disabled persons, and the rights of the mentally disabled individuals, the presentation of testimony at Congressional hearings on rights of persons who are mentally disabled, and generally attempting to involve the nation's lawyers with other individuals in addressing and solving critical issues in this field.

DEVELOPMENTAL DISABILITIES
STATE LEGISLATIVE PROJECT

Advisory Board

Judge Joseph Schneider, Chairperson
Commission on the Mentally Disabled
American Bar Association

Josephine Barresi
Council for Exceptional Children

James D. Clements
American Association
on Mental Deficiency

Thomas L. Conlan
Member-at-Large

David Deitz
United Cerebral Palsy
Associations

Gunnar Dybwad
Commission on the
Mentally Disabled
American Bar
Assocation

Paul R. Friedman
Mental Health Law Project

Dennis Haggerty
Family Law Section
American Bar Association

Lawrence A. Kane
National Center for
Law and the Handicapped

Michael Kindred
Member-at-Large

Sandy Kortschmaryk
Epilepsy Foundation of
America

Brian McCann
National Association
for Retarded Citizens

Sheridan Neimark
National Society for
Autistic Children

Dolores Norley van Dyk
President's Committee
on Mental Retardation

Serena Stier
American Psychological
Association

Ann H. Britton (Epilepsy Foundation of America), and Alan Abeson and Scottie Torres Higgins (Council for Exceptional Children), also served on the Board during part of the Project's tenure.

SPECIAL CONSULTANTS

Josephine Barresi (Education of Exceptional Persons — Chapter 5)

James W. Ellis (Right to and Admission to Services — Chapter 6)

Steven S. Goldberg (Marriage, Annulment and Divorce — Chapter 1; Sterilization — Chapter 1; Abortion — Chapter 1; Education of Exceptional Persons — Chapter 5)

Robert J. Hopperton (Zoning for Community Homes — Chapter 4)

Michael Kindred (Criminal and Juvenile Justice — Chapter 8)

Ross Dolloff (Emergency Medical Care and Treatment for Infants — Chapter 1)

We also wish to thank H. Rutherford Turnbull III, Marianne Bennett, Kathleen Watson and Ruth Luckasson who commented on drafts of Chapter 5 on the Education of Exceptional Persons, David B. Wexler who commented on a draft of Chapter 8, and the following law student research assistants: Randi G. Lewis, K.L. Becker, J. Christopher Ide, Barbara Joseph, Diana Ceresi, Patricia Truscelli, Joanne Lichtenstein, and Jeanne Bringgold.

TABLE OF CONTENTS

1
PERSONAL AND CIVIL RIGHTS

Table of Contents

INTRODUCTION

This chapter deals with legislation impacting upon the exercise by mentally disabled persons of a number of important, often "fundamental", personal and civil rights. These rights range from the provision of life-preserving care and treatment for infants to the personal disposition of property after one's death. They fall into several related categories. The first are those rights which are highly personal, such as the right to marry, to have and raise children, to be brought up in a family and to dispose of one's property upon death as he or she wishes. These rights are translated into publicly sanctioned procedures such as marriage, termination of parental rights, adoption, and testamentary dispostion. The second involves the medically related areas of sterilization, abortion and medical care and treatment for infants, while the third involves the right to vote, which arises out of our system of government, and driver licensing, important to most of us for the mobility it provides.

All of these rights are fundamental in two senses. When taken as a whole, they define the essence of a functional human being within our society and spell out the individual's obligations as a member of society. Indeed, the ability to be a contributing member of society is in large part dependent upon the legal ability to exercise these rights. Conversely, it is doubtful that any of us would be a contributing member of society if the exercise of these rights were denied us.[1] Several of these rights are also "fundamental" in a legal or constitutional sense, *viz.* they are among the rights so basic and so important that they receive added protection from the courts based upon the equal protection and due process clauses of the United States Constitution. Where a state imposes restrictions on the exercise of these fundamental rights, these restrictions are subject to the "strict-scrutiny" or "compelling state interest" test, *i.e.*, state restrictions must be "necessary" to further a "compelling state interest" when challenged or they must be struck down as unconstitutional.[2] Restrictions may be found "necessary" if they are "closely tailored" to affectuate the asserted state interest, thus restricting the rights of persons only as necessary to serve the state's interests.[3]

Despite the importance of these fundamental and other basic rights, mentally disabled persons in our society are systematically denied the exer-

[1] *See* Wald, *Basic Personal and Civil Rights, Principal Paper* in THE MENTALLY RETARDED CITIZEN AND THE LAW 4 (M. Kindred ed. 1976).

[2] *See, e.g.,* Loving v. Virginia, 388 U.S. 1 (1967) (invalidating a state miscegenation statute on the basis of the lack of a compelling state interest); Meyer v. Nebraska, 262 U.S. 390 (1923) (invalidating a mandatory public school attendance law as interferring with parents' right to direct the education and upbringing of children under their control); Roe v. Wade, 410 U.S. 113 (1973) (holding that the fundamental right to privacy encompassed "a woman's decision whether or not to terminate her pregnancy"); Griswold v. Connecticut, 381 U.S. 479 (1965) (holding that a state may not constitutionally prohibit the use of contraceptives by married persons); Home Bldg. and Loan Ass'n v. Blaisdell, 290 U.S. 398 (1934) (discussing the importance of the right to contract). In Meyer the majority wrote that the liberty interest protected by due process "denotes not merely freedom from bodily restraint but also the right of an individual to contract, to engage in any of the common occupations of life, to acquire useful knowledge, to marry, establish a home and bring up children, to worship God according to the dictates of [ones] own conscience, and generally to enjoy those privileges long recognized at common law as essential to the orderly pursuit of happiness by free men." 262 U.S. at 399.

Strict judicial scrutiny can be triggered either by the impact of a law on the exercise of a "fundamental right," or on the basis of the use of a "suspect" classification. The elements of a suspect classification were outlined by the Court in United States v. Carolene Prod. Co., 304 U.S. 144, 152-153 (1938), and emphasize the concept of political underrepresentation. The court noted in its well known footnote 4 that "... prejudice against discrete and insular minorities may be a special condition, which tends seriously to curtail the operation of those political processes ordinarily to be relied upon to protect minorities, and which may call for a correspondingly more searching judicial inquiry." For an argument that disabled persons should be treated as a suspect classfication, *see* Burgdorf and Burgdorf, *A History of Unequal Treatment: The Qualifications of Handicapped Persons as a Suspect Class Under the Equal Protection Clause,* 15 SANTA CLARA LAW. 855 (1975).

[3] *See, e.g.,* Shapiro v. Thompson, 394 U.S. 618 (1969) (voting restrictions invalidated because "less drastic means" were available and could be employed); Kusper v. Pontikes, 414 U.S. 15 (1973) (voting restrictions invalidated because they "unnecessarily" burdened the exercise of a protected interest and were not the least drastic means available.)

cise of many of them through statutory restrictions in many states. These restrictions are referred to here as "state-imposed disabilities" because they may or may not reflect actual disabilities. These "state-imposed disabilities" and the resulting inability to exercise fundamental personal rights are an indication that an individual is not a full, functional participant in society, and thus is, to some extent, to be looked on as an outsider.[4] As society establishes the goals of integration and community-based habilitation for disabled persons, it becomes critical to define the extent to which we will afford these individuals the tools to participate effectively within the society to which we have finally allowed them at least conditional admittance. Just as the movement to deinstitutionalization heightened the interest in advocacy, in zoning legislation and in limited guardianship, so, too, must it question current restrictions on the personal and civil rights of mentally disabled persons, for we now know that mentally disabled persons, if given appropriate services and support, can achieve at a level far above that once believed. New theories of education and habilitation are being utilized with success. Our factual knowledge of genetics, mental retardation, epilepsy, educational potential and employability, for example, has greatly improved. And mentally disabled persons are proving themselves capable of self-sufficiency and of responsible exercise of rights in the community. Yet our statutes have not changed to reflect this, nor in many cases have the public policies behind them.

This introduction to the topic of personal and civil rights will examine some of the philosophical and legal bases for the statutory restrictions on the exercise of these rights by disabled persons, challenge these bases as inconsistent with basic legal principles and with our moral and ethical ideals upon which they are based, and suggest several new bases from which to begin an analysis of restrictions on the exercise of these rights by disabled persons. Finally, this introduction will suggest the bases for determining what restrictions, if any, should be placed upon the exercise of personal and civil rights by disabled persons.

Philosophical and Legal Bases for Current Restrictions

The "state-imposed disabilities" described above are, among other things, a reflection of the social philosophy and the legal assumptions from which we operate. They appear to reflect a strong element of social-darwinist theory, which was a prevalent part of American political and social thought from about 1890 to the First World War[5] and which supported the eugenic movement in the United States.[6] That movement was based upon the idea that some persons are more *socially* desirable than others and that future generations can be improved *biologically* by increasing the proportion of "desirable" types of individuals and by decreasing the rate of propagation of "inferior" individuals. Proponents of the movement relied on scientific claims that mental disabilities among the other "undesirable qualities" were inheritable traits and therefore mentally disabled persons should be sterilized to prevent transmission of the disability.[7]

For example, it was argued that individuals who could not compete successfully should not be allowed to participate on an equal basis with others. Certainly, the argument went on, mentally disabled persons were naturally inferior to others and were, therefore, at the bottom of the social order. Thus, society had the right to protect itself from their proliferation. Out of this philosophy came restrictions on their marriage and especially their forced sterilization. This attitude was also reflected in the notion that some individuals were "unfit" to be parents, to vote, or to dispose of their property by will and that society must be protected against their folly. Most strikingly, unfortunately, it was reflected in the notion that the "quality of life" of some individuals did not measure up to the minimum standard necessary to warrant life itself. While social-darwinism no longer exists as a conscious philosophy and its rhetoric is seldom heard in any formal discussion, it has persisted as part of political and social folklore and is not without its adherents today.[8] In fact, many of today's statutes (*e.g.* marriage and sterilization) still reflect the need to protect society from the supposedly "unfit."

In addition to social philosophy, legal theories or assumptions about human behavior have greatly shaped the law related to mentally disabled persons. For example, two assumptions—that people know and understand what they do and the consequences which may arise therefrom, and that it is morally right that they suffer the consequences of their actions—are the bases for granting the exercise of rights and imposing legal obligations upon persons. When categories of persons are *perceived* as being unable to function in this manner, they have been traditionally denied the legal ability to

[4] R. BARRON & H. FABING, EPILEPSY AND THE LAW 35 (1956).

[5] *See* R. HOFSTADTER, SOCIAL DARWINISM IN AMERICAN THOUGHT 201-204 (1955).

[6] The eugenics movement resulted in the adoption of numerous state sterilization statutes beginning in 1907. The United States became a pioneer in the field of sterilization, though not without considerable legislative and judicial opposition. While many of the statutes were declared unconstitutional by state courts, the U.S. Supreme Court upheld a statute authorizing sterilization of "mentally deficient" persons in *Buck v. Bell,* 274 U.S. 200 (1927). *See* S. BRAKEL and R. ROCK, THE MENTALLY DISABLED AND THE LAW 207-209 (1971).

[7] BRAKEL and ROCK *supra* note 6, at 206.

[8] See *Emergency Medical Care and Treatment For Infants infra.*

exercise a broad range of choices associated with everyday life[9]—that is, the law frequently assumes one is either totally competent or totally incompetent. It thereby expands a person's incapacity in one area to include other areas as well, whether or not the person is actually incapable in those other areas.

The Inadequacies of State Statutes on Personal and Civil Rights

Current state statutes governing personal and civil rights for disabled persons suffer from a number of inadequacies. These statutes:

1. Assume an inability on the part of mentally disabled persons to participate in normal activities;
2. Oversimplify by treating all as if they are part of a homogeneous group regardless of individual characteristics and differences and by providing no means to establish that a given individual does not fit within the category. This approach tends to reduce the individual to the lowest common denominator, the lowest level of functioning, rather than emphasizing individual strengths and abilities;
3. Embody a gross overreaction to disabling conditions, resulting in more restrictive treatment of disabled persons than is warranted by the circumstances;
4. Condone the concept of "levels of humanity" — that some persons are somehow "better," more deserving of the enjoyment of basic rights than are others. Yet, such notions fly in the face of the guarantees of the Bill of Rights and of equal protection and due process or "fundamental fairness," as well as the basic moral and religious tenents upon which our social structure and the law are purportedly based; and
5. Deny the reality of state intervention by hiding behind the legal facades of "voluntariness" and "parens patriae." For instance, the law frequently allows substitute consent by relatives or others for those who are deemed to be incompetent to exercise their own choice. Because these substitute decisionmakers are presumed to have the person's best interests at heart, this exercise of choice is said to be "voluntary" on the part of the disabled individual whether the individual had any say in the matter or not. Thus, for example, the individual may find himself or herself "voluntarily" sterilized. In fact the choice may fit the needs of the substitute decisionmaker, not the disabled individual. Likewise, the state, acting as *parens patriae,* may also cause the individual to be sterilized; in addition, because it is said to be acting in the interests of the individual, the legislature has often assumed that no particular safeguards are required to protect the individual. The rationale of *parens patriae* is that the state must intervene in order to protect an individual who is not able to make decisions in his or her own best interest. Yet, the question of when this power should be exercised in a given situation is quite difficult to answer. In fact, the real basis for such action may be some generalized public interest which the state undertakes pursuant to its police powers and not the individual's interest at all. Under its police powers, a state may exert its coercive force on an individual if he or she represents a significant danger to other persons or to society. Yet, the application of the state's police power *vis-a-vis* the individual generally calls for greater scrutiny of the state's action and more protection of individual rights. The fiction that the state acts as *parens patriae* when its real motives lie in its police power can lead to abuse of individual rights.

Major Premises

The analyses, recommended statutory revisions, and model statutes which follow are based upon several major premises which are largely the counterpart of the inadequacies described above and challenge the traditional social and legal assumptions upon which most current statutes on the subject are based. Initially, we start from the premise that disabled persons are entitled to the same rights as others or as simply put in our Declaration of Independence, "All men are created equal." Certainly, all persons do not share equally in intelligence, physical ability, self-sufficiency or the capacity to contribute economically to society. But these factors do not warrant the denial of equality before the law. Thus, we specifically reject revival of the social-darwinist notion that some persons are "inferior" to others and may be so treated. For example, some have argued that mentally retarded persons should be treated differently from "normal" persons and that they should be allowed to die rather than being given medical treatment. (See *Right to Emergency Medical Care and Treatment for Infants infra.*) At least one writer has strongly warned against this trend:

> At stake is a basic question of equal protection. Medical literature, legal literature, and the theological literature reflect the encroachment of a new kind of ethnic which is being called by its proponents a "quality of life" ethnic. The "quality of life" ethic, in essence, rests on the principle that "some are more equal than others," that unequal capacity or potentiality provides grounds for discriminatory treatment. At stake here is the fundamen-

[9] *See* Wald *supra* note 1, at 6.

tal moral, and hence legal, equality of human beings as human beings.[10]

She goes on to point out that these issues are issues of social justice. But social justice is more than legal protection or justice in the social order; more than just a matter of external relations and regulation of behavior:

> It is also an ordering of the various values, attributes and tendencies within the inner person. One of the issues which should be raised, therefore, has to do with the kind of people we want to be: What does it do to us as persons to violate the rights of other people, including mentally retarded people, and to create a world in which the moral community excludes whole categories of people? Do we not diminish as people when we participate in the myths and stereotypes that support this kind of unequal treatment?[11]

The writings which follow are also based upon the premise that restrictions on "fundamental" rights should be minimized. Since the exercise of personal and civil rights is so fundamental to an individual's growth, development and well-being as a functional member of society, restrictions on these rights are appropriate only where their exercise would result in a wholly unacceptable set of adverse consequences, either for the state or for the individual. Freedom to make fundamental personal decisions should be presumed unless it can be shown, clearly and objectively, that the individual involved is so critically in need of added protection that this need overrides the multiplicity of benefits from personal autonomy or unless it can be illustrated convincingly that the state as a whole cannot tolerate the great risk to its citizens or its institutions that free exercise of a right by certain individuals is almost certain to entail. In determining when such restrictions are appropriate, the state's interests must be sharply defined and narrowly drawn *vis-a-vis* the individual interests of disabled persons.

More than anything else, disabled persons need the right to try, to learn by experience and to succeed or fail on their own. Yet, some disabled persons will not be capable of fending entirely for themselves and cannot be left to their own devioes. Thus, there will be widely differing and perhaps competing interests which must be considered when the state acts on behalf of the individual. At the same time, the state is responsible to ensure the well-being of all its citizens. In some instances, the free exercise of a right by one individual may have a direct impact on the freedom with which other individuals can safely and effectively exercise the same or another right.[12] Thus, some restrictions are necessary for the public welfare.

Since the restrictions imposed by the state are required by the Constitution to be "closely tailored" to serve the state's interest, the state's interest should be clearly defined at the earliest point in the legislative process and be the central point from which substantive restrictions grow.[13] Confronted with a diverse range of goals for each legislative decision, the states face a real dilemma in reconciling them all through the legislative process. However, only after the interests involved have been defined and analyzed and the interests favoring some restrictions found adequate, can the process of developing substantive restrictions and procedural safeguards profitably begin.

This step in the process necessarily involves the balancing of conflicting public and individual interests and may well be the most difficult since it involves highly subjective elements. The Supreme Court, by characterizing certain rights as "fundamental" and requiring that states deprive persons of their exercise only upon a showing that these restrictions on individual rights are "necessary" to further a "compelling state interest," has at least provided a starting point, but the court's analysis defines only what is minimally acceptable under Constitutional guarantees. Legislators, on the other hand, while not free to take a less critical approach than the Constitution mandates, may certainly go beyond its mandate by taking a more critical stance. In the course of the

[10] Lebacqz, *Nonconsensual Medical Procedures and the Right to Privacy, Reaction Comment* in THE MENTALLY RETARDED CITIZEN AND THE LAW *supra* note 1, at 113.

[11] *Id.*, at 114.

[12] A number of Supreme Court cases have recognized that the safeguards afforded one individual's exercise of rights may be restricted when his or her exercise of the right conflicts with the rights of others, particularly when free exercise may have an impact on health and safety. *See* Smith v. Organization of Foster Families, 431 U.S. 816 (1977) (procedures for removal of children from foster families were adequate where foster parents rights conflicted with those of biological parents); North Am. Cold Storage Co. v. City of Chicago, 211 U.S. 306 (1908) (summary seizure and destruction of individual inventory of poultry upheld in light of the public health hazards presented).

[13] In applying an equal protection analysis to state statutes the Supreme Court has shown a tendency to examine more closely the state's purpose in enacting the statute rather than some purpose "hypothesized" in an attempt to support the legislation after-the-fact. Summing up the line of Supreme Court cases which illustrates this tendency, Justice Brennan wrote in his dissent in Schlesinger v. Ballard, 419 U.S. 498, 520 (1975):

> While we have in the past exercised our imaginations to conceive of possible rational justifications for statutory classifications... we have recently declined to manufacture justifications in order to save an apparently invalid statute. *Cf., James v. Strange,* 407 U.S. 128 (1972); *Weber v. Aetna Casualty and Surety Co.,* 406 U.S. 164 (1972). Moreover, we have analyzed asserted governmental interests to determine whether they were in fact the legislative purpose of a statutory classification, *Eisenstadt v. Baird,* 405 U.S. 438, 442-443 (1972), and we have limited our inquiry to the legislature's stated purpose when these purposes are clearly set out in the statute or its legislative history. *Johnson v. Robinson,* 415 U.S. 361, 376 (1974).

discussion of individual rights, this section on personal and civil rights will illustrate both the balance struck by the judiciary and the balance that we, as legislators, would strike. Both can act as useful guidelines in their own right. With this in mind we turn to criteria for the development of standards for determining those individuals whose free exercise of personal rights must be restricted and for developing an administrative system to screen individuals in a non-discriminatory, individualized manner.

The development of standards for restricting individual rights begins from the premise that persons traditionally singled out for special treatment with respect to these rights do not require a special system of laws. Rather, there is a presumption in favor of a single system of laws with respect to these rights.[14] The general standard of capacity to exercise a right shall be applicable to all persons and not to any particular category of persons. A statutory standard is both too broad and too narrow if the standard places categorical restrictions or disqualfications on persons based upon a label. The writings which follow ignore, for most purposes, classifications of individuals developed in other contexts and for other purposes, *i.e.,* mental retardation, civil incompetency and the like. Rather, these categories must be recognized as encompassing persons of diverse background, ability and learning who do not all function equivalently in any given context.[15] Legislatures must look within and beyond these categories for the real indicia of inability to exercise a given right, the goal being always to restrict the smallest number of persons necessary to reconcile the competing interests involved.[16] In place of these broadly sweeping categorical definitions to determine the exercise of rights, it is proposed that each standard which restricts individuals from exercising certain rights be specially tailored to a determination for each particular right. Thus, the standard is to be a functional rather than a categorical one, which is directly related to the interest the state is attempting to achieve through its restrictions. Given the precision with which the state's purpose should, at this point, have been defined, the development of a standard is primarily a matter of identifying specific, objective, and administratively verifiable criteria which relate as precisely as possible to an individual's ability to function in the *specific context involved.* Two rules should guide the search for these criteria. First, the state must assume the burden of proving the direct relationship between the criteria utilized and the inability to properly exercise a given right.[17] If objective facts do not support the utility of a given restriction to further the state's interest, it should not be imposed. Intuition and supposition are inadequate grounds for the deprivation of fundamental rights.

Second, any standard should reduce administrative discretion to a minimum. It should contain criteria which are concrete, understandable and objective, allowing an administrator to decide, with a degree of certainly, whether or not an individual should be identified as one of the class of persons to be restricted.[18]

In many instances the development of a suitable standard for determining which individuals should be allowed to freely exercise a personal right will be nothing more than a determination of a person's capacity to adequately understand and make the choices which exercise of the right entails. Thus, the basic tests in a number of situations will be quite similar, reflecting an individual's ability to understand the nature and consequences of the decision to be made. A distinction must be recognized, however, which is often overlooked: simply because the overt test is the same does not mean that the specific objective criteria involved will also be the same or even similar. The character of the decision to be made will in each case dictate what specific elements of personal make-up,

---

[14] *See* Wald, *supra* note 1, at 3-6.

[15] *See* Murdock, *Civil Rights of the Mentally Retarded — Some Critical Issues,* 7 FAM. L.Q. 1 n.2; J. MERCER, LABELING THE MENTALLY RETARDED (London 1973). One rehabilitation counselor for retarded persons expressing his own experience with traditional labels said... "it soon became apparent to me that knowing that the young person sitting in my office was labelled [educable mentally retarded] told me very little about him....I began to view the...label as of benefit only to educators... and before long began questioning that assumption as well... The one useful function of the label was that it allowed the funds necessary to provide the... services needed by these students." Foss, *Rehabilitation Counseling with the Mentally Retarded: Profession and Practice,* in MENTAL RETARDATION: REHABILITATION AND COUNSELING 249, 260 (1974).

[16] This approach is the basis of the constitutional requirement that restrictions imposed on the exercise of a fundamental right be "necessary," *i.e.,* be "closely tailored" to effectuate the asserted state goal, thus restricting the smallest number of persons possible while still serving the state's important interests.

[17] This requirement follows from the basic premise that a presumption of competency should operate in all instances. Much as the judiciary imposed the burden of proof on the party claiming incompetency, the legislature, which by classifying persons declares a person's incompetence, should in fairness accept the same burden of justification before imposing like restrictions on entire classes of individuals. For a discussion of the presumption of competence *see AMERICAN ASSOCIATION ON MENTAL DEFICIENCY, CONSENT HANDBOOK 14 (H. Turnbull ed. 1976);* Neuwirth, Heisler and Goldrick, *Capacity, Competence, Consent: Voluntary Sterilization of the Mentally Retarded,* 6 COLUM. HUMAN RIGHTS L. REV. 447, 448 (1975).

[18] The Supreme Court has recognized that some laws, owing to their extreme vagueness, may be too imprecise to sanction a denial of personal liberty. Statutes which restrict persons' exercise of individual rights on the basis of their being "feebleminded," "insane" or the like, rather than on the basis of meaningful criteria, may be subject to a vagueness challenge. The Court's test for vagueness is whether a person "of common intelligence must necessarily guess at [the classification's] meaning and differ as to its application." Connally v. General Constr. Co., 269 U.S. 385, 391 (1926). *See generally,* Note, *The Void-for-Vagueness Doctrine in the Supreme Court,* 109 U. PA. L. REV. 67 (1960).

physical ability, experience and intelligence are relevant to a determination that an individual is capable of understanding the nature and consequences of the particular type of decision to be made.

In addition, the writings which follow are based upon the premise that all standards of incapacity should be applied through individual determinations of a person's capacity to exercise a particular right before its exercise can be restricted. Determinations of capacity must recognize the importance of personal experience, familiarity with the type of decision involved, maturity, educational and intellectual progress, psychological indicators, personal determination and all other relevant factors, not simply I.Q. scores and other limited indicators of personal ability. Only an individual's capacity to do specific things can be judged. Thus, judicial determinations based on unproven assumptions and so-called "common knowledge" are inappropriate.[19] Judgments must be related to the specific narrow and relevant behavior specified in the statutory standard through an appropriate system of administrative implementation. This system should be (1) individualistic, allowing each person an opportunity to demonstrate the inappropriateness of the contemplated state action before any substantive restrictions are imposed,[20] (2) non-discriminatory, assuring that every individual is scrutinized based on the same standards and according to the same administrative procedures, and (3) procedurally adequate, providing the basic due process rights of notice and hearing. The allegedly incompetent person or his or her representative has a right to notice of the specific disability alleged and the basis for that allegation, and a right to challenge the allegation before a neutral decisionmaker prior to deprivation of his or her rights. The administrative burden created by such an application of the laws, and the state's willingness or hesitancy to assume it, may be the best indicator of how important the legislature actually views the public purposes for which the restriction is being imposed. Arguably, if the purpose is not important enough to warrant the burden of universal application of the law, then it is likewise not important enough to warrant restrictive measures. Drawing the analytical process full circle, such hesitancy may require a return to the earliest elements of the analysis and a careful rethinking of the relative importance of the state and private interests involved.

Finally, the fact that a person is incapable today does not indicate his or her incapacity is permanent. In recent years it has been illustrated that in many cases the inability to understand and effectively make personal decisions is often a result of the lack of prior opportunities to become familiar with these decisions rather than the result of immutable mental disabilities.[21] We owe each individual the opportunity to demonstrate that, given the same experimental tools that the majority of people take for granted, he or she can rise to the level of personal competency that is found minimally necessary to protect both his or her private interest and the specific state interest which survived the foregoing analysis. Thus, while restrictions will in some cases remain necessary, the frequency can be minimized through innovative use of counselling and education designed to aid individuals to acquire the basic skills necessary to exercise rights rationally and effectively.[22]

Disabled persons have largely been left out of efforts in recent years to guarantee the rights of minority groups. These efforts are therefore necessarily incomplete. Basic, fundamental rights are essential for all persons. The sections within this chapter are offered with the intent of assisting in efforts to maximize the rights of disabled persons within the context of competing state and private or individual interests.

---

[19] *See generally,* MERCER, LABELING THE MENTALLY RETARDED 53-66 (Berkeley 1973) for a study of a single community's response to retarded persons, including an in-depth study of how administrative agencies go about deciding, with vastly differing results, who is and is not a "retarded" person.

[20] This is a basic requirement of procedural due process. The Court will determine how much individual scrutiny will be applied to any administrative decision, and at what point it will be applied based on a balancing of three factors: (1) the private interest that will be affected by the official action; (2) the likelihood or risk of an erroneous decision without the additional scrutiny and (3) the governmental interest involved. *See* Matthews v. Eldridge, 424 U.S. 319, 335 (1976). Clearly then, when an administrative decision impacts on the ability to exercise fundamental personal rights, the need for additional individualized procedures is great. Ordinarily, when the personal interest is especially strong, due process will require notice and a hearing *prior* to any actual deprivation. *See* Bell v. Bursen, 402 U.S. 535 (1971).

A similar concept of individualized determination has been developed under the guise of equal protection analysis. Under this approach the Court has seen fit to invalidate laws which create an "irrebuttable presumption" which is neither "necessarily or universally true in fact." Where this is so and the law impacts on the exercise of an important right, the Court has sometimes required the development of an administrative apparatus whereby individual determinations may be reviewed. *See, e.g.,* Cleveland Bd. of Educ. v. La Fleur, 414 U.S. 632 (1974); Vlandes v. Kline, 412 U.S. 441 (1973).

[21] Freedom to make personal choices is a major source of personal growth and development. *See* Shaman, *Persons Who Are Mentally Retarded: Their Right to Marry and Have Children,* 12 FAM. L.Q. 62-65 (1978); J. MERCER, LABELING THE MENTALLY RETARDED 127-40 (1973).

[22] For a good overview of facilitative counseling and its importance *see,* Halpern and Berard, *Counseling the Mentally Retarded: A Review for Practice,* in MENTAL RETARDATION: REHABILITATION AND COUNSELING 269 (P. Browning ed. 1974).

1.1 MARRIAGE, ANNULMENT AND DIVORCE

Table of Contents

INTRODUCTION

It has long been assumed that the act of marriage, or the ability to terminate a marital relationship are both moral and constitutional rights to be freely exercised by all citizens. For most persons, such is the case. But for millions of mentally disabled Americans, there have traditionally existed different domestic relation laws which include a widespread series of state prohibitions on the right to independently choose whether to get married, divorced or to seek an annulment of a marriage. These prohibitions have discriminated against handicapped persons, based solely on handicap. For instance, virtually all state laws historically attached labels to people with disabilities, referring to them as "idiots", "feebleminded" or worse, then promulgated unfair and arbitrary prohibitions against their right to engage in domestic relations matters on a level equal to that of their peers who were not thought to be handicapped. These laws, which still exist in some jurisdictions, have caused great hardship, do not benefit society and have been arbitrarily enforced.[1]

In the area of marital relationships, as in virtually all other areas where persons must be able to exercise their basic human rights free of state interference, laws which invidiously discriminate must be eliminated. As one noted commentator has observed: "It is essential to our constitutional scheme that stereotyping of men by government is to be prohibited, and that stereotyping of men by men is not to be encouraged by government, but affirmatively to be discouraged."[2] If this caveat is observed, respect will be fostered for all persons, regardless of whether they are thought by some to be "different." The laws relating to a marital relationship which have been used to discriminate thus deserve close scrutiny. This section will provide such an analysis.

Marriage

Marriage has been characterized by the United States Supreme Court as "one of the basic rights of man, fundamental to our very existence and survival" and "essential to the orderly pursuit of happiness by free men."[3] Despite this basic language of constitutional law, there are still some states which claim an interest in restricting the right of mentally disabled persons to marry. The individuals affected by these prohibitions are variously classified as idiots, insane, feeble-minded, lunatic, under guardianship, retarded and institutionalized. For these persons, the formation of a marital relationship is not only proscribed, but when the marriage does occur, however successful it may be, the union may later be held invalid.[4]

When a fundamental right such as marriage is the subject of legislation, the Supreme Court has held that "any statutory classifications [which] significantly interferes with the exercise of [the right to marry] cannot be upheld unless it is supported by sufficiently important state interests and is closely tailored to effectuate those interests."[5] If such a right is in any way impaired, the courts must conduct a "critical examination of the state interests advanced in support of the classification scheme employed."[6]

But what are the interests claimed by those state legislatures which restrict mentally handicaped persons in this basic right? First, because marriage is considered a condition precedent to having children[7], eugenic considerations have been involved.[8]

[1]*See, e.g.*, Wald, "Basic Personal and Civil Rights," in THE MENTALLY RETARDED CITIZEN AND THE LAW (M. Kindred, J. Cohen, D. Penrod and D. Shaffer, eds., 1976) (hereinafter cited as THE MENTALLY RETARDED).

[2]Gilhool, "The Right to Community Services," in THE MENTALLY RETARDED 172, 174.

[3]Skinner vs. Oklahoma, 316 U.S. 535, 541 (1942); *see also* Meyer vs. Nebraska, 262 U.S. 390 (1923) and Loving vs. Virginia, 388 U.S. 1, 12(1967).

[4]*See, e.g.*, THE MENTALLY DISABLED AND THE LAW (2nd ed. S. Brakel & R. Rock, eds. 1971) at 266; Wald *supra* note 1, at 1; and, cases and materials cited in M. Burgdorf, "Marital and Family Rights," in LEGAL RIGHTS OF MENTALLY DISABLED PERSONS (P. Friedman ed. 1979).

[5]Zablocki vs. Redhail, 434 U.S. 374, 388 (1977) citing, *inter alia*, Carey vs. Population Services International, 431, U.S., at 686 Memorial Hospital vs. Maricopa County, 415 U.S. at 262-263, San Antonio Independent School District vs. Rodriguez, 411 U.S. at 16-17.

[6]*Id.* at 383.

[7]*See, e.g.*, Maynard vs. Hill, 125 U.S. 190 (1888); and, Skinner, *supra* note 3, at 541 ("Marriage and procreation are fundamental to the very existence and survival of the human race.")

[8]*See, e.g.*, the introduction to Subchapter 1.4 entitled Sterilization and the cases and materials cited therein (hereinafter cited as the "Introduction").

The statutes of a few states removed prohibitions on marriage if handicapped persons were unable to procreate. These laws were obviously based upon eugenic grounds, including the state's alleged interest in protecting the public from the perceived burdens of caring for the disabled issue of the marriages of disabled individuals. The presumptions implicit in this rationale are that mental disabilities are genetically controlled, that most disabled children are the issue of disabled parents, and that the disabled parents will not be financially able to care for their offspring. The presumptions and thus the interests they advance, are unsupportable by the evidence. While a few disabilities are inheritable, it is clear that environmental, nutritional, and a variety of similar non-genetic determinants are more likely to result in disabilities.[9] In fact, there are over 250 known causes of retardation, the vast majority involving non-genetic factors. Even today, in approximately 75% of retardation cases no specific instrumental cause can be identified.[10] Thus, broadly drawn prohibitions are unsupportable when based solely on eugenic grounds. Even when the evidence does support inheritability of disabilities in certain limited cases, only a statute specifying such disabilities would conceivably be a justification for deprivation of a fundamental right. Finally, there is no evidence to support the assumption that disabled parents will be an economic burden on society. It is impossible "to make confident predictions in the cases of retarded persons any more than in other cases."[11]

A second and related facet of the eugenic rationale contemplates a state desire to protect the children of, and prevent children being born to, potentially unfit parents.[12] Yet, prohibiting marriage is surely not the most efficient or least intrusive method of encouraging birth control, nor is it the preferable way of protecting children. Restrictions should, at least theoretically, be drafted to eliminate only "unfit" parents; however, even for those few who may require some assistance from the state — such as child care and training for parents, whether disabled or not — such aid is preferable and less restrictive than denying the right to marry. The strongest attack on the unfit parent rationale, however, is that it is unsupportable. Experts now believe that disabled individuals make quite adequate parents, at least as adequate as persons not thought of as handicapped.[13]

A third rationalization for restrictions which requires scrutiny is the claim that marital prohibitions protect disabled individuals from entering into legally binding contractual arrangements, the consequences and obligations of which they are incapable of fully appreciating. Statutes to the effect that "marriage is a civil contract which may be entered into only by persons with sufficient capacity" are best explained by this rationale. On its face, the concept might appear logical, since an agreement cannot literally be said to exist unless both parties understand its basis and assent to its component parts. Problems arise, however, concerning the application of this principle in practice because there are many different tests for capacity even though the capacity to enter into a marriage contract has usually been seen as requiring less intellectual ability than capacity to contract commercially. In fact, a prior adjudication of incompetency is usually not even controlling.[14] In addition, while a finding of incapacity to contract is supposedly based on concern for the individual's welfare and is designed to prevent others from taking advantage of disabled individuals, it is difficult to apply this rationale to marriage. It hardly comports with our societal ideal of marriage as a supportive institution based on mutual trust to assume that every marriage is potentially detrimental to the disabled persons or to society. To the contrary, most mentally handicapped individuals who have married find the situation beneficial, providing them with the support and assistance necessary to mature and function effectively.[15]

Annulment

Annulment has been defined as the invalidation through court declaration that a marriage never came into being.[16] Traditionally, mental handicaps of a spouse have served as grounds for this action. For mentally handicapped persons, statutes, have allowed annulment for spouses of persons classified as having insanity, lunacy, idiocy, feeblemindedness, imbecility and the like.

These classifications have not been nor are they now valid reasons for granting an annulment. The real issue is whether the individual at the time of the marriage had the capacity to consent to forming the relationship. There is no data available to prove that

---

[9] Wald, *supra* note 1 at 1-7, 16-17; Kindregan, "Sixty Years of Compulsory Eugenic Sterilization: Three Generations of Imbeciles and the Constitution of the United States, 43 CHI-KENT L. REV. 123, 139 (1966); Brakel and Rock *supra* note 4, at 226; and, introduction generally.

[10] Shaman, "Persons who are Mentally Retarded: Their Right to Marry and Have Children," 12 FAM. L.Q. 61, 63 (1978).

[11] Wald, *supra* note 1 at 8.

[12] *See*. "The Right of the Mentally Retarded to Marry: A Statutory Evaluation," 15 J. FAM. L 463, 467 (1976-77).

[13] This belief has been documented for years. *See, e.g.*, "The Feeble-minded Parent: A Study of Ninety Family Cases," 51 AM. J. MENTAL DEFICIENTY 644 (1947).

[14] *Supra* note 12 at 468; *See generally*, Clark, LAW OF DOMESTIC RELATIONS (1968), § 2.11 and 2.15.

[15] *Supra* note 10.

[16] H. Bass and M. L. Rein, DIVORCE OR MARRIAGE: A LEGAL GUIDE 24 (1976).

mentally retarded or other handicapped persons are any more or less capable than other persons of having this capacity. Thus, the handicapped label is irrelevant to making a rational decision on whether to grant an annulment. Further, there are no studies available to indicate that persons deemed "handicapped" are any less able to cope with the pressures of marriage than persons not so labelled. Indeed, for the same reasons generally advanced and delineated above that marriage involving handicapped partners should not be prohibited, neither should a marriage once entered into be invalidated.

Another problem in this area is that in addition to the prohibitions being based on unsubstantiated assumptions, the legal restrictions are arbitrarily enforced. As Judge Wald has noted, there is no way of determining the number of persons who are mentally handicapped who indicate on marriage applications that they are not handicapped, or deemed handicapped, and thus marry without state intrusion. In fact, enforcement of these exclusionary and discriminatory laws if often left to clerks who happened to notice that a person had answered "yes" in response to the question relating to disability on a marriage application. Even when persons have answered "yes" to such questions, and annulment proceedings were then brought, retarded couples have been granted the right to marry.[17]

A final area where these laws may operate to discriminate against handicapped persons relates to who may bring the annulment action. At common law, marriages entered into by a person who was subject to annulment have been void *ab initio* — from the start — and no court order was usually necessary to set aside a marriage deemed never to have existed. The modern trend is to consider such marriages voidable if a party seeks to annul, but not necessarily void from the outset.[18] If a marriage is sought to be adjudged void, usually either party to the relationship may bring the action. If a marriage is considered voidable the question of who may bring an annulment action varies in the jurisdictions among the husband, wife, and third parties. These statutes cause unnecessary problems for handicapped persons. For example, there would appear to be no basis for denying the right of the spouse or of the disabled person to seek an annulment if that is his or her desire, unless the person has been declared incompetent in domestic relations matters — an unlikely event. But statutes permitting third parties the ability to destroy another's marriage relationship are based on antiquated notions of paternalism toward persons thought to be handicapped, as well as a desire on the part of society to control those who are different. There is just no basis for voiding a marriage involving handicapped persons on the basis of third party allegations when such interference would not be tolerated for all other persons.

Divorce

Divorce is the judicial termination of a marriage relationship that has been in legal existence.[19] Divorce laws in some jurisdictions, as in marriage and annulment, have allowed for one type of handicap — the antiquated generic classification of "insanity," to be used as the basis for terminating a marriage if one spouse is deemed to be insane.[20] There are several standards which must be met and which place a heavy burden on the person seeking a divorce pursuant to this ground. The insanity must be incurable; continue for a number of years; and, the person thought to be insane must be confined to an institution for the care of the mentally ill, or adjudicated insane, or at least be examined by a specified number of doctors who concur with the "diagnosis" of insanity. This heavy burden, however, still does not obviate the invidious and stigmatizing effect of such a law.

The real issue in determining whether to grant a divorce, at least in states which use the concept of fault in their divorce laws, is the behavior of the spouses toward each other. The label "insane", however, is a legal term and is far too amorphous to correlate with any specific behavioral handicap. Thus, it is inappropriate to use it as the grounds for granting a divorce. The same logic holds true if the term "mentally ill" was substituted for insane. The variety of behavioral symptoms due to mental illness are sufficiently numerous that it is impossible to rationally grant a divorce based upon such a label — especially considering that many forms of mental illness (*e.g.*, neurosis) would, at most, only strain a marriage but not break the relationship. Moreover, the ultimate issue, the behavior of the spouses, should be addressed directly as it would be for all other types of persons. By doing so, we are providing a direct assessment of the facts which are rationally related to granting a divorce decision while at the same time reducing the potential stigma to any party by not requiring them to be labelled insane. In states which have adopted no-fault divorce laws, the use of the label insane is equally inappropriate. One of the justifications for no-fault is that it reduces the need to stigmatize one of the parties as being at fault. Hence, a determination of insanity is unnecessary for granting a divorce under these types of laws.

[17]Wald, *supra* note 1 at 7-17.

[18]*See generally* REVIEW OF STATE LEGISLATION *infra*. The question of whether a marriage is void or voidable has implications for property settlements. Because these implications apply equally to all persons, however, they will not be considered here. *See generally*. Clark *supra* note 13.

[19]Bass and Rein, *supra* note 15.

[20]Clark, *supra* note 12; *see generally* Mattinson, MARRIAGE AND MENTAL HANDICAP (1971).

REVIEW OF STATE LEGISLATION

In order to more fully understand the level of discrimination that has occurred in this area, this section will review relevant state and territorial legislation on marriage and annulment. Divorce has been intentionally excluded because of the very specific way it discriminates against handicapped persons — i.e. applying to insane persons only. Thus, reviewing that state legislation would not add to what has already been noted in the Introduction.

Restriction Governing the Formation of Marriages

Thirty-three states prohibit certain subclasses of the developmentally disabled persons from entering into a marriage.[1] These provisions generally take one of two forms depending on the state's view of the legal foundation of the marital relationship. One group of restrictions, based on a notion of marriage as a civil contract between spouses, prohibits persons legally incapable of contracting from entering into a marital agreement.[2] States utilizing contract approach[3] derive their legal authority from common law doctrine[4] and some evidence indicates that the contractual approach is followed by the courts in most states whether or not any specific provision exists in support of it.[5]

The effect of the "contractual capacity" type statute will depend, to some extent, on the case law of the particular state concerning contract matters. Some states follow the modern trend and their courts will determine capacity based only on the individual's ability in relation to the specific contractual act to be performed;[6] this approach requires only that a person be capable of understanding the nature of the particular contract at hand and the obligations arising under it. Other states allow a single adjudication of capacity for contractual matters generally and capacity to marry will presumably be included in such rulings.[7] The commentary accompanying the Uniform Marriage and Divorce Act indicates that its authors contemplated use of the former method in states enacting this legislation.[8]

Also applicable to consideration of contractual capacity based disabilities is the possible existence of a judicial presumption in favor of adequate capacity and the existence in other statutory code sections of general categorical prohibitions against contracts entered into with persons of "unsound mind" or the like.[9] In many states case law and statutory law may overlap, presumptions concerning capacity may conflict; almost an infinite variety of interpretive difficulties may arise. One attempting to discover the true meaning and application of the simple statement that "marriage is a personal relationship...arising out of a civil contract to which the consent of the parties is essential,"[10] will do well to look far beyond the language of the provision itself.

In order to eliminate some of the interpretive difficulties and in a misguided attempt to protect the developmentally disabled from themselves, 23 states have enacted labelling statues which prohibit named classes of people from entering into marriage contracts.[11] The terminology employed ranges from practicaly meaningless (imbeciles, insane persons, idiots, lunatics and feeble-minded persons) to

[1]California, Delaware, Georgia, Idaho, Iowa, Kentucky, Maine, Massachusetts, Michigan, Minnesota, Mississippi, Missouri, Nebraska, Nevada, New Jersey, New Mexico, New York, North Carolina, North Dakota, Ohio, Oklahoma, Oregon, Pennsylvania, Rhode Island, South Carolina, South Dakota, Tennessee, Vermont, Virginia, Washington, West Virginia, Wisconsin, Wyoming.

[2]*See, e.g.*, IDAHO CODE ANN. §32-201 (1963).

[3]Idaho, Michigan, Nebraska, Nevada, New Mexico, New York, North Dakota, Oklahoma, Oregon, Wisconsin, Wyoming; *see also* Guam, Puerto Rico, Virgin Islands.

[4]*See* Jacobs, "The Right of the Mentally Disabled to Marry: A Statutory Evaluation," 15 J. OF FAM. L. 483, 465 (1976-77).

[5]*See* McCurdy, "Insanity as a Ground for Annulment or Divorce in English and American Law," 29 VIR. L. REV. 771, 772 (1943).

[6]*See* Clark, LAW OF DOMESTIC RELATIONS (1968), Section 2.15.

[7]*Id.*

[8]*See* UNIFORM MARRIAGE AND DIVORCE ACT 298 (1974 version).

[9]*See, e.g.*, IDAHO CODE ANN. §29-101 (1967).

[10]*Supra* note 8, at § 201.

[11]California, Delaware, Georgia, Idaho, Iowa, Kentucky, Maine, Massachusetts, Michigan, Minnesota, Mississippi, Missouri, New Jersey, North Carolina, North Dakota, Ohio, Pennsylvania, Rhode Island, South Carolina, Tennessee, Vermont, Virginia, West Virginia.

considerably more precise.[12] An intermediate group prohibits the marriage of institutionalized persons, persons with guardianship, mentally retarded persons and persons adjudged of unsound mind.[13] The significant distinction is between categories that imply a former judicial determination of capacity and those that call for an *ad hoc* decision on the part of the government functionary responsible for the issuance of a marriage license or some other necessary formality.

A much smaller subgroup of the labelling statutes employ a classification scheme which indicates a predominantly eugenic purpose. Statutes of this type remove marital prohibitions if the disabled person is rendered incapable of procreation by virtue of his or her prior sterilization.[14]

In what manner the states hope or intend to enforce the above provisions is in many instances a mystery. A number of approaches have been enacted into law, most appearing haphazard and inefficient. The most frequently encountered are those providing criminal or civil penalties against clerks who issue marriage licenses to persons otherwise prohibited or against the person who performs the ceremony.[15] Some of these provisions only apply if the person had actual knowledge of the incapacity,[16] others presumably apply if the party knew or should have known.[17] Another approach directly penalizes a party attempting to be married to an "idiot or lunatic" by making the person liable for a misdemeanor.[18] Additionally, some states penalize "any person" who violates any laws relating to marriage.[19] Still another approach requires a health certificate attesting to necessary capacity before a marriage license can be issued.[20] Finally 30 states simply have no enforcement provisions at all.[21] Clerks in these states, required to enforce and abide by the state's statutory requirements, are presumably free to withhold licenses should they be of the opinion that an individual is covered by some statutory disability. On what basis these decisions are to be made (considering the vague disabling terminology) and what expertise clerks possess in this context are administrative mysteries.

Provisions Allowing Annulment On The Basis of Developmental Disability

Thirteen states which have enacted no outright prohibitions concerning the right to marry, instead allow the judicial dissolution or annulment of a marriage on the grounds of mentally disability.[22] In addition, the states which do prohibit certain marriages, with the exception of 5,[23] also explicitly provide for annulment. Two distinctions are important with regard to these statutes. First, they can be separated, as were the prohibitive statutes, into those premised on contractual capacity and those employing a labelling approach. The states are about evenly split in this regard and a number of states employ both types. The charts indicate which form operates in each state.

A second, and more difficult, distinction concerns the effect of the legal incapacity on the married parties; some states declare the marriage void[24] while others terms it only voidable.[25] A third group declare that the marriage is void "from the time its nullity shall be declared by a court"[26] — a somewhat intermediate position. Early American case law addressing the effect of these statutes is inconsistent though generally the effect of the term void was construed to be a complete nullification of the marriage with or without court order.[27] A voidable marriage, on the other hand, presupposes that at least one party to the transaction has the power to nullify its effect but that in the absence of their intent to do so the contract will be successful. Where a party, with power to do so, does choose to rescind the contractual agreement a further question remains whether a judicial decree, as opposed to an information election, is necessary to do so. Case law is mixed on this point.[28] It may be in reaction to the validity of informal annulment that the

[12]Minnesota prohibits only persons under the guardianship or conservatorship of the Commission of Public Welfare, and only if the terms of the appointment specifically limit the right to marry. MINN. STAT. ANN. §517-.03 (Supp. 1978).

[13]*See, e.g.*, IOWA CODE ANN. §595.3(5) ("Mentally retarded persons or persons under guardianship as incompetents").

[14]*See, e.g.*, N.C. GEN. STAT. §51-12 (1976); N.D. CENT. CODE §14-03-07 (Supp. 1977).

[15]*See, e.g.*, GA. CODE ANN. §53-9901 (1974); MO. REV. STAT. §451.020 (1977).

[16]Georgia, Idaho, Maine, Nebraska, Oklahoma, Wyoming.

[17]*See* MICH STAT ANN. §25.6 (1974).

[18]*See, e.g.*, GA. STATE. ANN. §53-9901 (1974).

[19]*See, e.g.*, N.D. CENT. CODE §14-03-28 (1971).

[20]N.C. GEN. STAT. §51-12 (1976).

[21]Alabama, Alaska, Arizona, Arkansas, California, Colorado, Connecticut, Hawaii, Illinois, Indiana, Iowa, Kansas, Kentucky, Maryland, Massachusetts, Minnesota, Mississippi, Missouri, Montana, Nevada, New Hampshire, New Jersey, Ohio, Oregon, Rhode Island, South Carolina, South Dakota, Tennessee, Texas, West Virginia.

[22]Alaska, Arkansas, Colorado, Connecticut, District of Columbia, Hawaii, Illinois, Indiana, Montana, Rhode Island, South Dakota, Texas, West Virginia.

[23]New Mexico, Pennsylvania, South Carolina, Tennessee.

[24]*See, e.g.*, MICH. STAT. ANN. §25.81 (1974).

[25]*See, e.g.*, ALASKA STAT. §09.55.090(2) (1973).

[26]*See, e.g.*, ARK. STAT. ANN. §55-106 (1971).

[27]*See. e.g.*, Wightman, v. Wightman, 4 Johns. Ch. 343 (N.Y. 1820); Clark, *supra* note 6, at 97, 119-121.

[28]*Id.*, at §3.1.

intermediate type statute arose which, in essence, allows a marriage to be voided only by a court decree. The legal effect of the words "void when declared a nullity by a court" thus only acts to create a voidable marriage in most jurisdictions, though once the proper court decree is obtained its effect will relate back and the marriage will become void.[29]

As if this were not all complicated enough, two other statutory provisions muddle the area even further. The first are the, so-called, ratification provisions, which reduce the ability to avoid an otherwise prohibited or annulable marriage. The provisions provide that should the spouses cohabitate voluntarily after the disabled person "comes to reason" or, in Georgia, if a child is born to the couple, the marriage will no longer be voidable. Nine American jurisdictions have statutory ratification procedures.[30]

The final important question concerns which party or parties are authorized to petition for annulment. Traditionally, under general contract principles, only the incompetent party to a transaction was in a position to have it declared void.[31] Only mistake, failure of consideration or fraudulent inducement were available to the competent party as grounds for termination, not the other party's incapacity to consent at the time of formation.[32] In terms of the marital contract the issue has received differing treatment. Most states which have legislated in the area provide a remedy for either party; incompetents may be variously represented by parents, guardians, relatives, *guardians ad litem,* or may bring the action themselves.[33] A few jurisdictions deny a remedy to the competent spouse if he or she knew or should have known of the disability at the time of the marriage or if he or she subsequently learned of it and nevertheless voluntarily cohabitated.[34] Four states explicitly deny the competent party any remedy.[35] Finally, Puerto Rico provides that any interested person may petition for annulment.[36]

---

[29]*See* Am. Surety Co. v. Connor, 251 N.Y. 1, 166 N.W. 783 (1929).

[30]California, Georgia, Hawaii, Idaho, Nevada, North Dakota, South Dakota, Texas, Guam.

[31]*See* J. Calamere & J. Perillo, CONTRACTS 252 (2d ed. 1977).

[32]*See* McCurdy, *supra* note 21, at 789-90.

[33]*See, e.g.*, ALASKA STAT. ANN. §09.55.070 (1973).

[34]Delaware, District of Columbia, Kentucky, New York, Texas, Vermont, Virgin Islands.

[35]Indiana, Oklahoma, Vermont, Washington. In addition the language of the Mississippi statute might be construed to deny a remedy. *See* Miss. Code Ann. §93-7-3 (1972).

[36]P.R. LAWS ANN, tit. 31, §235 (Supp. 1977).

MARRIAGE AND ANNULMENT

State	Description of Persons Prohibited from Marrying	Exceptions to Prohibition	Enforcement of Prohibition	Grounds for Annulment Affecting the Developmentally Disabled	Who is Authorized to Petition for Judicial Dissolution
Alabama Code (1975)					
Alaska Stat. (1977)				Marriage voidable if either party was of unsound mind; §09.55.090(2) or incapable of consenting for want of sufficient understanding §25.05.031	Either party §09.55.070
Arizona					
Arkansas Stat. Ann. (1971)				When either party is incapable, from want of understanding, of consenting to the marriage it shall be void from the time its nullity shall be declared by a court §55-106	
California Civ. Code (West 1970)	Imbecile or insane person §4201 (Supp. 1978)			Incurable insanity[1] §4506, 4510 Marriage voidable if either party of unsound mind, unless such party, after coming to reason, freely cohabited with the other §4425 (c)	The party injured or the relative or conservator of the person of unsound mind §4426 (c)
Colorado Rev. Stat. (1973)				Party lacked capacity to consent due to mental incapacity or infirmity §14-10-111(1)(a)	Aggrieved party or the legal representative of the party who lacked capacity to consent §14-10-111 (IV)(2)(a)
Connecticut Gen. Stat. Ann. (West 1977)				Legal confinement in an institution because of mental illness for a total of 5 years within the 6 year period immediately preceding the complaint. Sec. 46b-40[2]	
Delaware Code (1974)	Person of any degree of unsoundness of mind or a present or former patient of an insane asylum unless certified as fit to marry by the asylum superintendent tit. 13, §101 (b)(1), (2)		$100 fine or imprisonment for 30 days tit. 13, §102	Insanity of either party tit. 13, §1551(5)[3]	Either party except the competent party must have been ignorant of the insanity at the time of marriage tit. 13, §1551(5)
District of Columbia Code (1973)				Party is an idiot or adjudged to be a lunatic §30-103 (Supp.V 1978)	Next friend of incompetent but not any competent party who knowingly and willfully contracted the illegal marriage §30-104
Florida Stat. Ann.					
Georgia Code Ann. (1974)	Person not of sound mind §53-102(Supp. 1977) idiot or lunatic §53-9901		Anyone who marries a party known to him to be an idiot or lunatic is guilty of a misdemeanor §53-9901	Inability to contract, §53-104 but only if no children result from marriage §53-601	Either party §53-602 Next friend for persons of unsound mind §53-603
Hawaii Rev. Stat. (1976)				One of the parties was an idiot or lunatic; §580-21(4) One of the parties is an idiot or insane person. (But not if the parties freely cohabited after restoration to sound mind) §580-26	Sane party or a relative of the idiot or lunatic or a person permitted to act as his or her next friend, or, if restored to reason, the lunatic himself. §580-26

State	Description of Persons Prohibited from Marrying	Exceptions to Prohibition	Enforcement of Prohibition	Grounds for Annulment Affecting the Developmentally Disabled	Who is Authorized to Petition for Judicial Dissolution
Idaho Code (1963)	Person incapable of making a civil contract §32-201 Includes person of unsound mind §29-101 (1967)		Anyone who shall presume to marry persons with knowledge of the incompetency of either party is guilty of a misdemeanor §32-406	If either party was of unsound mind unless that party, after coming to reason, freely cohabited with the other. §32-501 (3)	The party injured or the relative or guardian of the person of unsound mind §32-502 (3)
Illinois Ann. Stat. (Smith-Hurd. Supp. 1978)				Party lacked capacity to consent due to mental incapacity of infirmity ch. 40, §301(1)	Either party or the legal representative of the person who lacked capacity ch. 40, §302
Indiana Code Ann. (Burns 1973)				Either party is insane or idiotic, at the time of marriage §31-1-1-2[4] Party incapable of contracting through want of understanding §31-1-7-6	The incapable party §31-1-7-6
Iowa Code Ann. (West Supp. 1978-1979)	Mentally retarded persons or persons under guardianship as incompetents §595.3(5)			Either party was mentally ill or a mental retardate at the time of marriage §598.28(4)	Either party §598.30
Kansas					
Kentucky Rev. Stat. Ann. (Baldwin 1973)	Idiots or lunatics §402.020 (Supp. 1977)			Either party was an adjudicated incompetent at the time of marriage §402.020(Supp. 1977) Party lacked capacity to consent due to mental incapacity or deformity §403.120 (1)(a)	The party lacking capacity or legal representative or the capable party if they did not know of the incapacity §403.120 (2)(a)
Louisiana					
Maine Rev. Stat. (1964)	Persons who are mentally ill, feebleminded or idiots tit. 19§32		Willfully joining persons in marriage contrary to law punishable by a fine of $100 tit. 19, §3 (Supp. 1978-1979)	Marriages among persons prohibited from marrying are void tit. 19, §631 (Supp. 1978-1979)	Either party tit. 19, §632 (Supp. 1978-1979)
Maryland					
Massachusetts Ann. Laws (Michie/Law Co-op 1969)	Idiots, insane or feeble minded persons under commitment to an institution ch. 207, §5			Either party is an idiot, feeble-minded or insane person ch. 207, §5	Either party ch. 207, §14 (Supp. 1978)
Michigan Stat. Ann. (1974)	Person legally incapable of contracting §25.2 Includes insane persons, idiots and persons who have been confined to a public institution by a court unless certified as cured by two physicians §25.6		Any competent person who marries or aids or abets the marriage of any person prohibited from marrying is guilty of a felony §25.6	When either of the parties was insane or an idiot, marriage is void §25.81	Either party §25.83
Minnesota Stat. Ann. (West Supp. 1978)	Person under the guardianship or conservatorship of the commissioner of public welfare if the terms of the appointment limit the right to marry §517.03[5]	If commissioner of public welfare consents. §517.03(c)		Marriage voidable if a party was incapable of assenting to the marriage for want of capacity to consent and other party unaware of incapacity at time of marriage §518.02 Commitment of party pursuant to court adjudication under §§253A, 518.06	Either party or by legal representative of party who lacks capacity to consent §518.05
Mississippi Code Ann. (1972)	Persons insane or imbeciles §93-1-5			Insanity or idiocy of either or both parties §93-7-3 Either party incapable, for want of understanding, or consenting §93-7-3(d)	Guardian or next friend of insane's spouse, provided action is brought within 6 months after marriage §93-7-3(b)

State	Description of Persons Prohibited from Marrying	Exceptions to Prohibition	Enforcement of Prohibition	Grounds for Annulment Affecting the Developmentally Disabled	Who is Authorized to Petition for Judicial Dissolution
Missouri Rev. Stat. (1977)	Persons insane, mentally imbecile or feeble minded §451.020		Person who issues license is guilty of a misdemeanor §451.020	Marriage is void if either party is insane, mentally imbecile or feeble-minded §451.020	
Montana Rev. Codes Ann. (Supp. 1977)				Party lacked capacity to consent due to mental incapacity or infirmity §48-311(1)(a)	Either party, but action must be commenced within one year of obtaining knowledge of the condition §48-311(2)(a)
Nebraska Rev. Stat. (1974)	Persons lacking contractual capacity §42-101		Any person who solemnizes a marriage with knowledge of a legal impediment is guilty of a misdemeanor §42-113 (Supp. 1977)	Marriage void when either party is insane or mentally incompetent to enter into the marriage relation §42-103	Either party §42-119
Nevada Rev. Stat. (1977)	Persons lacking contractual capacity §122.010			Party lacks capacity to consent due to want of understanding unless after restoration of reason of the incapacitated party the parties freely cohabited §125.330	
New Hampshire					
New Jersey Stat. Ann. (West 1971)	Imbeciles, persons of unsound mind, inmates of insane asylum who have not been satisfactorily discharged §37:1-9			Either party lacked capacity to consent due to want of understanding §2A:34-1(d)	
New Mexico Stat. Ann. (1962)	Persons incapable of contracting §57-1-1		Person who performs marriage may be fined §57-1-8 Deceiving or attempting to deceive officials for purposes of unlawfully marrying punishable by $100 fine or 60 days in jail §57-1-17		
New York Dom. Rel. Law (McKinney 1977)	Persons incapable of contracting §10		Town or city clerk who violates any statutory provision concerning marriage is guilty of a misdemeanor §22	Either party is mentally retarded or mentally ill §140 Party incapable for want of understanding §7	By mentally illl person after restoration of sound mind, or relative or next friend, or other party if unaware of mental illness, but not allowed if cohabitation after restored to sound mind; §140 Either party if incurable insanity for 5 years §140(f)
North Carolina Gen. Stat. (1976)	Person adjudicated an idiot, imbecile, mental defective or of unsound mind §51-12	If sterilized §51-12	Health certificate required showing sound mind §51-9 or Sterilization §51-12	Marriage void if either party incapable of consent for want of understanding §51-3 (Supp. 1977)	Either party §51-4
North Dakota Cent. Code (1971)	Person incapable of consenting §14-03-01(Supp. 1977) Any woman under 45, and man of any age if mentally deficient, insane, or afflicted with hereditary insanity §14-03-07 (Supp. 1977)	Woman over 45 may marry, regardless of mental condition §14-03-07	Any person violating any laws relating to marriage is guilty of a misdemeanor §14-03-28 (Supp. 1977)	Either party was of unsound mind unless, after coming to reason, they freely cohabited with the other §14-04-01(3)	Injured party or the relative or guardian of the person of unsound mind §14-04-02(3)
Ohio Rev. Code. Ann. (Page Supp. 1977)	Imbecile or insane person §3101.06			Either party adjudicated to be mentally incompetent unless cohabitation after restoration to competency §3105.31	Aggrieved party or relative or guardian of incompetent person §3105.32(D)

State	Description of Persons Prohibited from Marrying	Exceptions to Prohibition	Enforcement of Prohibition	Grounds for Annulment Affecting the Developmentally Disabled	Who is Authorized to Petition for Judicial Dissolution
Oklahoma Stat. Ann. (West 1954)	Persons incapable of contracting tit. 43, §1		Person knowingly solemnizing a prohibited marriage is guilty of a felony tit. 43, §14	Either of the parties to the marriage are incapable, for want of understanding of contracting such marriage, but cohabitation after restoration of reason is a defense tit. 12, §1283 (1961)	The incapable party or his or her parent or guardian tit. 12, §1283 (1961)
Oregon Rev. Stat. (1977)	Persons incapable of contracting §106.010			Either party is incapable of contracting due to lack of sufficient understanding §106.030, 107.015	
Pennsylvania Stat. Ann. (Purdon Supp. 1978-1979)	Persons who are weak-minded, insane or under guardianship as a person of unsound mind tit. 48, §1-5(d)[a]	If a judge of orphans' court shall decide it is in best interest of public to issue license tit. 48, §1-5(d)	If license issued in violation of §1-5(d), $50 fine for each conviction. To $100 fine and imprisonment for applicants misrepresenting facts tit. 48, §1-21		
Rhode Island Gen. Laws (1970)	Persons under the control of a guardian §15-2-11	Consent of guardian or court order §§15-2-10; 15-2-11		If either party is an idiot or lunatic the marriage is void §15-1-5	
South Carolina Code (1976)	Mentally incompetent persons §20-1-10				
South Dakota Compiled Laws Ann. (1977)	Incapable of making civil contract §25-1-1			Either party was of unsound mind, except if after coming to reason the parties freely cohabited §25-3-2	The injured party, or the relative or guardian of the person of unsound mind §25-3-2
Tennessee Code Ann. (1977)	Insane persons or imbeciles §36-411				
Texas Fam. Code Ann. (Vernon 1974)				As a result of mental disease or defect a party did not have ability to consent to the marriage or understand the nature of the marriage ceremony and has not since regained reason and freely cohabited §2.45(a)	Either party, guardian or next friend unless competent party knew or should have known of the defect and freely cohabitated §2.45(a)
Utah					
Vermont Stat. Ann. (1974)	Persons under guardianship or *non compos mentis* tit. 18, §5142 (Supp. 1978)	Marriage allowed if guardian provides written consent tit. 18, §5142	Clerk who issues license may be fined tit. 18, §5141 (Supp. 1978)	Either party is an idiot or lunatic tit. 15, §512	A relative of the idiot or lunatic or the person, after restoration of reason, provided that the parties have not freely cohabited after restoration tit. 15, §514
Virginia Code (1976)	Feeble minded or insane persons §20-46	Unless a woman over 45 or a man and marrying a woman over 45 §20-46	Fine of $100; jail term to 90 days or both for violating this law §20-46	Either party lacked capacity to consent due to mental incapacity or infirmity §20-45.1	Either party §20-45.1
Washington Rev. Code Ann. (1961)	Persons incapable of making civil contract §26.04.010		Any person knowingly violating the act shall be punished by a fine of up to $1000 or imprisonment for up to 3 years §26.04.230	Either party was incapable of consenting for want of sufficient knowledge §26.04.130	The party laboring under the disability §26.04.130
West Virginia Code (1976)	Insane person, feebleminded person, idiot, imbecile §48-2-1			Either of the parties was an insane person, feebleminded person, idiot or imbecile §48-2-1	Either party §48-2-2
Wisconsin Stat. Ann. (West Supp. 1978-1979)	Person who for want of understanding is incapable of assenting to marriage §245.03(1)		Fines §245.30	A party who lacks capacity to consent because of mental incapacity or infirmity §247.03(1)	Either party or the legal representative of the party lacking capacity §247.03(1)

State	Description of Persons Prohibited from Marrying	Exceptions to Prohibition	Enforcement of Prohibition	Grounds for Annulment Affecting the Developmentally Disabled	Who is Authorized to Petition for Judicial Dissolution
Wyoming Stat. (1977)	Persons incapable of contracting §20-1-102		Performing marriage ceremony with knowledge of a legal impediment is a misdemeanor §20-1-108	Either party is mentally incompetent at time of marriage §20-2-101 (a)(ii)	Either party §20-2-101(c)
American Somoa					
Guam Civ. Code (1970)	Persons incapable of contracting §55			Either party was of unsound mind, unless the party, after coming to reason, freely cohabited §82(3)	The party injured or a relative or guardian of the person of unsound mind §83(3)
Puerto Rico Laws Ann. (1967)	Persons incapable of contracting tit. 31, §221. Includes persons of unsound mind tit. 31, §232 (2); and persons suffering from insanity or idiocy tit. 31, §235 (Supp. 1977)			Either party is suffering from insanity or idiocy tit. 31, §235 (Supp. 1977)	Prosecuting attorney or any interested person tit. 31, §235 (Supp. 1977)
Virgin Islands Code Ann. (1964)	Person incapable of contracting tit. 16, §31 (1967)			Either party is an idiot or person adjudged a lunatic tit. 16, §2	The next friend of an idiot or lunatic or the competent party if they did not know of the condition at the time of contracting tit. 16, §3

[1]Spouse not relieved from any obligation imposed by law for support of the spouse who is incurably insane. CAL. CIV. CODE §4510 (West 1970).

[2]See CONN. GEN. STAT. §46-40 (1977) for procedures. Court may order support payments for the benefit of the mentally ill spouse after dissolution. CONN. GEN. STAT. §46-53 (1977).

[3]No decree shall relieve a spouse from any obligation imposed by law as a result of the marriage for the support or maintenance of a spouse adjudicated to be mentally incompetent. DEL. CODE tit. 13, §1518 (Supp. 1977).

[4]These marriages are declared void. It is unclear whether any proceeding is required. Compare language with §31-1-7-1 (1973).

[5]Commissioner must approve unless the marriage is not in the best interest of the ward and the general public. MINN. STAT. ANN. §517.03 (West Supp. 1978).

[6]Judge of the Orphan's Court may allow marriage if in the best interests of the person and the public.

CONCLUSION

As demonstrated in the Introduction, the facts do not support the assertion that there is a compelling state interest that justifies legislating discrimination against handicapped persons in the areas of the fundamental rights of marriage, annulment, and divorce. Thus, those states which still have discriminatory classifications should reconsider their proscriptions and revise them in light of the following principles:

1. The same state domestic relations law applicable to the population generally should apply to persons deemed handicapped. Thus, no longer should statutes allow disparate treatment of an individual because of a label such as "mentally ill," "mentally retarded," "insane," "persons in institutions," etc. In this manner, the freedom to make family related decisions will not be unfairly denied to a particular class of persons who are, in fact, able to make such decisions for themselves.
2. Implicit in equal treatment for all persons in this area of the law is the recognition that handicapped persons must not be over-protected either. Thus, for example, a non-handicapped person wishing to divorce a spouse who happens to be handicapped should be able to do so. Since, however, divorce laws predicated on a theory of fault may require the labeling and hence stigmatizing of the handicapped person in the course of the divorce proceeding, states are urged to adopt a no-fault divorce law. If this is not feasible, states should eliminate mental disability as a ground for or consideration in granting a divorce. The issue should never be a label but rather should be the behavior of one partner toward the other.
3. Since labels should never be used as a basis for determining whether to grant a request to marry, annul or divorce, states should rely on the behavior of the two individuals. Since this would entail contractual considerations, it is important that the test of capacity to enter into a marriage contract be explicitly defined in the statute to avoid ambiguity and narrowly drawn to exclude those capacities that are irrelevant to determining the capacity to enter into a contract to marry. Such a definition would comport with the intent of the drafters of the Uniform Marriage and Divorce Act and would eliminate the confusion that exists in many state laws in this area today. In addition, because a capacity to consent standard, no matter how narrowly drawn, may prohibit a marriage that would benefit both parties, there should be written into the law a presumption of competence that would have to be affirmatively disproven.
4. In addition to eliminating labeling from these laws, states should also eliminate sterilization as a condition precedent to the granting of a marriage license. As already noted in the Introduction, eugenic considerations have not been shown to be valid. Thus, their use invidiously discriminates against handicapped persons.
5. Since marriage is a supportive institution and ultimately benefits both partners and society, laws which allow third parties the right to bring an annulment action should be repealed unless the third party is the guardian of a partner who is a minor. Relatedly, because all laws in this area are to be evenly applied to all individuals without discrimination expressed or implied, both partners should be allowed to bring an annulment action pursuant to the statutory criteria. As noted in point 3, the criterion relating to mental capacity should be the capacity of the individual to have entered into a marriage contract.
6. Finally, because enforcement of marriage laws is often arbitrary and in some cases capricious, it is important to insure that the discretion of government officials to deny a license be structured as much as possible. Thus, the law should establish a presumption that an applicant for a marriage license has the necessary mental capacity to enter into a marriage contract. When the official does not believe that the applicant has the requisite capacity, the official may require the individual to have an examination by a qualified psychologist or psychiatrist of that individual's choice and have that professional sign an affidavit attesting to that individual's capacity. If the professional attests to the applicant's capacity, the government official must grant the license. Such a procedure insures that a government official's denial of a marriage license is supported by the judgment of an expert.

In sum, the laws pertaining to marriage must have a neutral effect on all citizens, regardless of their perceived handicap. Statutes which are not applicable equally to all must be repealed forthwith.

1.2 RELINQUISHMENT AND TERMINATION OF PARENTAL RIGHTS

Table of Contents

INTRODUCTION

The right of parents to raise their children is protected by both American tradition and the Constitution.[1]

> It is plain that the interest of a parent in the companionship, care, custody, and management of his or her children "come[s] to this Court with a momentum of respect lacking when appeal is made to liberties which derive merely from shifting economic arrangements. . . ." The rights to conceive and to raise one's children have been deemed "essential," Meyer v. Nebraska, 262 U.S. 390, 399 (1923), "basic civil rights of man," *[sic]* Skinner v. Oklahoma, 316 U.S. 535, 541 (1942), and "[r]ights far more precious . . . than property rights," May v. Anderson, 345 U.S. 528, 533 (1953) . . . . The integrity of the family unit has found protection in the Due Process Clause of the Fourteenth Amendment, Meyer v. Nebraska, *supra,* at 399, the Equal Protection Clause of the Fourteenth Amendment, Skinner v. Oklahoma, *supra,* at 541, and the Ninth Amendment, Griswold v. Connecticut, 381 U.S. 479, 496 (1965) (Goldberg J. concurring).[2]

Increasingly recognized as well, are the rights of a child:

> 1. To receive parental love and affection, discipline and guidance, and to grow to maturity in a home environment which enables him [or her] to develop into a mature and responsible adult;
> 2. To be supported, maintained, and educated to the best of parental ability . . .;
> 3. To be regarded as a *person,* within the family, at school and before the law;
> 4. To receive fair treatment from all authority; [and]
> 5. To be heard and listened to . . . .[3]

Despite the weight which this interlocking web of rights and responsibilities has been accorded, the standards and procedures to be applied when a parent-child relationship is severed by force of law[4] have only recently begun to receive serious attention. The procedural safeguards and other protections generally accorded the waiver or denial of comparable rights have not generally been included in the process required for a parent to relinquish his or her parental rights or for a court to terminate involuntarily a parent's responsibility for raising his or her child.[5] As is discussed in greater detail below, for parents with a developmental disability, or mental illness, the impact of the vagueness of the current standards for termination and the absence of a full explanation of the ramifications of relinquishment or termination is intensified by provisions which establish mental disability as a ground for termination and which make the consent of a mentally disabled parent unnecessary before his or her child is released for adoption.

Adoption

Since at least the time of Moses, biological parents have decided to relinquish their children and place them so that they might be nurtured and loved by others.[6] Yet until little more than a century ago, these procedures were apparently wholly unrecognized by the legal system of common law countries.[7] Even today, biological parents in many American states are able to cast off their parental rights and responsibilities with far less effort than it took Moses' mother to weave and seal the basket and place it in the bullrushes. In many jurisdictions, all that has been required is the signature of one or two witnesses. In some in-

---

1. See *e.g.,* Ginsberg vs. New York, 390 U.S. 629 (1968); Wisconsin vs. Yoder, 406 U.S. 205 (1971); Quilloin vs. Walcott, 434 U.S. 246 (1978).
2. Stanley vs. Illinois, 405 U.S. 645, 651 (1972).
3. Foster and Freed, *A Bill of Rights for Children,* 6 FAMILY LAW QUARTERLY 318 (1972).
4. Parental rights may also be abrogated through natural causes — *e.g.,* the child attains majority — and when a child meets the criteria for emancipation — e.g., by getting married. See V. DEFRANCIS, TERMINATION OF PARENTAL RIGHTS 11 (American Humane Association 1971). H. BEASER, RUNAWAY CHILDREN — FROM WHAT TO WHERE: THE LEGAL STATUS OF RUNAWAY CHILDREN, 55-71 (1975).
5. S. BRAKEL and R. ROCK, THE MENTALLY DISABLED AND THE LAW, 238 (Rev. Ed. 1971); DEFRANCIS, *id.,* at 13.
6. EXODUS 2.
7. BRAKEL and ROCK, *supra* note 5 at 234; Katz, *The Adoption of Baby Lenore: Problems of Consent and the Role of Lawyers,* 5 FAMILY LAW QUARTERLY 405, 406 (1971).

stances the signature must be in the presence of a notary public.[8] Often this procedure has been performed in the hospital within a few days after the birth of the child with the adoption agency representative or caseworker serving as a witness.[9] While this practice serves to free the child quickly for permanent place, it does little to assure that the relinquishing parents understand the full implications of their actions. Consequently, there have been an increasing number of legal challenges to adoption focusing on whether the biological parents' consent was voluntary and informed.

> The recent challenges which have been sustained were predicated on finding that the parent was too weighed down by the emotional pressures and tensions surrounding the decision to surrender the child to know what she was doing. These mothers claimed they were too distraught by the crisis, or too pressured by relatives and others, to give genuine consent.[10]

The result has been legal court proceedings which have often been time-consuming for the courts and traumatic for the biological parents, the foster parents and the child.[11]

For parents with a mental disability, particularly those who have mental retardation, the inadequacy of the procedures for assuring a knowing and voluntary waiver of their parental rights is even more threatening than it is for other parents. Not only are they even more susceptible to pressure from authority figures, but they are less likely to understand the long-term ramifications of relinquishing their parental rights absent a patient explanation in the language, terms, and mode of communication which they are able to understand.[12] However, in many states, mentally disabled parents are not even given an opportunity to decide since their consent to the relinquishment of their parental rights may be waived, or the consent of a guardian or other third party accepted instead, without a formal termination proceeding.[13] These provisions are probably designed to protect the rights of adopting parents when the biological parent is incapable of consenting.[14] But, neither the definition of competency nor the procedures for determining competency in most adoption and guardianship statutes provide assurance that a court or third party decision to relinquish a mentally disabled person's parental rights will be limited to only those persons who are truly unable to provide an informed consent.[15] Hence, the adoption remains open to challenge unless other criteria justifying termination of parental rights have been proven.

For children with a developmental disability or mental illness who are without parents, a different problem arises. Current estimates indicate approximately 15% of the children in foster homes and institutions have a developmental disability.[16] These children have often been considered unadoptable.[17]

> The result for many of them has been a precarious existence without their parents or a barren childhood in detention centers, institutions and temporary foster homes, where they grow older but remain psychologically immature in the absence of the stable family support they need.[18]

In addition, a few states permit the adoption of a developmentally disabled child to be annulled if the condition existed prior to the adoption and the adopting parents were not informed of it. The period during which annulment can be obtained may be as long as 5 years.[19] Most of these statutes pay little attention to the needs and rights of the child either during or following the annulment proceeding.[20] However, over the last dozen years, most states have enacted some form of adoption subsidy program[21] designed to:

> Make possible through public subsidy the appropriate adoption of each child . . . who is legally free for adoption and who otherwise, because of special circumstances, might not be adopted.[22]

In general, these laws seek to compensate adopting parents for the additional cost of the medical, mental health, educational and other habilitative services which may be required by the disabilities

8. See Review of State Statutes, *infra*.
9. *DEFRANCIS, supra* note 4 at 14.
10. *Id.; see e.g.,* Scarpetta vs. Spence Chapin Adoption Service, 28 N.Y. 2d 185, 265 N.E. 2d 787, 321 N.Y.S. 2d 65 (1971); In re Adoption of P, Morris County Court — Probate Division No. M-14-379 (1972), as reported in Polow, THE LAWYER IN THE ADOPTION PROCESS, 6 FAMILY LAW QUARTERLY 72 (1972).
11. *Id.*
12. *See* AMERICAN ASSOCIATION ON MENTAL DEFICIENCY, (AAMD), CONSENT HANDBOOK, 24-27 (1977).
13. See Review of State Statutes, *infra;* P. Wald, *Personal and Civil Rights: Principle Paper,* in PRESIDENTS COMMITTEE ON MENTAL RETARDATION, (PCMR), THE MENTALLY RETARDED AND THE LAW 2, 14 (1976).
14. BRAKEL and ROCK, *supra* note 5, at 235.
15. *Id.;* and Chapter 7: Guardianship and Conservatorship, *infra,* Charts I and III.
16. Clement, *Adoptive Families Sought for Children with Developmental Disabilities,* 1 D.D. POLESTAR No. 8 (1980); see Katz and Gallagher, *Subsidized Adoption in America,* 10 FAMILY LAW QUARTERLY 3 (1976).
17. *Id.*
18. Katz and Gallagher, *supra* note 16.
19. BRAKEL and ROCK, *supra* note 5, at 237, 248-249.
20. *Id.,* at 237; *see also* Katz, *supra* note 7, at 412.
21. Chart II *infra;* and Katz and Gallagher, *supra* note 16, at 7.
22. Katz and Gallagher, *supra* note 16, at 4.

which the child has.[23] While thousands of mentally disabled children continue to be in permanent out-of-home placements, these statutes at least remove a major stumbling block to their adoption.

Termination of Parental Rights

All American jurisdictions authorize their juvenile or family courts to terminate parental rights over the parents objections.[24] The statutes arise from the *parens patriae* philosophy under which the state has the responsibility to intervene "into private family life when necessary to assure that the welfare of the child . . . [is] being served."[25] Most commonly, criteria are set forth which focus on parental conduct or status including abandonment, mental deficiency, continuing neglect, nonsupport, and immorality or drunkenness.[26] While the court's decisions are to be governed strictly by the "best interests of the child," current statutes generally fail to require proof of harm or the imminent threat of harm to the child. Neither do they require proof that termination of parental rights is likely to improve the child's chances for a permanent or at least long-term placement much less the opportunity to develop a psychological parent-child relationship.[27] Courts are left to make the delicate and difficult decision whether to terminate parental rights with little statutory guidance and few alternatives. As a result, parental rights have been abrogated on the basis of lifestyle, values, poverty, or lack of available services,[28] and children relegated to a foster care system in which they are more likely to suffer a series of short-term placements until they attain majority, than to find a permanent home.[29]

The effect of these vague statutes on parents who have a mental disability is illustrated by the saga of the Alsager family. In June, 1969, following complaints from neighbors, an earlier finding of neglect concerning their eldest child, and a 20-minute visit to their home by a juvenile court probation officer, the six children of Charles and Darlene Alsager were removed to the Polk County, Iowa Juvenile Home. The Alsagers were a poor, apparently mentally retarded couple.[30] Within a week after the children had been initially removed, they were found to be neglected and ordered to remain in county custody pending placement in a foster home or institution. Less than a month later, a termination petition was filed by the child probation officer alleging that the Alsagers:

> Substantially and continuously and repeatedly refused to give their children necessary parental care and protection; and were unfit parents by reason of conduct detrimental to the physical or mental health or morals of their children.[31]

In September, 1969, the court found "adequate and sufficient cause" to terminate the Alsager's parental rights but declined to do so as "long as there is any substantial hope that the parents will be able to improve to the extent that they can provide even minimal care." The two eldest children were then returned to their parents custody. Following an interim review hearing, a final termination was held on May 22, 1970, at which the Alsager's parental relationship to five of their six children was severed. There is no mention of what if any services were provided to the family to help the Alsagers improve as parents. The termination order was appealed and affirmed by the Iowa Supreme Court.[32]

A Federal civil rights proceedings was initiated, and in 1975 the U.S. District Court ruled that the Iowa termination statute was unconstitutionally vague.[33] In addition, it found that the Alsagers were denied adequate notice and that the preponderance of the evidence standard of proof was too low. The Court of Appeals affirmed the decision but limited its basis to the inadequacy of the evidence and the lack of notice.[34] Between the initiation of the case and the end of the proceedings, the four

23. For more detailed descriptions, see Chart II *infra* and Katz and Gallagher, *supra* note *16*.
24. NATIONAL TASK FORCE TO DEVELOP STANDARDS AND GOALS ON JUVENILE JUSTICE AND DELINQUENCY PREVENTION, A COMPARATIVE ANALYSIS OF STANDARDS AND STATE PRACTICES: ABUSE AND NEGLECT, 115 (1977).
25. Ketcham and Babcock, *Statutory Standards for the Involuntary Termination of Parental Rights,* 29 RUTGERS LAW REVIEW 530, 534 (1976).
26. NATIONAL TASK FORCE, *supra* note 24, at 184-185.
27. Ketcham and Babcock, *supra* note 25, at 543; NATIONAL ADVISORY COMMITTEE ON JUVENILE JUSTICE AND DELINQUENCY PREVENTION (NACJJDP), REPORT TO THE ADMINISTRATOR ON STANDARDS FOR JUVENILE JUSTICE AND DELINQUENCY PREVENTION, §3.185 (1976).
28. *See* NACJJDP, *id.* at Commentary to §3.113; *see e.g.,* Bross, *Termination of the Parent Child Legal Relationship in Colorado,* 7 COLORADO LAWYER, 363 (1978); Roe vs. Conn., 417 F. Supp. 769 (N.D. Ala. 1976); Alsager vs. District Court, Polk County, Iowa, 406 F. Supp. 10 (D. Iowa 1975).
29. Ketcham and Babcock, *supra* note 25, at 543; V. PIKE, S. DOWNS, A. EMLEN, G. DOWNS, D. CASE, PERMANENT PLANNING FOR CHILDREN IN FOSTER CARE: A HANDBOOK FOR SOCIAL WORKERS (1975).
30. The Iowa Supreme Court stated that material facts in the Alsager Case were identical to those in In re Interest of McDonald 201 N.W. 2d 447 (Ia. 1972) which involved a couple with mental retardation. In re Interest of Franklin, 201 N.W. 727 (Ia. 1972). But, the U.S. District Court characterized the Alsagers as "affectionate parents of below average but by no means inadequate intelligence," and states that the record of the case "reveals no intelligence testing of the Alsagers." Alsager vs. District Court of Polk County Iowa, 406 F. Supp. 10, 21023 (S.D. Ia. 1975), *aff'd in part* 545 F.2d 1137 (8th Cir. 1976).
31. See IOWA CODE §§ 232 41(2)(b) and (e) (1973). This history of the case is drawn from the opinion of the U.S. District Court. *Id.*
32. In re Interest of Franklin, 201 N.W. 2d. 727 (Ia. 1972).
33. 406 F. Supp. at 22.
34. 545 F.2d 1137 (8th Cir. 1976).

younger Alsager boys were placed in at least 15 foster homes and had eight juvenile home placements. In light of this record the federal District Court concluded:

> Termination . . . thus failed to provide the Alsager children with either stable or improved lives. . . . One lesson emerges clearly from this sad testament, termination is a drastic final step which when improvidently employed can be fraught with danger. Accordingly, to preserve the best interest of both parents and children. . . terminations must only occur where more harm is likely to befall the child by staying with his [or] her parents than by being permanently separated from them.[35]

Unfortunately, the Alsagers have not been the only developmentally disabled parents to lose custody of their children on the basis of questionable evidence.[36] As noted by one commentator, whether or not the facts justified termination, the general observations of some courts regarding the incapacity of retarded persons to act as parents is not justified.

> The general principle is that neglect by retarded parents must be proven in the same way as neglect by other parents. It may not be presumed from mental retardation.[37]

While this principle has found favor with some courts[38] and in recent sets of standards concerning termination of parental rights,[39] it is not reflected in the current state legislation governing voluntary or involuntary termination proceedings.

REVIEW OF STATE LEGISLATION

This statutory survey is based on the analysis of the statutes of each of the 50 states, the District of Columbia, American Samoa, Guam, Puerto Rico and the Virgin Islands regarding relinquishment and termination of parental rights and adoption subsidies. It is not an examination of the entire juvenile and civil codes of these jurisdictions. Hence, an indication that there is no provision requiring a particular procedure in a jurisdiction should be understood to refer only to the adoption and termination laws, since the matter in question may be included in another chapter, article, or set of rules which were not examined. It should be noted that the provisions on relinquishment and termination are not always distinct. In some states, the provisions on adoption govern both voluntary and involuntary severance of the parent-child relationship.[40] In most, however, the provisions on voluntary relinquishment are limited to those for adoption, and the authority to terminate parental rights involuntarily is included within or in close proximity to the chapter on child neglect and abuse.

In this review and throughout this report, voluntary and involuntary abrogation of a parent-child relationship are dealt with separately. Chart I concerns the procedures for giving up a child for adoption which most affects parents with a developmental disability. Chart II indicates the agency responsible and the children eligible for adoption subsidy programs. Finally, Chart III lists the statutory grounds for terminating parental rights which specifically refer to mental disabilities, and the basic due process safeguards included in termination statutes.

Relinquishment of Parental Rights (Chart I)

This chart addresses two questions which particularly affect parents with a developmental disability or mental illness. The first concerns the procedures which are required to assure that the consent to give up a child is knowing and voluntary. The statutes of 36 jurisdictions include at least some mechanism responding to this question. Ten states (Indiana, Maine, Montana, Nebraska, New Mexico, North Carolina, Oklahoma, Oregon, Pennsylvania, and Wyoming) require only that the consent be in writing. Twenty-two require that the consent be signed before a judge (Connecticut, Colorado, Michigan, South Dakota, Utah, Washington, Wisconsin), a notary (Arizona, Arkansas, Delaware, District of Columbia, Florida, Hawaii, Idaho, Iowa, Massachusetts, Nevada, Vermont, West Virginia), or either a judge or notary (Alaska, California, Ohio). Of these jurisdictions, Arizona, District of Columbia, Florida, and Massachusetts plus Missouri require, in addition, that the consent documents be signed in front of witnesses. Arizona and Ohio also forbid execution of consent forms within 72 hours of the time

---

35. 406 F. Supp. at 23-24.
36. *E.g.,* State ex rel Paul vs. Department of Public Welfare 170 So. 2d 549 (La. Ct. of App. 1965); In re Rathbun, 128 Vt. 429, 266 A.2d 423 (1970); In re McDonald 201 N.W. 2d 447 (Ia. 1972); Matter of C.M., 556 P.2d 514 (Wyo. 1976); State of Utah in E. and B. vs. J.T. No. 15140 (Utah, April 1978) as reported in MENTAL DISABILITY LAW REPORTER 707 (1978). *See also* Malmquist, *The Role of Parental Illness in Custody Proceedings,* 2 FAMILY LAW QUARTERLY 360, 367-368 (1968).
37. P. Wald, *supra* note 13, at 15.
38. *See e.g.,* Jewish Child Care Association vs. Benjamin and Rhoda G., as reported in 4 MENTAL DISABILITY LAW REPORTER 104 (1980); State of Utah in the Interest of E. and B. vs. J.T., as reported in 2 MENTAL DISABILITY LAW REPORTER 707 (1978).
39. NACJJDP, *supra* note 27; NATIONAL ADVISORY COMMITTEE ON CRIMINAL JUSTICE STANDARDS AND GOALS, JUVENILE JUSTICE AND DELINQUENCY PREVENTION, §14.32 (1976); INSTITUTE OF JUDICIAL ADMINISTRATION/AMERICAN BAR ASSOCIATION JOINT COMMISSION (IJA/ABA), STANDARDS RELATING TO ABUSE AND NEGLECT, §8.2 (Tentative Draft, 1977).
40. *E.g.,* RHODE ISLAND GENERAL LAWS §15-7-7 (Cum. Supp. 1978).

the child is born in order to give the mother at least some time to recover before making the relinquishment decision. Of the states which require a judicial proceeding, 3 (Colorado, South Dakota, Wisconsin) require the Court to satisfy itself that the parent understands the consequences of relinquishing his or her parental rights, Michigan requires the court to explain to the parents what their rights are and the consequences of signing a release, and Washington state requires an investigation by a guardian *ad litem* or "next friend" appointed by the Court.[41]

The second issue concerns the circumstances in which the consent to the relinquishing of the parental rights of a developmentally disabled or mentally ill parent may be waived or given by another person. The statutes of 31 jurisdictions address this issue. More than half of these jurisdictions permit personal consent to be waived only when a court has declared a parent to be incompetent (Alaska, Arizona, Kentucky, Missouri, New Mexico, Ohio, South Dakota, Vermont, and Washington), mentally retarded or incompetent (Guam, Hawaii, Illinois, Indiana, North Dakota), developmentally disabled (California), or incompetent and unlikely to become competent (Florida, Nevada, New York). The statutes of 12 states simply require a person to be insane or incapacitated (Alabama, Arkansas, Georgia, Maine, Massachusetts), incompetent (Rhode Island and Tennessee), mentally ill or deficient (Oregon), or unfit (Minnesota, Nebraska, North Carolina, and Virginia). The remaining state (Michigan) requires that before a guardian of a mentally incompetent person may waive that person's parental rights, the guardian must obtain the specific authorization of the court to do so.

Adoption Subsidy (Chart II)

Thirty-four American jurisdictions now have adoption subsidy provisions. Most of these statutes place the authority to operate the program in the agency responsible for providing foster care.[42] The statutes generally list several factors which may make a child eligible for an adoption. Although developmental disabilities are not mentioned *per se*, a child with a developmental disability may fit within one or more of these factors. Twenty-nine of the statutes include physical disabilities in the list of factors; 28 include mental deficiency, and 20 cite "emotional disturbance." Although the emphasis is on the child, about half of the adoption subsidy statutes list the low income of the adopting parents as an eligibility criteria. A wide array of intended uses for subsidy funds are cited. However, three-quarters of the statutes specify "maintenance care," about half include medical and surgical services, and about a third list "special services."[43]

Termination of Parental Rights (Chart III)

All American jurisdictions have statutes authorizing the involuntary termination of parental rights. The grounds for severing the parent-child relationship vary widely, though in general, the court is given broad discretion in determining what course will best serve the interests of the child. This chart sets forth the statutory provisions which list mental disability as a basis for terminating parental rights, the procedures specified for determining whether a parent has a developmental disability or mental illness and its impact on the child, and the due process protections prescribed for the parents and the child.

Thirty-four jurisdictions mention or allude to a mental disability. The most common type of provision empowers the court to terminate parental rights when because of a mental deficiency, disability or condition the parents are unable to provide proper care for their child or discharge their parental disabilities, and that the condition is irremediable or permanent (Arkansas, Idaho, Illinois, North Dakota), or will continue for an extended period of time/cannot be cured within a reasonable period of time (Arizona, Colorado, Connecticut, Massachusetts, Michigan, Nebraska, Oregon, Pennsylvania, Rhode Island, Virginia, Wisconsin). The statutes in nine states include the "inability to provide proper care" language but contain no reference to a period for recovery or habilitation (California, Delaware, Hawaii, Iowa, Montana, New Hampshire, Ohio, South Carolina, Wyoming). Two jurisdictions (Guam and Utah) cite a condition of the parents which is "seriously detrimental to the child," and two others (Indiana and New Jersey) require only that the parent be "incompetent." Finally, the statutes of three jurisdictions (District of Columbia, Kentucky, Maryland) take a somewhat different tack by instructing courts to consider the mental health of all parties in determining the best interests of the child,[44] and Missouri requires that the child be repeatedly neglected and the parent be so mentally deficient as to be unable to act knowingly. It should be noted that the absence of a specific mention of a mental disability does not prevent its consideration under such broad terms as "unfit," without "proper parental care or control," or "best interest of the child"

41. Illinois permits a judicial investigation when there are allegations of fraud or duress.
42. Indiana, Michigan, Missouri and Nevada on the other hand invest this authority in the juvenile court.
43. For a detailed comparative analysis of adoption subsidy statutes and a discussion of the proposed Model State Subsidized Adoption Act, see Katz and Gallagher, *supra* note 16.
44. D.C. also requires consideration of the child's mental health.

Fifteen termination statutes specify procedures for determining whether a parent has a mental disability and its effect on his or her parenting. Eight of these states require medical or mental health examinations and the testimony by 2 doctors (California, Illinois), 2 psychiatrists (Delaware and Rhode Island), 2 examiners (Idaho), 2 psychiatrists or psychologists (New Hampshire), a doctor and a psychologist (New York), or a person(s) qualified to present psychological or psychiatric evidence (Kentucky). Six other jurisdictions (Alabama, Arizona, Connecticut, Hawaii, Massachusetts, Missouri) require only a social study. Wisconsin requires a determination of incompetence at an independent proceeding.

The first two due process protections, notice and a hearing, are required by all jurisdictions. With regard to the level of proof which the court must find before terminating a person's parental rights, 6 jurisdictions (Arizona, Idaho, Kentucky, Massachusetts, Oregon, Texas) require only a preponderance of the evidence, 9 require clear and convincing evidence (Alabama, Alaska, Iowa, Missouri, New Hampshire, New Mexico, North Carolina, Tennessee, Virginia), and one (Louisiana) requires proof beyond a reasonable doubt. Two states (New York, Washington) set different standards of proof depending on the grounds on which the termination action is brought. New York requires clear and convincing proof rather than a prepondernace of the evidence for termination of the parental rights of a mentally retarded parent whose child has been in an out-of-home placement. The Washington statute does not make a distinction on the basis of a mental disability.

Most jurisdictions make some provision for assuring representation either by an attorney or a guardian *ad litem*. About a third of the jurisdictions require appointment of counsel for parents subject to a termination proceeding who are indigent. Six others (Arkansas, Montana, Nevada, New Mexico, North Dakota, Texas) leave the appointment of counsel for a parent within the discretion of the court. Nine jurisdictions (Arizona, Connecticut, Delaware, Idaho, Kansas, Nebraska, New Hampshire, New York, Wisconsin) require appointment of a guardian *ad litem* for parents alleged to be incompetent. Three others (Illinois, North Dakota, Virginia) allow but do not require such appointments.

With regard to children involved in termination proceedings, 13 jurisdictions require appointment of counsel (Alabama,[45] District of Columbia, Illinois, Iowa, Kentucky, Louisiana, Maine, Minnesota, South Dakota, Utah, Washington, West Virginia), and 12 (Arizona, Arkansas, California, Colorado, Kentucky, Massachusetts, Michigan, Montana, Nevada, New Mexico, North Dakota, Texas) leave the appointment to the court's discretion. The pattern is reversea for the appointment of a guardian *ad litem* for the child. In 10 jurisdictions (District of Columbia, Illinois, Iowa, Kansas, Kentucky, Minnesota, New Hampshire, Tennessee,[45] Vermont,[45] Virginia) such appointments are mandatory, and in 14 jurisdictions (Alaska, Arizona, California, Georgia, Hawaii, Idaho, Indiana, Mississippi, Montana, Nebraska, North Dakota, Rhode Island, Texas[46] Washington) it is discretionary. In 3 states it is possible to appoint both counsel and a guardian *ad litem* for an allegedly incompetent parent (Idaho, Kansas, New Hampshire). Twelve jurisdictions permit or require dual appointments for the child (Arizona, California, District of Columbia, Illinois, Iowa, Kansas, Kentucky, Minnesota, Montana, Texas, Virginia, Washington).

Finally, the termination statutes of 37 jurisdictions authorize appeal of the court's decision.[47]

---

45. Only when the child's interests conflict with those of his or her parents.
46. Only if the interests of the child are not adequately served.
47. Appeals in the remaining jurisdictions may be permitted by other provisions in their juvenile codes or rules of civil procedure.

	I RELINQUISHMENT OF PARENTAL RIGHTS		II ADOPTION SUBSIDY	
STATE	**PROCEDURE FOR ASSURING THAT CONSENT TO ADOPTION IS VOLUNTARY AND INFORMED**	**WHEN MAY A THIRD PARTY CONSENT OR MAY CONSENT BE WAIVED**	**ADOPTION SUBSIDY PROGRAM**	**ELIGIBILITY FACTORS FOR SUBSIDY**
ALABAMA Code (1977)		Person insane or otherwise incapacitated from giving consent. §26-10-3		
ALASKA Stat. (1975 & Cum. Supp. 1979)	Consent executed in presence of court or one authorized to take acknowledgements. §20.15.606	Person judicially declared incompetent or mentally defective. §20.15.121(a)(2)(6)	Department of Health and Social Services §20.15.190	Physical or mental disability, emotional disturbance, recognized high risk of physical or mental disease, age, membership in a sibling group, racial or ethnic factors or any combination of these conditions. §20.15.240
ARIZONA Rev. Stat. Ann. (1974 & Cum. Supp. 1979)	Consent must be acknowledged before two witnesses at least 72 hours after the birth of the child.	Person declared incompetent. §8-106(A)(1)(a)	Department of Economic Security §8-142	Physical or mental disability, emotional disturbance, recognized high risk of physical or mental disease, age, sibling relationship, racial or ethnic factors, or any combination. §8-141
ARKANSAS Stat. Ann. (Cum. Supp. 1979)	Written consent must be executed. §56-206	Person is incompetent or mentally defective. §56-207	Administered by the Division of Social Services of the Department of Human Services. §56-132	Physical or mental disability, emotional disturbance, recognized high risk of physical or mental disease, age, sibling relationship, racial or ethnic factors; any combination of these conditions. §56-131
CALIFORNIA Civ. Code (West Cum. Supp. 1979); Welfare and Inst. Code (Cum. Supp. 1979)	Consent executed before court or authorized person. Civ. §226.1	Person declared developmentally disabled or mentally ill. Civ. §§232(a)(5) & (6)	Department of Social Welfare W&I §16118	Adverse parental background or a handicapped child, or a child age 3 or more. W&I §16116
COLORADO Rev. Stat. (1978)	Court must be satisfied relinquishing parent knows consequences of act. §19-4-101		Department of Social Services §26-7-102	Special, unusual or significant physical or mental handicap, or emotional disturbance, or such other conditon which acts as a serious barrier to adoption. §26-7-102
CONNECTICUT Gen. Stat. Ann. (West Supp. 1979)	Court must hold a hearing. If parent is found to be incompetent the consent must be cosigned by guardian *ad litem*. §§45-61c thru 45-61f		Commissioner of Human Resources §17-44B	Race, a physical or mental handicap, age, or sibling group and has been designated as hard to place by the Commissioner. §17-44A
DELAWARE Code (Cum. Supp. 1978)	Must be acknowledged. tit. 13, §906			
DISTRICT OF COLUMBIA Code (1973 and Supp. 1978)	Consent must be acknowledged and at least one witness. §16-304 §32-786		The Mayor §3-115(b)(1)	Age, race, ethnic background, physical or mental condition, or membership in a sibling group, or who has not been placed within 6 months. §3-115(b)(2)
FLORIDA Stat. Ann (West 1976 & Cum. Supp. 1978)	Written consent must be executed before a notary public and 2 witnesses. §63.082	Judicially declared incompetent person for whom restoration of competency is medically improbable. §63.072		
GEORGIA Code Ann. (Cum. Supp. 1979)		Person who is insane or otherwise incapacitated from surrendering rights. §74-405		
HAWAII Rev. Stat. (1976)	Written consent must be executed §578-2	Judicially delcared mentally incompetent or mentally retarded person. §578-2		
IDAHO Code (1979)	Consent must be acknowledged. §16-1506		Department of Health and Welfare §56-803	Ethnic background, race, color, age, sibling grouping or physical, mental, emotional, or medical handicap. §56-802(A)(1)

STATE	PROCEDURE FOR ASSURING THAT CONSENT TO ADOPTION IS VOLUNTARY AND INFORMED	WHEN MAY A THIRD PARTY CONSENT OR MAY CONSENT BE WAIVED	ADOPTION SUBSIDY PROGRAM	ELIGIBILITY FACTORS FOR SUBSIDY
ILLINOIS Ann. Stat. (Smith-Hurd Cum. Supp. 1979)	The consent is valid unless a court finds fraud or duress. ch. 4. §9.1-10	A person who is found mentally impaired, mentally ill, in need of mental treatment or mentally retarded, after examination by two (2) qualified physicians. ch. 4. §9.1-8	Department of Children and Family Services ch. 23, §505	Physically or mentally handicapped or older children who immediately prior to their adoption, were wards of the Department. ch. 23 §505
INDIANA Code Ann. (Burns Cum. Supp. 1979)	Consent must be in writing. §31-3-1-3	A person who has been judicially declared to be incompetent or mentally defective. §31-3-1-6(g)(5)	Department of Public Welfare of the county responsible for foster care of the adoptive child or children. §31-3-3-1	Ethnic background, race, color, language, physical, mental or medical handicaps, age, or because he or she is a member of a sibling group which should be placed in the same home. §31-3-3-1
IOWA Code Ann. (West Cum. Supp. 1978)	Consent must be signed before a notary public. §600.7		Department of Social Services §600.17	Physically or mentally handicapped, older or otherwise hard to place child. §600-17.
KANSAS Stat. (Supp. 1978)	Consent must be acknowledged before a notary public or a court of record. §59-2102-(5)		Department of Social Welfare §38-321	Age, racial or ethnic background, mental, emotional, or physical handicap or membership in a sibling group. §38-322-(a)
KENTUCKY Rev. Stat. Ann. (Baldwin. 1975 & Cum. Supp. 1978)		Person adjudged incompetent. §199.500(1)(a)	Department of Human Resources §199.555(2)	Physical handicap, a mental condition, or the age of the child. §199.555
LOUISIANA				
MAINE Rev. Stat. (Cum. Supp. 1979-1980)	Written consent required. tit. 19. §532	Person who is hopelessly mentally ill or intemperate. tit. 19. §532	Department of Human Services tit. 19, §541	Physically or mentally handicapped, emotionally disturbed, or who by virtue of age, or sibling relationship otherwise may not be adopted. tit. 19, §541
MARYLAND Ann. Codde (Cum. Supp 1979)			Local Department of Social Services art. 16§88B §88C	Physical or mental disability, emotional disturbance, recognized high risk of physical or mental disease, age, sibling relationship, racial or ethnic factors, any combination of these special circumstances. art. 16, §88A
MASSACHUSETTS Ann. Law (Michie Law Co-op Cum. Supp. 1978)	Written consent attested and subscribed before notary public in presence of two witnesses. ch. 210. §16	Person who is hopelessly insane. ch. 210. §3		
MICHIGAN Comp. Laws (Cum. Supp. 1979)	Release must be executed before a judge or referee who has explained to the parents their rights and the effect of signing the release. §710.29	Person who is mentally incompetent for whom a guardian has been appointed by a court and given the power to execute a release. §710.28	Probate Court §710.48	Child who was in foster care for not less than 4 months, physical, mental or emotional condition which existed before the adoption. §710.48
MINNESOTA Stat. Ann. (West Cum. Supp. 1979)		If parent or guardian is not fit to give consent. §259.24	Commissioner of Public Welfare §259.40(1)	Child's characteristics or circumstances make it difficult to provide an adoptive home without subsidy. §259.40(4).
MISSISSIPPI Code Ann. (1972 & Cum. Supp. 1979)			State Department of Public Welfare §93-17-57	Severe physical or mental disability, severe emotional disturbance, recognized high risk of physical or mental disease, any combination of these handicaps. §93-17-55
MISSOURI Ann. Stat. (Vernon 1977)	Written consent must be witnessed by two persons. §453.030	Persons adjudged incompetent. §453.040	Juvenile court §453.073	Those children previously considered suffering from physical handicaps or mental retardation, or those children belonging to minority racial and ethnic groups for whom adoptive homes are not readily available. §453.073
MONTANA Rev. Codes Ann. (Cum. Supp. 1977)	Written consent required. §61-205			

STATE	PROCEDURE FOR ASSURING THAT CONSENT TO ADOPTION IS VOLUNTARY AND INFORMED	WHEN MAY A THIRD PARTY CONSENT OR MAY CONSENT BE WAIVED	ADOPTION SUBSIDY PROGRAM	ELIGIBILITY FACTORS FOR SUBSIDY
NEBRASKA Rev. Stat. (1978)	Written consent required. §43-104	Person is incapable of consenting. §43-104 §43-105	Department of Public Welfare §43-117	Ward of the department with special needs. §43-117
NEVADA Rev. Stat. (1975)	Written consent acknowledged by person or persons consenting. §127.040	Person adjudged insane for two years and felt by court to be incurable. §127.040	Welfare Division of De- Department of Human Resources §127.186	Race, age, physical or mental problems where difficult to find adoptive parents capable of bearing full costs. §127.186
NEW HAMPSHIRE				
NEW JERSEY Stat. Ann. (Cum. Supp. 1979-1980)			Department of Youth and Family Service §30:4C-46	Physical or mental condition, race, age or membership in a sibling group, or for any other reason falls into the category of hard-to-place. §30:4C-46(4)
NEW MEXICO Stat. Ann. (Supp. 1975)	Written consent required. §22-2-25	Person judicially declared incompetent. §22-2-26	Health and Social Services Board §40-7-28(A)	Physically or mentall handicapped, emotionally disturbed, or who is in special circumstances by virtue of age, sibling relationship or racial background. §40-7-27(B)
NEW YORK Dom. Rel. Law (McKinney 1977)		Person presumed presently and forseeably unable to properly care for child due to mental illness or retardation. §111	Social Services §452 §453	Physical, mental, or emotional condition of such severity or kind which in accordance with department regulations would constitute a significant obstacle to adoption, and child who has not been placed within 6 months §451-2 §452-3
NORTH CAROLINA Gen. Stat. (Supp. 1977)	Written consent required. §48-9	If person is not qualified to consent.	Department of Human Resources §48-39.2	Physically or mentally handicapped, older or otherwise hard to place for adoption. §48-39.1
OHIO Rev. Code Ann. (Page Supp. 1978)	Consent must be executed before court or notary no less than 72 hours after birth of the child. §3107.08	Person who is judicially declared incompetent. §3107.07		
OKLAHOMA Stat. Ann. (West Cum. Supp. 1979-1980)	Written consent. tit. 10, §60.5		Department of Public Welfare tit. 10, §60.13a	Physical or mental condition which makes child hard to place. tit. 10, §60.13a
OREGON Rev. Stat. (1977)	Written consent required. §109.312	Person who is mentally ill or deficient. §109.322	Children's Service Division §418.330	Physical or mental condition, race, age or membership in a sibling group, or the financial needs of the child are beyond the economic resources of the family. §418.330
PENNSYLVANIA Stat. Ann. (Purdon Supp. Pamph. 1964 to 1978)	Written consent required. tit. 1, §1411 tit. 1, §1411		Department of Public Welfare tit. 1, §773	Child has not been adopted for at least 6 months or has a physical and/or mental handicap, emotional disturbance, or is difficult to place because of age, sibling relationship or ethnicity. §772
RHODE ISLAND Gen. Law (Cum. 1978)		Person who is incompetent. §15-7-7	Department of Social and Rehabilitative Services §15-7-25	Handicapped or hard to place. §15-7-25 §15-7-25
SOUTH CAROLINA Code (Cum. Supp. 1978)			Department of Social Services and the Children's Bureau §15-45-310	Physical or mental disability, emotional disturbance, recognized high risk of mental disease, age, sibling relationship, racial or ethnic factors, or any combination §15-45-320
SOUTH DAKOTA Codified Laws (1976)	Petition must be verified and court must determine that the parent is aware of the consequences of the proceeding. §25-5A-6 §25-5A-16	Person adjudged mentally incompetent. §25-6-4.3	Division of Social Welfare or other public agency §28-7-31	Needs of the children beyond the economic resources of the adopting family. §28-7-31

STATE	PROCEDURE FOR ASSURING THAT CONSENT TO ADOPTION IS VOLUNTARY AND INFORMED	WHEN MAY A THIRD PARTY CONSENT OR MAY CONSENT BE WAIVED	ADOPTION SUBSIDY PROGRAM	ELIGIBILITY FACTORS FOR SUBSIDY
TENNESSEE Code Ann. (1977)		Person who is incompetent. §36-08	Department of Human Services §14-105	Physical handicap, mental or emotional condition, or other serious impediments to adoption. §14-105
TEXAS Human Resources Code (1978)			Human Resources Department §47.002	Child who is difficult to place because of age (over 3), color, ethnic background, language, physical, mental or emotional handicap, or sibling relationship. §47.001
UTAH Code (1976)	Consent in writing before the court. §78-30-8			
VERMONT Stat. Ann. Cum. Supp. 1979)	Consent must be executed. tit. 15 §435	Person who is incompetent to have care and custody of the minor in the opinion of the probate court. tit. 15 §435		
VIRGINIA (Cum. Supp. 1979)	Consent must be acknowledged. §63.1-225	Person for whom valid consent is unattainable. §63.1-225	Local board of public welfare or social services may make subsidy payments. §63.1-238.2	Physical, mental, or emotional conditions existing prior to adoption, hereditary tendency, congenital problem, or birth injury leading to substantial risk of future disability or individual circumstances of the child related to age, racial, or ethnic background or close relationship with one or more siblings. §63.1-238.1
WASHINGTON Rev. Code Ann. (1961)	Investigation by guardian *ad litem* or "best friend" appointed by the court. §26.32.070	Person adjudged incompetent and has remained incompetent for at least a year prior to the adoption proceeding. §26.32.040		
WEST VIRGINIA Code (1976 & Cum. Supp. 1979)	Consent must be acknowledged. §48-4-1			
WISCONSIN Stat. Ann. (West 1979)	Consent must be in writing and the court must explain that consent is irrevocable and satisfy itself that the consent was given voluntarily. §48.84(2)			
WYOMING Stat. (1977)	Written relinquishment of child and written consent to adoption. §1-22-109			
AMERICAN SAMOA				
GUAM Civ. Code (1970)		Person adjudged feeble-minded. §224		
PUERTO RICO				
VIRGIN ISLANDS				

III TERMINATION OF PARENTAL RIGHTS

STATE	GROUNDS FOR TERMINATION REFERRING TO DEVELOPMENTAL DISABILITIES	PROCEDURE FOR DETERMINING INCOMPETENCE	NOTICE TO PARENT	HEARING REQUIRED	STANDARD OF PROOF	APPOINTED COUNSEL FOR CHILD? FOR PARENT?	GUARDIAN *AD LITEM* FOR CHILD? FOR PARENT?	APPEAL
ALABAMA Code (1975)		Social study prior to disposition tit. 12, §15-65	tit. 12, §15-53 §15-54	tit. 12, §15-65	Clear and convincing evidence tit. 12, §15-65(e)	For child-mandatory if conflict of interest tit. 12, §15-63 For parent-mandatory if request counsel tit. 12, §15-63(b)		tit. 12, §15-12
ALASKA Stat. (1975)			§47.10.030	§47.10.070	Clear and convincing evidence §47 10.080 (Supp. 1978)		For child-discretionary §47.10.050	§47.10.080
ARIZONA Rev.Stat.Ann. (1974)	Parent's mental deficiency makes parent unable to discharge parental responsibilities and there are reasonable grounds to believe the condition will continue for an indeterminate period. §8-533(3)	Social study prior to disposition §8-536	§8-535	§8-537	Preponderance of the evidence §8-537(b)	For child-discretionary §8-535(e)	For child-discretionary §8-535(d) For alleged incompetent parent-mandatory §8-535(d)	§8-543
ARKANSAS Stat.Ann. (Supp. 1977)	Physical or mental incapacity which renders parent unable to provide necessary care and the condition is irremediable such that the child is suffering or probably will suffer harm. §56-220(c)(2)		§56-220(f)	§56-220(f)		For child-discretionary §56-220(f) For parent-discretionary §56-220(f)		
CALIFORNIA Civ.Code (West-Supp. 1978)	Mental deficiency of parents results in their inability to care for their child. §232(b)	Hearing is required with testimony of two physicians. §232(b)	§234 §235	§237.5		For child-discretionary §237.5 For parent-mandatory §237.5	For child-discretionary §237	§238
COLORADO Rev.Stat. (1973)	Mental deficiency of the parent is of such duration or nature as to render the parent unlikely within a reasonable time to care for the ongoing physical. mental and emotional needs of the child. §19-1-105(2)(a)			§19-1-103		For child-discretionary §19-1-106 For parent-mandatory §19-1-106(1)(c)		§19-1-112 §19-11-109
CONNECTICUT Gen.Stat.Ann. (West Supp. 1978)	Parents by reaon of continuing mental deficiency have and for such time as will be detrimental to the best interests of the child will be unable to provide the care, guidance and control necessary for the child's well-being. §17-43a	Social study §45-61(f)	§45-61(d)	§45-61(d)			For alleged incompetent parent-mandatory §45-61(e)	§17-70
DELAWARE Code (Supp. 1978)	Parent is incompetent or retarded and is unable to discharge his/her responsibilities. tit. 13, §1103(3)	Evidence from two qualified psychiatrists tit. 13, §1103(3)	tit. 13, §1107	tit. 13, §1107			For alleged incompetent parent-mandatory tit. 13, §1103(3)	tit. 13, §1109 (1974)

STATE	GROUNDS FOR TERMINATION REFERRING TO DEVELOPMENTAL DISABILITIES	PROCEDURE FOR DETERMINING INCOMPETENCE	NOTICE TO PARENT	HEARING REQUIRED	STANDARD OF PROOF	APPOINTED COUNSEL FOR CHILD? FOR PARENT?	GUARDIAN *AD LITEM* FOR CHILD? FOR PARENT?	APPEAL
DISTRICT OF COLUMBIA Code (Supp. 1978)	Mental health of all individuals involved is considered in determining best interests of the child. §16-2353		§16-2357	§16-2358 §16-2359		For child-mandatory §16-2304 For parent-mandatory §16-2304	For child-mandatory §16-2304	§16-2362
FLORIDA Stat.Ann. (West Supp. 1978)			§39.11	§39.09				§39.14
GEORGIA Code Ann. (1976)			§24A-1701	§24A-1701			For child-discretionary §24A-3301	§24A-3801
HAWAII Rev.Stat. (1976)	Parent is mentally retarded and incapacitated from providing adequate care to child. §571-62	Social study §571-62	§571-61	§571-62			For child-discretionary §571-62	§571-63
IDAHO Code (1979)	Parent is unable to discharge parental responsibilities because of mental deficiency and the condition is permanent §16-2005(d)	Personal exam by 2 examiners §16-2009 §66-329	§16-2007	§16-2009	Fair preponderance of the evidence §16-2009	For parent-mandatory §16-2009	For child-discretionary §16-2007 For alleged incompetent parent-mandatory §16-2007	§16-2014
ILLINOIS Ann.Stat. (Smith-Hurd 1972 and Cum. Supp. 1979)	Parent is mentally impaired to the extent of being incompetent to discharge parental responsibilities and will continue to be so. ch.37, §705-9	The court hearing and testimony of two qualified physicians ch.37, §705-9	ch.37, §704-2 §704-4	ch.37, §704-2		For child-mandatory ch.37, §704-5(3)	For child-mandatory ch.37, §704-5(b) For alleged incompetent parent-discretionary ch.4, §9.1-8(e)	ch.37, §701-20(3)
INDIANA Code Ann. (Burns 1979)	Parent is incompetent. §31-6-5-2		§31-6-5-3	§31-6-5-3		For parent-mandatory §31-6-3-2 §31-6-5-3	For child-discretionary §31-6-3-1	
IOWA Code Ann. (West Supp. 1978)	Child is without proper parental care because of his/her parents' mental disability. §232.2(13)(8)		§600A.6	§600A.7	Clear and convincing evidence §600A.8	For child-mandatory §232.28 For parent-mandatory §232.28	For child-mandatory §600A.6(2)	§600A.9(2)
KANSAS Stat. (1973 & Supp. 1978)			§38-810	§38-817		For child-mandatory §38-821 For parent-mandatory §38-820	For child-mandatory §38-820(a) For alleged incompetent parent-mandatory §38-821(b)	§38-834
KENTUCKY Rev.Stat.Ann. (Baldwin Supp. 1978)	Mental deficiency of parent is considered in determining the best interests of the child. §199.603	Court requires psychiatric or psychological evidence that the parent is consistently unable to care for the child for extended periods of time §199.603	§199.603(3)	§199.603(6)	Preponderance of the evidence §199.603	For child-discretionary §199.603 For parent-on request §199.603	For child-mandatory §199.603	§199.617
LOUISIANA Rev.Stat.Ann. (West Supp. 1978)			§13:1602	§13:1602(a)	Proof beyond a reasonable doubt §13:1603(a)	For child-mandatory §13:1602 For parent-mandatory §13:1602(e)		§13:1603

STATE	GROUNDS FOR TERMINATION REFERRING TO DEVELOPMENTAL DISABILITIES	PROCEDURE FOR DETERMINING INCOMPETENCE	NOTICE TO PARENT	HEARING REQUIRED	STANDARD OF PROOF	APPOINTED COUNSEL FOR CHILD? FOR PARENT?	GUARDIAN AD LITEM FOR CHILD? FOR PARENT?	APPEAL
MAINE Rev.Stat. (Supp. 1978)			If child is committed to Department of Health and Welfare tit. 22, §3792	For protective custody issue only tit. 22, §3792		For parent-mandatory tit. 22, §3792		tit. 22, §3792
MARYLAND Ann.Code (Supp. 1976)	The mental and physical health of the parent is considered in determining the best interests of the child. art. 16, §75		art. 16, §72	Discretionary art. 16, §72				art. 16, §84
MASSACHUSETTS Gen.Laws Ann. (West Supp. 1978)	Parents are incompetent to provide proper care. ch. 119, §24	Social study ch. 119, §24	ch. 119, §24	ch. 119, §24	Preponderance of the evidence ch. 119, §26	For child-disretionary ch. 119, §29 For parent-mandatory ch. 119, §29		ch. 119, §27
MICHIGAN Comp.Laws Ann. (Supp. 1978-1979)	Parent has failed to provide proper care for his/her child for more than 2 years because of a mental disability and there is no reasonable expectation that the situation will change. §712A.19A		§712A.13	§712A.12 §712A.17 §712A.19		For child-discretionary §712A.17		§712A.22
MINNESOTA Stat.Ann. (West Supp. 1978)			§260.135 §260.141 §260.158	§260.231(1) §260.155		For child-mandatory §260.155 For parent-mandatory §260.155	For child-mandatory §260.155	§260.291
MISSISSIPPI Code Ann. (1972)	Parent is mentally, morally or otherwise unfit. §93-15-1		§93-15-3	§93-15-7			For child-discretionary §93-15-5	§93-15-11
MISSOURI Ann.Stat. (Vernon Cum. Supp. 1979)	Parent is incompetent or so mentally deficient as to be unable to form an intent or act knowingly, and has substantially and repeatedly neglected the child or failed to provide care and protection. §211.447	Investigation by juvenile officer, state division of family services, or other public or private agency. §211.472	§211.457	§211.467	Clear, cogent and convincing evidence §211.447	For parent-mandatory §211.562		
MONTANA Rev.Codes Ann. (Supp. 1977)	Parent who by reason of physical or mental incapacity refuses or is unable to discharge parental responsibilities. §10-1301(2(b)		§§10-1310(4), (5)	§10-1312		For child-discretionary §10-1310(12) For parent-discretionary §10-1310(12)	For child-discretionary §10-1310(12)	
NEBRASKA Rev.Stat. (Supp. 1977)	Parent is unable to discharge responsibilities because of a mental deficiency and there are reasonable grounds to believe the condition will continue. §43-209(5)		§43-209	§43-209 §43-306.03			For child-discretionary §43-209 For alleged incompetent parent-mandatory §43-209	§43-209

STATE	GROUNDS FOR TERMINATION REFERRING TO DEVELOPMENTAL DISABILITIES	PROCEDURE FOR DETERMINING INCOMPETENCE	NOTICE TO PARENT	HEARING REQUIRED	STANDARD OF PROOF	APPOINTED COUNSEL FOR CHILD? FOR PARENT?	GUARDIAN AD LITEM FOR CHILD? FOR PARENT?	APPEAL
NEVADA Rev.Stat. (1975)			§128.060 §128.070 §128.080	§128.090		For child-discretionary §128.100 For parent-discretionary §128.100		§128.120
NEW HAMPSHIRE Rev.Stat.Ann. (Supp. 1977)	Because of mental deficiency parent is incapable of giving child proper care. §170-C:5	Testimony of either two licensed psychiatrists or psychologists or one of each; social study §170-C:10	§170-C:7	§170-C:9	Clear and convincing §170-C:10	For parent-mandatory §170-C:10	For child-mandatory §170-C:8 For alleged incompetent parent-mandatory §170-C:8	§170-C:15
NEW JERSEY Stat.Ann. (West 1976)	Parent is mentally incompetent. §9:2-19		§9:2-18	§9:2-18				
NEW MEXICO Stat.Ann. (Supp. 1975)			§22-2-23(d)	§§22-2-23(e), (f)	Clear and convincing evidence §22-2-23(f)	For child-discretionary §22-2-23 For parent-discretionary §22-2-23		§22-2-41
NEW YORK Mental Hyg. Law (McKinney Supp. 1978)		Judge orders exam and testing of an alleged mentally retarded parent by a physician and certified psychologist. §§384-b(6)(c), (d), (e)	§617 §384-b	§623 §384-b(3)	Fair preponderance of evidence if child is permanently neglected; clear & convincing evidence if parent is mentally retarded & the child has been in the care of an agency for more than 1 yr. §§384-b(4)(c), (d)		For parent-mandatory §262 §384-b	§1112
NORTH CAROLINA Gen.Stat. (Cum. Supp. 1979)			§7A-289.27 §7A-289.28	§7A-289.30	Clear, cogent and convincing evidence §7A-289.30		For child-discretionary §7A-289.27(2)	§7A-289.34
NORTH DAKOTA Cent. Code (Supp. 1977)	By reason of mental incapacity, a parent is unable to provide care for his/her child, the condition is irremediable, and the child is suffering. §14-15-19(3)(b)		§14-15-19(6)	§14-15-19(6)		For child-discretionary §14-15-19(6) For parent-discretionary §14-15-19(6)		§14-15-15(1)
OHIO Rev.Code Ann. (Page 1973 & Supp. 1977)	Child lacks care because of mental or physical condition of parent. §2151.04(B)		§2151.353	§2151.353				
OKLAHOMA Stat.Ann. (West Supp. 1978)			tit. 10, §1131	tit. 10, §1134				tit. 10, §1130
OREGON Rev.Stat. (1977)	Parent is unable to care for his/her child for an extended period of time due to a mental deficiency. §419.523(2)(a)		§419.488	§419.525	Fair preponderance of the evidence §419.525(2)			
PENNSYLVANIA Stat.Ann. (Purdon Supp. Pamph. 1978)	Repeated and continued incapacity of a parent has caused his/her child to be without proper care and the causes of the incapacity can not be remedied by the parent. tit. 1, §311(2)		tit. 1, §313	tit. 1, §313				

STATE	GROUNDS FOR TERMINATION REFERRING TO DEVELOPMENTAL DISABILITIES	PROCEDURE FOR DETERMINING INCOMPETENCE	NOTICE TO PARENT	HEARING REQUIRED	STANDARD OF PROOF	APPOINTED COUNSEL FOR CHILD? FOR PARENT?	GUARDIAN AD LITEM FOR CHILD? FOR PARENT?	APPEAL
RHODE ISLAND Gen. Laws (Supp. 1978)	Incapacity to provide care for a prolonged period. §15-7-7	Testimony of not less than two licensed psychiatrists §15-7-7	§15-7-8	§15-7-8			For child-discretionary	
SOUTH CAROLINA Code (1976)	Parent is mentally retarded to such an extent that he/she cannot function as a parent. §20-11-20(2)(b)		§20-11-30	§20-11-30				§20-11-50
SOUTH DAKOTA Compiled Laws Ann. (1976)			§26-8-13	§26-8-22		For child-mandatory §26-8-22-2 For parent-mandatory §26-8-22-2		
TENNESSEE Code Ann. (Supp. 1978)	Child has been removed from the home for 1 year because parent is unfit and there is little likelihood that these conditions will be remedied soon. §37-202(6) §37-246(d)(1)		§37-221 §37-246(a)	§37-246(d)	Clear and convincing evidence §37-246(d)		For child-mandatory if interests conflict with parent's interest §37-248	
TEXAS Fam.Code Ann. (Vernon Supp. 1978)			tit. 2, §11.09	tit. 2, §11.13 §11.19	Fair preponderance of the evidence tit. 2, §11.15	For child-discretionary tit. 2, §11.10(c) For parent-discretionary tit. 2, §11.10(c)	For child-mandatory if interests are not adequately served tit. 2, §11.10(a)	tit. 2, §11.19
UTAH Code Ann. (Supp. 1979)	Parent is unfit or incompetent by reason of conduct or a condition which is seriously detrimental to the child. §78-3a-48(1)(a)		§78-3a-48(2)	§78-3a-48(2)		For child-mandatory §78-3a-48(2) For parent-mandatory §78-3a-48		§78-3a-48(3) §78-3a-51
VERMONT Stat. Ann. (Supp. 1978)			tit. 33, §647 §648 §649	tit. 33, §651			For child-mandatory if interest conflicts with his/her parent's interests tit. 33, §653	
VIRGINIA Code (Supp. 1979)	Parent is suffering from a mental deficiency of such severity that there is no reasonable expectation that the parent will be able to undertake responsibility for the care needed by the child. §16.1-283(B)		§16.1-263 §16.1-264 §16.1-283	§16.1-283	Clear and convincing evidence §16.1-283(b)	For child-mandatory §16.1-266(a) For parent-mandatory §16.1-260	For child-mandatory §16.1-266(a) For parent-discretionary §16.1-266(d)	§16.1-276
WASHINGTON Rev.Code Ann. (Supp. 1978)			§13.34.090	§13.34.190	Clear, cogent & convincing evidence if child has been in an out-of-home placement for 6 months, or beyond a reasonable doubt if the child has not been placed. §13.34.190	For child-mandatory §13.34.090 For parent-mandatory §13.34.090	For child-discretionary §13.34.100	
WEST VIRGINIA Code (Supp. 1978)			§49-6-1	§49-6-2		For child-mandatory §49-6-2(a) For parent-mandatory §49-6-2(a)		§49-6-2(c)

STATE	GROUNDS FOR TERMINATION REFERRING TO DEVELOPMENTAL DISABILITIES	PROCEDURE FOR DETERMINING INCOMPETENCE	NOTICE TO PARENT	HEARING REQUIRED	STANDARD OF PROOF	APPOINTED COUNSEL FOR CHILD? FOR PARENT?	GUARDIAN AD LITEM FOR CHILD? FOR PARENT?	APPEAL
WISCONSIN Stat. Ann. (West Supp. 1978)	Parent is mentally deficient and because of that deficiency the parent is and will continue to be incapable of providing proper parental care for a prolonged, indeterminate period. §48.40(3)	Reference to independent procedure §48.40(3)	§48.42	§48.42 §48.425			For alleged incompetent parent-mandatory §48.42(2)	
WYOMING Stat. (1977)	Parent unable to discharge responsibilities to the child because of hospitalization or other physical or mental incapacity. §14-6-201(xvi) §14-3-101		§14-2-305	§14-2-305				
AMERICAN SAMOA Code (1973)			Discretionary tit. 602. §653(b)	Discretionary tit. 602. §653(b)				
GUAM (1970)	Parent is unfit because of a condition deemed seriously detrimental to the child. §268		§268	§268				
PUERTO RICO Laws Ann. (Supp. 1977)			tit. 31. §634					
VIRGIN ISLANDS Code Ann. (Cum. Supp. 1979)			tit. 5. §2509	tit. 5. §2509				

SUMMARY OF MODEL STATUTE

The above analysis of existing legislation reveals that the law of most American jurisdictions fails to recognize the parental rights of mentally disabled persons to the same extent as those of persons who do not have a developmental disability or mental illness. Except with regard to adoption subsidies, many of the provisions which deal specifically with mentally disabled persons tend to facilitate the abrogation of their rights rather than protect them. Accordingly, portions of existing and proposed legislation has been melded with original provisions into a Model Protection of Parental Rights Act which seeks to erase the synonymity between being mentally disabled and a bad parent.[48]

The model act is intended as amendatory legislation. It addresses the problems and imbalances affecting the parental rights of developmentally disabled and mentally ill persons within the framework of current relinquishment and termination practices. While a number of commentators have pointed out procedural, substantive, and philosophical shortcomings of the termination of parental rights process in particular,[49] the drafting of a more comprehensive statute which addresses each of these issues is far beyond the scope and resources of this project.

The proposed model act focuses solely on the relinquishment (sections 4-5) and termination (sections 6-10) of parental rights. The recent widespread enactment of adoption subsidy statutes and the existence of a model act and comparative analysis which seeks to remedy the shortcomings of some of these existing statutes,[50] make it unnecessary to include adoption subsidy in this series of proposals. Unlike some of the model legislation in this volume, the Model Protection of Parental Rights Act covers both mentally ill and developmentally disabled persons. Most current relinquishment and termination statutes cover both groups of individuals identically, and the problems faced by persons with a developmental disability and those who are mentally ill when their parental rights are at stake are highly similar.

The provisions of the model act are designed to strike current sections which permit courts to disregard the rights, desires, and abilities of mentally disabled parents. It would also prevent the severing of the relationship between mentally disabled parents and their children in circumstances in which the parental rights of a person without a developmental disability or mental illness would not be abrogated.[51] The act also includes procedures which will provide the court with the information necessary to determine whether a mentally disabled parent is able to care for his or her child[52] or has provided voluntary and informed consent to that child's adoption.[53] Finally, it seeks to provide mentally disabled parents with assistance both in the court and at home to protect their rights and meet their responsibilities.[54]

It is anticipated that if enacted, the changes incorporated in the model statute will better protect not only the interests of mentally disabled parents but also those of their children. As observed by one commentator writing for the Children's Division of the American Humane Association:

> Because it is easier to place a child than it is to pour into his (or her) home the social services which could stabilize family life, far more children are separated than should be. In the process, the right of a child to be with his [or her] natural parents siblings is overlooked, or it is disregarded as unimportant . . . .
>
> Termination of parental rights erases all rights, not only the rights of the parents but, also, the rights of their children . . . .
>
> While procedural changes [may] improve . . . additional responsibilities on child welfare services, it is a small price to pay for the peace of mind which results in terms of assuring full protections of substantial legal rights.[55]

The text of the model act appears in **bold type.** Optional language appears [within brackets/ separated by a slash]. Comments on statutory provisions appear in standard type. The comments explain, elaborate upon or provide a rationale for the text of the act. They are intended to assist the user in interpreting the act and in drafting a statute for a particular state from this model.

48. P. Wald, *supra* note 13, at 15.
49. See DE FRANCIS, *supra* note 4; NACJJDP, *supra* note 27, at §§ 3.113, 3.171; 3.184, 3.185, 3.1812 and 3.1813; Ketcham and Babcock, *supra* note 25; Areen, *Intervention Between Parent and Child: A Reappraisal of the State's Role in Child Abuse and Neglect Cases,* 63 GEORGETOWN LAW REVIEW 887 (1975); M. Wald, *State Intervention on Behalf of "Neglected" Children: A Search for Realistic Standards,* 27 STANFORD LAW REVIEW 985 (1975).
50. Katz and Gallagher, *supra* note 16.
51. See sections 4 and 6.
52. See sections 6(2) and 7.
53. See section 5.
54. See sections 5, 6(2), 9 and 10.
55. DEFRANCIS, *supra* note 4 at 9, 10 and 15.

MODEL STATUTE

A BILL FOR

AN ACT TO PROTECT THE RIGHTS OF PARENTS WITH A DEVELOPMENTAL DISABILITY; TO ASSURE THAT RELINQUISHMENTS OF PARENTAL RIGHTS BY PERSONS WITH A DEVELOPMENTAL DISABILITY OR MENTAL ILLNESS ARE MADE KNOWINGLY AND VOLUNTARILY; TO REVISE THE STANDARD FOR TERMINATING THE PARENTAL RIGHTS OF PARENTS WHO ARE DEVELOPMENTALLY DISABLED OR MENTALLY ILL; TO PROVIDE FOR EVALUATIONS OF SUCH PARENTS; AND TO REQUIRE THAT INDIVIDUALS WHOSE PARENTAL RIGHTS ARE SUBJECT TO TERMINATION RECEIVE ADEQUATE NOTICE AND ARE ENTITLED TO REPRESENTATION BY COUNSEL.

Section 1. Title

This act shall be known as the Protection of Parental Rights Act.

COMMENT. None Necessary.

Section 2. Purpose

Recognizing that the custody, care and nurture of a child resides first in his or her parents, that the authority to abrogate the relationship between a parent and child should be exercised with great care and with the utmost concern for both the parent and the child, that the parental rights of a developmentally disabled or mentally ill person should be terminated involuntarily only when there is no less drastic alternative, that every individual has unique needs and abilities, and that many individuals with developmental disabilities or a mental illness desire to and, with appropriate assistance, are capable of properly caring for and nurturing their children, the (name of legislature) declares that it is the purpose of this act to promote the general welfare by assuring that the relinquishment by a developmentally disabled or mentally ill person of their parental rights is made knowingly and voluntarily; by limiting the grounds on which the parental rights of individuals who have a developmental disability or mental illness may be terminated without their consent to those which apply to persons not suffering from such a disability, and by providing all parties with an opportunity to prepare for and protect the rights and interests affected by a relinquishment or termination proceeding. This act shall be liberally construed to achieve this purpose.

COMMENT. None necessary.

Section 3. Definitions

(1) "Developmental disability" means a chronic disability of a person which:

(a) Is attributable to a mental or physical impairment or a combination of mental and physical impairments;

(b) Is likely to continue indefinitely;

(c) Results in substantial functional limitations in two or more of the following areas of major life activity:

(i) Self-care;
(ii) Receptive and expressive language;
(iii) Learning;
(iv) Mobility;
(v) Self-direction;
(vi) Capacity for independent living; or
(vii) Economic self-sufficiency; and

(d) Reflects the need for a combination and sequence of special, interdisciplinary or generic care, treatment or other services which are of lifelong or extended duration and individually planned and coordinated.

COMMENT. This definition, like that in the Developmental Disabilities Assistance and Bill of Rights Act,[1] adopts a functional rather than a categorical approach. Thus, it places greater significance on the effect of a disability than on its particular cause. The definition differs from the one contained in the recently amended federal act in that there is no requirement that the disability be "severe" or that it appear before the afflicted person is age 22, and in that two rather than three areas of major life activity must be detrimentally affected. Whereas the federal definition is designed to target certain populations for intensive federal support, the purpose of this act's definition is to assure that the right of developmentally disabled persons to raise their

1. 42 UNITED STATES CODE §6001(7) (Cum. Supp. 1978) (P.L. 94-103 as amended by P.L. 95-1780).

children and the right of the children of developmentally disabled parents to receive natural parental love and affection is protected to the same extent as the rights of persons and children of parents who are not disabled. Since relatively few of the more severely handicapped persons within the federal definition are likely to have children, the meaning of the term developmental disability has been broadened so as to extend the protections afforded by this act to the wider range of people adversely affected by current laws on the relinquishment and termination of parental rights.

(2) "Mental illness" means a substantial disorder of a person's emotional processes, thought or cognition which grossly impairs judgment, behavior or capacity to recognize reality. It does not include a developmental disability, brief periods of intoxication caused by substances such as alcohol or drugs, or dependence upon or addiction to any such substances.

COMMENT. This definition seeks to describe mental disorders in terms comprehensible to non-mental health professionals. It is based upon the definition used in the "suggested statutes" developed by the Mental Health Law Project;[2] and covers each of the three components of mental functioning — emotion, volition, and cognition.[3] Like the definition of developmental disabilities in subsection (1), it focuses on the effects of the illness rather than on diagnostic labels. However, it makes clear that mental illness is not intended to be a generic term covering all forms of mental dysfunction.[4]

Persons who may be mentally ill are included in this statute for two reasons. First, most current statutes lump all mental disabilities together.[5] Second, the impact of and problems faced by developmentally disabled parents, mentally ill parents, and their children in termination and relinquishment proceedings are substantially the same, despite the differing conditions. Thus, it would make little sense to provide the protections offered by this statute only to persons with a developmental disability.

(3) "Counsel" means an attorney admitted to the practice of law in this state.

COMMENT. While a lay advocate can provide valuable assistance in a variety of contexts, given the importance and complexity of the matters at issue in relinquishing parental rights or in a termination proceeding, it is essential that the parents be represented by a licensed attorney. A distinction is made in this statute between guardians *ad litem* and counsel. The role of counsel is to serve as a zealous advocate of the legal interests of his or her client but not to determine those interests.[6] The function of the guardian *ad litem* is to assist individuals to determine their interests and, if they are incapable of doing so, of acting in their stead. Just as lawyers are discouraged from representing themselves in important litigation, the attorney who is compelled to represent an individual and serve as that individual's guardian *ad litem* may be caught between conflicting interests and have difficulty maintaining objectivity and avoiding the subrogation of one role to the other. A number of states currently recognize that the rights of individuals in termination proceedings are better protected when separate persons serve as counsel and guardian *ad litem*.[7]

(4) "Guardian *ad Litem*" means a person appointed by the court to assist a parent or a child to determine his or her interests in regard to a termination proceeding, or to make that determination if the parent or the child is unconscious or otherwise wholly incapable of determining his or her interests in that proceeding even with assistance.

COMMENT. See Comment to section 3(3).

(5) "Multidisciplinary evaluation team" means a group of experts appointed by the court to evaluate a parent in conjunction with a termination proceeding, if the parent is asserted to have a developmental disability or mental illness. When the parent is asserted to have a developmental disability, the multidisciplinary evaluation team shall include a physician, a social worker with a graduate degree in social work and field training or experience in working with persons with developmental disabilities, and an expert with knowledge of the particular disability which an individual is alleged or has been found to have. When the parent is asserted to have a mental illness, the multidisciplinary team shall include a clinical psychologist or psychiatrist and a social worker with a graduate degree in social work and field training or experience in working with mentally ill persons.

COMMENT. As is indicated on Chart III, a few jurisdictions provide for some type of examina-

---

2. Mental Health Law Project (MHLP), *Legal Issues in State Mental Health Care: Proposals for Change — Suggested Statute on Mental Health Treatment for Minors* §1(a), 2 MENTAL DISABILITY LAW REPORTER 473, 474, (1978).
3. MHLP, *Legal Issues in State Mental Health Care: Proposals for Change — Civil Commitment,* 2 MENTAL DISABILITY LAW REPORTER 89 (1977).
4. In re Rathbun, 128 Vt. 429, 266 A.2d 423 (1970).
5. AMERICAN BAR ASSOCIATION (ABA), CODE OF PROFESSIONAL RESPONSIBILITY, EC 7-1, EC 7-7, and EC 7-12 (1976).

6. See Charts I and III, *supra.*
7. *See Chart III, supra;* see also sections 9 and 10.

tion when a person is threatened with termination of his or her parental rights because of a mental disability. However, with two exceptions,[8] these statutes require only a unidimensional examination — i.e., either an assessment of the parent's mental or physical condition or an assessment of the home situation. Because many persons who have a developmental disability suffer from a combination of physical, mental and adaptive impairments, such a unidimensional approach is inadequate. Hence, the model statute provides for a multidisciplinary team of experts to conduct evaluations. A physician is included to examine the physical condition of the individual. The social worker is to focus on the environment, adaptive behavior and parenting skills of the individual. The additional member or members of the team should have the expertise to evaluate the particular type or types of disability which the individual is alleged or has been found to have in order to provide the court with information regarding the degree and speed with which the parent will be able to improve his or her parenting skills if they have been found to be inadequate. For example, if the person is alleged to be mentally retarded, a clinical psychologist and/or an expert in the education of retarded persons should be placed on the team. For parents alleged to be mentally ill, the multidisciplinary evaluation team includes both a clinical psychologist or psychiatrist and a social worker since both the parent's and the home's condition must be known if the court is to make an intelligent decision.

Care should be taken to assure that multidisciplinary team members have the requisite expertise in developmental disabilities or mental illness, and that they are able to perform an objective evaluation without conflicts of interests which may influence their assessments. Enacting jurisdictions which have established an Evaluation Oversight Board or similar mechanism to identify qualified experts to perform evaluations and examinations of persons with mental disabilities[9] should assure that the Board's authority extends to evaluations under this act and require that multidisciplinary evaluation team members be selected from the list published by the Board.

(6) "Evaluation" means a determination whether a parent has a developmental disability or mental illness; an assessment of the effect of a developmental disability or mental illness on a parent's ability to (provide necessary/proper care/discharge his or her parental responsibilities), and an opinion on whether the parent will be able to (provide necessary/proper care/discharge his or her parental responsibilities) within (one (1) year/a reasonable time/the foreseeable future) as well as on the training and services, if any, which would assist the parent to be able to (provide such care/discharge such responsibilities).

COMMENT: The evaluation is intended to be a comprehensive investigation of the abilities and disabilities of an individual and of the types of habilitative or rehabilitative services required to assist a developmentally disabled or mentally ill person, who has been found to have endangered his or her child as a result of the disability or illness, to become a competent parent.[10] Since this model act is intended to be amendatory legislation, the most common standards for termination of parental rights are listed as alternatives. Enacting jurisdictions should employ the terms which appear elsewhere in their termination provisions.

(7) "Termination proceeding" means any proceeding at which the parental rights of an individual may be involuntarily abrogated.

COMMENT. None necessary.

Section 4. Repeal of Authority to Waive Consent to Relinquish Parental Rights

(Citation to the section permitting the court to waive or a third party to provide consent to the relinquishment of parental rights) is repealed and the following provision is inserted in its place:

The consent of a person to the relinquishment of his or her parental rights shall not be waived nor shall the consent of a guardian or another third party be substituted therefor on the grounds that the parent has a development disability or mental illness. The parental rights of a person who is unable to provide informed voluntary consent pursuant to section 5 of this act as a result of a developmental disability or mental illness shall not be abrogated except following a termination proceeding in which the criteria set forth in section 6 have been satisfied, the evaluation requirements set forth in section 7 have been met, and the person has been afforded the rights set forth in sections 8 through 10.

COMMENT. In accordance with the law of approximately 40% of American jurisdictions, this section permits the severance of a parental-child relationship only if the parents consent or grounds for involuntary termination of parental rights have been met.[11] It prohibits the court

8. NEW HAMPSHIRE REVISED STATUTE ANNOTATED §170C:10 (Supp. 1979) (psychiatrists, psychologists or both, and a social study); NEW YORK MENTAL HYGIENE LAW §384-b(6)(McKinney Supp. 1978) (a physician and a psychologist).
9. See DEVELOPMENTALLY DISABLED OFFENDERS ACT, §14, in Chapter 8: Criminal and Juvenile Justice, *infra*.
10. See section 6.
11. See Chart I, *supra*.

from waiving the consent of a parent who is developmentally disabled or mentally ill or from accepting substitute consent by a guardian or other third party.

As noted in the introduction to this statute, the provisions which permit waiver of consent to relinquish parental rights are designed to facilitate adoption and protect the rights of the adopting parents.[12] However, little protection is offered to either the natural parents or to the children. The definitions of incompetence to provide consent are vague. There is often no requirement for an evaluation or other procedure for determining competence. And, there is usually no provision offering the natural parents an opportunity to make their views, feelings, abilities and disabilities known.[13]

This section addresses these problems in three ways. First, it repeals the existing provision authorizing the court to waive consent or accept substitute consent for parents who have a mental disability. Second, it expressly prohibits acceptance of a waiver or substitute consent to relinquish parental rights, in order to make the intent of the repeal clear and to establish an explicit policy in states which do not have an express provision permitting waiver. Third, it provides alternatives to the waiver procedure through references to the sections to be followed when a parent with a developmental disability or mental illness wishes to and is able to provide informed, voluntary consent to the relinquishment of parental rights, and when a mentally disabled parent is unable to provide such consent and his or her child is endangered. Both the consent provisions in section 5, and the provisions on termination of parental rights in section 6-10, are designed to assure that those mentally disabled parents who are able and willing to do so, may exercise their constitutionally protected right to raise their children, but to permit abrogation of that parental right for those developmentally disabled or mentally ill persons who endanger their children or who wish to relinquish them for adoption.

A number of states permit the court to waive the consent of parents who have abandoned their children. While this section would prohibit such waiver as well, it is anticipated that this will not significantly impede the process for freeing abandoned children for adoption, since almost all states include abandonment as a ground for termination of parental rights.

Section 5. Relinquishment of Parental Rights

(1) The relinquishment of parental rights by a person who has a developmental disability or a mental illness shall not be effective unless:

12. S. BRAKEL and R. ROCK, THE MENTALLY DISABLED AND THE LAW, 235 (Rev. Ed. 1971).
13. See *id.*, at 236 and Chart I, *supra.*

(a) The person has had the meaning and consequences of relinquishing his or her parental rights explained in the manner set forth in subsection (2) of this section;

(b) The person has signed a consent form meeting the requirements set forth in subsection (3) in addition to the requirements prescribed by (citation to the relevant section of the State code); and

(c) The consent has been found to be both informed and voluntary by the (name of appropriate court) pursuant to subsection (4).

COMMENT. In order for the waiver of a constitutionally protected right to be valid and enforceable, it must be found to have been made voluntarily and with knowledge of its consequences.[14] As discussed in the introduction to this statute, the right of a person to raise his or her child is protected by several clauses of the Bill of Rights.[15] Hence, it is necessary to assure that the relinquishment of that right by a parent giving up his or her child for adoption is made knowingly and voluntarily.[16] This section establishes procedures which are intended to accomplish this objective for persons with a developmental disability or mental illness. A three-step process is required. The first step is an explanation to a mentally disabled person who is considering the relinquishment of his or her parental rights of the implications of the relinquishment and the alternatives available. The second step is a clarification of the consent agreement itself. The final and most important step is a colloquy between the relinquishing parent and the judge to assure that the consent is not the result of coercion, duress or misapprehension.

In recent years, there have been an increasing number of adoptions that have been vacated on the grounds that the natural parent did not fully appreciate the consequences of relinquishing his or her parental rights.[17] The damage to the parties and to the child when an adoption is vacated under these circumstances, would dictate full support of a move to strengthen the relinquishment process so as to assure full protection of all parties to the transaction.[18] It is anticipated that if the above steps are followed in the manner set forth in subsections (2)-(4), there will be few challenges and even fewer successful ones.

14. See Johnson vs. Zerbst, 304 U.S. 458 (1938); Boykin vs. Alabama, 395 U.S. 238 (1969); *see also* AMERICAN ASSOCIATION ON MENTAL DEFICIENCY (AAMD), CONSENT HANDBOOK (1977).
15. *See e.g.*, Stanley vs. Illinois, 405 U.S. 645, 651 (1972).
16. NATIONAL ADVISORY COMMITTEE ON JUVENILE JUSTICE AND DELINQUENCY PREVENTION (NACJJDP), STANDARDS FOR THE ADMINISTRATION OF JUVENILE JUSTICE, §3.185 (1980).
17. V. DEFRANCIS, TERMINATION OF PARENTAL RIGHTS: BALANCING THE EQUITIES, 13 (The American Humane Association, 1971).
18. *Id.*

(2) (a) Not less than three (3) days after the birth of the child to be relinquished and not more than five (5) days before the hearing required by subsection (4), the following matters shall be explained to person with a developmental disability or mental illness who wishes to relinquish his or her parental rights:

(i) By consenting to the relinquishment of parental rights, he or she is irrevocably giving up the right to raise the child;

(ii) Consent must be given voluntarily and the person is free to provide or withhold consent; and,

(iii) There are services available which can help the person to care for the child.

(b) The explanations shall be provided by a social worker or counselor employed by the (name of the state or local agency(ies) responsible for providing services to persons with a developmental disability or a mental illness) or an advocate employed by the (name of the state developmental disabilities advocacy office,)

(c) Following the explanation, the individual providing it shall file with the agency which will be responsible for the care, custody and adoption of the child who may be relinquished, a certificate indicating that he or she has explained the matters set forth in paragraph (2)(a) of this subsection in the manner prescribed by paragraph (2)(d). A copy shall be provided to the parent.

(d) The explanation shall be in the language, terms, and mode of communication which the person is most likely to understand and shall be accompanied by a description of the services available.

COMMENT. As noted above, the first step in the relinquishment process is a detailed explanation of the implications and the alternatives to relinquishing parental rights. While in the past, mothers considering giving up their baby for adoption have received counseling, this has usually occurred prior to the birth of the child and has often been provided by the adoption agency.[19] The counseling called for by this section differs in that it takes place after the immediate effects of the natal process are over, and in that the explanation is provided by a person not directly involved in the adoption process who is experienced in working with developmentally disabled or mentally ill persons. It is intended as a supplement to, not as a substitute for pre-natal adoption counseling.

Specifically, paragraph (2)(a), prohibits a mentally disabled parent from relinquishing his or her parental rights until at least three days after the child has been born.[20] In reality, it is likely that a few more days will pass before a hearing can be scheduled and the mother is strong enough for the trip to the courthouse. This gives the mother particularly, an opportunity to recover from the immediate emotional and physical stresses of birth and to think through her decision without a delay that is likely to affect the infant. The paragraph also requires that the explanation be provided relatively close to the hearing so that it can reflect the circumstances and conditions which exist at the time that the relinquishment decision must be made. Moreover, this assures that the relinquishing parent will still have the information freshly in mind when he or she is presented with the consent form even when the relinquishment occurs when the child is no longer an infant.

It is intended that the individual providing the explanation will go beyond the words of the statute to convey the listed concepts and information, and that he or she will do so in a manner consistent with the parent's abilities and disabilities.[21] The certificate required to be filed by paragraph (2)(c) is to facilitate the court's inquiry into whether the procedures called for by this act have been followed. It is to be filed with the adoption or child custody agency since a court file is unlikely to exist at this stage of the proceedings. An alternative would be to require the individual who provided the explanation to be present at the relinquishment hearing. However, the certificate presents a much less costly method which in most cases, will provide adequate assurance that the act's requirements have been satisfied.

(3) (a) The consent form shall state clearly that by signing, the person is irrevocably giving up his or her right to raise the child to be relinquished, and shall contain all the terms and conditions on which the relinquishment is based.

(b)(i) Prior to signature, the contents of the consent form shall be explained to the person relinquishing his or her parental rights in the language, terms and mode of communication which he or she is most likely to understand.

(ii) The individual providing this explanation shall certify on the consent form that a full explanation was provided in the specified manner.

COMMENT. As indicated on Chart 1, *supra*, most jurisdictions require written consent to the relinquishment of parental rights. In light of this requirement, subsection (3) is intended to assure that when a developmentally disabled or mentally ill parent seeks to relinquish his or her parental rights, the consent forms are written as plainly as possible, contain all the promises and conditions agreed to or set by the child care or adoption

19. *Id.*
20. At least two states currently proscribe relinquishment of parental rights until the mother had had at least a 72-hour recovery period. ARIZONA REVISED STATUTES ANNOTATED §8-16 (Cum. Supp. 1979); OHIO REVISED CODE ANNOTATED § 3107.08 (Page Supp. 1978).

21. For a discussion of the phrase "Language, terms and mode of communication which the person is most likely to understand", see the comment to section 8, *supra*.

agency, and are fully explained before they are presented for signature.

Enacting jurisdictions which do not require persons to sign a written consent to relinquish their parental rights may wish to omit this procedure. If this subsection is omitted, a clause should be added to subsection (2) requiring the explanation called for by that section to include the terms and conditions on which the relinquishment is based.

(4) (a) The consent of a person with a developmental disability or mental illness to the relinquishment of his or her parental rights shall be reviewed by the (name of the appropriate court) at a hearing at which the relinquishing parent and a representative of the agency(cies) responsible for the care, custody and adoption of the child to be relinquished are present.

(b) Before accepting or rejecting the consent of a parent with a developmental disability or mental illness, the court shall determine, in addition to any other matters prescribed by law, whether the procedures prescribed in subsections (2) and (3) of this section have been followed, and whether the consent is informed and voluntary. In making these determinations, the court shall ascertain through questioning the parent personally, whether:

(i) The parent understands that by consenting to relinquish his or her parental rights the parent is irrevocably giving up the opportunity to raise his or her child;

(ii) The consent is not a product of force or threats or of promises other than those in the consent form; and

(iii) The parent understands and accepts the terms and conditions on the consent form.

(c) At the conclusion of the hearing, the court shall state its findings and whether the parent's consent has been accepted or rejected.

(d) A stenographic and/or mechanical record shall be made of the hearing.

COMMENT. This subsection requires the court to make a detailed inquiry before accepting the consent of a parent with developmental disability or mental illness, and to make certain that the parent has not been coerced into giving up his or her child and that the parent understands the consequences of his or her actions. A similar though less comprehensive inquiry is required by the adoption laws of Colorado, South Dakota, and Wisconsin and the Illinois code permits a judicial inquiry where there are allegations of fraud and duress.[22] The inquiry may be made at a formal courtroom hearing or informally in the judge's chamber depending on the practice in other adoption proceedings in the state. However, it must be on the record, and both the relinquishing parent and a representative of the agency which will take responsibility for the child must be present "so that the judge may make a personal assessment of the kind and degree of counseling and interpretation available to... (the parent), and the genuineness of her (or his) consent."[23] The statute calls for a direct colloquy between the judge and the parent.[24] However, it is not intended to prohibit the use of an interpreter when the parent is not fluent in English or when the parent's disability affects his or her ability or means of communication. The court may also question the agency employee regarding the representations made to the parent and the terms and conditions of the adoption.

Since this is not an adversary proceeding, there is no requirement that the parent be represented by counsel.[25] However, nothing in the provision is intended to prevent the parent from obtaining the assistance of counsel or a lay advocate. The provision in paragraph 5(4)(c) requiring the court to state whether it finds the consent to be informed and voluntary and whether in light of this finding, and the others which may be required by law, the consent has been accepted or rejected, is to prevent misunderstandings among the parties and to facilitate review.

Section 6. Amendment to the Grounds for Terminating Parental Rights

(Citation to relevant provision(s) on termination of parental rights) is (are) amended as follows:

(1) The (developmental disability and/or mental illness/mental deficiency/mental incapacity/mental incompetence/mental condition) of a parent shall not constitute a ground on which to base a finding (that the parent has been unable to provide necesssary and proper care/to discharge his or her parental responsibilities/or that a child has been neglected).

(2) The (developmental disability and/or mental illness/mental deficiency/mental incapacity/mental incompetence/mental condition) of a parent may be considered in determining whether or not a parent will be able to (provide necessary and proper care/discharge his or her parental responsibilities) within (one (1) year/a reasonable time/the foreseeable future). However, no finding shall be made that a parent who (has a developmental disability and/or mental illness/mental deficiency/mental condition/is mentally incompetent/mentally incapacitated) will not be able to

22. COLORADO REVISED STATUTES §19-4-101 (1978); SOUTH DAKOTA CODIFIED LAWS §§25-5A-6 and 25-5A-16 (1976): ILLINOIS ANNOTATED STATUTES ch. 4, §9.1-10 (Smith-Hurd Cum. Supp. 1979). See also MICHIGAN COMPILED LAWS §710.29 (Cum. Supp. 1979) which requires the judge or referee to explain to relinquishing parents their rights and the effect of signing the release form.

23. DEFRANCIS, *supra*, note 17, at 15.

24. For a similar requirement in another context, see Boykin vs. Alabama, 395 U.S. 238 (1969).

25. *Cf.* Section 6.

(provide necessary and proper care/discharge his or her parental responsibilities) within (one (1) year/a reasonable time/the foreseeable future) unless that individual has been evaluated in the manner prescribed in section 7 of this act and either:

(a) (i) A program of homemaking, child care, social, rehabilitative, advocacy, counseling, medical, mental health or other training services which is designed to teach or assist that parent to (provide necessary and proper care/discharge his or her parental responsibilities) has been provided to the parent for one (1) year; and

(ii) It is proven at a hearing that the child is still (neglected/dependent); or

(b) It is proven beyond a reasonable doubt that the provision of such training and services will not enable the parent to (provide necessary and proper care/discharge his or her parental responsibilities) within (one (1) year/a reasonable time/the foreseeable future).

COMMENT. As indicated in the Review of Statutes, *supra,* the most common criteria for the termination of parental rights of mentally disabled persons include three elements: an allusion to the mental disability, a statement of inadequate parenting, and some time period within which the problems must be resolved. This section addresses each of these three elements separately. It prohibits consideration of a mental illness or developmental disability as an independent ground for termination of parental rights either in and of itself or in conjunction with one or more of the parenting deficiency standards listed in the section. As noted in the introduction to this statute, "the relationship between parent and child is constitutionally protected."[26] While other decisions,[27] current statutes,[28] and the doctrine of *parents patriae*[29] make clear that the state has the power to intervene and limit or terminate parental rights when the health or safety of a child is in jeopardy, the mere fact that a parent has a developmental disability or mental illness does not prove that a child is in jeopardy of being neglected or harmed.

> A general finding of "incompetence" based on intellectual deficiencies is inadequate as a criterion for taking away parental rights. There are no data to support a general finding that retarded parents are generally bad parents. ". . . [W]hen one looks around and sees the number of children who are in the care of non-retarded parents who are incompetent for the task, . . . one wonders whether the principles outlined . . . for dealing with the retarded do not apply equally as well to us all.[30]

The standard for parental insufficiency is left unaltered. The most common phrases are listed as alternatives. Enacting jurisdictions should insert the formulation used in their termination provisions. While there are strong arguments for focusing both the definition of neglect and the criteria for termination of parental rights "on specific harms to the child rather than on broadly drawn descriptions of parental behavior,"[31] such a comprehensive restructuring of termination statutes goes far beyond the purposes and resources of this Project. Thus, under subsection (1) of this section, a court will still have to find that a parent is, for example, "unable to discharge his or her parental responsibilities," but it must do so on the basis of the parent's actions and omissions and not simply infer this inability from the fact that the parent has a developmental disability or is mentally ill.[32]

> Because parents are poor or stupid or have speech defects, can be no reason in themselves to take a child from its mother. It may well be that a given child might have had a better chance in life had it been born in a different home, but that by itself does not give the state the power to deprive a mother of her child.[33]

The third element — the recovery period — is governed by subsection (2). This subsection expressly permits consideration of a parent's mental disability in determining whether the threat to the child's welfare proven under subsection (1) can be remedied within a reasonable time.[34] However, it prohibits the court from issuing an order terminating a mentally disabled person's

26. Quilloin vs. Walcott, 434 U.S. 246 (1978).
27. Wisconsin vs. Yoder, 406 U.S. 205, 233-234 (1971).
28. See chart III, *supra.*
29. A legal doctrine which encourages the government to serve in the role of a benevolent parent caring for his or her poor, rich, or wayward children. See *e.g.,* Prince vs. Massachusetts 321 U.S. 158, 166 (1944). For a critique of this doctrine, see *e.g.,* Rothman, *The State as Parent: Social Policy in the Progressive Era,* in W. GAYLIN, I. GLASSER, S. MARCUS, & D. ROTHMAN, DOING GOOD: THE LIMITS OF BENEVOLVENCE, 67 (1978).
30. P. Wald, *Personal and Civil Rights* in PRESIDENT'S COMMISSION ON MENTAL RETARDATION (PCMR), THE MENTALLY RETARDED CITIZEN AND THE LAW, 2, 14 (1976).
31. See *e.g.,* NACJJDP, *supra* note 16, at Commentary to § 3.113, and §3.185 (1976); NATIONAL ADVISORY COMMITTEE ON CRIMINAL JUSTICE STANDARDS AND GOALS, REPORT OF THE TASK FORCE ON JUVENILE JUSTICE AND DELINQUENCY PREVENTION, §14.32 (1977); M. Wald, *State Intervention on Behalf of "Neglected" Children: A Search for Realistic Standards,* 27 STANFORD LAW REVIEW 985 (1975).
32. Ketcham and Babcock, *Statutory Standards for Involuntary Termination of Parental Rights,* 29 RUTGERS LAW REVIEW 530 (1976).
33. State *ex rel.,* Paul vs. Department of Public Welfare, 170 So. 2d. 549 (La. Ct. of App. 1965) (Tate, J. concurring).
34. Various time formulations drawn from current statutes are listed as alternatives.

parental rights unless certain conditions have been satisfied.

The first condition is that the parent must undergo a thorough evaluation, provided in accordance with section 7 of this act. As is indicated on Chart III, *supra,* few jurisdictions provide for the multidisciplinary evaluation necessary to accurately assess the affects of a developmental disability or mental illness, the nature of the home situation and the prospects for change. Second, on the basis of the evaluation and other evidence presented by the parties, the court must either order that appropriate habilitative or rehabilitative services be provided to the parent for at least one year or find that it has been proven beyond a reasonable doubt that the parent will not meet the appropriate statutory standard after the services have been provided. If this finding is made or if after the services have been provided, the child continues to be endangered when left in the parent's care, the court may, but is not required by this statute, to abrogate the parental-child relationship. This formulation seeks to balançe the interests of the parent and those of the child. If a parent wishes to care for his or her child and is able with the help of training, and social, medical, mental health, homemaking, and/or financial services to overcome the problems which led to the termination proceeding, it is clearly not in the parent's interest to terminate his or her parental rights until there has been an adequate opportunity for that assistance to work.[35] Neither is it in the interest of the child to have the bonds with his or her parents severed summarily and to be plunged into the uncertain world of foster care[36] without a reasonable attempt to correct the endangering conditions.[37] On the other hand, the importance of a stable environment for a child and a child's heightened sense of time[38] demand that the period during which children may be shuttling between their parental home and a foster placement be limited.

> It is true that the court should not be concerned with questions of the passage of arbitrary periods of time, but with whether the child's relationship with his natural parent is irreparably damaged, or whether the child has entered into a new psychological parent-child relationship. However, until behavioral scientists can provide more objective criteria for the determination of such factors, more traditional time concepts acceptable to the judicial system may be necessary. For the sake of both the parent and the child, there should be some outer limit to the period of time during which a decision to terminate can be effected. While durational "limits" of this sort too frequently become timetables for acting rather than the cut-off points they were intended to be, such limits are necesary either to encourage the responsible social service or placement agency to act promptly in those cases in which termination may be warranted, or to remove the threat of termination from the natural parent whose child has been adjudicated neglected.[39]

As suggested in the above quotation, the one year length of the habilitation period is somewhat arbitrary. Some commentators have suggested that when young children are involved, no more than 6 months should pass betwen the court's interim and final decree,[40] and at least one state has imposed a 2-year interim period.[41] One year appears to be a reasonable compromise which accommodates the training or treatement needs of mentally disabled persons while protecting against unwarranted delay. At the end of this period it is anticipated that in many cases, the family will be able to continue as a unit. In some, supporting services will continue to be necessary; in others they will not. And, in a few cases, the court will have to sever the parental-child relationship and order permanent placement for the child. But regardless of the outcome, the court will have more complete information and all parties will have a better opportunity to assure that the best interests of the child are furthered.

Alternative. *In enacting states which do not have a time requirement in their termination statute, subsection (2) should be changed to read:*

> *(2) No finding shall be made that a parent who (has a developmental disability and/or mental illness/mental deficiency/mental condition/is mentally incompetent/mentally in-*

35. *Accord* In re Appeal in Pima County, Juvenile Action No. 5-111, 25 Ariz. App. 380, 583 P.2d 809 (1975); *cf.* L. and L. vs. Big Horn County Department of Public Assistance and Social Services, 556 P.2d 514, 519 (Wyo. 1976); Illinois vs. Brendendick, 393 N.E.2d 675 (Ill. Ct. of App. 1979); *but see* In re Rathbun, 128 Vt. 429, 266 A.2d 423 (1970).
36. *See e.g.,* V. PIKE, S. DOWNS, A. EMLEN, G. DOWNS, D. CASE, PERMANENT PLANNING FOR CHILDREN IN FOSTER CARE: A HANDBOOK FOR SOCIAL WORKERS (1977); A. EMLEN, J. LAHTI, G. DOWNS, A. MCKAY, S. DOWNS, OVERCOMING BARRIERS TO PLANNING FOR CHILDREN IN FOSTER CARE (1978).
37. *N.B.* If the parent is not willing to care for his or her children and is unable to provide informed, voluntary consent to relinquish parental rights pursuant to section 5, a finding that the services would be futile will be in order in most instances.
38. *See* J. GOLDSTEIN, A FREUD, and A. SOLNIT, BEYOND THE BEST INTERESTS OF THE CHILD 40-45 (1973); NATIONAL COUNCIL OF JUVENILE AND FAMILY COURT JUDGES (NCJFCJ), MODEL STATUTE FOR TERMINATION OF PARENTAL RIGHTS, Prologue (1977).
39. Ketcham and Babcock, *supra* note 18, at 554.
40. See *e.g., id.* at 554-555. INSTITUTE OF JUDICIAL ADMINISTRATION/AMERICAN BAR ASSOCIATION (IJA/ABA), STANDARDS RELATING TO ABUSE AND NEGLECT, §8.3 (Tentative Draft 1977).
41. MICHIGAN COMPILED LAWS ANNOTATED §712A.19A (Supp. 1978-1979).

capacitated) will not be able to (provide necessary and proper care/discharge his or her parental responsibilities) unless that individual has been evaluated in the manner prescribed in section 7 of this act and either:

(a) . . . ;
(b) . . . .

Section 7. Evaluation.

(1) (a) When it is asserted that a parent will be unable to (provide necessary and proper care/discharge his or her parental responsibilities) within (one (1) year/a reasonable time/the forseeable future) because of a (developmental disability and/or mental illness/mental deficiency/mental incapacity/mental incompetence/mental condition), the court shall appoint a multidisciplinary evaluation team to evaluate the parent. Each party may submit names of qualified experts for the court's consideration.

(b) Each expert appointed to a multidisciplinary team pursuant to this act shall receive at the time of appointment a written description of his or her duties and of the matters to be addressed in the evaluation report.

COMMENT. Despite the acknowledged importance and difficulty of decisions to terminate parental rights, few states provide for thorough evaluation when a termination action has been initiated against a developmentally disabled or mentally ill parent.[42] While there have been no studies of the validity of the evaluation procedures used in termination cases, the appellate court cases which touch on the question reveal highly questionable assessment evidence and apparently little interest by the courts in obtaining more reliable evaluative information.[43] Whether or not these cases were correctly decided, the quality of the information available raises serious questions regarding the process for abrogating fundamental rights. Accordingly, this provision, together with sections 3(5), 3(6), and 6(1) require a multdisciplinary evaluation of mentally ill or developmentally disabled parents before a court can consider whether to terminate their parental rights. As discussed in more detail below, the examiners must assess both the abilities and disabilities of the parents, the home environment, the possibilities of improvement, and the services required. The obligation of obtaining and bearing the cost of the evaluation is placed on the court in order to avoid imposing an inordinate burden on the parent. Section 3(5) sets forth the fields of expertise which should be represented on a multidisciplinary evaluation team and the qualifications required of its members. In selecting the team, the court should accommodate the preferences of the parties to the extent that it is possible to do so without appointing more than one expert with the same specialty. In order to facilitate and assure the fairness of the identification and selection of qualified experts for evaluations in termination, guardianship, criminal and delinquency proceedings, enacting jurisdictions should consider establishing the evaluation overnight board described in section 14 of the Model Developmentally Disabled Offender Act.[44]

Paragraph (b) of this subsection addresses a problem which has hampered the working relationship between the courts and the medical and mental health professions. All too often, appointed experts do not understand the legal doctrines and standards which must be applied and lawyers and judges misinterpret technical terms and professional caution.[45] The written explanation to be furnished to each member of a multidisciplinary team is seen as one means of remedying this problem.

(2) The evaluation shall be conducted so as to minimize interference with the parent's activities and intrusion into the parent's privacy.

COMMENT. Except in extraordinary circumstances, evaluations should be conducted on an outpatient basis at a time which is mutually convenient to the parent(s) and the expert performing the evaluation. Only that information which is essential should be collected, and great care should be taken so as not to stigmatize the family among their neighbors and friends.

---

42. See Chart III, *supra*.
43. For example, in State vs. McDonald, 201 NW.2d 447 (Iowa 1972), the primary evidence regarding the degree of the parent's mental retardation were I.Q. tests administered 5 to 6 years earlier while the parents were in high school. The record cited by the United States District Court in Alsager vs. District Court of Polk County, Iowa 406 F. Supp., 10 (S.D. Ia. 1975) contained no evaluation evidence. In In re Rathbun, 128 Vt. 429, 266 A.2d 423 (1970), the court relied solely on the testimony of a general practitioner who when asked if the mother's "condition would be characterized as falling within the category of mental illness . . ." replied 'Yes. It is a personality defect and they are closely related. Her chief mental illness is retardation and a tendency to develop psychosis under stressful situations.' " (*sic*). Similarly in L. and L. vs. Big Horn County Department of Public Assistance and Social Services, 556 P.2d 514 (Wyo. 1976), the court found that a family doctor who had run drug studies at a school for persons with mental retardation was qualified to testify on the mental age of the parents purely on the basis of his personal knowledge of them. See also In re Appeal in Pima County, Juvenile Action No. S-111 (Ariz. Ct. of App. 1975); *but see* Overturf vs. District Court of the 20th District, 602 P.2d 850 (Colo. 1979).
44. *Supra*, at note 9.
45. Bennett, *Competency to Stand Trial: A Call for Reform*, 59 JOURNAL OF CRIMINAL LAW, CRIMINOLOGY AND POLICE SCIENCE 569 (1968); A. STONE, MENTAL HEALTH AND LAW: A SYSTEM IN TRANSITION, 202 and 224 (1975); Ennis, *CLU Client Crazy But Competent*, 17 CIVIL LIBERTIES IN NEW YORK 2 (1969).

(3) The report of the multidisciplinary evaluation team shall include:

(a) A description of the nature and extent of the parent's developmental disabilities and/or mental illness, if any;

(b) A description of the parent's mental, emotional, physical, and educational condition, adaptive behavior and social skills;

(c) A description of the particular services, if any, being used by the parent to (provide necessary and proper care/discharge his or her parental responsibilities);

(d) An opinion regarding the probability that the parent will be able to (provide the necessary and proper care/discharge his or her parental responsibilities) within (one (1) year/a reasonable time/the foreseeable future) and the type of training and services which will facilitate improvement in the parent's ability to (provide necessary and proper care/discharge his or her parental responsibilities);

(e) A description of the techniques or tests on which the above mentioned descriptions and opinions are based;

(f) The names and addresses of all individuals who examined, interviewed or investigated the parent in conjunction with the evaluation;

(g) The signatures of each member of the multidisciplinary evaluation team; and

(h) Any dissenting opinions or other comments by members of the multidisciplinary evaluation team.

COMMENT. The format for the report is designed to assure that the evaluation directly addresses the questions at issue in the proceeding,[46] and to provide the court and the parties with the information necessary to assess the conclusions, facts and opinions presented in the report. As a consequence, the statutory requirements are more detailed than those in current termination legislation.

(4)(a) Upon submission, the evaluation report shall be promptly sent by the court to the attorney of record for each of the parties, the guardian *ad litem*, if any, for each of the parties, and directly to any adult party who is not represented by counsel and who does not have a guardian *ad litem*.

(b) The attorney or guardian *ad litem* for a parent asserted to (have a developmental disability and/or mental illness/mental deficiency/mental condition/be mentally incompetent/mentally incapacitated) shall have the evaluation report explained to the parent in the language, mode of communication and terms which the parent is most likely to understand. Portions of reports which are likely to cause severe harm to the parent may be disclosed through a psychiatrist, psychologist or other appropriate professional.

COMMENT. This model act, like the others in this volume, adheres to the precept that the subject of a proceeding should participate in all important decisions to the maximum extent possible. Therefore, this section provides for the disclosure of the evaluation report to the persons whose parental rights are in jeopardy, as well as to the court and counsel, and guardian *ad litem*. While some reports may include highly sensitive information which might disturb the parent, his or her parental rights appear, on balance, to be of greater importance than his or her "interest in not being perturbed."[47] However, provision is made for disclosure of potentially disturbing portions of the reports through an individual with the professional skills to deal with the questions and emotions involved and thereby, lessen the likelihood of harm.

(5) The multidisciplinary evaluation team may, subject to court approval, incur such expenses as are reasonable and necessary for conducting the evaluation.

COMMENT. None necessary.

Section 8. Notice.

(1) Parents whose parental rights are subject to termination, their children, all other parties to the proceeding, the attorney of record of each parent, child, and other party, and the guardian *ad litem*, if any, of each parent or child are entitled to notice of all pleadings filed, hearings to be held, and actions taken relating to the termination proceeding.

(2) Notices shall set forth:

(a) The rights of the parties;

(b) The procedures and time limits for exercising those rights in the particular proceeding to which the notice refers; and

(c) The date, time, place, purpose and possible consequences of the hearing, if any, to which the notice refers.

(3) (a) Attached to any notice served in conjunction with this act shall be a copy of the pleading or order which gave rise to the notice.

(b) Personal service shall be made on each of the individuals or entities receiving notice in the manner prescribed by (citation to the relevant Rules of Civil Procedure).

(c) The attorney or guardian *ad litem* for a person whose parental rights are subject to termination shall have the contents of all documents served pursuant to this section explained to the parent in the language, mode of communication,

46. See section 6 for a discussion of this issue in another context, *supra*.

47. Suzuki vs. Quisenberry. 411 F. Supp. 1113, 1127 (D. Hawaii 1976).

and terms which that individual is most likely to understand.

COMMENT. As indicated in Chart III, *supra,* all American jurisdictions currently require some form of notice. However, the content of the notice is often unspecified. In jurisdictions in which a termination order is one of several dispositional alternatives following a finding of neglect, the notice of the neglect proceeding may not alert the parent that his or her right to raise her child may be in jeopardy. Full, detailed notice is particularly important to devleopmentally disabled and mentally ill parents who may be less likely to be aware of their rights and possible ramifications of court proceedings, and less likely to ask questions about these matters. However, since the shortcomings of current notice procedures affect all respondents in termination proceedings and since administratively, it is simpler to have a single notice form and procedure for such proceedings, this provision, unlike the preceding sections of the model act, encompass all persons subject to involuntary termination of their parental rights, not just those who are mentally disabled.

Specifically, it requires that all the interested parties receive notice and that notices must describe the purposes, procedures and significance of the pleading, hearing or order which they concern as well as the rights to which the parties are entitled. Notice under this section may be served in any manner consistent with the enacting jurisdiction's rules of civil procedure. Finally in order to increase the likelihood that the subjects of the termination proceeding will understand what is being alleged, what the proceedings entail, what the consequences may be, and what their rights are, all notices must be explained personally. While some individuals will not be able to grasp the meaning of the notice no matter how well it is explained, most will be able to understand if it is explained properly. For example, an explanation of the notice in English to a respondent who understands only Spanish, Vietnamese, or Navajo is of little use whether or not that individual has a developmental disability or mental illness. Attempting to communicate verbally or through a standard printed form to someone who can only understand sign language or read Braille, or couching an explanation in technical legal terms is equally futile. Yet, in too many instances, the subjects of legal proceedings fail to receive proper notice, not because they are unable to understand but because no one has made the effort to present the information in a manner that is comprehensible to them. By specifying that the notice must be in the language, mode of communication and terms which the subject of the intervention proceedings is most likely to understand, this provision mandates that such an effort be made. Responsibility for personally explaining or having the notice and accompanying materials explained is assigned to the attorney or guardian *ad litem* for a parent. It is anticipated that attorneys and guardians *ad litem* are likely to have the best access to the parent, to have an understanding o the proceedings, and to be able and willing to identify and obtain any special services necessary to provide the explanation. As an alternative, some jurisdictions may wish to assign this duty to the office or agency which normally serves notices and to require verification on the form demonstrating receipt of service that the explanation was given.

Section 9. Right to Counsel; Duties of Counsel; Compensation

(1) (a) All parties to a termination proceeding shall be entitled to be represented by counsel.

(b) An attorney shall be appointed to represent a parent who is subject to having his or her parental rights terminated if:

(i) The parent is unable to retain counsel for financial reasons; or

(ii) The parent is asserted to (have a developmental disability and/or mental illness/ mental condition/mentally incapacitated/mentally incompetent) and independent counsel has not been retained by or on behalf of that individual.

(c) Attorneys appointed to represent a parent in a termination proceeding shall be entitled to reasonable compensation. If the parent is unable for financial reasons to pay such compensation in whole or in part, the court shall award to a private attorney appointed pursuant to this subsection, an amount representing reasonable compensation for the services performed, or shall appoint (the name of the Public Defender or Legal Services Office) to represent the parent. In setting the amount of compensation, the court shall take into consideration any sum paid by or on behalf of the parent.

COMMENT. This subsection gives all parties to a termination proceeding — the parent(s), the child, and the petitioning agency, a right to counsel. Furthermore, it mandates appointment of counsel for at least those parents who are indigent or who are alleged to have a developmental disability or mental illness. Currently, about half of the states provide for appointment of counsel for persons subject to termination of their parental rights,[48] and the importance of counsel in termination proceedings has been stressed in most recent sets of standards or recommended statutes.[49] Like the provision on notice, this section covers all

48. See Chart III, *supra;* see *e.g.* DISTRICT OF COLUMBIA CODE §16-2304 (1973) and DISTRICT OF COLUMBIA SUPERIOR COURT RULES — NEGLECT RULE 20.

49. See NACJJDP, *supra* note 16, at §§3.132, and 3.185; NATIONAL ADVISORY COMMITTEE ON CRIMINAL JUSTICE STANDARDS AND GOALS, *supra* note 31, at §16.6; IJA/ABA, *supra* note 25, at §5.1; NCJFCJ, *supra* note 23, at §7(2); MODEL ACT TO FREE CHILDREN FOR PERMANENT PLACEMENT §§12 and 14 (1978).

parents subject to a termination proceeding, not just those with a mental disability, since the presence of a developmental disability or mental illness may not be apparent at the initial stages of the proceeding when the availability of counsel is critical.[50]

The subsection also makes clear that attorneys are entitled to compensation commensurate with the services they perform. When payment of some of all of the compensation would impair the financial independence of the subject of the proceeding, the cost of counsel should be borne by the state. This can be accomplished by compensating private attorneys appointed by the court or appointing the agency normally responsible for providing legal representation to individuals who are indigent. Any additional expense to the public resulting from the appointment of counsel should be viewed as a necessary cost of maintaining the integrity of the judicial intervention process when fundamental rights are at stake.

(2) The principal duty of an attorney representing a parent in a termination proceeding is to represent zealously that individual's legitimate interests. At a minimum, this shall include:

(a) Personal interviews with the parent;

(b) Explaining or having explained to the parent in the language, mode of communication and terms which the parent is most likely to understand, the nature and possible consequences of the proceeding, the alternatives which are available, and the rights to which the parent is entitled; and

(c) Securing and presenting evidence and testimony and offering arguments to protect the rights of the parent and further that individual's interests.

COMMENT. Reflecting the canons of the ABA Code of Professional Responsibility, this section requires retained or appointed attorneys for the subjects of termination proceedings to be zealous advocates for the lawful interests of their clients.[51]

> In our government of laws and not of men (and women), each member of our society is entitled to have his (or her) conduct judged and regulated in accordance with the law; to seek any lawful objective through legally permissible means; and to present for adjudication any lawful claims, issue or defense.[52]

(3) To the maximum extent possible, a parent alleged or found to have a developmental disability or a mental illness shall remain remain responsible for determining his or her interests. However, an attorney may seek appointment of a separate guardian *ad litem* when the circumstances set forth in section 10 of this act apply.

COMMENT. Consistent with the position adopted throughout these model statutes that, to the maximum extent possible, persons who have a mental disability should act in their own behalf and participate in all important decisions which affect them, this subsection provides that an attorney should not automatically assume the role of a guardian *ad litem.* It is for the client to decide what his or her interests are and what course will best promote those interests.

> In certain areas of legal representation not affecting the merits of the case or substantially prejudicing the rights of a client, a lawyer is entitled to make decisions on his (or her) own. But otherwise the authority to make decisions is exclusively that of the client and, if made within the framework of the law, such decisions are binding on his (or her) lawyer.[53]

It is anticipated that in most cases, the client will be the one to make those decisions. Attorneys may seek the appointment of a guardian *ad litem* only when they have reason to believe that the two sets of conditions set forth in section 10 of this act have been met. The first is that the subject of an intervention proceeding is unable to determine his or her interests without assistance because of youth or because his or her ability to receive and evaluate information or to communicate decisions is impaired. The second is that a client who is unable to determine his interests without assistance either has no limited personal guardian or personal guardian who can provide such assistance or has interests which conflict with those of the individual serving in a guardianship capacity. In seeking appointment of a guardian *ad litem* the attorney will have to weigh the possible prejudicial effects on the client's assertion that he or she is able to meet the standard for retaining his or her parental rights.

Section 10. Procedures and Criteria for Appointment of a Guardian *ad Litem;* Duties and Powers

(1) (a) A parent subject to termination of his or her parental rights, the attorney for that individual, the limited personal guardian, personal guardian, if any, for that individual, and/or anyone interested in the welfare of that individual, may file an application to have a guardian *ad litem* appointed.

(b) The application shall set forth:

(i) The name and address of the parent;

---

50. Although beyond the scope of this statute, the mandatory provision of counsel to the children affected by a termination proceeding is not inconsistent with its objectives. See *e.g.*, NACJJDP, *supra* note 16, at §3.132.

51. ABA, *supra,* note 6 at Canon 7.

52. *Id.*, at EC 7-1.

53. *Id.*, at EC 7-7.

(ii) The name, address and interest of the applicant;

(iii) The name and address of the attorney of record of the parent;

(iv) The name and address of the limited personal guardian, personal guardian, limited conservator or conservator, if any, for the parent;

(v) The facts and reasons supporting the request; and

(vi) The name, address and qualifications of the individual proposed for appointment as guardian *ad litem.*

COMMENT. Currently, about a dozen states provide for appointment of a guardian *ad litem* for an allegedly mentally disabled parent who is the subject of a termination proceeding. However, in most of these jurisdictions, there is no separate provision for the appointment of counsel[54] even though the role of an attorney is quite different from that of a guardian *ad litem.*[55] This section seeks to clarify the distinction between these two roles and to assure that a guardian *ad litem* is appointed only when the parents are so severely disabled that they are unable to determine their legal interests for themselves. Subsection (1) permits anyone interested in the well-being of the parent to request the court to appoint a guardian *ad litem* when there is reason to believe that the criteria set forth in subsection (4) are met. It is anticipated that such requests will ordinarily be made by the attorney for the parents pursuant to section 9(3). The form of the application sets forth the basic information required by the court and the parties, including the names and addresses of the persons most likely to have an interest in the application, the specific relief and the basis therefor.

(2) Any party may file an objection to the appointment of a guardian *ad litem* or to the individual proposed for appointment. Objections by a parent asserted to (have a developmental disabiity and/or mental illness/mental deficiency/mental condition/be mentally incapacitated/mentally incompetent) may be communicated to the court by any means, including but not limited to oral communication or informal letter. Objections by other individuals or entities shall be in writing.

COMMENT. Any party may challenge the need for appointing a guardian *ad litem* or the eligibility of the individual nominated for appointment in the petition. The objection should state the name of the objecting party and the facts and reasons on which it is based. In order to facilitate direct participation by the subject of the termination proceedings, objections by the subject of the proceeding need not be in writing nor comply with the usual formal requirements for written pleadings.

(3) (a) The court may grant or deny an application without a hearing unless a timely objection is submitted.

(b) If an objection is filed, the court shall hold an informal, *in camera* hearing.

COMMENT. If no objection is filed, the court may make its decision on the basis of the information contained in the application. If an objection is filed, the court must hold an informal hearing. The question of what procedural safeguards should be required when there is an objection to a request for appointment of a guardian *ad litem* raises some difficult issues. Even though the criteria for appointment of a guardian *ad litem* are different from those which must be met in order to terminate parental rights,[56] in some cases the appointment of a guardian *ad litem* could increase the likelihood that the parent-child relationship may be severed. On the other hand, if a formal hearing were required following the filing of a request by the parent's attorney, it would follow that a separate attorney would have to be appointed to represent the parent at that hearing. If that attorney also had reason to believe that appointment of a guardian *ad litem* was necessary, the cycle would begin anew. Accordingly, this section provides solely for an informal hearing in the judge's chambers where each party can present his or her views and the basis therefor, and the matter can be disposed of without delaying determination of the primary issues. Although the hearing is informal, the notice requirments set forth in section 8 apply and a record of it should be made.

(4) The court shall appoint guardians *ad litem* for a parent subject to termination of his or her parental rights if it is shown that because of impaired ability to receive and evaluate information effectively regarding the proceeding, or because of impaired ability to communicate decisions regarding the proceeding, they cannot determine their own interests without assistance, and either:

(a) No limited personal guardian or personal guardian has been appointed;

(b) Their interests and those of their parent, limited personal guardian or personal guardian conflict; or

(c) The appointment is otherwise required in the interests of justice.

COMMENT. This provision sets forth the criteria which the court should use in reviewing both uncontested and contested applications for appointment of a guardian *ad litem.* It requires the judge

54. See Chart III, *supra.*

55. For a discussion of this difference and a description of the duties of counsel and guardians *ad litem* in termination proceedings, see sections 3(3) and (4), 9(3) and 10(6).

56. See section 6.

to make two determinations. The first is that the parent is unable because of age and inexperience or because of a mental disability to weigh the alternatives and determine, without assistance, the course of action which best promotes his or her welfare. The second is that there is no natural or appointed guardian who is able and willing to provide this assistance. The criteria are explicitly limited to a determination of interests in the termination proceeding.

(5) The court may appoint as guardian *ad litem* any competent adult other than the parent's attorney whose interests do not conflict with those of the parent. Preference shall be given to a qualified individual nominated by the parent.

COMMENT. Any adult who is not providing or responsible for providing some service or treatment to the parent or child involved in the termination proceeding, or who is not a creditor of either the parent or child, or whose interests do not otherwise conflict with those of the parent may be appointed guardian *ad litem*. The parent's attorney is explicitly prohibited from being appointed. The reason for this was succinctly stated by the Supreme Court of Vermont in a case involving a minor.

> [A] lawyer attempting to function as both guardian *ad litem* and legal counsel is cast in the quandary of acting as both attorney and client, to the detriment of both capacities and the possible jeopardizing infant's [client's] interests.[57]

(6) Guardians *ad litem* shall assist the individuals for whom they have been appointed to determine their interests in regard to the legal proceedings in which they are involved. If an individual is wholly incapable of determining his or her own interests, the guardian *ad litem* shall make that determination and advise counsel accordingly. In so doing, guardians *ad litem* shall:

(a) Inquire thoroughly into all the circumstances that a prudent individual in the position of the person for whom the guardian *ad litem* has been appointed would consider in determining his or her interests in the proceedings; and

(b) Encourage the individual whom they are serving to participate, to the maximum extent of that individual's ability, in all decisions and to act on his or her own behalf on all matters in which he or she is able.

COMMENT. This subsection sets forth the duties and powers of a guardian *ad litem*. It stresses that the subject of the termination proceeding should be encouraged to take as active a role as he or she is able. The extent to which a guardian *ad litem* may act as a substitute decision-maker rather than as an advisor should be indicated by the court in its appointment order.

(7) The court should inform guardians *ad litem*, upon appointment, of their duties and powers.

COMMENT. Since few people are likely to be familiar with the role of a guardian *ad litem* prior to being appointed, the court is required to explain the duties and powers assigned to a guardian *ad litem* at the time of appointment. This explanation should be in non-technical language to the greatest degree possible.

Section 11. Authorization of Appropriations

To carry out the purposes of this act, there is authorized to be appropriated $______ for the fiscal year ending _____, $______ for the fiscal year ending _____, and $______ for the fiscal year ending _____.

COMMENT. Every effort has been made to limit the cost of implementing this act. However, many states will need to appropriate or transfer some additional funds in order to provide for the comprehensive evaluations required by section 7. Some enacting states will also have to provide funds to compensate attorneys pursuant to section 9(1). However, in relation to the fundamental rights at stake and the significance of the decisions for the parent, for the child, and for the public (which may have to bear the cost of long-term foster care for the child or services to the parents), this is a small price to pay for a signficantly improved decision-making process.

Section 12. Severability of Sections

If any section, subsection, paragraph, sentence, or any other part of this act is adjudged unconstitutional or invalid, such judgment shall not affect, impair or invalidate the remainder of this act, but shall be confined to the section, subsection, paragraph, sentence or other part of this act directly involved in the controversy in which the judgment was rendered.

COMMENT. None necessary.

Section 13. Effective date

This act takes effect ____________________

COMMENT. None necessary.

---

57. In re Dobson, 125 Vt. 165, 168, 212 A.2d 620, 622, (1965); *cf.* ABA, *supra* note 34 at EC-12.

1.3 TESTAMENTARY CAPACITY*

Table of Contents

INTRODUCTION

Testamentary disposition is the passing of property upon the death of the owner to his or her beneficiaries or heirs in accordance with the terms of a will or last testament.[1] State statutes list certain requirements which must be satisfied if a will is to be considered valid. Among these is the requirement, which virtually all state statutes impose,[2] that the testator, the writer of the will, must possess testamentary capacity, i.e. be mentally competent at the time of executing the will. This requirement seems to reflect two public policy objectives. First, the state, acting as parens patrie, is said to have an interest in protecting persons from testamentary acts which would not be undertaken had the person a full understanding of those acts and their consequences. Thus, the requirement of testamentary capacity reflects the belief that the testator should know and understand what he or she is doing and what the act will accomplish after his or her death. Second, the requirement attempts to insure that individuals with moral and legal "claims" on the testator are not lightly passed over in the testamentary process.[3] Since oftentimes those persons who appear deserving of testamentary consideration are relatives of the deceased, it is sometimes asserted that the law of wills is in part directed toward family harmony[4] and, more broadly, toward preservation of a societal structure based on the nuclear family.[5] To this end the law looks with suspicion on significant departures from normal intra-family distributions and clearly disfavors those departures which appear to have been made in an arbitrary and capricious manner.[6]

The law governing testamentary capacity recognizes various degrees of mental unsoundness, and not every degree of unsoundness is sufficient to destroy testamentary capacity; or to state it another way, absolute soundness of mind and memory in every respect is not essential to testamentary capacity. There may be testamentary incapacity due to mental disability, but mental illness, weakness or retardation alone does not establish incapacity, nor does one need average intelligence since the level of mental capacity required is low.[7]

Few cases deal with the particular type or level of mental impairment necessary to establish incapacity. One Pennsylvania court attempted to deal with the issue in the context of mental retardation and found that a person possessing the intellectual abilities of a five-year-old was incapable of devising an estate of 150,000 dollars in one lump sum to a first cousin,[8] a result which is reconcilable with the terms of the general test outlined below. Most courts, however, have taken the position that if a person suffers from a mental disability, its particular etiology or nature is not of great impor-

*This section surveys existing statutory law and judicial standards on the capacity of disabled individuals to make a valid will. It does not attempt to deal with questions of estate planning and will drafting involving disabled persons, whether as maker of or beneficiary under a will, since these two topics, while of great interest to persons involved in the field, are beyond both the scope and the resources of this project. For those interested in these topics see Effland, *Trusts and Estate Planning* in THE MENTALLY RETARDED CITIZEN AND THE LAW 116 (M. Kindred, ed., 1976); Gorman, *Planning for the Physically and Mentally Handicapped*, 11 INST. ESTATE PLANNING 14.1-47 (1977); and Comment, *Legal Planning for the Mentally Retarded*, 10 IDAHO L. REV. 245 (1974).

[1] This passing of property occurs through the probate process. Probate refers to the "proving" of a will, i.e., the establishing in probate court that the will meets all requirements for a valid will. The term probate is popularly used to refer to the administration and distribution of a testator's estate as well. Where there is no will or the will is declared invalid, state intestacy (meaning without a will) statutes govern the distribution of the property.

[2] *See* State Statute Chart *infra*. *See also* Weihoffen, *Mental Incompetency to Make a Will*, 7 NAT. RESOURCES J. 89 (1967). This article contains a thorough discussion of testamentary capacity.

[3] *See* Green, *Public Policies Underlying the Law of Mental Incompetency*, 38 MICH. L. REV. 1189, 1219 (1940).

[4] *Id.*

[5] Epstein, *Testamentary Capacity, Reasonableness and Family Maintenance: A Proposal for Meaningful Reform*, 35 TEMP. L.Q. 231, 232 (1962).

[6] *Id.* at 241.

[7] *See* Weihoffen, *supra* note 2 and cases cited therein.

[8] In re Estate of Glesenkamp, 378 Pa. 635, 107 A.2d 731 (1954). *But see* In re Estate of Teel, 14 Ariz. App. 371, 483 P.2d 603 (1971).

tance from the testamentary standpoint.[9] The vital requirement is only whether the testator's abilities at the time of execution were sufficient to satisfy the requirements of the general situational test.

The law also has not confused other tests of legal incompetency with testamentary incapacity. Thus, every sane person is generally presumed to have testamentary capacity, but "sanity", a criminal responsibility test, and "testamentary capacity" are not synonymous. Testamentary capacity is not the same as the ability to transact ordinary and common business,[10] nor, according to most authorities and jurisdictions, is it the same as capacity to execute a deed or contract.[11] Indeed, the solid weight of authority indicates that testamentary capacity is unrelated to any other form of legal capacity and should not be equated with, for example, contractual capacity.[12] It is especially enlightening in this area that courts in Maryland and the District of Columbia, even though their statutes have required contractual capacity, usually ignored this element and applied the common law requirements, which will be discussed in the following section.[13]

Similarly, a prior adjudication of incompetency or appointment of a guardian is never conclusive.[14] The factors considered in a hearing for the appointment of a guardian are different than the factors used to determine testamentary capacity.[15] The fact that the testator was under guardianship at the time of the will's execution, however, is relevant and admissible on the issue of testamentary capacity. In a number of states such evidence will raise a rebuttable presumption of incapacity.[16]

It has often been said that institutionalization is of even less probative value than an adjudication of incompetency and thus it does not raise a presumption of any sort.[17] At most, the testator's institutionalization at the time of execution of the will, will be allowed as probative evidence and, in rare cases, will be held to make out a prima facie case of testamentary incapacity.

THE GENERAL TEST OF CAPACITY

Five elements are generally recognized as necessary for a finding of testamentary capacity. These elements reflect the public policies mentioned above. First, the testator must understand the nature of the act in question; that is, he or she must be aware that the instrument being executed is in fact a legally operative will.[18] Second, the testator must have a general understanding of the nature and extent of his or her property.[19] This does not require a comprehensive knowledge of all items owned by the testator but only a reasonably complete general understanding is envisioned. Third, the testator must have a general recognition of those persons who are "the natural objects of his bounty." By this is meant those persons who by virtue of their familial relationships or service to the testator can be considered to have some equitable or moral claim to his or her property.[20] It is this requirement which introduces the public policy in favor of family harmony into the law of testamentary capacity. It is not, however, true that all family members can be considered the natural objects of the testator's bounty solely on the basis of their blood relationship.[21] Each individual's relationship with and service to the testator are considered in determining who the testator should have considered when deciding on the distribution of the property.[22] Fourth, courts often require that the testator have a reasonable comprehension of the distributive scheme to be utilized.[23] This requirement essentially constitutes an intent element since it cannot be said that an individual intends to devise property in a manner he or she never actually understood. Finally, an additional requirement, that the testator be able to appreciate all of the above elements in relation to each other is imposed, either implicitly or explicitly, by a number of courts.[24]

Statutes concerning testamentary capacity exist in every state but they add little of significance to the common law test just outlined. Nearly every statute requires that the testator be of "sound" or "sound and disposing" mind in order to execute a valid will[25] and these terms have, in turn, been uniformly construed by the courts to sanction the continued use of the common law analysis.[26]

---

[9] Epstein, *supra* note 5, at 235.

[10] *Id.* This is a minority view. *See* cases cited therein.

[11] Maryland and the District of Columbia have required that the testator be of sound mind and have the ability to execute a valid deed or contract. Maryland's statute was recently revised, omitting the above language and substituting "legally competent to make a will."

[12] *See* Weihoffen, supra note 2, at 91 and cases cited therein.

[13] *Id.* *See supra* note 11.

[14] *See* Note, 16 S. CAL. L. REV. 355 (1943). The District of Columbia is an exception in this regard, requiring that one be competent to execute a valid contract in order to execute a will. D.C. CODE §18-102 (1970).

[15] **Compare the test of testamentary capacity outlined below with the materials in Chapter 7: Guardianship and Conservatorship.**

[16] *See* Weihoffen, *supra* note 2, at 93.

[17] *Id.* at 94.

[18] *Id.* at 90.

[19] *Id.*

[20] *Id.*

[21] ATHINSON, HANDBOOK OF THE LAW OF WILLS 51 (ed. 1953).

[22] *Id.*

[23] Weihoffen, *supra* note 2.

[24] Note, *Psychiatric Assistance in the Determination of Testamentary Capacity*, 66 HARV. L. REV. 1116, 1116-17 (1953).

[25] See State Statute Chart *infra*.

[26] Epstein, *supra* note 5, at 236.

THE GENERAL DOCTRINE IN PRACTICE—EFFECT ON MENTALLY DISABLED PERSONS

In the reality of probate practice the tests for testamentary capacity have been shown to be quite lenient. Of all the various competency requirements imposed on mentally disabled persons throughout our laws, the requirements for execution of a valid will are generally conceded to be the least restrictive of all.[27] This is because the test is to some extent variable, with no specific level of intelligence, aptitude or understanding being sufficient in every case. Yet, the tests are applicable to all testators and provide quite specific guidance in making individual determinations. Furthermore, they possess a very legitimate and rational relationship to the state interests they are designed to further. Lastly, the probate process provides a mechanism for properly making individual determinations. Mental disability advocates recognize that a flexible competency standard based upon functional characteristics will in the long run be far less onerous to disabled persons than the stigma ladden, categorical disqualifications (i.e., those that prohibit "idiots", "insane persons" and the like) which are still prevalent in other areas of the law. Certainly the standard as applied meets the criteria outlined in the introduction to these materials on Personal and Civil Rights better than existing standards in other areas within this topic.

This is not to say that assessing testamentary capacity is without difficulty. Unlike other questions of capacity, this one is decided after the fact, i.e., after the death of the maker of the will. This can make assessment of capacity more difficult.

PROBLEMS IN ASSESSING TESTAMENTARY CAPACITY

Perhaps the single most unsatisfactory aspect of probate is the usual lack of reliable information on which the court can base an opinion as to capacity. Because capacity cases are often decided on little more than conjecture, a growing tendency has developed for dissatisfied relatives to contest testamentary capacity on a "what-do-we-have-to-lose" basis, oftentimes without any real basis in fact.[28] The nature of the proceeding, occurring as it does after the testator's death, and the rules of evidence which prevent the introduction of some relevant information can combine to create an unsatisfactory climate for proper presentation of the issue.[29] For example, testimony is frequently presented by psychiatrists or clinical psychologists who had no contact whatsoever with the testator while alive and are asked to make an evaluation solely on the basis of anecdotal evidence accumulated by the parties and presented to the expert in the form of protracted hypothetical questions.[30] Each party in questioning the expert will include only evidence favorable to their position, and rarely are the courts or the experts themselves satisfied with the reliability of testamony elicited in this manner.[31]

One commentator has suggested that a more exacting look at psychiatric and psychological classifications and clinical signs might alleviate some of the arbitrariness in contests based on capacity and might provide precedent which could aid lawyers in determining whether a projected contest is meritorious enough to be pursued.[32] The result of psychiatric and psychological examinations may be roughly transferable, in some instances, into the legal cognitive tests outlined above and so may deserve greater consideration in probate proceedings and practice. However, the results of such examinations are useful only insofar as they assist in applying the tests for testamentary capacity outlined above and in avoiding the problems identified here.[33]

SOME POSSIBLE SOLUTIONS

The most effective solution to the problem is not a matter for statutory reform but rather one to be addressed through improvements in probate practice. It has been suggested that lawyers can, and probably should, arrange as a matter of course for adequate professional evaluation of a testator's mental capacity contemporaneous with the execution of any will.[34] At least it ought to be done where there is any question concerning the testator's capacity. Evaluations may be geared toward the effect of specific disabilities upon capacity and provide the probate court, acting after the fact, with reliable information on which to base its findings. Records of diagnostic and evaluative results of such testing are ordinarily considered business records and as such are admissable as an exception to the hearsay rule in most jurisdictions.[35] Even if the physician or psychologist who performed the original evaluation cannot testify at the later probate hearing, his or her records will be

---

[27] *See e.g.*, In re Safer's Will, 19 App. Div. 2d 725, 242 N.Y.S. 2d 445 (1963).

[28] Epstein, *supra* note 5, at 231.

[29] Note, *supra* note 24, at 1118-19.

[30] *Id.*

[31] Hulbert, *Probate Psychiatry — A Neuro-Psychiatric Examination of Testator from the Psychiatric Viewpoint*, 25 ILL. L. REV. 288 (1930). The author was a practicing psychiatrist dissatisfied with the probate procedures of his time, most of which are still operative today.

[32] *Id. See also* Stephens, *Probate Psychiatry — Examination of Testamentary Capacity by a Psychiatrist as a Subscribing Witness*, 25 ILL. L. REV. 276 (1930).

[33] As previously noted, the tests utilized for determining testamentary capacity are well suited for carrying out the public policies upon which they are based. It is not suggested that these be replaced de facto by medical categorizations of mental disability whose purpose is different. Indeed, it is this type of decision-making by classification which is to be avoided.

[34] Hulbert, *supra* note 31; Stephens, *supra* note 32; Note, *supra* note 24.

[35] Stephens, *supra* note 32, at 286.

of inestimable help to experts unfamiliar with the testator who are asked to testify after the fact.

It has been asserted that lawyers have tended to avoid such evaluations because they may suggest the lawyer's own doubts as to a clients competency and consequently, may serve to promote will contests.[36] If, however, these examinations were conducted as a matter of course, no single evaluation would indicate anything about the lawyer's own views.[37] In addition, a contemporaneous examination would allow the lawyer to more effectively evaluate the client's capacity, prevent the drafting of wills for clients who are incompetent to execute them,[38] and allow for the development of trust devices or other alternative distribution schemes to assure that the disabled clients wishes are, to the greatest extent possible, carried out. Remember that by the time a will is challenged in court on the basis of capacity there remains little that either a lawyer or a judge can do to preserve the testator's plan if lack of testamentary capacity is found.

A related reform, judicial in character, may also contribute to a more equitable determination of testamentary capacity. One commentator has noted that faced with inadequate factual material on which to base a competency determination courts sometimes appear to decide cases on the basis of the objective reasonableness of the dispositive scheme embodied in the contested will.[39] This practice is completely at odds with the long-recognized doctrine that a will may be "as eccentric, as injudicious or as unjust as caprice, frivolity or revenge can dictate" provided that the testator has the mental capabilities to execute a valid will under the tests already described.[40] Since the entire rationale for personal wills, as opposed to statutorily mandated disposition schemes, is to afford each individual an element of sovereignty with respect to personally held property, any test which puts a premium on conformity is in conflict with the law of wills generally. It would be more appropriate for courts to concentrate exclusively on the testator's capacity rather than on the fairness of his or her desired ends as illustrated by his or her will.[41] The additional data collected by evaluations like those already suggested would remove the need for courts to resort to this latter, inconsistent form of analysis.

Moreover, if necessary reforms in probate practice were carried out, attorneys at the time a will is contemplated could reasonably assess their client's capacity and make alternative arrangements should capacity be found wanting. Adequate evidence of capacity would also serve to deter the bringing of will contests based on mere speculation and would aid courts in reaching appropriate decisions when meritorious contests are brought.

---

36 *See* SINGER & KROHN, INSANITY AND LAW 343 (1924).

37 Note, *supra* note 24, at 1118.

38 *Id.*

39 Epstein, *supra* note 5, at 231-32. Georgia has even codified this "reasonableness" evaluation. GEORGIA CODE ANN. §113-205 (1975) states in part, "In case of doubt as to the extent of (the testator's disability), the reasonable or unreasonable disposition of his estate should have much weight. . . :"

40 *See e.g.*, Schneider v. Vosburgh, 143 Mich. 467, 106 N.W. 1129 (1906). *See also* Athinson, *supra* note 21.

41 Epstein, *supra* note 5, at 248.

CHART OF STATE TESTAMENTARY CAPACITY STATUTES

State	Sound Mind	Other Standards
ALABAMA Code (1975)	tit. 43, §43-1-1	
ALASKA Stat. (1972)	§13.06.010	
ARIZONA Rev. Stat. Ann. (1975)	§14.2501	
ARKANSAS Stat. Ann. (1971)	§60-401	
CALIFORNIA Prob. Code (West Supp. 1978)	§20	
COLORADO Rev. Stat. Ann. (1973)	§15-11-501	
CONNECTICUT Gen. Stat. Ann. (West 1960)	§45-160	
DELAWARE Code Ann. (Supp. 1978)		Sound and disposing mind and memory tit. 12,§201
DISTRICT OF COLUMBIA Code (1973)		Sound and disposing mind and capable of executing a valid deed or contract. §18-102
FLORIDA Stat. Ann. (West 1976)	§732.501	
GEORGIA Code Ann. (1975)		Every person may make a will, unless laboring under some legal disability arising from a want of capacity.[1] §113-201
HAWAII Rev. Stat. Ann. (1976)	§560:2-501	
IDAHO Code Ann. (1979)	§15-2-501	
ILLINOIS Ann. Stat. (Smith-Hurd 1961)		Sound mind and memory ch.3,§42

State	Sound Mind	Other Standards
INDIANA Code Ann. (Burns Supp. 1978)	§32-1-2-2	
IOWA Code Ann. (West 1964)	§633.264	
KANSAS Stat. Ann. (1976)	§59-601	
KENTUCKY Rev. Stat. Ann. (Baldwin 1978)	§394.020	
LOUISIANA Civ. Code Ann. (West 1952)	art.1475	
MAINE Rev. Stat. Ann. (Supp. 1978)	tit.18,§1.3607	
MARYLAND Est. & Trusts Code Ann. (1974)		Legally competent [2] §4-101
MASSACHUSETTS Ann. Laws (Michie/Law. Co-op Supp. 1978)	ch.191,§1	
MICHIGAN Comp. Laws Ann. (1976)	§702.1	
MINNESOTA Stat. Ann. (West 1975)	§524.2-501	
MISSISSIPPI Code Ann. (Supp. 1978)		Sound and disposing mind §91-5-1
MISSOURI Rev. Stat. (Vernon 1956)	§474.310	
MONTANA Rev. Codes Ann. (1976)	§91A-2-501	
NEBRASKA Rev. Stat. (1975)	§30-201	
NEVADA Rev. Stat. (1973)	§133.020	

State	Sound Mind	Other Standards
NEW HAMPSHIRE Rev. Stat. Ann. (1974)		Sane mind §551:1
NEW JERSEY Stat. Ann. (West Supp. 1979)		Invalid if made by idiot, lunatic or person of unsound mind or memory §3A:2A-3
NEW MEXICO Stat. Ann. (1978)	§45-2-501	
NEW YORK Est. Powers & Trusts Law (McKinney 1967)		Sound mind and memory §3-1.1
NORTH CAROLINA Gen. Stat. (1976)	§31-1	
NORTH DAKOTA Cent. Code (1976)	§30.1-08-01	
OHIO Rev. Code Ann. (Page 1975)		Sound mind and memory §2107.02
OKLAHOMA Stat. Ann. (West 1970)	tit.84,§41	
OREGON Rev. Stat. (1972)	§114.020	
PENNSYLVANIA Cons. Stat. Ann. (Purdon Supp. 1979-1980)	tit.20,§2501	
RHODE ISLAND Gen. Laws Ann. (1969)		Sane mind §33-5-2
SOUTH CAROLINA Code Ann. (1976)	§21-7-10	
SOUTH DAKOTA Compiled Laws Ann. (1976)	§29-2-3	
TENNESSEE Code Ann. (1977)	§32-102	
TEXAS Prob. Code Ann. (Vernon 1980)	§57	

State	Sound Mind	Other Standards
UTAH Code Ann. (1978)	§75-2-501	
VERMONT Stat. Ann. (1958)	tit.14,§1	
VIRGINIA Code Ann. (1973)	§64.1-47	
WASHINGTON Rev. Code Ann. (Supp. 1978)	§11.12.010	
WEST VIRGINIA Code Ann. (1966)	§41-1-2	
WISCONSIN Stat. Ann. (West 1971)	§853.01	
WYOMING Stat. Ann. (1977)	§2-4-101	

1 See also §§113-202 through 113-205.

2 This statutory language replaces the earlier language requiring sound mind and the ability to execute a valid deed or contract. The old language was identical to the statute of the District of Columbia.

1.4 STERILIZATION

Table of Contents

INTRODUCTION

Developmentally disabled persons have in the last decade successfully overcome many legal obstacles which traditionally excluded them from exercising their full rights as citizens. State laws prohibiting equal access to decent educational programs[1] were challenged in the courts and in many cases eliminated; unfair and inappropriate confinements and procedures relating to confinements were attacked and eliminated;[2] and rights in community settings were developed and defined.[3]

As public awareness substantially increased, Congress began to legislate in the area of disability law. The Rehabilitation Act of 1973[4] prohibited discrimination against handicapped persons in any program receiving federal assistance; the Education for All Handicapped Children Act[5] set up funding for special education programs; and the Developmentally Disabled Assistance and Bill of Rights Act[6] established requirements for habilitation plans for institutionalized persons, along with an advocacy system to protect their rights.

But a most personal individual right — to make one's own decision in the area of procreation — is still frequently denied handicapped individuals.[7] Based upon a fear that mentally disabled persons would lower the level of intelligence of the general society or commit crimes out of proportion to their number, states have utilized unproven scientific theories as a rationale for developing compulsory sterilization laws for these persons. While societal attitudes and the courts have both served to diminish and in some cases eliminate these sterilization laws, the remnants of a century of legislation and ignorance remain.[8] (see Review of State Legislation, *infra.*).

[1]See, e.g., Frederick L. vs. Thomas, 419 F. Supp. 960 (E.D. Pa. 1976), *aff'd*, 557 F.2d 373 (3rd Cir. 1977); Fialkowski vs. Shapp, 405 F. Supp. 946 (E.D. Pa. 1975); Mills vs. Board of Education of the District of Columbia, 348 F. Supp. 866 (D.D.C. 1972); Pennsylvania Association for Retarded Children (PARC) vs. Commonwealth of Pennsylvania, 334 F. Supp. 1257 (E.D. Pa. 1971). For general legal commentary and discussion of right to education litigation, see Haggerty and Sacks, *Education of the Handicapped: Towards a Definition of an Appropriate Education,* 50 TEMPLE L.Q. 961 (1977); Note, *Enforcing the Right to an 'Appropriate' Education: The Education for All Handicapped Children Act of 1975,* 92 HARV. L. REV. 1103 (1979).

[2]See, e.g., Rouse vs. Cameron, 373 F.2d 451 (1966), "involuntary confinement without treatment is shocking," *id.* at 455; civil commitment entails a "massive curtailment of liberty," Humphrey vs. Cady, 405 U.S. 504, 509; the only permissible justifications for committing the mentally ill are: (1) danger to the individual, (2) danger to others, and (3) need for treatment, Jackson vs. Indiana, 406 U.S. 715, 737; Wyatt vs. Aderholt, 503 F.2d 1305, 1312 (5th Cir. 1974); Halderman vs. Pennhurst State School and Hospital, 446 F. Supp. 1295 (E.D. Pa. 1977), which continued by stating that "although this court entertains serious doubts as to whether retarded individuals should ever be subjected to 'commitment,' there is no doubt that under the present case law, the only possible justification for committing the retarded to an institution . . . is to provide them with habilitation, i.e., education, training and care. Failure to provide adequate habilitation may well mean commitment for the life of the retarded individual. Welch vs. Likins, 373 F. Supp. 487, 497 (D. Minn. 1974), *aff'd in part and vacated and remanded in part,* 550 F.2d 1122 (8th Cir. 1977); see also O'Connor vs. Donaldson, 422 U.S. 563 (1975).

[3]See, e.g., Halderman vs. Pennhurst State School and Hospital, Nos. 78-1490, 78-1564, 78-1602 (3rd Cir. 1979); Parham vs. J.L., ____ U.S. ____ (1979); Lake vs. Cameron, 364 F.2d 657, 663 (D.C. Cir. 1966); Eubanks vs. Clarke, 434 F. Supp. 1022, 1027-28 (E.D. Pa. 1977); Lynch vs. Baxley, 386 F. Supp. 378, 392 (M.D. Ala. 1974); Lessard vs. Schmidt, 349 F. Supp. 1078 (E.D. Wis. 1972), *vacated and remanded for a more specific order,* 414 U.S. 473, *order on remand,* 379 F. Supp. 1376 (E.D. Wis. 1974), *vacated and remanded on other grounds,* 421 U.S. 957 (1975), *order reinstated on remand,* 413 F. Supp. 1318 (E.D. WIs. 1976); Wyatt vs. Stickney, 344 F. Supp. 373 (M.D. Ala. 1972); 344 F. Supp. 387 (M.D. Ala. 1972), *enforcing* 325 F. Supp. 781 (M.D. Ala. 1971), *affirmed in relevant part, and decision reversed in part,* Wyatt vs. Aderholt, 503 F.2d 1305 (5th Cir. 1974), cited in Halderman, slip op. at 64.

[4]See 29 U.S.C. §504; 45 C.F.R. Part 84.

[5]20 U.S.C. §§1401 *et seq.;* 45 C.F.R. Part 121a.

[6]42 U.S.C. §§6000 *et seq.;* 45 C.F.R. Parts 1385-7.

[7]*The Legal Rights of People With Mental Disabilities In the Community: A Plea for Laissez Faire,* in LEGAL RIGHTS OF THE MENTALLY HANDICAPPED (Practising Law Institute, 1973) and reprinted in *The Mentally Retarded Citizen and the Law* (1974).

[8]Landman, *The History of Human Sterilization in the United States — Theory, Statute, Adjudication,* 23 ILL. L. REV. 463 (1929); see generally, Note, *Human Sterilization,* 35 IOWA L. REV. 251 (1950); Challener, *The Law of Sexual Sterilization in Pennsylvania,* 57 DICK. L. REV. 298 (1952); O'Hara and Sanks, *Eugenic Sterilization,* 45 GEO. L. J. 30 (1956); Ferster, *Eliminating the Unfit — Is Sterilization the Answer?* 27 OHIO ST. L. REV. 591 (1966); Murdock, *Sterilization of the Retarded: A Problem or a Solution?,* 62 CAL. L. REV. 917 (1974); Burgdorf and Burgdorf, *The Wicked Witch is Almost Dead — Buck v. Bell and the Sterilization of Handicapped Persons,* 50 TEMP. L Q. 995 (1977).

> Sterilization is the surgical means by which both male and female are rendered incapable of reproduction. The operation is much more serious in women than in men. In females the surgeon must do an abdominal operation, removing segments of the fallopian tubes (salpingectomy) and tying the cut ends. In men the operation is relatively simple. Small scrotal skin incisions are made, segments of the *vas deferens* are removed (vasectomy) and the proximal ends of the *vas* are tied. Neither interferes with the desire for sexual intercourse or with its gratification.

Deutsch, *The Mentally Ill in America,* cited in Lindman and McIntyre, *infra* note 11.

Eugenic Sterilization

According to an early critic of compulsory sterilization laws, this country "has the longest legislative history in this . . . field and enjoys the distinction . . . of being the most successful in the execution of the statutes."[9] This observation is not surprising given the well known but controversial studies of such famous families as the Jukes, the Kallikaks, and the Ishmaels which supposedly proved that "feeble-mindedness" was inherited through successive generations and which served to support proponents of the eugenic sterilization movement around the turn of the century.[10]

This movement saw four justifications for compulsory sterilization laws. Sterilizations were supposedly valuable in treating certain diseases which would result in danger to the health of the woman undergoing pregnancy or childbirth, in limiting the population for socio-economic reasons, in punishing criminals and sex offenders, and in creating what some people thought would be a more socially acceptable group of persons.[11] For example, a "Model Eugenic Sterilization Law" of the 1920s described as "unacceptable" and therefore subject to sterilization the following persons: (1) feeble-minded; (2) insane (including psychopathic); (3) criminalistic (including the delinquent and wayward); (4) epileptic; (5) inebriate (including drug-habituates); (6) diseased (including the tuberculous, the syphilitic, and leprous, and others with chronic infections and legally segregable diseases); (7) blind (including those with seriously impaired vision); (8) deaf (including those with seriously impaired hearing); (9) deformed (including the crippled); and (10) dependent (including orphans, ne'er-do-wells, the homeless, tramps, and paupers).[12]

The first attempt to pass such a law occurred in 1897 in Michigan,[13] but the bill was defeated. The Pennsylvania legislature, eight years later, did pass a bill "for the prevention of idiocy," but it was vetoed by the governor, who ridiculed this attempt by stating that

> this bill has, what may be called with propriety, an attractive title. If idiocy could be prevented by an act of assembly, we may be quite sure that such an act would have long been passed and approved in this state, and . . . in all civilized countries. . . . The nature of the operation is not described, but it is such an operation as they shall decide to be safest and most effective. It is plain that the safest and most effective methods of preventing procreation would be cut the heads off the inmates, and such authority is given by the bill to this staff of scientific experts. . . .[14]

Eventually, however, laws were enacted by state legislatures providing for involuntary sterilization.[15] While legislation in some states, including Connecticut and Illinois, went unchallenged in the courts, suits were brought to invalidate laws in a number of other jurisdictions. Indeed, various courts declared the sterilization statutes unconstitutional based upon prohibitions against cruel and unusual punishment, or because they violated the right to equal protection of the law. In other cases, judges held that the procedures resulting in sterilization did not comport with the right to fair hearings and appropriate procedural safeguards consistent with due process of law guaranteed by the Fourteenth Amendment to the U.S. Constitution.[16] Yet, advocates of compulsory sterilization were not deterred, and organizations such as the American

---

[9]Landman, *id.*, at 472.

[10]See, e.g., H. GODDARD, THE KALLIKAK FAMILY (1912), cited in Landman, *id.* at 471; holding of the New Jersey Supreme Court in Smith vs. Board of Examiners of Feeble-minded, *infra* note 16; O'Hara and Sanks, *supra* note 8, at 20; see generally, Note, *Eugenic Sterilization Statutes: A Constitutional Reevaluation,* 14 J. FAM. L. 281 (1975) citing E. GOSNEY and P. POPNOE, STERILIZATION FOR HUMAN BETTERMENT, 70 (1931), A. DEUTCH, THE MENTALLY ILL IN AMERICA (2d ed., 1949) and L. CAVILLI-STORZA and W. BODMER, THE GENETICS OF HUMAN POPULATION (1971).

[11]O'Hara and Sanks, *supra* note 8, at 21; S. BRAKEL and R. ROCK, THE MENTALLY DISABLED AND THE LAW, 183 (1971) and cases and materials cited therein.

[12]Note, *supra* note 10, at 283.

[13]O'Hara and Sanks, *supra* note 8, at 22.

[14]Brakel and Rock, *supra* note 11, at 184, citing Challener, *The Law of Sexual Sterilization in Pennsylvania,* 57 DICK L. REV. 298 (1952).

[15]In 1907, Indiana became the first state to enact a compulsory sterilization statute. Washington State followed in 1909 and California in the same year became the third state to pass a law in this field.

[16] The challenges have relied primarily on three parts of the United States Constitution: U.S. CONST. amend. VIII, which states that "excessive bail shall not be required, nor excessive fines imposed, nor cruel and unusual punishment inflicted"; U.S. CONST. amend. XIV, §1, which states that no state shall "deny any person within its jurisdiction the equal protection of the law"; and U.S. CONST. amend. XIV, §1, which states that no state shall "deprive any person of life, liberty, or property without due process of law".

Note, *supra* note 10, at 281; See, Landman, *supra* note 8 at 472-80 and O'Hara and Sanks, *supra* note 8 generally for an extensive discussion of case law and citations concerning the early eugenic sterilization statutes. See also Smith vs. Board of Examiners of Feeble-minded, 85 N.J.L. 46, 88 A. 963 (S. Ct. N.J. 1913) in which the court was particularly cognizant of the extremes to which the "social darwinism" of the eugenic sterilization movement could be carried. This court emphasized that the majority could easily characterize an unpopular minority or otherwise "different" group as an undesirable whose population must be sterilized:

> There are other things besides physical or mental diseases that may render persons undesirable citizens or might do so in the opinion of a majority of a prevailing legislature. Racial differences, for instance might afford

(continued on next page)

Breeders Association were organized to advocate for such laws.[17]

The Supreme Court Rules on Sterilization

Although enforced sterilization remained controversial and appeared to be generally in disfavor among courts and the public, a U.S. Supreme Court case[18] was to serve as the impetus for the legalization of sterilization laws across the nation. The Virginia legislature's sterilization act, which was the subject of the case, *Buck vs. Bell,* was enacted in 1924 and concluded that a law was necessary because

> . . . both the health of the individual patient and the welfare of society may be promoted in certain cases by the sterilization of mental defectives under careful safeguards and by competent and conscientious authority, . . . and
>
> . . . human experience has demonstrated that heredity plays an important part in the transmission of insanity, idiocy, imbecility, epilepsy, and crime. . . ."[19]

The legislature allowed the superintendent of any state institution which had custody over one of the persons specified to perform a sterilization when "it is for the best interest of the patients and of society that an inmate of the institution under his care should be sexually sterilized."

The act required the service of a copy of the sterilization petition on the inmate and his or her guardian or parents if the inmate was a minor, including notice of the time and place when a hearing would be held, as well as representation by counsel. The act also allowed for the committee, guardian, next friend, the special board considering the sterilization, or the superintendent to appeal to a court.

---

a basis for such an opinion in commuities where that question is unfortunately a permanent and paramount issue. Even beyond all such considerations it might be logically consistent to bring the philosophic theory of Malthus to bear upon the police power, to the end that that tendency of population to outgrow its means of subsistence should be counteracted by surgical interference of the sort we are now considering. 88 A. at 966.

[17]Testimony of Dr. I. Ignacy Goldberg in Pennsylvania Association for Retarded Children vs. Commonwealth of Pennsylvania, 343 F. Supp. 279, 294 (E.D. Pa. 1972). He stated that in about 1900:

> "[T]he wonderful idea of adjusting the individuals to our society became the dumping grounds for children who could not manage in other classes and started to be called classes for the feebleminded, classes for idiots, and so on . . . .
>
> And then the Eugenic Association in the United States started to raise quite of lot of cry that the American Society is going to pieces, mental retardation is hereditary, mentally retardeds are criminals, are prostitutes as the [I.Q.] tests proved. Therefore, something very drastic has to be done.
>
> And in 1912, the Eugenic Society, the Research Section of the Eugenic Society, namely, the American Breeders Association suggested that drastic measures be taken to prevent the Americans from becoming all feebleminded [such as] segregation or segregation during the reproductive period, for women, . . . organizing institutions for feebleminded women of child-bearing age in order to prevent them from having children, . . . compulsory sterilization law for women, and castration for men. . . . Another recommendation was euthanasia. This, of course, just introduced and I hope was not implemented. . . . I really want to point out that the days we are talking about are not so far removed, that the stigma attached to mental retardation is still with us, with the general public."
>
> (citation omitted)

[18]Buck vs. Bell, 274 U.S. 200 (1927).

[19]The Virginia Sterilization Act (Acts 1924, c. 394, p. 569) reads in part as follows:

> "Whereas, both the health of the individual patient and the welfare of society may be promoted in certain cases by the sterilization of mental defectives under careful safeguard and by competent and conscientious authority, and
>
> "Whereas, such sterilization may be effected in males by the operation of vasectomy and in females by the operation of salpingectomy, both of which said operations may be performed without serious pain or substantial danger to the life of the patient, and
>
> "Whereas, the commonwealth has in custodial care and is supporting in various state institutions many defective persons who if now discharged or paroled would likely become by the propagation of their kind a menace to society but who if incapable of procreating might properly and safely be discharged or paroled and become self-supporting with benefit both to themselves and to society, and
>
> "Whereas, human experience has demonstrated that heredity plays an important part in the transmission of insanity, idiocy, imbecility, epilepsy and crime, now, therefore
>
> "1. Be it enacted by the General Assembly of Virginia, that whenever the superintendent of the Western State Hospital, or of the Eastern State Hospital, or of the Southwestern State Hospital, or of the Central State Hospital, or of the State Colony for Epileptics and Feeble-Minded, shall be of opinion that it is for the best interests of the patients and of society that any inmate of the institution under his care should be sexually sterilized, such superintendent is hereby authorized to perform, or cause to be performed by some capable physician or surgeon, the operation of sterilization on any such patient confined in such institution afflicted with hereditary forms of insanity that are recurrent, idiocy, imbecility, feeblemindedness or epilepsy; provided that such superintendent shall have first complied with the requirements of this act.
>
> "2. Such superintendent shall first present to the special board of directors of his hospital or colony a petition stating the facts of the case and the grounds of his opinion, verified by his affidavit to the best of his knowledge and belief, and praying that an order may be entered by said board requiring him to perform or to have performed by some competent physician to be designated by him in his said petition or by said board in its order, upon the inmate of his institution named in such petition, the operation of vasectomy if upon a male and of salpingectomy if upon a female."
>
> 130 S.E. 516, 517 (S. Ct. App. Va. 1925).

In January of 1924, Carrie Buck was adjudicated to be "feeble-minded," and was committed to an institution called the State Colony for Epileptics and Feeble-minded. In September of that year, the Colony's superintendent presented a sterilization petition to the board and a hearing in accordance with the requirements of the Virginia law was held. An appeal was taken to the courts by her guardian and eventually to the Supreme Court of Appeals of Virginia, which ruled in 1925, as had the lower court, that the sterilization statute was constitutional.[20] The record indicated that Carrie Buck was committed to the Colony when her chronological age was 17, and her mental age was nine. Her mother had previously been committed to the Colony, and Carrie was said to have had a child who had "defective mentality." The Virginia high court, in upholding the constitutionality of the statute, held that "unless sterilized by surgical operation, she must be kept in the custodial care of the colony for thirty years, until she is sterilized by nature, during which time she will be a charge upon the state. If sterilized under the law, she could be given her liberty and secure a good home, under supervision, without injury to society. Her welfare and that of society would be promoted by such sterilization."[21]

The challenge to the law alleged that the Constitution of the United States was violated because the statute did not provide for due process of law, because it imposed cruel and unusual punishment, and because it denied Carrié Buck and other inmates of the Colony equal protection under the law. The court did not accept any of the arguments made on her behalf. It stated that all of the requirements of procedural due process, including notice and a hearing, were offered. It further stated that since the statute was not penal in nature, the question of cruel and unusual punishment did not apply.

Perhaps the most convoluted logic of the court surrounded the equal protection argument. Miss Buck's representative had claimed that two classes were created by the passage of the sterilization act, those judged to be feeble-minded inside the institution and those not yet committed to state custody. In upholding the statute, the court held that equal protection of the laws was not violated because "there can be no discrimination against the inmates of the colony, since the women on the outside, if in fact feeble-minded, can, by the process of commitment, and afterwards a sterilization hearing, be sterilized under the act."[22] The court summarily dismissed New Jersey and Indiana supreme court decisions which held similar statutes to be unconstitutional.

The case was appealed to the U.S. Supreme Court. In the high court, Buck's counsel argued that the Virginia law violated due process of law as protected by the Fourteenth Amendment to the U.S. Constitution. While not challenging the actual procedures involved, counsel for the appellant claimed that "substantive due process" was violated by the very subject matter of the law, because it illegally deprived her of her "bodily integrity," and the right to enjoy life with "whatever god has given to everyone." Moreover, he asserted that "the inherent right of mankind to go through life without mutilation of organs of generation needs no constitutional declaration," and that the law violated equal protection by infringing upon the rights only of the class of persons already committed to state institutions. He concluded that if the Virginia law was valid, then the limits on the power of the state to sterilize persons thought to be undesirable would be unlimited. He stated that "a reign of doctors will be inaugurated, and in the name of science new classes will be added, even races may be brought within the power of such regulation, and the worst form of tyranny practiced. In the place of the constitutional democracy of the [founding] fathers, we shall have set up a Plato's republic."[23] In rebuttal, counsel for the state repeated the lower court's holding that due process of law had been followed and that the statute created a "reasonable classification" which must necessarily be upheld by the Supreme Court.[24]

In a much quoted and extremely controversial opinion written by Mr. Justice Holmes, the Supreme Court affirmed the constitutionality of the Virginia law. Holmes, basing his opinion upon the unproven eugenic theories of some, wrote that

> we have seen more than once that the public welfare may call upon the best citizens for their lives [in wartime]. It would be strange if it could not call upon those who already sap the strength of the state for these lesser sacrifices, often not felt to be such by those concerned, in order to prevent our being swamped with incompetence. It is better for all the world, if instead of waiting to execute degenerate offspring for the crime, or to let them starve for their imbecility, society can prevent those who are manifestly unfit from continuing their kind. The principle that sustains compulsory vaccination is broad enough to cover cutting the fallopian tubes . . . . Three generations of imbeciles are enough.[25]

[20] *Id.*

[21] *Id.*, at 517-518.

[22] 130 S.E. 516, 520.

[23] 274 U.S. 200, 201-3.

[24] *Id.*, at 203-4.

[25] *Id.*, at 207.

There was only one dissent, by the little known Justice Butler, who did not write an opinion in this case. The court included such luminaries as the civil libertarian Louis Brandeis, Chief Justice and former President William Howard Taft, and the well-regarded former Columbia law dean, Harlen Fisk Stone.

The Holmes opinion appeared to be based on his own elitist view of society and was not even the prevailing opinion among the scientific community at the time.[26] In addition, the facts in the case were not closely examined. Review by later experts, in fact, proved that Carrie Buck's child who was supposedly "feeble-minded" was declared to be so at the age of one month by a nurse. The girl actually attended school and did well until her death in the second grade.[27] The opinion which took less than four pages was bitterly attacked by commentators, including the leading writer on sterilization at the time, Professor Landman. He called the opinion "astoundingly brief, and unsually platitudinous. The jurist is disconcerted by the absence of citations to support its legal principles and the psychiatrist and sociologist are equally surprised by the lack of a thorough understanding of the field of eugenics."[28] Another writer noted that Justice Holmes "dispenses with rules of logic," while others deemed the holding to be "felacious absurdity" and "inconsistent with and dangerous to our American heritage."[29] In spite of these attacks, the opinion, rendered in 1927, resulted in a number of laws being passed for compulsory eugenic sterilization.

In 1937 the American Neurological Association charged a committee with the responsibility of looking into eugenic sterilization matters. This committee came forth with an influential report which recommended that no involuntary sterilization take place since they wanted to avoid potential abuse of individual rights; rather they recommended that consent of the person faced with sterilization, or his or her guardian, should be obtained prior to any operation. For those serious diseases which could definitely be proven to be hereditary in nature, sterilization was regarded by this committee as acceptable. Even in these selected cases, however, the consent of the patient or responsible guardian was felt to be needed. According to one of the leading legal commentaries relating to eugenic sterilization, the Association's report "undoubtedly has had great influence in bringing about the gradual decline of the eugenic sterilization movement."[30] The report cautioned that "society did not have to hurry into a program based on fear and propaganda. Although the problem of mental disease and defectiveness is enormous, there exists no new social or biological emergency." As a result, this "go slow" attitude led to the decline of the eugenic sterilization movement.[31]

This attitude was reflected in subsequent court cases such as *Skinner vs. Oklahoma*[32] which the U.S. Supreme Court heard in 1942. In that case, an Oklahoma statute provided for the sterilization of "habitual criminals" who were defined as persons who had been convicted at least three times in Oklahoma of felonies involving moral turpitude. Certain offenses such as violations of revenue acts, embezzlement and political offenses were not covered by the law.[33] The appellant, Skinner, was a person who had been convicted of three crimes, including stealing chickens and twice for robbery, so the state attorney general instituted proceedings against him under the sterilization law. The trial court ordered that a vasectomy be performed, and the Supreme Court of Oklahoma affirmed by a 5-4 decision.[34] The law was challenged because of the vague nature of scientific authority in the area of inheritability of criminal traits and because the state improperly relied on that inheritability when it exercised its police power to provide laws protecting the public safety and welfare. Unlike the statute upheld in *Buck vs. Bell,* the potential person to be sterilized in the Oklahoma law was given no opportunity to be heard to determine if he or she was potentially

[26]See generally Landman, *supra* note 8. For a later attack on Holmes' legal theory in Buck vs. Bell, see J.B. Gest, *Eugenic Sterilization: Justice Holmes v. Natural Law,* 23 TEMP. L. Q. 306 (1950). Compare, Van Roden, *Sterilization of Abnormal Persons as Punishment for and Prevention of Crimes,* 23 TEMP. L. Q. 99 (1950).

[27]O'Hara and Sanks, *supra* note 8, at 31.

[28]Landman, *supra* note 8, at 465.

[29]See Burgdorf and Burgdorf, *supra* note 8, at 1006 and studies cited therein; including Gest, *supra* note 26, at 306; and O'Hara and Sanks, *supra* note 8, at 32.

[30]O'Hara and Sanks, *supra* note 8, at 36.

[31]*Cf.* In re M.K.R., 515 S.W.2d 467 (Mo. 1974); See generally Burgdorf, *Marital and Family Rights,* in LEGAL RIGHTS OF MENTALLY DISABLED PERSONS (Practising Law Institute 1979), and cases and materials cited therein; see generally, Note, *Sterilization, Retardation and Parental Authority,* B.Y.U. L. REV. 379 (1978); and Gaurney and Shugar, *The Permissibility of Involuntary Sterilization under the Parents Patriae and Police Power Authority of the State: In Re Sterilization of Moore,* 6 MD. L. FORUM 109 (1976).

A distinct line of cases began to develop regarding the judiciary's role in sterilization matters. See, e.g., Wade vs. Bethesda Hospital, 337 F. Supp. 671 (S.D. Ohio 1971) (judge could be held liable for damages when sterilization ordered was inappropriate; court has judicial immunity against liability if it had statutory authority to order the operation. The dissent argued that a general grant of authority to hear "all matters" did not give specific authority to order sterilizations. According to Stewart, J., "What Judge Stump did . . . was beyond the pale of anything that could sensibly be called a judicial act." 435 U.S. at 365.

[32]316 U.S. 535 (1942).

[33]*Id.*

[34]189 Okla. 235, 115 P.2d 123.

the parent of socially undesirable children. Cruel and unusual punishment arguments were also advanced.

The U.S. Supreme Court declared the law unconstitutional, holding that it violated the equal protection clause of the Fourteenth Amendment to the U.S. Constitution. Essentially, the Court held that since some criminals were covered by the laws while others were not, an equal protection violation resulted. For example, a person who is three times convicted of stealing chickens may be sterilized, but an embezzler may not be so treated. The Court did not reverse *Buck vs. Bell*, but appeared to establish new questions concerning sterilization statutes which put *Buck's* holding in question.

More recently, the landmark case of *Wyatt vs. Aderholt*[35] was a wide-ranging attack on conditions in state institutions for the retarded in Alabama, including the practice of involuntary sterilization. Judge Johnson's order[36] in the case set up standards and procedures relating to sterilization, after a three-judge panel found Alabama's compuslory sterilization law to be un-

[35] 368 F. Supp. 1383 (M.D. Ala. 1973).

[36] ORDER

On December 20, 1973, the three-judge court in Wyatt v. Aderholt, D.C., 368 F. Supp. 1382, declared that Tit. 45, § 243, Code of Alabama, is unconstitutional. As a consequence of that action, the three-judge feature of this case is now complete and that court has been dissolved. However, since it appears that sterilization continues to be performed in certain instances by the state health authorities, it is necessary that this Court promulgate adequate standards and procedural safeguards to insure that all future sterilizations be performed only where the full panoply of constitutional protections has been accorded to the individuals involved. Accordingly, it is ordered that the following standards be and they are hereby adopted and ordered implemented from this date for the sterilization of mentally retarded residents of the state retardation facilities:

1.(a) "Sterilization," as used in these standards, means any medical or surgical operation or procedure which results in a patient's permanent inability to reproduce.

(b) A determination that a proposed sterilization is in the best interest of a resident, as referred to in these standards, must include a determination that no temporary measure for birth control or contraception will adequately meet the needs of such resident, and shall not be made on the basis of institutional convenience or purely administrative considerations.

2. No resident who has not attained the chronological age of 21 years shall be sterilized except in cases of medical necessity as determined in accordance with the procedures set forth below. No other resident shall be sterilized except in accordance with procedures set forth below.

3. No resident shall be sterilized unless such resident has consented in writing to such sterilization. Except as set forth below, such consent must be informed, in that it is (a) based upon an understanding of the nature and consequences of sterilization, (b) given by a person competent to make such a decision, and (c) wholly voluntary and free from any coercion, express or implied. It shall be the responsibility of the Director of the Partlow State School (with the assistance of employees or officials designated by him) to provide the resident with complete information concerning the nature and consequences of sterilization, to assist the resident in comprehending such information, and to identify any barriers to such comprehension.

4. The Director shall prepare a report evaluating the resident's understanding of the proposed sterilization and describing the steps taken to inform the resident of the nature and consequences of sterilization. If the resident has been determined by a court of competent jurisdiction to be legally incompetent, or if the Director cannot certify without reservation that the resident understands the nature and consequences of sterilization, the sterilization shall not be performed unless (a) the Director sets forth reasonable grounds for believing that such sterilization is in the best interest of the resident; (b) the Review Committee described below approves such sterilization; and (c) it is determined by a court of competent jurisdiction that such sterilization is in the best interest of the resident.

5. No sterilization shall be performed without the prior approval of a Review Committee formed in accordance with this paragraph. The Review Committee shall consist of five members, and shall be selected by the Partlow Human Rights Committee and approved by the Court. The members shall be so selected that the Committee will be competent to deal with the medical, legal, social, and ethical issues involved in sterilization; to this end, at least one member shall be a licensed physician, at least one shall be a licensed attorney, at least two shall be women, at least two shall be minority group members, and at least one shall be a resident of the Partlow State School (the foregoing categories are not mutually exclusive). No member shall be an officer, employee, or agent of the Partlow State School, nor may any member be otherwise involved in the proposed sterilization.

Any fees or costs incurred by reason of services performed by the Review Committee, including reasonable fees for the physician and the attorney, shall be paid by the Alabama Department of Mental Health upon a certification of reasonableness by the Partlow Human Rights Committee.

6. Prior to approving the proposed sterilization of any resident, the Review Committee shall:

(a) Review appropriate medical, social, and psychological information concerning the resident, including the report of the Director prepared pursuant to paragraph 4;

(b) Interview the resident to be sterilized;

(c) Interview concerned individuals, relatives, and others who in its judgment will contribute pertinent information;

(d) Determine whether the resident has given his or her informed consent to the sterilization, or, if the resident is legally incompetent or the Director cannot certify without reservation that the resident understands the nature and consequences of sterilization, whether the resident has formed, without coercion, a genuine desire to be sterilized. In making such determination, the Review Committee shall take into consideration, *inter alia*, the report prepared by the Director pursuant to paragraph 4 and the interview required by paragraph 6(b).

(e) Determine whether the proposed sterilization is in the best interest of the resident.

If the Review Committee does not reach an affirmative determination as to the matters set forth in paragraphs 6(d) and (e), it shall not approve the proposed sterilization. Any doubts as to such matters shall be resolved against proceeding with sterilization.

7. Residents shall be represented throughout all the procedures described above by legal counsel appointed by the Review Committee from a list of such counsel drawn up by the Partlow Human Rights Committee and approved by the Court. Such counsel shall, *inter alia*, ensure that all considerations militating against the proposed sterilization have been adequately explored and resolved. No such counsel shall be an officer, employee, or agent of the Partlow State School, or agent of the Partlow State School, nor may such counsel be otherwise involved in the proposed sterilization.

(continued on next page)

constitutional.[37] The court's order[38] requires that no resident of a state instituion may be sterilized if he or she is under the age of 21, unless a medical emergency exists. Further, no adult resident may be sterilized unless the resident consents in writing, including the statement that the resident desires such an operation based upon his or her understanding of its nature and consequences, and that this decision is wholly voluntary and free of coercion. The director of Partlow State School, the institution in question, is required to provide any resident with information that will allow any person consenting to sterilization to understand the matter they are undertaking.

In addition, a review committee was set up, the purpose of which was to allow for prior approval of sterilizations. The committee consists of five members selected by the Human Rights Committee of the school, but no school or state employee can be involved in a sterilization decision. The committee is responsible for reviewing existing medical and psychological information concerning a person who is about to undergo sterilization; for interviewing the resident and other concerned individuals, including relatives, who have pertinent information; for determining whether the resident has given consent for the sterilization free of coercion; and for determining whether a proposed sterilization is in the best interests of the resident. Throughout the entire procedure, residents must be represented by independent legal counsel.

These guidelines set the stage for future state statutes and reflect a progressive and modern view of the sterilization which includes prohibiting sterilization of institutionalized minors and prohibiting involuntary sterilization of institutionalized developmentally disabled adults.[39]

In a challenge similar to that of *Wyatt* in Alabama, a three-judge federal court held that a North Carolina statute,[40] allowing for involuntary sterilization, was overly broad. The court noted that modern thought destroyed the assumptions made by the court in *Buck vs. Bell,* and that eugenic sterilization was no longer a viable legal or social concept. Only in extreme cases could sterilization be legally permissible.

Also relatively recently, a federal district court in Washington, D.C. decided *Relf vs. Weinberger.*[41] This case involved two sisters who were 12 and 14 years old and who were sterilized under a federally authorized and funded family planning program. The federal regulations challenged permitted the involuntary sterilization of minors and incompetent persons. In March, 1974, Judge Gesell enjoined the U.S. Department of Health, Education and Welfare from providing federal funds for the sterilization of such persons because federal regulations were inconsistent with the legal requirements that sterilizations funded be voluntary in nature. Since the sisters could not give consent, their submission to sterilization was not, of course, voluntary. In addition, according to this Court, sterilization regulations promulgated by the Department of Health, Education and Welfare to implement these laws were arbitrary and unreasonable because they authorized sterilizations of legaily competent persons without advising them that federal bvenefits could not be withheld from them if they refused to be sterilized. Prior to this order, persons were frequently told that they could be deprived of participation in federal programs if they refused an operation. Judge Gesell admonished government officials to provide full and competent notice and to safeguard the rights of persons sterilized pursuant to these programs.

---

[36](continued from preceding page)

8. The Review Committee shall maintain written records of its determinations and the reasons therefor, with supporting documentation. Such records shall be available for examination by the Partlow Human Rights Committee, the Court, and counsel of record in this cause. The Review Committee shall report in writing at least monthly to the Human Rights Committee, the Court, and counsel of record in this cause as to the number and nature of sterilizations approved and disapproved the procedures employed in approving or disapproving such sterilizations, the reasons for determining that such sterilizations were in the best interest of the residents involved, the number and nature of proposed sterilizations referred to courts of competent jurisdiction, and all other relevant information. The identity of residents sterilized or to be sterilized shall not be disclosed in such reports.

9. There shall be no coercion in any form with regard to sterilization of any resident. Consent to sterilization shall not be made a condition for receiving any form of public assistance, nor may it be a prerequisite for any other health or social service, or for admission to or release from the Partlow State School. Any individual having knowledge of coercion of any resident with regard to sterilization shall immediately bring such matter to the attention of the Partlow Human Rights Committee, the Court, or counsel of record in this cause.

It is further ordered that the defendants, their agents, employees and those acting in concert with them be and each is hereby enjoined from failing to implement the standards hereinabove set out for the sterilization of mentally retarded residents of the Alabama retardation facilities.

[37]368 F. Supp. 1382 (M.D. Ala. 1973).

[38]*Supra* note 35, at 1384.

[39]The courts have begun to accept the concept that bodily intrusions upon mentally handicapped persons must not occur without the person's consent. See, e.g., Kaimowitz vs. Michigan Department of Mental Health, No. 73-19434-AW (Cir. Ct. Wayne County, Mich., July 10, 1973) (psychotherapy not permissible if mentally ill adult lacked ability to understand the nature and consequences of the therapy), cited in Burgdorf, *supra* note 31, at 1019, comparing In re Cavitt, 182 Neb. 712, 157 N.W.2d 171, ". . . in the years between 1940 and 1970, Cavitt is the only decision relying on Buck v. Bell to uphold the constitutionality of an involuntary sterilization statute." *Id.* at 1016.

[40]North Carolina Association for Retarded Children vs. North Carolina, 420 F. Supp. 45, (M.D. N.C. 1976); for a lengthy discussion of the case, see, Burgdorf, *supra* note 31, at 1025-33.

[41]372 F. Supp. 1197 (D.D.C. 1974), *aff'd* 565 F.2d 722 (D.C. Cir. 1977).

After *Relf,* HEW issued final regulations in November, 1978 covering federal financial support for sexual sterilizations performed under the Public Health Service, Medicaid, and Social Services Programs.[42] The most significant aspect of these regulations is that HEW gave up attempting to construct a consent and review procedure and hence terminated financial support for sterilization for instituionalized persons;[43] for persons under the age of 21;[44] and for persons adjudicated incompetent unless these individuals also had been adjudicated competent to consent to sterilization.[45] Furthermore, sterilization on the basis of substituted consent is not permitted.[46]

Moreover, case law and federal regulations are recognizing the rights of handicapped persons against involuntary sterilization, and this trend is consistent with recent U.S. Supreme Court holdings dealing with abortion which recognize that all persons have the fundamental right to make decisions for themselves in childbearing and family matters.[47] Despite this fact, however, several states still maintain anachronistic statutes in this area. We now turn to an examination of this law.

REVIEW OF STATE LEGISLATION

The review covers the following matters relating to involuntary sterilization legislation. The first section describes the class of persons who are affected by these laws. The second section sets forth the standards in the various jurisdictions which officials must follow prior to approval of a sterilization. A third section describes the persons who may initiate a request for an involuntary sterilization. The procedural protections, if any, granted by the various jurisdictions to persons faced with sterilizations against their will are included in a fourth section. Finally, those jurisdictions where substituted consent is permitted to allow for involuntary sterilizations are indicated. Detailed citations are provided in the accompanying chart.

Persons Addressed

Twenty-one states maintain statutes providing for the involuntary sterilization of handicapped persons. The categories of disabilities covered by these laws is broad and varied, but the most common disability included is mental retardation. In Delaware, Mississippi, Oklahoma, and South Carolina, epileptics may be subject to involuntary procedures, while persons classified as "feebleminded" are subject to similar treatment in Maine, Mississippi, Oklahoma and South Carolina."Mental Incompetents" are included in the laws of Arkansas, Utah and West Virginia; brain damaged individuals may be sterilized by third party action in Georgia. Antiquated but still effective state statutes in Mississippi, Oklahoma and South Carolina provide for involuntary sterilization of "idiots."

Eleven states permit involuntary sterilization of persons who are institutionalized, as well as those in the community. Of the jurisdictions where handicapped persons in both settings are affected, Arkansas law covers "mental incompetents"; retarded persons are impacted negatively in Colorado, Delawaree, Georgia, Idaho, Minnesota, North Carolina, Oregon and Virginia. Persons classified by a state as mentally ill or epileptic may also be subject to involuntary proceedings in several other states.

Ten state statutes permit involuntary sterilizations of institutionalized persons only. Covered by these laws are, for example, "Inmates of Homes for the Mentally Deficient" in Alabama; residents of state hospitals for the mentally retarded in California; patients in institutions for the retarded in Connecticut and Ohio, as well as residents of state schools in North Dakota. Persons deemed to have "recurrent" insanity, idiocy, imbecility, feeble-mindedness and epilepsy in Mississippi, Oklahoma and South Carolina can be sterilized when they are institutionalized. Utah law covers persons adjudicated mentally incompetent, while Wisconsin officials may have criminal, mentally ill and mentally deficient persons sterilized.

Standards for Sterilization

The jurisdictions often differ over the standards that must be met for involuntary sterilizations to be sought. In Alabama, an operation is available

[42]Programs funded by the Public Health Service, Title V of the Social Security Act, §505(a)(8) of the Act, 42 U.S.C. 705(a)(8); The Medicaid Program (medical assistance) under Title XIX of the Social Security Act, 42 U.S.C. §§1396, *et seq.;* §§1902(a)(13) and 1905(a)(4)(c) of the Act, 42 U.S.C. §§1396a(a)(13) and 1396d(4)(c); Title XX (Social Services Program) of the Social Security Act, 42 U.S.C. §1397(a)(1) and Title IV-A of the Act (Aid to Families with Dependent Children).

[43]42 C.F.R. §§50.202, 50.203(b), 50.206; 441.253(b), 441.254.

[44]42 C.F.R. §§50.202, 50.206, 441.254.

[45]42 C.F.R. §§50.203a, 441.253.

[46]42 C.F.R. §§50.203(c), 441.253(c).

[47]See Eisenstadt vs. Baird, 405 U.S. 438, 453 (1972); *cf.* Roe vs. Wade, 410 U.S. 113 (1973).

"If . . . advisable" and in Wisconsin and Delaware if "Procreation is inadvisable." If a handicapped person is thought to be unable to care for or support a family, he or she may be sterilized in Georgia, Idaho, North Carolina and Oregon. The "best interest" of the person can subject handicapped persons to involuntary treatment in Minnesota, North Carolina, North Dakota, Oklahoma, South Carolina and Virginia, while the best interests of society at large is a statutory standard in Mississippi, North Carolina, North Dakota, Oklahoma, South Carolina and Virginia. In several states, including California, Maine, Mississippi, North Carolina, Oklahoma, South Carolina and West Virginia, persons can be subject to involuntary sterilization if it is thought that such an operation will prevent the propagation of genetic defects in society.

Initiator of Sterilization

In those states which permit only institutionalized persons to be involuntarily sterilized, usually the institution's superintendent is the official who is required to initiate the request for sterilization. In Wisconsin, the Department of Public Welfare is responsible for this action, while in Maine the medical staff of the institution is designated as the initiator. In a number of states, parents are vested with the authority to request sterilizations, including Arkansas, Colorado, Georgia, Idaho, North Dakota, Virginia and West Virginia. Legal guardians have this authority in Arkansas, Colorado, Georgia, Idaho, Virginia and West Virginia. In Georgia and Idaho, the next of kin may initially file for a sterilization, and in Virginia a spouse has the ability to initiate a sterilization.

Due Process Procedures

The protections offered to mentally handicapped persons cover the range of full procedural protections to extremely limited or non-existent hearing rights. Seven states do not require a hearing prior to a sterilization. In these states, a statement by the superintendent or director of the state institution or a physician that a sterilization is required is sufficient to accomplish the treatment. In four of these states, Delaware, Maine, North Dakota and Wisconsin, however, notice to a third party is provided for prior to the operation None of these states allow the person subject to the sterilization to confront the person who initiates the request for an operation; as a result, these procedures are certainly constitutionally suspect.

Five states provide for administrative hearings before sterilization. California requires a hearing only if an objection to a sterilization is entered. This hearing is conducted by the director of the state department with jurisdiction over institutions. The person to be sterilized may attend this hearing, as may others interested in the person concerned. No right to counsel is specifically granted. Mississippi, Oklahoma, Oregon and South Carolina all require full administrative hearings before sterilization. They allow a handicapped person to be present and to be represented by counsel during the hearing, and require that full records of all written and oral testimony be maintained.

Nine states require full judicial hearings before sterilization can be performed. The procedure followed is similar to the commencement of other civil cases in the state: filing of a petition in the appropriate court, service of process on the person to be sterilized and his or her next of kin, spouse or legal guardian; a hearing; the right to counsel; the right to be present and cross-examine witnesses; and the right to a full and complete record of all oral and written testimony.

Substituted Consent

In eleven states, substituted consent for involuntary sterilization is permitted pursuant to statute. For example, California, Maine, North Carolina and Oregon allow the next of kin to provide consent. A spouse may act in this capacity in California, North Dakota and Oregon. A legal guardian in California, Connecticut, Idaho, Maine, North Carolina, North Dakota, Ohio and Oregon are authorized to give substituted consent. If they are not filing the initial petition, parents may offer this form of consent in Georgia. A public department of welfare official is vested with this authority in Minnesota and Idaho, while any department having custody of a handicapped person may perform this function in Delaware.

INVOLUNTARY STERILIZATION

STATE	PERSONS ADDRESSED		STANDARD	INITIATOR	RIGHT TO A HEARING	RIGHT TO AN ATTORNEY	DECISION-MAKER	RIGHT OF APPEAL	SUBSTITUTE CONSENT
	Institutionalized	Non-institutionalized and Institutionalized							
ALABAMA Code (1959)	Inmates (Homes for the Mentally Deficient) tit.45, §243		"If . . . Advisable" tit.45, §243	Asst. Supt. with advice & consent of Supt. tit.45, §243			Asst. Supt. and Supt. tit.45, §243		
ALASKA									
ARIZONA									
ARKANSAS Stat. Ann. (1971)		Mental Incompetents §59-501	Person is "incompetent" §59-501	Parent or guardian §59-502	§59-501	Guardian *ad litem* §59-501	Probate Court §59-501	§59-501	
CALIFORNA Welf. & Inst. Code (West Supp. 1978)	Resident (State hospital for mentally disordered or mentally retarded) §7254		Hereditary mental disease, mental retardation, or marked departure from normal mentality §§7254 (a)-(c)	Superintendent of hospital §7254	§7254		Director of Dept. of Mental Health; Director of Dept. of Developmental Services §7254	§7254	Spouse, guardian, or next of kin §7254
COLORADO Rev. Stat. (Supp. 1978)		Mentally retarded persons over 18 years of age §27-10.5-128	Consent required of person, and parent or guardian §27-10.5-128[1]	Mentally retarded person, parent, legal guardian, or custodian §27-10.5-128	§27-10.5-130	§27-10.5-130	Court determines competency to give consent §27-10.5-130		
CONNECTICUT Gen.Stat.Ann. (West 1977)	Patients of training schools for mentally retarded §19-569g		Person incapable of comprehending consequences of action §19-569g	Superintendent and Medical Board §19-569g			Probate court §19-569g	§19-569g	Next of kin, guardian, Board of Trustees §19-569g
DELAWARE Code Ann. (1974)	Mentally ill, mentally retarded or epileptic in state institutions ------ habitual criminal tit.16, §5701	Mentally retarded, epileptic, or chronic mentally ill tit.16, §5701	"Procreation is inadvisable" tit.16, §5701	Institutional Board and Superintendent ------ Div. of Mental Health and Mental Retardation tit.16, §§5701, 5702			Institutional Board and Superintendent ------ Dept. of Health & Social Services tit.16, §5702	tit.16, §5702	Dept. having custody of person tit.16, §§5701, 5702
DISTRICT OF COLUMBIA									
FLORIDA									
GEORGIA Code Ann. (1975)		Mentally retarded or brain-damaged persons §84-933	Inability to care for and support children §84-933	Parent, legal guardian, next of kin, or state official §84-933	§84-933	§84-933	Probate court §84-933	§84-933	Parents (if not filing petition), Guardian *ad litem* §84-933
HAWAII									
IDAHO Code (1977)		Persons with mental retardation or organical retardation §39-3901	Inability to care for and support children §39-3901	Parent, legal guardian, next of kin, Director of State Dept. of Health & Welfare, or prosecuting Attorney §39-3903	§39-3903	§39-3093[2]	District Court §39-3903	§39-3904	Parent, Legal guardian, or Director of Health & Welfare §39-3903
ILLINOIS									

STATE	PERSONS ADDRESSED: Institutionalized	PERSONS ADDRESSED: Non-institutionalized and Institutionalized	STANDARD	INITIATOR	RIGHT TO A HEARING	RIGHT TO AN ATTORNEY	DECISION-MAKER	RIGHT OF APPEAL	SUBSTITUTE CONSENT
INDIANA									
IOWA									
KANSAS									
KENTUCKY									
LOUISIANA									
MAINE Rev.Stat. (1978)	Mentally ill or feeble-minded persons in state institutions tit.34, §2452	Non-institutionalized tit.34, §2461	Prevent further propagation of mental deficiency; or therapeutic treatment tit.34, §2462	Physician tit.34, §2462 Medical Staff of institution tit. 34, §§2462, 2463		[3]	Council of Physicians	For institutionalized persons only tit.34, §2465	Nearest relative or guardian if person is incompetent to consent
MARYLAND									
MASSACHUSETTS									
MICHIGAN									
MINNESOTA Stat.Ann. (West Supp. 1978)		Mentally retarded persons under guardianship of conservatorship of Commissioner of Public Welfare §252A.13	"Best interests of ward" §252A.13		[4]	§252A.13	Probate or county court §252A.13	§252A.13	Commissioner of Public Welfare, Parental consent required if ward is a minor §252A.13
MISSISSIPPI Code Ann. (1972)	Recurrent insanity, idiocy, imbecility, feeble-mindedness, or epilepsy §41-45-1		Prevent propagation of "socially inadequate offspring" and promote welfare of inmate and society §41-45-1	Director of Institution §41-45-1	§41-45-7	§41-45-7	Board of Trustees of §41-45-9	§41-45-11	
MISSOURI									
MONTANA									
NEBRASKA									
NEVADA									
NEW HAMPSHIRE									
NEW JERSEY									
NEW MEXICO									
NEW YORK									
NORTH CAROLINA Gen.Stat. (1976)		Mentally ill or retarded persons §35-36	Best interest of person, for public good, prevent propagation of serious deficiency or person's inability to care for child §35-36	County Director of Social Services; Director of Institution §35-37	§35-43	§35-45	District Court §35-46	Yes §35-44	Next of kin, Legal Guardian, or Guardian *ad litem* §35-39
NORTH DAKOTA Cent.Code (1977)	Resident of state school §25-04.1-06		Protection of society or improvement of person's well-being §25-04.1-06	Supt. of Institution or Parent §25-04.1-06 to §25-04.1-08	§25-04.1-06 to §25-04.1-08		Board of Physicians §25-04.1-06 to §25-04.1-08		Consent of Parent, Spouse, or Guardian required

STATE	PERSONS ADDRESSED		STANDARD	INITIATOR	RIGHT TO A HEARING	RIGHT TO AN ATTORNEY	DECISION-MAKER	RIGHT OF APPEAL	SUBSTITUTE CONSENT
	Institutionalized	Non-institutionalized and Institutionalized							
OHIO Rev.Code Ann. (Page Supp. 1977)	Resident of Institution for mentally retarded §5123.86		Informed consent of person §5123.86						If person is incompetent to consent, guardian may consent §5123.86
OKLAHOMA Stat. Ann. (West 1954)	Recurrent insanity, idiocy, imbecility, feeblemindedness, or epilepsy tit.43A, §341		Prevent propagation of "socially inadequate offspring" or promote welfare of patient & society, & patient will continue to be public charge tit.43A, §341	Superintendent of Institution tit.43A, §341	tit.43A, §342	tit.43A, §342	Board of Affairs tit.43A, §342	tit.43A, §343	
OREGON Rev.Stat. (1973)		Persons with mental retardation or mental illness §436.070	Inability to provide adequate care for children or prevent propagation of mental retardation or mental illness §436.070	Surgeon §436.100	§436.061 §436.080 §436.090 §436.100	§436.056[5]	State Board of Social Protection §436.020	§436.110	Consent of Guardian, spouse, or next of kin also required where person is mentally ill or mentally retarded §436.100
PENNSYLVANIA									
RHODE ISLAND									
SOUTH CAROLINA Code (1976)	Recurrent insanity, idiocy, imbecility, feeblemindedness, or epilepsy §44-47-10		Prevent propagation of "socially inadequate offspring," health of inmate and promote welfare of inmate & society §44-47-10	Superintendent of Institution §44-47-10	§44-47-30	Guardian *ad litem* §44-47-30	Dept. of Health & Environmental Control §44-47-10	§44-47-60	
SOUTH DAKOTA									
TENNESSEE									
TEXAS									
UTAH Code Ann. (1978)	Adjudicated mentally incompetent persons §64-10-1		"Compelling State Interest" §64-10-1	Superintendent of Mental Retardation Facility §64-10-1	§64-10-3	§64-10-3[7]	Court §64-10-4	§64-10-10	
VERMONT									
VIRGINIA Code (1976)	Patients with mental retardation or hereditary mental illness §37.1-171.1 (Supp. 1978)	Minors with recurrent mental illness or mental retardation, persons adjudicated legally incompetent §32-424	Best interest of person and society §32-423	Director of hospital, Parent, guardian, spouse, next friend or committee §32-423, §32-424.1	§32-424.1	Guardian *ad litem* §32-424.1	Circuit court §32-424, §32-424.1		
WASHINGTON									
WEST VIRGINIA Code (1976)		Mentally incompetent persons §27-16-1	Prevent propagation of genetic defect or inability of person to care for child; and, no feasible alternative method of birth control §27-16-1	Parent, guardian, committee or responsible authority §27-16-1	§27-16-1	§27-16-1[8]	Court §27-16-1	§27-16-2	

STATE	PERSONS ADDRESSED: Institutionalized	PERSONS ADDRESSED: Non-institutionalized and Institutionalized	STANDARD	INITIATOR	RIGHT TO A HEARING	RIGHT TO AN ATTORNEY	DECISION-MAKER	RIGHT OF APPEAL	SUBSTITUTE CONSENT
WISCONSIN Stat.Ann. (West 1978)	Criminal, mentally ill and mentally deficient persons		"Procreation is inadvisable"	Dept. of Public Welfare			Medical Experts & Supt.		
WYOMING									
AMERICAN SAMOA									
GUAM									
PUERTO RICO									
VIRGIN ISLANDS									

[1]The statute provides that any mentally retarded individual over eighteen years of age who has given consent and whose lawful parent or legal guardian has given consent may be sterilized if the following consultants also give consent: ". . . (1) a psychiatrist or psychologist who consults with and interviews the mentally retarded person; and (2) a person who works in the field of mental retardation, who is knowledgeable in it, and who consults with and interviews the mentally retarded person." COLORADO REV. STAT. §27-10.5-128 (Supp. 1978).

[2]The statute provides that the "person alleged to be subject to the provisions of this act shall have counsel at all stages of the proceedings provided for herein." IDAHO CODE §39-3903 (1977).

[3]The statute provides for notice to be given to designated relatives or the legal guardian. In the event that none of the relatives are known and no legal guardian has been appointed, the statute provides: "the department shall request a Justice of the Superior Court to appoint some attorney to protect the rights of the inmate and such notices and copies shall be sent to such attorney." MAINE REV. STAT. tit.34, §2464 (1977).

[4]The statute provides that "in every case a probate or county court shall determine if such operation is in the best interest of the ward." MINNESOTA STAT. ANN. §252A.13 (West Supp. 1978).

[5]The statute provides for the right to counsel and further provides that: "The Public Defender shall appear at all hearings conducted by the board pursuant to ORS 436.050 unless notified prior to any hearing that the person to be examined is represented by other counsel." OREGON REV. STAT. §436.056 (1973).

[6]The statute provides for the appointment of a guardian *ad litem* to defend the rights and interests of each inmate. SOUTH CAROLINA CODE §44-47-30 (1976).

[7]The statute provides: "The court shall appoint counsel to represent any resident or guardian upon the filing of a verified petition of impecuniosity." UTAH CODE ANN. §64-10-3 (1978).

[8]If an incompetent person is not represented by counsel, the court will appoint legal counsel and allow the incompetent to be present at hearings and to present evidence and cross-examine witnessses. WEST VIRGINIA CODE §27-16-1 (1976).

SUMMARY OF THE MODEL STATUTE

As can be seen from the review of state legislation, disabled persons are not protected in a number of jurisdictions against involuntary sterilization. Procedural safeguards are often vague or non-existent, and consent of third parties for a sterilization is acceptable in a variety of states. Thus, a model act was developed to assist state officials in revising or replacing their law. This act is based upon certain premises derived from existing legislation and legal caes, as well as a desire to correct the inadequacies of that legislation:

(1) Since states that have legislation allowing for involuntary sterilizations are relying on outmoded legal and genetic principles, a model act must prohibit such involuntary procedures whether or not the minor or adult is institutionalized;

(2) On the other hand, adults, pursuant to their right to make decisions in the area of procreation and family planning, should be able to obtain a voluntary sterilization if they wish. Because institutions are a potentially coercive environment, however, the act must incorporate a review procedure for requests from institutionalized adults so that independent observers may certify that the requested sterilization is truly voluntary. Independent means that these evaluators must not be employees or agents of the institutions.

(3) Since sterilization is an irreversible process in virtually all women and most men, persons under 21 years of age should be prohibited from consenting to such a procedure. A decision of a minor may be based upon immature judgment and insufficient life experiences. In addition, since the finality of the procedure would prevent the person from reversing their decision in later years and eventually having children, requiring the person to wait until they were 21 is not an unreasonable limitation on their rights — especially when compared to the protection against erroneous judgments that this limitation may provide.

The text of the model act appears in **bold type.** Comments on statutory provisions appear in standard type. The comments explain, elaborate upon or provide a rationale for the text of the act. They are intended to assist the user in interpreting the act, in revising their existing legislation and/or in drafting an act for a particular state based upon this model act.

MODEL STATUTE

A BILL FOR AN ACT TO PROVIDE FOR AND PROTECT THE RIGHT TO REPRODUCTIVE FREEDOM OF HANDICAPPED PERSONS.

Section 1. Title

This Act shall be known as the "Handicapped Persons Reproductive Rights Act."

COMMENT. None necessary.

Section 2. Purpose

The legislature of this state finds that all persons, including those with a handicap should have the right to make personal decisions regarding family matters. Therefore, it is the purpose of this act to allow handicapped persons to make decisions for themselves regarding sterilizations, free of the intervention of other parties, except in the case of medical emergencies. This Act shall be liberally construed toward this end.

COMMENT. At the turn of this century, the theory of eugenic sterilization was scientifically unproven, and few states enacted laws requiring compulsory elimination of childbearing capabilities. Even in those jurisdictions which enacted compulsory sterilization laws for persons classified "feeble-minded" or criminals, state courts frequently rejected their enforcement as cruel or unusual punishment, or as violations of due process or equal protection under the law.[1] Beginning with the U.S. Supreme Court case of **Buck vs. Bell**[2] in 1927, which legitimized eugenic sterilization laws, the majority of state legislatures began to pass laws in this field. After the same court decided **Skinner vs. Oklahoma** in 1942, however, enforcement of compulsory sterilization laws either declined or they were entirely repealed.[3] In later cases, including **Griswold vs. Connecticut**,[4] the court began to write of a right to privacy in constitutional law covering procreation and marital rights. The abortion cases, commencing with **Roe vs. Wade**[5] in 1973, further developed a right of individuals "to be let alone" to decide matters regarding procreation, and other family and personal matters.

Yet, a number of states[6] still maintain statutes allowing involuntary sterilizations of various citizens ranging, for example, from epileptics to institutionalized and mentally ill persons. It is the purpose of this model act to create uniform standards relating to the reproductive freedom of all handicapped persons in the nation. The increasing recognition that handicapped citizens have the same constitutional rights as others in all areas of human conduct, including the right to receive appropriate educational opportunities[7] and to live in the community[8] applies as well to childbearing and family matters. The model statute recognizes these latter rights and mandates them for the class of citizens it covers.

Section 3. Definitions

(1) "Sterilization" means any medical or surgical operation or procedure which results in a patient's permanent inability to reproduce.

---

[1]Legal histories of eugenic sterilization in the United States include J. Landman, *The History of Human Sterilization in the United States — Theory, Statute, Adjudication,* 23 ILL. L. REV. 463 (1929) (cases and statutes until 1929); O'Hara & Sanks, *Eugenic Sterilization,* 45 GEO. L. J. 30 (1956); Burgdorf and Burgdorf, *The Wicked Witch is Almost Dead — Buck vs., Bell* and *The Sterilization of Handicapped Persons,* 50 TEMP. L. Q. 995 (1977); Note, *Sterilization, Retardation, and Parental Authority,* 2 B.Y.U. L. REV. 38 (1978). Recent abortion law histories include Dellapenna, *History of Abortion: Technology, Morality and Law,* 40 U. PITT. L. REV. 359 (1979) and F. Susman, *Recent Developments in Women's Rights: A Symposium,* 22 ST. LOUIS U. L. J. 566 (1979).

[2]274 U.S. 200 (1927). For discussions on the implications of *Buck vs. Bell,* see articles in note 1, *id.,* as well as F. Lindman and D. McIntyre, *The Mentally Disabled and the Law (1961);* Paul, *State Eugenic Sterilization History: A Brief Overview* in EUGENIC STERILIZATION (Robitscher ed. 1973) and Gest, *Eugenic Sterilization, Justice Holmes vs. Natural Law,* 23 TEM. L. Q. 306 (1950).

[3]316 U.S. 535 (1942); O'Hara and Sanks, *supra* note 1, at 36.

[4]381 U.S. 479 (1965).

[5]410 U.S. 113 (1973).

[6]See Review of State Statutes, *supra.*

[7]See. e.g., Pennsylvania Association for Retarded Children vs. Commonwealth of Pennsylvania, 334 F. Supp. 1257 (E.D. Pa. 1971); Mills vs. Board of Education of the District of Columbia, 348 F. Supp. 866 (D.D.C. 1972); Frederick L. vs. Thomas, 419 F. Supp. 960 (E.D. Pa. 1976), *aff'd,* 557 F.2d 373 (3rd Cir. 1977); and Armstrong vs. Kline, 476 F. Supp. 583 (E.D. Pa. 1979), and cases and materials cited at n. 1, Introduction, *supra.*

[8]Halderman vs. Pennhurst State School and Hospital, 446 F. Supp. 1295 (E.D. Pa. 1977), *aff'd in part* Nos. 78-1490, 78-1564, 78-1602 (3rd Cir. 1979); and cases and materials cited at Notes 2 and 3, Introduction, *supra.*

COMMENT. This definition is in accord with the standards developed in court cases and are generally accepted as basic in the field.[9] For example, the definition is incorporated in regulations developed by the U.S. Department of Health, Education and Welfare (HEW) covering federal funding of sterilizations under Public Health Service, Medicaid (Title XIX of the Social Security Act), and social services (primarily Title IV-A and Title XX) programs.[10]

(2) "Handicapped Person" means any person who (i) has a physical or mental impairment which substantially limits one or more major life activities, (ii) has a record of such impairment, or (iii) is regarded as having such an impairment.

COMMENT. This language is included in Section 504 of the Rehabilitation Act of 1973,[11] and is included herein as the most broadly accepted definition of "handicapped person" presently in use. It is distinguishable from the much more limiting definition of "developmental disability" contained in the Developmentally Disabled Assistance and Bill of Rights Act,[12] which requires that a disability be "one which is severe, chronic . . . is manifested before the person attains age twenty-two, is likely to continue indefinitely [and which] results in substantial functional limitations in three or more of the following areas of major life activity: (i) self-care, (ii) receptive and expressive languages, (iii) learning, (iv) mobility, (v) self-direction, (vi) capacity for independent living, and (vii) economic self sufficiency. . . ."[13]

The great majority of persons traditionally subject to involuntary limitations of procreative freedom have been those who are retarded or mentally ill. Mentally ill persons, however, would not necessarily fall within the narrow definition of "developmentally disabled," but certainly must be included within a model act. Thus, a broad definition that will provide for the rights of the greatest number of handicapped persons has been selected for use herein.

(3) "Institutionalized Person" means a handicapped adult or minor person who is (i) involuntarily confined or detained, under a civil or criminal statute, in a correctional, care or rehabilitative residential facility including a mental hospital or school, or (ii) confined under a voluntary admission/commitment, in a mental hospital or other residential facility for care and treatment.

COMMENT. None necessary.

(4) "Non-institutionalized Person" means a handicapped adult or minor person who is not an "Institutionalized Person."

COMMENT. This language is similar to that incorporated in the H.E.W. sterilization regulations described above, except that the model act language includes all handicapped persons rather than being limited to mentally ill and mentally retarded persons. This expansion is warranted since the eugenic sterilization movement advocated sterilizing persons not included solely within these latter two classifications.[14]

Because constitutional protections are not waived by persons entering institutions, they are vested with the same fundamental rights accorded to individuals who have not been so placed. Thus, the right to decide matters relating to procreation is protected for persons in an institution., See, e.g., **Wyatt vs. Aderholt.**[15] In addition, handicapped persons who are in the general community or who live in "halfway" houses or group homes are entitled to the same constitutional rights regarding procreation matters as other citizens.[16] In order to ensure that these rights are not undermined, "non-institutionalized persons" are included within the model act.

(5) "Consent" means making a voluntary decision or undertaking a voluntary action after being fully informed and after fully understanding its nature and consequences.

COMMENT. None necessary.

(6) "Voluntary" means performed while competent to make the decision or undertake the action, and as a matter of free choice and will and not in response to coercion, duress, or undue influence.

COMMENT. None necessary.

(7) "Involuntary" means performed not as a matter of free choice and will, but in response to coercion, duress, undue influence, or while incompetent to make the decision or undertake the action.

---

[9] Wyatt vs. Aderholt, 368 F. Supp. 1383 (M.D. ALa. 1973).
[10] 42 C.F.R. §§50.201, *et seq.* See generally, 43 Fed. Reg. 52146 (1978).
[11] 29 U.S.C. §794; 45 C.F.R. §84.3(j) (1979)
[12] 42 U.S.C. §§6000, *et seq.*
[13] 42 U.S.C. §60001.

[14] See generally materials cited in note 2, *supra.*
[15] 344 F. Supp. 373 (M.D. Ala. 1972); 344 F. Supp. 387 (M.D. Ala. 1972), *enforcing* 325 S. Supp. 781 (M.D. Ala. 1971), *affirmed in relevant part, and decision reversed in part,* Wyatt vs. Aderholt, 503 F.2d 1305 (5th Cir. 1974); see also cases and materials cited in Notes 2 and 3, Introduction, *supra.* For a general discussion of early institutionalization cases and problems with enforcement, see generally, President's Committee on Mental Retardation, THE MENTALLY RETARDED CITIZEN AND THE LAW (Kindred ed., 1978) and Lottman, *Enforcement of Judicial Decrees: Now Comes the Hard Part,* 1 M.D. L.R. 69 (1976).
[16] See, e.g., Ruby vs. Massey, 452 F. Supp. 361 (D. Conn. 1978); North Carolina Association for Retarded Citizens vs. North Carolina, 420 F. Supp. 451 (M.D. N.C. 1976); In re Moore, 289 N.C. 95 (1976); and, Wyatt, *supra,* at note 15.

COMMENT. In regards to the possible imposition of a medical procedure, consent is only valid when a handicapped person has the ability to understand and make voluntary informed choices.[17] Retardation or mental illness are not dispositive on the issue of ability to comprehend. Even though a person is retarded, he or she still may possess the mental capacity necessary to understand and make personal decisions in procreation matters. Similarly, an individual may be declared incompetent to act in other matters, but can still be in a position to act in this area. The critical issue is that any information that is conveyed is fully understood. Assuming comprehension ability, the physician is required to "give the patient all information relevant to the patient's decisional process. This includes a reasonable familiarity with the medical procedures, their alternatives, and their benefits and hazards."[18] Finally, consent given must be voluntary.[19] Included in this concept is the requirement that all consent must be given in an atmosphere free of coercion.

(8) "Involuntary Emergency Treatment" means medical action taken without a person's consent which is immediately necessary to protect the life of an institutionalized or non-institutionalized person.

COMMENT. None necessary.

(9) "Records" includes, but are not limited to, all written clinical information, observations and reports or fiscal documents relating to an institutionalized or non-institutionalized person who has undergone or is about to undergo sterilization and which relate to the sterilization.

COMMENT. None necessary.

(10) "Minor" means an individual who has not attained the chronological age of 21.

COMMENT. None necessary.

Section 4. Involuntary Sterilization

In no event shall an institutionalized or non-institutionalized person be involuntarily sterilized unless it occurs as, or as part of, involuntary emergency treatment.

COMMENT. Sterilization is not only a bodily intrusion, but it is also a permanent deprivation of a fundamental right.[20] Traditionally, this type of treatment has been imposed upon disabled adult and minor persons — particularly those who are mentally handicapped — without consulting the person or obtaining his or her consent.[21] Because of the drastic nature and irreversible consequences of a sterilization procedure, it must be performed only when a person consents to it, or when necesssary to save the person's life. Thus, sterilizations requested by third persons including guardians and parents, for whatever reasons, including administrative convenience, eugenic reasons, or assumed inability to raise children, are precluded in all cases, absent a medical emergency. (See Section 3(8).)[22] Some may argue that such an approach ignores the reality of severely mentally handicapped women who are unable to consent but who might be subject to rape and unwanted pregnancy and child-bearing in the future. The solution to such a possibility, however, should lie in better protective services and not in a third party enforced sterilization.

Section 5. Voluntary Sterilization of Non-Institutionalized Persons

Non-institutionalized persons who are minors may not voluntarily consent to a sterilization. Non-institutionalized persons who have attained the chronological age of 21 have the right to be sterilized upon their request, when the person has given his or her consent pursuant to the procedures set forth in Section 8 of this act; and when at least thirty (30) days, but not more than one hundred and eighty (180) days have passed between the date of request and the date of the sterilization.

COMMENT. This provision allows for and underscores the right of handicapped persons in the community to make personal decisions regarding sterilizations.[23] Yet, this right is not unlimited. Consent must be voluntarily given after the person is fully informed and comprehends the nature and consequences of a sterilization, its risks and benefits, and potential alternatives to it; a person must be 21 years old; a waiting period of 30 days between the date of the request for sterilization and the operation is required; and the sterilization must occur before 180 days has passed from the date of the consent.

The age limitation has been set because experts generally agree that a form of "tender years" doctrine exists, in that persons under the age of 21 may not necessarily realize the nature or

[17]See generally, AMERICAN ASSOCIATION ON MENTAL DEFICIENCY (AAMD) CONSENT HANDBOOK, Special Publication No. 3 (1977); Wyatt, *supra* note 7 at 1384; Canterbury vs. Spence, 464 F.2d 772, 780 (D.C. Cir. 1972), *cert. den.* 409 U.S. 1064 (1972).

[18]AAMD, *id.*, at 6-10.

[19]*Id.*, at 10-11, 75-80.

[20]See Skinner, *supra* note 2.

[21]See generally, Introduction to the model act and cases and materials cited therein.

[22]See, e.g. Relf vs. Weinberger, 372 F. Supp. 1196 (D.D.C. 1974), 403 F. Supp. 1235 (D.D.C. 1975), *vacated as moot* 565 F.2d 722 (D.D.C. 1977).

[23]See, e.g., Wald, *The Legal Rights of People with Mental Disabilities in the Community: A Plea for Laissez Faire,* in LEGAL RIGHTS OF THE MENTALLY HANDICAPPED, 1035 (Practicing Law Institute, 1973) and cases cited therein; and M. Burgdorf, *Marital and Family Rights, in* LEGAL RIGHTS OF MENTALLY DISABLED PERSON (Practicing Law Institute 1979) and cases and materials cited therein.

consequences of sterilization. This concept is not related to an inability to comprehend an action based upon mental disability, but rather on a consideration that young persons are inexperienced in many ways and thus frequently regret sterilizations years later.[24] This age is also consistent with federal family planning regulations which prohibit funding of sterilizations for any persons under 21.[25]

A waiting period, also consistent with the same federal rules is incorporated in the model act.[26] The time allows a person to consider family planning information or to receive counselling which may lead the individual to reverse his or her decision. The time provided is reasonable, but not unlimited. No consent is valid 180 days after it is given. Thus, a physician will not be able to sterilize a person long after a person authorizes but has not undergone a sterilization. (See Section 8.) After 180 days, any consent is invalid, and the consent procedure must be initiated again. Therefore, the 180-day limitation serves as an additional check to insure that the sterilization is truly voluntary.

Section 6. Voluntary Sterilization for Institutionalized Persons

(1) If an institutionalized person is a minor, a sterilization shall not be performed even if voluntarily requested by the minor unless it occurs as, or as part of, involuntary emergency treatment.

COMMENT. See comment to Section 5.

(2) If an institutionalized person has attained a chronological age of 21 years, a voluntary sterilization may be performed if the following conditions are met:

(a) The person consents to the sterilization in accordance with Section 8 of this act;

(b) The Human Rights Committee described in Section 7 of this act approves of the sterilization; and

(c) At least thirty (30) days but not more than one hundred and eighty (180) days have passed between the date of the request and the date of the sterilization.

COMMENT. The mere fact of institutionalization should not preclude the sterilization of an adult if he or she voluntarily requests it. However, due to the potentially coercive nature of a mental health facility, hospital, or related center, a consent and review procedure and a time limit on the person's consent should be created to protect the person's constitutional rights.[27]

Section 7. Human Rights Committee

(1) There is hereby created a Human Rights Committee (hereinafter "Committee").

(2) The Committee shall consist of five members, and shall be selected by (name of the state protection and advocacy agency). The members shall be so selected that the Committee will be competent to deal with the medical, legal, social and ethical issues involved in sterilization; to this end, at least one member shall be an attorney, at least one shall be a psychologist, at least one shall be a physician, at least two shall be women, at least two shall be minority group members, and at least one shall be a resident of the institution. The foregoing categories are not mutually exclusive. No member shall be an officer, employee, or agent of the institution, nor may any member be otherwise involved in the proposed sterilization.

(3) Prior to approving the requested sterilization of any institutionalized person, the Committee shall:

(a) Review appropriate medical, social, and psychological information concerning the person;

(b) Interview the person;

(c) Interview concerned individuals, relatives, and others, who in its judgment will contribute pertinent information; and

(d) Determine whether the person has given his or her consent to the sterilization.

COMMENT. In a potentially coercive setting such as an institution, mentally handicapped persons may arrive at a sterilization decision involuntarily.[28] Therefore, the model act establishes a Human Rights Committee to ensure that all sterilizations are performed only after consent is voluntarily given by an institutionalized adult person in accordance with the provisions of this section and section 8. The concept of a Review Committee is similar to that ordered by the court in **Wyatt vs. Aderholt.**[29] In this landmark challenge to Alabama's treatment of institutionalized retarded persons, the court acted to end coercive sterilizations at Partlow State School and Hospital. The statute had permitted sterilizations of inmates based solely upon approval of the institution's director. After the law was declared unconstitutional based upon its failure to provide inmates basic procedural safeguards, the court ordered standards to protect the rights of persons subject to sterilizations. Included among these protections was a Review Committee consisting of independent citizens which examined all aspects relating to the proposed operation and which can ensure that an operation is based solely upon an institutionalized person's consent. Without the approval of its members, a sterilization could not be performed. Thus, sterilizations may no longer

[24]Wyatt, *supra* note 9, at 1384, *et seq.*
[25]42 C.F.R. 50.203(a)(i) *cf.* 42 C.F.R. 441.253(a).
[26]42 C.F.R. 50.203(i) *cf.* 42 C.F.R. 441.256(b), (c).
[27]Wyatt, *supra* note 9.

[28]See cases cited in footnote 16, *supra*.
[29]*Supra* note 9, at 1384-6.

be accomplished only after approval of an institution official.

This provision also may be compared with HEW comments relating to its refusal to fund sterilizations or abortions of institutionalized persons.[30] HEW regulations prohibit such funding partly because of the "inherently coercive nature" of institutions, and because of its apparent failure to develop a workable consent and review procedure in this area — something the provisions of this act solves; in addition, it is important to note that only federal funding is prohibited, not the right of institutionalized persons to obtain sterilizations through other means. Thus, this act is consistent with federal law.

The State Protection and Advocacy System is suggested as the Board having power to appoint Review Committee members, because it is usually the only statewide advocacy system available in every state. If an institution is currently under court order to desegregate its facilities and place persons in the community, then the court could, of course, take precedence in appointing members of the Review Committee, similar to Review Panels appointed in **Wyatt,**[31] **New York State Association for Retarded Children and Parisi vs. Rockefeller,**[32] and **Pennhurst.**[33]

(4) During the Committee's deliberations, the institutionalized person may be represented by an individual of his or her choice, including legal counsel, except that this individual may not be an officer, employee, or agent of the institution, nor otherwise involved in the proposed sterilization.

COMMENT. It is now universally accepted that due process requires representation for handicapped persons faced with any action potentially having negative impact on their person.,[34] Thus such representation should be accorded institutionalized persons whose request for a sterilization is being considered by the Committee. The individual representing the person should examine all matters relating to the proposed sterilization and thus provide a further check against a request that was not truly voluntary.

(5) If the Committee does not reach an affirmative determination as to the matters set forth in paragraphs 3(d) and (e), it shall not approve the proposed sterilization. Any doubts as to such matters shall be resolved against proceeding with sterilization.

COMMENT. None necessary.

(6) The Committee shall maintain written records of its determinations and the reasons therefore, with supporting documentation.

COMMENT. None necessary.

(7) Where the Committee does not approve the requested sterilization the institutionalized person shall have a right to appeal the decision to the (name of state trial court). This court shall conduct the trial de novo.

COMMENT. The purpose of a Review Committee is to ensure that a person in an institutional setting who requests a sterilization is acting in a voluntary manner. If the Committee opposes a requested sterilization because they do not believe that it was voluntarily requested, a handicapped person should have a right to seek judicial review. Thus, the model act specifically allows a person to appeal a Committee decision with which he or she disagrees.[35]

(8) The members of the Committee shall receive compensation at the rate not to exceed $ for each day, including the travel time, that they are engaged in the performance of their duties. In addition, all members shall be entitled to reimbursement for travel, subsistence and other reasonable expenses incurred by them in carrying out their duties.

COMMENT. None necessary.

Section 8. Consent Required

(1) Consent does not exist unless a consent form is completed voluntarily and in accordance with the following requirements:

(a) A physician must obtain consent for a sterilization from the institutionalized or non-institutionalized person;

(b) The physician must provide the person with a copy of the consent form which the person is to sign, written in the person's native language. A full and detailed explanation of the contents of the consent form is to be given to the person in the mode of communication that the person is most likely to understand, and prior to the person's signing of the form;

COMMENT. In order to preclude abuses, a consent form signed by the person requesting sterilization must be obtained prior to any operation.[36] Consent is ineffective without appropriate communication between a physician

[30]42 C.F.R. §50.206; for regulatory history, see 43 Fed. Reg. 52156 (1978).

[31]*Supra* note 9.

[32]357 F. Supp. 752 (E.D. N.Y. 1973); NYSARC and Parisi vs. Carey, No. 72-C-356/357 (E.D. N.Y. April 30, 1975), *approved* 393 F. Supp. 715 (E.D. N.Y. 1975).

[33]*Supra* note 8.

[34]Wyatt, *supra* note 9.

[35]Requirements for court proceedings in this area are described in Position Paper of the American Association on Mental Deficiency, Washington, D.C. (1975); Comer, *Sterilization of Mental Defectives: Compulsion and Consent,* 27 BAYLOR L. REV. 174 (1973) as well as in Wyatt, *supra* note 9, and NCARC, *supra* note 16.

[36]42 C.F.R. §§50.204-205; *cf.*, 42 C.F.R. §§441.257-258.

and patient undergoing such a procedure. This includes the necessity for obtaining consent in the primary language or mode of communication a handicapped person is most likely to comprehend. (See Section 3(5).)

(c) The consent form is to contain the following information:

(i) Advice that the person is free to withhold or withdraw consent to the procedure any time before the sterilization without affecting his or her right to future care or treatment and without loss or withdrawal of any state or federally funded program benefits to which the individual otherwise might be entitled;

COMMENT: This provision is included to prohibit forced sterilizaticn resulting from fear of loss of benefits for failure to accept a sterilization procedure. For years involuntary sterilizations have resulted from such institutional coercion. This act, however, mandates that all sterilizations, absent a medical emergency, shall be voluntary. Moreover, pursuant to Title XIX of the Social Security Act ("Medicaid") family planning services must be provided to "eligible individuals who desire such services." The Social Security Act further requires that "acceptance of family planning services . . . shall be voluntary on the part of the individual to whom such services are offered. Title X of the Public Health Service Act authorizes support for "voluntary family planning projects."[37] In addition, it is a federal felony, punishable by a fine up to $1000, or by imprisonment of up to one year, or both, to coerce any person to undergo an abortion or sterilization through the threat of withholding any service or benefits under a program receiving federal financial assistance.[38]

(ii) A description of available alternative methods of family planning and birth control;

(iii) Advice that the sterilization procedure is considered to be irreversible;

(iv) A thorough explanation of the specific sterilization procedure to be performed;

(v) A full description of the discomforts and risks that may accompany or follow the performing of the procedure, including an explanation of the type and possible effects of any anesthetic to be used;

(vi) A full description of the benefits or advantages that may be expected as a result of the sterilization;

COMMENT. Consent is effective only when an individual with the capacity to understand the nature and consequences of a sterilization or abortion procedure agrees voluntarily to its performance, after full information regarding the procedure, its risks and benefits, and any alternatives are disclosed. The American Association on Mental Deficiency in its *Consent Handbook* agrees with this position when they emphasized that in the area of medical treatment, all "information relevant to the patient's decisional process must be supplied,"[39] including a discussion of the procedures to be used, their hazards and benefits, as well as alternatives available. The model act language, extracted from H.E.W. regulations providing for funding of abortions and sterilizations, includes all of these indispensable requirements.[40]

(vii) An affirmation that the person has attained the chronological age of 21 years;

COMMENT. See comment to Section 5.

(viii) Advice that the sterilization will not be performed for at least thirty (30) days, but not more than one hundred and eighty (180) days from the date of the request for sterilization; and

COMMENT. Because it is irreversible, a waiting period of 30 days is included herein, so that a handicapped person may have the opportunity to consider relevant information concerning sterilizations. During this period a person may wish to seek legal or medical advice, including counselling or therapy, in considering whether to proceed with the sterilization. In addition, a person will be in a position during any waiting period to consider the procedures involved free of potentially coercive professionals or family members.[41] It may, of course, be argued that non-handicapped persons may obtain an immediate sterilization free of a mandatory waiting period and thus this provision is discriminatory. The model act suggests such a time limitation especially in the case of institutionalized persons who are subject to coercion by the nature of their placement because sterilizations are not generally immediately required and because it is advisable for persons in any event to have time to consider the nature and consequences of such treatment. The waiting period is inapplicable in the event of a medical emergency. (See Section 4.)

(ix) Advice that the physician will answer any questions that the person may have;

COMMENT. None necessary.

(d) The consent form must be signed and dated by the individual to be sterilized; the interpreter, if one was employed in explaining the contents of the form, including a sign language inter-

[37] 43 Fed. Reg. 52146 (1978), n. 1.
[38] *Id.*, p. 52147, n. 2.
[39] Notes 17 and 18, *supra*.
[40] See, *supra* note 36.
[41] *Id.;* see also, 43 Fed. Reg. 52158-9.

preter; and the physician who will perform the sterilization procedure;

(i) The physician must certify by signing the consent from that he or she complied with the provisions of this section, and that to the best of his or her knowledge and belief, the person to be sterilized knowingly and voluntarily consented to be sterilized and has attained the chronological age of 21 years;

(ii) If an interpreter is provided, the interpreter must certify by signing the consent form that he or she provided a full and detailed explanation of the contents of the consent form orally or in sign language or the person's other mode of communication, and that to the best of the interpreter's knowledge and belief, the individual to be sterilized understood what the interpreter told him or her; and

COMMENT. These provisions generally reflect the requirements of federal law relating to funding of sterilizations, and further ensure that consent is voluntary and knowing.[42]

(e) Twenty-four hours prior to sterilization, the physician, and the interpreter if needed, must once again provide to the institutionalized or non-institutionalized person the information and advice contained in subsection (c) of this Section.

COMMENT. This provision is included in the model act to allow handicapped persons a final opportunity to consider the seriousness of a sterilization, and to consider the option of less drastic alternative measures of birth control.[43]

(2) Consent may not be obtained if the institutionalized or non-institutionalized person who requests the sterilization is in labor or childbirth, or is under the influence of alcohol or other substances that may affect the individual's state of awareness.

COMMENT. Because consent cannot be given if a person is unable to comprehend the nature of the information provided, the provision is included to ensure that persons under physical or mental duress would not consent involuntarily to a sterilization. Thus, the model act does not permit any consent to be valid in circumstances, unrelated to mental disability, when a person may be in an altered state of consciousness. This provision is also consistent with H.E.W. regulations.[44]

[42] 42 C.F.R. §§50.204(b); 50.205(c).

[43] See, e.g., LaVeck and de la Cruz, *Contraception for the Mentally Retarded: Current Methods and Future Prospects*, HUMAN SEXUALITY AND THE MENTALLY RETARDED (1973).

[44] 42 C.F.R. §50.204(e).

Section 9. Confidentiality of and Access to Records

(1) All handicapped persons affected by this Act shall be informed of their right of and be entitled to access to all portions of any records relating to his or her sterilization or proposed sterilization.

COMMENT. The provisions relating to access to records is consistent with and representative of the modern policy that personal records should be made available to the person they affect. Access provisions are common in other areas of law. For example, the Family Educational Rights and Privacy Act,[45] requires parent — or in appropriate cases, student — access to all records relating to the student in question. The provisions of this law, known as "FERPA" or the "Buckley Amendment," have been incorporated in the Education for All Handicapped Children Act,[46] also known as P.L. 94-142, which applies to all programs funded under this law for mentally and physically disabled students in primary and secondary education programs. The Federal Freedom of Information Act,[47] which generally allows access to federal agency records, numerous state "sunshine" laws, and a growing number of state welfare department regulations covering programs for handicapped persons also allow such access to records. Thus, the model act adopts the right of the affected person to examine the information contained in official hospital and institution records.

(2) All records relating to a sterilization or proposed sterilization of a handicapped person shall be confidential and may not be released to third parties, except in the following circumstances:

(a) (the name of the state protection and advocacy agency) shall have complete access to all records without the consent of the person, notwithstanding any other provisions of this act;

(b) When an institutionalized or non-institutionalized person designates a third party as either a payor or co-payor for the sterilization, this designation will be assumed to carry with it his or her consent to release information to representatives of that payor which is necessary to establish reimbursement eligibility. Unless otherwise consented to by the patient, information released to the third-party payors shall be limited to that necessary to establish the claims for which reimbursement is sought;

(c) When state or federal agencies have a legitimate interest in obtaining information regarding sterilizations performed, records may be released without consent of the institutionalized or non-institutionalized person, providing that (i) the

[45] 20 U.S.C. 1232g; 45 C.F.R. Part 99.

[46] 45 C.F.R. §§121a.560, *et seq.*

[47] 5 U.S.C. §552.

person is informed of the need for these records; and (ii) all personally identifiable information is deleted from these records prior to disclosure to the state or federal agency; and

(d) Records may be disclosed to third parties upon written consent of the handicapped person who is the subject of the records.

COMMENT. None necessary.

(3) A charge, not to exceed the actual cost of duplication, may be assessed for copies of any records that need to be reproduced.

COMMENT. None necesssary.

(4) When records are released to a third party, it shall be unlawful for that party to redisclose any part of the record that contains personally identifiable information unless the person who is the subject of the record consents in writing. The release of the record to the third party shall be accompanied by the following statement: "This information has been disclosed to you from records whose confidentiality is protected by State law which limits redisclosure of personally identifiabale information to additional parties without the express consent of the subject of the record."

COMMENT. None necessary.

Section 10. Remedies

Any handicapped person who is aggrieved by a violation of this act shall have the right to bring a civil action in (name of the state trial court) to enforce its provisions and/or to receive monetary damages for a violation of it.

COMMENT. None necessary.

Section 11. Appropriations

To carry out the purposes of this act, there is authorized to be appropriated $______ for the fiscal year ending ______; $______ for the fiscal year ending ______; and $______ for the fiscal year ending ______.

COMMENT. None necessary.

Section 12. Severability Clause

If any provision of this act or the application thereof to any person or circumstance is held invalid, such invalidity shall not affect any other provision or application of the act which can be given effect without the invalid provision or application, and to this end the provisions of the act are declared severable.

COMMENT. None necessary.

Section 13. Effective Date

This act shall take effect __________.

COMMENT. None necessary.

1.5 ABORTION

As noted in the previous section, eugenic sterilization is a permanent deprivation of the fundamental rights to "privacy, marriage, sexual intercourse and procreation."[1] It is usually irreversible and is, of course, the most restrictive form of birth control available. For developmentally disabled persons, this procedure has traditionally been administered involuntarily by state hospital or penal authorities under the guise of treatment, and a number of state laws still permit sterilization without the express and sole consent of the handicapped person who is affected by the operation.[2] While abortion is not a permanent deprivation, it does, if performed involuntarily, impinge upon these same fundamental rights.

The United States Supreme Court has delineated the right to make personal decisions relating to abortion and family planning. This right began to be developed and defined by the Court in *Skinner vs. Oklahoma*,[3] a sterilization case, where the Court held unconstitutional as a violation of equal protection under the law a state statute requiring certain "habitual criminals" to be sterilized. Usually, the rights developed under this line of cases were based upon the concept of privacy. The Court subsequently found sufficient basis for this concept in the First, Fourth, Fifth, Ninth and Fourteenth Amendments to the U.S. Constitution, and has held that it is included within the meaning of the Bill of Rights.[4] Essentially, privacy includes those personal rights which are fundamental to "the very essence of a scheme of ordered liberty"[5]—those rights consisting of personal activities, including the sanctity of the marriage relationship and sexual freedom.

The right of married persons to practice contraception was affirmed in ***Griswold vs. Connecticut***,[6] which declared unconstitutional a state law prohibiting the sale, distribution and use of contraceptives. Justice Douglas wrote in his majority opinion that the government may not intrude in individual freedoms to undertake or refrain from procreation when the privacy of the marital relationship is involved.[7]

The Court extended this privacy right beyond the marriage relationship in ***Eisenstadt vs. Baird***,[8] which struck down a law banning the distribution of contraceptives to unmarried persons. In this 1972 opinion of the Court, Mr. Justice Brennan firmly asserted the scope of the privacy right: "If the right of privacy means anything, it is the right of the individual, married or single, to be free from unwarranted governmental intrusion into matters so fundamentally affecting a person as the decision whether to bear or beget a child."[9]

Perhaps the most significant case in the area of privacy and procreation is ***Roe vs. Wade***,[10] in which the Court held that a woman's right to her personal privacy includes the decision to make personal choices in abortion matters. The Court based this right on both the Ninth and Fourteenth Amendments, and held that its decision was a natural extension of earlier cases concerning contraception.[11] According to the Court, the state may not intrude in a decision to have an abortion during the first trimester of pregnancy. The state's interest is limited to safeguarding the woman's health only during the second trimester; its officials may only move to protect "fetal life" during the third trimester. Thus, the decision to have an abortion, while not absolute, is a private and in-

[1] P. Wald, *The Legal Rights of People with Mental Disabilities in the Community: A Plea for Laissez Faire*, in PRACTISING LAW INSTITUTE, LEGAL RIGHTS OF THE MENTALLY HANDICAPPED 1044 (1973).

[2] See Introduction to Subchapter 1.4: Sterilization

[3] 316 U.S. 535 (1942); *see generally*, Introduction, *id.*, for discussion of Skinner, as well as cases and materials cited therein.

[4] *See e.g.*, Griswold vs. Connecticut, 381 U.S. 479 (1965).

[5] Palko vs. Connecticut, 302 U.S. 319, 325 (1937).

[6] *Supra*, note 4.

[7] *Id.*, at 495-6.

[8] 405 U.S. 438 (1972).

[9] *Id.*, at 453.

[10] 410 U.S. 113 (1973).

[11] *Id.*, at 152-3.

dividual matter during the first pregnancy trimester.[12]

Abortion rights were further extended in *Planned Parenthood of Missouri vs. Danforth.*[13] The Court held by a 6 to 3 vote that spousal consent could not be required prior to an abortion, because of the privacy right extant during the first trimester.[14] By a separate vote of 5 to 4, the Court also held that parents of an unmarried minor could not prevent an abortion requested by the minor during the first trimester because "the abortion decision and its effectuation must be left to the medical judgment of the attending woman's physician [and not to third parties]".[15] The clear holding in *Danforth* was supported in *Belotti vs. Baird.* In this case, a Massachusetts law required unmarried minors to seek parental consent for all non-emergency abortions. If a parent refused permission, the minor could petition the superior court, which could grant authorization for the abortion for "good cause shown". Parents were to be provided with notice and consulted whenever such a judicial proceeding was instituted. The Supreme Court affirmed a District Court's holding that this law was unconstitutional because it unduly burdened the right to obtain an abortion by allowing a court to withhold consent, even if a minor was competent to make independent judgments concerning abortions. Unfortunately, however, this holding contains *dicta* which somewhat muddles the Court's position. Although eight of the nine justices agreed that the law was unconstitutional, they evenly split over whether a similar law could be enacted which would be constitutional.[16]

Although the clear mandate of these opinions is that there is a right to an abortion, there is still a question regarding the right to have the government fund such medical procedures. In a series of decisions, including *Maher vs. Roe*[17] and *Poelker vs. Doe,*[18] the Court held that equal protection is not violated when states refuse to fund "non-therapeutic" abortions[19] with Medicaid funds. The cases involved indigent women who challenged various state laws prohibiting the use of state medicaid money for such abortions, but the impact upon disabled persons is obvious and harsh. Frequently, disabled women are supported by Supplemental Security Income for the Aged, Blind and Disabled,[20] or by Social Security disability benefits,[21] or are receiving Medicaid for other medical procedures. As a result, women who may qualify for other medical services may not receive abortions funded by Medicaid. While persons sufficiently able to afford abortions are, of course, not affected by these decisions, it is apparent that a number of handicapped women, already relying on insufficient disability payments, are disproportionately discriminated against as a result. On June 30, 1980, the Supreme Court extended the *Maher* concept in *Harris vs. McRae*[22] to medically necessary abortions. Although the Court's opinion states that indigent women are not precluded for obtaining medically necessary abortions with their own funds, they are barred from securing federal Medicaid funds for such procedures.

The major problem created for women by the Maher and Harris prohibitions could be remedied through effective state legislation to ensure that persons dependent on medical assistance could obtain a full range of services. State funding would underscore the rights, especially of disabled citizens, to obtain equality with their non-handicapped peers and to ensure that all women may exercise their right to choice and privacy guaranteed by previous United State Supreme Court decisions.

A state statute on this topic also would be valuable for other reasons. First, it could include prohibitions against involuntary abortions performed on disabled women, whether or not the minor or adult is institutionalized. A similar prohibition and the rationale for it is included in the Model Sterilization Act presented in the previous section and should be examined for comparison.[23] Second, it could establish a consent and review procedure for women in institutions requesting abortions. Because institutions are potentially coercive environments, safeguards, including a review committee of independent citizens, should

[12] *Id.*, at 154. In Doe vs. Bolton, decided on the same day as **Roe** (January 22, 1973), the Court struck down provisions of a Georgia statute which impinged upon a woman's right to have an abortion. This law required, *inter alia*, approval of two physicians other than the woman's own to certify that an abortion was appropriate prior to the operation; approval of a hospital abortion committee; and performance of the operation in a state licensed and accredited hospital. The Court held that the acquiescence of two co-practicioners impinged on a physician's right to practice; that the hospital committee was unduly restrictive of the woman's personal rights; and the committee requirement which fails to take account of the woman's rights during the first trimester pursuant to Roe is invalid. 410 U.S. 179 (1973). In Colautti vs. Franklin, 439 U.S. 379 (1979), the Court emphasized that the physician, rather than the state, has the discretion to determine when a fetus becomes "viable" or capable of maintaining itself outside of the mother. Critical decisions relating to abortion matters are therefore left up to the woman and her physician.

[13] 428 U.S. 52 (1976).

[14] *Id.*, at 67-72.

[15] *Id.*, at 72-75.

[16] 61 L. Ed. 2d 797, 816-821 (1979).

[17] 432 U.S. 964 (1976).

[18] 432 U.S. 319 (1976).

[19] Although the Court did not specifically define "non-therapeutic", this phrase is generally regarded to refer to abortions that are not "medically necessary"—another amorphous term.

[20] 42 U.S.C. §§ 1381, *et seq.*

[21] 42 U.S.C. §§ 40, *et seq.*

[22] Harris vs. McRae, ___ U.S. ___ (1980).

[23] Model Sterilization Act, Section 4 precludes involuntary sterilizations in all instances, in the absence of a medical emergency, whether a developmentally disabled person is an institutionalized or non-institutionalized person.

be established to allow for total freedom of choice for disabled women so that abortions performed are truly voluntary. Such a procedure is incorporated in the Model Sterilization Act[24] and could, if modified to account for the immediate nature of abortions, serve as a guide for states wishing to draft an act in this field. Third, and finally, given the Supreme Court's specific holding that married women and unmarried minors need not necessarily obtain third party consent prior to seeking an abortion, a state statute should make it clear that this right applies equally to mentally handicapped women. Unless specifically adjudicated incompetent to make an abortion decision, these women should not be precluded from obtaining abortions on the same basis as their non-handicapped peers. To do so would be to discriminate against them solely on the basis of their handicap.

A model statute is not presented since it would be highly similar to the Model Sterilization Act. Thus, state officials should look to that act for guidance while specifically incorporating the points described above.[25]

[24] Section 7 of the Model Sterilization Act, *supra* Subchapter 1.4, describes the Human Rights Committee which is to act as an independent analyst to ensure that a sterilization is indeed voluntary. Section 8 provides for consent procedures and forms to further protect against involuntary sterilizations.

[25] For a discussion and legal case history of the states' involvement in abortion and developmentally disabled women, see "Abortion for the Severely Retarded: A Search for Authorization", 1 M.D.L.R. 484 (May-June 1977); and, M. Burgdorf, "Marital and Family Rights", an outline of cases and materials in the field set forth in PRACTISING LAW INSTITUTE, LEGAL RIGHTS OF MENTALLY DISABLED PERSONS (1979).

1.6 EMERGENCY MEDICAL CARE AND TREATMENT FOR INFANTS

Table of Contents

INTRODUCTION

With increasing frequency parents of infants born with a variety of physiological "defects" are, with the consent or on the advice of a physician, refusing to allow the treatment necessary to ensure the child's survival.[1] Among the conditions sometimes deemed to justify non-treatment are myelomeningocele, Down's syndrome, anencephaly, hydrocephaly and a variety of other central nervous system defects.[2]

Frequently a child born with one of the listed conditions will, in addition, suffer from related but independent "defects", most frequently a form of intestinal blockage or atresia, which, if uncorrected, will prevent the infant's receiving the nourishment necessary to survive.[3] In these cases the physician's refusal or inability to perform surgery will result in death by starvation. In other cases the infant, weakened by the primary disabling condition, will be ultra-sensitive to a variety of viral or bacterial infections, particularly pneumonia, which, if untreated, may ultimately prove fatal.[4] In still other cases the primary disability itself will be fatal if left untreated or uncorrected.[5]

Ironically, in some cases, the primary defect itself is far from life endangering. In fact, if promptly treated, infants currently allowed to die would in many cases develop into adults with only moderate permanent physical or mental disability.[6]

A variety of justifications are advanced for non-treatment. In extreme cases the most frequently advanced argument is a philosophical one: that the affected infant's condition renders him or her sub-human or a non-person.[7] While philosophers and theologians have on occasion attempted to define the critical indicia of "humanhood"[8] the argument is difficult to consider seriously in a legal context. The divergence of opinion as to the necessary characteristics illustrates the subjectivity of any such analysis. In addition, it seems clear that the ability to predict the extent of disability a child will suffer as an adult, if properly treated, is currently suspect[9] since at the time when a non-treatment decision is made it is frequently impossible to develop a satisfactory prognosis for the child's later life.[10] Critics of non-treatment decisions also point to the historical use of such arguments to justify discrimination and atrocity. In their view the use of life quality criteria is too susceptible to abuse to warrant serious consideration as a legal or ethical standard.[11]

An argument advanced less frequently is that the infant's right to life is outweighted by the psychological and economic impact on the infant's family if treatment is provided.[12] Clearly there is a "quality of life" aspect to this argument as well since no similar rationale has ever been advanced

---

[1] See Duff and Campbell, ***Moral and Ethical Dilemmas in the Special-Care Nursery***, 289 NEW ENG. J. MED. 890 (1973).

[2] Robertson, *Involuntary Euthanasia of Defective Newborns: A Legal Analysis*, 27 STAN. L. REV. 213-14 (1975). That the practice of withholding care is widespread is evidenced by entries in recent medical textbooks and journals outlining specific clinical signs suggesting non-treatment as the appropriate response. *See, e.g.*, F. INGRAHAM AND D. MATSON, NEUROSURGERY OF INFANCY AND CHILDHOOD, 35-39 (1954); Lorber, *Results of Treatment of Myelomeningocele*, 13 DEVELOP. MED. AND CHILD NEUROL. 279-303 (1971).

[3] Robertson, *supra* note 2, at 214 n.8.

[4] *Id.*

[5] Severe cases of anencephaly and hydrocephaly are the primary conditions which will, in and of themselves, prove fatal.

[6] Evidence suggests that at the time a non-treatment decision must be made current diagnostic procedures cannot accurately evaluate the extent of later disability the individual will show. Particularly in the case of infants with mongolism and myelomeningocele the condition may result in only moderate disability with or without proper treatment. *See* Freeman, *The Shortsighted Treatment of Myelomeningocele: A Long-Term Case Report*, 53 PEDIATRICS 311 (1974); Ames and Schut, *Results of Treatment of 171 Consecutive Myelomeningoceles—1963-1968*, 50 PEDIATRICS 466 (1972); *but see* Lorber, *Selective Treatment of Myelomeningocele: To Treat or Not to Treat*, 53 PEDIATRICS 307 (1974). *See also* Gustafson, *Mongolism, Parental Desires and the Right to Life*, 16 PERSPECTIVES IN BIOLOGY AND MEDICINE 529 (1973).

[7] See, *e.g.*, Fletcher, J. *Indicators of Humanhood: A Tentative Profile of Man*, HASTINGS CENTER REPORT (November 1972).

[8] *Id. See also*, Carney, *Medical Standards for Deformed Newborn*, WASH. POST, (Mar. 20, 1974), *A 15, col. 1.*

[9] See text accompanying note 6, *supra*.

[10] *Id.*

[11] See Kamisar, *Some Non-Religious Views Against Proposed "Mercy Killing" Legislation*, 42 MINN. L. REV. 969, 1012-37 (1958).

[12] *See* Robertson, *supra* note 2, at 255-62. For a discussion of the psychological effect on parents of the birth of a disabled infant, *see also* Giannini and Goodman, *Counseling Families During the Crisis Reaction to Mongolism*, 67 AM. J. MENTAL DEFICIENCY 740 (1962); Cohen, *The Impact of the Handicapped Child on the Family*, 43 SOCIAL CASEWORK 137 (1962).

concerning chronically ill individuals unless they also had a mental disability. Thus the argument is discriminatory in the same sense as those discussed previously and it is equally subject to abuse. In addition, if the state demonstrates a willingness to provide financial assistance and family counselling or, more drastically, if the state is willing to assume full responsibility for the care and custody of an infant, then the argument loses any force it might otherwise have. Finally, a decision not to treat which is justified by economic considerations is squarely at odds with fundamental constitutional principles. Economic considerations have never been thought sufficient to warrant deprivation of a fundamental right, of which the right to life stands as the most fundamental.[13]

Given the growing acceptance of non-treatment decisions within the medical community, the state legislative body may be the only feasible forum for a resolution of this issue. While a decision not to treat an infant implicates a host of criminal prohibitions—from homicide to child neglect—prosecutions are rare and convictions almost non-existent.[14] Given the manner of decision-making, it is unlikely that most decisions will receive the visibility necessary to set the state's criminal machinery in gear.[15]

Thus the non-treatment of infants is an issue ripe for legislative attention and dialogue. When brought to light and openly debated, it seems clear that non-treatment decisions are not justifiable in a society which values, above all else, the right to life itself, and which has the resources to aid parents with the financial and psychological adjustments they may incur.

REVIEW OF STATE LEGISLATION

There are currently only three states with any legislative initiative affecting the treatment of infants. Arkansas allows a parent or legal guardian to sign a declaration, on behalf of an infant or incompetent, prohibiting the provision of extraordinary means of care and treatment to prolong the infant's life.[16] On the basis of this Act, it appears that the parents or guardian may make an autonomous and unreviewable choice to allow an infant's death so long as the care necessary to prolong life may be characterized as extraordinary. As noted above, the actual life-threatening aspect of a disabled infant's condition, most commonly an intestinal or esophagial blockage, may be remedied by relatively routine surgery.[17] It remains to be seen whether the Arkansas statute will be construed to apply to these cases. Extraordinary means are nowhere defined in the Act.

A similar statute exists in New Mexico which allows parents or guardians to direct the withholding of care from terminally ill minors.[18] The New Mexico Act is circumscribed to a greater degree than the Arkansas statute because care may only be withheld from the terminally ill, defined as individuals who, regardless of the use of extraordinary medical procedures, will die as a result of their illness.[19] This would clearly not cover the class of infants who would survive in the event that standard corrective surgery was performed. This Act will, however, probably be applicable in cases of hydrocephaly or anencephaly since infants suffering from these conditions will, in all likelihood, eventually die as a result of the condition. Unlike the Arkansas law, the parental decision not to treat is reviewable under the New Mexico statute, and provision is made for appointment of a guardian ad litem to represent the interests of the child. It is not clear, however, that an appointed guardian will have an effective role in the proceeding since no requirement exists that the decision made represent the infant's best interests. On the face of the statute it would appear that the parental decision must stand if all legal formalities are complied with.

In contrast to these laws stands an Indiana statute[20] which declares that once a fetus is born alive it must be treated thereafter as a person under the laws of the state regardless of its physical or mental condition. In keeping with this philosophy the statute states that "failure to take all reasonable steps, in keeping with good medical practice, to preserve the life and health of said live born person shall subject the responsible persons to Indiana laws governing homicide, manslaughter and civil liability for wrongful death and medical malpractice".[21] Though the "good medical practice" proviso may allow non-treatment of infants in some instances, it is clear that this statute demonstrates a strong legislative intent to protect infants from non-treatment deci-

[13] An infant's right to life is a fundamental right under constitutional analysis. See Roe vs. Wade, 410 U.S. 113, 157-67 (1973). As such it would be inconsistent with basic principles to allow deprivation of the right on the basis of purely economic justifications. Memorial Hospital vs. Maricopa County, 415 U.S. 250, 263 (1974). Essentially a decision to allow non-treatment based on economic cost would be tantamount to sanctioning a rule requiring one individual to sacrifice his life for the good of society; a precedent with almost unimaginable implications.

[14] Robertson, *supra* note 2, at 217-44.

[15] Physicians and parents ordinarily make the non-treatment decision autonomously without peer review or any administrative process. See Duff and Campbell, *supra* note 7. Only when the physician or hospital disagrees with the parental decision is a judicial proceeding likely to be initiated. See Maine Medical Center vs. Houle, No. 74-145 (Maine Super. Ct. Feb. 14, 1974).

[16] ARK. STAT. ANN. §§82-3801 to 3804 (Supp. 1978).

[17] See text accompanying note 3, *supra.*

[18] N.M. STAT. ANN. §§24-7-1 to 11 (1978).

[19] *Id.* at §24-7-3.

[20] IND. CODE §35-1-58.5-7 (1975).

[21] *Id.*

sions and to enforce already existing criminal and civil penalties which impact on treatment decision-making.

SUMMARY OF MODEL STATUTE

Because of the paucity of legislation on this topic and because of clear inadequacies in two of those laws, a model act was developed. It adopts the philosophy that life-preserving care and treatment should be provided on a nondiscriminatory basis to all infants regardless of present or potential disability, and explicitly rejects the quality of life and economic hardship rationales that have been advanced in favor of non-treatment. It is recognized, however, that the burden placed on parents by the birth of a disabled infant is substantial. Consequently, a framework is provided within which the state attempts to play a supportive role: providing psychological counselling, financial assistance and referral services—all designed to minimize parental trauma and hardship associated with the infant's birth.

The Act is structured so as to provide the state a constructive rather than coercive role. To the extent state coercion is allowed, the coercion itself is to be used constructively to compel the attendance of parents or guardians at counselling sessions. Further, such coercion may only be employed after informal means have failed to produce results.

The Act seeks to maintain the integrity of the family unit and to preserve traditional parental control over the care and custody of an infant. Rather than usurp the parental role, states, under the Act, provide services designed to aid parents to successfully undertake their traditional role. Only in one instance—when failure to provide treatment will endanger an infant's life—may treatment be provided without parental approval and then only to the extent necessary to assure that the infant's life is preserved. In all other instances the parents are free to remain in control of the infant's care if they choose to do so.

Thus, given the overriding premise that non-treatment which endangers an infant's life is never justified, the Act strikes a balance between the legitimate needs of parents faced with the birth of a disabled child and the infant's compelling human right to non-discriminatory life-preserving medical care. Substantial attempts have been made to preserve the traditional power of parents and guardians. The Act attempts to err, if at all, in favor of parental rights, while providing the essential protection needed to prevent an infant's death through conscious non-treatment.

MODEL STATUTE*

A BILL FOR

AN ACT TO ESTABLISH THE RIGHT OF DEVELOPMENTALLY DISABLED INFANTS TO RECEIVE EMERGENCY MEDICAL CARE AND TREATMENT; TO PROHIBIT PARENTS OR GUARDIANS FROM DENYING THESE INFANTS THEIR RIGHT; AND TO ESTABLISH COUNSELING, REFERRAL AND FINANCIAL ASSISTANCE FOR PARENTS OF SUCH INFANTS WHEN NEEDED.

Section 1. Title

This act shall be known as the "Infant Emergency Medical Care Act."

COMMENT: None Necessary.

Section 2. Purpose

(1) The General Assembly declares that it is the purpose of this act to recognize and protect the right of a developmentally disabled infant to receive all medical care and treatment necessary to cure, alleviate or prevent any physical condition which, according to prevailing medical standards, involves a substantial risk of death to such infant.

(2) The General Assembly intends by this act to prohibit the withholding of medical care or treatment from a developmentally disabled infant on the basis of the predicted quality of such infant's life, when the withholding of such care or treatment involves, according to prevailing medical standards, a substantial risk of death to the infant.

(3) The General Assembly declares that the increased economic burden resulting from the provision of such care and treatment is, in all cases, outweighted by the state's interest in the sanctity of an infant's life.

COMMENT. It has become increasingly clear in recent years that a significant number of newborn infants are, by agreement between parent and physician, allowed to expire through the withholding of necessary medical care and treatment. The infants for whom non-treatment is most frequently suggested are those with multiple or chronic physical or mental impairments which are likely to result in later adult dysfunction and to require continuing treatment.

The act's statement of purpose delineates the most common rationales for withholding care and rejects them unequivocally. Specifically, commentators have suggested that certain conditions are so severe that an afflicted infant cannot be considered truly "human" and thus should not be afforded the normal incidents of personhood, primary among them the right to remain alive.[1] This is the typical "quality of life" argument rejected in subsection (2). The act takes the position that an objective definition of "personhood" is irrelevant in the context of infants. Regardless of what we view objectively as "normal" and what we feel our own desires would be if reduced to the condition of some infants, we must take cognizance of the fact that the infants in question know no other cognitive state and thus cannot be "dissatisfied" with their existence in the manner often attributed to them. Thus, substituted consent should never be permitted to terminate an infant's life.

Similarly, some commentators take what they consider the more pragmatic view that the economic burden imposed by treatment cannot be justified in terms of the concomitant benefits conferred on the infant.[2] Again the statute rejects this premise. It simply cannot be squared with our societal view of the sanctity of human life. There is no other situation in which we, as a society, have seriously contemplated economic justifications for the extinguishment of human life and there is no plausible reason for an exception in this situation.

*Ross Dolloff had primary responsibility for drafting this model statute.

[1] *See* Tooley, *Abortion and Infanticide,* 2 PHIL. AND PUB. AFFAIRS 37 (1972).

[2] Matson, *Surgical Treatment of Myelomeningocele,* 42 PEDIATRICS 225, 226 (1968).

Section 3. Definitions

(1) "Infant" means a developmentally disabled child, born alive, during the period commencing with birth and continuing up to and including the child's second birthday.
COMMENT. This definition controls the period of a child's life during which the act operates to protect the child. While, in most cases, the critical period when non-treatment decisions are made, will be during the first weeks after birth, some conditions currently believed to justify non-treatment require routine surgical or medical attention well beyond the first few weeks.[3] Thus, the definition is framed to extend protection and assistance throughout the child's first two years. Ordinarily, further state protection beyond this period is unnecessary[4] since a parent or guardian's initial negative reaction to the birth of a disabled infant should subside if counselling and other assistance is provided during the first two years.

(2) "Developmentally disabled" means having a disability which:

(a) (i) Is attributable to autism, cerebral palsy, epilepsy, or mental retardation;

(ii) Is attributable to any other condition found to be closely related to autism, or cerebral palsy, epilepsy or mental retardation or requires treatment and services similar to those required for such persons; and

(b) Can be expected to continue indefinitely.
COMMENT. The model act employs the categorical definition of developmental disability previously utilized in federal law. Its use is deemed preferable over potentially broader but more conceptual definitions because it is reasonably easy to apply on a case-by-case basis. A definition premised on later life-function impairment would, in contrast, be difficult to apply to newborn infants.

All the conditions which are currently considered suggestive of later retardation, and thus are implicated in non-treatment decisions, are covered. Among the most frequently involved conditions are Down's syndrome, spina bifida, anencephaly, hydrocephaly, and myelomeningocele.[5] It is important to recognize that as used in this act this definition will denote which *individuals* are covered, not which *diseases*. In other words, an individual who is developmentally disabled under the above definition must be treated for any condition which endangers his or her life whether or not the condition requiring treatment is directly related to the condition enumerated in this definition. Often times the life-endangering condition is incidental to the condition causing disability as when an infant with Down's syndrome is born with an intestinal blockage which, if left uncorrected, prevents normal feeding.[6] The act intends that such incidental conditions will be treated.

(3) "Necessary medical care and treatment" means all available medical care and treatment, including but not limited to, medical care, nursing care, supplies, medication and other assistance that according to prevailing medical standards is required to cure, alleviate, or prevent any physical condition which involves a substantial risk of death to a developmentally disabled infant.
COMMENT. This definition specifies the amount and kinds of medical care and treatment which must be provided when a developmentally disabled infant is afflicted with a life-endangering condition. If the withholding of any form of care or treatment holds a substantial risk of death to the infant, that treatment must be provided. Further, the act makes no distinction between ordinary and extraordinary treatment measures. Both must be provided if reasonably necessary.

(4) "Department" means the state agency which is responsible for health care policy implementation.
COMMENT. None necessary.

(5) "Support services" means counselling, psychological or psychiatric services, home care services, special educational services, and other services as may be of assistance to a parent or guardian responsible for the care of an infant.
COMMENT. The act requires parents or guardians to be informed of support services available to assist them in the care of a disabled infant. Thus, the counselling, psychological or psychiatric services mentioned in the definition refer to the provision of such services to the parents or guardians themselves. This definition indicates that such services are to be broadly construed to include any services which may be of assistance to such individuals.

Section 4. Provision of Medical Care and Treatment

(1) Every developmentally disabled infant shall be provided necessary medical care and treatment at the earliest time reasonably possible.
COMMENT. This section provides the basic enforceable duty under the statute. The limits of this duty are defined by subsections (1), (2), and (3) of section 3 which define the terms "developmentally disabled", "infant", and "necessary medical care and treatment". Taken together these definitions establish that the duty involved is to provide to a developmentally disabled individual, up to two years of age, all medical care and treatment reasonably necessary to preserve that individual's life.

---

[3] Robertson, *Involuntary Euthanasia of Defective Newborns: A Legal Analysis*, 27 STAN. L. REV. 213, 214 (1975).

[4] *Id.*, at 258.

[5] *See* Duff and Campbell, *Moral and Ethical Dilemmas in the Special Care Nursery*, 289 NEW ENG. J. MED. 890 (1973).

[6] *See* J. Warkany, CONGENITAL MALFORMATIONS 324-31 (1971).

Some medical authorities suggest that certain identifiable conditions, particularly spina bifida, may currently be more disabling than necessary because of delays in initiating treatment.[7] Thus, this provision also mandates that treatment begin as early as is reasonably possible.

(2) Responsibility for the provision of such care and treatment shall rest with the person or persons having responsibility for the care and custody of the infant.

COMMENT. The person responsible for the provision of necessary care and treatment is the person otherwise responsible, under the laws of the state, for the infant's care and custody; most often the infant's parent or parents. The model act attempts to maintain and support the family unit rather than to strip the infant's custodian of responsibility for his or her care. By clearly identifying the party responsible for care, this subsection also establishes a duty which if violated may trigger other aspects of state child-protection law including proceedings for termination of parental rights and child neglect proceedings.

Section 5. Failure to Give Consent; Emergency Treatment

(1) A physician shall provide necessary medical care and treatment, without consent, if delay would jeopardize the life of the infant.

COMMENT. If delay in the introduction of treatment while parental consent was being sought would jeopardize the infant's chances of survival, the physician has a duty to treat without consent.[8] This provision is essential to protect the life of the infant whether the difficulty in procuring consent is due to happenstance or deliberate intent on the parent's part. Conversely, however, if delay would not jeopardize the infant's chances of survival, parental consent must be secured as is typical in medical practice. This provision thus attempts to strike a balance between the traditional rights of parents and the needs of the affected infant, overriding the parental right only when strictly necessary. Even when non-consensual treatment is authorized, however, it is desirable to involve parents in the treatment process to the greatest extent possible. In keeping with this philosophy attempts should be made to secure parental approval even if treatment has been previously initiated under this section.

(2) Notwithstanding any other provision of law, a physician who provides necessary medical care and treatment without consent, in a reasonable good faith belief that the provision of such care and treatment is consistent with the requirements of this section, shall not be subject to civil or criminal liability for his or her actions in this regard.

COMMENT. This subsection immunizes physicians from legal liability for acts performed in the belief that they are required by this section. Where an emergency situation exists and life or death lies in the balance, it is to everyone's advantage that the physician act promptly without hesitating due to considerations of possible liability. The physician's belief must, however, be reasonable and consequently this subsection should not provide an overly broad immunity for non-consensual treatment decisions.

Section 6. Reporting of Violations; Investigation

(1) Whenever a parent intentionally refuses to consent to the provision of necessary medical care and treatment, the attending physician or physicians shall report such refusal to the Department.

COMMENT. Since a physician, in most cases, will be in a good position to determine whether a parent is complying with his or her responsibilities under this act, the physician is given the additional duty of reporting to the state administrative authority parental refusal to give consent.[9] This provision is essential to ensure that the state can provide counselling, referral services and financial assistance to those parents who are in need of them.

(2) Any persons believing themselves to have knowledge of a violation of section 4 of this act shall promptly notify the Department.

COMMENT. Primary reporting responsibility under the act will probably be shouldered by physicians pursuant to subsection 6 (1). This subsection provides an additional safeguard by making it the legal duty of all individuals believing themselves to have knowledge of a parental violation to report the same to the administrative authority. (See Section 9 (2) for violations by physicians).

(3) The Department, upon receipt of notification, as provided for by subsections (1) or (2) of this section shall promptly conduct an investigation to determine the existence or non-existence of a violation.

COMMENT. Investigation is essential to insure that actions are not taken based upon incorrect reports. But such investigations must be conducted as expediously as possible because of the emergency nature of many of the situations which will arise under this act.

[7] *See* Freeman, *The Shortsighted Treatment of Myelomeningocele: A Long-Term Case Report,* 53 PEDIATRICS 311 (1974).

[8] Treatment without consent, in emergency situations, has long been authorized under the common law. *See Physicians and Surgeons* 70 C.J.S. §48 (1951).

[9] This aspect of the physician's duty is analagous to that contained in many state child abuse and neglect laws.

Section 7. Parental Counselling

(1) When, after investigation, the Department determines that there exists reasonable grounds to believe that an infant is not receiving necessary medical care and treatment the Department shall commence parental counselling services.

COMMENT. This subsection sets out the standard of certainty required before the administrator may act to rectify a violation and also identifies the type of action the administrator may take. The standard employed is that of "reasonable grounds to believe", a lower standard than probable cause. This standard is adopted because the actual existence of a violation of the act's care and treatment requirements require difficult medical determinations involving the severity of an infant's condition. During the short period available to an investigator, and assuming investigators are not trained medical personnel, it would be improbable that a violation could be established with any greater degree of certainty. The "reasonable grounds to believe" standard is high enough, however, to insure that parents and guardians will not be subjected to frivolous intrusions by the Department. Furthermore, discussions between the parties should easily clear up any reasonable but unfounded concerns that the infant is not receiving necessary medical care and treatment.

Once a violation is established the Department is to initiate parental counselling. The act opts initially for a cooperative, rather than a coercive approach. Oftentimes the initial shock and grief experienced by a parent who bears a disabled child is the motivating factor for a non-treatment decision.[10] Such decisions are typically made in haste with little opportunity for full consideration of the repercussions.[11] Parental counselling serviices are designed to assure that no such hasty decisions are made and that, instead, parents or guardians are provided with an opportunity to discuss their fears and assess their alternatives. It is anticipated that the vast majority of these situations can be resolved without resorting to more coercive strategies.

(2) The Department shall commence parental counselling services by causing written notice to be given, to the person or persons responsible for the care and custody of the affected infant. The notice shall contain:

(a) A description of the acts or omissions believed by the Department to establish a violation of this act;

(b) An explanation of the Department's authority, pursuant to this section, to initiate parental counselling services;

(c) A request that the recipient of the notice contact the Department within 24 hours and arrange for a meeting with counselling personnel; and

(d) Any other matters established by regulation of the Department.

COMMENT. The counselling process begins with the dispatch of notice, which, since not specified in the act, may be provided by formal service or may consist of a visit by counselling personnel to the parents or guardian of the infant. In whatever manner service is provided, however, it must include presentation in writing of the information specified in this provision in order to provide evidence of proper notice and a tangible reference for the individual(s) notified.

(3) Counselling meetings may be conducted at such time and place as is convenient to the parent, parents, or guardian and to counselling personnel. Where conflicts arise the convenience of a parent of guardian shall take precedence over the convenience of counselling personnel.

COMMENT. The birth of a developmentally disabled child is often a traumatic experience for the parent and decisions regarding treatment are often difficult to make. Thus, the Department should respect the parents' situation and allow the convenience of the parents to take precedence over the convenience of administrative personnel. The requirements for the time and place of a counselling meeting are designed to provide administrative personnel with maximum discretion and flexibility.

(4) If, within twenty-four (24) hours of receipt of notice pursuant to this section, the person or persons having the care and custody of an infant do not contact the Department, the Department shall make reasonable efforts to contact such person or persons to arrange a time and place for a counselling meeting. If such efforts fail or if the person or persons refuse to agree to a meeting, the Department shall file a petition in the (name of court of general jurisdiction) in the county of residence of the infant seeking an order requiring the attendance at a counselling meeting of such person or persons. Such order shall be granted if the court finds, on the basis of affidavits provided by the Department, that probable cause exists to believe that such person or persons have violated the provisions of this act.

COMMENT. This subsection is the first to sanction any coercive efforts by the Department. The Department must first, however, use informal means to arrange for a counselling meeting which can include telephone calls, visits or any other method the administrator deems appropriate. If these efforts are unsuccessful, however, the Department is left no choice but to use the coercive powers of the court. Clearly, counselling is likely to be more successful if it is arranged cooperatively. Thus seeking a court order is available only after voluntary arrangements fail. Frank discussion between parents and counselling per-

---

[10] *See* Cohen, *The Impact of the Handicapped Child on the Family*, 43 SOCIAL CASEWORK 137 (1962); Fletcher, *Attitudes Toward Defective Newborns*, HASTINGS CENTER STUDIES, Vol. 2, No. 1 (1974).

[11] *See* Duff and Campbell, *Supra* note 5.

sonnel, even in a coercive atmosphere, however, will be preferable to an approach which would allow the state at this point to terminate parental responsibilities for the well-being of an infant.

This provision allows the Department to petition the court for an *ex parte* order so that a parent's non-appearance at the court hearing would not delay further the counselling meeting and the infant's receipt of needed medical care and treatment.

(5) The goals of a counselling meeting shall be to:

(a) Maintain, where possible, the integrity of the family unit;

(b) Suggest alternatives to non-treatment of the affected infant and inform the parties of available support services in their community or region;

(c) Inform the parties of the availability of state financial assistance pursuant to section 8 of the act and the qualifications therefor;

(d) Describe to the parties their legal rights and obligations and the probable consequences of their acts or omissions, including, in the counsellor's discretion, possible civil liability and criminal penalties to which they may be subject; and

(e) Discuss any other goals established by regulation of the Department which are consistent with those enumerated in paragraphs (a) through (d).

COMMENT. This provision enumerates some of the matters to be discussed in counselling. It is intended that counselling personnel will go beyond these issues to conduct a full and frank discussion of the situation, allowing the individuals counselled to fully air their personal feelings. It is hoped that this counselling meeting will culminate in the individuals counselled demonstrating a personal resolve to act in a manner consistent with the requirements of this act. This is the overriding goal of the counselling process.

Also, as is suggested throughout, the act favors maintenance of the family unit over other possible dispositions. Counselling should be conducted so as to maximize the likelihood that the family unit will remain intact. It is believed, however, that counselled individuals deserve a full description of available alternatives so that they may evaluate them with the help of counselling personnel. To this end, counsellors in informing the parties of their legal rights and obligations should inform parents or guardians of their right to relinquish their parental responsibilities. Adverse consequences of such a choice, from the infant's perspective, should also be discussed.

It must be emphasized that the list enumerated here is not comprehensive and it is the intent of this subsection to provide only a guideline for counselling personnel. It is intended that counselling personnel be provided some autonomy in their conduct of meetings and that the discussion which takes place be relatively unconstrained. The personalities of the individuals involved and the characteristics of the situation(s) which trigger counselling will be extremely varied and, thus, no single set of guidelines can be expected to respond adequately to every situation.

Section 8. State Assistance to Parents or Guardians of Infants in Need of Medical Care and Treatment

(1) The Department shall either provide necessary medical care and treatment or cause it to be provided to infants of eligible parents or guardians, or, where the parents or guardians have already had the necessary medical care and treatment provided or where they choose to do so through their own private arrangements, the Department shall provide the eligible parents or guardians financial reimbursement for the reasonable costs of the necessary medical care and services.

COMMENT. It is a fundamental tenet of the act, expressed in the Introduction, that the sanctity of human life outweighs any economic considerations which might be advanced as a justification for withholding care. Yet, such economic considerations are still important, especially when the fected individuals may well be too distraught or too the medical care and treatment the infant requires, or when such payments would work an extreme hardship on the family. This provision recognizes both the state's commitment to the sanctity of human life and the need of some parents or guardians for financial aid by authorizing the Department to aid eligible parents or guardians by one of three means: (1) providing the services directly to infants, (2) contracting for the services to be provided, or (3) reimbursing the parents or guardians for the reasonable costs incurred in having the necessary medical care and treatment provided. The options are provided to allow the Department to make best use of its resources while still assuring that the necessary medical care and treatment is provided.

Given that states will vary in their existing requirements for eligibility for other aid or financial assistance, in funds available to provide such aid and assistance, and in the rapidly changing cost of living this act does not define the term "eligible". Rather it is left to the legislature to specify who will be eligible or to delegate this responsibility to the Department.

(2) Any individual believing him or herself eligible for assistance under this section may make application for such assistance on forms provided by the Department. In addition, any person having knowledge of an individual believed to be eligible for assistance under this section may bring such facts as they believe establish eligibility to the attention of the Department.

COMMENT. The specifics of application procedures are left to the discretion of the Department. This section, however, makes clear that an application process may be initiated by any interested person, not solely the affected individual or individuals. Considering the nature of the situation which triggers assistance under the act, affected individuals may well be too distraught or too busy to deal with the application formalities. By allowing surrogates to act in their behalf, this subsection will help to facilitate the prompt receipt of assistance by those who need it most.

(3) The Department, upon receipt of an application or notice of an eligible person pursuant to subsection (2) shall promptly investigate the validity of such application or notice and shall determine eligibility for assistance pursuant to this section. If the Department finds an individual eligible it shall direct the payment or provision of assistance in the amount required by subsection (1) of this section.

COMMENT. This subsection, while again leaving specifics to Departmental regulations, requires the Department to act expeditously in processing applications for assistance. By placing investigative responsibility on the Department, pursuant to this subsection, and requiring only a threshold application process, pursuant to subsection (2), this section creates a system which places affirmative responsibility on the Department to help eligible individuals receive prompt assistance.

(4) If the Department, after investigation, finds an individual to be ineligible for assistance pursuant to this section it shall promptly notify the individual of its determination in writing stating the reasons for such determination and notifying the individual of his or her right to resubmit such application if he or she believes it to have been improperly denied. Any reapplication for assistance, after initial denial by the Department, must state such facts, errors, or changed circumstances which the applicant believes establish eligibility for assistance.

COMMENT. This subsection is intended to provide liberal rules for reapplication while eliminating clearly frivolous claims. Thus there are no barriers to prevent individuals from reapplying for assistance, but only those applications which state new facts or previous administrative error will receive substantive attention.

Section 9. Court Ordered Treatment

(1) If after counselling has proceeded for a reasonable period of time the parents or guardian of an infant still do not agree to consent to necessary medical care and treatment, the Department shall file a petition in the district court in the county of residence of the infant pursuant to (cite to the state's child abuse and neglect laws) seeking an order mandating that the infant be provided necessary medical care and treatment, and such other relief as warranted by the facts of the case.

COMMENT. Where counselling fails to induce the parents to consent to provide the infant with necessary medical care and treatment, the Department has the responsibility to seek a court order mandating that the child be provided these services. Since this action will be brought pursuant to the state's child abuse and neglect laws, the court has substantial flexibility in fashioning its order, including ordering the parent to bring the child to a physician or hospital, ordering a third party to take this responsibility, and temporarily or permanently terminating parental rights.

(2) When after investigation, the Department determines that there exists reasonable grounds to believe that an infant is not receiving necessary medical care and treatment, and that the lack of these services presents an immediate and substantial threat to the infant's life, the Department may seek a court order as provided in subsection (1) of this section, without engaging in counselling as provided in section 8.

COMMENT. It is expected that in most cases, counselling will generate the needed consent to medical services for the infant. Where this is not the case, subsection (1) of this section will insure that the infant receives the services in spite of the lack of parental consent. However, subsection (1) requires that counselling proceed for a reasonable time, time which some infants may not have if their life is to be saved. This provision provides for these special circumstances.

Section 10. Confidentiality of Counselling Meetings, Communications and Records

(1) All counselling meetings shall be held in private and all persons other than counselling personnel and individuals to be counselled shall be excluded unless their inclusion has been expressly requested by a parent or guardian.

(2) All communications, verbals or written, between participants during the course of or in furtherance of counselling shall be confidential.

(3) All records held by the Department involving counselling meetings shall be confidential.

(4) Counselled parties shall have the right to inspect and copy all records held by the Department that relate to the counselling meetings.

COMMENT. Since confidentiality is a prerequisite to the integrity of the cooperative approach embodied in the act, this section imposes strict guidelines as to privacy and on the use of documents and records accumulated during the investigative and counselling stages. This requirement is broadly consistent with modern state statutes dealing with custody of mental health information and records.[12]

Section 11. Enforcement; Penalties

(1) Any physician who knowingly fails to provide necessary medical care and treatment to an infant in accordance with the provisions of this Act shall be subject to loss of his or her licence to practice medicine in this state for a period not to exceed ____ years and shall be subject to such other civil

[12] *E.g.*, MD. ANN. CODE art. 59, §19 (Supp. 1979-80).

or criminal liability as is elsewhere provided by the laws of this state.

COMMENT. This provision provides sanctions against physicians who fail to carry out their responsibilities under this Act. It should be read together with section 5 (2), which provides immunity for physicians who treat an infant without consent, in a reasonable good faith belief that such treatment is necessary to preserve the infant's life.

The professional sanction adopted by this section is in addition to, not in lieu of, other potential civil or criminal liability. Failure to provide care and failure to report an infant in need of care may, in some jurisdictions constitute negligent or involuntary homicide, misdemeanor-manslaughter, and may violate child abuse or neglect laws.[13]

(2) Any parent or guardian who fails to comply with a court order issued pursuant to this act may be held in contempt of court. Contempt proceedings may be initiated by the Department or by the court on its own motion.

COMMENT. A parent or guardian is responsible under the act to cooperate with counselling personnel and to arrange for and attend a counselling session. If a court order is granted compelling attendance at a counselling meeting and the party or parties ordered fail to attend, this section authorizes the Department to seek a contempt order from the appropriate court. This sanction may also be used if a parent or guardian does not comply with a court order mandating that the infant be provided with necessary medical care and treatment. Section 9, however, would allow the court, pursuant to the child abuse and neglect laws of the state, to have a sheriff or other peace officer remove the child from the parent's home, thereby avoiding a parent's or guardian's potential disobedience of a court order.

It should be noted that far more serious charges may be brought against parents or guardians who fail to provide proper medical care to an infant.[14] This provision does not affect the right of state agencies to bring appropriate actions otherwise provided by state law.

Section 12. Appropriations

To carry out the purposes of this act, there is authorized to be appropriated $__________for the fiscal year ending __________; $__________for the fiscal year ending__________; and $__________for the fiscal year ending __________.

COMMENT. None necessary.

Section 13. Severability of Sections

If any provision of this act or the application thereof to any person or circumstance is held invalid, such invalidity shall not affect any other provision or application of the act which can be given effect without the invalid provision or application, and to this end the provisions of this act are declared severable.

COMMENT. None necessary.

Section 14. Effective Date

This act shall take effect __________________

COMMENT. None necessary.

---

[13] Robertson, *supra* note 3, at 224-35.

[14] *Id.*, at 217-24.

1.7 VOTING RIGHTS

Table of Contents

INTRODUCTION

Most of us take for granted our right to vote, whether we exercise it or not. Even if we do not vote, we know that if something makes us mad enough or if someone happens to inspire us sufficiently, we can exercise our right to vote. Yet, it has not always been so. For example, it was not so long ago that women were not allowed to vote in most jurisdictions. Even more recently, blacks were prevented from voting in many states by law or administrative practices. Despite our progress in this area, however, restrictions on the voting rights of disabled persons still persist either because some are seen as too unintelligent and irresponsible to cast a ballot, as women once were; because in some communities where large numbers have been institutionalized, their political power is feared; or because the state has shown little interest in providing the means to overcome administrative, physical or structural barriers to voting.

For disabled persons the right to effectively cast a ballot has both practical and symbolic significance. As a practical matter the right to vote is a prerequisite to being represented in the political process, to having a voice in selecting those who will adopt and administer the public policies which to a great degree govern our lives. This concern is one which is fundamental to persons in a democracy. Indeed, the United States Supreme Court has recognized this in describing the right to vote as a fundamental political right "because (it is) preservative of all rights,"[1] and as "a fundamental matter in a free democratic society.".[2]

This is certainly no less true of disabled persons and perhaps more so given the extent of state control, described in other chapters, which has been traditionally exercised over the lives of disabled citizens. More recently governmental efforts to provide needed services and to integrate disabled persons into the mainstream of society have also become a primary focus of political interest. Disabled persons need, deserve and have a right to the same political access as other persons and groups in asserting their interests. Certainly no documentation is needed to establish that politicians are most responsive to those voters and interest groups who can best exert political pressure at the polls. This potential voter impact appears to be one reason behind the refusal in some cases to register the residents of institutions as voters. (See text at note 11 *infra*.) The continued attempts to keep such voters politically impotent cuts off a vital avenue to pursuit of their interests—one that is open to all other groups and individuals in our political system. This tends to perpetuate both paternal attitudes and repressive legislation directed at disabled persons and constitutes a fundamentally unfair aspect of the political system.

There is a symbolic importance to the right to vote as well. Restrictive voting provisions are at odds with the "state of the art" approach being slowly applied in other areas—one of normalization, integration, reduced control and enhanced individual autonomy. Granting the right to effectively cast a ballot challenges old notions that disabled persons are capable of only minimum control over their lives. New medical and psychological literature indicates that mentally disabled persons dramatically respond to the challenge of greater personal responsibility in ways that benefit both society and themselves. The recent trends toward community care facilities and least restrictive treatment and habilitation alternatives are a response to this new awareness of the benefits of reduced parental and institutional control. These recent expansions of the personal and civil rights of disabled persons, while welcome, are necessarily incomplete so long as voting prohibitions remain. Granting the right to effectively vote would symbolize the recent movement away from legislative paternalism and toward greater personal responsibility and control. As such it is an appropriate cornerstone for any legislative effort designed to enhance the well-being of disabled persons.

The following sections will review existing state legislation restricting the right of mentally disabled persons to vote and examine their consistency with constitutional guarantees of equal protection and due process. Provisions for encouraging voter participation by removing physical and structural barriers to the balloting process will also be examined. Finally, alternative voting rights provisions are presented along with suggested legislative language where appropriate for reforming state voting rights statutes.

---

[1] Yick Wo v. Hopkins, 118 U.S. 356, 370 (1886).

[2] Reynolds v. Sims, 377 U.S. 533, 561-62 (1964).

REVIEW OF STATE LEGISLATION

Forty-five states and territories now disenfranchise some mentally disabled individuals. (See chart *infra.*) The provisions employed vary in their terminology, but are all class-based. The frequency of use (some states use more than one term) of specific terms is as follows:

Insane persons	17 states
Idiots	13 states
Mentally incompetent persons	13 states
Persons under guardianship	9 states
Persons non compos mentis	7 states
Incompetent institutional residents	3 states
Persons of unsound mind	3 states
Persons committed to an institution for the mentally ill or insane	2 states
Mentally diseased persons	1 states
Mentally retarded persons	1 states

In addition, states have, until recently, often employed some form of literacy or comprehension test, but many of these have been temporarily banned under the Voting Right Act of 1965 as amended by the Voting Rights Act Amendments of 1970.[3] Many of the prohibitions appear in state constitutions. "Idiots" and "insane persons," while still the most frequent terms, appear to be slowly being replaced by the terms "mentally incompetent persons" and "persons under guardianship" which seem to be favored in recent amendments.[4] A few states have recently amended their constitutions by eliminating disqualifications that had previously been in force and substituting a provision which leaves mental disability disqualifications to the discretion of the legislature.[5] One state, California, has recently amended its constitution to eliminate disability-based voter prohibitions entirely. A few states expressly require an adjudication of incompetency before a person may be disqualified as a voter.[6] In some states with guardianship-based prohibitions the procedural requirements in the guardianship statute may have the same effect.[7] Still other states may have reached the same result by enacting "bill of rights" provisions for institutionalized persons which include the retention of all personal and civil rights until a formal adjudication of incompetency.[8]

The extent to which statutory voting prohibitions are, in fact, enforceable is an open question. Only eight jurisdictions have specific provisions providing for enforcement of the substantive measure. Five of these take the form of requirements that some official notify the registrar of voters of any adjudications of incompetence and that the registrar then remove the individual's name from the voter registration lists.[9] One state, Wisconsin, provides civil penalties which may be assessed against anyone who votes in an election while subject to a statutory disqualification.[10]

It would seem that in the vast majority of states the enforcement of statutory prohibitions is left to the ad hoc judgment of the official responsible for voter registration. It is not known to what extent these officials do or do not attempt enforcement. However, the majority of judicial decisions involving voter qualifications arise as a result of challenges to election results, implying that no attempt was made to enforce the provision at an earlier stage in the voting process. Thus, the need for disability-based voting restrictions seems questionable, since the states apparently do not deem them important enough to provide any systematic enforcement mechanism.

A few court cases have arisen when large groups of institutional residents attempted to register together and were turned away en masse by the official responsible for voter registration.[11] While a few of these cases involve unique political situations in which a referendum issue affecting the institution had been placed on the ballot, creating an adversary situation between the community and the institutional populations, as a whole these cases imply that voting restrictions, if enforced at all, may have a particularly severe effect on institutionalized persons. Yet, most of these latter cases arise in states where no institutional/noninstitutional distinction appears contemplated in the statutory provision itself. Thus, this ad hoc enforcement, where it occurs, often results in unequal application of the voting laws.

Finally, a significant number of jurisdictions (10) have no disability-based voter disqualifications. Legislatures in these jurisdictions apparently have perceived no threat to the voting process in allowing all persons otherwise eligible to exercise the right to vote.

---

[3] 42 U.S.C.A. §1973 through 1973aa-5 (1974 and Supp. 1980).

[4] *See, e.g.,* UTAH CONST. art IV, §6, amended in 1977 by dropping the terms idiots and insane persons and substituting mentally incompetent persons.

[5] *See* KAN. CONST. art. V, §2 (amended in 1974); MICH. CONST. art II, §3 (amended in 1963); OKLA. CONST. art. 3, §1 (amended 1979).

[6] *See, e.g.,* FLA. STAT. ANN. §97.041(3) (West Supp. 1980).

[7] *See* Boyd v. Board of Registrars of Voters of Belchertown, 334 N.E.2d 629 (Mass. 1975).

[8] *See, e.g.,* WYO. STAT. §25-3-124(d) (1977).

[9] *See, e.g.,* ARIZ. REV. STAT. ANN. §16-165 (Supp. 1979-1980).

[10] *See* WIS. STAT. ANN. §12.59 (West 1967).

[11] *See, e.g.,* Boyd v. Board of Registrars of Voters of Belchertown, 334 N.E.2d 629 (Mass. 1975). Often these cases are generated by provisions prohibiting voting by "persons under guardianship."

CHART OF STATE VOTING PROVISIONS

State	Persons Disqualified	Adjudication Required	Enforcement
Alabama Code (1975)	Idiots and insane persons Const. art. VIII, §182		
Alaska Stat. (1979)	Persons of unsound mind unless the disability has been removed Const. art. V, §2; §15.05.040	§15.05.040	
Arizona Rev. Stat. Ann. (Supp. 1979-1980)	Persons under guardianship, non compos mentis or insane Const. art. 7, §2; §16-101(6)		When proceedings are had in superior court resulting in a person being declared incapable of taking care of himself . . . the recorder shall cancel the name on the register §16-165 May be challenged by any qualified elector. §16-591
Arkansas Stat. Ann. (1947)	Idiots and insane persons Const. art. 3, §5		
California			
Colorado			
Connecticut Gen. Stat. Ann. (1979)	Mentally incompetent persons §9-12		
Delaware Code (1974)	Idiots and insane persons Const. art. V, §2; tit. 15, §1701		
District of Columbia Code (1973)	Mentally incompetent §1-1102	§1-1102	
Florida Stat. Ann. (West Supp. 1980)	Persons adjudicated mentally incompetent in this or any other state and who have not had their competency restored pursuant to law §97.041 (3) (a)	§97.041 (3) (a)	
Georgia Code Ann. (1977)	Idiots and insane persons Const. art. 2, §501		
Hawaii Rev. Stat. (1976)	Non compos mentis Const. art. II, §2		
Idaho Code (Supp. 1979)	Those adjudicated incompetent and not restored to legal capacity §66-346 (a) (6)	§66-346 (a) (6)	
Illinois			
Indiana			
Iowa Code Ann. (1949)	Idiots and insane persons Const. art. 2, §5		

State	Persons Disqualified	Adjudication Required	Enforcement
Kansas Stat. Ann. (1969)	Persons under guardianship, non compos mentis, and insane unless restored to civil rights Const. art. 5, §2		
Kentucky Rev. Stat. (1977)	Incompetent institutional residents §§202A.170; 202B.050	§202A.170	
Louisiana			
Maine			
Maryland Ann. Code (Supp. 1979)	General Assembly may regulate or prohibit right to vote of persons under guardianship Const. art. I, §2		
Massachusetts Gen. Laws Ann. (West 1978)	Persons under guardianship Const. amend. Art. III, §105		
Michigan Comp. Laws Ann. (1977)	Mentally incompetent persons may be excluded by legislature Const. art. II, §3		
Minnesota Stat. Ann. (West Supp. 1978)	Persons under guardianship, insane and not mentally competent Const. art. VII, §1; §201.15		
Mississippi Code Ann. (1972 and Supp. 1978)	Idiots and insane persons Const. art. 12, §241; §23-5-85		
Missouri Ann. Stat. (Vernon Supp. 1978)	Idiots and persons under guardianship Const. art, VIII, §2 Adjudged incompetent §111.021 Institutionalized and adjudicated incompetent §202.847	§§111.021; 202.847	
Montana Rev. Codes Ann. (Supp. 1977)	Persons of unsound mind Const. art. IV, §2; §23-2701	§23-2701	
Nebraska Rev. Stat. (1975 and 1978)	Non compos mentis Const. art. VI, §2; §32-1048		
Nevada Rev. Stat. (1979)	Idiots and insane persons Const. art. 2, §1		§293.540
New Hampshire			
New Jersey Stat. Ann. (West 1971 and Supp. 1980-1981)	Idiots and insane persons Const. art. 2, §6; §19:4-1	[1]	
New Mexico Stat. Ann. (1978)	Idiots and insane persons Const. art. VII §1	§1-4-26	Determination of insanity reported §1-4-26

State	Persons Disqualified	Adjudication Required	Enforcement
New York Elec. Law (McKinney 1978)	Persons adjudicated incompetent or committed to an institution for the mentally ill. §5-106	2 §5-106	
North Carolina Gen. Stat. (1970)	Persons who cannot read and write sections of Constitution in English Const. art. VI, §4		
North Dakota Cent. Code (Supp. 1979)	Persons who have been declared mentally incompetent Const. art. V, §122	Const. art. V, §122	
Ohio Rev. Code Ann. (Page 1979 and Supp. 1979)	Idiots and insane persons Const. art. V, §6	§5122. 30.1	Probate judge shall file with Board of elections names of those adjudicated incompetent to vote §3503.18
Oklahoma Stat. Ann. (West Supp. 1979)	Determined by the Legislature Const. art. 3, §1 Persons adjudged mentally incompetent; mentally retarded persons tit. 26, §4-101	tit. 26, §4-101	tit. 26, §16-102
Oregon Rev. Stat. (1979)	Idiots and mentally diseased persons Const. art. II, §3		
Rhode Island Gen. Laws (1876)	Non compos mentis Const. art. XXXVIII, §1	Const. art. XXXVIII, §1	
South Carolina Code (1976)	Mental incompetence Const. art. II, §7; §7-5-120 General Assembly may require each person to demonstrate a reasonable ability, except for physical disability, to read and write the English language Const. art. II, §6	§7-5-120	
South Dakota Compiled Laws Ann. (Supp. 1979)	Mentally incompetent persons Const. art. VII, §2		Clerk of court to provide names of persons declared mentally incompetent §12-4-18
Tennessee			
Texas Code Ann. (Vernon Supp. 1980)	Idiots and lunatics Const. art. VI, §1; Elec. Code art. 5.01(2)		
Utah Code Ann. (1978 and Supp. 1979)	Mentally incompetent person Const. art. IV, §6	Institutional residents §64-7-48(1) (c)	
Vermont Stat. Ann. (1968)	Incompetent institutional residents tit. 18, §7705(a) (3)	tit. 18, §7705(a) (3)	
Virginia Code (1973)	Mentally incompetent persons §24.1-42	§24.1-42	

State	Persons Disqualified	Adjudication Required	Enforcement
Washington Rev. Code (1979)	Idiots and insane persons Const. art. VI, §3		
West Virginia Code (1978)	Persons of unsound mind Const. art. IV, §1		
Wisconsin Stat. Ann. (West 1957, 1967 and Supp. 1979-1980)	Persons under guardianship, non compos mentis, and insane Const. art III, §2; §6.03	Court may determine that the person is competent to exercise the right to vote when a person is under limited guardianship §6.03	Penalties §12.59
Wyoming Stat. (1977)	Idiots and insane persons Const. art. 6, §6	Institutional residents §25-3-124(d)	
American Samoa Code (1973)	Persons under guardianship, non compos mentis, or insane Const. art. II, §7		
Guam			
Puerto Rico Laws Ann. (Supp. 1980)	Persons judicially declared as unqualified tit. 16, §§3053, 3055	tit. 16, §3055	Court to send list of persons judicially declared mentally incompetent tit. 16, §3076
Virgin Islands Code Ann. (1976)	Inmates of institutions for insane and persons under guardianship tit. 18, §263(c)		

[1] *See* N.J. STAT. ANN. §30:4-24.2(a) (Supp. 1980-1981).
[2] *See also* N.Y. MENTAL HYG. LAW §33.01 (McKinney Supp. 1978).

THE CONSTITUTIONALITY OF VOTING RESTRICTIONS

The state statutory provisions restricting voting rights of disabled persons, as outlined above, are subject to a number of constitutional challenges as violative of both the Equal Protection and Due Process clauses of the 14th Amendment to the United States Constitution. We will examine these arguments in detail in the succeeding sections.

Voting as a Fundamental Right

The U.S. Constitution in art. 1, sec. 2, clause 1 allows the states to set voter qualification standards in both state and federal elections. This power has been explicitly recognized by the United States Supreme Court in *Lassiter v. Northampton County Board of Education,*[12] in which the Court stated that "the states have long been held to have broad powers to determine the conditions under which the right of sufferage may be exercised."[13] But the Court has limited the state's power by stating that "once the franchise is granted to the electorate, lines may not be drawn which are inconsistent with Equal Protection clause of the Fourteenth Amendment."[14] Thus, like other restrictions on voting rights, those provisions restricting "idiots," "insane persons," those "non compos mentis" and those "under guardianship," whether imposed by statute or state constitution, are subject to the Fourteenth Amendment requirements of due process and equal protection.

The Equal Protection Clause Applied

In establishing voter qualification standards, as in most other "state action," the state must necessarily utilize classifications which include some persons and exclude others. The Equal Protection clause forbids the states from utilizing classifications which constitute arbitrary and invidious discrimination.[15] Traditionally, the classifications established by the state have been required to bear some reasonable relationship to a legitimate state objective.[16] More recently the Supreme Court has employed a stricter level of scrutiny and a higher level of justification of state interests where the state action sufficiently impairs a fundamental right or is based upon an inherently suspect class.[17] This higher level of scrutiny has been generally referred to as the "compelling state interest" test.

The compelling state interest test involves two parts.[18] First, where a fundamental right is at stake, state regulations restricting the right may be justified only by a "compelling state interest." Second, the means utilized in the legislative enactment must be neither over nor under inclusive and must accomplish the state's interest with precision, *i.e.,* they must be narrowly drawn to express only the legitimate interest at stake.[19]

In numerous cases in recent years the Supreme Court has held that the right to vote is a fundamental right and has applied the compelling state interest test to strike down state statutes restricting the right to vote in various contexts.[20] The Court, however, has never decided a challenge to voting restrictions based upon mental disability. Yet, a look at the voting cases which have applied constitutional principles in the context of durational residency requirements and literacy tests is instructive. Presumably, the state's interest in excluding mentally disabled persons from voting is the maintenance of an "intelligent" and "informed" electorate. The burden is on the state to show that its interest is compelling in the face of the fundamental right to vote when it seeks to impose such restrictions. The state must show that the harm to its interest is substantially more severe than the consequences which flow to those individuals affected by the mechanism chosen to further that state interest, *i.e.,* preventing the exercise of the fundamental right to express one's personal and group interests through the political process. Current means to implement the state's interest make that showing difficult indeed.

---

[12] 360 U.S. 45 (1959).

[13] *Id.* at 50-51.

[14] Harper v. Virginia State Bd. of Elections, 383 U.S. 663, 665 (1966).

[15] *See, e.g.,* Loving v. Virginia, 388 U.S. 1 (1966).

[16] *See generally* Tussman and ten Broeck, *The Equal Protection of the Laws,* 37 CAL. L. REV. 341 (1949).

[17] *See generally* Gunther, *In Search of Evolving Doctrine of a Changing Court: A Model for a Newer Equal Protection,* 86 HARV. L. REV. 1 (1972). The court has recognized only three classifications as suspect—race, alienage and illegitimacy. Therefore, the remainder of this discussion will focus on fundamental rights as the basis for strict scrutiny of state action. For an argument that disability should be recognized as a suspect classification, based upon the criteria set out by the Court in *United States v. Caroline Prod. Co.,* 304 U.S. 144, 152-53, n. 4 (1938), *see* Burgdorf and Burgdorf, *A History of Unequal Treatment: The Qualifications of Handicapped Persons as a "Suspect Class" Under the Equal Protection Clause,* 15 SANTA CLARA LAW. 855 (1975).

[18] *See* Gunther, *supra,* note 17.

[19] Roe v. Wade, 410 U.S. 113 (1973).

[20] *See, e.g.,* Bullock v. Carter, 405 U.S. 134 (1972); Evans v. Cornman, 398 U.S. 419 (1970); Williams v. Rhodes, 393 U.S. 23 (1968); Harper v. Virginia Bd. of Elections, 383 U.S. 663 (1966); Reynolds v. Sims, 377 U.S. 533 (1964).

In fact, the Supreme Court has not held under the compelling state interest test that the state's interest in maintaining an intelligent electorate is compelling.[21] As the Court noted in *Dunn v. Blumstein,*[22] "the criterion of 'intelligent' voting is an illusive one and susceptible to abuse,"[23] and in a footnote the Court suggested that by passing the Voting Rights Act of 1965 Congress declared it federal policy that people should be allowed to vote even if not informed about the issues.[24] State restrictions on voting by mentally disabled persons raise the same problems identified in *Dunn v. Blumstein.* Any criteria which purport to insure "intelligent" voting are readily susceptible to the same type of abuse which resulted in the passage of the 1965 Voting Rights Act, as amended, which is directed at racial discrimination in voting. Certainly the number of mentally disabled persons who are disenfranchised by voting officials is lower and the methods of doing so are less systematic. Yet, the attitudes of these officials toward both groups is somewhat similar. Recent cases interpreting disability-based voting restrictions provide evidence of a continuing hostility toward the exercise of voting rights by mentally disabled persons.[25] Thus, one may assert that the state interest, if any, should be overridden by the individual's right to vote when consistent patterns of discrimination appear to be displayed among those who administer the voting laws, just as was found necessary with respect to racial discrimination.[26]

In addition to the questionable state interests, the voter restrictions suffer from the fact that they are both over and under-inclusive vis-a-vis the state interest. They are under-inclusive because they allow registration by a vastly greater number of voters who are arguably as uninformed and incompetent to cast an intelligent, considered vote as are those who are restricted from voting.[27] It is well known that a substantial number of qualified electors exercise the right to vote in an arbitrary and capricious fashion without any substantial knowledge of or regard for the intricacies of the electoral process or the particular issues involved. They are also over-inclusive because they catagorically prohibit many qualified persons from voting. Interestingly enough, an illuminating study on voting patterns among institutional residents during a New York mayorial election revealed that no significant difference existed between the distributed vote of the patient population and that of the surrounding community.[28] A subsequent study supported this result and concluded that hospitalized mental patients are capable of voting in an informed and thoughtful manner.[29] Thus, where allowed to vote, mentally disabled persons show voting patterns which display no significant differences from that of the general population.

Provisions restricting persons under guardianship are no better. Since guardianship is not a self-initiating process, usually being instituted only upon the petition of an interested party or by the voluntary action of the proposed ward, a classification based on it is also both over and under-inclusive. It is over-inclusive in that persons volunarily committed or placed under guardianship may have independent motivations for seeking a guardianship appointment which have little or no bearing on factors relevant to the state's interest in an intelligent electorate. Yet, they will still be disenfranchised. It is under-inclusive in that those persons who remain under the watchful eye of their families or friends but who do not have a formal guardian will be allowed to vote, even if they are less capable than those with a guardian. Also, a classification of this type may help to defeat a countervailing state interest, seeing that persons

---

[21]A number of early cases upheld such restrictions but were decided before the advent of strict scrutiny and so are no longer reliable precedent. *See, e.g.,* Pope v. Williams, 193 U.S. 621 (1904); Lassiter v. Northampton Bd. of Elections, 360 U.S. 45 (1959).

[22]405 U.S. 330 (1972).

[23]*Id.* at 356.

[24]*Id.* at 357 n. 29. The 1965 Act generally prohibits the denial of the right to vote on the basis of race or color because of the failure to comply with any test or device and defines "test or device" to mean, among other things, any requirement that a person demonstrate the ability to read, write, understand or interpret any matter, or demonstrate any educational achievement or his knowledge of any particular subject. 42 U.S.C.A. §1973(b) (1974).

[25]*See, e.g.,* Carroll v. Cobb, 139 N.J. Super. 439, 354 A.2d 355 (1976); Boyd v. Board of Registrars of Voters of Belchertown, 334 N.E.2d 629 (Mass. 1975). In *Carroll* v. *Cobb* the court quotes the following statement of the trial judge:

> From what I have heard here, I am firmly convinced that the reason why those men were not permitted to vote was the hostility that the Municipal Fathers and even the County Board of Election had against people who were confinea to such a school. I don't think that they really were concerned about their idiocy or their insanity because it hasn't been satisfactorily proven that these persons were in that category at all. 345 A.2d at 357.

[26]The Supreme Court appears somewhat hostile to restrictions which purport to establish "intelligence" criteria for voting, but has shown a reluctance to address this issue. Rather, the Court has decided the question of constitutionality of such restrictions using the second part of the equal protection analysis (see text at note 19 *supra*) and finding that the means chosen to further the asserted state interest are insufficiently narrow and impinge upon right to vote.

[27]*See* Dunn v. Blumstein, 405 U.S. 330 (1972), in which the Court struck down a statute requiring new residents to reside in the state for a fairly lengthy period of time before they were eligible to vote, ostensibly to insure they were knowledgable about state and local issues before they voted. However, the Court noted that this "classification obviously permit(s) any longtime resident to vote regardless of his knowledge of the issues—and obviously many (longtime residents) do not have any." *Id.* at 358.

[28]Klein and Grossman, *Voting Patterns of Mental Patients in a Community State Hospital,* 3 COMMUNITY MENTAL HEALTH J. 149-152 (Summer 1967).

[29]Howard and Anthony, *The Right to Vote and Voting Patterns of Hospitalized Psychiatric Patients,* 49 PSYCHIATRIC QUARTERLY 124 (1977).

in need of supervision receive it, by discouraging persons who otherwise might seek to have a guardian appointed from doing so. Thus, utilizing guardianship as a criterion suffers from all the infirmities recognized by an equal protection analysis and seems to be a particularly inappropriate standard. Unfortunately, as statutes and state constitutions are amended, there appears to be a developing trend toward its utilization.

Any consideration of the importance to the state of a statutory or constitutional classification must in the end give substantial weight to the *real* possibility of injury to a state concern. As the Court noted in *Storer v. Brown,*[30] a decision in this context is very much a matter of considering the facts and circumstances behind the law, the interests which the state claims to be protecting, and the interests of those who are disadvantaged by the classifications. In the context of the voting rights of mentally disabled persons the facts and circumstances behind the law are, for the most part, archaic stereotypes, and the state's interest is meager when compared to the real impact that voting restrictions have on the unquestionably fundamental right to participate in the political process. In fact, the impact on the electoral process of allowing mentally disabled persons the right to vote would be far less than that feared by the states.[31]

Due Process—Void for Vagueness

The Fourteenth Amendment also protects against state deprivation of liberty and property by means which do not meet minimum standards of due process. Among the Due Process or "fundamental fairness" standards is the requirement that a statute which infringes upon a liberty be sufficiently specific to give notice of what is prohibited by the statute. Statutes which contain vague standards are subject to a vagueness attack which stems from "the exaction of obedience to a rule or standard which (is) so vague and indefinite as really to be no rule or standard at all."[32] This "void for vagueness" test is normally applied in the criminal context but applies as well to civil statutes.[33]

The United States Supreme Court has identified several dangers inherent in vague laws:

> It is a basic principle of due process that an enactment is void for vagueness if its prohibitions are not clearly defined. Vague laws offend several important values. First, because we assume that man is free to steer between lawful and unlawful conduct, we insist that laws give the person of ordinary intelligence a reasonable opportunity to know what is prohibited, so that he may act accordingly. Vague laws may trap the innocent by not providing fair warning. Second, if arbitrary and discriminatory enforcement is to be prevented, laws must provide explicit standards for those who apply them. A vague law impermissibly delegates basic policy matters to policemen, judges, and juries for resolution on an *ad hoc* and subjective basis, with the attendant dangers of arbitrary and discriminatory application. Third, but related, where a vague statute "abut(s) upon sensitive areas of basic First Amendment freedoms," it "operates to inhibit the exercise of (those) freedoms." Uncertain meanings inevitably lead citizens to "steer far wider of the unlawful zone ... than if the boundaries of the forbidden areas were clearly marked."[34]

Thus, statutes which are vague give rise to three dangers: no understandable basis for denial of a right and, thus, no fair warning; impermissable delegation of discretion; and undue inhibition of the exercise of a Constitutional right. When applied to the standards utilized to prohibit disabled persons from voting, those standards must be found unconstitutionally vague.

The initial danger of these statutes stems from the fact that they use terms like "idiot," "imbecile," and "insane person" which provide no understandable basis for determining who is and who is not prohibited from voting. Even the term "under guardianship" has been the subject of litigation over its meaning in this context. In *Boyd v. Board of Registrars of Voters of Belchertown,* and in *Town of Lafayette v. City of Cheppewa Falls,*[36] residents of state facilities for the mentally disabled sought judicial review of registrars' decisions that they were ineligible to vote by virtue of the states' "under guardianship" prohibitions. The *Boyd* court recognized a distinction between commitment and guardianship because only the latter implied a judicial determination of incompetence to manage one's own affairs. The court ruled, therefore, that commitment alone was insufficient to deny an individual the right to vote, since it was not necessarily related to competency. As a practical matter, many individuals are committed to institutions because they are unable to be served by the social service system outside the institution, not because they are incompetent. In *Lafayette,* however, the court decided that, since a person committed to a state institution was considered, albeit informally, to be a ward of the state, institutional residents were "under guardianship" and within the purview of the state's voter restriction. These irreconcilable opinions point up the deficiencies of provisions prohibiting persons under

---

[30] 415 U.S. 724, 730 (1974).

[31] *See* text at *supra,* notes 28 and 29.

[32] A. B. Small Co. v. American Sugar Ref. Co. 267 U.S. 233, 239 (1925).

[33] Giacciuo v. Pennsylvania, 382 U.S. 399, 402 (1966).

[34] Graynel v. City of Rockford, 408 U.S. 104, 108-09 (1972) (footnotes omitted).

[35] 334 N.E.2d 629 (Mass. 1975).

[36] 70 Wis.2d 610, 235 N.W.2d 435 (1975).

guardianship from exercising the franchise. The same adverse and unfair result occurred when persons exercised a statutory right to receive residential services from the state only to find that voter registration officials subsequently denied them the right to vote because they were thereby found to be "idiots" or "under guardianship."[37]

Related to this first danger is the second danger, impermissible delegation of discretion from the state legislature to administering officials. Vague standards such as these afford such officials so much discretion in their interpretation and application that arbitrary and discriminatory denials are inevitable. An appellate court in deciding a voting rights case, sharply outlined this problem:

> The municipal clerk in this case insists, however, that she is vested with discretion to determine whether to reject an applicant for registration who lacks sufficient mental capacity to vote. We deem the contention untenable. While it is true that persons applying for registration who meet the voter requirements of age, citizenship and residence may still not be entitled to register if they are *otherwise disqualified*, N.J.S.A. 19:31-5, and "idiots" and "insane persons" are disqualified from voting, it should be abundantly evident that a lay person is completely unequipped to determine whether an applicant is either an "idiot" or an "insane person," as those terms are used in the Constitution and the statute, and thus disenfranchised. Indeed, we suspect that those imprecise terms may be troublesome to experts in the fields of psychiatry or psychology.[38]

One cannot expect the registrar to be any more consistent applying these standards than are the courts who must review the initial decision.[39] This is exactly the type of situation feared by the Court in *Dunn v. Blumstein,* which, as noted above, found intelligence to be an illusive criterion and "susceptible of abuse."[40]

The denial of registration and the right to vote may thus turn upon which official is deciding, rather than upon any explicit standards reflecting legislative intent. Given the degree of misinformation and fear which many persons have concerning disabled individuals, such arbitrary decision-making is not only possible but probable.

This kind of arbitrary denial necessarily discourages the exercise of the right to vote. Faced with the prospect of being labelled an "idiot," being denied the right to register and vote and having to sue to realize the right, many disabled persons, otherwise qualified to vote, but who are often unsophisticated in dealing with the bureaucracy, will simply forgo their right.

Procedural Due Process

Also included in the due process guarantees of the Fourteenth Amendment are the procedural due process requirements of notice and opportunity to be heard. State voting rights restrictions based upon disability also infringe upon the procedural due process guarantees in several ways.

Initially the statutes categorize individuals as incompetent to vote based upon the labels attached to them by voter registrars, without providing procedures for challenging that categorical determination or for showing that they are not suitable for inclusion in that class of persons.[41] And in doing so the statutes raise an irrefutable presumption which violates basic due process guarantees.[42] In addition, no notice is provided to those who are allegedly incompetent as to what constitutes incapacity and how they specifically have been found to demonstrate that incapacity. Thus, the statutes allow arbitrary classification, a presumption of incompetency based upon that classification, no notice of the basis for the classification of any given individual and no means for challenging the classification as erroneous. Such a denial of a fundamental right is at odds with the concept of "fundamental fairness," allows capricious decision-making and thus violates the basic right to due process.

Summary

The preceding indicates that present state disability-based voting restrictions may present serious constitutional problems, and may constitute invidious discrimination under the Equal Protection clause as well as violate the Due Process clause of the Fourteenth Amendment. Whether or not the court would actually find these provisions unconstitutional is, of course, still an open question. However, Fourteenth Amendment strict scrutiny is really nothing more than a complex judicial formula by which the court hopes to arrive at a result that is, if nothing else, essentially fair to both the disadvantaged group and the state. It is not then particularly different from the concepts of fairness, equity and morality by which we as individuals try to guide ourselves. The real thrust of this analysis is to demonstrate, within the context of constitutional guarantees, the inequities of many of the legislative choices in question. In the case of these voting prohibitions, the analysis illustrates that in order to accomplish a marginal purpose the states have singled out one highly visible group for disadvantageous treatment and by

[37]*See* Carroll v. Cobb, 139 N.J. Super, 439, 354 A.2d 355 (1976); Boyd v. Board of Registrars of Voters of Belchertown, 334 N.E.2d 629 (Mass. 1975); Town of Lafayette v. City of Chippawa Falls, 70.Wis.2d 610, 235 N.W.2d 435 (1975).

[38]Carroll v. Cobb, 139 N.J. Super. 439, 354 A.2d 355, 359 (1976).

[39]*See* text at *supra,* notes 35 and 36.

[40]405 U.S. 330, 356 (1972).

[41]*See* Skinner v. Oklahoma, 316 U.S. 535 (1942).

[42]*See* Dunn v. Blumstein, 405 U.S. 330 (1972); Carrington v. Rash 380 U.S. 89 (1965).

doing so have accomplished little that furthers the purpose itself. Instead, these laws serve to remove the last vestige of political power from a group that has historically been the target of restrictive legislation on a broad scale and is to this day consistently misunderstood.[43] Fundamental fairness as much as constitutional doctrine demands that a new approach be taken. Developmentally disabled persons should become active participants in the political process which effects their lives, which in turn may make legislators better understand their particular concerns and needs.

ACCESS TO THE BALLOT

Some persons with disabilities face, not prohibitions on their voting rights, but an inability to physically participate in the voting process. These limitations on access to the ballot may present problems in at least three areas: travelling to the polling place to register and vote, physical access to the polling place itself and marking the paper ballot or using the voting machine.

Absentee Ballots

Provisions for absentee ballots have long been a means for providing access to the balloting process for those who for one reason or another are unable to attend the polls and personally cast their ballot on election day. Absentee voting has been characterized by state courts as a privilege, not a right.[44] Likewise, federal courts have found that, unlike the right to vote, the right to cast an absentee ballot is not a constitutionally protected right. In *Whalen v. Heiman*,[45] for example, the court held that a person is not denied the constitutional right to vote when the only means permitted is physically impossible for the person, at least when there is no state action preventing attendance at the polls. Since it is the voter's own personal circumstances, rather than any state action, which prevents attendance, the court applied the rational basis test rather than the strict scrutiny test. Using this test the court found that a requirement that persons physically attend the polls is not so unreasonable so as to violate due process guarantees. In so finding, the Court stated that "(It is not) the province of courts to weigh the relative ease or difficulty which the state could accommodate its voting procedures to meet the needs of various handicapped voters. These are policy questions to be resolved by legislators."[46] Thus, the legislatures have a great deal of discretion in whether and in what form to adopt absentee ballot provisions.

Despite the fact that absentee voting is a privilege and not a right, every jurisdiction has adopted some form of absentee ballot provisions.[47] However, absentee ballot provisions frequently suffer from some major deficiencies:

1. Absentee ballots are not provided in all elections in many jurisdictions, restricting the vote of persons unable to personally attend the polls to only general or general primary elections.[48] Unfortunately, such limited provions have been upheld as reasonable based upon the "inordinate time, effort and expense" involved in providing absentee ballots.[49]
2. Absentee ballots must be cast too far in advance of the election to allow the voter to take advantage of last minute information about candidates and issues.
3. Procedures for obtaining and returning an absentee ballot vary considerably, are frequently complex and confusing, and are not publicized.
4. There are frequently problems with timely transmittal of the absentee ballot after it is requested.
5. Not all states adequately provide for absentee ballots for everyone, including physically disabled persons.
6. Some jurisdictions have no provisions for absentee registration.

These are problems common to all persons who commonly use absentee ballots, including military personnel, those traveling or living abroad, students and persons away on business as well as disabled persons. They are raised here as a further impetus to general reform of the ballot process which is heard from the other quarters as well. At a time when voter participation continues to fall, further effort is needed to remove the obstacles to absentee voting.

---

[43]For a concise history of the unequal treatment of disabled persons, particularly the mentally disabled, see Burgdorf and Burgdorf, *supra*, note 17.

[44]97 A.L.R.2d 264, §3.

[45]373 F. Supp. 353 (D. Conn. 1974).

[46]*Id.* at 357.

[47]For a summary of state registration and absentee ballot provisions *see* 1 MENTAL DISABILITY LAW REPORTER 237 (November-December 1976).

[48]*Id. See also* McDonald v. Board of Election, 394 U.S. 802, 810-811 n. 9 (1969) in which the court discusses state absentee ballot provisions.

[49]*See* Fidell v. Board of Election, 343 F. Supp. 913, 916 (E.D.N.Y. 1972); *See also* Whalen v. Heimann, 373 F. Supp. 353 (D. Conn. 1974).

Access to the Voting Booth

Some disabled voters are quite capable of getting to the polling place, but because of physical barriers are unable to enter the polling place to cast their vote. In a challenge to inaccessible polling places, a federal court held in *Selph v. Council of the City of Los Angeles*[50] that a handicapped person has a constitutional right to vote, but has no right to insist that polling places be modified to eliminate architectural barriers. It also held that some burden upon the right to vote was permissible as long as the right to vote was not denied, relying as the federal court had in other absentee ballot cases upon *McDonald v. Board of Election.*[51] "Whereas here, the right is not totally denied and there are reasonable alternatives provided to the person who finds that his polling place is inaccessible to him, the traditional standard of rational relationship to a legitimate state objective is proper to follow."[52] The court was particularly concerned about the cost and administrative burden of making all polling places accessible. It noted that at trial several suggestions involving minimal public expenditures were offered, including using centrally located, accessible polling places where disabled persons could vote, changing the state's law forbidding the removal of voting material from the polling place to allow disabled voters to vote outside, and providing absentee ballots, which, instead of being returned by mail, could be handed to election officials to allow for casting of one's vote on election day.[53] Each of these suggestions involves potential administrative problems and some have potential for voter fraud, but each is certainy worth study.

Constitutional right or not, states could be doing more to provide accessible voting places without expending large sums in doing so. Iowa, for example, requires that in the selection of polling places, consideration shall be given to the use of buildings accessible to elderly and physically disabled persons.[54] Idaho goes further in providing that to the maximum extent deemed possible polling places shall conform to the state's accessibility standards contained in the state's environmental barriers statute.[55] Kansas requires that all polling places meet specified accessibility standards unless there is no acceptable and accessible voting place available, unless it is anticipated that the polling place will be brought into compliance in the near future or unless the polling place will be made to temporarily comply with the standards during the time which the polls are open.[56] Thus, some states are making an effort to improve the physical accessibility of polling places.

Assistance in Casting a Ballot

Some voters may get to the polls, have no difficulty gaining physical access to the polling place, but have difficulty either in marking a ballot or utilizing a voting machine so as to effectively register their choice. There is some authority for the proposition that the constitutional right to cast a vote also includes the right to assistance which would effectuate the voter's choice. In *Garza v. Smith*[57] a federal court applied the strict scrutiny test to find a state statute which provided assistance to physically disabled but not illiterate voters violated the Equal Protection and Due Process clauses of the Fourteenth Amendment. However, where the state had adopted a voter assistance statute which applied to plaintiffs, a federal court has found that the rational basis test applies. Thus, a state may have a great deal of discretion in adopting a method of providing assistance, so long as it provides it to all voters who need it.

In practice, the provision of assistance presents no difficult administrative problems but does raise the possibility of voter fraud by improperly marking the ballot or influencing the voter's choice in the process. For this reason such assistance is closely regulated in most jurisdictions such that if assistance is not provided for by statute for a particular class of persons it will not be given by election officials. The result may be that some individuals may not be able to effectively make their choice at the polls, a result which is fundamentally unfair.

SUGGESTED LEGISLATION FOR VOTING RIGHTS REFORM

In the light of the previous discussions the changes in voting laws which follow are suggested to effectuate the goals of fundamental fairness and maximum voter participation. Fundamental fairness demands that all individuals who desire to exercise their fundamental right to vote be allowed to effectively do so. Such participation is in the in-

[50] 390 F. Supp. 58 (C.D. Cal. 1975).

[51] 394 U.S. 802 (1969).

[52] 390 F. Supp. 58, 62 (C.D. Cal. 1975).

[53] *Id.*

[54] IOWA CODE ANN. §49.21 (Supp. 1980-81).

[55] IDAHO CODE §34-302 (Supp. 1979).

[56] KAN. STAT. ANN. §25-2710 (Supp. 1979).

[57] 320 F. Supp. 131 (W.D. Tex. 1970), vacated on other grounds 401 U.S. 1006 (1971). *See also* Smith v. Dunn, 381 F. Supp. 822, 826 n. 3 (M.D. Tenn. 1974). *Contra* State v. Sweeney, 154 Ohio St. 223, 94 N.W.2d 785 (1950). (The Ohio Secretary of State had no authority to issue regulations providing voting assistance for illiterate voters.)

terest, not only of the individual voter, but of our democratic society as well. Democracy demands that all individuals and groups be allowed to participate in the democratic process, if it is to function as a means of expression of interests, of authoritative decision-making and of conflict resolution. Toward these goals the following legislative reforms are suggested:

1. Repeal of all disability-based restrictions on voting rights.

   States should repeal all provisions denying the right to vote to "idiots, imbeciles, persons non compos mentis, under guardianship," etc.

2. For those states which desire some test of competency, adopt the following:

   Any person who is able to provide the information, whether orally, in writing, through an interpreter or interpretive device or otherwise, which is reasonably required of all persons seeking to register to vote, shall be considered a qualified voter of this state and shall be registered to vote and allowed to cast a ballot in any election held in the election precinct in which the voter is registered.

   Such a provision imposes a minimum test of competency which is objective rather than vague and arbitrary. It thus overcomes the constitutional objections outlined *supra.* At the same time it provides for those persons to register and vote who are unable because of illiteracy, blindness, or physical inability to write or speak, including those who use "talking boards" or other interpretive devices.

3. Adopt a provision guaranteeing that no person shall be denied any civil right because he or she is receiving services for a mental disability or is a resident of a state or private facility for the housing, care or treatment of persons with mental or physical disabilities. (See Chart 1 of *Rights of Disabled Persons in Residential Facilities* in the appendix.) In addition, such persons should be specifically permitted to declare the residential facility their domicile for purposes of voting.

   The following statutory language is suggested:

   Voter residence in special cases.

   No person shall gain or lose a residence for voting purposes while in a hospital, nursing home, or facility for the housing, care or treatment of persons with physical or mental disabilities. Any person who is a resident of a nursing home or facility for the housing, care or treatment of persons with physical or mental disabilities may at the time he or she registers to vote, select the nursing home or facility, or his or her residence immediately prior to entering the home or facility as his or her voting residence.

4. Expand the coverage of, and simplify the procedures for obtaining and returning absentee ballots.

   It is not the intent of this study to present a comprehensive model act for reform of state absentee ballot laws for it would be far beyond its scope to do so. Rather, the intent is to point out the importance of such reform efforts to disabled persons as well as those usually considered most affected such as military personnel.

   Such proposals for reform have been made by others as well. For example, the Special Committee on Election Law and Voter Participation of the American Bar Association has recommended that the ABA support (1) the comprehensive recodification of the three federal statutes governing the right of United States military personnel stationed outside their voting residence and of U.S. civilians living abroad to vote absentee (Federal Voting Assistance Act of 1955, 42 U.S.C. 1973; Overseas Citizens Voting Assistance Act of 1977, 42 U.S.C. 1973; Overseas Citizens Voting Rights Act of 1975, 42 U.S.C. 1973; and Section 202 of the 1970 amendments to the Voting Rights Act of 1965, 42 U.S.C. 1973); (2) the designation of one federal official or agency to implement the recodified absentee balloting procedures; and (3) the granting of federal assistance to states that adopt for their state and local elections the comprehensive federal election absentee balloting provisions.[58] States would do well to examine these examples of absentee balloting procedures toward reform of their own procedures.

5. Adopt the following provision promoting physical access to polling places:

   Standards of accessibility for voting places; exceptions. (1) Except as hereinafter provided, each voting place provided by the (election official), in which a national, state, county, township, city and school primary and general election, referendum or bond issue is to be held, shall be accessible to and usable by elderly persons and by physically handicapped persons by complying with the following standards of accessibility:

   (1) Doors, entrances and exits used to gain access to or egress from the voting place shall have a minimum width of thirty (30) inches;

---

[58] Special Committee on Election Law and Voter Participation of the American Bar Association, *Report to the House of Delegates* (August 1980). The specific sections referred to in the Report are: 42 U.S.C, §§1973 cc-11, 1973 aa-1, 1973dd through 1973dd-5.

[59] *See* Smith v. Dunn, 381 F. Supp. 822 (M.D. Tenn. 1974).

(2) any curb adjacent to the main entrance to the voting place shall have curb cuts or temporary ramps;

(3) any stairs necessarily used to enter the voting place shall have temporary handrail and ramp; and

(4) at the voting place, no barrier shall impede the path of physically handicapped persons to the voting booth.

(b) When no voting place within an election precinct meets the accessibility standards in subsection (a) the (election official) shall select, wherever possible, a voting place which will be:

(1) temporarily made to comply with the standards during the time the polls are open or;

(2) brought into compliance with the standards at some future date.

(c) Any (election officer) who selects a voting place which does not meet the standards prescribed in subsection (a) shall report such selection to the Board of County (Supervisors/Commissioners etc.). The Board shall receive and respond to all suggestions for improving accessibility of voting places.

One must question the wisdom of spending scarce public resources to make all voting places accessible when other important needs of disabled persons such as transportation, housing and education go unmet. Yet, more can be done with available resources.

This provision calls for use of facilities which are accessible or can temporarily be made accessible or will be made accessible in the future. Thus, while it does not require that large amounts of public funds be used to make all voting places accessible, it does insure that efforts are made to provide accessibility where it can be accomplished without excessive expenditures of money and effort.

6. Provide by statute for assistance in the marking of ballots or use of voting machines to all voters who need it. In doing so the state should insure:
   a. Maximum privacy of the voter in casting his or her ballot;
   b. Maximum choice by the voter in who shall provide assistance, including relatives and friends; and
   c. Minimum opportunity for fraud or influence in casting such ballots.[59]

The following language is suggested:

Assistance to voter

(a) Whenever a voter who declares to the presiding election official that he or she cannot read or write or by reason of physical or mental disability is unable to mark his or her ballot or register his or her vote on a machine, as the case may be, the presiding election official shall determine whether the voter needs assistance. If the presiding official determines that assistance is needed, the presiding official shall appoint two election officials of different parties to assist the voter in the marking or registering of the voter's ballot. If, however, a voter desires assistance by someone other than the two election officials, he or she may be assisted by that person, provided that no person who is not an election official shall assist more than two voters in any election.

(b) No official or other person shall assist or offer to assist any elector in marking or registering of the voter's ballot, except as provided in this section.

(c) No official or other person at any election shall, while rendering assistance to a voter, suggest to the voter the name of any political party or candidate for any office.

(d) Any person who gives assistance to a voter in the marking or registering of his or her ballot shall not in any way divulge any information regarding the choice of the voter or the manner in which the voter's ballot was cast.

1.8 DRIVER LICENSING

Table of Contents

INTRODUCTION

All American jurisdictions place substantial restrictions on the ability of some disabled individuals to procure a driver's license.[1] Two strong competing interests are at stake when the states impose such limitations.

On the one hand, these restrictions reflect the need to protect the public from the consequences of the acts of persons who are incompetent to drive, and to protect these persons from themselves. The state has a responsibility to insure the safety of all its citizens and thus must be careful to scrutinize those persons to whom it grants the privilege of operating a motor vehicle.[2] Irresponsible or unfit drivers create a significant safety hazard both to themselves and to those with whom they share the roads. The number of traffic fatalities on our highways each year is a testament to the need for continuing vigilance in this area.[3]

On the other hand, driving in our society has become the predominant means toward fulfillment of a number of significant personal ends. By providing individuals with the freedom to work, travel, shop and seek recreational opportunities that is often not otherwise available, the ability to drive allows people to expand and diversify their experience. This growth process is urgently needed by developmentally disabled persons because of their history of isolation and deprivation. In fact, the right to drive has risen to a preeminent position in the scheme of personal privileges granted by the state to its citizens and is one which many individuals hold to be of the highest importance in their daily lives. As one noted commentator has stated, of all deprivations the inability to drive has the most immediate and practical effect.[4] Thus, it should not be surprising that persons who are institutionalized on the basis of mental disability are reported to inquire more frequently about their status as a driver than about the effect of their commitment on any other right.[5]

In addition to restricting opportunity for growth and experience, loss of one's driving privileges

---

[1] *See* State Statute Charts, *infra.*

[2] *See* Dixon v. Love, 431 U.S. 105, 114 (1977).

[3] Traffic deaths commonly exceed 50,000 annually in the United States. *See* U.S. DEPT. OF TRANSPORTATION, 1977 HIGHWAY SAFETY ACT REPORT.

[4] R. ALLEN, E. FERSTER & H. WEIHOFEN, MENTAL IMPAIRMENT AND LEGAL INCOMPETENCY 345 (1968).

[5] *Id.*

can also have a profound stigmatizing effect on the individual. Driving has become so pervasive in our society that an individual's inability is a highly visible characteristic which labels the person unfit, in some respect, to fully participate in our normal way of life.[6] Most of the reforms in the mental and developmental disabilities fields in recent years have in part been designed to avoid this type of stigmatization which prevents full social acceptance and integration into the community. Current driving restrictions pose a significant obstacle to this integrationist policy.

In weighing the two competing interests of public safety and individual opportunity, the proper balance must be struck to allow the state to impose those restrictions which are calculated toward, and suitable to, the achievement of its public safety goals without unduly restricting the issuance of licenses. Thus, in order to deprive an individual of the driving privilege, it is incumbent on the state to illustrate a real, not conjectural, link between the individual's particular disability and an impaired ability to safely operate a vehicle. The following sections will summarize the existing state statutes on driver licensing, point out some of the inadequacies in existing statutes and suggest basic premises upon which amended licensing statutes should be based.

REVIEW OF STATE LEGISLATION

State driver licensing statutes vary considerably with respect to provisions directed at disabled applicants. Despite their variations, however, nearly all statutes contain some form of provisions for prohibiting the licensing of certain individuals, for suspending or revoking licenses, for restricting licenses, and for implementing these restrictions.

Persons Not to Be Licensed (Chart I)

Provisions prohibiting the licensing of certain individuals take a variety of forms and seem to be based on criteria with no demonstrable link to driving ability. Perhaps the most onerous are the familiar types of class-based prohibitions which variously exclude persons characterized as idiots, insane persons, imbeciles and feebleminded.[7] More frequently the prohibited class is defined to include persons "adjudged to be afflicted with or suffering from any mental disability or disease and. . .not. . . restored to competency by the methods provided by law."[8]

In addition, most states have discretionary criteria which allow the commissioner of motor vehicles or the equivalent to refuse to license anyone who "by reason of physical or mental disability" is believed unable "to drive a motor vehicle with safety upon the highways."[9] Other common provisions prohibit the issuance of a license to persons who are unable to understand highway warning and traffic signs;[10] are unable to exercise reasonable and ordinary control over a vehicle[11]; are mentally or physically unqualified to operate a motor vehicle;[12] or whose driving would be inimical to public safety.[13]

Oftentimes states include specific provisions directed toward persons with epilepsy.[14] A small number categorically prohibit these persons from acquiring licenses, though the majority include some type of certification procedure whereby temporary or provisional licenses may be issued to persons with epilepsy whose seizures are controlled by drug therapy as attested to by one or more physicians. Generally these licenses will remain in effect so long as the person continues to acquire certifications of seizure control on a periodic basis.[15]

Persons Subject to Loss of License: Revocation and Suspension of Driving Privileges (Chart I)

Almost every state provides for the suspension or revocation of a previously issued license when the department of motor vehicles determines that a licensee is "incompetent to drive."[16] Suspension ordinarily means that one's license automatically becomes valid again after a given period of time elapses. After revocation, however, the licensee is usually required to reapply for a license and submit to all examinations. In all likelihood a person applying to be relicensed after a prior revocation will be subjected to far greater scrutiny the second time around.

Restricted Licenses (Chart I)

States commonly have statutes which allow licensing officials to impose any reasonable license restrictions which they have good cause to believe are warranted by an applicant's physical or mental condition.[17] The most common restriction

---

6 R. BARRON & H. FABING, EPILEPSY AND THE LAW 35 (1956).

7 *See e.g.*, OHIO REV. CODE ANN. §4507.08(B) (Page Supp. 1978); ALA. CODE tit. 32, §6-7 (1977).

8 *See, e.g.*, ARIZ. REV. STAT. §28-413(5) (1976); FLA. STAT. ANN. §322.05 (West 1975).

9 *See, e.g.*, MISS. CODE ANN. §63-1-9 (1973); R.I. GEN. LAWS §31-10-3 (1969).

10 *See, e.g.*, MINN. STAT. ANN. §171.004 (West Supp. 1979); TEXAS HIGH. CODE ANN. art. 6687b, §4 (Vernon 1977).

11 *See, e.g.*, ALA. CODE tit. 32, §6-7 (1977); IND. CODE §9-1-4-30 (1976).

12 *See, e.g.*, D.C. CODE ANN. §40-301(a)(1) (1973); N.Y. VEH. & TRAF. LAW §502 (1) (McKinney Supp. 1978-1979).

13 *See, e.g.*, NEV. REV. STAT. §483.250(7) (1977); N.M. STAT. ANN. §66-5-5 (1978).

14 *See* State Statute Chart, *infra*. *See also*, THE LEGAL RIGHTS OF PERSONS WITH EPILEPSY (Epilepsy Foundation of America, 1976). (Herein cited as Epilepsy Study.) This publication surveys state legislation restricting the rights of persons with epilepsy and includes a discussion of non-statutory administrative procedures beyond the scope of the present study.

15 *See, e.g.*, OHIO REV. CODE ANN. §4507.08 (Page Supp. 1978).

16 *See* State Statute Chart I, column 2, *infra*.

17 *See* State Statute Chart I, column 3, *infra*.

of this nature limits the licensee to drive while wearing corrective lenses as a result of the detection of a visual impairment. However, restrictions may be imposed as a result of a wide variety of other disabling conditions. Persons with epilepsy, for instance, may be issued licenses that remain valid only if the individual submits to repeated physical examinations and other disabled individuals may also be granted only temporary licenses.[18] Other restrictions imposed in practice are limitations on the type of vehicles a licensee may drive, the hours of the day during which the license is valid and restrictions requiring the licensee to drive only with the aid of specific mechanical devices.

Implementation of Restrictions (Chart II)

In order to implement its restrictions on licensing of disabled drivers a state must have some sort of review procedure to identify and evaluate applicants for a license. This screening process is complicated by the fact that many disabilities which may adversely affect driving ability are not readily apparent. On the other hand, many disabilities which are easily identifiable do not necessarily preclude the disabled individual from driving safely. In addition, we do not have as yet any generally agreed upon, empirically established criteria which may be utilized to determine who can and cannot safely drive a motor vehicle.[19] These factors point up the difficulty involved in identifying and evaluating applicants as well as the need for careful, individualized review of applicants identified as potentially hazardous drivers.

Given both the importance and the difficulty of adequate screening mechanisms to identify and evaluate disabled applicants, it is surprising that few states have adopted special review procedures.[20] Of those that do, some are applied generally to persons with a disability,[21] some are limited to persons with epilepsy or similar conditions,[22] while others consist only of a special right of appeal.[23] The better review procedures allow for independent examination of the applicant, careful, individual review and an opportunity for administrative review and appeal to the courts.

To aid in the development of better licensing criteria and review procedures the National Highway Traffic Safety Administration (NHTSA) has proposed the creation, at the state level, of medical advisory boards made up of practitioners in various medical specialties, as well as administrators involved with highway and traffic safety.[24] These boards would be charged with the responsibility for analyzing existing data and with developing medical criteria which are thought to bear a significant relationship to the state's safety concerns. NHTSA believes that by formulating objective criteria the states can remove much of the objectionable arbitrariness from the administrative process. Unfortunately, experts in the field are unable to agree on the relationship of specific disabilities to driving competence, and thus it is difficult to imagine that such boards would be able to establish valid criteria. (See Comment to Section 5 of the model act *infra*.)

Some states have established medical advisory boards by statute to advise on or establish guidelines for licensing of persons alleged to be disabled,[26] for selecting screening procedures for use by licensing examiners and for assisting in the making of individual determinations as to fitness to drive. Others use the boards only as medical review boards to hear appeals of denials of licenses on the basis of disability.[27] It should be noted that the medical advisory boards referred to in Chart II *infra*, have been established by statute; thus, the chart does not include the many jurisdictions which have established such boards by administrative agreement.[28]

The difficulty in identifying and evaluating applicants who have impaired driving ability raises the probability that some incompetent drivers will initially be granted licenses. However, all licensees in nearly every state may be subjected to reexamination.[29] In addition, such drivers are subject to suspension or revocation of their license in accordance with due process guarantees. Such guarantees require the states to provide for some sort of opportunity for a hearing on suspension or revocation.

A majority of states allow summary suspension or revocation, i.e., without a prior hearing, thus affording the individual no opportunity to contest

[18] *See* State Statute Chart II, column 2 and discussion *infra*.

[19] R. ALLEN, E. FERSTER & H. WEIHOFEN, *supra* note 4, at 345 n. 10.

[20] *See* State Statute Chart II, column 2, *infra*.

[21] *See, e.g.*, N.C. GEN. STAT. §20-9(g) (Supp. 1979); PA. CONS. STAT. ANN. 75, §§1518, 1519 (Purdon 1977).

[22] *See, e.g.*, OHIO REV. CODE ANN. §4507.08 (Page Supp. 1978); WIS. STAT. ANN. §343.09 (West Supp. 1978-1979).

[23] OKLA. STAT. ANN. tit. 47, §§6-207, 6-211 (West Supp. 1978).

[24] NATIONAL HIGHWAY TRAFFIC SAFETY ADMINISTRATION, U.S. DEPARTMENT OF TRANSPORTATION, THE ROLE OF MEDICAL ADVISORY BOARDS IN DRIVER'S LICENSING (March 1977).

[25] Evidence indicates that arbitrary decision-making in this area is pervasive. Researchers have found, for instance, that in states which do not categorically deny licenses to persons with epilepsy, but instead leave the decision to administrators on the basis of a general "competency to drive" standard, some administrators nonetheless insist on denying licenses to all persons with epilepsy, fearing perhaps, that they will be criticized if one of "their" drivers is subsequently responsible for an accident. R. Allen, E. Ferster & H. Weihofen, *supra* note 4, at 351.

[26] *See, e.g.*, DEL. CODE ANN. tit. 21, §2723 (1974).

[27] *See, e.g.*, N.C. GEN. STAT. §20-9(g) (4) (Supp. 1979); OHIO REV. CODE ANN. §4507.08(c) (3) (Page Supp. 1978).

[28] For a more complete discussion of medical advisory boards and their existing functions refer to U.S. DEPARTMENT OF TRANSPORTATION, NATIONAL HIGHWAY TRAFFIC SAFETY ADMINISTRATION, THE ROLE OF MEDICAL ADVISORY BOARDS IN DRIVER'S LICENSING. This publication is available for sale by the Superintendent of Documents, U.S. Government Printing Office, Washington, D.C. 20402 for $1.20.

[29] *See* State Statute Chart II, column 1, *infra*.

the basis for the administrator's decision until after the fact. Most of these statutes require the licensee to request a hearing within a specified time of the notice of suspension or revocation.[30] Some states, however, do require a hearing to be held prior to suspension or revocation unless the licensee poses an immediate threat to the public safety.[31]

Nearly every statute explicitly provides for judicial review of administrative decisions.[32] Even if not explicit in the licensing statutes, it may be provided by a state's administrative procedure act.[33]

While common judicial understanding holds that no citizen is "entitled" to a driver's license and thus that rejected applicants do not have a constitutional right to review of an initial license denial,[34] the importance of the right to drive to the deprived individual does not countenance summary denials. Still a number of states do not, at least explicitly, permit judicial review of this significant administrative decision.[35]

[30] *See* State Statute Chart II, column 4, *infra*.

[31] *See, e.g.*, MASS. GEN. LAWS ANN. ch. 90, §22, 28 (1975). Others are cited in State Statute Chart II, column 4, *infra*.

[32] *See* State Statute Chart II, column 5, *infra*.

[33] *See, e.g.*, IOWA CODE ANN. §17A.19 (1978).

[34] *Cf.* Bell v. Bursen, 402 U.S. 535 (1971).

[35] *See, e.g.*, IOWA CODE ANN. §§321.210, 321.215, 17A.18 (1978).

CHARTS OF STATE DRIVER LICENSING STATUTES

I. PERSONS SUBJECT TO RESTRICTION

State	Persons Not to Be Licensed*	Persons Subject to Loss of License	Persons Whose License May Be Restricted
ALABAMA Code (1977)	[1] and [3] tit. 32,§6-7	Persons incompetent to drive under tit. 32, §§6-7, 6-16	Physically disabled only tit. 32, §6-12
ALASKA Stat. (1978)	[2] and [3] §§28.15.031(4),(5)	Anyone who is determined to be incompetent to drive §28.15.161	Anyone for good cause §28.15.121
ARIZONA Rev. Stat. Ann. (1976)	[2] and [5] §28-413(5),(8)	Incompetent to drive a motor vehicle §28-446(A)(5) (Supp. 1978-1979)	Anyone for good cause §28-424 (Supp. 1978-1979)
ARKANSAS Stat. Ann. (1979)	[2] and [5] §75-309(a)(5)(9)	Anyone without hearing if showing of incompetence to drive §75-324(6)(1)	Anyone for good cause §75-324
CALIFORNIA Veh. Code (West Supp. 1978)	[3] §§12805,12806	Upon any of the grounds which authorize refusal §13359	Anyone for good cause §12813
COLORADO Rev. Stat. (1973)	[2] §42-2-103(d)	Persons specifically determined by a court to be mentally incompetent to such a degree that the person is incapable of safely operating a vehicle §42-2-122 (Supp. 1978)	Anyone for good cause §42-2-114(1) Mentally ill and developmentally disabled persons §42-2-114(5) (Supp. 1978)
CONNECTICUT Gen. Stat. Ann. (West 1970)	[3] and [6] §§14-36(b),14-46		May be issued for any reason commissioner deems advisable §14-36(b)
DELAWARE Code (1979)	[1] and [2] tit. 21,§2707(b) (4) [3] tit.21,§2707(b)(7)[8]	Upon information that the driver is incompetent to drive or afflicted with physical or mental infirmities or disabilities rendering driving unsafe tit.21,§2733(a) (3)	Physically disabled only tit. 21,§2722
DISTRICT OF COLUMBIA Code Ann. (Supp. 1978)	[4] §40-301(a) (1)	Anyone for any cause without hearing; effective 5 days after revocation or suspension order §40-302(a)	
FLORIDA Stat. Ann. (West 1975)	[2] §322.05(5)	If incompetent to drive §322.27(1) (c)[9] (West Supp. 1979)	Anyone for good cause §322.16(1) (a)

State	Persons Not to Be Licensed	Persons Subject to Loss of License	Persons Whose License May be Restricted
GEORGIA[1] Code Ann. (1980)	[2] and [3] §§68B-203(b) (4), (6)	Incapacitated by reason of disease, mental or physical disability to the extent that he is incompetent to drive §68B-307 (a) (2)[10] (Supp. 1978)	Anyone for good cause §68B-211(a)
HAWAII Rev. Stat. (1976)	[3] §286-104(5)	Anyone reasonably believed to be afflicted with mental or physical infirmities or disabilities which would make it unsafe for the licensee to operate a motor vehicle §286-119[11]	May impose such limitations appropriate for licensee's abilities §286-103
IDAHO Code (Supp. 1979)	[5] and [2] §§49-309,49-330	Licensee incompetent to drive; any persons subject to any condition resulting in lapses of consciousness or control or suffering from a physical or mental disability or disease; without preliminary hearing §49-330 (a) (3) (1)	Anyone for good cause §49-320 (1967)
ILLINOIS Ann. Stat. (Smith-Hurd Supp. 1979)	[3] and [6] ch. 95½, §6-103(5), (8)	One adjudged to be afflicted with a mental disease or disability ch. 95½,§6-206(18) Cancellation if person was not entitled to issuance ch. 95½,§6-201	Anyone for good cause ch. 95½,§§6-113(a), 6-206(c) (3)
INDIANA Code Ann. (Burns 1973)	[2], [3], [5] and [6] §§9-1-4-30	Incompetent to drive §9-1-4-34	Anyone for good cause §9-1-4-37 (Supp. 1979)
IOWA Code Ann. (West Supp. 1979)	[2] and [3] §331.177(5), (7)	One incompetent to drive §321.210(4) Cancellation §321.201	Anyone for good cause §321.193
KANSAS Stat. (Supp. 1978)	[2] §8-237(e)	Incompetent to drive §8-255(a) (3) (1975)	Anyone for good cause §§8-245,8-241
KENTUCKY Rev. Stat. Ann. (Baldwin Supp. 1978)	[1], [3] and [6] §§186.440(5),(6)[12]	Person whose physical or mental infirmities render it unsafe for him to drive §186.570	

State	Persons Not to Be Licensed	Persons Subject to Loss of License	Persons Whose License May be Restricted
LOUISIANA Rev. Stat. Ann. (West Supp. 1979)	[7] §32.419(1963)	Licensees incompetent to drive or afflicted with such mental or physical infirmities or disabilities as would constitute grounds for refusal of a license §§32.414(D) (5),(8)	Anyone for good cause §32.423 (1963)
MAINE Rev. Stat. (1978)	[7] tit. 29,§530 (Supp. 1979-1980)	One incompetent to drive tit. 29,§2241(1) (D) For any cause which Secretary of State deems sufficient tit.29,§2241(1)	Anyone as deemed to be in the interest of highway safety tit. 29, §534
MARYLAND Transp. Code Ann. (Supp. 1978)	[3] and [5] §16-103.1 (1977)	Unfit or unsafe drivers §16-206(a) (2) Cancellation §16-201 (1977)	Anyone for good cause §16-113(a) (2)
MASSACHUSETTS Gen. Laws Ann. (1975)	[7] ch. 90,§8		
MICHIGAN Stat. Ann. (Supp. 1979-1980)	[2] and [3] §9.2003(e),(f)	Upon admission to hospital for care and treatment of mental illness §9.2003(1) Incompetent to drive §9.2020	
MINNESOTA Stat. Ann. (West Supp. 1979)	[2], [3] and [5] §§171.04(5),(8),(9)	Persons incompetent to drive as determined and adjudged in a judicial proceeding §171.18(5) Cancellation §171.14	Anyone for good cause §171.09 (1960)
MISSISSIPPI Code Ann. (1972)	[2] and [3] §§63-1-9(d),(f)	Persons incompetent to drive §63-1-53	
MISSOURI Ann. Stat. (Vernon (1972))	[2] §302.060	Incompetent drivers: those who have become physically incapable of meeting requirements of exam [or] adjudged of unsound mind §§302.010(8), 302.291 (Vernon Supp. 1979)	Anyone for good cause §302.301
MONTANA Rev. Codes Ann. (Supp. 1977)	[3], [6] and [2] §§31-127(5),(8) (1961 & Supp. 1977)	Persons incompetent to drive §31-147(a),(4) Cancellation §31-142	Probationary license may be issued to one incompetent to drive §31-147(b) (1961) Anyone for good cause §31-137 (1961)

State	Persons Not to Be Licensed*	Persons Subject to Loss of License	Persons Whose License May Be Restricted
NEBRASKA Rev. Stat. (Supp. 1978)	[3] §60-407	Institutionalized persons; but only if superintendent certifies they are incompetent to drive §60-419(1) Persons mentally and physically incompetent to operate a vehicle; license may be cancelled indefinitely §60-407	
NEVADA Rev. Stat. (1979)	[2], [3] and [5] §§483.250(4),(6),(7)	Without hearing where holder is physically or mentally incompetent to drive §483.470(1)(e) Cancellation §483.420	Anyone for good cause §483.360
NEW HAMPSHIRE Rev. Stat. Ann. (1977)	[7] §261:3	Without hearing, when reason to believe holder is physically or mentally incompetent to drive §262:40 Director may revoke or suspend after due hearing for any cause deems sufficient §262:39	Physically handicapped only §261:5
NEW JERSEY Stat. Ann. (West 1973)	[4] and [6] §§39:3-10,39:3-10.5 (Supp. 1979)	May suspend or revoke for any reasonable ground after due notice §39:5-30	Director may for good cause issue anyone a restricted license §39:3-11
NEW MEXICO Stat. Ann. (1978)	[3] and [5] §66-5-5(F),(I)	Without hearing, if person is incompetent to drive §66-5-30(A)(5) Cancellation §66-5-24	Anyone for good cause §66-5-19
NEW YORK Veh. & Traf. Law (McKinney- Supp.) 1978-1979)	[7] §502(1)	Physically or mentally disabled and those committed to mental institution §510(3)(b)(1970)	Anyone for good cause §502(5)(b)
NORTH CAROLINA Gen. Stat. (Supp. 1979)	[1] and [3] §§20-9(d), (e)[13]	Those incompetent to drive §20-16(a), (4) Those adjudged mentally incompetent §20-17.1 Cancellation §20-15	May restrict license to a specific vehicle or class of vehicles §20-29.1
NORTH DAKOTA Cent. Code (1979)	[2], [3] and [5] §§39-06-03(4), (7), (8)	Those incompetent to drive §39-06-32(2) Cancellation §39-06-24	Anyone for good cause §39-06-17

State	Persons Not to Be Licensed*	Persons Subject to Loss of License	Persons Whose License May Be Restricted
OHIO Rev. Code Ann. (Page Supp. 1978)	1, 2 and 3 §§4507.08(B), (C), (D)[14]	Mentally ill persons after an adjudication specifically finding the individual incompetent for purposes of holding a license §4507.161	For 6 months, to persons subject to condition resulting in episodic impairment of consciousness or loss of muscular control, upon evidence that the infirmity is under medical control §4507.08(C)(1) Good cause §4507.14
OKLAHOMA Stat. Ann. (West Supp. 1978)	2 3 5 tit.47, §6-103(5), (8) (1962) tit.47, §6-207	Cancellation where afflicted with mental disability or physical condition with history of seizures, lapses of consciousness or loss of control tit.47, §6-207	Anyone for good cause tit.47, §6-113
OREGON Rev. Stat. (1977)	2 §482.120(2) 3 §482.130 6 §482.141	Those incompetent to drive or afflicted with mental or physical infirmities or disabilities rendering it unsafe for them to drive a motor vehicle upon the highways §482.450(1), (c)	Anyone for good cause §482.280 §482.283
PENNSYLVANIA Cons. Stat. Ann. (Purdon 1977)	2 75,§1503	Anyone mentally or physically unqualified 75,§1519	Anyone for good cause 75,§1512
RHODE ISLAND Gen. Laws (1968)	2, 3 and 5 §31-10-3	Incompetent due to physical or mental disability to safely operate a vehicle §31-11-7(a)(5) Cancellation §31-11-1	Anyone for good cause §31-10-28
SOUTH CAROLINA Code (1976)	2 §56-1-40	Those incompetent or not qualified to drive because of mental or physical disability §56-1-270 Cancellation §56-1-240	Anyone for good cause; no discrimination against handicapped person §56-1-170 (Supp. 1979)
SOUTH DAKOTA Complied Laws Ann. (1976)	2 §32-12-32 5 §32-12-35 6 §32-12-5.1 (Supp. 1979)	Those incompetent or otherwise not qualified to be licensed §32-12-46 Physically or mentally incompetent to drive §32-12-49	Anyone for good cause §32-12-36
TENNESSEE Code Ann. (1968)	2 and 3 §59-705(e), (g) (Supp. 1979)	Those incompetent to drive §59-713(a)(5)	Anyone for good cause §59-708(b)

State	Persons Not to Be Licensed	Persons Subject to Loss of License	Persons Whose License May be Restricted
TEXAS High. Code Ann. (Vernon 1977)	[1] and [2] art.6687b,§4(6) [3] and [5] art.6687b,§4(8),(9)[15]	Those who are epileptic or adjudged insane, imbecile, or feebleminded and not restored to competency art.6687b,§30 Those incapable of driving art.6687b,§22(b)	Anyone for good cause art.6687b,§12
UTAH Code Ann. (1970)	2 §41-2-5(4) (Supp. 1979) 3 §§41-2-5(5), (6)	Incompetent to drive or afflicted with mental or physical infirmities rendering it unsafe for a person to drive §41-2-19(a)(3) (Supp. 1979)	Certificates of seizure control required of otherwise qualified epileptics §§41-2-33 to 35 Anyone for good cause §41-2-9.
VERMONT State. Ann. (1978)	4 tit.23,§603 7 tit.23,§631	Improper or incompetent persons tit.23,§671	Anyone for good cause tit.23,§612
VIRGINIA Code (1974 and Supp. 1978)	2 §46.1-360 3 §46.1-361(a)	Legally adjudged insane or feebleminded, mentally ill, mentally retarded or epileptic §46.1-427 Physically or mentally incompetent to operate a vehicle; or afflicted with a disease or disability rendering it unsafe for him to drive §§46.1-383,430	Anyone for good cause §46.1-378
WASHINGTON Rev. Code Ann. (1979)	[2] and [3] §§46.20.031(5),(8)[16]	Persons incompetent to drive §46.20.291 Cancellation §46.20.207	Persons suffering from physical or mental disability §46.20.041
WEST VIRGINIA Code (1974)	[2] and [5] §§17B-2-3(5),(8)	Incompetent to drive §17B-3-6(5)	Anyone for good cause §17B-2-10
WISCONSIN Stat. Ann. (West Supp. 1978-1979)	[2], [5] and [6] §343.09[17] §§343.06(5),(6),(10)	Incompetent to drive §343.16(3)(a) Cancellation §343.25(4)	Epileptics otherwise qualified to obtain a license §343.09 (1973) Anyone for good cause §343.13
WYOMING Stat. (1977)	[2] and [3] §31-7-108(b)	Incompetent to drive §31-7-128 Cancellation §31-7-122	Anyone for good cause §31-7-117
AMERICAN SAMOA			
GUAM			

State	Persons Not to Be Licensed*	Persons Subject to Loss of License	Persons Whose License May Be Restricted
PUERTO RICO Laws Ann. (1976)	[4] tit.9, §653(a)(1)	Physically or mentally disabled for purposes of driving tit.9,§691(2)	Special or restricted licenses for disabled persons tit.6,§657
VIRGIN ISLANDS Code Ann. (1976)	[4] tit.20,§375	Those whose physical or mental infirmities render them unsafe to drive tit.20,§548	

FOOTNOTES TO CHART I

* "Persons not to be licensed" can be categorized as follows:
1 Anyone adjudged insane, imbecile, epileptic, or feebleminded, until restored to competency.
2 Anyone adjudged to be afflicted with mental disability or disease and not restored to competency.
3 Anyone subject to lapse of consciousness, or with any disease, disorder, or disability which affects ability to exercise reasonable and ordinary control, or to operate a motor vehicle with safety.
4 Persons not mentally and physically qualified.
5 Those inimical to public safety.
6 Restrictions for epileptic persons.
7 Any cause satisfactory to the department; not "proper person".
8 Must present a new physician's certificate each year to retain license.
9 License may be cancelled if determined that licensee was not entitled to it originally. §322.22 (West Supp. 1979).
10 Cancellation if licensee was not entitled to license. §6813-301.
11 Cancellation. §286-120
12 Epileptic person seeking a license must present certification that his condition is controlled by drug therapy, a description of the drugs taken and a personal certification that he is free of seizures and has taken the prescribed medication. §186.
13 Division may in its discretion issue a license to such persons if they present a certificate stating that the applicant possesses the requisite skills to safely operate a motor vehicle. §20-9(a).
14 Except a six-month restricted license, followed by a regular license may be issued to persons subject to conditions causing episodic impairments to consciousness, if such a person presents a physician's statement that the condition is under control.
15 Statute provides that no one shall be refused a license because of any physical defect, unless it is shown by common experience that such defect incapacitates him from driving safely.
16 Department must allow such persons an opportunity to demonstrate that they are proper persons to drive. §46.20.041.
17 Epileptic person may be issued a temporary license which remains in force as long as licensee submits a doctor's certificate at six month intervals. §343.09.

II. IMPLEMENTATION OF RESTRICTIONS

State	Re-Examination By Licensing Agency	Special Review Procedure Where Disability Alleged	Use of Medical Advisory Board	Notice and/or Hearing	Judicial Review
ALABAMA Code (1977)	At director's discretion tit.32,§6-16(h)	tit.32,§6-16		Immediate notice: hearing within 30 days of a written request tit.32,§6-16(h)	tit.32,§6-16(l)
ALASKA Stat. (1978)	May be required §28.15.091			Notice of the opportunity for an administrative hearing prior to suspension unless immediate action is necessary for the protection of health and safety §28.05.131	§28.05.14(d)
ARIZONA Rev. Stat. Ann. (1976)	May compel §28-447		Advises department on medical or any vision standards for licensing §§28-431,28-432,28-433	Immediate notice; hearing within 30 days of the request §28-446(B) (Supp. 1978-1979)	§28-451
ARKANSAS Stat. Ann. (1979)				Notice after suspension/ Opportunity for a hearing within 20 days after request §75-334	§75-338
CALIFORNIA Veh. Code (West Supp. 1978)	May compel §13801			Notice 30 days prior to action unless public safety requires immediate action §13950 Hearing §14100 et. seq.	§14400
COLORADO Rev. Stat. (1973)	§42-2-110(2)		§42-2-301	Immediate notice; hearing 30 days after request §42-2-122(2),(3)	§42-2-122(3)
CONNECTICUT Gen. Stat. Ann. (West 1970)		§14.36(b)			

State	Re-Examination By Licensing Agency	Special Review Procedure Where Disability Alleged	Use of Medical Advisory Board	Notice and/or Hearing	Judicial Review
DELAWARE Code (1979)	At person's own expense tit.21,§2724	tit.21,§2707(b)(7) tit.21,§2724	To advise on medical or any vision standards relating to the licensing of driver tit.21,§2723	Notice tit.21,§2736	Denial and revocation tit.21,§2734
DISTRICT OF COLUMBIA Code Ann. (Supp. 1978)				Appeal for hearing does not stay order to revoke or suspend license §40-302(a)	§40-302(a)
FLORIDA Stat. Ann. (West Supp. 1979)	May compel §322.221(c)	Special report §322.126	§§322.125,.126		§322.31
GEORGIA Code Ann. (1980)	May compel §68B-309	§§68B-216, 68B-217	§§68B-216, 68B-217	Hearing within 30 days of request §68B-307(d)	Applicant may appeal §68B-315
HAWAII Rev. Stat. (1976)			§286-4.1	Immediate notice and subsequent hearing §286-119	Applicant may appeal §286-129
IDAHO Code (Supp. 1979)				Immediate notice; hearing within 20 days of request §49-330(d)	Applicant may appeal §49-334
ILLINOIS Ann. Stat. (Smith-Hurd Supp. 1979)	ch.95½, §6-109	ch.95½,§§506-5 to 506-10	ch.95½,§506-3	Immediate notice; hearing within 20 days of request ch.95½,§§2-118, 2-206(c),6-209	ch.95½,§6-212
INDIANA Code (Burns 1973)	May compel §9-1-4-34	§9-1-8-8	See detailed provisions §§9-1-8-1 through 9-1-8-11	§9-1-8-8	§§9-1-4-34(f), 9-1-8-8
IOWA Code Ann. (West Supp. 1979)	§321.186			Notice 20 days prior to suspension 321.210 Hearing with 30 days of request §321.211	Suspensions stayed while review is pending §321.210, 321.215
KANSAS Stat. (1975)	§8-241		§8-255(b)	Immediate notice; hearing within 30 days of request §8-255(b)	§8-259

State	Re-Examination By Licensing Agency	Special Review Procedure Where Disability Alleged	Use of Medical Advisory Board	Notice and/or Hearing	Judicial Review
KENTUCKY Rev. Stat. Ann. (Baldwin Supp. 1978)		Epileptic persons §186.411		Notice and hearing within 20 days of a request §186.570(2)	Applicant may appeal §186.580 (1971)
LOUISIANA Rev. Stat. Ann. (West Supp. 1979)	Department may compel exam §32:424 (1963)	§32:403.2		Notice and hearing §32:414(E)	Applicant may appeal §§32:414(E), 32:419 (1963)
MAINE Rev. Stat. (Supp. 1978)	May compel tit.29,§2241		tit. 29,§547, tit.29,§2246	Without undue delay tit.29,§2214(3)	Applicant may appeal tit.29,§2242
MARYLAND Transp. Code Ann. (1977)	Department can compel exam of one who suffers from disorders characterized by lapses of consciousness §§16-119(c)(1), 16-207	§16-119	§16-118	Notice and hearing prior to suspension unless licensee poses an immediate threat to safety §16-206(b)(1)	
MASSACHUSETTS Gen. Laws Ann. (1975)			ch.90,§8	Prior to revocation or suspension ch.90,§§22,28	
MICHIGAN Stat. Ann. (Supp. 1979-1980)	May compel §9.2020			§9.2022	Applicant may appeal §9.2023
MINNESOTA Stat. Ann. (West Supp. 1979)	May compel §171.13(3) (1960)		§171.13	Immediate notice; hearing within 20 days of request §171.18	Applicant may appeal §171.19
MISSISSIPPI Code Ann. (1972)				Immediate notice; hearing within 20 days of request §63-1-53	Applicants §63-1-31
MISSOURI Ann. Stat (Vernon 1972)	Department may compel §302.291			Notice of desire to re-examine only §302.291	Applicant may appeal §302.311
MONTANA Rev. Codes Ann. (Supp. 1977)	May compel §31-148 (1961)			Immediate notice; hearing within 20 days of request §31-147(c)	Applicant may appeal §31-152 (1961)

State	Re-Examination By Licensing Agency	Special Review Procedure Where Disability Alleged	Use of Medical Advisory Board	Notice and/or Hearing	Judicial Review
NEBRASKA Rev. Stat. (Supp. 1978)	May compel §60-407	§§60-403,60-407, 60-408		§60-408	Applicant may appeal §60-420 (1974)
NEVADA Rev. Stat. (1979)	May compel §483.480		May be appointed to review specific cases §483.205	Immediate notice; hearing within 20 days of request §483.470	Applicant may appeal §483.520
NEW HAMPSHIRE Rev. Stat. Ann. (1977)	May compel §261:3-a			Hearing upon holder's written application within 10 days of request §262:40	Applicant may appeal §262:41
NEW JERSEY Stat. Ann (West 1973)				Notice prior to revocation hearing in administrator's discretion §39:5-30	§39:5-2
NEW MEXICO Stat. Ann. (1978)	May compel §66-5-31		§66-5-6	Immediate notice; hearing within 20 days of request §66-5-30(B)	Applicant may appeal §66-5-36
NEW YORK Veh. & Traf. Law (McKinney Supp. 1978-1979)	May compel §506			Right to hearing unless court commitment is basis for suspension or revocation §510(3)	§510(7)
NORTH CAROLINA Gen. Stat. (Supp. 1979)	May compel §20-29.1	§20-9(g)	Commissioner selects a board to develop licensing criteria §20-9(g)(2),(4)	Immediate notice for suspensions; hearing within 20 days of a request §20-16(d) For revocation on grounds of mental incompetency; hearing upon request prior to revocation §20-17.1 Review of denial upon request §20-9(g)(4)	Applicant may appeal §§20-25, 20-9(g)(4)(f)

State	Re-Examination By Licensing Agency	Special Review Procedure Where Disability Alleged	Use of Medical Advisory Board	Notice and/or Hearing	Judicial Review
NORTH DAKOTA Cent. Code (1979)	May compel §39-06-34			Notice and hearing prior to suspension §39-06-33	Applicant may appeal §39-06-39 (1972)
OHIO Rev. Code Ann. (Page Supp. 1978)	May compel §4507.20 (1973)	§4507.08	§4507.08(C)(2)	Hearing for refusal of restricted license §4507.08(C)(2)	Review boards appointed solely to hear appeals §4507.08(C)(3)
OKLAHOMA Stat. Ann. (West Supp. 1978)	May compel tit.47,§6-119	tit.47,§6-119	tit. 47,§6-118		tit.47,§6-211 tit.47,§6-118(d)
OREGON Rev. Stat. (1977)	May compel reexamination §§482.260(4),482.240(4)	May require additional examinations initially §482.240		Immediate notice and hearing §§482.450, 482.570	Applicant may appeal §482.490
PENNSYLVANIA Cons. Stat. Ann. (Purdon 1977)	May compel 75,§1519	75,§1519	75,§1517		Applicant may appeal 75,§1550
RHODE ISLAND Gen. Laws (1968)	May compel §31-11-9		§31-10-44	Immediate notice; hearing within 20 days of request §31-11-7	Applicant may appeal §31-11-15
SOUTH CAROLINA Code (1976)	May compel §56-1-270			Immediate notice §56-1-350 Hearing within 20 days of request §56-1-370	Applicant may appeal §56-1-410
SOUTH DAKOTA Compiled Laws Ann. (1976)	§32-12-46	§§32-12-4.3,32-12-5.1 (Supp. 1979)	§§32-12-4.2, 32-12-4.7	Prior to suspension §§32-12-46, 32-12-49	Applicant may appeal §32-12-59
TENNESSEE Code Ann. (1968)	§§59-713(b),59,707 (Supp. 1979)			Notice and hearing within 20 days of request §59-713(b) (Supp. 1979)	By discretionary criteria §59-713(f)
TEXAS High. Code Ann. (Vernon 1977)			To hear appeals only art. 6687b,§21A	Prior to revocation art.6687b,§22a	Applicant may appeal art.6687b,§31

State	Re-Examination By Licensing Agency	Special Review Procedure Where Disability Alleged	Use of Medical Advisory Board	Notice and/or Hearing	Judicial Review
UTAH Code Ann. (Supp. 1979)	May compel §41-2-19(g)	Board may be convened and recommendations made §41-2-39	§41-2-40	Immediate notice; hearing within 20 days of request §41-2-19(b)	May appeal §41-2-20
VERMONT Stat. Ann. (1978)	May compel[1] tit.23,§636	tit.23,§637	tit.23,§637	Prior to suspension tit.23,§671	
VIRGINIA Code (1974 and Supp. 1978)	May compel §46.1-383		§46.1-26.1	Prior to suspension unless suspension is based on legal adjudication of incompetency §§46.1-430 to 46.1-434	Applicant may appeal §§46.1-367, 46.1-437
WASHINGTON Rev. Code Ann. (1979)	§46.20.305	§46.20.041		If "driver improvement interview" prior to suspension §46.20.322 Formal hearing §§46.20.328,46.20.334	Applicant may appeal §46.20.334
WEST VIRGINIA Code (1974)	§17B-3-7		§17B-2-7a	Immediate notice; hearing within 20 days of request §17B-3-6	
WISCONSIN Stat. Ann. (West Supp. 1978-1979)	§343.16(3) (a)	One of following conditions: 1. legally declared recovered; 2. certificate and determination of competency 3. Medical exam §343.06 Persons with epilepsy §343.09		Hearing for denial of epileptics' temporary license §343.09(2)	Applicant may appeal §343.40
WYOMING Stat. (1977)	§31-7-128	Examination of records §31-7-121	§31-7-121	§31-7-127(c)	May appeal §31-7-133
AMERICAN SAMOA					
GUAM					
PUERTO RICO Laws Ann. (1976)	tit.9,§693	tit.9,§654	tit.9,§654	tit.9§692	tit.9,§692
VIRGIN ISLANDS Code Ann. (1976)	tit.20,§375				

1. If applicant is not satisfied with results of examination, may demand a second examination at own expense. tit.23, §638.

THE ADEQUACY OF EXISTING STATUTORY LAW

A review of existing licensing statutes reveals that many still restrict the granting of licenses on a catagorical basis. Yet despite these widespread restrictions in state statutes, there is almost no empirical evidence concerning the factors which contribute to a decreased ability to control a motor vehicle and even less to support the criteria most frequently used by states to justify their current prohibitions.[36] For example, one study concluded that no reliable data exists linking any known disease, other than alcoholism, with an increased incidence of motor vehicle accidents.[37] Thus, it is clear that, considering the extent of our knowledge at present, the driving prohibitions currently imposed are based more on intuition than on concrete evidence of a legitimate causal relationship.

In addition, screening procedures are frequently inadequate and not well defined. They may lack procedures for delineating factors to be utilized in examining the abilities of individual applicants and sometimes do not fully consider the data available while emphasizing disabilities over remaining abilities. For example, some helpful medical data currently exists concerning epilepsy. Modern drug therapies can successfully control seizures in the majority of cases and states must recognize this factor in their licensing procedures.[38] The doctor of a person with epilepsy is ordinarily in a better position to evaluate the individual's fitness as a safe driver than are licensing administrators. Thus, the evaluation procedures established should provide for deference to a physician's considered opinion. If an individual is in fact seizure free, as attested to by a knowledgeable physician, that person can pose no extraordinary threat as a motor vehicle operator. Thus, restrictions should be limited to those persons whose seizures cannot confidently be termed under control.[39] Likewise, requirements for periodic, indefinite review of the condition which are contained in some statutes are unwarranted if medical evidence indicates that, in all probability, the control is permanent. It should be remembered that the physical examination procedures and certifications are undertaken at the disabled individual's expense and thus create a financial burden as well as an inconvenience for the prospective driver. If the burden is needless in the face of medical or other evidence, it should not be imposed.

In addition to inadequately delinating evaluation factors, the screening procedures may also not be sufficiently individualized. Given the margin of error involved where no definative standards exist, it becomes particularly important that decisions be made on the basis of *individual* abilities and disabilities.

Many statutes do not provide adequate administrative review procedures. Few provide explicitly for review and appeal of license denials[40] or restrictions, though they do ordinarily allow for a hearing before a license is revoked on the basis of a driver's failure to abide by the restrictions imposed.[41] This dichotomy could, if employed in practice, have the anomolous result of forcing drivers to violate their restrictions in order to have their appropriateness reviewed. In addition, most do not provide for prior review of license suspensions or revocations despite the fact that few of these licensees present any significant threat to public safety.

While some states have relatively good licensing provisions affecting applicants with disabilities, many are in need of substantial amendments if these applicants are to be treated equitably. Such treatment need not compromise public safety and will greatly enhance the lives of those who might otherwise be denied the privilege to drive and the many opportunities it brings to them. The model driver licensing act which follows is intended to assist in this effort.

MAJOR PREMISES OF THE MODEL ACT

The model act which follows is based upon several premises which are consistent with those which guide all the subjects in the personal and civil rights area. These premises are:

1. Prohibitions on licensing should be functionally based, not categorically based. All epithetical terms like idiot and imbecile should be removed from statutes. Such terms do not provide a valid basis for determining who can and cannot be licensed. Likewise, involuntary commitment or guardianship proceedings which are based upon standards which do not necessarily have any bearing on the ability to drive should not, per se, be grounds for a finding of incompetency to operate a motor vehicle.[42] Rather, decisions on issuance, suspension, revocation or cancellation of drivers' licenses should be made on the basis of factors which are functionally based and as objectively related to safe driving requirements as possible. At

[36] R. ALLEN, E. FERSTER & H. WEIHOFEN, *supra* note 4, at 345 n. 10.

[37] *Id.*

[38] R. BARRON and H. FABING, *supra* note 6.

[39] To this end a number of authors have proposed the issuance of restricted licenses for a one to two year probationary period followed by an unrestricted license if no seizures have occurred during the probationary period. See *Epilepsy Study, supra* note 14, at 5-6. Fabing and Barron, *Restricted Driver's License To Controlled Epileptics: A Realistic Approach to a Problem of Highway Safety*, 2 UCLA L. REV. 500 (1955); Keys, Marten, Barron & Fabing, *The Epileptic Automobile Driver in Ohio*, 35 OHIO BAR 63 (1962).

[40] *See, e.g.*, IOWA CODE ANN. §§321.210, 321.215, 17A.18 (1978).

[41] *See, e.g.*, NEV. REV. STAT. §483.360 (1977).

[42] R. ALLEN, E. FERSTER and H. WEIHOFEN, *supra* note 4, at 257.

the same time it must be recognized that truly objective criteria are impossible to develop given our current level of information. Such criteria do not lend themselves to precise, predictable cut-off points which clearly indicate who can and cannot safely drive.[43] Thus, any "criteria" developed must be recognized only as factors to be considered and not definitive guidelines, since these "criteria" will have a significant possibility of error in their application.

2. Determinations must be made individually through established procedures. Individual applications of relevant factors should concentrate on the person's functional abilities, not disabilities. Application procedures should include appropriate input by the applicant's physician and other professionals and consideration of remedial or compensatory steps to overcome disabilities consistent with public safety.

3. Given the margin of error possible in individual cases, an opportunity to challenge final determinations, including denials and restrictions on licenses should be provided. This should include administrative hearings as the most efficient procedure. The possible loss of employment and other personal and economic hardships that may flow from the suspension of driving privileges can never be completely compensated for if the state's initial decision is found erroneous on review.[44] For this reason, it is particularly important that the review process occur prior to the actual administrative action in all but emergency situations.[45] (See Chart II, column 4 and discussion in subsection 10(3) of the model act *infra*.) As previously noted, the link between disabilities and driving competence is so tentative that it is extremely unlikely that an emergency justification can be supported for a summary revocation on disability-based grounds, without specific facts in support of each particular case. Therefore, this determination should be made on an individual not a catagorical basis. All such administrative determinations should in turn be subject to judicial review or appeal.

SUMMARY OF THE ACT

Statutory Provisions

The model act is presented as a resource from which to draw in amending existing licensing statutes. It has as its goals the protection of the public safety as well as the individual driver, while at the same time insuring that no individuals are unduly denied driving privileges. It rejects categorical prohibitions on driver licensing and establishes physical or mental disabilities or alcohol or drug dependency which might reasonably prevent safe driving as a starting point for further inquiry rather than as the sole determinative factors.

The model act provides for special application and individual review procedures, including medical input from the applicant's physician, where a disqualifying disability or dependency is reasonably suspected. Restricted licenses may be issued to deal with limiting conditions.

The administrator is assisted in this process by an Advisory Board representing various specialities. An Advisory Board aids in the process by establishing factors for identifying and evaluating individuals with potentially serious driving impairments, providing individual evaluations in difficult cases, and assisting in educating examiners, physicians and the public.

The act also provides for reexamination of licensees and for cancellation, suspension or revocation of licenses under certain circumstances. A special administrative review mechanism is provided before a Review Panel where action is taken on the basis of physical or mental disability or alcohol or drug dependency. The Review Panel is to include experts on the disability in question. A right of judicial appeal is provided for in all cases.

The text of the model act appears in **bold type**. Comments on statutory provisions appear in standard type. The comments explain, elaborate upon or provide a rationale for the text of the act. They are intended to assist the user in interpreting the act and in drafting an act for a particular state based upon this model act.

Adapting the Model Act

In using this model act, drafters amending existing codes should refer to the "major premises" stated *supra*. At a minimum, the amendments should provide for repeal of existing categorical prohibitions. Rather, a "good cause" standard for suspected disabling conditions should trigger a careful, individualized review procedure, focused on functional abilities of the applicant. An Advisory Board should be established to recommend: (1) factors to aid in identifying and examining applicants; (2) review procedures; and (3) training and educational programs. Administrative review and appeal to the courts should be available. If the proper provisions are adopted, the twin goals of maximizing public safety and minimizing the undue denial of driving privileges can be realized.

---

[43] *See* NATIONAL HIGHWAY TRAFFIC SAFETY ADMINISTRATION, *supra* note 24 at 41-42.

[44] *See*, Dixon v. Love 431 U.S. 105, 113 (1977).

[45] *Cf.* Bell v. Bursen, 402 U.S. 535, 542 (1971).

MODEL STATUTE*

A BILL FOR

AN ACT TO AMEND PROVISIONS OF THE LAW RELATING TO EXAMINATION OF APPLICANTS FOR DRIVER LICENSES AND THE CANCELLATION, SUSPENSION OR REVOCATION OF LICENSES AND TO ESTABLISH AN ADVISORY BOARD TO ASSIST IN EXAMINING APPLICANTS WITH DISABLING CONDITIONS.

Section 1. Additional definitions.

(1) "Commissioner" means the Commissioner of Motor Vehicles.

COMMENT. The term "Commissioner" is used throughout the model act to refer to the Chief Administrative Officer. It should be replaced as necessary with the title of the appropriate state official having authority over driver licenses.

(2) "Department" means the Department of Motor Vehicles.

COMMENT. The term "Department" is used throughout this model act to refer to the driver licensing agency. It should be replaced as necessary with the title of the appropriate state administrative agency having authority over driver licenses.

(3) "Physical or mental disability" means visual impairment; hearing impairment; cardiovascular condition; neurological condition, including epilepsy, multiple sclerosis, parkinsonism or muscular distrophy; metabolic condition, including diabetes; mental or emotional condition, including mental illness, mental retardation or autism; musculoskeletal condition, including amputation or skeletal deficit, paralysis, severe arthritis or cerebral palsy; and any other physical or mental condition which may adversely affect the ability to drive a motor vehicle.

COMMENT. This term is defined to include major diseases or disabilities which may adversely affect the ability to drive. It is not intended to be all inclusive, nor is the definition designed to create a categorical prohibition. Rather, the existence of a disability, when coupled with good cause to believe it prevents the reasonable and ordinary control over a motor vehicle (see subsection 5(2)) triggers the special review procedures contained in section 6 of the act.

(4) "Alcohol or other drug dependency" means addiction, habituation or abuse of alcohol or other drugs.

COMMENT This term is defined to include any persistent misuse or abuse of drugs including alcohol (See subsection 5(1)). It does not include the taking of prescription drugs for treatment of a disability as defined in subsection 1(3) above. In these cases it is one factor to be considered in evaluating the impact of the disability upon driving ability. Indeed, in some cases, such as high blood pressure, parkinsonism or epilepsy, the taking of a prescribed medication may be a condition of licensing.

(5) "Board" means the Driver License Advisory Board.

COMMENT. None necessary.

(6) "Panel" means the Driver License Review Panel.

COMMENT. None necessary.

(7) "Review Board" means the Driver License Review Board.

COMMENT. None necessary.

Section 2. Application procedures.

(1) The general application form shall ask no questions relating to medical, psychiatric or psychological information except, "Do you have any mental, physical or other condition which might affect your ability to operate a motor vehicle with safety upon the highways?" Supplementary forms designed by the Board are to be provided to applicants who answer this question positively. These completed supplementary forms shall be stored separately from other Department files and the Commissioner shall adopt special procedures in accordance with sections of the code (cite to sections on rulemaking procedures) in order to protect the confidentiality of these records, [not withstanding/consistent with] section of the code (cite the state's public records law).

COMMENT. This provision is intended as an addition to the other application procedures in existing licensing statutes. Such things as eye examinations, for example, are assumed to be included in the general application and licensing procedures.

*D. Matthew Powell, Associate Director of the project, had primary responsibility for drafting this model statute.

The subsection provides that application is to be made on a form which is to be used only for screening purposes. Those who answer the questions on this form affirmatively are to be given supplementary forms to complete. This two step process has two advantages. It permits the supplemental form to be tailored to specific types of disabilities so as to elicit maximum relevant information from the applicant. It also allows for separate treatment of the information so that confidentiality may be maintained. Such personal information should be available only to those licensing personnel who require it to perform their duties. (See section 14, *infra.*)

The act relies upon the reporting of disabilities by applicants. It does not require mandatory reporting of certain diseases, such as epilepsy, by physicians to the licensing authority. The imposition of such a requirement interferes with the doctor-patient privilege of confidentiality and makes it less likely that persons who could benefit from medical treatment will seek it, knowing it may result in the denial or loss of a driver license.

Yet those who forgo treatment may present a significant hazard on the road[1], whereas with such treatment many persons would be as capable of driving as safely as anybody. Thus, the model act encourages reporting of disabling conditions by providing equitable licensing provisions for disabled applicants. These persons will naturally cooperate with states which provide standards and procedures for review of applicants which are fair and impose reasonable burdens upon them consistent with considerations of public safety. On the other hand, persons who can do so will frequently attempt to hide their disability, if they know that the state has a policy of categorically denying licenses to certain disabled persons. Those applicants who fail to voluntarily disclose disabling conditions, whether they are in fact competent to drive or not, may have their licenses cancelled. (See section 8, *infra.*)

Other means of disclosure are also available. In addition to questions on the application form, the examinations given to applicants include written, driving and eye examinations, which provide an opportunity for screening people to readily detect disabilities. Other potentially hazardous drivers may come to the attention of the Department through observations of law enforcement officials and through reports of accidents in which a disabling condition may have been a contributing factor. Finally, information may be volunteered by other persons or in conjunction with hearings to suspend or revoke licenses for other reasons.[2] These identified individuals may then be re-examined pursuant to section 9, *infra.*

[1] *See* ALLEN, FERSTER AND WEIHOFEN, MENTAL IMPAIRMENT AND LEGAL INCOMPETENCY 351 (1968).

[2] *Id.* at 247.

Some states also provide for notice of adjudication of incompetency or civil commitment to be given to the licensing agency. Such notice frequently requires automatic revocation. Yet the determining factors for these two decisions are vastly different than those used to determine driving ability. Thus, the model act rejects such categorical revocations. (See section 5, *infra.*) If a state wishes to have such provisions, persons so identified should be re-examined pursuant to section 9, *infra.*

(2) Driver fitness examinations may not include any further physical or mental examinations except as specifically required by section 6.

COMMENT. Disabled applicants, like other applicants, must pass the written and driving demonstration exams, and the general eye examination. In addition, special assessment procedures are provided in section 6 for those individuals who are identified as having disabilities which are potentially hazardous to driving ability. The assessment procedures provided in the act are not to be circumvented or negated by other ad hoc examinations which may be devised by licensing personnel. Such procedural specificity is necessary in order to protect applicants who have a physical or mental disability from potentially arbitrary or discriminatory examinations.

Section 3. Driver License Advisory Board.

(1) There is hereby created within the Department a Driver License Advisory Board for the purpose of advising the Commissioner concerning the licensure of persons with a physical or mental disability or alcohol or other drug dependency.

COMMENT. The Advisory Board is intended to advise the Commissioner rather than having the policymaking and administrative authority itself. (See text of Introduction at notes 26 through 28. See also section 4, *infra,* for a listing of specific Board duties.)

(2) The Board shall be composed of eleven members, appointed by the Commissioner, and shall consist of a neurologist, a family practitioner, a psychiatrist, an opthamologist, a cardiologist, an orthopedist, an optometrist, a psychologist specializing in mental health or mental retardation, a person active in alcohol abuse programs, a person active in drug abuse programs, and a person active in physical rehabilitation.

COMMENT. Since the Commissioner is responsible for administration and enforcement of the state's licensing program, it is suggested that the Board members be appointed by the Commissioner rather than the Governor, who normally appoints persons to policymaking Boards. (Contrast this with subsection 4 (4) of the *Environmental Barriers* model act.) The size of the Board is considered the minimum necessary to provide adequate numbers of specialists in medicine, mental health, mental retardation, physical rehabilitation, alcohol abuse and drug abuse.

At a minimum the Board should include at least one neurologist, one cardiovascular expert and one opthomological specialist. These three specialists are considered crucial since most of the driver disabilities that would require medical input will involve one of these three areas. Emphasis on rehabilitation in selection of Board membership is also crucial in carrying out the Act's mandate to emphasize abilities over disabilities. For a state who wishes to expand the number of the members of the Board, additional specialists may be added to increase the level of expertise in these areas.

(3) The Board shall be appointed initially as follows: three members to serve two-year terms, four members to serve three-year terms, and four members to serve four-year terms; thereafter, appointments shall be four-year terms, and vacancies shall be filled by appointment for the unexpired portion of the term. The Board members shall annually elect the Chairman of the Board.

COMMENT. The act provides for staggered terms for Board members to insure that some experienced members always serve on the Board.

(4) Board members shall serve without compensation but shall be reimbursed for necessary expenses incurred in the performance of their duties. Such payments shall be made from funds appropriated to the Department.

COMMENT. It is not considered necessary to provide compensation in order to attract well-qualified members to the Board. The interest of individual Board members in serving should be sufficient motivation to insure the quality of the membership.

(5) The Board shall meet at least six times a year and may hold such additional meetings as are necessary to fulfill its responsibilities described under section 4 of this act. A majority of the Board shall constitute a quorum.

COMMENT. None necessary.

Section 4. Duties of the Driver License Advisory Board.

The Board shall:

(1) Advise the Commissioner on vision guidelines relating to the safe operation of motor vehicles;

(2) Advise the Commissioner on factors to be considered in identifying and evaluating for licensing, individuals with physical or mental disabilities or alcohol or drug dependency;

(3) Initiate the development of acceptable report forms to be prescribed by the Commissioner in accordance with section 6;

(4) Recommend a training course for driver examiners in the aspects of licensure provided in this act and initiate efforts to orient the general physician population as well as the public in these issues; and

(5) Evaluate individual problem cases as requested by the Commissioner.

COMMENT. The Driver License Advisory Board is given two major roles in driver licensing: to advise on the factors to be considered in evaluating and identifying individuals with disabilities or dependency and to assist in the evaluating of individual problem cases as requested by the Commissioner. The other duties of the Board are supportive of these two major duties.

The act has selected the word "factors" to be considered rather than criteria or guidelines. Even the National Highway Traffic Safety Administration (NHTSA), which has promoted the use of medical advisory boards and the establishment of valid medical criteria to be used in determining who should and should not be licensed to drive, admits that most medical conditions do not readily lend themselves to precise, predictable cutoff points.[3] The NHTSA notes that even in the area of vision, which seems most amenable to being measured, criteria used are more or less arbitrary but have been handed down through the years. Most criteria tend to be completely arbitrary, through seemingly logical, and there is little evidence to indicate whether these arbitrary standards or criteria should be raised or lowered.[4] Given the state of the art, it would be extremely difficult, if not impossible, to establish valid medical criteria.

The NHTSA distinguishes between guidelines and criteria, criteria being established by scientific data, whereas guidelines are established on the basis of experience, common sense and interpolation from allied information and statistical evidence.[5] While this distinction may seem logical given the difficulty in establishing criteria, guidelines have a way of becoming the same thing as criteria. As an alternative approach the model act has used the term "factors" to be considered. This approach appears to better reflect the current reality of the absence of any real empirical evidence to establish relationships between disabilities and unsafe driving.[6] Thus, the act has deliberately adopted a rather vague standard of public safety as a criteria for who should or should not be allowed to drive. And the Board is to establish par-

[3] NATIONAL HIGHWAY TRAFFIC SAFETY ADMINISTRATION, THE ROLE OF MEDICAL ADVISORY BOARDS IN DRIVER LICENSING 41 (1977).

[4] *Id.*

[5] *Id.*

[6] NHTSA is seeking to develop emperical data for various disabilities. The results of a study done for NHTSA entitled *The Feasibility of Developing a Medical Condition Data Collection System for Driver Licensing* are expected to be released in 1980. This study may lead to further studies which will attempt to develop emperical data on specific disabling conditions.

ticular factors which may be examined in determining who may be a potentially unsafe driver.[7]

Section 5. Persons not eligible for driver licenses.

The department shall not issue or renew any license to any person:

(1) Who is subject to alcohol or drug dependency except as he or she may qualify under section 6;

(2) Who has a physical or mental disability which the Commissioner has good cause to believe will prevent the person from exercising reasonable and ordinary control over a motor vehicle while operating it upon the highways, except as he or she may qualify under section 6;

(3) Who is unable to read and understand official signs regulating, warning and directing traffic; or

COMMENT. The difficulty in making individual determinations on driving ability is exacerbated by the fact that we currently have no empirical data linking a particular disability or disease to impaired driving ability. Some studies have purported to show a statistical link between the existence of certain physical disabilities and a higher accident incidence.[8] Such studies, however, are inconclusive and subject to a high degree of unreliability, and thus are not useful guides in establishing general standards or in making determinations in individual cases. (See comment to section 4 *supra*.) One study has summed up our current state of knowledge:

> No scientific data are available to indicate that drivers with any particular disease, including diabetes, epilepsy, color blindness, cardiovascular disease, or deafness have higher accident rates than comparable groups not having the condition. The identification of persons with certain diseases is of little value in preventing accidents.[9]

Only drinking and alcoholism have been shown to be clearly linked to higher accident rates.[10] Yet this condition is not addressed in existing licensing legislation.

This section reflects our existing state of knowledge on disability and accident prevention. It avoids categorical determination for which there is no evidence. Rather, identified disabilities, including alcoholism, merely trigger further inquiry into the applicant's driving ability pursuant to section 6.

(4) When the Commissioner has good cause to believe from the Department's records and other sufficient evidence that the person is unable or unwilling to drive a motor vehicle with safety upon the highways as evidenced by:

(a) ______ or more convictions of driving while under the influence of alcohol or other drugs within a ______ year period;

(b) ______ or more convictions for other moving violations within a ______ year period; or

(c) the accumulation of ______ points under the point system established by section ______ .

COMMENT. Some licensees demonstrate an inability to drive safely as evidenced by convictions or an accumulation of points, where the state employs such a system. Such "demonstrated" inability is treated separately from the alleged inability under subsections 1 and 2 of this section. Since the inability to drive safely has already been demonstrated, no further inquiry under section 6 is necessary.

Section 6. Special Licensing Procedures-Disabled or Dependent persons.

(1) The Department may issue an operator's or chauffeur's license to any person whom it reasonably believes is subject to alcohol or drug dependency or has a physical or mental disability described in subsections 1(3) and 1(4) but who

---

[7] For example, the following factors have been suggested for evaluation of potential drivers with epilepsy:

1. Frequency of the Attack. The less frequent the attacks, the less likely is a recurrence. A consistent record of one or fewer seizures per year over a period of time would appear to represent a low level of risk. Thus, in the instance of a previously seizure free individual who experiences a single seizure, especially during change of medication or other unusual circumstance, the overall pattern and previous history should be important determinants.

2. The Nature of the Onset of the Attacks. In some persons each attack is preceded by a warning or "aura", which may last for periods up to several minutes. Persons consistently experiencing a warning of such a nature as to permit them to take protective action are at reduced risk.

3. The Nature of the Attacks. The most serious attacks are those characterized by sudden loss of consciousness with or without generalized muscle spasm. Attacks characterized by local muscle spasms or sensory phenomena and without loss of consciousness may involve little or no risk of loss of control.

4. Customary Time of Occurrence of Attacks. Some persons are subject to attacks only while sleeping — for others only during the initial waking hours. Such persons are at reduced risk of seizures during the waking hours.

5. Individual Responsibility. The data indicate that the personal characteristics of the individual are a very significant factor in risk. Thus, epileptic women have fewer accidents than do all men. Epileptic men over age 35 have fewer accidents than non-epileptic men age 18-26. A history of erratic behavior and especially of noncompliance in the utilization of anticonvulsant drugs, excessive use of alcohol, or other evidences of unreliability are serious risks for persons with epilepsy. On the other hand the conservative epileptic driver who consciensciously monitors his or her own condition will avoid undue risks and maintain a safe driving record.

Thus, careful evaluation of each individual will make it possible to select those for whom the privilege of driving carries little additional risk.

Affidavit of Richard Mastand, M.D., on file with the Epilepsy Foundation of America, Washington, D. C.

[8] *See* Negri and Ibison, *Accidents Involving Handicapped Drivers*, 40 REHABILITATION LITERATURE 149 (May-June 1979) and *Allen*, *supra* note 1, at 345 n. 10.

[9] *Medical Aspects of Driver Limitation-Special Report*, 187 J. AM. MED. ASSN. 376 (1964).

[10] *Allen, supra* note 1, at 345.

is otherwise qualified to obtain a license, provided such person complies with subsection 2 of this section.

COMMENT. This section establishes a procedure for individual assessment of applicants who have been identified through application procedures outlined in section 2 as having a physical or mental disability or dependency which may affect their ability to safely operate a motor vehicle.[11]

(2) (a) An applicant for issuance of an operator's or chauffeur's license under subsection (1) of this section may be required by the Department to submit a certificate in the form prescribed by the Commisioner. This certificate shall include specific information concerning the applicant's condition and treatment.

(b) Such applicant also may be required to submit to a physical examination by a physician of the applicant's choice, or to a mental examination by a psychiatrist or clinical psychologist of the applicant's choice, who shall complete and sign a certificate in the form prescribed by the Commissioner reporting the results of the examination. In addition, the applicant may submit a written evaluation from other qualified professionals.

(c) All certificates to be submitted under this section shall be devised by the Commissioner with the advice of the Driver License Advisory Board. The certificates shall be designed to elicit the information reasonably necessary to assist in determining whether or not it would be a hazard to the public safety to permit the applicant to operate a motor vehicle on the public highways. Where applicable, it shall include a statement that the applicant is under the care of a physician, psychiatrist or clinical psychologist and is under treatment or medication, that the applicant's physical or mental disability or dependency is under control, that the applicant has been free of any seizures, loss of consciousness or control for a minimum of six months and that the applicant experiences no effects from any medication which the applicant may take which may affect his or her ability to operate a motor vehicle with safety on the highways.

COMMENT. Information concerning applications for licenses under this section is to be initially gathered from the applicants on forms prescribed by the Commissioner. It is anticipated that different forms will be developed for different disabilites or dependencies with the advice of the Advisory Board in accordance with subsection 4 (3). This will allow the forms to be tailored to specific disabilities in order to maximize the degree of information obtained.

In addition to information submitted by the applicant, the applicant may also be required to undergo a physical examination by a physician of the applicant's choice in order to elicit medical information, or undergo a mental examination by the psychiatrist or clinical psychologist of the applicant's choice in order to elicit psychological information, required to make a determination on the applicant's driving ability. In addition, the applicant is given the option of submitting other information which he or she may feel is relevant for determining driving ability. This will allow the submission of information by persons with expertise in the symptoms or etiology of a particular disability or dependency related to the particular case in question. (See the discussion of factors to be considered in licensing in Comment to section 4 *supra*.) Thus, this subsection provides for the submission of information beyond that which may be required in the existing statutes.[12]

(3) The Commissioner is not bound by the recommendation of an examining physician, psychologist, social worker or other professional but shall give full and fair consideration to such recommendation. The Commissioner shall issue a license unless, upon all the evidence, it appears that it is unsafe to permit the applicant to operate a motor vehicle. In deciding whether to issue or deny a license, the Commissioner may request an opinion from the Driver License Advisory Board. (See section 4 *supra*.)

COMMENT. The act makes it clear that the Commissioner is the final determinant of an applicant's ability to drive a motor vehicle. However, the Commissioner must have some evidence that the applicant is an unsafe driver in order to deny the applicant a license.[13] Where the Commissioner is in doubt about the driving ability of the particular applicant, he or she may request an opinion from the Driver License Advisory Board. However, like the opinion of the applicant's examining professional, the opinion of the Advisory Board is not binding on the Commissioner.

(4) In evaluating individual cases the Board may formulate its recommendation from records and reports previously submitted or may cause an additional examination and written report to be made by an expert in diagnosing and treating the particular disability or dependency which the applicant is said to possess. Such experts shall be compensated on an equitable basis by the Department. The written report shall be forwarded to the Board by the expert and such report shall be given due consideration by the Board.

COMMENT. The Advisory Board would normally make its recommendation to the Commissioner

[11] Section 6 is based upon similar provisions in existing state statutes. *See, e.g.* N.C. GEN. STAT. §20-9 (1978) and OHIO REV. CODE §4507.08 (1976).

[12] *See, e.g.*; N.C. GEN. STAT. §20-9(9) (2) (1978).

[13] *See* Ormond v. Garrett, 8 N.C. App. 662, 175 SE 2d 371 (1970).

based upon existing records and reports by the applicant's examining professional. The Board, however, is given the authority in cases where it is necessary, to have the applicant re-examined by an expert in a particular disability or dependency in question. Where the Board chooses to do this, the experts will be compensated by the Department, since the applicant has already personally paid for one examination and the second examination is undertaken solely at the behest of the Advisory Board. Any reports submitted by the experts shall be duly considered by the Board and its recommendation reported to the Commissioner.

(5) Any person under review who refuses to submit to a reasonable examination or to consent to provide information, or both, shall as a matter of law be considered unqualified to operate a motor vehicle until such time as the individual complies with the Board's requests and the Board makes its findings and recommendations to the Commissioner.

COMMENT. The examination by experts at the request of the Advisory Board is, like the initial examination, a precondition to granting a license under this section, since the information solicited is necessary to make a rational assessment of the applicant's driving ability. Failure to comply with this requirement will logically result in the withholding of an assessment, and thus the issuance of a license, until such time as the applicant does comply.

(6) Notwithstanding section ______ (cite to state's open meetings law), the portions of meetings of the Board in which reports are received for the purpose of determining the driving ability of an applicant are closed sessions, since those reports are confidential under section 14 of this act.

COMMENT. The Board will be considering in part information which is normally accorded privacy. Thus, any meetings at which the Board considers such information are to be closed to the public. (See comment to subsection 14 *infra*.)

(7) The Board, any of its members, the Commissioner or its employees, or any physician or optometrist shall not be subject to civil or criminal liability, absent a showing of bad faith, for providing any reports, records, examinations, opinions, or recommendations pursuant to this act.

COMMENT. Persons submitting or utilizing information pursuant to this act for the purpose of making individual determinations on licensing of applicants must be free to make judgments without the threat of law suits, civil or criminal, which may arise out of those judgments. For this reason these persons are given immunity from liability absent a showing of bad faith on their part in carrying out their duties under this act. Where an individual plaintiff can show bad faith in submitting erroneous or misleading reports, records, opinions, or recommendations pursuant to the act, such individuals are rightly subjected to liability for such actions. Therefore, the act does not exempt such bad faith submissions.

(8) In issuing a license under this section the Commissioner may restrict the license in accordance with the provisions of section 7, including a provision that the licensee annually submit the certificates required by subsection (2) of the section. However, where the applicant's or licensee's physician has certified that the person's physical disability, or the applicant's or licensee's psychiatrist or clinical psychologist has certified that the person's mental disability is not subject to change, the Commissioner shall require no more than one subsequent annual certificate to be submitted. If no change has occurred during the one year period, no further certificates shall be required. Where the applicant or licensee has a physical or mental disability which is subject to change or requires medication or has an alcohol or drug dependency, the Commissioner may require a certificate to be submitted at six month intervals for up to two years and annually thereafter.

COMMENT. Many conditions such as failing eyesight, diabetes, epilepsy, or alcohol or drug dependency may need to be monitored over time to insure that a licensee still presents no undue danger to the public safety. Thus, the Commissioner may restrict the license by requiring periodic review of the applicant or licensee's physical and mental disability. However, where the disability is not subject to change such as deafness, cerebral palsy or skeletal deficit, the licensee should not be required to submit certificates beyond that required to show that no change is occurring. It must be remembered that the provision of certificates by one's physician is a financial burden which must be borne by the licensee. Therefore, these should be kept to the minimum and only be required to establish the continued ability of the licensee to safely operate a motor vehicle. Indeed, even those licensees whose condition may be subject to change should not be required to indefinitely submit certificates where they show over a period of time a stability in their condition.

Section 7. Restricted licenses.

(1) The Commissioner, upon issuing any license pursuant to this chapter, may, whenever good cause appears, impose restrictions suitable to the licensee's operating ability with respect to the type of special mechanical control devices required on a motor vehicle which the licensee may operate, to the time of day or to the distance from licensee's residence that the licensee may operate a motor vehicle, or such other restrictions applicable to the licensee as the Commissioner may determine to be appropriate to as-

sure the safe operation of a motor vehicle by the licensee.

(2) Such a limitation or restriction shall be noted on the face of the license, and it shall be unlawful for the holder of such limited or restricted license to operate any motor vehicle in violation of the restriction. To do so shall be deemed the equivalent of operating a motor vehicle without a chauffeur's or operator's license. Any such restricted or limited licensee may at any time apply for and receive an unrestricted operator's or chauffeur's license upon meeting the requirements therefor.

COMMENT. Many drivers have various disabilities which affect driving ability but which may be compensated for by various devices or by limiting driving to certain conditions. Perhaps use of corrective lenses is the most common example of this. Other drivers perform quite safely with the use of prosthetic devices or special mechanical controls or by taking certain drugs which prevent the adverse symptoms of a disease from limiting driving ability. Licensing procedures must recognize such functional approaches for overcoming certain handicaps. This subsection allows certain restrictions to be imposed in granting licenses which are intended to assure the safe operation of motor vehicles despite the existence of disabling conditions. This approach should allow the license to be tailored to the individual licensee, rather than granting or denying a license in an "all or nothing" manner. It is intended in this way to maximize the granting of licenses while protecting the public safety.

Section 8. Cancellation of Licenses.

(1) The Department is hereby authorized to cancel any driver's license upon determining that the licensee was not entitled to the issuance thereof under this chapter or that the licensee failed in a material way to give the required or correct information in his or her application or committed any fraud in making such application. The licensee shall be afforded an opportunity for a hearing under section 12 prior to cancellation.

(2) Upon such cancellation, the licensee must surrender the license to the Department.

COMMENT. When deciding whether to grant a license, the Commissioner relies heavily on voluntary disclosure to discover the existence of disabling conditions which may impair driving ability. (See section 2 *supra*.) Thus, the issuance of a license is based upon and necessarily assumes the accuracy of information submitted by the applicant. To help assure such accuracy, the Department is authorized to cancel a license where it is determined that the applicant failed to give correct information which would have made a material difference in whether or not a license would have been granted. Cancellation on the basis of disability or dependency is reviewable pursuant to section 12 to protect the rights of the licensee.

Section 9. Reexamination of licensees.

Whenever the Commissioner has good cause to believe that a licensed operator or chauffeur is incompetent or otherwise not qualified to be licensed, the Commissioner may, upon written notice of at least ten days, require the licensee to be reexamined in accordance with section 6. Upon conclusion of the examination, the Commissioner shall take such action as may be appropriate, and may permit such person to retain such license, may issue a license subject to restrictions, or may suspend or revoke the license of such person in accordance with section 10.

COMMENT. Dependency and many of the disabilities which give rise to questions concerning driving ability are subject to change over time. For this reason periodic review is required in some cases. (See subsection 6(8).) Changing conditions may also cause a licensee who was competent at the time he or she was licensed or reviewed, to become incompetent at a later time. These persons come to the attention of the Department through a variety of means already discussed in the comment to section 2. Where this occurs, the Commissioner may reexamine the licensee upon good cause. This reexamination procedure is to be conducted in the same manner as initial examinations.

Section 10. Suspension or Revocation of Licenses for Driver Incompetency.

(1) The Commission may suspend or revoke the license of a person who fails reexamination or who refuses or neglects to submit to reexamination pursuant to section 9 until the person satisfactorily complies with that section.

(2) The Commission shall not suspend or revoke a license under this section without first providing an opportunity for a hearing pursuant to section 12, provided, however, that notwithstanding any other provision in this act, the Commissioner is authorized to suspend the license of a driver without a hearing when there is a clear and convincing showing by the Department's records or other competent evidence that the licensee is so incompetent to drive a motor vehicle because of a physical or mental disability or alcohol or drug dependency that he or she will present a clear and immediate danger to the public safety. The licensee may seek review of such a suspension pursuant to section 12. The Review Board may in its discretion order the Department to reimburse the licensee for any and all expenses incurred as a result of the suspension of the license by the Commissioner where the Commissioner fails to sustain the burden of evidence upon review. The licensee shall be reimbursed from funds appropriated to the Department.

COMMENT. Suspension or revocation of a license under this section is subject to an opportunity for hearing unless the licensee presents a clear and present danger to the public safety. Such an approach is consistent with constitutional principles of due process which are to be observed prior to depriving the licensee of his or her property, i.e. his or her license. (See section 12(1).) When the licensee clearly endangers the public safety, the state interest in public safety on the highways allows the due process balance to be struck in favor of the state, thus permitting a suspension or revocation without a prior opportunity for a hearing. (See discussion in Comment to subsection 12(1).) This approach is intended to be the clear exception, not the norm. To ensure that this action is taken only where the evidence is clear and convincing that there is an immediate danger to public safety, the Review Panel may order the Department to reimburse the licensee for expenses incurred as a result of the Commissioner's action where the Commissioner has not sustained the burden of evidence upon review. Reimbursement will also help to compensate for the personal and economic hardships which may flow from the state's erroneous decision.[14]

(3) When a license has been suspended, revoked, or denied, and the licensee's livelihood depends upon the use of the driver's license, the Commissioner may upon application, and shall if the Review Board or a court so recommends, issue a limited license. If by application, the Commissioner shall issue or deny a limited license within 30 days of the application. If by recommendation, the body so recommending may impose such limits on the license as it deems necessary in the interests of public safety and welfare, including limitations to operation of particular vehicles, to particular times of operation, to particular routes and to particular conditions of traffic. In making such a determination, the entire driving record of the licensee and the number of miles driven annually by the licensee must be considered.

COMMENT. This is a general provision applicable to all drivers which has been slightly adapted to incorporate the provisions for special review contained in section 12. It is intended to alleviate the severe adverse impacts which loss of a license might cause, such as loss of employment, without unduly compromising the public safety and welfare. Such provisions are consistent with the state's dual interests in maximizing public safety and in minimizing public assistance payments.

Section 11. Driver License Review Panel and Boards.

(1) The (Director of the Public Health Department) shall establish a Driver License Review Panel for the purpose of selecting members of individual Review Boards to hear reviews under section 12 of this act. The Panel shall consist of a sufficient number of physicians, optometrists, mental health, mental retardation, epilepsy, alcohol, drug abuse and physical rehabilitation specialists to accomplish this task.

(2) Hearings held pursuant to section 12 shall be heard by a Driver License Review Board. Such a Board shall be created each time the (Director of the Public Health Department) receives notice from the Commissioner that a request for review has been made pursuant to Section 12. Each Review Board shall consist of three members, one of whom shall be a hearing officer employed by the (Public Health Department) for this purpose and the other members shall be selected from the Driver Review Panel by the (Director of the Public Health Department). At least one member of the Review Board shall be a physician and at least one shall be an expert in alcohol or drug dependency or in the physical or mental disability at issue in the review.

(3) Members designated by the Director to serve on a Review Board shall receive $_________ per diem and expenses while serving on a case which shall be paid for from funds appropriated to the (Department of Public Health).

COMMENT. The section establishes Review Panels and Boards to review the Commissioner's actions with respect to licensees with disabilities or dependency. This approach is adopted from existing statutes in North Carolina and Ohio.[15] It has the advantage of tailoring the reviewers to the disability or dependency which gave rise to the Commissioner's action on a license, thereby assuring greater expertise in the review process, while separating the review process from the initial investigation and determination by the Commissioner. (See discussion in "Organization of the Office of Civil Rights" contained in the *Prohibiting Discrimination* chapter.) It would make little sense to expect the Commissioner to objectively review his or her own decision. Consistent with this separation, the Director of the Department of Health or its equivalent is to select the Panel and Board members and to control the appropriations made to support them.

Section 12. Driver License Review Hearings.

(1) Any applicant or licensee whose license is denied, limited, restricted, cancelled, suspended or revoked because of a physical or mental disability or alcohol or drug dependency shall be afforded an opportunity for a hearing under this section.

COMMENT. This subsection provides for an opportunity for a hearing whenever action on a license application is taken on account of a dis-

[14] *Cf.* Dixon v. Love, 431 U.S. 105, 113 (1977).

[15] N.C. GEN. STAT. §20-9 (1978); OHIO REV. CODE §4507.08 (1976).

ability or dependency. Where a license already has been granted, a prior opportunity for hearing is to be provided before suspensions or revocation occurs, except where there is imminent danger to the public safety. (See subsections 8(1) and 10(2) *supra*.)

Whether constitutional due process mandates a prior hearing is still an open question. Three United States Supreme Court cases which frame the law in this area are of importance to an understanding of the issues involved. First, in *Bell v. Bursen*[16] the Court found a prior hearing to be required prior to revocation of a license for failure to establish financial responsibility because the state's interest involved only the payment of judgments by the licensee *not* any issues impacting directly on matters of public safety.[17]

In *Dixon v. Love*[18] the Court held that where the state's interest does involve the safety of persons travelling on its highways, the due process balance may be struck differently, allowing the state to revoke licenses summarily. In that case the Court held constitutional a summary revocation based on a statutorily defined number of past traffic violations under a point system. In part the Court's decision was premised on the limited amount of administrative discretion involved with the revocation action, since the number and type of violations which warranted and required revocation were set out by statute and regulation.[19] Thus, there was nothing to contest except whether or not the licensee had committed the number and type of violations specified in the statute. In addition, since the revocation criteria involved past court adjudications of fault, the Court found that licensees had actually received a prior opportunity to contest each violation on the merits, and thus a reviewing court would have little to review, except perhaps the administrator's arithmetic in determining that a revocation was called for.[20]

Finally, in *Mackey v. Montrym*[21] the Court held that states could constitutionally suspend a driver's license without any prior hearing on the basis of a signed police report stating that the licensee had refused to submit to a breathalyser test. While the Court, however, claimed that the state's purpose in summary revocations was the promotion of public safety, the dissent pointed out that since drivers who did take the test and were found intoxicated did *not* have their licenses revoked prior to a hearing, the effect of the law was simply to punish drivers who did not cooperate with police.[22] In addition, the driver in *Mackey* had no prior opportunity to contest the factual basis for a revocation as had been the case in *Love*.[23] Thus *Mackey* may cut back on the due process safeguards that *must* be afforded before a driver's license is revoked.

However, there are reasons to believe that in the context of disability-based restrictions, a prior hearing may still be constitutionally required. Both cases which upheld summary revocations involved administrative decisions that permitted officials to exercise very little discretion. Thus, the Court may well have reasoned that a hearing could rarely be of any use to the deprived individual. On the other hand, the necessity for a prior hearing has long been recognized as most vital when the decision to be made is complex and controversial.[24] The decision to suspend or revoke on the basis of a disability or dependency is surely of this nature for two reasons.

First, the necessarily vague statutory criteria used, the "ability to drive safely", does nothing more than describe the problem. It does little to aid the administrative decision-maker in determining who among disabled persons is or is not competent. Thus, the administrator is vested with a great deal of discretion to decide what does and doesn't constitute incompetence even when regulations and the Advisory Board provide some further guidance.

Second, it is not clear what factors actually adversely affect driving ability or at precisely what point this occurs. In addition, each individual varies in the way a particular disability may manifest itself and thereby affect performance. The uncertainty of such a multitude of factors makes the administrative decision a difficult one to make and results in a significant possibility of erroneous decisions. Thus, the interjection of a due process hearing could be of significant value in the process and would constitute much more than the rubber stamp approval of the administrative decision-making as was involved in *Love* and *Mackey*.

Furthermore, since our present understanding is inadequate to make any conclusive links between driving ability and any but the most profound disabilities, there is no reason to be overly concerned with the threat posed by most drivers in the period prior to initiation of a hearing. Where there is an immediate threat to public safety, the Commissioner may act pursuant to subsection 10 (2) to immediately suspend the license. When one recognizes that states commonly do *not* summarily revoke the licenses of drivers found intoxicated by

16 402 U.S. 535 (1971).

17 *Id.* at 540.

18 431 U.S. 105 (1977).

19 *Id.* at 113-15.

20 *Id.*

21 443 U.S. 1 (1979).

22 *Id.* at 20.

23 *Id.* at 22-23.

24 The Constitutional test most frequently employed by the court to determine the amount of process due and explicitly stated in Matthews v. Eldridge, 424 U.S. 319, 334-35 (1976) included three elements: 1) the private interest involved, 2) the government's interest and 3) the *risk of an erroneous deprivation through the procedures used and the probable value of additional or substitute procedural safeguards.* (Emphasis added.)

breathalyser examination,[25] even though drinking and alcoholism have been shown to contribute to higher accident rates,[26] it becomes even harder to accept that any legitimate need exists which warrants summary treatment of disabled persons.

(2) In taking any adverse action with respect to a license or application for a license, the Commissioner shall give notice to the licensee or applicant of the right to a review of the decision. The licensee or applicant then has 30 days from the date of notification in which to request a review unless he or she can show good cause for a delay. The request must be submitted to the Commissioner in writing.

COMMENT. To insure that persons are apprised of their rights, the Commissioner is required to give notice of the right to review under this section. All requests are required to be made within 30 days unless good cause is shown in order to prevent the bringing of stale review petitions.

(3) The review hearing shall be held within 30 days of the request for review unless the applicant or licensee agrees otherwise. At least ten days prior to the review hearing the applicant or licensee shall be given notice stating the time, place and subject of the hearing. The notice shall be in writing and shall be delivered in person or sent by certified mail, return receipt requested.

COMMENT. Given the importance of a driver's license, the applicant or licensee is entitled to a prompt determination on review with adequate notice to allow sufficient time to prepare for the hearing. The applicant may, of course, agree to a different time for the hearing.

(4) All hearings under this section shall be conducted in accordance with (cite to state's administrative procedure act or appropriate section in existing license statutes). The hearing may be closed at the request of the applicant or licensee as necessary to protect the privacy of the applicant or licensee.

COMMENT. The hearings are intended to be conducted in a manner consistent with the state's general administrative procedures and those outlined in this section. Subsequent drafters should examine the state's administrative procedure act for consistency with this section. Also see generally sections 22 and 23 of the *Prohibiting Discrimination* model act.

Much of the information considered by the Panel, however, is of a private nature and is supplied to the state only as necessary to meet the specific needs for information that will be used in determining driving ability. The public need to know is clearly outweighed by the privacy interests of the individual, since there is no significant public need for access of the information beyond that which is already satisfied by the state's access in the interest of public safety. Therefore, the act allows review hearings to be closed at the request of the applicant or licensee whose privacy is threatened. (See also the Comment to section 14.)

(5) The Review Board may compel the attendance of witnesses and the production of such books, records and papers, consistent with section (cite to state's administrative procedure act or other appropriate section). Upon request, a subpoena to compel the attendance of any witness or a subpoena duces tecum to compel the production of any books, records, or papers shall be issued by the hearing officer.

COMMENT. None necessary.

(6) The Review Board may continue a hearing upon motion where there is good cause shown.

COMMENT. None necessary.

(7) The hearing officer shall preside at the hearing and shall pass upon the admissibility of evidence at the hearing. Incompetent, immaterial, irrelevant or unduly repetitious evidence may be excluded. The party adversely affected by the exclusion may at the time object to the ruling, and, if evidence offered is rejected, the party may proffer the evidence, and such proffer shall be made a part of the record.

COMMENT. The hearing officer, who is trained in administrative procedure, is to preside at the hearings and rule on evidentiary matters since he or she will be best suited amongst Board members to assure that proper procedures are observed.

(8) The Review Board shall not be bound by common law or statutory rules of evidence which prevail in courts of law or equity and may give probative value to evidence commonly accepted by reasonably prudent men in the conduct of their affairs. Uncontested facts may be stipulated by agreement. All evidence, including records and documents in the possession of the Department or the Review Board of which the Review Board desires to avail itself, shall be made a part of the record. Documentary evidence may be received in the form of copies or excerpts, or by incorporation by reference.

COMMENT. None necessary.

(9) The hearing officer shall prepare an official record, which shall include testimony and exhibits. A record of the testimony and other evidence submitted shall be taken, but it shall not be necessary to transcribe shorthand notes or electronic recordings unless requested for purposes of court review.

COMMENT. None necessary.

(10) Every decision and order adverse to an aggrieved party shall be in writing or stated in

[25] *Cf.* Mackey v. Mantrym, 443 U.S. 1, (1979).

[26] *See* U.S. DEPT. OF TRANSPORTATION, 1977 HIGHWAY SAFETY ACT REPORT APP. A-9; U.S. DEPT. OF HEALTH, EDUCATION AND WELFARE, THIRD SPECIAL REPORT ON ALCOHOL AND HEALTH 61 (1978). *See also* ALLEN supra note 1.

the record and shall be accompanied by findings of fact and conclusions of law. The findings of fact shall consist of a concise statement of the Review Board's conclusions on each contested issue of fact and its application of the law to the facts. Counsel for the aggrieved party, or the aggrieved party, if he or she has no counsel, shall be notified of the Review Board's decision in person or by certified mail with return receipt requested.

COMMENT. None necessary.

(11) Actions of the Review Board are subject to judicial appeal as provided under section 13 of this act.

COMMENT. None necessary.

Section 13. Right of appeal to court.

Any person aggrieved by a decision of the Review Board after hearing pursuant to section 12 may file a petition within 30 days thereafter for a hearing in the matter in the ________ court of the county wherein such person shall reside. The court shall set the matter for hearing upon 30 days written notice to the Department.

The hearing shall be *de novo*. The court shall determine whether the petitioner is entitled to a license or is subject to limitation, restriction, suspension, cancellation or revocation of a license under the provisions of this chapter.

COMMENT. The act adopts the position of most state statutes in allowing appeal of actions of the Review Board to be heard *de novo*, i.e. as if being heard for the first time. States may also wish to limit judicial action to review of the administrative action, especially in view of the expertise of the Review Board, thereby limiting the reviewing court to an examination of the law and the reasonableness of the factual findings of the Review Board. For an example of a model section on such judicial review see section 24 of the *Prohibiting Discrimination* model act.

Section 14. Confidentiality of reports and records.

Notwithstanding section _____ of the code (cite to state's public records law), reports or records, received or made by the Board or any of its members, or by the Department, pursuant to this act for the purpose of assisting the Commissioner in determining whether a person qualifies for a license to drive are for the confidential use of the Board and the Commissioner and, except as may otherwise be specifically required by Federal statute, such reports or records shall not be divulged to any other person, the Federal, State, or local government and/or private entity, or used as evidence in any trial. Copies of such reports or records shall be provided, however, upon demand, to the licensed driver or applicant to whom they relate or to his or her authorized representative, and may be submitted in proceedings under sections 12 and 13 of this acct, and any person conducting an examination or submitting a report or copy of records pursuant to this act may be compelled to testify relative to it in such proceedings.

COMMENT. Decisions concerning driving ability made pursuant to section 6 must necessarily be made in part on the basis of personal, medical or other information which must be provided by the applicant or licensee. This section weighs the need of the state for such information against the right to privacy normally enjoyed by the individual. The information submitted is to be held confidential except as it is necessary for the purpose of determining driving ability, whether initially (see subsection 6(2) and 6(3)), upon examination by the Advisory Board (see subsection 6(4)), upon review by the Review Board (see section 12), or on judicial appeal (see section 13). Thus, the Advisory Board's (subsection 6(6)) or the Review Board's (subsection 12(4)) considerations of this confidential information may be closed to the public to protect the individual's right to privacy. This approach insures that the need of the state is met for such information while maximizing individual privacy. It will also tend to encourage the submission of full and accurate information concerning the condition of the applicant and thereby contribute to the realization of the public safety goal of the act.

The subsection also contributes to further disclosure by making clear that the doctor-patient privilege does not apply with respect to written reports of examinations submitted by an applicant or licensee; hence, a doctor may be compelled to testify about the examination in a subsequent hearing and is not to be held liable for doing so. (See subsection 6(7).) The effect of this requirement of disclosure is mitigated, however, by the provisions for closed meetings and closed hearings on such information.

Section 15. Appropriation.

The General Assembly is authorized to appropriate _____ dollars for the immediate fiscal year to cover the initial expenses incurred in implementing this act.

COMMENT. None necessary.

Section 16. Severability Clause.

If any provision of this act or the application thereof to any person or circumstance is held invalid, such invalidity shall not affect any other provision or application of the act which can be given effect without the invalid provision or application, and to this end the provisions of the act are declared severable.

COMMENT. None necessary.

Section 17. Effective Date of the Act.

This act shall take effect ________________ .

COMMENT. None necessary.

2
PROHIBITING DISCRIMINATION

Table of Contents

INTRODUCTION

"The heart of the question is whether all Americans are to be afforded equal rights and opportunities . . ."[1]

During the fifteen years that have passed since President John F. Kennedy expressed this challenge, gains have been made on behalf of blacks, women, the poor, and the powerless toward assuring equal justice for all. The Civil Rights Acts of 1964[2] and 1968[3] and the Voting Rights Act of 1965[4] have brought us as a nation closer to a positive resolution of the question posed. Under the impetus of such legislation millions of blacks were enfranchised to vote, segregation in public accommodations eliminated, and employment discrimination based on "race, color, religion, sex or national origin"[5] proscribed. In fact, there have been great changes in the last ten years in the generally accepted social roles of women and blacks.

Unfortunately, the movement toward equality of opportunity and rights has failed to include Americans with disabilities who confront attitudinal, legal, and physical barriers on a daily basis. In relation to other minority groups, disabled persons are years behind in their struggle for equal rights. For example, negative attitudes about the proper social roles for disabled Americans are very strong, with much of the community behaving on the basis of certain assumptions, whether those assumptions are verifiable or not.[6] Thus, many jobs are considered inappropriate for disabled persons and many disabled children are considered uneducable, even though they are not.

Discrimination against the individual with a disability is no more apparent than in the employment area. It has been conservatively estimated that in the United States there are 7 to 12 million disabled persons of work age and able to work, though other estimates are much higher.[7] Of this number, only 800,000 persons are employed.[8] This extraordinarily high rate of unemployment is directly related to employer attitudes toward the disabled person.[9] Employers fear that hiring the worker with a disability will impair the productivity of the business and will result in higher insurance rates.[10]

Such misconceptions[11] and assumptions are not the only impediments confronted by the disabled individual in finding gainful employment. Paternalism has been a frequent criticism of rehabilitation programs which have limited conceptions of the disabled person's job capabilities.[12] Other obstacles in the form of design barriers in mass transportation systems and job locations also present formidable barriers.

Certainly, the economic costs of discrimination against disabled people are serious; however, the human costs are even more distressing.[13] Although approximately 14% of the population in the United States live below the poverty level, more than 21% of non-institutionalized disabled persons are part of this group.[14] The statistics in the field of education are equally dismal, with 60% of an estimated 7 million disabled children being denied special educational assistance needed for full equality of opportunity, and one million being excluded entirely from public school systems.[15]

Negative attitudes and assumptions are not only manifested in the behavior of employers and edu-

[1] Speech of President John F. Kennedy, June 11, 1963, N.Y. Times, June 12, 1963, p. 6, Col. 8 (cited in Bonfield, *State Civil Rights Statutes: Some Proposals*, 49 IOWA L. REV. 1067, 1078-1079 (1964)).

[2] 42 U.S.C. §§2000e *et seq.* (1976).

[3] 42 U.S.C. §§3601 *et. seq.* (1976).

[4] 42 U.S.C. §§1971 *et. seq.* (1976).

[5] *See, e.g.*, 42 U.S.C. §2000 e-2 (1976).

[6] Kriegel, *Uncle Tom and Tiny Tim: Some Reflections on the Cripple as Negro*, 35 AM. SCHOLAR 412,423 (1969).

[7] Wright, *Equal Treatment of the Handicapped By Federal Contractors*, 26 EMORY L. J. 65, 68 n. 13 (1977).

[8] Weiss, *Equal Employment and the Disabled: A Proposal*, 10 COLUM. J. LAW SOC. PROB. 457 (1974).

[9] Note, *Abroad in the Land: Legal Strategies to Effectuate the Rights of the Physically Disabled*, 61 GEO. L.J. 1501, 1513 (1973).

[10] Note, *Potluck Protections for Handicapped Discriminatees: The Need to Amend Title VII to Prohibit Discrimination on the Basis of Disability*, 8 LOY. CHI. L.J. 814, 816-817 (1977).

[11] *See* Note, *supra* note 9, at 1513. *But see* Wright, *supra* note 7, at 77 (handicapped generally have a higher risk of injury; therefore the cost of their insurance is correspondingly higher).

[12] Weiss, *supra* note 8, at 458.

[13] Note, *supra* note 9, at 1501.

[14] Note, *supra* note 10, at 816.

[15] Note, *supra* note 9, at 1503 nn. 12 & 13.

cators among others, but are also given legal sanction in a large number of areas affecting legal rights. As a result the context in which disabled persons are discriminated against is probably broader and farther-reaching than that faced by any minority in America today. Thus, in addition to discrimination in employment, education and transportation, a disabled individual faces legally imposed restraints upon marriage, voting, contracting, testamentary dispositions, and driver's licenses and is not uncommonly subjected to involuntary sterilization and termination of parental rights.[16] Moreover, many persons including legislators see nothing wrong with such categorical restrictions, despite their similarity to laws which once categorically restricted women and blacks from voting and declared women incompetent to handle their own financial affairs.

This is not to say that discrimination occurs or restrictive laws are passed out of malevolence; quite the contrary. They reflect paternalistic attitudes and a belief in the need to protect disabled persons from themselves and others and to protect society from their folly, just as restrictions on women once did. Regardless of the motivation, just as such restrictions are not appropriate for women, they are not appropriate for disabled persons.

All disabled Americans should have an equal opportunity to realize their potential to the fullest extent and to share in the "fruits of our society."[17] By achieving this goal our nation remains faithful to the American ideal "that all men are created equal."[18] Although not every person has "equal talent, or an equal ability or equal motivation . . . they should have the equal right to develop their talent and their ability and their motivation . . . "[19] Protection of disabled persons from discrimination in every aspect of life would represent a fuller realization of "the principled system of morality"[20] embodied in the Civil Rights Act of 1964[21] and would be a reaffirmation of the individual's dignity and worth — a reaffirmation that one can make it one one's own.[22]

Eradication of discrimination against disabled people would also greatly benefit society. The high incidence of unemployment and poverty among disabled persons in the United States is reflective of the "drain on the productive resources of our economy" and, more importantly, the "waste of human resources" fostered by the stereotypes of the prejudices toward the disabled individual.[23] Equal employment opportunity would provide for greater utilization of the disabled labor force,[24] increase the GNP and tax revenues, increase the purchase of goods and services in the market place[25] and decrease the burden on social welfare programs.[26] Finally, equality of opportunity for the individual with a disability would enable society to make greater use of and receive contributions from untapped human resources in activities from which disabled persons had been excluded.[27]

The ramifications of discrimination in the United States reach beyond our shorelines. In the eyes of world opinion, discrimination against a minority group detracts from "our image as a land of justice, equal opportunity, and democratic principles."[28] Only by providing equality of rights for all minorities do we signal to the world that our national commitment to human rights is a reality, not merely verbal window-dressing.

How does a nation go about effecting the change in attitudes and behavior necessary for assuring equality of opportunity and rights for disabled people? Several possibilities are suggested: education, litigation, and legislation.

Education of a people would be a massive, time-consuming effort with no assurance of success. Indeed, it will never be totally effective. And to allow discrimination against disabled persons to continue during an educational process would be a callous disregard of the rights of a minority that already has suffered much indignity.

Litigation to achieve one's rights is generally viewed as a last resort. One goes to court when legislation protecting one's rights is not fully implemented or is non-existent,[29] since litigation is expensive and time-consuming and there is a paucity of legal theories on which to proceed.[30] Further, it does not allow a systemic approach to change since litigation is largely a piecemeal approach, though it has been used in some cases with success.

Due process arguments under the Fifth and Fourteenth Amendments have been successfully

---

[16] These legal restrictions on rights are dealt with as a separate topic in this series of model acts. However, they are mentioned here because they demonstrate how pervasively negative attitudes and assumptions have affected our society, including our legal system. *See* Chapter 1: Personal and Civil Rights, *supra*.

[17] Bonfield, *supra* note 1, at 1079.

[18] *Id.* at 1078 n. 51.

[19] *Id.* at 1078 n. 53.

[20] Note, *supra* note 10, at 844.

[21] 42 U.S.C. §2000e *et seq.* (1976).

[22] *See* Kriegel, *supra* note 6, at 414.

[23] Note, *supra* note 10, at 815-816.

[24] *See id.*

[25] *See* Bonfield, *supra* note 1, at 1079.

[26] *See id.*, at 1080. *But cf.* Wright, *supra* note 7, at 99 (the cost burden of employing disabled persons will fall arbitrarily upon groups of employers, employees, shareholders and consumers, not upon the general public).

[27] *See id.*, at 1079.

[28] *See id.*, at 1080.

[29] *See* Note, *supra* note 9, at 1502.

[30] *See generally id.*

used as a basis for a constitutional claim of discrimination against disabled persons in public employment[31] and education. For example, the U.S. Supreme Court, in *Cleveland Board of Education v. La Fleur*,[32] found that the mandatory maternity leave for pregnant teachers established an irrebuttable presumption of the physical incompetence of these teachers — a presumption that was "neither necessarily [nor] universally true" and therefore violative of the due process clause.[33]

The court in *Mills v. Board of Education*[34] stated that the right to a publicly-supported education extended to physically handicapped as well as to mentally retarded children[35] and that denial of rights violated the due process clause of the Fifth Amendment.[36] A similar due process reasoning was apparent in *Gurmankin v. Costanzo*[37] where the third circuit, relying on *La Fleur*, held unconstitutional a school district policy barring blind persons from jobs as teachers of sighted students. By not allowing Gurmankin the opportunity to take a competency examination to demonstrate her competency as a teacher, the school district violated her due process rights. The denial of the opportunity, according to the court, subjected Gurmankin to an irrebuttable presumption that blindness renders her incompetent to teach sighted students.[38]

Use of the due process theory, however, is necessarily limited by the fact that some state or federal governmental action must be implicated if there is to be a due process claim.[39] A constitutional claim based on denial of equal protection is not likely to meet with much success because "disabled persons" has not been found to be a "suspect classification."[40] Accordingly, matters affecting disabled persons are subjected to the "rational basis" test rather than the more demanding "compelling state interest" standard.

The most promising remedy appears to be legislation and education in a comprehensive approach to the problem, backed up by litigation when necessary. Although the law may not be able to legislate morality and attitudes, legislation proscribing discriminatory behavior can provide the environment within which attitudes incubate.[41]

Federal legislation in the area of civil rights has been quite prominent during the last fifteen years. For example, Title VI of the Civil Rights Act of 1964 proscribes discrimination on the basis of "race, color, or national origin" in any activity or program receiving federal financial assistance.[42] Title VII is a comprehensive prohibition of private acts of employment discrimination based on "race, color, religion, sex or national origin."[43] The omission in the federal civil rights legislation is glaring — discrimination against disabled persons is not prohibited.

In 1973 Congress enacted legislation, the Rehabilitation Act of 1973,[44] that has been called "the greatest step taken by the federal government to guarantee equal employment to the disabled in both federal agencies and private employment,"[45] though its scope is considerably narrower than that of Title VII since it does not cover private employment. The most notable provisions are Section 503,[46] which requires private contractors doing business in excess of $2,500 with the federal government to take affirmative steps to employ the handicapped, and Section 504,[47] which prohibits exclusion from participation, denial of benefits, or discrimination against a qualified disabled person by any program or activity which receives federal financial assistance. Addressing the needs of disabled persons in federal employment, Section 501[48] established the Interagency Committee on Handicapped Employees to review "the hiring, placement, and advancement practices" of executive branch agencies with regard to handicapped persons. An affirmative action program from each department, agency, and instrumentality in the executive branch is also required.

Although the Rehabilitation Act of 1973 may very well be the "greatest step" taken by the federal government to assist disabled persons, it has inherent weaknesses in its enforcement mechanism. There is no specific provision in Sections 503 and 504 authorizing initiation of a civil action by the Office of Federal Contract Compliance Program[49] or establishing a private cause of action by the aggrieved individual.[50] The limited right of a complainant to a formal hearing[51] and the emphasis in

31 *See* Lang, *Employment Rights of the Handicapped*, 11 CLEARINGHOUSE REV. 703, 705 (Dec. 1977).

32 414 U.S. 632 (1974).

33 *Id.* at 645-646.

34 348 F. Supp. 866 (D.D.C. 1972).

35 *Id.* at 878.

36 *Id.* at 875.

37 556 F. 2d 184 (3d. Cir. 1977), *aff'g.* 411 F. Supp. 982 (E.D. Pa. 1976).

38 *Id.* at 187.

39 Note, *supra* note 9, at 1508 n. 53.

40 Weiss, *supra* note 8, at 468-470. *See also* Note, *supra* note 10, at 827.

41 Note, *supra* note 10, at 817.

42 42 U.S.C. §2000d (1976).

43 42 U.S.C. §2000e (1976).

44 29 U.S.C. §§701 *et seq.* (1976).

45 Weiss, *supra* note 8, at 466.

46 29 U.S.C. §793 (1976).

47 29 U.S.C. §794 (1976).

48 29 U.S.C. §791 (1976).

49 Note, *supra* note 10, at 834.

50 *But cf.* Lloyd v. Regional Transportation Authority, 548 F2d 1277 (7th Cir. 1977) (implied cause of action found to exist under §504 until promulgation of effective enforcement mechanisms). *See* Wright, *supra* note 7, at 89-97 for a discussion of private right of action under 29 U.S.C. §793.

51 *See* Note, *supra* note 10 at 834.

Section 503 regulations on informality in complaint resolution may delay vindication of the complainant's rights.[52] Moreover, federal legislation in general and the Rehabilitation Act in particular are commendable but inadequate attempts to protect the rights of the disabled person. That such legislation is confined in scope to federal employment practices and certain private businesses contracting with the federal government demonstrates the need for a more comprehensive legislative program to protect disabled people in all aspects of life.[53] Furthermore, it is unrealistic to expect discrimination to be eliminated solely from the federal level. Even if Title VII of the federal Civil Rights Act were amended to include disabled persons, experience shows that state and local involvement is needed as well to effectuate the policy of the law.[54]

Many states, in varying degrees, have moved to provide greater protection of disabled individuals' rights through legislation. Such legislation can be flexible and easily adapted to the unique needs of a state's citizenry and since state government is "closer" to the people, the opportunities for effective implementation and enforcement of the statutory protections are enhanced.

---

[52] *Id.*

[53] Proposals to amend Title VII to prohibit discrimination against disabled persons in private employment have not met with congressional approval, despite strong support for the amendments among the disabled population. Note, *supra* note 9, at 1502. One commentator, stating that Congress "has chosen not to amend Title VII to prohibit handicap discrimination," lists thirty-three amendments to the Civil Rights Act of 1964 that have been proposed but not enacted during the years of 1972 to 1977. Wright, *supra* note 7, at 65-66 n. 2. For a discussion of the advantages and potential problems of amending Title VII. to include the disabled *see* Weiss, *supra* note 8. *See also* Note, *supra* note 10.

[54] Of course, in any scheme designed to remedy discrimination and to provide equality of opportunity the most important element is not the apparatus chosen, but rather the commitment of the political leadership to make the scheme work. *See* Note, *supra* note 10, at 817.

REVIEW OF STATE LEGISLATION

Introduction

Since state legislation is a desirable approach, it is important to first consider what already exists in this area. State legislation prohibiting discrimination has now been enacted in various forms in forty states and the District of Columbia.[55] Fifteen of these states[56] have utilized a "comprehensive" approach covering at least the three basic areas of employment, housing and public accommodations. These statutes may also cover other substantive areas (see Extent of Coverage *infra*) and include specialized administrative and enforcement mechanisms, usually under a civil rights or human rights commission, though sometimes under a single commissioner (see Administrative and Enforcement Mechanism *infra)*. Where not previously included, these existing statutes have been amended by adding handicapped or disabled persons as a protected class.

Some states have adopted "limited coverage" statutes prohibiting discrimination in only one area, usually employment,[57] although they may also cover housing, public accommodations or insurance.[58] In addition, a state may have more than one limited coverage statute and thus prohibit discrimination in several different areas. These acts have frequently been amended to include handicapped or disabled persons. (See Chart I *infra.)* Administration and enforcement mechanisms vary considerably, though most have an administrative mechanism with some combination of enforcement powers. Some are administered by a human rights commission, but these statutes are frequently administered by an agency not specialized in antidiscrimination, e.g. a department of labor administering an employment statute. (See Administrative and Enforcement Mechanisms *infra.*)

Finally, other states have adopted so-called "White Cane Laws" or similar legislation covering handicapped or disabled persons specifically. These are brief, general statutes guaranteeing blind and otherwise physically disabled persons full and equal access to housing and public facilities.[59] The "White Cane Law" type of legislation does not list prohibited discriminatory practices. Rather, they simply provide that disabled persons are entitled to equal treatment in the specifically protected area. Statutory language is often imprecise and terms are not defined.[60] There are seldom administrative mechanisms provided. Enforcement mechanisms, where included, consist only of misdemeanor provisions with small monetary penalties. (See Administrative and Enforcement Mechanisms *infra.*)

This section will review the existing legislation, highlighting the various approaches, the classes of disabled persons covered, the extent of substantive coverage and the administrative and enforcement mechanisms provided and thus should serve as a useful guide for gauging the level of effective protection from discrimination afforded disabled persons in the various states.

Extent of Coverage

Existing state statutes cover a broad range of areas, including employment, housing, public accommodations, education, credit, insurance, public services, and public contracts, though some areas may appear in only a few states. Employment, housing and public accommodations are the only areas covered by a significant number of states.

A common feature of the comprehensive legislation and the better-developed, limited coverage statutes is a provision setting forth with some particularity the nature of unlawful discriminatory practices in employment, housing, and public accommodations. In the employment field, employers, employment agencies and labor organizations are often prohibited from: using applications

---

55 *See* CHART I: Persons Covered by Statutes.

56 Conn., Ind., Iowa, Me., Md., Mich., Minn., Mont., N.H., N.J., N.M., N.Y., Ohio, Ore., and Pa.

57 *See, e.g.,* HAWAII REV. STAT. §378-2 (1976); NEV. REV. STAT. §613.330 (1973); and R.I. GEN. LAWS §28-5-7 (1968 & Supp. 1977).

58 *See, e.g.,* MASS. GEN. LAWS ANN. Ch. 149 §24k, Ch. 272 §98 (West Supp. 1977-1978).

59 *See, e.g.,* LA. REV. STAT. ANN. §§46:1951 *et seq.* (West Supp. 1978); NEB. REV. STAT. §20-126 *et seq.* (1977); N.C. GEN. STAT. §168-1 *et seq.* (1976); S.C. CODE §§43-33-10 *et seq.* (1976).

60 An example of this criticism can be found in the South Dakota statute which entitles blind and otherwise physically disabled persons to "reasonably equal accommodations" of a place to which the general public is invited. S.D. COMPILED LAWS ANN. §20-13-23.1 (Supp. 1977). The meaning of "reasonably equal accommodations" is not given in the statute.

which seek to elicit information concerning a person's disability;[61] discriminating against a disabled individual in the hire, apprenticeship, tenure, conditions or privileges of employment;[62] and publishing any notice of employment which expresses a limitation or preference as to "handicap" of a prospective employee or member.[63] The more prevalent prohibited discriminatory practices in housing include: refusal to rent, sell, lease or transfer any real property to a person because of a disability;[64] discrimination in terms, conditions or privileges of the housing transaction;[65] and publication of a housing notice which indicates a limitation or preference as to a person with a disability.[66] Several states go further and prohibit discrimination by banks, credit unions, and other lending institutions against a disabled person seeking financial assistance for the purchase, construction or rehabilitation of any real property.[67] Public accommodations provisions prohibit an establishment serving the general public from refusing, withholding from or denying disabled persons full and equal enjoyment of the services, facilities, and privileges of the place of public accommodation.[68]

Exemptions from the legislative mandate against discrimination are included in nearly all statutes covering employment, housing and public accommodations. Several are commonly found in most statutes. In the area of employment, the most frequent type exempts those employers who hire below a specified number of employees. Other exemptions permit an employer to differentiate on the basis of a physical or mental disabliity when based on a "bona fide occupational qualification" needed for the normal operation of business.[69] The more common housing exemptions are for housing accommodations for two families living independently with the owner residing in one, rentals in a single-family dwelling if the lessor resides therein,[70] and certain sales by individual owners of single-family housing.[71] In addition, nearly all public accommodations statutes exclude private clubs from their provisions. Not infrequently the antidiscrimination statutes have a clause stating that these housing or public accommodations provisions do not require modification of the property or the exercise of a higher degree of care for a disabled person.[72] One state has in effect provided a general exemption for all prohibited discriminatory practices where "reasonable grounds" exist.[73]

Those states with a "comprehensive" approach are more likely than the other states to have extended protection beyond the three principal areas of employment, housing, and public accommodations. "Limited coverage" stautues have also been adopted covering these other areas. However, the number of states with such legislation is still quite small.

Provisions protecting disabled persons from discrimination in credit transactions have been enacted in at least seven states.[74] Generally, the statutes have made it unlawful for a creditor to grant, withhold, modify or fix the terms of an extension of credit solely on the basis of a physical or mental disability. A more detailed law, such as Ohio's,[75] precludes a creditor from inquiring as to a disability where such an inquiry will be used to limit or specify those persons to whom credit will

---

[61] *See, e.g.,* MINN. STAT. ANN. §363.03 [Subd. 1 (4)(a)] (West Supp. 1978); N.M. STAT. ANN. §4-33-7(D) (1974 & Supp. 1975); PA. STAT. ANN. tit. 43 §955(h)(6) (Purdon Supp. 1978-1979).

[62] *See, e.g.,* OHIO REV. CODE ANN. §4112.02(A) (Page Supp. 1977); PA. STAT. ANN. tit. 43 §955(a) (Purdon Supp. 1978-1979).

[63] *See, e.g.,* KAN. STAT. §44-1009(3) (Supp. 1975); N.H. REV. STAT. ANN. §354-A:8 (III) (Supp. 1977); OHIO REV. CODE ANN. §4112.02 (E)(4) (Page Supp. 1977).

[64] *See, e.g.,* MD. ANN. CODE art. 49B §20. (Supp. 1978); MONT. REV. CODES ANN. §64-306(4)(a) (Supp. 1977); ORE. REV. STAT. §659.430(1)(a) (1973).

[65] *See, e.g.,* MICH. COMP. LAWS ANN. §37.1502(b) (Supp. 1978-1979); N.M. STAT. ANN.§4-33-7(G)(2) (Supp. 1975); PA. STAT. ANN. tit. 43 §955(h)(3) (Purdon Supp. 1978-1979).

[66] *See, e.g.,* IOWA CODE ANN. §601A.8(3) (1975); MINN. STAT. ANN. §363.03[Subd. 2(2)(c)] (West. Supp. 1978); N.J. STAT. ANN. §10:5-12(f) (West 1976).

[67] *See, e.g.,* KY. REV. STAT. §207.190 (Supp. 1976); MICH. COMP. LAWS ANN. §37.1504 (Supp. 1978-1979); N.J. STAT. ANN. §10:5-12(i) (West 1976); N.M. STAT. ANN. §4-33-7(H) (Supp. 1975); PA. STAT. ANN. tit. 43 §955(h)(2) (Purdon Supp. 1978-1979).

[68] *See, e.g.,* LA. REV. STAT. ANN. §46:1952(B) (West Supp. 1978), MONT. REV. CODES ANN. §65-306(13) (Supp. 1977).

[69] *See, e.g.,* KAN. STAT. §44-1009(a)(3) (Supp. 1975); ORE. REV. STAT. 659.425(1)(1973); R.I. GEN. LAWS §28-5-7(D) (Supp. 1977).

[70] *See, e.g.,* CONN. GEN. STAT. ANN.§53-35 (West Supp. 1978); MICH. COMP. LAWS ANN. §37.1503 (Supp. 1978-1979); N.H. REV. STAT. ANN. §354-A:8(V-b) (Supp. 1977).

[71] In New Mexico a significant portion of the housing market is not covered by the anti-discrimination statute. Sales or rental of single-family housing is exempted if the owner has not engaged in proscribed advertising expressing a limitation or preference for specified minority group persons and he or she does not own more than three single-family dwellings. N.M. STAT. ANN. §4-33-8 (A) (1974 & Supp.1975). A similar exemption exists in New Hampshire with the additional requirement that the owner does not use the sale or rental serivce of a real estate broker in the transaction. N.H. REV. STAT. ANN. §354-A:8 (V-b)(a) (Supp. 1977).

[72] *See, e.g.,* IND. CODE ANN. §22-9-1-13(c) (Burns Supp. 1977) (employer not required to modify physical plant or administrative procedures to accommodate a handicapped person); LA. REV. STAT. ANN. §46:1952(E) (West Supp. 1978) (owner of public accommodations not required to modify property or provide higher degree of care for disabled person); NEB. REV. STAT. §20-131.03(1977) (person renting housing not required to modify property to accommodate the special needs of any lessee). *But see* MICH. COMP. LAWS ANN. §37.1207 (Supp. 1978-1979)(employer not exempted from obligation to accommodate employee or applicant for employment).

[73] Montana will exempt a person from complying with a provision if the commission for human rights finds that "reasonable grounds" exist for an exemption. MONT. REV. CODES ANN. §64-306.1 (Supp. 1977).

[74] IND. CODE ANN. §22-9-1-3(1) (Burns Supp. 1977); IOWA CODE ANN. §601A.9(1975); MONT. REV. CODES ANN. §§64-306(5) (Supp. 1977); N.J. STAT. ANN. §10:5-12(i) (West 1976); N.M. STAT. ANN. §4-33-7(H) (Supp. 1975); N.Y. EXEC. LAW §296-a(McKinney Supp. 1972 to 1977); OHIO REV. CODE ANN. §4112.021 (Page Supp. 1977).

[75] OHIO REV. CODE ANN. §4112.02.1 (Page Supp. 1977).

be granted. Credit worthiness must be based on established criteria used in the normal course of business.

Statutory prohibitions of discrimination in education are not widespread.[76] Where legislatures have acted in this area, educational institutions are not permitted to discriminate against an individual with a disability in the admission to and in the full utilization of benefits and service from the institution. Coverage of the statutes varies from both private and public institutions in Michigan[77] to just publicly-funded educational programs and schools in Oregon.[78]

A distinct minority of states protect disabled persons in insurance transactions.[79] In what may in part be a reflection of the field's complexity, the insurance statutes are conspicuously different in the types of disability protected and insurance covered. The provisions may extend to physical handicap,[80] or mental retardation,[81] or both.[82] Very narrowly, North Carolina included either physical or mental disability, but only of minor children of insured persons.[83] Coverage is sometimes confined to one or two types of insurance, such as auto and health insurance. Only Ohio[84] and Washington[85] prohibit discrimination in transactions involving any type of insurance.

Currently, there are several statutes which require that all public contracts to which the state or any of its political subdivisions is a party must contain a non-discrimination clause.[86] In Minnesota, a public contract shall not be awarded to any firm unless the business has received a certificate of compliance from the commissioner of human rights.[87] The Illinois legislation permits the state to avoid a contract with any firm that has violated the Fair Employment Practices Commission's rules prohibiting unlawful discrimination.[88] In addition, affirmative action programs are required of public contractors in New Jersey[89] and Illinois.[90] This requirement goes beyond remedying a discriminatory practice after it has occurred. Also, it gives the public contractor clear notice of his obligations and establishes a discernible standard for gauging compliance. The failure to include an affirmative action plan in a public contract renders the contract null and avoid under New Jersey law.[91]

Minnesota[92] and Montana[93] have enacted statutes prohibiting discrimination against disabled persons in access to, full utilization of, or benefit from any public service or program. And lastly, in the area of state licensing, Montana prohibits state and local agencies from granting, denying or revoking the license of a person on the grounds of physical or mental handicap.[94] Maryland makes it unlawful for any state-licensed person or firm to discriminate on the basis of a person's physical or mental handicap.[95] Upon a finding of unlawful discrimination by a state licensee, the licensing authority may take such action against that licensee as it deems "appropriate," according to the Pennsylvania statute.[96]

Protected Class

There is a great deal of variability among the class of disabled persons protected by existing statutes, though they fall into the physical and mentally disabled subcategories. However, legislative protection against discrimination does not apply to both the physically disabled and the mentally disabled in all 40 states with such statutes. (*See* Chart I: Persons Covered by Statutes.) Although the two classes are covered in half of these states, the other jurisdictions, with the exception of Arizona, have restricted protection to the physically disabled.[97] A summary of the coverage in the major areas follows.

76 *See* IND. CODE ANN. §22-9-1-3(1) (Burns Supp. 1977); MICH. COMP. LAWS ANN. §37.1402 (Supp. 1978-1979); MINN. STAT. ANN. §363.03 (Subd. 5) (West Supp. 1978); MONT. REV. CODES ANN. §64-306(7) (Supp. 1977); ORE. REV. STAT. §659.150(2) (1973) WIS. STAT. ANN. §101.223 (West Supp. 1978-1979).

77 MICH. COMP. LAWS ANN. §§37.1401-37.1402 (Supp. 1978-1979).

78 ORE. REV. STAT.§659.150 (2) (1973).

79 *See, e.g.*, FLA. STAT. ANN. §§627.644, 627.6576 (West Supp. 1978); MASS. GEN. LAWS ANN. Ch.175 -§§120A (West Supp. 1977-1978); MINN. STAT. ANN. §72A.20 (West Supp. 1978); N.C. GEN. STAT. §58-251.5(1976); OHIO REV. CODE ANN. §3999.16 (Page Supp. 1977); ORE. REV. STAT. §746.225 (1973); R.I. GEN. LAWS §42-62-14 (Supp. 1977); WASH. REV. CODE ANN. §48.30.300 (Supp. 1977).

80 *See, e.g.*, ILL. ANN. STAT. Ch. 73 §1031(2) (Smith-Hurd 1974 & Supp. 1978).

81 *See, e.g.*, MASS. GEN. LAWS ANN. Ch. 175 §120A (West Supp. 1977-1978).

82 *See, e.g.*, FLA. STAT. ANN. §§627.644, 627.6576 (West Supp. 1978).

83 N.C. GEN. STAT. §58-251.5 (1975)

84 OHIO REV. CODE ANN. §3999.16 (Page Supp.1977).

85 WASH. REV. CODE ANN. §48.30.300 (Supp. 1977).

86 *See, e.g.*, IND. CODE ANN. §22-9-1-10 (Burns 1974 & Supp. 1977); MONT. REV. CODES ANN. §64-319 (Supp.1977).

87 MINN. STAT. ANN. §363.073 (West Supp. 1978).

88 ILL. ANN. STAT. Ch. 48 §854 A (Smith-Hurd Supp.1978).

89 N.J. STAT. ANN. §10:5-32 (West 1976).

90 *See supra* note 85.

91 N.J. STAT. ANN. §10:5-35 (West 1976).

92 MINN. STAT. ANN. §363.03 (Subd. 4) (West Supp. 1978).

93 MONT. REV. CODES ANN. §64-306(6)(a) (Supp. 1977).

94 MD. ANN. CODE art. 49B §11c (Supp. 1978).

95 MONT. REV. CODES ANN. §64-321 (Supp. 1977).

96 PA. STAT. ANN. tit. 43 §959 (Purdon Supp.1978-1979).

97 *See* CHART 1: Persons Covered By Statutes (The District of Columbia housing statute covers the physically and the mentally disabled but the employment provision is limited to the physically disabled).

STATUTORY COVERAGE BY AREA

Class	Employment	Housing	Public Accommodations
Physically Disabled	15	9	12
Mentally Disabled	1	2	1
Both	20	17	16
Total # of States	36	28	29

Only those states which cover both physically disabled and mentally disabled persons will cover all developmentally disabled persons in some form in one or more areas. Thus, in some states only persons with cerebral palsy, for example, will be covered while another state will only cover persons with mental retardation or autism. Only 40% of the states currently protect in some form all developmentally disabled persons from employment discrimination. The figures for housing and public accommodations are even lower. Where given protection, it is not always equal. Some states having extensive prohibitions of discrimination against blacks, women and religious and ethnic minorities have given lesser protection to disabled persons.[98]

While most classes of persons protected by antidiscrimination statutes are relatively easy to determine, e.g. race, sex, age or marital status, the class of disabled persons raises definitional questions which relate to the scope of coverage of the class and may raise problems in determining which individuals fall within the protected class.

The types of definitions of the class of disabled or handicapped persons tend to fall into three categories. Some state statutes contain no specific definition of the class, but merely use the terms "mental or physical handicap" throughout the statute.[99] Other states have defined "handicap" or "disability" in various ways, ranging from the circular[100] to the all-encompassing.[101]

Examples of circular definitions include the Hawaii statute which defines "physical handicap" as a substantial physical impairment . . . "[102] and the Iowa statute which defines "disability" to mean a physical or mental condition of a person which constitutes a "substantial handicap."[103] The Iowa definition has some rationality behind it since the word "handicap" is more specific than the word "disability". A disability may or may not be handicapping. However, it seems to add little to the meaning of the definition. Those states which fail to adequately define the protected class may create notice problems in terms of an employer's or proprietor's obligation under the statute.[104]

Those states with all-encompassing definitions are more specific in their definition of the disabled class. These more expansive definitions describe "handicap" as a physical disability caused by bodily injury, birth defect or illness, including epilepsy, which includes but is not limited to paralysis, amputation, lack of physical coordination, blindness, visual, hearing or speech impediment or physical reliance on a seeing eye dog, wheel chair, or other remedial device.[105] A potential problem with this type of definition is that by trying to be all-inclusive it may actually preclude coverage of a disability not specifically included by the legislature.

Some states have gone beyond basic definitions of disability or handicap, apparently with the intent to limit the protected class to persons with more than superficial disabilities. These states require that additional conditions be met, such as the handicap "is expected to continue for a considerable length of time" and "can reasonably be expected to limit the person's functional ability . . ."[106]

[98] Washington, for example, prohibits discrimination on the basis of disability only in the areas of employment and insurance but gives additional protection to other minorities in housing, public accommodations, and credit transactions. WASH. REV. CODE ANN. §§49.60.176 (Supp. 1977) (prohibits discrimination based on race, creed, color, national origin, sex or marital status in credit transactions); §49.60.215 (1962) (prohibits discrimination based on race, creed, color, or national origin in public accommodations); §49.60.222 (Supp. 1977) (prohibits discrimination based on sex, marital status, race, creed, color or national origin in real estate transactions).

[99] *See, e.g.*, WASH. REV. CODE ANN.§49.60.040 (Supp.1977).

[100] *See, e.g.*, IND. CODE ANN.§22-9-1-3(q) (Burns Supp. 1977).

[101] *See, e.g.*, MD. ANN. CODE art. 49B §19(b)(u) (Supp. 1978); MONT. REV. CODES ANN. §64-305(13) (Supp.1977); N.J. STAT. ANN. §10:5-5(q) (West 1976); R.I. GEN. LAWS §28-5-6 (H) (Supp.1977).

[102] HAWAII REV. STAT. §378-1 (1976).

[103] IOWA CODE ANN. §601A.2 (1975).

[104] Note, *Potluck Protections for Handicapped Discriminatees: The Need to Amend Title VII to Prohibit Discrimination on the Basis of Disability*, 8 LOY. CHI. L.J. 814, 838 (1977). *But See* Chicago, Milwaukee, St. Paul & Poc. R.R. Co. v. Wash. State Human Rights Comm'n., 14 Fair Empl. Proc. Case 130, 132 (1976). After vacating a prior ruling that the anti-discrimination statute was void because it lacked a definition of the term "handicapped" the Washington Supreme Court adopted the definition found in Webster's Third New International Dictionary.

[105] *See supra* note 101.

[106] *See, e.g.*, OHIO REV. CODE ANN. §4112.01 (M) (Page Supp. 1977).

In defining "disability" or "handicap" several state legislatures have added a special qualifier applicable only to the class of disabled or handicapped persons. They qualify "handicap" as one being unrelated to a person's ability to perform the duties of a particular job or as one being unrelated to a person's ability to acquire or maintain real property.[107] They may also describe "handicap" as a determinable physical or mental characteristic which is "unrelated to the individual's ability to utilize and benefit" from public accommodations and educational opportunities.[108] These definitional qualifications, like the "bona fide occupational qualification" exemptions to the statutory prohibitions, among others, serve to narrow the class of disabled persons protected by these laws. However, unlike the other qualifiers and exemptions, this one places an additional condition only upon the disabled class.[109]

Administrative and Enforcement Mechanisms

One of the more important features of a statutory scheme to vindicate the rights of a minority class is the enforcement mechanism. Statements of policy and laws against discrimination are at most symbolism if there are no provisions for an administrative procedure and enforcement remedies through which disabled persons may vindicate their rights.

Some state statutes have no administrative mechanism and rely solely upon criminal penalties to enforce their provisions. This is particularly true of the "White Cane Laws." (*See* Approaches to Prohibiting Discrimination *supra*.) Criminal penalties also supplement administrative mechanisms in some statutes as noted *infra*. (*See* Chart II: Enforcement Procedure.)

Quite a large number of state statutes have administrative enforcement procedures of some sort. In the employment area nearly 70% of the state statutes prohibiting discrimination have complete administrative enforcement procedures, including authority to order compliance. In the areas of housing and public accommodations the percentage drops to approximately 55% of the 29 state statutes in each of those areas. (*See* Chart II: Enforcement Procedures for individual state provisions.) A summary table follows.

ADMINISTRATIVE PROCEDURE BY AREA

Administrative Procedure	Employment	Housing	Public Accomodations
Investigation	29	20	18
Administrative Hearing	26	17	17
Ordering Compliance	25	16	16
Total # of States	36	29	29

107 *See, e.g.,* ORE. REV. STAT. §659.030 (1)(a) (1973); IND. CODE ANN. §22-9-1-3 (Burns Supp. 1977).

108 MICH. COMP. LAWS ANN. §37.1103 (Supp.1978-1979).

109 Indiana seems to carry this narrowing trend to an extreme by providing that before a person can bring a valid complaint of employment discrimination that individual first must be "properly certified as a handicapped person." Thus a disabled individual must be classified and officially certified as a member of the protected class before he or she can claim any protection under the statute. This is not required of persons in any of the other protected classes. IND. CODE ANN. §22-9-1-13(c) (Burns Supp.1977).

The most prevalent form of administrative agency is a commission with a Civil Rights, Human Relations or Human Rights designation. However, some agencies are administered by individuals rather than commissions. In Minnesota[110] and New York[111] the responsible agencies are headed by a Commissioner of Human Rights. New Jersey's law against discrimination is administered by the Attorney General through the Division of Civil Rights within the state's Department of Law and Public Safety.[112] This form of administrative agency is most often associated with "comprehensive statutes," but may also be included in "limited coverage" statutes. (*See* Approaches to Prohibiting Discrimination *supra*.) Those statutes which limit their scope to a particular area such as employment or housing frequently do not have administrative enforcement procedures but where provided may also include a state department of labor[113] as the designated agency in the case of employment discrimination.

Uniquely, the Oregon legislation delegates to the Commissioner of the Bureau of Labor responsibility for enforcement of discrimination prohibitions in the areas of employment, housing, public accommodations, and educational programs.[114] This statutory scheme is an example of an administrative enforcement mechanism in which responsibility is given to an agency whose primary function is other than the vindication of the civil rights of disabled persons and other minorities.

Civil rights or human rights commissions or commissioners bring specialization and expertise to bear on the problem of discrimination. They are generally given the powers to receive and investigate complaints of discrimination, eliminate unlawful discriminatory practices through informal means of conference, conciliation, and persuasion, and, failing this, to institute a formal hearing.[115] (For individual state provisions see Chart II: Enforcement Provisions.) In addition, delegated duties include promulgation of regulations for carrying out the prohibitions and development of educational programs and reports. These duties emphasize the commission's or commissioner's specialization role.

Upon receiving a complaint the common practice is to investigate to determine if probable cause exists for crediting the allegations of unlawful discrimination. If there is probable cause, the commission will attempt resolution through informal means. Should these informal efforts fail, a formal hearing is held in which both parties are called upon to present evidence. Obviously, upon a finding that no discrimination has occurred the complaint is dismissed. Where the respondent has been found to have engaged in unlawful discrimination, 27 states empower the commission to order the respondent to cease and desist from the practice and to take such affirmative action as "will effectuate the purposes" of the statute.[116] (*See* Chart II: Enforcement Procedure.) Some states[117] have further developed "affirmative action" to specifically include recovery of back pay. Maine and Vermont are the only states having administrative mechanisms which are not empowered to order compliance.[118] Judicial review of or the right of appeal from a commission order is a standard provision of the statutes, with the exceptions of the Hawaii, Kentucky, Montana, Nevada, and Oregon legislation. (*See* Chart II: Enforcement Prodedure.)

Twenty states expressly provide for penalties upon a commission or court finding that the respondent has discriminated against a disabled person. (*See* Chart II: Enforcement Procedure.) The penalties, usually of the misdemeanor class, are minor, from a minimum of a $25 fine[119] to maximums of a $1,000 fine [120] and 90 days imprisonment.[121] Criminal penalties of this nature have been termed ineffective — a greater need being for "positive civil remedies" such as injunctive relief or damages.[122]

There are certain statutes which allow the commission to bring suit for a temporary injunction where it appears likely that the complainant may suffer irreparable injury as a result of the alleged discrimination.[123] At the other end of the administrative process, the New York law,[124] for example, empowers its Division of Human Rights to seek injunctive relief from the courts where the respon-

110 MINN. STAT. ANN. §363.04 (West Supp. 1978).

111 N.Y. EXEC. LAW §293 (McKinney 1972).

112 N.J. STAT. ANN. §10.5-7 (West 1976) (state attorney general and a commission compose the Division of Civil Rights).

113 *See, e.g.,* KY. REV. STAT. §207.200 (Supp.1976).

114 ORE. REV. STAT. §§659.040-659.045 (1973).

115 *See, e.g.,* PA. STAT. ANN. tit. 43 §959 (Purdon Supp. 1978-1979).

116 Note that the number here is greater than that given in the summary table *supra* since some states provide for such powers in only one area. This total figure draws from all three areas listed.

117 See, e.g., N.J., N.Y., Ohio, Pa., and R.I.

118 However, commissions may seek judicial action to remedy discriminatory practices.

119 CONN. GEN. STAT. ANN. §53-35(d) (West Supp. 1978) (penalty for discrimination in housing and public accommodations).

120 ALASKA STAT. §18.06.040 (1974).

121 D.C. CODE ANN. §6-1506 (1973).

122 Note, *supra* note 10, at 837. *See also* Weiss, *Equal Employment and the Disabled: A Proposal,* 10 COLUM. J. LAW & SOC. PROB. 457, 465 (1974) and Bonfield, *State Civil Rights Statutes: Some Proposals,* 49 IOWA L. REV. 1067, 1114-1119(1964). *But cf.* Wright, *Equal Treatment of the Handicapped by Federal Contractors,* 26 EMORY L.J. 65, 89-97(1977) (no private right of action should be implied under 29 U.S.C. §793).

123 *See, e.g.,* IOWA CODE ANN. §601A.5(4) (1975); ME. REV. STAT. tit. 5 §4612 (4) (Supp. 1978).

124 N.Y. EXEC. LAW §297(6) (McKinney Supp. 1972 to 1977).

dent is engaging in an act that renders ineffectual an order of the Commissioner.

One of the more controversial remedies for discrimination is a private cause of action, as opposed to administrative enforcement procedures. The advantages of a private right of action have been cited as having a greater deterrent effect than criminal penalties and complementing the administrative remedy so as to allow the aggrieved individual the opportunity to seek pecuniary redress where it is appropriate.[125] Kentucky, Maine, Michigan and New York give a person who has been discriminated against a cause of action for injunctive relief and damages. Arizona, California, Montana, Oregon, and Texas provide a private right of action only for recovery of damages, with the Arizona, California and Oregon statutes specifically entitling the individual to recover punitive or exemplary damages. (*See* Chart II: Enforcement Procedures.) In New Jersey standing in the courts to institute actions to enforce the antidiscrimination law is given not only to the person who has been discriminated against but also to "any organization which represents or acts to further the interests of individuals who have been discriminated against . . ."[126] This provision is an important recognition of the need for flexibility of enforcement remedies for the vindication of disabled persons' rights.

125 *See* Bonfield, *supra* note 122, at 1113.

126 N.J. STAT. ANN. 10:5-38 (West 1976).

STATE STATUTE CHARTS
ANTI-DISCRIMINATION STATUTES
Chart I: Persons Covered By Statutes

STATE	EMPLOYMENT		HOUSING		PUBLIC ACCOMMODATIONS	
	Physically Disabled	Mentally Disabled	Physically Disabled	Mentally Disabled	Physically Disabled[1]	Mentally Disabled
ALABAMA						
ALASKA Stat. (1974)	§18.80.220(a)(1)				§18.06.020	
ARIZONA Rev. Stat. Ann. (1974)		§36-506(B) (1)		§36-506(B) (3)		
ARKANSAS Stat. Ann. (1976)				§82-2905		§82-2902
CALIFORNIA	Cal. Lab. Code §1420 (West Supp. 1978)		Cal. Civ. Code §54.1(b) (West Supp. 1978)		Cal. Civ. Code §54.1(a) (West Supp. 1978)	
COLORADO Rev. Stat.(1974)	§24-34-801[2]				§24-34-801	
CONNECTICUT Gen. Stat. Ann. (West Supp. 1978)	§31-126		§53-35		§53-35	
DELAWARE						
DISTRICT OF COLUMBIA Code Ann.(1973)	§6-1504		§6-1505	§§6-1501, 6-1502		
FLORIDA Stat. Ann. (West Supp. 1978)	§413.08(3)		§413.08(4)[3]		§413.08(1)[3]	

STATE	EMPLOYMENT		HOUSING		PUBLIC ACCOMMODATIONS	
	Physically Disabled	Mentally Disabled	Physically Disabled	Mentally Disabled	Physically Disabled[1]	Mentally Disabled
GEORGIA						
HAWAII Rev. Stat. (1976)	§378-2					
IDAHO						
ILLINOIS Ann. Stat. (Smith-Hurd)	Ch. 48, §853 (1966 & Supp. 1978)		Ill. Const. of 1970, Art. 1, §19[4]		Ch. 38, §13-2 (1972 & Supp. 1978)	
INDIANA Code Ann. (Burns 1974 & Supp. 1977)	§§22-9-1-2, 22-9-1-3 (L)		§16-7-5.5-2(1973 & Supp. 1976), §22-9-1-2, §22-9-1-3(L)		§22-9-1-2, §22-9-1-3-(L)	
IOWA Code Ann. (1975)	§601A.6		§601A.8		§601A.7	
KANSAS Stat. (1973 & Supp. 1975)	§§44-1009(a), (b)		[5]		§44-1009(c)	
KENTUCKY Rev. Stat. Ann. (Supp. 1978)	§§207.150 - 207.170[6]		§207.180			
LOUISIANA Rev. Stat. Ann., (West Supp. 1978)			§46:1953		§46:1952	
MAINE Rev. Stat. (Supp. 1978)	Tit. 5 §4572		Tit.5 §4582		Tit.5 §4592	
MARYLAND Ann Code (1972 & Supp. 1978)	Art. 49B §19		Art. 49B §22		Art. 49B §11	

STATE	EMPLOYMENT Physically Disabled	EMPLOYMENT Mentally Disabled	HOUSING Physically Disabled	HOUSING Mentally Disabled	PUBLIC ACCOMMODATIONS Physically Disabled[1]	PUBLIC ACCOMMODATIONS Mentally Disabled
MASSACHUSETTS Gen. Laws Ann. (West Supp. 1978)	Ch. 149 §24K				Ch. 272 §98[7]	
MICHIGAN Comp. Laws Ann. (Supp. 1978-1979)	§§37.1202 - 37.1206[8]		§37.1502		§37.1302	
MINNESOTA Stat. Ann. (West Supp. 1978)	§363.03 (Subd. 1)		§§363.03 (Subd. 2), 256c.025	§363.03 (Subd. 2)	§§363.03 (Subd. 3), 256c.02	§363.03 (Subd. 3)
MISSISSIPPI Code Ann. (Supp. 1978)					§§43-6-3, 43-6-5	
MISSOURI						
MONTANA Rev. Codes Ann. (1970 & Supp. 1977)	§§64-306(1), 64-304		§64-306(4)		§64-306(3)	
NEBRASKA Rev. Stat. (1974)	§§48-1104 - 48-1107		§20-131.01 (1977)		§20-127, §20-129 (1977)	
NEVADA Rev. Stat. (1973)	§613.330					
NEW HAMPSHIRE Rev. Stat. Ann. (1966 & Supp. 1977)	§354-A:8 (I,II,III)		§354-A:8(V)		§354-A:8(IV)	
NEW JERSEY Stat. Ann. (West 1976 & Supp. 1978)	§§10:5-4.1, 10:5-12		§§10:5-4.1, 10:5-12		§§10:5-4.1, 10:5-12	
NEW MEXICO Stat. Ann. (1974 & Supp. 1975)	§§4-33-7(A-E)		§4-33-7(G)		§4-33-7(F)	

STATE	EMPLOYMENT		HOUSING		PUBLIC ACCOMMODATIONS	
	Physically Disabled	Mentally Disabled	Physically Disabled	Mentally Disabled	Physically Disabled[1]	Mentally Disabled
NEW YORK Exec. Law (McKinney Supp. 1972 to 1977)	§296 (1, 1a)		§296 (2a,5)		§296(2)	
NORTH CAROLINA Gen. Stat. (1976)	§168-6		§168-9		§§168-2,3	
NORTH DAKOTA						
OHIO Rev. Code Ann. (Page 1973 & Supp. 1977)	§§4112.02(A-F)		§4112.02(H)		§4112.02(G)	
OKLAHOMA						
OREGON Rev. Stat. (1973)	§659.425(1)		§659.430		§659.425(2)	
PENNSYLVANIA Stat. Ann. (Purdon 1964 & Supp. 1978 - 1979)	Tit. 43 §955		Tit. 43 §955(h)		Tit. 43 §955(i)	
RHODE ISLAND Gen Laws (1968 & Supp. 1977)	§28-5-7					
SOUTH CAROLINA Code (1976)			§43-33-70		§43-33-20	
SOUTH DAKOTA Compiled Laws Ann. (Supp. 1977)					§20-13-23.1[9]	
TENNESSEE Code Ann. (Supp. 1977)	§8-4131					
TEXAS Rev. City Stat. Ann. (Vernon 1976)	Art. 4419e(3) (f)		Art. 4419e(3) (h) (Supp. 1978)		Art. 4419e(3)	

STATE	EMPLOYMENT Physically Disabled	EMPLOYMENT Mentally Disabled	HOUSING Physically Disabled	HOUSING Mentally Disabled	PUBLIC ACCOMMODATIONS Physically Disabled[1]	PUBLIC ACCOMMODATIONS Mentally Disabled
UTAH						
VERMONT Stat. Ann. (Supp. 1978)	Tit. 21 §498					
VIRGINIA Code (1976)	§40.1-28.7					
WASHINGTON Rev. Code Ann. (Supp. 1977)	§§49.60.180 - .200					
WEST VIRGINIA Code (Supp. 1978)	§§5-11-9 (a-e)[10]		§5-11-9(g)[10]		§5-11-9(f)[10]	
WISCONSIN Stat. Ann. (West)	§111.32 (1973 & Supp. 1978-1979)		§101.22 (1973 & Supp. 1978-1979)			
WYOMING						
AMERICAN SAMOA						
GUAM						
PUERTO RICO						
VIRGIN ISLANDS						

[1] See also the provisions cited in the survey of envirommental barriers, Chart I, Column I, entitled Provide a Right of Equal Access to Public Facilities in ABA, Commission on the Mentally Disabled, *Environmental Barriers: Statutory Survey and Model Statute.*
[2] Covers public employment only.
[3] Limited to deaf, blind and visually handicapped persons.
[4] Municipalities may enact ordinances defining unfair housing practices, establishing administrative and enforcement mechanisms for the prohibition of discrimination based on, *inter alia,* physical or mental handicap in the sale, rental or financing of real property. ILL.II ANN. STAT. Ch. 11 §11.1-1 (Smith-Hurd Supp. 1978).
[5] *Unlike the anti-discrimination sections on employment and public accommodations, the housing provision has not been amended to include* "physical handicap" as a prohibited basis for discrimination. See KAN. STAT. §§44-1016, 44-1017 (1973)
[6] Cities and counties may adopt and enforce measures prohibiting unfair treatment of individuals on the basis of physical handicap and may prescribe penalities therefor. KY. REV. STAT. §207.220 (Supp. 1976).
[7] Limited to deaf and blind persons.
[8] The employer is *not* exempted from the obligation to accommodate an employee or applicant with a handicap unless such accommodation would cause undue hardship on the business. MICH. COMP. LAWS ANN. §37.1207 (Supp. 1978-1979) With Commission approval, an employer may adopt a plan to eliminate the present effects of past discriminatory practices or assure equal opportunity with respect to "individuals who *have handicaps." MICH. COMP. LAWS ANN. §37.1208 (Supp. 1978-1979).*
[9] According to S.D. COMPILED LAWS ANN. §20-13-23 (Supp. 1977) the blind, visually handicapped and otherwise physically disabled are entitled to "reasonably equal accommodations," advantages, facilities and privileges of place to which the general public is invited.
[10] Limited to blind persons.

Chart II: Enforcement Procedure

STATE	TYPE OF DISCRIMINATION	INVESTIGATION	ADMINISTRATIVE HEARING	AUTHORITY TO ORDER COMPLIANCE	APPEAL	PENALTY
ALABAMA						
ALASKA Stat. (1974)	**Employment**	§18.80.110	§18.80.120	§18.80.130	§18.80.135	Misdemeanor §18.80.270
	Public Accommodations					Misdemeanor §18.06.040
ARIZONA Rev. Stat. Ann. (1974)	**Employment Housing**					§36-516(B)[1]
ARKANSAS Stat. Ann. (1976)	**Housing Public Accommodations**					Misdemeanor §82-2906
CALIFORNIA Lab. Code (West Supp. 1978)	**Employment**	§§1418, 1422	§§1422.2, 1424	§1426	§1427, Cal. Govt. Code §11523	
	Housing Public Accommodations					[2]
COLORADO Rev. Stat. (1974)	**Public Employment Public Accommodations**					Misdemeanor §24-34-802 $100 fine, 60 days or both.
CONNECTICUT Gen. Stat. Ann. (West Supp. 1978)	**Employment**	§31-127	§31-127	§31-127	§31-128	
	Housing Public Accommodations	§53-36	§53-36	§53-36	§53-36b	$25 - $100 fine and/or up to 30 days imprisonment. §§53-35(d), 53-34
DELAWARE						

STATE	TYPE OF DISCRIMINATION	INVESTIGATION	ADMINISTRATIVE HEARING	AUTHORITY TO ORDER COMPLIANCE	APPEAL	PENALTY
DISTRICT OF COLUMBIA Code Ann. (1973)	**Employment Housing Public Accommodations**					$300 fine and/or maximum 90 days imprisonment §6-1506
FLORIDA Stat. Ann. (West Supp. 1978)	**Employment Housing**					2nd Degree Misdemeanor §413.08(5)
GEORGIA						
HAWAII Rev. Stat. (1976 & Supp. 1977)	**Employment**	§378-3(a)		§§378-3(b), 3(c), 3(d), 3(f)		
IDAHO						
ILLINOIS Ann. Stat. (Smith-Hurd 1966 & Supp. 1978)	**Employment**	Ch. 48 §858	Ch. 48 §858.01	Ch. 48 §858.01	§§858.02, 860	
	Housing					
	Public Accommodations					Ch. 38 §13-3 (1972 & Supp. 1978) Misdemeanor
INDIANA Code Ann. (Burns 1974 & Supp. 1977)	**Employment Housing Public Accommodations**	§22-9-1-11[3]	§22-9-1-6(j) (1)	§22-9-1-6(k) (1)	§22-9-1-6(k) (2)	
IOWA Code Ann. (1975)	**Employment Housing Public Accommodations**	§601A.5	§§601A.5, 601A.14	§§601A.14(12), 601A.15(2)	§601 A.15 (1)	§601D.7 Misdemeanor
KANSAS Stat. (1973 & Supp. 1975)	**Employment Public Accommodations**	§44-1005	§44-1005	§44-1005	§§44-1010, 44-1011	

STATE	TYPE OF DISCRIMINATION	INVESTIGATION	ADMINISTRATIVE HEARING	AUTHORITY TO ORDER COMPLIANCE	APPEAL	PENALTY
KENTUCKY Rev. Stat. (Supp. 1976)	**Employment**	§207.200[4]	§207.210(1)	§207.210(3)		
	Housing					[5]
LOUISIANA	**Housing Public Accommodations**					
MAINE Rev. Stat. (Supp. 1978)	**Employment Housing Public Accommodations**	Tit. 5 §4612(1)	Tit.5 §4612(3) Tit.5 §4612(4)[6]			[7]
MARYLAND Ann. Code (1972 & Supp. 1978)	**Employment Housing Public Accommodations**	Art. 49B, §13	Art. 49B, §14	Art. 49B, §14(e), Art. 49B, §15		
MASSACHUSETTS Gen. Laws Ann. (West Supp. 1978)	**Employment**					$25-$200 fine Ch. 149 §24k
MICHIGAN Comp. Laws Ann. (Supp. 1978-1979)	**Employment Housing Public Accommodations**	§37.2602	§37.2602	§37.2605	§37.2606	[8]
MINNESOTA Stat. Ann. (West Supp. 1978)	**Employment Housing Public Accommodations**	§363.05 (Subd. 1) (9)	§363.05 (Subd. 1) (18)	§§363.06 (Subd. 4) (3), 363.091	§§363.06 (Subd. 4) (1), 363.072	Misdemeanor §363.101
MISSISSIPPI Code Ann. (Supp. 1978)	**Public Accommodations**					§46-6-11 Misdemeanor
MISSOURI						
MONTANA Rev. Codes Ann. (1970 & Supp. 1977)	**Employment Housing Public Accommodations**	§64-308(4)	§§64-308(2), (5)	§64-310		§64-304 Misdemeanor[9] §64-312(3)

STATE	TYPE OF DISCRIMINATION	INVESTIGATION	ADMINISTRATIVE HEARING	AUTHORITY TO ORDER COMPLIANCE	APPEAL	PENALTY
NEBRASKA Rev. Stat. (1974)	**Employment**	§§48-1117(1), 48-1118	§§48-1117(2), 48-1119	§48-1119(3)	§48-1120	
	Housing **Public Accommodations**					Misdemeanor §20-129 (1977)
NEVADA Rev. Stat.	**Employment**	§613.405 (1975) §613.415 (1969)	§613.415 (1969)	§613.415 (:1969)		
NEW HAMPSHIRE Rev. Stat. Ann. (1966)	**Employment** **Housing** **Public Accommodations**	§§354-A:7(VI), 354-A:9(II)	§§354-A:7(VIII), 354-A:9(II)	§§354-A:9(II), 354-A:10	§354-A:10	§354-A:12 (Supp. 1977) Misdemeanor
NEW JERSEY Stat. Ann. (West 1976 & Supp. 1978)	**Employment** **Housing** **Public Accommodations**	§10:5-14	§10:5-16	§§10:5-14.1, 10:5-17, 10:5-19	§10:5-21	10
NEW MEXICO Stat. Ann. (1974)	**Employment** **Housing** **Public Accommodations**	§4-33-4(B)	§4-33-10 (Supp. 1975)	§§4-33-4(B), (D)	§4-33-12 (Supp. 1975)	§4-33-10(E) (Supp. 1975)
NEW YORK Exec. Law (McKinney Supp. 1972 to 1977)	**Employment** **Housing** **Public Accommodations**	§297(2)	§297(4)	§§297(4.c), 297(6)	§§297-a, 298	11
NORTH CAROLINA	**Employment** **Housing** **Public Accommodations**					
NORTH DAKOTA						
OHIO Rev. Code Ann. (Page Supp. 1977)	**Employment** **Housing** **Public Accommodations**	§4112.05(B)	§4112.05(B)	§4112.05(G)	§4112.06 (A) (1973)	§4112.99 Misdemeanor
OKLAHOMA						
OREGON Rev. Stat. (1973)	**Employment** **Housing** **Public Accommodations**	§659.050	§659.060	§§659.060(3), 659.070		12

STATE	TYPE OF DISCRIMINATION	INVESTIGATION	ADMINISTRATIVE HEARING	AUTHORITY TO ORDER COMPLIANCE	APPEAL	PENALTY
PENNSYLVANIA Stat. Ann. (Purdon 1964 & Supp. 1978-1979)	**Employment Housing Public Accommodations**	Tit. 43 §§957(f), 959	Tit. 43 §§957(g), 959	Tit. 43, §§959, 960	Tit. 43 §960	
RHODE ISLAND Gen Laws (1968)	**Employment**	§28-5-13	§28-5-18	§28-5-24	§28-5-28	
SOUTH CAROLINA Code (1976)	**Housing**					
	Public Accommodations					Misdemeanor §43-33-40
SOUTH DAKOTA Compiled Laws Ann. (Supp. 1977)	**Public Accommodations**	§20-13-28	§20-13-35	§§20-13-42, 20-13-49	§20-13-47	Misdemeanor §20-13-23.3
TENNESSEE Code Ann. (1973 & Supp. 1977)	**Employment**					Misdemeanor §8-4131
TEXAS Rev. Civ. Stat. Ann. (Vernon 1976)	**Employment Housing Public Accommodations**					Misdemeanor $100-$300 fine Art. 4419e (6)[13]
UTAH						
VERMONT Stat. Ann. (Supp. 1978)	**Employment**	Tit. 21 §495b	Tit. 21 §495b (Judicial Hearing)			
VIRGINIA Code (1976)	**Employment**					
WASHINGTON Rev. Code Ann. (Supp. 1977)	**Employment**	§§49.60.120, 49.60.240	§49.60.140	§49.60.250, §49.60.260	§49.60.270 (1962)	[14]

STATE	TYPE OF DISCRIMINATION	INVESTIGATION	ADMINISTRATIVE HEARING	AUTHORITY TO ORDER COMPLIANCE	APPEAL	PENALTY
WEST VIRGINIA Code (Supp. 1978)	**Employment Housing**	§5-11-10	§5-11-10	§§5-11-10, 5-11-11	§29A-5-4	
WISCONSIN (West Supp. 1978-1979)	**Employment**	§111.36(1)	§111.36(3) (1974)	§111.36(3) (b)	§111.37	
	Housing	§101.22(4) (a)	§§101.22(4) (b), (4) (c) (1973)	§101.22(4) (c)	§101.22(5)	
WYOMING						
AMERICAN SAMOA						
GUAM						
PUERTO RICO						
VIRGIN ISLANDS						

[1] Violation of person's right not to be discriminated against in employment and housing gives that person a cause of action for "greater of either $1000 or 3 times the amount of actual damages." ARIZ. REV. STAT. ANN. §36-516(b) (1974).

[2] Aggrieved individual may sue for actual damages and no more than $1000 in punitive damages. CAL. CIV. CODE §54.3 (West Supp. 1978).

[3] No discrimination in employment complaint shall be considered valid unless complainant has been properly certified as a handicapped person by the Indiana rehabilitation services board. IND. CODE ANN. §22-9-1-13(c) (Burns 1974 & Supp. 1977).

[4] Commissioner of labor makes a preliminary determination as to whether the aggrieved individual has a statutorily protected "physical handicap" and is fully capable of carrying out the duties of the job which he had been denied.

[5] Any person deeming himself injured by an unlawful discriminatory act shall have a civil cause of action to enjoin further violations and to recover actual damages. KY. REV. STAT. §207.230 (Supp. 1978).

[6] There is no provision for a formal hearing; informal resolution is sought through conference, conciliation and persuasion. If conciliation efforts fail the Human Rights Commission shall file a civil action seeking appropriate relief.

[7] A person subjected to unlawful discrimination may file a civil action against the person who committed the unlawful discrimination. ME. REV. STAT. tit. 5 §4621 (Supp. 1978).

[8] Person alleging a violation may bring a civil action for appropraite injunctive relief or damages, or both. MICH. COMP. LAWS ANN. §37.2801(Supp. 1978-1979).

[9] Aggrieved individual may seek appropriate relief, injunctive or otherwise, from the district court. §64-329 MONT. REV. CODES ANN. §64-329 (Supp. 1977). Person discriminated against in employment because of a physical handicap may bring suit for civil damages. MONT. REV. CODES ANN. §64-304 (Supp.1977).

[10] Any individual who has been discriminated against and any organization which represents the interests of individuals who have been discriminated against shall have standing in courts of law to institute actions to enforce the law against discrimination. N.J. STAT. ANN. §10:5-38 (West 1976).

[11] Any person claiming to be aggrieved by an unlawful discriminatory practice shall have a cause of action for damages and other appropriate remedies. N.Y. EXEC. LAW §297(9) (McKinney Supp. 1972 to 1977).

[12] As made applicable to violations of the civil rights of the physically and mentally handicapped by §659.435(1973), §659.105(1973) gives an aggrieved person a cause of action against the violator for damages sustained and for exemplory damages as may be reasonable.

[13] "Handicapped person" who has also been deprived of his civil liberties by a person, firm or other organization may maintain a civil cause of damages with a conclusive presumption of at least $100 to the handicapped person. TEX. CIV. CODE ANN. Art. 4419e(6)(b) (Vernon 1976).

[14] Any aggrieved person shall have a civil action in a court to enjoin further violations, to recover actual damages, or both together with cost of suit and any other remedy authorized by the United States Civil Rights Act of 1964. WASH. REV. CODE ANN. §49.60.030(2) (Supp. 1976).

SUMMARY OF THE MODEL ACT

As can be seen from the review of existing legislation, no state has a comprehensive and adequate statute to protect disabled persons. Thus, the next section presents a model act whose goal is to promote the greatest possible participation of disabled persons in the social and economic life of the state commensurate with their abilities and desires. This goal is achieved through a general statute outlawing discrimination in specific areas and by providing investigation and enforcement procedures to assure compliance with its provisions.

Approach

The act does not create a special anti-discrimination statute for disabled persons. Rather, disabled persons are included as one among a number of commonly included minority groups covered by such laws. This approach, reflecting the belief that disabled persons confront the same problems of exclusion faced by other minority groups, sometimes on an even wider and more pervasive scale, has several benefits. First, it should assist in raising public consciousness of the minority group status of disabled persons. Second, it is more realistic, since it is not expected that a state will adopt a separate statute with separate enforcement mechanisms for every minority group discriminated against. Third, it is also consistent with the general trend to amend existing anti-discrimination statutes to include disabled persons among those protected. (See Review of State Legislation *supra*.)

Statutory Provisions

The model statute is presented in a comprehensive form. It protects persons with all types of disabilities, with the other minority groups included being those most commonly protected by comprehensive anti-discrimination statutes. Yet, the listing is certainly not dispositive and may be viewed as under- or over-inclusive by a legislature. Only the categories of *disability* and *use of adaptive devices* are specifically recommended for inclusion and are "musts" for protection of the interests of developmentally disabled and other mentally or physically impaired persons.

The act has two major parts: the substantive provisions outlawing discrimination and the procedural provisions providing complaint, investigation and enforcement mechanisms. Both are necessary to sound and effective legislation.

The substantive provisions of the act prohibit discrimination on specified bases in seven major areas. Each area lists prohibitions and exceptions separately. Employment discrimination by employers, employment agencies, and labor organizations is prohibited. Prohibitions include provisions on hiring, promotion, discharge, compensation, training programs and membership in labor organizations, and on solicitation of information and occupational licensing. Employers are required to make reasonable accommodations for disabled persons, except where accommodation would impose undue hardship on a business.

Discrimination in real estate transactions including sale, rental, construction or repair of housing or commercial property is prohibited. Provisions include financing of transactions and restrictive covenants which could otherwise be used to circumvent statutory prohibitions. Similarly, creditors may not grant, withhold, modify, of fix the terms of an extension of credit on the basis of a disability, without violating the act.

Public accommodations provisions prohibit any entity serving the public from refusing or withholding from a person because of disability the full use and enjoyment of the services, facilities, or privileges of a place of public accommodation. Other provisions forbid discrimination against disabled individuals in the admission to and full utilization of the benefits of public and private educational institutions.

The act also prohibits discrimination in insurance transactions based solely upon disability but does not affect distinctions based upon actuarial or claim differentials. Other provisions prohibit discrimination in public services and the granting of public assistance or licenses, and discrimination by public contractors. The act also provides for affirmative action programs where public funds are involved. Finally, there are prohibitions against aiding and abetting discriminatory practices and engaging in reprisals against those who comply or seek compliance with the act.

These provisions are implemented and enforced by a state regulatory civil rights agency consisting of a Director and staff and an advisory Civil Rights

Commission which approves appointments of hearing examiners. Persons aggrieved by a violation of the act may file a complaint with the Director. The Director makes an initial determination of validity on its face and then issues a notice of violation to the alleged violator or respondent. Each complaint is then investigated to determine if there is probable cause to find that a violation has occurred. If so, an informal settlement is attempted. If this fails, the Director may initiate a hearing on the matter before a hearing examiner. The Director may also seek a temporary injunction in appropriate situations. Where the hearing examiner finds that a discriminatory practice has occurred, the examiner issues an order containing appropriate relief. This order may be appealed by an aggrieved person and is subject to enforcement by the Director. A private right of action is granted to aggrieved persons where compliance through administrative action has first been sought.

The portions of the text relating specifically to disabled persons are <u>underlined</u>. The text of the act appears in **bold type**. Comments on statutory provisons appear in standard type. Alternatives to the text appear in *italics*. The comments explain, elaborate upon or provide a rationale for the text of the act. They are intended to assist the user in interpreting the act, in drafting an act for a particular state based upon this model act and in operating an advocacy agency. The alternatives suggest other approaches which might be considered.

Adapting the Model Act

The model act is presented in comprehensive form. It can be used by states which have few or no provisions prohibiting discrimination. States which already have anti-discrimination statutes, on the other hand, can use the model to amend their existing code by including the model's provisions concerning disabled persons and adopting any general provisions deemed preferable to existing legislative content. Thus, the act is meant to have application, provide guidance, and offer drafting assistance in areas beyond the preparation of a "long form," all-inclusive anti-discrimination act covering all kinds of discrimination and all minority sub-groups requiring protection. The Model Legislative Project is well aware that not every state may be ready or prepared for such an omnibus enactment (although deemed preferable from the viewpoint of developmentally disabled and other handicapped persons). It is, of course, quite legitimate for advocacy groups to seek special purpose statutory protection or to add the developmentally disabled and other physically and mentally handicapped persons to existing antidiscrimination legislation in the state if that is the only feasible legislative course. In such cases, the model act offers a wealth of ideas, reminders and guidance for these less comprehensive efforts, both in terms of drafting and critical review of proposed bills. For example:

(1) The careful definitions of "disability" and "use of adaptive devices" as a handicapped category will help insure proper coverage of disabled persons. Even where shorter definitions are expedient in adding this group to existing legislation, the act's concepts and thinking will help the drafter insure clarity and comprehensiveness in coverage.

(2) Where disabled persons are being added to existing special purpose legislation or more comprehensive anti-discrimination statutes, it is important to keep in mind, as the Model Act cautions, those needs that have special relevance to disabled persons as opposed to other minority groups. For example, in employment discrimination guarantees, it is particularly important to avoid singling out disabled or handicapped persons with provisions that disabled candidates be "qualified," to insure that medical requirements be strictly "job related," to allow for compensatory accommodation for disabled persons, and to include all forms of employee benefits and remuneration within the discrimination ban. Similarly, with regard to insurance discrimination, it is essential that differential insurance coverages and rates be mandatorily tied to actual risks as measured by hard claims and actuarial data rather than misconceptions about the "high risk" character of disabled groups.

(3) Where a legislative thrust involves the addition of developmentally disabled and other handicapped persons to existing legislation, this would also be a good time to take a look at the "teeth" and effectiveness of existing enforcement apparatus. It may well be possible to tighten up an existing statute by incorporating a few new enforcement features without necessitating a major legislative overhaul. Broad investigatory authority with subpoena powers, an appropriate array of remedies (administrative fines and private suits for injunctive relief or damages) and sound procedures for dealing with complaints (administrative hearings, mediation, regulatory power, administrative orders enforceable in court, etc.) are the kinds of features that might be considered and these are covered in detail in the act's text and commentary.

One further possibility which advocates may have to confront is that state philosophy in this kind of social legislation may favor "short form" enactments with broad regulatory power and significant detail left to the rulemaking and regulatory process. This is often a legitimate legislative approach, and where drafters are faced with this political reality much of the model act's detail can be utilized in fashioning regulations. Here, however, care should be taken to specify in the short form statute that detailed implementing regulations must be formulated, consistent with the statute's objectives and issued with reasonable promptness. In such cases, the short form bill will require close attention to clarity and tightness and an "eye" to accommodation of those provisions of the comprehensive statute which are to be preserved through regulation.

MODEL STATUTE *

A BILL FOR

AN ACT TO CREATE A HUMAN RIGHTS AGENCY TO PROHIBIT DISCRIMINATORY PRACTICES IN EMPLOYMENT, REAL ESTATE TRANSACTIONS, PUBLIC ACCOMMODATIONS, PUBLIC SERVICES, EDUCATION, CREDIT, AND INSURANCE, TO PROVIDE FOR ADMINISTRATIVE ENFORCEMENT OF PROHIBITED PRACTICES AND TO PROVIDE PENALTIES FOR VIOLATION OF PROVISIONS OF THIS ACT.

Section 1. Title

This act shall be known as the (name of state) Civil Rights Act.

COMMENT. None necessary.

Section 2. Statement of Purpose

(1) The purpose of this act is to encourage and enable all persons to fully participate to the maximum extent of their abilities and desires in the social and economic life of the state, to engage in remunerative employment, to obtain suitable housing and commercial property in the real estate market, to utilize available public accommodations, and public services, to utilize educational facilities and to otherwise pursue their general welfare and exercise their rights and privileges as inhabitants of this state.

COMMENT. Our political system is based upon the belief in freedom to pursue one's own happiness and equality of opportunity to do so. The areas covered by the anti-discrimination portions of this act are essential to these bases. Discriminatory practices threaten these rights of freedom and equality, deny the fullest participation of all persons and ultimately injure society by interfering with the full utilization of productive capacities and by contributing to our social problems. Thus, the ultimate goal of this act is to allow persons the widest possible participation in the state's social and economic life by eliminating the artificial barriers imposed through acts of discrimination.

(2) The general assembly finds that the practice or policy of discrimination based upon race, color, religion, national origin, disability, age, marital status, sex or use of adaptive devices[1] are unwarranted interferences with the purposes of this act and are a matter of concern to this state. Such discrimination is contrary to the principles of freedom and equality of opportunity, threatens the rights and proper privileges of the inhabitants of this state and threatens the institutions and foundations of a free democratic society.

The general assembly further finds that such discrimination results in a failure to utilize the productive capacities of individuals to their fullest extent, deprives large segments of the population of the state of earnings necessary to maintain or contribute to a decent standard of living and necessitates their resort to public support, compels many individuals to live in dwellings which are substandard, unhealthful and overcrowded, and thereby threatens the peace, health, safety and general welfare of the state and its inhabitants.

(3) This act shall be deemed to be an exercise of the state's police power for the protection of the peace, public health, safety and general welfare of all inhabitants of the state, for the fulfillment of the provisions of the state's laws guaranteeing civil rights, for the maintenance of business and good government, and for the promotion of the state's trade and commerce. This act shall be liberally construed to accomplish its purpose and any law inconsistent with any provision of this act shall not apply.

COMMENT. Interference with the rights and exercise of functions so basic to our social, economic and political system is a legitimate public concern and its amelioration a legitimate reason for exercise of the state's police power. As an exercise of this power, it is intended that the act be liberally construed and implemented to achieve its purposes. Thus, although the act is not in-

---

* D. Matthew Powell, Associate Director of the project, had primary responsibility for drafting this model statute.

[1] The bases for discrimination given in the model statute are those most commonly found in state civil rights statutes. Other possible bases include status with regard to public assistance (Minnesota), liability for service in the armed forces (New Jersey), alcoholism, drug addiction, and sexual preference. Only *disability* and *use of adaptive devices* are specifically recommended for inclusion in the statute. (See Summary of the Act *supra*.)

tended to repeal any existing provisions of state law relating to discrimination, it is intended to take precedence over any conflicting state law. Therefore, other state laws which may conflict should be checked for consistency.

Section 3. Civil Rights

It is the public policy of this state to assure equal opportunities to all individuals, to foster their employment in accordance with their fullest capacities and protect their right to be free from discrimination based upon race, color, religion, national origin, disability, age, marital status, sex or use of adaptive devices. This right is declared to be a civil right and shall include, but not be limited to:

(1) Obtaining and holding employment without such discrimination;

(2) Engaging in real estate transactions without such discrimination;

(3) Utilizing any of the public accommodations without such discrimination;

(4) Utilizing available educational institutions without such discrimination;

(5) Utilizing public services and assistance without such discrimination;

(6) Obtaining credit and insurance without such discrimination.

COMMENT. The section specifies certain rights and prohibits certain discriminatory actions, thereby helping to achieve the stated goal of the act and the public policy of the state. The right to assurance of equal opportunity and protection from discrimination are given the status of civil rights and are enforceable through the substantive and procedural provisions of the act.

Section 4. Definitions

As used in this act:

(1) "Person" means one or more individuals, partnerships, associations, unincorporated organizations, corporations, cooperatives, labor unions or organizations, legal representatives, trustees and receivers, agents or any group of persons, the state of (name of state) or any political or civil subdivision or instrumentality thereof.

COMMENT. The term is defined to include the widest application of the word. As used in the act, it gives broad application to provisions prohibiting discrimination by "any person," except where more specific provisions are necessary, e.g. section 5 on employment discrimination. Review and enforcement provisions allow a "person who has a substantial personal interest" and a "person aggrieved" certain rights which are also intended to be broadly applied.

(2) "Discriminate" or "discrimination" means to segregate or unreasonably differentiate in treatment, whether intended or unintended or to act in a manner fair in form but discriminatory in operation based upon race, color, religion, national origin, disability, age, marital status, sex or use of adaptive devices.

COMMENT. The act outlaws certain acts of discrimination on specified bases. The definition of these terms focuses upon behavior or actions and their result, rather than upon motive or intent. If through discrimination in employment, for example, a person is denied a job, it is irrelevant whether the employer's intentions were deliberate or benign for the result is the same for those discriminated against. The action is still prohibited even if unintended. Motive or intent is, of course, a legitimate consideration in fashioning relief and in considering the assessment of any penalty therefor.

(3) "Age" means over forty (40) years of age.

COMMENT. This term is defined in accordance with federal provisions on age discrimination contained in the Age Discrimination Act, 29 U.S.C. §631 (1976).

(4) "Marital status" means the state of being married, single, divorced, widowed or separated.

COMMENT. None necessary.

(5) "National origin" means both national origin and ancestry.

COMMENT. None necessary.

(6) "Religion" is defined to include all aspects of religious observances, practices, and beliefs.

COMMENT. None necessary.

(7) "Disability" means any condition or characteristic that renders a person a disabled person as defined in subsection 8 of this section.

(8) "Disabled person" means any person who:

(a) Has a physical or mental impairment which substantially limits one or more major life activities;

(b) Has a history of, or has been misclassified as having an impairment which substantially limits one or more major life activities;

(c) Has a physical or mental impairment that does not substantially limit major life activities but that is treated by others as constituting such a limitation;

(d) Has a physical or mental impairment that substantially limits major life activities only as a result of the attitudes of others toward such impairment;

(e) Has none of the impairments defined in this subsection but is treated by others as having such an impairment;

(i) "Physical or mental impairment" means:

[a] Any physiological disorder or condition, cosmetic disfigurement or anatomical loss affecting one or more of the following body systems: neurological; musculoskeletal; special

sense organs; respiratory, including speech organs; cardiovascular; reproductive; digestive; genito-urinary; hemic and lymphatic; skin; and endocrine; or

[b] **Any mental or physiological disorder, such as mental retardation, organic brain syndrome, emotional or mental illness, and specific learning disabilities.**

(ii) "Major life activities" means functions such as, but not limited to, caring for one's self, performing manual tasks, walking, seeing, hearing, speaking, breathing, learning, and working.

COMMENT. This definition is based upon the definition of handicapped person contained in section 706(6) of the Rehabilitation Act of 1973 as amended[2] and the Department of Health, Education and Welfare regulations.[3] To be covered a person must either have a disability which substantially impairs one or more major life activities, have a record of such disability or be treated by others as if the person had a disability. Most persons covered by the act will be in the first category.

There are some existing statutes which have been criticized for their vagueness.[4] The definition of physical or mental impairment in the act is quite specific without being so specific in its listings as to leave out disabling conditions which should be covered. This impairment must substantially limit major life activities. Thus, to come within the terms of the statute a limitation does not constitute a handicap unless it is severe enough to have a substantial impact on such things as sight, communication, mobility, learning or working. The act is not intended to cover minor impairments which are not handicapping, nor does it cover conditions arising from cultural, environmental, or economic disadvantages. There must be a physical or mental basis.

Note that section 706(6) has been interpreted by the United States Attorney General in a legal opinion to the HEW Secretary to include alcoholism and drug addiction since they are generally believed to be diseases, though it is uncertain whether they are mental or physical.[5] Since both conditions may substantially impair major life activities, persons with alcoholism or drug addiction may be protected by this act, if qualified, as that term is used in various portions of the act. Thus, as with any other disability, these conditions can be taken into account as they affect eligibility for employment or services.

Persons are also considered disabled if they have a record of a disability. This category includes persons who had a disability at one time but are no longer disabled or those who have been misclassified as having an impairment. Common examples of the first category include mental illness and cancer, while a common example of the second category is persons who have been misclassified as mentally retarded. Although persons in these two categories would normally be covered by paragraphs (8)(c), (8)(d), or (8)(e) as well, respondents are likely to allege as a defense to a complaint that a complainant is not covered by the act. Thus, the effort may be made to preclude consideration of the merits of a complaint. This section protects against the success of such an argument.

A person may also come within the definition of disabled if that person is treated as if he or she is disabled, whether or not the person has a physical or mental impairment. Included are persons who have an impairment but are not substantially limited in any major life activity and those who have no history of such disability, but are treated as if they are disabled. Examples include a person who has controlled epilepsy or mild cerebral palsy and is not precluded in any way from engaging in major life activities, but is refused employment because an employer considers them disabled and unable to work. Such persons are ordinarily considered disabled but do not technically come within the first two paragraphs. Others not ordinarily considered disabled such as those with disfiguring scars, and those who have no impairment at all are included, if treated as if disabled. Thus, the act covers discrimination in the eye of the beholder, whether or not it is factually based.[6] The act is concerned with behavior which has a discriminatory effect, not intent.

This section, of course, decides only an initial question of membership in a protected class. There are still several major issues to be decided before a violation of the act can be found, including whether or not the complainant is a "qualified individual" within the meaning of the act and whether the discrimination occurred on the basis of a prohibited criterion.

(9) "Use of adaptive device" means the utilization of any item to compensate for a physical or mental impairment, including, but not limited to, braces or other supports, wheel chairs, talking boards, hearing aids, corrective lenses, or seeing eye dogs.

COMMENT. This category is added to more traditional bases for discrimination to prevent a kind of "secondary discrimination." A common example is the refusal to serve a blind person in a restaurant because that person has a guide dog.[7]

---

[2] 29 U.S.C. §706(6) (1976).

[3] 45 C.F.R. §84.3(j) (1977).

[4] *See* discussion in Review of State Legislation, Protected Class *supra*.

[5] Analysis of Final Regulation, §504, 45 C.F.R. pt. 84 (1977).

[6] *See* Wright, *supra* note 4, at 70.

[7] Some States have prohibited discrimination because of seeing eye dogs. *See e.g.* LA. REV. STAT. ANN. §46:1951-1953 (West Supp. 1978); NEB. REV. STAT. §20-129 to 131 (1977); N.C. GEN. STAT. §168 (1976); S.C. CODE §§43-33-10 to -70 (1976).

(10) "Unfair practice" or "discriminatory practice" means those practices specified as unfair or discriminatory and prohibited by sections 5 through 13 of this act.

COMMENT. None necessary.

(11) "Affirmative action program" means any bona fide effort to advertise, seek out, recruit, attract, or otherwise indicate that persons who are disabled, over forty (40) years of age, non-white or female are welcome or desirable with respect to employment, real estate transactions, public accommodations, public services, educational institutions, credit transactions, insurance transactions or other actions covered by this act, whether voluntary or required by statute or rule or by a consent, administrative or court order. The term includes the making or keeping of records on race, color, disability, age or sex so long as they are not made, kept or used for reasons contrary to the provisions or purposes of this act.

COMMENT. Affirmative action has become a major part of the effort to eliminate the effects of discrimination, whether it is voluntary or ordered to be undertaken, The act makes clear that these efforts are not to be adversely affected by the terms of the act so long as they are consistent with its provisions and purposes. Thus, affirmative action programs are exempted from the otherwise restrictive provisions of the act primarily covering solicitation of information and record-keeping practices.

(12) "Office" means the Office for Civil Rights.

COMMENT. None necessary.

(13) "Director" means the Director of the Office.

COMMENT. None necessary.

(14) "Commission" means the (name of state) State Civil Rights Commission created by this act.

COMMENT. None necessary.

(15) "Local commission" means an agency of a city or county, created pursuant to law, city charter, or municipal ordinance, for the purpose of dealing with discrimination on the basis of race, color, religion, national origin, disability, age marital status, sex, or use of adaptive devices.

COMMENT. None necessary.

(16) "Complaint" means a charge of discrimination filed by an individual alleging an unfair or discriminatory practice by a person.

COMMENT. None necessary.

(17) "Complainant" means an individual who files a complaint.

COMMENT. None necessary.

(18) "Petition" means a pleading filed by the director with a hearing examiner or the district court alleging the occurrence of an unfair or discriminatory practice, seeking relief, or both.

COMMENT. None necessary.

(19) "Respondent" means a person against whom a complaint or petition is filed.

COMMENT. None necessary.

(20) "Consent order" means a formal written agreement entered into in lieu of adjudication.

COMMENT. None necessary.

Introduction to Section 5

The applicability of section 5 on employment is somewhat different than the other substantive provisions of the act. They apply to any person who discriminates, whereas this section is limited to employers, employment agencies, and labor organizations. Thus, the specific provisions which apply to these latter three categories are listed separately below.

Section 5. Employment Discrimination Prohibited

(1) Additional definitions as used in this act:

(a) "Employer" means any person employing individuals within the state and includes any agent of such person;

COMMENT. This definition covers any person who employs another within the state either directly or through an agent, including the state and its subdivisions, unless excluded by subsection 5(9). Some employers using public funds are also subject to affirmative action programs in addition to the discrimination prohibitions of this section. These include the state and all recipients of state financial assistance or parties to a public contract. (See subsections 11(3) and 11(4).)

Alternative. *Some states may wish to exclude smaller employers, e.g. those employing less than four employees, for two reasons. Working relationships will tend to be closer where there are only a few employees and a state may feel that freedom of association even in an employer-employee context outweighs the right to freedom of equality in very small employer situations. In addition the number of employers covered will be greatly reduced and, presumably, so will the number of complaints filed. Civil rights agencies are notorious for consistently being behind on processing of complaints. Limited resources may be better spent on larger employers with a larger impact on the employment market.*

(b) "Employment agency" means any person undertaking to procure employees or opportunities to work for any other person or any person holding himself or herself or itself to be equipped to do so, and includes any agent of such person;

COMMENT. Any person undertaking to bring employer and employee together necessarily stands in the same relationship as would an employer with respect to the act. Therefore, any person who seeks employees for an employer or jobs for persons seeking employment is covered by this section.

(c) "Labor organization" means:

(i) An organization of any kind, such as an agency or employee representation committee, group, association, or plan in which employees participate and which exists for the purpose, in whole or in part, of dealing with employers concerning grievances, labor disputes, wages, rates of pay, hours, or other terms or conditions of employment;

(ii) A conference, general committee, joint or system board, or joint council which is subordinate to a national or international labor organization;

(iii) An agent of a labor organization;

COMMENT. Any entity to which employees belong or are required to join and which represents them as employees vis-a-vis an employer is covered by this section, since membership in such an entity is often necessary for employment.

(d) "Qualified individual" means an individual who can perform the essential functions of the job in question. As applied to a disabled individual, "qualified individual" means an individual who, with reasonable accommodation, can perform the essential functions of the job in question. Receipt or alleged receipt of treatment for a disability, whether physical or mental, shall not constitute evidence of a person's inability to perform the essential functions of a particular job or position. In addition, uninsurability or increased cost of insurance under a group or employee insurance plan does not render a disabled person unqualified.

If a disabled person is qualified to perform a particular occupation by reason of training or experience, the nature of that occupation shall not be the basis for exception to the unfair or discriminatory practices prohibited by this section.

COMMENT. A number of state statutes go beyond inclusion of disability as a protected class by defining this class in terms of the member's qualifications to perform job functions.[8] Since this is not required of any other protected class, it is another example of discrimination against disabled individuals by singling them out for separate treatment. It also displays a lack of faith or understanding of the exception for a bona fide occupational qualification as a defense to claims of discrimination.[9] (See paragraph 5(9)(b) *infra*.)

The requirement that disabled persons be able to perform job functions suffers from overbreadth as well. Where handicap or disability is defined as a physical condition unrelated to the person's ability to engage in a particular job or occupation, employers often compare the abilities of the *category of persons* with a disability to requirements of a particular *occupation,* rather than *an individual's abilities* with the requirements of a *specific job.* This creates the danger that a person fully qualified for a particular job could be rejected.[10] Since this requirement, however, is such a prevalent part of existing statutes[11] and is likely to be insisted upon by many employers and some legislators, it is unrealistic to simply remove it. The definition proposed avoids the discriminatory aspects and overbreadth of many existing statutes and deals with disabilities and qualifications directly. That is, this definition applies to *all* applicants, not just disabled ones. And it matches the applicant's individual abilities to the essential functions of the job in question. The only employment criteria allowed to be determinative are those which have substantial predictive value as to an individual's capacity to perform the particular job in question.[12] Evidence that certain persons with a disability cannot perform a job is not dispositive absent a showing that the *applicant* cannot perform adequately, except where a bona fide occupational qualification applies. The standard for selection of employees must be neutral, job-related, and reliable in prediction.

Furthermore, many disabled applicants are able to perform adequately if artifical barriers to their performance are removed or certain adaptations to the work environment are made. Therefore, the qualifications of disabled applicants are to be measured after accommodation to the disability of the applicant is made. (See subsection 5(7).) Note that it is not necessary that the individual perform every task involved in a job. The fact that an individual has difficulty in performing tasks which are marginally related to the job should not bar his or her employment.

Employers, like other persons, are prone to make decisions based upon previously established generalizations or fears which often have no applicability to individuals. Thus, generalizations about abilities based upon treatment received, the employer's fears about increased insurance costs, or an alleged uninsurability are not allowed.[13] No

[8] *See e.g.* Indiana, New Jersey, Michigan, New Mexico, Oregon and Pennsylvania.

[9] *See* Note, *Potluck Protections for Handicapped Discriminatees: The Need to Amend Title VII to Prohibit Discrimination on the Basis of Disability,* 8 LOY. CHI. L.J., 814, 839 (1977).

[10] Note, *supra* note 9, at 840.

[11] *See supra* note 8.

[12] *See* Note *supra* note 9, at 840.

[13] The Wisconsin anti-discrimination statute is noteworthy in that it forbids an employer "to contribute a lesser amount to the fringe benefits, including life or disability insurance coverage, of any employee because of a handicap." WIS. STAT. ANN. §111.32(5)(f)(2) (West Supp. 1978-1979).

Prior to 1976 the statute provided that an employer's exclusion of a handicapped employee from life or disability coverage

factual justification is given for such exclusions from insurance benefit programs (see comment to 5(2)(b) *infra*), nor is there justification for a rationalization that workmen's compensation insurance rates will be increased by hiring disabled workers. Rates are based upon the relative hazard of the employer's industry and the employer's actual accident experience.[14] In fact, studies show that the accident rates of properly placed disabled individuals is at least equal, if not superior, to nondisabled workers and there is no clear evidence that employment of disabled persons results in higher workmen's compensation costs.[15]

(2) Employer prohibitions.

It shall be an unlawful discriminatory practice for an employer because of race, color, religion, national origin, disability, age, marital status, sex or use of adaptive devices to:

(a) Fail or refuse to hire, recruit, or promote a qualified individual;

COMMENT. Any failure or refusal to hire, recruit or promote a qualified individual is specifically prohibited. However, the real issue will center around whether the individual is a qualified individual. (See comment to 5(1)(d) *supra*.)

(b) Discharge or otherwise discriminate against qualified individuals with respect to compensation, fringe benefits or the terms, conditions, or privileges of employment;

COMMENT. Disputes over the discharge of employees will also center around whether the discharged employee was qualified. Other issues covered by this paragraph include equal pay,[16] work environment,[17] sick leave, vacation time and life and health insurance benefits.[18] These issues may arise *after* employment. Of these, insurance coverage is perhaps the most controversial. The cost of insuring disabled persons is often cited as a reason for not hiring such persons. (See discussion within comment to 5(1)(d) *supra*.) The act prohibits this as well as discriminating with respect to employment benefits, including insurance, once hired.

The inclusion of insurance in compensation and fringe benefits has been a source of concern and criticism from employer interests. However, criticism of this coverage of insurance benefits recites assumptions about increased costs but offers no authoritative support for such statements.[19] Proposed regulations by HEW implementing section 504 as applied to employee benefit programs had allowed for differences in contributions to or benefits from employer insurance where justified by actuarial figures and an actual, substantial increase in insurance costs.[20] However, this section was dropped from the regulations on the belief that currently available data and experience do not demonstrate a basis for promulgating a regulation specifically allowing for differences in benefits or contributions.[21] Therefore, the act adopts the stance taken by HEW and prohibits discrimination in insurance benefits.

(c) Limit, segregate, or classify an employee or applicant for employment in a way which deprives or tends to deprive a qualified individual of employment opportunities or otherwise adversely affects the status of an employee;

COMMENT. Applicants and employees must be treated equally and no separate consideration or treatment which adversely affects the employment status of qualified individuals is permitted. This does not affect differential treatment of applicants under affirmative action programs, since these would aid employment opportunities of protected classes. Such programs are exempted by paragraph 5(9)(c).

(d) Fail or refuse to hire, recruit, or promote a qualified individual on the basis of physical, mental or other examinations that are not directly related to the requirements of the specific job;

(e) Discharge or take other discriminatory action against a qualified individual on the basis of physical, mental or other examinations that are not directly related to the essential functions of the job;

---

did not constitute discrimination. However, in Chrysler Outboard Corp. v. Dept. of Indus., Labor & Human Relations (14 Fair Empl. Prac. Cas. 344) (1976) the court affirmed a finding of discrimination for refusal to hire a person with leukemia because of "the risk of future absenteeism and the higher insurance costs." *Id.* at 345. The court stated that although some discrimination in insurance coverage of a handicapped employee was permitted, an employer was not permitted "to discriminate in hiring on the basis of increased insurance costs." *Id.*

[14] Note, *supra* note 9, at 820.

[15] *Id.* at 821.

[16] While equal pay provisions cover sex, 29 U.S.C. §206 (d) (1976), age (40-65 years of age), 29 U.S.C. §§621-634 (1976) and race, color, religion, sex, or national origin, 42 U.S.C. §2000e-2 (1976), disabled persons have only limited coverage under 29 U.S.C. §793 (1976) applicable to federal contracts over $2500.

The Fair Labor Standards Amendments of 1974, 29 U.S.C. §214-16 (1976) authorize the Secretary of Labor to provide employment under special certificates of individuals whose earning or productive capacity is impaired by physical or mental deficiency or injury at wages which are substantially less than the minimum wage established by Section 206 of the Fair Labor Standards Act, 29 U.S.C. 206 (1976). For a criticism of this practice *see* Bernstein, *The Right to an Adequate Income and Employment, Principal Paper* in THE MENTALLY RETARDED CITIZEN AND THE LAW 272-295 (M. Kindred ed. 1976). These persons would not be covered by the act, since they would not be "qualified."

[17] Katz, *Sexual Harassment, Ethnic Slurs and other Breaches of the Employer's Obligation to Maintain a Discrimination-Free Work Environment*, 13 ARIZ. B.J. 42 (April, 1978).

[18] Follmann, *Insurance, Principal Paper* in THE MENTALLY RETARDED CITIZEN AND THE LAW 144, 149, 160 (M. Kindred ed. 1976).

[19] *See* Wright, *supra* note 4, at 77, 86.

[20] *See* 41 Fed. Reg. 29,564 (1976) (to be codified in 45 C.F.R. §84.18).

[21] *See supra* note 5. *But see* Wright, *supra* note 4, at 84-89.

COMMENT. Most larger employers use various examinations to screen applicants for employment. Since these examinations may be used for discriminatory purposes, they are required to be directly related to job requirements. Thus, examinations which tend to screen out protected classes of persons must be shown to be job-related to be allowable. What may appear to be neutral examinations or criteria are suspect if they in fact have an adverse impact upon a protected class of persons. Such examinations or criteria may then be used only if they qualify as business necessities which override their discriminatory impact.[22] Thus, they must be shown to be necessary to the employer's operations and that no acceptable alternative method of screening is available.[23] This provision is intended to be interpreted like Title VII of the federal Civil Rights Act.[24] (Contrast this with the bona fide occupational qualification exception in paragraph 5(9)(b).)

With respect to disabled persons, employers must take into account that some examinations or criteria depend upon speaking, manual or sensory skills which are themselves not essential to the job in question but which may prevent the individual from passing the examination. Thus, the employer must select and administer tests so as to measure a disabled person's ability to perform the specific job in question and not the ability to see, hear, speak or perform manual tasks, except where those skills are factors which the test is designed to measure and where they are necessary to the performance of the job. (See 5(9)(b) for discussion of bona fide occupational qualification exception.) This section, however, does not prohibit the use of examinations so long as they meet the requirements of paragraphs 5(9)(d) and (e).

(f) Fail or refuse to hire, recruit, or promote a qualified individual when accommodation or adaptive devices or aids may be utilized, thereby enabling that individual to perform the essential functions of the job, except as noted in subsection 7 of this section; or

(g) Discharge or take other discriminatory action against a qualified individual when accommodation or adaptive devices or aids may be utilized thereby enabling that individual to perform the essential functions of the job, except as noted in subsection 7 of this section.

COMMENT. See comment to subsection 5(7) *infra*.

(3) Employment agency prohibitions.

It shall be an unlawful discriminatory practice for an employment agency, because of race, color, religion, national origin, disability, age marital status, sex or use of adaptive devices to:

(a) Refuse or fail to accept, register, classify properly, or refer for employment or otherwise to discriminate against a person; or

(b) Comply with a request from an employer for referral of applicants for employment if the request indicated directly or indirectly that the employer fails to comply with the provisions of this act.

COMMENT. Employment agencies stand at the threshold of employment. Their compliance with the requirements of the act is as essential as any employer's. To allow otherwise would be to exempt a large portion of the employment market from the efforts of the act. These agencies, if anyone, should be expected to conform to both the letter and spirit of the law by treating applicants for employment equally and by making sure that employers who use their services comply with the act as well. Licensed agencies which violate the act, in addition to being subject to other provisions of the act, may also be subject to suspension or revocation of their license. (See subsection 23(5)).

(4) Labor organization prohibitions.

It shall be an unlawful discriminatory practice for a labor organization, because of race, color, religion, national origin, disability, age, marital status, sex or use of adaptive devices to:

(a) Deny full and equal membership rights to a member or to a qualified individual seeking membership;

(b) Expel a member from membership; or

(c) Discriminate against a qualified individual seeking membership or a member by adversely affecting his or her classification, hire, tenure, compensation, terms, upgrading, conditions, facilities, or privileges of employment.

COMMENT. Labor union membership is often desirable and sometimes mandatory, such as in closed union shops, in order to hold many jobs. Therefore, it is essential to the viability of the act that labor organizations be prohibited from discriminating in their membership or in the way they represent or affect the employment of their members. (See also subsection 5(7)).

(5) Prohibition — training programs.

It shall be an unlawful discriminatory practice for an employer, employment agency, or labor organization to discriminate against a qualified individual because of race, color, religion, national origin, disability, age, marital status, sex or use of adaptive devices in admission to, or employment or continuation in, a program established to provide apprenticeship or other training.

COMMENT. Training programs run by em-

---

22 Note, *Equal Employment and the Disabled: A Proposal*, 10 COLUM. J.L. SOC. PROB. 457, 472-76 (1974).

23 *Id.* at 476.

24 42 U.S.C. §§2000e to 2000e-17 (1976).

ployers, employment agencies or labor organizations are often used as entry positions for later jobs. Since precluding entry into such training programs may prevent or hinder subsequent employment opportunities, it is essential to prohibit discrimination against qualified individuals who apply for apprenticeship or training.

(6) Other prohibitions.

(a) Except where based on a bona fide occuptional qualification certified in advance by the director pursuant to paragraph 5(9)(b) of the section or where done in accordance with paragraph 5(9)(c) of this section, it shall be an unlawful discriminatory practice for any employer, employment agency or labor organization to:

(i) Elicit or attempt to elicit any information concerning the race, color, religion, national origin, disability, age, marital status, sex or use of adaptive devices of an applicant for employment or membership;

(ii) Make or keep a record of the race, color, religion, national, origin, disability, age, marital status, sex or use of adaptive devices of any applicant for employment or membership;

(iii) Use any form of application for employment or personnel or membership blank which seeks to elicit information regarding race, color, religion, national origin, disability, age, marital status, sex, or use of adaptive devices unless required by the United States, this state, or a political subdivision or agency of the United States or this state;

COMMENT. Pre-employment inquiries are generally prohibited. They are allowed only in two contexts: where a protected class is also shown to be a bona fide occupational qualification (paragraph 5(9)(b)) or where the inquiry is made as part of a bona fide affirmative action program and complies with paragraph 5(9)(c). Otherwise pre-employment inquiries must be stated in terms of the applicant's ability to perform job-related functions. (See paragraph 5(9)(d)).

The general prohibition on pre-employment inquires concerning race, sex, disability, etc. was adopted because of the tremendous potential for their use in discriminating against persons in the protected classes without serving any legitimate interests of employers. This is especially true of disabled applicants for employment.

Employers may argue that pre-employment inquiries concerning disabilities are necessary to determine qualifications, identify safety hazards caused by a disability, determine accommodations that may be required and identify individuals for consideration under affirmative action programs. All of these can be done by direct inquiry. An employer may inquire into the applicant's ability to perform job-related tasks without the need to inquire about disabilities. (See paragraph 5(9)(d)). For example, where a driver's license is necessary for a job, the employer should ask whether the applicant possesses one, not whether the applicant has epilepsy.

(iv) Print or publish or cause to be printed or published any notice or advertisement relating to employment or membership indicating any preference, limitation, specification, or discrimination, based upon race, religion, national origin, disability, age, marital status, sex, or use of adaptive devices;

(v) Announce or follow a policy of denying or limiting, through a quota system or otherwise, the employment or membership opportunities of any group because of the race, color, religion, national origin, disability, age, marital status, sex or use of adaptive devices.

COMMENT. No employer, employment agency or labor organization may take steps to discourage potential applicants from applying for employment or membership, even though they may not otherwise discriminate. Such acts are contrary to the goal of the act. Furthermore, it is hoped that employers, employment agencies and labor organizations will utilize the affirmative action provisions of paragraph 5(9)(c).

(vi) Utilize in the recruitment or hiring of persons any employment agency, placement service or training school or center, labor organization, or any other employee-referring source known to discriminate against persons because of their race, color, religion, national origin, disability, age, marital status, sex, or use of adaptive devices;

COMMENT. The services of an employment referral source which is known to discriminate cannot be used to circumvent the obligations of employers, employment agencies, or labor organizations under the act.

(vii) Attempt any of the unfair practices prohibited by this section.

COMMENT. It is not necessary that an employer, employment agency or labor organization succeed in discriminating to come within the proscriptions of the act. The Director may act to prevent or terminate attempts at employment discrimination as well.

(b) It shall also be an unlawful discriminatory practice for any person seeking employment to publish any advertisement which specifies or in any manner indicates a limitation or preference as to the race, color, religion, national origin, disability, age, marital status, sex or use of adaptive devices of any prospective employer, except as provided in subsection 9 of this section.

COMMENT. While a person cannot be required to work for an employer, if the person objects to the employer on the basis of membership in a protected class or for any other reason, that person cannot publicly advertise or publish the biased beliefs as the basis for accepting or rejecting employment. Such advertising of bias has no place in the employment market.

(7) Obligation to accommodate employees and applicants with a disability; exception.

An employer shall make accommodation to the known disability of an otherwise disabled employee or applicant for employment unless the employer demonstrates that the accommodation would impose an undue hardship in the conduct of business. A labor organization shall cooperate in making such accommodations to the extent they influence or control job structure and other employment conditions.

COMMENT. Where an employee or applicant for employment has a known disability which hinders employment, the employer is required to make a reasonable accommodation to the disability and remove impediments. Examples include physical modifications or relocation of an office or job so that facilities are accessible to disabled persons, job restructuring such as transferring nonessential duties to other employees, modifying work schedules, including part-time employment, and acquisition or modification of equipment.[25]

Labor organizations frequently have an impact upon the means of accommodation such as job structures which can effectively negate employer efforts at accommodation. Accordingly, labor organizations are required to cooperate with employer efforts at accommodation.

Employers. however, are exempted from making accommodations where they would entail undue hardship on the conduct of the employer's business. Factors to be considered include the overall size of an employer as measured by the number of employees and facilities, the type of facilities and the level of annual budgets, the type of operation including the composition and structure of the work force and the nature and cost of the accommodation.[26] Larger employers tend to have greater flexibility in scheduling, place of work and structuring of job duties as well as a greater ability to afford modifications. Small employers may not be able to afford even a modest modification and may have no flexibility in scheduling or work place. Each is different. None is intended to be unduly burdened.

An employer has no obligation to employ or retain a disabled person when accommodation cannot overcome the effects of the person's disability so that the person can perform the essential functions.

(8) Nondiscrimination in occupational licensing.

If it is required that a person receive a license, certification, or other credential, or become a member or an associate of any club, association, or other organization, or pass any examination in order to engage in a profession, occupation or trade, it is an unlawful employment practice for any person to discriminate on the basis of race, color, religion, national origin, disability, age, marital status, sex or use of adaptive devices, against any other person seeking or holding such license, certification, or other credential, or seeking to become a member or associate of or holding membership or associate status in such club, association, or other organization, or seeking to take or pass such examination.

COMMENT. Many jobs, particularly professional and trade positions, require a certificate, license or membership in order to qualify. To allow discrimination to occur in the granting or holding of such certification would make other protections of the act meaningless. While some cases may be challenged on other bases such as 42 U.S.C. §1983, where state action is involved and race is the basis, the coverage of §1983 is limited. Thus, all persons covered by this act are protected by this subsection so that other protections within this act maintain their vitality. (See also paragraph 11(2)(e).)

(9) Exceptions.

(a) This section shall not apply to the employment of any individual by a parent, spouse or child or of an individual in domestic service, or where, as part of his or her employment, the individual resides in the personal residence of the employer or renders a personal service to the employer or members of his or her family.

COMMENT. Employment between family members is exempted in recognition of the right to privacy in family matters. Intervention in this context is considered beyond the proper scope of state authority. The exemptions in this paragraph also give recognition to the right of privacy and freedom of association. Where an employee is to work or live in one's home as part of the employment, these interests take priority over an interest in equal treatment. The sanctity of one's home which is the final retreat from the outside world, should not be disturbed in this context.[27] (Compare this with the exemption applicable to the renting of rooms in one's home in paragraph 6(7)(d).) Similarly, the exemption for personal services recognizes the intimate nature of that association. Examples include baby sitters, private nurses, and tutors.[28]

(b) This section shall not apply to the use of qualifications based upon race, color, religion,

---

[25] *See* 45 C.F.R. §84.12 (1977).

[26] *Id.*

[27] Bonfield, *State Civil Rights Statutes: Some Proposals*, 49 IOWA L. REV. 1067, 1109 (1964).

[28] *Id.*

national origin, disability, age, marital status, or sex when the qualification is a bona fide occupational qualification, certified by the director in advance of the use. The exceptions permitted in this subsection based on bona fide occupational qualifications shall be strictly construed.

COMMENT. Unlike the neutral hiring practices which may have a discriminatory effect, discussed in the comment to 5(2)(d) and (e) *supra*, the bona fide occupational exception is a defense to overt discrimination when job qualifications allegedly preclude hiring members of protected classes.[29] The employer alleges that substantially all members of a class are incapable of performing the job-related functions of the specific job in question. Thus all persons who are of a certain sex, religion, disability, race, or national origin would be precluded from holding the job. Such a requirement has obvious potential for discrimination and thus is required to be certified by the Director in advance of any use. This allows the Director to weed out inappropriate uses of the bona fide occupational qualification exception and to suggest other alternatives which have a lesser adverse impact upon protected classes.

Employers utilizing this exception must justify it by proving that there is a reasonable cause to believe that substantially all persons within the class are incapable of performing the job involved. This necessitates the showing of a factual basis for such a belief. The bare assumption that a member of a class is incapable of doing a job is insufficent since that assumption is often based upon the employer's misconception of what constitutes a physical risk and upon ignorance of the problems that disabled persons can handle. Absent a factual showing, individual members of the class must be allowed to demonstrate their ability to perform the essential functions of the job in question.[30]

Adverse reactions from peers, fellow workers, or customers would not justify utilizing a bona fide occupational qualification to exclude disabled individuals. The fact that fellow workers may be uncomfortable around disabled persons does not justify their exclusion from employment. These are the prejudices that the act is intended to overcome.[31] Nor would an employer be able to exclude disabled individuals based upon the employer's notions about proper roles or occupations for disabled individuals. (See statutory language in statute 5(1)(d).)

(c) Nothing in this section shall prohibit the operation of a bona fide affirmative action program which complies with this exception. An employer, employment agency, or labor organization may as part of a remedial affirmative action program to correct the effects of past discrimination pursuant to a consent order, hearing examiner's order or court order or as part of a bona fide voluntary affirmative action program, invite applicants for employment, promotion or membership to indicate their race, color, age, sex or any disability, provided that the employer, employment agency or labor organization:

(i) States clearly on any written form or questionnaire used for this purpose or makes clear orally if no written form or questionnaire is used that the information requested is intended for use solely in connection with its remedial action obligations or affirmative action efforts;

(ii) States clearly that the information is being requested on a voluntary basis, that it will be kept confidential, that refusal to provide it will not affect employment, promotion or membership or subject the applicant or employee to any adverse treatment, and that it will be used only in accordance with this act; and

(iii) Attaches a summary of the plan and how it works for the applicant's information.

COMMENT. Affirmative action programs provide an exception to the prohibition against pre-employment inquiries. However, since information about membership in protected classes can be easily used for discriminatory purposes, the bona fide affirmative action programs must comply with the requirement of this exception. These requirements are quite simple. The program must be voluntary on the part of the applicant for employment. The applicant must be clearly informed that the information requested is to be given voluntarily, that it will be kept confidential if given, that it will be used solely for affirmative action purposes, and that failure to provide the information will not affect any employment action taken as a result. The form must also include a summary of the employer's affirmative action program so that the applicant has a meaningful understanding of what is being asked and of what use will be made of the information. Any affirmative action program which does not comply with these basic requirements is in violation of prohibitions on pre-employment inquiries.

(d) Nothing in this section shall prohibit pre-employment or pre-promotional inquiries which are directly related to an applicant's ability to perform essential job-related functions.

COMMENT. An employer is not prohibited from using tests and making inquiries, provided that they are directly related to the ability of the applicant to perform essential job-related functions and are not designed to elicit information about membership in a class. The employer is to compare applicants' abilities with essential job-related functions. (See discussion in comment to paragraphs 5(1)(d) and 5(2)(d) and (e).) Such questions may inquire into various aspects of educational background, experience, job-related performance and

[29] *See* Note, *supra* note 22, at 476.

[30] *Id.* at 478-479.

[31] *Id.* at 480.

the ability to perform safely where this is a job-related function, as well as seek other information useful for determining the applicant's ability to perform essential functions.

(e) Nothing in this section shall prohibit an employer from conducting a comprehensive medical examination solely for the purpose of determining the person's capability to perform available employment, provided that all entering or promoted employees are subjected to examination and investigation and provided that the results of such an examination shall be used only in accordance with the requirements of this section. Whenever an employer inquires into an applicant's or employee's physical or mental condition or conducts a medical examination prior to employment or change in employment status, information obtained in response to such inquiries or examination shall be collected and maintained on separate forms and shall be kept confidential as medical records except that supervisors and managers may be informed of accommodations for and restrictions on the work or duties of disabled persons. First aid and safety personnel may be informed to the extent appropriate, if the condition might require emergency treatment, and government officials investigating compliance with this act shall be provided relevant information upon request.

COMMENT. Medical examinations are a valid means of employment screening for particular jobs. Thus, nothing in this act is intended to interfere with an employer's examination of an applicant's physical health and abilities where this is an indicator of the applicant's ability to perform essential job functions and provided that all applicants are subjected to such examinations. If certain physical attributes are required for a particular job then all applicants should be screened for such physical attributes.

Any information arising out of these medical examinations is to be collected and maintained separately from other information and be kept confidential. Such information may be made available to necessary personnel but otherwise is not to be made public, since no other individual should have any interest in receiving such information. Government officials are excepted from this requirement in order to carry out their enforcement functions under the act.

(f) The employment of one person instead of, or in place of, another, standing by itself, shall not be evidence of an unfair or discriminatory practice.

(g) Nothing in this section shall be interpreted to prevent the termination or change of the employment of any person who is unable to adequately perform his or her duties, nor to preclude discrimination among individuals on the basis of competence, performance, conduct or any other reasonable standard, if the employer has complied with subsection 7 of this section.

COMMENT. These exceptions are included to allay fears about unjustified charges of discrimination. Any such charges should be screened out during the investigation process provided for in subsections 21(4) and 21(5).

(h) The provisions of this act relating to discrimination because of age shall not be construed to apply to any retirement plan or benefit system of any employer unless such plan or system is a subterfuge adopted for the purpose of evading the provisions of this act.

COMMENT. Retirement plans or other benefit systems are, of course, designed to take age into account based upon actuarial data. Nothing in the act is intended to affect them. Such plans, however, are subject to other provisions of the act such that they may not, for example, discriminate on the basis of sex.

(i) The provisions of this section shall not apply to any action taken on the basis of disability or age, pursuant to law or regulation governing any employment or training program, which is designed to benefit persons of a particular disability or age group.

COMMENT. Many governmental programs are designed to meet the special needs of particular groups, such as farmers, consumers, dependent children and veterans. The fact that some of these groups are also covered by the act should not preclude their special needs being met. Thus, governmental employment or training programs designed to meet the special employment needs of disabled or senior citizens are explicitly excluded.

(j) Nothing in this section shall apply to a bona fide religious institution as defined in subsection 8(1) with respect to any qualifications for employment based upon religion when those qualifications are for a bona fide religious purpose.

COMMENT. This exception recognizes and is likely mandated by the right to freedom of religion. The exception applies only to religious criteria for employment and only where the criteria have a religious purpose. Thus, for example, where a religious institution engages in a secular activity for which it hires employees, it may not discriminate with respect to such hiring.

(k) The provisions of this section relating to religion shall not apply if an employer demonstrates that he is unable to reasonably accommodate an employee's or prospective employee's religious observances or practices without undue hardship on the conduct of the employer's business.

COMMENT. Employers are exempted from honoring employee religious practices only if the employer can show that the practices create an undue hardship on the employer's business. This normally entails a change in work days or hours. Whether these changes create an undue hardship is largely dependent upon the number, composition, and structure of the employer's work force, number and type of facilities, and especially the number of hours of normal operation.

(l) An employer, employment agency, their employees, or agents may advertise and offer employment to only disabled persons when other employment compatible with their ability would not be available to disabled persons because of their disability. Any such employment or offer of employment shall not discriminate among disabled persons on the basis of race, color, religion, national origin, age, marital status or sex.

COMMENT. The purpose of the act is to assure equal access to the job market without the artificial barriers created by stereotypes and prejudice. For those who cannot qualify or compete in the regular job market, specialized work programs can provide an oportunity to utilize or develop available skills. Thus, sheltered workshops for disabled persons are exempted from the act.

(10) Posting of statutory provisions.

Every employer employing over five persons, employment agency, and labor organization subject to this act shall post in a conspicuous place or places on his or her premises a notice to be prepared or approved by the director, which shall set forth excerpts of this act and such other relevant information which the director deems necessary to explain the act. Any employer, employment agency, or labor organization, refusing to comply with the provisions of this section shall be punished by a fine of not less than one hundred ($100) dollars nor more than five hundred ($500) dollars for each violation.

COMMENT. Employees must know about their rights if they are to have any meaning. Posting a notice of rights at places of work is an inexpensive way of providing such information. The notice must be prepared by or approved by the Director to insure that it is accurate and understandable.

Unlike other portions of the act which merely proscribe certain discriminatory actions, this subsection requires a larger employer to act affirmatively to give notice. This section, however, is not applicable to employers of less than five employees since their inclusion would make the enforcement of the provision impractical.

(11) Evidence of predetermined pattern.

The director or hearing examiner, in ascertaining the practices followed by a respondent, shall take into account all evidence, statistical or otherwise, which may tend to prove the existence of a predetermined pattern of discriminatory employment or membership.

COMMENT. Complaints or investigations may include or give rise to allegations of widespread predetermined or systematic patterns of discrimination. Evidence tending to support these allegations may be taken into account in the investigation, conciliation or hearing phases.

(12) Maintenance of employment records.

The director may by rule require employers, employment agencies, and labor organizations to maintain such records on age, sex, disability and race as are consistent with this act and with any requirements of federal law to aid in the administration of the civil rights law and regulations. These records are confidential and shall be available only to federal and state personnel legally charged with administering such laws and regulations. However, the director shall compile statistical information summarizing these records and make it available to the general public.

COMMENT. The Director may require certain employers and others to maintain records on employment to aid in administration and enforcement, consistent with the other provisions of this act and federal law. Thus, an employer may be required to maintain records by race or sex on employees hired, but could not be required to list the race or sex of all applicants since that information is at most given at the applicants' option. These requirements must be set by rule which normally provides for public comment and notice of the requirements. It is anticipated that this provision will be limited to larger employers, employment agencies and labor organizations.

These personnel records are confidential, as are any other personnel records, and are thus not open to public inspection. They are to be made available to state and federal personnel charged with enforcing anti-discrimination laws for whose use they are made. The public's right to know is satisfied by the Director making available statistical summaries of the progress that has been made in eliminating employment discrimination.

Section 6. Discrimination in Real Estate Transactions Prohibited

(1) Additional definitions.

As used in this act:

(a) "Real property" means buildings, structures, real estate, lands, tenements, leaseholds, interests in real estate cooperatives, condominiums, and hereditaments, corporeal and incorporeal, or any interest therein;

(b) "Real estate transaction' means the sale, exchange, purchase or rental, construction, rehabilitation, repair or maintenance of real property or offer to sell, exchange, purchase, rent or lease real property.

(c) "Rent or rental" means to lease, to sublease, to let or otherwise to grant for a consideration the right to occupy premises not owned by the occupant.

COMMENT. Real estate transactions can take many forms and cover many different types of interests. These definitions are intended to cover all types of transactions and interests whatever their form in order to make housing and commercial properties available to all without discrimination. Thus, the act, unlike some existing statutes which are limited to housing only, [32] covers all types of real estate transactions whether they involve a building or other tangible property or an interest in land, whether by sale or rental and whether for personal housing or for commercial endeavors or investment purposes.[33] These transactions also include rehabilitation, repair and maintenance.

(2) Discrimination in real estate transactions prohibited.

On the basis of race, color, religion, national origin, disability, age, marital status, sex or use of adaptive devices, it is an unlawful discriminatory practice for any person having the right to sell, rent, lease, control or manage any real property, whether acting for himself, herself, or another, to:

(a) Refuse to engage in a real estate transaction with a person;

(b) Discriminate against a person in the terms, conditions or privileges of a real estate transaction or in the furnishing of facilities or services in connection therewith;

(c) Refuse to receive or to fail to transmit a bona fide offer to engage in a real estate transaction from a person;

(d) Refuse to negotiate or execute a real estate transaction with a person;

(e) Represent to a person that real property is not available for inspection, sale, rental, or lease when in fact it is so available, or to fail to bring a property listing to his or her attention, or to refuse to permit him or her to inspect real property;

(f) Print, circulate, post or mail or cause to be so published, a statement, advertisement or sign, or to use a form of application for a real estate transaction, or to make a record or inquiry in connection with a prospective real estate transaction, which indicates, directly or indirectly, an intent to make a limitation, specification, or discrimination with respect thereto;

(g) Offer, solicit, accept, use or retain a listing of real property with the understanding that a person may be discriminated against in a real estate transaction or in the furnishing of facilities or services in connection therewith;

(h) Expel a person from occupancy of real property;

(i) Deny any person access to or membership or participation in any multiple-listing service, real estate broker's organization, or other service organization or facility relating to the business of selling or renting housing accommodations, or to discriminate against any person in the terms or conditions of such access, membership, or participation;

(j) Attempt to do any of the unfair practices prohibited by this section.

COMMENT. Although housing is a basic human need, many minority persons, including those with disabilities, have inadequate housing. One reason has been their inability to access the housing market, due to discrimination in real estate transactions. Prohibitions on discrimination cannot remove the economic barriers to inadequate housing but they can assist in removing the artificial barriers of prejudice and discrimination which result in greater differentials in the quantity and quality of housing available to minority groups. Prohibitions on discrimination in real estate transactions apply to all persons who have the right to engage in a real estate transaction, whether as an owner or an agent of the owner. Thus, the prohibitions also apply to an apartment manager who rents units as an agent of an owner. The provisions also cover all aspects of real estate transactions including advertising, listings, solicitation, negotiations, offers and executions of the transaction, or expulsion from occupancy.

(3) Discrimination in financing of real estate transactions.

It is an unlawful discriminatory practice for any person to discriminate on the basis of race, color, religion, disability, marital status, sex or use of adaptive devices in the course of financing a real estate transaction, whether it be in its granting, withholding, extension, modification or renewal or in its rates, terms, conditions, or privileges, or in the negotiation or execution of any item of service related thereto, including issuance of title insurance, mortgage insurance, loan guarantee, or other aspect of the financing, and whether that financing be by mortgage, deed of trust, contract or other instrument imposing a lien or other security in real property.

COMMENT. Few people can afford to buy real estate outright. As a practical matter, financing is as important to obtaining or improving real estate as its actual availability. Therefore, all aspects of financing are covered by anti-discrimination provisions.[34]

---

[32] *See e.g.,* MICH. COMP. LAWS ANN. §37.1502(b) (Supp. 1978-1979); N.M. STAT. ANN. §4-33-7 G(2) (Supp. 1975); PA. STAT. ANN. §955(h)(3) (Purdon Supp. 1978-1979).

[33] For examples in existing state statutes *see* MD. ANN. CODE art. 49B §22 (Supp. 1978); MONT. REV. CODES ANN. §64-306(4)(a) (Supp. 1977); ORE. REV. STAT. §659.430(1)(a) (1973).

[34] For examples in existing state statutes *see* KY. REV. STAT. §207.190 (Supp. 1976); MICH. COMP. LAWS ANN. §37.1504

(4) Discrimination in soliciting information prohibited.

In connection with the negotiation, execution or financing of a real estate transaction, no person shall use or make or intend to use or make any form or record or inquiry which expresses, directly or indirectly, any limitation, specification, or discrimination as to race, color, religion, national origin, disability, age, marital status, sex or use of adaptive devices. Nothing in this section shall limit the effect of section 9 relating to unfair practices in credit transactions.

COMMENT. The prohibitions on discrimination are carried through to the formalization of offers, acceptances and agreements throughout various stages of real estate transactions. Thus, no forms or records used or made relating to a real estate transaction may be used if they contain references to the prohibited bases of discrimination. In addition, this subsection makes it clear that the provisions on credit are also applicable to financing where credit is extended. So too are the exemptions relating to age and marital status in the credit transactions and the use of forms and records relevant to the transaction. (See subsection 9(4).)

(5) Inducing a sale or rental by representing a change in neighborhood prohibited.

For the purpose of inducing a real estate transaction, it shall be an unlawful discriminatory practice for any person who may financially benefit from that transaction, to:

(a) Represent that a change has occurred or will or may occur with respect to race, color, religion, national origin, disability, age, marital status, sex or use of adaptive devices in the composition of the owners or occupants in the block, neighborhood or area in which the real property is located; or

(b) Represent, directly or indirectly, that this change will or may result in undersirable consequences in the block, neighborhood, or area in which the real property is located, including but not limited to the lowering of property values, an increase in criminal or antisocial behavior, or a decline in the quality of schools or other public facilities.

COMMENT. As a corollary to prohibitions on discrimination in real estate transactions this subsection prohibits "scare tactics" and "block busting" by those who may stand to profit from such practices by buying property at reduced prices and selling to purchasers at a high rate of profit. Such practices are detrimental to the integrity of neighborhoods, relations among neighbors and groups, and the financial interests of both buyers and sellers of the property and are therefore prohibited.

(6) Discrimination through restrictive covenants prohibited.

(a) Every provision in a written instrument relating to real property which purports to forbid or restrict the conveyance, encumbrance, occupancy or lease thereof to individuals of a specified race, color, religion, national origin, or disability, and every condition, restriction or prohibition, including a right of entry or possibility of reverter, which directly or indirectly limits the use or occupancy of real property on the basis of race, color, religion, national origin or disability is void as against the public policy of this state and shall be given no legal or equitable force or effect.

COMMENT. This subsection prohibits doing by private agreement through restrictive convenants or other deed restrictions what is otherwise forbidden by the act. These private agreements are against the express public policy of the state and are therefore declared void whether they are for existing or future restrictions.

Similar statutory provisons have been adopted in at least two states voiding private agreements prohibiting the sale, lease or use of real property based upon race, religion or national origin.[35]

(7) Exceptions.

(a) Notwithstanding any other provision of this act it shall not be an unfair practice or a denial of civil rights for any public or private educational institution to separate the sexes or give preference to or limit use of dormitories, residence halls or other student housing to persons of one sex or to make distinctions on the basis of marital or family status.

(Supp. 1978-1979); N.J. STAT. ANN. §10:5-12(i) (West 1976); N.M. STAT. ANN. §4-33-7(H) (Supp. 1975); PA. STAT. ANN. tit. 43 §955(h)(2) (Purdon Supp. 1978-1979). *See also* Bonfield, *State Civil Rights Statutes: Some Propsals*, 49 IOWA L. REV. 1067, 1106-07.(1964).

[35] *See* N.Y. GEN. OBLIG. LAW §5-331 (McKinney 1978) and TEXAS REV. CIV. STAT. ANN. art. 1293a (Vernon Supp. 1978). While no cases have yet interpreted the statutes in these two states, there is case law upholding provisions voiding existing agreements for the rental of property. *See* Industrial Development & Land Company v. Goldschmidt, 56 Cal. App. 507, 206 P. 134 (1922). *See also* Home Building and Loan Ass'n. v. Blaisdell, 290 U.S. 398 (1934).

Judicial enforcement of restrictive racial covenants in land deeds is unconstitutional. Shelley v. Kraemer, 334 U.S. 1 (1948). But while such covenants are judicially unenforceable, it requres legislative action to make the covenants themselves illegal. Title VIII of the Fair Housing Act of 1968, §804(c), 42 U.S.C. §3604(c) (1970) contains language very similar to subsection 6(4) of this act. Based largely on this language at least one court has found it unlawful to include such covenants in deeds even if no effort is made to enforce them. Mayers v. Ridley, 465 F. 2d 603, n. 1 (D.C. Cir. 1972), *rev'g* 330 F. Supp. 417 (D.D.C. 1971). In *Mayers v. Ridley* homeowners whose deeds contained restrictive racial covenants successfully brought suit enjoining the recorder of deeds from accepting for filing any instrument containing a restrictive racial covenant and requiring the recorder of deeds to mark such instruments in the volumes in which they are recorded to indicate that such covenants are void and unenforceable.

COMMENT. This is a standard provision in many statutes and it is included solely for the convenience of subsequent drafters.

(b) Notwithstanding the other provisions of this act, a person may sell, rent, or advertise any dwelling to aged or elderly persons only, or to disabled persons only or to persons of one sex only, if the dwelling is planned exclusively for or occupied exclusively by such persons.

COMMENT. Certain groups of persons sometimes need or desire special housing to serve their needs. Examples are homes for elderly persons which may provide special facilities or services or where older persons may choose to live without the noise of children or the lifestyles of younger persons, housing for disabled persons designed to be totally accessible and perhaps with special services available to residents, or housing available to battered women in the community. The act is not intended to interfere with meeting these special housing needs.

(c) Nothing in this act shall prohibit the renting of rooms by a nonprofit organization, if the discrimination is by sex.

COMMENT. This paragraph is a standard provision in many existing statutes and is included solely for the convenience of subsequent drafters. It exempts the renting of rooms to one sex by non-profit organizations like the YMCA and YWCA or youth hostels. Other prohibitions on discrimination still apply.

(d) Nothing in this act shall apply to the renting of less than five single rooms in a single dwelling by an occupant of such dwelling nor to the renting of a housing unit in a building which contains housing units for no more than two individuals or families living independently of one another if the owner resides in one of such housing units.

COMMENT. This exemption gives recognition to the sanctity of one's home and freedom of association. It provides a balance between society's interest in equal opportunity and the individual's interest in freedom of association. Where the unusual intimacy of opening one's home is involved there should be a bias in favor of freedom of association. Therefore, when one rents less than five rooms in one's home whether one owns it or not, the act permits selection of tenants on any basis the lessor sees fit to use. The exemption also allows this choice where an owner rents the other half of a duplex in which he or she resides. No exemption, however, is allowed where the housing provides three or more separate housing units, e.g. an apartment house having 3 or more units or an older house which has been divided up into separate apartments where facilities other than common access ways are not shared. In these circumstances the unusual intimacy between landlord and tenant which justifies the exemptions is no longer present.[36]

(e) Nothing in this act shall be construed to:

(i) Require any person or group of persons selling, renting or leasing property to modify the property in any way, or exercise a higher degree of care for a person having a disability than for a person who does not have a disability;

(ii) Relieve any person or persons of any lawful obligations generally imposed on all persons in a written lease, rental agreement, or contract of purchase or sale;

(iii) Forbid distinctions based on the inability to fulfill the terms and conditions, including financial obligations, of such lease, agreement or contract; or

(iv) Prohibit an owner, lender or his agent from requiring that a person who seeks to engage in, or obtain financial assistance for, a real estate transaction supply information concerning the applicant's financial, business, or employment status or other information designed solely to determine the applicant's credit worthiness.

COMMENT. Nothing in this section is intended to affect the tortious or contractual liability of landlord and tenant or seller and buyer except as they may arise out of discriminatory practices in violation of the act. For example, a seller or landlord is not required to make any changes in the property to meet the needs of a disabled buyer or tenant. Nor is a landlord subject to any higher degree of care in a tort action by a disabled tenant. This provision of the act is intended to allay the fears of landlords about potential liability above that normally experienced. These fears are otherwise likley to discourage renting to disabled persons.

Other exemptions are intended to make clear that normal obligations imposed in leases and contracts are unaffected by this section so long as these obligations are imposed on all persons indiscriminately. Finally, no restriction is placed on efforts to determine credit worthiness. Inability to meet the obligations of a rental or contractual agreement remain a valid basis for refusing to engage in a real estate transaction.

Section 7. Discrimination in Public Accommodations Prohibited

(1) Additional definitions.

(a) "Full and equal enjoyment of" means the right to purchase without discrimination any service, commodity or article of personal property offered or sold on, or by, any establishment to the public, to admission without discrimination of any person to accommodations, facilities or privileges of any place of public accommodation

---

[36] *See* Bonfield, *supra* note 27, at 1105-06.

and to be free of acts directly or indirectly causing persons to be treated as not welcome, accepted or desired.

COMMENT. Persons have the right to enter a public accommodation and to conduct business or engage in appropriate activities on the same basis as any other person. Thus, for example, persons are entitled to be served or treated on an equal basis with other members of the public.[37]

(b) "Public accommodation" means, but is not limited to, any place, property, vehicle or facility, licensed or unlicensed, which is made available for public use or enjoyment with or without charge, or where charges are made for admission, service, occupancy or use. The term shall include, but is not limited to, places, facilities, vehicles or other property used for the entertainment, housing or lodging of transient guests or for the benefit, use or accommodation of those seeking health, recreation or rest, or for the burial or other disposition of human remains, or for the sale of goods, merchandise, services, or personal property, or for the rendering of personal services, or for public conveyance or transportation on land, water, or in the air, including the stations and terminals thereof and the garaging of vehicles, or where food or beverages of any kind are sold for consumption on the premises, or where public amusement, entertainment, sports or recreation of any kind is offered with or without charge, or where medical service or care is made available, or where the public gathers, congregates, or assembles for amusement, recreation or public purposes, or any public halls, public elevators and public washrooms of buildings and structures occupied by two or more tenants, or by the owner and one or more tenants, or any public library.

COMMENT. All places, property, buildings, vehicles or other facilities which are available for public use and enjoyment are covered by this section regardless of whether a charge is made for use or enjoyment or whether they are licensed by the state or a subdivision. The definition does not attempt a complete listing, since such a listing would be impossible without leaving something out. However, examples include all retail stores, restaurants, theaters, barber shops, dry cleaners and laundries, taxis, buses, trains, planes, cemeteries, swimming pools, hospitals and doctors' offices or any public parks, recreational facilities or places for spectator sports.

(2) Unfair practices.

It is an unfair and unlawful practice for any person, because of race, color, religion, national origin, disability, age, marital status, sex, or use of adaptive devices to:

(a) Refuse or deny an individual the full and equal enjoyment of the goods, services, facilities, privileges and accommodations of a public accommodation; or

(b) Print, circulate, post, mail, or otherwise cause to be published a statement, advertisement, or sign which indicates that the full and equal enjoyment of the goods, services, facilities, privileges and accommodations of a public accommodation will be refused, withheld from, or denied an individual, or that an individual's patronage of, or presence at, a place of public accommodation is objectionable, unwelcome, unacceptable or undesirable.

COMMENT. The conscious decision to exploit the opportunities available by serving the general public necessitates acceptance of the obligation to provide equal opportunity and enjoyment to all who wish to patronize that endeavor. Precisely because such endeavors cater to the general public, the contacts involved are likely to be brief, sporadic and non-social. Consequently, the right to freedom of association in this context is greatly overshadowed by the right to equal opportunity. Having invited the patronage of the public, one cannot justifiably bar members of minority groups, especially when the aggregate of such exclusions works such an unjust and unwarranted injury on such groups and society as a whole.[38]

(3) Exceptions.

(a) Nothing herein contained shall be construed to include or apply to:

(i) Any institute, bona fide club or accommodation, which is by its nature distinctly private, including fraternal organizations, except that the use shall be covered to the extent that:

[a] **Public use is permitted or that the goods, services, facilities, privileges or accommodations of the private club or establishment are made available to the customers or patrons of another establishment that is a place of public accommodation; or**

[b] **It is licensed, chartered, or certified to do business by the state or any of its political subdivisions.**

COMMENT. Any accommodation which is distinctly private and does not cater to the general public is exempted in recognition of the right to freedom of association. Under these circumstances the interest in freedom of association outweighs the interest in equal opportunity since the interaction is largely social and the accommodation is not open to the public, but caters only to particular individuals.

[37] For a general discussion of statutory provisions prohibiting discrimination in public accommodations *see* Bonfield, *supra* note 27, at 1095-1107.

[38] *Id.* at 1100.

This exemption does not apply where the accommodation is in fact open to the public at large during the time and to the extent it is open. For example, a club that opened its facilities to the public when not in use by its members must comply with the act during the time public use is permitted. The exemption also does not apply where the accommodation is licensed to do business by the state or its subdivisions.

(ii) Any bona fide religious institution as defined in subsection 8(1) with respect to any qualifications the institution may impose based on religion when such qualifications are related to a bona fide religious purpose; or

COMMENT. In the interests of freedom of religion, religious institutions are exempted from the provisions of the section so long as the discrimination is based solely on religion and is done for a bona fide religious purpose.

(iii) The rental or leasing to transient individuals of less than four rooms within a single dwelling by the occupant of such dwelling.

COMMENT. Where a person lives in a single dwelling and rents to transient individuals, whether as an owner or as a lessee who sublets, the interest in freedom of association outweighs the public interest in equal opportunity to the accommodations. (See comment to paragraph 6(7)(d).) This is the case even though there is a less extended contact in this situation than there would be in rentals by the month[39] for there is less opportunity for screening those whom one admits to one's home. This exemption, however, applies only where less than four rooms are rented. The larger the number rented, the less personalized are the relationships and the less equity does freedom of association have vis-a-vis equal opportunity.

(b) The provisions of this section relating to sex shall not apply to such facilities as restrooms, locker rooms, and other similar places.

COMMENT. None necessary.

(c) Nothing in this section shall be construed or interpreted to prohibit the proprietor of any public accommodation or its employees from denying service to any person for failure to conform to the usual and regular requirements, standards, and regulations for the public accommodation as long as the denial is not based upon discrimination on the basis of race, color, religion, national origin, disability, age, marital status, sex or use of adaptive devices.

COMMENT. None necessary.

Section 8. Unfair Practices by Educational Institutions

(1) Additional definitions.

As used in this act:

(a) "Educational institution" means a public or private institution, including an academy, college, elementary or secondary school, extension course, kindergarten, nursery school system, school district, or university, and a business, nursing, professional, secretarial, technical or vocational school or an agent thereof.

COMMENT. The broad definition covers all types of educational facilities or programs at all levels of general, technical, vocational or professional education.

(b) "Qualified individual" means, for purposes of this section, an individual who meets or can adapt to the nondiscriminatory academic and technical standards of an educational institution.

COMMENT. An individual is qualified for purposes of this section if the individual can meet or adapt to the admission standards of an institution, provided those standards are nondiscriminatory. These include all academic and non-academic criteria which are essential to effective participation in the educational program. Thus, an individual with blindness, cerebral palsy, or epilepsy could not be excluded if he or she could adjust to the institution's programs with reasonable adaptation. However, an individual who is severely retarded or physically disabled *and* needs special programming would not be a qualified individual since that person could not adapt to regular programming.

Note that the scope of this definition and of this section is narrower than that of the Education for All Handicapped Children Act, Pub. L. 94-142, 20 U.S.C. §§1401 et seq. (1976) and the Rehabilitation Act of 1973, §504, 29 U.S.C. 794 (1976). This section deals only with discrimination in educational institutions. Issues such as a free appropriate public education, identification and location of unserved handicapped children and evaluation procedures are dealt with in the model education statute in this series. A state's version of Section 504 provisions covering nondiscrimination by recipients of state funds is included in section 11 of this act.

This section is intended to supplement those provisions. It covers such things as private preschool programs and trade and occupational schools not receiving federal or state funds. Qualified individuals could not be excluded from these programs. For those programs or institutions covered by federal legislation, individuals may wish to look to federal requirements which are more extensive in their protections.

(c) "Religious institution" means an institution which is operated, supervised, controlled or sustained primarily by a religious organization, or is one which is stated by the parent religious body to be and is, in fact, officially related to that

[39] *Id.* at 1105-06.

religion by being represented on the board of the institution and by providing substantial financial assistance, and which has certified, in writing, to the director that it is a religious educational institution.

COMMENT. Religious institutions are limited to those institutions operated by or officially related to a religious organization. Those institutions include parochial schools, seminaries or colleges.

(2) Unfair practices in admission and benefits.

It shall be an unfair and unlawful practice for any educational institution on the basis of race, color, religion, national origin, disability, age, marital status, sex or use of adaptive devices to discriminate against any qualified individual's use of or benefit from the institution, or its services or exclude, expel, limit, or otherwise discriminate against a qualified individual seeking admission, being admitted or enrolled as a student.

The obligation to comply with this section shall not be obviated or alleviated because employment opportunities in any profession or occupation are or may be more limited for disabled persons than for non-disabled persons.

COMMENT. Discrimination is prohibited in the admission to the institution, in the use of its services, or in the treatment of students. Students cannot be discriminated against in academic programs or in nonacademic or extracurricular programs or services when these are provided by the institution. These include counseling, health, financial assistance, transportation, housing or athletics. In addition, the fact that educational personnel feel that employment opportunities are not promising does not justify excluding a disabled person, a black, a woman, or a person over 40 from the program. The decision to attend should be made by the applicant. To allow such discrimination perpetuates the self-fulfilling prophecy that certain people cannot enter certain jobs.

(3) Unfair practices in records and forms.

It shall be an unfair and unlawful practice for any educational institution to make or use a written or oral inquiry or form of application for admission that elicits or attempts to elicit information, or make or keep a record, concerning the race, color, religion, national origin, disability, age, marital status, sex, or use of adaptive devices by an applicant for admission for reasons contrary to the provisions or purposes of this act or to indicate a preference, limitation, specification or discrimination in a printed or published catalog or other notice or advertisement.

COMMENT. Indications of preference or limitation can function to discourage applicants and provide an opportunity for discrimination. Thus, educational institutions are prohibited from preadmission inquiries concerning race, religion, disability and so forth other than for the purpose of correcting past discrimination, keeping health records or other purposes not contrary to the purpose of the act.

(4) Exceptions.

(a) It is not an unfair discriminatory practice for a religious institution to limit admission or give preference to applicants of the same religion;

COMMENT. Consistent with the guarantee of freedom of religion, religious institutions are exempted from the provisions of this section where discrimination is based solely upon religion. However, other provisions of the act do apply.

(b) The provisions of this section relating to sex shall not apply to a private educational or religious institution, or one of its branches or divisions, in which students of only one sex are permitted to enroll;

COMMENT. This section is included since it is a common feature of statutes allowing the continuation of institutions which are private in nature and optional in attendance.

(c) Nothing in this section shall limit the functioning of an affirmative action program.

COMMENT. Educational institutions are allowed, indeed encouraged, to engage in bona fide affirmative action programs. Nothing in this section is intended to prevent such efforts. Thus, they may advertise, recruit and give preference to protected classes consistent with federal and state laws.

Section 9. Unfair Practices in Credit Prohibited

(1) Additional definitions. As used in this section only:

(a) "Credit" means the right granted by a creditor to a person to defer payment of a debt, or to incur debt and defer its payment, or to purchase property or services and to defer payment therefor.

(b) "Creditor" means any person who regularly extends, renews, or continues credit, any person who regularly arranges for the extension, renewal, or continuation of credit, or any assignee of an original creditor who participates in the decision to extend, renew, or continue credit, whether or not any interest or finance charge is required.

(c) "Credit reporting agency" means any person who for monetary fees, dues, or on a cooperative nonprofit basis, regularly assembles or evaluates credit information for the purpose of furnishing credit reports to creditors.

(d) "Age" means age of majority.

(2) Unfair practices by creditors.

(a) It shall be an unlawful discriminatory practice for any creditor on the basis of race,

color, religion, national origin, disability, age, marital status, sex, or use of adaptive devices to:

(i) Discriminate against any applicant for credit in the granting, withholding, extending, or renewing of credit, or in the fixing of the rates, terms, or conditions of any form of credit;

(ii) Use or make any inquiry for the purpose of limiting or specifying those persons to whom credit will be granted;

(iii) Refuse to consider the sources of income of an applicant for credit, or disregard or ignore the income of an applicant, in whole or in part;

(iv) Impose any special requirements or conditions, including, but not limited to, a requirement for co-obligators or reapplication, upon any applicant or class of applicants in circumstances where similar requirements or conditions are not imposed on other applicants similarly situated, unless the special requirements or conditions that are imposed with respect to age are in accordance with paragraph 4(e) of this section or are the result of programs that grant preferences to certain age groups administered by instrumentalities or agencies of the United States, a state, or a political subdivision of a state;

(v) Fail or refuse to maintain records on any account established after (date) or to furnish information on the accounts to credit reporting agencies in a manner that clearly designates the contractual liability for repayment as indicated on the application for the account, and, if more than one individual is contractually liable for repayment, to maintain records and furnish information in the name of each individual. This section does not apply to individuals who are contractually liable only if the primary party defaults on the account.

(b) It shall be an unlawful discriminatory practice for any creditor to:

(i) Refuse to grant credit to an individual in any name that individual customarily uses, if it has been determined that the individual is otherwise qualified for credit, or to maintain, upon the request of the individual, a separate account for that individual to whom credit is extended;

(ii) Fail or refuse to provide an applicant for credit a written statement of the specific reasons for rejection of the application if requested in writing by the applicant within sixty days of the rejection. Such a statement must be provided within thirty days after receipt of such request;

(iii) Fail or refuse to print on or firmly attach to each application for credit, in a type size no smaller than that used throughout most of the application form, the following notice: "The laws against discrimination require that all creditors make credit equally available to all credit-worthy persons, and that credit reporting agencies maintain separate credit histories on each individual upon request. The (name of state) office for civil rights is responsible for enforcing this law."

(3) Unfair practices by credit agencies.

It shall be an unfair practice for any credit reporting agency to fail or refuse to:

(a) Maintain, upon the request of the individual, a separate file on each individual about whom information is assembled or evaluated; or

(b) Clearly note, maintain and report any information furnished it under paragraph (2)(a)(v) of this section.

(4) Exceptions.

(a) This section does not prohibit a creditor from requesting the signature of both spouses to create a valid lien, pass clear title, or waive inchoate rights to property.

(b) Nothing contained in this section shall bar a creditor from reviewing an applicant's credit history, and application for credit on the basis of established criteria used in the normal course of business for the determination of credit worthiness.

(c) An inquiry concerning marital status does not constitute discrimination for the purposes of this section, if the inquiry is made for the purpose of ascertaining the creditor's rights and remedies applicable to the particular extension of credit.

(d) This section shall not apply with respect to any inquiry, elicitation of information, record or form of application required of such creditor by an instrumentality or agency of the United States.

(e) This section shall not apply with respect to age in the financing of any real estate transaction between a financial institution, insurance company or similar institution and its customers.

COMMENT. The credit provisions contained in this section[40] are consistent with the intent of the act as a whole, i.e., credit is to be extended based upon credit worthiness and not one of the prohibited bases.

The section contains general provisions covering extension of credit, fair reporting of credit decisions and record-keeping. Special provisions relating to age and marital status are also included. Age for purposes of this section is the age of majority since discrimination against younger persons is a possibility here. Other age provisions allow an exception to the section's application for real estate transactions with financial institutions

---

[40] This section is based upon the Ohio credit practices provisions administered by the Ohio Civil Rights Office. OHIO REV. CODE ANN. §4112.021 (Page Supp. 1977).

when advanced age and reduced life expectancy and income make payment more uncertain. The section also contains provisions relating to marital status, maintenance of separate accounts, consideration of all sources of income and exceptions to allow creditors to protect their interests based upon contractual or property law provisions relating to marriage.

Unlike most of the other substantive sections of the act, this one has no provisions relating directly to disabled persons. They are to be treated as all other persons in the extension of credit. Yet, this provision is, of course, important to disabled persons since credit is necessary to finance educational or business opportunities, purchase real estate for commercial endeavors or personal housing, or buy durable goods. Thus, it is an important part of the act's protections.

Section 10. Unfair Practices in Insurance Prohibited

(1) As used in this section:

"Insurance transaction" means any soliciting, negotiating, or executing of an insurance contract, or insuring or transacting of matters subsequent to the executing of the contract and arising out of it.

COMMENT. This section covers all phases of insurance transactions between the insurer and the insured party, including the initial solicitation, negotiation or execution of the contract, its terms, and any transaction arising out of the contract. In addition, all types of insurance are covered, including health, auto, home and life, whether they be bought by or for an individual or group. (Distinguish coverage under this section with prohibitions on discrimination in employee benefits, including insurance under group or other policies, provided by an employer, contained in paragraph 5(2)(b)).

(2) Unfair practices in insurance transactions.

No person engaged in the business of insurance in this state, whether acting for himself or herself or another in connection with an insurance transaction, shall fail or refuse to issue any contract of insurance or cancel or decline to renew such contract because of race, color, religion, national origin, disability, marital status, sex or use of adaptive devices by the insured or prospective insured. The amount of benefits payable, or any term, rate, condition, or type of coverage shall not be restricted, modified, excluded, increased or reduced on the basis of race, color, religion, national origin, disability, marital status, sex, or use of adaptive devices by the insured or prospective insured.

COMMENT. This section does not require mandatory or compulsory insurance.[41] Rather, it requires that all insurance transactions be based upon legitimate underwriting standards without discrimination. Thus, the section prohibits discrimination in an insurance transaction solely on the basis of race, color, religion, national origin, disability, marital status, sex or use of adaptive devices. Age is not included, since it is a well-established risk factor in insurance transactions and thus is a permissible basis for differential treatment.

(3) Exceptions.

(a) The provisions of this section shall not prohibit fair discrimination on the basis of disability, marital status, sex or use of adaptive devices when bona fide statistical differences in risk have been substantiated. Such discrimination must be justified by claims experience and actuarial projections and other data sufficient to establish significant and substantial differences in class rates or coverage.

(b) Nothing in this section shall be construed to require an insurer to provide coverage against a person's disability which the applicant or policyholder has already sustained.

COMMENT. Some bases for discrimination may also be factors which affect risk and therefore can affect coverage and rates. Nothing prohibits the use of such factors where they can be shown to create substantial differences. Insurers, however, cannot base decisions on assumptions or unreliable data which does not establish differences in risk. These assumptions, like those made in other areas covered by this act, can result in denial of insurance coverage, increased rates or reduced coverage.

In addition, nothing in the section requires an insurer to insure against a pre-existing disability. Thus, a person may be covered for all health problems except pre-existing conditions which are excluded from coverage by the policy. This is a standard and obviously fair practice of many insurance companies.

Alternative. *This section merely contains a general prohibition against discrimination in insurance transactions by private insurers. Yet, the insurance industry is different from other areas covered by this act, since it is regulated in the public interest. However, the regulatory controls in most states have not addressed discrimination on the basis of disability. While many disabled persons have been able to obtain insurance coverage, they have often had to pay higher rates and absorb greater risks than non-disabled persons even though there has been little actuarial support for*

[41] References to mandatory insurance programs may be found in Friday, *Insurance: Reaction Comments*, in THE MENTALLY RETARDED CITIZEN AND THE LAW 169 (M. Kindred, ed., 1976); Goldberg, *State Laws Mandating Mental Health Insurance Coverage*, 28 HOSPITAL AND COMMUNITY PSYCHIATRY 759 (October 1977).

this differentation. The Office is unlikely to have the expertise to challenge this form of discrimination. Similarly, it is unlikely that individuals already at an economic disadvantage, as many disabled persons are, will have the resources to establish that an insurance rate is excessive. Hence, this statute encourages the state insurance regulatory agency to close this gap in its control of the insurance coverage and rates offered.

Such an approach does have a danger, however, aside from the broad problem of the regulatees controlling regulatory agencies. That is, many disabled persons, who are now insured under individual health, life, or auto policies at normal rates, may end up with higher rates when classes or categories are used to establish ratings and rates. Since little data or experience exists on creation of classes of disabilities as a rating mechanism and since what does exist is of questionable reliability, it makes the establishment of a valid and reliable rate system very difficult.[42]

Other insurance issues beyond the scope of this act include whether and when certain individuals have the ability to contract for insurance, whether certain health-related problems such as custodial care, catastrophic medical expenses, mental illness or other disablilities are insurable by private companies or public programs, and whether mandatory or compulsory coverage legislation should be passed.[43]

One writer, although speaking only of mentally retarded persons, has summarized well the difficulties involved in this area:

> *Insurance for mentally retarded citizens is a complex subject. It does not lend itself to simple answers, applicable in all cases. The viewpoints of insurers are continually qualified, depending upon the degree of mental retardation, the socioeconomic circumstances present, the forms of care being received, and the legal status of the individual. The challenges for insurers are (1) to seek practical and workable solutions to the unusual problems of mentally retarded persons and (2) to avoid assuming that special problems exist in the absence of adequate data to support this assumption.*[44]

Section 11. Nondiscrimination in Public Services and Assistance; Licensing; Public Contracts

(1) Additional definitions.

As used in this act:

(a) "Public service" means any facility, program, or service, operated or managed by or on behalf of the state of (name of state) or any subdivision thereof.

COMMENT. This definition includes grounds, parks, buildings, pools, recreational or other facilities and any services, information, advice, etc., which are available to some or all members of the public. It is intended to be broad enough to cover all areas and activities of the state and its subdivisions which serve the public.

(b) "Subdivision" means any county, city, town, township, village, independent district or school district in this state.

COMMENT. None necessary.

(c) "Qualified individual" means any person who meets the essential nondiscriminatory eligibility requirements for the receipt of a public service, financial assistance or a license or charter.

COMMENT. None necessary.

(d) "Public contract" means any contract to be performed for or on behalf of the state or its subdivisions.

COMMENT. None necessary.

(e) "Financial assistance" means any grant of funds, loans, services of public personnel, or interest in or use of real or personal property from the state or any subdivision thereof.

COMMENT. None necessary.

(2) Discrimination prohibited.

(a) It is an unlawful discriminatory practice for any person on the basis of race, color, religion, national origin, disability, age, sex, marital status, or use of adaptive devices to exclude a qualified individual from participation in, deny a qualified individual the benefits of or subject a qualifed individual to discrimination under or as a part of any public service.

COMMENT. While public services have different target groups and deal with different problems which naturally necessitate different eligibility requirements for different programs or services, there is no justification for discrimination on the bases listed. Citizens have the right to expect and receive equal treatment in the delivery of public services since these services are supported by public funds.

(b) No agency of the state or its subdivision shall discriminate on the basis of race, color, religion, national origin, disability, age, sex, marital status, or use of adaptive devices, in providing financial assistance to any person.

COMMENT. Public funds also support financial assistance programs. Thus, any discrimination by the state or its subdivisions in the distribution of financial assistance is also prohibited.

(c) It is an unlawful discriminatory practice

---

[42] For discussion of insurance coverage for the mentally retarded *see* Follman, *Insurance, Principal Paper* in THE MENTALLY RETARDED CITIZEN AND THE LAW 144-70 (M. Kindred ed. 1976).

[43] *See id.* and Goldberg, *supra* note 41.

[44] Follman, *supra*, note 42, at 146.

for any person on the basis of race, color, religion, national origin, disability, age, sex, marital status, or use of adaptive devices to exclude a qualified individual from participation in, deny a qualified individual the benefits of or subject a qualified individual to discrimination under any program or activity receiving state financial assistance.

COMMENT. No recipient of state finanical assistance shall discriminate in any program or activity receiving such funds. Just as the state is precluded from discrimination so are persons using state funds to carry out a program. This is a condition upon which financial assistance is accepted.

This paragraph is similiar to Section 504 of the Rehabilitation Act of 1973[45] which prohibits discrimination on the basis of handicap in programs receiving federal financial assistance. It is intended to be implemented in a manner similar to the federal law.[46]

(d) No facility owned, operated or managed by the state or its subdivisions shall be used in the furtherance of any discriminatory practice, nor shall any agency of the state or its subdivisions become a party to any agreement, arrangement or plan which has the effect of sanctioning discriminatory practices. Each such agency shall analyze all of its operations to identify possible instances of noncompliance with this act and shall initiate programs to remedy any noncompliance found to exist.

COMMENT. No public facilities are to be used by others to engage in discrimination. In addition, the state or a subdivision cannot sanction discrimination in any way and must examine their practices for instances of discrimination of which they may not be aware.

(e) No department, board, commission or agency of the state or its subdivisions shall grant, deny, suspend or revoke the license or charter of any qualified person on the basis of race, color, religion, national origin, disability, age, sex, marital status or use of adapative devices. Each such agency shall take appropriate action in the exercise of its licensing or regulatory power to assure equal treatment of all persons similarly situated, to eliminate discrimination, and to enforce compliance with the provisions of this act.

COMMENT. All licensing is done by the state which acts for the public welfare pursuant to its police power. This power must be exercised in conformity with the goal of equality which also contributes to the public welfare. Thus, determinations on licensing are to be made solely on the basis of the licensing or regulatory purpose. (Contrast this paragraph with subsection 22(5) on suspension or revocation of licenses because of discrimination by licensees).

(3) Public contracts.

(a) Any contract in excess of $_____for procurement of personal property and nonpersonal services, including construction, to which the state or its subdivisions is a party shall contain a provision stating that the contractor or the subcontractor shall not commit an unlawful discriminatory practice and shall take affirmative action in the performance of the contract in accordance with paragraph 11(4)(c).

(b) Expenses incurred by a public body or private firm or individual in furthering equal employment opportunities relating to a public contract may be included in the costs of the project to be performed under the public contract. Such expenses may be paid for or financed by any method used to pay or finance other costs of the project.

COMMENT. In accepting the benefits of public contracts the contractor or subcontractor agrees as part of the contract not to engage in discrimination and to take affirmative action in hiring. This subsection is similar to section 503 of the Rehabilitation Act of 1973[47] requiring affirmative action in employment under federal contracts.

Some may question the financial impact of these requirements. There should be none, however, since contractors will include any costs attendant to affirmative action in the bid for the cost of contract performance. Such bidding should also provide a cost control mechanism to hold down such costs.

(4) Affirmative action.

(a) Each department, board, commission or agency of the state, within 180 days after ______, 19_, shall submit to the director for approval an affirmative action program for the hiring, placement, and advancement of persons

---

[45] 29 U.S.C. 794 (1976). Section 504 states that "no otherwise qualified handicapped individual . . . shall, solely by reason of handicap, be excluded from participation in, be denied the benefits of, or be subjected to discrimination under any program or activity receiving Federal financial assistance."

[46] It is intended that the state's governor designate the Office of Civil Rights to develop rules for implementing this section as the Department of Health, Education and Welfare has done. *See* 45 C.F.R. pt. 84 (1977). These should then be used by other state agencies which oversee recipients of state financial assistance to develop rules to enforce the provisions of this section.

[47] 29 U.S.C. 793 (a) (1976). Section 503(a) states: "Any contract in excess of $2,500 entered into by any Federal department or agency for the procurement of personal property and nonpersonal services (including construction) for the United States shall contain a provision requiring that, in employing persons to carry out such contract the party contracting with the United States shall take affirmative action to employ and advance in employment qualified handicapped individuals . . . The provisions of this section shall apply to any subcontract in excess of $2,500 entered into by a prime contractor in carrying out any contract for the procurement of personal property and nonpersonal services (including construction) for the United States . . ." *See* N.J. STAT. ANN. §§10:5-31 to 37 (West 1976) for an example of an existing comprehensive state statute covering public contracts.

who are disabled, over forty (40) years of age, nonwhite or female. Such a program shall be updated annually and shall be reviewed annually and approved by the director.

(b) Any program, activity, organization, or private institution receiving state financial assistance in excess of $______, or having more than ______ employees shall submit to the director for approval an affirmative action program. Programs submitted shall be evaluated and approved or disapproved in accordance with rules adopted by the director.

(c) Each department, board, commission, or agency of the state or its subdivision having contract-letting authority shall submit to the director for approval an affirmative action program to be followed by those parties entering into a public contract in excess of $______. Included in this program shall be a procedure for agency monitoring and the enforcement of compliance with the affirmative action plan by the contractor and subcontractor.

COMMENT. The provisions of the act are primarily directed toward eliminating or redressing current acts of discrimination. Affirmative action is required in the private sector only where necessary to remedy the effects of past discrimination after a pattern of discrimination has been snown and an order issued to eliminate it. The public sector, however, can be expected to go beyond eliminating discrimination to implementing the public policy expressed in the act throughout the activities of state government. Thus, where the state acts directly, or the private sector receives the benefits of public funds through a grant or as party to a contract, affirmative action efforts are required to achieve the act's goals. It is intended to be limited to larger contractors, e.g. over $2500 as required by §503, or recipients of state financial assistance above a selected minimum to reduce the administrative burdens. Those who do not comply with these requirements can be denied further dealings with the state or have the grant or contract terminated. (See subsections 23(7) and 23(8)).

Administrative oversight of affirmative action programs is provided in two ways. State agencies and recipients of state financial assistance must submit their plans to the Director for approval. Thus, they are approved directly by the Office. Contractees, on the other hand, are monitored by the contract-letting agencies in accordance with plans previously approved by the Director, thereby reducing the administrative load on the Director and Office.

(5) Violations; enforcement.

(a) Any person alleging a violation of this section may file a verified complaint with the Director, as provided in section 21.

(b) The Director shall promulgate rules and regulations necessary for the effective implementation of the affirmative action programs mandated herein. It shall be the responsibility of the Director to review periodically the adequacy of such programs and make appropriate recommendations.

COMMENT. In addition to monitoring and enforcement of this section by the state, this provision allows individuals to make use of the standard complaint, investigation, and enforcement procedures of sections 21 and 23. Violators in addition to the general sanctions of the act are subject to loss of benefits under public financial assistance or contracts. (See subsections 23(7) and 23(8).)

(6) Exceptions.

(a) Nothing in this section shall limit the functioning of an affirmative action program.

(b) Nothing in this section shall prohibit or limit any public service whose benefits are limited by federal, state or local statute, ordinance or executive order to persons of a specified race, national origin, age, disability or sex.

COMMENT. Affirmative action programs are to be encouraged and in some cases required. They are clearly not to be affected by the act. In addition, this act should not affect programs which have been established to meet the needs of particular groups. Examples are cultural centers for Spanish-speaking persons or American Indians, recreation centers or programs providing meals for elderly persons, transportation programs for elderly or disabled persons, or rape crisis centers for women.

Section 12. Aiding, Abetting and Obstruction

It is an unfair discriminatory practice for any person to intentionally:

(1) Aid, abet, incite, compel, or coerce a person to engage in any of the practices forbidden by this act; or attempt to do so; or

(2) Obstruct or prevent any person from complying or attempting to comply, or discriminate against any person who is complying with provisions of this act or with any order issued thereunder; or

(3) Resist, prevent, impede, or interfere with the Director or any Office employees or designees in the performance of their duties under this act.

COMMENT. This section covers persons who assist or force others to discriminate or attempt to discriminate, who interfere with or discriminate against others who are complying with the act, or who interfere with the administration or enforcement of the act. These actions are as detrimental to the purpose and spirit of the act as direct discrimination and are, therefore, prohibited.

Section 13. Reprisals Prohibited

It is an unlawful discriminatory practice for any person covered by this act to intentionally engage in any reprisal against any person because that person:

(1) Opposed a practice forbidden under this act, or has filed a complaint, testified, assisted, or participated in any matter in an investigation, proceeding or hearing under this act; or

(2) Associated with a person or group of persons of a different race, color, religion, national origin, disability, age, marital status, sex or using adaptive devices.

COMMENT. Persons who exercise their rights under the act or who assist in enforcement actions pursuant to its provisions are protected from any actions taken in retribution. Without such protection it is likely that persons will be discouraged from enforcing their rights under the act as well as being discouraged from participating as witnesses. For these reasons, this section is vital to the integrity of the act.

Section 14. Programs for Disabled Persons; Conflicting Provisions Not Applicable

(1) Nothing in this act shall be construed to prohibit any program, service, facility or privilege afforded to a person with a disability which is intended to habilitate, rehabilitate or accommodate that person.

COMMENT. Although having common problems, the types of persons covered by this act are a diverse group with different or special needs. Many of these needs are recognized by the exceptions throughout the substantive provisions of the act.

In addition, it must be kept in mind that it is neither necessary nor desirable, and frequently impossible, to meet a broad array of needs in one public program. For example, many programs are realistically designed to meet the special needs of one group of persons. The act is not intended to interfere with the operation of these programs or services. Indeed, without them many disabled persons would never be able to exercise their rights under the act because they could not become "qualified persons," could not live independently or could not become sufficiently mobile to utilize public accommodations. Therefore, these programs are specifically exempted.

(2) The provisions of this act shall apply notwithstanding any provisions of the code or any local law, ordinance or other requirement that imposes on the basis of disability prohibitions or limitations upon the eligibility of qualified disabled individuals to receive services or to practice any occupation or profession.

COMMENT. The provisions of the act are not to be frustrated or circumvented by state or local law which impose restrictions solely on the basis of a disability rather than the qualifications of an individual. Yet, this subsection does not prohibit excluding persons with certain types of disabilities where the disability per se would render the person unqualified.

(3) The obligation to comply with the provisions of this act is not obviated because employment opportunities in any occupation or profession are or may be more limited for disabled persons than for nondisabled persons.

COMMENT. Existing attitudes or potential market conditions should not preclude qualified individuals from pursuing a desired career. That is an individual decision. Thus, this subsection precludes the practice of prohibiting entry into educational institutions or training programs or the use of employment services because "handicapped people don't do that" or because it is believed the person won't be able to get a job.

Note that this subsection does not apply to habilitation programs, vocational rehabilitation or other programs designed to prepare persons for other educational or employment opportunities. The provisions of the act apply only to individuals who are already qualified for admission to educational institutions, for job training programs or for employment.

Section 15. Education Program

In order to eliminate prejudice among the various groups in this state and to further good will among such groups, the Office and the state department of education are jointly directed to prepare a comprehensive educational program, designed for the students of the public schools and for all other residents of this state, calculated to emphasize the origin of prejudice against minority groups, its harmful effects, and its incompatibility with American principles of equality and fair play.

COMMENT. Educational efforts can be aided by the preparation and dissemination of educational packets which can be used with a minimum of preparation by teachers and others in their educational programs. This can be done without great expenditure of money and effort and should yield positive results.

Organization of the Office for Civil Rights[48]

Many state statutes create an Office for Civil Rights with a governing commission having primary responsibility for the functioning of the Office, and often with a Director carrying out the administrative duties.[49] This statutory approach,

[48] *See also* the discussion of the organization of a state advocacy office in Chapter 9: Advocacy, *infra*.

[49] *See e.g.* IND. CODE ANN. §22-9-1-6 (Burns Supp. 1977), IOWA CODE ANN. §601A.3 (Supp. 1978-1979); OHIO REV. CODE ANN. §4112.04 (Page Supp. 1977); WASH. REV. CODE ANN. §§49.60.051, 49.60.120 (Supp. 1977).

however, seems to suffer from several recurring problems: (1) The major activity of a civil rights office is the receipt and resolution of complaints. Yet, a major failing of most civil rights commissions has been their inability to resolve complaints in a timely fashion; (2) Many statutes do not focus responsibility for prompt investigation so that those complainants without sufficient evidence are dropped and the remaining complaints processed; (3) Multiple administrative appeals are sometimes provided, encouraging such appeals and delaying resolution of a complaint; (4) Many statutes provide that a commission shall be responsible for investigating, hearing cases and issuing orders, while others allow individual commissioners to conduct the investigations. This makes it virtually impossible for a respondent's case to be heard by an impartial decision maker; and (5) Disharmony sometimes develops between commission members or between the commission and director. This might result in an inability to focus the efforts of the office and to agree upon and carry out a means of reducing a backlog of cases.

In view of these problems, several considerations need to be addressed in deciding upon an appropriate alternative approach:

How can complaints be most efficiently and effectively handled by the Office?

How can settlement be encouraged at earlier stages in the complaint/enforcement process?

How can the conflict inherent in having the same person or body be both investigator/prosecutor and judge best be avoided?

How can an adversarial relationship between Commission and Director best be avoided?

The model act utilizes the familiar Office/Commissioner/Director organization. However, the allocation of duties between the Director and Commission is different from that found in many state civil rights statutes. This model act gives the Director primary responsibility for the functioning of the Office,[50] including the authority to investigate, make initial determination of probable cause for complaints, negotiate and enter into enforceable consent orders, bring cases to hearing and seek enforcement. Thus, the Director carries out a prosecutorial function. Cases are then heard by hearing examiners insulated from the Director with the hearing examiner's decision being final for administrative purposes. Further action on appeal is to the state's trial courts.

The strong Director approach gives recognition to the importance of the Director's administrative ability in the efficient management of the Office as well as the day-to-day experience and expertise of the Director and staff. This is not likely to be developed by part-time commissioners. In addition, giving the Director the investigative and prosecutorial functions consolidates decision-making and avoids the possible inability of diverse commissioners agreeing on a comprehensive enforcement program for the Office. Finally, a Director and staff can more easily negotiate and settle cases through use of consent orders than can a full commission since commissioners dependent upon staff are less likely to be able to give informed approval.

Having cases heard by semi-autonomous hearing examiners avoids the conflict of interest inherent in having commissioners investigate and decide cases. (See subsections 19(14), 20(2), and 24(1).) This system also streamlines enforcement procedures and reduces delay by limiting formal administrative procedures to one hearing in which the determination is final for administrative purposes. (See paragraphs 23(3)(a) and 23(4)(a) and (d).) And several hearing examiners can hear many more cases than can a commission.

It might be argued that giving a commission the power to hear and decide cases while the Director acts as investigator/prosecutor would also eliminate any conflict of interest. That is true but such an approach has several disadvantages. It would put the Director and commission in opposing roles, would make cooperation between them more difficult and would preclude the giving of advice by the commissioners, if they are to maintain their detached decision-making status. Furthermore, it is unlikely that a part-time commission can hear all the cases and appeals in a timely fashion which will delay resolution of complaints and create a backlog of cases. Also, creating an additional level of administrative appeal to a commission will tend to encourage more appeals and delays by respondents wishing to avoid a final determination or complainants who do not like the result.

The Director in the proposed model act still receives citizen and other input through the advisory Commission. The Commissioners are free to comment, criticize, and recommend changes. The Commission can also carry out studies as needed to aid in the administration of the Office.

Alternative. *Some states may still wish to adopt a citizen governing commission or to retain the existing one. These states may have a value preference for citizen boards over individual administrators. Where this is the case the duties of the Director and Commission can be rather easily reallocated between the two. States with existing commissions, however, should examine the model act for possible additions or amendments to their statutes based upon the duties enumerated in section 19 below.*

---

[50] This approach is used in Minnesota, MINN. STAT. ANN. §363.05 (West Supp. 1978), in New Jersey, N.J. STAT. ANN. §§10:5-7 through 10:5-10 (West 1976), and in New York, N.Y. EXEC. LAW §293 (McKinney 1972).

Section 16. Office for Civil Rights Created

The Office for Civil Rights is created as an executive agency of state government. The Office shall consist of the Director of Civil Rights and his or her staff and the state Civil Rights Commission.

COMMENT. See Organization of the Office *supra*.

Section 17. Director of Civil Rights

(1) The Office shall be headed by the Director of Civil Rights and shall be administered under his or her supervision and direction.

COMMENT. See Organization of the Office *supra*.

(2) The Director shall be appointed by the governor, with the advice and consent of the state senate for a term of four (4) years beginning______ and until a successor is appointed and approved by the senate. The governor may remove the Director for good cause shown after notice and hearing.

In case of vacancy, the governor shall appoint an acting Director who shall take office and carry out all duties of the Director until his or her appointment is confirmed or denied by the state senate or withdrawn by the governor. His or her appointment shall be submitted to the senate at the next session of the legislature.

COMMENT. The Office is a regulatory agency. Thus, the Governor, who is responsible for leadership, administration and enforcement of the state's policies and programs is given authority to appoint the administrative head of the Office. As with most appointments at this level, it is subject to confirmation by the state senate. This gives a clear focal point of responsibility for appointment of a well-qualified person and of accountability for a bad choice.

The Director is appointed for a specified term of years to help insulate him or her from disagreements over the conduct of individual cases and from any undue pressures which may otherwise threaten his or her removal. It should also give the Director greater freedom to administer the Office as he or she sees fit consistent with established policies. It is not intended that the Governor attempt to influence the outcome of cases, although the Governor is authorized to remove the Director during his or her term for incompetent performance. In that event, the Director is to be given a hearing with full due process rights attached. No appeal from the judicial decision is contemplated unless some mechanism for appeal is provided by some other state statute.

Alternative. *The Commission could be given authority to appoint the Director. This may help insulate the selection from the broader political processes. However, strong interests are likely to be actively represented on the Commission. Selection would be made by those with a real interest, but the process could divide the Commission and create disharmony. On the other hand, it is possible that such selection may be made on a consensus of all interests represented on the Commission and foster a more cooperative relationship between Commissioners and the Director.*

If selection is to be made by the Commission, section 18 may be made more specific to insure that designated interests are represented on the Commission. A broad representation of interests becomes more important as Commission powers and duties increase.

Finally, the Director might serve at the pleasure of the appointing authority rather than serve a specified term subject to removal for cause. This would give greater flexibility to remove a Director whose performance was not satisfactory.

(3) The Director shall have demonstrated an interest in the civil rights field and shall be selected and confirmed or denied appointment based upon his or her ability to efficiently and effectively perform the duties and functions specified by this act.

COMMENT. None necessary.

(4) The Director shall be in unclassified service and shall be paid a salary provided by law.

COMMENT. Since the Director is appointed, it is intended that he or she be exempted from civil service requirements.

Section 18. Civil Rights Commission

(1) There is established a Civil Rights Commission.

COMMENT. None necessary.

(2) The Commission shall consist of fourteen (14) members appointed by the governor, with the advice and consent of the state senate, for a term of four (4) years and until his or her successor is appointed and qualified, except that of those first appointed, three (3) shall be appointed for a term of one (1) year, three (3) for a term of two (2) years, four (4) for a term of three (3) years and four (4) for a term of four (4) years. The governor shall strive to ensure representation of minority groups. Vacancies caused other than by expiration of term shall be filled in the same manner, but for the unexpired term only. Members of the Commission shall serve without compensation, but shall be reimbursed for necessary expenses incurred in the performance of their duties. The chairman shall be elected annually by the members.

COMMENT. The Commission's major duties are to provide review and advice on Office operations. Thus, it can be larger than a governing commission whose size should be kept more manageable. The large size enables a broader range of interests to be represented and provides greater input. The four year term should allow commissioners to develop sufficient expertise to effectively contribute to its mission. Terms are staggered to insure that a wholesale turnover in membership will not occur and that some members will always have the experience of serving previously. It is anticipated that the Governor will consult interested groups before making appointments.

Alternatives. *A state may wish to reduce the size of the Commission or specify groups from which members may be appointed. These two items are more important if the Commission is given appointment or governing authority. In addition, Commissioners may be given per diem, especially where members of other boards and commissions performing similar duties within the state receive such compensation.*

Section 19. Powers and Duties of Director

The Director shall have authority to:

(1) Exercise leadership in the development of human rights policies and programs, and make recommendations to the governor and the general assembly for their consideration;

COMMENT. The Director, as head of the Office, is expected to provide leadership in the development and implementation of a civil rights program for the state. Where he or she sees the need for changes in existing law and procedures he or she is given specific authority to make recommendations to the Governor and General Assembly.

(2) Establish and maintain such offices as are necessary to efficiently and effectively carry out the duties and powers of the Director;

COMMENT. None necessary.

(3) Meet and function at any place within the state;

COMMENT. None necessary.

(4) Adopt rules and regulations as appropriate to effectuate the purposes of this act;

COMMENT. None necessary.

(5) Cooperate and consult with other state agencies as appropriate in developing plans and programs to effectuate the purposes of this act;

(6) Comment upon or recommend to any state agency procedures or regulations for the purpose of safeguarding the civil rights of individuals within the state. A state agency shall notify the Director within 60 days of the receipt of such comments or recommendations of the action taken thereon and the reasons therefor;

(7) Obtain upon request and utilize the services of all state governmental departments and agencies;

COMMENT. Other state agencies are responsible for carrying out portions of the state's civil rights program, especially the provisions of section 11 of the act prohibiting discrimination in public services, assistance and licensing, and mandating affirmative action programs where public funds or hiring are involved. The state and its subdivisions are also covered by other portions of the act, including employment and public accommodations. These agencies and the Director should cooperatively develop affirmative action programs and adopt procedures to uncover and correct those actions which may be discriminatory. The Director is given the authority to both recommend regulations and procedures to state agencies, and to require that he or she be informed of the action taken by a state agency and the reason therefor. This provision is not intended to give the Director the authority to enforce its proposed regulations and procedures vis-a-vis other agencies of the state, but merely to require that the receiving state agency give the suggestions serious consideration and respond to the Director with a reason for its actions.

(8) To the extent permitted by federal law and regulation, utilize the records (of the department of employment services of the state) when necessary to effectuate the purposes of this act;

COMMENT. Records on population, minority group numbers and employment statistics are extremely useful in presenting statistical evidence in support of allegations of pattens of discrimination. For this reason the Director is given access to records of the state's employment agency.

(9) Develop and disseminate technical assistance to persons subject to the provisions of this act;

COMMENT. Each person subject to the act is responsible for compliance with its terms. However, development of guidelines and provision of technical assistance by the Director should assist in achieving better compliance with the act's provisions.

(10) Receive complaints, issue notices of complaint, investigate complaints, initiate investigations in his or her discretion, with or without the filing of a complaint, determine whether or not probable cause exists for a hearing, file petitions for hearings, and represent complainants or the Office at hearings and in the courts of the state directly or through attorneys employed by the Office;

COMMENT. The Director is given an investigative and prosecutorial role by the act. Thus, the Director is responsible for all phases from receipt of complaints to enforcement of the orders of the hearing examiner in the courts of the state. (See Organization of the Office for Civil Rights *supra*).

(11) In connection with any investigation held pursuant to the provisions of this act, subpoena witnesses, compel their attendance, administer oaths, take the testimony of any person under oath, and, in connection therewith, require the

production of any books, papers or materials relating to any subject matter under investigation or in question by the Office and conduct such discovery procedures which may include the taking of interrogatories and oral depositions as shall be deemed necessary by the Director in any investigation. The Director shall adopt rules as to the issuance of subpoenas pursuant to this subsection. The Director may seek enforcement of this subsection in accordance with subsection 22(2) of this act;

COMMENT. In connection with his or her investigative duties the Director is given subpoena authority, a common provision of regulatory agency law. This allows the Director to fully investigate a complaint prior to a hearing and should contribute to the likelihood of a settlement without a hearing. In addition, without the power to subpoena persons and records not otherwise obtainable during an investigation, the Director would be forced to forgo the relevant information or be required to proceed to hearing and have the hearing examiner issue a subpoena to get it. This encourages resort to formal hearing procedures where none may be necessary. The Director, however, must adopt by rule procedures for issuance of subpoenas so that there is a uniform and reasonable procedure with notice of it being given to the public.

Note that the Director's action to issue a subpoena would be an agency action within the meaning of most state administrative procedure acts and would be appealable, if the subpoenaed person wished to challenge it. If this is not the case in a particular state, some appeal mechanism should be provided for. In addition, enforcement of subpoenas is not automatic. They must be enforced through the court pursuant to subsection 22(2).

(12) Attempt, by means of education, conference, conciliation, and persuasion to eliminate unfair or discriminatory practices as being contrary to the public policy of the state;

COMMENT. None necessary.

(13) Employ personnel in accordance with the state's merit system as may be necessary to effectuate the purposes of this act;

COMMENT. The Director is given wide discretion to employ staff members as appropriate. All employees of the Office except the Director are to be hired through the state merit or civil service system.

(14) Appoint with the approval of the Commission, a chief hearing examiner for the Office;

COMMENT. The Director has authority to appoint a chief hearing examiner subject to the approval of the Commission. (See subsection 20(2).) The hearing examiners are to function semi-autonomously without direction from the Director in any case. (See subsection 24(1).)

(15) Delegate such duties to Office employees as appropriate;

COMMENT. The Director is given the authority to delegate his or her duties under the act to such Office employees as appropriate to carry out those duties most efficiently and effectively.

(16) Contract or enter into agreements as appropriate to effectuate the purposes of this act;

COMMENT. None necessary.

(17) Develop such programs as will aid in determining compliance with the provisions of this act and in the furtherance of such duties, conduct research and study discriminatory practices;

COMMENT. None necessary.

(18) Publish and distribute the results of research and study when in the judgment of the Director the purposes of this act will be served thereby;

COMMENT. None necessary.

(19) Create such local and statewide advisory committees as will in the judgment of the Director aid in effectuating the purposes of this act;

COMMENT. It may at times be desirable to appoint special advisory committees to deal with problems in a particular subject or geographic area or to study the development of programs or guidelines. These committees could be composed of persons affected by the problems or persons with relevant expertise, such as sociologists, psychologists, social workers or lawyers. These committees can then make recommendations to the Director.

(20) Provide appropriate staff services to such advisory committees as may be created;

COMMENT. None necessary.

(21) Develop and conduct programs of formal and informal education designed to eliminate discrimination and inter-group conflict;

COMMENT. None necessary.

(22) Make a written report each year of the activities of the Office to the governor and to the general assembly;

COMMENT. None necessary.

(23) Accept public and private gifts, bequests, grants or other payments to help finance the activities of the Office; and

COMMENT. None necessary.

(24) Prepare an annual budget for the Office for submission to the governor and general assembly.

COMMENT. None necessary.

Section 20. Powers and Duties of Commission

The Commission shall:

(1) Consult with and advise the Director with respect to the work of the Office;

COMMENT. The Commission is to provide a periodic source of advice and oversight for the operations of the Office and provide internal accountability for such operations. The Commission is not to interfere with the conduct of individual complaints or cases. Rather, it should assist in the development or review of priorities and programs.

(2) Approve the appointment of hearing examiners pursuant to subsections 19(14) and 24(1).

COMMENT. As a check on who is appointed and to assist in insuring the independence of hearing examiners, all examiners whether appointed by the Director or the chief hearing examiner are to be approved by the Commission.

(3) Survey and study the operations of the Office and report to the Director, governor and the general assembly with respect to such matters and at such times at it deems in the public interest; and

(4) Contract with an outside person to evaluate the functions and effectiveness of the Office at least once every three years and distribute the resulting report to the governor, general assembly and the public.

COMMENT. The Commission also acts in an oversight role, providing internal evaluation of Office operations and reporting findings and suggestions to the Director, Governor and General Assembly when appropriate. In addition to its internal evaluation function the Commission is responsible for arranging periodic external evaluations from an outside, unbiased person. Ongoing evaluation can contribute greatly to keeping programs dynamic and responsive to current needs.

Enforcement Provisions in Anti-discrimination Statutes[51]

Anti-discrimination statutes have used three main approaches to enforce their substantive provisions: criminal prosecution, private civil suits for damages or injunctive relief, and administrative enforcement. (See the Review of State Legislation, especially Chart II on enforcement procedures, for a summary of existing statutory enforcement procedures.)

Many early statutes relied exclusively on criminal prosecution for enforcement. However, criminal or penal approaches have inherent shortcomings when compared to the other two approaches. First, criminal statutes must be unusually specific and be construed narrowly in order to meet due process requirements; due process demands that criminal statutes contain clear and specific notice of their exact proscriptions. Yet discriminatory acts are rarely overt. They tend to be carried out subtly and may take a vast variety of forms which may not be readily obvious. Consequently, specific anti-discrimination statutes of a criminal nature are likely to leave a great deal of room for evasive conduct outside the precise letter of the particular statute involved.

Second, criminal statutes also have a number of procedural requirements, making it difficult to achieve a criminal conviction under them. They rely upon prosecutors who must initially be persuaded to act. Prosecutors may be unsympathetic with the goal of the statutes or the penalties imposed, which may lead to prosecutions being undertaken less vociferously than would otherwise be the case. In addition, proof in a criminal case must be beyond a reasonable doubt. This makes conviction extremely difficult, especially since the precise motivation of any defendant is difficult to prove beyond a reasonable doubt. Further, trial by jury is always available in criminal prosecutions, which permits the defendant to avail him or herself of a body that may be as prejudiced as the defendant is or may be very unsympathetic to the objectives or approach of the statute.

Third, and finally, perhaps the biggest drawback of the criminal statute approach is that it does not meet the objective of the anti-discrimination statute, namely to eliminate discrimination of the kind prohibited. Anti-discrimination acts are not geared to be punitive or to work retribution upon those who violate their terms. Rather the purpose is to terminate discriminatory action. Since criminal statutes penalize without necessarily eliminating the discriminatory acts, persons discriminated against may continue to be denied the rights which other members of the general public share. The relatively small penalties assessed under some statutes and the reluctance of judges to imprison people in such cases make criminal statutes a poor deterrent even when violations are found.

Private civil actions for damages have certain advantages over criminal statutes. Unlike the criminal prosecutions, injured parties can initiate their own suits to vindicate their legal rights rather than having to rely upon a prosecutor. Further, in civil suits proof usually need only be by a preponderance of the evidence. In addition, if substantial exemplary damages were awarded, the deterrent effect is likely to be greater than that achieved under criminal statutes.

However, civil suits for damages also have a number of severe defects which render them undesirable as a mode for enforcing anti-discrimination statutes. Like criminal pros-

[51] The following text is based largely upon an article by Bonfield discussing state civil rights statutes. *See* Bonfield, *State Civil Rights Statutes: Some Proposals*, 49 IOWA L. REV. 1067, 1110-1119 (1964).

ecutions, they give the defendant a right to a jury trial in most cases. As noted above, this permits the defendant to try his case before a body which may be as prejudiced as the defendant and out of sympathy with the anti-discrimination statute. In addition, the private action is likely to involve considerable expense in the form of legal fees and other litigation costs. Since the amounts awarded under such suits may be minimal, the expenses involved in such suits will normally far exceed any damages received as a result of the action. Additionally, the burden of going forward with such suits remains on the person discriminated against. These people are likely to be reticent about litigating their rights, less knowledgeable on how to do so and have fewer resources for carrying out such suits. Finally, private damage suits may also share the most significant weakness of criminal statutes; that is, they may not help eliminate discriminatory acts. If the assessment of damages is small, the best they would do is penalize the perpetrator of such acts.

Private suits for injunctive relief do not have this disadvantage. In addition, they avoid the jury aspects of damage actions. Thus, they are a considerably better means of enforcing anti-discrimination provisions. However these private suits also suffer the disadvantages of considerable expense in engaging counsel and maintaining the litigation and also necessitate placing the burden on persons discriminated against to go forward with the injunctive suits. These persons are often not prepared to do so.

Administrative enforcement mechanisms have a number of superior advantages over both criminal statutes and private suits. Such administrative mechanisms generally provide more effective enforcement in a shorter period of time, at a lesser cost than litigation, and more readily achieve the legislation's real objective — the termination of discriminatory actions. In addition, these mechanisms overcome a number of the disadvantages of other approaches. First, the complainant no longer must bear the expense of investigation and going forward with a private suit. Second, the administrative agency becomes expert at discerning bias, and its orders will frequently be upheld by the court, if they are supported by substantial evidence. Third, no jury is required in the enforcement process. Fourth, the administrative agency is frequently authorized to seek out the discrimination or to take complaints from organizations. Thus, the agency is in a position to keep a constant check on recalcitrant individuals. Fifth, the administrative enforcement mechanisms normally stress informal, and thus faster and cheaper resolution of alleged violations. Conciliation and mediation are often used, coupled with broad educational campaigns conducted by the administrative agency. Sixth, when informal means of resolution fail, the agency may proceed to an administrative hearing on the matter. Administrative hearings are generally conceded to be a faster and less expensive means of resolution and have several distinct advantages. Proof is satisfied by a preponderance of the evidence and the truth is generally easier to ascertain since the technical rules of evidence do no apply. Seventh, where violations of the act are found, the administrative approach normally includes the power to issue enforceable cease and desist orders and other flexible remedies designed to assure achievement of the legislative purpose. Again the real purpose of these laws is to terminate discriminatory practices and not to penalize offenders. Flexible cease and desist orders issued by vigilant administrative agencies are considerably more effective in achieving this objective. Eighth, because such statutes are considerably broader in coverage of proscribed acts than criminal statutes, they may be made sufficiently general to catch novel attempts at evasion or subterfuge of the anti-discrimination statutes. And, because these provisions are not penal, they can also be construed broadly to effect greater purpose. Thus, as a means of enforcement, administrative remedies are vastly superior to the other two approaches.

This is not to say that private actions cannot supplement administrative enforcement. In the long run the effect of the enforcement of anti-discrimination statutes is more important than the particular approach used. Weak administrative enforcement would be unproductive in eliminating discrimination regardless of how comprehensive the statute may be. Thus, private civil actions can provide an excellent means of supplementing administrative enforcement procedures. For example, private suits afford injured parties a means of vindication of their rights where the administrative agency has wrongfully refused to act. Thus, it tends to keep the administrative agency on its toes.[52]

The proposed act adopts an administrative enforcement mechanism supplemented by private civil suit. No criminal action is provided for. In addition, the usual powers of a regulatory agency are included: investigation, subpoena, issuance of orders and enforcement of orders in the state's courts. (Contrast this to the purposes and powers

[52] A number of states have supplemented administrative enforcement mechanisms with provisions for private civil suits. Most of these states require a complainant to elect the remedy by demanding that the aggrieved party pursue either a private suit or administrative enforcement. This eliminates needless badgering of the defendant. *See e.g.* ARIZ. REV. STAT. §36-516(b) (1974); CAL. CIV. CODE §54.3 (West Supp. 1978); KY. REV. STAT. §207.230 (Supp. 1978); ME. REV. STAT. tit. 5 §4621 (Supp. 1978); MICH. COMP. LAWS ANN. §37.2801 (Supp. 1978-79); MONT. REV. CODES ANN. §§64-304, 64-392 (Supp. 1977); N.J. STAT. ANN. §10:5-38 (West 1976); N.Y. EXEC. LAW §297(9) (Supp. 1972-1977); ORE. REV. STAT. §§659.105, 659.435 (1973); TEX. CIV. CODE ANN. art. 4419e(6)(b) (Vernon 1976); WASH. REV. CODE ANN.§49.60.030(2) (Supp. 1976).

of the Advocacy Office proposed in the model advocacy act.)

Section 21. Grievance Procedures

(1) Filing of complaints.

(a) Any person aggrieved by a violation of this act may file a verified complaint with the Director or designated agent. The original shall be signed and verified before a notary public or other person duly authorized by law to administer oaths and take acknowledgements. Notorial service shall be furnished by the Office without charge. To be accepted by the Director, a complaint shall be sufficiently complete so as to reflect properly the full name and address of the complainant; the name and address of the respondent against whom the complaint is made; the alleged discriminatory practice and a statement of particulars thereof; the date or dates and places of the alleged discriminatory practice and if the alleged discriminatory practice is of a continuing nature, the dates between which said continuing acts of discrimination are alleged to have occurred; and a statement as to any other action, civil or criminal, instituted in any other form based upon the same grievance as is alleged in the complaint, together with a statement as to the status or disposition of such other action.

(b) Any person who shall knowingly make a false allegation in a complaint shall be guilty of a misdemeanor and shall be subject to a penalty not to exceed $100 for each offense.

COMMENT. Any person aggrieved by a violation of the act may file a complaint. It must be sufficiently specific for the Director to determine whether a prima facie violation of the act has occurred and whether the complainant has taken any previous action relating to the matter. By signing the complaint the complainant is deemed to have attested to the truth of the allegations. A penalty is provided for those who abuse the complaint procedure by making false accusations.

(2) Referral to local commissions.

(a) Whether or not a complaint has been filed under this act, the Director may refer a matter involving discrimination because of race, color, religion, national origin, age, disability, marital status, sex or use of adaptive devices to a local commission for study and report.

(b) Upon referral by the Director, the local commission shall make a report and make recommendations to the Director and take other appropriate action within the scope of its powers.

(c) A local commission may refer a matter under its jurisdiction to the Office.

(d) The complainant has the option of filing a complaint either with a local commission or the Office. Filing a complaint with one agency shall preclude filing it with the other. At the time a complaint comes to the attention of a local agency, the agency or its representative shall inform the complainant of this option and of his or her rights under this act.

COMMENT. Some local jurisdictions may have created civil rights or human rights commissions. Frequently, larger cities in a state have established such commissions before the state has done so. The Office should establish cooperative relationships with these commissions and defer to them in certain cases. Handling cases at the local level may well be the best approach. These commissions may also wish to use the authority or resources of the Office to handle an investigation or hold a hearing. Either is authorized to act so long as efforts are not duplicated. Local commissions are required to inform complainants of their rights under this act since they may be more extensive than those provided at the local level. A complainant should have knowledge of the options available before selecting which to use.

(3) Time for filing.

A complaint of an unfair or discriminatory practice must be filed within 180 days after the complainant knew or should have known of the occurrence of the alleged practice, or within 90 days after the termination of a grievance procedure provided by a respondent employer or labor organization, whichever is longer.

COMMENT. This section establishes a statute of limitations for complaints running from the time a complainant gained knowledge of the occurrence of a discriminatory act or from the time the complainant should have known from available information. This protects respondents against dilatory complaints while still allowing the complainant to raise the issue where he or she had no means of knowing of the discriminatory act. It is unlikely that a respondent would supply information that would reveal such an act and thus it would be unfair to preclude a potential complainant from filing because of a successful coverup.

Where a complainant has previously participated in an internal grievance procedure, the time period is shorter. In this case the complainant will already have had time to gather information to prepare a complaint so further time is not needed for this purpose.

(4) Notice of complaint issued by Director.

(a) The Director shall examine a complaint immediately after its receipt and determine on its face, taking into account any other information the Director may currently have in his or her possession, whether there is reason to believe that a person is engaging in an unfair or discriminatory practice.

COMMENT. The Director is required to determine initially whether the complaint states a prima facie case of discrimination. This minimum showing must be made before an allegation will be made to a respondent. This is the first stage of screeening out false complaints.

(b) If the Director has reason to believe that a person is engaging in an unfair or discriminatory practice, the Director shall issue a notice of complaint listing any alleged violations of this act and include a summary of the procedures in this section for the information of the respondent. It shall be served upon the respondent personally or mailed to the respondent by registered or certified mail within ten (10) days of the filing of the complaint.

COMMENT. Once an initial determination of validity has been made, the respondent is informed of the complaint, the nature of the allegations and what procedures will be used to resolve the complaint. Thus, the respondent will know what is being alleged, how the investigation will proceed and what enforcement procedures may be invoked. To minimize delay in handling complaints, this determination must be made and notice given where appropriate within ten days of filing of the complaint.

(c) Periodically after the filing of a complaint, but at intervals of no more than 60 days, the Director shall, in writing, inform the complainant of the status of his or her complaint.

COMMENT. Since the investigation which normally follows a complaint may take place over a lengthy period of time, a complainant should be informed of the status of the complaint at periodic intervals. This subsection puts the burden upon the Director to do so.

(5) Investigation of complaints.

When a notice of complaint has been issued, the Director shall promptly inquire into the truth of the allegations of the complaint to prevent a complainant from suffering irreparable loss in the absence of immediate action. On each charge the Director shall make a determination as to whether or not there is probable cause for an allegation of an unfair or discriminatory practice.

COMMENT. This section provides the second screening for unsupported or false complaints and insures that allegations are followed up in a timely manner.

(6) Where no probable cause found; reconsideration.

If the Director shall determine after investigation that no probable cause exists for the allegation of an unfair or discriminatory practice, the Director shall, within ten (10) days of such determination, serve upon the complainant and respondent written notice of such determination. Within ten (10) days after receipt of this notice, the complainant may request in writing that the Director reconsider the determination. The request shall contain a brief statement of the reasons for, and any new evidence in support of, the request for reconsideration. At the time of submission of the request to the Director, the complainant shall deliver or mail to the respondent a copy of the request. The Director shall either reaffirm or reverse the determination of no probable cause within 20 days after receipt of the request, and shall within ten (10) days thereafter notify the complainant and respondent in writing of the decision to reaffirm or reverse. A decision by the Director that no probable cause exists for an allegation of an unfair or discriminatory practice shall not be appealed to the district court pursuant to subsection 24(1) or section (cite to state's administrative procedure act where appropriate) of the code.

COMMENT. With a minimum of formality, complainants are accorded a right to request a redetermination and may include arguments and any additional evidence to support their position. No appeal from a decision of no probable cause is allowed, however, since the opportunity to provide evidence during the investigation and obtain a reconsideration of an adverse decision should be adequate to establish probable cause, if indeed there is any. Resources of the Office should not be wasted in defense of spurious appeals.

(7) Where probable cause found; attempts to eliminate unfair practices; consent orders.

(a) If the Director determines after investigation that probable cause exists for an allegation of and unfair or discriminatory practice, the Director shall endeavor to eliminate the unfair or discriminatory practice through education, conference, conciliation and persuasion. Agreements reached by this means shall be entered into as consent orders.

COMMENT. The act encourages settlement as early as possible without the need for formal hearing procedures. All conciliation agreements must be entered into in writing and signed by the respondent and Director in the form of a consent order so that all parties clearly understand what is being agreed to and what is expected.

(b) Each conciliation agreement shall include provisions requiring the respondent to refrain from the commission of unlawful discriminatory practices in the future and may contain such further provisions as may be agreed upon by the Director and the respondent, including a provision for the entry of a consent decree embodying the terms of the conciliation agreement in the district court in any county in the judicial district where the alleged unlawful discriminatory practice was committed, or where any respondent resides or maintains an office for the transaction of business.

COMMENT. Conciliation agreements must, at a

minimum, include a provision requiring the respondent to refrain from further violations of the act. Other provisions may include an affirmative action program to rectify past discrimination, periodic compliance reports, or payment of damages to the complainant.

Where compliance by a respondent is doubtful, it is desirable for the terms of the consent order to be filed with the trial court of the state in the form of a consent decree. This should allow the Director to seek a contempt of court citation if the respondent continues to refuse or fails to comply with the act and the terms of the agreement. Without entry of a decree the Director will normally be required to file an action in the trial court alleging noncompliance with the consent order, and show that the order is lawful and reasonable. The court could then issue an order against the respondent requiring compliance. Only if this court order is not met will the respondent be subject to contempt proceedings. Thus, this provision can protect against a recalcitrant respondent further delaying compliance.

(c) If the respondent and the Director agree upon conciliation terms, the Director shall serve upon the complainant personally or by certified mail, a copy of the proposed conciliation agreement. If the complainant agrees to the terms of the agreement or fails to object to such terms within fifteen (15) days after its service, the Director shall issue a consent order embodying the conciliation agreement. If the complainant objects to the agreement, he or she shall serve a specification of his or her objections upon the Director within such period. Unless such objections are met or withdrawn within ten (10) days after service thereof, the Director shall proceed to a hearing in accordance with subsection 21(8).

(d) Notwithstanding any other provisions of this section, the Director, where he or she finds the terms of a conciliation agreement to be in the public interest, may issue a consent order embodying the conciliation agreement, and limit the hearing to the objections of the complainant. If, however, the Director finds that the complainant's objections to the proposed conciliation agreement are without substance or that initiating a hearing on the complaint would be otherwise undesirable, the Director may in his or her unreviewable discretion, at any time prior to a hearing, dismiss the complaint on the grounds of administrative convenience.

COMMENT. The Director acts in the public interest to eliminate discrimination, but does so within the context of the interests of a complainant. Therefore, the complainant is given an opportunity to object to the terms of a conciliation agreement. If the complainant objects and his or her objections are not met, the Director will normally proceed to a hearing on the objections.

Where the public interest is affected by the broader impact of a case, the Director is authorized to enter into a consent order on the broader issues, while proceeding to a hearing on the complainant's objections. For example, where complainant has been discriminated against by a company which has admittedly engaged in a pattern of past discrimination, the company and the Director could enter into a consent order covering affirmative action in hiring and promotion without settling the complainant's individual claims on back pay and damages. These issues could go to hearing separately.

Note that the Director is not obliged to proceed on objections which are frivolous or without grounds. The Director has an obligation to protect respondents from unwarranted actions and to use the limited resources of the Office wisely. Thus, the Director may dismiss the complaint at any time prior to the hearing.

(e) With the approval of the hearing examiner, the Director and the respondent may enter into a consent order in resolution of a complaint at any time prior to commencement of a hearing, or at any time subsequent to commencement of a hearing, but prior to issuance of an order.

COMMENT. Once a hearing is commenced a case proceeds under the jurisdiction of the hearing examiner, who must approve a consent order as a disposition of the case. However, such informal settlements are to be encouraged and should be approved unless there is some compelling reason for not doing so.

(f) If a consent order is entered into, the Director shall serve a copy of the order upon all parties to the proceeding.

COMMENT. None necessary.

(g) A consent order entered into pursuant to this subsection is a final decision of the Office within the meaning of section 24. The Director may seek enforcement of the consent order in accordance with section 24.

COMMENT. A consent order is a final administrative action in a case. Since it is done by consent, it is not subject to judicial review (i.e. would not be appealed) but it is subject to the Director's enforcement powers.

(h) Not later than one (1) year from the date of an order issued under this section, and at any other times in his or her discretion, the Director shall investigate whether the respondent is complying with the terms of the order. Upon a finding of noncompliance, the Director shall take appropriate action to assure compliance.

COMMENT. None necessary.

(8) Initiating hearing where probable cause is found.

If the Director determines after investigation

that probable cause exists for an allegation of an unfair or discriminatory practice, and if attempts to settle the complaint by conference, conciliation and persuasion have not succeeded, the Director may file a petition with the chief hearing examiner alleging the occurrence of an unfair or discriminatory practice by the respondent, stating the specific facts upon which the allegation is made, and praying for relief. The chief hearing examiner shall assign the case to a hearing examiner in accordance with subsection 23(1) of the act.

COMMENT. The Director has the sole discretion to initiate a hearing, but only after informal attempts at settlement have failed. The petition should be specific in its allegations including statements of fact and a prayer. Once the petition is filed the hearing examiner obtains jurisdiction over the case.

(9) Temporary injunction.

At any time after the Director has determined that there is probable cause to believe that a respondent has engaged in an unfair or discriminatory practice, the Director may file a petition in the district court in the county in which the alleged practice occurred or in the county in which a respondent resides or transacts business, and seek appropriate temporary relief against the respondent, including an order or decree restraining respondent from doing or procuring any act tending to render ineffectual any order a hearing examiner may enter with respect to the complaint. The court shall have the power to grant such temporary relief or restraining order as it deems just and proper, but no such relief or order extending beyond ten (10) days shall be granted except by consent of the respondent or after hearing upon notice to the respondent and a finding by the court that there is reasonable cause to believe that the respondent has engaged in a discriminatory practice. The (name of state) rules of civil procedure shall apply to such application and the district court shall have authority to grant or deny such relief sought on such conditions as it deems just and equitable. All hearings under this section shall be given precedence as nearly as practicable over all other pending civil actions.

COMMENT. The Director has authority to seek a temporary injunction to protect the interests of a complainant until the issues can be heard through the administrative process. The procedure outlined in this subsection should be checked against a state's existing procedures for consistency.

Section 22. Subpoena of Witnesses

(1) Subpoenas.

After the issuance of a notice of hearing pursuant to subsection 23(2), a respondent, the Director or other admitted party may request that the hearing examiner issue a subpoena requiring the presence of witnesses or the production of relevant books, papers or materials.

COMMENT. As regulatory agency, the Office has been given the subpoena power to facilitate a fair hearing and an informed finding by the hearing examiner. Without a subpoena to compel attendance or production of evidence the parties would have no way of getting a reluctant witness to testify or bring forward important information. Also, the subpoena serves to equalize the ability of the parties to obtain witnesses and evidence to support their respective positions. Were there no subpoena power it is likely that the party with greater resources, and hence greater powers of persuasion, would have the advantage. Such equalization is especially appropriate in the context of this statute which specifically protects the rights of minorities who tend to have fewer resources.

Note that the Director has the power of subpoena during an investigation (see subsection 19(11)). Once the hearing examiner issues a notice of hearing, all subpoenas must be issued by the examiner since the examiner now has jurisdiction over the case.

(2) Enforcement.

Disobedience of a subpoena issued by the Director pursuant to section 19(11) or by a hearing examiner pursuant to this section shall be punishable in like manner as a contempt of the district court upon application of the Director or the person who requested its issuance. Such application shall be to the district court of the county where the hearing was held or where the subpoenaed witness resides or has a principal place of business.

COMMENT. The power to subpoena would be an empty one if there were no provision for the punishment of persons who disobeyed it. For this reason, disregard of the subpoena and obstruction of the Office's administrative and regulatory functions is to be treated in the same manner as contempt of court. Disobedience of a subpoena, however, is not automatically punished since the Office has no direct enforcement authority. Rather, the Director or party requesting the issuance of the subpoena must apply to the district court for a contempt citation and appropriate sanctions.

(3) Evidence produced in obedience to summons; immunity of witnesses.

No person shall be excused from attending and testifying or from producing records, correspondence, documents or other evidence in obedience to the subpoena of the Director or hearing examiner on the ground that the required testimony or evidence may tend to incriminate or subject him or her to a penalty or forfeiture. But no person shall be prosecuted or subjected to any penalty or forfeiture on the basis of any testimony or evidence compelled by the subpoena

or on the basis of any information directly or indirectly derived from such testimony or evidence after having claimed his or her privilege against self-incrimination. Such person may, nevertheless, be prosecuted or subjected to penalty or forfeiture for any perjury or contempt committed in answering, or failing to answer, or in producing or failing to produce evidence in accordance with the subpoena, and any such testimony given or evidence produced shall be admissible against him or her in any proceeding concerning such perjury or contempt. The immunity herein provided shall extend only to natural persons so compelled to testify or produce evidence.

COMMENT. Following the example of the federal immunity statute,[53] this provision confers "use" immunity[54] rather than "transaction" immunity[55] upon witnesses testifying before the Director or examiner. This means that no testimony given or evidence produced or any information derived from such testimony or evidence can be used against the witness in a civil or criminal proceeding.

To receive immunity the witness first must assert the privilege against self-incrimination.[56] If the witness refuses to testify after having been informed of the immunity he or she may be cited for contempt. The immunity given by this section, however, does not protect the witness from perjury or contempt for failure to comply with the subpoena.

(4) Witness and deposition fees.

Witnesses before a hearing examiner shall be paid the same fees and mileage that are paid witnesses in the courts of the state. Witnesses whose depositions are taken shall be entitled to the same fees as are paid for like services in the courts of the state.

COMMENT. None necessary.

Section 23. Hearings

(1) Hearing examiners.

The chief hearing examiner shall, in addition to hearing contested cases under this act, appoint such additional hearing examiners with the approval of the Commission and subject to budgetary limitations as may be necessary to efficiently hear contested cases arising under this act. The chief hearing examiner shall assign cases among the hearing examiners from the Office and, when necessary and available, from other state agencies, so as to efficiently handle the case load.

COMMENT. All hearing examiners are to function semi-autonomously from the rest of the Office. Only the chief hearing examiner is appointed by the Director. All others are appointed by the chief hearing examiner who also oversees the functioning of the hearing examiners and assigns cases among them. As an extra check on their independence, all hearing examiners must be approved by the Commission before their appointment is effective.

These measures are designed to separate the investigative and prosecutorial functions of the Director from the decision-making function of the hearing examiners. As noted earlier, these functions are frequently vested in the same entity, usually a commission. This creates an inherent conflict of interest and at least the appearance of bias to those who are regulated by it. Thus, the act separates those functions and should give respondents a fairer hearing.

(2) Initiating hearings, petition and answer.

Upon being assigned a case pursuant to subsection 23(1), the hearing examiner shall cause a written notice of the hearing to be served on the Director and to be served together with a copy of the petition upon the complainant and respondent within fifteen (15) days of the filing of the petition pursuant to subsection 21(8). The notice shall require the respondent to file a written answer to the petition either admitting or denying each paragraph of the petition and stating any defenses relied upon and to appear at a hearing before a hearing examiner at a time and place specified in the notice, but not less than fifteen (15) nor more than ninety (90) days after service of the notice unless the parties agree to a shorter or longer period of time before the hearing. The notice shall inform the respondent that the failure to answer will result in default being entered against the respondent.

COMMENT. After the filing of a petition and assignment of the case, the hearing examiner serves a notice of hearing upon the parties. The notice informs the complainant and respondent that the Director has initiated a hearing by filing a petition and that the petition attached has made specific allegations of violations of the act, pleaded facts supporting those allegations and requested relief. The respondent thus knows what is alleged and what relief is requested. The respondent is required to respond to each allegation by admitting or denying it and to plead any affirmative defenses to be relied upon at the hearing. In this way, the issues to be decided at hearing are clearly drawn. The hearing can thus proceed more efficiently and the parties can more easily prepare their cases, obtain witnesses and prepare evidence on only those issues which are still contested. While it puts a burden upon the respondent to prepare an answer, it will save time, effort and expense for all parties.

The case is required to be moved along rapidly.

---

[53] 18 U.S.C. §6002 (1976).

[54] *See* Murphy v. Waterfront Commission, 378 U.S. 52 (1964).

[55] *See* Counselman v. Hitchcock, 142 U.S. 547 (1892).

[56] *See* U.S. CONST. amend. V.

The hearing examiner must be appointed and the notice of hearing issued within fifteen days of the filing of the petition. The hearing must then be held within ninety days of service of the notice. Since there is a tendency for proceedings to drag on beyond original expectations, these time limits will insure that some resolution is reached within a reasonable time. In the interest of fairness, however, the hearing cannot be held prior to fifteen days after service of the notice to insure that the parties have sufficient time to prepare. The parties may agree to a different time, if they so wish.

(3) Conduct of hearings.

(a) A petition filed by the Director shall be heard as a contested case, except that the report of the hearing examiner shall be binding on all parties to the proceeding and, if appropriate, shall be implemented by an order as provided for in subsection 23(4). The hearing shall be conducted at a place designated by the hearing examiner within the county where the unfair or discriminatory practice occurred or where the respondent resides or has his or her principal place of business. The hearing shall be conducted in accordance with sections ______ of (cite the state's administrative procedure act or other appropriate sections) and is subject to appeal in accordance with section 24 of this act.

COMMENT. This act differs from most existing state statutes in that it separates the hearing functions from other Office functions and thereby eliminates potential conflicts of interest. (See subsection 23(1)). It also streamlines the administrative procedure by providing for one proceeding before an independent hearing examiner with his or her decision being final. No further administrative review, appeal or rehearing is allowed. This simplified procedure will provide earlier, more eficient resolution of cases. (See discussion in Organization of the Office of Civil Rights *supra*.) Persons aggrieved by an order of the hearing examiner may still appeal the decisions directly to the district court. This should reduce spurious appeals that might occur if administrative review was provided, and result in an earlier definitive decision for those parties who intend to appeal a decision. Hearings are to be conducted in accordance with the state's administrative procedure act or its equivalent and thus that act should be checked against this one for consistency.

(b) When a respondent has failed to answer a petition as provided by this section, the hearing examiner may enter a default order or may proceed to hear testimony adduced on behalf of the complainant. After hearing such testimony, the hearing examiner may enter such order as in the examiner's opinion the evidence warrants. For good cause shown, the hearing examiner may set aside an entry of default within ten (10) days after the date of such entry.

COMMENT. Where a respondent fails to answer, the hearing examiner is authorized to enter a default order, though it may be set aside where a good reason for the failure is shown. In many cases, the hearing examiner will still wish to hear evidence for the purpose of fashioning affirmative relief or assessing damages or a penalty in the order.

(c) The case in support of a complaint shall be presented at the hearing by the Director or by one of the Office's attorneys or agents.

COMMENT. The interests of the complainant are represented by the Director or, more typically, by an Office attorney or agent. In cases where broader interests are involved, such as when a pattern of discrimination is alleged, the Director also represents the interests of all those similarly situated in the sense that he or she will attempt to eliminate the pattern of discrimination as well as resolve the complainant's problem.

(d) A respondent who has filed an answer, or whose default in answering has been set aside for good cause shown may appear at the hearing in person or otherwise, with or without counsel.

COMMENT. None necessary.

(e) The hearing examiner may in his or her discretion allow the Director to amend a petition and the respondent to amend an answer.

COMMENT. None necessary.

(f) The hearing examiner may in his or her discretion permit any person who has a substantial personal interest to intervene as a party and may require that necessary persons not already parties be joined.

COMMENT. Parties to the hearing may extend beyond the Director and respondent if the hearing examiner believes that it is necessary in the interest of justice.

(g) The hearing examiner shall not be bound by the strict rules of evidence prevailing in courts of law or equity but all parties shall have the right of cross-examination. The Director or the complainant shall bear the burden of establishing a prima facie case. Once established, the burden of going forward with the evidence and of persuasion shifts to the respondent. The testimony taken at a hearing shall be under oath, reported, and, if necessary, transcribed.

COMMENT. A prima facie case must be established on behalf of the complainant. Thus, it must first be shown that the respondent is engaged in conduct which constitutes discrimination under the act and is thereby prohibited, that the conduct was based on a prohibited ground (e.g. disability), that the conduct was taken with respect to complainant, and that damages occurred.

Once established, the respondent must negate the case made for complainant by controverting proof of at least one element of the case: that the conduct alleged was not engaged in, that the complainant was not covered by the act (i.e. was not a "qualified individual"), that the ground for discriminatory treatment was not one statutorily proscribed (e.g. another individual was hired because he or she was a longtime friend of the employer), that the discriminatory treatment comes within one of the exceptions in the act, or that there was no damage.

(4) Determination of discriminatory practice.

(a) The hearing examiner shall make findings of fact and conclusions of law, and if the hearing examiner finds that the respondent has engaged in an unfair or discriminatory practice, the hearing examiner shall issue an order directing the respondent to cease and desist from such practice, to take such affirmative action as will effectuate the purposes of this act, and to periodically report on the manner of compliance. Such order shall be a final decision of the Office. In all cases, the examiner may order the respondent to pay compensatory damages, including but not limited to back pay or other loss, to an aggrieved party who has suffered discrimination, and punitive damages in an amount not less than $500 nor more than $5000, in addition to any other relief the examiner deems just and equitable.

COMMENT. Early civil rights statutes often contained only misdemeanor provisions which discouraged discrimination by penalizing those who engaged in it. While criminal penalties may be appropriate in certain cases where the discrimination is willful and malicious, the purpose of the act is not to penalize persons who violate it, but to eliminate discrimination and its effects. Thus, for example, the examiner may order hiring, reinstatement, upgrading, restoration of union membership, or completion of a real estate or insurance transaction.

In addition to eliminating discriminatory practices the examiner may order payment of damages to compensate for losses resulting from the discrimination. And where the respondent has willfully or maliciously engaged in discriminatory practices, punitive damages may be assessed. This penalty is appropriate under such circumstances and should act as a deterrent to such behavior.

(b) If any lessor, after he or she has engaged in an unfair or discriminatory practice, shall lease or rent such real property to a person who has no knowledge of such practice, or of the existence of any charge with respect thereto, such lessor shall, in addition to such other relief ordered by the hearing examiner, be liable for actual damages sustained by such person by reason of any final order requiring such person to be evicted from such real property.

COMMENT. In addition to injuring a complainant by discriminating in a real estate transaction, a respondent may also indirectly cause a loss to a lessee who is required to move as a result of corrective action ordered by the Office. Since the lessee is an innocent party, it is only fair that the respondent who is responsible for the lessee's having to move be liable for any actual loss incurred. Without such a provision, an examiner may be reluctant to order the needed remedial action. It is, of course, expected that other remedial arrangements will be fully explored first and that such a provision in an order will be a final resort in settling the complaint.

(c) The examiner shall cause the findings of fact, conclusions of law, and order to be served on the respondent and complainant personally or by registered or certified mail, and shall furnish copies to the Director within 60 days of the close of the hearing.

COMMENT. None necessary.

(d) If the hearing examiner makes findings of fact, conclusions of law, and an order in favor of the respondent, the order shall be a final decision of the Office for purposes of section 24.

COMMENT. An order in favor of the respondent is a final administrative action and may be appealed by the Director or other aggrieved person.

(e) Not later than one year from the date of an order issued under this section, other than an order in favor of a respondent, and at any other times in its discretion, the Office shall investigate whether the respondent is complying with the terms of the order. Upon a finding of noncompliance, the Office shall take appropriate action to assure compliance.

COMMENT. None necessary.

(5) Respondents subject to state licensing or regulatory power.

If the hearing examiner determines that the respondent has engaged in a discriminatory practice, and if the respondent does not cease to engage in such discriminatory practice, the Director may certify that fact to the licensing or regulatory agency in the case of a respondent who is subject to the licensing or regulatory power of the state or any political subdivision or agency thereof. Unless the determination of unfair or discriminatory practice is reversed in the course of judicial review, a final determination is binding on the licensing or regulatory agency. Every licensing or regulatory agency of the state or any political subdivision or agency thereof shall have authority to take appropriate administrative action, including a suspension or revocation of the respondent's license or certificate of public conveyance and necessity.

COMMENT. The public has the right to expect that licensees operating under the authority of the state or its subdivisions should obey the law in their operations. Thus, it is appropriate that such persons be subjected to suspension or revocation of their licenses for violations of this act. Suspension or revocation is not automatic, however. Although the final determination of discrimination is conclusive, most administrative procedure acts or other provisions require that licenses not be suspended or revoked without a hearing. In this case, the hearing would be limited to whether or not the license should be suspended or revoked in view of the violation of the act.

(6) Governor may act on orders against state or political subdivisions.

In any case in which the hearing examiner shall issue an order against any political or civil subdivision, any agency, or instrumentality of the state, or any officer or employee thereof, the Director shall transmit a copy of the order to the governor who shall take such action as he or she deems appropriate to secure compliance.

COMMENT. The Governor is ultimately responsible for the administration and upholding of state policy and law. Thus, for example, in the case of a state agency which is violating the act, the Governor may take informal action, such as consultation with the agency head, or formal action, such as the issuance of an executive order to eliminate the unlawful practices.

(7) Public contracts.

In the case of a respondent who is a party to a public contract, if the hearing examiner determines that the respondent has engaged in a discriminatory practice, the Director may certify that fact to the contract-letting agency. Unless the findings of a discriminatory practice are reversed in the course of judicial review, such agency shall have the power to take appropriate administrative action, including imposition of financial penalties of termination of the contract, in whole or in part, to effectuate the purposes of this act.

(8) Respondents receiving state financial assistance.

In the case of a nongovernmental respondent who is receiving state financial assistance, if the hearing examiner determines that the respondent has engaged in a discriminatory practice, the Director may certify that fact to the state agency authorizing the financial assistance. Unless the findings of a discriminatory practice are reversed in the course of judicial review, such agency shall have the power to take appropriate administrative action, including suspension or termination of, or refusal to grant or continue state financial assistance, or imposition of terms and conditions upon further state financial assistance, to effectuate the purposes of this act.

COMMENT. It is appropriate that public funds not be expended through or by persons who engage in discriminatory acts. Thus, in addition to other remedies under this act, holders of public contracts and recipients of state financial assistance shall be subject to penalties and termination of their contract or suspension or termination or refusal to grant or continue state financial assistance by the contract-letting agency or funding or granting agency. The contracting or funding agency is given authority to take such action in accordance with applicable due process proceedings before the agency.

Section 24. Judicial Review; Enforcement

(1) Judicial review.

Any person aggrieved by a final decision of the Office as rendered by the hearing examiner, including the Director, may seek judicial review pursuant to section (cite to state's administrative procedure act). The district court review proceedings shall conform to section (cite to state's administrative procedure act) on the scope of judicial review of agency decisions. The petition for judicial review must be filed within thirty (30) days of the issuance of a final decision by the Office.

COMMENT. Any person aggrieved by the final decision of the hearing examiner may seek review of the decision. This includes a complainant directly, the Director representing a complainant, one of a class of persons affected by the decision or a respondent. The procedures for review and the scope of review are intended to conform to the standard administrative procedure act provisions. The review is intended to be limited to questions of law, or issues of lawfulness or reasonableness of the order.[57]

(2) Transcript of hearing; availability.

The Director shall provide to the respondent, upon request, and for cost, a transcript of any hearing in which the respondent is a party in interest for the purposes of judicial review or of defending against an enforcement action.

COMMENT. None necessary.

(3) Enforcement.

(a) If no proceeding to obtain judicial review is instituted within thirty (30) days from the service of an order of the Office under subsection 23(3) or if an order is issued after a review and if the respondent fails or refuses to comply with a final decision of the Office, the Director may seek enforcement in the district court of the county in which the unfair practice occurred or in which the respondent resides or maintains his or her principal place of business. The Office may appear in court by its own attorney (notwithstanding section____of the code).

(b) Such an enforcement proceeding shall

---

[57] *See e.g.* IOWA CODE ANN. §17A.19(7),(8) (1978).

be initiated by the filing of a petition in such court and by the service of a copy upon the respondent. The Director shall certify to the court the fact of respondent's failure to comply and also shall file with the court a transcript of the record of the hearing before the hearing examiner.

After such filing, the court shall issue an order directed to the respondent to show cause why an order directing compliance should not be issued.

(c) Petitions filed under this subsection shall be heard expeditiously and the case determined upon the transcript filed, without requirement for printing.

An objection that has not been urged before the hearing examiner shall not be considered by the court in an enforcement proceeding, unless the failure or neglect to urge such objection is excused because of extraordinary circumstances.

(d) Any party to the enforcement proceeding may move the court to remit the case to the hearing examiner for the purpose of adducing additional specified and material evidence and findings thereof, if the party shows reasonable grounds for the failure to adduce that evidence before the hearing examiner.

(f) The court shall have the power to grant such temporary relief or issue a restraining order as it deems just and proper, and to make and enter upon the pleadings, testimony, and proceedings set forth in the transcript an order enforcing, modifying, or setting aside the order of the Office in whole or in part. If the hearing examiner has ordered an award of damages pursuant to subsection 23(4) and if the court sustains or modifies the award, it shall enter a judgment on the order or modified order in the same manner as in the case of an order of the district court.

(g) If a respondent fails to comply with a court order issued pursuant to this act, that respondent shall be guilty of contempt and shall be fined not more than one thousand (1,000) dollars for each day he or she fails to comply with the order. Conviction under this section shall not bar prosecution under any other penal statute.

COMMENT. Orders of the Office are not self-enforcing, and thus the Director must seek enforcement through the courts. Respondents are given the opportunity to appeal the order by seeking review prior to enforcement while the Director may seek enforcement at any time the terms of the examiner's order are not being met, if judicial review is completed or if it is not sought within the specified time.

Office attorneys are given the authority to appear in court on behalf of the Office, since they will have far more expertise than an outside attorney and are likely to be more dedicated to fulfilling the purposes of the act. In some states this will require an exception to be stated in the act since the Attorney General's Office is usually given authority to represent state agencies.

Enforcement is initiated by filing a petition certifying the failure of the respondent to comply and seeking relief, and by filing a transcript of the record of the hearing in the case. This establishes a prima facie case of violation with the court having the authority to determine, based solely upon the transcript, whether the order of the Office was lawful and reasonable. If so, the only question to be determined is whether the respondent has complied with the order. If not, the respondent shall be ordered by the court to comply.

The court is given broad authority to grant orders, including temporary injunctive relief pending a hearing on an enforcement order. The order shall include a previous award of damages unless such award is clearly unlawful or unreasonable. Any respondent failing to comply with the court's order shall be guilty of contempt and subject to a fine.

(4) Right to appeal adverse judgment.

The Director or a respondent may at his or her discretion appeal from an adverse judgment rendered by a court. Such appeal shall proceed as in the case of appeals in civil actions as provided in (cite to appropriate code sections and rules of appellate procedure).

COMMENT. None necessary.

(5) Remedies not exclusive.

Nothing in this section shall adversely affect provisions in section 26 of this act.

COMMENT. This subsection makes it clear that the private right of action granted by section 26 is in addition to any remedies granted by this section.

Section 25. Confidentiality of Information; Publication of Cases

(1) Nothwithstanding section (cite to state's public records statute) of the code, all complaints filed with the Director under this act and all records and documents in the custody of the Office which relate to and identify a particular person shall be confidential during an investigation, and shall not be disclosed by the Director or staff except to the parties or as necessary in the course of an investigation or a hearing or proceeding under this act. However, access to any record or document which is part of the record of any hearing or court proceeding shall not be restricted.

(2) The Director may publish an account of a case in which the complaint has been dismissed, the terms of the settlement of a case that has been voluntarily adjusted, or the terms of an order which has been issued.

COMMENT. Since the act encourages informal

settlement prior to formal proceedings, it will be easier to accomplish this if the parties can do so without publicity during the investigation and negotiation stages. Such confidentiality will also protect respondents from adverse publicity arising out of false or fraudulent claims. Thus, this section provides an exception to many public records statutes which make most public records open to the public. This exception applies to records and documents while the investigation is being conducted.

Confidentiality ends when these materials become part of the hearing record or when the case is settled by an agreement or order. In this way, the public's right to know is still protected, since records of the case will always become available during or after the settlement process. The Director may also publicize cases which have been terminated by dismissal, settlement or order.

Section 26. Suits by Private Parties; Intervention; District Court Jurisdiction; Attorney's Fees and Costs

(1) Suits by private parties; intervention.

(a) Any person aggrieved by a violation of this act, including any organization which represents or acts to further the interests of individuals so aggrieved, may bring a civil action seeking redress, including but not limited to equitable relief and damages for an unfair or discriminatory practice, upon withdrawal of the complaint from the Office and within the following times:

(i) Within ninety (90) days after the Director has determined that there is no probable cause for an allegation contained in a complaint or, if the complainant requested a reconsideration, within ninety (90) days after the Director has reaffirmed his determination of no probable cause;

(ii) After ninety (90) days but within one (1) year after the filing of a complaint if at or prior to the time of bringing the civil action a hearing has not been held pursuant to section 23; or

(iii) At any time after the Director fails or refuses to enforce an order relating to the person pursuant to subsection 24(3).

COMMENT. A private right of action for any aggrieved person is allowed as a supplement to other rights granted by the statute, but only where the Office fails to act. In addition, since many individuals protected by this act will not have the resources and expertise to initiate an action, organizations representing these people (e.g., the NAACP and NARC) may also bring actions on their behalf. Such representation can benefit the broader interests of persons similarly situated by eliminating systematic discrimination. Note, however, that since all persons are required to utilize the admininstrative procedures prior to initiating suit and since most cases will be handled by the Office, there should not be many of these private actions.

(b) A complainant bringing a civil action shall send by registered or certified mail a copy of the notice and the petition to the Director. Upon his or her receipt thereof, the Director shall cause all proceedings in the Office relating to an original complaint to terminate. No complaint shall be filed or reinstituted with the Director after a civil action relating to the same unfair or discriminatory practice has been brought, unless the civil action has been dismissed without prejudice.

COMMENT. All persons wishing to bring a private action must elect between administrative remedies and the private remedy. If the private action is elected, all administrative action on the original complaint must be terminated. This avoids multiple litigation of issues. It should also discourage private actions, since the complainant must begin anew on a crowded civil docket. Thus, there may be little gained and much expended where delay is the only complaint against administrative action. And unlike using the administrative procedure, the complainant must normally finance his or her own suit. Finally, since suits are primarily equitable, with few damages awarded, the financial incentive to continue to pursue administrative remedies is also strong.

(c) Upon application by an indigent complainant, the district court may appoint an attorney for such person and may authorize the commencement of the action without payment of fees, costs, or security.

COMMENT. Private actions are expensive to bring even for persons of moderate income. And since the monetary return from such suits is slight, if any, contingency fee arrangements are not a viable option. Justice should not depend upon one's financial assets. Those discriminated against are likely to be the least able to afford the cost of a suit. For this reason the court is given authority to appoint an attorney and waive costs for indigent complainants where they appear to have just cause for a suit.

(d) Upon timely application, the court may, in its discretion, permit the Office to intervene in a civil action brought pursuant to this section upon certification that the case is of general public importance.

COMMENT. In some instances where a complainant elects to bring a private suit, there may still be substantial public interest involved. The Office is allowed to intervene in these cases to protect those public interests which may be different from or be broader than the private interests.

(e) Upon request, the court may, in its discretion, stay further proceedings for not more than sixty (60) days pending further efforts of the Office to obtain voluntary compliance.

COMMENT. None necessary.

(2) District court jurdisdiction.

(a) Any action brought pursuant to this section shall be filed in the district court of the county wherein the unfair or discriminatory practice is alleged to have been committed or where the respondent resides or has his or her principal place of business.

(b) Any action brought pursuant to this act shall be heard and determined by a judge sitting without a jury.

(c) If the court finds that the respondent has engaged in an unfair or discriminatory practice, it shall issue an order directing such relief as it deems appropriate and which effectuates the purpose of this act. Such relief may include any relief permitted by section 23 of the act.

(d) Such relief shall be in addition to any remedy authorized by the United States Civil Rights Act of 1964, 42 U.S.C. §§2000a - 2000h (1976) or the Rehabilitation Act of 1973, 29 U.S.C. §701-794 (Supp. V 1975).

COMMENT. The private right of action is essentially an equitable action and is therefore heard by a judge with no right to a jury trial. In addition, the judge is given broad discretion in fashioning relief. This right of action is not intended to alter any right under federal law outlawing discrimination on the basis of race, color, religion, sex or disability. However, it is expected that at least some complainants will proceed under the more accessible state proceedings.

(3) Attorney's fees and costs.

In any action or proceeding brought pursuant to this section, the court, in its discretion, may allow the prevailing party other than the Office, reasonable attorney's fees in addition to costs.

COMMENT. Potential litigants should not be discouraged from asserting their rights because of the need to pay legal fees. Such fees are part of the cost of a legal action. They should be assessed between the parties along with other costs of the suit. Note that this provision cuts both ways. Complainants who lose may also be assessed for the reasonable attorney's fees of the respondent. This should help to deter unfounded suits. The allocation of fees is to be made based upon financial abilities and the relative merits of the respective cases.

Section 27. Severability Clause

If any provision of this act or the application thereof to any person or circumstance is held invalid, such invalidity shall not affect any other provision or application of the act which can be given effect without the invalid provision or application, and to this end the provisions of the act are declared severable.

COMMENT. None necessary.

Section 28. Effective Date of Act

This act shall take effect ________.

3
ELIMINATING ENVIRONMENTAL BARRIERS

TABLE OF CONTENTS

INTRODUCTION

The built environment in which we live, work and play is pervasive. We rarely escape it. Architecture is man's attempt to tame this environment, render it functional,[1] and make it aesthetically pleasing. For most people it contributes greatly to improving the quality of their lives. But for a significant portion of our population, the built environment tames their abilities and is dysfunctional for it is filled with barriers. Thus, there is a disparity between the physical environment as we have erected it about ourselves and the environment as it should be constructed to permit the large numbers of physically disabled persons[2] to live and work in equality with their fellow citizens.[3]

Environmental barriers restrict potential places of residence and opportunities for education, employment, shopping, recreation, entertainment and worship. They contribute to segregation of environmentally limited persons, thereby limiting their societal interaction and denying them full and independent lives.[4] Some figures on the effects of inaccessible mass transportation are illustrative. It is estimated that with the present transportation system the disabled travel about half as much as the rest of the population,[5] with the largest difference being in social/recreational, work and shopping excursions. If an accessible transportation system were available the number of trips made by disabled persons would increase significantly.[6]

The impact of barriers on employment is no less negative. Large numbers of disabled persons are unemployed or underemployed. One source notes that only 47.8% of the handicapped population between the ages of 17 and 64 are members of the labor force as against 65% of the general population in this age group.[7] Of those who are employed

---

[1] Farber, *The Handicapped Plead for Entrance – Will Anyone Answer?*, 64 KY. L. J. 99, 99-100 (1975).

[2] The number of disabled persons in the United States is difficult to ascertain. The figures vary from 18 to 68 million nationally, depending upon who is being counted as disabled and for what purpose. Note, *Access to Buildings and Equal Employment Opportunity for the Disabled: Survey of State Statutes*, 50 TEMP. L.Q. 1067 n. 7 (1977). Other estimates are as low as 12 million for physically disabled persons. Farber, *supra* note 1, at 99 n. 2. Congress has found there are several million children and at least 28 million adults with physical or mental handicaps. Achtenberg, *Law and the Physically Disabled: An Update with Constitutional Implications*, 8 SW. U.L. REV. 847 n. 1 (1976).

The difficulty in obtaining accurate and meaningful statistics is attributable to the inability of statisticians to measure the effect of a defined handicap on the capacity of the handicapped person to function normally in society. For example, the epileptic person may not be handicapped in his or her capacity to use public transportation; however, he or she is severely limited in ability to secure and maintain employment. Similarly, an individual with a spinal cord injury may be able to obtain employment but incapable of utilizing public transportation in order to seek and maintain employment. Thus, numerical statistics must be evaluated in terms of the resultant effect of a specific disability on participation in normal activity. Note, *Abroad in the Land: Legal Strategies to Effectuate the Rights of the Physically Disabled*, 61 GEO. L.J. 1501 n. 2 (1973) (citations omitted).

Yet, figures vary widely even when estimates are made for a particular purpose. The Department of Health, Education and Welfare, using a broad definition of disability, has estimated that 29 million persons in America suffer from conditions which would be less handicapping without transportation and architectural barriers. Reed, *Equal Access to Mass Transportation for the Handicapped*, 9 TRANSP. L.J. 167, 169 n. 15 (1977). Another estimate places at one in ten the number of persons in this country who have some disability which prevents them from using buildings and facilities designed only for the physically fit. This number does not include the many persons who become temporarily disabled. Note, *The Forgotten Minority: The Physically Disabled and Improving Their Physical Environment*, 48 CHI.-KENT L. REV. 215, 220 n. 31 (1971).

The Department of Transportation made the highest estimate of those who would benefit from removal of transportation obstacles — nearly 44 million people. Reed, *supra*, at 169 n. 12. The Urban Mass Transportation Administration has estimated that there are 13.3 million disabled and elderly persons who cannot use or experience difficulty in using transit systems. *Id.* at 170 n. 17. Of these, 6 million are disabled and 7 million are elderly persons. Comment, *Mass Transportation for the Handicapped and the Elderly*, 1976 DET. C.L. REV. 277 n. 1. No matter which estimate is utilized, it is obvious that the number of persons affected by environmental barriers is substantial.

[3] Note, *The Forgotten Minority: The Physically Disabled and Improving Their Physical Environment*, 48 CHI.-KENT L. REV. 215, 216 (1971).

[4] *See* Reed, *Equal Access To Mass Transportation for the Handicapped*, 9 TRANS. L.J. 167, 169-71 (1977); Achtenberg, *Law and the Physically Disabled: An Update with Constitutional Implications*, 8 SW. U.L. REV. 847, 855-75 (1976).

[5] Reed, *supra* note 4, at 170 n. 18. Transportation accessibility is only one component of total accessibility, and the area is largely preempted by federal legislation. (See section 13 of the model act *infra*.) However, figures on transportation accessibility are available because the area has been extensively studied by the federal government, while comparable figures on the impact of barriers in other components are not readily available. Thus, these figures are offered to illustrate the larger problem of accessibility.

[6] *Id.* nn. 19 & 21.

[7] Achtenberg, *supra* note 4, at 860 n. 39.

53% used public transportation to get to work, indicating a much higher dependency on public transportation than the general public. Of those who are unemployed, but looking for jobs, 30% indicated that lack of transportation was a factor in their unemployed status.[8] Another source indicates that lack of transportation is the single most important factor preventing 13% of disabled persons from working.[9] When coupled with the impact of the inaccessibility of many places of employment, the lack of accessible housing near places of employment, and the presence of other barriers to pedestrian travel (e.g., curbs and stairways), the adverse effect of barriers on employment of disabled persons is clearly felt.

The cost of removing barriers can certainly be significant, especially in transportation systems. But the economic benefits are significant as well. For example, the removal of travel barriers should allow 200,000 handicapped persons to enter the work force,[10] and estimates indicate that a benefit of $300 to $500 million would result from each 100,000 handicapped persons returned to the work force.[11] In addition, the Department of Transportation found that a net economic benefit of approximately $800 million would result from elimination of transportation barriers.[12] Since the cost of eliminating structural barriers is considerably less than in transportation systems (*see* Comment to section 8 of the model act *infra*), the potential cost benefit ratio is likely to be even greater in that area.

But dollars and cents hardly tell the whole story. The human factors must also be considered. Environmental barriers which segregate and isolate contribute to lower societal expectations and evaluations. These in turn contribute to a social stigma projected upon disabled persons. Moreover, barriers have a debilitating, denigrating impact upon the mental as well as physical well being of a disabled indivdual by instilling in that individual a sense of inferiority commensurate with the status society awards to him or her.[13]

True integration and equality will not be had for disabled persons until they are free to move about their neighborhoods, their communities, and their country. Where once equality meant only that the disabled persons be protected, cared for and allowed to live in seclusion, they are now going out to live as normally as their personal limitations will permit.[14] Yet, society is not adequately accommodating them and their desires. Thus, public policy must be developed and implemented to allow disabled persons to be free, independent, and self-sustaining to the maximum extent of their abilities; every individual must be allowed to make his or her contribution to society, as well as to enrich his or her own life.[15]

> (S)ociety has a distinct interest in utilizing every possible source of human skill and ingenuity, including the skills and talents of mobility-handicapped individuals. When effectively confined to a single floor, building or city block, not only are the handicapped deprived of the myriad benefits of society, but society is deprived of the valuable contributions of these otherwise normal human beings. And this deprivation is compounded by the fact that, when unable to fend for themselves, the handicapped must depend upon the public purse for their sustenance. This is a burden that should be inflicted upon neither the mobility-handicapped nor society in general.[16]

The following sections will emphasize the importance of mobility, point out three general restrictions on mobility and discuss the combined adverse impact which the three have on disabled persons.[17] It is then suggested that the law needs to take an affirmative approach in eliminating this adverse impact. Federal and state statutes which were passed to do so are reviewed and both are evaluated for their adequacy and effectiveness. Finally, the factors which should be considered when developing alternative legislation are presented and a model act is summarized, with suggestions for adapting it to individual state's needs.

---

8 Reed, *supra* note 4, at 170.

9 *Id.* at n. 22.

10 *Id.*

11 *Id.* at 171 n. 24.

12 *Id.*

13 Comment, *Mass Transportation for the Handicapped and the Elderly,* 1976 DET. C.L. REV. 277.

14 Note, *supra* note 3, at 219.

15 *See* Hull, *Forward – The Specter of Equality: Reflections on the Civil Rights of Physically Handicapped Persons,* 50 TEMP. L.Q. 944, 952 (1977).

16 Bartels v. Biernat, 405 F. Supp. 1012, 1017-18 (E.D. Wis. 1975).

17 Since many elderly persons are also restricted in their mobility, much of the discussion concerning these restrictions in the text *infra* applies to them as well. They too are environmentally limited and would benefit from the removal of barriers.

RESTRICTIONS ON MOBILITY

Introduction

> Movement, we are told, is a law of animal life. As to man, in any event, nothing could be more essential to personality, social existence, economic opportunity — in short, to individual well-being and integration into the life of the community — than the physical capacity, the public approval, and the legal right to be abroad in the land.[18]

Most people take for granted their mobility without considering its impact upon their lives, including their educational, employment, recreational and entertainment opportunities. Only when it is lost does its importance become evident. Indeed, movement is an integral part of our life-style and culture, without which we cannot participate. Yet most members of society are or have been either temporarily or permanently restricted in their movement. For example, nearly all of us have at one time or another been temporarily physically incapacitated due to illness such as influenza or a broken bone. For some, a heart attack or stroke may dictate an extended period of incapacity. Public attitudes or mores, whether conscious or unconscious, may also restrict our movement. For example, some areas of the community may be viewed as appropriate for some groups of persons but not for others and vice versa. These "socially determined" areas can contribute to "ghettoization" of both groups and neighborhoods. Finally, movement can also be restricted as a result of restrictions on the legal rights of persons. The most common examples are incarceration of persons convicted of a criminal offense or the civil commitment of persons with mental or physical disabilities.

The Impact of Restrictions for Disabled Persons

Developmentally disabled persons,[19] particularly persons with cerebral palsy or epilepsy, and those with other physical disabilities are affected by all three of these factors. To truly understand their combined impact, we must ask to what extent do physical capacity, public attitudes and legal rights coincide and what are the consequences of this combination?[20]

Physical Capacity

No one can deny that a person's physical disability can frequently restrict mobility and that this, in turn, may have devastating impact upon one's life. Specifically, we must move to be educated, we must move to be employed, we must move to recreate. We no longer live and work in small self-sufficient communities or neighborhoods. But equally problematic is the fact that the severity of a physical handicap is frequently a function of the physical demands of society. For example, the built environment, including buildings and transportation systems, has been designed and constructed for the average person. In fact, since handicaps have traditionally been considered to be the product of something within the person as opposed to being the product of external factors, consideration for the disabled has been completely absent from the design and construction of the built environment. While new techniques have

---

[18] tenBroek, *The Right to Live in the World: The Disabled in the Law of Torts,* 54 CALIF. L. REV. 841 (1966).

[19] The term "developmental disability" is a federally created term referring to a disability which interferes with the normal development of functional abilities. The original definition contained in the Rehabilitation Act of 1973 as amended was a categorical one which read:

> The term "developmental disability" means a disability of a person which —
> (A)(i) is attributable to mental retardation, cerebral palsy, epilepsy, or autism; (ii) is attributable to any other condition of a person found to be closely related to mental retardation because such condition results in similar impairment of general intellectual functioning or adaptive behavior to that of mentally retarded persons or requires treatment and services similar to those required for such persons; or (iii) is attributable to dyslexia resulting from a disability described in clause (i) or (ii) of this subparagraph;
> (B) originates before such person attains age eighteen;
> (C) has continued or can be expected to continue indefinitely; and
> (D) constitutes a substantial handicap to such person's ability to function normally in society.

Pub. L. No. 94-103, §125, 89 Stat. 496 (1975).

The Act has since been amended to replace the original categorical definition with a functional definition which reads:

> The term "developmental disability" means a severe, chronic disability of a person which —
> (A) is attributable to a mental or physical impairment or combination of mental and physical impairments;
> (B) is manifested before the person attains age twenty-two;
> (C) is likely to continue indefinitely;
> (D) results in substantial functional limitations in three or more of the following areas of major life activity: (i) self-care, (ii) receptive and expressive language, (iii) learning, (iv) mobility, (v) self-direction, (vi) capacity for independent living, and (vii) economic sufficiency; and
> (E) reflects the person's need for a combination and sequence of special, interdisciplinary, or generic care, treatment, or other services which are of lifelong or extended duration and are individually planned and coordinated.

42 U.S.C. §6001 as amended by Rehabilitation, Comprehensive Services, and Developmental Disabilities Amendments of 1978, Pub. L. No. 95-602, §503, 92 Stat. 3004-06 (1978).

[20] tenBroek, *supra* note 18, at 842.

been developed to minimize the impact of many kinds of physical disabilities, the handicaps imposed by the built environment continue to proliferate.[21] The time has come to consider rehabilitation, education, housing, employment and the exercise of rights of physically disabled persons within the context of the built environment in which these activities must be pursued.[22] "The greatest single obstacle to employment (and college educations and just plain mobility) for the handicapped is the physical design of the buildings they must use."[23]

Public Attitudes

Public attitudes toward disabled persons have also contributed to their lack of mobility. As one author has pointed out:

> The actual physical limitations resulting from the disability more often than not play little role in determining whether the physically disabled are allowed to move about and be in public places. Rather, that judgment for the most part results from a variety of considerations related to public attitudes, attitudes which not infrequently are quite erroneous and misconceived. These include public imaginings about what the inherent physical limitations must be; public solicitude about the safety to be achieved by keeping the disabled out of harm's way; public feelings of protective care and custodial security; public doubts about why the disabled should want to be abroad anyway; and public aversion to the sight of them and the conspicuous reminder of their plight. For our purposes, there is no reason to judge these attitudes as to whether they do credit or discredit to the human head and heart. Our concern is with their existence and their consequences.[24]

At best these public attitudes have resulted in benign negligence. While the public seems to be aware of the existence of environmental barriers, the public seems substantially unaware of the real meaning of these barriers and are seemingly unconcerned.[25] This lack of concern, in turn, contributes to the continued presence of environmental barriers and indeed becomes a barrier in itself.[26]

In addition, significant numbers of individuals who cannot engage in normal activities without some external adaptation or assistance are forced to remain inconspicuous to the general public.[27] This circumstance contributes to the "out of sight, out of mind" societal attitude which has had two primary effects. First, the "average" architect, engineer or builder has designed and built facilities for the "average" person. While the less severely disabled persons may pass as "average," more severely disabled persons are substantially excluded from society. Second, inaccessibility reinforces the social stigma which is projected on disabled persons by further isolating or segregating them.[28] Thus, these persons are subjected to sufficient ego-destroying pressures to deter the militancy necessary for demanding legally protected rights,[29] and "(m)ore people are forced into limited lives and made to suffer by these manmade obstacles than by any specific physical or mental disability."[30]

The Legal Right

The law itself has been a third factor contributing to lack of mobility, either directly, or more frequently, indirectly. The historical treatment of mentally disabled persons in state statutes is one of inequality.[31] These persons have been and still are subjected to restrictions on marriage, voting, contracting, and testamentary disposition, and have been subjected to sterilization, termination of parental rights, involuntary as well as "voluntary" commitment to custodial institutions and exclusion from educational facilities.[32] Physically disabled persons have fared considerably better with regard to these kinds of restrictions; but there are state statutes which have restricted their movement.

Perhaps the most bizarre of these laws are those which have been referred to as the "ugly laws."[33] For example, until repealed in 1974, the Chicago Municipal Code provided:

> No person who is diseased, maimed, mutilated or in any way deformed so as

---

[21] Note, *supra* note 3, at 220-21.

[22] *Id.* at 219.

[23] *Id.* (quoting Report of the National Commission on Architectural Barriers to the Rehabilitation of the Handicapped, at 1 (Dec. 1976)).

[24] tenBroek, *supra* note 18, at 842.

[25] Note, *supra* note 3, at 221.

[26] *Id.* at 221-22.

[27] *Id.* at 221.

[28] Hull, *supra* note 15, at 948.

[29] Achtenberg, *supra* note 4, at 852. *See* Kriegel, *Uncle Tom and Tiny Tim: Some Reflections on the Cripple as Negro*, 38 AM. SCHOLAR 412 (1969) for a discussion of the effects of societal stereotyping of disabled persons upon the ego.

[30] Achtenberg, *supra* note 4, at 848.

[31] *See* Burgdorf and Burgdorf, *A History of Unequal Treatment: The Qualifications of Handicapped Persons as a "Suspect Class" under the Equal Protection Clause*, 15 SANTA CLARA LAW. 855, 861-99 (1975) for an overview of the unequal treatment of handicapped persons past and present.

[32] *See* Chapter 1: Personal and Civil Rights, *supra;* Chapter 5: Education of Exceptional Persons, *infra;* Chapter 6: Right to Services, *infra*.

[33] Burgdorf, *supra* note 31, at 863.

> to be an unsightly or disgusting object or improper person to be allowed in or on the public ways or other public places in this city, shall therein or thereon expose himself to public view, under a penalty of not less than one dollar nor more than fifty dollars for each offense.[34]

The authors also cite to similar laws in Columbus, Ohio, Omaha, Nebraska and other cities and note that as recently as 1974 a person was arrested for violating the Omaha ordinance.[35]

However, except for instances such as these, physically disabled persons have not been subjected to outright prohibitions in the law. Indeed, Professor tenBroek has asserted that integration of the physically disabled is the policy of the nation.[36] The law contains provisions for rehabilitation of physically disabled persons and prohibits discrimination in employment, housing and public accommodations. But of what use are opportunities for education, rehabilitation, employment, better housing, recreation and entertainment if these facilities or the transportation to reach them are inaccessible? Simply declaring the right to access and use does little if physical barriers exist.[37] The legal rights of disabled persons must include a constructive and affirmative approach to the elimination of physical barriers in the environment if these rights are to be truly meaningful. The right to freedom from employment discrimination must include the right to enter places of employment without being blocked by unnecessary barriers, just as a right to public accommodations necessarily entails a rights to effective access and use. Facilities must be erected in accordance with a design which takes into account the needs of disabled persons so that facilities are accessible to and functional for everyone.

Both the federal and state governments have responded in varying degrees to the need for affirmative approaches to the elimination of environmental barriers. The next sections will summarize those provisions and examine their effectiveness.

FEDERAL LAWS PROHIBITING ENVIRONMENTAL BARRIERS

Introduction

Of the three impediments to full accessibility noted above, inadequate architectural design and lack of legal recognition of the rights of disabled persons are the two major ones.[38] Yet, the first should not be a problem since physical barriers can be effectively eliminated by modern technology and proper planning. Thus, the lack of legal principles to implement the right to full access and utilization of facilities[39] and the commitment to carry them out is of prime concern.

Various constitutional arguments have been asserted to establish the right of access for disabled persons. These include the right to travel,[40] the right to petition the government,[41] and equal protection.[42] However, litigation relying on these theories has been largely unsuccessful to date.[43] In contrast, judicial enforcement of legislatively created rights has been more effective.[44] Since the federal government and nearly all of the states have adopted statutes requiring access to varying degrees (see *infra*), it is to these statutorily created rights we look.

Federal Statutory Law

The first federal legislation aimed at eliminating barriers was the Architectural Barriers Act of 1968.[45] It requires that most buildings constructed, altered, leased or financed by or on behalf of the United States government after the effective date of the act shall comply with standards adopted to

---

34 Chicago, Ill., Mun. Code §36-34 (1966) (repealed 1974).

35 Burgdorf, *supra* note 31, at 863-64.

36 tenBroek *supra* note 18, at 843-47.

37 *Id.* at 859.

38 *See* Note, *Abroad in the Land: Legal Strategies to Effectuate the Rights of the Physically Disabled,* 61 GEO. L.J. 1501, 1506 (1973).

39 *Id.*

40 *Id.* at 1507-09.

41 *Id.* at 1509-12.

42 *See* Burgdorf and Burgdorf, *supra* note 31, at 899-908; Reed, *supra* note 4, at 181-85; Comment, *supra* note 13, at 287-89.

43 *See* United Handicapped Federation v. Andre, 409 F. Supp. 1297 (D. Minn. 1976) (transportation); Snowden v. Birmingham-Jefferson County Transit Authority, 407 F. Supp. 394 (N.D. Ala. 1975) (transportation); Selph v. Council of the City of Los Angeles, 390 F. Supp. 58 (C.D. Cal. 1975) (voting).

44 *See* Marsh v. Edwards Theater Circuit, No. C 55, 376 (L.A. County Super. Ct., June 16, 1975); People v. Smuggler's Inn, Associated Hosts, Inc., No. 251428 (Sacramento County Super. Ct., March 1, 1975). *Cf.* Lloyd v. Regional Transp. Auth., 548 F.2d 1277 (7th Cir. 1977); Bartels v. Biernat, 427 F. Supp. 226 (E.D. Wis. 1977); Bohlke v. Golden Gate Bridge, Highway & Transp. Dist., Memorandum and Minute Order No. 73362 (Marin County, Cal. Super. Ct., May 9, 1975). *See also* Achtenberg, *supra* note 4 at 855-75 and Reed, *supra* note 4 at 174-85 for a discussion of accessibility-related cases.

45 42 U.S.C. §§4151-56 (1976).

insure ready access to and use of these buildings wherever possible by physically handicapped persons.[46] Modification or waiver of such standards is authorized on a case by case basis upon a determination that it is clearly necessary.[47]

Standards generally applicable to federal buildings have been adopted by the Administrator of the General Services Administration (GSA).[48] Other agencies such as the Department of Housing and Urban Development have also adopted standards as required by the act.[49] These standards were not developed specifically for implementing the act, but rather they were first developed in 1961 by the American National Standards Institute (ANSI)[50] at the request of the President's Committee on Employment of the Handicapped.[51] A large number of organizations were represented on the development committee, including many representing disabled persons,[52] with the resultant standards being accepted as the best available.[53] However, since they have also been the subject of criticism,[54] new standards are to be issued in 1979. (See Comment to section 8 *infra.*)

The Architectural Barriers Act of 1968 contained no enforcement provisions, thus making its effectiveness extremely questionable. It has been noted that voluntarism in the areas of accessibility, as in civil rights issues, is essentially nonexistent;[55] change cannot be achieved by voluntary efforts alone.[56] Congress gave recognition to this reality by including enforcement provisions in the Rehabilitation Act Amendments of 1973.[57] Specifically, Section 502 of the 1973 Amendments established the Architectural and Transportation Barriers Compliance Board[58] which was made up of the heads of nine different federal departments or agencies headed by the Secretary of Health, Education and Welfare.[59] A Consumer Advisory Panel, a majority of which must be made up of handicapped persons, was to provide guidance, advice and recommendations to the Board.[60] One of the Board's functions is to insure compliance with the standards adopted pursuant to the 1968 act, as amended,[61] and in carrying out its responsibilities, the Board is empowered to conduct investigations, hold public hearings and issue orders as it deems necessary.[62]

Unfortunately, both the standards provisions and the enforcement provisions have been noted for their ineffectiveness. (See Criticisms of Federal and State Legislation *infra*) Some of the major criticisms may be summarized as follows:

1. The statute establishes no minimum standards and no ultimate compliance date;

2. The statute allows the granting of waivers and modifications of standards without providing strict criteria for granting them and without an adequate showing of the factual basis for granting them. This can result in the granting of waivers and modifications which are contrary to the intent of the statute;

3. The Board is made up of the heads of the agencies which it is to regulate, creating an inherent conflict of interest. Disabled persons are given only an advisory role;

4. The statute's scope, enforcement authority and procedures are inadequate and therefore not effective.

In response, Congress amended Section 502 with the Comprehensive Rehabilitation Amendments of 1978.[63] According to its provisions:

1. The Board is empowered to "establish minimum guidelines and requirements" for the standards issued pursuant to the Architectural Barriers Act.[64] The current federal minimum standard is the ANSI Standard. Since revision of the ANSI Standard is near completion as noted *supra*,

---

46 *Id.* §§4151-55.

47 *Id.* §4156.

48 41 C.F.R. §§101-19.600-607 (1977). Certain changes have been made in the regulations relating to the applicability of standards to federally leased buildings. *See* 43 Fed. Reg. 16478-80 (1978).

49 *See, e.g.,* 24 C.F.R. §§40.1-40.6 (1978) (Department of Housing and Urban Development).

50 Formerly known as the American Standards Association. American National Standards Institute, Inc., AMERICAN NATIONAL STANDARD SPECIFICATIONS FOR MAKING BUILDINGS AND FACILITIES ACCESSIBLE TO, AND USABLE BY, THE PHYSICALLY HANDICAPPED, No. A117.1-1961 (Reaffirmed 1971).

51 Farber, *supra* note 1, at 108 n. 31.

52 The organizations represented on this Committee are listed in the Foreward to the ANSI Standards, *supra* note 50.

53 Farber, *supra* note 1, at 108.

54 *See id.* at 108-11; *Hearings on the Effectiveness of the Architectural Barriers Act of 1968 (Pub. L. 90-480) Before the Subcommittee on Investigations and Review of the House Committee on Public Works and Transportation,* 94th Cong., 1st Sess. 12 (1975).

55 Farber, *supra* note 1, at 112.

56 Note, *supra* note 3, at 225 n. 78.

57 *See* §502, 29 U.S.C. §792 (1976).

58 *Id.*

59 *Id.* §792(a).

60 *Id.*

61 *Id.* §792(b).

62 *Id.* §792(d). Regulations of the Board covering practice and procedures may be found at 36 C.F.R. §§1150.1-1150.115 (1978). These procedures will quite likely be amended somewhat as a result of changes in the statute noted in the text which follows.

63 *See* Rehabilitation, Comprehensive Services, and Developmental Disabilities Amendments of 1978, Pub. L. No. 95-602, §118, 92 Stat. 2979-82 (1978).

64 *Id.* §118(b)(3).

these may be adopted by the Board in place of the current standards.

2. The Board is empowered to insure that "all waivers and modifications of standards are based upon findings of fact and are not inconsistent with the provisions of the Act."[65]

3. The composition of the Board was increased by twelve members with the addition of a representative from the Department of Justice and eleven public members (five of whom shall be handicapped individuals) appointed by the President. At the same time, the Board's National Advisory Committee was abolished.[66]

4. The scope of the statute has been widened. The Board is now empowered to "insure that public conveyances, including rolling stock, are readily accessible to, and usable by, physically handicapped persons."[67] Previously only fixed facilities were covered.

The Board's enforcement authority and the procedures used have also been strengthened. The Board is empowered to insure compliance with the Architectural Barriers Act of 1968 (including the application of that Act to the U.S. Postal Service), including but not limited to enforcing all standards under that Act.[68] The Board's executive director has final authority to investigate instances of noncompliance and to issue formal complaints, codifying current administrative practice.[69] The Board no longer performs as an appellate body. For judicial review purposes, an administrative law judge's order is a final order of the Board.[70] The executive director may also, with Board approval, bring a civil action in U.S. district court to enforce the Board's orders and intervene, appear and participate as friend-of-the-court in federal or state court actions relating to Section 502 of the Architectural Barriers Act.[71]

Moreover, these amendments should improve the effectiveness of the federal accessibility statutes in two important ways. First, the authority to set minimum standards should improve the quality of applicable standards by insuring that they reflect the comprehensive needs of disabled persons, while the authority to monitor the granting of waivers should insure that these standards are not rendered ineffective through the granting of waivers, modifications or exemptions. Second, improved enforcement procedures should increase the effectiveness of the standards by helping to insure that they are implemented where applicable.

There are two other federal mechanisms for promoting accessibility. The first is contained in Section 504 of the Rehabilitation Amendments of 1973.[72] This section, which is applicable to all recipients of federal funds, requires among other things that no recipient shall discriminate against otherwise qualified handicapped persons in any program or activity for which it uses the federal funds.[73] Executive Order 11914 issued April 28, 1976[74] designated the Secretary of Health, Education and Welfare to coordinate the implementation of Section 504 by all federal departments and agencies empowered to extend federal financial assistance to any program or activity. The Secretary of HEW was to adopt rules and regulations and issue orders as necessary to carry out this responsibility. Each department or agency empowered to extend federal financial assistance is to issue rules and regulations for implementing Section 504 which are consistent with those adopted by the Secretary.[75]

Pursuant to the mandate, the Secretary has issued HEW regulations to implement Section 504 and regulations to implement the terms of the Executive Order.[76] These latter regulations contain essentially the same provisions as the former.[77] Other departments and agencies are in various stages in the process of adopting regulations.

The Department of Health, Education and Welfare's implementing regulations contain provisions requiring program accessibility for both new construction and existing facilities.[78] In general, Subpart C of the regulations prohibits the exclusion of qualified handicapped persons from federally assisted programs or activities because a recipient's facilities are inaccessible or unusable by them. It requires that all new construction commencing after June 3, 1977 must be designed so that the facility will be accessible to and usable by handicapped persons.[79] Commencement of construction

65 *Id.* §118(b)(1).

66 *Id.* §118 (a).

67 *Id.* §118(b)(3).

68 *Id.* §118 (b)(1).

69 *Id.* §118(d)(2).

70 *Id.* §118(d)(3).

71 *Id.* §118(c)(5).

72 29 U.S.C. §794 (1976).

73 The text of §504 reads: "No otherwise qualified handicapped individual in the United States, as defined in section 706(6) of this title, shall, solely by reason of his handicap, be excluded from the participation in, be denied the benefits of, or be subjected to discrimination under any program or activity receiving Federal financial assistance." *Id.*

74 *See* Appendix A, 43 Fed. Reg. 2139 (1978).

75 *Id.*

76 43 Fed. Reg. 2132-39 (1977) (to be codified in 45 C.F.R. §§85.1-.58).

77 They do provide for certain limited extensions of time for compliance with respect to certain modes of transportation. *See* 43 Fed. Reg. 2139 (1978) (to be codified in 45 C.F.R. §§85.57(c), 85.58).

78 *See* 45 C.F.R. §§84.21-.23 (1977).

79 *Id.* §84.23(a).

means groundbreaking. All alterations after June 3, 1977 which affect or could affect usability of a facility must to the maximum extent feasible, be altered so as to make the altered portion of the facility readily accessible to and usable by handicapped persons.[80]

Accessibility requirements also apply to existing programs.[81] Generally, recipients operating existing programs are required to make their programs accessible to all handicapped persons so that the program or activity, when viewed in its entirety, is readily accessible to handicapped persons. Accessibility may be achieved by a variety of methods, including redesign of equipment, reassignment of classes or other services to accessible buildings, or by making aids available to the handicapped. According to HEW's interpretation, structural changes in existing facilities are required only where there is no other feasible way to make the program accessible. For example, a university is not required to make all of its existing classroom buildings accessible to handicapped students if some of its buildings are already accessible, and if it is possible to reschedule or relocate classes in those facilities.[82]

For existing programs, all non-structural adjustments must be made within 60 days of the June 3, 1977 effective date. If structural changes are required, recipients will be given up to three years to attain program accessibility, but they must submit a transition plan within six months detailing how the program plans to make its facilities accessible.[83]

Finally, according to Section 504, all new construction, alterations, and structural changes to existing programs must comply with the ANSI Standard[84] on building accessibility with a departure from them being permitted when it can be demonstrated that equivalent access to the recipient's facilities is provided.[85] Thus, a wide range of options for compliance could exist. However, the Secretary of HEW has stated in his analysis of the final regulations that enough flexibility exists in the ANSI Standards to permit recipients to devise ways to make their programs accessible short of extremely expensive or impractical physical changes in facilities. Accordingly, no waivers are provided for in the regulations.[86]

The implementation of 504 regulations by all grantors of federal financial assistance should help to improve accessibility in a number of areas, including education, housing, health care services and transportation. Thus, while limited in coverage, it can still have a significant impact.

The second mechanism of federal law applicable to accessibility involves the planning of and federal financial assistance to mass transportation and highway systems primarily and other modes of federally regulated transportation secondarily. Since the area of transportation is largely preempted by federal standards and regulations, the state role in transportation issues is more limited. Federal requirements and their relation to state responsibility and authority are discussed in the Comment to section 13 of the model act, *infra*.

In conclusion, federal initiatives in the area of accessibility have been significant. However, even if tighter standards were adopted and the enforcement of standards completely effective, the impact on the lives of most disabled persons would be negligible because of its very limited coverage. These federal requirements apply only to federally constructed, altered, leased or financed buildings and to programs operated by recipients of federal financial assistance. To be effective, accessibility requirements must extend to the environment in which most of us conduct our daily activities. For this reason we look next to state-created statutory rights.

REVIEW OF STATE LEGISLATION

Every state has adopted some type of legislation covering access to certain buildings and facilities.[87] However, not all of these statutes provide affirmative measures to insure that buildings and facilities are actually accessible and usable.

Two distinct types of legislation have been enacted by the states to assist disabled persons gain access to public facilities. The first type is merely declaratory, rather than affirmative. Commonly referred to as "White Cane Safety Acts," most of these statutes were originally enacted to prevent discrimination against the blind and were only later amended to include other physical disabilities. Their purpose is to declare a right of equal access to public facilities for disabled per-

80 45 C.F.R. §84.23(b) (1977).

81 *Id.* §84.22.

82 Analysis of Final Regulation, Appendix A, subpart C, 45 C.F.R. pt. 84 (1977).

83 *See* 45 C.F.R. §§84.22(d), (e) (1977).

84 ANSI Inc., *supra* note 50.

85 *Id.* §84.23(c).

86 Analysis, *supra* note 82.

87 *See* "State Statute Charts" *infra* and those in Chapter 2: Prohibiting Discrimination, *supra*.

sons.[88] While these acts provide that "the physically disabled (shall) have the same right as the able-bodied to the full and free use of the streets, highways, sidewalks, walkways, public buildings, public facilities and other public places,"[89] they have consistently been held by the courts to be insufficient manifestations of legislative intent to require any environmental modifications to facilitate such use.[90] As such they have been of little benefit to persons with severe mobility-limiting handicaps. This is not to say that these statutes are not useful in cases where outright discrimination occurs. When this is the case, these statutes may provide the most effective remedy. For a listing of the states with this type of law see the "State Statute" Charts in Chapter 2: Prohibiting Discrimination.

The second type of statute more directly addresses the problems of those persons who are mobility-limited and cannot pass as "normal" within the built environment. These statutes provide for positive efforts to make new or existing buildings accessible. Statutory provisions vary considerably from simple delegations of authority to promulgate rules[91] to lengthy acts incorporating minimum standards, definitive scope provisions and specific administrative and enforcement methodologies.[92] It is this second type of statute with which we are concerned and they are listed in the charts accompanying this section.

Policy and Purpose (Chart I)

Twenty-four states now preface their acts with a statement that the legislation is designed to make buildings and/or facilities accessible to, and functional for, the physically handicapped.[93] Three states also include among their policies and purposes maximized integration into society.[94] Statements to this effect are particularly useful in those states where actual design and construction criteria are left to be developed by an administrative agency or board in that the "legislative declaration" may be held to create an affirmative duty for the board to act expeditiously. They also give the agency or board legislative guidance in the adoption and implementation of criteria. In addition, statements which indicate that legislation is intended to promote safety, now present in the environmental barriers statutes of six states,[95] help to justify use of the state's police power and insulate the accompanying legislation from constitutional attack.[96]

Scope — Coverage of the Total Environment (Chart II)

The "scope" of environmental barriers legislation refers to those statutory sections which describe the various elements of the man-made or built environment to which barrier-free design standards are applicable. As design standards described *infra* define the degree of accessibility required within a given facility, scope provisions determine the degree of accessibility throughout the total environment.

The states have taken widely divergent approaches in formulating these critically important sections. A majority (27) limit the applicability of their statutes to buildings and facilities constructed in whole or in part with state or municipal funds. Many also apply to publicly leased facilities.[97] By so limiting coverage to publicly funded facilities, these states effectively exempt the majority of facilities to which physically limited persons might reasonably have a need for access. Both constitutional and economic justifications have been advanced for these limitations but a substantial number of other state legislatures, which have enacted significantly broader provisions, found neither rationale compelling.

Presently twenty-two states do provide for regulation of some facilities designed and constructed exclusively with private funds but used by the public.[98] While some provisions of this type are narrowly drafted to include only specific types of buildings over a particular size,[99] others are broad enactments encompassing all buildings open to, and used by, the public.[100] Both the statutes lim-

[88] *See* ALA. CODE tit. 21, §§7-2, 7-3 (1977); ARK. STAT. ANN. §82-2901 (1976); COLO. REV. STAT. §24-34-801 (1974); DEL. CODE tit. 16, §9502 (1975); D.C. CODE ANN. §§6-1501, 6-1502 (1973); IDAHO CODE §§56-701 to 03 (1976); ILL. ANN. STAT. ch. 23, §3363 (Supp. 1978); IND. CODE ANN. §§16-7-5-2, 16-7-5.5-2 (Burns 1973); IOWA CODE ANN. §§601D.3, .4 (1975); KAN. STAT. §46:1952 (West Supp. 1978); MINN. STAT. ANN. §256C.02 (West Supp. 1977); MISS. CODE ANN. §§43-6-3, 43-6-5 (Supp. 1977); MONT. REV. CODES ANN. §64.301 (Supp. 1978); NEB. REV. STAT. §§20-127, 20-131.01 (1977); NEV. REV. STAT. §651.070 (1977); N.H. REV. STAT. ANN. §167-6:2 (1964); N.M. STAT. ANN. §12-26-3 (1976); N.C. GEN. STAT. §§168-2, 168-3 (1976); ORE. REV. STAT. §659.405 (1977); PA. STAT. ANN. tit. 43, §955 (1964); S.C. CODE §43-33-20 (1976); TEX. REV. CIV. STAT. ANN. art.4419e(3)(a) (1976); WASH. REV. CODE §49.60.030 (Supp. 1976); WYO. STAT. §42-1-126 (1977).

[89] *See, e.g.*, N.M. STAT. ANN. §12-26-3 (1976).

[90] *See* Marsh v. Edwards Theaters Circuit, Inc., 64 Cal. App. 3d 881, 134 Cal. Rptr. 844 (Ct. App. 1976).

[91] *See, e.g.*, UTAH CODE ANN. §§26-27-1 *et. seq.* (1976 & Supp. 1977).

[92] *See, e.g.*, LA. REV. STAT. ANN. §§40:1731 *et seq.* (West 1977); §§49:148 *et seq.* (West Supp. 1978).

[93] See "State Statute Charts," Chart I, column 1 *infra*.

[94] See Chart I, column 2 *infra*.

[95] See Chart I, column 3 *infra*.

[96] *See* Bibb v. Navajo Freight Lines, Inc., 359 U.S. 520, 524 (1959).

[97] See Chart II, column 1 *infra*.

[98] See Chart II, column 2 *infra*.

[99] *See* PA. STAT. ANN. tit. 71, §1455 (Purdon Supp. 1978-1979).

[100] *See, e.g.*, R.I. GEN. LAWS §37-17-1 (1977).

ited to publicly funded facilities and those which include privately funded facilities used by the public sometimes list exceptions to their coverage.[101] Most frequent exclusions include: private one and two family residences, buildings designated as hazardous occupancies, warehouses and storage areas, and agricultural facilities.

While many states use the terms "buildings and facilities" to define the general scope of their provisions, the breadth of this scope is dependent upon how these terms are actually defined in the statute. Frequently the definitions, when they appear, are ambiguous at best. It is not always clear whether publicly funded buildings also mean privately owned buildings leased with public funds. Nor is it always clear whether privately owned facilities include places of employment, housing, business or accommodation available to the public, nor to what extent ancillary facilities necessary for their effective use are covered.

Scope — Specific Types of Facilities Covered (Chart III)

Some statutes include specific types of facilities in their coverage, many of which are necessary to the effective use of general facilities. For example, streets, sidewalks, terminals, and transportation facilities may be included. Accessible buildings are of little benefit to the handicapped individual unable to negotiate the streets and sidewalks adjacent to them. At present 30 states [102] have enacted provisions that require that either curb cuts or ramps be installed at intersections to allow access by those in wheelchairs.[103] Usually these alterations need only be made, however, when some other structural work necessitates repairs to the particular curb; thus, a substantial period of time may pass before localities covered by such a provision will make these changes. However, this specific inclusion can avoid subsequent problems in interpretation of coverage.[104] Transportation must also be accessible and usable if the environment in which disabled persons live and work is to be barrier free. Some states have partially addressed this problem by including transportation terminals and stations in their specific provisions.[105] These terms do not, as a general matter, mandate accessible vehicles and equipment; thus, they provide only limited immediate assistance. However, a federal statute[106] now requires that "special efforts" be made to ensure that any new equipment or facilities purchased or constructed with federal assistance may be "effectively utilized" by handicapped persons. (See the Comment to section 13 of the model act *infra.*)

Other provisions in some state statutes specifically cover temporary and emergency buildings.[107] For example, consider the case where services and employment are offered in temporary structures prior to the opening of permanent facilities. These provisions help to insure that disabled persons can receive benefits on an equal basis with other persons during the interim period. In addition, it is uncertain whether general provisions cover outdoor facilities such as parks and playing fields. Thus, some states have specific provisions which include such recreational facilities[108] (*e.g.,* outdoor athletic facilities and parks). Finally, many states have also specifically included educational facilities,[109] presumably because of the importance of accessibility for attaining an education. However, even where such facilities are not specifically included within a statute, they will generally be included within the general provisions.[110]

Phases of Construction Covered (Chart IV)

While the scope provisions of the statutes determine the extent to which particular facilities within the total environment are made accessible, the provisions discussed in this subsection apply to different stages of construction or to the modification of facilities. Nearly every jurisdiction requires accessibility to some degree in newly constructed facilities,[111] while fourteen states have enacted provisions for the inclusion of accessibility features in buildings already under construction on the legislation's effective date.[112] The majority of statutes call for full compliance with barrier-free design standards unless construction has reached a stage where compliance would result in a substantial increase in cost or delay in

---

[101] See Chart II, column 3 *infra.*

[102] See Chart III, column 2 *infra.*

[103] The critical importance of this provision to the safety of disabled persons was dramatically emphasized recently by the tragic death of Jack Achtenberg, a leading commentator on the legal problems of handicapped persons. Professor Achtenberg, a quadriplegic, was killed by a motor vehicle while attempting to negotiate the streets of Los Angeles where curb-ramping has, as yet, not been provided. See, *Remarks of Roberta Achtenberg,* 50 TEMP. L.Q. 941, 943 (1977).

[104] An illustrative case occurred recently in California, where the legislature was forced to amend that state's act because a Los Angeles city official, charged with enforcement responsibility for his municipality, refused to interpret the phrase as including sidewalks adjacent to state buildings. *See* Achtenberg, *supra* note 4, at 855.

[105] See Chart III, column 1 *infra.*

[106] Urban Mass Transportation Act of 1964, §16, 49 U.S.C. §1612 (1976).

[107] See Chart III, column 4 *infra.*

[108] See Chart III, column 3 *infra.*

[109] See Chart III, column 5 *infra.*

[110] *See* Op. Ill. Att'y Gen. No. S-606 (1973).

[111] See Chart IV, column 3 *infra.*

[112] See Chart IV, column 2 *infra.*

construction.[113] Of these, Mississippi requires only entrance ramps[114] and Nevada only accessible toilet facilities.[115]

A good many states also cover, at least in part, existing buildings or facilities.[116] A substantial number (29) require that renovated areas in existing buildings conform,[117] with nearly all of these states specifically mentioning additions.[118] At present only seven states require alterations to existing buildings solely to improve accessibility when no other construction or remodeling is planned. Of these, one requires only entrance ramps [119] and two others require alterations only when a change from private to public use is made.[120]

Design and Construction Standards (Chart V)

The degree of accessibility and usability of a particular building covered by a statute will be determined largely by the design and construction standards adopted (and the waivers permitted, see *infra*). Thus, they are extremely important. New architectural design concepts are at the heart of any attempt to improve the physical environment. Since their promulgation in 1961, the ANSI Standards,[121] have come closest to adoption as a national standard for barrier-free design. At present they have been incorporated in whole or in part into the legislation of 27 states while an additional ten statutes require that they be "consulted" or "fully considered" by the agency responsible for the production of state standards.[122] Most states that have adopted the ANSI Standards include provisions automatically adopting any revisions or updates.

Approximately half of the state legislatures, in an attempt to provide greater adaptability to changing circumstances and new technologies, have placed the responsibility for standard development in the hands of an administrative agency.[123] Of these, ten have done so without providing any significant legislative guidance while the remainder make reference to ANSI Standards or to another state's published code.[124]

Because the vast majority of states do not require that existing buildings conform to accessibility standards, it is essential that those facilities which are accessible be easily identified. A graphic representation of an individual confined to a wheelchair has been adopted as the international symbol to be used for this purpose. This symbol is employed to denote usable entrances, toilet facilities, ramps and elevators, as well as accessible buildings themselves. Since the ANSI Standards do not include provisions requiring that these facilities be marked, a separate statutory section is required to enact this feature. Fourteen states presently have such a provision.[125]

Waiver Provisions (Chart VI)

The effectiveness of legislative efforts may well be determined by the availability of waivers or exceptions to the requirements of an environmental barriers act. Even in states with detailed and exhaustive statutory sections, easily obtainable waivers may allow a substantial number of facilities to remain unaffected. Given the demonstrable inclination of some administrative officials to favor lenient enforcement and narrow construction, it is likely that waiver provisions will be utilized in many states to the fullest extent possible.[126]

Seventeen statutes now include provisions which allow an administrative agency to exempt buildings and facilities from the states' barrier-free design provisions as a whole. Generally a waiver is allowed under these statutes when the authority finds that compliance would be "impractical." [127] Another seventeen states have provided that only individual standards may be waived.[128] Of these, ten require either "substantial compliance" or proof that equivalent facilities and/or safety features have been otherwise provided. New Hampshire allows no waivers when state funds are involved in the construction project[129] and New York allows an exemption only if compliance will impair the structural stability of the building or facility in question.[130] Finally, fifteen states provide no statutory authorization for waivers, although a number of these may allow the authority responsible for enforcement to exercise discretion in requiring full compliance with applicable standards.

Administrative Authority (Chart VII)

In order to fully implement the policies and procedures embodied in environmental barriers legislation, an efficient system must be developed to

[113] *See, e.g.*, IOWA CODE §104A.5 (Supp. 1978-1979).

[114] MISS. CODE ANN. §43-6-101 (Supp. 1977).

[115] NEV. REV. STAT. §338.100 (1977).

[116] See Chart IV, column 1 *infra*.

[117] See Chart IV, column 4 *infra*.

[118] See Chart IV, column 5 *infra*.

[119] MISS. CODE ANN. §43-6-101 (Supp. 1977).

[120] MASS. GEN. LAWS ANN. ch. 22, §13A (West Supp. 1978); MICH. COMP. LAWS ANN. §125.1352 (1976).

[121] *See* ANSI, *supra* note 50.

[122] See Chart V, columns 1 and 3 *infra*.

[123] See Chart V, column 2 *infra*.

[124] See Chart V, columns 1 and 2 *infra*.

[125] See Chart V, column 4 *infra*.

[126] *See* Achtenberg, *supra* note 4, at 885-63, and Farber, *supra* note 1, at 105-08.

[127] See Chart VI, column 2 *infra*.

[128] *Id.*

[129] N.H. REV. STAT. ANN. §275-C:15 (Supp. 1977).

[130] N.Y. PUB. BLDGS. LAW §51 (McKinney Supp. 1977).

administer the act. Effective administrative procedures should ensure compliance at an early stage, thereby substantially reducing the need for utilization of more drastic enforcement mechanisms.

All but five states now supply statutory measures which describe both the administrative duties considered necessary to carry out the intent of the legislators and the particular agency or individual responsible for the performance of each.[131] For the purposes of this survey, administrative responsibilities are defined as including development of design standards, promulgation of administrative rules and regulations, approval of plans and specifications, issuance of construction and use permits and granting of exceptions and waivers. Enforcement authority is considered separately since enforcement is viewed as a vehicle of last resort and not as a normal administrative procedure. Provisions pertaining to it are shown in Charts VIII and IX.

The states in enacting administrative guidelines have taken two basic approaches. Eighteen states now name an existing single authority to perform all enumerated duties while 28 others delegate specific duties piecemeal to a number of different agencies and/or individuals.[132] The majority of states taking the latter approach delegate authority alternatively to state or local officials depending upon which governmental unit has primary design and construction responsibility for the individual project. In addition most provide that where school funds are utilized the State Board of Education will have primary administrative responsibility.[133]

While the majority of states now delegate primary administrative responsibility to existing departments or officials, the acts of five states [134] create a special board to carry out this function. The legislators in these states see an advantage in a centralized authority capable of developing singular expertise in this conceptually and procedurally difficult field. In addition, such an approach allows the legislature to insure that handicapped individuals may participate directly in the decision-making process, since the board membership provisions in all five states mandate that at least one physically disabled individual be present as a voting member.[135] New Hampshire's statute goes farthest in this respect by providing that a majority of the state's 11 member board must be physically limited.[136] An additional two states have created boards with similar membership characteristics to serve in an advisory capacity under the administrative authority[137] and Delaware has created a special commission to advise on waiver requests only. [138] Finally, other states have given preexisting state building code commissions or other boards administrative responsibility. However, these preexisting commissions do not share the advantages of the specialized boards, since they do not have special expertise and do not provide direct participation by disabled persons in the decision-making process.

Enforcement Provisions (Charts VIII and IX)

Adequate enforcement has been a recurrent problem with environmental barriers legislation.[139] Both public contracting agencies and private individuals undertaking construction projects have shown a marked resistance to voluntary compliance in some areas.[140] Therefore, effective enforcement provisions are also a vital part of a good statute.

Forty-five states do have provisions which delegate enforcement responsibility to specific agencies or individuals.[141] Typical statutes place power alternatively in the hands of a state or municipal authority depending upon which political division is primarily responsible for the project in question. Common recipients at the state level are: the

---

[131] See Chart VII, columns 1 and 2 *infra*. Columns 1 and 2 of Chart VII provide a summary of the present administrative provisions. The names of agencies or individuals responsible for some aspect of administration are given in column 1 and directly adjacent to them in column 2 are a list of the particular duties delegated to each. Since enforcement provisions are generally treated separately, they are only included in Chart VII when the statutory provision makes no distinction and names a single authority responsible to "administer and enforce" the act. *See, e.g.*, IDAHO CODE §39-3204 (1977).

Because the states employ a confusing array of terms, specific phrases are used in Chart VII and repeatedly throughout the charts to denote certain general types of provisions. The term "standards" refers to the actual design and construction criteria utilized by the state to promote accessibility and to which facilities covered by the act must conform. "Rules and regulations" are procedures developed by the administrative authority to carry out its responsibilities under the act. The term "complaints" refers to both official grievances and those of private citizens. "Waivers" are administrative judgments exempting facilities from some or all of the requirements of the act and are also referred to as exceptions, variances, exemptions and releases within the various state statutes.

[132] See Chart VII, column 1 *infra*.

[133] *Id*.

[134] Massachusetts, Michigan, New Hampshire, South Carolina and Vermont.

[135] See listing of these states in Chart VII, column 3 *infra*.

[136] N.H. REV. STAT. ANN. §275-C:12 (Supp. 1977).

[137] NEB. REV. STAT. §72-1121 (Supp. 1976); OR. REV. STAT. §447.230 (1977).

[138] DEL. CODE tit. 29, §6917 (Supp. 1977).

[139] See "Adequacy of Existing Statutory Law" *infra*, at notes 20-28.

[140] The District of Columbia's new transit system was finally made completely accessible only after a federal judge enjoined use of certain stations until relevant criteria were met. *See* Washington Urban League, Inc. v. Washington Metropolitan Transit Authority, Inc., Civil No. 776-72 (D.D.C. 1973).

[141] See Chart VIII *infra*.

agency contracting for construction of the project, the state building code council, the state department of commerce or its equivalent and, when school funds are utilized, the state board of education. Six states assign responsibility to the state fire marshall, presumably because inspection and enforcement procedures are already operating under his or her authority.[142] Recipients at the local level include the governing body of a local unit of government or its building official or department.[143]

In addition to delegations of authority, twenty-eight states have included specific enforcement strategies within their legislation.[144] All but one of these allow the responsible authority to issue orders and enter court to "compel compliance" with applicable standards. Only eight of these states, however, include provisions allowing the authority to enter buildings to investigate reports of noncompliance and so the effectiveness of a number of these statutes may be open to question.[145] Eleven states also allow construction to be enjoined if architectural design standards are not being met.[146] Wisconsin's act is unique in this regard, in that it provides any private individual a statutory right to enter court seeking such an injunction and provides for reimbursement of all costs should they be successful.[147] A number of states provide criminal sanctions for violating their laws.[148]

Tax Incentives (Chart X)

As noted previously, the economic burdens placed on private individuals is often a principal justification given for limiting the applicability of barrier-free design regulations and for the enactment of expansive waiver provisions. Though authorities in the field have found the additional expense incurred by including accessibility features in new buildings to be as low as $^1/_{10}$ of 1 percent of the total cost of construction,[149] there remains a general reluctance to place even this encumbrance on private parties.[150] Thus, nine states have attempted to alleviate the burden in some cases through incentives included within the state's taxing mechanisms. Since each of these provisions differs substantially, they have been included nearly verbatim in Chart X.

142 See Chart VIII, column 1 *infra*.

143 See Chart VIII, column 2 *infra*.

144 See Chart IX *infra*. Citations to all code sections which provide for these various enforcement mechanisms are given in Chart IX. Where an enforcement authority is listed in Chart VIII but no citations appear in Chart IX, the state in question has delegated authority but provided only a general mandate to "enforce." There are presently fifteen state acts of this sort.

145 See Chart IX, column 1 *infra*.

146 See Chart IX, column 3 *infra*.

147 *See* WIS. STAT. ANN. §101.13(6)(g) (West Supp. 1978-1979).

148 See Chart IX, column 4 *infra*.

149 See Comment to subsection 8(2) of the model act *infra*.

150 Case law suggests that imposing the burden is firmly within the state's authority under its police power. For a decision holding that the state of California could constitutionally mandate an *existing* building's compliance with newly enacted building code provisions at a cost of $55,000 dollars against a total assessed value for the structure of $250,000, *see*, City of Bakersfield v. Miller, 64 Cal. 2d 120, 48 Cal. Rptr. 869, 410 P.2d 393 (1966), *cert. denied*, 384 U.S. 988 (1966).

CHARTS OF STATE ENVIRONMENTAL BARRIERS STATUTES

I. STATEMENT OF POLICY AND PURPOSE

State	Make Facilities Accessible to and Functional For the Physically Handicapped	Maximize Integration into Society	Promote Safety and/or Independence
ALABAMA Code (1977)	tit.21, §4-1		
ALASKA Stat. (1976)	§35.10.015		
ARIZONA Rev. Stat. (1974)	§34-401		
ARKANSAS			
CALIFORNIA Gov't. Code (Deering Supp. 1977)	§4450		
COLORADO Rev. Stat. (1973)	§9-5-103		
CONNECTICUT			
DELAWARE			
DISTRICT OF COLUMBIA			
FLORIDA			
GEORGIA Code Ann. (Supp. 1977)	§91-1104	§91-1104	
HAWAII Rev. Stat. (1977)	§103-50		
IDAHO Code (1976)	§39-3201		
ILLINOIS Ann. Stat. (Smith-Hurd 1977)			ch.111, §13
INDIANA			
IOWA Code (Supp. 1978)	§104A.1		
KANSAS Stat. (1973)	§58-1303		
KENTUCKY			

State	Make Facilities Accessible to and Functional For the Physically Handicapped	Maximize Integration Into Society	Promote Safety and/or Independence
LOUISIANA Rev. Stat. Ann. (West 1977 & Supp. 1978)	§49:148 §40:1731	§40:1731	§40:1731
MAINE			
MARYLAND Ann. Code (1975)			art. 78A,§51
MASSACHUSETTS Gen. Laws Ann. (West Supp. 1978)			ch. 22, §13A
MICHIGAN Comp. Laws Ann. (1976)	§125.1351		
MINNESOTA			
MISSISSIPPI			
MISSOURI			
MONTANA Rev. Codes Ann. (Supp. 1977)	§69.2110		
NEBRASKA Rev. Stat. (1976)	§72-1102		
NEVADA Rev. Stat. (1977)	§338.180		
NEW HAMPSHIRE Rev. Stat. Ann. (Supp. 1977)	§275-C:14		
NEW JERSEY			
NEW MEXICO Stat. Ann. (Supp. 1975)	§67-35-52		
NEW YORK			
NORTH CAROLINA			
NORTH DAKOTA			
OHIO Rev. Code Ann. (Page Supp. 1977)	§3781.111		
OKLAHOMA			

State	Make Facilities Accessible to and Functional For the Physically Handicapped	Maximize Integration Into Society	Promote Safety and/or Independence
OREGON Rev. Stat. (1977)	§447.220		
PENNSYLVANIA			
RHODE ISLAND Gen. Laws (1977)	§37-17-2		
SOUTH CAROLINA Code (1976)		§10-5-210	§10-5-210
SOUTH DAKOTA			
TENNESSEE Code Ann. (1977)	§53-2545		
TEXAS Rev. Civ. Stat. Ann. (Vernon Supp. 1978)	art. 678g,§3(b)		art. 678g,§1
UTAH Code Ann. (1976)	§26-27-2		
VERMONT			
VIRGINIA			
WASHINGTON			
WEST VIRGINIA Code (1977)	§18-10F-1		
WISCONSIN			
WYOMING			
AMERICAN SAMOA			
GUAM			
PUERTO RICO			
VIRGIN ISLANDS			

II. SCOPE — COVERAGE OF THE TOTAL ENVIRONMENT

State	Public Buildings	Privately Owned Buildings	Facilities Specifically Excluded
ALABAMA Code (1977)	Used by the public and built in whole or in part with the use of state, county or municipal funds, or the funds of any political subdivision of the state. tit. 21, §4-4		
ALASKA Stat. (1976)	Built by or for the state and its political subdivisions. §35.10.015		
ARIZONA Rev. Stat. (1974 & Supp. 1977)	Used by the public and built in whole or in part with the use of state funds or the funds of any political subdivision. §34-403	All establishments which cater or offer their services to or solicit patronage from the general public. §34-402,§34-403	Places of public accommodation which have less than three floor levels including basements. §34-403
ARKANSAS Stat. Ann. (1968)	All buildings of assembly, educational institutions and office buildings which are constructed in whole or in part with the use of state funds or the funds of any instrumentality of the state. §14-627		
CALIFORNIA Gov't. Code (Deering Supp. 1977)	Constructed with the use of state, county or municipal funds or the funds of any political subdivision. §4450 Leased, contracted, sublet or hired for periods in excess of 2 years by any of the above. §4451		Portable buildings leased or owned by a school district which are not used by the physically handicapped. §4457
COLORADO Rev. Stat. (Supp. 1976)	Used by the public and built in whole or in part with the use of state, county or municipal funds, or the funds of any political subdivision of the state. §9-5-102	Same as "Public" but constructed with wholly private Funds. §9-5-102	Privately funded projects for the construction of separate houses designed as single family residences or other types of residential property containing less than seven residential units. §9-5-111
CONNECTICUT Gen. Stat. Ann. (West 1977 & Supp. 1978)	All buildings and building elements constructed under permits issued after effective date. §19-395a	See Under "Public"	Buildings which have been approved by the Department of Housing & Urban Development as being in conformance with Federal standards for housing for the elderly and physically handicapped and for which a permit was issued prior to June 9, 1976. §19-395a. The following use groups in the state building code: A, high hazard; B-1, storage, moderate; B-2, storage, low; L-3 residential, one & two family. §19-395L

State	Public Buildings	Privately Owned Buildings	Facilities Specifically Excluded
DELAWARE Code (Supp. 1977)	Every public works contract awarded by the state or any political subdivision thereof . . . for any public building in which public funds are involved. Also state leased. tit.29, §6917		
DISTRICT OF COLUMBIA			
FLORIDA Stat. Ann. (West 1975 & Supp. 1978)	Buildings or facilities intended for use by the general public built or altered or operated as a lessee by or on behalf of the state or any subdivision, municipality, or special district thereof. §255.21	All new buildings except single family dwellings and duplexes, which the general public may frequent, live in, or work at. §553.48	Single-Family dwellings §533-48(f). Building maintenance & storage areas where only employees have occasion to enter and within which the work cannot reasonably be performed by the handicapped unless such areas provide the only path between areas normally used by the handicapped. §553-48(a)
GEORGIA Code Ann. (Supp. 1977)	All government buildings, public buildings and facilities constructed or substantially renovated. §91-1105 Only buildings open to the public. §91-1125	All buildings, structures, streets, sidewalks, walkways and access thereto used by the public or in which handicapped or elderly persons may be employed. §91-1107	Rental apartment complexes and temporary lodging facilities need only provide entrances & doors wide enough to accommodate wheelchairs. §91-1105 Private single family residences, duplexes, triplexes & condominiums. §91-1107 Public schools exempted from some standards. See §91-1124
HAWAII Rev. Stat. (1977)	Construction of public buildings and facilities by the state or any political subdivision thereof. §103-50		
IDAHO Code (1976)	Constructed by the state, any county, city, district, authority, board, or any public corporation or entity which has any reasonable availability to, or usage by, physically handicapped persons. §39-3202, §39-3203		
ILLINOIS Ann. Stat. (Smith-Hurd Supp. 1978)	A building, structure or improved area owned or leased by this state or its political subdivisions. ch.111, §12	A building, structure or improved area used primarily by the public as a place of gathering or amusement. Office buildings & factories. ch.111, §12	The second floor of any two story office building or factory having less than 15,000 square feet at the second floor level or any occupied levels below grade having less than 15,000 square feet unless the building is publicly owned. ch.111, §12

State	Public Buildings	Privately Owned Buildings	Facilities Specifically Excluded
INDIANA			
IOWA Code (1975)	All public and private buildings and facilities, temporary and permanent, used by the general public. §104A.2	See under "Public"	
KANSAS Stat. (1976)	All buildings and facilities used by or open to the public which are constructed in whole or in part with moneys appropriated by the state or any political or taxing subdivision thereof. §58-1301		See §58-1308
KENTUCKY Rev. Stat. (1976 & Supp. 1978)	All public buildings and public accommodations.[1] See def. §227.305	See under "Public"	
LOUISIANA Rev. Stat. Ann. (West 1977)	All state owned buildings, educational institutions, and office buildings which are constructed, renovated or remodeled in whole or in part with the use of state funds or the funds of any board, commission, agency or department of the state (except school boards). §49:148 (Supp. 1978) See also §40:1732	Building, structure or improved area to which the general public customarily has access or utilizes for a number of specified purposes. See §40:1732	Privately owned one and two family dwellings. §40:1732 Building maintenance, freight loading & storage areas where only employees have occasion to enter and within which the work cannot reasonably be performed by any handicapped person; public facilities which are utilized for the purchase, rental, or acquisition of goods & services with five thousand square feet or less per floor, which are accessible at habitable grade levels; public facilities for accomodation which are two & three story buildings with less than 49 units which are accessible at habitable grade levels. §40:1734 Unleased portions of buildings partially leased by the State or its political subdivisions. §40:1739
MAINE Rev. Stat. (1977)	Structure to which the public customarily has access and utilizes and which is constructed in whole or in part with funds of the state or its political subdivisions. tit. 25, §2701	A place where 5 persons or more will be employed. tit.25, §2701	
MARYLAND Ann. Code (1975)	All public buildings erected by the state or any political subdivision thereof, except penal institutions. art. 78A, §51		

State	Public Buildings	Privately Owned Buildings	Facilities Specifically Excluded
MASSACHUSETTS Gen. Laws Ann. (West Supp. 1978)	Buildings constructed by the commonwealth or any political subdivision thereof with public funds and open to public use. ch.22, §13A		
MICHIGAN Comp. Laws Ann. (1976)	A building, structure or improved area owned, leased, financed, rented by or on behalf of the state or its political subdivisions or with federal funds. See def. §125.1351	A facility used by the public. See §125.1351	
MINNESOTA Stat. Ann. (West Supp. 1977)	Any building and grounds appurtenant thereto, the cost of which is paid for by the state of Minnesota or any governmental subdivision or school district. §299F.41		
MISSISSIPPI Code Ann. (Supp. 1977)	All buildings of assembly, educational institutions, office buildings and other public buildings which are constructed in whole or in part with the use of state, county, or municipal funds, or the funds of any instrumentality of the state. §43-6-101		
MISSOURI Ann. Stat. (Vernon Supp. 1978)	All buildings and facilities for public use and assembly which are constructed in whole or in part with the use of state funds or the funds of any political subdivision of the state. §8.610		
MONTANA Rev. Codes Ann. (Supp. 1978)	Any place which a municipality or state maintains for the use of the public, or a place where the public has a right to be. (May be further limited by an administrative agency.) §69-2105		
NEBRASKA Rev. Stat. (1976)	Buildings and facilities used by the public, constructed or remodeled in whole or in part with the use of state, county or municipal funds, or the funds of any political subdivision of the state. §72-1101	All buildings where the public is invited to enter or remain upon the premises as business invitees. §72-1101	Buildings structures or installations, or portions thereof, used for agricultural purposes or heavy industry and any other business or industrial building where the presence of handicapped persons could or would tend to place them in a position of potential hazard to their person. §72-1123

State	Public Buildings	Privately Owned Buildings	Facilities Specifically Excluded
NORTH DAKOTA Cent..Code	All public buildings and facilities constructed, in whole or in part, from funds of the state or of its political subdivisions. §48-02-19 (Supp. 1977)	All buildings and facilities held open to the public and used for public assemblages of any character must have accessible toilet facilities only. §23-13-13 (1978)	
OHIO Rev. Code Ann. (Page Supp. 1977)	Any building owned by the state or any political subdivision thereof which is open to and used by the public. §3781.111		
OKLAHOMA Stat. Ann. (West Supp. 1977)	Any public buildings erected by the state or any agency or political subdivision thereof or erected with the use of public funds. tit.61, §11		
OREGON Rev. Stat. (1977)	All buildings and structures used by the public that are constructed, purchased, leased or rented in whole or in part with the use of state, county or municipal funds or the funds of any political subdivision of the state, and, to the extent lawful, federal funds. §447.210	All buildings and structures used by the public having a ground area of more than 4,000 feet and more than 20 feet high. §447.210	
PENNSYLVANIA Stat. Ann. (Purdon Supp. 1978)	All buildings of assembly, educational institutions and office buildings constructed in whole or in part with the use of commonwealth funds or the funds of any instrumentality of the commonwealth or leased by the commonwealth or any instrumentality thereof. tit.71, §1455.1	Department stores, theaters, retail stores, sport arenas, and restaurants with sitdown dining with 2,800 feet or more of usable floor space. tit.71, §1455.1	
RHODE ISLAND Gen. Laws (1977)	Public buildings constructed by the state or any municipality of the state. §37-8-15 Buildings and facilities used by the public which are constructed in whole or in part with the use of state or municipal funds or the funds of any political subdivision of the state. §37-17-1	Any privately financed buildings that are open to and used by the public. §37-17-1	Apartment houses. §37-17-1

State	Public Buildings	Privately Owned Buildings	Facilities Specifically Excluded
NEVADA Rev. Stat. (1977)	Public buildings and facilities constructed by the state or by a political subdivision, district, authority, board, public corporation or entity of the state. §338.180	Public buildings and facilities constructed by a public corporation. §338.180	
NEW HAMPSHIRE Rev. Stat. Ann. (1977)	Buildings, facilities, appurtenant grounds and curbs which are used or to be used by the public and the cost of the construction, rehabilitation, or substantial remodeling of which is to be paid for, in whole or in part with federal, state, county or municipal funds. §275-C:10		See §275-C:15
NEW JERSEY Stat. Ann. (West Supp. 1977)	Any building, structure, facility or complex used by the general public and constructed by any state, county or municipal government agency or instrumentality. §52:32-6	Any building, structure, facility or complex used by the general public and constructed by any private individual, partnership, association or corporation. §52:32-6	One-to-four-family private residences; warehouse storage areas; and all buildings classified as hazardous occupancies. §52:32-6 (A Hazardous occupancy is a building or structure which has inherent characteristics that constitute a special fire hazard.)
NEW MEXICO Stat. Ann. (Supp. 1975)	Public buildings constructed in the state through expenditure of state, county or municipal funds. §67-35-52 Leased or rented facilities. §6-2-26.1		
NEW YORK Pub. Bldgs. Law (McKinney Supp. 1977)	Any building or portion thereof constructed wholly or partially with state or municipal funds which is likely to be used by physically handicapped persons. §50		
NORTH CAROLINA Gen. Stat. (1978)	[3]		

State	Public Buildings	Privately Owned Buildings	Facilities Specifically Excluded
SOUTH CAROLINA Code (1976)	Buildings, structures, streets and sidewalks and access thereto used by the public or in which physically handicapped may be employed that are constructed, purchased, leased or rented in whole or in part with the use of state, county or municipal funds or the funds of any political subdivision of the state. §10-5-220	Same as public but constructed, purchased, rented or leased with the use of private funds. §10-5-220	See §10-5-220
SOUTH DAKOTA Compiled Laws Ann. (Supp. 1977)	Buildings and facilities used by the public which are constructed in whole or in part with the use of state, county, or municipal funds, or the funds of any political subdivision of the state. §5-14-12		
TENNESSEE Code Ann. (1977)	Any building, structure, or improved area owned or leased by the state or its political subdivisions. §53-2546	Any building, structure or improved area used primarily by the general public as a place of gathering or amusement. §53-2546	
TEXAS Rev. Civ. Stat. Ann. (Vernon Supp. 1978)	All buildings and facilities used by the public which are constructed in whole or in part with the use of state, county or municipal funds, or the funds of any political subdivision of the state. Also leased or rented by same. art. 678g, §2	Privately financed buildings, building elements, and improved areas which are open to public use for education, employment, transportation, or acquisition of goods and services in counties with a population of 50,000 or more. art. 678g, §2(d)	
UTAH Code Ann. (1976 & Supp. 1977)	All buildings and facilities contructed or remodeled in whole or in part with the use of state, county or municipal funds or the funds of any political subdivision of the state. §26-27-1	Private individuals encouraged to apply the standards. §26-27-1	Caretaker dwellings, service buildings & heating plants as well as other buildings not intended for public use. §26-27-1
VERMONT Stat. Ann. (Supp. 1977)	Buildings to be used by the public for which the federal, state, county or municipal government pays any part of the construction cost. tit.18, §1322		

State	Public Buildings	Privately Owned Buildings	Facilities Specifically Excluded
VIRGINIA Code (Supp. 1978)	Building or facility used by the public which is constructed in whole or in part or altered with the use of state, county or municipal funds or the funds of any political subdivision of the state. §2.1-514		
WASHINGTON Rev. Code (Supp. 1976)	Buildings, structures or portions thereof of used primarily for Group A through Group H occupancies as defined in state building code. §70.92.110	See under "Public"	Any building or structure used solely for dwelling purposes and which contains not more than two dwelling units; apartment buildings with 10 or fewer units, any building not used primarily for Group A through Group H occupancies as set forth in the Washington state building code. §70.92.110
WEST VIRGINIA Code (1977)	Building or facility the public has a general right of access to and the ways of travel to and from the same. §18-10F-1	See, under "Public"	Apartment houses with less than twenty units, row houses or rooming houses; convents or monasteries; jails or other places of detention; garages, hangars or boat houses; hazardous occupancies, warehouses; building specifically built for field service purposes; residence halls at colleges or universities which have at least two others accessible for each sex. §18-10F-1
WISCONSIN Stat. Ann. (West Supp. 1978)	Any place of employment or public building. §101.13(2) Defined as any structure, including exterior parts of such buildings, used in whole or in part as a place of resort, assemblage, lodging, trade, traffic, occupancy or use by the public or 3 or more tenants. §101.01	See under "Public"	Same as West Virginia except 3 residence halls for each sex at colleges & universities must be accessible. §101.13(2)

State	Public Buildings	Privately Owned Buildings	Facilities Specifically Excluded
WYOMING Stat. (1977)	Buildings for general public use built by the state or any governmental subdivision or any school district or other public administrative body within the state. §35-13-101		

[1] Public building defined as "any building constructed, owned or leased by the Commonwealth or any political subdivision." Public accommodations defined as those covered by Title II of the Federal Civil Rights Act of 1964. §277.220.

[2] Policy statement included urging that private enterprise comply voluntarily with accessibility standards. §49:148.3.

[3] To be determined by building code council. §143-138.

III. SCOPE — SPECIFIC TYPES OF FACILITIES COVERED

State	Transportation Terminals and Stations	Streets and Sidewalks (Ramps)	Recreational Facilities	Temporary and Emergency Buildings	Education Facilities
ALABAMA					
ALASKA Stat. (1976)	§35.10.015[1]				§35.10.051
ARIZONA Rev. Stat.		§9-499.02 (1977)	§34-402 (1974)	§34-403 (Supp. 1977)	§34-410 (1974)
ARKANSAS					
CALIFORNIA Gov't. Code (Deering Supp. 1977)		§4450		§4451	§4451
COLORADO					

State	Transportation Terminals and Stations	Streets and Sidewalks (Ramps)	Recreational Facilities	Temporary and Emergency Buildings	Education Facilities
CONNECTICUT Gen Stat. Ann.	§13b-11a (West Supp. 1978)[2] Pub. Acts. 77-207[3]	§7-118a (West 1977)			
DELAWARE					
DISTRICT OF COLUMBIA Code (1973)	§1-442[4]				
FLORIDA Stat. Ann. (West Supp. 1978)		§335.075 (1975)	§553.47		§553.48
GEORGIA Code Ann. (Supp. 1977)		§91-1107 §69-1609 (1976)		§91-1105	
HAWAII					
IDAHO Code (1976)		§39-3203,§40-527			
ILLINOIS Ann. Stat. (Smith-Hurd Supp. 1978)	ch. 127, §49.21[5] ch. 127, §49.19(3)	ch. 24, §11-80-11			
INDIANA Code Ann. (Burns Supp. 1977)		§8-11-11.1-2			
IOWA Code (1975)	§601J.1 et seq.	§601D.9		§104A.2	
KANSAS Stat.		§12-697 et seq. (1975)			§31-150 (1973)
KENTUCKY Rev. Stat. (Supp. 1978)		§66-660			

State	Transportation Terminals and Stations	Streets and Sidewalks (Ramps)	Recreational Facilities	Temporary and Emergency Buildings	Education Facilities
LOUISIANA Rev. Stat. Ann. (West Supp. 1978)	§40:1732	§40:1732	§40:1732	§40:1732	§40:1732
MAINE Rev. Stat. (Supp. 1975)			tit. 23, §106		
MARYLAND Ann. Code (1975 & Supp. 1977)	art. 78A, §51	art. 78A, §51	art. 78A, §51		art. 78A, §51
MASSACHUSETTS Gen. Laws Ann. (West Supp. 1978)	ch. 22, §13A	ch.22, §13A			ch. 22, §13A
MICHIGAN Comp. Laws Ann. (1976)	§125.1351(d)	§125.1351	§125.1351(d)		§125.1351(d)
MINNESOTA Stat. Ann. (West Supp. 1977)		§471.464			
MISSISSIPPI Code Ann. (Supp. 1977)		§21-37-6			§43-6-101
MISSOURI Ann. Stat. (Vernon Supp. 1978)			§8.621		
MONTANA					
NEBRASKA Rev. Stat. (1976)				§72-1101	
NEVADA					
NEW HAMPSHIRE					
NEW JERSEY Stat. Ann. (West Supp. 1977)	§52:32-6	§52:32-14, §52:32-15	§52:32-6		§52:32-6, §52:32-16 §18A:18A-17

State	Transportation Terminals and Stations	Streets and Sidewalks (Ramps)	Recreational Facilities	Temporary and Emergency Buildings	Education Facilities
NEW MEXICO Stat. Ann. (Supp. 1975)		§55-2-53			
NEW YORK (McKinney Supp. 1977)	Pub. Bldgs. §50	High. §330	Pub. Bldgs. §50 Envir. Conserv. §3-0301 note		Pub. Bldgs. §50
NORTH CAROLINA					
NORTH DAKOTA Cent. Code (Supp. 1977)		§48-02-19, §40-31-01.1			
OHIO Rev. Code Ann. (Page 1976)		§729.12			
OKLAHOMA					
OREGON Rev. Stat. (1977)	§267.240[6]	§447.310, §447.210		§447.260	
PENNSYLVANIA Stat. Ann. (Purdon Supp. 1978)		tit. 53, §1898, tit. 53, §1899			tit. 7.1, §1455.1
RHODE ISLAND Gen. Laws (1977)	§37-17-1		§32-1-16	§37-17-1	
SOUTH CAROLINA Code (1976)		§10-5-220		§13-11-60	
SOUTH DAKOTA Compiled Laws Ann. (Supp. 1977)		§9-46-1.1, §9-46-1.2		§5-14-12	
TENNESSEE Code Ann. (1977)				§53-2548	

State	Transportation Terminals and Stations	Streets and Sidewalks (Ramps)	Recreational Facilities	Temporary and Emergency Buildings	Education Facilities
TEXAS Rev. Civ. Stat. Ann. (Vernon Supp. 1978)	art. 678g, §2(d)			art. 678g, §2(b)	
UTAH Code Ann. (1976)					§26-27-1
VERMONT					
VIRGINIA Code (Supp. 1978)		§15.1-381		§2.1-521	§2.1-518
WASHINGTON Rev. Code (Supp. 1976)		§35.68.075, §35.68.076			
WEST VIRGINIA Code (1977)		§18-10F-1, §18-10F-2		§18-10F-2	§18-10F-1
WISCONSIN Stat. Ann. (West Supp. 1978)		§66.616	§30.90 (1973)		§101.13
WYOMING					

[1] Includes ferries.
[2] Commisslon established to develop guidelines for accessibility of transportation facilities.
[3] Enabling act to allow funding for non-profit corporations providing accessible transportation.
[4] Enabling act to receive funding to meet accessibility standards of 42 U.S.C. §4151.
[5] Provision allocating state funds to municipalities to upgrade transportation for the handicapped.
[6] Both terminals and vehicles must be accessible when such vehicles are commercially available.

IV. PHASES OF CONSTRUCTION COVERED

State	Existing Facilities	Facilities Under Construction	New Construction	Renovations and Remodeling	Additions	Work Requiring A Building Permit Generally
ALABAMA Code (1977)		tit. 21, §4-4	tit. 21, §4-4			
ALASKA Stat. (1976)	§35.10.015[1]		§35.10.105			
ARIZONA Rev. Stat. Ann. (Supp. 1977)			§34-403			
ARKANSAS Stat. Ann. (1968)			§14-627			
CALIFORNIA Gov't. Code (Deering Supp. 1977)			§4451			
COLORADO Rev. Stat. (Supp. 1976)			§9-5-102	§9-5-102	§9-5-102	
CONNECTICUT Gen. Stat. Ann. (West 1977 & Supp. 1978)			§19-395a	§19-395a	§19-395a	§19-395a
DELAWARE Code (Supp. 1977)			tit. 29, §6917	tit. 29, §6917	tit. 29, §6917	
DISTRICT OF COLUMBIA						
FLORIDA Stat. Ann. (West 1975)		§255.21	§255.21 §553.46 (Supp. 1978)	§255.21	§255.21	
GEORGIA Code Ann. (Supp. 1977)		§91-1105	§91-1105	§91-1105		

State	Existing Facilities	Facilities Under Construction	New Construction	Renovations and Remodeling	Additions	Work Requiring A Building Permit Generally
HAWAII Rev. Stat. (1977)			§103-50			
IDAHO Code (1976)			§39-3201			
ILLINOIS Ann. Stat. (Smith-Hurd 1977)			ch.111, §15	ch. 111, §15 ch. 111, §12 (Supp. 1978)	ch. 111, §15 ch. 111, §12 (Supp. 1978)	ch. 111, §16
INDIANA						
IOWA Code (1975)		§104A.5	§104A.2			
KANSAS Stat. (1976)			§58-1305	§58-1305	§58-1305	
KENTUCKY Rev. Stat. (1976 & Supp. 1978)			§227.305	§227.305	§227.305	§227.305
LOUISIANA Rev. Stat. Ann. (West Supp. 1978)	§40:1738, §40:1745		§49:148, §40:1732	§49:148, §40:1732	§40:1732	§40:1740
MAINE Rev. Stat. (1977)			tit. 25, §2703	tit. 25, §2703[2]	tit. 25, §2703	
MARYLAND Ann. Code (1975)			art. 78A, §51			
MASSACHUSETTS Gen. Laws Ann. (West Supp. 1978)	ch. 22, §13A[3]		ch. 22, §13A	ch. 22, §13A	ch. 22, §13A	ch. 22, §13A
MICHIGAN Comp. Laws Ann. (1976)	§125.1352(2)		§125.1352(1)	§125.1352(2)	§125.1352(2)	
MINNESOTA Stat. Ann. (West Supp. 1977)			§299F.44	§299F.44		

State	Existing Facilities	Facilities Under Construction	New Construction	Renovations and Remodeling	Additions	Work Requiring A Building Permit Generally
MISSISSIPPI Code Ann. (Supp. 1977)	§43-6-101[4]	§43-6-101	§43-6-101			
MISSOURI Ann. Stat. (Vernon Supp. 1978)			§8.610	§8.623[5]		
MONTANA Rev. Codes Ann. (Supp. 1978)			§69-2120			
NEBRASKA Rev. Stat. (1976)		§72-1101	§72-1101	§72-1101		
NEVADA Rev. Stat. (1977)		§338.100[6]	§338.180			
NEW HAMPSHIRE Rev. Stat. Ann. (1977)			§275-C:14	§275-C:14	§275-C:14	
NEW JERSEY Stat. Ann. (West Supp. 1977)			§52:32-6	§52:32-6	§52:32-6	
NEW MEXICO Stat. Ann. (Supp. 1975)			§67-35-52			
NEW YORK Pub. Bldgs. Law (McKinney Supp. 1977)			§51	§50, §51	§50, §51	
NORTH CAROLINA						
NORTH DAKOTA Cent. Code (1978)			§48-02-19 (Supp. 1977), §23-13-13	§23-13-13[7]		
OHIO Rev. Code Ann. (Page Supp. 1977)			§3781.111			
OKLAHOMA Stat. Ann. (West Supp. 1977)			tit. 61, §11	tit. 61, §11	tit. 61, §11	

State	Existing Facilities	Facilities Under Construction	New Construction	Renovations and Remodeling	Additions	Work Requiring A Building Permit Generally
OREGON Rev. Stat. (1977)	§447.235	§447.230	§447.230	§447.230	§447.230	
PENNSYLVANIA Stat. Ann. (Purdon Supp. 1978)	tit. 71, §1455.1	tit. 71, §1455.1	tit. 71, §1455.1			
RHODE ISLAND Gen. Laws (1977)		§37-17-1	§37-17-1			
SOUTH CAROLINA Code (1976)			§10-5-260	§10-5-260	§10-5-260	
SOUTH DAKOTA Compiled Laws Ann. (Supp. 1977)			§5-14-12	§5-14-12	§5-14-12	
TENNESSEE Code Ann. (1977)			§53-2548, §53-2546	§53-2548, §53-2546	§53-2548 §53-2546	
TEXAS Rev. Civ. Stat. Ann. (Vernon Supp. 1978)		art. 678g, §2(b)	art. 678g, §2(a)	art. 678g, §2(a)	art. 678g, §2(a)	
UTAH Code Ann. (1976 & Supp. 1977)		§26-27-1	§26-27-1	§26-27-1		
VERMONT Stat. Ann. (Supp. 1977)			tit. 18, §1322			
VIRGINIA Code (Supp. 1978)			§2.1-519	§2.1-519		
WASHINGTON Rev. Code (Supp. 1976)			§70.92.110	§70.92.130, §70.92.110	§70.92.130, §70.92.110	
WEST VIRGINIA Code (1977)		§18-10F-2	§18-10F-2			

State	Existing Facilities	Facilities Under Construction	New Construction	Renovations and Remodeling	Additions	Work Requiring A Building Permit Generally
WISCONSIN Stat. Ann. (West Supp. 1978)		§101.13(2),(5)	§101.13(2)	§101.13(6)	§101.13(6)	
WYOMING Stat. (1977)			§35-13-105		§35-13-101	

[1] Administrative agency recommends priorities for upgrading existing public buildings.
[2] Estimated total cost must exceed 100,000 dollars.
[3] Standards applicable if use of an existing facility changes from private to public.
[4] Any existing building or one under construction must provide entrance ramps only.
[5] Covers repairs executed in whole or in part by the use of federal funds.
[6] Must provide accessible toilet facilities only.
[7] Must provide accessible toilet facilities only.

V. DESIGN AND CONSTRUCTION STANDARDS

State	Use of ANSI Standards	Promulgates Standards Independently[1]	Standards Prescribed in Statute	Accessible Buildings Display International Symbol
ALABAMA Code (1977)	Standards "may be based" on ANSI at fire marshall's discretion. tit. 21, §4-3	tit. 21, §4-3		
ALASKA Stat. (1976)	Standards must be "comparable" to ANSI. §35.10.015	§35.10.015		
ARIZONA Rev. Stat. (1974 & Supp. 1977)	Adopts in toto. §34-404 et seq.		§34-404 et seq.	
ARKANSAS Stat. Ann. (1968)	Adopts in toto. §14-627		§14-627	
CALIFORNIA Gov't. Code (Deering Supp. 1977)	Adopts as minimum standards to be revised and expanded. §4450	§4450		§7251[2]
COLORADO Rev. Stat. (1973 & Supp. 1976)	Adopts in toto. §9-5-104 et seq.		§9-5-104 et seq.	
CONNECTICUT Gen. Stat. Ann. (West 1977 & Supp. 1978)	Adopts as minimum standards. §19-395a			§19-395a
DELAWARE Code (Supp. 1977)	Adopts in toto. tit. 29, §6917		tit. 29, §6917	
DISTRICT OF COLUMBIA				
FLORIDA Stat. Ann. (West)	Adopts in toto. §255.21(1975), §553.48 (Supp. 1978)		§553.48 (Supp. 1978)	§255.211(1975),
GEORGIA Code Ann. (Supp. 1977)	Adopts in toto. §91-1108 et seq.[3]		§91-1108 et seq.	
HAWAII Rev. Stat. (1977)	Adopts in toto. §103-50			
IDAHO Code (1976)	Adopts in toto. §39-3203			§39-3209, §39-3208[4]
ILLINOIS Ann. Stat. (Smith-Hurd 1977)		ch. 111, §14		

State	Use of ANSI Standards	Promulgates Standards Independently[1]	Standards Prescribed in Statute	Accessible Buildings Display International Symbol
INDIANA Code Ann. (Burns 1973)		5		
IOWA Code (1975)	Adopts partially. §104A.3 et seq.	§104A.6	§104A.3 et seq.	
KANSAS Stat. (1976)	Adopts in toto but may be modified by state agency. §58-1301	§58-1301		
KENTUCKY Rev. Stat. (1976 & Supp. 1978)	Adopts in toto. §277.305			
LOUISIANA Rev. Stat. Ann. (West Supp. 1978)	Adopts in toto. §49:148.1, §40:1733		§49:148.1	§40:1735(1977)
MAINE Rev. Stat. (1977)	Adopts in toto. tit. 25, §2702		tit. 25, §2702 tit. 5, §4593	
MARYLAND Ann. Code (1975)	May be modified by department of public improvements. art. 78A, §51	art. 41, §257JK (1978)		art. 78A, §51
MASSACHUSETTS Gen. Laws Ann. (West Supp. 1978)		ch. 22, §13A		
MICHIGAN Comp. Laws Ann. (Supp. 1978-1979)		§125.1355(7)		
MINNESOTA Stat. Ann. (West Supp. 1977)		§299F.43		§299G.12
MISSISSIPPI Code Ann. (Supp. 1977)	Adopts in toto. §43-6-103 et seq.		§43-6-101 et seq.	§43-6-125
MISSOURI Ann. Stat. (Vernon Supp. 1978)	Adopts in part. §8.620		§8.620	§8.655
MONTANA Rev. Codes Ann. (Supp. 1978)	May be considered or adopted. §69-2111	§69-2111		
NEBRASKA Rev. Stat. (1976)	Adopts in toto. §72-1103 et seq.	§72-1122	§72-1103 et seq.	
NEVADA Rev. Stat. (1977)	Adopts in toto. §338.180			

State	Use of ANSI Standards	Promulgates Standards Independently[1]	Standards Prescribed in Statute	Accessible Buildings Display International Symbol
NEW HAMPSHIRE Rev. Stat. Ann. (1977)	Minimum standards to be expanded by barrier-free design board. §275-C:14	§275-C:14		§275-C:9
NEW JERSEY Stat. Ann. (West Supp. 1977)		§52:32-6		
NEW MEXICO Stat. Ann. (Supp. 1975)	May consult or adopt. §67-35-52	§67-35-52		
NEW YORK (McKinney Supp. 1977)		Pub. Bldgs. §51 Exec. §373 et. seq.		
NORTH CAROLINA Gen. Stat. (1978)	§143-138[6]	§143-138		
NORTH DAKOTA Cent. Code (Supp. 1977)	Must be given full consideration. §48-02-19	§48-02-19	§23-13-12,[7] §23-13-13	
OHIO Rev. Code Ann. (Page Supp. 1977)		§3781.111		
OKLAHOMA Stat. Ann. (West Supp. 1977)	Adopts in part (excludes section on elevators). tit. 61, §11			
OREGON Rev. Stat. (1977)	Must be given full consideration. §447.230	§447.230		§477.240, §447.245
PENNSYLVANIA Stat. Ann. (Purdon Supp. 1978)	Adopts in toto. tit. 71, §1455.2		tit. 71, §1455.2 Also see 34 Pa. Code §47.111 et seq.	
RHODE ISLAND Gen. Laws (1977)	Taken into consideration. §37-8-15 Adopted. §37-17-3		§37-17-3 et seq.	
SOUTH CAROLINA Code (1976)	Given full consideration. §10-5-250	§10-5-250		§10-5-280
SOUTH DAKOTA Compiled Laws Ann. (Supp. 1977)	Adopts in toto. §5-14-13			§5-14-13.1
TENNESSEE Code Ann. (1977)		§53-2547[8]		§53-2547

State	Use of ANSI Standards	Promulgates Standards Independently[1]	Standards Prescribed in Statute	Accessible Buildings Display International Symbol
TEXAS Rev. Civ. Stat. Ann. (Vernon Supp. 1978)	Adopts in toto. art. 678g, §2(a)		art.678g,§3 et seq.	
UTAH Code Ann. (1976 & Supp. 1977)		§26-27-3		
VERMONT Stat. Ann. (Supp. 1977)	Adopted — may be revised or amended by architectural barriers compliance board. tit. 18, §1322			
VIRGINIA Code (Supp. 1978)		§2.1-516, §2.1-517, §2.1-518		
WASHINGTON Rev. Code (Supp. 1976)	Considered. §70.92.150	§70.92.140		§70.92.120
WEST VIRGINIA Code (1977)		§18-10F-3		
WISCONSIN Stat. Ann. (West Supp. 1978)	Considered. §101.13(5)	§101.13(5)	§101.13(7)	
WYOMING Stat. (1977)	Adopts in toto. §35-13-101			

1 See chart VII, *infra*, for the title of the particular agency or individual responsible for the promulgation of standards.
2 Must display signs showing availability and location of accessible toilets and ramps.
3 Several state agencies are empowered to waive any of the standards presently set forth and to substitute others consistent with any new ANSI standards adopted.
4 Must post signs showing the locations of accessible toilets and entrances but need not use international symbol.
5 The state building council may adopt standards at its discretion. *See*, 22-11-1-1 et seq. (1974).
6 Statute mandates that the state building code council must "follow" standards promulgated by the United States Standards Institute. It is unclear whether all such standards must be adopted. *See* §143-138.
7 Standards prescribed for toilet facilities only.
8 Adopts the provisions of the North Carolina state building code.

VI. WAIVER PROVISIONS[1]

State	General Waiver at Discretion of Administrative Authority	Specific Provisions
ALABAMA Code (1977)		May waive particular standard if any alternative facility is available or compliance would be unreasonably costly or impractical. tit. 21, §4-5
ALASKA		
ARIZONA		
ARKANSAS Stat. Ann. (1968)	§14-627	
CALIFORNIA Gov't. Code (Deering Supp. 1977)		Must be clearly evident that equivalent facilitation and protection are secured. §4451
COLORADO Rev. Stat. (Supp. 1976)	§9-5-102	
CONNECTICUT Gen. Stat. Ann. (1977)		Municipal board of appeals may review under several circumstances. See §19-402
DELAWARE Code (Supp. 1977)	tit. 29, §6917	
DISTRICT OF COLUMBIA		
FLORIDA Stat. Ann. (West 1975)	§255.21	
GEORGIA Code Ann. (Supp. 1977)		State fire marshall may declare full compliance with a particular standard to be impractical whereupon there shall be substantial compliance to the maximum extent practical. §91-1105
HAWAII		
IDAHO		
ILLINOIS		
INDIANA Code Ann. (Burns 1973)	§22-11-1-16	
IOWA		

State	General Waiver at Discretion of Administrative Authority	Specific Provisions
KANSAS		
KENTUCKY		
LOUISIANA Rev. Stat. Ann. (West)	§49:148.2 (Supp. 1978)	Only if evident that equivalent facilitation and protection are secured. §40:1737 (Supp. 1977)
MAINE		
MARYLAND		
MASSACHUSETTS Gen. Laws Ann. (West Supp. 1978)		Board may make modifications of or substitutions for particular impractical standards. ch.22, §13A
MICHIGAN Comp. Laws Ann. (Supp. 1978-1979)	§125.1355(6)(a)	
MINNESOTA		
MISSISSIPPI Code Ann. (Supp. 1977)	§43-6-106	
MISSOURI Ann. Stat. (Vernon Supp. 1978)	§8.650	
MONTANA Rev. Codes Ann. (Supp. 1978)	§69.2114	
NEBRASKA Rev. Stat. (1976)		Particular standards may be waived at discretion of authority if found impractical. §72-1101
NEVADA		
NEW HAMPSHIRE Rev. Stat. Ann. (1977)	§275-C:15	No waiver allowed when state funds are used. §275-C:15
NEW JERSEY Stat. Ann. (West Supp. 1977)		Must demonstrate that equivalent facilitation and protection are secured. §52:32-8 May waive only specific provisions. §52:32-9
NEW MEXICO Stat. Ann. (Supp. 1975)	Applies only to leased or rented facilities. §5-2-26.2	
NEW YORK Pub. Bldgs. Law (McKinney Supp. 1977)		Not applicable if compliance will impair the structural stability of the public building or its facilities. §51

State	General Waiver at Discretion of Administrative Authority	Specific Provisions
NORTH CAROLINA Gen. Stat.(1978)		May substitute comparable measures for those required by the code. §143-141
NORTH DAKOTA Cent. Code (Supp. 1977)	§48-02-19	
OHIO Rev. Code Ann. (Page Supp. 1977)	§3781.111	
OKLAHOMA		
OREGON Rev. Stat. (1977)		General waiver if cost of compliance exceeds 25% of total cost of construction. §447.230 Otherwise specific waiver if compliance would defeat the purpose of the project. §447.250
PENNSYLVANIA Stat. Ann. (Purdon Supp. 1978)	tit.71, §1455.3a	
RHODE ISLAND Gen. Laws (1977)		Specific standards may be waived if impractical. §37-17-1
SOUTH CAROLINA Code (1976)		Waiver only if compliance would defeat the purpose of the project and (1) an acceptable alternative is substituted; or (2) the additional cost to conform exceeds 7% of the total construction cost. §10-5-270
SOUTH DAKOTA Compiled Laws Ann. (Supp. 1977)		Specific standards may be waived if impractical. §5-14-12
TENNESSEE Code Ann. (1977)		Must demonstrate that equivalent facilitation and protection are thereby secured. §53-2548
TEXAS Rev. Civ. Stat. Ann. (Vernon Supp. 1978)		There must be substantial compliance with the standard or specification to the maximum extent practical. art.678g, §2(a)
UTAH		
VERMONT Stat. Ann. (Supp. 1977)	tit.18, §1322	
VIRGINIA Code (Supp. 1978)	§2.1-520	
WASHINGTON Rev. Code (Supp. 1976)	§70.92.160	

State	General Waiver at Discretion of Administrative Authority	Specific Provisions
WEST VIRGINIA Code (1977)	§18-10F-3	
WISCONSIN		
WYOMING Stat. (1977)	§35-13-101	

[1] See Chart VII, *infra*, for the title of the particular agency or individual responsible for granting waivers.

VII. ADMINISTRATIVE AUTHORITY

State	Who Administers the Act	Duties and Responsibilities of Administrative Authority	Makeup, Length of Term and Frequency of Meeting If a Board or Commission is Created to Administer the Act	Who Appoints Board	Board Staffing Provisions
ALABAMA Code (1977)	State fire marshall.	Inspect for compliance. tit.21, §4-7 Prescribe and publish standards. tit.21, §4-3			
	State fire marshall and the director of the state building commission.	Construe and interpret standards, grant waivers. tit.21, §4-5			
	State board of education; state building commission; the governing bodies of governmental subdivisions.	Review and approve plans and specifications for buildings under their jurisdiction, tit.21, §4-6			
ALASKA Stat. (1976)	Department of public works.	Prepare & promulgate regulations, develop & maintain an inventory of facilities with respect to their compliance, recommend priorities for upgrading existing facilities. §35.10.015 Prepare & adopt plans & specifications for each public work. §35.10.010			
ARIZONA Rev. Stat. (1974)	Administered as part of all state & local building codes. §34-411				
ARKANSAS					
CALIFORNIA Gov't. Code (Deering Supp. 1977)	State architect.	Adopt standards. §4450			
	Department of general services.	Approve plans before contract is awarded (for state funded projects only). §4454			
COLORADO Rev. Stat. (Supp. 1976)	Governmental unit responsible for enforcement.	Grant waivers. §9-5-102			

State	Who Administers the Act	Duties and Responsibilities of Administrative Authority	Makeup, Length of Term and Frequency of Meeting If a Board or Commission is Created to Administer the Act	Who Appoints Board	Board Staffing Provisions
CONNECTICUT Gen. Stat. Ann. (West 1977 & Supp. 1978)	State building code standards committee & state building inspector.	Adopt, promulgate & administer a state building code. §19-395 Hold hearing on standards. §19-395. Approve plans. §19-398a	9 members; none handicapped; 3 year terms. §19-395f	Public works commissioner. §19-395f	May employ such assistance as may be necessary. §19-395f
	Local building inspector.	Pass upon questions concerning construction; inspect. §19-397. Approve plans & issue permits. §19-398			
DELAWARE Code (Supp. 1977)	Secretary of the department of administrative service in consultation with review committee.	Grant waivers. tit.29, §6917.	5 members; 1 disabled; indefinite terms. tit.29, §6917	various state agencies. See tit.29, §6917	
DISTRICT OF COLUMBIA					
FLORIDA Stat. Ann. (West 1975)	Department of general services.	Provide standards. Conduct surveys & investigations. Grant waivers. §255.21			
GEORGIA Code Ann. (Supp. 1977)	State fire marshall, board of regents of the university system, or the local governing authority.	Grant waivers for projects within their jurisdiction. §91-1105 Promulgate rules, regulations & procedures, substitute for & amend prescribed standards, enforce the act. §91-1123			
HAWAII					
IDAHO Code (1976)	Administrator for the department of public works.	Prescribe rules & regulations. §39-3205 "Administer & enforce. §39-3204. Approve plans before permit issued. §39-3206			

State	Who Administers the Act	Duties and Responsibilities of Administrative Authority	Makeup, Length of Term and Frequency of Meeting If a Board or Commission is Created to Administer the Act	Who Appoints Board	Board Staffing Provisions
ILLINOIS Ann. Stat. (Smith-Hurd 1977)	Capital development board.	Publish tentative & final standards. ch.111, §14			
	Contracting authority.	Determine that plans & spefications meet the standards. ch.111,§15			
INDIANA					
IOWA Code (1975)	State building code commissioner.	Promulgate rules &, regulations & additional standards. §103A.7			
	Each governmental subdivision.	Examine & approve plans; issue & revoke permits; inspect. §103A.19			
KANSAS Stat. (1976)	Architectural services division of the department of administration.	Modify & revise standards. §58-1301 Approve plans for school buildings only, pursuant to KSA 72-4604. §58-1304			
KENTUCKY Rev. Stat. (1976 & Supp. 1978)	State fire marshall[1].	Issue regulations & standards. §227.305			See §227.230
LOUISIANA Rev. Stat. Ann. (West 1977)	State fire marshall.	Inspect for compliance before work may be accepted by contracting authority (state buildings only). §49:148.2 (Supp. 1978) Grant waivers. §40:1737 Issue certificate of approval before any construction contract awarded. §40:1740 Grant extensions of compliance deadlines §40:1737 Educate the public. §40:1744			

State	Who Administers the Act	Duties and Responsibilities of Administrative Authority	Makeup, Length of Term and Frequency of Meeting If a Board or Commission is Created to Administer the Act	Who Appoints Board	Board Staffing Provisions
MAINE Rev. Stat. (1977)	Director of public improvements commissioner of educational & cultural services & the governing bodies of political subdivisions.	Administer & enforce the act with respect to facilities under their respective jurisdictions. tit.25, §2703.			
	State, county or municipal authority responsible for reviewing plans.	Require on site inspections; approve plans & construction. tit.25, §2703			
MARYLAND Ann. Code (1975)	Board of public works & department of public improvements.	Publish & revise standards. art.78A, §51(d)			
	Contracting authority.	Must determine that all plans are in compliance. art.78A, §51(e)			
MASSACHUSETTS Gen. Laws Ann. (West Supp. 1978)	Supervisor of plans.	Issue a certificate of approval for the erection or alteration of a public building. ch.143, §3W			
	Architectural barriers board.	Promulgate standards, rules & regulations; provide waivers; receive & review complaints. ch.22, §13A	Seven members; 3 handicapped; 4 year terms; frequency of meetings not specified. ch.22, §13A	Secretary of the executive office of public safety. ch.22, §13A	
MICHIGAN Comp. Laws Ann. (1976 & Supp. 1978-1979)	Construction code commission.	Promulgate standards; interpret standards. §125.1355			
	Department of management & budget, department of education & local administrative authority with responsibility & duty of issuing building permits.	Administration & enforcement with respect to those facilities within their respective jurisdictions. §125.1353			
	Barrier free design board.	Review & grant waivers; review & act on complaints; advise the construction code commission on standards. §125.1355	9 members; 4 physically limited; 3 year terms; not less than 6 meetings annually. §125.1355	Governor §125.1355	Staff provided by department of labor. §125.1355
MINNESOTA Stat. Ann. (West Supp. 1977)	State fire marshall.	Administer & enforce. §299F.42 Promulgate standards, rules & regulations. §299F.43 Approves plans for state buildings & make recommendations for other public buildings. §299F.44			

State	Who Administers the Act	Duties and Responsibilities of Administrative Authority	Makeup, Length of Term and Frequency of Meeting If a Board or Commission is Created to Administer the Act	Who Appoints Board	Board Staffing Provisions
MISSISSIPPI Code Ann. (Supp. 1977)	State building commission, the board of supervisors of each county & the governing authorities of towns & political subdivisions of the state.	Cause the construction of entrance ramps in all public buildings under their supervision. §43-6-101			
	State board of health.	Grant waivers. §43-6-101 Review & approve plans & specifications; promulgate necessary rules. §43-6-123			
MISSOURI Ann. Stat. (Vernon Supp. 1978)	Commissioner of administration.	Grant waivers & keep records of those granted. §8.650			
MONTANA Rev. Codes Ann. (Supp. 1978)	Department of administration	Issue orders; appoint technical advisors & experts to assist in code formulation; enter & inspect buildings. §69-2109 Adopt a state building code & hold hearings pursuant to it's adoption. §69-2111 Grant waivers or modifications; review permit refusals. §69-2114			
	Municipal authority.	Approve plans; issue permits, licenses & certificates; inspect buildings & promulgate administrative rules. §69-2117			
NEBRASKA Rev. Stat. (1976)	Public buildings safety advisory committee.	Grant waivers. §72-1101 Advise authorities. §72-1120 Promulgates codes & standards. §72-1122, §72-1114 (Supp. 1977)	5 members; 1 handicapped; 4 year terms. Meet at least 3 times annually. §72-1121	Governor. §72-1120	
	State building division and the appropriate officials of political subdivisions.	Issue use & occupancy permits; enforce. §72-1119 (Supp. 1977)			

State	Who Administers the Act	Duties and Responsibilities of Administrative Authority	Makeup, Length of Term and Frequency of Meeting If a Board or Commission is Created to Administer the Act	Who Appoints Board	Board Staffing Provisions
NEVADA					
NEW HAMPSHIRE Rev. Stat. Ann. (1977)	Governor's commission for the disabled permanent subcommittee on barrier-free design.	Establish publish & enforce standards; promulgate necessary rules & regulations; hold public hearings. §275-C:14 Grant waivers; investigate complaints; issue cease & desist orders. §275-C:15	11 Members; at least 6 handicapped. 275-C:11 Serve at chairman's pleasure. Meet at least 6 times annually. §275-C:12	Chairman of the commission for the disabled. §275-C:11	May utilize staff of any state agency. §275-C:13
NEW JERSEY Stat. Ann. (West Supp. 1977)	Department of the treasury.	Promulgate standards. §52:32-5			
	State division of building & construction for the department or the treasury; building departments of municipalities having jurisdiction over particular projects.	Grant waivers. §52:32-8 Issue certificates of use or oçcupancy only if buildings comply. §52:32-10			
NEW MEXICO Stat. Ann. (Supp. 1975)	General construction board.	Promulgate standards. §67-35-52(D) Review all plans for compliance. §67-35-52(E)			
	Trade board inspectors.	Inspect for compliance. §67-35-50			
NEW YORK Pub. Bldg. Law (McKinney Supp. 1977)	The official, governing body or board having design approval for the state or local building construction.	Ensure that the design of any such building complies. §52			
NORTH CAROLINA Gen. Stat. (1978)	Building code council.	Prepare state building code which includes architectural barriers standards. §143-138 Take appeals of orders by enforcement agencies. §143-140 Recommend legislation. §143-142	11 Members; 6 year appointments. At least two meetings annually. §143-136, §143-137	Governor. §143-136	May use personnel of the division of engineering. §143-137

State	Who Administers the Act	Duties and Responsibilities of Administrative Authority	Makeup, Length of Term and Frequency of Meeting If a Board or Commission is Created to Administer the Act	Who Appoints Board	Board Staffing Provisions
NORTH DAKOTA Cent. Code (Supp. 1977)	State construction superintendent after consultation with the state board of architecture.	Promulgate standards, rules & regulations; revise annually; approve plans & specifications; grant exemptions. §48-02-19			
	Governing bodies of political subdivisions.	Require statements of persons preparing plans that such are in conformance. §48-02-19			
OHIO Rev. Code Ann. (Page Supp. 1977)	Board of building standards.	Adopt standards & rules. §3781.111 Certify county & municipal building departments for enforcement authority. §3781.10	11 members; 4 year terms. §3781.07	Governor. §3781.07	May employ additional staff. §3781.07
	Board of building appeals.	Hear appeals. §3781.111	3 members; 4 years terms. §3781.19	Governor. §3781.19	
	Local building departments.	Accept & approve plans & specifications. §3781.10			
OKLAHOMA Stat. Ann. (West Supp. 1977)	The governing body of the political entity controlling the funds involved (where State funds involved the appropriate authority is the state board of affairs).	Approve plans & specifications before construction contract awarded. tit.61, §11			
OREGON Rev. Stat. (1977)	Director of commerce.	Promulgate rules. §447.260 Grant waivers. §447.250 Establish standards. §447.230			
	Advisory committee.	Assist director. §447.230	8 Members; at least 2 handicapped. Serve at the pleasure of the governor. §447.230	Governor. §447.230	

State	Who Administers the Act	Duties and Responsibilities of Administrative Authority	Makeup, Length of Term and Frequency of Meeting If a Board or Commission is Created to Administer the Act	Who Appoints Board	Board Staffing Provisions
PENNSYLVANIA Stat. Ann. (Purdon Supp. 1978)	Advisory board.	Develop guidelines, make recommendations for waivers. tit.71, §1455.3a	5 members; 2 year terms. tit.71, §1455.3a	Secretary of labor & industry. tit.71, §1455.3a.	
	Secretary of labor & industry. tit.71, §1455.3a	Grant waivers. tit.71, §1455.3a			
RHODE ISLAND Gen. Laws (1977)	Authority responsible for the construction.	Grant waivers. §37-17-1			
SOUTH CAROLINA Code (1976)	Board for barrier-free design.	Establish & publish a code; promulgate rules & regulations. grant waivers. §10-5-250 Grant waivers. §10-5-270	6 appointed members, at least 2 handicapped; 3 ex officio members; 4 year terms. At least one meeting annually. §10-5-230, §10-5-240	Governor. §10-5-230	Division of general service provides staff. §10-5-240
SOUTH DAKOTA Compiled Laws Ann. (Supp. 1977)	The authority responsible for proper construction for the governmental unit concerned.	Grant waivers. §5-14-12			
TENNESSEE Code Ann. (1977)	Local building inspector. See def., §53-2546	Approve specifications. §53-2547 Grant waivers. §53-2548.			
TEXAS Rev. Civ. Stat. Ann. (Vernon Supp. 1978)	State building commission[2] with the assistance of appropriate state rehabilitation agencies.	Approve plans & specifications; inspect & issue certificates of compliance for privately owned buildings; grant extensions of time allowed for compliance; modify & revise standards; grant waivers; promulgate rules & regulations. art.678g, §20			
	University of Texas board of regents.	Same as above but only with regard to buildings under its jurisdiction. art.678g, §20			

State	Who Administers the Act	Duties and Responsibilities of Administrative Authority	Makeup, Length of Term and Frequency of Meeting If a Board or Commission is Created to Administer the Act	Who Appoints Board	Board Staffing Provisions
UTAH Code Ann. (1976 & Supp. 1977)	State building board.	Promulgate standards. §26-27-3 Advise & review plans upon request. §26-27-1	5 members; 6 year terms. §63-10-1 (1978) Meetings as called by governor. §63-10-3 (1978)	Governor. §63-10-1 (1978)	May procure such services as it deems necessary. §63-10-7(1978)
	State buildings board, state board of education, & the governing bodies of political subdivisions.	Adopt standards. §26-27-4			
VERMONT Stat. Ann. (Supp. 1977)	Architectural barriers compliance board.	Adopt, amend & repeal rules & standards. tit.18, §1312 Grant waivers. tit.18, §1322	8 members; 2 handicapped; 6 year terms. Meetings when necessary. tit.18, §1321	Governor. tit.18, §1321	
VIRGINIA Code (Supp. 1978)	Division of engineering & buildings.	Prescribe standards for state funded facilities. §2.1-516			
	Governing body of local governmental unit.	Prescribe standards for facilities under its jurisdiction. §2.1-517			
	State board of education.	Prescribe standards for public school facilities. §2.1-518 Authority responsible for standards is also authorized to grant waivers & conduct surveys & investigations. §2.1-520			
WASHINGTON Rev. Code (Supp. 1976)	Building department of each county, city or town. §70.92.130	Grant waivers. §70.92.160			
	Board of building code appeals.	Hear appeals from waiver denials. §70.92.160			
	State building code advisory council.	Adopt standards. §70.92.140			
WEST VIRGINIA Code (1977)	Director of the division of vocational rehabilitation with the approval of the state board of public buildings.	Promulgate reasonable rules and regulations. §18-10F-3	5 members, serve at the will & pleasure of the governor. §18-10F-4	Governor. §18-10F-4	

State	Who Administers the Act	Duties and Responsibilities of Administrative Authority	Makeup, Length of Term and Frequency of Meeting If a Board or Commission is Created to Administer the Act	Who Appoints Board	Board Staffing Provisions
WISCONSIN Stat. Ann. (West Supp. 1978)	Department of industry, labor & human relations.	Provide standards. §101.13(1)(m) Grant waivers. §101.13(2)(d)			
WYOMING Stat. (1977)	State fire marshall.	Approve construction. §35-13-105 Grant waivers. §35-13-101 Review & approve plans & specifications. §35-13-102			
	Department of fire prevention.	Hear appeals from orders of the fire marshall. §35-13-104			

[1] Other administrative procedures are included in the chapter but as written are applicable to fire regulations only.

[2] Art. 678M.1 abolishes the State Building Commission and transfers its responsibilities to the State Board of Control upon passage of a constitutional amendment. For general provisions relating to the structure and function of the State Board of Control. *See* art. 601 et seq. (1964 & Supp. 1978).

VIII. ENFORCEMENT — RESPONSIBLE AUTHORITY

State	When State Funds are Utilized	When Funds of Counties, Municipalities or Other Political Subdivisions are Utilized	When Wholly Private Funds are Utilized
ALABAMA Code (1977)	State fire marshall. tit.21, §4-7	Same	
ALASKA Stat. (1976)	Public works department. §35.10.010	Same	
ARIZONA[1] Rev. Stat. (1974)	State agency responsible for planning & supervising the construction of state buildings. §34-410 Arizona state board of regents responsible for state university or college buildings. §34-410	Governing body of the governmental unit whose fund are being utilized for construction. §34-410 Governing body of school district where public school funds utilized. §34-410	
ARKANSAS Stat. Ann. (1968)	The department of health. §14-629	Same	
CALIFORNIA Gov't. Code (Deering Supp. 1977)	Director of the department of general services.[2] §4453	Governing body of the political subdivision §4453	
COLORADO Rev. Stat. (Supp. 1976)	Office of state planning & budget. §9-5-110	Governing body of the political subdivision. §9-5-110	The building department of the political subdivision having jurisdiction. §9-5-110
CONNECTICUT Gen. Stat. Ann. West (1977)	Building code standards committee & state building inspector. 19-395	Local building official. §19-398	Local building official. §19-398
DELAWARE Code (Supp. 1977)	Secretary of the department of administrative services. tit.29, §6917	Same	
DISTRICT OF COLUMBIA			
FLORIDA Stat. Ann. (West 1975)	Department of general Services. Department of education when school funds utilized. §255.21	Governing body of the political subdivision. §255.21	
GEORGIA Code Ann. (Supp. 1977)	See under "Administrative Authority".		
HAWAII			
IDAHO Code (1976)	Administrator for the department of public works. §39-3204	Same	

State	When State Funds are Utilized	When Funds of Counties, Municipalities or Other Political Subdivisions are Utilized	When Wholly Private Funds are Utilized
ILLINOIS Ann. Stat. (Smith-Hurd 1977)	Attorney general ch.111, §16.1	Same	Same
INDIANA Code Ann. (Burns 1973)	Administrative building council. §22-11-1-18	Same in cooperation with local building inspectors. §22-11-1-19	
IOWA Code (1975)	See under "Administrative Authority."		
KANSAS Stat. (1976)	The architectural services division of the state department of administration. §58-1304	The governing body of the political subdivision. §58-1304	
KENTUCKY Rev. Stat. (1976 & Supp. 1978)	State fire marshall. §227.230	Same	Same
LOUISIANA Rev. Stat. Ann. (West 1977)	State fire marshall. §49:148.2 (Supp. 1978)	Local building code authorities & health authorities. §40:1742	Local building code authorities & health authorities. §40:1742
MAINE Rev. Stat. (1977)	See under "Administrative Authority." Attorney general enforces all civil violations & fines. tit.49, §2704		
MARYLAND Ann. Code (1975)	Department of public improvements or, when school funds utilized, the state department of education. art.78A, §52(j)	The governing body of the political subdivision. art.78A, §52(j)	
MASSACHUSETTS Gen. Laws Ann. (West Supp. 1978)	Architectural barriers board. ch.22, §13A		
MICHIGAN Comp. Laws Ann. (1976)	Department of management and budget or, if school funds utilized The department of education. §125.1353	Local administrative authority responsible for the issuance of building permits. §125.1353	Local administrative authority responsible for the issuance of building permits. §125.1353
MINNESOTA Stat. Ann. (West Supp. 1977)	State fire marshall. §299F.42	Same	
MISSISSIPPI Code Ann. (Supp. 1977)	State board of health. §43-6-123	Same	

State	When State Funds are Utilized	When Funds of Counties, Municipalities or Other Political Subdivisions are Utilized	When Wholly Private Funds are Utilized
MISSOURI Ann. Stat (Vernon Supp. 1978)	The agency, division or instrumentality having jurisdictional control of the design function of the work or when school fund utilized, the department of education. §8.630	The governing body of the political subdivision. §8.630	
MONTANA Rev. Codes Ann. (Supp. 1978)	Department of administration. §69-2109	Municipal authority. §69-2117 Department of administration. §69-2109	
NEBRASKA Rev. Stat. (1976)	See Under "Administrative Authority".		
NEVADA			
NEW HAMPSHIRE Rev. Stat. Ann. (1977)	See under "Administrative Authority."		
NEW JERSEY Stat. Ann. (West Supp. 1977)	Division of building & construction for department of the treasury. §52:32-7	Building department of the governmental unit having jurisdiction. §52:32-7	Building department of the governmental unit having jurisdiction. §52:32-7
NEW MEXICO Stat. Ann. (Supp. 1975)	General construction board. §67-35-52(e)	Same	
NEW YORK (McKinney)	See under "Administrative Authority."		
NORTH CAROLINA Gen. Stat. (1978)	The insurance commissioner. §143-139	Same	
NORTH DAKOTA			
OHIO Rev. Code Ann. (Page Supp. 1977)	Detailed provisions. See §3781.10	Same	
OKLAHOMA			
OREGON Rev. Stat. (1977)	Director of commerce. §447.280	Same	Same
PENNSYLVANIA Stat. Ann. (Purdon Supp. 1978)	Department of labor & industry. tit.71, §1455.3	Same	Same
RHODE ISLAND Gen. Laws (1977)	Director of administration or, if school funds utilized, the commissioner of education. §37-17-19	The governing body of the governmental unit. §37-17-19	

State	When State Funds are Utilized	When Funds of Counties, Municipalities or Other Political Subdivisions are Utilized	When Wholly Private Funds are Utilized
SOUTH CAROLINA Code (1976)	Board for barrier-free design. §10-5-250 Local building code inspector, fire chief or, if neither available, the state fire marshall. §10-5-300	Same	Same
SOUTH DAKOTA Compiled Laws Ann. (Supp. 1977)	The authority responsible for proper construction for the governmental unit concerned. §5-14-14	Same	
TENNESSEE Code Ann. (1977)	Local building inspector. §53-2548	Same	Same
TEXAS Rev. Civ. Stat. Ann. (Vernon Supp. 1978)	State building commission. art. 678g, §20	Same	Same
UTAH Code Ann. (1976 & Supp. 1977)	Utah state building board or, if school fund utilized, the state board of education. §26-27-4	The governing board of the governmental unit. §26-27-4	
VERMONT Stat. Ann. (Supp. 1977)	The department of labor & industry and the governor's committee on employment of the handicapped or, if school funds used, the state board of education & the governor's commission. tit.18, §1322	The department of labor & industry & the governor's commission on employment of the handicapped. tit.18, §1322	
VIRGINIA Code (Supp. 1978)	The division of engineering & building. §2.1-521.1		
WASHINGTON			
WEST VIRGINIA Code (1977)	Director of the division of vocational rehabilitation. §18-10F-5	Same	Same
WISCONSIN Stat. Ann. (West Supp. 1978)	Any interested person may bring suit to enforce. §101.13(6g)	Same	Same
WYOMING			

[1] Also enforced as part of all state and local building codes. §34-411
[2] The District Attorney, Attorney General and City Attorney are empowered to enjoin violations.

IX. ENFORCEMENT — POSSIBLE ACTIONS

State	Investigate Reports of Non-Compliance	Compel Compliance	Seek Injunctive Relief	Criminal Penalties
ALABAMA Code (1977)		tit.21, §4-7		
ALASKA				
ARIZONA				
ARKANSAS				
CALIFORNIA Gov't. Code (Deering Supp. 1977)		§4452	§4458	
COLORADO				
CONNECTICUT Gen. Stat. Ann. (West 1977 & Supp. 1978)	§19-397	§19-397		
DELAWARE Code (1975 & Supp. 1977)		tit.29, §6917		
DISTRICT OF COLUMBIA				
FLORIDA Stat. Ann. (West 1975)	§255.21	§255.21		Misdemeanor §553.46 (Supp.1978)
GEORGIA Code Ann. (Supp. 1977)		§91-1123		
HAWAII				
IDAHO Code (1976)		§39-3206		Misdemeanor. §39-3210
ILLINOIS Ann. Stat. (Smith-Hurd 1977)	ch.111, §16.1	ch.111, §16.1	ch.111, §16.1	
INDIANA				
IOWA				
KANSAS				
KENTUCKY Rev. Stat. (1976 & Supp. 1978)				Misdemeanor. §227.990

State	Investigate Reports of Non-Compliance	Compel Compliance	Seek Injunctive Relief	Criminal Penalties
LOUISIANA Rev. Stat. Ann. (West 1977)		§49:148.2 (Supp. 1978) §40:1740	§40:1742	Misdeameanor. §40:1621
MAINE Rev. Stat. (1977)		tit.25, §2703		Civil Fine. tit.25, §2704
MARYLAND				
MASSACHUSETTS Gen. Laws Ann. (West Supp. 1978)	ch.22, §13A	ch.22, §13A; ch.143, §3W		Misdemeanor. ch.22, §13A, ch.112, §60G(d)
MICHIGAN Comp. Laws Ann. (1976)		§125.1356	§125.1356	
MINNESOTA				
MISSISSIPPI				
MISSOURI				
MONTANA Rev. Codes Ann. (Supp. 1978)		§69-2117	§69-2118	Misdemeanor. §69-2119
NEBRASKA Rev. Stat. (1976)		§72-1119 (Supp. 1977)		Misdemeanor. §72-1124
NEVADA				
NEW HAMPSHIRE Rev. Stat. Ann. (Supp. 1977)		§275-C:17	§275-C:17	Misdeameanor. §275-C:17
NEW JERSEY Stat. Ann. (West Supp. 1977)		§52:32-7, §52:32-10		
NEW MEXICO Stat. Ann. (Supp. 1975)	§67-35-50	§67-35-52	§67-35-60	Petty Misdemeanor. §67-35-59
NEW YORK				
NORTH CAROLINA Gen. Stat. (1978)		§143-138	§143-138	Misdeameanor. §143-138
NORTH DAKOTA				
OHIO Rev. Code Ann. (Page Supp. 1977)	§3781.10	§3781.10		

State	Investigate Reports of Non-Compliance	Compel Compliance	Seek Injunctive Relief	Criminal Penalties
OKLAHOMA Stat. Ann. (West Supp. 1977)		tit.61, §12		
OREGON Rev. Stat. (1977)		§447.280		
PENNSYLVANIA				
RHODE ISLAND				
SOUTH CAROLINA Code (1976)		§10-5-290	§10-5-290, §10-5-320	Misdemeanor §10-5-310
SOUTH DAKOTA				
TENNESSEE Code Ann. (1977)		§53-2548		Misdemeanor §53-2548
TEXAS Rev. Civ. Stat. Ann. (Vernon Supp. 1978)		art.678g §20(b)	art.678g §20(b)	
UTAH				
VERMONT				
VIRGINIA Code (Supp. 1978)	§2.1-520	§2.1-520		
WASHINGTON				
WEST VIRGINIA Code (1977)		§18-10F-5		
WISCONSIN Stat. Ann. (West 1974)			§101.13(6)[1]	
WYOMING Stat. (1977)	§35-13-101	§35-13-105		

[1] Any private citizen may bring suit.

X. TAX INCENTIVES

ARIZONA
Rev. Stat. (1977)

Property used for the operation of a residential apartment housing facility which is not used or held for profit and is structured for handicapped persons or persons sixty-two years of age or older and for which a substantial subsidy or payment is given by federal, state or local government or by non-profit organizations, is tax-exempt.

Property used for the operation of such a facility which is located adjacent to a health care institution providing services to the elderly & handicapped is also exempt. §42-271.

CALIFORNIA
Rev. & Tax. Code
(Deering 1978)

Property used exclusively for housing and related facilities for elderly or handicapped families, financed by the federal government, and owned and operated by religious, hospital, scientific or charitable funds shall be tax exempt.§214

CONNECTICUT
Gen. Stat. Ann.
(West Supp. 1978)

The legislative body of any municipality may, by ordinance, authorize such municipality to enter into a written agreement with an owner of any real property who agrees to improve, rehabilitate or renovate any building thereon to meet the standards of design & construction of the state building code to accomodate physically disabled persons, to fix the assessment on such property for a period of not longer than five years and to defer any increase in assessment attributable to such improvement during such period. §12-65g.

FLORIDA
Stat. Ann.
(West 1977)

Any taxpayer who renovates an existing building or facility owned by such taxpayer in order to permit physically handicapped persons to enter and leave such building or facility or to have effective use of the accomodations and facilities therein shall be deemed not to have increased the value of such building more than the market value of the materials used in such renovation, valued as salvage materials. §193.623.

IOWA
Code (1978)

The property owned and operated by a nonprofit organization providing low-rent housing for the elderly and the physically and mentally handicapped shall not be taxed §427.1(34).

NEVADA
Rev. Stat. (1977)

No increase may be made to the assessed valuation of a residence owned and occupied by a handicapped person for improvements made to the premises for the purpose of removing architectural barriers. §36.087.

NEW HAMPSHIRE
Rev. Stat. Ann. (1977)

Every owner of residential real estate upon which he resides, and to which he has made improvements for the purpose of assisting a physically handicapped person who also resides on such real estate, is entitled to an exemption from taxation upon such residential real estate determined by deducting the value of such improvements from the assessed value of the real estate before determining the taxes upon such real estate. §72-37a.

NORTH CAROLINA
Gen. Stat. (1977)

There shall be allowed to resident owners of multifamily rental units an amount equal to five hundred fifty dollars as a credit against the tax imposed by the division for each dwelling unit constructed which conforms to the recommendations of the state Building Code for the taxable years within which the construction of such dwelling unit is completed up to 5% of the total number of such units completed during the year. §105-151.1.

OREGON
Rev. Stat. (1977)

There shall be subtracted from federal taxable income (which is used as the taxable income of state residents for state taxation purposes) the entire cost of a renovation project designed to eliminate architectural barriers. §316.066, §316.067, §317.330, §317.335.

THE ADEQUACY OF EXISTING STATUTORY LAW

Despite the existence of both federal and state statutes, environmental barriers continue to abound. Existing legislation has been referred to as "illusory"[1] and as "an ineffectual policy statement rather than a master plan for a mandatory program."[2] A similar conclusion was reached by the Comptroller General in a report to Congress.[3] And a 1974 study in the state of Iowa found that six years after the passage of the Architectural Barriers Act of 1968 the stated intent of Congress to insure that federal buildings are accessible to the physically handicapped had not been met.[4] Incredibly, that study found that even most of the new federal construction did not meet the requirements of the Act.[5] Unfortunately, state legislation has been criticized as no more effective, with its overall effectiveness rated as poor.[6]

Given the allegations of ineffectiveness of the current statutes, it will be instructive to look at the inadequacies of existing federal and state statutes and determine what might be learned about drafting new and more effective legislation. The major inadequacies can be summarized as follows:[7]

1. The statutes are frequently too limited in scope to be effective, applying only to publicly funded buildings.[8] Many of these do not clearly apply to publicly leased buildings,[9] and few provide for an ultimate compliance date for the publicly funded buildings.[10] Many of the statutes do not apply to privately owned buildings which have the greatest impact upon the built environment.[11] Those that do are not clear whether they apply to all places of business open to the public, to public accommodations and to places of employment.[12] While it is true that the cost of construction of new accessible facilities is somewhat higher, studies have shown the incremental costs to be minimal[13] and some states have provided tax incentives to offset these costs.[14]

2. In addition, most statutes are limited in scope in a second way, applying only to new structures.[15] This effectively

[1] Note, *The Forgotten Minority: The Physically Disabled and Improving Their Physical Environment*, 48 CHI.-KENT L. REV. 215, 222 (1971).

[2] Farber, *The Handicapped Plead for Entrance – Will Anyone Answer?*, 64 KY. L.J. 99, 112 (1975).

[3] *See* Comptroller General of the United States, REPORT TO CONGRESS: FURTHER ACTION NEEDED TO MAKE ALL PUBLIC BUILDINGS ACCESSIBLE TO THE PHYSICALLY HANDICAPPED 6 (1975).

[4] Iowa Governor's Committee on Employment of the Handicapped, ACCESSIBILITY — THE LAW AND REALITY: A SURVEY TO TEST THE APPLICATION AND EFFECTIVENESS OF PUBLIC LAW 90-480 IN IOWA 15 (1974).

[5] *See id.* at 9-12.

[6] National League of Cities, for HEW Rehabilitation Services Administration, REPORT ON STATE AND LOCAL EFFORTS TO ELIMINATE ARCHITECTURAL BARRIERS TO THE HANDICAPPED (Nov., 1967) cited in Note, *supra* note 1, at 223 n. 54.

[7] For a critique of the federal statute see Farber, *supra* note 2, at 102-13. For a critique of state statutes see Note, *Access to Buildings and Equal Employment Opportunity for the Disabled: Survey of State Statutes*, 50 TEMP. L.Q. 1067, 1069-78 (1977) and Note, *supra* note 1, at 222-24. Specific examples of state statutes exhibiting these inadequacies can be found by consulting the "State Statute Charts" *supra*.

[8] *See, e.g.*, ARK. STAT. ANN. §14-627 (1968); MD. ANN. CODE art. 78A, §51 (1975). This is also true of the federal statute. *See* 42 U.S.C. §§4151-4156 (1976).

[9] *Compare* CAL. GOV'T CODE §4451 (Deering 1975) (covering publicly leased facilities) *with* HAW. REV. STAT. §103-50 (1977).

[10] *But see* MISS. CODE ANN. §43-6-101 (Supp. 1977); PA. STAT. ANN. tit. 71, §1455.1 (Purdon Supp. 1978-1979).

[11] *See, e.g.*, DEL. CODE tit. 29, §6917 (1975); KAN. STAT. §58-1301 (1976).

[12] Provisions covering privately owned buildings take a variety of forms. The most ambiguous are those which include any facilities "used by the public" or "held open to the public." *See, e.g.*, N.J. STAT. ANN. §52:32-6 (West Supp. 1978-1979). Others which gain in terms of clarity lose considerably more in terms of scope, carefully circumscribing their reach to include only enumerated classes of privately owned facilities. *See, e.g.*, ME. REV. STAT. tit. 25, §2701 (1977) (covering only buildings where 5 or more persons will be employed); PA. STAT. ANN. tit. 71, §1455.1 (Purdon Supp. 1978-1979) (covering stores, theaters, sports arenas and certain restaurants).

[13] For estimates of the incremental cost increase ranging from ½ of one percent to 1/10 of one percent of total construction costs *see* Note, *Access to Buildings and Equal Employment Opportunity for the Disabled: Survey of State Statutes*, 50 TEMP. L.Q. 1067, 1068 (1977); U.S. Dept. of Housing and Urban Development, *Barrier Free Site Design* 10 (April 1975).

[14] See Chart X *supra*.

[15] For an example of the few statutes which do cover existing facilities see OR. REV. STAT. §447.235 (1977) and PA. STAT. ANN. tit. 71, §1455.1 (Purdon Supp. 1978-1979).

forecloses the possibility of comprehensive accessibility for many decades into the future.

3. Sometimes the statutory language does not reflect the stated intent of the law.[16] For example, the provisions on standards and enforcement mechanisms often do not, as adopted, insure that the built environment is accessible to all people.

4. Some statutes do not set minimum standards, leaving wide discretion in an administrative body, which may not exercise its authority effectively.[17] In addition, compliance with standards is sometimes not mandatory,[18] while in other instances, broad escape clauses or equally broad waiver authority with a lack of specific statutory guidance for the granting of waivers effectively negates the practical effect of the standards.[19]

5. Many statutes lack adequate administrative mechanisms for implementation and enforcement. Specifically, the following problems appear: definite, fixed responsibility is not set or is divided among several agencies, making effective implementation difficult at best; [20] the administrative body is made up only of those who are to be regulated by it,[21] with disabled persons rarely being included,[22] although a few states do give them an advisory role;[23] the administrative body is placed in an agency geared to traditional building code requirements;[24] there is no requirement for the review of plans and specifications prior to construction;[25] and finally, some states lack full or even partial administrative enforcement mechanisms,[26] or the specific authority to seek judicial enforcement.[27]

"It seems apparent that the failure to designate definitive standards and enforcement responsibility has undermined the effectiveness of architectural barriers legislation in many states."[28] It has been suggested that access problems of handicapped persons could be remedied by a major federal response.[29] However, experience across a broad range of federal programs would indicate that the effective implementation of broad scale efforts of this nature is more easily and effectively accomplished through state and local entities coordinated with federal efforts.

While a number of states have credible programs to eliminate environmental barriers, it is readily evident from an examination of existing statutes that overall they represent a patchwork of provisions of widely varying impact. The goal of accessibility to education, employment, housing, recreation, goods and services and of general social interaction is not achievable with existing mechanisms.

Thus, new initiatives are required[30] and several writers have suggested the need for model legislation.[31] A uniform act to this effect would guarantee that all states require adherence to the same minimum standards, and most importantly, if it was adopted, funded and enforced, it would provide a uniform environment free from handicapping barriers and would promote equality and normalcy for physically disabled persons.[32] The model environmental design act suggested in the following pages is intended to assist in this initiative.

---

[16] *See* Farber, *supra* note 2, at 100-05.

[17] *See, e.g.,* N.J. STAT. ANN. §52:32-5 (West Supp. 1978-1979).

[18] *See* UTAH CODE ANN. §26-27-1 (1976).

[19] *See, e.g.,* ARK. STAT. ANN. §14-627 (1968); IND. CODE ANN. §22-11-1-16 (Burns 1973).

[20] *See, e.g.,* MICH. COMP. LAWS ANN. §125.1353 (1976); OHIO REV. CODE ANN. §3781.03 (Page Supp. 1977).

[21] *See, e.g.,* IDAHO CODE §39-3204 (1976); ALASKA STAT. §§35.10.010-.015 (1976). *See also,* Achtenberg, *Law and the Physically Disabled: An Update with Constitutional Implications,* 8 SW. U.L. REV. 847, 855 n. 24 (1976).

[22] *But see* MASS. GEN. LAWS. ANN. ch. 22, §513A (West Supp. 1978); MICH. COMP. LAWS ANN. §125.1355 (West Supp. 1977); N.H. REV. STAT. ANN. §275-C:12 (Supp. 1977); S.C. CODE §10-5-230 (1976); VT. STAT. ANN. tit. 18, §1321 (Supp. 1977).

[23] *See* NEB. REV. STAT. §72-1120 (Supp. 1976); OR. REV. STAT. §447.230 (1977).

[24] *See, e.q.,* ARIZ. REV. STAT. §34-411 (1974).

[25] *See, e.g.,* MO. ANN. STAT. §8.650 (Vernon Supp. 1978); N.C. GEN. STAT. §§143-138 to 142 (1978).

[26] See Chart IX *supra.*

[27] *Id.*

[28] National League of Cities, *supra* note 6.

[29] Farber, *supra* note 2, at 100-01.

[30] The National Commission on Architectural Barriers to Rehabilitation of the Handicapped made several recommendations "to speed the Nation toward" the elimination of those barriers: (1) federal legislation for application to new public buildings if any federal funds are used in their construction; (2) state legislation for public buildings constructed with state funds; (3) building code revisions for application to privately owned public buildings; (4) assignment of enforcement responsibilities and resources for implementation; and, (5) public information programs. From the foregoing discussion, the first and, in some instances, the second of these recommendations have since been adopted. Noticeably lacking, however, has been any action regarding the third, fourth and fifth points.
Note, *supra* note 1, at 225 (footnotes omitted).

[31] *See* Farber, *supra* note 2, at 100; Note, *supra* note 1, at 229.

[32] Note, *supra* note 1, at 229.

MAJOR PREMISES OF THE MODEL ACT

The model act which follows is based upon certain premises derived from existing legislation, a desire to correct the inadequacies of that legislation, the realities of state administrative structures, and the need for more efficient and effective implementation. They are:

1. The act should specifically identify as goals the fullest access to and use of the built environment and the fullest integration of disabled persons into the mainstream of society. These goals should be strongly reflected in statutory language which requires affirmative efforts to achieve them, and the resulting act should be coordinated with legislation promoting employment, housing, and access to public accommodations and educational facilities to achieve maximum impact.

2. The coverage of the act should be as broad as possible when balanced against cost factors. Thus, new construction which is cheaper to make accessible[33] merits broader coverage than existing facilities, while use of public funds in construction or alteration justifies greater coverage than when private efforts are undertaken. Relatedly, broader coverage is merited where increased public use occurs with strictly private areas being excluded. Finally, there should be economic incentives in the form of tax provisions to induce compliance and voluntary private initiatives.

3. The act should specify minimum standards of accessibility with flexible means of updating them periodically. The standards should reflect an environmental approach, not just a structural "brick and mortar" approach, and should stress maximum integration of disabled and non-disabled persons within the built environment. Disabled persons should have strong input in the standard-setting process to insure that standards fit their needs, and any standards adopted should be consistent with but not limited by applicable federal standards. The act should include strict standards for the granting of any variances or exemptions, permitting them only where the standard cannot be met and equivalent access and use is provided to the maximum extent possible.

4. The act should contain a well defined and complete implementation mechanism. Authority and responsibility as well as procedures should be clearly stated rather than left to uncertainty, with the authority being backed by a professional staff. In addition, major affected interests, i.e., handicapped persons, should be included within the composition of the authority. The implementation mechanism should include "before the fact" review of a design to insure compliance and to avoid construction of nonconforming facilities, as well as procedures for quick response to questions and requests for assistance.

5. Effective administrative enforcement mechanisms should be provided with resort to judicial enforcement only if necessary. The mechanism should take account of existing state mechanisms, with responsibility for standard setting, waivers and enforcement retained in one agency.

SUMMARY OF THE ACT

Statutory Provisions

The next section presents a model act whose goal is to promote accessibility and use of public facilities by all environmentally limited persons, i.e., those who are restricted by barriers in their use of the built environment, so that they may enjoy the same opportunities for education, employment, housing, public accommodations and social interaction as do others. Public facilities are defined as buildings, structures or improved areas utilized or held out for use by the public. This goal is sought through an agency of the state to insure that affirmative steps are undertaken to remove existing barriers and prevent the creation of new ones.

The act applies to all new public facilities and at least partially to certain existing public facilities which are to undergo alteration. Its provisions are carried out through an Environmental Design Board on which major affected interests are represented.

33 See Comment to subsection 8(2) of the model act *infra*.

The Environmental Design Board is responsible for the setting of agency policy and priorities, the adoption of environmental design standards and the granting of variances and exemptions. The Board's policies and priorities are carried out by a professional staff headed by a Secretary appointed by and responsible to the Board.

The primary implementation mechanism in the act is the requirement that persons constructing or altering facilities covered by the act submit plans and specifications and receive a permit prior to construction or alteration. This "before the fact" review is designed to minimize the need for use of enforcement procedures against those facilities which might otherwise later be found not to comply and to eliminate the potential cost of subsequent structural changes.

The act gives recognition to the fact that other state and local agencies have concurrent jurisdiction over the facilities covered, and seeks to utilize these existing resources by allowing for delegation of plan review and permit authority where it is shown that the second agency has the authority and capability to carry out these tasks. Authority to grant variances and exemptions and enforce the act, however, is retained by the Board. In addition, the Secretary must review and evaluate the work of these delegated agencies to insure that it complies with the act and the standards and rules adopted pursuant to it.

Pursuant to administrative enforcement, the Secretary is given authority to issue compliance orders where there is clear evidence of a violation of the act. These orders may be appealed to the Board with the appellant being given an administrative hearing on the Secretary's order. Judicial review of all agency actions is then available. The act also authorizes the Secretary to seek judicial enforcement of its substantive provisions where necessary and provides penalties for violations of the act. In certain cases private actions are authorized.

Finally, compliance with environmental design standards is encouraged by tax provisions which grant preferential treatment of those facilities which are altered to comply. The act suggests several alternative tax provisions from which states may choose.

The text of the model act appears in **bold type.** Comments on statutory provisions appear in standard type. The comments explain, elaborate upon or provide a rationale for the text of the act. They are intended to assist the user in interpreting the act, in drafting an act for a particular state based upon this model act and in operating an agency.

Adopting the Model Act

The model act presented here is recommended for adoption in its entirety. However, it is recognized that many states already have legislation which is some form addresses the elimination of environmental barriers. Some of these states may wish to retain portions of existing statutes for administrative, political or other reasons. They can use the model as a resource from which to draw in amending their existing statutes, since it offers ideas, reminders and guidance for redrafting these less comprehensive efforts.

Subsequent drafters should pay particular attention to the "Major Premises" section *supra*. Any adaptation should, at a minimum, retain the environmental design concept rather than a strictly structural approach and should reflect an integrationist policy with respect to disabled persons. It should have broad coverage, including privately owned public facilities.

Minimum standards should be specified with an ultimate compliance date for publicly funded public facilities. It should insure that further standards formulated by an administrative agency are consistent with the goals of the act and that waiver provisions are based upon strict criteria to prevent circumvention of the standards. There should be direct input from environmentally limited persons, both in the adoption of standards and in the granting of waivers; they should not be relegated to an advisory role. Strong implementation mechanisms should be adopted, including authority for review of plans and specifications and issuance of permits as well as a strong enforcement authority.

"The successful implementation of state laws and regulations relative to environmental barriers depends upon a conviction of their value, the development of systematic enforcement procedures to ensure compliance, and financial and administrative support by state legislatures and officials."[34] "Indeed, the only true barrier to the solution of accessibility problems seems to be governmental inertia."[35]

---

[34] Note, *supra* note 1, at 223 n. 54 (quoting National League of Cities, *supra* note 6, ch. II at 35).

[35] Farber, *supra* note 2, at 100.

MODEL STATUTE*

A BILL FOR

AN ACT TO ESTABLISH AN ENVIRONMENTAL DESIGN BOARD WHICH WILL ADOPT AND IMPLEMENT DESIGN STANDARDS FOR THE ELIMINATION OF ENVIRONMENTAL BARRIERS TO THE FULL PARTICIPATION OF ALL PERSONS IN THE SOCIETY, TO PROVIDE PENALTIES FOR VIOLATION OF THE PROVISIONS OF THE ACT AND TO PROVIDE TAX INCENTIVES TO ENCOURAGE THE REMOVAL OF ENVIRONMENTAL BARRIERS.

Section 1. Title

This act shall be known as the "Environmental Design Act."

COMMENT. None necessary.

Section 2. Statement of Findings and Purpose

(1) The general assembly finds that:

(a) Environmental barriers and the lack of suitable accommodations within the built environment constitute a serious threat to the safety and welfare of all members of society.

(b) Environmentally limited persons are often denied access to and use of the built environment as the result of environmental barriers and the lack of suitable accommodations, preventing them from exercising many of their rights and privileges as citizens, including, but not limited to, education, employment, voting, and use of public transportation.

(c) The integration of environmentally limited persons into the mainstream of society furthers the goals and policies of this state to assure the right of all persons to live as independently as possible and to participate in the life of the community as fully as possible.

(2) It is the purpose of this act to integrate environmentally limited persons into the mainstream of society, to eliminate accommodations for environmentally limited persons separate from those used by other persons when possible, and to enable members of society to interact with the built environment with maximum safety and independence and at each person's full potential by providing for the establishment and implementation of standards for the elimination of environmental barriers to such integration and interaction. This act shall be liberally construed toward these ends.

COMMENT. The act specifically recognizes that our society is based in large part upon the manmade changes in our environment, i.e., the built environment, that both necessitate and contribute to the need for physical mobility. It also recognizes that this environment contains a significant number of barriers to disabled persons, thus making them environmentally limited. The act challenges the existence of these environmental barriers and the results they produce.

Environmental barriers have the effect of limiting the lives of millions of Americans. They interfere with the use of public services and benefits, including education, health care, transportation, and recreation, and with the exercise of civil rights and of free citizenship, including the right to petition and to vote. They restrict employment opportunities and access to private goods and services. They make adequate housing difficult to find, and they deny environmentally limited persons their independence and participation in society by de facto segregating them.

The act endorses the goals of integration and independence for all persons, including those who are environmentally limited, in all spheres of society. It rejects the old notion that disabled persons must be "protected" from society; that keeping them "busy" is enough. It promotes the notion that disabled persons need and deserve to participate in society, to learn, to work, to live as others do, though not necessarily at the same speed or in exactly the same way. Separate but equal is no longer applicable.

Interference with one's integration and independence, and with the exercise of functions so basic to our social, economic and political systems for such a large proportion of our population is a legitimate matter for public concern and justifies the exercise of the state's police power for the benefit of the safety and welfare of its citizens. At the same time the act rejects the view that these goals can be achieved by voluntary efforts alone. It is clear from past experiences that legally mandated imperatives are necessary for their

---

* D. Matthew Powell, Associate Director of the project had primary responsibility for drafting this model statute.

achievement.[1] Accordingly, as an exercise of the state's police power, the act seeks to eliminate environmental barriers by providing for the establishment and implementation of appropriate accessibility standards. The act is to be liberally construed toward this end.

Section 3. Definitions

As used in this act:

(1) "Built environment" means those parts of the physical environment which are designed, constructed or altered by people, including all public facilities.

COMMENT. The model act is intended to emphasize the impact of the physical environment as constructed and controlled by man upon the functioning of certain persons who do not fit the normal assumptions about users of this environment. (See Comment to subsection 3(6).) Thus, as used in the statement of findings and purpose (section 2 *supra)*, the term "built environment" is broadly defined to include the total man-made physical environment.

(2) "Public facility" means any building, structure or improved area utilized or held out for use or intended to be utilized or held out for use by the public or employees for:

(a) Purposes of education, employment, entertainment or amusement, lodging, recreation, transportation, housing, other than single or two-family dwellings; or

(b) The purchase, rental, sale, or acquisition of any goods, personal property or services.

The term includes public telephones, drinking fountains and restrooms.

COMMENT. Only a portion of the built environment is intended to be regulated by this act. That portion is defined as any structure which is used or intended to be used by persons in general, whether they be members of the public or employees of an individual or company. This use is intended to cover a broad range of purposes, including education, employment, entertainment, housing and transportation, as well as the sale of goods and services. Thus, facilities such as stores, shopping centers, office buildings, factories, theaters, restaurants, libraries, gas stations, bus terminals, doctors' offices, hospitals, barber shops and so forth are intended to be covered by the act. This definition also makes clear that telephones, drinking fountains, and rest rooms intended for use by the public are covered even though they may not be attached to or associated with a building or other structure.

A public facility is subject to the environmental design standards set forth in section 8 of the act and to the plan review and permit requirements of section 10 of the act. Note that public facilities are covered only to the extent that they are new facilities or are being altered to a substantial extent. (See subsection 8(2).)

(3) "Improved area" means:

(a) Parking areas;

(b) Sidewalks, ramps, or other means of ingress or egress; and

(c) Landscaped areas, parks, beaches, harbors, and similar areas.

COMMENT. This term is included to eliminate the uncertainty as to the inclusion of parking areas and sidewalks adjacent to buildings or provided as public access ways and to eliminate any argument that they are not intended to be covered.[2] The term also includes some areas which may not popularly be thought of as public facilities, including parks and beaches which are frequently constructed with public funds and can easily be made accessible.[3]

(4) "Publicly owned facility" means a public facility which is:

(a) Owned by, or on behalf of, the state or its political subdivisions;

(b) Leased or rented in whole or in part by the state or its political subdivisions after the effective date of this act; or

(c) Financed in whole or in part by a grant or a loan made or guaranteed by the state or its political subdivisions after the effective date of this act.

COMMENT. The act adopts the premise that public facilities which are constructed or maintained with public funds rightfully may be subjected to broader application of standards of accessibility and use than privately owned facilities. Accordingly, the act is applied to currently existing publicly owned facilities in addition to those which are newly constructed or altered. It is intended that these facilities be assigned a compliance plan and be the subject of an ultimate compliance date after which all such facilities shall be fully accessible. (See subsection 5(11).)

(5) "Publicly available housing" means any housing development of single or two-family dwellings intended for sale to the public, includ-

[1] *See* "The Adequacy of Existing Statutory Law" *supra*, and "Enforcement Provisions" *infra*.

[2] An illustrative case occurred recently in California, where the legislature was forced to amend that state's act because a Los Angeles city official, charged with enforcement responsibility for his municipality, refused to interpret the phrase as including sidewalks adjacent to state buildings. Achtenberg, *Law and the Physically Disabled: An Update with Constitutional Implications*, 8 SW. L. REV. 847,855 (1976).

[3] Recently a number of communities have made great progress in this area and have shown the incremental cost increase for accessible outdoor facilities to be minimal. *See* Zucker, *Accessibility: How One Department Is Making It Possible*, 11 PARKS AND RECREATION 24 (June, 1976).

ing but not limited to detached or semidetached dwellings, townhouses, or duplexes, and dwellings sold individually or as condominiums.

COMMENT. Privately owned private facilities are generally excluded from inclusion in the regulatory provisions of the act. This applies to both single and two-family dwellings. (See subsection 3(2).) However, there is a need for accessible housing for a portion of our citizens. This definition is intended to carve out a portion of the private single and two-family dwelling market to make a portion of these homes accessible to environmentally limited persons. (See subsection 8(2).)

(6) "Environmentally limited person" means a person with a particular physical, mental, or communicative condition or characteristic, regardless of its cause, nature, extent, permanence or temporariness, and who, because of the condition or characteristic, is restricted within or in the use of the built environment.

COMMENT. The act uses the term "environmentally limited" rather than physically handicapped in order to emphasize the effect of environmental barriers on the functioning of disabled (and elderly or temporarily injured) persons. The intent is to alter the view of handicapping or limiting conditions as being caused only by the presence of the disability. Handicaps are in fact caused by a combination of the disability and the environment in which the disabled person must operate. Thus, disabilities are considered in light of the impact of the environment upon the functioning of the individual. It specifically recognizes that many disabled persons are limited because of the choices made by society in the construction of buildings and other facilities which, due to their design, exclude disabled persons from using them. The act challenges the traditional notions of design by looking at how the environment can be changed to accommodate the needs of diabled individuals rather than always attempting to change disabled individuals to fit as best they can in an environment which is clearly ill-prepared to receive them.

(7) "Environmental design" means the planning and design of all public facilities and publicly available housing to provide safe, independent and efficient use of the built environment by all people regardless of their physical, mental or communicative conditions or characteristics.

COMMENT. The act strongly endorses the concept of environmental design, also known as barrier-free design, as opposed to the traditional ramp and elevator accommodations. Thus, the objective is to make the total environment and public facilities more accessible and functionally usable. This requires that interiors and furnishings also be considered in addition to exterior modifications permitting ingress and egress.

This concept is based upon the premise that the needs of environmentally limited persons must be pragmatically met. Environmental design must consider maximum independence and effective use by all persons despite the existence of disabilities, the need for environmental accommodations both inside and outside the facility, the maximum safety of all users of the facility, the cost involved in operation, maintenance and energy consumption, and the aesthetic appeal of the facility. Application of the concept means, for example, that entrance to a building be achieved through landscaped access areas used by all persons rather than separate ramped areas intended only for use by environmentally limited persons. Proper application of the concept should insure not only maximum accessibility and functional use, but facilities which are aesthetically pleasing as well.

(8) "Environmental barrier" means a physical design feature which restricts access to or use of the built environment by environmentally limited persons.

COMMENT. Use of the term "environmental barrier" is intended to emphasize the total environment and the impact of barriers upon its accessibility and use. This replaces the term "architectural barriers" which tends to be limited to "brick and mortar" types of barriers. The term "environmental barriers" is intended to include such things as counter and display case heights, directional signals, information devices and the like.

(9) "Accommodations" means those physical features of a public facility or publicly available housing which make it usable by or serve the needs of its occupants or users, including, but not limited to, adaptions to parking areas, means of ingress and egress, elevators, rest rooms and phone booths.

COMMENT. Certain physical design features and adaptions can be included in the environmental design process to meet the needs of environmentally limited persons and insure that public facilities are maximally accessible and usable. Consistent with the goal of maximum integration, accommodations should be designed and adapted so that they may be used by both environmentally limited persons and other persons, thereby keeping separate accommodations to a minimum.

(10) "Independence" means the ability to move about or utilize the built environment without the direct assistance of other persons.

COMMENT. A constant need to rely upon the assistance of others to function within the built environment is not only a nuisance but demeaning to a person's ego and self-concept. The need to be independent and self-sustaining is as important for disabled persons as it is for anyone else. Accordingly, one of the goals of this act is to promote independence for environmentally limited persons. This independence must include the ability to gain access to and utilize the built environment without the assistance of other persons.

(11) "Planning" means the development or preparation of architectural or engineering designs or plans, technical or other specifications, landscaping plans or other preconstruction plans or specifications.

(12) "Construction" means any erection, building, installation or reconstruction.

(13) "Alteration" means any modification, renovation or addition involving a structural change which exceeds five percent of the full market value of a public facility.

COMMENT. Environmental design standards adopted under the act apply to all new construction of public facilities and to a portion of existing facilities, i.e., those scheduled for alteration. Due to the higher cost involved in modification of existing facilities, only larger alterations involving renovation or additions which (1) involve a structural change and (2) exceed 5% of the full market value of the public facilities are intended to be covered. This should insure that the cost of alterations to meet accessibility standards is imposed only where the alterations involve substantial changes, thereby reducing the cost of such accessibility requirements. (See subsection 8(2).) These standards are intended to be applied during the development of engineering plans and technical specifications for the construction or alteration of any facility covered by the act. The plans and specifications must then be reviewed and approved by the Secretary. (See section 10.)

(14) "Variance" means the use of an accommodation for environmentally limited persons which is different from that specified in an environmental design standard.

COMMENT. None necessary.

(15) "Exemption" means an exclusion from an environmental design standard.

COMMENT. None necessary.

(16) "Person" means one or more individuals, partnerships, associations, unincorporated organizations, corporations, cooperatives, legal representatives, trustees and receivers, agents or any group of persons, the state of (name of state) or any political or civil subdivision or instrumentality thereof.

COMMENT. The term is defined to include the largest application of the word as used in the act.

(17) "Subdivision" means any county, city, town, village, independent district or school district in the state, or any combination thereof.

COMMENT. None necessary.

(18) "Board" means the environmental design board.

COMMENT. None necessary.

(19) "Secretary" means the secretary of the environmental design board.

COMMENT. None necessary.

Administrative Organization of the Environmental Design Board

Many existing statutes addressing the removal of environmental barriers have placed administration of the law in traditional building code or fire safety agencies or have divided the authority for administration among a number of agencies.[4] They seldom provide formal mechanisms for input by affected groups, especially physically disabled persons in standard-setting and implementation.

These approaches suffer from several problems. Many of the implementing agencies are tied to existing concepts of structural and safety standards which tend toward a "brick and mortar" approach to environmental barriers instead of the environmental design concepts promoted by this model act. The model act's philosophy is based on the fullest access and integration within the total environment and not just the traditional ramps and elevators. Agencies geared to building and fire codes are unlikely to have or soon develop this philosophy and accompanying expertise.

In addition, the absence of formal mechanisms for input in the standard setting and the granting of variances makes it less likely that the needs of those persons whom the statute is meant to serve will be adequately represented. Environmentally limited persons are a valuable source of information and their input should be formally sought. Finally, coordination of policymaking and implementation is frequently made more difficult by the fragmented administration of many statutes. This can easily produce a wide degree of variation in the effective implementation of the act's requirements.

The administrative organization proposed in this model act differs from most current approaches. It utilizes a board which is responsible for policymaking, standard-setting, rule-making and the granting of waivers. The Board-established policies, standards and administrative rules are carried out by a Secretary appointed by and answerable to the Board. Implementation occurs through the review of plans and specifications for the construction of facilities covered by the act and the inspection of facilities to insure that construction is completed in accordance with approved plans and specifications. Enforcement provisions are utilized only where this procedure fails or is inapplicable. The Secretary and his or her staff performs these more technical functions.

This Board/Secretary dual approach gives recognition to the need for the broad policymaking which must go on and the need for technical ex-

[4] See Chart VII, Administrative Authority *supra*.

pertise in implementing this policy as established. The real policymaking decisions that are made in the setting and revision of standards, adoption and revision of rules and granting of waivers are part of a continuous process in the model act. Therefore, there is much greater need for a corporate body with multiple interests represented to establish policy. This provides input from all affected persons. Thus, the approach suggested here is different from that suggested in Chapter 2: Prohibiting Discrimination, in which all major policies are established by the act and merely carried out by the Director. The focus here is on both policymaking and implementation.

The implementation function involves both plan review, inspection and enforcement. These tasks require the technical expertise of trained and experienced professionals to review plans and specifications, answer day-to-day technical questions, conduct inspections of built facilities, conduct investigations and testify as expert witnesses in enforcement proceedings. Thus, these tasks are given to a Secretary who can provide the required technical expertise directly or through his or her staff that is necessary to carry out the policy of the Board.

Another advantage of this approach is that it combines the responsibility for policy and implementation in a single entity. This should greatly facilitate the coordination of policymaking, standards development and implementation. It should also avoid the fragmentation of these functions which occurs in states which allocate authority among several agencies. This fragmentation can easily result in a patchwork of standards and inconsistent implementation of them which makes the achievement of accessible, integrated public facilities unlikely.

At the same time the model act recognizes the existence of current programs. Thus, the act provides for delegation of the authority to carry out the requirements of the act by cities, counties and other state agencies which conduct plan review and building inspection programs for public health and safety purposes. They must, however, apply the same standards adopted by the Board, have adequate authority and provide adequate organizational capability, qualified staff and resources to effectively implement the intent of the act. Enforcement authority and mandatory review of the implementation of this delegated authority is retained to insure compliance with the act where necessary.

Section 4. Environmental Design Board Created

(1) There is hereby created an Environmental Design Board as the sole agency of the state having responsibility for carrying out the purposes of this act. It shall be located for administrative purposes within the department of ________________. Other agencies of the state or its subdivisions shall assist where appropriate in the implementation of this act.

COMMENT. The Environmental Design Board is created to be the exclusive agency of the state for carrying out the purposes of the act. This provides a single focus for policymaking and implementation of the act, while specifically rejecting the division between various state agencies of authority for implementation of access and integration provisions. This integrated approach should provide for a more effective implementation program. The act does, however, provide for participation of other agencies in this process and specifically requires other agencies of the state and its subdivisions to assist the Board in carrying out this mandate. (See subsection 4(3) and section 12.)

(2) The Board shall be composed of nine voting members who shall be responsible for carrying out its duties and shall include:

(a) One registered architect;

(b) One registered professional engineer in the field of mechanical or structural engineering;

(c) One general contractor with experience in residential and commerical construction;

(d) One licensed attorney;

(e) Four members who are environmentally limited, at least one of whom must be a wheelchair user; and

(f) One member of the general public who has a demonstrated interest in eliminating environmental barriers.

Nothing in this act shall be construed to prevent an environmentally limited person from serving in one of the positions in paragraphs (a) through (d) of this subsection.

COMMENT. Because the Board will be developing major policy and passing on requests for variances and exemptions from the act's requirements, it is important that affected interests be represented on the Board. The act specifically provides for the inclusion of an architect, an engineer and a general contractor who can provide technical expertise in the development of standards and on the feasibility of standards in particular instances when variances or exemptions are applied for. It is also extremely important that standards are developed and implemented which are responsive to the actual day-to-day needs of environmentally limited persons. Therefore, a substantial portion of the Board is required to be made up of persons who are environmentally limited, at least one of whom must be a wheelchair user to insure that the particular needs of this group are met.

(3) The following persons or their designees shall be ex-officio, nonvoting members of the

Board, who shall act as liaisons between the board and their respective entities:

(a) The state building inspector;

(b) The director of transportation;

(c) The director of public safety;

(d) The director of the commission on the aging;

(e) The director of the governor's commission on employment of the handicapped;

(f) The chairman of the developmental disabilities council;

(g) The director of administration/general services administrator/or other person responsible for construction, purchase or lease of state buildings;

(h) The director of the department of education;

(i) The chairman of the board of regents or other entity responsible for institutions of higher education;

(j) The director of the office for civil or human rights; and

(k) A building inspector from a local unit of government who shall be selected by the state league of municipalities or similar organization for a three-year term.

COMMENT. The Board must coordinate its activities with the programs of a number of other state and local agencies in order to carry out its mandate. The model act provides a formal mechanism for carrying out this coordination by including as ex officio, nonvoting members of the Board, the heads of various agencies whose programs will be affected by the efforts of the Board. This formal mechanism should ensure that input is received from these agencies and that they assist in the implementation of the mandate of the act. It should be noted, however, that the listing suggested here is not necessarily complete and the names of the affected agencies should be modified to reflect the existing counterpart in each state.

(4) Voting members of the Board shall be appointed by the Governor with the advice and consent of the State Senate.

The voting members shall serve for a term of three years, except that of those initially appointed, three shall be appointed for a term of one year, three for a term of two years, and three for a term of three years. A vacancy shall be filled in the same manner as the original appointment for the balance of the unexpired term. Members shall serve until replaced by a new appointment.

COMMENT. The Board carries on a regulatory function and thus the Governor, who is responsible for leadership, administration and enforcement of the state's policies and programs, is given authority to appoint the Board members. As is typical with appointments at this level, they are subject to confirmation by the state senate. Appointment by the Governor gives a clear focal point for responsibility for appointment of well qualified persons and provides accountability for bad choices. Staggered terms are provided for Board members to ensure that some experienced members will always serve on the Board.

(5) The Board shall meet as often as deemed necessary by the majority of its members, but no less than ________.

COMMENT. The Board has considerable responsibility under the model act and must necessarily meet very frequently to carry out that responsibility. It is suggested that the Board meet at least once a month, though more frequent meetings may be required because of its responsibilities.

(6) Five voting members of the Board shall constitute a quorum, with a concurrence of a majority vote of the quorum being necessary to take an action.

COMMENT. Note that only the voting members of the Board have the authority to carry out the responsibilities of the Board. All ex-officio members may act as liaisons only.

(7) The Board shall annually elect from its voting membership a chairperson and such other officers as it determines to be appropriate.

COMMENT. None necessary.

(8) Voting and ex-officio members of the Board not employed by the state shall receive compensation at a rate not to exceed $ for each day, including the travel time, they are engaged in the performance of their duties as members of the board. In addition, all members of the board shall be entitled to reimbursement for travel, subsistence and other reasonable expenses incurred by them in carrying out their duties.

COMMENT. None necessary.

Section 5. Duties of the Board

The Board shall:

(1) Organize itself and adopt procedures as it deems appropriate.

COMMENT. None necessary.

(2) Appoint a secretary to the Board who shall carry out the duties specified in section 6 of the act. The Secretary shall be selected on the basis of his or her interest, knowledge of the area and administrative abilities. The salary of the Secretary shall be initially established by the Board, but it shall not exceed ____________ dollars per annum and, thereafter, it shall be determined by the General Assembly. The appointment or removal of the Secretary shall not be subject to the provisions of (cite to state's civil service code).

COMMENT. While the Secretary performs substantial implementation functions pursuant to section 6 of the act, the Secretary is appointed by and is responsible to the Board. The appointment of the Secretary is not subject to civil service requirements and he or she therefore serves at the pleasure of the Board. This should assure that ultimate authority to implement the act is retained in the Board as the responsible policymaking body.

(3) Develop comprehensive plans and programs for the elimination of environmental barriers in the state, and set priorities for their implementation.

COMMENT. The Board is intended to be responsible for the development of comprehensive plans and programs to carry out the purposes of the act. Since it will obviously not be able to perform all functions assigned to it in any given year, it is specifically directed to set priorities for implementation of plans and programs.

(4) Formulate, adopt and modify, as appropriate, environmental design standards in accordance with section 8 and (cite to rule-making section of state's administrative procedure act or similar provisions).

COMMENT. The Board is given the major responsibility for adopting environmental design standards which will largely determine the success of access and integration efforts. These standards must be adopted in accordance with a state's applicable rule-making procedures under its administrative procedure act or other statute. These should include, as a minimum, notice and provision for public hearing and comment.

(5) Establish, modify or repeal rules relating to:

(a) Submission and review of plans and specifications and specifying the conditions under which the Secretary shall issue, revoke, suspend, modify or deny permits in accordance with section 10;

(b) Variances and exemptions in accordance with section 11;

(c) Delegation of plan review, inspection and permit issuance authority pursuant to section 12; and

(d) Inspection and monitoring of public facilities and publicly available housing, as necessary to implement the provisions of the act.

COMMENT. As the policymaking authority, the Board is responsible for the promulgation of all rules governing the implementation of the act. Note that the scope of applicability of the plan review and permit requirements of section 10 will be determined by the rules adopted by the Board.

(6) Approve or deny applications for variances and exemptions from environmental design standards in accordance with section 11.

COMMENT. Accessibility will be determined largely by the scope of the act and the content of the standards adopted by the Board. Yet, accessibility requirements can be vitiated by the indiscriminate granting of variances and exemptions. Therefore, the ultimate authority to approve or deny all applications for variances or exemptions is retained in the Board itself.

(7) Receive complaints of alleged violations of this act of any rule adopted or permit issued pursuant thereto and refer them to the Secretary for investigation.

COMMENT. While policymaking by committee is frequently more effective than policymaking by a single individual, especially where multiple interests are to be represented, a committee approach is generally less effective in enforcement proceedings. Experience with boards and commissions generally supports the notion that a director or secretary and staff is nearly always out in front of the commission or board in its desire to carry out an assertive program of enforcement. Thus, the model act has opted for assigning the investigation and enforcement authority to the Secretary and staff.

The Board is still given the authority to hear complaints, and indeed it should, to help it improve the standards which it has adopted and to evaluate their effectiveness, scope and applicability. Beyond this, investigation and enforcement are more easily handled by the Secretary. The Board, of course, does have ultimate authority over the Secretary through its hiring and firing powers and could exercise it where the Board feels the Secretary is being too aggressive or not aggressive enough in his or her enforcement activity.

This approach also has another advantage in that it separates the investigation and enforcement functions from the ultimate decision-making functions on appeal. That is, the Secretary is assigned a prosecutorial role while the Board is assigned a judicial role by the act. This should help alleviate the frequent criticism of administrative agencies when the same entity acts as both prosecutor, judge and jury. (See sections 14 and 16.)

(8) Hold public hearings and issue and enforce subpoenas as necessary to accomplish the purposes of this act.

COMMENT. The Board must hold public hearings in a number of instances in order to carry out the duties assigned by this act. These instances include public hearings to receive comment during the rule-making process; hearings on appeal from the denial of permits, variances and exemptions, and delegations of authority pursuant to section 12; and hearings on appeal of administrative orders. As a regulatory agency the Board has been given subpoena authority to facilitate its hearing function with respect to appeals. (See section 22 in the Prohibiting Discrimination model statute, Chapter 2, *supra.*)

(9) Undertake efforts to create a positive awareness of the problems faced by environmentally limited persons and disseminate information about the benefits of environmental barrier-free design.

COMMENT. None necessary.

(10) Create such local and statewide advisory committees as will in the judgment of the Board aid in implementing this act and provide appropriate staff services to such committees.

COMMENT. The Board may occasionally benefit from the expertise of various individuals or groups in the development of standards applicable to particular types of environmental limitations or to particular types of facilities. Where this is the case, the Board has the authority to appoint appropriate advisory committees and to provide staff services to such committees as necessary.

(11) (a) Develop and maintain a record of the degree to which each existing publicly owned facility is in compliance with this act and any rule or permit issued pursuant thereto; and shall inspect and approve a transition plan for all existing publicly owned facilities which shall describe the changes necessary to meet environmental design standards and contain cost estimates for such changes. All publicly owned facilities shall comply with environmental design standards adopted by the Board no later than ______________, notwithstanding any applicable federal requirements.

(b) The Board shall prepare an annual report to the Governor and General Assembly describing the work of the Board during the preceding year to eliminate environmental barriers. The report shall also contain the Board's recommended priorities for changes in existing publicly owned facilities to meet environmental design standards consistent with the priorities developed pursuant to subsection 5(2), and the Board's cost estimates for such changes.

COMMENT. Facilities which are both utilized by the public and built with public funds deserve special attention with respect to compliance with the requirements of this act. The Board is required to inspect and approve transition plans for all such existing facilities and specify changes necessary to meet environmental design standards. This information should be transmitted along with the Board's annual reports to the Governor and General Assembly, together with suggested priorities for upgrading these facilities to comply with the act. Funds would then be allocated on a periodic basis to gradually alter existing publicly owned facilities to insure their compliance.

In order to facilitate this process, an ultimate compliance date is provided for in the act. This date should be set as appropriate in each state. One should bear in mind that too short a compliance date will be totally unrealistic and will therefore promote noncompliance. On the other hand, setting a date too distant in the future tends to lead to procrastination of work which could occur at an earlier date and therefore defeats the purpose of providing an ultimate compliance date. It is suggested that errors on the side of an early compliance date can always be corrected by a subsequent amendment should it prove to be unrealistic.

Note that some of these facilities may be required to comply with the program accessibility requirements of Section 504[5] and accompanying regulations.[6] For instance, recipients of federal funds housed in publicly owned facilities must comply with the requirements of that section and the regulations regarding program accessibility[7] regardless of the requirements of this statute. On the other hand, facilities not required by federal statute to be accessible may be required by this subsection to meet access requirements.

Section 6. Duties of the Secretary of the Board

(1) The Secretary shall recommend to the Board:

(a) Plans, programs and priorities for administration of this act; and

(b) Environmental design standards and other administrative rules.

COMMENT. While the Board retains the ultimate authority to establish programs and priorities for the administration of this act, the Secretary is given the authority to make recommendations based upon his or her perception of the needs and problems that the agency will encounter during its day-to-day operation. The Secretary and staff are also expected to provide technical assistance and information to the Board to aid in its development of plans, programs and priorities.

It is also expected that the Secretary and staff will provide drafting assistance in the development and modification of standards and rules. For example, much of the research and information gathering could be conducted by the Secretary and staff. They could also provide actual drafts of standards and rules based upon directions from the Board. The Board, of course, always retains final decision-making power.

(2) Subject to the approval of the Board, the Secretary shall:

---

5 29 U.S.C. §794 (1976). Section 504 states that "no otherwise qualified handicapped individual . . . shall, solely by reason of handicap, be excluded from participation in, be denied the benefits of, or be subjected to discrimination under any program or activity receiving Federal financial assistance."

6 45 C.F.R. §84.21 (1977).

7 See discussion in "Federal Laws Prohibiting Environmental Barriers" *supra*.

(a) Establish or reorganize the administrative structure;

(b) Appoint the technical, professional, secretarial and clerical staff necessary to accomplish the purposes of this act, subject to the provisions of (cite to state's civil service law).

The Secretary and staff shall receive, in addition to salary, their necessary traveling and related expenses when engaging in the performance of official business;

(c) Prepare a budget request for submission to the Governor and General Assembly; and

(d) Contract or enter into other agreements with public or private persons as necessary to implement the purposes of the act.

COMMENT. The Secretary, pursuant to his or her role as implementor of the act, is given the administrative authority over the day-to-day operation of the agency, always subject to the ultimate approval of the Board. All employees of the agency are to be selected by the Secretary through the state's merit or civil service system.

(3) In accordance with the rules approved by the Board, the Secretary shall:

(a) Direct and administer the plans, programs and services of the Board; and

(b) Review and approve or disapprove plans and specifications and grant, modify, deny, suspend or revoke permits in accordance with section 10.

COMMENT. Also consistent with the implementation role of the Secretary, he or she is to direct and administer all plans and programs developed and approved by the Board. In addition, the Secretary is given the express authority over plan review and permit issuance pursuant to section 10 of the act. These activities involve technical knowledge and experience which will be provided through the professional staff of the agency, although the review will always be conducted in accordance with the standards and rules adopted by the Board.

Most states' administrative procedure acts will require the Secretary to grant a right to a hearing on the suspension or revocation of permits. The act also extends this requirement to the denial of permits, as provided in subsection 10(3).

(4) The Secretary shall also:

(a) Conduct inspections and monitor public facilities or publicly available housing as appropriate to implement provisions of this act;

(b) Conduct investigations of complaints of alleged violations of this act or any standard, rule or permit issued pursuant thereto whether received directly or referred by the Board, and conduct such other investigations as deemed necessary;

(c) Periodically inform complainants of the status of their unresolved complaints;

(d) Take any action or actions allowed by law which are necessary to enforce or secure compliance with the provisions of this act or with any rule or standard established or permit issued pursuant thereto;

(e) Provide technical, scientific or other services to the Board as needed;

(f) Provide technical information and advice to persons covered by the provisions of the act to facilitate the implementation of the standards and rules issued pursuant thereto;

(g) Perform any other duties specified by this act or as directed by the Board; and

(h) Delegate to staff members such duties as appropriate to carry out the purpose of this act.

COMMENT. In performing both the plan review and permit functions, the Secretary is to provide technical advice and information to persons subject to the provisions of the act in order to facilitate their compliance. It is important in dealing with the construction industry, where costs can rise daily, that architects, engineers, and builders get prompt answers to questions they may have in the development of plans and specifications or in the construction of public facilities. Allowing the Secretary and staff to answer such technical questions should considerably speed up this process. In addition, it is unlikely that nontechnically oriented Board members will be able to decide specific questions of technical application of the rules.

If the Secretary has some difficulty in interpreting the meaning or intent of a rule, he or she could and should request an interpretation from the Board. In addition, any person receiving a determination which he or she considered incorrect could request, in most states with an administrative procedure act,[8] that the Board issue a declaratory ruling clarifying its standard or rule. As a check on the Secretary's interpretation of standards and rules, the Board always has the option of directing the Secretary to alter his administration of the rule, changing the rule or standard, or ultimately replacing the Secretary.

In addition to plan review and permit issuance the Secretary must inspect facilities to insure compliance with the act and investigate complaints alleging noncompliance. The Secretary also acts in a prosecutorial role to enforce the requirements of the act. As noted above, this allows for the separation of the prosecutorial function from the judicial function, which is allocated to the Board. This also has the advantage of consolidating investigative and prosecutorial functions in one decision-maker and avoids the possible inability of the diverse Board members to agree on a

---

[8] *See, e.g.*, IOWA CODE ANN. §17A.9 (1978).

comprehensive enforcement program. The Secretary is likely to be more assertive in carrying out an enforcement program and the procedure provided for in section 14 of the act is likely to be more efficient and effective.

Obivously the Secretary can not perform all functions specified by the act or directed by the Board. Accordingly, the Secretary is given authority to delegate his or her duties under the act to employees of the agency as appropriate to carry out those duties most efficiently and effectively.

Section 7. Authority to Inspect or Investigate

While conducting an inspection or investigation, the Secretary may enter at any reasonable time in and upon any private or public property, except private dwellings, to investigate any actual or possible violation of the provisions of this act or the rules, standards or permits adopted or issued under this act.

(1) If the owner or occupant of any property refuses admittance thereto, or if prior to such refusal the Secretary demonstrates the necessity for a warrant, the Secretary may make application under oath or affirmation to the ____________ court of the county in which the property is located for the issuance of a search warrant.

(2) In the application the Secretary shall state that an inspection of the premises is mandated by the laws of this state and a general administrative plan for enforcement of the law or that a search of certain premises, areas, or things designated in the application may result in evidence tending to reveal the existence of violations of this act or any standard, permit or rule issued pursuant to this act. The application shall describe the area, premises, or thing to be searched, give the date of the last inspection if known, declare the need for such inspection, recite that notice of desire to make an inspection has been given to affected persons and that admission was refused if that be the fact, and state that the inspection has no purpose other than to carry out the purpose of the statute or rule pursuant to which inspection is to be made. If an item of property is sought by the Secretary it shall be identified in the application.

(3) If the court is satisfied from the examination of the application, the applicant and other witnesses, if any, that there is probable cause to believe the existence of the grounds of the application, the court may issue such search warrant.

(4) In making inspections and searches pursuant to the authority of this section, the Secretary must execute the warrant in a reasonable manner within ten days after it is issued.

COMMENT. In order to carry out his or her regulatory functions under the act, including the inspection of completed facilities pursuant to subsection 10(4) and the investigation of complaints of alleged violations pursuant to paragraph 6(4)(b), the Secretary must have specific authority to conduct administrative inspections and investigations. Administrative inspections may be conducted where consent is given, where the areas to be inspected are open to the general public, or where a warrant has been issued in accordance with Fourth Amendment protections.

It is expected that in the vast majority of cases inspection will be authorized by the consent of the property owner, thereby precluding the necessity for a warrant.[9] In addition, no warrant will be necessary for simple view inspections of areas observable by or open to the general public.[10] What is observable by the public is observable without a warrant by a government inspector.[11]

However, the act covers areas which are not open to the public, especially places of employment, which require a warrant for inspection if consent is refused. Thus, legislative authority is needed for an administrative warrant in order to adequately implement the provisions of the act. The warrant procedure would be utilized only when necessary. Administrative expedience dictates that a warrant be sought only after consent has previously been denied or when evidence exists indicating the likelihood of denial. The model act mandates this approach, and requires, in accordance with recent Supreme Court decisions,[12] that an administrative search warrant be obtained by the Secretary before conducting an inspection over the objections of a property owner.

Warrants are obtainable for either of two purposes. First, a warrant may be issued pursuant to a general administrative inspection plan designed to further adequate enforcement of the model act or any other laws of the state.[13] Second, a warrant may be obtained on the basis of a reasonable belief that evidence may be obtained which will tend

---

[9] Traditionally a warrant is not required to conduct a search pursuant to an owner's consent. Consent to an administrative inspection may be evidenced by any manifestation of assent, no matter how casual, according to the majority of courts which have addressed the issue.

See United States v. Thriftimart, Inc., 429 F.2d 1006 (9th Cir. 1970), *cert. denied,* 400 U.S. 926 (1970), *reh. denied,* 400 U.S. 1002 (1971); United States v. Hammond Mining, 413 F.2d 608 (5th Cir. 1969).

[10] Marshall v. Barlow's, Inc., 436 U.S. 307, 315 (1978). *Cf.* Air Pollution Variance Bd. v. Western Alfalfa Corp., 416 U.S. 861 (1974).

[11] Marshall v. Barlow's, Inc., 436 U.S. 307, 315 (1978).

[12] Marshall v. Barlow's, Inc., 436 U.S. 307, 325 (1978); Camara v. Municipal Court, 387 U.S. 523 (1967); See v. City of Seattle, 387 U.S. 541 (1967).

[13] The constitutionality of administrative search warrants issued on the basis of the existence of "reasonable legislative or administrative standards" was established in Camara v. Municipal Court, 387 U.S. 523, 538 (1967).

to show a violation of the provisions of the act. These dual standards provide the Secretary the necessary authority both to develop a broad general enforcement strategy and to respond directly to individual enforcement problems as they arise. The warrant requirement itself will insure that a detached and disinterested individual reviews the reasonableness of any administrative inspection scheme before it is placed in operation.

The Secretary is required to execute a warrant within ten days after its issuance. While the type of violations that may arise under the provisions of the model act are ordinarily not of a type that will disappear over time, the ten day requirement insures that the Secretary will carry out the administrative plan properly and expeditiously. It prevents the storing of warrants for possible future use. In addition, the model act requires that searches be conducted in a reasonable manner. Ordinarily this will require that inspections be carried out during daylight hours and by means which are not overly intrusive.

Finally, the act exempts private dwellings from the inspection and investigation authority. This restriction is intended to preserve the long-standing distinction between a person's greater expectation of privacy in the home and lesser expectation in other environs.[14] The exemption for private dwellings responds to "the competing public and private interests" at stake which must be the touchstone for an analysis of the reasonableness of any inspection or investigatory scheme.[15]

Section 8. Environmental Design Standards; Applicability

(1) The environmental design standards adopted pursuant to subsection 5(4) shall be formulated and adopted so as to eliminate environmental barriers and encourage and facilitate maximum integration, independence and safety for environmentally limited persons within the built environment. Separate but equal accommodations shall be prescribed only when integrated accommodations are not possible or when separate accommodations are necessary to meet the special needs of certain environmentally limited persons.

In adopting standards the Board shall consider:

(a) The criteria contained in the "American Standard Specifications for Making Buildings Accessible to and Usable by the Physically Handicapped" and any amendments thereto as approved by the American Standards Association, Incorporated (ANSI Standards);

(b) Design manuals;

(c) Applicable federal guidelines;

(d) Current literature in the field;

(e) Applicable safety standards; and

(f) Any applicable environmental impact statements.

The standards shall be at least as stringent as the ANSI Standards and any applicable federal requirements.

COMMENT. The heart of the environmental barriers act is the adoption of environmental design standards, for it is these standards that will determine the degree of accessibility of individual facilities. Therefore, particular attention should be paid to the standards section of the act. To be enforceable, the standards must either be specified in the act or adopted as a part of the agency's official rules through the state's prescribed rule-making procedure. This presents several options:

1. Standards can be fully prescribed in the act. This approach provides a uniform set of standards but is extremely inflexible. Any required changes or new innovations will necessitate an amendment to the statute. This is a serious disadvantage.

2. The act can prescribe no standards, giving the administrative authority wide discretion to adopt standards. This approach has maximum flexibility and could result in more detailed, workable and effective standards which could be changed to respond to new needs. On the other hand, this approach lacks any uniformity and gives almost total administrative discretion without any legislative guidance. This could easily result in an ineffective act, since the effectiveness is totally dependent upon the promulgation of effective standards.

3. The act can grant administrative discretion to adopt standards while providing guiding principles and specifying standards, usually by reference, which must be considered or must serve as minimum standards. This approach provides at least minimum uniformity and a minimum level of accessibility in any standards adopted. At the same time it allows the adoption of new or revised standards, as needed.

This latter approach is adopted by the model act because it provides both specificity and flexibility. Any standard adopted must be specific enough to provide practical guidance to architects, engineers and contractors and to give adequate notice of what is required to those persons covered by the act so as to satisfy due process requirements. At the same time the standards must be flexible enough to meet the needs of persons with dif-

[14] *Cf.* Cady v. Dombrowski, 413 U.S. 433 (1973); Chambers v. Maroney, 399 U.S. 42 (1970).

[15] Camara v. Municipal Court, 387 U.S. 523, 539 (1967).

ferent types of disabilities[16] and to allow architects, engineers and contractors to utilize new and innovative approaches to accessibility without being hindered by narrow standards. In addition, the provision of guidelines specifying maximum integration, safety and independence and mandating integrated facilities should help insure that standards adopted are consistent with the goals of the act. This, along with the provision of minimum standards in the statute, gives strong legislative guidance to administrative discretion.

Many states have utilized the standards adopted by the American National Standards Institute (ANSI) in their statutes.[17] The ANSI standards were adopted in 1961. Since then they have been widely accepted as the best available, though they have been the subject of criticism.[18] The ANSI standards are cited in the act as minimum standards, as they were intended to be by the drafters,[19] i.e., any standards adopted by the Board must be at least as stringent in their effect. The ANSI standards are a good starting point, but are not as comprehensive or as aggressive as they need to be to assure accessibility and achieve the purposes of the model act. In addition, they are not specific enough as drafted for use in a mandatory, regulatory scheme of broad applicability. This is not to say that the ANSI standards fail to accomplish their purpose, but rather that they were not drafted, and cannot be properly incorporated by reference, as regulations under a specific law designed to establish a definite program for upgrading access to public facilities.[20] The agency must use its discretion in adopting specific rules.

In doing so, the agency should note that the ANSI standards are being completely revised. A Department of Housing and Urban Development (HUD)-sponsored research study has formed the basis for revised national accessibility standards. The study conducted by Syracuse University involved organizing findings from existing literature conceptually.[21] These concepts reflect the impact of various disabling conditions upon thirteen areas in which disability concerns must be reflected in design, e.g., entrances, exits and doorways, parking, work surfaces, and communications. Laboratory testing was then carried out on the capabilities of persons with walking, reaching and handling limitations and severe visual impairment. The results were translated into new proposed standards. The economic impact analysis of the proposed new standards was estimated and a cost-benefit analysis was done based on the assumed increased economic activity of disabled persons if the standards are applied. Based upon the results proposed new standards were presented to ANSI for approval since the final revision of the standards is the responsibility of an ANSI committee. The new standards are due to be released sometime in 1979 and are likely to be considerably better in addressing the needs of disabled persons. They also directly consider the economic impact of the standards. As such, they should be an excellent beginning point from which to draft the standards prescribed by this act. Subsequent drafters should cite specifically to the new standards in any new legislation proposed.

(2) The standards shall be known as the (name of state) Environmental Design Standards and shall apply to:

(a) Any new public facility or portion thereof the planning or construction of which is begun after the effective date of this act;

(b) Any new or any portion of any new publicly available housing development of ten (10) or more units the planning or construction of which is begun after the effective date of this act. If ten percent (10) of the housing complies with the standards, then for purposes of the act, the entire housing development shall be deemed to comply with the standards; and

(c) Any existing public facility alteration which is proposed after the effective date of this act, provided that:

---

[16] One author, commenting upon this, clearly illustrates the difficulty involved:

> Although all the disabled are helped by eliminating stairs, the crippled are helped far more than the deaf. Manholes, access panels and excavations are of greatest peril for the blind but are also hazardous for all. The deaf require visual signals which are of no use for the blind and vice versa for auditory signals. The paraplegic must have special toilet and washroom facilities and arrangements, while the blind couldn't care less where the mirror is located. For the persons in the wheelchair and the mobile cripple, a site is best developed which is level and without curbs and other abrupt changes.For the blind, large, level, open plazas and other areas around and among buildings, without discernible landmarks such as curbs and well-defined walks, can be traversed only by dead reckoning.

ten Broek, *The Right to Live in the World: The Disabled and the Law of Torts,* 54 CALIF. L. REV. 841, 861 (1966).

[17] See "Review of State Legislation, Design and Construction Standards" *supra* and Chart V *supra*.

[18] *See* Farber, *The Handicapped Plead for Entrance – Will Anyone Answer?*, 64 KY. L.J. 99, 108-11 (1975); *Hearings on the Effectiveness of the Architectural Barriers Act of 1968 (Pub. L. 90-480) Before the Subcommittee on Investigations and Review of the House Committee on Public Works and Transportation,* 94th Cong., 1st Sess. 12 (1975).

[19] *See* American National Standards Institute, Inc., AMERICAN NATIONAL STANDARD SPECIFICATIONS FOR MAKING BUILDINGS AND FACILITIES ACCESSIBLE TO, AND USABLE BY, THE PHYSICALLY HANDICAPPED, Foreward, 1961 (Reaffirmed 1971).

[20] *See* Farber, *supra* note 18, at 111.

[21] The researchers on the study won an award for their outstanding work. *See* PROGRESSIVE ARCHITECTURE 100 (January, 1979).

(i) If the alteration involves less than 50 percent (50) of the floor area of the public facility, only the area affected, and the area necessary to provide a continuous and unobstructed route of travel to and from the affected areas from and including the nearest entrance, shall be required to meet the standards, and

(ii) If the alteration involves fifty percent (50) or more of the floor area of the public facility, the entire public facility shall meet the standards.

COMMENT. The standards adopted by the Board are applicable to all new public facilities and publicly available housing and to some existing public facilities. This approach is intended to reflect the small cost of making new buildings accessible and the larger costs of making existing buildings accessible.

The cost of making new buildings accessible is minimal. Cost estimates by the National League of Cities based on seven hypothetical buildings show that the additional cost involved in making them barrier-free would be less than one-half of one percent.[22] Studies based on three buildings actually constructed indicated that the cost was increased by only one-tenth of one percent.[23] The requirement of plan review and permit issuance of section 10 seeks to insure that these costs are kept low by attempting to insure that accessibility is considered at the design stage. The cost of compliance by modifying a facility after construction and a finding of noncompliance will be often considerably higher than the figures cited.

On the other hand, the cost of modifying older structures may be very high. Therefore, the act does not require modification of these facilities, except where they are undergoing major renovation. Where this is true, estimates of the cost of accessibility range from .66 to 2.4 per cent of the cost of renovation.[24] At this cost, accessibility is also feasible for all major renovations. Thus, some alterations of facilities are required to meet the standards adopted by the Board.

To be covered by the act, the alteration must involve a structural change and the cost must exceed five percent of the market value of the facility.[25] Thus, mere redecorating or minor repairs are not covered. The degree of compliance with standards for those alterations covered is determined by the extent of the facility affected by the alteration as measured by floor area.[26] This recognizes the effect that the size of the alteration has on the cost of accessibility. The greater the alteration the lower the percentage of cost is for compliance with the standard. Thus, greater compliance with standards is mandated for the more extensive alterations.

Since barrier-free housing has traditionally been a problem for disabled persons, the act attempts to promote the accessibility of a portion of the single family or two-family detached or semidetached housing available for sale to the general public, in addition to promoting accessibility in public facilities, including apartment buildings. The act's requirements apply only to larger housing developments of ten units or more of which ten percent must be made accessible. This allows for economy of scale in designing accessible homes and for site selection where this is a factor in accessibility. It should produce a significant number of homes for use by disabled persons. Although ideally, each home should be barrier-free to promote maximum integration of environmentally limited persons, the cost appears too high to impose it on every new homeowner, and design factors and cost would preclude multiple-story homes in many instances.

Finally, the standards apply both to the planning (see subsection 3(11)) where applicable and construction (see subsection 3(12)) of public facilities and publicly available housing. Plans are to be submitted and a permit issued prior to construction when required by section 10 of the act. The standards are applicable to federal buildings and facilities only to the extent allowable under federal law. Since federal architectural barriers legislation[27] vests authority in a number of separate executive departments to set independent standards for facilities constructed by or for each department, a conflict may arise between the applicable federal design criteria and those promulgated by the Board under the model act. Where conflicts do arise, the constitutional doctrine of preemption may act to exempt the federal facility from compliance with state standards. Generally, preemption will be found whenever a substantial conflict exists between the state and federal laws,[28] though the most recent Supreme Court cases evidence a growing reluctance to find preemption without a clear congressional state-

---

[22] *See* National Commission on Architectural Barriers, DESIGN FOR ALL AMERICANS 7 (1967).

[23] *See id. See also* Achtenberg, *supra* note 2, at 856 n. 29; Note, *Access to Buildings and Equal Employment Opportunity for the Disabled: Survey of State Statutes,* 50 TEMP. L.Q. 1067, 1068 n. 12.

[24] Hearings, *supra* note 18, at 17.

[25] *See* MASS. GEN. LAWS ANN. ch. 22, §13A (West Supp. 1978).

[26] *See* MICH. COMP. LAWS ANN. §125.1352 (1976). The act uses a combination of cost and floor space. Other criteria are used in some state statutes. *See* Note, *supra* note 23, at 1071-72 n. 29.

[27] Architectural Barriers Act of 1968, 42 U.S.C. §§4151-56 (1976).

[28] *See* Burbank v. Lockheed Air Terminal Inc., 411 U.S. 624 (1973).

ment of intent.[29] The courts have never faced the issue in the precise context of barrier-free legislation. Even if preemption is not determinative, the doctrine of sovereign immunity may prevent effective enforcement of state standards against noncomplying federal entities.[30] Thus, unanswered questions exist as to the authority of states to adopt standards for federal facilities in the absence of a clear judicial statement. However, federal buildings would probably be found exempt from the design standards promulgated by the Board.

(3) The Board shall adopt standards within one (1) year of the effective date of this act, shall review them at least annually thereafter, and may amend or replace them with new standards at any time.

COMMENT. There will be no enforceable standards until the Board adopts them. Thus, the effectiveness of the act could be delayed indefinitely by attempts to delay the adoption of standards by rule. For this reason the act requires the Board to adopt some standards (based upon subsection 8(2)) within one year of the effective date of the act. They may, of course, be supplemented or amended at any time through the rule-making procedure. To insure that the standards are reviewed periodically for potential updating, the Board is required to review them at least once per year.

(4) No exceptions shall be made to the standards except in accordance with section 11.

COMMENT. Since the effectiveness of the best standards could easily be emaciated by indiscriminate granting of variances, exemptions or other exceptions to the rules, all exceptions are required to be granted through the procedures provided in section 11 of the act. That section contemplates the granting of exceptions for individual standards in individual cases. Blanket exceptions are not to be granted by variance or exemption or by rule since standards are intended to apply to all facilities covered by the act unless a specific exemption is granted. Note that this is different from permit requirements in section 10 of the act. The Board has discretion in determining which of the facilities covered by the act must receive a permit prior to construction.

(5) The Board shall publish the standards and shall make copies available to architects, engineers, contractors, state and local building inspectors and commissioners, state and local public works officials, chief executives of subdivisions, environmentally limited persons and other interested individuals or groups.

[29] *See* Exxon Corp. v. Governor of Maryland, 437 U.S. 117 (1978).

[30] *See* Massachusetts v. Hills, 437 F. Supp. 351 (D.C. Mass. 1977).

(6) All public newspaper solicitations for bids on projects for construction or alteration of public facilities or publicly available housing regulated by this act shall state that conformity to the (name of state) Environmental Design Standards shall be required.

All contract documents for public facilities or publicly available housing covered by this act and the rules of the Board shall provide that the construction or alteration covered by the contract shall conform to the requirements of this act.

COMMENT. The success of the agency's implementation efforts will be greatly enhanced by promoting the use of the standards among those people in particular who control the design and construction of most public facilities or who are most interested in accessible facilities. Thus, the Board is directed to insure the availability of the standards to these groups for their use. For the same reason all bid solicitations and contracts for public facilities should contain a notice of the need to conform to the act so that all parties involved are reminded of this and act accordingly.

Section 9. International Symbol for the Environmentally Disabled

(1) In facilities covered by this act, the symbol described in subsection 9(2) shall be posted at or near all entrances which meet environmental design standards and must be visible from the adjacent public sidewalk or way. This symbol shall also be displayed on or adjacent to the facilities' accommodations which meet environmental design standards. Where the total facility meets the environmental design standards, the symbol shall also be included in all public advertisements of it.

(2) The symbol for the environmentally limited shall be white on a blue background and shall look as follows:

COMMENT. Since not all public facilities will be accessible at any time in the near future, it is important that those facilities and accommodations within them which are accessible be readily identifiable. This requirement is not a part of existing ANSI standards and it is uncertain whether it will be included in new standards. Yet, because of its importance to environmentally limited persons, the act specifically requires that notice be given by displaying the generally recognized accessibility symbol.

Section 10. Written Permits Required

(1) Without first obtaining a written permit as required by this section and the rules of the Board, it shall be unlawful for any person to undertake the construction of any new public facility or publicly available housing or the alteration of any existing public facility to which environmental standards apply pursuant to section 8 and the rules of the Board.

COMMENT. Successful implementation of the act will depend upon the proper application of the standards to individual public facilities. While enforcement of penalty provisions might be used to promote compliance, this approach does nothing to insure adequate incorporation of the standards during the crucial design stage. It is strictly after the fact.

A large number of states have provided for the submission and review of plans and the issuance of construction permits.[31] Such administrative implementation provides the means for continuous review and surveillance by experts qualified to make determinations as to the adequacy of plans, recommendations for improvements[32] and inspections of built facilities for conformance with approved plans. This preconstruction review, along with post construction inspection, should help to insure that standards are effectively utilized.

This approach is adopted in the model act while at the same time allowing for a flexible application of permit requirements. Specifically, although all public facilities and publicly available housing are covered by the standards adopted by the Board, they are only covered by permit requirements to the extent required by the Board in its rules. Thus, the Board can exempt some types of facilities such as sidewalks, curbs and parking lots which do not logically need to be reviewed by the agency prior to construction even though the standards apply to them. It also allows the Board to exempt certain facilities, for example those under a specified size, where the agency cannot hope to handle the volume of submissions were all facilities covered. This means that facilities which have a greater impact on accessibility can receive greater attention. The number and type of facilities included can be altered rather easily by rule to respond to needed changes. It should also help maintain the program's credibility in two ways. It limits the time and expense involved in plan development and review to those facilities which really merit this investment and it insures that response time is kept to a minimum because the volume of plans to be reviewed is kept manageable by concentrating on more important facilities.

But to emphasize again, public facilities covered by the act are still covered by the standards. They may merely be excluded from plan review and permit requirements. Where they are covered by permit requirements, it is unlawful to proceed with construction or alteration without the required permit. To do so subjects the responsible person to the enforcement provisions of sections 14 and 16 of the act.

(2) (a) Plans and specifications for all construction or alteration for which a permit is required shall be filed with the Secretary by the responsible design architect or engineer whose seal shall appear on each page. The architect or engineer shall be registered to practice such profession in this state. However, such plans and specifications may be prepared and filed by responsible persons other then registered architects or engineers for:

(i) The construction or alteration of public facilities or publicly available housing which have a volume of thirty-thousand (30,000) cubic feet or less,

(ii) The construction of additions to existing public facilities, which have a volume of thirty-thousand (30,000) cubic feet or less, or

(iii) Alterations to existing public facilities which do not involve changes affecting the structural safety thereof.

COMMENT. In order to allow the Secretary to meaningfully review and approve plans and specifications, they must be thoroughly, accurately and competently developed to show the technical application of standards to the proposed facility. Accordingly, the act requires that the plans be prepared by a professional architect or engineer. This avoids the repeated submission of inadequate plans by untrained persons which cannot, because of their inadequacy, be reviewed and approved by the Secretary. This is a rather frequent requirement in existing statutes.[33]

The model act does allow an exception for facilities under a certain size. This recognizes that for some small projects the cost of such professional services may simply not be justified. Although the act presents one approach to exemption, a state may wish to adopt a different basis, e.g., square feet. However, in some states general provisions in other statutes may override the specific exemptions suggested here and specifically require that all plans received by a state agency be submitted under professional seal.[34] Thus, specific existing statutory provisions should be examined by subsequent drafters. But whatever approach is

---

[31] See Chart VII *supra*.

[32] Note, *The Forgotten Minority: The Physically Disabled and Improving Their Physical Environment*, 48 CHI.-KENT L. REV. 215, 227 (1971).

[33] These statutes generally require that plans be submitted under professional seal. *See, e.g.*, IOWA CODE §§114.16, .26 (1979).

[34] *Id.*

ultimately chosen, it is important to note that submission of plans by other than an architect or engineer does not exempt the submittor from other requirements of this section.

(b) Such plans and specifications shall be accompanied by a properly executed certificate on an application form made available by the Board attesting that said plans and specifications were prepared by the submitting architect, engineer or other person or under his or her immediate supervision, that they were prepared in accordance with the applicable standards and other rules promulgated by the Board and that the construction or alteration shall be reasonably and periodically inspected by a named responsible person. The inspection shall determine generally that the construction or alteration is proceeding in accordance with the plans and specifications as approved by the Secretary. Said certificates shall also provide an estimate of the cost of construction, the total square footage area, and cubic feet volume, and such other information as may be deemed necessary by the Board.

COMMENT. The act seeks accountability for plan preparation by requiring the preparer of plans to certify his or her responsibility for preparation and his or her use of applicable standards. The act also seeks accountability for construction by requiring the naming of a responsible person to inspect the construction periodically and insure it is proceeding in accordance with approved plans. This gives the Secretary a focal point for inquiry and responsibility.

Submitted plans must include information necessary to adequately review them. At a minimum, easily obtainable information such as cost, square footage, and volume is necessary to administer this section and section 8.

(c) The Secretary may require the owner of an existing public facility to file with the Secretary complete plans and specifications of the whole or any part of such facility and any other information concerning its construction or alteration which is necessary to administer this act.

COMMENT. In some instances plans of existing facilities may be necessary. For example, where a public facility has been constructed without plans, where required, or not in accordance with approved plans, it may be necessary to determine what does exist in order to determine how to properly alter the facility to comply. Thus, plans of the existing facility may be necessary to properly evaluate proposed changes. For this reason the Secretary may require submission of such plans.

(3) Once an application has been filed, the Secretary shall promptly notify the applicant in writing of his or her action and, if the permit is denied, state the reasons for denial. The applicant may appeal to the Board the denial of a permit or the imposition of any condition on the permit if he or she files a request for review with the Secretary within thirty (30) days after notice of denial is given or the permit is issued. The Secretary shall notify the applicant within thirty (30) days of the appeal request of the time and place for the review hearing. A decision of the Board shall be a final decision for purposes of section 15.

COMMENT. A successful review and permit program must insure prompt response time. Failure to do so will result in construction delays and increased costs for builders which they are not likely to tolerate gracefully. Accordingly, the Secretary is required to respond promptly to applications.

Fundamental fairness inherent in due process requires that applicants be informed of the reasons for denial where this occurs and an opportunity to challenge the denial of a permit or a condition imposed in a permit. Thus, applicants are required to receive notice and an opportunity for hearing and appeal. The review hearing should generally conform to the hearing on appeal of an order pursuant to subsection 14(3), since it is an administrative appeal of an agency action.

(4) Upon completion of the permitted construction or alteration, the Secretary shall inspect the work done and issue a certificate of compliance, if the public facility or publicly available housing meets the requirements of this act and the standards, rules and permit conditions and any variance or exemption applicable to it. The owner or lessee in possession of a facility or housing which does not comply shall be subject to sections 14 and 16.

COMMENT. If plan review and approval is to be effective, it must result in a constructed facility which complies with approved plans. To insure this, the Secretary must inspect and approve the construction or alteration of each permitted facility. Failure to comply will subject the responsible person to the enforcement provisions of the act.

(5) If it is necessary or desirable to make material changes in approved plans or specifications, a new application including the revised plans or specifications together with reasons for the proposed changes must be submitted to the Secretary or state agency or subdivision for a supplemental written permit. A decision on the application shall be reached in accordance with the requirements of subsection 10(3).

COMMENT. If plan review is to be meaningful, all facilities must be built in accordance with approved plans. A change in plans which might affect the application of standards must naturally be reviewed for continued compliance. To allow such changes would negate the value of both the review of plans and the post-construction inspection based upon the approved plans. Thus, no such changes are allowed without a subsequent permit.

(6) Upon the adoption of standards and permit rules by the Board, the Board may delegate plan review and permit authority to state agencies and subdivisions pursuant to section 12. Authority may be delegated if the agency or subdivision employs a qualified registered architect or engineer whose qualifications are approved by the Board. The qualified architect or engineer shall review plans and specifications using standards known as the "(name of state) Environmental Design Standards" adopted by the Board.

After delegation of authority is made, all plans and specifications within its jurisdiction shall be submitted to the authorized state agency or subdivision which shall issue permits for construction or alteration and certificates of compliance, if the submitted plans and specifications and the resulting public facility or publicly available housing comply with the standards and rules adopted by the Board or a variance or exemption has been granted pursuant to section 11. After issuing a permit, the agency or subdivision shall forward a copy to the Secretary. If, after inspection, the agency or subdivision cannot issue the certificate of compliance, it shall notify the Secretary, who shall proceed in accordance with sections 14 and 16. The Secretary may request additional information, if necessary.

COMMENT. The model act seeks to utilize existing resources and avoid duplication and multiple reviews where feasible. (See Comment to subsection 12(1).) However, authority to administer a program for elimination of barriers is always to be retained in the Board. Thus, while division of labor is acceptable, division of authority is not. (See "Administrative Organization of the Environmental Design Board" *supra*.) Delegation of plan review and permit authority is allowed only if the applicant shows the availability of employees capable of conducting the reviews. All reviews must be conducted using the Board approved Environmental Design Standards to insure uniform accessibility and use throughout the state and avoid a patchwork of standards which might otherwise be applied. Note that although permit authority may be delegated, this subsection specifically acknowledges that exemption authority cannot. Only the Board can grant an exemption pursuant to section 11.

After delegation the Secretary is to receive copies of all permits and any additional information required. This will facilitate the oversight of delegated activity and help to insure proper implementation of standards. (See subsection 12(4).) In addition, where a certificate of compliance cannot be issued, the Secretary will seek compliance and enforce the act pursuant to sections 14 and 16 since this authority also cannot be delegated to another state agency or subdivision. (See subsection 12(1).)

Section 11. Variance and Exemption

(1) If a person undertaking the construction or alteration of any public facility determines that full compliance with the environmental design standards is impossible or impracticable, such person may apply to the Board for a variance or exemption from a particular standard. The application shall be accompanied by reasons for the request and a proposal setting forth the maximum extent to which it is believed practical to comply with the Environmental Design Standards. The application shall include all relevant data required by the Board that may be helpful in evaluating the request for variance or exemption, including the cost of the total project, the cost of construction and accommodations necessary to comply with the Environmental Design Standards and one copy of the building plans for the applicable areas of the construction or alteration. The architect, engineer or builder in charge of planning the project shall certify as valid all data submitted and shall certify as reasonable and proper the proposal for variance or exemption of the Environmental Design Standards.

COMMENT. Regardless of how carefully the standards are drawn, there will always be some instances in which they cannot feasibly be applied or in which their application would cause extreme hardship. Thus, some mechanism is needed to grant exceptions to the application of standards. For these reasons the model act provides a specific procedure. However, the discretion to grant exceptions is limited by subsection 11(4).[35]

(2) The Secretary shall investigate the application, determine the estimated building costs and the incremental costs of construction to conform to the Environmental Design Standards and recommend to the Board the disposition of the application.

COMMENT. The Secretary is to undertake a technical review of the application, including costs and feasibility of alternative means of compliance, and make a recommendation to the Board for disposition. This may include recommended conditions suggested in subsection 11(5). The final decision on any variance or exemption is, of course, up to the discretion of the Board.

(3) A public hearing shall be held, if the Board concludes that a hearing is advisable, or if requested by an interested person.

COMMENT. Because the granting or denial of a variance or exemption can have such an adverse impact upon both the users and the owners of a facility, interested persons are given the opportu-

[35] *See* Farber, *supra* note 18, at 105-08. *See also* Note, *supra* note 23, at 1076.

nity to comment before the Board on the pending application upon request. Such a public hearing is in the nature of a rule-making hearing, rather than an appeal, and is held for the purpose of gathering information and comments on the application.

(4) Upon the basis of the findings of the investigation and other relevant evidence, the Board may grant in whole or in part a variance or exemption from a particular standard for that applicant if:

(a) The purpose of this act can be fulfilled by an acceptable alternative to the particular standard which will provide equivalent access and use; or

(b) The incremental construction or alteration costs of conforming to the particular standard exceed ____________ percent of the total construction or alteration costs.

COMMENT. The indiscriminate granting of variances and exemptions can emaciate the effectiveness of standards.[36] Thus, it is important that the bases for granting them be specified, and that they be consistent with the goals of the act and with economic realities.

The act provides for the granting of exceptions to a standard in two circumstances. An exception may be granted if an acceptable alternative is shown to the standard which provides roughly equivalent access or if the cost of meeting a particular standard is shown to exceed a specified percentage of the total construction or alteration costs. This assumes that the cost of any accommodation which exceeds this percentage is per se a disproportionate and unreasonable expense as to that particular standard. Although no figure is inserted in the act, drafters should consider using five percent, since each exception must only refer to a specific standard and must be considered separately.

(5) In granting a variance or an exemption, the Board may require the applicant to effect a partial elimination of environmental barriers within a reasonable period of time, may require a schedule of compliance, may prescribe interim measures to minimize environmental barriers and may prescribe other requirements with which such applicant shall comply.

The Board may grant a variance or exemption for a specified period of time, provided that no variance or exemption shall exceed five (5) years. The Board may further specify that the applicant make periodic reports specifying the progress that has been made toward compliance with any standard for which the variance or exemption was granted. A variance or exemption may be extended by affirmative action of the Board.

[36] *Id.*

COMMENT. While an applicant may not fully comply with a standard, at least partial compliance is frequently feasible. In other instances compliance may be achieved over time or barriers to current compliance removed at some anticipated date in the future. In such instances the Board may set certain conditions in the granting of exceptions to insure that standards are met to the maximum extent possible. Finally, since circumstances change over time the Board may not grant an indefinite exception.

(6) The Board shall promptly notify the applicant, in writing, of its action and, if the application is denied, state the reasons for denial. The applicant may request a review hearing if the application is denied in whole or in part by filing a request for review with the Secretary within thirty (30) days of the denial. The Secretary shall notify the applicant within thirty (30) days of the time and place of the review hearing. A decision by the Board after the hearing shall be a final decision for purposes of section 15.

COMMENT. Due process is granted where an application is denied in whole or in part by providing notice of the reasons for denial and an opportunity for hearing and appeal to challenge the denial. The review hearing should generally conform to the hearing on appeal of an order pursuant to subsection 14(3), since it is an administrative appeal of an agency action.

(7) Any person aggrieved by a decision of the Board may seek judicial review in accordance with section 15.

COMMENT. None necessary.

Section 12. Delegation of Authority to Review Plans and Issue Permits

(1) Any agency of the state or its subdivisions may apply for delegation of authority to review plans and issue permits within its jurisdiction. Such authority may be undertaken by a single agency or subdivision or jointly by two or more agencies or subdivisions. In reviewing plans and issuing permits pursuant to a granting of this authority, an agency or subdivision may adopt rules and standards to implement the purposes of this act within their respective authority or jurisdictions.

COMMENT. This section recognizes that some state agencies such as the Department of Transportation, the State Board of Regents or Department of Education, or the General Services Administration conduct their own plan review for facilities within their jurisdiction which may also be covered by the provisions of the act and that certain cities or counties conduct building permit programs similar to that mandated by the act. This section attempts to utilize existing programs of this nature to avoid duplication while insuring the requirements of the act are carried out under the standards and procedures adopted by the Board as the agency for implementation of barrier-free requirements. The section invites state agencies and local governments to apply for authority to conduct plan review and permit issuance and allows these agencies or subdivisions of the state to act either alone or jointly in carrying out these responsibilities. The act also specifically grants authority to the agencies and subdivisions to adopt appropriate rules and standards for carrying out the delegated authority. Note, however, that variance, exemption and enforcement authority cannot be so delegated (see subsection 11), since the delegation of these responsibilities to different administrative units would create a real danger of unequal and inadequate application and enforcement of the standards.

(2) In evaluating an application for a delegation of authority, consideration shall be given to whether the program provides for the following:

(a) Ordinances, or rules establishing the (name of state) Environmental Design Standards as those to be utilized in the program along with permit requirements and other procedures consistent with, or stricter than, those imposed by this act and the rules and standards adopted by the Board; and

(b) Administrative organization, staff, financial and other resources necessary to administer an efficient and effective program.

COMMENT. It is intended by the act that all public facilities within the state covered by its provisions be subjected to a single set of standards. This avoids the confusion which will likely occur between architects, engineers, and builders where multiple sets of standards are to be implemented. It also insures that the degree of accessibility within the state is not dependent upon the jurisdiction in which one resides or finds oneself. Therefore, any jurisdiction receiving delegation of authority must utilize the Environmental Design Standards adopted by the Board. In addition, it must show that it has the administrative organization and resources which will be necessary to properly administer the program. These criteria are required to be met prior to approval of the delegation of authority.

(3) Any agency or subdivision desiring a certificate of delegation shall apply on forms prescribed by the Board. The Secretary shall promptly investigate the application and recommend the disposition of such application to the Board. The Board may conduct a public hearing before action is taken on the recommendation.

Upon acceptance of a program, the Board shall issue a certificate of delegation to the appropriate agency or subdivision for the entire program or a portion thereof. If the recommendation is against issuing a certificate, or for issuing a certificate for only a portion of the requested program, the agency or subdivision shall be entitled to a review hearing. The burden of proof shall be upon the agency or subdivision raising the appeal. Any decision by the Board after review shall be a final decision for purposes of section 15.

COMMENT. The Secretary is directed to investigate factual matters to determine whether the criteria for delegation in subsection 12(2) have been met. Thus, the Secretary would inquire as to whether the Environmental Design Standards have been adopted by the applicant, and what resources the applicant proposes to allocate to the administration of the program. The Board then considers these factual matters in making a determination as to whether to grant or deny the application for certification of delegation. The Board is authorized to seek public comment at a public hearing should it be deemed appropriate.

Since some agencies or subdivisions may be interested in or only capable of performing a part of the program mandated under the act, the Board is authorized to delegate such portions of a program as may be appropriate to the situation and retain the remaining portion under the direct authority of the Board. For example, the Board may wish to delegate authority for all buildings under a certain specified number of cubic feet or square feet, retaining jurisdiction for larger buildings, or delegate authority only for certain types or classes of public facilities or publicly available housing.

Where the Board denies the application for delegation, or grants a certificate for only part of the requested program, due process is satisfied through the provision for a review hearing on the decision. Since such a denial would be an agency action under most state administrative procedure acts, it would also be subject to judicial review pursuant to those acts.

(4) The Secretary shall annually review and evaluate the work done by those holding certificates of delegation and report results to the Board. If the Board determines at any time that an agency or subdivision's program is being conducted in a manner inconsistent with the substantive provisions of this act or the rules or Environmental Design Standards adopted thereunder, the Board shall notify the agency or subdivision, citing the deviations from the acceptable standards and rules and specifying the corrective measures that must be completed within a reasonable amount of time in order to be in compliance. If the corrective measures are not implemented as prescribed, the Board shall suspend or revoke the certificate of delegation for the entire program or portion thereof. The responsibilities that were suspended or revoked shall be assumed by the Secretary until the certificate of delegation for those responsibilities is reissued pursuant to subsection 12(3). In cases where a certificate of acceptance or portion

thereof is proposed to be suspended or revoked the agency or subdivision is entitled to a hearing prior to suspension or revocation. Any decision by the Board after review shall be a final decision for purposes of section 15.

COMMENT. The granting of delegation authority should not end the Board's responsibility to insure proper implementation of the act through continuous oversight. It is needed to insure that delegation is being properly carried out. Accordingly, the Secretary is to perform investigations and evaluations at least annually to oversee the work of the agencies and subdivisions and report the findings to the Board for their action. The Board is authorized to suspend or revoke the delegation authority or a portion thereof where the agency or subdivision is not performing properly. In accordance with the requirements of most administrative procedure acts the subsection provides for a hearing prior to suspension or revocation of what amounts to a license. Such acts would also be subject to judicial review in accordance with a state's administrative procedure act.

(5) The Board or Secretary shall give advice or technical assistance to agencies or subdivisions upon request or as appropriate.

COMMENT. None necessary.

Section 13. Coordination of Transportation Programs

(1) Whenever an agency or subdivision of the state intends to develop or implement a federal grant program under the Urban Mass Transportation Act of 1964 (49 U.S.C. §§1601-1613) or Section 127 of the Federal-Aid Highway Act of 1970 (23 U.S.C. §142), it shall, at the earliest practicable time, consult with and obtain the views of the Board on matters concerning the provision of accessible public transportation for handicapped or environmentally limited persons.

(2) The state agency or subdivision shall give special deference to the views of the Board on matters within the Board's area of expertise.

COMMENT. Transportation planning and development is largely preempted by the federal government because of its jurisdiction over interstate commerce and because of the fact that states and their subdivisions have not had the necessary financial resources. Thus, when planning and development is carried out by a state or a subdivision it is done in accordance with federal requirements and using federal funds. This process provides a major access point for the input of needs of environmentally limited persons, especially with respect to mass transit projects, since a state or local agency is involved and since federal requirements mandate that special efforts be made to accommodate environmentally limited persons when designing transportation systems.

Specifically, federal funds for planning and implementing transportation programs are available under two federal acts. The Urban Mass Transportation Act of 1964[37] provides federal financial assistance for the planning, development, construction and operation of public transportation systems with local public bodies being eligible to receive up to 80 percent of construction, acquisition and planning expenses and 50 percent of operating costs for transit systems;[38] including the costs of fixed facilities,[39] buses,[40] rapid rail systems,[41] feasibility studies,[42] technical studies,[43] developmental studies[44] and demonstration projects.[45] Section 16 of the act[46] requires as a precondition to receiving financial aid, that "special efforts" be made in the planning and design of mass transportation facilities and services to assure the availability of mass transportation to the elderly and handicapped.

Funds are also available under section 165 of the Federal-Aid Highway Act of 1970[47] for the construction of preferential bus lanes, bus passenger loading areas and facilities, and fixed-rail rapid transit systems and equipment. In addition, local officials may apply highway funds to the purchase of buses and other public transit vehicles should the needs of local transportation planning indicate the preferability of a mass-transit alternative to highway construction. 1974 amendments to the section added a note incorporating the same requirement of "special efforts" set out above and requiring further that no program or project be approved for federal assistance which does not comply with the "special efforts" mandate.[48]

Regulations promulgated under the authority of both laws provide a sketchy protrait of the mean-

---

[37] 49 U.S.C. §§1601-13 (1976), as amended by the Federal Public Transportation Act of 1978, Pub. L. No. 95-599, §§301-13, 92 Stat. 2735.

[38] *Id.* §1604(e)(1976).

[39] *Id.* §1604(d)(1).

[40] *Id.*

[41] *Id.*

[42] *Id.* §1607.

[43] *Id.*

[44] *Id.* §1605.

[45] *Id.*

[46] *Id.* §1612.

[47] 23 U.S.C. §142 (1976), as amended by the Federal-Aid Highway Act of 1976, Pub. L. No. 94-280, §127, 90 Stat. 440.

[48] Federal-Aid Highway Amendments of 1974, Pub. L. No. 93-643, §105(a), 88 Stat. 2282.

ing of "special efforts,"[49] although some examples are provided: (1) any system which assures each handicapped person a minimum of 10 round-trips per week on public conveyances; (2) a program for the acquisition of mass transit vehicles until 50 percent of the entire transit fleet is accessible to the handicapped; or (3) the expenditure of 5 percent of the local unit's total federal apportionment for a program designed specifically to improve wheelchair users' and semi-ambulatory persons' access to public transportation.[50] In addition, the regulations that will benefit environmentally limited persons contain specific minimum design criteria for fixed transportation facilities (i.e., stations, platforms and terminals),[51] buses,[52] rapid rail cars[53] and light rail vehicles.[54]

Since a state agency or a subdivision of the state is responsible for making these special efforts, this section requires the agency or subdivision to consult with the Board during the planning process because of the Board's special expertise. The Board must be provided a timely opportunity to present its views concerning matters of importance to the handicapped community and the transportation authority must give special deference to the Board's views. "Special deference" as used here will ordinarily mean that the Board's suggestions should be considered unless the transportation authority can show good cause for overriding them. The model act contemplates a continuing consultation process with significant give and take in order to assure a coordinated approach to transportation planning from the earliest possible time. Consequently, the Board should be notified of an authority's intention to develop and implement a grant program.

The reasons for providing the Board only an advisory role are several. First, the federal grant programs mentioned, particularly those under the Urban Mass Transportation Act, require long range, comprehensive programs and planning.[55] Provisions in the transportation system to meet the needs of handicapped persons are to be fully coordinated and integrated into the larger mass transportation scheme. It is imperative, therefore, that the agency with general expertise in transportation planning retain its jurisdiction over all transportation matters in order for the state or local government to effectively participate in these federal programs. Second, if the Board were in more than an advisory role the model act would be applicable in full to transportation facilities and rolling stock, and a preemption issue would exist since fairly specific national design criteria are provided by regulation for most transportation equipment.[56] In addition, such uniform national design standards should assure the availability of accessible equipment at the earliest possible time and at the most economical price. Varying standards from state to state would make the manufacture of accessible equipment a losing proposition and might well cause interminable delays. Thus, it is in the best interests of all concerned to support the federal government lead in transportation equipment design.

Some transportation facilities will, however, be covered by the model act since the act is applicable to buildings and facilities held out for use by the public for purposes of transportation.[57] Most importantly, the act will cover *commercial* transportation facilities (e.g., bus depots, train stations, airports) which are not included in federal programs. Some borderline cases will undoubtedly arise, but these will be best dealt with on a case-by-case basis as federal design regulations are further articulated and expanded. As a general matter fixed commercial transportation facilities are covered by the model act while fixed public transportation facilities are not. The model act contains no provisions for rolling stock of any kind which with few exceptions will either be covered by the federal laws discussed above or, if travelling in interstate commerce, may be subject to other federal requirements.

Enforcement Provisions

One of the most frequent criticisms of existing state statutes, and the federal statute as well, is

---

49 The "special efforts" requirement has been narrowly construed by the courts, at least prior to the issuance of these implementing regulations. *See* Atlantis Community, Inc. v. Adams, 453 F. Supp. 825 (D. Colo. 1978) (Buses need not be equipped with hydraulic lifts); Vanho v. Finley, 440 F. Supp. 656 (N.D. Ohio 1977) (Separate paratransport system is sufficient — regular bus system need not be accessible); United Handicapped Federation v. Andre, 409 F. Supp. 1297 (D. Minn. 1976) (Provision of 10 small buses to transport wheelchair users is sufficient "special effort" for the Minneapolis-St. Paul area); Snowden v. Birmingham — Jefferson County Transit Authority, 407 F. Supp. 394 (N.D. Ala. 1975) (Vehicles accessible to wheelchair users need not be provided). A number of the newer decisions indicate that the recently promulgated regulations may require a more substantial commitment to accessible design. *See* United Handicapped Federation v. Andre, 558 F.2d 413 (8th Cir. 1978) (Vacating the lower court judgment cited *supra* on the basis of the new regulations); Michigan Paralysed Veterans of America v. Coleman, 451 F. Supp. 7 (E.D. Mich. 1977). In addition, as access technology improves and accessible vehicles become commercially available, the "special efforts" mandate may require their acquisition. *See* United Handicapped Federation v. Andre, 558 F.2d 413 (8th Cir. 1978); Snowden v. Birmingham-Jefferson County Transit Authority, 407 F. Supp. 394 (N.D. Ala. 1975).

50 49 C.F.R. §613.204 appendix (1978).

51 *Id.* §609.13.

52 *Id.* §609.15.

53 *Id.* §609.17.

54 *Id.* §609.19.

55 49 U.S.C. §1604(1) (1976).

56 *See* §2(1)(c).

57 See subsection 3(2) of the model act *supra*.

the inadequacy of the enforcement provisions.[58] The National Commission on Architectural Barriers to Rehabilitation of the Handicapped in its recommendations concluded that state legislation was needed which should include not only the ANSI standards but strong enforcement provisions. "Intentions, policies, and criteria are of little value if they are not supported by the force, funding and people necessary to implement them."[59] Even with maximum education the changes needed to achieve accessibility will not come about by voluntary efforts alone.

The model act attempts to remedy this problem by creating strong and effective enforcement mechanisms where the plan review and permit process has failed to implement the requirements of the act. As in the *Prohibiting Discrimination* model statute in this volume, the model act proposes administrative control of enforcement. This approach is generally believed to provide more effective enforcement with less expenditure of time and money than would litigation and to more readily achieve the legislation's objective of widespread accessibility and use of public facilities.

The administrative approach has a number of advantages which should contribute to effective enforcement. First, the agency develops an expertise in administering and enforcing the statute. Second, it is in a position to provide continuous surveillance by qualified experts of compliance with the act, as well as act upon complaints lodged by individuals or organizations concerning violations. Third, administrative enforcement mechanisms can stress informal, faster and cheaper resolution of alleged violations using conciliation and mediation techniques coupled with broad educational programs to avoid the need for litigation. Fourth, where informal meetings or resolutions fail, the agency may be authorized to issue flexible cease and desist orders or other flexible remedies which are designed to insure achievement of legislative purposes. These orders may be tailored to specific fact situations. Thus, the flexible cease and desist orders are considerably more effective in achieving the objectives of the act than proceeding directly to litigation.

These administrative mechanisms may be supplemented by a private right of action which can provide an excellent means of assuring that interested parties are vindicated in their pursuit of rights granted under the act. It also tends to provide a means of keeping the agency vigilant in its task.

Section 14. Violations; Notice to Offenders; Orders; Appeals; Hearings

(1) Whenever the Board or the Secretary has evidence that any person has violated or is violating any provision of this act or any rule or standard established or permit issued pursuant thereto, the Secretary shall notify the alleged violator and by informal negotiation attempt to resolve the problem. Agreements reached by this means shall be entered into as consent orders which shall prescribe the actions necessary to achieve compliance and may prescribe timetables for such action. The Secretary may seek enforcement of the consent order in accordance with section 16, if necessary.

COMMENT. The model act encourages use of informal negotiations where evidence of violations of the act are found. Resolutions of violations are required to be entered as consent orders between the Board and the alleged violator so that each of these parties understands their obligations and so that subsequent enforcement won't have to begin anew if the informal agreement is not carried out by the violator. The Secretary is given specific authority to enforce the consent order in accordance with section 16 after subsequent violations of the order by the violator.

(2) If such negotiations fail to resolve the problem within a reasonable period of time, the Secretary may issue an order directing the person to cease and desist from any action which constitutes a violation and to take such corrective action as is necessary to meet the requirements of this act and any rule or standard established or permit issued pursuant thereto.

COMMENT. Administrative orders are an effective means of promoting compliance with a minimum expenditure of time and resources especially where the Secretary is allowed to fashion them. Such orders can be tailored to the specific situations involved and contain both cease and desist provisions and affirmative action provisions necessary to carry out the purposes of the act. This allows for prompt response by the Secretary in those instances where informal negotiations fail to resolve violations of the act, and is considerably more efficient than immediate judicial action. In addition, since due process is still satisfied by allowing for subsequent appeal (see subsection 14(3)), this procedure eliminates the need for an administrative hearing in each situation and should achieve resolution of the problem without further resort to administrative mechanisms in a significant percentage of the cases.

Note also that this is different than the procedure proposed in the Prohibiting Discrimination model act. There the administrative procedure calls for hearings followed by issuance of an order should the violation be shown. However, that act deals with acts of discrimination which are more difficult to establish since their occurrence is based on intangible human interaction which must be proved after the fact. In the extant situation the

[58] See "The Adequacy of Existing Statutory Law" *supra*.

[59] Note, *supra* note 32, at 225.

occurrence of violations is shown by the existence of tangible evidence, *i.e.,* the structure of public facilities covered by the act. Thus, violations of this act are considerably easier to detect and prove. Therefore, there is less need for formal fact-finding mechanisms prior to the issuance of the order.

(3) Any person aggrieved by an order of the Secretary may appeal by filing a written notice with the Board within thirty (30) days of the issuance of the order. The Board shall adopt rules relating to the specificity of allegations made on appeal such that the Secretary can be adequately apprised of all issues raised.

COMMENT. Due process requires the provision of an opportunity to appeal the order of the Secretary, including an opportunity to challenge the finding of a violation, alleging it did not occur or challenging the Secretary's interpretation of the standards or rules adopted by the Board. The Board is directed to adopt rules to insure adequate notice to the Secretary of the issues being raised on appeal.

In order to insure that the appeals are carried out promptly, the appeal must be filed within 30 days of the issuance of the order. During the appeal the effectiveness of the order is stayed. Where irreparable harm would take place during the time required for administrative appeal, the Secretary has the authority under subsection 16(4) to take immediate judicial action.

(4) (a) The Secretary shall schedule a hearing within thirty (30) days of the filing of the notice of appeal. The hearing may be held before the Board or its designee. The hearing shall be conducted in accordance with sections __________ of (cite to state's administrative procedure act or other appropriate sections).

(b) The Board or its designee may issue a subpoena requiring the presence of witnesses or the production of books, papers or materials upon the request of the Secretary or appellant.

(c) The Secretary may appear personally or by a staff attorney or other counsel.

(d) A complete record of the hearing shall be made. A copy shall be furnished to appellant at his or her request and expense.

(e) On appeal the Board may affirm, modify or vacate the order of the Secretary. The Board shall issue its findings within forty-five (45) days of the conclusion of the hearing, which shall constitute a final decision of the Board. Judicial review of findings may be sought pursuant to section 15.

COMMENT. The administrative hearings are to be conducted in accordance with the procedures contained in the state's administrative procedure act.[60] To insure the hearing is held promptly, the act requires that the hearing be scheduled within 30 days of the filing of the notice of appeal. While the Secretary acts as a prosecutor, the Board acts as the trier of fact and law which is consistent with the Board's role as the ultimate authority in administering the act, since it allows the Board to interpret its standards and rules. The Board, however, may delegate its hearing authority to a hearing officer for the sake of administrative efficiency.

As with other administrative hearing provisions the Board or a hearing officer on its behalf is given the authority to issue subpoenas and require the production of appropriate evidence. Subpoena power insures that all parties have an equal opportunity to produce the evidence required to support their appeal.[61]

The Secretary has the option of appearing at an administrative hearing personally or by staff attorney or through use of other counsel, such as the attorney general of the state. Under normal practice it is anticipated that the staff attorney would represent the Secretary at such hearings.

To insure that there is a prompt decision following the hearing, the Board is required to issue a final decision within 45 days of its conclusion. Any decision by the Board is subject to judicial review in accordance with section 15 of the act.

Section 15. Judicial Review

Any person aggrieved by a final decision of the Board or of the Secretary may seek judicial review pursuant to section (cite to state's administrative procedure act). The district court review proceedings shall conform to section (cite to state's administrative procedure act) on the scope of judicial review of agency decisions. The petition for judicial review must be filed within thirty (30) days of the issuance of a final decision by the Board or Secretary.

COMMENT. Any person who is aggrieved by a final decision of the Board or Secretary may seek judicial review of the decision in accordance with the procedures specified in the state's administrative procedure act or its equivalent. This may include the Secretary (in the case of decisions by the Board), applicants for permits, applicants for variances or exemptions, or alleged violators against whom administrative orders have been issued. The procedures for review and the scope of review are intended to conform to standard administrative procedure act provisions. Thus, the review is intended to be limited to questions of law

[60] In adopting rules relating to hearings subsequent drafters may wish to consult the procedures outlined in sections 21 through 23 of the Prohibiting Discrimination model act, Chapter 2, *supra.*

[61] *See* section 22 of the Prohibiting Discrimination model act, Chapter 2, *supra.*

or issues of lawfulness or reasonableness rather than questions of fact. To insure timely exercise of this right, the petition for review must be filed within 30 days of the issuance of the final decision.

Section 16. Enforcement; Penalties; Burden of Proof; Private Actions

(1) Failure to obey any order issued by the Secretary with reference to a violation of this act or any rule promulgated or permit issued pursuant thereto shall constitute prima facie evidence of contempt. In such event the Secretary may certify to the __________ court of the county in which such alleged disobedience occurred the fact of such failure. The __________ court, after notice, as prescribed by the court, to the parties in interest, shall then proceed to hear the matter. At that hearing, any previous findings of fact after notice and hearing on an appeal shall be deemed conclusive, if supported by substantial evidence in the record when the record is viewed as a whole. If the court finds that the order was lawful and reasonable it shall order the party to comply with it. If the person fails to comply with the court order, that person shall be guilty of contempt and shall be subject to a fine not to exceed five hundred dollars ($500) for each day that he or she fails to comply with the court order. The penalties provided in this section shall be considered as additional to any penalty which may be imposed under this section or any other provision of state law.

COMMENT. Where the Secretary has issued an administrative order against an alleged violator, the violator has received notice and opportunity to appeal. In cases where the violator chooses to ignore the order, or in other ways fails to comply, the Secretary is authorized to seek a contempt citation in the state's trial court. The court is limited in its review of factual matters where the order has been appealed and the order upheld by the Board. In this situation any findings of fact are to be deemed conclusive, provided they are supported by substantial evidence in the record. The court is directed to review the order for its lawfulness and for its reasonableness based on these findings of fact. If the order is upheld, the court shall require the party to comply with it. In addition to violating the terms of the order, the person may also violate provisions of the act, rules adopted or permits issued pursuant thereto. Such person may be subjected to penalties for violation of these provisions as well.

(2) Any person who violates any provision of this act or any permit, rule or standard issued under this act shall be subject to a civil penalty not to exceed five hundred dollars ($500) for each day of such violation.

COMMENT. Whether or not an order has been issued by the Secretary, a person may violate the terms of the act or any permit, standard or rule issued under the act. The subsection provides for a civil penalty for each day of such violation. These violations are separate from and may be in addition to the contempt citation provided for in subsection 16(1).

(3) Any person who knowingly makes any false statement, representation or certification in any application, plan or other document filed or permit, variance or exemption issued under this act or who falsifies any information required by this act or by any permit, rule, or order issued under this act shall upon conviction be punished by a fine of not more than one thousand dollars ($1,000) or by imprisonment in the county jail for not more than six months or by both such fine and imprisonment.

COMMENT. Many of the administrative implementation activities of the Board and the Secretary rely upon the submission of adequate and factual information relating to proposed plans and technical specifications to be utilized in constructing public facilities. Because of the importance of adequate information to the administration of the act, a separate penalty is provided for persons who knowingly make false statements or representations in their submission of such information. Violation requires that this false information be submitted knowingly. Thus, the standard of proof goes considerably beyond mere negligence before a person could be convicted of such a violation. The terms of the subsection apply also to Board or staff members of the agency who knowingly grant a permit, variance or exemption where one is not justified under the law.

(4) The Secretary may institute at any time, either individually or at the request of the Board, any legal proceedings, including an action for an injunction or a temporary injunction, necessary to enforce the penalty provisions of this act or to obtain compliance with the provisions of this act or any rules promulgated or any provision of any permit issued under this act. In any such action, any previous findings of fact by the Board after notice and hearing shall be conclusive, if supported by substantial evidence in the record when the record is viewed as a whole.

COMMENT. The Secretary is given specific authority to enforce requirements of the act, both through injunctive relief and the penalty provisions. This would normally occur after the use of administrative orders. However, in those situations where irreparable harm may be done unless immediate action is taken, the Secretary has the authority to seek temporary injunctions without the use of administrative orders to prevent such harm until the matter may be litigated.

In any such enforcement proceedings it is not intended that matters litigated at the administrative level be relitigated at the judicial level. Thus, where previous findings of fact have been made, the court is bound to act upon them, provided that they are supported by substantial evidence in the record.

(5) The Secretary or any other interested party may apply to the ______________ court for a stay of a judicial order pending the outcome of an appeal. After a hearing the stay may be granted and may be conditioned upon the furnishing by the appellant of such reasonable security as the court may direct.

COMMENT. Where the enforcement of an order would cause irreparable harm prior to the final determination of the issues on appeal, the affected party may seek a stay order pending final determination. In such cases, the court may require such reasonable security as it deems necessary to insure compliance following a final determination.

(6) In any proceedings under this act the Board or the Secretary may be represented by the attorney general or by counsel employed or designated for such purpose (notwithstanding section ______________ of the code).

COMMENT. It is intended that the Board and Secretary be represented by its own counsel, who has developed an expertise with respect to the administration of the act. In states where this is precluded by existing statutory law, an exemption to such a statute is recommended. This would not preclude utilizing the assistance of the Attorney General where appropriate.

(7) In all proceedings with respect to any alleged violation of the provisions of this part or any rule established or permit issued under this act, the burden of proof shall be upon the Board or the Secretary except in any action for contempt as provided in subsection 16(1) or as otherwise provided in subsection 16(4).

COMMENT. As in all regulatory activities, the regulator naturally has the burden of proof in showing a violation of the law. However, where matters of fact have been previously litigated at the administrative level, this rule is intended to be reversed. In these situations the offending party must show that the previously issued order is unreasonable or in some manner unlawful to overcome the presumption raised by the previous determination.

(8) If the Secretary has instituted legal proceedings in accordance with this section, all related issues which could otherwise be raised by the alleged violator in a proceeding for judicial review under section 15 shall be raised in the legal proceedings instituted in accordance with this section.

COMMENT. Where the Secretary has instituted legal proceedings to enforce the provision of the act during the time when a person may seek judicial review of an agency action, it is possible that two separate actions, both involving the same or similar issues may be filed. To avoid the duplication of trying issues in two different cases, this section provides that the issues on judicial review are to be consolidated with the enforcement action brought by the Secretary.

(9) The owner, owner's agent or lessee in possession of a public facility which fails to meet the requirements of this act or any rule or standard adopted pursuant thereto may be required to alter or reconstruct the same by a court in a civil suit based upon such requirements. Such suit may be brought by any interested person without exhausting any administrative remedy which may be available. Such person shall be reimbursed, if successful, for all costs and such actual attorney fees as may be allowed by the court. Successful plaintiffs in such suits may also recover any actual and punitive damages for injuries resulting from failure to meet such requirements.

COMMENT. As exists in the Wisconsin statute,[62] this subsection allows private actions to supplement the enforcement actions of the Secretary. It is intended that such suits be separate from any proceedings which might be brought under this act and accordingly the subsection provides that the action may be brought without exhausting administrative enforcement provisions. Such rights of action not only keep the regulated persons on their toes, but the agency personnel as well.

Section 17. Taxation of Existing Facilities Altered to Comply

(1) Any taxpayer who alters an existing public facility which he or she owns to meet the Environmental Design Standards adopted pursuant to this act; and the owner of a residence in which the owner resides and to which the owner has made alterations to eliminate environmental barriers:

Alternative 1 — Shall not be subject to an increase in the assessed valuation of the facility based on that portion of the alteration necessary to meet Environmental Design Standards.

Alternative 2 — Is entitled to an exemption from property tax upon the facility to be determined by deducting the value of improvements necessary to meet Environmental Design Standards from the assessed value of the real estate.

Alternative 3 — May deduct from taxable income (the cost of the alterations made to eliminate environmental barriers/a credit of ______________). Such (deduction/credit) shall be

[62] WIS. STAT. ANN. §101.13(6) (West 1974).

taken in the tax year in which eligible expenses were paid.

(2) No taxpayer is eligible for an exemption, deduction or credit under this section for any public facility, the alteration or use of which is subject to the permit requirements of section 10 of this act or any rule adopted pursuant to this act unless the proper permit(s) were obtained.

COMMENT. Taxation provisions are frequently used at both the federal and state level to promote various public policies. It is strongly suggested that the state's taxing authority be utilized to promote barrier-free public facilities and public housing.[63] These provisions should provide an incentive for barrier-free construction or alteration by partially offsetting the cost involved.

This section presents a number of alternatives which are by no means all-inclusive. Tax incentives may involve property taxes, income taxes, or both. Each state should study its own tax provisions to determine which would be most appropriate for its situation.

It is strongly suggested that these provisions be utilized to promote compliance with the permit requirements of the act by providing that tax provisions are not applicable to those who do not receive the proper permits prior to alteration. However, the tax incentives are not intended to be limited to those covered by permit requirements or even those covered by standards. Thus, those individuals who are interested in making a facility barrier-free, even though it is not required under the act, should also be allowed to take advantage of the tax incentives provided.

Section 18. Appropriation

The general assembly is authorized to appropriate ____________ dollars for the immediate fiscal year to cover the initial expenses incurred in implementing this act.

COMMENT. None necessary.

Section 19. Severability Clause

If any provision of this act or the application thereof to any person or circumstance is held invalid, such invalidity shall not affect any other provision or application of the act which can be given effect without the invalid provision or application, and to this end the provisions of the act are declared severable.

COMMENT. None necessary.

Section 20. Effective Date of the Act

This act shall take effect ____________________.

COMMENT. None necessary.

[63] Some tax deductions or exemptions already exist for the removal of barriers. For existing state statutes see Chart X *supra*. And see 26 U.S.C. §190 (1976) for federal provisions.

4
ZONING FOR COMMUNITY HOMES

Table of Contents

INTRODUCTION

Normalization

Important progress has been achieved in recent years toward providing humane habilitative programming for developmentally disabled persons. This goal requires that these individuals not be warehoused in large, traditional institutions but rather be afforded the opportunity to live in community-based residential settings. This approach is called "normalization" — the principle of providing the "patterns of life and conditions of everyday living which are as close as possible to the regular circumstances and ways of life of society."[1] According to the principle, disabled persons, if unable to live with their families, should reside in homes of normal size, located in normal neighborhoods that provide opportunities for normal societal integration and interaction.[2] Such community living permits developmentally disabled persons to reach their human potential and to become contributing, productive members of society. It also enables them to participate in generic services, to receive training for employment, to be employed, and in many cases to become part of the taxpaying public rather than an enormous strain on the public treasury.[3]

The progress of recent years towards "normalization" has been the result of efforts on many fronts.[4] Developmental disability experts have increasingly advocated community-based care and habilitation. Equally important steps have been achieved through: (1) judicial decrees requiring that placement in less restrictive settings than institutionalization be considered prior to commitment;[5] (2) recent Federal legislation calling for increased community habilitation of developmentally disabled persons;[6] and (3) state legislative enactments providing for humane care for these persons.[7]

Notwithstanding the weight of professional, judicial, and legislative authority favoring community-based treatment, insufficient numbers of community homes[8] are currently available to serve developmentally disabled persons who are not in need of institutionalization. Although this lack of residential facilities is attributable to many causes, a very significant factor has been local zoning regulations which effectively exclude or restrict community homes from residential areas.

Zoning in General

Zoning is a type of land use control deriving from public legislative bodies and is most often implemented at the local level. It is the most important and prevalent land use control in the United States and serves the function of regulating land uses by separating commercial and industrial districts from residential districts of the community. The establishment of industry or places of business is normally prohibited in districts designated as residential.[9]

Types of Ordinances Restricting Community Homes

The activities of local communities in the last few years demonstrate beyond question that many, for whatever reasons, will do whatever they can by means of exclusionary zoning laws and practices to frustrate efforts to establish commu-

---

[1] B. Nirje, "The Normalization Principle," in *Changing Patterns in Residential Services for the Mentally Retarded* 231 (R. Kugel & A. Shearer eds. 1976).

[2] *Id.* at 232.

[3] *See generally* M. Kindred, Written Testimony submitted to Ohio General Assembly in support of Substitute Senate Bill 71 (June 15, 1977), and J. Chandler and S. Ross, Jr., "Zoning Restrictions and the Right to Live in the Community," in *The Mentally Retarded Citizen and the Law* 313 (M. Kindred ed. 1976).

[4] *See generally* ABA Commission on the Mentally Disabled, "Community-Based Mental Health Treatment: Impact of Zoning Development;" *Clearinghouse Rev.* 356 (Aug. 1977).

[5] *E.g., Wyatt* v. *Stickney,* 344 F.Supp. 373 (M.D. Ala. 1972), *aff'd sub nom. Wyatt* v. *Aderholt,* 503 F.2d 1305 (5th Cir. 1974); *Suzuki* v. *Quisenberry,* 411 F.Supp. 1113 (D. Hawaii 1976).

[6] *E.g.,* Developmentally Disabled Assistance and Bill of Rights Act, 42 U.S.C. §§6001-6081 (Supp. V 1975).

[7] *E.g.,* COLO. REV. STAT. §§27-10.5-101 to 123 (Supp. 1976); FLA. STAT. ANN. §393.13 (West Supp. 1978); NEB. REV. STAT. §§83-1, 141 (Reissue 1976); OHIO REV. CODE ANN. §§5123.67 to .99 (Page Supp. 1977).

[8] A variety of terms, such as "group homes," "foster homes," and "community residences" have been used to describe community-based residential facilities serving developmentally disabled persons. For purposes of this commentary, the generic term "community homes" has been used. The Act uses the term "family home" to define a particular type of community home covered by the Act. *See* the Model Act, Section 3, Definitions, *infra* at

[9] N. Williams, 1 *American Land Planning Law,* ch. 16, at 327-43 (1974) [hereinafter cited as N. Williams].

nity homes.[10] Some localities specifically exclude such homes from single-family residential districts which are the neighborhoods most desirable for community homes.[11] More frequently a political subdivision excludes these homes through its definition of "family." The most restrictive type of definition is similar to the one found in *Village of Belle Terre* v. *Boraas*[12] in which the village defined "family" by limiting the persons authorized to live on the premises to those who were related by blood, marriage, or adoption, with no more than a specified number of unrelated persons.[13] These local devices usually result in a political subdivision having no community homes.[14]

Elsewhere, political subdivisions designate community homes as a "conditionally permitted use"[15] in some residential areas, or as a business use or as a boarding house thus limiting them to commercial zones,[16] or, finally, as a use allowed only in areas where hospitals and nursing homes are permitted.[17] These types of restrictions frequently result in the creation of ghettos of community homes, particularly in larger center cities. Such a concentration occurs because the only districts open to community homes are transitional and politically weak residential neighborhoods or business or institutional zones. This ghettoizing leads to the creation of a new form of institutionalization — large numbers of community homes in certain areas of a city so that the homes become the dominant feature of a residential neighborhood.[18] These concentrations change the character of neighborhoods and undercut the very purposes behind "normalization." Further, they provoke justified, negative reactions on the part of neighborhoods where the community homes are impacted, and strengthen the resolve of other communities to avoid admitting community homes for fear that such concentrations will occur in their communities.[19]

Local Discretion and Exclusionary Zoning: The Remedy — State Legislation

As long as each political subdivision retains broad discretion to admit or exclude community homes, there is little incentive for admitting them. On the contrary, there is strong incentive for political subdivisions to exclude or restrict such homes because, under present circumstances, if one community acts in a progressive, constructive manner to permit community homes, there is a high probability that it will become a magnet for large numbers of homes. This occurs because operators have nowhere else to establish much needed residences. Once this rush to one city occurs and officials in other political subdivisions perceive the phenomenon, their conviction to exclude community homes is reinforced.

One thing then has become abundantly clear: local decision-making on the location of community homes allows for and potentially encourages exclusionary and undesirable results. If the community homes concept is to succeed, it must do so across an entire state as a result of state legislation that is not subject to the veto of political sub-

---

[10] Planning bodies in several political subdivisions have been sufficiently concerned about property values to conduct research on the question of the effect of community facilities on property values in residential neighborhoods. In none of these studies were the researchers able to verify any significant impact of such facilities upon property values. *See generally The Social Impact of Group Homes: A Study of Small Residential Service Programs in First Residential Areas,* prepared for the Green Bay, Wis. Plan Commission, prepared by Eric Knowles and Ronald K. Baba (June 1973); *The Influence of Halfway Houses and Foster Care Facilities Upon Property Values,* or City of Lansing, Mich., Planning Dept. (Oct. 1976); *The Effect of Siting Group Homes on the Surrounding Environs,* (White Plains, N.Y.) by Stuart Breslow (Nov. 1976); Michael Dear, "Impact of Mental Health Facilities on Property Values," 13(2) *Community Mental Health Journal* 150 (1977).

[11] R. Hopperton, *Zoning for Community Homes: A Handbook for Local Legislative Change,* 3 (Ohio State University Law Reform Project 1975).

[12] 416 U.S. 1 (1974).

[13] *See supra* note 11. Comment, "Exclusionary Zoning and its Effect on Group Homes in Areas Zoned for Single Family Dwellings," 24 *Kan. L. Rev.* 677, 683 (1976).

[14] In some areas apartment buildings and duplex homes are increasingly being utilized in the placement of developmentally disabled persons. One reason for this development is that the use of multiple family dwelling units by small numbers of developmentally disabled persons living independently can many times avoid zoning obstacles that face community homes located in free-standing single-family dwelling units because local definitions of "family" permit only up to three, four, or five unrelated persons to live together. A second reason may be that apartments and duplexes are a dominant form of housing in a given area and therefore provide a ready source for community living opportunities regardless of local zoning laws. No value judgment is offered on the question of the use of free-standing single-family homes versus apartments or duplexes. It should be said, nevertheless, that (1) a "family home" is defined for purposes of this Act as requiring "supervision in a family environment . . .," and (2) apartments are usually used for developmentally disabled persons capable of independent living without supervision in a family environment. In this latter case, as already indicated, zoning restrictions are a significantly lesser problem, and for this reason such apartment and duplex living arrangements are not treated in this Act.

[15] A "conditionally permitted use" is a land use authorized in a particular zoning district only if certain requirements or standards are met and only after approval is given by the local zoning appeals board or other public body. The terms "conditional use," "special use," "special use permit," and "special exception" are other terms often used by political subdivisions to designate a "conditionally permitted use." "Conditionally permitted uses" are sometimes confused with "variances" which are granted normally only to relieve a particular hardship arising from application of a zoning ordinance. Finally, a "permitted use" is to be contrasted with a "conditionally permitted use." The former is a use by right specifically authorized in a particular zoning district. *See* M. Meshenberg, *The Language of Zoning* (American Society of Planning Officials, Planning Advisory Service Report No. 322, Nov. 1976).

[16] *See* J. A. Chandler and S. Ross, Jr., "Zoning Restriction and the Right to Live in the Community," in *The Mentally Retarded Citizen and the Law* 313 (M. Kindred ed. 1976).

[17] *Id.* at 314.

[18] *See* "Crackdown Planned on Homes for Mentally Ill," *Dayton Journal Herald.* (Dayton, Ohio) April 13, 1978, at 9.

[19] *Id.*

divisions.[20] Uniform state requirements, superseding parochial, local legislative and administrative actions will open up desirable neighborhoods for placement of community homes on an equitable basis and will remove principal reasons (*e.g.*, overconcentration of such homes in one area) for excluding such homes.[21]

Significantly, state legislation limiting local zoning authority is not unprecedented. In fact, sixteen states have already enacted such statutes.[22] Each of these states recognized that zoning restrictions which exclude or restrict community homes are a state problem, not a local one, that must be addressed at the state level.[23]

REVIEW OF RECENT STATE LEGISLATION

As indicated in the previous section, state legislation limiting local discretion to exclude community homes is not novel. Sixteen states have enacted statutes mandating in various ways and to various degrees the location of community homes in residential areas. These recent state enactments exhibit several common characteristics: each statute identifies, and most define, the type of community home treated; each designates the type of population served and all but one specify the number of residents permitted in community homes; each identifies the type of zone/s in which they will be permitted; most require state licensing of the homes; most indicate whether local zoning authorities can impose any additional conditions not specified in the state law;[24] and approximately half list dispersal requirements. (*See* Chart.)

General Approaches to Limiting Local Discretion

Rhode Island's brief zoning statute designates the residents of community homes (all must have six or fewer) as a "family" and waives all local zoning requirements pertaining to them. Similarly, Tennessee establishes community homes (for eight or fewer) as single family residences for zoning purposes.[25] Thus, in both of these states

---

[20] The Ohio experience prior to passage of SB 71, legislation of the sort contemplated by this Act, is instructive. Major efforts aimed at breaking down zoning barriers facing community homes were undertaken by various reform groups at the local level in 1975 and 1976. The Ohio Association for Retarded Citizens formed a special zoning subcommittee of its Legal and Governmental Affairs Committee. The Ohio Developmental Disabilities Planning and Advisory Council and the Law Reform Project at the Ohio State University, College of Law, jointly undertook the development, publication, and distribution of zoning handbooks aimed at local legislative change. *See* R. Hopperton, *Zoning for Community Homes: A Handbook for Local Legislative Change*, and *Zoning for Community Homes: A Handbook for Municipal Officials*, (Ohio State University Law Reform Project 1976). In addition, various *ad hoc* zoning committees were established in communities around the state. At the time these efforts were undertaken, approximately six Ohio political subdivisions permitted community homes for developmentally disabled persons in some manner (usually conditionally). After approximately one year of intensive efforts, only five more communities had reformed their zoning regulations to permit community homes (again, usually on a conditional basis). Thus, notwithstanding a major investment of time and resources only *eleven* of Ohio's over 930 municipal corporations specifically allowed community homes for developmentally disabled persons. In numerous communities in the state, reform efforts were met by flat rejection or interminable delays. One community tightened its zoning ordinance to close a possible loophole that could have permitted a community home. Increasingly, the response given by cities for not reforming their local zoning regulations was, "If we allow one community home, we will be deluged by others." It became obvious to concerned groups and individuals in Ohio that no amount of time or effort would result in any significant number of Ohio's political subdivisions voluntarily admitting community homes. As a consequence, Ohio reform organizations led by the Ohio Association for Retarded Citizens and the Law Reform Project concluded that state-wide legislation constituted the only answer to local zoning barriers.

Substitute Senate Bill 71 was drafted, introduced, and passed overwhelmingly by the Ohio General Assembly (OHIO REV. CODE ANN. §5123.18 (Page Supp. 1977), effective October 31, 1977). For discussion *see* "Review of Recent State Legislation."

[21] *See* Chart on *State Zoning Laws Regulating Community Facilities for Developmentally Disabled Persons* [hereinafter cited as *Chart*], *infra.* at

[22] *Id.*

[23] Additionally, state courts have increasingly provided more active and informed judicial review of local legislation restricting or excluding community homes. These cases, which strike down exclusionary regulations and practices, are well summarized in the recent concurring opinion of Mr. Justice Stevens in *Moore* v. *East Cleveland,* 431 U.S. 494 (1977).

Mr. Justice Stevens states

> [I]n well-reasoned opinions, the courts of Illinois, New York, New Jersey, California, Connecticut, Wisconsin and other jurisdictions have permitted unrelated persons to occupy single family residences notwithstanding an ordinance prohibiting, either expressly or implicitly, such occupancy.
>
> The state decisions have upheld zoning ordinances which regulated the identity, as opposed to the number, of persons who may comprise a household only to the extent that the ordinances require such households to remain non-transient, single-housekeeping units. [Footnotes omitted.] *Id.* at 516-17, 519.

Both Mr. Justice Stevens and Mr. Justice Powell, writing for the court, indicate the legitimacy of zoning regulations which are directed specifically at typical zoning problems such as adequate parking and overcrowding of physical structures. Both indicate, however, that those problems can be addressed specifically and not directed toward restricting certain persons from living in a single-housekeeping unit. Mr. Justice Stevens also reviews state judicial opinions in which group living arrangements and community homes have not received favorable treatment. *See also* Comment, *supra* note 13, for an in-depth review of state case law regarding community homes.

[24] *E.g.*, architectural design and site layout.

[25] As will be seen in the Model Act, the approaches taken by Rhode Island and Tennessee are not recommended since licensure of such homes is needed to insure appropriate levels of care for residents. Moreover, the Rhode Island approach creates unjustifiable reverse discrimination by exempting community homes from zoning requirements (*e.g.*, height, parking restrictions) that apply to all other dwellings in the zone. The State of Maryland appears to have attempted to treat community homes very much like Rhode Island. The Maryland provision is contradictory, however, ("Although the public group home is exempt from any local zoning rule or regulation, public group homes may not be located in any area where specifically prohibited by the local zoning law.") and therefore the effect of this provision has yet to be determined.

smaller community homes are treated as "permitted uses"[26] which means that the operator of a home can locate it as a matter of right, without having to meet any special local conditions or standards. Thus local discretion regarding the zoning of these homes is removed.

Like the above two states, Arizona, Minnesota, Ohio and Vermont also establish the smaller community homes (in Arizona, Minnesota and Vermont six or fewer; in Ohio eight or fewer) as "permitted uses" in all residential zones. In addition, these four states require state licensing of community homes. Therefore, these statutory provisions create a most favorable setting for the establishment of smaller community homes because local discretion regarding these homes is removed and state licensing is required. Larger state-licensed community homes in Arizona, Minnesota and Ohio, however, are treated as "conditionally permitted uses" and may be required to meet certain requirements or standards imposed by a political subdivision. The Wisconsin approach is similar to Arizona, Minnesota, Ohio and Vermont except that community homes in that state are subjected to annual review by a political subdivision.

The statutes of California,[27] Colorado, Michigan, and Montana[28] declare that smaller state-licensed community homes for handicapped persons are residential uses for zoning purposes in all residential zones. Political subdivisions, however, may treat these state-licensed homes as "conditionally permitted uses" and require them to meet special requirements or standards. Thus, smaller state-licensed community homes in these states receive essentially the same treatment as the larger state-licensed community homes in Arizona, Minnesota and Ohio. This means that significant local control is retained by a political subdivision.

The statute of South Carolina declares that community homes are to be construed as natural families for local zoning purposes, but then establishes a very unusual procedure that gives political subdivisions the right to object to the location of a community home. The state agency may then appeal the political subdivision's objection to the State Budget and Control Board. If that Board does not decide the state agency appeal within a specified time, however, the objection of the political subdivision controls and the establishment of the community home is prevented.

In Virginia, a very general statute also allows significant local discretion, but does require that political subdivisions provide community homes in "appropriate residential zoning districts." The Virginia statute, however, does not define "appropriate residential zoning districts," but it does include a strong policy statement encouraging and promoting community homes. Another general statute has been passed in New Jersey prohibiting political subdivisions from discriminating between children of normal single-family homes and children living in community homes.[29] Unfortunately discrimination is not defined, thus leaving its full strength undetermined. Finally, New Mexico's statute, though similar to those of California, Colorado, Michigan, and Montana, differs in one very important respect — New Mexico does not require political subdivisions to treat community homes as residential uses. Instead, such treatment is left to the discretion of the local community.

---

[26] *See supra* note 15.

[27] The validity of the California statute (CALIF. WELF. AND INST. CODE §§5115-5116) was upheld in *City of Los Angeles* v. *California Department of Health,* No. 116571 (Cal. Super. Ct. October 24, 1975), 1 *Mental Disability L. Rep.* 26.

> In its complaint, filed March 5, 1975, the city [of Los Angeles] sought, inter alia, an injunction prohibiting the State Department of Health and other defendants from licensing community care facilities in areas where such facilities would violate the Los Angeles Planning and Zoning Code.
>
> Los Angeles maintained that as a chartered city, it was not subject to State regulation or control with regard to municipal affairs, and that its regulation of land-use through zoning requirements constituted "municipal affairs." Thus the city argued that it was entitled to exclude community care facilities from single family and certain other residential zones. In other words, the city contended that it was exempt from the California Welfare and Institutions Code (§§ 5115 and 5116), which requires that community care facilities serving six or fewer persons on a 24-hour non-medical basis be permitted in all residential zones, including single family areas.
>
> The California Department of Health argued in its answer, filed April 17, 1975, that the sections of the State code pertaining to community care facilities "embody a subject of statewide concern and said provisions must prevail over local ordinances which might otherwise be deemed appropriate municipal action."
>
> In granting the defendants' motion for summary judgment, the court declared that it was not necessary to decide the matter on constitutional grounds, since under the California code, "a state authorized, certified, or licensed family or group home serving six or fewer mentally disordered or otherwise handicapped persons [must] be considered a residential use of property" and such a use must be permitted in all residential zones in the city of Los Angeles. The court thus declared invalid and void those provisions of the Los Angeles Planning and Zoning Code prohibiting family and group homes in residential areas.

1 *Mental Disability L. Rep.* 26-27 (1976). *See infra.* note 34.

[28] The Supreme Court of Montana upheld that state's statute (MONT. REV. CODES ANN. §§11-2702.1-.2 (Supp. 1977)) that exempts community homes for developmentally disabled persons serving eight or fewer residents from local exclusionary zoning requirements. *State ex. rel. Thelen* v. *City of Missoula,* No. 13192 (Mont. Sup. Ct., Dec. 8, 1975), 543 P.2d 173, 1 *Mental Disability L. Rep.* 27 (1976).

[29] The validity of, the New Jersey Statute (N.J. STAT. ANN. §30:4C-26 and §40:5533.2 (West Supp. 1977)) was sustained in *Berger* v. *State of New Jersey,* 71 N.J. 206, 364 A.2d 993 (1976), 1 *Mental Disability L. Rep.* 214 (1976).

Characteristics of Residents

Fifteen states limit the types of residents who may live in community homes, with seven states (Arizona, Colorado, Montana, New Mexico, Ohio, Vermont, and Virginia) specifically including developmentally disabled persons. In addition, the following states define developmental disabilities: the Colorado statute defines developmentally disabled persons as those persons having cerebral palsy, multiple sclerosis, mental retardation, autism, or epilepsy; according to the New Mexico law, a developmental disability is a disability attributable to mental retardation, cerebral palsy, autism, or neurological dysfunction, that requires treatment or habilitation similar to that provided to persons with mental retardation. Ohio's legislation covers disabilities that originate before the person becomes 18 years old and can be expected to continue indefinitely, consitute a substantial handicap to the person's ability to function normally in society, or are attributable to mental retardation.

Persons having other kinds of disabilities are covered in four of these seven states' statutes, as well as in the nine remaining states. The categories in the various states include mentally disabled, mentally retarded, mentally ill, handicapped, physically handicapped, children and persons in need of supervision and care. (Obviously mental retardation is a type of developmental disability, but the states of Maryland, Minnesota, Rhode Island and Tennessee use that term apparently to exclude other types of developmental disabilities.) Unlike other states, Wisconsin does not limit the types of residents who may live in community homes.

Licensing and Dispersal Requirements

State licensing or other recognition of community homes is mandatory in thirteen of the sixteen statutes. The specific licensing authority is indicated in the chart whenever referred to by statute. Provisions in seven states prohibit the issuance of a license for operating a community home if the proposed location would result in an excessive concentration of such homes in an area. Also, six states specifically require that a certain distance be maintained between community homes in order to ensure dispersal of such homes within a political subdivision. These provisions attempt to ensure the effectiveness of the de-institutionalization process by avoiding the creation of a so-called community home ghetto or district.

STATE STATUTE CHART

State Zoning Laws Regulating Community Facilities For Developmentally Disabled Persons

STATE	TYPE OF COMMUNITY FACILITY	RESIDENTS		ZONE IN WHICH PERMITTED	CONDITIONAL USE PERMITS ALLOWED	STATE LICENSING OF FACILITY		DISPERSAL OF FACILITIES REQUIRED
		No.	Type			Required	Licensor	
ARIZONA Act of June 7, 1978, Ch. 198 §37 (H.B. 2426) (Effective Dec. 1, 1978)	Residential facility	6 or fewer 7 or more	Developmentally disabled persons	Single family Multiple family[1]	No Yes[2]	Yes	Dept. of Economic Security	1200 ft. required between residential facilities
CALIFORNIA Welf. & Inst. Code §§5115-5116 (Deering Supp. 1977)	Family care home Foster home Group home	6 or fewer	Mentally disordered or otherwise handicapped persons, or dependent and neglected children	Single family	Yes	Yes	Not specified	Not specified
COLORADO Rev. Stat. §§27-10.5-133 (Supp. 1976), 30-28-115 (Supp. 1976)	Group home	8	Developmentally disabled persons	Single family	Yes	Yes	Dept. of Health	750 ft. required between facilities
MARYLAND No. 1343, amending Art. 59A, Ann. Code (1972 Replacement Vol. and 1977 Supp.)	Public group home Private group home	4-8	Mentally retarded persons	[3] All Residential zones	No[4]	Yes Yes[5]	Sec. of Health and Mental Hygiene Director of Mental Retardation Administration	Not specified
MICHIGAN Acts of Jan. 3, 1977 Pub. L. Nos. 394-396	Residential facility	6 or fewer	Persons in need of supervision or care	Single family	Yes	Yes	Director of Dept. of Services	1,500 ft. (3,000 ft. in cities over one million population) required unless permitted by local ordinance
MINNESOTA Stat. Ann. §§252.28 (West Supp. 1977), 462.357 (West Supp. 1977)	Group home Foster home Residential facility	6 or fewer 7-16	Mentally retarded or physically handicapped persons	Single family Multiple family	Not specified Yes	Yes	Commissioner of Public Welfare	300 ft. required between facilities unless conditional use permit is granted

STATE	TYPE OF COMMUNITY FACILITY	RESIDENTS		ZONE IN WHICH PERMITTED	CONDITIONAL USE PERMITS ALLOWED	STATE LICENSING OF FACILITY		DISPERSAL OF FACILITIES REQUIRED
		No.	Type			Required	Licensor	
MONTANA Rev. Codes Ann. §§112702.1-.2 (Supp. 1977)	Community residential facility: group, foster or other home	8 or fewer	Developmentally disabled or handicapped persons	Single family	Yes	Yes	Dept. of Health and Environmental Science; Dept. of Social and Rehabilitative Services	Not specified
NEW JERSEY Stat. Ann. §§30:4C-2(m),-26 (West Supp. 1977)	Group home	12 or fewer	Children	[6]	Not specified	[7]	[7]	Not specified
NEW MEXICO Act of April 7, 1977, ch. 279 §20 (to be codified in Stat. Ann. §14-20-1)	Community residences	10 or fewer	Mentally ill or developmentally disabled persons	Single family	Not specified	Yes	Not Specified	Not specified
OHIO Act of Aug. 1, 1977, Amended Substitute S.B. No. 71 (to be codified in Rev. Code Ann. §5123.18 [Page])	Family home Group home	8 or fewer 9-16	Developmentally disabled persons	Single family Multiple family	Where such ordinance in effect prior to 6/15/77 Yes	Yes	Chief of Division of Mental Retardation and Developmental Disabilities	Not specified Limitation on excessive concentration of facilities permitted
RHODE ISLAND Act of May 13, 1977, ch. 257 (to be codified in Gen. Laws 45-24-22)	Any type of residence	6 or fewer	Retarded children or adults	[8]	Not specified	No	—	Not specified
SOUTH CAROLINA Act of April 4, 1978; H.B. 2121 Amendments to Act 653 of 1976 and 1976 Code §44-21-525, §44-17-10	Community residential facility	9 or fewer	Mentally handicapped persons	Single family[9]	Not specified	Yes	Dept. of Mental Health & Dept. of Mental Retardation	No excessive concentration of community residential facilities
TENNESSEE 1978 Tenn. Pub. P.C. 863 (H.B. 777)	Community home	8 or fewer	Mentally retarded or physically handicapped persons	Single family[10]	Not specified	Not specified	Not specified	Not specified

STATE	TYPE OF COMMUNITY FACILITY	RESIDENTS No.	RESIDENTS Type	ZONE IN WHICH PERMITTED	CONDITIONAL USE PERMITS ALLOWED	STATE LICENSING OF FACILITY Required	STATE LICENSING OF FACILITY Licensor	DISPERSAL OF FACILITIES REQUIRED
VERMONT Act of March 24, 1978, H. 698 (to be codified in Stat. Ann. tit. 24 §4409(d)	Community care home or group home	6 or fewer	Developmentally disabled or physically handicapped	Single family	Not specified	Yes	Not specified	1,000 ft. required between facilities
VIRGINIA Code §15.1-486.2 (Supp. 1977)	Family care home, foster home, group home,	Not specified	Mentally retarded and other developmentally disabled persons	Appropriate private residential districts	Yes	Not specified	—	Yes
WISCONSIN Signed into law March 21, 1978; Ch. 205, Laws of 1977	Child welfare agency Group foster home for children Adult residential facility	8 or fewer[11] 9-15 16 or more	All children or adults	All residential zones	[12] None Permit required for 1 or 2 family zones Permit required for all residential zones	Yes	Dept. of Health and Social Services	[13] 2,500 ft. distance, and density limit of the greater of 25 or 1% of population in municipality or aldermanic district for class 1-4 cities

[1] Residential facilities of 7 or more residents are a permitted use in any zone in which residential buildings of similar size, containing rooms provided for compensation, are a permitted use.

[2] Conditional use permits allowed only if no conditions are imposed on such facilities which are more restrictive than those imposed on similar dwellings in the same zone.

[3] The statute is somewhat confusing: "Zoning classifications. Although the public group home is exempt from any local zoning rule or regulation, public group homes may not be located in any area prohibited by the local zoning law. However, for the purposes of the mental retardation law, and zoning, the public group home conclusively shall be deemed a single family residential use, permitted in all residential zones. . . ." §19B(b)(5)

[4] Group homes shall not be subject to a special exception or conditional use permit or procedure different from those required for a single family dwelling in the same zone.

[5] Certificate of approval required.

[6] No municipality shall enact an ordinance governing single-family use of land which discriminates between children who are members of single families and children who are placed in group homes.

[7] Must be recognized as group home by Department of Institutions and Agencies in accordance with state rules and regulations.

[8] Residents are considered a family, and all requirements pertaining to local zoning are waived.

[9] While the South Carolina statute provides that community homes are to be treated as natural families for county or municipal zoning purposes, it also gives the political subdivision the right to object to a proposed location. This objection is made to the State Budget and Control Board, and that Board must make a decision on the objection, if appealed by the state agency, within a given time period. If no decision is made by the Board within the time period, then the political subdivision's objection controls and the location of the community home is prevented.

[10] Group homes of 8 or fewer residents are considered single family residences.

[11] A private, licensed foster home for four or fewer children is permitted, without space or density limits, in all residential areas.

[12] The municipality may annually review the effect of any community facility and may order it to close unless special zoning permission is obtained. The order is subject to judicial review.

[13] The municipality may, at its discretion, agree to increase the density limit or decrease the dispersal distance.

MUNICIPAL HOME RULE

The Problem

Although not specifically mentioned in the review of existing legislation, another issue warrants consideration — municipal home rule. It is a system of powers, including zoning powers, conferred upon municipalities by a state constitutional provision[30] in approximately forty states.[31] Home rule greatly insulates municipalities from state legislative action of the sort contemplated by this Act. For this reason, a crucial technical question must be correctly answered if this proposed Act is to avoid potential constitutional infirmities; that is: what language must be used in the Act to ensure that it does not violate a state's constitutional home rule provision?

Because various municipal home rule systems differ in content, interpretation and effect, no general discussion can adequately treat all of the individual provisions. For example, both Ohio[32] and California[33] are constitutional home rule states, and both have adopted, as indicated above, state legislation that limits local zoning authority with regard to community homes. Although the language of the constitutional home rule provisions of the two states is similar, judicial interpretation of these two provisions varies significantly.[34] Therefore, statutory language to satisfy home rule requirements in these two states had to be significantly different. In California, the Code[35] established a state-wide policy that the use of property as a community home for handicapped persons is to be considered a residential use. This simple pronouncement met California home rule requirements. In Ohio, however, the Revised Code[36] licenses individual operators of community homes for developmentally disabled persons in order to meet the tests set up by the Ohio Supreme Court's interpretation of the home rule provision. The important point to be made is that because of differing home rule requirements, that which is effective in California would have met early and complete failure in Ohio.

The Solution

What then is necessary to ensure a correct answer to the problem posed by municipal home rule? In each state where efforts are undertaken to gain passage of this proposed Act, proponents should do the following:

1. Determine if the state is a municipal home rule state (*See* note 31);

2. If it is, then obtain professional legal advice on what particular legislative language requirements must be met as a result of state judicial interpretation of the municipal home rule provision in the state constitution;

3. Ensure that the necessary language is incorporated in appropriate ways and at appropriate places into the proposed act;

4. Prepare answers and testimony for use in the legislative process that will provide a clear dem-

---

[30] *See generally* R. Anderson, 1 *American Law of Zoning* 2d, §§2.14-18 (1977).

[31] "Provisions of some sort for home rule are contained presently in constitutions of forty states. . . ."

See Alas. Const. art. X; Ariz. Const. art. XIII, §§2-3; Cal. Const. art. XI, §§35-37; Colo. Const. art XX; Conn. Const. art. X; Fla. Const. art. XIII 1(g), 2(b); Ga. Const. art. XV; Hawaii Const. art. VII, §2; Idaho Const. art. XII, §2; Ill. Const. art. VII, §6; Iowa Const. art. III, §40; Kan. Const. art. XII, §5; La. Const. art. VI; Me. Const. art. VIII-A; Md. Const. arts. XI-A, XI-E, XI-F; Mass. Const. art. of amend. II; Mich. Const. art. VII, §§2, 22; Minn. Const. art. XI, §3; Mo. Const. art. VI, §18(a)-(s), 19-19(a); Mont. Const. art. XI, §§5-6; Neb. Const. art. XI, §§2-5; Nev. Const. art. VIII, §8; N.H. Const. pt. I, art. XXXIX; N.M. Const. art. X, §6; N.Y. Const. art. IX; N.D. Const. art. VI, §130; Ohio Const. art. XVIII; Okla. Const. art. XVIII, §3-4; Ore. Const. art. XI, §2; Pa. Const. art. IX; §2; R.I. Const. art. of amend. XXVIII; S.C. Const. art. VIII, §11; S.D. Const. art. IX, §2; Tenn. Const. art. XI, §9; Tex. Const. art. XI, §5; Utah Const. art. XI, §5; Wash. Const. art. XI, §§10-11; W. Va. Const. art. VI, §39(a); Wis. Const. art. XI, §3; Wyo. Const. art. XIII, §1.

K. Vanlandingham, "Constitutional Municipal Home Rule Since the AMA (NLC) Model," 17 *William and Mary L. Rev.* 1, 4 n.9 (1975).

[32] Ohio Const. art. XVIII, §3.

[33] Cal. Const. art. XI, §7.

[34] For Ohio, *see Village of Struthers* v. *Sokol*, 108 Ohio St. 263, 140 N.E. 519 (1923); *Youngstown* v. *Evans*, 121 Ohio St. 342, 168 N.E. 844 (1929); and *Village of West Jefferson* v. *Robinson*, 1 Ohio St. 2d 113, 205 N.E.2d 382 (1965). For California, *see In Re Porterfield*, 28 Cal.2d 91, 168 P.2d 706 (1946); *Johnston* v. *Board of Sup'rs of Martin County*, 31 Cal. 2d 66, 187 P.2d 686 (1947); *Oakland Raiders* v. *City of Berkeley*, 65 Cal. App. A.3d 623, 137 Cal. Rptr. 648 (1976). *See also City of Los Angeles* v. *California Department of Health*, *supra* note 26.

[35] CAL. WELF. AND INST. CODE §§5115-5116 (Deering Supp. 1978).

[36] OHIO REV. CODE ANN., §5123.18 (Page Supp. 1977).

onstration of (a) an understanding of the home rule problem and (b) an appropriate approach to solving that problem.

In general, the home rule problem is one of form, not substance. Once the proper means of limiting local zoning power has been determined in a given state, the technical problem of home rule can be readily solved.

SUMMARY OF THE PROPOSED ACT

The overall goal of this Act is to improve the quality of life of all developmentally disabled persons by guaranteeing them the benefits of living in normal residential communities. The principal means to achieve this goal is a state-imposed statutory limitation upon the zoning power of a state's political subdivisions. This limitation prevents local communities from using their zoning power to exclude community homes for developmentally disabled persons[37] and is achieved by establishing the community home as a "permitted use" for zoning purposes in all residential districts throughout this state (*See* section 4 of the Act).

While the narrowly drawn limitation on local zoning power is the principal implementation device, this Act also utilizes four other important means to achieve its ends:

1. It ensures appropriate care and habilitation, as well as decent, safe and adequate housing for developmentally disabled persons through the establishment of a state licensing program for community homes in the state administrative department serving the developmentally disabled population. State administrative supervision based on licensing regulations can provide the necessary tools and sanctions in one agency with state level authority to ensure that community homes are well-run and residents well-treated. State licensing also insulates the regulatory process for community homes from strong local political pressures (*See* subsection 5(a) (i)-(iii));

2. The Act ensures that the present problem of over-concentration of community homes in certain types of areas will be avoided by giving the appropriate state agency responsibility for prevention. State-imposed dispersal requirements, achieved through state administration of the licensing system, can achieve direct, uniform, and coordinated supervision of community home locations. Various specific techniques are available to implement density control and avoid impaction and are appropriate for administrative regulations. It is important that these controls be exercised uniformly at the state level rather than at the local level to prevent attempts at indirect exclusion of large numbers of community homes by political subdivisions (*See* subsection 5(a) (i));

3. Appropriate safeguards and procedural rights are provided to political subdivisions to permit appropriate local regulation and to reduce or avoid local resistance or hostility to community homes (*See* subsections 5(a) (iii)-(iv), 5(b), 5(c));

4. The Act is also designed to prohibit exclusion of community homes by private agreement. This measure is included because evidence is accumulating in states that have already passed legislation similar to this Act that community homes now face a new obstacle: private restrictive covenants specifically prohibiting such homes[38] (*See* section 6).

In summary, the Act is a narrowly drawn effort to balance the need to facilitate location of community homes with legitimate local neighborhood concerns about health, fire and safety regulations, and the desire of local communities to be heard regarding the development of such homes in their neighborhoods.[39]

---

37 This limitation on local zoning powers is likely to be a controversial issue in most states. Yet, it is negligible compared to the overriding importance of implementing state policies and programs to guarantee the right to humane care and habilitation and the right to live in a community setting for developmentally disabled persons. In addition, the proposed limitation is minimal in comparison to the vast range of zoning powers enjoyed by cities, townships, villages, and boroughs of the various states throughout the country. Moreover, given the unique nature and significance of these rights it is also important to note that such state action regarding zoning should in no way be viewed as the precursor of any broad inroads into the area of local land use regulation.

Thus, the title of this Act accurately describes its thrust: first and foremost, it is state legislation designed to advance important rights of developmentally disabled persons and only secondarily is it a bill which imposes a limitation of local zoning power. This emphasis is important to remember when efforts are undertaken to achieve political support for the legislation.

38 *See Berger* v. *State of New Jersey* 71 N.J. 206, 364 A.2d 993 (1976) and *Adams* v. *The Toledo City Plan Commission*, No. 77-1198 (Lucas Co., Ohio Ct. of Common Pleas Dec. 7, 1977) holding that use of residential dwellings as community homes for mentally disabled children was not in conflict with private covenants. *See also* J. Morales, Missoula, MT. letter to M. Kindred, Sept. 28, 1977.

For a case which held that a foster home for mentally retarded children was in violation of a private restrictive covenant, *see Bellarmine Hills Association* v. *The Residential Systems Co.*, No. 77-156161 C2 (Oakland Co., Mich. Aug. 25, 1977).

39 Sections 5 and 6 of the Act are carefully and purposefully patterned after Sections 4 and 5 of the Mental Health Law Project's zoning legislation draft for homes serving the mentally handicapped. The MHLP materials are thoughtfully designed and well-drafted. While the coverage and approach of the MHLP legislation is somewhat different, its general purposes are virtually identical to those of this Act. Thus, it is recom-

The text of the Act appears in **bold type.** Comments on statutory provisions appear in standard type. Alternatives to the text appear in *italics.* The comments explain the Act. They are intended to assist the user in interpreting the Act and in drafting an act for a particular state from this model act. The alternatives suggest other approaches which might be considered.

---

mended that persons interested in zoning law reform with regard to community homes review both sets of materials carefully.

The differences between this Act and the MHLP draft legislation are:

(1) This Act includes a title clause; the MHLP draft does not.

(2) This Act is designed to serve the "developmentally disabled person"; the MHLP draft serves the "mentally handicapped person."

(3) This Act defines "developmental disability"; the MHLP draft does not define either this term or "mental handicap."

(4) In order to assign appropriate responsibilities to state officers, this Act defines "director" [of developmental disabilities]; the MHLP draft does not define any analogous term or official.

(5) This Act defines "permitted use" in order to make the zoning treatment for "family homes" as explicit and clear as possible.

(6) This Act defines "political subdivision" to provide a means of dealing with municipal home rule problems.

(7) This Act establishes a state licensing system for community homes. As indicated above, licensing is of fundamental importance to the protection of community home residents, to the effectiveness of density control provisions, and to the avoidance of local political pressures.

(8) This Act establishes a procedure for notifying the governing body of the political subdivision in which a family home is to be located.

MODEL STATUTE*

AN ACT TO ESTABLISH THE RIGHT TO LOCATE COMMUNITY HOMES FOR DEVELOPMENTALLY DISABLED PERSONS IN THE RESIDENTIAL NEIGHBORHOODS OF THIS STATE.

Section 1. Title

This act shall be known as the "Location Act for Community Homes for Developmentally Disabled Persons."

COMMENT. None necessary.

Section 2. Statement of Purpose

The general assembly declares that it is the goal of this act to improve the quality of life of all developmentally disabled persons and to integrate developmentally disabled persons into the mainstream of society by ensuring them the availability of community residential opportunities in the residential areas of this state. In order to implement this goal, this act should be liberally construed toward that end.

COMMENT. The goal of this Act is to assist in improving the quality of life of developmentally disabled persons within this state by fully integrating them into the community in which they live. The intent of the Act is to achieve this goal by removing local obstacles that prevent the establishment of community homes serving developmentally disabled persons in residential neighborhoods.

Section 3. Definitions

As used in this act:

(1) "Developmental Disability" means a disability of a person which:

(a)(i) is attributable to mental retardation, cerebral palsy, epilepsy, or autism;

(ii) is attributable to any other condition found to be closely related to mental retardation because such condition results in similar impairment of general intellectual functioning or adaptive behavior to that of mentally retarded persons or requires treatment and services similar to those required for such persons; or

(iii) is attributable to dyslexia resulting from a disability described in clause (i) or (ii) of this paragraph; and

(b) has continued or can be expected to continue indefinitely.

COMMENT. This Act adopts the definition of "developmental disability" formerly contained within the Developmental Disabilities Assistance and Bill of Rights Act, Pub. L. 94-103, 42 U.S.C. §§6001-6081 (Supp. V 1975), except that the requirements that the disability originate before the person reaches eighteen and that the disability constitute a substantial handicap have been deleted. The reason for the first of these two deletions is that the class of persons fitting within the category of developmentally disabled has been expanded since adults who become disabled and otherwise fit the definition of developmentally disabled may also need community-based residential services. Adults can develop epilepsy, mental retardation or cerebral palsy from a number of sources, including trauma and disease. The reason for the second deletion is that the purpose of the federal definition is to target certain populations for intensive federal support, whereas the purpose of this Act's definition is to enable all developmentally disabled persons who need community-based residential services to be able to obtain them regardless of whether they have a substantial handicap.

The definition adopts a categorical approach rather than a functional one [see 42 U.S.C. §6001(7) (Cum. Supp. 1978)]. Thus, the disability must be caused by one or more of the conditions described in paragraph (a). In addition, the condition contained in paragraph (b) must also be met for the disability to qualify as a developmental disability. States wishing to adopt a functional definition of this term should refer to Section 3(4) of the Model Right to Developmental Disabilities Services Act in Chapter 6, *supra*.

ALTERNATIVE. *Some states may wish to expand the coverage of this Act to include homes which serve other disabled persons since exclusionary measures against any type of community home should not be sanctioned. On the contrary, expansion of this legislation to encompass the licensing and operation of community homes for all groups receiving some form of treatment, care, habilitation, or supervision is desirable. Yet, expansion of the statute may bring increased resistance to the Act's passage. Where this is the case, the passage of the Act as currently drafted is strongly recommended. The Act will significantly enhance exist-*

---

* Robert J. Hopperton, Associate Professor of Law at the University of Toledo College of Law, had primary responsibility for drafting the model statute.

ing developmental disability programs and eliminate the serious and justifiably attacked problem of over-concentration of community homes in certain areas. Finally, it will provide valuable experience which will be beneficial to drafters and supporters of a broader statute.

(2) "Developmentally Disabled Person" means a person with a developmental disability.

COMMENT. None necessary.

(3) "Director" means the director of developmental disabilities (or appropriate state official).

COMMENT. As can be seen from the chart, various state departments and officials are given responsibility for licensing community homes in the states that have already passed legislation similar to this Act — in Colorado, the Department of Health; in Michigan, the Director of Social Services; in Minnesota, the Commissioner of Public Welfare; in Montana, the Department of Health and Environmental Science and the Department of Social and Rehabilitation Services; and in Ohio, the Chief of the Division of Mental Retardation and Developmental Disabilities. Because the most appropriate official to have licensing authority can vary from state to state, this determination is best made after a review of the administrative structure in each state.

(4) "Family Home" means a community-based residential home licensed by the director that provides room and board, personal care, habilitation services, and supervision in a family environment for not more than [six (6)] developmentally disabled persons.

COMMENT. The term "family home" is designed to include only smaller community-based residences as opposed to institutions such as hospitals and schools. By definition these homes are licensed by the state to ensure responsible supervision, density controls, and other requirements designed to protect both residents and neighbors. (*See* subsection 5(1) of the Act.) By including only smaller homes, the Act encourages community-based residences that are most nearly like normal family arrangements. Moreover, the limitation to six residents makes optimum living arrangements such as single bedrooms for some or all residents more possible. In addition, since the "family home" is treated as a "permitted use" in all residential zones, this limitation encourages maximum compatibility with residential densities.

ALTERNATIVE. *Some states may wish to consider raising the maximum size of the family home, for example, to not more than eight residents. The decision as to the upper limit of residents for the smaller community homes is best left to a determination based on financial (e.g., appropriate size to ensure economic feasibility of the home) programmatic and political considerations in each state.*

(5) "Permitted Use" means a use by right which is authorized in all residential zoning districts.

COMMENT. "Permitted use" is defined so as to make completely clear the intent of the Act, i.e., to eliminate the exclusion and restriction of family homes. Thus a political subdivision could no longer restrict a family home by treating it as a "conditionally permitted use." (*See supra* notes 15 and 26 and COMMENT to section 4 *infra.*)

(6) "Political Subdivision" means a municipal corporation, township, or county.

COMMENT. In order to implement the limitation on local zoning power that is the primary means of this Act, it is necessary to define the subordinate political units of a state. Since some variations in terminology occur from state to state, it is recommended that the official names of political subdivisions be identified, i.e., municipal corporations, townships, boroughs, villages, etc., and then used appropriately.

Section 4. Permitted Use for Family Homes

A family home is a residential use of property for the purposes of zoning and shall be treated as a permitted use in all residential zones or districts, including all single-family residential zones or districts, of all political subdivisions. No political subdivision may require that a family home, its owner, or operator obtain a conditional use permit, special use permit, special exception, or variance.[40]

COMMENT. Generally, under the zoning ordinances of political subdivisions, real estate development by private parties proceeds as a matter of right. There are exceptions, however, to this general framework; when a given use is treated as a conditional use, special use, or special exception, the owner is not entitled to proceed until permission is received from the local zoning authority. Frequently, it is difficult to determine what standards guide local discretion in such matters. For this reason, this Act establishes the family home as a "permitted use" in all residential districts of the state. (*See* note 15 *supra.*) This entitles an owner/operator to develop such a home as a matter of right, not subject to local discretion,[41] and thus this provision is the key section of the act.

Section 5. Licensing Regulations and Density Control for Family Homes

(1) For the purposes of safeguarding the health and safety of developmentally disabled persons and avoiding over-concentration of fam-

---

[40] *See supra* note 15.

[41] N. Williams, *supra* note 3, at 330.

ily homes, either along or in conjunction with similar community-based residences, the director or the director's designee shall inspect and license the operation of family homes and may renew and revoke such licenses. A license is valid for one year from the date it is issued or renewed although the director may inspect such homes more frequently, if needed. The director shall not issue or renew and may revoke the license of a family home not operating in compliance with this section and regulations adopted hereunder. Within one hundred eighty (180) days of the enactment of this act, the director shall promulgate regulations which shall encompass the following matters:

COMMENT. This section establishes a state licensing system for family homes which will help assure that developmentally disabled persons living in the community will have not only decent, safe, and adequate housing, but also sufficient care and habilitation. In addition, state licensing will help prevent existing neighborhoods from losing their essential residential characteristics because of over-concentration of family homes and similar facilities. The responsibility for promulgating regulations to ensure compliance with this section lies with the Director.

(a) Limits on the number of new family homes to be permitted on blocks, block faces, and other appropriate geographic areas taking into account the existing residential population density and the number, occupancy, and location of similar community residential facilities serving persons in drug, alcohol, juvenile, child, parole, and other programs of treatment, care, supervision, or rehabilitation in a community setting;

COMMENT. This paragraph requires that licensing and operation of family homes be conditioned on considerations of statistical and compositional factors in the existing community. Establishing guidelines for determining when the density of family homes in the community is reaching a critical or excessive level will be an essential, but difficult task. *See* D. Lauber and F. Bangs, *Zoning for Family and Group Care Facilities* (American Society of Planning Officials, Planning Advisory Service Report No. 300, March, 1974) which suggests reasonable limits on minimum floor area, minimum lot area, and number of facilities per block, block face, and neighborhood. *See also* R. Hopperton, *Zoning for Community Homes: a Handbook for Local Legislative Change* (Ohio State University Law Reform Project, 1975). These sources are recommended as a starting point for the Director.

(b) Assurance that adequate arrangements are made for the residents of family homes to receive such care and habilitation as is necessary and appropriate to their needs and to further their progress towards independent living;

COMMENT. A major criticism of many of the de-institutionalization programs has been the lack of services for persons discharged from institutions. This provision, together with paragraph (c) below, will afford developmentally disabled persons protection by requiring the Director to adopt regulations which will assure that developmentally disabled residents receive services that they need in the family home and in their advancement towards independent living.

These subsections should be worded consistently with any existing legislation which places licensing authority in a state department.

(c) Protection of the health and safety of the residents of family homes, provided that compliance with these regulations shall not relieve the owner or operator of any family home of the obligation to comply with the requirements or standards of a political subdivision pertaining to building, housing, health, fire, safety, and motor vehicle parking space that generally apply to single family residences in the zoning district; and provided further that no requirements for business licenses, gross receipt taxes, environmental impact studies or clearances may be imposed on such homes if such fees, taxes, or clearances are not imposed on all structures in the zoning district housing a like number of persons;

COMMENT. Paragraph (c) permits political subdivisions to impose ordinary, general requirements regarding building, housing, health, fire, and safety code provisions and off-street parking requirements on family homes so long as they do not exceed standards applied to single family homes in the same area. Similarly, the second provision prohibits local fees, taxes, and environmental impact clearance procedures on family homes from exceeding those imposed on residential structures of similar capacity in the same area. Political subdivisions could not, for example, demand payment of business taxes. They could, however, where not otherwise forbidden, demand payment of local property taxes at the same rate applied to single-family dwellings in the neighborhood.[42]

Conceivably, some political subdivisions may attempt to subvert the intent of this paragraph through discriminatory local non-zoning regulations. If such attempts succeed, it may be necessary to propose statutory limitations on those regulations.

(d) Procedures by which any resident of a residential zoning district or the governing body of a political subdivision in which a family home is, or is to be, located may petition the director to deny an application for a license to operate a

[42] *See* MHLP, *supra* note 39.

family home on the grounds that the operation of such a home would be in violation of the limits established under paragraph (1)(a) of this section;

COMMENT. The paragraph, together with subsection 2 below, assures that political subdivisions and local residents have the means to inform the Director of violations of the density controls in the state licensing regulations that might occur if an additional family home were to be established in a particular neighborhood. Local involvement will facilitate the identification and avoidance of potential problems. The only objections, however, which can be raised are those which would violate the Director's regulations regarding the number of other, similar facilities, their occupancy and the population of the area. To further ensure the quality of family homes and adherence to regulatory standards, subsection (1) mandates annual inspections and permits the Director to make additional ones.

(2) All applicants for a license to operate a family home shall apply to the director for the license and shall file a copy of the application with the governing body of the political subdivision having jurisdiction over the zoning of the land on which the family home is to be located. All applications must include population and occupancy statistics reflecting compliance with the limits established pursuant to paragraph (1)(a) of this section.

COMMENT. This section establishes a step-by-step application procedure and provides for notification to the political subdivision that a family home may be located within its jurisdiction. Notification allows the political subdivision to make comment under the procedures adopted by the Director under paragraph (1)(a).

(3) The Director may not issue a license for a family home until the applicant has submitted proof of filing with the governing body of the political subdivision having jurisdiction over the zoning of the land on which such a home is to be located, a copy of the application at least thirty (30) days prior to the granting of such a license, and any amendment of the application increasing the number of residents to be served at least fifteen (15) days prior to the granting of a license.

COMMENT. This section prevents licensure of a family home until timely and sufficient notice has been provided to the political subdivision.

(4) In order to facilitate the implementation of paragraph (1)(a), the director shall maintain a list of the location, capacity, and current occupancy of all family homes. The director shall ensure that this list shall not contain the names or other identifiable information about any residents of such home and that copies of this list shall be available to any resident of this state and any state agency or political subdivision upon request.

COMMENT. This section also facilitates the implementation of the provisions that prevent overconcentration of family homes within the neighborhoods of a state. The required list is limited to the population and other statistical information required in paragraph (1)(a), information relevant to the concerns of governmental entities and private citizens in preventing over concentration of homes.

Section 6. Exclusion by Private Agreement Void

Any restriction, reservation, condition, exception, or covenant in any subdivision plan, deed, or other instrument of or pertaining to the transfer, sale, lease, or use of property which would permit residential use of property but prohibit the use of such property as a family home for developmentally disabled persons shall, to the extent of such prohibition, be void as against the public policy of this state and shall be given no legal or equitable force or effect.[43]

COMMENT. This section is designed to void private agreements whenever they are used to deny developmentally disabled persons the benefits of living in normal residential units.[44] Because this section is designed to void already existing, as well as future, private agreements that exclude family homes, some opponents may raise questions of due process, retroactivity, or impairment of obligation of contract. Such objections are not well-founded. One state, Wisconsin, has already passed legislation similar to section 6 dealing with community homes.[45] Other states, e.g., New York and Texas, have enacted statutory provisions similar to section 6 voiding private agreements that prohibit the sale, lease, or use of real property on the basis of race, religion, or national origin.[46] While there is no case law as yet interpreting the Wisconsin, New York, or Texas statutes, case law in other states indicates that provisions voiding already existing contracts for the sale of stock[47] and for the rental of property[48] have been found valid.[49]

---

[43] *Id.* at §5 and Comment to §5.

[44] *See supra* note 38.

[45] 1977 Wis. Laws, Ch. 205, §2 [to be codified in WIS. STAT. §46.03 (22)(d)].

[46] *See* N.Y. General Obligations Law §5-331 (McKinney 1978) and TEX. REV. CIV. STAT. ANN. art. 1293a (Vernon Supp. 1978).

[47] *Massillon Savings and Loan Co.* v. *Imperial Finance Co.*, 114 Ohio St. 523, 151 N.E. 645 (1926).

[48] *Industrial Development & Land Co.* v. *Goldschmidt*, 56 Cal. App. 507, 206 P.134 (1922).

[49] *See also Home Building & Loan Ass'n.* v. *Blaisdell* 290 U.S. 398 (1934).

Care should be taken to conform the language of this section to existing subdivision control statutes.

Section 7. Severability of Sections

If any section, subsection, paragraph, sentence, or any other part of this act is adjudged unconstitutional or invalid, such judgment shall not affect, impair or invalidate the remainder of this act, but shall be confined to the section, subsection, paragraph, sentence, or any other part of this act directly involved in the controversy in which said judgment has been rendered.

COMMENT. None necessary.

5
EDUCATION OF EXCEPTIONAL PERSONS

Table of Contents

Introduction

Dramatic progress in establishing the right to appropriate education programs for exceptional children — those who are handicapped, as well as gifted or talented — has been accomplished in the past decade.[1] Although states have legislated on behalf of exceptional persons for the last one hundred and fifty years, state laws in the previous century focused on the establishment of state schools and institutions. Since 1960 state legislative attention to special education has considerably increased in comprehensiveness. Yet, until the very recent past, many children with special education needs still have been either underserved or ignored, misclassified, or excluded entirely from school. For example, some laws for years barred from an education those deemed "uneducable" by school officials who, for reasons of administrative convenience, ignorance or lack of funds, chose not to deal with special education students.

The first significant attack on an exclusionary state law was brought by the Pennsylvania Association of Retarded Children (*PARC*), [2] which sought to establish a constitutional right to education for retarded children. In two opinions, issued in 1971 and 1972, a three-judge federal court agreed with *PARC*'s contention that Pennsylvania's practice of excluding retarded children from the opportunity to participate in free public education violated their rights to equal treatment and due process of law required by the Fourteenth Amendment to the United States Constitution.

As a result of this case, "due process" procedures were incorporated in Pennsylvania's State Board of Education regulations so that parents could challenge any decision by school officials which affected the delivery of services to a retarded child.[3] When a dispute arises between parents and school district officials regarding any aspect of a retarded person's education, a conference must be held to resolve the dispute in an informal, yet legally mandated, fashion. If agreement cannot be reached at the conference, both parents and school officials have the option to request a special education due process hearing. An impartial hearing officer — a person familiar with special education matters, but who is not a lawyer — is appointed by the State Department of Education to hear evidence of both parties' concerns. A written transcript is taken and the hearing officer is required to write an opinion, based upon the evidence presented, detailing an appropriate special education program for the person who is the subject of the hearing. These opinions may be appealed to the State Secretary of Education and then to the courts.

PARC and a challenge to exclusions of all handicapped children in Washington, D.C., *Mills v. Board of Education of the District of Columbia,*[4] were landmarks used by other concerned groups to establish legal protections and a right to education for children with special needs.[5] For example, another Pennsylvania case, *Catherine D. v. Pittenger*[6] extended the *PARC* procedures to cover all exceptional children in that state, including a newly defined group of persons with special needs — those who are gifted or outstandingly talented. Experts found that children who are intellectually gifted or outstandingly talented in art, music or dance, suffer the same boredom, frustration and failure to learn when faced with a regular education program as handicapped persons. As a result, 24 states have included a right to education for these non-handicapped exceptional children. However, federal law does not yet guarantee a right to education for gifted and talented exceptional persons; thus, these persons only have the opportunity to contest any aspect of their education in states which are willing to provide these rights.[7] Even in states which have provided the right to education and due process protections for its gifted and talented persons, few would disagree that this population is grossly underserved in

---

[1] *See, e.g.*, Frederick L. v. Thomas, 419 F. Supp. 960 (E.D. Pa. 1976), *aff'd*, 557 F2d 374 (3d Cir. 1977); Fialkowski v. Shapp, 405 F. Supp. 946 (E.D. Pa. 1975); Mills v. Bd. of Educ., 348 F. Supp. 866 (D.D.C. 1972); Pennsylvania Ass'n for Retarded Children v. Commonwealth of Pennsylvania, 334 F. Supp. 1257 (E.D. Pa. 1971) and 343 F. Supp. 279 (E.D. Pa. 1972).

[2] *Id.*

[3] 22 PA. CODE ch. 13. *see also*, ch. 171, 341.

[4] *Supra*. Note 1.

[5] Panitch v. Wisconsin, 371 F. Supp. 935, 390 F. Supp. 611 (E.D. Wis. 1974), 444 F. Supp. 320, 76 F.R.D. 608 (E.D. Wis. 1977), per Curiam Order, 451 F. Supp. 132 (E.D. Wis. 1978); Harrison v. Michigan, 350 F. Supp. 846 (E.D. Mich. 1972); LeBanks v. Spears, 60 F.R.D. 135 (E.D. La. 1973); Maryland Ass'n of Retarded Children v. Maryland, Eq. No. 100-182-77676 (Cir. Ct. Baltimore Cnty., April 9, 1974). *See also* Note, *The Right of Handicapped Children to an Education: The Phoenix of Rodriguez,* 59 CORNELL L. REV. 519 (1974).

[6] No. 74-2435 (E.D. Pa., *Order*, June 27, 1975).

[7] *See* Review of State Statutes, *infra*.

every aspect of education. As in the case of other exceptional persons, advocacy on behalf of this group is increasing and programs are being considered in greater numbers.

Armed with the court cases described and increasing public awareness, new cases for previously unidentified and underserved populations were pursued in the courts. For example, *Frederick L. v. Thomas,*[8] was brought to provide appropriate programs for all learning disabled persons residing in the Philadelphia School District. In this case, a federal district court found that an effective plan for identifying and screening these persons did not exist, and appointed a master to supervise the development and implementation of a meaningful plan for serving them.[9]

Increasing advocacy on behalf of exceptional children gained the attention of Congress, and the result has been two major laws which are the genesis of a federal mandate to serve all exceptional children with appropriate educational programs and services. The Education for All Handicapped Children Act of 1975 (P.L. 94-142) and its implementing regulations require programs, services, and due process protections for all handicapped children residing in states accepting funding under this law.[10] The primary focus of this legislation is the requirement that each handicapped child be provided with an appropriate program of special education and related services for his or her unique needs. Section 504 of the Rehabilitation Act of 1973,[11] a civil rights act based upon previous federal laws barring discrimination on the basis of race and sex, prohibits the exclusion of persons on the basis of handicap who are "otherwise qualified" from participating in any program receiving federal financial assistance. The U.S. Department of Health, Education and Welfare issued regulations to this Act in 1977, which describe the persons and programs, including education, covered by the law.[12] In addition, Congress has recently passed Title IX of the Education Amendments of 1978[13] to increase federal funding to states which develop programs for gifted and talented persons. While the appropriations under this law are currently very limited and due process protections are not provided, this law is the first Congressional action of note on behalf of these exceptional persons. It is therefore significant and bodes well for additional federal legislative activity in this area. A description of these federal laws follows.

The Education for All Handicapped Children Act.

In response to advocacy on behalf of handicapped children throughout the nation, Congress passed The Education for All Handicapped Children Act of 1975, also known as P.L. 94-142. The purpose behind this law is to provide a nationwide system of protections for handicapped persons and their parents, as well as federal funds for states willing to provide programs of education that are appropriate to meet the needs of each of its handicapped children. As a condition for receiving this federal money, state educational agencies were required to insure that all programs would be provided at no cost to an exceptional person and at no cost to the person's parent.[14] In 1977, the U.S. Department of Health, Education and Welfare passed regulations for P.L. 94-142, the purpose of which is to define more fully the rights set forth under the Act and the obligations of recipients of federal funds.[15]

> P.L. 94-142 defines a "free appropriate public education" as a program of special education and related services which (a) have been provided at public expense, under public supervision and direction, and without charge, (b) meet the standards of the state educational agency, (c) include an appropriate preschool, elementary, or secondary school education in the state involved, and (d) are provided in conformity with the individualized education program. . . .[16]

Special education is defined more specifically as "specially designed instruction, at no cost to parents or guardians, to meet the unique needs of a handicapped child, including classroom instruc-

---

[8] *Supra.* Note 1.

[9] *See* Armstrong v. Kline, 476 F. Supp. 583 (E.D. Pa., 1979) (court ordered programs in excess of normal school year for severely handicapped persons); *see also* New York State Ass'n of Retarded Children v. Rockefeller, 357 F. Supp. 752 (E.D. N.Y. 1973) and 393 F. Supp. 714 (E.D. N.Y. 1975); New Jersey Ass'n for Retarded Citizens v. New Jersey Dep't of Human Resources, No. C2473-76 (N.J. Super. Ct., filed Mar. 14, 1977). *See also* McMillan v. Bd. of Educ., 430 F.2d 1145 (2d Cir. 1970) (mentally handicapped persons); Mattie T. v. Holladay, No. DC-73-31-S (N.D. Miss., Jan. 26, 1979) (consent decree affecting all handicapped persons unserved with appropriate educational programs in Mississippi); Lora v. Board of Education of the City of New York, 456 F. Supp. 1211 (E.D. N.Y. 1978).

[10] 20 U.S.C. §§1401 *et seq.* (1976); 45 C.F.R. pt. 121a (1978).

[11] 29 U.S.C. §794 (1976); *see,* Exec. Order No. 11,914, 41 Fed. Reg. 17,871 (1976).

[12] 45 C.F.R. §§84.1 *et seq.* (1978); *see,* Cherry v. Mathews, 419 F. Supp. 922 (D.D.C. 1976).

[13] 20 U.S.C. §§3311, *et seq.* (1976).

[14] 20 U.S.C. §§1412(1), 1414(a)(1)(C)(ii) (1976); 45 C.F.R. §§121a.2, 121 (1978).

[15] *See, supra* Note 10. Commentary regarding the developing P.L. 94-142 law may be found in the following law review articles: Note, *Enforcing the Right to an "Appropriate" Education: The Education for All Handicapped Children Act of 1975,* 92 HARV. L. REV. 1103 (1979); Haggerty & Sacks, *Education of the Handicapped: Towards a Definition of an Appropriate Education,* 50 TEMP. L.Q. 961 (1977); Note, *The Education of All Handicapped Children Act of 1975*, 19 U. MICH. J.L. REF. 110 (1976).

[16] 20 U.S.C. §1401(18) (1976); 45 C.F.R. 121a.4 (1978).

tion, instruction in physical education, home instruction and instruction in hospitals and institutions."[17] Related services include such services as transportation, physical and occupational therapy, recreation, speech pathology and audiology, parent and student counseling, and psychological counseling whenever they are required by a person to benefit from special education. Importantly, the law does not limit the type or number of services to be provided, and the lengthy list of potential services set forth in the regulations is specifically not all-inclusive. Thus, the mandate is for a program of specific services to serve the particular needs of the individual.[18]

The definition of "free appropriate education" has been a source of controversy between parents of handicapped persons and school district officials, which the courts have only recently been asked to resolve. In *Armstrong v. Kline,*[19] a case which challenged the refusal of state and local officials to provide more than 180 days of education (the "normal school year") for severely handicapped persons and the first case brought to trial under P.L. 94-142, a federal court agreed that "it is still sometimes difficult to determine precisely what the state is required to provide."[20] Citing the legislative history of P.L. 94-142, this Court held that, at the very least, an appropriate education means providing that type of educational opportunity which would allow individuals to achieve "self-sufficiency and independence from caretakers."[21] As cited by the court, Senator Williams, one of the sponsors of P.L. 94-142, stressed the need to provide a program of education and training which would allow handicapped persons to become independent citizens. He stated that:

> Failure to provide appropriate educational services for all handicapped children result in public agencies and taxpayers spending billions of dollars over the lifetime of these individuals to maintain them as dependents in minimally acceptable lifestyles. Yet, providing appropriate educational services now means that many of these individuals will be able to become a contributing part of our society, and they will not have to depend on subsistence payments from public funds.[22]

Appropriateness, according to the Court, is not necessarily the provision of similar services to handicapped, as well as non-handicapped persons. An individualized education program must be developed that incorporates the necessary special education and related services to meet the individual's unique needs.[23] This requires that a program be set forth in writing describing the goals and objectives for educating each child, and the specific educational services to be provided each person in order to meet those goals. This Individualized Education Program (IEP) was to have been in effect on October 1, 1977 for each child who had been identified as requiring special education prior to that date. After that time, IEPs must be in effect at the beginning of every school year for each handicapped person who is to receive services.[24] The IEP is to be developed at a meeting among the child's teacher, one of the parents, the person, if appropriate, and a representative of the educational agency (usually a principal).[25] Others with a legitimate educational interest in the person may be present at the discretion of either party. The educational agency must ensure that the parent is given sufficient notice of the time and date of the meeting where the IEP will be developed; importantly, the notice must be in the parent's native language or mbde of communication and provisions must be taken at the meeting to make certain that a parent fully understands the proceedings. When necessary, for example, a sign language interpreter for deaf parents, or a language interpreter for language minority parents must also be provided.[26]

A major concept incorporated into the federal law is the provision of special education and related services to each person in the least restrictive environment or most normal setting possible.[27] P.L. 94-142 sets forth this requirement as follows.

> Each public agency shall insure: (1) that to the maximum extent appropriate, handicapped children, including children in public or private institutions or other care facilities, are educated with children who are not handicapped, and (2) that special classes, separate schooling or other removal of handicapped children from the regular educational environment occurs only when the nature of severity of the handicap is such that education in regular classes with the use of supplementary aids and services cannot be achieved satisfactorily.[28]

---

[17] 20 U.S.C. §1401(16) (1976); 45 C.F.R. §121a.14 (1978).

[18] 20 U.S.C. §1401(17) (1976); 45 C.F.R. §121a.13 (1978).

[19] Armstrong v. Kline, 476 F. Supp. 583 (ED. Pa. 1979).

[20] Armstrong, *supra* at 603.

[21] *Id.* at 603.

[22] *Id.* at 604.

[23] 20 U.S.C. §1401(19) (1976); 45 C.F R. §§121a.340 *et seq.* (1978).

[24] 45 C.F.R. §121a.342 (1978).

[25] *Id.* §§121a. 343-345, 347-348.

[26] 20 U.S.C. §1412(5) (1976); 45 C.F.R. §121a.345 (1978).

[27] 20 U.S.C. §1412(5) (1976); 45 C.F.R. §§121a.550 *et seq.*, 121a.346(C) (1978).

[28] 45 C.F.R. §121a.550(b) (1978).

A person may be placed in a regular class for certain purposes, but may require special education resource rooms for other programs or services. The point of this requirement is that a handicapped child may no longer be segregated from his other non-handicapped peers, unless it is dictated by his or her individual needs. Such integration will sensitize each person to the unique qualities of others while allowing them to reach their full potential.

Due to the nature and severity of an individual's handicap, there may be no program available in a public school facility. Thus, that person may have to be educated, at least temporarily, in a private school at public expense. However, the goal of the law is to provide each child with appropriate programs as close to the home as possible, and private settings may not be used to ignore the responsibility of each school district to provide an education. When a person is placed in a private school or other institutional environment, an IEP must still be developed, annually reviewed,[29] and the person's local school district of residence remains responsible for its enforcement. In addition, all rights accorded to handicapped persons in a public setting must also be provided to the children placed by public agencies in private schools, including the provision of the education at no cost to the person or that person's parent.[30]

Another significant legal mandate of P.L. 94-142 is the special education due process procedures.[31]

> Parents have a right to receive notice at any time a public educational agency proposes to initiate or change the identification, evaluation, or educational placement of the child or the provision of a free appropriate public education to the child, or refuses to initiate or change the identification, evaluation, or educational placement of the child or the provision of a free appropriate public education to the child.[32]

This notice must be provided to the parent in a form easily understandable to the general public and in the parent's native language or other mode of communication. It must indicate why an action is proposed or refused, and give supporting evidence of why the educational agency is acting or refusing to act. In addition, parental consent must be obtained for an initial evaluation.[33]

If a parent objects to any proposed action or refusal to act, he or she has a right to contest the action or inaction by requesting a special education due process hearing.[34] This hearing must be conducted by an impartial hearing officer who cannot be an employee of the school district which is proposing the evaluation or program change.[35] It must be held in a location which is convenient for the parent, a written or recorded transcript must be maintained, and all evidence not described by either party at least five days before the hearing can be excluded. At the hearing, each side presents its reasons why the proposed action or refusal to act should or should not be accomplished. Either side has the right to be represented by legal counsel.[36]

The hearing officer is required to write an opinion, no later than 45 days after the receipt of a request for a hearing, and a copy of the decision must be mailed to each of the parties by that time. The state educational agency is mandated to provide a final decision on review no later than 30 days after the receipt of a request for a review. After the state educational agency's final decision is rendered, the party aggrieved by these findings may appeal all or part of the opinion to the appropriate state court, or to a federal district court.[37] During the course of all administrative and judicial proceedings, the child is to remain in the present educational placement unless the parent and school district agree otherwise.[38]

The confidentiality protections of the Family Educational Rights and Privacy Act (known as "FERPA" or the "Buckley Amendment") have been incorporated into P.L. 94-142. Parents are permitted to obtain access to all educational records maintained regarding the handicapped person by an educational agency.[39] If a parent believes that any information contained in a handicapped person's education records is "inaccurate or misleading or violates the privacy or other rights" of the person, the parent may request that

---

[29] 20 U.S.C. §1414(a) (1976); 45 C.F.R. §§121a.551-552 (1978).

[30] 20 U.S.C. §1413(a)(4) (1976); 45 C.F.R. §§121a.2, 400-403, 450 *et seq.* (1978). The refusal of education officials to provide non-public placements at no cost to parents of an exceptional person violates the equal protection clause of the 14th Amendment to the U.S. Constitution. *See* Halderman v. Pittenger, 391 F. Supp. 872 (E.D. Pa. 1975); Kruse v. Campbell, 431 F. Supp. 180 (E.D. Va. 1977), *vacated and remanded*, 98 S. Ct. 38 (1977), *on remand*, No. 75-0622-12 (E.D. Va., Jan. 5, 1978); Scavella v. School Bd., 363 So.2d 1095 (Fla. 1978). *But see* Cuyahoga County Ass'n for Retarded Adults v. Essex, 411 F. Supp. 46 (N.D. Ohio 1976); Levy v. City of New York, 345 N.E.2d 556 (N.Y. 1976).

[31] 20 U.S.C. §1415 (1976); 45 C.F.R. §§121a.500 *et seq.* (1978).

[32] 45 C.F.R. §121a.504 (1978).

[33] 45 C.F.R. §121a.505 (1978). In addition to the requirement that notice to parents be provided in the native language or mode of communication, evaluations conducted concerning the person must not be a manner which is racially or culturally discriminatory. 45 C.F.R. §121a. 532 (1978). *See also* Larry P. v. Riles,—F. Supp.—(N.D. Cal. 1979); Lora, at note 9, *supra*.

[34] 20 U.S.C. §1415 (1976); 45 C.F.R. §121a.506 (1978).

[35] 45 C.F.R. §121a.507 (1978).

[36] *Id.* §121a.508.

[37] *Id.* §§121a.509-512.

[38] *Id.* §121a.513.

[39] 20 U.S.C. §§1232g, 1412, 1417(C) (1976); 45 C.F.R. §§121a. 560 *et seq.* (1978).

the educational agency "amend" the records to exclude the inappropriate information.[40] If the educational agency refuses a request to amend, the opportunity for a hearing must be provided at the parent's request to contest the refusal. The hearing must be conducted by an impartial hearing officer who may be an employee of the local educational agency being challenged, but who cannot be directly involved in the specific controversy. A written record must be kept of the testimony presented at the hearing, and a written opinion must be rendered. If the opinion concludes that the records will not be amended, a parent has the opportunity to include a statement indicating his or her views in the person's permanent record file.[41]

Section 504 of The Rehabilitation Act of 1973

Section 504 provides in part as follows: 'No otherwise qualified handicapped individual in the United States . . . shall solely by reason of his handicap, be excluded from the participation in, be denied the benefits of, or be subjected to discrimination under any program or activity receiving federal financial assistance."[42] The 1977 implementing regulations for Section 504 require public elementary and secondary educational agencies receiving federal funding to provide a "free appropriate public education" to every "qualified" handicapped person who is within that recipient's jurisdiction.[43] These regulations define an "appropriate" education as: "[T]he provision of an appropriate program of regular or special education and related aids and services that (i) are designed to meet individual educational needs of handicapped persons as adequately as the needs of non-handicapped persons are met . . . .[44]

Section 504 requires similar protections and programs as P.L. 94-142 for all handicapped persons in need of special education programs and services. In fact, Section 504's regulations specifically state that implementation of an IEP in accordance with P.L. 94-142 meets the appropriate education mandate of §504.[45] As in P.L. 94-142, "qualified" handicapped students must be educated in the least restrictive setting, and integrated with non-handicapped students to the maximum extent possible.[46] Non-academic school services, including recess, lunch and similar activities must also provide for the maximum amount of integration possible.[47] In addition, a multidisciplinary team must determine the appropriate program of education and related services for each person only after reviewing the results of all necessary tests and evaluations administered to the person.[48] If a public education program cannot be located, a private special education facility must be used, including the provision of room and board at no cost to the person or his or her parent.[49]

For all practical purposes, each handicapped child protected by P.L. 94-142 will similarly benefit from the provision of Section 504. However, the definition of "handicapped person" is broader under Section 504, and is not limited by the P.L. 94-142 definitions of handicapping conditions.[50]

---

40 45 C.F.R. §121a.562 (1978).

41 *Id.* §§567-570.

42 29 U.S.C. §794 (1976). The United States Supreme Court has recently emphasized that "otherwise qualified" refers to a handicapped person's ability to participate in a program despite the existence of a disability. Southeastern Community College v. Davis, 99 S.Ct. 2361 (1979). Thus, a hearing impaired applicant for a registered nurse training program may properly be denied admission without violating §504, because she could not participate in the program or as a nurse without substantial physical and academic modifications. Those disagreeing with the decision have argued that auxiliary aids must be provided in order to assist an "otherwise [academically] qualified" person to enter a specific program and should have been in this case. *See* 45 C.F.R. §84.4 (1978). In any event, the *Davis* opinion, which was unanimous is limited in its application to the admission of the named plaintiff to a high technical post-secondary training program; thus, future matters should be decided on a case by case basis.

43 45 C.F.R. §§84.1 *et seq.* (1978).

44 45 C.F.R. §84.33(b) (1978). *See* Camenisch v. Univ. of Texas, No. A-78-CA-061 (S.D. Texas 1978); Barnes v. Converse College, 436 F. Supp. 635 (D.S.C. 1977); Crawford v. Univ. of North Carolina, 440 F. Supp. 1047 (M.D. N.C. 1977).

45 45 C.F.R. §84.33(b)(2) (1978).

46 *Id.* §84.34(a). *See, e.g.,* Hairston v. Drosick, 423 F. Supp. 180 (S.D. W.Va. 1976) (court ordered admission of spina bifida child to regular classroom). *See also* Pecunas v. Kline, No. 78-3133 (E.D. Pa., filed September 19, 1978) (action resulted in admission of quadraplegic child with cerebral palsy to regular classroom).

47 45 C.F.R. §84.37 (1978).

48 *Id.* §84.35.

49 *Id.* §§84.33(3), (4). *See* Howard S. v. Friendswood Independent School Dist., 454 F. Supp. 634 (S.D. Tex. 1978); Kruse v. Campbell, 431 F. Supp. 180 (E.D. Va. 1977), *vacated and remanded,* 98 S. Ct. 38 (1977), *on remand,* No. 75-0622-12 (E.D. Va., Jan. 5, 1978). *Compare* Halderman v. Pittenger, 391 F. Supp. 872 (E.D. Pa. 1975), *and* Scavella v. School Bd., 363 S.2d 1095 (Fla. 1978); with Cuyahoga Co. Ass'n for Retarded Adults v. Essex. 411 F. Supp. 46 (N.D. Ohio 1976) *and* Levy v. City of New York, 345 N.E.2d 556 (N.Y. 1976).

50 20 U.S.C. §1401(1),(15)(1976); 45 C.F.R. §121a.5 (1978) states:

(a) As used in this part, the term "handicapped children" means those children evaluated in accordance with §§121a.530-121a.534 as being mentally retarded, hard of hearing, deaf, speech impaired, visually handicapped, seriously emotionally disturbed, orthopedically impaired, other health impaired, deaf-blind, multi-handicapped, or as having specific learning disabilities, who because of those impairments need special education and related services.

(b) The terms used in this definition are defined as follows:

(1) "Deaf" means a hearing impairment which is so severe that the child is impaired in processing linguistic information through hearing, with or without amplification, which adversely affects educational performance.

(2) "Deaf-blind" means concomitant hearing and visual impairments, the combination of which causes such severe communication and other developmental and educational problems that they cannot be accommodated in special education programs solely for deaf or blind children.

Section 504 regulations define a "handicapped person" as "any person who (i) has a physical or mental impairment which substantially limits one or more major life activities, (ii) has a record of such an impairment, or (iii) is regarded as having such an impairment." "Physical or mental impairment" is defined in part as

> (A) any physiological disorder or condition, cosmetic disfigurement, or anatomical loss affecting one or more of the following body systems: neurological; musculoskeletal; special sense organs; respiratory, including speech organs; cardiovascular; reproductive; digestive; genito-urinary; hemic and lymphatic; skin; and endocrine; or (B) any mental or psychological disorder, such as mental retardation, organic brain syndrome, emotional or mental illness, and specific learning disabilities.

"Major life activities" means "functions such as caring for one's self, performing manual tasks, walking, seeing, hearing, speaking, breathing, learning and working."[51]

Thus, handicapped persons who have severe or unusual disabilities not covered by P.L. 94-142 or its implementing regulations may well be covered by Section 504 law. In addition, Section 504 requires an affirmative mandate that recipients provide architectural access to programs for all handicapped persons entering a school building or other educational building or program. The 504 regulations became effective on June 3, 1977. By August 2, 1977, access to programs or activities in existing facilities were to be made accessible to handicapped persons where structural changes were not required. By June 2, 1980 structural changes must be accomplished in existing facilities where such changes are necessary to obtain access to programs.[52]

The Gifted and Talented Children's Education Act

Realizing that gifted and talented children must be served, Congress passed the Gifted and Talented Children's Education Act of 1978.[53] This law provides limited funds for states willing to provide programs for gifted and talented persons and which apply for federal assistance; it does not require appropriate programs and procedural safeguards. Yet, its declaration of findings may very well set the stage for additional federal action in this area. Congress declared that-

> (1) the Nation's great resource for solving critical national problems in areas of national concern is its gifted and talented children,-
>
> (2) unless the special abilities of gifted and talented children are developed during their elementary and secondary school years, their special potentials for assisting the Nation may be lost, and-

---

(3) "Hard of Hearing" means a hearing impairment, whether permanent or fluctuating, which adversely affects a child's educational performance but which is not included under the definition of "deaf" in this section.

(4) "Mentally retarded" means significantly subaverage general intellectual functioning existing concurrently with deficits in adaptive behavior and manifested during the developmental period, which adversely affects a child's educational performance.

(5) "Multihandicapped" means concomitant impairments (such as mentally retarded-blind, mentally retarded-orthopedically impaired, etc.), the combination of which causes such severe educational problems that they cannot be accommodated in special education programs solely for one of the impairments. The term does not include deaf-blind children.

(6) "Orthopedically impaired", means a severe orthopedic impairment which adversely affects a child's educational performance. The term includes impairments caused by congenital anomaly (e.g., clubfoot, absence of some member, etc.), impairments caused by disease (e.g., poliomyelitis, bone tuberculosis, etc.), and impairments from other causes (e.g., cerebral palsy, amputations, and fractures or burns which cause contractures).

(7) "Other health impaired" means limited strength, vitality or alertness, due to chronic or acute health problems such as a heart condition, tuberculosis, rheumatic fever, nephritis, asthma, sickle cell anemia, hemophilia, epilepsy, lead poisoning, leukemia, or diabetes, which adversely affects a child's educational performance.

(8) "Seriously emotionally disturbed" is defined as follows:

(i) The term means a condition exhibiting one or more of the following characteristics over a long period of time and to a marked degree, which adversely affects educational performance:

(A) An inability to learn which cannot be explained by intellectual sensory, or health factors;

(B) An inability to build or maintain satisfactory interpersonal relationships with peers and teachers;

(C) Inappropriate types of behavior or feelings under normal circumstances;

(D) A general pervasive mood of unhappiness or depression; or

(E) A tendency to develop physical symptoms or fears associated with personal or school problems.

(ii) The term includes children who are schizophrenic or autistic. The term does not include children who are socially maladjusted, unless it is determined that they are seriously emotionally disturbed.

(9) "Specific learning disability" means a disorder in one or more of the basic psychological processes involved in understanding or in using language, spoken or written, which may manifest itself in an imperfect ability to listen, think, speak, read, write, spell, or to do mathematical calculations. The term includes such conditions as perceptual handicaps, brain injury, minimal brain disfunction, dyslexia, and developmental aphasia. The term does not include children who have learning problems which are primarily the result of visual hearing, or motor handicaps, of mental retardation, of emotional disturbance or of environmental cultural, or economic disadvantage.

(10) "Speech impaired" means a communication disorder such as stuttering, impaired articulation, a language impairment, or a voice impairment, which adversely affects a child's educational performance.

(ii) "Visually handicapped" means a visual impairment which, even with correction, adversely affects a child's educational performance. The term includes both partially seeing and blind children.

[51] 45 C.F.R. §84.3 (1978).

[52] *Id.* §84.21-33.

[53] 20 U.S.C. §§3311 *et seq.* (1976).

(3) gifted and talented children from economically disadvantaged families and areas often are not afforded the opportunity to fulfill their special and valuable potentials due to inadequate or inappropriate educational services.[54]

Further, this law defines "gifted and talented children" for federal purposes as persons-who are identified at the preschool, elementary, or secondary level as possessing demonstrated or potential abilities that give evidence of high performance capability in areas such as intellectual, creative, specific academic, or leadership ability, or in the performing and visual arts, and who by reason thereof, require services or activities not ordinarily provided by the school.[55]

A state education department wishing to apply for funding under the Act must submit an application, including an agreement to commit funding received for the purposes of statewide administration and planning designed to meet the needs of gifted and talented persons; the remainder is to be distributed to local educational agencies. A state's application must also certify that its programs will include a component for identification of and programming for low-income gifted and talented persons. While participation in the Act is discretionary at this date and funding is relatively low, this new law is significant in its recognition of the special educational needs of gifted and talented persons.[56] Moreover, it further points the way to the future acceptance of the exceptional child approach to education for gifted and talented persons.

Conclusion

It is evident that exceptional persons have obtained many victories. Cases have been won, and policies relating to the education of exceptional persons are now incorporated in federal statutes. Yet, case and federal statutory laws and regulations do not ensure that all children who are in need of services are presently obtaining them.[57] Children remain underserved, unidentified, or unjustifiably excluded,[58] and the cases brought years ago (*PARC, Mills, Frederick L.*) remain under the jurisdiction of courts convinced that their Orders have not been fully met. Thus the future requires continuing advocacy and effective state legislation to enforce the rights of exceptional persons. Toward this end, many states have adopted or amended legislation to comport to the trends of federal case law and statutes. We turn now to a review of state legislation.

---

[54] *Id.* §3311(b).

[55] *Id.* §3312.

[56] *Id.* §§3314, 3315. Litigation to develop and fund programs for gifted exceptional persons has only recently been progressing through the courts. *See, e.g.*, Central York School Dist. v. Pennsylvania Dep't of Educ., 349 A.2d 167 (Commw. Ct. 1979) (school districts must provide programs for gifted prior to state reimbursement). *cf.* Frederick L. v. Thomas, 419 F. Supp. 960 (E.D. Pa. 1976) (similar holding for programs for learning disabled students); *See also* EDUCATION DAILY, May 4, 1979 (describing Illinois lawsuit brought to provide appropriate education for gifted children).

[57] *See* Lottman, *Enforcement of Judicial Decrees: Now Comes the Hard Part,* 1 MENTAL DISABILITY L. REP. 69 (1976)

[58] For example, handicapped students have been known to be excluded for disruptive behavior that is the manifestation of the very handicap that the program of education and related services is designed to remediate. *See* Stuart v. Nappi, 443 F. Supp. 1235 (D. Conn. 1978); 22 PA CODE §§13.62 and 341.90. *cf.* Mrs. A.J. v. Special School District Number 1, 478 F. Supp. 418 (D. Minn. 1979).

REVIEW OF STATE LEGISLATION

Introduction

Due to states' own initiatives, the successful litigation over the past ten years and the passage of the Education for All Handicapped Children Act,[1] every state has some sort of provision in its statutes for the education of handicapped children. Therefore, it is important to examine what currently exists in this area.

Most of the state statutes were enacted or amended after the Education for All Handicapped Children Act of 1975. Yet it is surprising how diverse a set of statutes now exist in the states. Because the federal act has created a uniform mandate for special education, many state statutes require revision. This is necessary not only to comply with the federal requirements so that the state can receive federal funds under P.L. 94-142, but also because the federal law, while comprehensive, nevertheless establishes only a minimum policy base which state policy will need to reinforce.

State Policy (Chart I)

Federal law requires that in order to receive financial assistance for special education programs states must demonstrate the existence of a state policy to provide all handicapped children the right to a "free appropriate public education."[2] Undoubtedly the existence in state law of a policy statement to this effect would help evidence compliance with this requirement of federal law. Some states utilize the language of the federal act nearly verbatim as a statement of the state's educational policy toward exceptional children.[3] A number of other states use similar language. For instance, Alabama mandates free appropriate instruction[4] while New Jersey law seeks to provide for the handicapped the fullest possible opportunity to develop their intellectual capacities.[5] Tennessee calls for free public education to meet the needs and maximize the capabilities of handicapped children.[6] A few states have no general policy statements concerning special education but include a mandate to provide free appropriate educational opportunities in some other section of the state's act.[7] Often statutes will call for an "appropriate education" without specifying that it must be provided at no cost,[8] though this may be the implication of the act's provisions taken as a whole. Finally, several states merely mandate that the education and training of exceptional children shall be provided for[9] or that *all* children in the state will be provided an adequate education.[10] In addition, twenty-four states have expanded the federal mandate to include the gifted and/or talented.[11]

Children between the ages of 3 and 21 must be provided educational services in order for a state to receive federal assistance.[12] The age ranges provided for under state law vary considerably, but it is encouraging to note that a substantial number of state statutes exceed the federal requirement by providing services to children outside the federally set age ranges. Seventeen states provide for mandatory special education and related services from birth.[13] Still others provide for optional services at an age earlier than 3 in the discretion of the local school district providers.[14] A few states mandate services to children of "lawful school age," thereby allowing the state board of educa-

---

[1] P.L. 94-142, 89 STAT. 773 (1975).

[2] 20 U.S.C. §1412(1)(1976).

[3] *See e.g.*, LA. REV. STAT. ANN. §17:1941 (West Supp. 1978).

[4] ALA. CODE tit. 16 §39-3 (1975).

[5] N.J. STAT. ANN. §18A:46-19.1 (West Supp. 1978-1979).

[6] TENN. CODE ANN. §49-2912 (1977).

[7] *See, e.g.*, CONN. GEN. STAT. ANN. §10-76d(a) (West Supp. 1978-1979) (enumerating the duties of regional or town boards of education).

[8] *See, e.g.*, ALASKA STAT. §§14.30.180 (1975).

[9] *See, e.g.*, IDAHO CODE §33-2001 (Supp. 1978).

[10] *See, e.g.*, GA. CODE ANN. §32-603a (1976).

[11] See Chart I *infra*.

[12] 20 U.S.C. §1412(2)(B) (1976); *but see* the limitation within the definition; it states: a free appropriate public education will be available for all handicapped children between the ages of three and eighteen within the State not later than September 1, 1978, and for all handicapped children between the ages of three and twenty-one within the State not later than September 1, 1980, except that, with respect to handicapped children aged three to five and aged eighteen to twenty-one, inclusive the requirements of this clause shall not be applied in any State if the application of such requirements would be inconsistent with State law or practice, or the order of any court, respecting public education within such age groups in the State.

[13] *See* Chart I *infra*.

[14] *See e.g.*, ARK. STAT. ANN. §80-2124 (Supp. 1977).

tion, in most instances, to set age requirements by regulation.[15]

A critical element of any fully integrated effort to provide special educational services is an outreach system which searches out and identifies exceptional children in need of special services. The Education for All Handicapped Children Act of 1975 requires the state to accept this responsibility before receiving funds. Thirty-one states now include in their educational statutes provisions requiring the identification of exceptional children. Generally these provisions simply state the goal of identification and name the governmental unit responsible, usually the local school board.[16] Some, however, include a specific date by which all individuals must be identified and their names submitted to the state administrative authority.[17]

Evaluation (Chart II)

In order to provide a truly "appropriate" education for exceptional children with differing needs and educational requirements a state must develop methods by which individual children's exceptionalities can be identified. Forty-four states have provisions calling for personal evaluations to achieve these ends. The great majority of these statutes do not articulate the means of evaluation or the personnel who will be responsible for performing them. Rather, most statutes simply call for the state administrative authority to prescribe procedures by regulation.[18] Where personnel are described within the state statute the task of evaluation has been variously assigned to: school psychologists,[19] multidisciplinary teams,[20] a committee of qualified professionals,[21] "a specialist"[22] and other "qualified personnel."[23] Multidisciplinary teams in various states include among their members: school principals, teachers, special education experts, parents, physicians, psychologists, certified examiners and the child in question.[24]

Federal law requires that, regardless of who performs the evaluation, the techniques utilized in the evaluation process must be free of any racial, cultural or other biases based on handicap.[25] Only seven states mandate this as part of their special education legislation, though a number of states may provide for nondiscriminatory evaluation techniques by regulation.

Only a few states (9) provide explicitly for notice to parents of any impending evaluation[26] and even fewer (6) require a parent's consent before an evaluation may take place.[27] Notice in this context serves two important purposes. First it assures the parent's right to contest the results of an evaluation at the earliest feasible time, reducing the possibility of a mistaken placement and the ensuing detrimental effect on the exceptional child, and, second, it constitutes a step in the important goal of involving parents in educational decisions concerning their children. Also to this end, seven states provide parents the right to participate in the development of their child's educational program.[28] Finally, nine states give parents contesting an evaluation decision the right to seek an independent evaluation at state expense to be used as evidence at any hearing held to review the child's placement.[29] One other state leaves the hearing board with discretion to allow or deny an independent evaluation[30] and a final state allows the parents to seek an evaluation at their own expense.[31] Considering the relative infancy of special education training and substantive understanding and the diversity in makeup of the various state's evaluation boards a second independent evaluation may frequently turn up a major difference of opinion. Thus the right to an independent evaluation may be a significant contribution in the effort to assure each exceptional student the opportunity to develop his or her skills and the abilities in an optimal educational setting.

Delivery of Services (Chart III)

Where services are delivered to exceptional children is critical. The Education for All Handicapped Children Act of 1975 does not waiver in this regard. It demonstrates a clear preference for education in the least restrictive environment through participation by exceptional students in

---

[15] *See, e.g.,* KAN. STAT. §72-962(f) (Supp. 1978). In order to find all provisions relating to age requirements for special educational services the reader should consult both the column marked "age range" in Chart I and the column marked "preschool program" in Chart III. The latter will provide citations to optional preschool programs which may have the effect of extending the age range otherwise covered by the state act.

[16] *See, e.g.,* CAL. EDUC. CODE §56330(a) (West Supp. 1978).

[17] *See, e.g.,* WIS. STAT. ANN. §111.85(3) (West Supp. 1978-1979).

[18] *See, e.g.,* MO. ANN. STAT. §162.685(2) (Vernon Supp. 1978). *see* Chart II *infra.*

[19] PA. STAT. ANN. tit. 24, §13-1371 (Purdon Supp. 1978-1979).

[20] R.I. GEN. LAWS §16-24-7 (Supp. 1977); WIS. STAT. ANN. §115.80(3) (West Supp. 1978-1979).

[21] COLO. REV. STAT. §22-20-108 (Supp. 1977).

[22] ALA. CODE tit. 16, §39-4 (1975).

[23] ALASKA STAT. §14.30.330(c)(1975).

[24] *See* Chart II *infra.*

[25] 20 U.S.C. §1412(5)(C)(1976).

[26] *See* Chart IV *infra.*

[27] *See* Chart IV *infra.*

[28] *See* Chart II *infra.*

[29] *See* Chart II *infra.*

[30] MD. EDUC. CODE ANN. §8-415(c)(iii) (Supp. 1978).

[31] OHIO REV. CODE ANN. §3323.05(a) (Page Supp. 1978).

regular educational programs whenever appropriate to the needs of the individual.[32] In addition, since it is clear that not all exceptional children will benefit from regular education, a corollary to the concept is incorporated in federal law which requires that an individualized education program be developed for each child to insure that he or she is provided with the particular services which will maximize the development of his or her specific and unique potentialities.[33] In order for states to qualify under the act for federal aid they must accept and utilize these federal requirements.

Nevertheless only seven states require education in the least restrictive environment and only 14 require that services be provided in the context of regular educational programs when appropriate to the individual's needs.[34] Most states merely fail to deal with the specifics of this issue by statute, providing only the identity of the officials responsible for deciding where services will be delivered without any significant substantive guidelines. In this respect the policy statements discussed previously provide the only real guidance as to how administrators should structure substantive aspects of their special education programs.

Fifteen states specifically provide for individualized educational programming. A number of other states have similar requirements. Alaska, for instance, requires a "special learning program" designed to meet the needs of a child's specific exceptionality.[35] While New York provides that the child's individual needs will determine what program will ultimately be used.[36] Iowa law calls for each child to receive an appropriate educational program[37] and Connecticut requires the provision of services designed to meet the child's educational needs.[38] Again the reader should be aware that some states may have expanded their general policy statements mandating an "appropriate" educational program, by regulation, to provide for individualized educational programming.

Since not all exceptional children would benefit from attendance in regular education, a complete program for special education must make provision for alternative service settings. Thirty-three states now provide for such alternative settings. Usually these provisions are categorical, enumerating specific settings which are appropriate. Most frequently included are education at home, in hospitals and in institutions,[39] with services most often provided through visiting teacher programs.[40] In a number of states these categorical listings are not exclusive and the local school authorities are free to develop other necessary alternatives.[41] Frequently provisions concerning alternative educational settings can be found in the definitional section of the state act and are part of the state's general definition of special education.[42]

In addition to educational service settings, a few states include in their code a mandate to provide certain specific types of educational programming. Ten states specifically mention some form of occupational training or therapy, two states require training in basic life skills, and 20 states provide preschool programs for exceptional children.[43] These provisions are important because the types of programming mentioned are not ordinarily included as part of traditional educational programs and, consequently, may not be available unless the programs are specifically established by local service providers. Yet these programs or services may be necessary to the educational development of exceptional persons. While some local units may take the initiative and develop these and similar non-traditional programs independent of state requirements a specific mandate in the state code will help assure equality of educational opportunity throughout the state by the provision, in each local district, of uniform educational services.

That person or entity responsible for deciding where and what types of services will be delivered to each exceptional person is usually not identified in the state code and is instead to be designated by regulation. Most frequent among those actually named in the code are the local school board,[44] the superintendent of schools[45] or some multidisciplinary team.[46]

Federal law provides that during the pendency of any service delivery decision children must be allowed to remain where they are currently receiv-

---

[32] 20 U.S.C. §1412(5)(B)(1976).

[33] *Id.*, §1414(a)(5) (1976).

[34] *See* Chart III *infra.*

[35] ALASKA STAT. §14.30.350(1)(1975).

[36] N.Y. EDUC. LAW §4402(2)(a) (Mckinney Supp. 1978-1979).

[37] IOWA CODE ANN. §273.5(1)(West Supp. 1978-1979).

[38] CONN. GEN. STAT. ANN. §§10-76(a), 76d(a) (West Supp. 1978-1979).

[39] *See, e.g.*, OR. REV. STAT. §343.035 (1977).

[40] *See, e.g.*, W.VA. CODE §18-20-1 (Supp. 1978).

[41] *Id.*

[42] *See, e.g.*, OR. REV. STAT. §343.035(3) (1977); ILL. ANN. STAT. ch.-122, §14-1.08 (Smith-Hurd Supp. 1978). This is consistent with the federal act's treatment. *See* 20 U.S.C. §1401(16) (1976).

[43] *See* Chart III *infra.*

[44] MICH. COMP. LAWS ANN. §380.1711(g) (Supp. 1978-1979); WIS. STAT. ANN. §115.81(3) (West Supp. 1978-1979).

[45] ARIZ. REV. STAT. ANN. §15-1013 (Supp. 1978-1979); ILL. ANN. STAT. ch. 122, §14-8.01 (Smith-Hurd Supp. 1978).

[46] ALA. CODE tit. 16, §39-8 (1975); CONN. GEN. STAT. ANN. §10-76d(a) (West Supp. 1978-1979); COLO. REV. STAT. §22-20-108 (Supp. 1977); LA. REV. STAT. ANN. §17:1945 (West Supp. 1978).

ing services, or if they are applying to public school for the first time, be placed in an ordinary public school program during the interim.[47] Only 7 states address this issue by statute. Of these, 5 provide that the child should remain where previously situated, one provides that the child should remain in the regular classroom and one allows the child to be temporarily excluded from school.[48] Only one of these explicitly allows for temporary admission in the public school system during the pendency of a final service delivery decision.[49]

Procedural Guarantees (Chart IV)

P.L. 94-142 contains strict requirements safeguarding the due process rights of parents and exceptional persons in the education system.[50] Charts IVA through IVC survey the responsiveness of present state legislation to the procedural requirements of the federal law. Twenty states have included no procedural safeguards in their education statutes. A few of these, however, provide that due process procedures will be prescribed by regulation.[51] Of those procedures which are enumerated by statute the most frequently included are: the right to seek judicial review of an administrative decision (18 states), notice to parents before initiating or refusing a change in evaluation or educational program (19 states) and the right to an administrative hearing (23 states).[52] Nine states provide the right to have counsel present, to present evidence and to cross examine witnesses in an administrative hearing.[53] Of these, 5 allow parties to compel the attendance of witnesses.[54] In addition, 11 states mandate an impartial hearing officer; that is, one who was not previously involved with decisions relating to the delivery of services to the child.[55]

In order to safeguard the right to later judicial review, seven states require the administrative decisionmaker to hand down a written opinion, and 10 provide for the keeping of a written transcript of any administrative hearing, available to the parties, usually at no cost.[56]

Of those states which provide notice to parents of proposed changes in evaluation or educational program, 11 require that the notice contain an explanation of the party's due process rights, 6 require that the notice be in the recipient's native language and 4 require that it contain explanations as to why alternative programs were rejected.[57] Ten states go beyond federal procedural requirements by mandating parental consent before any change in program may be initiated.[58]

Eleven states allow parents full access to their children's educational evaluation, written record and, ordinarily, clinical history.[59] This feature ensures parents the information necessary to effectively contest one evaluation on the merits. Nine states provide assurances that these records will otherwise be held confidential and that only persons directly concerned with the case of the exceptional child will be allowed access to them for any purpose.[60]

While formal administrative procedures are a necessary and important safeguard for parents and children, the goal of parental involvement in the development of the child's educational program may be better served by informal non-adversary procedures through which a parent can object to or question any aspect of the child's education. Only 8 states, however, currently provide for these informal sessions by statute.[61]

Finally, federal law requires, in keeping with a philosophy of individualized educational programming, that program decisions be reviewed at least annually.[62] Ten states currently comply with this requirement by statute.[63] Two states, Louisiana and New York, provide for reevaluation but not as frequently as is required by the federal statute.[64] The remainder of Chart IV indicates the time requirements for hearing and deciding matters on appeal.

Administrative Responsibilities (Chart V)

The Education for All Handicapped Children Act of 1975 articulates a number of administrative responsibilities which state and local agencies must assume in order to receive federal grants for special education.[65] Chart V reviews state compliance

---

[47] 20 U.S.C. §1415(e)(3) (1976).

[48] *See* Chart III, *infra.*

[49] OR. REV. STAT. §343.077(7) (1977).

[50] 20 U.S.C. §1415 (1976).

[51] *See, e.g.,* ME. REV. STAT. tit. 20, §3131 (Supp. 1978); MONT. REV. CODES ANN. §75-7802(1)(b) (Supp. 1977).

[52] *See* Chart IV, *infra.*

[53] *Id.* Kansas, Louisiana, Missouri, Nebraska, North Carolina, Ohio, Oregon, Tennessee, Wisconsin

[54] *Id.* Kansas, Louisiana, Nebraska, Ohio, Oregon.

[55] *See, e.g.,* FLA. STAT. ANN. §230.23(4)(M)(4) (West Supp. 1979); MD. EDUC. CODE ANN. §8-415(a)(3) (1978).

[56] *See* Chart IV *infra.*

[57] *Id.*

[58] *Id.*

[59] *See, e.g.,* N.C. GEN. STAT. §§115-376(b), 179.1 (1978); TENN. CODE ANN. §49-2947 (1977).

[60] *See, e.g.,* LA. REV. STAT. ANN. §17:1955 (West Supp. 1978)

[61] *See* Chart IV *infra.*

[62] 20 U.S.C. §1414(a)(5) (1976).

[63] *See* Chart IV *infra.*

[64] LA. REV. STAT. ANN. §§17:1941, 1945 (West Supp. 1978) (every 3 years or upon request); N.Y. EDUC. LAW §4402(1)(b)(3)(d) (McKinney Supp. 1977-1978) (every 3 vears).

[65] 20 U.S.C. §1414 (1976).

with these requirements and other frequently used administrative provisions.

Nearly all states (41) place primary responsibility for the provision of special education on local school authorities, ordinarily at the school district level.[66] In 28 states, however, state administrators retain some power over local officials by virtue of a statutorily granted authority to withhold special education funds in the event of local non-compliance.[67] In addition, Maine and Alabama allow injunctive relief against non-complying districts.[68]

Since a local school district may not be able to provide educational programming to meet the unique educational needs of a particular exceptional person, some statutes provide for contractual arrangements with private organizations and inter-district cooperation. Toward this end, inter-district cooperative programs are provided by 43 states, interstate cooperation by 19 states and contracts with private service providers by 38 states.[69] Often these statutes allow for cooperative programs or programs in private facilities only when to do so is deemed the sole method of achieving the needed educational program for the child.[70] In addition, most statutes provide that such arrangements do not shift the primary responsibility for the adequacy of the educational program which remains on the school district of the child's residence.[71]

Pursuant to federal law, the state educational agency must monitor local compliance with federal requirements before allocating funds to the local unit.[72] All federal monies are given directly to the state agency which is responsible for receiving applications from, and allocating funds to, the local service providers.[73] In order to facilitate compliance with these mandates a number of administrative techniques are employed. Fourteen states require by statute that local school districts annually submit a program plan outlining the measures it proposes to take to assure that the needed special education is provided.[74] Many of these same states and a few others (17 total) require local district administrators to submit additional annual reports on their compliance with state and federal law during the preceding funding period,[75] particularly describing the extent and types of services that have been and are being provided. Similarly, 16 states require local districts and state administrators to compile and keep records of the number and types of exceptional children within each district who are currently in need of or are currently receiving special education services.[76] Other procedures for assuring compliance by local districts are provided in 16 states.[77] By far, the most frequently employed method provides for an annual review by a state or intermediate authority of each district's compliance with state and federal law.[78] In addition most of these same states mandate an annual or semiannual evaluation of the effectiveness of these local programs.[79]

Responding to the complexity of special education management, 9 states have created separate administrative divisions within the states' larger education departments.[80] These special divisions are responsible for all the state level administrative duties regarding special services. Twenty-five states have also created special advisory committees which suggest improvements in educational services for exceptional persons to state and local administrators.[81] P.L. 94-142 requires state level advisory panels[82] and these are the most frequent type. California and Oregon go beyond the federal mandate in this regard and also provide for local advisory panels to aid in district-level planning and compliance.[83]

Financing (Chart VI)

State financing structures for public education are extremely diverse. Most state aid for regular educational programming is computed on the basis of the number of classroom units or qualified students present in the local district with a fixed sum available from the state for each classroom or student. Since special education may be more expensive than traditional education, the general financing system alone cannot provide sums adequate for local units to maintain special education programming.

Chart VI provides citations to the state code provisions which set out the extent and availability of state aid for the excess costs incurred by local

[66] *See* Chart V *infra.*

[67] *Id.*

[68] ALA. CODE tit. 16, §39-5 (1975); ME. REV. STAT. tit. 20, §3127(4)(C) (Supp. 1978).

[69] *See* Chart V *infra.*

[70] *See, e.g.,* LA. REV. STAT. ANN. §17:1947(D) (West Supp. 1978).

[71] *See, e.g.,* MINN. STAT. ANN. §120.17(1) (West Supp 1979).

[72] 20 U.S.C. §1414 (1976).

[73] *Id.*

[74] *See* Chart V *infra.*

[75] *Id.*

[76] *Id.*

[77] *Id.*, see column entitled "Procedures for Monitoring Each District's Compliance with Law."

[78] *See, e.g.,* CAL. EDUC. CODE §46312(e) (West Supp. 1978); CONN. GEN. STAT. ANN. §10-76b(d) (West Supp. 1978-1979); MICH. COMP. LAWS ANN. §380.1711(h) (Supp. 1978-1979).

[79] *See* Chart V *infra.*

[80] *Id.*

[81] *Id.*

[82] 20 U.S.C. §1413(a)(12) (1976).

[83] CAL. EDUC. CODE §56330(r) (West Supp. 1978); OR. REV. STAT. §343.293 (1977).

units providing special education services. Although the specific systems and provisions are too complex to be discussed with any detail here, a few frequently encountered approaches can be described. Some states provide a detailed list of which specific costs will be reimbursed by the state and at what level.[84] Others provide simply that all excess costs will be reimbursed.[85] A number of statutes limit state aid to a certain dollar figure for each exceptional child being educated by the local unit[86] or to a percentage of the local unit's total cost of providing special educational services.[87] Arizona is an example of a state that provides aid to local units only in the event that local taxing efforts produce insufficient revenues to fully support special education programs.[88]

Twenty-four states include specific provisions concerning financial arrangements when exceptional children receive educational programs and services in private facilities.[89] Ordinarily the local school district responsible for the education of the child must initially contract and pay for the private services and will then be reimbursed for part or all of its costs.[90] Most states will reimburse for the full excess costs of the private service, but some limit state aid to a percentage of the full cost or to a specific dollar figure.[91] Provisions for funding educational programs and services that are delivered out-of-state are included in 12 states and follow essentially the same patterns as the provisions for delivering services in private facilities.[92]

---

[84] *See, e.g.*, N.J. STAT. ANN. §18A:58-6 (West Supp. 1978-1979); OHIO REV. CODE ANN. §3317.02.4 (Page Supp. 1978).

[85] *See, e.g.*, UTAH CODE ANN. §53-18-3 (1970).

[86] *See, e.g.*, CAL. EDUC. CODE §84850(a) (West Supp. 1978); ILL. STAT. ANN. §14-7.02a (Smith-Hurd Supp. 1978).

[87] MICH. COMP. LAWS ANN. §388.1452 (Supp. 1978-1979); NEB. REV. STAT. §43-648 (Supp. 1978).

[88] ARIZ. REV. STAT. ANN. §15-1018 (Supp. 1978-1979).

[89] *See* Chart VI *infra*.

[90] *See, e.g.*, ALA. CODE tit. 16, §39-3 (1975); MINN. STAT. ANN. §124.32(5) (1979).

[91] *See e.g.*, N.J. STAT. ANN. §18A:58-6 (West Supp. 1978-1979); VA. CODE §22-10.8 (Supp. 1978).

[92] *See* Chart VI *infra*.

CHARTS OF STATE SPECIAL EDUCATION STATUTES

I. STATE POLICY

State	Educational Policy	Person-Find Required	Gifted and Talented Persons Included	Age Range
ALABAMA Code (1975)	"Appropriate instruction" tit. 16, §39-3	tit. 16, §§39-4, 39-9	tit. 16, §39-2(1)	tit. 16, §39-2 (Supp. 1978)
ALASKA Stat. (1975)	"Competent education" (does not specify free) §§14.30.180(c), 14.30.191(c)[1]		§14.30.350(E)	At least 3 years §14.30.180
ARIZONA Rev. Stat. Ann. (Supp. 1978-1979)	"Commen-surate with their abilities and needs" §15-1010(b)		§15.1011(2)	Lawful school age §15-1011(3)
ARKANSAS Stat. Ann. (Supp. 1977)	F.A.P.E. §80-2117			6-21 yrs. §80-2124
CALIFORNIA Educ. Code (West Supp. 1978)	F.A.P.E. §§56033, 56301, 56304	§§56001, 56330(a)		School age-22 yrs. (Mentally retarded only) §56501 3-21 yrs. (All others) §56302
COLORADO Rev. Stat. (1974)	"To provide educational opportunities" (Does not specify free) §22-20-102	§22-20-103(4)		5-21 yrs. §22-20-103(4)
CONNECTICUT Gen. Stat. Ann. (West Supp. 1978-1979)	[2]	§10-76d(a)	10-76a(e)	Under 21 yrs. §10-76a (1977)
DELAWARE Code Ann. (Supp. 1977)	F.A.P.E. tit. 14, §3120	tit. 14, §3122	tit 14, §3101(3)	4-20 yrs. tit. 14, §3110
DISTRICT OF COLUMBIA				
FLORIDA Stat. Ann. (West Supp. 1979)	§230.23(4)(m)(1) (does not specify free)	§232.13 (1977)	§228.041(19)	3 yrs. §232.01(f) (1977)
GEORGIA Code Ann. (1976)	§32-603a[3]		§32-605a	School age §32-605a(a)
HAWAII Rev. Stat. (1976)	". . . education, therapy and training to enable them to live normal competitive lives." (does not specify free) §301-22			Under 20 yrs. §301-21
IDAHO Code (Supp. 1978)	". . . shall provide for the education and training of exceptional pupils.' §33-2001		§33-2002	0-21 yrs. §33-2001

State	Educational Policy	Person-Find Required	Gifted and Talented Persons Included	Age Range
ILLINOIS Ann. Stat. (Smith-Hurd Supp. 1978)			ch. 122, §§14A-1 *et seq.*	3-21 yrs. ch.122, §§14-1.02 *et seq.*
INDIANA Code Ann. (Burns)	F.A.P.E. §20-1-6-15 (1976)			3-21 yrs. §20-1-6-1 (Burns Supp. 1978)
IOWA Code Ann. (West Supp. 1978-1979)	"Special education opportunities sufficient to meet the needs and maximize the capabilities of children requiring special education." §281.2	§§281.9(5), 273.1		Under 21 yrs. §281.2
KANSAS Stat. (Supp. 1978)	Provide for the development of each exceptional child in accord with his abilities and capacities. §§72-961, 72-966[5]		§72-962(f)	School age §72-962(f)
KENTUCKY Rev. Stat. (1977)	"A comprehensive educational program" §157.224	§157.260	§157.200(1)	Under 21 yrs. §157.200(1)
LOUISIANA Rev. Stat. Ann. (West Supp. 1978)	F.A.P.E. §17:1941	§§17:1944(B)(1), 17:1945(D)	§17:1943	3-21 yrs. §17:1947(A)
MAINE Rev. Stat. (Supp. 1978)	"equal education opportunities for all exceptional children" tit. 20, §3121 "An appropriate program for exceptional children may be established by any administrative unit operating schools tit. 20, §3125	tit.20, §3124		5-20 yrs. tit. 20, §3123(1)
MARYLAND Educ. Code Ann. (1978)	F.A.P.E. §8-402	§§8-403(d), 8-411		Under 21 yrs. §8-403(a)
MASSACHUSETTS Gen. Laws Ann. (West Supp. 1978-1979)	". . . To assure the maximum possible development of a child with special needs . . ." ch. 71B, §2	ch. 71B, §3		3-21 yrs. ch. 71B, §1
MICHIGAN Comp. Laws Ann. (Supp. 1978-1979)	F.A.P.E. §§380.1701, 380.1714, 380.1751	§380.1711(f)		0-26 yrs. §§380.1711(f), 380.1711(j)
MINNESOTA Stat. Ann. (1979)	F.A.P.E. §120.17(3a)			4-21 yrs. §120.17(1)
MISSISSIPPI Code Ann. (Supp. 1978)	"Competent education" (does not specify free) §37-23-1	§37-23-129(c)	§37-23-3	0-21 yrs. §37-23-3
MISSOURI Ann. Stat. (Vernon Supp. 1978)	F.A.P.E §162.670	§162.695(1)	§162.720	0-21 yrs. §162.675(2)

State	Educational Policy	Person-Find Required	Gifted and Talented Persons Included	Age Range
MONTANA Rev. Codes Ann. (Supp. 1977)	F.A.P.E. §75-7805(1)	§75-7802(1)(c)		6-18 yrs. after 9/1/77; 3-21 after 9/1/80 §75-7805
NEBRASKA Rev. Stat. (Supp. 1978)	"Meaningful education" §43-641 However, §43-649 & §43-652 speak about an "appropriate educational program."	§43-643	§79-339	0-21 yrs. §43-646
NEVADA Rev. Stat. (1977)	". . . a reasonably equal educational opportunity" (does not specify free) §388.450		§388.490(4)	3-18 yrs. §§388.490, 388.440
NEW HAMPSHIRE Rev. Stat. Ann. (1977)	"Best and most effective" §186-A:1	§186-A:4		0-21 yrs. §186-A:6
NEW JERSEY Stat. Ann. (West Supp. 1978-1979)	"Fullest possible opportunity to develop their intellectual capacities" §18A:46-19.1	§18A:46-6		5-20 yrs. §18A:46-6
NEW MEXICO Stat. Ann. (1978)	Sufficient to meet the needs of exceptional children §22-13-5 Free §22-13-8			
NEW YORK Educ. Law (McKinney Supp. 1978-1979)	F.A.P.E. §4402(2)(a)	§4402(1)(a)		0-21 yrs. §4401
NORTH CAROLINA Gen. Stat. (1978)	F.A.P.E. §§115-363, 115-364	§§115-363, 115-367(j)	§115-366	5-18 yrs. §115-366
NORTH DAKOTA Cent. Code (Supp. 1977)	"To meet the needs of" §15-59-01	§15-59-04.1	§15-59-01	0-21 yrs. §15-59-01
OHIO Rev. Code Ann. (Page Supp. 1978)	F.A.P.E. §3323.02	§3323.03	§3301.07(L)	0-22 yrs. §3323.01
OKLAHOMA Stat. Ann. (West Supp. 1978-1979)	"Necessary education" tit. 70, §13-101	tit. 70, §13-111	tit. 70, §13-101	4 yrs.; at any time for severely handicapped tit. 70, §13-101
OREGON Rev. Stat. (1977)	Oregon has a general policy to provide a free education to all persons §§343.041, 343.085[6]		§343.391 *et seq.*	0-21 yrs. §343.221
PENNSYLVANIA Stat. Ann. (Purdon Supp. 1978-1979)	"Proper education and training" tit. 24, §13-1372(1)[7]	tit. 24, §13-1371(2)		6-21 yrs. tit. 24, §13-1376(a)
RHODE ISLAND Gen. Laws (Supp. 1977)	"Education that will best satisfy the needs of . . ." §16-24-1 (1970)[8]	§16-24-3	§16-42-1 *et seq.* (1970)	3-21 yrs. for mentally retarded & multi-handicapped §16-24-2[9]

State	Educational Policy	Person-Find Required	Gifted and Talented Persons Included	Age Range
SOUTH CAROLINA Code (1976)	F.A.P.E. §59-33-10	§59-33-40		6-21 yrs. §§59-33-10, 59-63-20
SOUTH DAKOTA Compiled Laws Ann. (1975)	F.A.P.E. §13-37-14.1			0-21 yrs. §13-37-1 (Supp. 78)
TENNESSEE Code Ann. (1977)	Meet the needs & maximize the capabilities of handicapped children §49-2912	§49-2941(c)		School age and under 21 for exceptional, 4-21 for handicapped §§49-2901, 49-2914[10]
TEXAS Educ. Code Ann. (Vernon Supp. 1978-1979)	F.A.P.E. §16.104(a)			3-21 yrs. §16.104(a)
UTAH Code Ann. (1970)	Appropriate program of special instruction §53-18-2	§53-18-1		5-21 yrs. §53-18-3
VERMONT Stat. Ann. (Supp. 1978)	"Equal education" tit. 16, §2941			0-21 yrs. tit. 16, §2942(1)
VIRGINIA Code (Supp. 1978)	F.A.P.E. §22-10.4(A)			2-21 yrs. §§22-10.3, 22-10.4
WASHINGTON Rev. Code Ann. (Supp. 1978)	F.A.P.E. §28A.13.005			Common school age §28A.13.005
WEST VIRGINIA Code (Supp. 1978)	"Special programs shall be maintained for exceptional children" (does not specify free) §18-20-1	§18.20.2	§18-20-1	5-23 yrs. §18-20-1
WISCONSIN Stat. Ann. (West Supp. 1978-1979)	F.A.P.E. §§115.76(10), 115.85(1)	§115.85(3)		3-21 yrs. §§115.85(1), 115.76.2
WYOMING Stat. (1977)	F.A.P.E. §21-14-101			School age §21-14-101

1 Although there is no stated policy to provide an "appropriate" education, §14.30.191(c) speaks of an "appropriate educational program placement."

2 The Connecticut statute has no general policy statement dealing with the education of exceptional children. However §10-76b(a) states that, "The state board of education shall provide for the development and supervision of the educational programs and services for children requiring special education. . . ." Also, §10-76d(a) states that each town or regional board of education shall prescribe suitable educational programs for eligible children.

3 There is a general policy in Georgia that all persons be educated. There is no provision specifically dealing with the right of exceptional persons to be educated.

4 There is no policy statement in the statute. There is no mention of the type of education that exceptional persons are to receive or if it is free. The statute simply begins with a definitional section. §14.801, however, speaks of placing the child in an "appropriate program."

5 The Kansas statute contains a general policy to educate exceptional children. The types of education that will be provided are to be set in accordance with standards set by the state board of education. The statute does not specify whether the education provided is free.

6 There is no state policy specifically regarding the education of exceptional persons.

7 Those certified as uneducable and untrainable are excluded from the provisions of the statute.

8 The statute only applies to mentally retarded and multiple handicapped children. Others to which the statute will apply are defined in state board of education regulations.

9 The eligibility age for other exceptional persons will be established in state board of education regulations.

10 §49-2901 — "exceptional children" — children of school age and under 21 years of age, of educable mind, whose bodily functions or members are so impaired that they cannot safely or adequately be educated in the regular classes of the public schools without special educational services, and to include children of school age and under 21 who are psychologically exceptional, and to include children of pre-school age down to and including the age of 3 years, who are deaf §49-2914 — "handicapped children" — children and youth between the ages of 4 and 21 inclusive who have been certified under regulations of the state board of education by a specialist as being unsuited for enrollment in regular classes of the public schools or who are unable to be educated or trained adequately in such regular programs without the provision of special classes, instruction, facilities, or related services.

II. EVALUATION

State	Nondiscriminatory Testing	Right Of Parents To An Independent Evaluation	Parent Involvement In Development Of Educational Program	Evaluated By
ALABAMA Code (1975)				A specialist for each type of exceptionality tit. 16, §§39-4, 39-5[1]
ALASKA Stat. (1975)		§14.30.191(c)		[2]
ARIZONA Rev. Stat. Ann. (Supp. 1978-1979)			§15-1013(d)(5)	§15-1013(b)[3]
ARKANSAS Stat. Ann. (Supp. 1977)				§§80-2111, 80-2119[4]
CALIFORNIA Educ. Code (West Supp. 1978)	§56301(i)	§56337(c)(2)	§56036(a)(4)	School appraisal team §56337[5]
COLORADO Rev. Stat. (1974)				Committee of qualified professionals §22-20-108 (Supp. 1978)[6]
CONNECTICUT Gen. Stat. Ann. (West Supp. 1978-1979)		At parents' own expense §10-76d(a)	§10-76d(a)	[7]
DELAWARE Code Ann. (Supp. 1977)				[8]
DISTRICT OF COLUMBIA				
FLORIDA Stat. Ann. (West Supp. 1979)				"School board will provide the necessary professional service" §230.23(4)(m)(1)
GEORGIA Code Ann. (1976)				
HAWAII Rev. Stat. (1976)				Licensed physician §301-21(3) Dept. of Education §301-26
IDAHO Code (Supp. 1978)				[9]
ILLINOIS Ann. Stat. (Smith-Hurd Supp. 1978)	ch. 122, §14-8.01			ch. 122, §14-8.01[10]
INDIANA				

State	Nondiscriminatory Testing	Right Of Parents To An Independent Evaluation	Parent Involvement In Development Of Educational Program	Evaluated By
IOWA Code Ann. (West Supp. 1978-1979)				[11]
KANSAS Stat. (Supp. 1978)	§72-963(d)			[12]
KENTUCKY Rev. Stat. (1977)				§157.240[13]
LOUISIANA Rev. Stat. Ann. (West Supp. 1978)	§§17:1941, 17:1945(A)	§17:1952(B)(1)	§§17:1944(10), 17:1952(A)	Multidisciplinary team §17:1945[14]
MAINE Rev. Stat. (Supp. 1978)				[15]
MARYLAND Educ. Code Ann. (1978)		At hearing board's discretion §8-415(c)(iii) (Supp. 1978)		[16]
MASSACHUSETTS Gen. Laws Ann. (West Supp. 1978-1979)	ch. 71B, §7	ch. 71B, §3	ch. 71B, §3	[17]
MICHIGAN Comp. Laws Ann. (Supp. 1978-1979)				An appropriate professional §380.1711(j)[18]
MINNESOTA Stat. Ann. (1979)	§120.17(3a)(d)			[19]
MISSISSIPPI Code Ann. (Supp. 1978)				[20]
MISSOURI An . Stat. (Vernon Supp. 1978)				[21]
MONTANA Rev. Codes Ann. (Supp. 1977)				Trustees of local school district §75-7811
NEBRASKA Rev. Stat. (Supp. 1978)				[22]
NEVADA Rev. Stat. (1977)				[23]
NEW HAMPSHIRE Rev. Stat. Ann. (1977)				[24]
NEW JERSEY Stat. Ann (West Supp. 1978-1979)				[25]
NEW MEXICO Stat. Ann. (1978)				[26]

State	Nondiscriminatory Testing	Right Of Parents To An Independent Evaluation	Parent Involvement In Development Of Educational Program	Evaluated By
NEW YORK Educ. Law (McKinney Supp. 1978-1979)		§4402(3)	§4402(1)(b)(3)	Committee on the handicapped §4402(1)(b)(1)[27]
NORTH CAROLINA Gen. Stat. (1978)	§115-375(a)			[28]
NORTH DAKOTA Cent. Code (Supp. 1977)				
OHIO Rev. Code Ann. (Page Supp. 1978)		At parents' expense §3323.05(A)		[29]
OKLAHOMA Stat. Ann. (West Supp. 1978-1979)				[30]
OREGON Rev. Stat. (1977)				Qualified educational & medical authorities §343.227
PENNSYLVANIA Stat. Ann. (Purdon Supp. 1978-1979)				Public school psychologist plus any other expert which the child's condition may necessitate tit. 24, §13-1371
RHODE ISLAND Gen. Laws (Supp. 1977)				Individual multidisciplinary evaluation §16-24-7
SOUTH CAROLINA Code (1976)				[31]
SOUTH DAKOTA Compiled Laws Ann. (Supp. 1978)				[32]
TENNESSEE Code Ann. (1977)		(at state expense only if parents are unable to afford it) §49-2947		[33]
TEXAS				
UTAH (Code Ann. (1970)				Public school psychologist §§53-18-1, 53-18-2, 53-18-7[34]
VERMONT Stat. Ann. (Supp. 1978)				
VIRGINIA Code (Supp. 1978)				[35]

State	Nondiscriminatory Testing	Right Of Parents To An Independent Evaluation	Parent Involvement In Development Of Educational Program	Evaluated By
WASHINGTON Rev Code Ann. (Supp. 1978)				
WEST VIRGINIA Code (Supp. 1978)				Appropriate specialist §18-20-4[36]
WISCONSIN Stat. Ann. (West Supp. 1978-1979)		(at state expense only if parents unable to afford it) §115.81(5)	§115.80(3)(c)	Multidisciplinary team §115.80(3)[37]
WYOMING Stat. (1977)				[38]

1 A specialist is defined as a physician, psychologist, psychometrist, or other professional personnel qualified pursuant to state board of education regulations.
2 Evaluation procedures are defined in state board of education regulations. §14.30.330(c)
3 Each child must be evaluated by at least one professional specialist in a field relevant to the child's handicap. If appropriate he shall be evaluated by a psychologist.
4 Children may be admitted to all such special educational services and/or discharged therefrom on the recommendation of physicians under whom they are being treated or by medical, psychological, psychiatric or therapeutic experts chosen or approved by the Board.
5 Mandatory minimum members: (1) school principal, (2) the special education teacher or teachers or specialist or specialists most appropriate to the needs of the pupil, (3) the parent or his selected representative or both. Other optional members: (1) a regular teacher of the pupil, (2) special education specialist or whoever conducted an assessment of the pupil, (3) the pupil, whenever he is capable of benefitting from the discussion.
6 The committee is composed of but not limited to: (1) the director of special education for the administrative unit, (2) a psychologist, (3) social worker, (4) physician, (5) a school administrator, (6) a teacher of the handicapped.
7 Evaluation procedures defined by regulations promulgated by the secretary of the state board of education. §10-76d
8 Evaluation procedures defined by state board of education regulations. tit. 14, §3122
9 Evaluation procedures defined by state board of education regulations. §33-2003(4)
10 A child is eligible for special education only upon "a carefully completed case study fully reviewed by professional personnel in a staff conference and only upon the recommendation of qualified specialists."
11 Evaluation procedures established by the Division of Special Education with the approval of the state board of education. §281.7
12 Evaluation procedures defined in state board of education regulations. §72-963d
13 "No physically handicapped child shall be allowed special education facilities except upon a certified diagnosis of a physical defect by competent and appropriate professional authorities acceptable to the superintendent of public education."
14 Statute does not specify who comprises the "multidisciplinary team." Reference is only to members who are "certified as qualified examiners by the Department of Education."
15 Evaluation procedures established by the Commissioner of Education. tit. 20, §3124
16 Evaluation procedures defined in state board of education regulations. §8-403(d)(2)
17 Evaluation is conducted by (1) a representative of the local school department, (2) a classroom teacher who has dealt with the child in the classroom, (3) a complete medical assessment by a physician, (4) a psychologist, (5) a nurse, (6) a social worker, and (7) specialists as may be required. ch. 71B, §3
18 Evaluation procedures defined in state board of education regulations.
19 Evaluation procedures defined in state board of education regulations. §120.17(3)
20 Evaluation procedures defined in state department of education regulations. §37-23-125 to 131
21 Evaluation procedures defined in state board of education regulations. §162.685
22 Evaluation procedures defined in state department of education regulations. §43-646.10
23 Evaluation procedures defined in school district regulations. §388.450
24 Evaluation procedures defined in department of education regulations. §186-A:6
25 Evaluation procedures established by the Commissioner of Education with the approval of the state board of education. §18A:46-8
26 Evaluation procedures defined in state board of education regulations. §22-13-7(B)
27 Committee is comprised of: (1) school psychologist, (2) a teacher or administrator of special education, (3) school physician, (4) a parent of a handicapped child residing in the school district, and (5) such other persons as the Board of Education shall designate. The committee shall invite the appropriate professionals most familiar with the child's handicap or handicaps to attend any meeting concerning the educational program for each child.
28 Evaluation procedures defined in state board of education regulations. §115-375(b)
29 Evaluation procedures defined in state board of education regulations. §3303.03
30 Evaluation procedures defined in state board of education regulations. tit. 70, §13-302
31 Evaluation procedures defined in state board of education regulations. §59-33-30
32 Evaluation procedures defined in state board of education regulations. §13-37-4
33 Evaluation procedures defined in state board of education regulations. §49-2941
34 §53-18-2 states, "The state board of education shall adopt standards and regulations relating to the diagnosis and evaluation of the handicapped children by competent professional personnel. . . ."
35 Evaluation procedures defined in state board of education regulations. §22-10.4
36 Each child shall be examined by an appropriate medical specialist, psychologist or educational specialist. The specialist's report shall carry recommendations for eligibility and placement.
37 Multidisciplinary team includes "two or more persons who are skilled in assessing exceptional educational needs that a child may have and who are skilled in programming for children with exceptional educational needs. Additional members may be appointed by the state superintendent."
38 Evaluation procedures defined in state board of education regulations.

III. DELIVERY OF SERVICES

State	Decision-Maker	I.E.P. Guaranteed	Non-Traditional Educational Programming: Vocational Skills Included	Non-Traditional Educational Programming: Life Skills Included	Non-Traditional Educational Programming: Preschool Program Available	Least Restrictive Environment	Provision To Educate With Non-Exceptional To Maximum Extent Possible	Provision For Educating Outside of School Facility	Pending Evaluation: Remain Where Situated	Pending Evaluation: In Regular Classroom
ALABAMA Code (1975)	Placement committee (medicine, education, and psychology) tit.16, §39-8		tit.16, §39-2(6) (Supp. 1978)				tit. 16, §39-8			
ALASKA Stat. (1975)	Commissioner of education §14.30.330	A special learning program designed to serve their particular exceptionality §14.30.350(1)						§14.30.305		
ARIZONA Rev. Stat. Ann. (Supp. 1978-1979)	Chief administrative official of school district §15-1013(E)[1]	Recommendation of specific goals and instructional objectives §15-1013(B)(7)					§15-1015(a) (3)			
ARKANSAS Stat. Ann. (Supp. 1977)	§§80-2111, 80-2119[2]	To meet the needs and maximize the capabilities of §80-2116			Local unit may provide §80-2124		§80-2118	§§80-2125, 80-2110		
CALIFORNIA Educ. Code (West Supp. 1978)	School appraisal team §56337[3]	§§56301(d), 56302(g), 56333, 56330(i)	§§56070, 56330(g)	§56332(a)(3)	§§56301(c), 56302, 56331	§56332(a)(1)	§§56301(f), 56330(j)			
COLORADO Rev. Stat (1974)	Committee of qualified professionals §22-20-108 (Supp. 1977)[4]						§22-20-102			

State	Decision-Maker	I.E.P. Guaranteed	Non-Traditional Educational Programming: Vocational Skills Included	Non-Traditional Educational Programming: Life Skills Included	Non-Traditional Educational Programming: Preschool Program Available	Least Restrictive Environment	Provision To Educate With Non-Exceptional To Maximum Extent Possible	Provision For Educating Outside of School Facility	Pending Evaluation: Remain Where Situated	Pending Evaluation: In Regular Classroom
CONNECTICUT Gen. Stat. Ann. (West Supp. 1978-1979)	Planning and placement team §10-76d(a)[5]	Services designed to meet the educational needs of... §10-76a, 10-76d(a)	§10-76d(d)		§10-76d(b)			§10-76d(d)		
DELAWARE Code Ann. (Supp. 1977)	[6]	tit. 14, §3101(2)(d)								
DISTRICT OF COLUMBIA										
FLORIDA Stat. Ann. (West 1977)	[7]	Only for kindergarten through grade 3 §230.23(4)(o)	§228.041	§228.041(20)	§232.01(f)		§230.23(4)(m)(5) (Supp. 1979)	§232.01(f)		
GEORGIA Code Ann. (Supp. 1977)					§32.607a					
HAWAII Rev. Stat. (1976)		Education, therapy & training to enable them to live normal, competitive lives §301-22	§301-22					§301-22		
IDAHO Code (Supp. 1978)	[8]							§§33-2002A, 33-2004		
ILLINOIS Ann. Stat. (Smith-Hurd Supp. 1978)	Superintendent of public instruction ch. 122, §14-8.01							ch. 122, §§14-1.08, 14-7.03, 14-8.01		

State	Decision-Maker	I.E.P. Guaranteed	Non-Traditional Educational Programming: Vocational Skills Included	Non-Traditional Educational Programming: Life Skills Included	Non-Traditional Educational Programming: Preschool Program Available	Least Restrictive Environment	Provision To Educate With Non-Exceptional To Maximum Extent Possible	Provision For Educating Outside of School Facility	Pending Evaluation: Remain Where Situated	Pending Evaluation: In Regular Classroom
INDIANA Code Ann. (Burns) (1976)		F.A.P.E. which meets their special needs §20-1-6-15						§20-1-6-3 (Supp. 1978)		
IOWA Code Ann. (West Supp. 1978-1979)	[9]	An appropriate program. §273.5(1)[38]			§281.2		§281.2	§§281.2(2), 281.3(8)		
KANSAS Stat. (Supp. 1978)	[10]	[39]						§72-967(2)		
KENTUCKY Rev. Stat. (1977)	§157.240[11]							§157.270		
LOUISIANA Rev. Stat. Ann. (West Supp. 1978)	Multidisciplinary team §17:1945[12]	§§17:1943(2) 17:1947(A), 17:1945(G)	§17:1943(3)		§17:1943(4)	§§17:1941, 17:1946(B)		§17:1946(A)		
MAINE Rev. Stat. (Supp. 1978)	[13]	[40]			tit. 20, §3131(2)	tit. 20, §3121(2)	tit. 20, §3121(2)			
MARYLAND Educ. Code Ann. (1978)	[14]				§8-413					
MASSACHUSETTS Gen. Laws Ann. (West Supp. 1978-1979)	ch. 71B, §3 [15]	ch. 71B, §3	ch. 71B, §2					ch. 71B, §2		ch. 71B, §3
MICHIGAN Comp. Laws Ann. (Supp. 1978-1979)	Intermediate school board §380.1711(q)[16]									
MINNESOTA Stat. Ann. (1979)	[17]						§120.17(3a)(c)	§120.17(2)(c)		

State	Decision-Maker	I.E.P. Guaranteed	Non-Traditional Educational Programming: Vocational Skills Included	Non-Traditional Educational Programming: Life Skills Included	Non-Traditional Educational Programming: Preschool Program Available	Least Restrictive Environment	Provision To Educate With Non-Exceptional To Maximum Extent Possible	Provision For Educating Outside of School Facility	Pending Evaluation: Remain Where Situated	Pending Evaluation: In Regular Classroom
MISSISSIPPI Code Ann. (Supp. 1978)	[18]									
MISSOURI Ann. Stat. (Vernon Supp. 1978)	[19]		§162.895		§162.700	§162.680(2)	§162.675(4)	§162.955		
MONTANA Rev. Codes Ann. (Supp. 1977)	[20]	§§75-7801(1), 75-7802(1)(c)	§75-7801(1)		§75-7806(2)	§75-7802(1)(a)	§75-7805(1)	§75-7801(1)		
NEBRASKA Rev. Stat. (Supp. 1978)	[21]		§43-646		§43-646.01					
NEVADA Rev. Stat. (1975)	§388.470 [22]				§388.490					
NEW HAMPSHIRE Rev. Stat. Ann. (1977)	[23]									
NEW JERSEY Stat. Ann. (West Supp. 1978-1979)	[24]				§18A:46-14			§18A:46-14		[44]
NEW MEXICO Stat. Ann. (1978)	[25]									
NEW YORK Educ. Law (McKinney Supp. 1978-1979)	§4402[26]	The needs of the child shall determine what program will be used §4402(2)(a)						§4401(2)(a)	§4404(4)	If initial evaluation §4404(4)
NORTH CAROLINA Gen. Stat. (1978)	[27]	§115-375(f)	§115-365			§115-367(m)		§115-365	§115-179.1(d)	

State	Decision-Maker	I.E.P. Guaranteed	Non-Traditional Educational Programming: Vocational Skills Included	Non-Traditional Educational Programming: Life Skills Included	Non-Traditional Educational Programming: Preschool Program Available	Least Restrictive Environment	Provision To Educate With Non-Exceptional To Maximum Extent Possible	Provision For Educating Outside of School Facility	Pending Evaluation: Remain Where Situated	Pending Evaluation: In Regular Classroom
NORTH DAKOTA Cent. Code (Supp. 1977)								§15-59-01(2)		
OHIO Rev. Code Ann. (Page Supp. 1978)	[28]	§3323.08(C)			§3323.07		§3323.04	§3323.12		
OKLAHOMA Stat. Ann. (1972)	[29]							§§13-103, 13-104		
OREGON Rev Stat. (1977)	Administrative officers of the school district after an appropriate evaluation §343.077	§343.035(3)[41]						§343.035(3)	§343.077(7)	Initial placement §343.077(7)
PENNSYLVANIA Stat. Ann. (Purdon Supp. 1978-1979)								tit. 24, §13-1372(3)		
RHODE ISLAND Gen. Laws (Supp. 1977)	[30]	§16-24-7[42]								
SOUTH CAROLINA Code (1976)								§59-21-540(11)		
SOUTH DAKOTA Compiled Laws Ann. (1975)	[31]	[43]								

			Non-Traditional Educational Programming						Pending Evaluation	
State	Decision-Maker	I.E.P. Guaranteed	Vocational Skills Included	Life Skills Included	Preschool Program Available	Least Restrictive Environment	Provision To Educate With Non-Exceptional To Maximum Extent Possible	Provision For Educating Outside of School Facility	Remain Where Situated	In Regular Classroom
TENNESSEE Code Ann. (1977)	[32]	To meet the needs and maximize the capabilities of. . . §49-2902(b)					§49-2913(b)	§49-2902(a)	§49-2947	
TEXAS										
UTAH Code Ann. (1970)		Proper education and training §53-18-2			If paid for by parents §§53-18-2, 53-18-3			§53-18-2		
VERMONT Stat. Ann. (Supp. 1978)	[33]				tit. 16, §2943(3)			tit. 16, §2944(a)		
VIRGINIA Code (Supp. 1978)	[34]	§22-10.3(b)	§22-10.3(b)		·§22-10.8(c)			§22-10.3(b)		
WASHINGTON Rev. Code Ann. (Supp. 1978)		Total school program to meet the needs of handicapped children §28A.13.070			§28A.13.050			§28A.13.040		
WEST VIRGINIA Code (Supp. 1978)	§18-20-4[35]							§18-20-1		
WISCONSIN Stat. Ann. (West Supp. 1978-1979)	Local school board §115.85(2)[36]	§115.80 (3)(e)			§115.85 (2)(b)	§115.85(2)(c)		§115.80(3)(e)		§115.81(3)
WYOMING Stat. (1977)	[37]									

SECTION III

[1] In consultation with: (1) school principal, (2) special education teacher, (3) special education administrator, (4) teacher who is currently instructing the child, (5) an appropriate professional adviser who may be a physician, psychologist, professional social worker or school nurse, and (6) a parent or guardian of the child.

[2] See Chart II Evaluation, footnote 17.

[3] See Chart II Evaluation, footnote 18.

[4] See Chart II Evaluation, footnote 19.

[5] The composition of the planning and placement team is determined according to state board of education regulations.

[6] Placement procedures are defined in state board of education rules and regulations. tit. 14, §3122

[7] Placement procedures are defined in state board of education regulations. §230.23(4)(m)(4)(Supp. 1979).

[8] Placement procedures are defined in state board of education rules and regulations. §33-2003(4)

[9] Placement procedures are promulgated by the division of special education, subject to approval of the state board of education. §281.7

[10] Placement procedures are defined in state board of education regulations. §72-972

[11] See Chart II Evaluation, footnote 30.

[12] The multidisciplinary team shall consist of professionals from two or more disciplines plus any other specialist as may be necessary.

[13] Placement procedures are defined in regulations established by the Commissioner of Education. tit. 20, §3124

[14] Placement procedures are defined in State Board of Education regulations. §8-403(d)

[15] See Chart II Evaluation, footnote 30.

[16] Further placement procedures defined in state board of education regulations.

[17] Placement procedures defined in state board of education regulations. §120.17(3)

[18] Placement procedures defined in state department of education regulations. §37.23.5

[19] Placement procedures defined in state department of education regulations. §162.685

[20] Placement procedures defined in regulations established by the Superintendent of Public Instruction. §75-7811

[21] Placement procedures defined by state department of education regulations. §43.646.10

[22] The child is placed by the board of trustees of the school district after a consultation with the child's parents or guardians and an examination of the child is conducted in accordance with standards prescribed by the state department of education. A psychiatrist may be consulted when the board of trustees of the school district deems it necessary.

[23] Placement procedures are defined in state department of education regulations. §186-A:6

[24] Placement procedures established by the Commissioner of Education subject to the approval of the state department of education. §18A:46-6

[25] Placement procedures defined in state board of education regulations. §22-13-7

[26] See Chart II Evaluation, footnote 39.

[27] Placement procedures defined in state board of education regulations. §115.375(b)

[28] Placement procedures defined in state board of education regulations. §3323.04

[29] Placement procedures defined in state board of education regulations. §13-102

[30] Placement procedures defined in state board of education regulations. §16-24-2

[31] Placement procedures defined in state board of education regulations. §13-37-4(1975)

[32] Placement procedures defined in state board of education regulations. §49-2932(b)

[33] Placement procedures defined in state board of education regulations. tit. 16, §2944(c)

[34] Placement procedures defined in state board of education regulations. §22-10.4

[35] See Chart II Evaluation, footnote 48.

[36] The school board places the child after consultation with the multidisciplinary evaluation team and after the consent (in writing) of the child's parents.

[37] Placement procedures are defined in state board of education regulations. §21-14-102

[38] Courses of study and curricula are prescribed in state department of education regulations. §281.3(6)

[39] Courses of study and curricula are prescribed in state board of education regulations.§72-963

[40] Tit. 20, §3125 states that, "An appropriate program for exceptional children may be established by any administrative unit operating schools."

[41] §343.035(3) only applies to mentally retarded minors, who require "specialized instruction appropriate to their individual capacity."

[42] "Special education" means specially designed instruction to meet the unique needs of a handicapped child.

[43] The educational program is set up in accordance with the state board of education's accreditation rules. §13-37-2 (Supp. 1977)

[44] Pending a child's examination and classification he may be refused admission to, or be excluded temporarily from the schools of any district for a reasonable time. §18A:46-17.

IV A. PROCEDURAL GUARANTEES

State	Periodic Review and Reevaluation of Program	Notice To Parents Before Initiating or Refusing Any Change in Evaluation or Program	Contents of Notice: Explanation of Procedural Safeguards	Contents of Notice: Written in Native Language	Contents of Notice: Description of Alternative Programs	In Evaluation or Program Consent of Parents Before Initiating or Refusing Any Change	Confidentiality of Records	Person and Parents' Right To Access To Records
ALABAMA Code (1975)		tit. 16, §39-8					tit. 16, §39-4	tit. 16, §39-4
ALASKA Stat. (1975)[1]						§14.30.330(a)		
ARIZONA Rev. Stat. Ann. (Supp. 1978-1979)	Every semester §15-1014					§15-1013(E)		
ARKANSAS Stat. Ann. (Supp. 1977)[2]		§80-2133					§80-2135	
CALIFORNIA Educ. Code (West Supp. 1978)[3]	§§56310(i), 56336.3(a)	§§56036(a), 56337(a)	§56036(a)(1)	§§56036(a)(1), 56337(a)	§56036(a)(6)	Placement §§56036(a)(5), 56338	§56337(a)	§§56341.1, 56341.2
COLORADO Rev. Stat. (1974)	Every year § 2-20-108(4)	§22-20-108(1) (Supp. 1977)				§22-20-108(2)		
CONNECTICUT Gen. Stat. Ann. (West Supp. 1978-1979)	§10-76d (g)(1) (1977)	§10-76d(a)						
DELAWARE								
DISTRICT OF COLUMBIA								
FLORIDA Stat. Ann. (West Supp. 1979)		§230.23(4)(m)(4)	§230.23(4)(m)(4)		§230.23(4)(m)(4)			
GEORGIA								
HAWAII								

State	Periodic Review and Reevaluation of Program	Notice To Parents Before Initiating or Refusing Any Change in Evaluation or Program	Contents of Notice: Explanation of Procedural Safeguards	Contents of Notice: Written in Native Language	Contents of Notice: Description of Alternative Programs	In Evaluation or Program Consent of Parents Before Initiating or Refusing Any Change	Confidentiality of Records	Person and Parents' Right To Access To Records
IDAHO								
ILLINOIS								
INDIANA								
IOWA Code Ann. (West Supp. 1978-1979)[4]								
KANSAS Stat. (Supp. 1978)		§72-972(b)	§72-972(b)	§72-972(b)		§72-972(c)		§72-973(b)
KENTUCKY								
LOUISIANA Rev. Stat. Ann (West Supp. 1978)	3 years or upon request §§17:1941, 17:1945(E)	§17:1952(B)(3)		§17:1932(B)(4)		§17:1945(A)	§17:1955	§17:1952(B)(1)
MAINE Rev. Stat. (Supp. 1978)[5]							tit. 20, §3128	
MARYLAND Educ. Code Ann. (1978)								
MASSACHUSETTS Gen. Laws Ann. (West Supp. 1978-1979)	ch. 71B, §§2, 3	ch. 71B, §3	ch. 71B, §3	ch. 71B, §3	ch. 71B, §3	ch. 71B, §3	ch. 71B, §3	ch. 71B, §3
MICHIGAN[6]								
MINNESOTA Stat. Ann. (West Supp. 1979)		§120.17(3b)(a)						
MISSISSIPPI								

State	Periodic Review and Reevaluation of Program	Notice To Parents Before Initiating or Refusing Any Change in Evaluation or Program	Contents of Notice: Explanation of Procedural Safeguards	Contents of Notice: Written in Native Language	Contents of Notice: Description of Alternative Programs	In Evaluation or Program Consent of Parents Before Initiating or Refusing Any Change	Confidentiality of Records	Person and Parents' Right To Access To Records
MISSOURI Ann. Stat. (Vernon Supp. 1978)	§§162.685(2), 162.950	§162.945	§162.945					§§162.945, 162.963(1)
MONTANA[7]								
NEBRASKA Rev. Stat. (Supp. 1978)								
NEVADA Rev. Stat. (1977)		§388.470(1)(a)						
NEW HAMPSHIRE Rev. Stat. Ann. (1977)						§186-A:13		
NEW JERSEY Stat. Ann. (West Supp. 1978-1979)[8]	Those classified as uneducable only §18A:46-17	§18A:46-8						
NEW MEXICO								
NEW YORK Educ. Law (McKinney Supp. 1977-1978)	3 years §4402(1)(b)(3)(d)	§4402(1)(b)(3)(c)	§§4402(1)(b)(3)(c), 4403(8)				§4403(9)	
NORTH CAROLINA Gen Stat. (1978)	Every year §§115-363, 115-375(c)	§115-375(c)	§115-179.1(b)		§115-179.1(b)	§115-375(c)	§115-376(a)	§§115-376(b), 115-179.1(e)
NORTH DAKOTA								
OHIO Rev. Code Ann. (Page Supp. 1978)	§3323.04	§3323.05(C)	§3323.05(C)				§3323.06(A)	§3323.05(A)
OKLAHOMA								
OREGON Rev. Stat. (1977)		§343.077(2)	§§343.077(2), 183.415	§343.077(2)		§343.227		§343.077(3)

State	Periodic Review and Reevaluation of Program	Notice To Parents Before Initiating or Refusing Any Change in Evaluation or Program	Contents of Notice: Explanation of Procedural Safeguards	Contents of Notice: Written in Native Language	Contents of Notice: Description of Alternative Programs	In Evaluation or Program Consent of Parents Before Initiating or Refusing Any Change	Confidentiality of Records	Person and Parents' Right To Access To Records
PENNSYLVANIA								
RHODE ISLAND								
SOUTH CAROLINA								
SOUTH DAKOTA Compiled Laws Ann. (1975)	§13-37-5.1							
TENNESSEE Code Ann. (1977)		§49-2947	§49-2947					§49-2947
TEXAS								
UTAH								
VERMONT Stat. Ann. (Supp. 1977)								
VIRGINIA[9]								
WASHINGTON Rev. Code Ann. (Supp. 1978)								
WEST VIRGINIA								
WISCONSIN Stat. Ann. (West Supp. 1978-1979)	§115.80(5)	§115.81(2)(a)	§115.81(2)			§115.85(2)		§115.81(4)
WYOMING								

[1] Additional procedural safeguards not defined by statute are defined in state board of education regulations.
[2] Additional procedural safeguards not defined by statute will be established by the "State." §80-2133
[3] Additional procedural rights not defined by statute are defined in state board of education regulations.
[4] Procedural safeguards are defined in state board of education regulations.
[5] Procedural safeguards not defined by statute are defined in regulations established by the Commissioner of Education.
[6] All procedural safeguards are defined in state board of education regulations.
[7] §75-7802 (1)(b) states that the policies recommended by the superintendent of public instruction and adopted by the board of public education shall assure due process for all handicapped children.
[8] The parents or guardian is provided an opportunity to review the classification of the child pursuant to rules of the state board of education. §18A:46-8
[9] All procedural safeguards are defined in state board of education regulations.

IV B. PROCEDURAL GUARANTEES

State	Right To Contest Evaluation or Program Informally	Right To An Administrative Hearing	Right To An Impartial Hearing Officer	Right To Counsel	Right To Present Evidence	Right To Cross-Examine	Right To Compel Witnesses	Right of Exceptional Person To Be Present
ALABAMA Code (1975)								
ALASKA Stat. (1975)	§14.30.191(b)	§14.30.191(c)						
ARIZONA Rev. Stat. Ann. (Supp. 1978-1979)								
ARKANSAS Stat. Ann. (Supp. 1977)		§80-2133						
CALIFORNIA Educ. Code (West Supp. 1978)	§56341.1	§56036 (a)(6)	§56036 (a)(6)					
COLORADO Rev. Stat. (1974)	§22-20-108 (Supp. 1977)							
CONNECTICUT Gen. Stat. Ann. (West Supp. 1978-1979)	§10-76h(a)(2), (b)(1)	§10-76h(a)(1) (1977)	§10-76h(b)					
DELAWARE Code Ann. (Supp. 1977)			Only for disputes arising from private placement tit. 14, §3124					
DISTRICT OF COLUMBIA								
FLORIDA Stat. Ann. (West Supp. 1979)		§230.23 (4)(m)(4)	§230.23 (4)(m)(4)					
GEORGIA								
HAWAII								

State	Right To Contest Evaluation or Program Informally	Right To An Administrative Hearing	Right To An Impartial Hearing Officer	Right To Counsel	Right To Present Evidence	Right To Cross-Examine	Right To Compel Witnesses	Right of Exceptional Person To Be Present
IDAHO								
ILLINOIS								
INDIANA								
IOWA Code Ann. (West Supp. 1978-1979)		§290.1						
KANSAS Stat. (Supp. 1978)		§72-972	§72-973(c)	§72-973(a)(1)	§72-973 (a)(5)	§72-973 (4)	§§72-973 (4), (5)	§72-973 (6)
KENTUCKY								
LOUISIANA Rev. Stat. Ann. (West Supp. 1978)		§§17:1952 (B)(5), 17:1952 (B)(6)(e)		§17:1952 (B)(6)(a)	§17:1952 (B)(6)(b)	§17:1952(B)(6)(b)	§17:1952(B)(6)(b)	
MAINE Rev. Stat. (Supp. 1978)		tit. 20, §3131(1-B)	tit.20,§3131(1-A)(b)					
MARYLAND Educ. Code Ann. (1978)		§8-415	§8-415 (a)(3)					
MASSACHUSETTS Gen. Laws Ann. (West Supp. 1978-1979)		ch. 71B, §3						
MICHIGAN								
MINNESOTA Stat..Ann. (West Supp. 1979)	§120.17 (3b)(b)	§120.17(3b)(c)						
MISSISSIPPI								
MISSOURI Ann. Stat. (Vernon Supp. 1978)	§§162.950, 162.961	§162.950		§162.963	§162.963	§§162.961, 162.963		

State	Right To Contest Evaluation or Program Informally	Right To An Administrative Hearing	Right To An Impartial Hearing Officer	Right To Counsel	Right To Present Evidence	Right To Cross-Examine	Right To Compel Witnesses	Right of Exceptional Person To Be Present
MONTANA								
NEBRASKA Rev. Stat. (Supp. 1978)		§43-661	§43-663	§43-664(1)	§43-664(2)	§43-664 (2)	§§43-652 (2), 43-664 (2)	
NEVADA Rev. Stat. (1977)								
NEW HAMPSHIRE Rev. Stat. Ann. (1977)								
NEW JERSEY Stat. Ann. (West Supp. 1978-1979)	§18A:46-8	§18A:46-8						
NEW MEXICO								
NEW YORK Educ. Law (McKinney Supp. 1977-1978)		§§4402(1)(b) (3)(c), 4404(1)	§4404(1)					
NORTH CAROLINA Gen. Stat. (1978)		§115-179.1(b)	§115-179.1(f)	§115-179.1(e)	§115-179.1(e)	§115-179.1 (e)		§115-179.1 (e)
NORTH DAKOTA								
OHIO Rev. Code Ann. (Page Supp. 1978)		§§3323.05(E), (F)	§3323.05(E)	§3323.05(E)(1)	§3323.05(E)(2)	§3323.05(E)(2)	§3323.05(E)(2)	§3323.05
OKLAHOMA								
OREGON Rev. Stat. (1977)		§343.077(2)		§183.415	§183.415	§183.450(3)	§183.440	
PENNSYLVANIA								
RHODE ISLAND								

State	Right To Contest Evaluation or Program Informally	Right To An Administrative Hearing	Right To An Impartial Hearing Officer	Right To Counsel	Right To Present Evidence	Right To Cross-Examine	Right To Compel Witnesses	Right of Exceptional Person To Be Present
SOUTH CAROLINA								
SOUTH DAKOTA								
TENNESSEE Code Ann. (1977)		§49-2947	§49-2947	§49-2947	§49-2947	§49-2947		§49-2947
TEXAS								
UTAH								
VERMONT Stat. Ann. (Supp. 1977)	tit.16, §2944(b)(2)							
VIRGINIA								
WASHINGTON Rev. Code Ann. (Supp. 1978)		§28a.13.060[1]						
WEST VIRGINIA								
WISCONSIN Stat. Ann. (West Supp. 1978-1979)		§115.81		§115.81(4)	§115.81(4)	§115.81 (4)		§115.81
WYOMING								

[1] The parents or guardian may appeal the original placement decision to the superintendent of public instruction pursuant to procedures established by him or her.

State	Written Decision	Written Transcript Of Hearing	Time Between: Request And Administrative Hearing	Time Between: Administrative Hearing And Decision	Time Between: Decision And Appeal	Time Between: Appeal And Decision	Right To Court Hearing	Right To Public Hearing
ALABAMA								
ALASKA								
ARIZONA								
ARKANSAS								
CALIFORNIA Educ. Code (West Supp. 1978)	§56341.4		Within 15 days—before a "fair-hearing panel" §§56341.1, 56341.3		Within 30 days—before the superintendent of public instruction §56341.4	Within 15 days §56341.4	Parents can request a writ of mandamus ordering a specific placement of the child §56037	
COLORADO								
CONNECTICUT Gen. Stat. Ann. (West Supp. 1978-1979)	§10-76h(a)(1)	§§10-76h(a)(1), 10-76h(d)	Within 10 days §10-76h	Within 10 days §10-76h	Within 30-60 days §10-76h(c)	Within 60-120 days §10-76h(c)	§10-76h(e)	
DELAWARE								
DISTRICT OF COLUMBIA								
FLORIDA Stat. Ann. (West Supp. 1978)							§230.23(4)(m)(4)	
GEORGIA								
HAWAII								
IDAHO								
ILLINOIS								
INDIANA								
IOWA Code Ann. (West Supp. 1978-1979)		§290.3						

State	Written Decision	Written Transcript Of Hearing	Time Between: Request And Administrative Hearing	Time Between: Administrative Hearing And Decision	Time Between: Decision And Appeal	Time Between: Appeal And Decision	Right To Court Hearing	Right To Public Hearing
KANSAS Stat. (Supp. 1978)	§72-974	§72-973(9)	Reasonable time §72-973	Within 5 days §72-973(b)	Within 20 days §72-974	Within 5 days §72-974	§72-975(d)	§72-973(a)
KENTUCKY								
LOUISIANA Rev. Stat. Ann. (West Supp. 1978)	§17:1952 (B)(6)(d)	§17:1952 (B)(6)(c)					§17:1952 (B)(6)(e)	
MAINE								
MARYLAND Educ. Code Ann. (Supp. 1978)			Within 45 days §8-415(a)(2)	§8-415(a)(4)			§8-415(f)	
MASSACHUSETTS Gen. Laws Ann. (West Supp. 1978-1979)						Within 30 days ch. 71B, §3	ch. 71B, §3	
MICHIGAN								
MINNESOTA Stat. Ann. (West Supp. 1979)	§120.17 (3b)(d)(1)	§120.17(3b)(c)		Within 5 days §120.17(3b)(d)	Within 15 days §120.17(3b)(e)	Within 30 days §120.17(3b)(e)	§120.17(3b)(f)	
MISSISSIPPI								
MISSOURI Ann. Stat. (Vernon Supp. 1978)	§162.962		Within 20 days §162.961		Within 30 days §162.962	Within 30 days §162.962	§162.965	
MONTANA								
NEBRASKA Rev. Stat. (Supp. 1978)	§43-664(5)	§43-664(4)					§§43-652.03, 43-666	
NEVADA								
NEW HAMPSHIRE								
NEW JERSEY								
NEW MEXICO								
NEW YORK Educ. Law (McKinney Supp. 1978-1979)		§4404(1)					§4404(3)	

NORTH CAROLINA Gen. Stat. (1978)			Within 30 days §115-179.1(b)				§115-179.1(g)	§115-179.1(c)
NORTH DAKOTA								
OHIO Rev. Code Ann. (Page Supp. 1978)							§3323.05(F)	
OKLAHOMA								
OREGON Rev. Stat. (1977)	§183.470	§183.415 (8)					§183.480	
PENNSYLVANIA								
RHODE ISLAND								
SOUTH CAROLINA								
SOUTH DAKOTA Compiled Laws Ann. (1975)							Any decision by a school board may be appealed to he circuit court §13-46-1	
TENNESSEE Code Ann. (1977)		§49-2947					§49-2947	
TEXAS								
UTAH								
VERMONT Stat. Ann. (Supp. 1977)							§22-10.4(D)	
VIRGINIA Code (Supp. 1978)							§22-10.4(b)	
WASHINGTON								
WEST VIRGINIA								
WISCONSIN Stat. Ann. (West Supp. 1978-1979)		§115.81(4)	Within 60 days §115.81(6)	Within 30 days §115.81(6)	Within 30 days §115.81(7)	Within 30 days §115.81(7)	§115.81(8)	§115.81(4)
WYOMING								

V. ADMINISTRATION

State	Local Administration				
	School District Responsible For Special Education	Two Or More Districts May Combine To Provide All Necessary Services	Annual Report By District On The Extent It Is Providing Special Education	Separate Division Established Within State Education Agency	Advisory Panel Established
ALABAMA Code (1975)	tit. 16, §39-4				
ALASKA Stat. (1975)	§14.30.186	§§14.30.186(d), 14.30.285			§14.30.231
ARIZONA Rev. Stat. Ann. (Supp. 1978-1979)	§15-1015	§15.1015(D)		§15-1012	§15-1012(F)
ARKANSAS Stat. Ann. (Supp. 1977)	§§80-2107, 80-2123	§§80-2107, 80-2132	§80-2131	§80-2126	§80-2129
CALIFORNIA Educ. Code (West Supp. 1978)		§§56313, 56315(b)			Community §56330(r) State §33590
COLORADO Rev. Stat. (1974)	§22-20-106	§22-20-106(4)	§22-20-113		§22-20-104(2)
CONNECTICUT Gen. Stat. Ann. (West Supp. 1978-1979)	§10-76d(b)	§10-76d(d)	§10-76(1)		§10-76i
DELAWARE Code Ann. (Supp. 1977)	tit. 14, §3121	tit. 14, §3121			tit. 14, §3111
DISTRICT OF COLUMBIA					
FLORIDA Stat. Ann. (West Supp. 1979)	§230.23(4)(m)(2)	§230.23(4)(m)(2)			

State Administration

Provision For Interstate Cooperation	Annual Program Plan To Be Submitted Dealing With Future Needs	Statistical Data Maintained	Provision For Contracting With Private Agencies	Procedures For Monitoring Each District's Compliance With Law	Periodic Evaluation Of Effectiveness Of Programs	To Receive State Aid Compliance With Law Is Required
	tit. 16, §39-6[1]	tit. 16, §39-4	tit. 16, §39-6	tit. 16, §39-5	tit. 16, §36-6	
§14.30.285(c)						
	§15-1017(C)		School district or county school superintendent §15-1015(F)	§15-1012(D)		§15-1017(F)
		§80-2135	§§80-2112, 80-2113	§80-2128		§80-2106
	§56312(j)	§56352	Local school district §56032	§56312 (e)	§§56301 (1), 56314(b), 56355	
		§22-20-104(4)	§22-20-107			§22-20-114
§10-76d(f)		§10-76j	Town or regional board of education §10-76d(d)	§10-76b(d)	§10-76b(d)	State reimburses school district 66⅔% of its net cost of special education for the preceeding fiscal year §10-76g
			Local school district — Only if "complex or rare" handicap and no suitable state program available tit. 14, §3124			
	§230.23(4)(m)(3)		Local school board §230.23(4)(m)(2)			§230.23(4)(m)(2)

State	Local Administration: School District Responsible For Special Education	Local Administration: Two Or More Districts May Combine To Provide All Necessary Services	Local Administration: Annual Report By District On The Extent It Is Providing Special Education	Separate Division Established Within State Education Agency	Advisory Panel Established
GEORGIA Code Ann. (1976)	§32-605a(b)				
HAWAII Rev. Stat. (1976)				§301-23	
IDAHO Code (Supp. 1978)	§33-2001	§33-2004		§33-2003	
ILLINOIS Ann. Stat. (Smith-Hurd Supp. 1978)	ch. 122, §14-4.01	ch. 122, §14-7.01			County ch. 122, §14-2.01 State ch. 122, §14-3.01
INDIANA Code Ann. (Burns Supp. 1978)		§§20-1-6-20(6)(b), 20-1-6-3		§20-1-6-2	§20-1-6-16
IOWA Code Ann. (West Supp. 1978-1979)	§281.2	§281.3(9)		§281.1	
KANSAS Stat. (Supp. 1977)	§72-933	§72-967(4)	§72-971		§72-964
KENTUCKY Rev. Stat. (1977)	§157.224 (2)	§157.280		§157.221	
LOUISIANA Rev. Stat. Ann. (West Supp. 1978)		§17:1947(D)	§§17:1947(A)(1), 17:1948	§17:1944	Local §17:1948(A) State §17:1954
MAINE Rev. Stat. (Supp. 1978)	tit. 20, §3124	tit. 20, §3125(3)			
MARYLAND Educ. Code Ann. (1978)	§8-402	§8-403(d)			
MASSACHUSETTS Gen. Laws Ann. (West Supp. 1978-1979)	ch. 71B, §3	ch. 71B, §4	ch. 71B, §6		

State Administration

Provision For Interstate Cooperation	Annual Program Plan To Be Submitted Dealing With Future Needs	Statistical Data Maintained	Provision For Contracting With Private Agencies	Procedures For Monitoring Each District's Compliance With Law	Periodic Evaluation Of Effectiveness Of Programs	To Receive State Aid Compliance With Law Is Required
§32-605a(c)(3)	§32-606a(b)		State board of education §32-605a(c)(2)	§32-650a		32-669a
			Trustees of the school district §33-2004	§33-2003(3)		
ch. 122, §14-7.02						
§20-1-6-19			Superintendent of public ins ruction §20-1-6-19			
	§281.11					
§72-968		§72-979(c)	Local school district board of education §72-967(5)			§72-978
			School district §157.280			§157 224(2)
§17:1944(B)(18)		§17:1944(B)(3)	Parish or city school board §17:1949	§17:1944(B)(4)	§17:1944(B)(9)	§17:1944(B)(16)
tit. 20, §3125(6)	tit. 20, §3128	tit. 20, §3128	Local administrative unit tit. 20, §3125(2)	tit. 20, §3126	tit. 20, §3126	tit. 20, §3127 (4)(c)(1)
			§8-409			§8-405
ch. 71B, §10		ch. 71B, §6	School committee of any city, town, or school district with approval of state dept. of education ch. 71B, §4			

State	Local Administration: School District Responsible For Special Education	Local Administration: Two Or More Districts May Combine To Provide All Necessary Services	Local Administration: Annual Report By District On The Extent It Is Providing Special Education	Separate Division Established Within State Education Agency	Advisory Panel Established
MICHIGAN Comp. Laws Ann. (Supp. 1978-1979)	§380.1751	§380.1751(1)(b)			§380.1716
MINNESOTA Stat. Ann. (1979)	§§120.17(1), 120.17 (3a)	§120.17(1)			
MISSISSIPPI Code Ann. (Supp. 1978)					§37-23-127
MISSOURI Ann. Stat. (Vernon Supp. 1978)	§162.700(1)	§162.705	§162.705(4)		§162.690
MONTANA Rev. Code Ann. (Supp. 1977)	§75-7805(1)	§§75-7805, 75-7807, 75-7808			
NEBRASKA Rev. Stat. (Supp. 1978)	§43-642	§43-642			
NEVADA Rev. Stat. (1977)	§388.450(2)				
NEW HAMPSHIRE Rev. Stat. Ann. (1977)	§186-A:7	§§186-A:7, 186-A:8			
NEW JERSEY Stat. Ann. (West Supp. 1978-1979)	§18A:46-13	§18A:46-14(b)			§18A:46-2
NEW MEXICO Stat. Ann. (1978)	§22-13-5				
NEW YORK Educ. Law (McKinney Supp. 1978-1979)	§4402(2)(a)	§4401(2)(b)	§4402(1)(b)(3)(f)		§4403(6)
NORTH CAROLINA Gen. Stat. (1978)	§115-367(i)	§115-367(a)	§115-367(e)		
NORTH DAKOTA Cent. Code (Supp. 1977)	§15-59-04	§15-59-04			§15-59-02
OHIO Rev. Code Ann. (Page Supp. 1978)		§3323.08(F)	§3323.02		§3323.10

State Administration

Provision For Interstate Cooperation	Annual Program Plan To Be Submitted Dealing With Future Needs	Statistical Data Maintained	Provision For Contracting With Private Agencies	Procedures For Monitoring Each District's Compliance With Law	Periodic Evaluation Of Effectiveness Of Programs	To Receive State Aid Compliance With Law Is Required
§380.1751(1)(b)		§380.1711(f)		§380.1711(h)	§§380.1711(a), 380.1701(a)	§380.1702(3)
§120.17(2)(a)			Local school district §120.17(2)(h)			
			§37-23-69			
		§162.695	Local district or state board of education §§162.705(1), 162.735			§162.705(4)
§75-7809			Local school district §75-7809.1		§75-7811	§75-7811
	§43-643 (Supp. 1976)			§43-642	Biennially §43-650	§43-642
						§388.520(3)
§186-A:8			§186-A:8[2]			
§18A:46-14(b)			Local board of education §18A:46-14(g)	§18A:46-15 (b)	§18A:46-15(b)	
			Local school board §22-13-8(b)		Only private facilities §22-13-8(c)	§22-13-7(c)
§4407(1)		§4403(1)	Local district §§4401(2)(e)-(h)	§4403(4)	§§4402(1)(b)(3)(e), 4403(4)	
	§115-367(h)		State or local educational agencies §115-384	§115-367(k)	§115-363	
			Local school district §15-59-07			
		§3323.02	Local school district §3323.08(F)	§3323.06(A)	§3323.02	§3323.02

State	Local Administration: School District Responsible For Special Education	Local Administration: Two Or More Districts May Combine To Provide All Necessary Services	Local Administration: Annual Report By District On The Extent It Is Providing Special Education	Separate Division Established Within State Education Agency	Advisory Panel Established
OKLAHOMA Stat. Ann. (West Supp. 1978-1979)	§13-101	§13-101			
OREGON Rev. Stat. (1977)	§343.221(2)	§343.221(3)			State §343.287 Local §343.293
PENNSYLVANIA Stat. Ann. (Purdon Supp. 1978-1979)	tit. 24, §13-1372(3)	tit. 24, §13-1372(1)			
RHODE ISLAND Gen. Laws (Supp. 1977)		§16-24-5			
SOUTH CAROLINA Code (1976)	§§59-33-40, 59-33-50	§59-33-50	§59-33-40		
SOUTH DAKOTA Compiled Laws Ann. (1975)		§13-37-14.1			
TENNESSEE Code Ann. (1977)	§49-2912	§§49-2919, 49-2938		§49-2915	§49-2916
TEXAS Educ. Code Ann. (Vernon Supp. 1978-1979)	§16.104(a)	§16.104(g)			
UTAH Code Ann. (1970)	§53-18-2	§53-18-2			§53-18-8
VERMONT Stat. Ann. (Supp. 1977)		tit. 16, §2948(d)			tit. 16, §2945
VIRGINIA Code (Supp. 1978)	§22-10.5	§22-10.6			§22-10.13
WASHINGTON Rev. Code Ann. (Supp. 1978)	§28A.13.010	§28A.13.030		§28A.13.010	§28A.14.050
WEST VIRGINIA Code (Supp. 1978)	County boards §18-20-1				§18-20-6
WISCONSIN Stat. Ann. (West Supp. 1978-1979)	§115.85(1)	§115.85(2)(c)	§§115.84, 115.85(3)	§115.77	
WYOMING Stat. (1977)	Jointly with state board §21-14-102	§21-14-103			

[1] An annual program plan must be submitted every two to seven years.
[2] §186-A:8 does not specify who has the authority to contract.
[3] Procedures for contracting with private agencies will be defined in state board of education regulations. §21-14-103

State Administration

Provision For Interstate Cooperation	Annual Program Plan To Be Submitted Dealing With Future Needs	Statistical Data Maintained	Provision For Contracting With Private Agencies	Procedures For Monitoring Each District's Compliance With Law	Periodic Evaluation Of Effectiveness Of Programs	To Receive State Aid Compliance With Law Is Required
		§13-111 (1972)	Local district §13-101			§§13-106, 13-111 (1972)
	§343.221(1)		District school board §343.221(4)			§343.281
		tit. 24, §13-1371(2)	tit. 24, §13-1376			tit. 24, §13-1373
§59-33-50	§59-33-40	§59-33-50	School district §59-33-50			§§59-21-540, 59-21-570
§§49-2938, 49-2943	§49-2933		School district and special education associations §49-2924		§49-2945	§49-2949
			School district §16.104(l)		§16.104(n)	
						§53-18-3
tit. 16, §2949						
	§22-10.5		Local school board §22-10 6			§22-10.4(E)
			School district §28A.13.045			§28A.13.08 (Other sanctions available for non-compliance)
§115.85(2)(c)	§115.85(3)	§115.85(3)(a)	Local school board 115.85(2)(d)	§115.77(4)	§115.78	§115.88
§21-14-103			[3]			

VI. FINANCING

State	Excess Costs: State Aid	Excess Costs: Limit	State Aid For Private Services	Funds For Out-of-State Services
ALABAMA Code (1975)	tit. 16, §39-3[5]		tit. 16, §39-6(5)[14]	tit. 16, §36-6(5)[24]
ALASKA Stat. (1975)	[6]			
ARIZONA Rev. Stat. Ann. (Supp. 1978-1979)	§15-1018[7]			
ARKANSAS Stat. Ann (Supp. 1977)	§§80-2106, 80-2122[8]			
CALIFORNIA Educ. Code (West Supp. 1978)	§84850(a)	$785 per pupil §84850(a), up to $1570 for severely disabled §84850(e)		
COLORADO Rev. Stat. (Supp. 1978)	§§22-50-101.5 *et seq.*			§22-50-104
CONNECTICUT Gen. Stat. Ann. (West Supp. 1978-1979)	§10-76g[9]		§§10-76p, 10-76q	
DELAWARE Code Ann. (1974 & Supp. 1977)	tit 14, §§1701 *et seq.*			tit. 14, §1703
DISTRICT OF COLUMBIA				
FLORIDA				
GEORGIA	§32-639a(b)			
HAWAII				
IDAHO Code (Supp. 1978)[1]				
ILLINOIS Stat. Ann. (Smith-Hurd Supp. 1978)	§§14-7.02, 14-7.02a	$2000 §14-7.02a	$2500 limit §14-7.02	$2500 limit §14-7.02
INDIANA Code Ann. (Burns 1976)	§20-1-6-9		§20-1-6-19	
IOWA Code Ann. (West Supp. 1978-1979)	§281.9			§282.8[25]
KANSAS Stat. (Supp. 1978)	§72-7043		§72-967(5)	§72-968
KENTUCKY Rev. Stat. (Supp. 1978)	§157.360		§157.360(2)	
LOUISIANA Rev. Stat. Ann. (West Supp. 1978)	§17:1956		§§17:1949, 17:1956[15]	

State	Excess Costs: State Aid	Excess Costs: Limit	State Aid For Private Services	Funds For Out-of-State Services
MAINE Rev. Stat. (Supp. 1978)	tit. 20, §3126		§3130	
MARYLAND Educ. Code Ann. (1978)	§§8-417.2 *et seq.*		§§8-417.2(d), 8-417.3(d)	
MASSACHUSETTS Gen. Laws Ann. (West Supp. 1978-1979)	ch. 70, §§1 *et seq.*		ch. 70, §2A	ch. 71B, §10
MICHIGAN Comp. Laws Ann. (Supp. 1978-1979)		§388.1452[13]		
MINNESOTA Stat. Ann. (1979)	§124.32		§124.32(5)[16]	
MISSISSIPPI Code Ann. (Supp. 1978)	§§37-19-1 *et seq.*		§37-23-69[17]	
MISSOURI Ann. Stat. (Vernon Supp. 1978)	§162.975	Total limit of $250,000 §162.995	§162.980	
MONTANA Rev. Codes Ann. (Supp. 1977)	§75-7813.1		§75-7813.1(b)(IV)(A)	§75-7809
NEBRASKA Rev. Stat. (Supp. 1978)	§§43-646.02, 43-648	90% of allowable excess costs		
NEVADA				
NEW HAMPSHIRE Rev. Stat. Ann. (1977)	§186-A:8		§186-A:8(VI)	§186-A:8(I)
NEW JERSEY Stat. Ann. (West Supp. 1978-1979)	§18A:58-6[10]		§18:46-19.8[18] §18A:58-6[19]	
NEW MEXICO				
NEW YORK Educ. Law (McKinney Supp. 1978-1979)	[11]		§4405(3)(a)	§4407(1)
NORTH CAROLINA Gen. Stat. (1978)			§115-377 Unless so placed against the advice of the state	
NORTH DAKOTA Cent. Code (Supp. 1977)	§15-59-06	1½ times the state per pupil cost	§15-59-07[20]	same as private placement
OHIO Rev. Code Ann. (Page Supp. 1978)	§3317.02.4[12]			
OKLAHOMA Stat. Ann. (West Supp. 1978-1979)	§18-109			
OREGON Rev. Stat. (1977)	§343.281(3)	30% of the district's approved costs of special education	§343.281(5)[21]	

State	Excess Costs: State Aid	Excess Costs: Limit	State Aid For Private Services	Funds For Out-of-State Services
PENNSYLVANIA Stat. Ann. (Purdon Supp. 1978-1979)	tit. 24, §§25-2501, 25-2509.1, 13-1373		tit. 24, §13-1376[22]	
RHODE ISLAND[2]				
SOUTH CAROLINA Code (1976)	§§59-33-70, 59-21-540		§59-33-50	§59-33-50
SOUTH DAKOTA Compiled Laws Ann. (Supp. 1978)			$2 per day more §13-37-8.7	
TENNESSEE Code Ann. (1977)	§49-2959(c)(2)			
TEXAS Educ. Code Ann. (Vernon Supp. 1978-1979)	Legislature sets limit §16.104(m)			
UTAH Code Ann. (1970)[3]				
VERMONT Stat. Ann. (Supp. 1978)	tit. 16, §2948	75% of salaries and wages of special education personnel		
VIRGINIA Code (Supp. 1978)			§22-10.8[23]	
WASHINGTON Rev. Code Ann. (Supp. 1978)	§28A.41.053			
WEST VIRGINIA				
WISCONSIN Stat. Ann. (West Supp. 1978-1979)	§115.88(i)	70% of amount expended by local school board		
WYOMING Stat. (1977 & Supp. 1979)[4]				

[1] §33-1006A states that the state board of education shall determine the allowable costs for providing approved programs of education for exceptional students.
[2] Effective July 1, 1978, $4,000,000 is appropriated to the department of education to be distributed to local and regional school districts on the basis of the ratio which the handicapped children in that district bears to the state total of such children, provided that no school district receives less than $12,500. §16-7-34.2
[3] §53-18-3 simply states, "School districts maintaining special classes will be reimbursed by the state board of education."
[4] §21-14-103 simply states, "Such programs and services shall be financed as provided by law."
[5] If not enough funds are available for the education of both exceptional and non-exceptional children then the Board must prorate all funds on a per capita basis.
[6] Funds for special education are computed, as is all public education, on an "instructional unit" formula pursuant to §§14.17.021 - 14.17.051.
[7] Only if the district's qualifying tax rate does not produce enough revenue to cover all excess costs will it be eligible for state aid. §§15-1018(c)(4), (5)(a-c).
[8] State policy is for equality in expenditures. However §80-2122 adds that, "Any additional effort necessary to provide supplemental aids and services, shall be the ultimate responsibility of the State, but shall, to the maximum extent practicable, be the responsibility of the local school districts." §80-2106 states the basic premise that, "If for any reason the number of approved applications exceed the available funds, the proration of funds to the districts will be the responsibility and the authority of the Board."
[9] The state reimburses the school districts 66⅔% of their net costs.
[10] §18A:58-6 contains an extensive list of what costs are reimbursable and at what percentage.
[11] The excess cost ratio is computed according to the terms of §4401(7).
[12] §3317.024(N) — An amount for each approved unit for a deaf, blind, emotionally disturbed, crippled, neurologically handicapped, or educable mentally retarded class in the district or in the institution. The amount shall be the sum of the unit's teacher's minimum salary calculated on the basis of the teacher's training level and years of experience plus 15% of the allowance, plus $4540 or 105% of the amount per pupil paid during the current fiscal year times the number of full-time equivalent pupils enrolled in such classes, whichever is greater. §3317.024(O) — An amount for each approved unit for the gifted, child study, occupational or physical therapy, speech and hearing, special education supervisors, and special education coordinators. The amount shall be the sum of the unit's teacher's training level and years of experience, plus 15% of such allowance, plus $870.

[13] State aid equals 75% of operating costs; 100% of the operating costs for those assigned through the courts or a state agency.
[14] The school district pays the tuition and other costs of semi-public or private schools, with payment per child not to exceed the average per pupil appropriation for all exceptional children in the school district.
[15] Funding is limited to the average gross costs per child. §17:1949
[16] The state reimburses the residential district 60% of the difference between the amount of the contract and the "foundation aid formula allowance," which is computed according to §124.17.
[17] State financial assistance shall not exceed $1,000 per child per year, to be paid as follows: a) 100% of the first $600; and b) 50% of the next $800. In no case may such allowance exceed $600 unless the state department of education finds that the program of instruction is not being provided in the public schools of the district of the applicant's domicile.
[18] State aid pays 100% of the cost of examination, classification and speech correction for non-public school children.
[19] State pays 50% of the tuition of the private school.
[20] Each district is authorized to pay up to 3 times the state average per pupil cost, but not to exceed 100% of the tuition costs. The state will reimburse the district 60% of the amount it expends.
[21] The state will reimburse the school district the amount of the excess over the regular per capita cost of the school district.
[22] The state will fund "reasonable" tuition costs.
[23] The local school board shall pay to or on behalf of the parent or guardian, the reasonable tuition cost and other reasonable charges as may be determined by the state board of education. The local school board shall then be reimbursed from state funds 60% of its costs.
[24] No funds are to be expended for out-of-state schools.
[25] The child is treated as a pupil of the district of his residence for the apportionment of the current school fund and the payment of state aid.

SUMMARY OF THE MODEL STATUTE

As can be seen from the review of the state legislation, it is difficult to pinpoint a state with a fully comprehensive act for the education of its exceptional persons. Thus, the model act was developed to assist state officials in establishing procedural protections, specific definitions and appropriate time limitations for implementing the right to education for all persons having special education requirements.

Most importantly, the model act is not limited to mentally and physically handicapped persons, but also covers persons identified as gifted or talented.[1] In addition, the act applies to all exceptional persons between birth and 21. The act grants such persons a right to a free appropriate public education and grants their parents the right to contest every aspect of a proposed educational program through appropriate due process procedures.[2] Further, parents of all exceptional persons are given the right to participate in the development of an IEP for the purpose of designating the special education and related services that will be provided to meet a person's individual needs over the course of a school year.[3]

All of these rights described above shift to a person when he or she becomes eighteen;[4] this new and useful provision establishes the right of exceptional persons to control their own educational program when they reach a generally accepted age of majority. Further, the model law specifically provides that a free appropriate program of special education and related services includes programs in excess of a school district or state required "normal school year." Thus, if a severely handicapped person would regress substantially if not provided with a summer program, that person's IEP must include programming in excess of the normal school year.[5]

Importantly, the model act sets specific time limitations within which educational agencies must complete evaluations, IEPs and other special education requirements that are either not established in current federal law, or are too loosely set forth to fully and effectively provide expeditious access to appropriate programs.[6] Required evaluations must be completed within 10 days after an exceptional person is deemed by his or her school district of residence to require them; IEPs must be finalized within 30 days after the evaluations are complete; and, hearings must be convened within 10 days and a written hearing officer's opinion be rendered within 30 days after the date a special education hearing is requested.[7]

Applications to inspect and review a person's records must be honored by school authorities within 30 days after the date of such a request. In addition, these records must be provided to parents either at a cost not to exceed the price of duplication,[8] or at no cost if the charge would prohibit the parents from gaining access.

Specific time limitations are also set for disciplinary exclusions of an exceptional person from school; these exclusions are limited to no longer than a period of five consecutive days. Any proposed exclusion in excess of this period is deemed to be a change in educational status, and a due process hearing must be convened to ascertain an appropriate program for the exceptional person in question; as a result, the person either must not be excluded for extensive periods of time, or if a disciplinary concern recurs, new educational alternatives must be located or developed.[9]

The model act specifically allows the use of an "interim" program if necessary while negotiations between a parent or person and school officials are ongoing.[10] Additionally, the cost of special education and related services must be free to the parent and child, including the cost of transportation, food and lodging when services are delivered in a residential setting.[11] A section is also included requiring affirmative action in hiring exceptional persons.[12] Other provisions in the model act de-

[1] Model act section 3(1)(a).

[2] Model act section 6.

[3] Model act section 5.

[4] Model act section 3(2)(c).

[5] Model act section 3(5)(e).

[6] Model act sections 4(4) and 5(1)(a).

[7] Model act sections 6(6)(b) and (c).

[8] Model act section 11(3).

[9] Model act section 5(6).

[10] Model act section 4(5)(b).

[11] Model act section 3(4).

[12] Model act section 10(4).

scribe a state education office and its responsibilities, including the authority to impose sanctions — such as loss of reimbursement — upon a local district for failure to comply with the act's requirements;[13] a section covering local school district obligations;[14] and a section allowing states to use intermediate education units when a local education agency is unable to program for an exceptional person.[15] Additional sections cover provisions for confidentiality and access to records and describe very basic requirements for financing special education programs.[16]

The text of the act appears in bold type. Comments on statutory provisions appear in standard type. The comments explain, elaborate upon or provide a rationale for the text of the act. They are intended to assist the user in interpreting the act, in drafting an act for a particular state based upon this model act.

[13] Model act section 7.

[14] Model act section 8.

[15] Model act section 9.

[16] Model act sections 11, 12 and 13.

MODEL STATUTE*

A BILL FOR

AN ACT TO REQUIRE SCHOOL DISTRICTS TO PROVIDE A FREE APPROPRIATE PUBLIC EDUCATION FOR EACH EXCEPTIONAL PERSON AND TO PROVIDE PROCEDURAL SAFEGUARDS DESIGNED TO PROTECT THEIR EDUCATIONAL RIGHTS.

Section l. Title

This act shall be known as the Education of Exceptional Persons Act.

COMMENT. None necessary.

Section 2. Statement of Purpose

(1) It is the policy of this state to provide all exceptional persons a free appropriate public education in the least restrictive environment. Such education will include all special education and related services necessary to meet the needs of each exceptional person in the state regardless of the schools, institutions or programs by which the person is served. Such education will be provided in conformance with the individualized education program developed for each exceptional person and in accordance with procedural safeguards designed to protect the educational rights of such persons. No person shall be excluded from participation in, denied the benefits of, or subjected to discrimination on the basis of such person's mental or physical handicap and/or exceptionality under any public or publicly supported or regulated educational program.

(2) All exceptional persons will be provided a free appropriate public education in accordance with the provisions of this act. This act shall be liberally construed toward this end.

COMMENT. None necessary.

Section 3. Definitions

(1) "Exceptional person" means an individual between birth and the age of 21 who has a disability attributable to: deafness and hearing impairment, blindness and visual impairment, speech and language impairment, orthopedic or other health impairment, emotional and behavioral disorder, autism and related syndromes, learning disability, mental retardation; who are talented or mentally gifted; any combination thereof; and who require special education and related services.

COMMENT. P.L. 94-142 requires the provision of special education to all handicapped persons between ages 3 and 21 by 1980.[1] Yet, experts agree that early intervention in providing appropriate programs of special education is necessary so that exceptional persons may reach their fullest potential. In addition, early screening and identification allow educational agencies to properly plan for future personnel and facilities. Thus, this subsection extends services from birth through 21.

Similarly, this definition goes beyond P.L. 94-142 and Section 504 in identification in that it includes persons who are gifted and/or talented. The intellectually gifted person, or one who is particularly talented in art or music, for example, can be equally frustrated by regular education programs as are handicapped persons. Thus, their special needs must be met by providing them with the programs to develop and increase their creative capacity or special interest, so that, like handicapped children, they will be able to deal with their exceptionality in a manner most productive to themselves and to society. However, programs for those who excel have only been slowly developing, and only a limited number of gifted and talented persons have been identified. It is therefore important that early screening be provided and that the programs and protections required for other exceptional persons be applied equally to non-handicapped exceptional persons.

(2) (a) "Parent" means a parent, guardian, a person acting as a parent, or a surrogate parent as defined in paragraph (b) of this subsection.

COMMENT. None necessary.

(b) "Surrogate Parent" means a person assigned to protect the rights of exceptional per-

* Josephine Barresi of the Policy Research Center, Council for Exceptional Children and Steven S. Goldberg of the Education Law Center had primary responsibility for developing this model statute. We also appreciate the advice and assistance of Scottie Torres Higgens and Fredrick J. Weintraub of the Council for Exceptional Children, and H. Rutherford Turnbull, III of the Institute of Government, University of North Carolina.

[1] 20 U.S.C. §1412(2)(B) (1976); 45 C.F.R. §121a.300 (1978).

sons as set forth in this act and who is assigned to act in this capacity by the (name of state education agency).

COMMENT. See comment to subsection 6(11).

(c) The rights granted to parents under this act are automatically transferred to the exceptional person when the person reaches eighteen years of age.

COMMENT. The act adopts a widely accepted age of majority, eighteen years old, as the date when rights are transferred from a parent to an exceptional person. This age is consistent with the provisions of the Family Educational Rights and Privacy Act incorporated into P.L. 94-142, which transfer from parents to exceptional persons the right to enforce confidentiality of the person's education records.[2] While this act mandates that the person upon reaching eighteen is the only individual who may exercise the rights granted under the act, a court's determination at the end of a guardianship proceeding that a parent must act on the person's behalf would supercede this requirement.

(3) "Special education" means specially designed instruction at no cost to the parent or to the person to meet the unique educational needs of an exceptional person in the least restrictive or most normal setting appropriate to those needs.

COMMENT. The entire focus of federal law and this act is to ensure that all exceptional persons receive an education that is appropriate to their unique educational needs. For many handicapped persons such instruction will also include training in those self-help and basic life skills which will allow a person to act as independently as possible. Given an appropriate program of special education, virtually all exceptional persons can acquire a degree of self-sufficiency free from institutionalization.

In addition, special education must be provided in the most normal setting appropriate to the exceptional person's needs since educators are now convinced that removal from regular education should only occur when the nature or severity of a person's exceptionality is such that education in a regular setting with supplementary aids and services cannot be achieved satisfactorily. Integration of exceptional persons at least to some extent with their non-exceptional peers provides all persons with an environment most comparable to society at large. Integration also increases the experiences of non-exceptional persons and encourages acceptance of exceptional persons. Thus, this subsection prohibits removing exceptional persons to isolated settings when there is a less restrictive alternative that is appropriate to the person's needs.[3]

(4) "Related services" means transportation and such developmental, corrective, therapeutic and other supportive services including, but not limited to, speech pathology and audiology, psychological, psychiatric and counseling services, physical and occupational therapy, and recreation services as are required to assist an exceptional person to benefit from special education.

COMMENT. P.L. 94-142 mandates that all services a child requires to benefit from a program of special education must be provided. For example, physical therapy is necessary in some cases so that the person would be able to participate to the best extent possible in the educational program. It should be noted that the list in this subsection is not meant to be all-inclusive; rather it is to be illustrative of related services which the state must provide.[4]

(5) "Free appropriate public education" means special education and related services which:

(a) Have been or are being provided at no cost and under public supervision and direction by school district or intermediate education unit personnel; or as ordered by a hearing officer, by the Secretary of Education, or by a court on appeal;

(b) Meet the standards of the state education agency;

(c) Include an appropriate preschool, elementary, secondary or vocational education;

(d) Include a residential educational placement where appropriate to the person's education needs and include the costs of room and board when services are delivered in this setting;

(e) Include educational programs in excess of the regular school year where appropriate to the person's education needs; and

(f) Are provided in conformity with the person's individualized education program.

COMMENT. Pursuant to P.L. 94-142 and Section 504, this definition mandates that the person's school district of residence has the responsibility for providing a free appropriate public education.[5] This education must meet the person's individual needs and be provided at no cost to the parents, guardian, or person regardless of where it is offered. In addition, free appropriate public education is not limited to the regular school year for those handicapped students who require extended

[2] 20 U.S.C. §§1232(g), 1417(c) (1976); 45 C.F.R. §§121a.560 *et seq.* (1978).

[3] 20 U.S.C. §1401(16) (1976); 45 C.F.R. §121a.14 (1978).

[4] 20 U.S.C. §1401(17) (1976); 45 C.F.R. §121a.13 (1978); *cf.* services required pursuant to Developmentally Disabled Assistance and Bill of Rights Act, 42 U.S.C. §§6001-6081 (1976).

[5] 20 U.S.C. §1401(18) (1976); 45 C.F.R. §§121a.4 and 84.33 (1978).

programs because of the severe nature of their impairments and the resulting regression in their progress over summer months. This provision is also consistent with federal law.[6]

(6) "Individualized education program" means a written statement for each exceptional person and shall include:

(a) A statement of the present level of educational performance of the person;

(b) A statement of annual goals;

(c) A statement of short-term instructional objectives;

(d) A statement of the special education and related services to be provided to the person;

(e) A description of the extent to which the child will be able to participate in regular educational programs;

(f) The projected date for initiation and anticipated duration of the special education and related services; and

(g) Appropriate objective criteria, evaluation procedures and schedules for determining, on at least an annual basis, whether instructional objectives are being achieved.

It shall be developed in a meeting among a representative of the school district who shall be qualified to provide, or supervise the provision of, specifically designed instruction to meet the unique needs of exceptional persons, the person's teacher(s), the parents or guardian of the person and, whenever appropriate, the person. Others, at the discretion of either party may be present, if they have a legitimate interest in the education of the person.

COMMENT. The IEP is a statement of what the person needs and what will be provided to meet those needs. Its purpose as formulated in P.L. 94-142 is not just to match the services which happen to be available in the school district to the needs of the person but to insure that the needs of the person are met by all necessary programs and services.[7] Thus, the IEP must include a detailed description of how the goals will be accomplished and list all components, including related services, that are required to achieve an appropriate program of special education for each exceptional person. Services could include physical education, occupational and physical therapy, transportation, psychological intervention and all other services which are required to help an exceptional person learn.

To insure its appropriateness, this provision provides that the IEP must be formulated by a multidisciplinary team, including those individuals most familiar with the person's special needs — the teacher(s), parent, school district representative (such as a principal), other similar professionals, and the person when feasible. The teachers should include professionals from private agencies and facilities where there is a likelihood that the I.E.P. will require that the person receive or continue to receive special education and related services in a private setting. While P.L. 94-142 allows this meeting to include participation of individuals "at the discretion" of either party,[8] the U.S. Department of Health, Education and Welfare has held that the Buckley Amendment and its regulations proscribe the presence of persons, such as teachers' union representatives, who do not have a "legitimate educational interest" in the exceptional person's program.[9] Thus, those lacking such "interest" must be excluded.

(7) "School district" means a public board of education which is legally constituted within a state to administer, direct or perform a service for the public preschool, elementary, secondary or vocational schools in a city, county, township, other political subdivision, or such combination of school districts or counties as are recognized in the state as an administrative agency for its public schools, including an intermediate education unit.

COMMENT. None necessary.

(8) "Intermediate education unit" means any public educational agency which provides special education and related services to exceptional persons on a regional basis and which is supervised by the (name of state education agency).

COMMENT. P.L. 94-142 provides for the establishment of intermediate education units.[10] Many states create such public agencies to provide special education and related services to exceptional persons when local districts are incapable of providing these services.[11]

(9) The following words and phrases, as used in relation to records, shall have the following meanings:

(a) "Destruction" means the physical destruction or permanent removal of identifying data from education records of an exceptional person so that the information in those records is no longer personally identifiable.

(b) "Education records or records" means education records, as defined in 20 U.S.C. §1232g

[6] Armstrong v. Kline, 476 F. Supp. 583 (E.D. Pa. 1979); Mahoney v Administrative School District No. 1, 601 P.2d 826 (Or. App. 1979).

[7] 20 U.S.C. §1401(19) (1976); 45 C.F.R. §121a.343 and 45 C.F.R. §121a.512(1977) (1978).

[8] 45 C.F.R. 121a.344(a)(5) (1978).

[9] 20 U.S.C. §1232g (1976); 45 C.F.R. Part 99 (1978).

[10] 20 U.S.C. §1401(22) (1976); 45 C.F.R. 121a.7 (1978).

[11] *See* section 9, *infra*.

(1976) and 45 C.F.R. Part 99 (1978), of an exceptional person who (i) receives or has received special education and related services from a school district or an intermediate education unit; (ii) is enrolled or has been enrolled, pursuant to this act, in an approved private school for exceptional persons; (iii) is enrolled or has been enrolled in a program of special education operated by the state education agency.

COMMENT. These definitions are adapted from the regulations to the Family Educational Rights and Privacy Act (the Buckley Amendment).[12]

(c) "Educational agency" means a private agency, the (name of state educational agency), a school district or an intermediate education unit and any component part thereof which collects, maintains, uses or releases an exceptional person's education records containing personally identifiable information.

COMMENT. The purpose of this all-inclusive definition is to ensure that all agencies which have been or are providing special education programs and services to an exceptional person will be covered by this act in general and, more specifically, by those sections of the act relating to confidentiality and access to records.[13] Again, the Buckley Amendment must be followed by all of the above educational agencies which receive funds pursuant to U.S. Office of Education programs, and a similar definition is included in that law.[14]

(d) "Personally identifiable" means information that includes the name or address of an exceptional person or any of the person's family members; an identifying piece of information such as the person's telephone number or social security number; and a list of personal characteristics or other information which would make the person's identity easily traceable.

COMMENT. Personally identifiable information as described in this subsection is that data which includes private and personal information relating to a person and his or her family. This term is also included in the Buckley Amendment regulations.[15]

(e) "Release" means giving access to or disclosing education records which include personally identifiable information.

COMMENT. None necessary.

(10) "Services," "service delivery" and "delivery of services" means delivering special education and related services.

COMMENT. None necessary.

[12] 20 U.S.C. §1232g (1976); 45 C.F.R. §99.3 (1978); 45 C.F.R. 121a.560 (1978).

[13] See section 11 *infra*.

[14] 45 C.F.R. §§99.1, 99.3 (1978).

[15] *Id.* §99.3

Section 4. Identification and Evaluation

(1) Each school district shall adopt a policy and develop and implement a plan to identify, locate and evaluate all persons who are exceptional and in need of special education.

COMMENT. Identification and evaluation is a requirement for receiving federal funding pursuant to P.L. 94-142.[16] Further, Section 504 and its implementing regulations require that no handicapped person, otherwise qualified to participate in a program receiving federal financial assistance, "shall, solely by reason of his handicap, be excluded from the participation in, be denied the benefits of, or be subjected to discrimination in . . ." any such program.[17] Since education is one of the key federal activities from which handicapped exceptional persons cannot be excluded,[18] §504 also necessitates identification and evaluation. Thus, proven person find models, such as Pennsylvania's Commonwealth Plan for the Identification, Location, and Evaluation of Retarded Children,[19] issued in response to the *PARC* case,[20] must be developed and implemented.

(2) The identification and evaluation plan shall incorporate procedures and instruments including, but not limited to:

(a) Group screening instruments, which may be used to identify exceptional persons; and

(b) A procedure for referrals of exceptional or thought-to-be exceptional persons for further evaluation and testing by parents, teachers, psychologists, nurses, counselors and other professionals who know the person. Parents must be given prior notice when their child is referred for such evaluation.

COMMENT. Because P.L. 94-142 is skeletal in its requirements for identification and evaluation of exceptional persons, a state statute should provide guidance to education professionals responsible for making service delivery decisions. Additionally, such legislatively mandated compliance will preclude the need for court-ordered plans, such as the one required in *Frederick L. v. Thomas*.[21] Hence, this subsection is designed to give state educational agencies and school districts of residence some guidance in promulgating procedures and carrying out screening and evaluation plans which will result in the proper identifi-

[16] 20 U.S.C. §1414(a)(1) (1976); 45 C.F.R. §121a.220 (1978).

[17] 29 U.S.C. §794 (1976).

[18] 45 C.F.R. §§84.31 *et seq.* (1978).

[19] Armstrong v. Kline, 476 F. Supp. 583 (E.D. Pa. 1979).

[20] Pennsylvania Ass'n for Retarded Children v. Pennsylvania, 334 F. Supp. 1257 (E.D. Pa. 1971); 343 F. Supp. 279 (E.D. Pa. 1972).

[21] Frederick L. and Delaware Valley Ass'n for Children with Learning Disabilities v. Thomas, 419 F. Supp. 960 (E.D. Pa. 1976), *aff'd*, 557 F.2d 373 (3rd Cir. 1977).

cation of and delivery of service to of all exceptional persons.[22]

(3) The (name of state education agency) shall assure that all testing and evaluation materials and procedures shall be selected and administered so as not to be discriminatory on the basis of race, color, religion, national origin, sex, disability or handicapping condition, and shall be administered in the exceptional person's native language or mode of communication. No single procedure shall be used as the sole criterion for identifying, evaluating or determining an appropriate educational program for an exceptional person.

COMMENT. Pursuant to P.L. 94-142, this requirement is necessary to prevent the misclassification and inappropriate evaluation of individuals as a result of biased testing procedures.[23] For example, testing people in English when that is not their primary language would yield inaccurate results. Similarly, those persons (such as deaf and hearing impaired) who require sign language interpreters must be provided them if accurate testing results are to be achieved. In some instances this provision may require that interpreters also be available for the parents. This section also requires using more than one procedure to insure the reliability of the testing by enabling data gathered from a variety of sources to be compared. Relying on one assessment technique or device for making a decision may yield inaccurate results, even if the test is administered in the person's native language or mode of communication.

Generally, the greater the severity of a person's exceptionality, the greater the amount of information that needs to be collected. However, prescribing a standard battery of tests to be administered to all candidates for special education should be avoided. The choice of assessment techniques should be determined on an individual basis to avoid unnecessary or inappropriate evaluations and may include but not be limited to the following:

(a) An educational history assessment including a history of the person's prior evaluations, an overview of the person's school progress, and a statement of the person's current academic standing;

(b) An educational skills assessment including an analysis of the person's specific behavioral abilities along a developmental continuum with a comparison of those abilities to the tasks which are contained in the regular education program; a statement of school readiness, functioning and achievement; and a statement of the person's attentional capacity, motor coordination, activity levels and patterns, communication skills, and memory and social relations with groups, peers and adults;

(c) A health assessment including taking the previous medical history, obtaining a current comprehensive physical examination, and determining current existing medical conditions and prescribed treatment(s);

(d) A psychological assessment including obtaining an individually appropriate psychological examination, observing the person in familiar surroundings such as classroom or home, and obtaining an inventory of sensory, motor, language, perceptual, attentional, cognitive, affective, attitudinal, self-image, interpersonal, behavioral interest and vocational factors, as appropriate in each case; and

(e) A social assessment including obtaining a description of pertinent family developmental history and estimates of adaptive behavior at home, in the neighborhood and in local peer groups. Estimates of adaptive behaviors should be based to the greatest possible degree on information obtained by direct observation of the person or direct interview of the person and parent in the neighborhood setting.

Additional types of assessments necessary to evaluate specific difficulties may be required and should be based upon a determination of each person's case characteristics.

(4) All required evaluations described in this section shall be completed prior to the meeting held to determine an individualized education program, but no later than ten (10) days after an exceptional person is deemed to require them by a school district, or a parent requests them for a person who is or is thought to be exceptional, unless the parents and school district agree to different time limitations.

COMMENT. The time limitations are included so that decisions about service delivery can be resolved in a prompt manner. The ten day period for completion of all evaluations will guarantee that an IEP can be developed[24] and appropriate services delivered within a reasonable amount of time after the person is identified. This section comports generally with the requirements of P.L. 94-142 and Section 504.[25]

(5) Each person is presumed to have no need for special education and related services until he or she has been identified and evaluated as an exceptional person pursuant to the procedures set forth in this act. Pending development of an individualized education program for a person determined to be exceptional, the following

[22] *Cf.* 20 U.S.C. §1412(5)(c) (1976); 45 C.F.R. §§121a.530 *et seq.*

[23] 20 U.S.C. §1415 (1976); 45 C.F.R. §121a.532 (1978). Lora v. Board of Education of the City of New York, 456 F. Supp. 1211 (E.D. NY. 1978).

[24] *See* subsection 5(1).

[25] 45 C.F.R. §§121a.340 *et seq.* and §84.33(b) (1978).

provisions regarding the delivery of educational services shall apply:

(a) Each person already enrolled in the school district's programs shall remain in his or her current program until such procedures are completed, unless the parent or educational agency agree to an interim program; and

(b) An exceptional person neither attending school nor receiving educational services shall be educated in the district's program for nonexceptional persons pending completion of the individualized education program, unless the parents and the school district agree to an interim program.

COMMENT. Since evaluation procedures are frequently time-consuming, the subject of the testing and diagnostic procedures is usually excluded from the school or current program (or some aspect of it) pending completion of the assessment and a challenge to the results if the parent disagrees with the outcome. Thus, to avoid creating a gap in the person's education, this provision allows the person to remain in his or her current program or be placed in the school district's program for nonexceptional persons pending the outcome.[26]

This subsection extends somewhat beyond the rights accorded students in P.L. 94-142's regulations, which require full testing but no provision for alternative placement prior to the initial evaluation. However, a state statute should extend the best possible protection to all exceptional or thought to be exceptional persons.[27] In addition, the failure to place a handicapped or thought to be handicapped person in a regular class, at least initially, would amount to an exclusion on the basis of handicap and would violate Section 504 and its implementing regulations.[28] Finally, it must be kept in mind that homebound instruction is the most restrictive environment available when education is considered, and it should only be used as a last resort in all cases.

(6) Where the parent disagrees with the results of the evaluation obtained by the school district, the parent shall have the right to obtain an independent evaluation of the exceptional person and have that evaluation considered by the school district when developing the individualized education program. Where the cost of the independent evaluation would effectively prevent access to it by the parent, the school district shall bear the expense.

COMMENT. Having the right to an independent evaluation is consistent with federal law.[29] However, providing it at no cost to parents who cannot afford to pay for it varies slightly from P.L. 94-142 in that the federal statute requires a due process hearing when a parent requests the school district to bear the expense of the independent evaluation.[30] For indigent parents, this is an unnecessary burden especially when it can be handled administratively and since there is a right to a due process hearing if the request for reimbursement is denied.[31]

Section 5. Delivery of Services

(1) (a) The (name of state education agency) shall insure that each school district or intermediate unit shall develop and implement an individualized education program for each exceptional person resident therein, regardless of whether the person is to receive services in a public or private school or facility. The individualized education program shall be developed within thirty (30) days after all evaluations of the person are completed pursuant to subsection 4(4) of this act.

(b) Decisions concerning the education program or termination of an educational program of an exceptional person shall be made solely on the basis of the person's needs for special education and related services as set forth in the person's individualized education program and as consistent with the requirements of section 5(2) of this act.

(c) Each school district shall periodically, but at least annually, review, and revise if appropriate, the person's individualized educational program. In addition, each school district shall reevaluate each exceptional person every three years, or within two years of the last evaluation if requested by the parent, and revise the individualized education program if appropriate. Reevaluations may not be performed without prior written notice to the exceptional person's parent.

COMMENT. The IEP is one of the cornerstones of P.L. 94-142 and its requirements are therefore included in this act.[32] Although the IEP is not a legally binding contract, since the passage of P.L. 94-142 teachers and administrators have feared "educational malpractice" lawsuits if all of the goals listed in the IEP were not achieved. This is an unlikely result if genuine efforts are made to comply with it. If such efforts were not made, however, or if needed services were intentionally excluded from the original IEP, school officials would be in serious danger of losing funding for

---

26 20 U.S.C. §1415(e)(3) (1976); 45 C.F.R. §121a.513 (1978).

27 *See* 45 C.F.R. §§121a.530-534 (1978).

28 *Id.* Part 84.

29 20 U.S.C. §1415(b)(1)(4) (1976); 45 C.F.R. §121a.503 (1978).

30 45 C.F.R. §121a.503(b) (1978).

31 See subsection 6(6).

32 See subsection 3(6).

their programs or subjecting themselves to personal liability.[33]

(2) The (name of state education agency) and each school district shall insure that;

(a) Unless an exceptional person's individualized education program requires other arrangements, the person is educated in the school he or she would have attended if not exceptional;

(b) All exceptional persons, including those in public or private institutions or other care facilities, are educated with persons who are not exceptional;

(c) Each exceptional person participates with nonexceptional persons in nonacademic and extracurricular services and activities to the maximum extent appropriate to the needs of the exceptional person; and

(d) Special education classes, separate schooling or other removal of exceptional persons from the regular educational environment occurs only when the nature or severity of the exceptionality is such that education in regular classes with the use of supplementary aids and services would not benefit the person.

COMMENT. Every exceptional person must be educated in the most normal setting appropriate to his or her needs. This statement embodies the concept of the least restrictive environment and is another of the major and frequently misunderstood requirements of P.L. 94-142. It does not mean that all exceptional persons will always be educated in regular education settings; rather, it requires that such persons be educated in regular classrooms when appropriate to the person's needs. This requirement applies not only to classroom instruction, but also to field trips and extracurricular activities, and other school programs and activities, including art, music, and gym.[34]

(3) Each public educational agency shall have service delivery options available to meet the needs of exceptional persons for special education and related services. These options shall include but are not limited to regular classes; special classes in regular schools; special public schools; preschool, elementary, secondary, vocational and private schools; instruction in hospitals and institutions or instruction in the person's home.

COMMENT. P.L. 94-142 and Section 504 require that alternative programs be available to serve the various needs of exceptional persons.[35] As stated in subsection (2) above, however, all placements must be in the most normal or least restrictive setting, as appropriate for each person's needs.

(4) If at any time the (name of state education agency) finds that a school district has failed to provide a free appropriate public education to any exceptional person entitled to receive the same from the school district, the (name of state education agency) may withhold all or any portion of the state aid to which the district is or may be entitled for the operation of all education programs and services until the district remedies its failure. Furthermore, the (name of state education agency) may provide the education directly whether or not it elects to withhold aid.

The (name of state education agency) shall take no action pursuant to this subsection except after a public hearing is held pursuant to (cite sections of code relating to the administrative agency law of this state) which establishes the failure of the school district to meet its obligations to the exceptional person.

COMMENT. This subsection sets out the sanction of withholding funding from any local educational agency which fails to meet the requirements of the act. It should be realized, however, that the state educational agency should withhold funding only as a last resort, since such action would result in denying opportunities to the very children who must be served. Yet, such a measure may be required if a hearing determines that local officials failed or refuse to comply with the requirement of a free appropriate education for each exceptional person and refuse to take action to remedy their refusal or denial.[36]

(5) If the (name of state education agency) finds that a school district has failed to provide a free appropriate public education to an exceptional person entitled to it under this act, and if it decides to provide the education directly, it may, without limitation:

(a) Hire and employ special education teachers and any necessary supporting professional and other personnel;

(b) Incorporate such personnel into the affected school system;

(c) Procure and deploy such supplies, equipment and facilities as may be necessary or appropriate;

(d) Furnish such administrative supervision and services as may be necessary and appropriate to make the special education program effective;

(e) Intervene in state institutions or facilities

---

33 45 C.F.R. §121a.340 *et seq.;* §84.33(b) (1978); *see also* 45 C.F.R. §121a.349 (1978) relating to personal accountability.

34 *See* 20 U.S.C. §1412 (1976); 45 C.F.R. §§121a.550 *et seq.* and §84.34 (1978). For a legal discussion of least restrictive environment, see Hairston v. Drosick, 423 F.Supp. 180 (S.D. W.Va. 1976) and Halderman v. Pennhurst State School and Hospital, Nos. 78-1490, 78-1564 and 78-1602 (3rd Cir. December 13, 1979).

35 20 U.S.C. §§1412, 1414 (1976); 45 C.F.R. §§121a.550 *et seq.* and 84.34 (1978).

36 *Cf.* 20 U.S.C. §1414(b)(2) (1976); 45 C.F.R. §121a.194 (1978).

solely to guarantee or provide exceptional persons therein their rights to a free appropriate public education except that no person shall be removed from the school district in which he or she would regularly be entitled to receive special education without the consent of such person's parent or guardian;

(f) Take all other action to assure that exceptional persons in the school district receive a free appropriate public education.

COMMENT. None necessary.

(6) If an exceptional person is or is to be excluded from any educational program for a period in excess of five (5) consecutive days, or for two or more periods of at least one day each within a three-month period, such exclusion will be deemed to be a change in educational program, and a due process hearing shall be held to determine what is the appropriate program for the person. The hearing shall be held in conformity with section 6 of this act. During such exclusion, the person shall be given the opportunity to make up any work missed as a result of the exclusion.

COMMENT. Excluding exceptional persons from school for any period of time is indeed a controversial issue. However, it is established law that removal in terms of suspension or expulsion may well be punishment for behavioral manifestations of the very handicap the program of special education is designed to remedy. For handicapped persons in particular, this type of removal may indicate an exclusion on the basis of handicap in violation of Section 504.[37] But for all exceptional persons, any interruption in educational programming could be harmful and thus it is deemed a change in educational program. As such, an immediate due process hearing must be convened to determine the appropriate program for the person.[38]

This provision also recognizes that the goal of all special education laws should be to provide equal treatment to all students in all aspects of the educational setting. Therefore, the opinion of some that an exceptional person should never be excluded from the classroom for any reason may be unrealistic and legally untenable. As a result, this subsection provides that an exclusion up to five days may be allowed. During any exclusion, however, exceptional persons must be given the opportunity to make up the work they have missed in class. Further, continued exclusions for more than five days or for at least two one-day periods in three months (known to some as "serial suspension") are prohibited.[39]

(7) Every exceptional person completing a special education program at the high school level shall be granted a high school diploma by the school district of residence. This diploma shall be same as the regular high school diploma unless:

(a) The district offers to non-exceptional persons more than one type of diploma (certificate) and the format or content of each is determined by a program that has a significantly different curriculum than that of other programs in which other students in the district are enrolled; and

(b) The curriculum of the program in which the exceptional person was enrolled differs significantly from that of other programs in which other students in the district are enrolled.

COMMENT. The fact that many exceptional persons have needed accommodations to help them benefit from a program of education should not preclude them from receiving a regular high school diploma upon completion of the required curriculum unless the school district differentiates between curriculum and types of diplomas awarded for the regular students, and the curriculum completed by the exceptional person substantially differs from that which regular students complete. In fact, if an exceptional person is qualified for a diploma, its denial would be based solely on handicap and therefore would violate Section 504.[40]

Section 6. Procedural Safeguards

(1) The (name of state education agency) and each school district and any other state or local agency providing or required to provide special education and related services to exceptional persons, or any private agency providing special education and related services pursuant to a publicly supported agreement, shall establish and maintain procedures to assure that exceptional persons' parents are guaranteed procedural safeguards with respect to the provision of a free appropriate public education by the agency or district.

COMMENT. Pursuant to P.L. 94-142 and Section 504, procedural safeguards are yet another major protection which exceptional persons and their parents must be given in every state which receives federal funding. A state statute should, at the very least, include and mandate these

37 29 U.S.C. §794 (1976).

38 *See* Stuart v. Nappi, 443 F.Supp. 1235 (D. Conn. 1978); 45 C.F.R. §84.36 (1978); 20 U.S.C. 1412(5)(B), 1415(b)(1)(C) (1976).

39 Pennsylvania Ass'n for Retarded Children v. Pennsylvania, *supra;* Mills v. Board of Education, 348 F.Supp. 866 (D.D.C. 1972).

40 *See generally,* 45 C.F.R. §§84.4 and 84.31 *et seq.* (1978).

safeguards, as well as provide protection beyond the outlines set forth under federal law.[41]

Note that the definition of "parents" includes exceptional persons when persons become old enough to exercise their rights under this act.[42] For younger exceptional persons, it is the parent who exercises the rights set forth in the act on behalf of the exceptional person. As a result, all references to the individual advocating the person's rights under this section will be to parents.

(2) Notice which meets the requirements of subsections (3) and (4) of this section shall be given to the parent of an exceptional person before an educational agency takes any action or refuses to take any action regarding the identification, referral, evaluation and individualized education program of, or delivery of services to an exceptional person. The notice shall include:

(a) A complete statement of all rights granted to parents under this act and by federal law relating to the education of exceptional persons, including but not limited to a description of all procedural safeguards;

(b) A complete statement of the action the educational agency proposes to take or refuses to take, including an explanation detailing the reasons its officials propose or refuse to take such action;

(c) A complete statement of all tests, reports, evaluations and all other documents used as a basis for the proposed action or refusal to act; and

(d) A complete statement of any other factors considered by officials of the educational agency which are relevant to their decision.

COMMENT. The right to be informed of all matters relating to an exceptional person's involvement in the special education system is perhaps the most important right afforded to a parent pursuant to this act and federal law;[43] without the notice described in this subsection, it would be impossible for parents to act intelligently in obtaining an appropriate program of special education. Not only is such full notice required under the legal concept of due process, but educational experts agree that full participation of a parent in all aspects of an exceptional person's education is beneficial to all participants in the system.

(3) All notices mandated by this section shall be written in language understandable to the general public and provided in the native language or mode of communication used by the parent.

COMMENT. If parents are to effectively exercise the person's rights under this act, it is essential that they be informed of their existence and understand their import. Thus, all notices must be provided in the parents' native language or mode of communication. Legal jargon must be avoided, which means that sending parents digests of statutory requirements does not meet the requirements of properly notifying parents of actions officials propose to take.[44]

(4) If the native language or other mode of communication of the parent is not a written language, or if the parent cannot read, the state or local education agency shall take steps to insure:

(a) That the notice is translated orally or by other means to the parent in his or her native language or other mode of communication;

(b) That the parent understand the content of the notice; and

(c) Written documentation exists to demonstrate evidence of meeting these requirements.

COMMENT. None necessary

(5) The school district must obtain parental consent before conducting an initial evaluation or making an initial determination of the program of special education and related services which the person is to receive.

COMMENT. Due process, if nothing else, means the opportunity both to be put on notice and have the right to respond when a government agency takes or proposes to take any action which impacts on an exceptional person's life. As such, parental consent is a necessary feature of due process when school authorities propose to initially evaluate their child or implement the initial individualized education program. All matters relating to an initial evaluation, including the types of tests and procedure to be used and the purpose of each testing instrument must therefore be explained and consent obtained prior to their being administered. It may be unreasonable to allow parents the opportunity to delay testing indefinitely by not granting consent, but in that event a due process hearing should be convened.[45]

(6) (a) Whenever parents or a school district disagree with respect to any matter relating to the identification, referral, evaluation, individualized education program, or delivery of services, or to any other aspect of the provision of a free appropriate public education to the exceptional person, either party shall have the opportunity to present complaints to the (name of state education agency), and to have a hearing presided over by an impartial hearing officer

---

[41] 20 U.S.C. §1415 (1976); 45 C.F.R. §§84.36, 121a.500 *et seq.* (1978).

[42] *See* subsection 3(2).

[43] *See* 20 U.S.C. §1415(b) (1976); 45 C.F.R. §§121a.504, 121a.505 (1978).

[44] 20 U.S.C. §1415 (1976); 45 C.F.R. §121a.505 (1978).

[45] *See* subsection 6(a), *infra;* 20 U.S.C. §1415 (1976); 45 C.F.R. §121a.504 (1978).

selected by the (name of state education agency) if they so choose.

COMMENT. Due process also requires that parents and school officials have the right to present a complaint, or to present a complaint and have a hearing presided over by an impartial hearing officer, if they so choose, when they disagree with any matter relating to the exceptional person's right to a free appropriate public education. At the hearing each party shall have the opportunity to fully present their reasons for requesting the hearing.[46]

(b) The hearing shall be conducted in the parents' school district of residence, at a location and time that is convenient to the exceptional person's parent, and within no more than ten (10) days after the request for a hearing is made with the school district of residence. A hearing or reviewing officer may grant reasonable extensions of time beyond the ten (10) day limit at the request of either party.

COMMENT. Parents faced with the financial, emotional and procedural obstacles of locating, preparing and presenting a number of varied experts at a hearing should not be required to transport all such individuals great distances and at their own expense. This is particularly important since some state educational agencies have held hearings in distant state capitals in the past. Such action may make the hearing a sham. Thus, this subsection protects against such an occurrence.

The ten-day time period within which a hearing must be held after it is requested is provided so that any disputed matter regarding an exceptional person's right to a free appropriate public education may be expeditiously resolved and an appropriate program of special education and related services be promptly provided. However, unforeseen difficulties do arise — a witness may become ill or be unexpectedly called out of town. Therefore, a hearing officer must be given the authority to grant "reasonable" extensions to the time limitation which must otherwise be strictly enforced.

(c) The (name of the state education agency shall insure that no later than thirty (30) days after the hearing is requested, a final decision is reached, and a copy of it is mailed to each of the parties. The decision shall be based on a preponderance of the evidence presented at a hearing and shall be final unless appealed pursuant to subsection 7 of this section.

COMMENT. The time requirement set out in this subsection is generally within the mandate of P.L. 94-142.[47] However, the federal law is expanded to require a final hearing officer's opinion within 30 days after the hearing is requested, while P.L. 94-142 permits 45 days — which is far top much time when an exceptional person's rights are at stake.

(7) When either party disagrees with respect to any matter decided by the hearing officer the decision may be appealed to the chief education officer of this state.

COMMENT. This section is consistent with P.L. 94-142 and its implementing regulations.[48]

(8) The (name of state education agency) shall insure that not later than thirty (30) days after a review is requested, it is held, a final decision is reached, and a copy of it is mailed to each of the parties.

COMMENT. The 30-day period for deciding an appeal is extracted from the federal law.[49]

(9) A due process hearing may not be conducted by a person who is an employee of the school district or other public agency involved in the education or care of the exceptional person or any person having a personal or professional interest that may conflict with his or her objectivity in the hearing. However, a person assigned as a hearing officer is not deemed to be an employee of the (name of state agency) or school district solely because he or she is paid to serve as a hearing officer.

COMMENT. The impartiality of a hearing officer is, in practice, the most controversial aspect of the hearing. For example, hearing officers frequently are persons in a similar job (albeit in another school district) as those officials who the parent of an exceptional person is required to challenge at a hearing. In order to avoid this problem, hearing officers should be chosen from the ranks of advocacy organizations and academia, in addition to the usual source of special education administrators. In addition, attempts must be made to avoid choosing persons who may be subject to political pressures from the school district in which they are sitting as a hearing officer.[50]

(10) (a) Any party in a due process hearing has the right to:

(i) Be accompanied and advised by a personal advocate, including legal counsel, and by individuals with special knowledge or training with respect to the problems of exceptional persons;

(ii) Present evidence and confront, cross-examine, and compel the attendance of witnesses;

(iii). Prohibit the introduction of any evidence at the hearing that has not been disclosed to that party at least five days before the hearing;

---

[46] 20 U.S.C. §1415 (1976); 45 C.F.R. §121a.506 (1978).

[47] 45 C.F.R. §121a.500 *et seq.* (1978).

[48] 20 U.S.C. §1415(c) (1976); 45 C.F.R. §121a.510 (1978).

[49] 45 C.F.R. §121a.512 (1978).

[50] *See* 20 U.S.C. §1415(B)(2) (1976); 45 C.F.R. §121a.507 (1978).

(iv) Obtain an electronic or written record of the hearing; and

(v) Obtain written findings of fact and decisions.

(b) Parents involved in a hearing must be given the right to open the hearing to the public and have the exceptional person who is the subject of the hearing present.

COMMENT. These hearing rights reflect the requirements of P.L. 94-142.[51]

(11) Any party disagreeing with the findings and decisions of any final state level decision may bring a civil action with respect to the complaint in any state court of competent jurisdiction. In any action brought under this subsection the court shall receive the records of the administrative proceedings, hear additional evidence at the request of either party, base its decision on a preponderance of the evidence, and grant such relief as the court determines is appropriate after a hearing. This section shall not preclude the right of any party to bring a civil action in a United States district court pursuant to federal law.

COMMENT. Pursuant to P.L. 94-142, the final action of the state administrative agency may be appealed either to the local federal district court or to the appropriate state court.[52] Yet, there is still a good deal of confusion regarding whether the court should conduct a *de novo* or original proceeding, or if the court should merely act after reviewing the record of the due process hearing (in most cases the written transcript). This subsection adopts the latter approach and avoids additional burdens to the parents of an exceptional person by not requiring a new trial of the facts. However, in accordance with the congressional mandate of P.L. 94-142, a court may hear any additional evidence, at the request of either party.[53]

(12) (a) The (name of state education agency) shall insure that an individual is assigned to act as a surrogate parent for an exceptional person whenever, and only whenever, the exceptional person's parents cannot be identified, located, or is unavailable, or the person is a ward of the state.

(b) The (name of state education agency) shall develop and apply a method for determining whether a person needs a surrogate parent and for assigning a surrogate parent to that person. It shall establish procedures for the selection of surrogates that will insure that individuals selected have the knowledge and skills to adequately represent the best interests of the exceptional person and to insure against any conflicting interests. A person assigned as a surrogate may not be an employee of a public agency which is involved in the education or care of the exceptional person. However, a person who otherwise qualifies to be a surrogate parent under this subsection is not an employee of the agency solely because he or she is paid by the agency to serve as a surrogate parent.

(c) The surrogate parent may represent the exceptional person in all matters relating to the identification, referral, evaluation, individualized education program, delivery of services, and all other matters relating to the provision of a free appropriate public education to the exceptional person.

(d) No surrogate may be held liable to the person whom he or she represents for taking or omitting to take any action as a surrogate if he or she has acted or omitted to act in good faith.

COMMENT. When a parent or guardian cannot be located or is unavailable due to such causes an institutionalization, the appointment of a formal legal guardian — a time-consuming and cumbersome process should not be required in order to represent exceptional persons throughout the procedures described in this act. If a responsible person genuinely concerned with the exceptional person's education is available or can be provided, that person may be appointed for the limited purposes described, so that the exceptional person may be served without delay. Pursuant to P.L. 94-142,[54] state education agencies must develop and apply a method for determining if an exceptional person requires a surrogate, and criteria for appointing a surrogate parent. These surrogates should be appointed only in the situations described in this section and not merely when parents are uncooperative with the efforts of school officials.

Section 7. State Administration

(1) There is hereby established in the (name of state education agency) a Division of Education for Exceptional Persons, hereafter referred to as "the Division."

COMMENT. A Division of Education for Exceptional Persons is created within the state education agency so that federal and state programs will be administered in an efficient manner by a single office. It should be directly responsible to the chief executive officer of the state education agency, in order to ensure equal access to policy-makers and fiscal resources, which in turn should ensure that both school officials and consumers will only have to go to one office for information and funding. The Division could also engage in such functions as continuing and compensatory education, thereby conveniently placing all services for the

[51] 45 C.F.R. §121a.508 (1978).

[52] 20 U.S.C. §1415(e)(3) (1976); 45 C.F.R. §121a.511 (1978).

[53] 20 U.S.C. §1415(e)(2) (1976).

[54] 20 U.S.C. §1415 (1976); 45 C.F.R. §§121a.10, 121a.514 (1978).

education of exceptional persons under a single jurisdiction.

(2) In addition to all powers granted by (cite other state enabling legislation), the (name of state education agency) through the Division has the following powers, duties and responsibilities:

(a) To make and keep current a plan for the implementation of this act and to require school districts, intermediate education units and state, local, and private agencies where they receive public funds, to comply with this plan. The plan shall include, but not be limited to:

(i) Assurances that each exceptional person has the right to a free appropriate public education;

(ii) An inventory of the kinds and numbers of personnel, facilities and services necessary to execute the provisions of this act;

(iii) Provisions to assure that federal and state financial aid for the education of exceptional persons is expended in a manner consistent with the applicable federal or state law governing the receipt, expenditure and accounting of that aid;

(iv) Provisions for the implementation of sections 10 and 12 of this act;

(v) Provisions for making reports to evidence compliance with the terms of this or other state or federal law affecting the education of exceptional persons;

(vi) Provisions to enable the (name of state education agency), through the Division, to monitor and exercise general supervision over all special education and related services provided or required to be provided by all school districts, intermediate education units and state, local and private agencies;

(vii) Procedures for evaluating at least annually the effectiveness of the identification, referral, evaluation, individualized education program, and delivery of services in meeting the educational needs of exceptional persons; and

(viii) Provisions to give the public an opportunity to review and comment on the annual plan before its adoption.

The Division shall submit the Plan to the Governor, the state legislature and (name of state advocacy office) and make it available for public distribution no later than (day, month) of each year. Thereafter, amendments to or revisions of the plan shall be submitted to the Governor and the legislature and made available for public distribution at least thirty (30) days prior to the convening of each regular session of the legislature (or, if the legislature meets biannually only, by (day, month). All such submissions, except for the initial one, shall detail progress made in fulfilling the plan and in implementing the policy of this act.

COMMENT. A state plan is one of the necessary and important administrative requirements for funding pursuant to P.L. 94-142.[55] Additionally, advocates of exceptional persons can use the plan to determine if their clients are receiving the full range of services mandated by the plan; to gain access to information required by the plan pursuant to the federal Freedom of Information Act (and/or state "Right to Know" laws); and to obtain federal court or administrative orders mandating compliance with it, if the state plan is not being strictly enforced by the state educational agency.

The provision relating to public comment is extracted from P.L. 94-142, which requires that state education agencies provide an opportunity for public comment on a state's Annual Program Plan. This process may include both public hearings and written comments addressed to the state education agency.[56]

(b) To establish standards and regulations for the identification, referral, evaluation and individualized education program of, and delivery of services to each exceptional person which must be provided by any school districts, intermediate education units and state, local and private agencies;

COMMENT. This paragraph should be interpreted broadly to require that standards be developed by the state education agency that will secure appropriate, efficient and consistent services for all eligible exceptional persons in the state. These standards must be detailed and narrowly construed. Further, the state must ensure that these standards carry out the requirements of P.L. 94-142.[57]

(c) To exercise general supervision over the identification, referral, evaluation and individualized education program of, and delivery of services to, each exceptional person which must be provided by all school districts, intermediate education units and state, local and private agencies;

(d) To monitor and compile and to require school districts, intermediate units and state, local, and private agencies to monitor and compile and report to the Division, data concerning the identification, referral, evaluation and individualized education program of, and delivery of services to, each exceptional person and that person's lack of and need therefor; and to review and analyze the data collected in order to evaluate these activities of the reporting entities;

COMMENT. These paragraphs provide general authority for the Division to supervise the provision of a free appropriate public education for all

---

55 20 U.S.C. §1411 (1976); 45 C.F.R. §§121a.750 *et seq.* (1978).

56 20 U.S.C. §1412(7) (1976); 45 C.F.R. §121a.280 *et seq.* (1978).

57 20 U.S.C. §1412(6) (1976); 45 C.F.R. §121a.600 (1978).

exceptional persons. It is both appropriate and necessary that the ultimate responsibility for providing this education be vested in the single state agency responsible for education for at least the following reasons: (1) to centralize accountability; (2) to encourage the best utilization of education resources; (3) to guarantee implementation of the comprehensive state plan for the education of all exceptional persons within the state; (4) to ensure day-by-day coordination of efforts among education expertis, thereby reducing the likelihood of inadequate educational programs in other delivering agencies.

Furthermore, the collection and analysis of data required in paragraph (d) will enable the state and such federal agencies as the United States Department of Education Civil Rights Office to assess the implementation of this Act and to determine areas of need throughout the state in order to more effectively plan for program development and improvement. In addition, such data must be accurately maintained for the state educational agency to be in compliance with P.L. 94-142, Section 504, and their implementing regulations.[58] The information collected is subject to the confidentiality provisions of section 11.

(e) To provide technical assistance and assist school districts, intermediate education units and state, local and private agencies in the identification, referral, evaluation and individualized education program of, and the delivery of services to, each exceptional person;

(f) To purchase, lease or enter into other contracts for the purchase or lease of, and to receive gifts of, equipment, materials, supplies, building, structures and facilities to aid in the identification, referral, evaluation and individualized education program of, and the delivery of services to, each exceptional person;

(g) To develop, test, demonstrate, maintain, purchase or otherwise acquire, store, produce and make available equipment, materials and special supplies and devices useful in identifying, referring, evaluating, creating an individualized education program, and delivering services to exceptional persons;

(h) To collect, evaluate, develop and disseminate research data and other information related to special equipment, materials, supplies, devices and techniques for teaching and training;

(i) To conduct or contract with any federal, state or private agency for research and development projects designed to improve the quality or efficiency of the identification, referral, evaluation and individualized education program of, and delivery of services to, exceptional persons;

[58] 20 U.S.C. §1411 (1976); 45 C.F.R. §§121a.751 and 80.6-.10, 84.61 (1978).

COMMENT. None necessary.

(j) To cooperate with and assist public and private colleges and universities in developing courses and programs designed to prepare their graduates to serve the educational needs of exceptional persons;

(k) To provide in-service training for teachers of exceptional persons and others requiring special skills or understanding in connection with the education of exceptional persons.

COMMENT. These requirements are intended to increase the availability of trained staff to work with exceptional persons. Often schools need specialized personnel such as resource room instructors, and little or no training is available in the state to prepare individuals in such areas. Thus, the state education agency is empowered to provide an ongoing program of inservice training to train new persons, keep the knowledge and skills of existing personnel up to date and sensitize such persons to the needs of consumers and to provide technical assistance to and cooperate with institutions of higher education in accomplishing these tasks.[59]

(l) To adopt certification, licensing, registration or similar qualification standards or requirements for all personnel in school districts, intermediate education units and state, local and private agencies who identify, refer, evaluate and create individualized education programs, and deliver services to exceptional persons;

COMMENT. A wide variety of professional personnel beyond those needed for traditional education must serve exceptional persons. The standards required by this paragraph are intended to insure that all such personnel are adequately trained and capable of meeting this need. The language of this provision also allows the state education agency to enter into reciprocal agreements with other states regarding certification and other types of standards, so that unnecessary funds and administrative hours need not be wasted on reclassifying personnel who are available within the state and have already proven themselves competent by training and experience within another state.[60]

(m) To take affirmative action to employ and advance in employment qualified exceptional persons;

COMMENT. Federal law regarding nondiscrimination and affirmative action in hiring handicapped citizens should be included and strictly enforced in a state statute relating to the education of exceptional persons.[61]

[59] 45 C.F.R. §121a.382 (1978).

[60] *See generally*, 45 C.F.R. §§121a.380 *et seq.* (1978).

[61] *See* 20 U.S.C. §1415 (1976), 45 C.F.R. §121a.150 (1978) and 29 U.S.C. §§793, 794 (1976); 45 C.F.R. §§84.4 *et seq.* (1978).

(n) To receive and investigate complaints concerning any actions taken by any agency that are contrary to the provisions of this act and/or a denial of the right of an exceptional person to a free appropriate education; to investigate school districts, intermediate units, state, local and private agencies and take appropriate action concerning the failure of school districts, intermediate education units, state, local, and private agencies receiving public support, to identify, refer, evaluate, create an individualized education program, or deliver services to exceptional persons; and to conduct due process hearings or hear appeals brought pursuant to section 6 of this act.

COMMENT. This paragraph provides the state education agency with the authority to investigate complaints regarding denials or refusals of local and state level education officials to comply with the provisions of this act. As a result, parents have a mechanism for bringing violations to the attention of state level authorities without requiring that cumbersome and time-consuming due process hearings, federal administrative complaints or lawsuits be filed in every instance. Further, this paragraph provides the authority for special education due process hearings required by P.L. 94-142 and Section 504 and implementing regulations in order that parents or school districts may resolve disputes concerning the provision of free appropriate public education.[62]

(o) To identify, refer, evaluate, create an individualized education program, or deliver services directly to exceptional persons in a manner consistent with the requirements of this act;

COMMENT. See subsection 5(4) and its comment.

(p) To receive and allocate federal and state funds for programs for persons requiring special education and related services subject to the provisions of this act;

(q) To provide for the apportionment of federal and state aid to school districts, intermediate education units and state, local and private agencies, pursuant to section 12;

(r) To require school districts, intermediate education units and state, local and private agencies receiving any funds from the state to establish fiscal and cost accounting and reporting procedures, forms, schedules, rates and audits in conformity with Division standards, and to make reports to the Division at such times, in such fashion and on such forms as the Division may require;

(s) To ensure that public funds, whether from federal, state or local governmental sources, are expended in a manner not inconsistent with the requirements of this act;

(t) To recommend to the (name of state board of education) and the state legislature and any of its committees or commissions amendments to state law, changes in administrative practices and organization, and changes in the level, source, nature and distribution of financial support from the state for special education and related services;

COMMENT. None necessary.

(u) To develop public information programs regarding the nature and extent of special education and related service needs of exceptional persons in the state and the availability of such education and services to meet those needs; and

COMMENT. Public information programs is one means whereby people learn of the kinds of services that are available throughout the state for their own children or other exceptional persons with whom they have contact. In addition to this benefit, however, public dissemination of information will identify legal rights and crystallize issues, so that the time of consumers and administrators is focused less on determining positions and more on serving exceptional persons within a framework of mutual respect and understanding.

(v) To establish standards relating to facilities and to issue rules and regulations and take any other action necessary or desirable to execute the Division's responsibilities under this act or to assist school districts, intermediate education units, and state, local and private agencies to execute their responsibilities under this act.

COMMENT. None necessary.

(3) The Division shall be administered by a Director who shall be qualified by reason of his or her education, training and experience to take responsibility for and give direction to the programs of the (name of state education agency) relating to exceptional persons.

COMMENT. None necessary.

(4) (a) The (name of the state education agency) may enter into agreements with other school districts, intermediate education units, private agencies, or public or private agencies in other states to provide special education and related services, provided that the provision of such education and related services shall continue to be the responsibility of the school district in which the parents of the exceptional person reside.

COMMENT. In certain circumstances, the delivery of services may have to occur either out of the district of residence or out of state. This is permitted by federal law.[63] The delivery of services out of

[62] 20 U.S.C. §1415 (1976); 45 C.F.R. §§121a.500-.514; §84.36 (1978); 45 C.F.R. §121a.602 (1978).

[63] *See generally*, 45 C.F.R. §§121a.14 and 550 *et seq.* (1978).

state, however, must be discouraged in most cases. In addition, this subsection must not be interpreted to ignore or limit the federal requirements mandating the least restrictive environment appropriate to the person's needs.[64] Precise regulations also must be promulgated to monitor and evaluate the delivery of these services in order to insure quality and strict enforcement of the individualized education program.[65]

(b) Agreements made pursuant to this subsection may include the provision of all special education and related services, and the payment of the reasonable costs thereof.

COMMENT. None necessary.

(c) The parents of any exceptional person provided special education or related services pursuant to this subsection shall have the same rights they would have if the person were receiving special education or related services from the school district where the exceptional person would normally attend public school.

COMMENT. See comment to subsections 8(1) and (2).

(5) The (name of state education agency) and a school district through the (name of state education agency) may enter into an agreement with the Secretary of the Interior, the Secretary of Health, Education and Welfare, or any other federal agency to implement state or federal law requiring that American Indian exceptional persons residing on or near Indian reservations be provided special education and related services and a free appropriate public education.

COMMENT. American Indian persons who are members of federally recognized Indian tribes and who reside on or near reservations (i.e., any Indian tribe's reservation, Pueblo, Colony or Rancheria, including former reservations in Oklahoma, Alaska Native regions, and Indian allotments) are eligible for special programs and services provided by the Secretary of the Interior. As citizens of the state in which they reside, such persons are also eligible for state programs and services, and it is the responsibility of the state to provide for the education of each exceptional person. This subsection provides authority for the state and/or school districts acting through the state to enter into agreements with federal agencies for the discharge of their responsibilities regarding the provision of special education and related services to these persons. Funds under this act may not, if prohibited by federal law, be used to supplant any federal funds authorized for these persons.

(6) There is hereby created within the state an Advisory Council for the Education of Exceptional Persons. The Council shall be responsible for advising the (name of state education agency and other state agencies concerned with exceptional persons) regarding the identification, referral, evaluation and individualized education program of, and delivery of services to all exceptional persons. The Council's membership shall be appointed by the Governor and shall consist of at least one (1) person who is representative of each of the following groups: community organizations interested in the education of exceptional persons, teachers and parents of exceptional persons, exceptional persons and school district education officials. The state shall appropriate such funds as are necessary for a Council staff and furnish other facilities for the Council to carry out its functions under this paragraph, and for the reimbursement of expenses incurred during the course of their service to the Council's membership. The members shall otherwise serve without pay.

COMMENT. The Advisory Council is established independent of the state education agency, as it is intended to represent and advocate for the interests of exceptional persons rather than those of the agency responsible for their education. The composition of the Council is intended to insure that the interests and points of view of consumers, parents and professionals are appropriately represented.[66]

Section 8. Local Administration

(1) The school district of which the parent(s) of any exceptional person is a resident, is responsible for providing the person a free appropriate public education as required by this act. The school district maintains responsibility notwithstanding its assigning or placing the person in a program in a school district other than the one of residence, or in an education unit, or in a private program of special education and related services.

COMMENT. The local school district of residence is responsible for providing each exceptional person a free appropriate program of special education and related services for his or her individual needs. This requirement is consistent with Section 504's prohibition against excluding exceptional persons from education programs which receive federal financial assistance.[67] Further, once placed, the district must maintain responsibility for monitoring each person, regardless of where the local district placed that person.

(2) (a) Each school district may satisfy its obligations under this act through any appropriate

---

[64] 20 U.S.C. §1412(5)(B) (1976); 45 C.F.R. §§121a.550-556, 346(c) (1978).

[65] See subsections 3(5), 5(1) and 5(3).

[66] *See* 20 U.S.C. §1413 (1976); 45 C.F.R. §§121a.650-.653 (1978).

[67] 45 C.F.R. §84.31-.39 (1978). *See also* Pennsylvania Ass'n of Retarded Children v. Pennsylvania, *supra*, and Frederick L. and Delaware Valley Ass'n for Children with Learning Disabilities v. Thomas, *supra*.

administrative arrangement, including but not limited to direct service, agreements with other school districts, agreements with other agencies, including but not limited to private agencies, interstate agreements pursuant to subsection 5, and agreements with intermediate units as described in section 9 of this act. A school district may use any of these administrative arrangements simultaneously.

COMMENT. School districts may meet their obligations under this act by purchasing services from educational, diagnostic, or other service providers. However, it must be the primary objective of all local educational agencies to educate children as close to home as possible, consistent with the least restrictive environment mandate of P.L. 94-142.[68]

(b) Agreements made pursuant to this subsection shall include the provision of all special education and related services, and the payment of the reasonable costs thereof.

COMMENT. Because of financial concerns, problems invariably arise when governmental jurisdictions are required to work together. For example, questions including how charges are to be assessed and who will own or manage facilities and equipment, can often delay or defeat needed cooperative efforts. Thus, legal provisions are included in the act to authorize appropriate financial arrangements.

(c) The parents of any exceptional person provided special education or related services pursuant to this subsection shall continue to have the same rights they would have had if the person were receiving special education or related services from the school district where the exceptional person would normally attend public school.

COMMENT. See comment to subsections 8(1) and (2) above.

(3) Each school district shall provide for the administration of its responsibilities and duties under this act by maintaining a separate administration which has primary responsibility for carrying out the requirements of this act. The director of that administrative unit shall be qualified by reason of education, training, and experience to supervise the administration of the district's special education and related services.

COMMENT. Due to the extensive state and federal responsibilities school districts have in providing each exceptional person a free appropriate public education, it becomes necessary for school districts to have a separate department of special education in the exceptional person's local district. This in turn will ensure easy access to persons who are responsible for programmatic and fiscal decision-making in this area.

[68] 20 U.S.C. §1413 (1976); 45 C.F.R. §121a.550 *et seq.* (1978).

(4) (a) On or before (day, month) each school district shall submit an application to the (name of state education agency) for state aid for the education of exceptional persons. The application shall demonstrate how the school district complies with this act and shall include but not be limited to:

(i) Assurances that each exceptional person in the district has the right to a free appropriate public education and procedural safeguards to obtain this right;

(ii) Certification in writing that each exceptional person in the district is receiving a free appropriate public education; this certification shall include information relating to the number, ages, and types of exceptional persons who are in need of services, and who are receiving such services; the number, nature, and types of special education programs and related services being provided; and, an identification of the administrative arrangements by which such programs and services are offered;

COMMENT. The purpose of this application is twofold: first, it assists a state in preparing the required information to ensure compliance with Section 504 and to obtain funding pursuant to P.L. 94-142[69] and this act. Second, the application assists state educational agency officials in determining the local district's requirements for technical assistance, training and other information.

(iii) An inventory of the kinds and numbers of personnel, facilities and services that are used and are necessary but not available to execute the provision of this act;

(iv) Evidence of compliance with section 10;

(v) A report on individualized education programs including information on the extent of parent participation in program development, the type of school district representatives who attend individualized education program meetings, the extent of school district and private agency cooperation, and the evaluation results;

(vi) A report on and accounting of the amount of federal, state and local funds that the district has received, spent, encumbered, or obligated during the immediately preceding fiscal year for the education of exceptional persons;

COMMENT. None necessary.

(vii) Reports certifying that the district has used state and federal financial aid for the education of exceptional persons in a manner consistent with the applicable state and federal law governing the use of such aid, including but not limited to evidence of compliance with section 12 of the act.

COMMENT. The Secretary of the U.S. Depart-

[69] *See also* 45 C.F.R. §121a.180 *et seq.* (1978).

ment of Education through the Office of Special Education and Rehabilitative Services is charged with monitoring state compliance with federally funded education programs relating to handicapped children, including P.L. 94-142. In addition, the Office for Civil Rights within the U.S. Department of Education is responsible for gathering information and investigating specific complaints regarding discrimination based on handicap in federally funded activities, pursuant to Section 504, including education programs. Thus, the data required to be collected pursuant to this subsection is of vital importance to the federal offices in accomplishing their tasks and in insuring state compliance with federal mandates.

(viii) A projection of budget expenses for the forthcoming fiscal year for the education of exceptional persons as required by this act; and

(ix) A report on all of the district's activities concerning the employment and advancement in employment of exceptional persons.

COMMENT. None necessary.

(b) In addition, each school district shall:

(i) Furnish such other information required by the (name of state education agency) as may be necessary to enable the agency to perform its duties under this act, including but not limited to information relating to the educational achievement of exceptional persons participating in the district's program for exceptional persons;

(ii) Keep such records and afford access to those records as the (name of state education agency) requires to ensure the correctness and verification of the information that the school district furnishes under subsection (4)(a);

(iii) Make its application, all supporting documents, and all reports and records required by subsections (4)(a) available to the public and inform parents of exceptional persons of their rights of access to those documents, reports, and records; and

(iv) Provide all citizens the opportunity to review and comment on its annual application at least sixty (60) days prior to its submission to the (name of state educational agency), at a public meeting of the school district convened for the sole purpose of commenting on this application.

COMMENT. None necessary.

(5) In addition to the powers granted under (cite applicable state law), each school district shall have the power to:

(a) Establish, operate, and maintain all programs, classes, and educational facilities necessary or desirable for the purposes of providing a free appropriate public education for all eligible exceptional persons;

(b) Acquire, construct, maintain, and operate educational facilities in which to provide a free appropriate public education for exceptional persons, including facilities and services for those exceptional persons who may not reasonably go to and from home daily to receive an appropriate education;

(c) Make arrangements with school districts participating in the intermediate education unit for the provision of a free appropriate public education to the exceptional persons of such school districts;

(d) Make arrangements with and payments to private and other public agencies for the identification, referral and evaluation of, development of an individualized education program for, and delivery of services to exceptional persons;

(e) Receive, administer, and expend federal and state funds upon approval of the application submitted pursuant to subsection 4(a);

(f) Employ personnel necessary to administer the school district's programs for eligible exceptional persons;

(g) Employ special education teachers and other personnel required to identify, refer, evaluate, create an individualized education program, and deliver services to exceptional persons;

(h) Acquire, hold, and convey real and personal property;

(i) Provide transportation for exceptional persons in connection with any of its services;

(j) Receive, administer, and expend the proceeds of any issue of school bonds or other bonds intended wholly or partly for its benefit;

(k) Apply for, accept, and use grants, gifts, or other assistance, and, if not contrary to law, comply with the conditions, if any, attached thereto;

(l) Participate in, and make its employees eligible to participate in, any retirement system group insurance system, or other program of employee benefits, including salary schedules and other job-related benefits;

(m) Develop, test, demonstrate, maintain, purchase, or otherwise acquire, store, produce, and make available equipment, materials, and special supplies and devices for delivering services to exceptional persons;

(n) Collect, evaluate, develop, and disseminate research data and other information relating to special equipment, materials, supplies, devices, and techniques for teaching and training;

(o) Provide in-service training for teachers of exceptional persons and other persons requiring special skills or understanding in connection with delivering services to exceptional persons; and

(p) Do such other things as are necessary and incidental or desirable for the execution of

any of the foregoing or any other powers conferred upon school districts by this act.

COMMENT. None necessary.

Section 9. Intermediate Education Unit

(1) Whenever a school district is unable to provide by direct service all of the required services necessary for each exceptional person in the district to receive a free appropriate public education, it may participate in an intermediate education unit with other school districts.

COMMENT. Although P.L. 94-142 and its regulations [70] and this subsection allow for the utilization of intermediate education units, the use of facilities outside the local school district of an exceptional person's residence should be discouraged. Delivering services closest to the person's home (preferably in the neighborhood school) more accurately conforms to the least restrictive environment mandate.[71]

(2) Intermediate education units may:

(a) Establish, operate, make financial payments to, and maintain programs of preschool, developmental day care, alternative school, vocational education, "after hours" education, summer compensatory education and transportation; and

(b) Participate in, and make its employees eligible to participate in, any retirement system, group insurance system, or other program of employee and job-related benefits, including salary and tenure on the same terms as those that govern school districts and their employees.

COMMENT. None necessary.

(3) Each intermediate education unit shall have a board of directors, composed of at least one member chosen for a term of three years, from each of the boards of school directors of school districts included within it. The intermediate education board of directors shall:

(a) Appoint an Executive Director and determine his or her salary. The Director shall be qualified by reason of his or her education, training and experience to supervise the administration of the unit's special education and related services;

(b) Appoint assistant executive directors upon recommendations of the Executive Director and determine their salaries;

(c) Approve professional staff appointments made by the Executive Director;

(d) Adopt employment policies for all intermediate unit staff; and

(e) Adopt a program of services to be provided.

COMMENT. None necessary.

(4) The Director of the intermediate education unit shall have the power and duty to:

(a) Administer the intermediate unit program of services;

(b) Appoint professional staff subject to the approval of the intermediate unit board of directors:

(c) Employ nonprofessional staff in accordance with employment policies of the intermediate unit board of directors;

(d) Prepare the budget for adoption by the intermediate unit board of directors;

(e) Direct expenditures of funds within the budget or other authorizations of the intermediate unit board of directors;

(f) Appoint such advisory groups as will assist the staff in providing programs of services for school districts;

(g) Provide the (name of state education agency) with information, reports and services;

(h) Perform such other duties as may be required by the intermediate unit board of directors and the regulations of the (name of state education agency).

COMMENT. None necessary.

(5) (a) Any intermediate education unit that is in the process of formation shall submit to the Director of the Division the agreement under which it proposes to function. Each participating school district shall submit simultaneously to the (name of state education agency, the application required of it by section 8 of this act. No intermediate education unit may receive state aid unless and until it has been approved by the Director of the Division.

(b) The Director of the Division shall approve the creation of an intermediate education unit and, if appropriate, state aid to the unit if he or she determines that the intermediate education unit complies with all provisions of this act or, if the intermediate education unit is not yet in operation, that it will have the resources and authority to comply.

COMMENT. None necessary.

(6) Each intermediate education unit is required to have a written agreement regarding the establishment and provision of services by the unit. The written document shall contain at least the following provisions:

(a) The names and mailing addresses of the school districts;

(b) The services to be provided by the unit;

(c) The internal management and control of the unit;

(d) The relationships between the school districts and the unit in regard to the respon-

---

[70] 20 U.S.C. §1401(22) (1976); 45 C.F.R. §121a.7 (1978).

[71] See section 5(3).

sibilities for the delivery of services to exceptional persons;

(e) The financial responsibilities of each school district to the unit setting forth formulas, procedures and other methods for the calculation thereof;

(f) The duration of the agreement;

(g) Provisions for amendment, renewal, withdrawal from, or termination of the agreement;

(h) Provisions for the disposition of the unit property upon dissolution of the unit;

(i) Provisions for financial settlement, if any, with a withdrawing school district; and;

(j) Any other necessary or appropriate provisions.

Before becoming effective, an agreement shall be submitted to and approved by the (name of state education agency). The (name of state education agency's) failure to object to an agreement within ninety (90) days after receiving it shall constitute approval.

COMMENT. None necessary.

(7) (a) A school district may withdraw from participation in any or all of the services of an intermediate education unit only with the approval of the Director of the Division. The Director shall grant his or her approval unless there is reasonable grounds to believe that the district will be unable to provide substantially the same services provided by the unit or that such services are no longer necessary in order to carry out the provisions of this act.

(b) In order to withdraw from the intermediate education unit, members must adhere to the following conditions so that the appropriate education of exceptional persons is not interrupted or delayed until the district provides the required service(s).

(i) A school district's withdrawal may not become effective earlier than the end of the next school year.

(ii) A school district that wishes to withdraw from an intermediate education unit shall file a written application with the Director, the governing board of the unit, and the governing boards of the participating school districts no later than one hundred-twenty (120) days before it seeks to withdraw. The application shall set forth the reasons why the district wishes to withdraw and evidence that it will be able to provide all exceptional persons in its district substantially the same services provided by the unit or that such services are no longer needed to carry out the provisions of this act.

(iii) The withdrawing school district shall be liable for its share of debt incurred while it was a participant.

(iv) A unit established under this subsection may be dissolved by action of its Governing Board, but such dissolution shall not take place until the end of the school year in which the action was taken. When a unit is dissolved, assets and liabilities shall be distributed pro rata to all districts that participated in the unit.

COMMENT. The state must insure that the needs of exceptional persons served by an intermediate education unit will continue to be met upon its dissolution. Therefore, such dissolution should only occur at the end of a school year when education will not be interrupted. The use of an application process for approval of the dissolution is a further safeguard to enable the state to insure that the appropriate education of the exceptional persons will be neither diminished nor denied upon such dissolution.

Section 10. Personnel

(1) The (name of state education agency), each school district and each intermediate education unit shall recruit, hire and supervise personnel necessary to carry out the purposes of this act. Neither the (name of state education agency) nor any school district shall hire any person who has not met the (name of state education agency) approved or recognized certification, licensing, registration or similar qualification requirements applicable to the professional specialities of the job.

COMMENT. Every person who is charged with the identification, referral, evaluation, development of an individualized education program, or delivery of services to an exceptional person must be qualified in their respective fields to ensure the quality of their work.[72]

(2) The (name of state education agency) shall require local agencies to include in their applications provisions detailing the type and frequency of in-service training for regular and special education teachers, building principals, administrators, paraprofessional and other support personnel, hearing officers and surrogate parents.

COMMENT. In-service personnel development programs based on assessed needs must be available to all personnel engaged in the education of exceptional persons.[73] Such training is needed since personnel must frequently be apprised of changes in the law, public policy and regulations in this dynamic field, as well as new developments and literature in their respective specialities.

(3) Paraprofessional personnel shall be employed only after receiving training in special education or in the related services which they

[72] *See* 45 C.F.R. §§121a.12, 121a.380-.387 (1978) and subsection 7(2)(1) of this act.

[73] 45 C.F.R. §121a.380 (1978); *see* Frederick L. v. Thomas, *supra*, Note 21.

will provide. Each local educational agency shall set policy guidelines for their employment and training.

COMMENT. Paraprofessionals may be used to supplement, but not replace, the services of qualified professional personnel who provide educational programs or services for exceptional persons.[74]

(4) All educational agencies in this state shall take affirmative action in hiring qualified exceptional persons. No exceptional person shall be denied the benefit of, or denied a position of any type because of his or her handicap or other exceptionality.

COMMENT. Affirmative action in hiring and employment practices is one of the cornerstones of the Rehabilitation Act of 1973 and its implementing regulations.[75] P.L. 94-142 has corresponding requirements.[76]

(5) The state education agency and each school district, in conformance with the state's comprehensive system of personnel development shall provide, approve, or cause to be provided the following:

(a) The use of incentives, including but not limited to release time, payment for participation, options for academic credit, salary step credit, and certification renewal, that insure that the general and special education administrative instructional, related and support personnel update their professional skills; and

(b) Options to assist personnel in developing and improving skills required in order for them to carry out the policies of this act, including but not limited to college courses, workshops, institutes, meetings, membership in professional organizations, and attendance at professional conferences.

COMMENT. None necessary.

Section 11. Confidentiality of and Access to Records

(1) The state education agency shall give notice to fully inform parents about the collection of records pursuant to its responsibility to identify and locate all exceptional persons, including a description of:

(a) The types of exceptional persons on whom personally identifiable information is maintained, the types of information sought, the methods the state intends to use in gathering the information (including the sources from whom information is gathered), and the uses to be made of the information;

(b) The policies and procedures participating agencies must follow regarding storage, disclosure to third parties, retention and destruction of personally identifiable information; and

(c) All of the rights of parents regarding this information, including the rights under the Family Educational Rights and Privacy Act, 20 U.S.C. 1232g and 45 C.F.R. Part 99.

Before any major identification, or evaluation activity, the notice must be published or announced in newspapers or other media, or both, with circulation adequate to nofity parents throughout the state of the proposed activity.

COMMENT. Confidentiality in record matters can only be preserved when the parents of the subject of the records know where records are maintained, when collection of information will take place and what rights are available. When this occurs, they may properly object to the contents of the records which currently exist, or to collection of new information. This is a requirement of the Buckley Amendment[77] and is incorporated in P.L. 94-142's regulations.[78]

(2) Each educational agency shall provide personal notice to parents describing its record policies which shall include all requirements of 20 U.S.C. §1232g and 45 C.F.R. Part 99, and this section, and a list of the types and location of education records maintained or used by the educational agency, when requested to do so by the parents of an exceptional person.

COMMENT. Pursuant to the Buckley Amendment, public education agencies must give parents personal notice of their record policies because records relating to exceptional persons are not only kept in the various agencies which evaluate or diagnose an exceptional person but are frequently maintained in several locations within the person's local school district. For example, school districts often provide duplicate copies of records to central offices, local schools, division of special education and persons responsible for attendance, transportation and discipline. Thus it is important that the parents of the person to whom the records relate receive notice of this fact.[79]

(3) Each educational agency shall permit parents to inspect and review any education records relating to their children which are collected, maintained or used by the agency under this act. The agency shall comply with a request within thirty (30) days or sooner, if a prehearing confer-

[74] See comment to subsection 10(1).

[75] 29 U.S.C. §§793, 794 (1976); 45 C.F.R. §§84.11 *et seq.* (1978); *see also* Gurmankin v. Costanzo, 411 F.Supp. 983, *aff'd* 556 F.2d 184 (3rd Cir. 1977) (blind english teacher held to be excluded from teaching based solely upon her handicap).

[76] 20 U.S.C. §1405 (1976); 45 C.F.R. §121a.150 (1978); see subsection 7(2)(m) of this act.

[77] 20 U.S.C. §1232g (1976); 45 C.F.R. §§99.1 *et seq.* (1978).

[78] 20 U.S.C. §§1412(2), 1417(c) (1976); 45 C.F.R. 121a.560 *et seq.* (1978).

[79] 45 C.F.R. §99.6 (1978).

ence, or a hearing relating to the identification or evaluation of, and delivery of services to the exceptional person has been scheduled or must be held due to an immediate need to hold any such meeting.

The right to inspect and review education records under this section includes: the right to a response from the participating agency to reasonable requests for explanations and interpretations of the records including raw scores of psychological tests, the right to request that the agency provide copies of the records containing the information, and the right to have a representative of the parent inspect and review the records.

Educational agencies may charge a reasonable fee for providing copies, which may not exceed the actual cost of duplication. No fee shall be charged if it would effectively deny access to records; neither may any fee be charged for search and/or retrieval of any records.

COMMENT. The opportunity for access in order to comply with the spirit and intent of P.L. 94-142 must reasonably include the right to an expeditious response by an educational agency to a parent's request to review records. These strict deadlines are needed so that unnecessary delay in providing materials will be eliminated; prompt access helps ensure that exceptional persons will continue to be educated in an appropriate environment and will not be excluded from receiving the services they require.[80]

This subsection also eliminates the cumbersome and inefficient practice of requiring parents to make a copy of records; the educational agency is given this responsibility. Further, consistent with P.L. 94-142, it provides that no fee may be charged if it would "effectively prevent" access to a parent who could not afford to pay such a fee; however, the federal requirement is extended to include a right to obtain records at a cost no greater than that actually spent by an educational agency for the cost of duplication of records. This extension should be included in state law to preclude prohibitive or nuisance fees being charged to parents who request records.[81]

(4) No agency which maintains education records shall disclose any such records to any other agency or third party without the written consent of the parent.

COMMENT. This provision is taken from the Buckley Amendment regulations.[82] In the interests of confidentiality, advocates representing exceptional persons and school officials must obtain written permission from a parent prior to redisclosure of education records. As a result, random redisclosure to third parties is precluded.

(5) Each educational agency shall keep a record of any individual who has obtained access to education records collected, maintained or used under this act, including the name of the party, the date access was given and the purpose for which the party is authorized to use the records.

COMMENT. This language is mandated by the Buckley Amendment regulations as incorporated in P.L. 94-142, 45 C.F.R. §99.32 and is designed to promote access to authorized persons only.

(6) If any education record includes information on more than one exceptional person, the parents of those persons shall have the right to inspect and review only the information relating to their child.

COMMENT. None necessary.

(7) (a) A parent who believes that information in the education records collected, maintained or used under this act is inaccurate or misleading or violates the privacy or other rights of the person, may request that the educational agency which maintains the information amend the information.

(b) The agency shall decide whether to amend the information in accordance with the request within thirty (30) days of receipt of the request.

COMMENT. The federal law relating to confidentiality of student records was enacted to give parents the opportunity to request that false or inaccurate information be deleted from the records in question or amended.[83] A corresponding section is incorporated in the regulations to P.L. 94-142.[84] For exceptional persons, who are continually being labelled and whose educational programs are subject to more comment and flux than regular education students, this provision is especially critical.[85]

Practice has shown the federal laws require only a reasonable period of time to elapse between a request and action, or "in no event later than forty-five (45) days from the date of receipt of the request." This is far too lengthy a period for implementing the rights of an exceptional person who may be currently excluded or threatened with an exclusion or change in educational placement or classification. Thus, the requirements of this subsection extend beyond those of either of the preceding federal laws, in that the period within which an agency has to decide whether to amend the information is restricted to thirty days from the date of receipt of the request.

[80] 45 C.F.R. §§99.11, 121a.562 (1978).

[81] 45 C.F.R. §§99.11(2), 121a.566 (1978).

[82] 45 C.F.R. §99.33 (1978).

[83] *See* 45 C.F.R. §§99.20 *et seq.* (1978).

[84] 45 C.F.R. §121a.567 (1978).

[85] *See* 45 C.F.R. §§121a.560 *et seq.* (1978).

(8) If the agency decides not to amend the information in accordance with the request made pursuant to subsection 8(7), it shall inform the parent of the refusal and advise the parent of the right to a hearing. The hearing shall be held within thirty (30) days of the date it is requested. When a hearing is requested, the parent shall be given full opportunity to present reasons for deleting information from or amending the exceptional person's records.

COMMENT. Pursuant to the Buckley Amendment and P.L. 94-142, an administrative hearing must be convened upon refusal of educational agency officials to amend records in accordance with a parent's request.[86] While both of these federal statutes require a hearing to be held within a "reasonable time" after it is requested, school officials and parents may well disagree regarding interpretation of the term "reasonable." Therefore, this act specifies that a hearing must be held within thirty (30) days after it is requested by a parent in order to avoid excessive delays in scheduling it. In this manner, both parents of exceptional persons and school district officials will be aware of the need to act expeditiously in order to resolve a records dispute.

(9) The hearing shall be conducted in accordance with the requirements of 45 C.F.R. §99.22 and shall be conducted by a hearing officer who is not involved in the matter in dispute. The officer shall write a written summary of the evidence presented and an opinion of his or her conclusions within twenty (20) days of the date that hearing is held.

COMMENT. Both P.L. 94-142[87] and the Buckley Amendment[88] require a records hearing to be held. However, unlike a special education due process hearing, the hearing officer need not be from another district according to federal law.[89] Instead, a hearing officer need only be a person who may be employed by the school district but who is not directly concerned with the concerns of the person involved in the request. The section also establishes a time limitation within which the hearing officer's opinion must be written — twenty (20) days from the date the hearing is held. In this way, the dispute can be resolved in a reasonably expeditious fashion, thereby serving the needs of the consumer and administrative convenience as well.

(10) If, as a result of the hearing, the agency decides that the information is inaccurate, misleading or otherwise in violation of the privacy or other rights of the person, its officials shall delete or amend the information and so inform the parent in writing.

COMMENT. None necessary.

(11) If, as a result of the hearing, the agency decides that the information is not inaccurate, misleading or otherwise in violation of the privacy or other rights of the exceptional person, it shall inform the parent of the right to place in the person's records a statement setting forth any reasons the parent has for disagreeing with the agency's decision.

Any parent's statement placed in the records of the person under this section must be maintained by the agency as part of the records of the person as long as the record or contested portion is maintained by the agency. If the records of the person or the contested portion is disclosed by the agency to any party, the statement of the parents must also be disclosed to the party.

COMMENT. This subsection mirrors the requirements of P.L. 94-142 and the Buckley Amendment,[90] and clearly states that a parent has the right to at least include a statement in the file indicating that there is disputed information in it, even if the school district employed hearing officer disagrees with the parent's request for an amendment.

(12) (a) Each agency shall protect the confidentiality of personally identifiable information at collection, storage, disclosure and destruction stages with one official at each agency assuming this responsibility.

(b) All persons collecting personally identifiable information must receive training regarding the state's policies and those mandated by 45 C.F.R. Part 99.

(c) Each agency shall maintain, for public inspection, a current listing of the names and positions of those employees within the agency who may have access to personally identifiable information.

COMMENT. This subsection complies with a provision of the regulations to P.L. 94-142, requiring states to promulgate safeguards to protect confidentiality of personally identifiable information.[91] Included among these protections is the requirement that one official be accountable as the custodian for protection and disclosure of records. The designation of such an official precludes the frequent burden placed upon parents of seeking the person's usually voluminous records from a number of different sources throughout a local school district and avoids lawsuits requiring disclosure of these numerous sources and custo-

[86] *See* 45 C.F.R. §§121a.568, 99.21 (1978)

[87] 20 U.S.C. §§1412(2); 1417(c) (1976); 45 C.F.R. §§121a.567 *et seq.* (1978).

[88] 20 U.S.C. §1232g (1976); 45 C.F.R. Part 99 (1978).

[89] See subsection 6(8).

[90] 20 U.S.C. §1412(2)(D), 1417(c) (1976); 45 C.F.R. §§121a.569, 99.21 (1978).

[91] 45 C.F.R. §121a.572 (1978).

dians.[92] Also included as a safeguard is the requirement that parents and the public be informed of the names and titles of those employees who have access to personally identifiable information. Thus, the parent knows who has knowledge of the information contained in his or her child's records.[93]

(13) The agency shall inform parents when personally identifiable information which is collected, maintained or used under this section is no longer needed to provide special education and related services to the exceptional person. This information shall then be destroyed at the request of the parent. However, a permanent record of the person's name, address and phone number, his or her grades, attendance record, classes attended, grade level and year completed may be maintained without time limitations.

COMMENT. None necessary.

Section 12. Finance

(1) The state shall provide financial aid in each school year to the (name of the state education agency), school districts and intermediate education units for the special education and related services that they provide to exceptional persons. Such aid shall be determined and allocated in accordance with this section and the rules and regulations of the (name of state education agency).

COMMENT. Laws regulating state reimbursement for the education of exceptional persons must be broad enough to allow for flexible programming to meet the unique needs of each exceptional person, and such funding should be tied to a state approved plan for which the state and the district can be held accountable. Such a plan and subsequent reimbursement should include provision for, but not be limited to, instructional services, administration, transportation, facilities and personnel to serve all exceptional persons whether they be located in public day schools, state schools or institutions, hospitals, homes or any other public or private facility. Since some state constitutions prohibit the legislature from directly paying a private agency, such payments go to the state education agency or local school which contracted for the service under this subsection. They, in turn, are responsible for paying the private agency.

(2) For purposes of entitlement to state aid, exceptional persons shall be counted in the same manner as nonexceptional persons and per pupil aid shall be determined on the same basis as for nonexceptional pupils pursuant to (cite appropriate section of state law). In addition to the state aid claimed and paid pursuant to this subsection, any school district that has maintained an approved program of education for exceptional persons during any school year shall be entitled to and receive reimbursement from the state for the excess cost of educating each exceptional person. Excess costs shall be determined in the following manner:

(a) Add all expenditures of the school district made during the preceding year, except capital outlay and debt service:

(i) For elementary school students, if the exceptional person is an elementary school student; or

(ii) For secondary school students, if the exceptional person is a secondary school student; or

(iii) For preschool and vocational school students, if the exceptional person is a preschool or vocational school student.

(b) From this amount, subtract amounts from state and local funds which the district spent in the preceding school year for programs for exceptional persons in elementary, secondary, preschool or vocational school programs respectively.

(c) Divide the amount remaining under paragraph-(b) of this section by the average number of all students enrolled in the school district in the preceding school year:

(i) In its elementary schools, if the exceptional person is an elementary school student; or

(ii) In its secondary schools, if the exceptional person is a secondary school student; or

(iii) In its preschool and vocational school programs, if the exceptional person is a preschool or vocational school student.

(d) Divide the amount the district spent in the preceding school year for programs for exceptional persons in elementary, secondary, preschool or vocational programs respectively, by the average number of exceptional persons enrolled in the district's special education programs in the preceding school year;

(i) In its elementary and secondary schools if the exceptional person is an elementary or secondary school student; or

(ii) In its preschool and vocational school program if the exceptional person is a preschool or vocational school student;

(e) Subtract the amount obtained under subparagraph (d) of this subsection from the amount obtained under subparagraph (c) of this subsection.

[92] *See* Watson v. Costanzo, Civ. No. 75-459 (Ed. Pa., *Order*, May 24, 1977).

[93] 20 U.S.C. §§1232g(e) and (f), 1417(c) (1976); 45 C.F.R. §§99.5, 121a.572 (1978); see also subsection 11(5).

COMMENT. The costs of educating an exceptional person beyond that of educating a nonexceptional person should be assumed by state government. However, the person's district of residence shall be required to assume an expenditure for the person equal to that expended for a nonexceptional person, regardless of where the person receives an education.

(3) If any of the special education or related services aided pursuant to this section are provided partly by one school district or intermediate education unit and partly by another such district or intermediate education unit, and if there is no valid agreement by which one of the districts or intermediate education units is the proper claimant for all the aid in question, each such district or intermediate education unit shall be entitled to claim and receive a proportionate share of state aid in accordance with its actual assumption of costs. The (name of state education agency) shall provide for the calculation and apportionment of state aid in cases covered by this subsection.

COMMENT. None necessary.

(4) The (name of state education agency) annually shall submit to the (name of appropriate state authority) a request for operating funds in an amount sufficient to enable it to perform all of the duties required of it by this act, including but not limited to the following:

(a) Paying all school districts, intermediate education units and private agencies all excess costs for providing special education and related services to all exceptional persons;

(b) Enabling the (name of state education agency) to make grants to school districts, intermediate education units and private agencies for the following purposes:

(i) To enable recipients to secure technical assistance in planning and designing facilities, equipment, buses or other transportation for exceptional persons, and to supplement otherwise available but inadequate funds for the acquisition or construction of facilities, equipment, buses or other transportation for exceptional persons; and

(ii) To enable the (name of state education agency) to make grants, traineeships, payments or other awards to individuals, agencies or other entities as a result of activities conducted directly by the state or through contracts, agreements or other arrangements made in the course of fulfilling its responsibilities under this act.

(c) Assist the (name of state education agency) in carrying out its own responsibilities under section 7 of this act.

COMMENT. None necessary.

Section 13. Appropriations

The General Assembly is authorized to appropriate $_______for the immediate fiscal year to cover the initial expenses incurred in implementing this act.

COMMENT. None necessary

Section 14. Severability of Sections.

If any section, subsection, paragraph, sentence or any other part of this act is adjudged unconstitutional or invalid such judgment shall not affect, impair or invalidate the remainder of this act but shall be confined to the section, subsection, paragraph, sentence or any other part of this act directly involved in the controversy in which said judgment shall have been rendered.

COMMENT. None necessary.

Section 15. Effective Date of Act.

Because an emergency exists, this act takes effect ______________________________.

COMMENT. None necessary.

6

RIGHT TO SERVICES

Table of Contents

INTRODUCTION

No area of law which affects developmentally disabled people has engendered so many controversies as the right to receive services and the procedures by which these services are provided. Debates over segregation, deinstitutionalization, commitment, consent, the right to refuse treatment, the doctrine of least restrictive alternatives, and parental rights have raged in recent years. These same issues have found their way into both courts and legislatures. Almost none can be said to have been resolved definitively.

The issue which initially appears to be least controversial involves the contention that developmentally disabled people have a right to receive services which will assist their adjustment to adult life in the community. But this proposition, which appears so unexceptionable on the surface, proves difficult upon closer inspection. While few will argue that developmentally disabled citizens are not entitled to such assistance, mechanisms for enforcement of the right, and questions of competing equities in the apportionment of public resources, engender major disagreements.

It is also apparent that the issue of the role of institutions in the care of developmentally disabled people is controversial. Theories that mentally retarded people constituted a public menace led to the creation of our large state institutions in the early years of this century.[1] These theories have been universally discredited in recent decades, but the abandonment of the theoretical justification for their creation has not produced abandonment of the institutions themselves. They have persisted through public neglect and convenience. Professionals and experts continue to disagree about the nature and degree of harm done to residents of large institutions,[2] but none can deny that unconscionable abuses have developed in state after state.

While a consensus has developed concerning the need to abandon or transform our old institutional models, disagreement continues concerning what is to replace them. There appears little disagreement that a large number of people who would have been (or actually were) institutionalized in earlier times can now be well served and housed in smaller community settings. Whether larger facilities even serve a valuable function in housing and caring for the minority of more severely disabled individuals remains unresolved. No state has yet totally abandoned larger facilities in favor of community care for all its developmentally disabled citizens, despite suggestions in the literature that this is both possible and desirable.[3] But the trend is moving away from reliance on institutions, and there is ample justification for a policy which limits the use of such facilities to cases of last resort, in which no other services would meet the individual's needs adequately.[4]

This presumption against the use of institutions is most frequently expressed in terms of the legal principle of "least restrictive alternatives" or "least drastic means."[5] This doctrine is based on the Constitution's limitations on the exercise of state power over individuals, and holds that the state cannot exercise its interest in a manner which is unnecessarily restrictive of individual liberty. Thus, where the state can meet its objectives in two ways, it must choose the option which intrudes least on individual freedom and choice. This principle has been used in numerous cases involving mentally ill and developmentally disabled people.[6] While the principle is well established in constitutional caselaw, its implications remain uncertain and controversial. What constitutes "greater" or "lesser" restriction is by no means clear in every case. In addition, many professionals view the

---

[1] W. WOLFENSBERGER, THE ORIGIN AND NATURE OF OUR INSTITUTIONAL MODELS (1975).

[2] *See, e.g.*, R. KING, N. RAYNES & J. TIZARD, PATTERNS OF RESIDENTIAL CARE: SOCIOLOGICAL STUDIES IN INSTITUTIONS FOR HANDICAPPED CHILDREN (1971); Begab, *The Mentally Retarded and Society: Trends and Issues*, in THE MENTALLY RETARDED AND SOCIETY: A SOCIAL SCIENCE PERSPECTIVE 3 (M. Begab & S. Richardson eds. 1975); Tizard, *Quality of Residential Care for Retarded Children*, in VARIETIES OF RESIDENTIAL EXPERIENCE 52 (J. Tizard, I. Sinclair & R. Clarke eds. 1975); Zigler & Balla, *Impact of Institutional Experience on the Behavior and Development of Retarded Persons*, 82 AM. J. MENTAL DEFICIENCY 1 (1977). *See also* R. SCHEERENBERGER, DEINSTITUTIONALIZATION AND INSTITUTIONAL REFORM (1976).

[3] F. MENOLASCINO, CHALLENGES IN MENTAL RETARDATION: PROGRESSIVE IDEOLOGY AND SERVICES (1977).

[4] R. SCHEERENBERGER, *supra* note 2.

[5] Chambers, *The Principle of the Least Restrictive Alternative: The Constitutional Issues* in PRESIDENT'S COMMISSION ON THE MENTALLY RETARDED (P.C.M.R.), THE MENTALLY RETARDED CITIZEN AND THE LAW 486 (1976) [hereinafter cited as P.C.M.R.].

[6] *E.g.*, Covington v. Harris, 419 F.2d 617, 623-25 (D.C. Cir. 1969); Lake v. Cameron, 364 F.2d 657, 659-63 (D.C. Cir. 1966); Halderman v. Pennhurst State School & Hospital, 446 F.Supp. 1295, 1319-20 (E.D. Pa. 1977), *aff'd in part, rev'd in part*, 612 F.2d 84 (3d. Cir. 1979); Dixon v. Weinberger, 405 F. Supp. 974, 977-79 (D.D.C. 1975); Wyatt v. Stickney, 344 F.Supp. 387, 396 (M.D. Ala. 1972), *aff'd sub nom.* Wyatt v. Aderholt, 503 F.2d 1305 (5th Cir. 1974).

doctrine as an unwarranted intrusion upon their prerogatives of professional and clinical judgment, and they are afraid that this kind of presumption enforced by courts and lawyers may deprive some individuals of high quality services from which they could benefit.

The proper role of lawyers and courts is also a prominent concern in the debate over the most appropriate way to provide services. Perhaps no area of developmental disabilities law has proven so controversial and yet received so little analytical attention. While courts have addressed the procedures which are constitutionally required in civil commitment cases involving the mentally ill in the last decade,[7] far fewer procedural cases have arisen involving the mentally retarded and other developmentally disabled persons.[8] One authority has argued that the mental health cases should not be applied to mental retardation hearings.[9] Others have called for the abolition of civil commitment of retarded persons, at least in the form it takes in current state statutes.[10] But since abolition will not greatly reduce the number of developmentally disabled people who need services for which they are incapable of giving or withholding consent, some mechanism of substitute consent will still be needed.

Current state commitment and admission laws, to the extent that they provide an easy means of placing individuals in institutional facilities unnecessarily, must be replaced.[11] In addition, both the Fourteenth Amendment to the U.S. Constitution and sound public policy require that adequate due process be provided to all persons who may have their liberty infringed by placement in residential facilities, whether large institutions or smaller community settings.[12] And since even voluntary admission of developmentally disabled people raises the possibility of abuse through inadequate consent,[13] procedures must also be devised to assure that voluntary recipients of services are truly voluntary. Thus, there will be a role for lawyers and courts — albeit controversial — in the development of both voluntary and involuntary admissions procedures.

Issues involving the adequacy of consent to the delivery of services — both residential and nonresidential — are also arising with increasing frequency.[14] Until recently there was an unchallenged presumption among service providers and others that the vast majority of mentally handicapped people were incapable of giving or withholding consent or of participating in decisions regarding services. Often those decisions were made by family members or guardians or even the service providers themselves. It has become clear in recent years that we need a more sophisticated approach to this problem — an approach which allows greater participation by developmentally disabled people in their own decisions, but which also provides protection against unwise and potentially harmful choices when the individual lacks the capacity to make his or her own judgments about services.[15] The Model Guardianship and Conservatorship Act[16] addresses many of these issues, but they must also be considered in conjunction with various service delivery settings.

Possibly the most difficult issue concerning consent is the question of parental decision-making for developmentally disabled minors. The United States Supreme Court has recently considered this issue in the context of parental commitment of their children to mental retardation institutions,[17] and concluded that states are free to either require due process hearings or to grant more informal procedural protections to the children involved.[18] This decision will certainly not end the

[7] *See* Goldy v. Beal, 429 F.Supp. 640 (M.D. Pa. 1976); Stamus v. Leonhardt, 414 F.Supp. 439 (S.D. Iowa 1976); Suzuki v. Quisenberry, 411 F.Supp. 1113 (D. Hawaii 1976), *declaratory judgment that amendatory legislation also unconstitutional sub nom.* Suzuki v. Alba, 438 F.Supp. 1106 (D. Hawaii 1977); Doremus v. Farrell, 407 F.Supp. 509 (D. Neb. 1975); Kendall v. True, 391 F.Supp. 413 (W.D. Ky. 1975); Lynch v. Baxley, 386 F.Supp. 378 (M.D. Ala. 1974); Bell v. Wayne County Gen. Hosp., 384 F.Supp. 1085 (E.D. Mich. 1974); Lessard v. Schmidt, 349 F.Supp. 1078 (E.D. Wis. 1972), *vacated and remanded,* 414 U.S. 473 (1974), *on remand,* 379 F.Supp. 1376 (E.D. Wis. 1974), *vacated and remanded,* 421 U.S. 957 (1975), *on remand,* 413 F.Supp. 1318 (E.D. Wis. 1976).

[8] *See* Cramer v. Tyars, 23 Cal. 3d 131, 588 P.2d 793, 151 Cal. Rptr. 653 (1979); O'Brien v. Superior Court, San Bern. County, 61 Cal. App. 3d 62, 132 Cal. Rptr. 13 (1976); People v. Reliford, 65 Ill. App. 3d 177, 382 N.E.2d 72 (1978).

[9] Strauss, *Due Process in Civil Commitment and Elsewhere,* in P.C.M.R., *supra* note 5, at 433.

[10] Dybwad & Herr, *Unnecessary Coercion: An End to Involuntary Civil Commitment of Retarded Persons,* 31 STAN. L. REV. 753 (1979).

[11] *Id.*

[12] The U.S. Supreme Court has referred to civil commitment as a "massive curtailment of liberty." Humphrey v. Cady, 405 U.S. 504, 509 (1972). The balancing test for answering the question of "what process is due" is found in Mathews v. Eldridge, 424 U.S. 319 (1976):

> [I]dentification of the specific dictates of due process generally requires consideration of three distinct factors: First, the private interest that will be affected by the official action; second, the risk of an erroneous deprivation of such interest through the procedures used, and the probable value, if any, of additional or substitute procedural safeguards; and finally, the government's interest, including the function involved and the fiscal and administrative burdens that the additional or substitute procedural requirement would entail.

Id. at 335. *See also* Friendly, *"Some Kind of Hearing",* 123 U. PA. L. REV. 1267 (1975).

[13] For discussions of these issues as they affect the mentally ill, *see* Gilboy & Schmidt, *"Voluntary" Hospitalization of the Mentally Ill,* 66 Nw. U. L. REV. 429 (1971); Owens, *When is a Voluntary Commitment Really Voluntary?,* 47 AM. J. ORTHOPSYCHIATRY 104 (1977).

[14] *See* CONSENT HANDBOOK 39-47 (H. R. Turnbull ed. 1977).

[15] *Id.*

[16] Chapter 7: Guardianship and Conservatorship, *infra.*

[17] Secretary of Public Welfare v. Institutionalized Juveniles, 442 U.S. 640 (1979).

[18] *Id.* at 646-50.

debate on the constitutional question involved.[19] But aside from constitutional considerations, there remain important public policy questions about the need to protect children from unnecessary residential placements, and about the proper role parents should play in selecting services for their developmentally disabled child.[20]

While each of these issues is critical, few have received comprehensive treatment in current state legislation. Many of these matters have been left to informal administrative and professional resolution. The one area in which almost all states have statutes involves commitment and admission to residential facilities. Their provisions are summarized in the next section.

---

[19] *See* Ellis & Carter, *Treating Children Under the New Mexico Mental Health and Developmental Disabilities Code,* 10 N.M.L. REV. 279 (1980).

[20] *See* Teitelbaum & Ellis, *The Liberty Interest of Children: Due Process Rights and Their Application,* 12 FAM. L. Q. 153 (1978), *reprinted in* 2 MENTAL DISABILITY L. RPTR. 582 (1978).

REVIEW OF STATUTES

States have enacted a variety of schemes for the admission of developmentally disabled people to residential services. Some of the statutes appear to be modeled after the provisions for the admission and commitment of mentally ill persons to institutions. Others treat mentally ill and developmentally disabled persons under different sets of procedures. Some of the states in this latter group appear to have left intact the mental retardation statutes enacted during the era of eugenic concern in the early decades of this century. Others have revised their statutes substantially in recent years, but have chosen to treat mentally ill and developmentally disabled people differently.

A large number of states differentiate "voluntary admission" from "involuntary commitment." The discussion of the statutes which follows uses this method of categorizing types of admissions of adults to residential facilities. Provisions for the placement of children in residential services are treated separately in many statutes, and are therefore considered separately in this summary.

While the following summary accurately describes the types of statutes now in effect in the states, it should not be taken as an accurate description of the actual practices and procedures in all of those states. There is reason to believe, from anecdotal accounts and experience, that in many states the procedures prescribed by the statute are not followed. In other states, compliance with the procedures may be of a token and formalistic nature — hearings held in wholesale fashion with no factual consideration given to the person's condition or needs in individual cases. Similarly, in some states which provide for both a formal "involuntary" hearing process and an informal "voluntary" admission process, the vast majority of individuals placed in residential facilities may go through the informal process with little or no regard to how "voluntary" their admission is in fact. Thus the formal procedural protections described here, while they are present in the states' statutes, may seldom or never be used. Once again, anecdotal experience suggests that very few developmentally disabled people who enter residential facilities in this country do so following either a full adversarial hearing on the appropriateness of the proposed placement or a careful inquiry into the adequacy of their "voluntary" consent.

It is not possible either to document or disprove this conclusion with any studies in the professional literature. There is a need for careful social science research into the issue of the actual procedures by which people are admitted to institutions and other residential facilities for the developmentally disabled. But in the absence of such evidence, there is reason to consider the statutory descriptions of procedural protections in a number of states with skepticism.

I. VOLUNTARY ADMISSION OF ADULTS

There is some indication that the vast majority of developmentally disabled individuals who enter residential facilities while they are adults do so under "voluntary admission" provisions of their state statutes. These procedures can be characterized as informal and unencumbered by procedural protections of the individual's rights. The extent to which these procedures produce appropriate admissions and services depends upon the performance of admitting officials at the residential facility in question rather than upon any system of external accountability or review.

The majority of states[21] provide for the voluntary admission of mentally retarded or developmentally disabled persons to state institutions.[22] At least four states have left the process of designing voluntary admissions procedures up to their department of developmental disability services (or equivalent agency), typically giving the agency little or no statutory guidance in drafting these regulations.[23] Among those states which prescribe procedures in their statutes, several create categories which closely resemble voluntary status. At least two states provide for "administrative admission" (separately delineated from judicial admission), which may be considered similar to voluntary admission since the client may initiate

---

[21] At least two states make no provision for the voluntary admission of adults. *See* GA. CODE ANN. §88-2504.1(a)(1979); N.M. STAT. ANN. §§43-1-14 through 43-1-16 (Repl. 1979).

[22] Although some states have procedures governing the admission of mentally retarded clients to non-residential services, *e.g.*, N.H. REV. STAT. ANN. §§171-A: 2(XVI) through 171-A:6 (1977), most states' voluntary admissions procedures govern the admission to a state institution. The most common terms for such institutions are "training schools" or "training centers." *E.g.*, HAW. REV. STAT. §333-21 (1976); N.D. CENT. CODE §25-04-1 (Supp. 1979); WYO. STAT. §9-6-603 (1977). Some states discuss admissions in terms of the mentally retarded, *e.g.*, *id.*, while others use the term developmentally disabled, *e.g.*, ILL. STAT. ANN. ch. 91-1/2, §4-302 (Smith-Hurd Supp. 1979).

[23] ALASKA STAT. §47.80.130(a)(2)-(3) (1979); DEL. CODE ANN. tit. 16, §5521 (1974); KAN. STAT. ANN. §76-12a07 (1977); N.C. GEN. STAT. §122-70 (Supp. 1979).

it.[24] Another state provides both for voluntary admission and "voluntary commitment."[25] At least one state provides, in addition to voluntary admission, a special "non-objecting" status.[26] These variations will be considered along with voluntary admission statutes in this summary.

A. Initiator

At least five jurisdictions provide that only the developmentally disabled person may apply for voluntary admission. These include the District of Columbia,[27] Minnesota,[28] New York,[29] West Virginia,[30] and Wyoming.[31] (Among those five states, there is a variety of guarantees of the applicant's voluntariness.)[32] At the opposite end of the spectrum, at least eight states make no provision for the client to apply. In some of these states, either the parent or guardian may apply,[33] in others some other adult may also apply,[34] and in one only the guardian may apply.[35] Generally, in the rest of the states, the client, a parent, the guardian, or other custodian may apply.[36] Some of these states include others who may also apply.[37]

As a general matter, there is little guarantee of voluntariness of the application, but a few states have made efforts in this area.[38] At least one statute outlines a procedure for objection to voluntary admission.[39]

As indicated above, some states allow one person to apply for the admission of another, designating this substitute consent as a "voluntary admission." Traditionally, few limits were placed on this power, once granted. A small but growing number of such statutes now include provisions requiring that the guardian have specific authorization to admit the individual voluntarily,[40] and Virginia's statute requires, when the individual is incapable of requesting voluntary admission, that a parent, guardian, or other responsible person initiate a judicial proceeding empowering him or her to admit the individual to the facility.[41]

B. Criteria for Admission

The criteria for voluntary admission fall into four general categories, with a roughly equal number of states in each. Some state statutes require only that the client be mentally retarded or developmentally disabled;[42] of those, a few require an evaluation to determine the client's condition.[43] A second group of states adds that the individual also must be in need of care and treatment.[44] Another group requires that the developmentally disabled individual be "suitable for admission."[45] The final group specifies that the developmentally

---

[24] ILL. STAT. ANN. ch. 91-1/2, §4-302 (Smith-Hurd Supp. 1979); MICH. COMP. LAWS ANN. §330.1509(2) (1975).

[25] PA. STAT. ANN. tit. 50, §§4402, 4403 (Purdon 1969).

[26] N.Y. MENTAL HYG. LAW §15.25 (McKinney 1978).

[27] D.C. 2-108, Act 2-297, §§302, 303 (Nov. 8, 1978).

[28] MINN. STAT. ANN. §253A.03(1) (West. Supp. 1980).

[29] N.Y. MENTAL HYG. LAW §15.13(a) (McKinney 1978).

[30] W. VA. CODE §27-4-1(a) (Supp. 1978).

[31] WYO. STAT. §9-6-616(a)(i) (1977).

[32] The Wyoming provision merely states that the individual must have "sufficient insight or capacity to make responsible application." *Id.* The Minnesota statute states that "[s]uch person shall not be admitted to the hospital if he objects thereto." MINN. STAT. ANN. §253A.03 (West Supp. 1980). The West Virginia statute requires that the applicant be told and be given a written statement of the consequences of voluntary admission. W. VA. CODE §27-4-4(a) (Supp. 1978). In New York, the applicant must be advised of and "have the ability to understand" the nature of the voluntary status; the voluntary client must also annually consent in writing to continued residence. N.Y. MENTAL HYG. LAW §§15.15,15.17 (McKinney 1978). Only the District of Columbia provides that a court-appointed officer shall make a determination of competency *and* voluntariness; if the officer finds that there is a substantial question regarding either, the court must conduct a hearing. D.C. CODE ANN. §§2-108, 2-297, 302(c) (Supp. 1978).

[33] ARK. STAT. ANN. §59-1104(a) (1971); HAW. REV. STAT. §333-35.5 (1976); MISS. CODE ANN. §41-19-11(a) (1972); UTAH CODE ANN. §64-8-15 (1977).

[34] In Alabama, when neither the parent nor guardian "is available and capable," an adult next of kin may apply. ALA. CODE tit. 22, §52-51 (1975). In Iowa, another person "responsible for any person believed to be mentally retarded" may apply. IOWA CODE ANN. §222.13 (West Supp. 1979-80). In South Carolina, the spouse may also apply. S.C. CODE §44-21-80 (1976).

[35] NEB. REV. STAT. §83-383 (1976).

[36] *E.g.*, N.D. CENT. CODE §25-04-05(1)(a) (Supp. 1979) ("by his parent or guardian or the person or agency having legal custody of him, or by the mentally deficient person himself"); S.D. COMPILED LAWS ANN. §27B-5-3 (1976) ("by the individual himself if he is competent to do so, or by his guardian if he is not competent to do so").

[37] *E.g.*, R. I. GEN. LAWS §23-43.1-7(1) (Supp. 1978) ("any individual of lawful age, either personally, or on the application of any relative, friend or attorney with said individual's consent").

[38] The provisions range from nothing, *e.g.*, N.D. CENT. CODE §25-04-05(1) (Supp. 1979), to broad statements of competence or capability of requesting admission. *E.g.*, ARIZ. REV. STAT. ANN. §36-560(D) (Supp. 1979-80); VA. CODE §37.1-65.1(A) (Supp. 1979).

[39] ILL. ANN. STAT. ch. 91-1/2, §4-305(a) (Smith-Hurd Supp. 1979) ("Any interested person on behalf of a client or a client himself if he is 12 years of age or older may object . . .").

[40] *E.g.*, ILL. ANN. STAT. ch. 91-1/2, §4-302 (Smith-Hurd Supp. 1979) ("if he is authorized by the guardianship order of the Circuit Court"); WASH. REV. CODE ANN. §72.33.125(1)(b) (Supp. 1978) ("by his or her guardian, or limited guardian where so authorized").

[41] VA. CODE §37.1-65.1 (Supp. 1979).

[42] *E.g.*, ALA. CODE tit. 22, §52-51 (1975); NEV. REV. STAT. §435.081(1) (1977); W. VA. CODE §27-4-1(a) (Supp. 1978).

[43] *E.g.*, ARIZ. REV. STAT. ANN. §36-559 (Supp. 1979-80); N. H. REV. STAT. ANN. §171-A:6(II)-(III) (1977) (the fixed point of referral's staff must do a screening evaluation and make a recommendation for placement). *Accord*, UTAH CODE ANN. §64-8-13 (1977):

> All mentally retarded persons . . . whose defects prevent them from properly taking care of themselves, or who are a social menace, may be admitted . . . .
>
> . . . A mentally retarded person shall mean a person in whom there has been found, by comprehensive evaluation, a condition of mental retardation of such a nature and degree as to constitute a substantial, continuing prospective, educational, vocational, and social handicap.

[44] *E.g.*, ARK. STAT. ANN. §59-1011 (1971); TENN. CODE ANN. §33-501(a) (1977). *Accord*, PA. STAT. ANN. tit. 50, §4402(b) (Purdon 1969) (facility director must initiate examination to determine need).

[45] *E.g*, COLO. REV. STAT. §§27-10.5-103(2), 27-10.5-104(2) (Supp. 1978); MICH. COMP. LAWS ANN. §330.1509(2) (1975); N.Y. MENTAL HYG. LAW §15.13(a) (McKinney 1978). *Accord*, ILL. ANN. STAT. ch. 91-½, §4-300(a) (Smith-Hurd Supp. 1979) (diagnostic evaluation required to determine suitability); S.D. COMPILED LAWS ANN. §§27B-5-2,5-3, & 5-7 (Rev. ed. 1976) (preadmission examination required; facility may receive individual for up to 60 days to conduct examination (§27B-5-7)).

disabled person be in need of services, as determined by a pre-admission examination.[46]

Possibly in response to recent court rulings, an increasing number of states require that voluntary placements conform to the least restrictive alternative doctrine.[47] Similarly, some statutes now provide that voluntary residents have the right to an individualized habilitation plan.[48] Some states now have both of these requirements,[49] but a roughly equal number have neither.[50]

C. Review and Discharge

More than half the jurisdictions provide for automatic review of either the individual or the habilitation plan[51] at certain specified intervals.[52] Another mechanism adopted by several states is an "objection" procedure to review the admission.[53] Under these provisions, periodic review is not automatic, but rather depends upon the individual's or another person's initiative in challenging continued confinement. The balance of the states make no provisions for review all all,[54] although at least one state sets a limit on the duration of the voluntary admission.[55]

Some states have discharge provisions specifying that the individual may apply for his or her own discharge,[56] although some of these require that the person was the original applicant for admission in order to have the right to make such a request.[57] At least two other jurisdictions permit only the client to initiate discharge procedures.[58] A considerable number of states, however, permit someone else, such as a parent or guardian, to apply for an individual's discharge;[59] several of them specify that such person must have originally applied for the individual's admission in order to have the right to seek the discharge.[60]

A number of states also provide for discretionary discharge, at the option of an institutional official.[61] At least three provide only for discretionary discharge.[62] Of the states with provision for discharge initiatiated by request of the individual or other interested person, about half provide that the discharge should take place automatically on receipt of the request and the giving of the required notice.[63] The other half leave the decision in the hands of the facility director or some other official; typically, he or she decides on the basis of whether the individual meets the requirements for involuntary commitment.[64]

In summary, the states' provisions for voluntary admission of adults include a variety of procedures. In some cases, the admission requires the voluntary consent of the individual. In other cases, other persons are authorized to "volunteer" the developmentally disabled adult into a facility. Provisions for periodic review of the individual's continuing need for residential services are not universal, and some of those which have been enacted require the individual to initiate the review process, thus eliminating the necessity of review when the individual is unable to initiate it or when he or she is unaware of the right to do so. Most statutes give little guidance regarding appropriate discharge procedures, and therefore administrators are given considerable discretion in shaping and implementing such procedures.

II. INVOLUNTARY COMMITMENT OF ADULTS

Although there are several states which do not have statutory provisions for involuntary commitment of developmentally disabled clients,[65] the vast majority of states have some type of involun-

[46] *E.g.*, IOWA CODE ANN. §222.13 (West Supp. 1979-80); TEX. REV. CIV. STAT. ANN. art. 5547-300, §33 (Vernon Supp. 1980); VA. CODE §37.1-65 (1976). *Accord*, N.J. STAT. ANN. §30:4-25.3 (West Supp. 1979-80) (no examination specified).

[47] *See, e.g.*, TENN. CODE ANN. §33-1601 (Art. IV) (1977).

[48] *E.g.*, MINN. STAT. ANN. §243A.17(9) (West Supp. 1980); OHIO REV. CODE ANN. §5123.85 (Page Supp. 1979); W. VA. CODE §27-5-9(d) (Supp. 1978).

[49] *E.g.*, ILL. ANN. STAT. ch. 91-½, §§4-301(a), 4-309 (Smith-Hurd Supp. 1979); NEV. REV. STAT. §433.494(l) (1975); N.J. STAT. ANN. §§30:4-24.2(e)(2), 30:6D-10 & -11 (West Supp. 1979-80).

[50] *E.g.*, Arkansas, Idaho, Iowa.

[51] *E.g.*, ALA. CODE tit. 22, §52-57 (1975); MASS. GEN. LAWS ANN. ch. 123, §4 (West Supp. 1979); TEX. REV. CIV. STAT. ANN. art. 5547-300, §17(a) (Vernon Supp. 1980).

[52] The intervals range from three months, *e.g.*, NEV. REV. STAT. §433.494(2) (1975), to six months, *e.g.*, ARIZ. REV. STAT. ANN. §36-565(A) (Supp. 1979-80), to twelve months, *e.g.*, PA. STAT. ANN. tit. 50, §4402(d) (Purdon 1969).

[53] *E.g.*, MICH. COMP. LAWS ANN. §330.1511 (1975); S.D. COMPILED LAWS ANN. §§27B-5-9 to 27B-5-12 (1976).

[54] *E.g.*, North Dakota, Tennessee, Wyoming.

[55] HAW. REV. STAT. §333-35.3 (1976) (60-day limit, with one additional period of 30 days if a court finds the extension is in the client's best interest).

[56] *E.g.*, COLO. REV. STAT. §27-10.5-108(2)(a) (Supp. 1978); WASH. REV. CODE ANN. §72.33.140 (Supp. 1978).

[57] *E.g.*, MICH. COMP. LAWS ANN. §330.1512(1) (1975); S.C. CODE §44-21-100 (1976); S.D. COMPILED LAWS ANN. §27B-5-16 (Rev. ed. 1976).

[58] *E.g.*, MINN. STAT. ANN. §253A.03(2) (West Supp. 1980); PA. STAT. ANN. tit. 50, §4402(c) (Purdon 1969).

[59] *E.g.*, N.H. REV. STAT. ANN. §171-A:7 (1977) (legal guardian); OHIO REV. CODE ANN. §5123.70(A) (Page Supp. 1979) (resident's counsel, legal guardian, parent, spouse, or adult next of kin); WASH. REV. CODE ANN. §72.33.140 (Supp. 1978) (guardian or court-appointed personal representative).

[60] *E.g.*, ALA. CODE tit. 22, §52-53 (1975); IOWA CODE ANN. §222.15 (West Supp. 1979-80).

[61] *E.g.*, MICH. COMP. LAWS ANN. §330.1525(1) (1975) (facility director); R.I. GEN. LAWS §23-43.1-16(1) (Supp. 1978) (facility director); W. VA. CODE §27-4-2 (1976) (facility chief medical officer).

[62] *E.g.*, MASS. GEN. LAWS ANN. ch. 123, §10(a) (West Supp. 1979) (facility superintendent); NEV. REV. STAT. §435.077(3) (1975) (associate administrator for mental retardation of the division); VA. CODE §37.1-98(1) (Supp. 1979) (facility director).

[63] *E.g*, ARIZ. REV. STAT. Ann. §36-566(A) (Supp. 1979-80); IOWA CODE ANN. §222.15 (West Supp. 1979-80).

[64] *E.g.*, ALA. CODE tit. 22, §52-53 (1975); OHIO REV. CODE ANN. §5123.70(A)(2) (Page Supp. 1979); W. VA. CODE §27-4-3(c) (1976).

[65] Those states include Arizona, Kansas, New Jersey, Tennessee, and Virginia.

tary admission. At least two have administrative (non-judicial) admissions,[66] and some have both judicial commitment and "medically certified admissions,"[67] but most states have statutory provisions for some type of judicial proceeding for involuntary commitment.[68]

A. The Admission

1. Persons Authorized to Seek Commitment

Several jurisdictions allow a relative (usually the parent) or a guardian of a developmentally disabled adult to initiate involuntary commitment to a residential facility.[69] Quite a few jurisdictions permit anyone to initiate the process,[70] and some specify that a public agency or official may do so.[71]

Typically, the person seeking the commitment will petition a court for an order placing the individual in a residential facility. Of the states which provide for judicial commitment, twice as many utilize a court of general jurisdiction[72] as use probate courts.[73]

2. The Criteria

The standards for involuntary commitment are widely divergent among the states. A growing number of states use some form of "dangerousness" standard; the majority of those define it as dangerous to self or others,[74] although at least one defines it only as dangerous to self.[75] A second group of states uses a standard involving the individual's welfare or the protection of the community;[76] some of them specify this as an alternate standard to dangerousness.[77] A third group of states merely specifies that the mentally retarded person be in need of care or treatment.[78] Other states use a variety of standards for admission.[79]

Only a few state statutes specifically require both that the placement ordered be the least restrictive alternative and that an individualized habilitation plan be presented to the court before admission.[80] Some states require that the placement be the least restrictive alternative, but do not require the development of an habilitation plan until after the individual's admission.[81]

3. Procedural Rights

About half the jurisdictions specify that the developmentally disabled person has the right to an attorney, but not all of them provide that the attorney is mandatory.[82] One jurisdiction's statute only states that the judge "may appoint an attorney or guardian ad litem to serve on the examining committee for the protection of all legal rights of the retarded individual."[83] Quite a few jurisdictions permit the client's presence at the hearing to be waived,[84] and some specify that the court[85] or the client's attorney may waive it.[86] At least two states' statutes provide that the client's presence shall not be waived.[87]

A few jurisdictions provide that the developmentally disabled individual is entitled to pre-

---

[66] DEL. CODE tit. 16, §5522 (1974) (by the institution superintendent, with no hearing unless commitment is appealed to Court of Chancery); S.D. COMPILED LAWS ANN. §27B-7-1 to 27B-7-16 (1976) (by a county board of mental retardation, with a hearing).

[67] *E.g.*, PA. STAT. ANN. tit. 50, §§4404, 4406 (Purdon 1969); W. VA. CODE §§27-5-2, 27-5-4 (Supp. 1978).

[68] *E.g.*, HAW. REV. STAT. §§333-26, 333-27 (1976); N.M. STAT. ANN. §§43-1-10 to 43-1-13 (Repl. 1979); OHIO REV. CODE ANN. §§5123.71 to 5123.76 (Page Supp. 1979).

[69] *E.g.*, ARK. STAT. ANN. §59-1104(a) (1971); HAW. REV. STAT. §333-26 (1976).

[70] *E.g.*, ALA. CODE tit. 22, §52-55(a) (1975); ILL. ANN. STAT. ch. 91-½, §4-501(a) (Smith-Hurd Supp. 1979); UTAH CODE ANN. §64-8-16 (1977).

[71] *E.g.*, IDAHO CODE §66-329(a) (Supp. 1979); NEB. REV. STAT. §83-220(2) (1976).

[72] *E.g.*, N.M. STAT. ANN. §43-1-3(F) (Repl. 1979); PA. STAT. ANN. tit. 50, §4406(a) (Purdon 1969) (civil court commitment); W. VA. CODE §27-5-4(a) (Supp. 1978) (legal proceeding).

[73] *E.g.*, MINN. STAT. ANN. §253A.07(1) (West 1971); OHIO REV. CODE ANN. §5123.76(A) (Page Supp. 1979); S.C. CODE §44-21-90 (1976). In Mississippi, the chancery court is the forum. MISS. CODE ANN. §41-21-75 (Supp. 1978).

[74] *E.g.*, IDAHO CODE §66-329(i) (Supp. 1979); MICH. COMP. LAWS ANN. §330.1515 (1975); W. VA. CODE §§27-5-2(a), 27-5-4(a) (Supp. 1978) (both medically certified admissions and legal proceedings).

[75] OHIO REV. CODE ANN. §5123.68(N)(1) (Page Supp. 1979).

[76] *E.g.*, IOWA CODE ANN. §222.31 (West Supp. 1979-80); MINN. STAT. ANN. §253A.07(17)(b) (West Supp. 1980); NEB. REV. STAT. §83-222 (1976).

[77] *E.g.*, NEV. REV. STAT. §435.081(3) (1977); N.M. STAT. ANN. §43-1-13(E) (Repl. 1979); TEX. Rev. CIV. STAT. ANN. art. 5546-300, §37(b) (Vernon Supp. 1980).

[78] *E.g.*, ALA. CODE tit. 22, §52-55 (1975); S.C. CODE §44-21-90 (1976); S.D. COMPILED LAWS ANN. §27B-7-1 (1976).

[79] *E.g.*, GA. CODE ANN. §88-2504(f) (Supp. 1979) (the client requires direct medical services and needs 24-hour training in a residential care facilitiy); WYO. STAT. §9-6-621(d) (1977) (the client is a fit subject for care and training).

[80] *E.g.*, GA. CODE ANN. §§88-2503.21, 88-2504(f) (1979, Supp. 1979); N.M. STAT. ANN. §43-1-13 (Repl. 1979); OHIO REV. CODE ANN. §5123.76(E) (Page Supp. 1979).

[81] *E.g.*, ILL. ANN. STAT. ch. 91-½, §§4-609(b), 4-612 (Smith-Hurd Supp. 1979); S.D. COMPILED LAWS ANN. §§27B-7-16, 27B-8-10 (1976).

[82] Those which, although granting the right to an attorney, do not require one include: MICH. COMP. LAWS ANN. §330.1454(3) (1975); MISS. CODE ANN. §41-21-73 (Supp. 1978); OHIO REV. CODE ANN. §5123.76(A)(3) (Page Supp. 1979).

[83] ALA. CODE tit. 22, §52-55(b) (1975).

[84] *E.g.*, GA. CODE ANN. §88-2502(p) (Supp. 1979); IDAHO CODE §66-329(h) (Supp. 1979); WYO. STAT. §9-6-621(c) (1977).

[85] *E.g.*, MINN. STAT. ANN. §253A.07(12) (West 1971); S.C. CODE §44-21-90 (1976); TEX. REV. CIV. STAT. ANN. art. 5547-300, §37(m)(3) (Vernon Supp. 1980).

[86] *E.g.*, ILL. ANN. STAT. ch. 91-½, §4-606 (Smith-Hurd Supp. 1979).

[87] WASH. REV. CODE ANN. §71.05.310 (Supp. 1978); W. VA. CODE §§27-5-2a, 27-5-4(c) (Supp. 1978) (for both medically certified admissions and legal proceedings).

sent his or her own experts and witnesses to testify. A number of states also give the client the right to confront and cross-examine opposing witnesses.[88]

Very few state statutes specifically provide for an appeal of the commitment order,[89] but only Mississippi's provides that the order is not appealable.[90]

4. Duration of Commitment Orders

About half the jurisdictions make no provision regarding the duration of the involuntary commitment.[91] The rest specify a period or periods of admission, ranging from two months to two years.[92]

B. Periodic Review

About half the jurisdictions with involuntary commitment procedures have no provisions for review of the individual's need for continued confinement and habilitation. Of those which do have such reviews, a majority provide that they shall occur automatically at certain intervals, ranging from three months to one year.[93] Other states provide that the client or some other person may initiate the review process.[94] The jurisdictions providing for review are split about evenly between those in which the review is done by staff within the facility[95] and those in which the review is done in court.[96] Very few states specifically provide for an appeal of the decision on review.[97]

C. Discharge

At least two states provide for automatic discharge when the commitment period expires.[98] About half the jurisdictions provide for discretionary discharge at the option of an institution official,[99] and a few provide that either the developmentally disabled individual or another person may request discharge.[100] Most of the states allowing a request for discharge specify that a court shall make the determination whether to grant the requested discharge.[101] Of the states with such request provisions, about half specify that the client shall be released if he or she no longer meets the standard for involuntary admission.[102] The rest specify a variety of criteria for discharge, including appropriateness of the discharge,[103] capability of the individual to care for him or herself,[104] and a finding that the individual no longer needs treatment.[105]

III. MINORS

A. "Voluntary" Admissions

A majority of the states permit parents to apply for the voluntary admission of their developmentally disabled minor children.[106] Although most do not, some jurisdictions permit the minor to apply for his or her own admission,[107] or require that he or she consent to the voluntary admission sought by the parents.[108] In the majority of jurisdictions, the discharge and review procedures are the same for both minors and adults. At least three states have separate provisions for minors, governing admission, review, and discharge.[109]

B. Involuntary Admission

At least one statute includes a separate provision for hearings on the necessity of residential

---

88 *E.g.*, IDAHO CODE §66-329(h) (Supp. 1979); MICH. COMP. LAWS ANN. §330.1517(3)(d)-(g) (1975); WYO. STAT. §9-6-621(c) (1977).

89 *E.g.*, HAW. REV. STAT. §333-27 (1976); N.M. STAT. ANN. §43-1-13(D) (Repl. 1979); TEX. REV. CIV. STAT. ANN. art. 5547-300, §37(p) (Vernon Supp. 1980).

90 MISS. CODE ANN. §41-21-75 (Supp. 1978).

91 *E.g.*, Arkansas, Iowa, Utah.

92 *E.g.*, HAW. REV. STAT. §333-30 (1976) (60 days, then indefinite thereafter); OHIO REV. CODE ANN. §5123.76(C)(1) (Page Supp. 1979) (up to 90 days; there is also an absolute five-year limit on some commitments under §5123.76(I) (Page Supp. 1979); WASH. REV. CODE ANN. §71.05.320 (Supp. 1978) (90 days, then 180 days); Ga. Code Ann. §88-2504(f) (Supp. 1979) (six months); MICH. COMP. LAWS ANN. §330.1519(3) (1975) (12 months); W. VA. CODE §27-5-4(d) (Supp. 1978) (two years for commitment via legal proceedings).

93 *E.g.*, NEV. REV. STAT. §433.494(2) (1975) (every three months); MICH. COMP. LAWS ANN. §330.1531(1) (1975) (every six months); TEX. REV. CIV. STAT. ANN. art. 5547-300, §17(a) (Vernon Supp. 1980) (annually).

94 *E.g.*, ILL. STAT. ANN. ch. 91-½, §4-612 (Smith-Hurd Supp. 1979) (the client or anyone on his behalf); N.M. STAT. ANN. §43-1-13(G) (Repl. 1979) (the client, legal guardian, parent, spouse, relative, or friend).

95 *E.g.*, ALA. CODE tit. 22, §52-57 (1975); MICH. COMP. LAWS ANN. §330.1531(1) (1975); W. VA. CODE §27-5-8(a) (Supp. 1978).

96 *E.g.*, N.M. STAT. ANN. §43-1-13(G) (Repl. 1979).

97 *E.g*, W. VA. CODE §27-5-8(b) (Supp. 1978).

98 OHIO REV. CODE ANN. §5123.76(H) (Page Supp. 1979); WASH. REV. CODE ANN. §71.05.320 (Supp. 1978).

99 *E.g.*, NEB. REV. STAT. §§83-218, 83-387 (1976); UTAH CODE ANN. §64-8-8(10) (1977).

100 *E.g.*, IOWA CODE ANN. §222.42 (West Supp. 1979-80) (client or any "reputable person").

101 *E.g.*, ILL. STAT. ANN. ch. 91-½, §4-706(b) (Smith-Hurd Supp. 1979).

102 *Id.*

103 TEX. REV. CIV. STAT ANN. art. 5547-300, §§39 & 40 (Vernon Supp. 1980).

104 IOWA CODE ANN. §222.43(2) (West Supp. 1979-80).

105 MINN. STAT. ANN. §253A.19(1) (West 1971).

106 *E.g.*, OHIO REV. CODE ANN. §5123.69(A) (Page Supp. 1979); TENN. CODE ANN. §33-501(a)(2) (Repl. 1977).

107 *E.g.*, D.C. Code Ann. §§2-108, 2-297, 302(a), 303 (Nov. 8, 1978); *see* MINN. STAT. ANN. §253A.03(1) (West Supp. 1980).

108 *E.g.*, ARIZ. REV. STAT. ANN. §36-560(D) (Supp. 1979-80) (client 14-18 years old, capable of giving voluntary informed consent, shall apply jointly with his parent or guardian); W. VA. CODE §27-4-1(b) (Supp. 1978) (voluntary admission shall be conditioned on the client's consent if he or she is a minor 12 years of age or over).

109 HAW. REV. STAT. §333-35 (1976); N.M. STAT. ANN. §43-1-16 (Repl. 1979); WIS. STAT. ANN. §51.13 (West Supp. 1979-80).

habilitation of minors.[110] The provisions of most other states regarding involuntary treatment of adults are also applicable to minors.[111]

C. Provisions on Reaching Majority

At least five states provide that a minor client shall be discharged when he or she reaches majority unless steps have been initiated to continue treatment, on either a voluntary or involuntary basis.[112]

IV. MISCELLANEOUS PROVISIONS

Almost half the states specifically guarantee the right to habeas corpus for developmentally disabled persons in institutional facilities.[113] A few states have provisions for emergency commitment of the developmentally disabled.[114] And at least four jurisdictions have made statutory provision for short-term respite care for developmentally disabled persons.[115]

---

[110] N.M. STAT. ANN. §43-1-16.1 (Repl. 1979). *See* Ellis & Carter, *Treating Children Under the New Mexico Mental Health and Developmental Disabilities Code*, 10 N.M.L. REV. 279 (1980).

[111] However, at least two states, Illinois and Michigan, specifically exempt minors from their statutes' coverage. ILL. ANN. STAT. ch. 91-½, §4-500 (Smith-Hurd Supp. 1979); MICH. COMP. LAWS ANN. §300.1503(1) (1975).

[112] ARIZ. REV. STAT. ANN. §36-566(B) (Supp. 1979-80); HAW. REV. STAT. §333-35 (1976); NEB. REV. STAT. §83-388 (1976); NEV. REV. STAT. §435.360(1) (1977); N.J. STAT. ANN. §30:4-107.3(4) (West Supp. 1979-80).

[113] *E.g.*, GA. CODE ANN. §88-2503.14(a) (1979); N.Y. MENTAL HYG. LAW §33.15(a) (McKinney 1978); WASH. REV. CODE ANN. §71.05.480 (1975).

[114] *E.g.*, ARIZ. REV. STAT. ANN. §36-560(N) (Supp. 1979-80); ILL. ANN. STAT. ch. 91-½, §§4-400 to -408 (Smith-Hurd Supp. 1979); OHIO REV. CODE ANN. §§5123.71(C) & (D) (Page Supp. 1979).

[115] COLO. REV. STAT. §27-10.5-103(1) (Supp. 1978); ILL. ANN. STAT. ch. 91-½, §§4-311 & -312 (Smith-Hurd Supp. 1979); S.D. COMPILED LAWS ANN. §27B-5-1 (1976); TEX. REV. CIV. STAT. ANN. art. 5547-300, §34(h) (Vernon Supp. 1980).

SUMMARY OF MODEL STATUTE

The proposed model statute differs greatly from existing state laws in both its scope and its substantive provisions. Virtually all the states consider admission to services in the context of the state's coercive power to commit individuals to residential facilities. To the extent that they have any laws dealing with other kinds of services and other kinds of consent mechanisms, those laws are not integrated so as to provide for consideration of the entire spectrum of available services and of appropriate consent procedures. And no state currently guarantees the right to appropriate services in the context of consent provisions for all services.

The statement of legislative purpose is explicit and detailed in its statement of the premises of the model act. With it, an adopting legislature would go on record as guaranteeing the right of all developmentally disabled individuals to services appropriate to their habilitation and other needs. This guarantee is not limited to those services which are available, but rather extends to all needed services even if the state has chosen not to provide them in adequate quantity in the past. The state further commits itself to limiting the intrusion into individual autonomy, and implements this commitment through the requirement of least drastic means, through enhancement of the role of direct individual consent to services, and through the provision of due process to those who are unable to make decisions regarding their own services. This same concern about individual liberty dictates the decision to limit severely (or, in the alternative, to eliminate) the use of institutional residential facilities.

The substantive provisions of the model statute begin with a guarantee of services appropriate to the individual's needs, and a mechanism for obtaining evaluation of those needs and services to meet them. The enforcement mechanism includes a recognition of the power of the state's courts to order the provision of appropriate services when the state is not currently making them available, and even to order the creation of new services where they do not currently exist. All services are to be designed and provided on an individualized basis, and toward that end, all services must be provided pursuant to an individualized program plan. The statute details the components of that plan and the process by which it is to be developed.

At the heart of the statute are the provisions for consent to services. The act's presumption is in favor of direct consent by the individual who is to receive the services, and consent by others who are authorized to consent on his or her behalf is permitted only when it is shown that the individual is not capable of making his or her own decisions. There are different procedures provided for different kinds of services, with the greater procedural protections reserved for those situations in which serious liberty interests are at issue.

Adults may receive non-residential services either by their own consent or by the consent of their guardians. The individual's or guardian's decision to withhold consent can be overridden only if a hearing officer determines that the services proposed are appropriate to the individual's needs and that the individual will suffer serious harm if the services are not provided. Residential services in small facilities follow the same pattern of consent procedures, except that an adult's guardian lacks the power to consent to the individual's placement in such a facility. Such a placement requires a court order, which may be sought by the individual's guardian. Procedural requirements are considerably tighter for large facilities (if the state has not chosen to abolish their use). All placements in such facilities — voluntary or involuntary — require a court order, and the substantive standard for such orders requires a showing of the necessity of such a placement.

Services for developmentally disabled minors are to be provided by a different set of consent mechanisms, which take into account the appropriate role of the minor's parents in selecting such services. A parent may consent to his or her minor child's participation in non-residential services, and the parent's decision can be overridden only upon a showing equivalent to a finding of neglect. But a parent may consent to the child's placement in a small residential facility only upon the order of a hearing officer who has found the proposed placement to be appropriate for the child's needs. A child may be placed in a large residential facility only upon a court finding that such a placement is absolutely necessary for the child's needs.

Provisions are made for the delivery of services on an emergency basis and for the provision of respite care services. The statute also specifies the procedures to be followed in both judicial and nonjudicial hearings in cases involving developmentally disabled persons and the services they require.

In conclusion, the proposed model statute bears little resemblance to the existing statute of any state. But since states should commit themselves and their resources to addressing the needs of their developmentally disabled citizens, it is appropriate to consider an integrated legislative approach which guarantees both substantive and procedural rights of the individuals affected.

The text of the act appears in **BOLD TYPE.** Optional language appears [within brackets/ separated by slashes]. Comments on statutory provisions appear in standard type. Alternatives to the text appear in *italics.* The comments explain, elaborate upon or provide a rationale for the text of the act. They are intended to assist the user in drafting a statute for a particular state from this model. An alternative suggests other approaches which might be considered.

PROPOSED ARTICLE FOR STATE CONSTITUTION*

The State shall provide appropriate services to meet the needs of physically and mentally handicapped people and to enhance their potential for independence and self-sufficiency. Such services shall be provided in a manner consistent with the individual needs and rights of each handicapped person.

COMMENT. Not every state will wish to enact a constitutional amendment, and no constitutional provision is needed to support the proposed model statute. However, some states currently have constitutional provisions regarding services for the handicapped, and these may include references to the maintenance of institutions.[1] In such cases, it may be desirable to replace such provisions with an article which more accurately reflects the range of services which handicapped people may need for their development. Another benefit to a constitutional provision such as the one proposed here is that it provides a cause of action, as does the model statute, when the state fails to provide necessary resources to fund services for the handicapped.

---

* James W. Ellis, Associate Professor of Law, University of New Mexico, had primary responsibility for drafting this proposed article and model statute. The assistance of Kathleen Watson and Ruth Luckasson and the comments of H. Rutherford Turnbull III and Marianne Bennett are gratefully acknowledged.

[1] *E.g.*, MICH. CONST. art. 8, §8.

MODEL STATUTE

AN ACT TO GUARANTEE THE RIGHT OF DEVELOPMENTALLY DISABLED PERSONS TO RECEIVE APPROPRIATE SERVICES AND TO ESTABLISH CONSENT PROCEDURES TO ALLOW THE PROVISION OF THOSE SERVICES.

Section 1. Title

This act shall be known as the "Right to Developmental Disabilities Services Act."

COMMENT. None necessary.

Introduction To Section 2.

This statement of legislative purpose is considerably more detailed and philosophical than that for most state legislation. This more elaborate approach may be warranted by two factors. First, the proposed model statute differs significantly from the statute of any state at the present time, and this kind of major departure from previous statutes should be explained fully to prevent misunderstanding. Second, the proposed model statute assumes that the Legislature has committed itself to a new model of service delivery for developmentally disabled people. If this kind of statute is to guarantee real rights and appropriate services to individuals with developmental disabilities, the Legislature must be committed as well to the task of implementing it with appropriate funding, particularly for community based services. By enacting a declaration of purpose like the one proposed here, the Legislature goes on record as supporting a move away from institution-based services and toward implementing the principle of normalization.[2]

Section 2. Statement of Legislative Purpose

(1) The Legislature recognizes the needs of developmentally disabled citizens for habilitation, and that all developmentally disabled people can benefit from habilitation and related services. The Legislature acknowledges the state's obligation to meet those needs, and guarantees each developmentally disabled person's right to obtain those habilitation services which appropriately meet his or her individual needs, regardless of the degree of his or her handicap or the cost of providing the services.

COMMENT. This subsection commits the state to a "zero-reject" policy, under which no individual is viewed as too handicapped to benefit from habilitation and other needed services. The acknowledgement of its obligation to meet service needs does not mean that all services will be provided by the state rather than by private service providers, nor does it mean that the state will be forced to pay for all such services irrespective of the individual's need for financial assistance. But it does commit the state to the role of "service provider of last resort," and promises that where financial costs are the only barrier to an individual's receiving needed services, the state will come forward to remove that barrier. The detailed formula for determining financial need is not included in this model statute. The state may either enact such a formula to accompany this statute, or can incorporate an already existing formula into this statute's guarantee of services.

(2) The Legislature recognizes that the liberty interest of developmentally disabled people is as fundamental as that of non-handicapped citizens, and is equally protected by the Constitutions of the United States and the state. Developmentally disabled people have the right to be free from unnecessary restrictions on the exercise of their right to liberty. Every developmentally disabled person who receives services pursuant to this act has the right to receive those services with no more restrictions on his or her liberty and freedom of choice than are absolutely required by the nature and severity of the person's individual disability and habilitation needs. Therefore, every developmentally disabled person is entitled to due process in determining whether he or she needs services and whether the services proposed or provided adequately meet the individual's needs and are consistent with the least drastic means principle.

---

* James W. Ellis, Associate Professor of Law, University of New Mexico, had primary responsibility for drafting this model statute. The assistance of Kathleen Watson and Ruth Luckasson and the comments of H. Rutherford Turnbull III and Marianne Bennett are gratefully acknowledged.

[2] *See, e.g.*, MONT. REV. CODES ANN. §53-20-101 (1979).

COMMENT. This statement is an affirmation of the basic citizenship rights of developmentally disabled people.[3] It acknowledges that the fact that a person has a developmental disability does not in any way diminish the value of that person's liberty, either to him or herself or to society. It further recognizes that liberty includes not only the right to freedom from unnecessary physical confinement, but also the right to make one's own choices. Neither of these liberty interests can be infringed without an adequate showing of necessity. The mechanism for protecting the liberty interests of developmentally disabled citizens is the same as that which protects the liberty of other citizens — due process of law. Later provisions of the statute adapt the concept of due process to the needs of developmentally disabled people, but it is essential that the statement of purpose recognize the importance of liberty to developmentally disabled people and the appropriateness of due process as an enforcement mechanism.

(3) The Legislature recognizes that each developmentally disabled person has the right to as much personal independence and self-sufficiency as possible. Therefore, each developmentally disabled person has the right to comprehensive services designed and administered to build and enhance the individual's independence, self-sufficiency, and ability to cope with his or her environment to the maximum extent possible. The Legislature also recognizes that, as part of that independence and self-sufficiency, every developmentally disabled person has the right to control, to the maximum extent possible, his or her own destiny and to make his or her own decisions about accepting or refusing services.

COMMENT. This provision recognizes the service needs of developmentally disabled people and emphasizes that those services should be designed to enhance the individual's abilities. The emphasis on independence, self-sufficiency, and the ability to cope with one's environment are found throughout the model statute. It stands in marked contrast to the dependency and "learned helplessness"[4] which have often accompanied services to developmentally disabled people, particularly when those services involved placement in large institutions. Similarly, the element of individual choice generally has been lacking in traditional programs of services to handicapped people, and thus, the state's commitment to freedom of choice is a major principle in the model statute.

(4) The Legislature, recognizing that the history of segregation in services has harmed developmentally disabled people, commits the state to the principle that the developmentally disabled person's needs shall be met with generic (nonspecialized) services whenever possible. When specialized services are required, they shall be provided with as little segregation and stigmatization as possible.

COMMENT. While recognizing that specialization and segregation will sometimes be required by the service needs of some developmentally disabled individuals, this provision seeks to minimize the possibility of stigmatization which may accompany such services.[5] Too frequently in the past, segregation and specialized services were the norm, with integration allowed only in extraordinary circumstances.[6] The model statute reverses that expectation and creates a presumption in favor of integration. The presumption should be rebuttable only upon a showing that the individual's service needs cannot be met successfully through generic or integrated services.

(5) The Legislature recognizes that institutions are frequently harmful to developmentally disabled people and their use is seldom warranted by the needs caused by developmental disabilities. Therefore, it is the intent of the Legislature to greatly reduce the use of residential facilities serving more than six developmentally disabled people, and to restrict the provision of residential services in larger facilities to those cases in which there is a showing that the individual requires intensive specialized care which can be provided only in a larger facility. In order to accomplish this limitation on the use of large facilities, the Legislature is committed to the creation of high quality, community-based services which will meet the individual needs of developmentally disabled persons of all degrees of disability. Although the state will attempt to minimize and mitigate the disruptive effects of this change upon those communities in which large institutions are now located and upon persons now employed in those institutions, the state shall give first priority to the rights and habilitation needs of developmentally disabled people to receive appropriate services in community-based settings.

COMMENT. The dispute over the role of large residential institutions has raged long and loud.[7] The evidence on the effect of the size of residential facilities is neither complete nor unambigu-

[3] *See generally* Wald, *Basic Personal and Civil Rights*, in PRESIDENT'S COMMITTEE ON MENTAL RETARDATION (P.C.M.R.), THE MENTALLY RETARDED CITIZEN AND THE LAW 3 (1976) [hereinafter cited as P.C.M.R.].

[4] DeVellis, *Learned Helplessness in Institutions*, 15 MENTAL RETARDATION 10 (October 1977).

[5] *See* R. EDGERTON, THE CLOAK OF COMPETENCE: STIGMA IN THE LIVES OF THE MENTALLY RETARDED 207 (1967).

[6] *See generally* W. WOLFENSBERGER, THE PRINCIPLE OF NORMALIZATION IN HUMAN SERVICES (1972).

[7] *See* R. SCHEERENBERGER, DEINSTITUTIONALIZATION AND INSTITUTIONAL REFORM (1976).

ous,[8] but a pattern of abuse, neglect, and individual regression has become quite apparent in many of our larger state institutions. In the last decade, many courts have observed this phenomenon, and have held the conditions in particular institutions to violate constitutional or statutory minimum standards.[9] Simultaneously, professionals have devised programs which deliver a wide variety of services in non-insitutional, community settings to individuals of all degrees of handicap.[10] The combination of these factors makes it advisable for states, at the very least, to reduce their dependence upon institutional models of service delivery for developmentally disabled people.

A legislature adopting this provision recognizes the potential harms of institutionalization, and commits the state to using this means of service delivery in only the most exceptional of cases. This includes only a small minority of the persons now housed in institutions in this country. Modern professional experience suggests that it is possible to provide high quality services to all other individuals in nonresidential settings or in smaller residential facilities.[11]

This provision does not suggest that reducing the size of the facility in which an individual is served will automatically assure the receipt of high quality services. The correlation between size and quality is neither that simple nor that direct.[12] But the correlation is impressive enough, and the dehumanization which has so often been found in large institutions is sufficiently appalling,[13] that states are justified in moving to other models of service delivery to meet the needs of their developmentally disabled citizens.

The selection of a number of residents to serve as a boundary between "large" and "small" facilities is necessarily arbitrary and imprecise. Considerations of human interrelationships and economic viability of residential units must inform this judgment. The number six has been selected as a generally acceptable compromise. Some states may wish to reduce or increase that number slightly. Any number greater than twelve, however, creates a serious risk that the facilities which develop will not maintain the personal character and individual habilitation perspective which this statute seeks to promote.

This provision also recognizes the difficulties which may be encountered in the transition from institutional services to smaller facilities in the community. Yet the desire to mitigate those difficulties cannot supercede the right of developmentally disabled individuals to receive high quality, individualized services.

Finally, some states may wish to go further and may abolish large institutions altogether. This model statute can be readily adapted to the needs of states which choose this alternative. If current trends toward community care continue, it may be that many states will choose this alternative in the relatively near future. In the belief that not all states will do so immediately, the model statute provides a more conservative interim approach.

(6) The Legislature recognizes that at the time of this enactment there does not yet exist within the state a sufficient quantity of high-quality community-based services to meet the needs of the state's developmentally disabled citizens. The Legislature commits itself to create an adequate number and variety of such services.

COMMENT. This provision commits the state to the expenditure of sufficient funds to create the services which developmentally disabled people need. This means that the state is committed to the creation of community-based services for those individuals who can profit from them but who would remain in institutions if these alternatives were not created in sufficient numbers. A legislative "guarantee" of individualized service delivery cannot be implemented without this kind of commitment.

(7) The provisions of this act shall be liberally interpreted and construed to carry out the purposes described in this section.

COMMENT. None necessary.

Alternative. *For those states wishing a briefer statement of legislative purpose, the following is appropriate:*

The Legislature declares that it is the purpose of this act to guarantee the right of developmentally disabled persons to receive those services which

---

[8] *See, e.g.,* R. KING, N. RAYNES & J. TIZARD, PATTERNS OF RESIDENTIAL CARE: SOCIOLOGICAL STUDIES IN INSTITUTIONS FOR HANDICAPPED CHILDREN (1971); Begab, *The Mentally Retarded and Society: Trends and Issues,* in THE MENTALLY RETARDED AND SOCIETY: A SOCIAL SCIENCE PERSPECTIVE 3 (M. Begab & S. Richardson eds. 1975); Tizard, *Quality of Residential Care for Retarded Children,* in VARIETIES OF RESIDENTIAL EXPERIENCE 52 (J. Tizard, I. Sinclair & R. Clarke eds. 1975); Zigler & Balla, *Impact of Institutional Experience on the Behavior and Development of Retarded Persons,* 82 AM. J. MENTAL DEFICIENCY 1 (1977).

[9] *E.g.,* Wuori v. Zitnay, No. 75-80-SD (D. Maine July 14, 1978), *reported in* 2 MENTAL DISABILITY L. RPTR. 693 (1978); Halderman v. Pennhurst State School & Hospital, 446 F. Supp. 1295 (E.D.Pa. 1977), *aff'd in part, rev'd in part,* 612 F.2d 84 (3d. Cir. 1979), Welsch v. Likins, 373 F. Supp. 487 (D. Minn. 1974), *consent decree approved,* No. 4-72 Civ. 451 (D. Minn. Dec. 7, 1977), *aff'd in part, vacated and remanded in part,* 550 F.2d 1122 (8th Cir. 1977); New York State Ass'n for Retarded Children, Inc. v. Rockefeller, 357 F. Supp. 752 (E.D.N.Y. 1973), *consent decree approved sub nom.,* New York State Ass'n for Retarded Children, Inc. v. Carey, 393 F. Supp. 715 (E.D.N.Y. 1975); Wyatt v. Stickney, 344 F. Supp. 387 (M.D. Ala. 1972), *aff'd sub nom.,* Wyatt v. Aderholt, 503 F.2d 1305 (5th Cir. 1974).

[10] *See* B. BAKER, G. SELTZER & M. SELTZER, AS CLOSE AS POSSIBLE: COMMUNITY RESIDENCES FOR RETARDED ADULTS (1977); F. MENOLASCINO, CHALLENGES IN MENTAL RETARDATION: PROGRESSIVE IDEOLOGY AND SERVICES (1977); G. O'CONNOR, HOME IS A GOOD PLACE: A NATIONAL PERSPECTIVE OF COMMUNITY RESIDENTIAL FACILITIES FOR DEVELOPMENTALLY DISABLED PERSONS (1976).

[11] *See* R. SCHEERENBERGER, *supra* note 7, at 168-81.

[12] Balla, *Relationship of Institution Size to Quality of Care: A Review of the Literature,* 81 AM. J. MENTAL DEFICIENCY 117 (1976).

[13] *See* note 9 *supra.*

will promote their independence and self-sufficiency in a manner consistent with the least drastic means principle, to establish consent procedures for such services which enhance the developmentally disabled person's involvement in decisions regarding his or her own services, and to provide procedures for substitute consent when the individual lacks the capacity to make his or her own decisions.

Section 3. Definitions

(1) "Consistent with the least drastic means principle" means that services and the conditions under which they are administered:

(a) are not more harsh, hazardous or intrusive than necessary to achieve the appropriate habilitation objectives of the individual;

(b) involve no more restrictions on the individual's physical freedom, social interaction, or individual choice than are necessary to achieve the appropriate habilitation objectives of the individual;

(c) are administered at a suitable location as close as possible to the individual's place of residence; and

(d) are administered in a manner which respects the individual's humanity, privacy, and dignity.

COMMENT. The "least drastic means principle" has been expressed in other legislation and in some judicial decrees as the right to treatment in "the least restrictive alternative."[14] Whatever formulation is chosen, the principle minimizes the state's intrusion into the individual's freedom of movement and freedom of choice. It also minimizes the disruption of the individual's relationships with his or her family and community, and protects the individual's humanity, privacy and dignity. This principle pervades the model statute, and governs the delivery of services to clients in all settings.

(2) "Court" means [the trial court of general jurisdiction of the state].

COMMENT. The court structures of the states vary widely. It is the intention of the model statute that judicial proceedings should be conducted in courts of general jurisdiction rather than courts, such as probate courts, which specialize in matters unrelated to the needs of developmentally disabled people. Since important questions of individual liberty will often be litigated under this statute, the state is not justified in relegating these cases to such inappropriate specialized courts.

(3) "Department" means [the state developmental disabilities service agency].

COMMENT. Bureaucratic organization will also vary from state to state. This provision seeks to identify the organization in state government which has primary responsibility for providing services to developmentally disabled people. The model statute will have to be adjusted in states which divide this responsibility between more than one agency or which delegate some or all of it to local governmental units.

(4) "Developmental disability" means a chronic disability of a person which:

(a) is attributable to a mental or physical impairment or a combination of mental and physical impairments;

(b) is likely to continue indefinitely;

(c) results in substantial functional limitations in two or more of the following areas of major life activity:

(i) Self-care;

(ii) Receptive and expressive language;

(iii) Learning;

(iv) Mobility;

(v) Self-direction;

(vi) Capacity for independent living; or

(vii) Economic self-sufficiency; and

(d) reflects the need for a combination and sequence of special, interdisciplinary or generic care, treatment or other services which are of lifelong or extended duration and individually planned and coordinated.

COMMENT. This definition, like that in the Developmental Disabilities Assistance and Bill of Rights Act, 42 U.S.C. §6001(7) (Cum. Supp. 1978) (Pub. L. 94-103, as amended by Pub. L. 95-1780), adopts a functional rather than a categorical approach. The effect of the disability, rather than its cause, is of prime importance for this statute. The definition differs from the one contained in the recently amended federal act in that there is no requirement that the disability be "severe" or that it appear before the afflicted person is 22 years old. Both these changes are intended to expand the group of persons who will be entitled to services under this act. The federal definition targets certain populations for intensive federal support, while this act's definition is to provide a right to all needed services and a consent mechanism for receiving them, regardless of whether the individual has a "severe" disability or whether his or her disability is the result of a genetic defect or post-natal trauma or disease.

In considering this act, care should be taken to assure that the definition of developmental disabil-

---

[14] *E.g.*, Welsch v. Likins, 373 F. Supp. 487, 501-02 (D.Minn. 1974), *aff'd in part, vacated and remanded in part*, 550 F.2d 1122 (8th Cir. 1977); *see also* Chambers, *The Principle of the Least Restrictive Alternative: The Constitutional Issues*, in P.C.M.R., *supra* note 3, at 486.

The model statute uses the "least drastic means" formulation rather than the more commonly used "least restrictive alternative." The reason for this preference is that important components of the principle, such as providing services close to the individual's home, are not easy to categorize in terms of greater or lesser "restriction." *See* N.M. STAT. ANN. §43-1-3(D) (Repl. 1979).

ity is consistent with that adopted in the statute establishing the state protection and advocacy system for persons with developmental disabilities.

(5) "Developmental disabilities professional" means an individual who by training or experience is qualified to evaluate or work with individuals with developmental disabilities.

COMMENT. This definition is exceptionally broad and inclusive so the act's provisions allowing access to expert testimony (e.g., section 15(3)) assure flexibility in selecting the witness best able to assist in the preparation and presentation of the case.

(6) "Evaluation" means a study of whether an individual is developmentally disabled and of the developmentally disabled individual's condition, situation and needs which leads to a recommendation of what services, if any, would benefit the individual.

COMMENT. None necessary.

(7) "Generic services" means those services which are designed to serve the general population rather than designed specifically to meet the needs of developmentally disabled people.

COMMENT. None necessary.

(8) "Guardian" means a limited personal guardian or personal guardian appointed by a court pursuant to [citation to relevant section(s) in the state guardianship statute/Model Guardianship and Conservatorship Act].

COMMENT. None necessary.

(9) "Habilitation" means the process by which service providers assist a developmentally disabled person to acquire, develop and maintain those skills and behaviors which enable him or her to cope more effectively with the demands of his or her environment and to raise the level of his or her physical, mental, and social functioning.

COMMENT. None necessary.

(10) "Hearing officer" means an attorney designated by the court to conduct nonjudicial hearings.

COMMENT. The requirement that hearing officers be licensed attorneys reflects the fact that matters involving individual liberty should be decided by persons trained in evaluating evidence and analyzing legal questions. Since the act recommends using nonjudicial hearings to decide some liberty issues, the hearing officer must be an attorney.[15]

(11) "Large residential facility" means a residential facility for the care, custody, and/or habilitation of developmentally disabled people at which more than six people are in residence.

[15] *See* comment on §17(2).

COMMENT. See comment on subsection (14) below.

(12) "Judicial hearing" means a court proceeding as provided in section 16 of this act.

COMMENT. None necessary.

(13) "Minor" means a person under the age of eighteen years.

COMMENT. None necessary.

(14) "Small residential facility" means a residential facility in which six individuals or fewer reside. A facility shall not be considered small if it is part of a campus of other residential facilities where the total number of residents exceeds six.

COMMENT. The definitions in this subsection and in subsection (11) above distinguish large facilities from smaller ones. The dividing line of six residents is somewhat arbitrary, and some states may wish to revise it.[16]

The provision classifying as large facilities those which are aggregations of smaller units is to prevent a large facility from designating its smaller "cottages" or "dormitories" as independent facilities, thus escaping the procedural requirements of this act. It is not intended to apply to an apartment house, for example, which offers generic services to both handicapped and non-handicapped people and which has more than six developmentally disabled individuals residing in individual apartments.[17]

(15) "Nonjudicial hearing" means a proceeding before a hearing officer to determine a client's rights and the appropriateness of proposed services, as provided in section 17 of this act.

COMMENT. None necessary.

(16) "Program plan" means an individualized plan for services prepared pursuant to section 6 of this act.

COMMENT. None necessary.

(17) "Residential services" means a program in which habilitation or other services are administered on the premises where a developmentally disabled individual resides. Such residential services do not include situations in which the only services which the individual receives in his or her place of residence are occasional visits by a social worker or training in household skills.

COMMENT. None necessary.

(18) "Respite care" means short-term assistance to the family or other non-professional

[16] *See* comment on §2(5).

[17] *See generally* B. BAKER, G. SELTZER & M. SELTZER, *supra* note 10, at 163-76 (semi-independent apartments).

caretaker of a developmentally disabled person in the form of temporary substitute care by a developmental disabilities professional or agency.

COMMENT. Respite care, as described in section 15 of this act, should be made available to family members caring for a developmentally disabled relative, or to non-relatives who serve the same caretaking function.[18] This provision is not meant to authorize respite assistance to those who care for developmentally disabled people on a professional basis (such as the operators of a group home).

(19) "Services" means publicly or privately provided assistance to a developmentally disabled person or his family, including, but not limited to, the following areas:

advocacy
behavior modification or management
communication assistance and training
counseling
education and training of adults
family assistance
financial management services
habilitation
homemaking services
medical diagnosis and treatment
occupational therapy
physical therapy
psychological or psychiatric treatment
recreation
residential services
respite care
social work
speech therapy
transportation
vocational assistance and training

COMMENT. It is important that a listing of services should not be seen as exhaustive of those services to which developmentally disabled people have a right. But a list such as the one in this subsection may help courts and others understand the breadth of the spectrum of services to which individuals may be entitled.

Two items which are not within the scope of this act are worth mentioning. First, rights of minors to educational services are established and enforced under separate state and federal legislation (see the Model Education for Exceptional Persons Act). It would only create confusion to attempt to use this act as another guarantee of the educational rights of developmentally disabled children.

The second kind of service which is not within the scope of the act includes certain kinds of medical treatment. This act provides a framework under which developmentally disabled people can obtain ordinary medical services with legally adequate consent. But there are some medical procedures which are so intrusive, hazardous, irreversible, and/or experimental that more elaborate consent procedures are required. Examples include psychosurgery[19] and sterilization.[20] This act should not be used to limit the right of individuals to more elaborate protections which the law requires in the case of such medical procedures.

(20) "Service provider" means a person who provides services to developmentally disabled persons pursuant to this act.

COMMENT. None necessary.

Section 4. Right to Services; Enforcement of Rights

(1) Every developmentally disabled person residing in this state has the right to receive such habilitation and other services as may reasonably be expected to build and enhance the individual's independence, self-sufficiency, and ability to function in society.

COMMENT. This provision establishes those services to which developmentally disabled people are entitled under the act. The entitlement is limited to those services which are likely to help the individual develop his or her skills and abilities. The individual may also be entitled to other support, such as assistance in meeting the cost of food and shelter, under other provisions of law, but this act seeks only to assure those services which will help the person deal more effectively with his or her disability.

(2) The services to which developmentally disabled persons are entitled may be provided by state or local government agencies or by private parties or agencies. But neither the individual's inability to pay for services, the nature or severity of his or her handicap, nor the cost involved in provision of the services shall limit the individual's right to receive appropriate services. When a developmentally disabled person is unable to pay for needed services, the state shall pay for their provision. When appropriate services are not available to meet an individual's needs, the state shall create such services or shall contract with other public or private service providers to meet the person's individual needs.

COMMENT. Developmentally disabled people are entitled to receive services, but this act only obligates the state to pay for them when the individual lacks sufficient resources to do so. In every

---

[18] *See* comment on §15(1).

[19] *See, e.g.,* Aden v. Younger, 57 Cal. App. 3d 662, 129 Cal. Rptr. 535 (1976); Kaimowitz v. Department of Mental Health, Civil Action No. 73-19434-AW (Mich. Cir. Ct. July 10, 1973), *reported in* 2 PRISON L. RPTR. 433 (1973).

[20] *See, e.g.,* North Carolina Ass'n for Retarded Children v. North Carolina, 420 F. Supp. 451 (M.D. N.C. 1976); Relf v. Weinberger, 372 F. Supp. 1196 (D.D.C. 1974); Wyatt v. Aderholt, 368 F. Supp. 1383 (M.D. Ala. 1974). *See generally* Murdock, *Sterilization of the Retarded: A Problem or a Solution?*, 62 CALIF. L. REV. 917 (1974).

state there are services which the state government chooses to provide to all disabled people whether they are indigent or not, and this provision will not prohibit the state from continuing this practice. But it does provide that for those services which the state has not traditionally provided free of charge, the individual's inability to pay cannot prevent him or her from receiving them.

Similarly, the nature or severity of the individual's disability cannot be a barrier to receiving those services which are appropriate to his or her needs. The statute envisions services being provided by state or local government in some instances, and by private service providers in others. But where adequate services do not exist either in the public or the private sector, the state is obligated to create them or see that they are provided by others.

(3) The services to which developmentally disabled persons are entitled and the method of providing such services shall be:

(a) individualized and appropriate to the person's developmental needs;

(b) pursuant to a program plan which is designed to meet the service needs of the individual;

(c) consistent with the least drastic means principle; and

(d) designed to use generic (non-specialized) services wherever possible. When specialized services are required, they shall be provided with as little segregation and stigmatization as possible.

COMMENT. This provision assures that those services which persons receive are individualized to their needs, provided pursuant to a program plan, and no more intrusive than is necessary.[21] These limitations prohibit the provision of services to large numbers of individuals on a "wholesale" basis with little or no regard to the individual needs of the disabled person. This wholesale approach too frequently characterized both institutional and community services in the past.

(4) Any developmentally disabled individual, or any interested person, may petition the department for the provision of services if the petitioner believes that the individual is not receiving needed services to which he or she is entitled under this act, that services are being provided in a manner inconsistent with the act, or that the individual's program plan should be modified. The petition shall specify the services which the petitioner believes the individual should receive, the reasons for believing that such services are needed, and a history of the actions, if any, by service providing agencies which have excluded the individual from the services in question.

COMMENT. The right of the individual to petition for services should engender no controversy. The key feature of this provision, however, is that any interested person can submit a petition to obtain services for a developmentally disabled person. The right is problematic since it creates the possibility of officious meddlers intruding into an individual's life.[22] But failing to allow interested persons to submit such petitions could prevent some developmentally disabled people from receiving needed services simply because they did not understand the availability or desirability of such services, or because they did not know how to bring such a petition. This would particularly disadvantage severely and profoundly handicapped individuals. Another possible approach would be to limit those who could petition to certain categories (such as close relatives). Such a limitation might prove to be an ineffective protection against inappropriate petitions, and would be quite likely to prevent the filing of meritorious petitions by legitimately concerned friends of disabled individuals who did not happen to be related to them. Therefore it would appear desirable for the legislature to allow petitions from any interested person, and for the department to be wary of ill-motivated petitions which, if granted, might unnecessarily intrude into the individual's privacy and which do not identify legitimate service needs.

(5) Upon receipt of such a petition, the department shall conduct an investigation. The department shall involve the developmentally disabled person and consider his or her wishes to the maximum extent possible. If it determines that the individual is not receiving services to which he or she is entitled under this act, it shall provide such services, or assure that such services will be obtained from another public or private agency. If the department determines that no violation of the act has occurred, it shall inform the petitioner of its decision and of the petitioner's right to obtain administrative or judicial review of the decision. In all cases, the department shall report its conclusions to the petitioner within thirty (30) days.

---

[21] *Cf.* Wyatt v. Stickney, 344 F. Supp. 387, 396 (M.D. Ala. 1972), *aff'd sub nom.*, Wyatt v. Aderholt, 503 F.2d 1305 (5th Cir. 1974) (institution resident has right to individual habilitation plan; no mentally retarded person shall be admitted to institution if community services and programs can afford him adequate habilitation; institution resident has right to least restrictive conditions necessary to achieve habilitation, and institution shall make every attempt to move resident from segregation to integration into community life).

[22] For example, the neighbor of a developmentally disabled person might decide that it was offensive that the disabled person did not work, and petition for his or her enrollment in a job training program. While this might or might not prove beneficial to the individual, it is important to remember that non-handicapped people are not subject to such interference—potentially coercive interference—with their personal affairs and decisions.

COMMENT. This provision describes the process by which the department is to consider the petition. Perhaps the most important aspect of it is the requirement that the developmentally disabled individual should be involved in the investigation to the maximum extent possible and that his or her wishes should be given great weight in determining whether the individual needs the services which the petitioner proposed. This involvement of the disabled individual should not be limited to those persons who are thought to be legally "competent" for one purpose or another, but rather should include all individuals who are capable of understanding all or part of what is being proposed and who can express some preference. It is not necessary that an individual have the legal capacity to make decisions before we should listen to his or her ideas and preferences regarding services.[23] The barrier of capacity is designed to protect the incompetent from ill-advised decisions, and the investigation by the department described in this section creates no such risk for the individual of limited skills or experience.

(6) If the petitioner is dissatisfied with the decision of the department, he or she may petition for review of the decision in a judicial or nonjudicial proceeding. If the review is before a court, and the court finds by a preponderance of the evidence that the developmentally disabled individual is entitled to services which he or she is not receiving, it may order the department to obtain or create appropriate services for the individual. If the review is before a hearing officer, and the officer finds by a preponderance of the evidence that the person is entitled to services which he or she is not receiving, the officer may order the department to obtain such services for the individual. If no appropriate services are available, the petitioner, the individual or his or her attorney or advocate may petition the court to order the creation of the services pursuant to section 5(11) of this act.

COMMENT. This subsection creates an appeals mechanism for those who are dissatisfied with the department's response to the petition for services. The petitioner has the option of seeking either judicial or nonjudicial review of the adverse ruling. In addition to deciding whether the department's response to the petition was proper, both the court and hearing officer have the authority to order that the individual receive needed services, either directly from the department or from another service provider to be located by the department. Only the court, however, has the power to order the creation of services not currently available. Therefore, if the hearing has been nonjudicial, the dissatisfied party must petition the court for an order for the creation of services.

(7) Nothing in this section shall be construed to reduce the burden of proof or to eliminate necessary findings of fact provided in other sections of this act for cases in which the developmentally disabled individual does not personally consent to admission to services.

COMMENT. The preceding subsection allows the court or hearing officer to order the client's admission to services when it finds by a preponderance of the evidence that he or she needs them. However, in sections 7, 8, 10, and 11 of this act, it is provided that an individual can only be admitted to residential services without his or her consent when certain findings are made by clear and convincing evidence.[24] This provision assures that the lower standard of proof found in this section cannot be used to dilute the showing required for placement of an individual in a residential facility.

Introduction To Section 5 This section details the consent procedures to be followed for participation in an evaluation of an individual's need for services. Because of the act's emphasis on the individualized nature of services, thorough evaluation of the individual's needs is essential, and therefore participation by the individual in the evaluation process is encouraged. But even the process of evaluation can be intrusive in that it may interfere with the individual's liberty and freedom of choice. Therefore, proper consent must be obtained for the evaluation of an individual. The consent procedures of this section are designed to balance the individual's need for proper evaluation with the interest in making his or her own decisions or in having them made properly by others.

The relationship between the evaluation discussed in this section and the "investigation" pursuant to a petition for services under section 4 should also be noted. The investigation described in section 4 responds to a petition requesting particular services; the evaluation discussed in this section may proceed with no preliminary assumption on anyone's part about what kind of services the individual may need. The two kinds of inquiry are quite similar, however, and, in some individual cases, may be combined. In response to a petition for services pursuant to section 4, the department may determine that a full evaluation of the individual's needs is warranted, or a court or hearing officer reviewing a decision by the department may wish to have an evaluation performed in order to determine the individual's need for services. In such cases, consent to the evaluation process must be obtained pursuant to the provisions of this section.

---

[23] American Association on Mental Deficiency, CONSENT HANDBOOK 18-19 (H.R. Turnbull ed. 1977).

[24] With regard to civil commitment of mentally ill adults, the "clear and convincing" standard has been held to be constitutionally required. Addington v. Texas, 441 U.S. 418 (1979).

The evaluations provided under this section may be performed either by the department or by another service provider. Permitting service providers to do such evaluations upon request allows developmentally disabled people to take advantage of the special expertise that service providers may possess, and also may avoid the delays which may accompany evaluations by the centralized state service providing agency. One alternative to this approach would be for the state to create a separate evaluation agency whose only responsibility would be to evaluate the need for services of developmentally disabled individuals. This alternative would prevent the conflict of interest involved when the same agency is preparing the evaluation of an individual's needs and then also providing the recommended services. However, it risks creation of a new, large service bureaucracy. Another possibility is the approach recommended in the Model Developmentally Disabled Offenders Act. That model act establishes an Evaluation Oversight Board responsible for identifying and publishing a list of persons in all parts of the state who are qualified to perform evaluations. It would also keep the list current and monitor the selection, appointment and compensation process. However, the Board and its small staff would not be authorized to perform individual evaluations. Thus the independence of the state agency model is retained without the cost to the public of employing staffs of expert evaluators. The most appropriate alternative depends to great extent on the size and demographics of the enacting jurisdiction. In some larger states, and in those states where the concern about the potential conflict of interest is strong, the alternative of a separate agency may be desirable.

Section 5. Evaluation of Need for Services

(1) Any developmentally disabled person who believes that he or she needs services may present himself or herself to the department or a service provider for the purpose of obtaining those services. The department or a service provider shall evaluate the individual's condition and situation and shall make a determination whether he or she is in need of services, and if so, what services would best meet his or her needs. The parent, guardian or other custodian of a developmentally disabled minor may present the minor to the department or to a service provider for evaluation.

COMMENT. This provision allows any developmentally disabled individual to request an evaluation of his or her own need for services. When the individual is willing to participate in such an evaluation, there is relatively little cause for concern about possible violations of liberty, and therefore no elaborate checks on the adequacy of consent are warranted.[25] The parents of a minor may consent to his or her evaluation. As is true for adults, the Model Act seeks to facilitate the evaluation process so that services, if they are to be provided, may be individualized to the person's needs.

(2) The court-appointed guardian of a developmentally disabled adult may propose that that individual's need for services be evaluated. If the developmentally disabled individual does not object to participating in the evaluation the department or service provider shall proceed as provided in this section. If the individual declines to consent to the evaluation, the guardian may consent to the evaluation on his or her behalf if the guardian determines that the individual's objections are outweighed by the possible benefits the individual would receive from the proposed services.

COMMENT. Developmentally disabled individuals who have a court-appointed guardian may receive an evaluation upon the consent of their guardian, but the guardian is to take into consideration any objections expressed by the individual. The purpose of this provision, as is true elsewhere in the act, is to allow individuals who lack full capacity to participate nevertheless in important decisions about services which they will receive.[26]

(3) A relative or friend, or any other person interested in the welfare of a developmentally disabled adult, may propose that the individual be evaluated if the proponent believes that the individual needs services which he or she is not currently receiving. If the developmentally disabled person does not object to participating in the evaluation, the department or service provider shall proceed as provided in subsection (1) of this section. If the individual objects to the evaluation, the evaluation may proceed over his or her objection only if a court or hearing officer, after a hearing, determines that the objections are outweighed by the possible benefits the individual would receive from the proposed services and orders his or her participation.

COMMENT. This provision recognizes that there will be some developmentally disabled adults who will not have court-appointed guardians and who will also be unable to request an evaluation themselves. Therefore, interested persons are permitted to request such an evaluation on the individual's behalf. But unlike the power granted to guardians in the previous subsection, here interested persons do not have the authority to override the objections of the individual. If there is such an objection, the individual's wishes can only be overcome

---

[25] For a general discussion of the relationship between the nature of a proposed service and the nature of consent for that service, *see* CONSENT HANDBOOK, *supra* note 23.

[26] *See* text accompanying note 23 *supra*.

through judicial or nonjudicial proceedings pursuant to this act. This requirement of a hearing is warranted because an evaluation conducted over the individual's objection involves a substantial interference with individual liberty and freedom of choice.

An alternative provision which would be more protective of individual liberty, and more restrictive of the right of interested persons to obtain evaluations, would require a hearing whenever the individual "does not consent" to the proposed evaluation. The difference between the possible alternative and the provision suggested above is that the alternative would require a hearing in cases in which the individual neither consents nor objects to the proposed evaluation. The main provision requires a hearing only when the individual actively objects. Since a considerable number of developmentally disabled individuals will be unable to give legally valid consent, the alternative provision would require a substantial number of hearings on privately initiated requests for evaluation.

The choice between the main provision and the alternative will depend upon the legislature's attitude toward intervenors in the individual's life who are not court-appointed guardians. Since these persons will often be relatives of the individual, many states may choose to grant them considerable deference in situations where the individual is not objecting to the evaluation they propose.

(4) When the department or service provider is authorized, pursuant to subsections (1), (2), or (3) of this section, to conduct a comprehensive evaluation of a client, it shall take great care to determine the actual needs, abilities, and potential of the individual. In selecting possible services and in determining the appropriateness of those services for the person's individual needs, the department or service provider shall select those services which will enhance the individual's abilities, independence, and self-sufficiency to the maximum degree possible. Therefore, services which assist the individual's habilitation and development shall be selected instead of services which merely serve a custodial function or which only maintain the person's health and safety. The proposed services must also be consistent with the least drastic means principle.

COMMENT. This provision delineates the factors which the evaluator must consider in determining what services to recommend. Particularly important is the statement that services which enhance the individual's skills should be chosen over those services which merely protect the individual's health and safety. This does not demean the importance of considering the individual's health and safety when choosing services, but rather emphasizes that those considerations are not sufficient when it is possible to do more for the individual by choosing higher quality and more ambitious services. Thus this provision raises a presumption against recommendations of services which are merely custodial in character.[27]

(5) The evaluation shall be conducted in a manner consistent with the least drastic means principle. The evaluation shall not take place in a residential facility, unless (a) the department or service provider determines that portions of the evaluation are impossible to complete except in such a facility and the individual or the person authorized to consent does not object, or (b) the department or service provider determines that the health or safety of the individual would be endangered unless he or she were placed in a residential facility on an emergency basis, pursuant to section 14 of this act. Without proper consent or determination of an emergency situation, the evaluation can take place in a residential facility only upon the order of a court or hearing officer.

COMMENT. This provision delineates the conditions under which the evaluation can be performed. A strong presumption is raised against evaluations taking place in a residential facility.[28] (Evaluation in the individual's private home would not be considered "residential" for these purposes.) Evaluations can take place in a residential facility only if there is proper consent, an order of a court or hearing officer, or if there exists an emergency situation. The reason for this presumption is that many evaluations can take place on a non-residential basis, and residential facilities should not be employed merely for the convenience of the evaluator. Other provisions of this act place important limits on the use of residential facilities for the delivery of services, and the presumption in favor of non-residential services should extend to evaluations as well, since each involves an attempt to eliminate unnecessary restrictions on individual liberty.

(6) If a developmentally disabled person, or another person acting on his or her behalf, wishes to receive services without undergoing a new evaluation, he or she may present to the department or service provider the complete results of another evaluation completed by any developmental disabilities professional or agency within the last three months, or any other appropriate proof of the individual's need for services. The department or service provider shall have the right to reject such offered proof and to insist upon a new evaluation if it believes this to be in the best interest of the developmentally disabled person.

---

[27] *See* Wyatt v. Stickney, 344 F. Supp. 387, 392 (M.D. Ala. 1972), *citing* Wyatt v. Stickney, 325 F. Supp. 781, 784-85 (M.D. Ala. 1971), *both aff'd sub nom.*, Wyatt v. Aderholt, 503 F.2d 1305 (5th Cir. 1974).

[28] *See* discussion of least drastic means principle in comment on §3(1).

COMMENT. This provision makes clear that nothing in the act requires a new evaluation when all parties agree that it is not necessary. But the department or service provider is also given the power to reject the results of a prior evaluation if a new evaluation would be in the individual's best interest. Therefore, no limits are placed on the qualifications of the preparer of the original evaluation, since the department or service provider has the discretion to order a new evaluation if the original one is inadequate.

(7) Once the evaluation is completed, the department or service provider shall prepare an evaluation report and shall refer the individual to an appropriate provider of the recommended services (if other than the evaluating department or service provider). The individual may receive the proposed services only upon adequate consent or upon the order of a court or hearing officer as provided in sections 7 through 12 of this act.

COMMENT. This provision makes clear that a mere recommendation that a particular service or group of services is desirable will not, by itself, authorize the individual's admission. Proper consent procedures must be followed, as provided in other sections of this model statute.

(8) If the individual, or the person who has obtained the individual's evaluation, is dissatisfied with the results of the evaluation, or is unable to obtain the recommended services, he or she may petition the department or seek a judicial or nonjudicial hearing as provided in section 4 of this act.

COMMENT. None necessary.

Section 6. Preparation of Individualized Program Plans

(1) Upon agreeing to provide services to a developmentally disabled person, the service provider shall prepare a program plan for the individual. The plan shall include the following items in detail:

(a) An evaluation of the nature of the specific strengths, problems, and needs of the individual;

(b) a description of the services which will meet the needs of the individual;

(c) a description of the agencies or individuals which are proposed to provide each of the services recommended and the conditions under which the services would be administered, including a description of alternative services which were considered in preparing the plan;

(d) a description of the intermediate and long-range objectives for the individual's development and habilitation, with each objective stated in terms of a singular behavioral outcome;

(e) a statement assigning priorities to the objectives for the individual, and describing proposed strategies, methods of measuring attainment and projected timetables for their attainment;

(f) a statement and rationale of how the proposed services will assist the individual in achieving these intermediate and long-range objectives;

(g) a statement of criteria for transition to less restrictive or more advanced or independent services, and for discharge from services, and a projected date for transition or discharge;

(h) a description of the tests and other evaluative devices used and their results;

(i) proposed criteria for monitoring and evaluating the success of the services in meeting the individual's needs; and

(j) the signatures of the preparers of the plan and the date.

COMMENT. This provision requires service providers to prepare an individual program plan for each person they serve. The elements of that plan, as specified in this subsection, are designed to assure that the plan is tailored to the individual's needs and designed to enhance his or her independence and self-sufficiency.[29] The specifications for the plan are also designed to facilitate subsequent review of the plan's adequacy when a party is dissatisfied with the plan or the services provided under it. Increasing the accountability of service providers is an important function of program plans.

One problem with a single set of specifications for program plans is that the nature of services which developmentally disabled people receive varies widely. Therefore the specificity of a plan for residential services or for a behavior modification program may not be appropriate for a plan for homemaking services. As an alternative, some states may wish to have different descriptions of the program plans for various kinds of services. One difficulty with such an alternative is that categorizing services for purposes of prescribing different kinds of program plans is not a simple task. The provision suggested in this model act may be sufficient in that the amount of detail within each item may vary with the complexity and intrusiveness of the service provided. Thus, the program plan for homemaking services could remain quite simple, while still including basic information on each item listed in this subsection, whereas an adequate program plan for residential services would have to include a significant amount of detail on most of the items.

---

[29] *Cf.* Wyatt v. Stickney, 344 F. Supp. 387, 397 (M.D. Ala. 1972), *aff'd sub nom.*, Wyatt v. Aderholt, 503 F.2d 1305 (5th Cir. 1974) (listing six items to be included in each individualized habilitation plan); *see also* ACCREDITATION COUNCIL FOR SERVICES FOR MENTALLY RETARDED AND OTHER DEVELOPMENTALLY DISABLED PERSONS (AC MRDD), STANDARDS FOR SERVICES FOR DEVELOPMENTALLY DISABLED INDIVIDUALS §1.3 (1980).

It is also true that a program plan will not be required for every "service" received by a developmentally disabled person. If the individual's transportation needs are met by riding the buses of the generic municipal bus company, the company need not, of course, prepare a program plan on each such rider.

The evaluation specified in subsection 6(1)(a) may be the same as the evaluation of an individual obtained pursuant to section 5. But that section's evaluation process is discretionary, while this section's program plan is required for all persons receiving services within the terms of this act.

(2) The program plan shall be an individualized document. Use of standard or repetitious provisions for a large group of clients without regard to their individual capacities or needs is not acceptable.

COMMENT. There is a considerable danger that the requirement of program plans could result in the processing of a great deal of paper with no appreciable improvement in the individualization or planning of services. This result will occur if service providers view the plans as mere paperwork, unrelated to service delivery. The demands of administrative economy and convenience may produce a tendency toward "boilerplate" provisions, with each program plan identical to the one which preceded it. No legitimate interest will be served by this section if the tendency is allowed to materialize.

This subsection's admonition against such standardized plans cannot be drafted in such a way to ensure that it is self-enforcing. Yet, its inclusion in the statute may serve two implementation functions. The first is to put service providers on notice as to what is expected of them as they attempt to comply with the law. The second is to assure consumers, advocates, judges and hearing officers that they need not, and should not, accept such standardized non-individualized plans. This refusal to accept such plans is the most effective means of enforcing the individual's right to individualized services.

(3) The developmentally disabled person shall be involved, to the maximum extent possible, in the preparation of his or her own program plan. The preparers of the plan shall make all feasible efforts to describe and explain the options under consideration in the language, mode of communication, and terms which that individual is most likely to understand, and to discern the individual's wishes. The efforts to involve the individual and the results of those efforts shall be documented in the plan.

COMMENT. This subsection requires attempts to obtain the participation of as many individuals in their own plans as is possible. This includes those individuals who are not capable of making their own decisions concerning consent to services and even those who have been declared incompetent by a court and who have guardians.[30] The legal definition of competence generally sets a threshold that is high enough that there remain "incompetent" people who can nevertheless benefit from explanations of alternatives and who can communicate their preferences. These preferences should be taken into account even when the individual lacks full capacity to ultimately make his or her own decisions.

It will be the rare individual for whom no explanation and no consultation is appropriate. Some profoundly retarded individuals will be unable to understand any communication, however simple. This provision does not require efforts which are certain to prove futile in such a case. But absent such extraordinary circumstances, each individual has a right to the service provider's best efforts to involve him or her in the planning process.

(4) The preparer of the plan shall also make diligent efforts to involve the individual's guardian, advocate, attorney, close relatives, close friends, and previous service providers in the process of evaluation and preparation of the plan. The preparer of the plan shall ask the developmentally disabled individual which persons he or she believes should participate in this process, and, if possible, his or her wishes should be honored.

COMMENT. Individuals other than the proposed recipient of services will also have pertinent information which should guide the preparer of the program plan. These individuals may know the person's abilities and needs better than the service provider does, and this valuable knowledge should be considered in preparing a plan. Any advice which these individuals may wish to offer should also be given full consideration. A particularly useful source for obtaining the names of those who may be helpful is the developmentally disabled person, and this is another way in which he or she can participate in the preparation of the plan.

(5) The service provider shall complete the preparation of the plan as expeditiously as possible. In no case shall the process extend longer than thirty (30) days from the date of the agreement to provide services to the individual unless an explanation of why the delay is necessary is given and unless consent is received from the developmentally disabled individual or other party who consented to his or her receipt of services pursuant to this act.

COMMENT. Especially with complex or intrusive services, the preparation of a fully adequate program plan may not be possible immediately. Evaluation, consultation with the individual and with others (as provided in the preceding subsections), and the consideration of alternatives will

---

[30] *See* text accompanying note 23 *supra*.

take some time. Nevertheless, it is important that the plan be prepared and approved expeditiously so that the individual's access to services is not unduly delayed. Where the individual already is receiving services during the preparation of the plan, it is equally important that the plan be completed quickly so that the services can be tailored to the individual's needs.

The time required to prepare a plan will vary according to the nature of the services and other variables. Any outer time limit will be somewhat arbitrary, but such a limit is necessary to prevent unnecessary delay. Thirty days would appear to be a generous limit, and some states might wish to choose a somewhat shorter time period.

The right to receive an explanation of a proposed delay and to give or withhold consent to an extension rests with the person whose consent is required for the services proposed.

(6) When the service provider has completed its proposed program plan, it shall present and explain the proposed program plan to the developmentally disabled person, making all feasible efforts to do so in the language, mode of communication, and terms that the individual is most likely to understand. The plan shall also be presented, and upon request, explained to the individual's guardian, advocate, and attorney. The plan shall not be disclosed to any other person without the developmentally disabled adult's consent. If he or she is incapable of giving or withholding consent to disclose the contents of the plan to other persons, only a court-appointed guardian shall have the power to make that decision on the individual's behalf. If the developmentally disabled person is a minor, the plan shall be explained to both the minor and his or her parents or guardian, and shall be presented to the minor's attorney or advocate, if any. The minor's parents or guardian shall decide which other persons, if any, shall receive copies of the plan.

COMMENT. The individual's right to access to his or her own program plan is vital,[31] and for some individuals to receive the benefit of this right, special communicative efforts will be required.

A more difficult problem is presented by questions of others' access to the plan. Often other individuals will have a deep and sincere interest in the individual and in the services he or she is to receive. This interest may lead to a desire to see the program plan, and access by these individuals may constitute a uniquely useful form of accountability. The value of having interested individuals participate in the preparation of the plan carries over to their ability to evaluate its appropriateness for the individual's needs.

These benefits of disclosure, however, must be weighed against the interests of the developmentally disabled individual in confidentiality and in controlling the access of others to his or her records. The program plan may include material of some sensitivity: diagnostic and evaluative material, evaluations of personal characteristics and handicaps, etc. Because promiscuous dissemination of this information may serve to stigmatize the individual,[32] he or she should have the right to determine who has access to this material.

Yet, it must be kept in mind that a considerable number of developmentally disabled people lack the capacity to make an informed decision regarding release of this information. The possible benefits of sharing the program plan with those who are legitimately concerned about such an individual are not reduced by the fact that the individual lacks the capacity to consent to their obtaining the information. In fact, individuals lacking this capacity may particularly benefit from the accountability which accompanies limited release since they will often be unable to articulate their own objections and views about a plan which may be inappropriate.

This provision allows the court-appointed guardian of such incapacitated individuals to determine whether others will receive copies of the program plan. This has the advantage of placing the decision in the hands of a person who has been designated as the individual's surrogate decision-maker and who presumably has no loyalties or interests other than those of the individual. The problem with this approach is that there will be a considerable number of individuals who lack the factual capacity to make such a decision but who do not have a court-appointed guardian. For the plan of such individuals to be released, a guardian would have to be appointed. This has the disadvantage of proliferating guardians for persons who might otherwise be able to get by without this imposition on their freedom. Those states which provide for limited guardians who do not have plenary powers may find this feature of the act more acceptable.

An alternative to this provision would be to allow someone else to make the decision about release of the program plan. An advocate or attorney might be an appropriate person to designate for this responsibility, but not every individual will have one of these persons. The service provider might be in a position to make this decision, but there is cause for some concern about potential

---

[31] For a discussion of this issue in the context of mental health treatment, *see* Mental Health Law Project (M.H.L.P.), *Legal Issues in State Mental Health Care: Proposals for Change–Therapeutic Confidentiality,* 2 MENTAL DISABILITY L. RPTR. 337, 351-52 (1977).

[32] *See generally* R. EDGERTON, *supra* note 5.

conflict of interest (e.g., the service provider has a particular interest in releasing information to insurance companies if insurance benefits will provide the compensation for the proposed services).

Also troubling is the question of access to the plan by the parents of developmentally disabled adults. The parents of a developmentally disabled minor are given access to the plan and are allowed to make decisions regarding disclosure to other persons. But the parents of a developmentally disabled adult stand in a different legal relationship to their son or daughter. The law's general presumption is that legal parental control expires when the individual attains majority, and this presumption should be extended to the developmentally disabled just as it is to others in society. But the parents of developmentally disabled adults, particularly those adults who are severely or profoundly handicapped, often assume, as a practical matter, a continuing responsibility for their children after majority. Some do so through appointment as their son or daughter's guardian, but many others do so informally. Many such parents will feel that they are entitled to access to their son or daughter's program plan even if they have not been appointed guardians.

This provision does not recognize such a right of access. A blanket grant of access would permit disclosure of information to parents irrespective of whether they in fact have a close relationship with their child, or of whether their son or daughter wishes them to receive the information. It would also have the unfortunate symbolic effect of treating developmentally disabled people as "perpetual children" by granting life-long power to their parents. Emphasizing the adulthood of severely and profoundly handicapped adults is particularly important to their proper acceptance in the community,[33] and the statute should not undercut that effort.

An alternative provision, which would constitute a compromise with the parental interest, would allow release of records to the parents of such an adult unless the individual, informed of the proposed release, objected to it. This would retain at least a measure of individual control over information, but would allow the parents access if the individual expressed no objection (or was unable to express a choice).

(7) The individual's program plan shall be reevaluated and revised as circumstances warrant, and at least fourteen (14) days prior to the date upon which the plan is to expire. Reevaluation and revision of the plan shall be accomplished in the same manner as occurred for the creation of the original program plan pursuant to this section.

COMMENT. The program plan should be reevaluated and revised as circumstances in the individual's life change — for example, as he or she gains skills, or possibly as his or her residential situation or involvement in other services changes. The plan cannot guide individualized services effectively if it is viewed as a once-and-for-all document. It is particularly important to conduct such a reevaluation and revision in conjunction with the periodic review of the individual's services as provided in other sections of this act, particularly section 13.

Section 7. Delivery of Non-residential Services to Adults

(1) If a developmentally disabled adult is evaluated as needing non-residential services, and if he or she understands what is being proposed, the person shall have the option of accepting some or all of the recommendations in the evaluation report and of voluntarily consenting to some or all of the non-residential services proposed. The individual's agreement shall be recorded by signature or other appropriate means, and he or she shall voluntarily receive those services which he or she has accepted.

COMMENT. This section is premised on the understanding that there are many individuals who, despite their disability, will be able to understand what kinds of non-residential services are being proposed, and who can knowingly and voluntarily consent to receive them.[34]

(2) If the individual declines to accept any of the proposed non-residential services, or is unable to give or withhold valid consent because of inability to understand what is being proposed or to express a preference, he or she shall not receive those services unless his or her guardian who has the power to consent to the services does so or unless an order is obtained pursuant to subsection (3) of this section. A guardian may consent for the individual only after consulting with the individual and determining that the individual's objections, if any, are outweighed by the benefits the individual would be likely to receive from the proposed services.

COMMENT. The court may appoint a guardian with the power to consent to services on behalf of the developmentally disabled individual. But even though the guardian has the authority to give or withhold consent for the individual, he or she is directed to consult with the individual and to evaluate his or her objections, if any, to the proposed services. The guardian is given the power to override these objections, but the consultation requirement is included in order to enhance the participation of even seriously disabled people in

---

33 Bronston, *Concepts and Theory of Normalization*, in THE MENTALLY RETARDED CHILD AND HIS FAMILY: A MULTIDISCIPLINARY HANDBOOK 490, 508-11 (R. Koch & J. Dobson eds., rev. ed. 1976).

34 CONSENT HANDBOOK, *supra* note 23, at 14-15.

their own decisions. Once again, the underlying presumption is that even when an individual lacks the capacity to make his or her own decision, his or her wishes should be taken into account by the surrogate decisionmaker.[35]

(3) When a developmentally disabled adult has not consented to proposed non-residential services, and has no guardian or the guardian has not consented to the proposed services, any interested person may request a nonjudicial hearing pursuant to subsection 16(3) to determine the necessity of the individual's participation in the proposed services as described in the program plan. The petition shall explain why the petitioner believes the wishes of the individual, and of the guardian if any, should be overridden. A copy of the petition shall be served on the individual and his or her attorney. The hearing officer may order the individual to participate in the proposed services only after a hearing and only upon a finding, by clear and convincing evidence, of all of the following elements:

(a) the individual lacks the ability to understand the services which are proposed in the program plan or is unable to express a preference for or against acceptance of the services;

(b) the proposed services are appropriate and well suited to enhance the individual's independence and self-sufficiency;

(c) the proposed services are consistent with the least drastic means principle and are less restrictive of the individual's liberty than any other possible services which would address his or her needs; and

(d) the individual will suffer serious harm unless he or she is ordered to participate in the proposed services.

When a hearing officer considers ordering the mandatory participation of an individual in services under this section, he or she shall have the power to order substantive modification of the proposed program plan if modification is necessary to meet the requirements of this act. The hearing officer shall have the power to order the individual's receipt of services from any program or combination of programs which meet these requirements. The hearing officer shall set a date of expiration for the order, which shall not be later than one year from the date the order is issued. Upon the expiration of the order, the individual may continue to receive the services only if valid consent is obtained pursuant to this section or if a new order is issued pursuant to the provisions of section 13 of this act.

COMMENT. This subsection allows the wishes of the individual, and even of a guardian, to be overridden upon a proper showing that the individual needs the proposed services. This kind of intrusion of state power into the life of an individual is not to be undertaken lightly, and therefore the substantive standard prescribed for overriding the private judgment is quite high. The hearing officer must be convinced by clear and convincing evidence that the individual is incapable of making his or her own decision, that the proposed services are appropriate and are not unduly intrusive into his or her liberty, and that serious harm will result if the individual does not receive the services.

The determination is to be made by a hearing officer after a nonjudicial hearing. The provisions of section 17 regarding such hearings provide considerable procedural protections for the individual, but, as an alternative, some states may wish to require that such an order be issued by a court after a judicial proceeding. This alternative could be preferred by some states because individual liberty is at stake in hearings under this subsection, and because of a belief that the full authority of a court is required to justify such a proposed deprivation of the individual's freedom of choice. The model statute places these hearings before a hearing officer because of a policy choice to reduce the number of issues which must go to court, and in the belief that somewhat less formal proceedings may be appropriate in cases such as these. Many of these cases will involve relatively simple issues, and states may conclude that use of the courts should be reserved for cases which are likely to involve greater deprivations of liberty.

The power of the hearing officer (or in the alternative, the court) to tailor the order to the individual's service needs is considerable. He or she may order substantive modifications of the proposed program plan, and may order the individual to be admitted to services other than those specified in the petition. These powers may appropriately be used, for example, when the officer determines that the individual does need some services to which he or she cannot or will not consent, but when the services proposed in the petition are unduly restrictive of his or her liberty.

The hearing officer is to set an expiration date for the order. This date should be set at the point at which the officer believes the individual's service needs should be reevaluated, but in no case may this period exceed one year. Even though there are some individuals whose service needs can be predicted with reasonable accuracy for a period more than a year in the future, the potential for abuse of indeterminate orders outweighs the inconvenience which may be caused in these relatively rare cases.

As is true with other orders issued under this act, any order to continue an individual in services beyond the expiration of a previous order must meet the same substantive and procedural re-

---

[35] *See* text accompanying note 23 *supra*.

quirements as that original order. Thus the proponent of services to which an individual or guardian does not consent must prove again that the services are needed, appropriate, and that harm will occur if they are not provided. The burden of persuasion does not revert to the individual to show that he or she no longer needs the services he or she has been receiving. The subsequent proceedings retain the presumption in favor of individual liberty and freedom of choice, and that presumption can only be overcome upon a new showing.

Section 8. Admission of Adults to Small Residential Facilities

Consent for admission of adults to small residential facilities shall be sought and obtained as provided in section 7, except that a guardian shall not have the power to override the objections of the individual to receiving services in a small residential facility. If the individual objects or lacks the capacity to consent or object to the proposed services, he or she shall not be admitted unless his or her guardian who has the power to consent to the proposed services or any other interested person files a petition with the court requesting that the individual be ordered to receive the proposed residential services. The court shall hold a judicial hearing on the issue of the individual's need for the proposed residential services, and shall order them only upon finding all of the elements described in section 7(3) by clear and convincing evidence. In addition, all of the requirements of section 7(3) shall apply to hearings conducted under this section. The court shall set a date of expiration for its order, which shall not be later than one year from the date the order is issued.

COMMENT. An individual can consent to residential services in a small facility in the same manner that he or she can consent to nonresidential services under section 7. Thus no procedural barrier is erected between an individual who is capable of making such a decision and the small residential facility (such as a group home) which he or she desires to enter.

While court-appointed guardians may be given the power to consent to non-residential services, they are not given the authority under this act to consent to residential services. This is because of the generally greater loss of liberty which may be involved in a residential placement to which the individual has not consented. This greater intrusion into the individual's life warrants greater procedural protection than is necessary for non-residential services. Thus, this provision requires that when a guardian or another interested person petitions for an individual's receipt of services in a small residential facility, the placement can take place only upon a court order following a judicial hearing conducted according to the provisions of section 16. As was true with the nonjudicial hearing for non-residential services under section 7, the decision requires a finding by clear and convincing evidence that the order to receive services in a small residential facility is appropriate, that it is not unnecessarily intrusive or restrictive, and that without the order serious harm would result.

These proceedings are to take place before a judge rather than before a hearing officer because of the generally greater deprivation of liberty involved when an individual is ordered to receive services in a residential facility.

One possible objection to this provision is that it requires a judicial hearing for placement of individuals who do not have the capacity to consent for themselves, but who do not protest the proposed placement. It may be argued that such hearings are a waste of judicial and other resources. It is true that not every such hearing will contain a real controversy to be litigated. But hearings provide a valuable check against ill-advised placements of individuals who do not need confinement or who can be served in less restrictive settings when the individuals are unable to evaluate the proposed placements or to articulate objections. This accountability function is enhanced by the opportunity of the court to review the proposed program plan and by the presence of the individual's counsel, who has the opportunity to explore alternatives which might be more appropriate for the individual or which might involve less restriction of his or her liberty.

Introduction to Section 9. Large residential facilities present particular problems of potential abuse to developmentally disabled people and particular risks that individuals will not gain skills necessary for independence and self-reliance.[36] This section treats such large facilities in a manner substantially different from smaller facilities and provides increased procedural protections against their unnecessary use. Thus large facilities become the placement of last resort, appropriate only when no other alternative will meet the individual's habilitation needs.

An alternative to this approach would be to abolish such facilities altogether. For states which choose this alternative, this section and section 12 (involving the institutionalization of minors) may be omitted and replaced with statutory language enforcing the facilities' abolition.

Another alternative would be to allow such facilities to exist, but to prohibit any but voluntary admission to them. States choosing this alternative would enact only subsection (1) of this section.

---

[36] *See* comment on §2(5).

Section 9. Admission of Adults to Large Residential Facilities

(1) Admission of an adult to a large residential facility can be accomplished only by order of the court. If the individual wishes to accept proposed services in a large residential facility, the court shall order him or her admitted as a voluntary resident only after a judicial hearing and upon finding by clear and convincing evidence (a) that the individual was capable of making the decision in question, (b) that he or she fully understood the nature of the proposed placement and the alternatives which were available to him or her, and (c) that the individual's decision was not the product of coercion or undue influence by any person. The court's order that the individual may be admitted as a voluntary resident shall expire on a date set by the court, which shall not be later than one year from the date the order is issued.

COMMENT. This provision allows developmentally disabled individuals to become voluntary residents of large residential facilities, but requires judicial ratification of the individual's voluntariness. The judicial check on the voluntariness of the admission decision reflects both a reluctance to use large facilities unless absolutely necessary, and a concern that individuals may be coerced into entering those facilities. Allowing unchecked "voluntary" admissions would permit the procedural and substantive protections of subsection (2) of this section to be evaded whenever the individual could be convinced to sign a voluntary admission document.[37] Without some form of check, there would be no assurance that the individual understood the alternatives which this act provided in the form of other, less restrictive, forms of services.

An alternative to this provision would allow voluntary admission only upon the judicial finding described in this subsection *plus* a finding of the four elements described in subsection (2) of this section. This would limit voluntary admissions to those individuals who truly wish to enter the facility *and* whose admission is required by particular service needs. This alternative would greatly reduce the number of people eligible for voluntary admission, and thus the total number of people who could reside in large facilities. The cost of this limitation is that it would also restrict the right of individuals who are capable of making their own choices from exercising that right.[38] It may well be that few developmentally disabled individuals who possess this capability will choose to live in large residential facilities,[39] but the main provision preserves that choice for them, while the alternative denies it to them.

(2) If the individual does not consent to receiving services in a large residential facility, or if the court determines pursuant to subsection (1) of this section that the individual is not capable of giving fully informed and voluntary consent, the individual's guardian or any interested person may petition the court to order the involuntary receipt of services in the facility. The individual may be admitted to the large residential facility only if the court, after a judicial hearing, finds by clear and convincing evidence that:

(a) the individual requires services which can *only* be provided in a large residential facility;

(b) the proposed placement is the best possible choice of services for enhancing the individual's independence and self-sufficiency, and is not likely to produce regression, loss of skills, or other harm to the individual;

(c) the proposed placement is consistent with the least drastic means principle and is no more restrictive of the individual's liberty than any other possible choice of services which would address his or her needs; and

(d) unless the individual is involuntarily admitted to the large residential facility he or she will suffer serious bodily harm.

All the requirements of section 7(3) shall apply to hearings conducted under this section. The order of the court must expire within a period not to exceed twelve months.

COMMENT. For those individuals who have not voluntarily chosen to enter large residential facilities, and for those whose voluntary admission is denied by the court under subsection (1) of this section, admission can take place only upon a court order after a judicial hearing on the necessity of the placement. As a general matter, this represents the greatest restriction of individual liberty of any provision in this act, and thus warrants the greatest procedural and substantive protections.

The drastic nature of the standard is warranted both by the loss of liberty in involuntary residential placement and by the likelihood that large facilities will be harmful for a considerable number of individuals.[40] It is anticipated that courts will only rarely find individuals for whom all of these

---

[37] For a discussion of this problem in the context of mental health treatment, *see* Gilboy & Schmidt, *"Voluntary" Hospitalization of the Mentally Ill*, 66 Nw. U. L. REV. 429 (1971); Owens, *When Is a Voluntary Commitment Really Voluntary?*, 47 AM. J. ORTHOPSYCHIATRY 104 (1977).

[38] *See* Aden v. Younger, 57 Cal. App. 3d 662, 684, 129 Cal. Rptr. 535, 549 (1976).

[39] Kindred, *Guardianship and Limitations Upon Capacity*, in P.C.M.R., *supra* note 3, at 63, 68. Any discomfort involved in granting this choice to developmentally disabled people should be reduced because of the model act's guarantee that a range of services will be available.

[40] *See* comment on §2(5). *See also* People v. Reliford, 65 Ill. App. 3d 177, 382 N.E.2d 72 (1978).

criteria can be established by clear and convincing evidence.

One possible objection to this subsection is the length of the maximum time limit for such an order. As an alternative, some states may wish to shorten that period from one year to six or nine months. The longer time period was chosen here because it is anticipated that some of the individuals who will fit the restrictive criteria of this subsection will be so severely handicapped, and so clearly in need of the services which a particular large facility may uniquely offer, that more frequent review would be inappropriate. Even if the state chooses the one year time limit as provided in the text, it is anticipated that courts will set a shorter time period for individuals whenever there is any likelihood that more frequent review might lead to movement to a less restrictive placement.

Section 10. Delivery of Non-residential Services to Minors

(1) If the evaluation of a developmentally disabled minor recommends non-residential services, the minor's parent or guardian shall have the option of accepting some or all of the recommendations for non-residential services and of consenting for their child to receive services. The minor may receive those non-residential services to which the parent or guardian has consented, except as provided in subsection (2) of this section.

COMMENT. This subsection grants to the parents or guardians of developmentally disabled minors the authority to consent to their participation in non-residential services. Absent any challenge as provided in subsection (2) of this section, the parental decision to consent or withhold consent shall be legally sufficient.

An alternative to this position would allow older minors (e.g., over the age of 14) the authority to consent to some services if they are capable of understanding what is proposed and of expressing a preference. Another alternative would create a system of concurrent consent of both parents and such older minors. These provisions may be appropriate for consent to treatment of mentally ill minors,[41] but fewer developmentally disabled minors will have the capacity to consent to such services, and therefore the provision in the text leaves the power to consent in the hands of the parents of all minors.

(2) Any interested person may request the department to petition the court to overturn the decision of a parent or guardian to accept or reject non-residential services for a minor of any age. If the department concludes that such a petition is warranted, the petition shall detail the reasons for the department's belief that the parent's or guardian's decision will be harmful to the child. The court may overturn the decision of the parent or guardian only after a judicial hearing on the petition, at which the parent or guardian is a party and is represented by counsel, and upon finding by clear and convincing evidence that the parental decision will produce serious harm to the child and that the facts surrounding the decision would be sufficient to sustain a petition alleging neglect under the [state's juvenile code]. Any order which overturns the decision of a minor's parent or guardian shall have an expiration date to be set by the court which shall not be later than one year from the date of the court's order.

COMMENT. This provision allows interested persons to petition the court to overturn parental decisions to consent (or withhold consent) to services for their minor child. Such petitions cannot be granted unless the petitioner has satisfied the substantive standard of the neglect jurisdiction of the juvenile court.[42] An alternative which some states may prefer would be to omit this section altogether, and to allow parental decisions to be overturned only when petitioner established the parents as neglectful under the state's juvenile laws. The substantive effect of this alternative would be the same, and a neglect order would presumably be available whenever the parental decision endangered the life or health of the child. The provision presented in the text may be preferable to the alternative because it allows such disputes to be resolved by the court which customarily handles developmental disability cases, and affords all parties the procedural protections of section 16, which were designed for cases involving developmental disabilities services.

Section 11. Admission of Minors to Small Residential Facilities

(1) If the evaluation of a developmentally disabled minor recommends placement in a small residential facility, and if the minor's parent or guardian agrees with the recommendation, he or she may request a nonjudicial hearing to seek approval of the delivery of services in that setting. The hearing officer shall order the placement of a minor in a small residential facility only after a nonjudicial hearing and upon finding by clear and convincing evidence that:

(a) the proposed placement constitutes an appropriate choice of services for enhancing the minor's independence and self-sufficiency;

[41] *See* Mental Health Law Project (M.H.L.P.), *Legal Issues in State Mental Health Care: Proposals for Change–Mental Health Treatment for Minors*, 2 MENTAL DISABILITY L. RPTR. 459 (1978); Ellis & Carter, *Treating Children Under the New Mexico Mental Health and Developmental Disabilities Code*, 10 N.M. L. REV. 279 (1980).

[42] *See generally* Alsager v. District Court of Polk County, 406 F. Supp. 10 (S.D. Iowa 1975), *aff'd* 545 F.2d 1137 (8th Cir. 1976).

(b) the proposed placement is consistent with the least drastic means principle, and that no non-residential services can adequately address the minor's needs; and

(c) the proposed placement is not likely to produce regression, loss of skills, or other harm to the minor.

An order placing a minor in a small residential facility with parental consent shall expire on a date set by the hearing officer, which shall not be later than one year from the date of the order.

COMMENT. While parents are given the authority to consent to their minor child's admission to non-residential services,[43] their authority regarding residential services is more limited. This provision will certainly be controversial, because many parents will believe that they should be allowed to make all such decisions for their minor child. Although a number of courts had held that it was unconstitutional for states to allow parents the power to place their children in residential facilities (or at least institutional residential facilities),[44] in 1979 the United States Supreme Court ruled that the states have the option of either allowing parents this authority or of requiring a hearing mechanism for residential placement.[45] The model statute chooses to provide such hearings as a policy matter because of the potential harm from unnecessary residential placement of children.

Unlike the situation which arises with most non-residential services, a proposed residential placement may place some parents in a conflict of interest.[46] This is not to suggest that such parents are unconcerned about the welfare or best interest of their developmentally disabled child. It merely observes that such parents also have other interests in the peaceful operation of their home and in the best interest of their other children (the disabled child's brothers and sisters). These legitimate and understandable concerns will lead some parents to conclude that their disabled child can no longer live in the family home. Whether this judgment is correct or not in terms of the family's needs, the possibility of exclusion from the child's family home and the possibility of an ill-advised residential placement calls for an objective determination of the disabled child's individual needs and interests. The nonjudicial hearing mechanism provided in this subsection is designed to ascertain the child's needs and tailor services to those needs.

The substantive standard for an order placing a child in a small residential facility is not as high as that for placing a child in a large facility.[47] It only assures that the proposed placement meets the child's needs and is not unnecessarily restrictive or damaging to the child. As in all cases under this act, the hearing officer has the power to modify the proposed plan or to order an alternative approach to delivering services or an alternative facility if the evidence suggests that the proposed facility is ill-suited to meet the individual child's needs. Therefore the officer is not faced with a choice between placing a child in an unsatisfactory placement or sending him or her back to a family home which is unable or unwilling to take the child back. Thus the hearing mechanism can provide an evaluation of the appropriateness of the child's placement even when return to the family home is not a possibility.[48]

(2) Any interested person may request the department to petition the court to overturn the decision of a parent or guardian to refuse to consent to residential services which are recommended for a minor in his or her program plan. If the department concludes that such a petition is warranted, the petition shall detail the reasons for the department's belief that the parent's or guardian's refusal will be damaging to the child. The court may overturn the decision of the parent or guardian to reject the delivery of services in a small residential facility only after a judicial hearing on the petition, at which the parent or guardian is a party and is represented by counsel, and upon finding by clear and convincing evidence that the parental decision will produce serious harm to the child and that the facts surrounding the decision would be sufficient to sustain a petition alleging neglect under the [state's juvenile code] and that each of the elements of subsection 11(1) is satisfied. An order placing a child in residential services over parental objection shall expire on a date set by the court, which shall not be later than six months from the date of the court's order.

COMMENT. This provision parallels the petition to overcome a parental decision regarding non-

---

[43] *See* §10.

[44] *E.g.*, Institutionalized Juveniles v. Secretary of Public Welfare, 459 F. Supp. 30 (E.D. Pa. 1978), *rev'd and remanded*, 442 U.S. 640 (1979); J.L. & J.R. v. Parham, 412 F. Supp. 112 (M.D. Ga. 1976), *rev'd and remanded*, 442 U.S. 584 (1979); Bartley v. Kremens, 402 F. Supp. 1039 (E.D. Pa. 1975), *rev'd on other grounds*, 431 U.S. 119 (1977); Kidd v. Schmidt, 399 F. Supp. 301 (E.D. Wis. 1975); Saville v. Treadway, 404 F. Supp. 430 (M.D. Tenn. 1974); *In re* Roger S., 19 Cal. 3d 921, 569 P.2d 1286, 141 Cal. Rptr. 298 (1977); *In re* Long, 25 N.C. App. 702, 214 S.E.2d 626, *cert. denied*, 288 N.C. 241, 217 S.E.2d 665 (1975).

[45] Secretary of Public Welfare v. Institutionalized Juveniles, 442 U.S. 640 (1979) (mentally retarded and mentally ill children); Parham v. J.R., 442 U.S. 584 (1979) (mentally ill children).

[46] For a more expanded discussion of the issue of hearings for children whose residential placement is proposed, *see* Teitelbaum & Ellis, *The Liberty Interest of Children: Due Process Rights and Their Application*, 12 FAM. L. Q. 153 (1978), *reprinted in* 2 MENTAL DISABILITY L. RPTR. 582 (1978). *See generally* Bezanson, *Toward Revision of Iowa's Juvenile Commitment Laws: Thoughts on the Limits of Effective Government Intervention*, 63 IOWA L. REV. 561 (1978); Ellis, *Volunteering Children: Parental Commitment of Minors to Mental Institutions*, 62 CALIF. L. REV. 840 (1974).

[47] *See* comment on §12(2).

[48] *See* Ellis, *supra* note 46, at 890-94.

residential services.[49] The only substantive difference between these provisions is that an order for the delivery of services in a small residential facility over parental objections cannot exceed six months in duration. Since these parents presumably wish to keep the child at home, a more frequent review of any order which takes the child out of their home is warranted.

Section 12. Admission of Minors to Large Residential Facilities

(1) A minor may be placed in a large residential facility only by court order.

COMMENT. The drastic nature of placement in a large facility and its possibly damaging effects[50] on the child warrant the most rigorous procedural safeguards, and therefore a court hearing is required. The reasons for rejecting the Supreme Court's conclusions regarding such hearings are discussed in the commentary to subsection 11(1).

(2) The court shall order the delivery of services to a minor in a large residential facility only if, after a judicial hearing, it finds by clear and convincing evidence that:

(a) the proposed placement constitutes the best possible choice of services for enhancing the minor's independence and self-sufficiency;

(b) the proposed placement is consistent with the least drastic means principle, and that no non-residential services can adequately address the minor's needs;

(c) the proposed placement is not likely to produce regression, loss of skills, or other harm to the minor;

(d) unless the minor is admitted to the proposed facility he or she will suffer serious bodily harm; and

(e) that no small residential facility can adequately address the minor's needs.

COMMENT. The substantive standard for evaluating the decision to deliver services in a large residential facility is as high as any other in the proposed statute. The court must find by clear and convincing evidence that the proposed placement is the best possible placement for the minor, that it is not unnecessarily restrictive or damaging, and that no non-residential services or placement in a small facility can meet the minor's needs. This standard places severe limitations on the use of large facilities for minors.

As an alternative, some states may wish to abolish the use of large facilities for minors altogether. This option could be chosen whether or not the state also chooses to abolish such placements for adults.[51] The text suggested here rejects this alternative in the belief that there are some children for whom services can best be provided in settings which house more than six individuals, and in the belief that the substantive and procedural protections provided are sufficient to prevent unnecessary placement.

(3) An order placing a minor in a large residential facility shall expire on a date set by the court, which shall not be later than one year from the date of the court's order.

COMMENT. As was the case of orders placing adults in large facilities, this section sets the outer time limit at one year for orders involving children. An alternative to this provision would shorten that limit to six or nine months. The alternative could be justified on the basis of the greater likelihood that children may be harmed by unnecessarily long periods of institutionalization.

Section 13. Periodic Review

(1) Every developmentally disabled person has the right to periodic review of the services he or she receives.

COMMENT. Periodic review is one of the most important features of the proposed model statute.[52] Regular reevaluation of the individual's need for services and the appropriateness of the services being provided to meet his or her needs promotes both accountability and individualization. This is particularly important in cases of out-of-state placement. (See section 16(10).) The intensity of review which is appropriate will obviously vary from situation to situation, but in every case in which an individual is receiving services under this statute, some form of review should prove beneficial.

(2) Every individual's program plan shall expire on a date set by the service provider, or by the court or hearing officer ordering the delivery of services. The expiration date shall be the date at which a reevaluation of the individual's needs is appropriate, but shall not be more than twelve months after the completion of the program plan, or other periods established by this act.

COMMENT. Preceding sections[53] have provided that every court order or order of a hearing officer authorizing delivery of services to an individual shall have an expiration date. These sections set an outer time limit for each type of service situation at either one year or six months. This subsection provides that even when services are provided by consent mechanisms other than judicial or nonjudicial order, an expiration date must be provided

[49] *See* comment on §10(2).

[50] *See* comment on §11(5).

[51] *See* introduction to §9.

[52] *See* P. FRIEDMAN, THE RIGHTS OF MENTALLY RETARDED PERSONS 39 (1976); Fasulo v. Arafeh, 173 Conn. 473, 378 A.2d 553 (1977); *Developments in the Law–Civil Commitment of the Mentally Ill*, 87 HARV. L. REV. 1190, 1376-98 (1974) [hereinafter cited as *Developments in the Law–Civil Commitment*].

[53] *See* §§7-12.

in every individual program plan. In cases in which no court or hearing officer is involved (and this will undoubtedly encompass a majority of cases), the expiration date will be set by the service provider at the time the program plan is prepared. Since in these cases the service provider is setting the frequency of its own reevaluation, there may be a tendency to invariably apply the statutory maximum period of one year. But conscientious service providers will identify cases in which a shorter review period is appropriate. In addition, each of these cases will require that either the individual or parent or guardian consent to the proposed services and to the proposed program plan. Therefore, if the individual, parent, or guardian believes that a shorter evaluation period is appropriate, he or she can condition consent upon a mutually agreeable expiration date for the plan.

(3) Prior to the expiration date of the program plan, the service provider shall evaluate the individual's performance under the plan, progress or lack of progress in attaining the goals set forth in the plan, and shall recommend that either:

(a) the program be modified to meet the individual's current and anticipated needs during the next program plan period; or

(b) the program be renewed without modification; or

(c) the individual be discharged from services because he or she no longer needs them.

The service provider shall then consult with the individual and any other persons interested in the individual's program, including the parents of a minor, and propose a new program plan if appropriate. Consent to the new or renewed program plan and admission to the services proposed shall be accomplished as provided in sections 7 through 12 of this act.

COMMENT. This provision requires that the service provider begin the process of reevaluating the individual's service needs prior to the expiration of the program plan. The service provider is directed to consult with the individual and with others interested in the individual in the course of this process.

(4) If at any time during the period that an individual is receiving services under a program plan, the individual, the parent of a minor, the individual's attorney or advocate, or any other interested person believes that the individual's needs have changed or that he or she is not receiving adequate services, that person may petition for a modification or dissolution of the program plan. The petition shall state the reasons for the proposed change, and the petitioner shall have the option of requesting a judicial or nonjudicial proceeding. After the hearing, the court or hearing officer shall order modification, dissolution or continuation of the plan as required by the needs of the individual and the provisions of this act. This subsection shall not be interpreted as limiting an individual's right to petition for a writ of habeas corpus or to seek any other legal or equitable remedy otherwise available.

COMMENT. This subsection allows the individual receiving services or others interested in the individual to initiate the review process prior to the expiration of the program plan. This reevaluation is accomplished by the petitioner's proposing either a modification in the program plan or dissolution of the plan and discharge of the individual from services. The petitioner has the option of requesting either a judicial or nonjudicial hearing to consider the petition. This option is made available because many such reevaluations will require only a nonjudicial proceeding, but some (including some cases involving residential services, for example) will involve such important issues of liberty that the petitioner may prefer to have them decided by a court.

Section 14. Emergency Services

(1) Services may be provided to a developmentally disabled person on an emergency basis without complying with the consent provisions of this act when a service provider determines that there is an imminent danger that the physical health or safety of the individual will be seriously impaired if the services are not provided, and that the consent requirements of this act cannot be accomplished in time to avoid the danger.

COMMENT. There will be occasions on which an individual needs services before the consent requirements of this act can be met. These are situations in which the physical health or safety of the individual will be impaired if the services are not provided immediately. Examples might include residential services for an individual who is unable to obtain shelter and for whom no less restrictive short-term alternative can be arranged, or some form of restraint or intervention for an individual who is impairing his or her own physical health and safety by serious self-mutilating behavior. In such situations, some mechanism must be provided for the individual to obtain services on an emergency basis.

The requirement that there be "an imminent danger that the physical health or safety of the individual will be seriously impaired" is a relatively high substantive standard which must be present before these emergency provisions are invoked.[54] Service providers and others might prefer that other circumstances also be treated as emergencies, but limitations are placed on the use of this

---

[54] *Cf.* Mental Health Law Project (M.H.L.P.), *Legal Issues in State Mental Health Care: Proposals for Change–Civil Commitment*, 2 MENTAL DISABILITY L. RPTR. 73, 132 (1977) [hereinafter cited as M.H.L.P., *Legal Issues–Civil Commitment*] (definitions of "likelihood of serious harm to oneself" and "likelihood of serious harm to others").

section in order to minimize the number of circumstances in which proper consent is not obtained.

(2) When services are provided on an emergency basis pursuant to this section, the service provider shall document the nature of the emergency and the reason for failing to comply with the consent provisions of this act, and copies of this document shall be placed in the individual's records and shall also be sent to the individual, his or her attorney, advocate, guardian, and if the individual is a minor, his or her parents.

COMMENT. This provision establishes the requirement that the service provider document the nature of the emergency and inform interested parties of the fact that emergency services were provided and the reason they were provided.

(3) When an individual is receiving services on an emergency basis, the service provider shall make diligent efforts to comply with the consent provisions of this act as soon as possible. In no case shall an individual receive services on an emergency basis for a period longer than seven (7) days, and no person shall receive residential services on an emergency basis for longer than three (3) days.

COMMENT. This subsection emphasizes the temporary nature of the authority to provide services on an emergency basis. Often the consent provisions can be met within a few hours or days, and in such cases consent should be obtained as expeditiously as possible. An outer time limit of seven days is established for non-residential services, but since residential services usually involve a more serious deprivation of liberty, the shorter period of three days is established.

If the individual is unconscious or otherwise incapable of giving or withholding direct consent for longer than the time limits established by this subsection, substitute consent should be sought pursuant to sections 7 through 12 of this act.

Section 15. Respite Care

(1) The parent of a developmentally disabled minor, or the caretaker of a developmentally disabled person of any age, may seek and obtain respite care for that individual without complying with the evaluation and consent provisions of other sections of this act.

COMMENT. This section establishes respite care[55] as a general exception to the consent and evaluation requirements of the other sections of the model statute. This exception is justified by the temporary nature of respite care and by the benefits it can confer on developmentally disabled people and on their families and others who care for them.

Families who keep a developmentally disabled individual in the family home often face serious difficulties since the care which the individual requires may involve major physical, emotional, and other kinds of strains for the family.[56] Allowing relatively easy access to respite care will help alleviate some of these burdens by allowing the family members to have short periods of time during which their responsibility is relieved. This may allow them to take a vacation, or have a weekend to themselves, or to make other kinds of uses of short periods of time free from the responsibility of caring for the developmentally disabled family member.

This freedom will also benefit developmentally disabled people. By allowing their families to obtain relief through respite services, the state reduces the stress on family members and therefore makes it less likely that they will seek full-time residential services for the individual. Thus respite services may increase the number of families which are willing and able to keep the developmentally disabled family member at home. Even if the setting in which respite care is provided is more restrictive than the family home, allowing limited access to respite care will produce an overall reduction in the number of individuals who must enter full-time residential care, and thus will result in a net reduction in the deprivation of individuals' liberty.

The term caretaker does not easily admit of definition. The intention is to include anyone who cares for a developmentally disabled individual in the same fashion that a parent might care for a developmentally disabled son or daughter. This may include relatives, friends who are not related to the individual, or foster families. It should not include those who care for developmentally disabled people on a professional basis, such as the operators of a group home.

(2) The individual seeking respite care shall apply to the department for referral to an appropriate provider of respite services. The department, giving due consideration to the wishes of the developmentally disabled person, shall designate a governmental or private service provider whose services are appropriate to the needs of the person. Large residential facilities may not be used for respite care unless no other facilities are available.

COMMENT. The process of selecting a provider of respite services is left relatively informal and

[55] *See* Wolfensberger, *Will There Always Be an Institution? The Impact of New Service Models: Residential Alternatives to Institutions*, 9 MENTAL RETARDATION 31 (Dec. 1971).

[56] *See generally* PARENTS SPEAK OUT: VIEWS FROM THE OTHER SIDE OF THE TWO-WAY MIRROR (A. Turnbull & H.R. Turnbull eds. 1978).

unencumbered by procedures. The department is given considerable discretion in the selection process. The one prohibition is that large residential facilities may not be chosen when other services are available and would meet the individual's needs. This provision reflects the judgment that smaller settings are less likely to be disruptive to the individual who is going to be a resident for a short period of time.

The other important provision of this subsection is the requirement that the individual's wishes be ascertained and honored as much as possible. This does not give the developmentally disabled person a veto power over the decision to obtain respite services or over the particular selection of a service provider. But the individual's wishes, particularly his or her preference about different service providers, should be honored whenever that is practical.

(3) Before the individual is placed in respite care, the parent or caretaker and the provider of respite services shall reach written agreement about the duration and conditions of respite care, and both parties shall be bound by that agreement. A copy of the agreement shall be sent to the department.

COMMENT. The requirement of a contract between the caretaker and the service provider is designed to assure that respite care does not extend into a longer or permanent period of services, particularly residential services, for which adequate consent has not been obtained.

(4) No individual shall be placed in respite care for longer than thirty (30) days at a time, or for longer than a total of sixty (60) days during any 12 month period. Services may be provided for longer periods of time than allowed by this section only if the requirements of the rest of this act are met.

COMMENT. The time limits on respite care are, like the contract in the previous subsection, designed to prevent respite care from becoming an indefinite provision of services for which adequate consent has not been obtained. There may be legitimate reasons which may lead a caretaker to request more than thirty days respite care at one time, or more than sixty total days during one year, but in such cases the consent provisions of sections 7 through 12 of this act must be followed.

Section 16. Judicial Proceedings

(1) In every court proceeding under this act, the provisions of this section shall apply.

COMMENT. None necessary.

(2) Every developmentally disabled person whose need for services or right to services is at issue in a judicial proceeding shall be represented by counsel and both the individual and counsel shall receive timely and clear notice of the nature of the issue and proceedings. The developmentally disabled person shall have the right to be present, to present evidence on his or her own behalf, to cross-examine witnesses, and to testify or decline to testify on his or her own behalf.

COMMENT. This provision establishes the basic procedural rights of individuals involved in judicial proceedings under the model statute. Perhaps the most important among them is the right to counsel.[57] The language "shall be represented by counsel" indicates that counsel is not waivable in judicial proceedings, although waiver is possible in nonjudicial hearings.[58] Since judicial proceedings will generally deal with more weighty matters, many of which involve substantial questions of individual liberty, it is important that the individual be represented by an attorney.[59]

In addition, the individual has the right to be present at the hearing but that right can be waived as provided in subsection (9) of this section.

The right to decline to testify at the hearing is a matter of some controversy. There is disagreement among appellate courts on the question of whether the constitutional privilege against self-incrimination applies in cases involving mental health treatment or developmental disabilities services. Recently, the California Supreme Court ruled that the privilege does not apply in commitment proceedings involving mentally retarded persons,[60] but there is a substantial possibility of abuse if developmentally disabled people are forced to testify against themselves, and the facts of the California case itself suggest reasons for extending the privilege.[61] Therefore, whether or not the privilege against self-incrimination is eventually required in such proceedings as a constitutional right, it is advisable to extend it as a statutory right in all proceedings under this act.

(3) The right to present evidence shall include the right to the assistance of a developmental disabilities professional or professionals of the developmentally disabled person's choosing. If he or she is unable to pay for the services of such a professional, state funds will be provided for this purpose by order of the court.

---

[57] *See* Herr, *The New Clients: Legal Services for Mentally Retarded Persons*, 31 STAN. L. REV. 553 (1979).

[58] *See* comment to §17(4).

[59] *See Developments in the Law — Civil Commitment, supra* note 52, at 1283-88.

[60] Cramer v. Tyars, 23 Cal. 2d 131, 588 P.2d 793, 151 Cal. Rptr. 653 (1979).

[61] At trial, Mr. Tyars' claim to the privilege against self-incrimination was denied, and the district attorney called him as a witness.

> [T]he judge asked him whether he had been involved in any fights. Apparently unaware of the consequences to himself, appellant dramatized his affirmative reply by swinging his fists like a fighter and by uttering emphatic "pows" in a child-like effort to simulate the sound of his fists making contact. All of this occurred in front of the jury.

Id. at 143, 588 P.2d at 800, 151 Cal. Rptr. at 660 (Bird, C.J., dissenting).

COMMENT. For many, if not most, of the hearings conducted pursuant to this act, expert testimony about the developmentally disabled individual and his or her needs will be crucial to the outcome. Fair hearings and accurate results cannot be obtained if only one side has access to such testimony. Therefore, the right to an expert of the individual's own choosing is an essential element of due process in such hearings.[62] The expert will be able to offer not only his or her own testimony but can also assist counsel in evaluating the client's needs and other possible testimony. This right should not be denied because of the individual's inability to pay for the services of the expert,[63] and therefore the expenditure of state funds for this purpose is authorized. In most cases, one professional will be sufficient to assist counsel, but in some complex cases more than one may be needed. The court is empowered to appoint more than one expert when such cases arise.

(4) If the developmentally disabled person is unable to obtain counsel, the court shall appoint counsel, giving preference to non-profit organizations offering representation to developmentally disabled clients. If the individual cannot afford to pay counsel, state funds will be provided for this purpose by order of the court.

COMMENT. The right to counsel cannot be denied because of the individual's inability to obtain a lawyer or the inability to pay for his or her services. Therefore a mechanism for court appointment of counsel is established. This provision's preference for non-profit organizations specializing in representing developmentally disabled clients is designed to utilize the expertise of such specialists whenever it is available. Appointing lawyers from a general list on a random basis for one case at a time may not produce representation with the degree of sophistication about developmental disabilities issues that is desirable. The advocacy agency established by a companion model statute in this series[64] is one way of organizing such specialized representation.

(5) The hearing shall be conducted as clearly and as comprehensibly to the developmentally disabled person as possible. The rules of procedure and evidence applicable in civil proceedings shall apply.

COMMENT. It is important that unnecessary formalities not interfere with the individual's ability to comprehend and participate in the proceedings. Courts and counsel, over time, will be able to devise procedures which enhance the individual's understanding of what is happening during the hearing. But there are limits to the desirability of informality. It is even more important that the hearing be fair, and therefore the rules of evidence and civil procedure should be applied.[65] A desire for simplification and informality cannot be allowed to deprive the individual of important procedural rights, such as the prohibition against the introduction of hearsay evidence.

(6) The developmentally disabled person shall have the right to a trial by a six-person jury, if requested.

COMMENT. As with some other procedural rights, it is not completely clear whether the individual has a constitutional right to a jury trial in proceedings like those held pursuant to this act. A strong argument can be made that the Constitution guarantees a right to a jury trial at least in those hearings in which a substantial issue of physical liberty is in question.[66] But whether the right is constitutionally protected or not, it seems desirable to guarantee it in the proposed model statute, because in some cases it may enhance the fairness of the proceedings.[67]

(7) A complete record of the hearing shall be taken and shall be made available upon the request of any party.

COMMENT. The availability of a complete record is crucial to the possibility of an appeal.[68] It is not necessary under this provision to make a transcript of every hearing; it will suffice to make a tape recording of the type which can later be transcribed if a written record is required.

(8) Any party dissatisfied with the decision of the court shall have the right to an expeditious appeal to [name of appellate court] pursuant to [appellate rules].

COMMENT. States will differ in the structure of their appellate courts. In some states, appeal will lie with an intermediate court, and in others it will go directly to the state's highest court.

The availability of appeal serves two functions in these cases. The first is to assure a fair result when a party is dissatisfied with the ruling of the trial court. The second is to create a body of case law for the guidance of trial courts in future cases.

(9) A developmentally disabled person who is competent to do so may waive any rights except

---

62 Several courts have ordered the provision of such experts in mental health civil commitment cases. *E.g.*, Dixon v. Attorney General, 325 F. Supp. 966 (M.D. Pa. 1971). *See also* M.H.L.P., *Legal Issues–Civil Commitment, supra* note 54, at 104.

63 Farrell, *The Right of an Indigent Civil Commitment Defendant to Psychiatric Assistance of His Own Choice at State Expense*, 11 IDAHO L. REV. 141 (1975).

64 Chapter 9: Advocacy, *infra*.

65 Shuman, *The Road to Bedlam: Evidentiary Guideposts in Civil Commitment Proceedings*, 55 NOTRE DAME LAWYER 53 (1979).

66 O'Brien v. Superior Court, San Bern. County, 61 Cal. App. 3d 62, 132 Cal. Rptr. 13 (1976).

67 Experience in the area of civil commitment suggests that even when the right is explicitly guaranteed by state law, the right is not often exercised, and that may also prove true in these proceedings.

68 *E.g.*, Lynch v. Baxley, 386 F. Supp. 378 (M.D. Ala. 1974).

the rights to notice and counsel. With the permission of the court, counsel may waive the right of the client to be present at the hearing if the individual is not capable of making his or her own decision regarding presence and if presence would not be in his or her best interest.

COMMENT. Waiver of rights is a most difficult issue under this kind of statute,[69] since a considerable number of individuals whose cases are before the court will not have the capacity to waive any rights. Specific provision is made in this subsection for the authority of counsel to waive the presence of his or her client when that individual cannot make the decision for him or herself and when presence would not be in his or her best interest. For most individuals, presence at the hearing will not be harmful, and will often prove beneficial. But a small number of individuals may be so extraordinarily disruptive and so totally incapable of comprehending the proceedings that their presence should be waivable by counsel.

Two rights are not waivable, even by individuals who possess the ability to waive other rights. These are notice of the nature of the proceedings, and the assistance of counsel. Authorizing either the individual or counsel to waive either of these rights would create considerable potential for abuse, since they form the core of procedural fairness.

(10) In any judicial proceeding under this act, if the court concludes that a developmentally disabled person needs particular services which no identifiable public or private service provider is able or willing to provide to that individual, the court shall order the department to obtain such services for the individual. If such services are not currently available, the court shall order the department to furnish or to cause the needed services to be furnished for the benefit of that individual and of other individuals, if any, with similar needs. If the court determines that such services are not available for the individual, and that there are so few individuals with similar needs who are similarly situated that it would be impractical to create equivalent services within the state, the court may order the department to pay for equivalent services to be provided by a suitable facility or service provider in another state, if the court finds such an out-of-state placement to be in the individual's best interest and consistent with his or her habilitation needs and with the least drastic means principle. Any person receiving services out of state pursuant to this subsection shall retain all rights guaranteed under this act.

COMMENT. This provision authorizes the court to order the creation of new services where adequate services are not otherwise available for a developmentally disabled individual. This is a considerable grant of power to the courts, but the current deficiencies in the service delivery systems of all of the states warrant its adoption. Without this power, the courts are required to choose among those services which the state has chosen to make available. At the present time, this would certainly mean inappropriate services and unnecessary restriction of liberty for a considerable number of individuals. States which have committed themselves to the principles set forth in section 2 of this act, have recognized the need to reshape the service delivery system and to move away from unnecessary reliance upon residential, and particularly institutional, facilities. The state will also have declared the importance of individualized service and of eliminating unnecessary deprivations of liberty. Without this provision, there would be no effective way to enforce an individual's right to appropriate services and to receive those services in a manner consistent with the least drastic means principle.

There will be occasions, particularly in less populous states, when it is not appropriate to order the creation of new services because there are so few individuals within the state who could benefit from them. In these rare circumstances, the court is authorized to order the individual's admission to services in another state. This provision is included with some reluctance, because in the past a pattern of abuse has developed around inadvisable out-of-state placements.[70] But the need for such placements will occasionally arise, and the provision includes substantive safeguards, and these, plus the active vigilance of counsel, should serve as a deterrent to future abuses.

Section 17. Nonjudicial Proceedings

(1) In every nonjudicial proceeding in which a hearing officer considers the rights of a developmentally disabled person under this act, the provisions of this section shall apply.

COMMENT. This section creates a hearing mechanism for those cases which do not require a court hearing. Generally those cases are established categorically in other sections of this act, although some provisions allow parties to select between judicial and nonjudicial proceedings. The shared characteristic of cases which are appropriate for nonjudicial hearings is that they involve a somewhat lesser deprivation of liberty than those cases which must go to court. For example, judicial hearings are provided for large residential facilities, while nonjudicial hearings are employed in cases involving non-residential services for adults.

---

[69] *Compare* B. ENNIS & R. EMERY, THE RIGHTS OF MENTAL PATIENTS 80-82 (1978) *with* Wexler, *The Waivability of Recommitment Hearings*, 20 ARIZ. L. REV. 175 (1978).

[70] *See* Gary W. v. Louisiana, 437 F. Supp. 1209 (E.D. La. 1976).

(2) The court of each [judicial district] shall designate an attorney or attorneys to serve as hearing officers for the purpose of adjudicating cases under this act. No hearing officer may also serve as counsel or employee of a provider of developmental disabilities services. The rules for disqualification or recusal of a hearing officer shall be the same as those which govern judges.

COMMENT. The fact-finder in nonjudicial hearings under this statute will be a hearing officer who is an attorney. This provision is crucial to the satisfactory operation of the statute. While they involve somewhat less drastic issues than those involved in some judicial hearings, cases coming before the hearing officer will decide important questions of individual liberty. The hearing officer cannot perform his or her duties properly without the ability to weigh evidence, rule on objections, and consider interpretations of the law. Together these considerations require legal training.

An alternative which some states may choose to adopt would create interdisciplinary hearing boards to consider the cases which this act identifies as appropriate for nonjudicial resolution. As long as such boards each contained at least one member with legal training, they would probably withstand constitutional challenge and would meet the policy concerns expressed in the preceding paragraph. The model statute utilizes only single hearing officers because of the simplicity and reduced cost as compared to interdisciplinary boards. If a state believes that the additional costs are warranted by the broader experience that such boards would bring to nonjudicial cases, it may wish to replace the hearing officer with such a board.

The mechanics of compensation and the statement of duties of hearing officers (or, in the alternative, members of interdisciplinary boards) will usually be created by court rules. However, some states may wish to include them in the statute.

This subsection also provides protections against conflict of interest on the part of hearing officers. These would be particularly important to apply to non-attorney members of hearing boards, if established, since it would be essential to assure that staff members of the service provider in question would not participate in a case involving their facility or agency.

(3) The clerk of the court shall maintain a list of hearing officers appointed by the court. When any party seeks a nonjudicial hearing under the provisions of this act, the party shall petition the clerk for the designation of a hearing officer and the setting of a hearing date. The clerk shall, upon receipt of such a petition, designate a hearing officer and with his or her concurrence set a hearing date, which shall not be later than ten (10) days after the filing of the petition. The clerk shall then notify all parties of the identity of the hearing officer and the time and place of the hearing.

COMMENT. This provision provides the mechanism for selecting the hearing officer for each case.

If a state adopts the alternative of creating hearing boards, a mechanism similar to this one should be created to select the members of the board for each case.

(4) Every developmentally disabled person whose need for services or right to services is at issue in a nonjudicial proceeding shall have the right to be represented by counsel and shall be represented by counsel, or by an advocate of his or her choice who also is acceptable to the hearing officer. The developmentally disabled person and his counsel or advocate shall receive timely and clear notice of the nature of the issue and proceedings. The developmentally disabled person also shall have the right to be present, to present evidence on his or her own behalf, to cross-examine witnesses, and to testify or decline to testify on his or her own behalf. The hearing shall be conducted as clearly and as comprehensibly to the developmentally disabled person as possible.

COMMENT. This subsection establishes many of the procedural rights of individuals whose cases come before a hearing officer. With two important exceptions, these rights parallel those afforded in judicial proceedings.[71]

The first difference concerns the right to counsel. In judicial proceedings, the individual must be represented by an attorney. In nonjudicial proceedings, the individual has the right to be represented by an attorney, but he or she may choose to be represented by an advocate[72] who is not a lawyer. However, the individual must be represented by one or the other; he or she cannot be unrepresented. The provision allows representation by an advocate because there will be some cases which will not require the services of a lawyer. But any advocate must be acceptable to the hearing officer, and if the officer believes that a particular advocate lacks the skills necessary to handle a particular case he or she may insist that another advocate or an attorney be selected or appointed.

The other major difference is that the rules of evidence and procedure need not be followed. This allows for a more informal hearing where such informality is appropriate. However, the hearing officer always has the authority to reject any evidence or to insist upon any procedure when fairness dictates.

---

[71] *See* comments to §§16(2), (5).

[72] *See* Chapter 9: Advocacy, *infra*.

As an alternative, some states may wish to require that individuals be represented by an attorney, or that the rules of evidence and procedure should apply. If the state chooses to apply the rules of evidence and procedure, it may be appropriate, and possibly necessary, to require that the individual be represented by a lawyer.

(5) The right to present evidence shall include the right to the assistance of a developmental disabilities professional or professionals of the developmentally disabled person's choosing. If he or she is unable to pay for the services of such a professional, state funds will be provided for this purpose by order of the court, upon request of the hearing officer.

COMMENT. See comment on section 16(3). The statute does not authorize hearing officers to order the expenditure of state funds, so payment for the expert must come through a court order.

(6) If the developmentally disabled person wishes to be represented by an attorney and does not have one, or if the advocate wishes to be assisted by counsel, the hearing officer shall request the court to appoint counsel as provided in section 16(4).

COMMENT. See comment on section 16(4). The hearing officer lacks power to appoint counsel, so a court order is required.

(7) A complete record of the hearing shall be taken and shall be made available upon the request of any party.

COMMENT. See comment on section 16(7). The availability of a record is particularly important here since appeal from nonjudicial orders lies with the trial court, which will require a record for review.[73]

(8) Any party dissatisfied with the decision of the hearing officer may appeal to the court for review of the decision on the basis of the nonjudicial hearing record. The court shall resolve the matter as expeditiously as possible.

COMMENT. This provision allows appeal of nonjudicial orders to a trial court for review of the hearing officer's record of the proceedings. Further appeal from the decision of this court can be taken to an appellate court pursuant to section 16(8). This "dual appeal" mechanism assures that the appellate court will always have a trial court record to review; it should not be asked to review directly the actions of a nonjudicial officer.

Some states may wish to adopt an alternative to this provision which would give the dissatisfied party a right to a trial *de novo* rather than a review of the hearing record. This would give the dissatisfied party the opportunity to present its case directly to the court. Under such an alternative provision, testimony at the nonjudicial hearing should be admissible from the nonjudicial record upon the consent of all parties and the approval of the court, so that admissible testimony will not have to be repeated unnecessarily. Note, however, that the alternative would be more costly than the subsection in the text, since the new trials would be more elaborate than judicial review of the record.

(9) A developmentally disabled person who is competent to do so may waive any rights except the right to notice and to be represented by an attorney or advocate. With the permission of the hearing officer, the attorney or advocate may waive the right of the client to be present at the hearing if the individual is not capable of making his or her own decision regarding presence and if his presence would not be in his or her best interest.

COMMENT. See comment on section 16(9).

(10) The hearing officer shall be compensated with funds from the budget of the court.

COMMENT. It is important that the expenses of the nonjudicial hearing should be met through the court's budget rather than through the budget of the department, since the latter arrangement would create a serious potential for conflict of interest. Obviously, the department should not have the opportunity to influence the outcome of cases through budgetary pressures or threats.

(11) In any nonjudicial proceeding under this act, if the hearing officer determines that a developmentally disabled person needs particular services which no identifiable public or private service provider is able or willing to provide to that individual, the officer shall order the department to obtain such services for the individual. If the officer concludes that such services are not currently available, the individual or his or her attorney or advocate may petition the court to order the department to furnish or cause the needed services to be furnished for the benefit of that individual and of other individuals, if any, with similar needs, or to order the out-of-state placement of the individual pursuant to the provisions of section 16(10).

COMMENT. See comment on section 16(10). Since the hearing officer lacks the authority to order the state to expend funds, a court order must be obtained for such purposes, and upon a proper finding by the officer, a party may request such an order.

Section 18. Authorization of Appropriations

To carry out the purposes of this act, there is authorized to be appropriated $ __________ for the fiscal year ending ____________________, $ ____________________ for the fiscal year ending __________, and $ __________ for the fiscal year ending __________ .

[73] *See* §17(8).

COMMENT. The cost of establishing the consent and hearing mechanisms of this act will be substantial but not extraordinary. But a much greater expenditure of state funds will be required to provide the substantive services to which individuals are entitled. The move away from an institution-based service delivery system will not be cost-free, even though institutions are extremely expensive and wasteful. The start-up costs of creating new services will be substantial, as will the commitment to meet individuals' service needs which the current system ignores.

Section 19. Severability of Sections

If any section, subsection, paragraph, sentence, or any other part of this act is adjudged unconstitutional or invalid, such judgment shall not affect, impair or invalidate the remainder of this act, but shall be confined to the section, subsection, paragraph, sentence or other part of this act directly involved in the controversy in which that judgment was rendered.

COMMENT. None necessary.

Section 20. Effective Date

This act takes effect ____________________.

COMMENT. None necessary.

7
GUARDIANSHIP AND CONSERVATORSHIP

Table of Contents

INTRODUCTION

Guardianship is a legal relationship which authorizes one individual to become a substitute decision-maker for another. Its most common form is the "natural guardianship" relationship between parents and their minor children. A guardianship is established by court order when because of age, illness, or disability, a person is determined to be incapable of managing some or all of his or her personal and/or financial affairs. A guardian may be given partial or total authority to determine whether the disabled person will live in the community or an institution, and what type of medical, mental health and other services the disabled person will receive (personal guardianship), and/or partial or total power to manage and control that person's property and income (conservatorship). Correspondingly, the individuals for whom a guardianship has been established may lose the right to decide whether to remain in their own home, to make contracts for goods and services, to go to court to enforce their rights, to hold or convey property, and in some instances to marry, to have children, to vote and to make a will.

The roots of guardianship are deeply embedded in both Roman and English law.[1]

> . . . [I]n Rome at the time of Cicero elaborate provisions were made for the protection of the property of the mentally disabled while none at all existed for the person. The pattern was followed in England and also in colonial America. . . . .[2]

In England, guardianship was part of first the feudal lords' and then the Crown's duty to look after their subjects. In this country, the doctrine of *parens patriae* — the state serving as the benevolent parent caring for his or her poor, sick or wayward children — provided not only the legal basis but also the dominant philosophy for guardianship. Although this permitted considerable flexibility in guardianship proceedings, it presented substantial problems as well.

> The disadvantaged were the objects of care, they were to be done for. They did not require protection against the well-meaning parent, [or] rights to be exercised against the paternalistic state.[3]

As Lionel Trilling once commented: "Some paradox in our nature leads us, once we have made our fellow men [or women] the objects of our enlightened interest, to go on to make them the objects of our pity, then of our wisdom, ultimately of our coercion."[4] The criteria for establishing a guardianship are often broad and vague, permitting the imposition of restrictions on persons who are "different" as well as on those who are disabled.[5] The procedures often omit the safeguards we have come to expect when restrictions on liberty are imposed or fundamental rights threatened in other contexts. And perhaps most importantly, even today in many jurisdictions, guardianship orders and guardians have failed to recognize that individuals with disabilities are often capable of doing many things for themselves.

> [A] . . . serious difficulty arises because the law usually represents incompetence in simple black and white terms, with the result that most guardianships of the person are looked on as plenary guardianships. The effect is well expressed by the word *interdiction* (prohibition) which has until recently been in use in the statutes of many of the Latin countries. The person interdicted (declared incompetent) is deprived of the legal capacity to act in any way on his own behalf. Even though he [or she] may have a guardian appointed to exercise some of his [or her] rights, the emphasis usually is on the deprivation of rights rather than on implementing rights constructively through informed

---

[1] *See* S. BRAKEL & R. ROCK, THE MENTALLY DISABLED AND THE LAW, 250 (Rev. Ed. 1971); Regan, *Protective Services for the Elderly: Commitment, Guardianship, and Alternatives*, 13 WM. & MARY L. REV. 569, 570-573 (1972).

[2] BRAKEL & ROCK, *supra* note 1, at 250.

[3] Rothman, *The State as Parent: Social Policy in the Progressive Era*, in W. GAYLIN, I. GLASSER, S. MARCUS, & D. ROTHMAN, DOING GOOD: THE LIMITS OF BENEVOLENCE, 67, 70 (1978).

[4] *Id.* at 72.

[5] N. KITTRIE, THE RIGHT TO BE DIFFERENT: DEVIANCE AND ENFORCED THERAPY (1971).

> representation. Moreover, the idea that the person himself [or herself] can properly retain and exercise some personal and even property rights, selectively, according to his [or her] individual capacity, is not adequately expressed in most existing statutes pertaining to guardianship.[6]

Over the past two decades, a growing list of organizations and governmental commissions which have examined guardianship have called for correction of these problems.[7] For example, in 1962, the President's Panel on Mental Retardation stated:

> For some, of course, a comprehensive guardianship will be needed. But we urge that, as far as possible, mentally retarded adults be allowed freedom — even freedom to make their own mistakes. We suggest the development of limited guardianships of the adult person, with the scope of the guardianship specified in the judicial order.[8]

The 1969 Report of the International League of Societies for the Mentally Handicapped recommended:

> The retarded adult should be permitted to act for himself [or herself] in those matters which he [or she] has competence. The limitations of legal capacity inherent in guardianship should not extend to these matters. It follows that a person whose mental retardation is characterized by impairments of social competence which are partial should enjoy a partial guardianship specifically adapted to his [or her] strengths and weaknesses.[9]

The 1975 edition of the Uniform Probate Code makes a clear distinction between guardianship of the person and conservatorship of the estate, and establishes a number of less restrictive alternatives for the protection of the property of a disabled individual (though not the person). It also provides for the execution of durable powers of attorney as a means of obviating the need for a guardianship or conservatorship.[10] Finally, the 1976 Report of the President's Committee on Mental Retardation, urged that:

> Statutes and court procedures bearing on competency should be clarified and revised (a) to recognize gradations of competence, (b) to recognize that areas of competency may be quite varied and therefore should be separable in law, (c) to assure full and explicit due process safeguards on any and all areas of competency, and that the scope of any judgment of incompetence is made fully explicit, and (d) to ensure that restrictions of competency be limited to a specific period of time or subject to periodic review.[11]

Despite these concerns and recommendations, there has been relatively little litigation directly addressing guardianship other than the cases now pending before the U.S. Supreme Court on the power of parents as natural guardians to voluntarily commit their child to an institution for the mentally ill or mentally disabled.[12] The legislative response is considered below in the following examination of the structure and provisions of current guardianship statutes.

---

[6] INTERNATIONAL LEAGUE OF SOCIETIES FOR THE MENTALLY HANDICAPPED (ILSMH), SYMPOSIUM ON GUARDIANSHIP OF THE MENTALLY RETARDED, 11(1969).

[7] *E.g.*, THE PRESIDENT'S PANEL ON MENTAL RETARDATION (PPMR), REPORT OF THE TASK FORCE ON LAW (1963); PRESIDENT'S COMMITTEE ON MENTAL RETARDATION (PCMR), REPORT TO THE PRESIDENT — MENTAL RETARDATION: CENTURY OF DECISION (1976); THE PRESIDENT'S COMMISSION ON MENTAL HEALTH (PCMH), REPORT TO THE PRESIDENT (1978); ILSMH, *supra* note 6; COUNCIL OF THE AMERICAN ASSOCIATION ON MENTAL DEFICIENCY (AAMD), POSITION PAPER ON GUARDIANSHIP FOR MENTALLY RETARDED PERSONS (1973); National Center For Senior Citizens, *Model Guardianship, Conservatorship and Power of Attorney Legislation*, in U.S. SENATE SPECIAL COMMITTEE ON AGING, PROTECTIVE SERVICES FOR THE ELDERLY: A WORKING PAPER, 75-110 (1977); GOVERNOR'S COMMISSION FOR REVISION OF THE MENTAL HEALTH CODE OF ILLINOIS, REPORT (1976).

[8] PPMR, *supra* note 7, at 25.

[9] ILSMH, *supra* note 6, at 18-19.

[10] NATIONAL CONFERENCE OF COMMISSIONERS ON UNIFORM STATE LAW (NCCUSL), UNIFORM PROBATE CODE (U.P.C.) Article V (4th Ed. 1975).

[11] PCMR, *supra* note 7, at 65-66.

[12] *E.g.*, J.L. & J.R. v. Parham, 412 F. Supp. 112 (M.D. Ga. 1976) probable jurisdiction noted, 431 U.S. 936 (1977); Bartley v. Kremens, 402 F. Supp. 1039 *rev'd on other grounds*, 431 U.S. 119 (1977); Saville v. Treadway, 404 F. Supp. 430 (M.D. Tenn. 1974).

REVIEW OF STATUTES

This statutory survey is based on an analysis of the statutes of each of the 50 states, the District of Columbia, American Samoa, Guam, Puerto Rico and the Virgin Islands regarding guardianship and conservatorship and not on an examination of the entire civil code of these jurisdictions. Hence, an indication that there is no provision addressing a particular question in a jurisdiction should be understood to refer only to the guardianship and conservatorship laws, since the procedure, right or definition may be included in another chapter, article or set of rules which were not examined. The terminology relating to guardianship varies from state to state. To facilitate comparison, a single set of terms is used throughout the review. Hence, the word "conservator" incorporates curators, guardians of the estate, and guardians of property. Similarly, the phrase "partially disabled or disabled person" includes such terms as ward, incompetent, and incapacitated person.

The review is divided into two sections. The first covers guardianship of the person and conservatorship of the property of partially disabled and disabled adults. The second covers guardianship and conservatorship for minors. Each section describes the individuals subject to guardianship and conservatorship, the scope of the guardianship and conservatorship relationship, the rights and procedures in guardianship and conservatorship proceedings, the powers and duties of personal guardians and conservators, and the review, modification and termination of guardianships and conservatorships.

GUARDIANSHIP AND CONSERVATORSHIP FOR PARTIALLY DISABLED OR DISABLED ADULTS

Adults Subject to Guardianship and Conservatorship (Chart 1)

All American jurisdictions provide statutorily for some form of guardianship of adult persons and conservatorship of their property if they are not able to care for themselves without assistance. Most statutes employ a two part test for determining whether guardianship or conservatorship is required. The first identifies the type of disability affecting an individual — e.g. mental illness, mental retardation, physical disabilities, senility, mental deficiency, mental and physical weakness and in some older statutes,[13] "idiocy," "lunacy," "habitual drunkeness," or "having an unsound mind." The second requires that the disability prevent individuals affected from caring for themselves or managing their property. The effect of these standards on developmentally disabled persons varies. Fourteen states (Georgia, Indiana, Iowa, Massachusetts, Michigan, New York, North Carolina, Ohio, Pennsylvania, Tennessee, Texas, Washington, Wisconsin, and Wyoming) specifically include individuals with mental retardation as appropriate subjects for guardianship or conservatorship. Five states (Idaho, Illinois, North Carolina, Washington, and Wisconsin) also list individuals with developmental disabilities, or with autism, cerebral palsy or epilepsy. In most other states, inclusion of persons with developmental disabilities depends on the interpretation of such terms as mental or physical disability, mental or physical weakness, or mentally deficient.

Scope of Guardianship and Conservatorship (Chart II)

The extent to which and manner in which the legal capacity of a partially disabled or disabled individual is restricted varies along a number of important dimensions. The first of these is whether protection of an individual's property entails restrictions on the capacity of that individual to handle his or her personal affairs and conversely, whether the appointment of a personal guardian necessitates the appointment of a conservator as well. While most jurisdictions permit appointment of a personal guardian, a conservator or both, 6 jurisdictions (Alabama, Kentucky, Louisiana, New Jersey, Oklahoma, and the Virgin Islands) either do not distinguish between personal guardians and conservators, or do not permit legal capacity to be restricted in one area but not the other. Two others, the District of Columbia and Massachusetts, make no provision for an individual to serve solely as personal guardian.

The second major dimension is whether personal guardians and conservators are empowered to make nearly all important decisions on behalf of the individual whom they are serving (a plenary or

[13] *E.g.*, ALA. CODE §§26-2-1 (1975); ARK. STAT. ANN. §§57-601 (1975); and R.I. GEN. LAWS §§33-15-8 (1969).

full guardianship or conservatorship), or whether they are assigned only those duties and powers which the individual is incapable of exercising (a limited or partial guardianship or conservatorship). Currently, 17 states explicitly permit limited guardianship and/or conservatorship. Eleven of these (Connecticut,[14] Florida, Idaho, Illinois, Kentucky, New York, South Carolina, Texas, Washington, West Virginia, Wisconsin) have enacted a separate provision or set of provisions outlining the scope of and procedures applicable to limited guardianships and conservatorships. Others simply require that the appointing body set forth the powers and duties of a personal guardian or conservator and suggest that the assigned powers and duties need not include all those listed in the statute (Maine, Michigan, North Carolina, South Dakota, Virginia). Finally, Maryland has a statement of policy favoring exploration and use of less restrictive alternatives.[15]

The third dimension concerns who may be appointed as a personal guardian or conservator. Many of the states which distinguish between guardianship and conservatorship encourage the appointment of close relatives or friends to care for the personal needs of partially disabled or disabled persons, and permit banks, trust companies and similar institutions to look after their financial resources. Thirteen states also permit appointment of a public official as a personal guardian or conservator for partially disabled and disabled persons who have no close relatives or friends able and willing to serve in that capacity or whose assets may be exhausted trying to compensate a guardian or conservator.[16] Generally, public guardians have the same powers and duties as do private guardians and conservators. Six of the states with public guardian statutes (Colorado, Georgia, Maine, Minnesota, North Carolina, and Ohio) assign public guardianship duties to the state agency responsible for providing services to one or more of the classes of individuals subject to guardianship. The other states provide more independence for public guardians by making them state (Delware, Illinois), county (Alabama, Arizona, California) or court appointees (Oregon, Tennessee). When public guardianship provisions are significantly different from those on private guardianship, they are listed separately at the end of each chart. In addition, several topics which are unique to public guardianship are covered in Chart VII.

Finally, states differ with regard to the permissibility of testamentary, standby and emergency appointments. Statutes authorizing testamentary and standby appointments provide a means of easing transitional difficulties presented by the death or incapacity of the remaining parent of a partially disabled or disabled person, or by the death, resignation or incapacitation of the personal guardian or conservator. These statutes permit automatic transfer of guardianship and/or conservatorship authority to an individual named in the parent's will (testamentary appointment) or in the appointing authority's order (standby appointment). Currently, 34 juridictions permit testamentary appointments, 4 allow standby appointments (Connecticut,[17] Iowa, South Carolina, and Washington) and 3 provide for both[18] (Florida, Illinois and Michigan). A total of 31 jurisdictions permit appointment of a personal guardian or conservator on either a temporary or emergency basis. These guardianships are used when a remaining parent, personal guardian or conservator dies or resigns and no provision for a standby appointment has been made, or when an individual is incapable of giving an informed consent to specific medical treatment. Because of the time constraints usually associated with the need for an emergency or temporary appointment, notice provisions which apply in other guardianship and conservatorship proceedings (see Rights and Procedures, *infra*) either are dispensed with or shortened, and a hearing may not be a statutory requirement. The right to appeal is also often precluded. Most statutes set a definite limit on the duration of temporary appointments although 11 jurisdictions (District of Columbia, Iowa, Louisiana, Maine, Massachusetts, Michigan, Minnesota, Oregon, Pennsylvania, Rhode Island and Texas) either do not specify a limit or leave the matter entirely within the court's discretion. The time limits which are set statutorily vary from 6 days (Maryland) to 6 months (Alaska, Arizona, Colorado, Hawaii, Idaho, Montana, Nebraska, Nevada, New Mexico, North Dakota and Utah).

Rights and Procedures (Charts III and IV)

Except when there is a testamentary appointment, proceedings are generally initiated by the fil-

[14] Limited conservatorship only.

[15] Even in states denominated "Plenary" on Chart I, courts may in fact order less than total guardianship or conservatorship. However, the statute does not mandate or give preference to such action. Additionally, there is some authority in states which have adopted the Uniform Probate Code for courts to order less than plenary conservatorships. *See* U.P.C., *supra* note 10, at §§5-409 and 5-426.

[16] States which provide for the appointment of the director of the department of mental health and/or mental retardation or an institution superintendent to act as guardian for persons who are residents of institutions have not been included, since such appointments are generally made to enable the guardian to accept monetary entitlements of such persons.

[17] For persons with mental retardation only.

[18] Because only guardianship statutes were searched, even for states where neither an "S" nor a "T" appears on the charts, statutory provisions permitting use of these kinds of appointments may exist. This is particularly possible for testamentary appointment provisions, which may appear in the part of the state code dealing with wills.

ing of a petition. The contents of the petition and the individuals who may file a petition vary from jurisdiction to jurisdiction. Many statutes provide that any interested person may file. Following the filing of a petition, all jurisdictions except Alabama, South Dakota and in some instances Texas, require that the individual allegedly in need of a personal guardian or conservator be given notice of the date of the hearing on the petition. Notice is usually served on the family of the subject of the petition and on other persons who know the subject well. The notice period ranges from 3 to 20 days, with most jurisdictions requiring notice 7 to 14 days prior to the hearing. A few statutes require only that notice be reasonable (Delaware, Florida, Missouri, and Virginia), or conform to the notice requirements prescribed for other civil proceedings (e.g., Iowa and Maryland). Prior to the hearing, 33 jurisdictions provide for an examination of the subject with most requiring an examination by at least one physician. Five states (Connecticut, Kansas, Kentucky, New York and West Virginia) permit an examination by a psychiatrist or psychologist. Three other states (Georgia, North Carolina, and Texas) provide for an examination at an evaluation facility or by a multidisciplinary evaluation team. At the hearing, which except in North Carolina is before a judge, respondents are statutorily entitled in the overwhelming number of jurisdictions to be represented by a lawyer or to have a guardian *ad litem* — i.e. an individual appointed to act on the respondent's behalf in making the decisions regarding the proceedings. Specifically, 15 statutes provide for representation by counsel (California, Connecticut, Delaware, Florida, Kansas, Kentucky, Louisiana, Maryland, Massachusetts, New Jersey, Ohio, Oregon, South Carolina, West Virginia and Wyoming), 7 provide for appointment of a guardian *ad litem* (Alabama, District of Columbia, Michigan, New York, Rhode Island, Virginia, Puerto Rico), 5 permit a respondent to have both an attorney and a guardian *ad litem* (Illinois, Indiana, Texas, Washington and Wisconsin), two permit one or the other (Georgia and North Carolina), and 13 direct an attorney to take on both roles (Alaska, Arizona, Colorado, Hawaii, Idaho, Illinois, Missouri, Montana, Nebraska, New Mexico, North Dakota, Tennessee and Utah). Thirty-eight statutes entitle respondents to be present at the hearing if they are able. Twenty-two states entitle the respondent to a jury on request. In addition, Georgia has enacted a unique provision which empanels a "commission" of two physicians and one attorney to determine whether guardianship is required. Finally, 17 states permit respondents to present evidence at the hearing and 15 states permit the cross-examination of witnesses and closing the hearing to the public.

Powers and Duties of Guardians (Chart V)

All guardianship statutes provide some guidance to guardians and conservators regarding their responsibilities, and some protections to persons subject to guardianship or conservatorship, although, as noted earlier, the guidance is usually general in nature and the protection limited. Nearly every jurisdiction requires conservators to file a surety bond to indemnify against waste or misappropriation by the conservator of the financial resources over which the conservator has supervision or control. A bond requirement is imposed on public guardians in 10 states (Alabama. California, Colorado, Delaware, Georgia, Maine, North Carolina, Ohio, Oregon, Tennessee). To further assure accountability, 21 states impose the duties of a fiduciary on private conservators. With regard to personal guardians, the statutes of 22 states plus Guam and Puerto Rico require or at least establish a policy encouraging personal guardians to provide or take advantage of services which will maximize the ability of the partially disabled or disabled person whom they are serving to function in society. A similar habilitation requirement is imposed on public guardians in Colorado, Maine and Minnesota. The services encompassed within the habilitation requirement vary from state to state but may include education, vocational training, mental health counseling or treatment, physical therapy, or job placement assistance.

Finally, the guardianship statutes in all but 10 states and the District of Columbia empower personal guardians and conservators to collect a fee for services rendered from the financial resources of the individual whom they are serving. Most of these statutes impose a standard of reasonableness on the fee or leave the amount of compensation entirely to the discretion of the court. Arizona, California, Delaware, Maine,[19] North Carolina, Oregon and Tennessee also permit public guardians to charge a reasonable fee to recover costs.

Review, Modification and Termination (Chart VI)

In recognition of the fact that the disabilities of an individual subject to guardianship or conservatorship may lessen or increase over time, most jurisdictions provide some means for modifying the duties and powers of the guardian or conservator, or terminating the guardianship or conservatorship altogether and restoring to the former subject of that guardianship or conservatorship, the legal capacity to act in his or her own behalf. Ten states (Arizona, California, Connecticut, Florida, Idaho, Indiana, Iowa, Michigan, Nevada[20] and North Carolina) require regular court review of private guardianships and conservatorships. In addition, Colorado, Delaware, Maine, Minnesota

[19] The Maine Bureau of Mental Retardation can charge a fee. ME. REV. STAT. tit. 18, §3629 (Supp. 1978). The Maine Department of Health and Welfare cannot. ME. REV. STAT. tit. 18, §3647 (Supp. 1978).

[20] Personal guardian only.

and Ohio require such a periodic review when the public guardian is appointed. The majority of these provisions require at least an annual review, although some jurisdictions, such as California, impose a biennial requirement or leave the review period entirely within the discretion of the court (e.g., Indiana). The guardianship statute of every jurisdiction except American Samoa, Puerto Rico, and Virginia provides for a proceeding to terminate the guardianship or conservatorship or to limit or reduce the powers of the personal guardian or conservator. Thirty-nine jurisdictions have provisions explicitly authorizing the partially disabled or disabled person to initiate a termination proceeding. Most jurisdictions also permit other individuals interested in that person, including close relatives, to initiate such proceedings.

GUARDIANSHIP AND CONSERVATORSHIP FOR MINORS

Minors Subject to Guardianship and/or Conservatorship (Chart VIII)

The definitions of the minors subject to guardianship and/or conservatorship provided in Chart VIII often do not appear in the codes themselves but are derived from all the applicable provisions within the state probate statutes. All but 7 jurisdictions set 18 as the age of majority for guardianship purposes. Four jurisdictions (Missouri, New Jersey, Puerto Rico and Virgin Islands) retain the common law age of 21. Alabama, Alaska and Wyoming have adopted age 19 as the threshold for adulthood. The statutes in 4 jurisdictions (Nevada, Pennsylvania, Puerto Rico, and Rhode Island) list age as the sole eligibility criterion for the appointment of a personal guardian and conservator for a minor. Indiana lists age as the sole criterion for appointment of a conservator. The statutes in 19 jurisdictions specify that a minor must be orphaned before a personal guardian may be appointed; 32 provide that a minor who is married is ineligible for appointment of a personal guardian;[21] 14 include termination of parental rights among the criteria; 10 statutes specify that appointment must be necessary and/or convenient; 6 include that the minor have no guardian or conservator; and all include that appointment is necessary to protect the interests or rights of the minor.

Many of these same grounds apply to eligibility for appointment of a conservator. Among those grounds listed most often are: that the minor be unmarried (11 states),[22] that the minor must be entitled to property (9 states), that the property cannot otherwise be protected (11 states) and that the appointment must be necessary and/or convenient (9 states).[23] Three statutes (Missouri, North Carolina, and Virginia) specify that the parents of a minor must be dead before a conservator may be appointed to protect the property of a minor.

In addition, the juvenile codes of most states provide that custody of a minor who falls within the definition of a dependent, abandoned, neglected and/or abused child, may be transferred to an appropriate individual or agency.

Scope of Guardianship and/or Conservatorship (Chart IX)

While the statutes concerning guardianships and conservatorships for minors often parallel the provisions for partially disabled and disabled adults, there remain a number of important differences. These differences are attributable, in large part, to the fact that unlike adults, minors in all American jurisdictions are assumed to be legally incompetent until they come of age unless one of the events discussed below in conjunction with Chart XII occurs or they are declared by a court to be emancipated. Even following emancipation, certain disabilities may continue until the minor attains majority.[24]

As is true of the adult provisions, a substantial majority of states allow an individual to act as either personal guardian, conservator or both. However, 4 states have no probate mechanism to establish a separate guardianship of the person (Alabama, Massachusetts, Vermont and Virginia). These states, however, do allow juvenile courts to appoint custodians who have substantially the same powers as a guardian[25] for those children whose natural parents have shown an inability to provide suitable care.

One significant difference between provisions concerning adults and those concerning children is the nearly total absence of limited guardianships for minors. Almost all jurisdictions have established a mechanism for providing consent on behalf of a minor to receive medical care when the minor's parent refuses to provide such consent on religious grounds, but only Ohio and Washington state provide explicitly for limited guardianship for minors. Most states provide a list of acceptable appointees for minors similar to those provided for adults. Preference is given to parents, relatives, close friends, and parental or testamentary designees. Of the 12 state provisions which provide for public guardianship for adults, 5 permit the public

[21] This applies only to women in Delaware, Montana, Vermont, and the Virgin Islands and only to those persons between ages 18 and 21 in Mississippi.

[22] In Massachusetts this provision applies only to males; in the Virgin Islands it applies only to females; and in Mississippi it applies only to persons 18-21.

[23] In Massachusetts this applies to females only.

[24] H. BEASER, RUNAWAY YOUTH FROM WHAT TO WHERE: THE LEGAL STATUS OF RUNAWAY CHILDREN, 55-70 (1975).

[25] *See* Katz, *Legal Aspects of Foster Care*, in THE YOUNGEST MINORITY, 72 (S. Katz ed. 1974).

guardian to be appointed for minors. (See Chart VII.) In addition, a number of states loosely designate adoption agencies, state schools and homes, social services agencies and a variety of other state and local bodies involved in child welfare as eligible to be custodians of children who come under their care and supervision. As noted above, custodians in these circumstances generally have nearly the same powers and duties as guardians.

Nearly every American jurisdiction statutorily provides for testamentary appointments. Only Florida and Washington also permit appointment of a personal guardian or conservator for a minor on a standby basis. Iowa permits standby appointments but has no provision on testamentary appointments. Twenty-eight jurisdictions provide for temporary or emergency appointments. Half of these terminate after a specified amount of time (most often 6 months) or upon appointment of a permanent guardian. The remainder continue in effect for a period set by the court. The uses of and procedural requirements for temporary and emergency appointments are substantially the same as those for adults discussed above.

Rights and Procedures (Chart X)

As is true with the provisions for adults, a proceeding for the non-testamentary appointment of a personal guardian or conservator for a minor usually begins by the filing of a petition, typically by any interested person. However, beyond that point, the probate court appointment procedure for minors is substantially abbreviated in most states. Minors under age 14 are generally not entitled to notice of the proceedings. Only 26 jurisdictions require notice of the initiation of a guardianship or conservatorship proceeding to be sent to a minor over age 14. In three additional states (Connecticut, Iowa and Kansas) notice to minors over age 14 is required only in particular situations or left to the court's discretion. With regard to representation, 12 jurisdictions (Alaska, Arizona, California, Colorado, Idaho, Kansas, Missouri, Montana, Nebraska, New Mexico, North Dakota, Utah) provide that a minor may be represented by counsel either as a matter of right or at the discretion of the probate court in proceedings to appoint a personal guardian or conservator. Twenty other jurisdictions provide for appointment of a guardian *ad litem*. Washington State permits appointment of both an attorney and a guardian *ad litem*. Oregon courts can appoint an attorney in proceedings prior to the establishment of a guardianship for a minor; but Oregon law permits only the appointment of a guardian *ad litem* in proceedings after a guardianship for a minor has been established.

Generally, the right of a minor to be present at a guardianship proceeding, to present evidence and cross-examine witnesses at the hearing, and to have the matters at issue determined by a jury are not statutorily specified. Hence, these points are not included in the Chart. However, most state codes permit a minor over age 14 to nominate an individual to act as guardian. Forty-three states now provide that such a nominee, if found by the court to be a suitable person, must be appointed. An additional 4 states require that the nominee be at least considered by the court (Arkansas, California, Indiana, and New Hampshire).

Testamentary designees need not be appointed by the probate court in 29 of the 50 jurisdictions which statutorily permit testamentary appointments or designations. Thus, in these states, many if not most guardianships and conservatorships for minors become effective without any formal hearing. The impact of this provision, however, may be lessened by a companion section in 17 of these states which allows the minor, if over 14, to object to the appointment of the testamentary nominee and to force a court hearing. An additional 9 states which require a hearing before appointment of the testamentary nominee also allow objections by minors age 14 or older.

In addition to the probate court procedures outlined above, Chart X includes references to the provisions on notice to and representation of a minor in proceedings held pursuant to the juvenile or family court's jurisdiction over dependent, abandoned, neglected and abused children.

Powers and Duties in Guardianships and/or Conservatorships (Chart XI)

Most statutes set forth some duties for personal guardians and conservators of minors although the provisions on the duties of conservators are usually far more detailed than those on the duties of personal guardians. All but 6 jurisdictions (Iowa, Maryland, North Carolina, Pennsylvania, Tennessee and American Samoa) provide explicitly for custody of a minor by that minor's personal guardian, and all but 11 jurisdictions (Connecticut, Iowa, Kansas, Maryland, Minnesota, New Jersey, North Carolina, Pennsylvania, Tennessee, Wisconsin, and American Samoa) make the personal guardian of a minor responsible for the minor's education.[26] Fourteen jurisdictions specifically authorize personal guardians of minors to consent to medical treatment. Furthermore, the juvenile codes of 46 jurisdictions include a provision which authorizes the court to assume temporary custody of a child in need of emergency treatment. Since assumption of responsiblity by the court is usually the first step in a neglect proceeding, 9 of these jurisdictions (Arizona, Arkansas, Colorado, Idaho, New Jersey, Oklahoma, Rhode Island, Utah, Wyoming) have enacted an additional provision which specifies that in most instances in which medical

[26] In Connecticut, however, the responsibility for the minor's education is subsumed under the guardian's general authority to exercise the powers of a parent.

care is in conflict with the religious beliefs of the child's parents, the failure to provide medical services does not constitute neglect.

All jurisdictions except American Samoa specifically provide for management of a minor's estate. The statutes in each jurisdiction require conservators to file a bond, although most jurisdictions which provide for testamentary appointments permit the bond requirement to be waived if a waiver is consistent with the wishes of the testator. In addition, 38 jurisdictions include conservators for minors within the probate code definition of fiduciaries. Forty-six statutes require the personal guardian or conservator to file a periodic accounting of their activities with the court. The most common provision requires an annual financial statement, including all income and disbursements from the minor's estate. Finally, 43 jurisdictions provide for compensation of personal guardians and conservators for minors, with the standard most often used being "just and reasonable." A few jurisdictions (Connecticut, Louisiana, Mississippi, American Samoa) employ more specific criteria, setting a percentage figure of the estate income as the maximum allowable payment.

Termination of Guardianship and/or Conservatorship and Periodic Review (Chart XII)

Forty-six jurisdictions provide for automatic termination of a guardianship or conservatorship for a minor when the minor attains majority. An additional 24 jurisdictions provide for automatic termination upon the marriage of a minor. Six other jurisdictions provide more limited forms of automatic termination. In Guam, South Dakota and Wyoming conservatorships are not terminated by a marriage. In Illinois, the marriage of a girl for whom a conservator has been appointed does not automatically terminate the conservatorship. In South Dakota, the date of automatic termination falls one year after a minor attains majority. Eight jurisdictions (California, Florida, Georgia, Kentucky, Maryland, New Jersey, South Carolina, and the Virgin Islands) have no automatic termination provisions. Twenty-six states allow a minor, age 14 or over, to petition personally or through a guardian *ad litem* for termination prior to majority. If the court finds a continuing need for supervision, a petition for termination under these statutes may result only in the replacement of the current guardian by a nominee of the minor or another suitable person chosen by the court. In addition, these statutes generally allow petitions to be initiated by other interested persons as well as by the minor. Most states also provide for removal and replacement for cause of a personal guardian or conservator for a minor on motion of the minor or a person interested in the welfare of the minor, or on the court's own initiative. Grounds for removal include waste, failure to account or post bond, failure to obey a court order concerning the guardianship or conservatorship, habitual drunkenness, mental disability, and general neglect of duty.

In keeping with the traditional presumption in favor of guardianship and conservatorship for minors, only Florida provides for periodic review.

STATE STATUTES USED IN PREPARING CHARTS

ALA. CODE §§12-15-1 *et seq.* (1975); §§26-1-1 *et seq.* (1975); §§26-4-3 *et seq.* (1977).

ALASKA STAT. §§13.26.005 *et seq.* (1977); §§47.10.010 *et seq.* (1977).

ARIZ. REV. STAT. §§8-201.01 *et seq.* (1974 & Supp. 1977); §§14-5101 *et seq.* (1975); §36-547 (1974).

ARK. STAT. ANN. §§45-501 *et seq.* (1977); §§57-601 *et seq.* (Supp. 1977).

CAL. PROB. CODE §§1400 *et seq.* (1974 & Supp. 1977); WELF. & INST. CODE §§301 *et seq.* (Supp. 1978); §8006 (Supp. 1977); CIV. CODE §25 (Supp. 1977).

COLO. REV. STAT. §§15-14-101 *et seq.* (1973) & Supp. 1976); §§19-3-101 *et seq.* (1973 & Supp. 1977).

CONN. GEN. STAT. §17-59 (1977); §§45-45 *et seq.* (1977); §§45-70 *et seq.* (Supp. 1977); §§45-70a *et seq.* (Cum. Supp. 1979); §§45-78a *et seq.* (Cum. Supp. 1979).

DEL. CODE tit. 10 §§935 *et seq.* (1974); tit. 12 §§3901 *et seq.* (1975 & Supp. 1977).

D.C. CODE §§21-102 *et seq.* (1977); §§21-1501 *et seq.* (1973).

FLA. STAT. ANN. §39.06 *et seq.* (West 1976 & Supp. 1977); §§744.101 *et seq.* (West Supp. 1976); §§744.304 *et seq.* (West 1976 & Supp. 1977).

GA. CODE ANN. §§49-101 *et seq.* (1974 & Supp. 1977); §§74-101 *et seq.* (1964); §§88-2501 *et seq.* (1974 & Supp. 1977).

HAW. REV. STAT. §§560:5-101 *et seq.* (1976); §§571-31 *et seq.* (1976).

IDAHO CODE §§15-5-101 *et seq.* (Supp. 1978); §16-1610 *et seq.* (1978); §56-239 (Supp. 1977).

ILL. ANN. STAT. ch. 3 §§112 *et seq.* (Smith-Hurd Supp. 1977); ch. 3 §§139 *et seq.* (Smith-Hurd 1976 & Supp. 1978); ch. 37 §704-1 *et seq.* (Smith-Hurd 1972 & Supp. 1978); ch. 91½ §§701 *et seq.* (Smith-Hurd Supp. 1979); ch. 110½ §§11-3 *et seq.* (Smith-Hurd Supp. 1979); ch. 110½ §§11a-1 *et seq.* (Smith-Hurd Supp. 1979).

IND. CODE §§29-1-18-1 *et seq.* (1972 & Supp. 1977); §31-5-7-15 (Supp. 1977).

IOWA CODE §§232.28 and 232.33 (1969); §§633.552 *et seq.* (1964 & Supp. 1977).

KAN. STAT. §38-817 *et seq.* (Supp. 1977); §§59-3001 *et seq.* (1976).

KY. REV. STAT. §§203.010 *et seq.* (1975); §§208.010 *et seq.* (1975 & Supp. 1977); §§387.020 *et seq.* (1969 & Supp. 1977); H.B. No. 531 (1978).

LA. CIV. CODE ANN. art. 248 *et seq.* (West Supp. 1978); art. 354 *et seq.* (West Supp. 1977); art. 389 *et seq.* (1952); CODE CIV. PRO. ANN. art. 4006 *et seq.* (West 1961 & Supp. 1978); LA. REV. STAT. ANN. §§13:1575 and 13:1580 (West Supp. 1978).

ME. REV. STAT. tit. 18 §§3551 *et seq.* (1965 & Supp. 1977); tit. 18 §§3601 *et seq.* (1964 & Supp. 1977); tit. 18 §§3640 *et seq.* (1964 & Supp. 1977); tit.22 §§3792 *et seq.* (1965 & Supp. 1977).

MD. EST. & TRUSTS CODE ANN. §§13-101 *et seq.* (1974 & Supp. 1977).

MASS. ANN. LAWS ch. 201 §§1 *et seq* (Michie Law Coop 1969 & Supp. 1978).

MICH. COMP. LAWS ANN. §§330.1600 *et seq.* (1976); §§598.17 *et seq.* (1962 & Supp. 1977); §§703.1 *et seq.* (1976).

MINN. STAT. §§252A.01 *et seq.* (Supp. 1977); §260.111 (1971 & Supp. 1978); §525.54 *et seq.* (1975 & Supp. 1977).

MISS. CODE ANN. §43-23-17 (1972); §§93-13-1 *et seq.* (1973 & Supp. 1977).

MO. REV. STAT. §§210.125 *et seq.* (1962 & Supp. 1978); §§475.010 *et seq.* (1956 & Supp. 1978).

MONT. REV. CODE ANN. §§10-1301 *et seq.* (Supp. 1977); §§91A-5-101 *et seq.* (Special Pamphlet 1977).

NEB. REV. STAT. §§30-2601 *et seq.* (1975); §§43.205.01 *et seq.* (1974 & Supp. 1977).

NEV. REV. STAT. §§62.085 *et seq.* (1977); §159.017 *et seq.* (1973 & Supp. 1975).

N.H. REV. STAT. ANN. §169:13A (Supp. 1977); §§462:1 *et seq.* (1968 & Supp. 1975); §§463:1 *et seq.* (Supp. 1977).

N.J. STAT. ANN. §§3A:1-1 *et seq.* (West 1953 & Supp. 1977); §§3A:6-17 *et seq.* (West 1963 & Supp. 1978); §§9:6-8.1 *et seq.* (West 1976 & Supp. 1978).

N.M. STAT. ANN. §§13-14-9 *et seq.* (1976); §§32A-5-101 *et seq.* (1975 & Supp. 1976).

N.Y. CIV. PRAC. LAW §§1201 *et seq.* (McKinney 1977); DOM. REL. LAW §§81 *et seq.* (McKinney 1977); FAM. CT. ACT. §§249 and 634 (McKinney Supp. 1978); MENTAL HYG. LAW §§77.01 *et seq.* (McKinney 1976 & Supp. 1977); §§78.01 *et seq.* (McKinney 1976 & Supp. 1977); SOC. SERV. LAW §384-b (McKinney Supp. 1977); SURR. CT. PROÇ. ACT §§1701 *et seq.* (McKinney 1967 & Supp. 1977).

N.C. GEN. STAT. §7A-283 (Supp. 1977); §§33-1 *et seq.* (1976); §§35.1-1 *et seq.* (1966 & Supp. 1977).

N.D. CENT. CODE §§27-20-01 *et seq.* (1974); §§30.1-26-01 *et seq.* (1976; §§30.1-27-01 *et seq.* (1976).

OHIO REV. CODE ANN. §§2111.01 *et seq.* (Page 1976); §§2151.28 *et seq.* (Page 1976); §§5119.85 *et seq.* (Page Supp. 1976).

OKLA. STAT. ANN. tit. 10 §1101 (Supp. 1977); tit. 30 §§1 *et seq.* (1967); tit. 58 §§776 *et seq.* (1965); §§851 *et seq.* (1965 & Supp. 1977).

OR. REV. STAT. §§126.001 *et seq.* (1977); §§419.482 *et seq.* (1977).

PA. STAT. ANN. tit. 20 §2519 (Purdon 1975); §§5121 *et seq.* (Purdon 1975); §§5501 *et seq.* (Purdon 1975 & Supp. 1977); tit. 42 §§6224 *et seq.* (Purdon Pamphlet 1977).

R.I. GEN. LAWS §§14-1-16 *et seq.* (1970); §§33-15-1 *et seq.* (1969 & Supp. 1977); §§40-11-5 *et seq.* (1977).

S.C. CODE §14-21-590 (1970) (1977); §§21-19-40 *et seq.* (1976) & Supp. 1977); §§21-19-200 *et seq.* (1977).

S.D. COMPILED LAWS ANN. §§26-8-19 *et seq.* (1977);

	§§27B-6-1 *et seq.* (1976); §§30-27-1 *et seq.* (1976).
TENN.	CODE ANN. §§30-1501 *et seq.* (1977); §§33-313 *et seq.* (1977); §34-201 (1977); §§34-1008 *et seq.* (1977); §§37-101 *et seq.* (1977 & Supp. 1978).
TEX.	FAM. CODE tit. 2 §§11.01 *et seq.* (Vernon 1975 & Supp. 1978); PROB. CODE tit. 5 §§117 *et seq.* (Vernon 1956 & Supp. 1978); §§130 *et seq.* (Vernon Supp. 1978); §§415 *et seq.* (Vernon Supp. 1978).
UTAH	CODE ANN. §§75-1-101 *et seq.* (Pamphlet 1977); §78-3A-19.5 *et seq.* (1977 & Supp. 1978).
VT.	STAT. ANN. tit. 14 §§2601 *et seq.* (1974 & Supp. 1978); §§2683 *et seq.* (1974); tit. 33 §§639 *et seq.* (Supp. 1978).
VA.	CODE §16.1-246 (Supp. 1978); §§31-2 *et seq.* (1973); §§37.1-128 *et seq.* (1976).
WASH.	REV. CODE ANN. §§11.88-005 *et seq.* (1963 & Supp. 1977); §§13.04.050 *et seq.* (1963 & Supp. 1978).
W.VA.	CODE §§27-11-1 *et seq.* (1976 & Supp. 1977); §§44-10-1 *et seq.* (1966 & Supp. 1978); §49-6-1 *et seq.* (1976 & Supp. 1978).
WIS.	STAT. ANN. §§880.01 *et seq.* (West Pamphlet 1978); §§48.14 *et seq.* (West 1957 & Supp. 1978).
WYO.	STAT. §§3-2-201 *et seq.* (1977); §§14-8-101 *et seq.* (1977).
A.S.	CODE tit. 16 §§701 *et seq.* (1973).
GUAM	PROB. CODE §1401 *et seq.* (1970).
P.R.	LAWS ANN. tit. 31 §§661 *et seq.* (1967); §§681 *et seq.* (1971); tit. 34 App. I.R. 8.3 *et. seq.* (1967).
V.I.	CODE ANN. tit. 5 §§2503 *et seq.* (1967); tit.15 §§801 *et seq.* (1964).

I. ADULTS SUBJECT TO GUARDIANSHIP AND/OR CONSERVATORSHIP

State and Citation	Statutory Provision
Alabama §26-2-1	"persons of unsound mind" who are defined as "idiots, lunatics, or the insane"
Alaska §13.26.005 (Uniform Probate Code State)	"incapacitated person — any person who is impaired by reason of mental illness, mental deficiency, physical illness or disability, advanced age . . . or other cause (except minority) to the extent that he lacks sufficient understanding or capacity to make or communicate responsible decisions concerning his person."
§13.26.110	personal guardianship: incapacitated person for whom appointment is necessary or desirable as a means of providing continuing care and supervision of the person of the proposed ward.
§13.26.165	property guardianship: a person who is unable to manage his property and affairs effectively for reasons such as mental illness, mental deficiency . . . and who has property which will be wasted or dissipated unless proper management is provided, or that funds are needed for the support, care and welfare of the proposed ward or those entitled to be supported by him.
Arizona §36-547(A)	any person who is gravely disabled as a result of mental disorder.
§14-5301 §14-5401	(UPC State — see Alaska)
Arkansas §57-601	incompetence: incapability by reason of insanity, mental illness, imbecility, idiocy, senility . . . or other mental incapacity, either of managing property or caring for self.
California Prob. Code §1435.2 Prob. Code §1460	incompetence: legal, not medical, disability to be measured by substantial functional inabilities, not to be evaluated solely by isolated incidents of negligence or improvidence. personal guardianship: inability of adult to care for personal needs for physical health, food, clothing or shelter. property guardianship: substantial inability to manage financial resources.
Colorado §15-14-304 §15-14-401	(UPC State, see Alaska)
Connecticut §45-70a(g) §45-70a(h) (see also §45-78a)	"incapable of managing his own affairs" means a mental, emotional or physical condition resulting from mental illness, mental deficiency, physical illness or disability . . . which prevents such person from performing the functions inherent in managing his or her own affairs, and the person has property which will be wasted or dissipated unless proper management is provided or that funds are needed for the support, care or welfare of the person or those entitled to be supported by such person and that the person is unable to take the necessary steps to obtain or provide such funds. "incapable of caring for one's self" means a mental, emotional or physical condition resulting from mental illness, mental deficiency, physical illness or disability . . . which results in the person's inability to provide medical care for physical and mental health needs, nutritious meals, clothing, safe and adequately heated and ventilated shelter, personal hygiene and protection from physical abuse or harm and which results in endangerment to such person's health.

State and Citation	Statutory Provision
Delaware 12 §3914	person, not mentally ill, who by reason of advanced age, or mental infirmity or physical incapacity is unable properly to manage and care for his person or property and in consequence thereof is in danger of dissipating or losing such property or becoming the victim of designing persons, or where a guardian of the person is sought, such person is in danger of substantially endangering his health or of becoming subject to abuse by other persons or of becoming the victim of designing persons.
District of Columbia §21-1501	adult unable by reason of advanced age, mental weakness not amounting to unsoundness of mind, mental illness, or physical incapacity properly to care for his property. (Court may, in its discretion, invest conservator, with powers over the person, also §21-1506.)
Florida §744.102	person incapable of managing either person or property or both, because of minority, senility, lunacy, insanity, imbecility idiocy . . . or other physical or mental incapacity.
Georgia §88-2502	mental retardation: state of sub-average general intellectual functioning which originates during the developmental period and is associated with impairment in adaptive behavior; (mentally retarded person not accepted for hospitalization may have a guardian appointed over person and estate §88-2514, 1964.)
Hawaii §560:5-101 §560:5-301 §560:5-401	(UPC State see Alaska)
Idaho §15-5-101	(UPC State, see Alaska)
§56-239	mental retardation and developmental disabilities: court shall 1) inquire into the nature and extent of the general intellectual functioning of the individual asserted to need a guardian, 2) evaluate the extent of impairment in his adaptive behavior and 3) ascertain his capacity to care for himself and his estate.
Illinois ch. 110½ §11a-2	person who because of mental illness, developmental disability, mental deterioration or physical incapacity, is not fully able to manage his person or estate or who because of gambling, debauchery or excessive use of spirits or drugs so spends or wastes his estate as to reduce himself or his family to want
Indiana §29-1-18-1	"incompetent" or "disabled person": person incapable by reason of insanity, mental illness, mental retardation, senility, habitual drunkenness, excessive use of drugs, old age, infirmity, disappearance, or other incapacity, of either managing his property or caring for himself or both.
Iowa §633.552 §633.566	personal guardianship: mental retardate, mentally ill, senile, person incapable of caring for own person property guardianship: person incapable of managing own property.
Kansas §59-3002	person impaired by reason of mental illness, mental deficiency, physical illness or disability, advanced age . . . or other cause to the extent that he lacks sufficient understanding or capacity to make or communicate responsible decisions concerning either his person or his estate.

State and Citation	Statutory Provision
Kentucky §203.010	person of unsound mind who from confirmed bodily infirmity is unable to make known to others by speech, sign or otherwise his thoughts and desires and by reason thereof is unable to manage his estate, or one whose mind because of mental infirmity or old age has become so disabled as to render him unable to manage his estate.
Louisiana art. 355	person above age 15 possessing less than 2/3 of average mental ability of normal person of same age, evidenced by standard testing procedures administered by competent persons.
art. 389	no person . . . who is subject to an habitual state of imbecility, insanity or madness, shall be allowed to take care of his own person or administer his estate.
Maine 18 §3601	person over 18 and impaired by reason of advanced age, physical or mental illness or incapacity or other cause to the extent that he lacks sufficient understanding or capacity to make, communicate or implement responsible decisions concerning his person or property.
Maryland Estates and Trusts §13-101(d)	property guardianship: person other than a minor who is unable to manage his property because of physical or mental disability, senility, or other mental weakness or disease . . . and who has or may be entitled to property or benefits which require management. personal guardianship: persons unable to provide for his daily needs sufficiently to protect his health or safety because he lacks sufficient understanding or capacitv to make or communicate responsible decisions concerning his person, including provisions for health care, food, clothing, or shelter, because of any mental disability, senility or other mental weakness, disease.
Massachusetts 201 §6A	property guardianship: persons of advanced age, mental weakness or mental retardation or physical incapacity.
201 §16B	personal and property guardianship: person is mentally retarded to the degree that he is incapable of making informed decisions with respect to his personal and financial affairs, and failure to appoint a guardian would create an unreasonable risk to his health, welfare, and property, and appointment of a conservator would not eliminate such risk. property guardianship: person is mentally retarded to the degree that he is incapable of making informed decisions with respect to the conduct of his financial affairs and failure to appoint a conservator would create an unreasonable risk to his property.
Michigan §330.1600 §330.1618	mentally retarded person: person having significantly subaverage general intellectual functioning which originates during the developmental period and is associated with impairment in adaptive behavior; partial guardianship: mentally retarded individual who lacks the capacity to do some, but not all, of the tasks necessary to care for himself or his estate. plenary guardianship: a mentally retarded person who is totally without capacity to care for himself or his estate.
Minnesota §525.54	any person who is incompetent to manage his person or estate because of old age, or imperfection or deterioration of mentality.

State and Citation	Statutory Provision
Mississippi §93-13-111	mentally ill person who is reasonably expected in the near future to unintentionally or intentionally injure himself or others and is incapable of taking care of his person and property (does not apply to mentally retarded persons).
§93-13-251	person who, by reason of advanced age, physical incapacity or mental weakness is incapable of managing his own estate.
Missouri §475.010	any person who is incapable by reason of insanity, mental illness, imbecility, idiocy, senility . . . or other incapacity, of either managing his property or caring for himself or both.
Montana §91A-5-101 §91A-5-301 §91A-5-401	(UPC state, see Alaska)
Nebraska §30-2601 §30-2620 §30-2630	(UPC state, see Alaska)
Nevada §159.019	incompetent: person who, by reason of mental illness, mental deficiency, advanced age, disease, weakness of mind or any other cause is unable without assistance, properly to manage and take care of self or property.
New Hampshire §464:2	mentally incompetent person (not defined in statute). property guardianship: wherever person shall deem himself unfitted by reason of infirmities of age, or by other mental or physical disability for the management of his affairs with prudence and understanding, he may apply to the probate court for the appointment of a conservator of his property.
New Jersey §3A:1-1	mental incompetent: person who, as result of idiocy, insanity, lunacy, unsoundness of mind, or habitual drunkenness, is incapable of governing himself and managing his affairs.

State and Citation	Statutory Provision
New Mexico §32A-5-101 §32A-5-301 §32A-5-401	(UPC state, see Alaska)
New York Mental Hyg. L. §77.01	person not declared incompetent who, by reason of advanced age, illness, infirmity, mental weakness, intemperance, addiction to drugs or other cause, has suffered substantial impairment of his ability to care for his property or has become unable to provide for himself or others dependent upon him for support.
§78.01	incompetent: person unable to manage self or affairs by reason of age, drunkenness, mental illness or other cause. patient: person lawfully committed or admitted to any facility for the mentally retarded or mentally ill in the Department of Correction or Department of Mental Hygiene unable adequately to conduct his personal or business affairs.
NY Surr. Ct. Proc. Act §1750	certified as incapable to manage himself and/or affairs by reason of mental retardation.
North Carolina §35-1.7(11)	incompetent: person who lacks sufficient capacity to make or communicate important decisions concerning his person, family or property because of mental illness, mental retardation, epilepsy, cerebral palsy, or autism.
North Dakota §30.1-26-04 §30.1-29-01	(UPC state, see Alaska)
Ohio §2111.01	incompetent: person who by reason of advanced age, improvidence, mental or physical disability or infirmity . . . mental retardation or mental illness, is incapable of taking proper care of himself or his property or fails to provide for his family or other persons for whom he is charged by law to provide.

State and Citation	Statutory Provision
Oklahoma 58 §851 58 §852	person who is insane, or who from any cause is mentally incompetent to manage property, and is incapable of taking care of himself and managing his property.
Oregon §126.003	(UPC state, see Alaska)
Pennsylvania tit.20 §5501	person who, because of old age, mental illness, mental deficiency,.or retardation, drug addiction or inebriety 1) is unable to manage his property or is liable to dissipate it or become the victim of designing persons, or 2) lacks sufficient capacity to make or communicate responsible decisions concerning his person.
Rhode Island §33-15-8	personal or property guardianship: any idiot, lunatic, or person of unsound mind . . . who . . . from want of discretion in managing his estate, so spends, wastes or lessens his estate, or is likely to do so, that he may bring himself or his family to want. property guardianship: aged or incompetent person incapable of properly caring for his property.
South Carolina §21-19-200	any person other than a mentally ill person primarily in need of mental health services, whose intellectual deficit and adaptive level of behavior require for his benefit, or that of the public, special training, education, supervision, treatment, care or control.
South Dakota §30-27-6	incompetent: person . . . mentally ill or for any cause mentally or physically incompetent to manage his own property. person is mentally retarded and not receiving proper care or education.
Tennessee §33-313	mentally ill and mentally retarded persons undergoing hospital proceedings: clear, unequivocal and convincing evidence that the individual is incompetent to manage himself and his estate; the question of incompetency (to manage self and estate) is separate and apart from the hearing and consideration of mental illness or the need for judicial commitment or hospitalization.
§34-1008	person of advanced age, physical incapacity or mental weakness.
Texas Prob. Code 5 §130H	mentally retarded person who lacks capacity to do some, but not all, of the tasks necessary to care for himself or his estate.
5§415	person of unsound mind.
Utah §75-1-201 §75-5-301 §75-5-401	(UPC state, see Alaska)
Vermont 14 §2683	an insane person or . . . person mentally incapable of taking care of himself or his property.
Virginia §37.1-128	any person who by reason of mental or physical condition has become incapable either wholly or partially, of taking proper care of his person or properly handling and managing his estate.
Washington §11.88.010	person incompetent by reason of insanity, mental illness, mental retardation . . . or other mental incapacity of managing his property, caring for self, or both; disabled person who by reason of his disability has need of assistance but cannot be found to be fully incompetent.

State and Citation	Statutory Provision
West Virginia §27-11-2	person unable to care for his physical well-being: substantial risk of physical harm to self as evidenced by conduct demonstrating dangerousness to self, notwithstanding poor judgment; inability to manage business affairs; inability to know and appreciate the nature and effect of business transactions, notwithstanding poor judgment.
Wisconsin §880.01	incompetent: person adjudged to be substantially incapable of managing his property or caring for himself by reason of aging, developmental disabilities or like incapacities. developmentally disabled person: person having disability attributable to mental retardation, cerebral palsy, epilepsy, autism, or other neurological condition closely related to mental retardation or requiring similar treatment to that required for mentally retarded individuals, which has continued or can be expected to continue indefinitely, substantially impairs the individual from adequately providing for his own care or custody and is a substantial handicap to him; term does not include senility.
Wyoming §3-2-301	individual unable, unassisted, to properly manage and take care of self or property or both as the result of mental illness, mental deficiency, or mental retardation.
American Samoa 16 §701	mental or physical incompetence to manage one's own property.
Guam §1460	any person, whether insane or not, who by reason of old age, disease, weakness of mind, or other cause, is unable, unassisted, properly to manage to take care of himself or his property, and thereby is likely to be deceived or imposed upon by artful or designing persons.
Puerto Rico 31 §661	insane or demented person, including minors, deaf and dumb, prodigals, and habitual drunkards, persons under sentence of civil interdiction.
Virgin Islands 15 §841	insane person, idiot and other person incapable of conducting his own affairs.

PUBLIC GUARDIANSHIP

State and Citation	Statutory Provision
Alabama §26-2-50	persons of unsound mind
Arizona §14-5602	persons in need of guardianship, conservatorship or public administration for whom there is no person or corporation qualified and willing to act.
California §8006	persons requiring guardianship or conservatorship for whom there is no person or corporation qualified and willing to act in such capacity (includes patients in county-owned facilities).
Colorado §26-3-107	an incapacitated person.
Delaware 12 §3992	aged, mentally infirm or physically incapacitated persons in need of guardian under private guardianship. statute (§3914).
Georgia §49-604(d) (2)	person mentally ill, mentally retarded or mentally incompetent to the extent he is incapable of managing his estate . . . and is receiving or eligible for services from the Department of Human Resources.

State and Citation	Statutory Provision
Maine 18 §3623	mentally retarded persons incapable of managing self and affairs independently, requiring supervision and care of person and estate.
18 §3640	person 18 years or over who is impaired by reason of advanced age, physical or mental illness or incapacity or other cause to extent he lacks sufficient understanding or capacity to make, communicate or implement responsible decisions concerning his person or property (excludes mentally retarded persons).
Minnesota §252A.02	(mentally retarded) persons having significantly subaverage intellectual functioning existing concurrently with demonstrated deficits in adaptive behavior such as to require supervision and protection for his welfare or the public welfare.
North Carolina §35-1.6 §35-1.28	adult who lacks sufficient capacity to make or communicate important decision concerning his person, family or property because of mental illness, mental retardation, epilepsy, cerebral palsy or autism.
Ohio §5119.86(c)	see private guardianship — standard for appointment of public guardian is the same, although only for "mentally retarded and other developmentally disabled persons."
Oregon §126.905	see private guardianship — standard for appointment of public guardian is the same.
Tennessee §30-1501	see private guardianship — standard for appointment of public guardian is the same.

II. TYPES OF GUARDIANSHIP AND/OR CONSERVATORSHIP (ADULTS)

State	Power over Person, Property or Both	Plenary or Limited Powers	Testamentary and Standby Appointment	Temporary Appointment	Emergency Appointment
Alabama	person/property distinction not addressed	Plenary	T		
Alaska	X	Plenary	T	Court may appoint a temporary guardian for 6 months when a permanent guardian is discharged for misconduct §13.26.140	Pending notice and hearing for permanent guardianship the court may exercise the powers of a guardian §13.26.140
Arizona	X	Plenary	T	§14-5310 see Alaska 30 days if individual is gravely disabled § 36-547.02	§14-5310 see Alaska
Arkansas	X	Plenary	T	90 days §57-620	
California	X	Plenary	T	30 days or until permanent guardian appointed Prob. Code §1640; §1645	
Colorado	X	Plenary	T	15-14-310 see Alaska	see Alaska
Connecticut	X	Plenary (Limited conservator for M.R. §45-78(c))	S for M.R.	30 days § 45-72	
Delaware	X	Plenary			
District of Columbia	may be conservator or personal guardian and conservator (no provision for personal guardian only)	Plenary	T	until permanent guardian appointed §21-1505	
Florida	X	Limited §744.303	T. S		
Georgia	X	Plenary	T	90 days or until permanent guardian appointed. whichever is shorter § 49-604	

State	Power over Person, Propety or Both	Plenary or Limited Powers	Testamentary and Standby Appointment	Temporary Appointment	Emergency Appointment
Hawaii	X	Plenary	T	§560:5-310 see Alaska	see Alaska
Idaho	X	Limited §56-239	T	§ 15-5-310 see Alaska	see Alaska
Illinois	X	Limited ch.110½§11a-12	T	60 days ch. 110½§11a-4	
Indiana	X	Plenary	T S	60 days §29-1-18-24	
Iowa	X	Plenary	S	no time limit is specified §633.558	
Kansas	X	Plenary			
Kentucky	one individual to serve as both personal guardian and conservator	Limited §203.022(3)	T		
Louisiana	one individual to serve as both personal guardian and conservator	Plenary	T	no time limit is specified La. Civ. Code art. 394 (Conservator only)	
Maine	X	Limited 18§3512		no time limit is specified 18§3510 18 §3648	
Maryland	X	Limited §13-705(b)	T		6 days §13-709
Massachusetts	may be conservator or personal guardian and conservator (no provision for personal guardian only)	Plenary	T		until permanent guardian appointed 201§14
Michigan	X	Limited §330.1618	T S	60 days §330.1608	no time limit is specified
Minnesota	X	Plenary		until permanent guardian appointed §525.591	
Mississippi	X	Plenary	T		
Missouri	X	Plenary			

State	Power over Person, Property or Both	Plenary or Limited Powers	Testamentary and Standby Appointment	Temporary Appointment	Emergency Appointment
Montana	X	Plenary	T	§91A-5-310 see Alaska	see Alaska
Nebraska	X	Plenary	T	§30-2626 see Alaska	see Alaska
Nevada	X	Plenary		6 months or until permanent guardian appointed, whichever is shorter §159.051	
New Hampshire	X	Plenary		60 days §464:4	
New Jersey	one individual to serve as both personal guardian and conservator	Plenary	T		
New Mexico	X	Plenary	T	6 months §32A-5-310	
New York	X	Limited: Mental Hygiene Law §77.25, §78.02 Surr.Ct.Proc.Act §1751			
North Carolina	X	Limited §35-1.6	T		
North Dakota	X	Plenary	T	§30.1-28-10 see Alaska	see Alaska
Ohio	X	Plenary	T		
Oklahoma	one individual to serve as both personal guardian and conservator	Plenary	T		
Oregon	X	Plenary	T	no time limit is specified §126.133	no time limit is specified
Pennsylvania	X	Plenary		no time limit is specified 20§5513	
Rhode Island	X	Plenary	T	until permanent guardian appointed §33-15-10 §33-15-11	
South Carolina	X	Limited §21-19-210	S		
South Dakota	X	Limited — court discretion §30-28-1	T		

State	Power over Person, Propety or Both	Plenary or Limited Powers	Testamentary and Standby Appointment	Temporary Appointment	Emergency Appointment
Tennessee	X	Plenary	T		
Texas		Limited Prob. Code 5 §130H (mentally retarded persons only)	T	at court's discretion (may be made permanent) Prob. Code 5 §131	
Utah	X	Plenary	T	§75-5-310 see Alaska	see Alaska
Vermont	X	Plenary	T		
Virginia	X	Limited §37.1-128.1(A)	T		
Washington	X	Limited §11.88.005 §11.88.125	S		
West Virginia	X	Limited §44-10A-2	T		
Wisconsin	X	Limited §880.37	T	60 days §880.15	
Wyoming	X	Plenary	T		
American Samoa	X	Plenary			
Guam	X	Plenary			
Puerto Rico	X	Plenary			
Virgin Islands	one individual to serve as both personal guardian and conservator	Plenary			

PUBLIC GUARDIANSHIP

State	Power over Person, Property or Both	Plenary or Limited Powers	State	Power over Person, Property or Both	Plenary or Limited Powers
Arizona	X	Plenary	Maine	X	Limited
California	X	Plenary	Minnesota	X	Limited
Colorado	X	Limited	North Carolina	X	Limited
Delaware	X	Limited	Ohio	X	Plenary
Georgia		Plenary	Oregon	X	Plenary
Illinois	X	Limited	Tennessee		Plenary

III. PROCEDURES PRECEDING APPOINTMENT (ADULTS)[1]

State	Notice of Proceeding		Evaluation prior to hearing by				Judicial hearing
	to alleged partially disabled or disabled person	how many days	physician	psychologist	visitor (UPC)	other	
Alabama							yes
Alaska	§13.26.135 §13.26.185	14 days	X		X		
Arizona	§36-547.03(2) §14-5309(B) §14-5405	14 days	by two physicians §34-547.01		X		yes
Arkansas	§57-611(b)	3 days	by "one or more medical witnesses" §57-615(b)				yes
California	Cal. Prob. Code §1461 §1754	10 days				court appointed investigator (lawyer) Prob. Code §1461.14	yes
Colorado	§15-14-309(2) §15-14-405	10 days	X		X		yes
Connecticut	§45-70(c) §45-78(g)	7 days	by two physicians or one physician and one psychologist §45.78a				yes
Delaware	12§3914(b)	"reasonable notice"					yes
District of Columbia	§21-1502(a)	14 days					yes
Florida	§744.331 §744.335	"reasonable notice"	by two physicians §744.331(5) (a)			by one citizen §744.331(5) (a)	yes
Georgia	§49-604 §88-507.3(c)	within 5 days	X			may go to evaluation facility §88-507.3	yes
Hawaii	§560-5-309 §560-5-405	14 days	X		X		yes
Idaho	§15-5-309 §15-5-405	14 days	X		X		yes

State	Notice of Proceeding		Evaluation prior to hearing by				Judicial hearing
	to alleged partially disabled or disabled person	how many days	physician	psychologist	visitor (UPC)	other	
Illinois	ch.110½§11a-10(d)	14 days				qualified persons ch.110½§11a-9	yes
Indiana	§29-1-18-14	10 days; 3 days for good cause	no, but see §29-1-18-20				yes
Iowa	§633.554&568 §633.568	rules of civil procedure govern					yes
Kansas	§59-3010(A) (1) §59-3012	between 7-14 days	physician or psychiatrist §59-3010(A) (6)				yes
Kentucky	§203.016	3 days	2 physicians or 1 physician and 1 clinical psychologist §203.015				yes
Louisiana	notice "shall be made as in ordinary proceedings" art. 4544		"by physician or other skillful persons" art. 393				yes
Maine	18§3602	14 days				Dept. of Health & Welfare 18§3601	yes
Maryland	§13-201(a) §13-211	as prescribed by law or statute rules					yes
Massachusetts	201§7 201§17	7 days	X 201§6A, 201§16B	X		by one social worker	yes
Michigan	§330:1614(1)&(3)		X or psychologist §330:1612	X or physician §330:1612			yes
Minnesota	§525.55	14 days					yes
Mississippi	§93-13-111	5 days					yes
Missouri	§475.075(2)	reasonable time					yes
Montana	§91A-5-405	14 days	X		X		yes

State	Notice of Proceeding		Evaluation prior to hearing by				Judicial hearing
	to alleged partially disabled or disabled person	how many days	physician	psychologist	visitor (UPC)	other	
Nebraska	§30-2625(b) §30-2634(a).	14 days	X		X		yes
Nevada	§159.047(2) (a) §159.053	20 days					yes
New Hampshire	§464:1	14 days				3 suitable persons §464:1	yes
New Jersey	Rule 4:83-4(a)	20 days	two physicians Rule 4:83-2				yes
New Mexico	§32-5-309 §32A-5-405	14 days	X		X		yes
New York	M. Hyg. Law §78.03 M. Hyg. Law §77.07		two licensed physicians or one licensed physician & one licensed psychologist; SCPA §1750				yes
North Carolina	§35-1 14&16(b)	10 days				multi-disciplinary evaluation §35-1.16(b)	no, to hearing before clerk
North Dakota	§30.1-28-09 §30.1-29-05		X		X		yes
Ohio	§2111.04 (B)	3 days					yes
Oklahoma	58§851	5 days					yes
Oregon	§126.007, §126.103 §126.187	10 days	X		X		yes
Pennsylvania	20§5511(a)	in such manner as the court shall direct	X				yes
Rhode Island	§33-15-17, §33-15-44	14 days					yes
South Carolina	§44-17-520 §44-17-540 §21-19-220	10 days if possible				two examiners, one licensed physician; §44-17-540 see §21-19-220	yes

State	Notice of Proceeding			Evaluation prior to hearing by				Judicial hearing
	to alleged partially disabled or disabled person		how many days	physician	psychologist	visitor (UPC)	other	
South Dakota	no	§30-27-15						yes
Tennessee	yes	§34-1009	7 days	two physicians §34-1010				yes
Texas	no yes	Prob. Code 5 §416 Prob. Code 5 §130E					at evaluation center Prob. Code 5 §130F	yes
Utah	yes	§75-5-405(1)	10 days	§75-5-303(2) §75-5-308				yes
Vermont	yes	14§2684	6 days if in state, 10 days if out of state					yes
Virginia	yes	§37.1-128(A)	"reasonable notice"					yes
Washington	yes	§11.88.040	10 days				"sworn medical report" §11.88.045	yes
West Virginia	yes	§27-11-1(b)	10 days	two licensed physicians; §44-1-A-1	one licensed psychologist; §44-1-A-1			yes
Wisconsin	yes	§880.08(1)	10 days	X §880.33	X §880.33			yes
Wyoming	yes	§3-2-304	10 days					yes
American Samoa								
Guam	yes	§1461	5 days					yes
Puerto Rico							experts 31§706	yes
Virgin Islands	yes	§842	10 days					yes

[1] The public guardianship statutes in California, Delaware, Georgia, North Carolina, Ohio, Oregon and Tennessee explicity provide that the procedures which apply to the appointment of a private personal guardian or conservator also apply to the appointment of the public guardian. Colorado [§26-3-104], Maine [18§3626 and 18§3644] and Minnesota [§252A.08], provide for somewhat shorter notice periods in their public guardianship statutes and in the case of Colorado and Minnesota, more detailed requirements for the membership of the evaluation team. Alabama, Arizona and Illinois do not address specifically the procedures which apply when appointment of the public guardian is sought.

IV. RIGHTS OF THE RESPONDENT (Adults)[1]

State	Representation: by attorney	Representation: by guardian *ad litem*	Be present	Jury trial	Cross-examine witnesses	Present evidence	Closed hearing
Alabama		§26-2-42(B)	if consistent with health or safety §26-2-43	§26-2-43 §26-2-44			
Alaska		appropriate official or attorney with powers & duties of guardian *ad litem* §13.26.105 §13.26.195	§13.26.105	upon request §13-26.110 does not apply to conservatorship proceedings	yes	yes	yes
Arizona		attorney with powers & duties of guardian *ad litem* §14-5303(B) §14-5407(B) §36-547.03(4)	§14-5303(B)	upon request §14-5303 does not apply to conservatorship proceedings	yes	yes	yes
Arkansas		no; §57-611(c)	at court's discretion §57-615(b)				
California	Prob. Code §1606 §1754	Prob. Code §1607	Prob. Code §1461 §1754	if demanded Prob. Code §1461			
Colorado		appropriate attorney or official with powers & duties of guardian *ad litem* §15-14-303(2), §15-14-407	§15-14-303(2)	§15-14-303 does not apply to conservatorship proceedings	yes	yes	yes
Connecticut	§45-70c		§45-70c				
Delaware	12§3914(b)						
District of Columbia		§21-1502(b)					
Florida	if indigent, counsel appointed §744.331(4)						

State	Representation		Be present	Jury trial	Cross-examine witness	Present evidence	Closed hearing
	by attorney	by guardian *ad litem*					
Georgia	§88-507.3(f)	if no other representative §88-507.3(a)	§88-507.3	decided by a commission of two physicians and one attorney §88-507.3			
Hawaii		attorney with powers & duties of guardian *ad litem* §560:5-303(b) §560:5-407(b)	§560:5-303(b)		yes	yes	yes
Idaho		attorney with powers & duties of guardian *ad litem* §15-5-303(b) §15-5-407(b) §56-239	§56-239 §15-5-303(b)		yes	yes	yes
Illinois	attorney & guardian *ad litem* ch.110½§11a-10	attorney & guardian *ad litem* ch.110½§11a-10		if requested ch. 110½§11-1			
Indiana	court may appoint §29-1-18-19	at court's discretion §29-1-18-14	unless injurious to health §29-1-18-19	if requested §29-1-18-19			
Iowa				§633.555			
Kansas	§59-3010		§59-3013	if requested at least 48 hours prior to hearing §59-3013	yes	yes	yes
Kentucky	§203.014		§203.018	§203.022			
Louisiana	art. 4544						
Maine							
Maryland	Maryland Rules Rule R76			no; §13-211 but see Maryland Rule R77(b)			

State	Representation: by attorney	Representation: by guardian *ad litem*	Be present	Jury trial	Cross-examine witness	Present evidence	Closed hearing
Massachusetts							
Michigan		at court's discretion §330.1616					
Minnesota							
Mississippi			§93-13-111				
Missouri		§475.075(3) incapacitated persons: appropriate official or attorney with powers & duties of guardian *ad litem*	§475.075(2)	if requested §475.075(1)			
Montana		protected persons; attorney with powers & duties of guardian *ad litem* §91A-5-303 §91A-5-407(2)	§91A-5-303	upon request 91A-5-303 does not apply to conservatorship proceedings	yes	yes	yes
Nebraska		appropriate official or attorney with powers & duties of guardian *ad litem* §30-2619(b) §30-2636(b)	§30-2619(b)		yes	yes	yes
Nevada			§159.047				
New Hampshire			§464:2				
New Jersey	within court's discretion to appoint Rule 4:83-4(a)		Rule 4:83-4(a)	if demanded §3A:6-35 Rule 4:83-4(a)			
New Mexico		attorney with powers & duties of guardian *ad litem* §32A-5-303(B) 32A-5-407(B)	§32A-5-303(B)	upon request §32A-5-303 does not apply in conservatorship proceedings	yes	yes	yes

State	Representation		Be present	Jury trial	Cross-examine witness	Present evidence	Closed hearing
	by attorney	by guardian *ad litem*					
New York		Mental Hyg. Law §77.07(b) §78.03(e)	unless inability is established to satisfaction of court Mental Hyg. Law §77.07(b)	Mental Hyg. Law §77.07(c) §78.03(4)			
North Carolina	§35-1.16(a)	for persons who do not want counsel	on request §35-1.16(c)	yes	yes	yes	on request §35-1.16(d)
North Dakota		attorney with powers & duties of guardian *ad litem* §30.1-28-03 §30.1-29-07(2)	§30.1-28-03		yes	yes	yes
Ohio	§2111.02						
Oklahoma			if able to attend 58§851				
Oregon	§126.103(5) §126.197(3)		§126.103(5)		yes	yes	yes
Pennsylvania		no 20§5511(a) (ii)	unless testimony satisfies court that physical or mental welfare will not be promoted 20§5511(a) (ii)	20§5511(a)			
Rhode Island		§33-15-9					
South Carolina	appointed within 3 days after notice §44-17-530 see §21-19-220		if over 21 §21-19-220				
South Dakota			at court's discretion §30-27-17			"any person may be heard" §30-27-16	

State	Representation		Be present	Jury trial	Cross-examine witness	Present evidence	Closed hearing
	by attorney	by guardian *ad litem*					
Tennessee	court may designate counsel guardian *ad litem* in proposed ward's best interest §33-313	§34-1010	mentally ill & mentally retarded persons undergoing hospitalization proceedings: unless waived by counsel §33-313				
Texas	Prob. Code 5§130G	Prob. Code 5§417	Prob. Code 5 §130G Prob. Code 5 §115	on request Prob. Code 5 §130G Prob. Code 5 §417			on request Prob. Code 5 §130G
Utah		incapacitated persons: appropriate offical or attorney with powers & duties of guardian *ad litem* §75-5-303(2) protected persons: attorney with powers & duties of guardian *ad litem* §75-5-407(2)	§75-5-303(2)	upon request §75-5-303 does not apply in conservatorship proceedings	yes	yes	yes
Vermont							
Virginia		§37.1-128.1(A)					
Washington	§11.88.045	§11.88.090	guardian *ad litem* may waive, or court may conduct hearing at proposed ward's residence if physically unable to be present §11.88.040	upon request §11.88.045			
West Virginia	retained or appointed §27-11-1(b)		§27-11-1(b) §44-10A-3				
Wisconsin	right to counsel whether or not present §880.33(2)(a)	guardian *ad litem* must be appointed in any event §880.33(2) (a)	if able to attend §880.08(1)	on request §880.33(a)	yes	yes	

State	Representation		Be present	Jury trial	Cross-examine witness	Present evidence	Closed hearing
	by attorney	by guardian *ad litem*					
Wyoming	at court's discretion §3-2-305(b)		where presence cannot be expected because of health or because proposed ward is patient in facility, licensed physician & court-appointed examiner or head of institution must state reasons for inability to attend §3-2-304	in court's discretion §3-2-306			
American Samoa							
Guam			if able to attend §1261				
Puerto Rico		next friend appointed for persons presumed to be incapable who do not wish or are unable to defend themselves. 31§705					
Virgin Islands							

[1] The public guardianship provisions in California, Delaware, Georgia, North Carolina, Ohio, Oregon and Tennessee afford the same rights to respondents as the provisions for the appointment of a private personal guardian or conservator. In Colorado the public guardianship statute unlike the provisions on private guardianship or conservatorship does not expressly guarantee respondents the rights to a jury, to cross-examine opposing witnesses, and to a closed hearing. The Minnesota public guardianship statute on the other hand, does explicitly provide for these rights even though the private guardianship and conservatorship provisions do not. The Alabama, Arizona, Illinois and Maine codes do not address the question.

V. DUTIES AND POWERS IN GUARDIANSHIPS AND/OR CONSERVATORSHIPS FOR ADULTS

State	A bond must be filed by conservators	Conservators are made fiduciaries by law	Compensation	Personal guardians required to secure habilitation for the individuals they are serving
Alabama	§26-3-1		2½% of receipts and disbursements §26-5-16	
Alaska	§13.26.215	§13.26.245	reasonable §13.26.230	whenever appropriate §13.26.150
Arizona	§14-5411	§14-5417	reasonable §14-5314	whenever appropriate §14-5312
Arkansas	§57-616		reasonable §57-641	
California	Prob. Code §1480	Prob Code §1400	reasonable Probate Code §1556. §1908	
Colorado	§15-14-411	§15-14-424	reasonable §15-14-312(2) §15-14-414	whenever appropriate §15-14-312(1) (b)
Connecticut	§45-70d			
Delaware	12§3907			
District of Columbia	§21-1503	§21-1701(a)		
Florida	unless waived by the subject of the conservatorship §744.351(1)		reasonable §744 108	
Georgia	§49-603			§49-604(b)
Hawaii	§560:5.411		reasonable §560 5-312 §560 5-414	whenever appropriate §560:5-312
Idaho	§15-5-411		reasonable §15-5-312 §15-5-414	whenever appropriate §15-5-312(2) (a)
Illinois	ch.110½§12-2		reasonable ch.110½§27-1	ch.110½§11a-17

State	A bond must be filed by conservators	Conservators are made fiduciaries by law	Compensation	Personal guardians required to secure habilitation for the individuals they are serving
Indiana	§29-1-18-22	as determined by court §29-1-18-45		
Iowa	§633.634	§633.634	§633.673	
Kansas	§59-3014(2)		reasonable §59-3032	
Kentucky	§387.070		same compensation as trustee of estate §387.230(3)	
Louisiana	art. 4131			
Maine	18§3801			
Maryland	13§208	§13-213	Same as for trustee of a trust §13-218	§13-214(b) (2)
Massachusetts	201§19		compensation as court allows 206§16	201§6A
Michigan		§330:1632		
Minnesota	§525.551			
Mississippi	§93-13-17	§93-13-38		
Missouri	§475.100		reasonable §475.265	
Montana	§91A-5-411	§91A-5-417	reasonable §91A-5-312, §91A-5-414	whenever appropriate §91A-5-312(b)
Nebraska	§30-2640	§30-2646	reasonable §30-2628(b), §30-2643	whenever appropriate §30-2628(2)
Nevada	§159.065			§159.079
New Hampshire	§462:3		reasonable §462:8	
New Jersey	Rule 4:83-6(c)			
New Mexico	§32A-5-411	§32A-5-417	reasonable §32A-5-312, §32A-5-414	whenever appropriate §32A-5-312(A) (2)

State	A bond must be filed by conservators	Conservators are made fiduciaries by law	Compensation	Personal guardians required to secure habilitation for the individuals they are serving
New York	Mental Hyg. Law §78.09, §77.13		fixed by court Mental Hyg. Law §78.21, §77.27	when required Mental Hyg. Law §78.01, §77.19
North Carolina	§35-1.19		expenses incurred §35-1.34(b)	§35-1.34(a) (2)
North Dakota	§30.1-29-11	§30.1-29-17	reasonable §30.1-28-12-2 §30.1-29-14	whenever appropriate §30.1-28-12-1(b)
Ohio	§2109.12	§2109.01	as court directs §2109.23	
Oklahoma	58§853	58§776	reasonable 58§884	
Oregon	§126.237	§126.237	reasonable §126.263	whenever appropriate §126.137(b)
Pennsylvania	20§5515, 20§5516	20§5516		
Rhode Island	§33-15-45 §33-17-1		reasonable §33-15-27	
South Carolina	§21-19-40		same as executors, administrators & other trustees §21-19-180	§21-19-200
South Dakota	§30-27-29	§30-29-2		§30-28-4
Tennessee	§34-1013		reasonable §34-810	
Texas	Prob. Code 5 §130K	Prob. Code 5 §108	governed by law on decendents' estates Prob. Code 5 §108	
Utah	§75-5-411	§75-5-417	reasonable §75-5-312(2) §75-5-414	whenever appropriate §75-5-312(1) (b)
Vermont	14§2751		reasonable 14§2847	
Virginia	§37.1-128.1(B)	§37.1-128.1(B)		

State	A bond must be filed by conservators	Conservators are made fiduciaries by law	Compensation	Personal guardians required to secure habilitation for the individuals they are serving
Washington	unless estate less than $1500-then waivable §11.88.100	§11.92.040(4)	reasonable §11.92.180	§11.92.040(3)
West Virginia	§27-11-2 §44-10 5			
Wisconsin	may be waived §880.13		reasonable §880.24(1)	as far as may be necessary §880.21(1)
Wyoming	may be waived §3-2-108 §3-2-111		reasonable §3-2-121	
American Samoa	16§704		not exceeding 2½% of amounts received & dis-bursed minimum fee of $40 16§709	
Guam	§1480		as determined by court §1556	as far as necessary §1502
Puerto Rico	unless parent or grandparent 31§767		4%-10% of net income or proceeds of property 31§790	§31-783(2)
Virgin Islands	§881(b) (1)		as determined by court §888	
		PUBLIC GUARDIANSHIP		
Alabama	§26-3-5 §26-3-6			
Arizona			reasonable §14-5604	
California	Welf. & Inst. Code §8008		cost incurred plus reasonable fee of $25-$100 Welf. & Inst. Code §8006.5	
Colorado	§26-3-109			§26-3-105(b)
Delaware	if required 12§3991		costs incurred may be charged 12§3993	
Georgia	§49-502			

State	A bond must be filed by conservators	Conservators are made fiduciaries by law	Compensation	Personal guardians required to secure habilitation for the individuals they are serving
Illinois			in accordance with set standards following notice at the time of appointment ch.91½§709	
Maine	18§3629 18§3647		18§3629 reasonable compensation to Bureau of Mental Retardation 18§3647 no compensation to Dept. of Health and Welfare	insofar as practicable 18§3628
Minnesota				§252A.11(2)
North Carolina	§35-1.19		for time and trouble §35-1.28	§35-1.34 (a) (2)
Ohio	§5119.86		Public Guardian not entitled to compensation §5119.86	
Oregon	§126.935		reasonable §126.955	
Tennessee	§30-1501		reasonable §30-1507	

VI. TERMINATION AND PERIODIC REVIEW

State	Action to terminate guardianship or conservatorship		Periodic review
	statute provides a termination procedure	subject of guardianship or conservatorship may petition	
Alabama	§26-2-51	yes	
Alaska	§13.26.125(b)	yes	
Arizona	§14-5307(B)	yes	annually §36-547 (B) (1974)
Arkansas	§57-643		
California	Prob. Code §1470	yes	one year after appointment and bienially thereafter Prob. Code §1500.1 §1851.1
Colorado	§15-14-307(2) §15-14-430	yes	
Connecticut	§45-77		annually §45-70
Delaware	12§3914(e)	yes	
District of Columbia	§21-1504	yes	
Florida	§744.371 See also §744.464		annually. §744.364
Georgia	§49.605(b)	yes	
Hawaii	§560-5-307 §560-5-415	yes	
Idaho	§15-5-307 §15-5-415	yes	at interval ordered by court, but at least annually. §56-241
Illinois	ch.110½§11a-19 ch.110½§11a-20	yes	
Indiana	§29-1-18-47	yes	department of public welfare investigates whenever ordered by court §29-1-18-13 see also §29-1-18-28
Iowa	§633.679	yes	at such times as court orders, guardian must report §633.669

State	Action to terminate guardianship or conservatorship: statute provides a termination procedure	Action to terminate guardianship or conservatorship: subject of guardianship or conservatorship may petition	Periodic review
Kansas	§59-3027	yes	
Kentucky	§203.024	yes	
Louisiana	La. Civ. Code art. 361 art. 420	yes	
Maine	18§3607	yes	
Maryland	§13-221(1974) Rule R.80	yes	
Massachusetts	201 §13	yes	
Michigan	§330.1636	yes	annual report §330:1630
Minnesota	§525.61	yes	
Mississippi	§93-13-151	yes	
Missouri	§475.360	yes	
Montana	91A-5-430	yes	
Nebraska	§30-2623 §30-2659	yes	
Nevada	§159.191	yes	for personal guardianships only 159.081(1)
New Hampshire	§462:29. §462:30		
New Jersey	§3A:6-43	yes	
New Mexico	§32A-5-307 §32A-5-430	yes	
New York	Mental Hyg. Law §77.35 §78.27(b) Surr. Ct. Proc. Act §1752	yes	
North Carolina	§35-1.38	yes	six months after appointment. then annually §35-1.31
North Dakota	§30.1-28-07 §30.1-29-30	yes	
Ohio	§2111.47	yes	

State	Action to terminate guardianship or conservatorship		Periodic review
	statute provides a termination procedure	subject of guardianship or conservatorship may petition	
Oklahoma	58§854	yes	
Oregon	§126.123	yes	
Pennsylvania	20§5517		
Rhode Island	§33-15-44	yes	
South Carolina	§21-19-220	yes	
South Dakota	§30-28-3		
Tennessee	§33-315	yes	
Texas	Prob. Code 5 §426* 5 §130(M)	yes	
Utah	§75-5-307	yes	
Vermont	14§3007		
Virginia			
Washington	§11.88.140		
West Virginia	§44-10A-3		
Wisconsin	§880.26(c)		
Wyoming	§3-2-310	yes	
American Samoa			
Guam	§1470		
Puerto Rico			
Virgin Islands	§884		

PUBLIC GUARDIANSHIP

Alabama			
Arizona			
California			

State	Action to terminate guardianship or conservatorship		Periodic review
	statute provides a termination procedure	subject of guardianship or conservatorship may petition	
Colorado	§26-3-112	yes	annually, §26-3-11
Delaware			12§3996
Georgia			
Maine	18§3633(1)	yes	annually and whenever ordered by court 18§3634 18§3650-C
Minnesota	§252A.19	yes	annually §252A.16
North Carolina	§35-4	yes	
Ohio	by reference §5119.88(D)	by implication	at least annually §5119.87
Oregon			
Tennessee			

VII. PUBLIC GUARDIANSHIP

State	Identity of Public Guardian	Funding Provisions	Provide Guardianship/Conservatorship Services to Minors as well as Adults
Alabama	general guardian of county or sheriff §26-2-50		yes
Arizona	county appointee §14-5601	county funds §14-5601B	yes
California	county appointee Welf. & Inst. Code §8000		yes
Colorado	State Dept. of Social Services §26-3-107		
Delaware	appointee of Chancellor 12§3991	salary paid from general fund 12§3991	
Georgia	Commissioner of Human Resources §49-604(d)(3)		
Illinois	Division of Guardianship and Mental Health Advocacy Commission ch.91½§703	State funds ch.91½§735	
Maine	Bureau of Mental Retardation 18§3621 Department of Health & Welfare 18§3638		
Minnesota	Commissioner of Public Welfare or his designee §252A.03(1)		
North Carolina	officer, agent or employee of state human resources agency §35.1-28		
Ohio	division of mental retardation and developmental disabilities §5119.85		
Oregon	public guardian created by county §126.905(1)	county funds §126.905(2)	yes
Tennessee	county court appointee §30-1501		yes

VIII. MINORS SUBJECT TO GUARDIANSHIP and CONSERVATORSHIP[1]

State	Guardianship	Conservatorship
Alabama	A person under the age of 19 with no living parent or between 14 and 19 whether a parent is alive or not. §26-2-22(1977), §26-1-1(1977)	A person under the age of 19 with an estate in his or her own right. §26-2-22(1977), §26-1-1(1977)
Alaska	A person under 19 and unmarried for whom all parental rights of custody have been terminated by circumstances or prior court order. §13.06.050(1972), §13.26.045(1972)	A person under the age of 19 who owns money or property that requires management or protection which cannot otherwise be provided. §13.06.050(1972), §13.26.165(1972)
Arizona	A person under 18 years of age and unmarried for whom all parental rights of custody have been terminated by circumstances or prior court order. §1-215 (1974) §14-5204(1974)	A person under the age of 18 who owns money or property that requires management or protection which cannot otherwise be provided. §1-215 (1974) §14-5401(1974)
Arkansas	A person under the age of 18 and unmarried whose disabilities have not been removed and for whom a guardianship is desirable to protect his or her interests. §57-103(1975), §57-601(Supp. 1977), §57-614(1975), §57-643(1975)	Same, except marriage does not terminate.
California	A person under the age of 18 and unmarried for whom an appointment appears necessary or convenient. Prob. Code §1440, §1500, & §1590 (Supp. 1977) Civ. Code §25 (Supp. 1977)	Same, except marriage does not terminate.
Colorado	See Arizona §15-14-204(1973), §15-10-201(1973)	See Arizona §15-14-401(1973), §15-10-201(1973)
Connecticut	A person under the age of 18 who has no parent or guardian or for whom all parental rights have been terminated. §45-43(1977), §45-44(1977)	A person under the age of 18 who is entitled to an estate and has no guardian. §45-47(1977), §45-43(1977)
Delaware	A male under the age of 18 or a female under the age of 18 and unmarried, or a male or female under 18 whose parents have been convicted of a specified offense. 12§3911 (1975)	Same 12§3903(1975) 12§3905(1975) 12§3911(1975)
District of Columbia	A person under the age of 18 and unmarried who has no parents or guardian or whose welfare requires an appointment. §21-101(Supp. 1978), §21-104(Supp. 1978)	A person under the age of 18 and unmarried who is entitled to a distributive share of the personal estate of an intestate or to a legacy or bequest under a last will or personal property by gift or purchase. §21-104(Supp. 1978), §21-106(Supp. 1978)
Florida	A person under the age of 18 whose disabilities have not been removed by marriage or otherwise. §744.102 (1976)	Same
Georgia	A person under the age of 18 who has no guardian. §49-105(1974), §74-104(Supp. 1977)	Same

State	Guardianship	Conservatorship
Hawaii	See Arizona §560:1-201 (1976). §560:5-204(1976)	See Arizona §560:1-201(1976). §560:5-401(1976)
Idaho	See Arizona §15-1-201 (Supp. 1978) §15-5-204 (Supp. 1978)	See Arizona §15-1-201(Supp. 1978) §15-5-401(Supp. 1978)
Illinois	A person under the age of 18 for whom an appointment appears necessary or convenient. ch.3§133(Supp. 1978) ch.3§131(Supp. 1978)	Same
Indiana	A person under the age of 18 and unmarried who has no natural guardian in the state properly performing his duties. §29-1-18-6(Supp. 1977) §29-1-1-3 (Supp. 1977)	Any person under 18 years of age. §29-1-1-3(Supp. 1977) §29-1-18-6(Supp. 1977)
Iowa	A person under the age of 18 and unmarried whose best interests require the appointment of a guardian. §633.3(Supp. 1978) §633.552(1964)	Same
Kansas	A person under the age of 18 and unmarried. §59-3028(1976)	A person under the age of 18 unless marriage has conferred the rights of majority. §59-3028 (1976)
Kentucky	Any person who has not reached the age of 18 and is a resident of the state or requires consent for some act in the state. §387.010 (Supp. 1977) §387.020(Supp. 1977)	Any resident or person having property or real estate in the state who is under 18 years of age. §387.010(Supp. 1977) §387.020(Supp. 1977)
Louisiana	A person under 18 years of age and unemancipated after the dissolution of the marriage of his father and mother or the separation from bed and board of either one of them from the other. R.S. §13:1569 (Supp. 1978) CC art. 246 (1952) A person under the age of 18 who is an orphan. CC art. 270 (Supp. 1978)	Same
Maine	A person under 18 and unmarried 18§3509 (Cum. Supp. 1978) 18§3553 (Cum. Supp. 1978)	Same
Maryland	An unmarried person under 18 years of age with no parent serving as guardian. §13-101 (Supp. 1977) §13-702 (Supp. 1977)	A person under 18 years of age who owns or is entitled to property or is in need of funds for his or her support, care, welfare or education. Est. & Trusts §13-101 (Supp. 1977) Est. & Trusts §13-201 (1974)
Massachusetts	A person under the age of 18 and unmarried for whom an appointment is necessary or convenient. 201§1 (Supp. 1978) 201§4 (Supp. 1978) 201§5 (1969)	A male under the age of 18 and unmarried or a female under the age of 18 for whom an appointment is necessary or convenient. 201§1 (Supp. 1978) 201§4 (Supp. 1978) 201§5 (1969)

State	Guardianship	Conservatorship
Michigan	A person under 18 and unmarried having no father or mother living, competent and suitable to have the custody and care of the education of such minor or a minor who has one parent competent and suitable and another who though able to support and care for the education of such child, has nevertheless failed and neglected to do so, for a period of one year. §703.1 (1976) §554.451 (1976)	A person under 18 with an estate within Michigan. §703.1 (1976) §554.451 (1976) Marriage terminates guardianship of the person of a female minor. §703.23 (1976)
Minnesota	Any male under 18 years of age or any female under 18 and unmarried who has no suitable or competent parent. §524.54 (1975) §525.60 (1975)	Any person under 18 years of age with no suitable or competent parent. §525.24 (1975)
Mississippi	A person under the age of 18 or between the ages of 18 and 21 and unmarried who has no parent suitable to discharge the duties of guardian-ship. §93-13-1 (1972) §93-13-75 (1972) §93-13-5 (Supp. 1977)	A person under the age of 18 or between 18 and 21 and unmarried with an estate larger than 1000 dollars §93-13-1 (1972) §93-13-75 (1972)
Missouri	A person under the age of 21 and unmarried who has no living parent suitable to discharge the duties of guardianship or whose father is imprisoned in the state peniten-tiary. §475.030 (1956) §475.010 (Supp. 1978)	A minor under the age of 21 who has an estate which is not derived from a living parent who is acting as natural guardian. §475.010 (1956) §475.030 (Supp. 1978)
Montana	See Arizona §91A-1-201 (Special Pamphlet 1977) §91A-5-204 (Special Pamphlet 1977)	See Arizona §91A-1-201 (Special Pamphlet 1977) §91A-1-401 (Special Pamphlet 1977)
Nebraska	See Arizona §30-2209 (1975) §30-2608 (1975)	See Arizona §30-2209 (1975) §30-2630 (1975)
Nevada	A person under 18 years of age. §129.010 (1977) §159.035 (1973)	Same
New Hampshire	A person under the age of 18 whose parents are unfit or living apart and when it is in the interest of that person for an appointment to be made. §463:18a (Supp. 1977) §463:1 (1968) §463:6 (1968)	Same
New Jersey	A person under the age of 21 and unmarried with no living mother. §3A:6-18 (1953) §3A:6-28(1953)	Same
New Mexico	See Arizona §13-13-1 (1976) §32A-5-204 (Pamphlet 1976)	See Arizona §13-13-1 (1976) §32A-5-401 (Pamphlet 1976)

State	Guardianship	Conservatorship
New York	Any person under the age of 18 years who has no guardian whether or not the parent or parents of such person are living. SCPA §103(26) (Supp. 1977) SCPA §1701 (Supp. 1977)	Same
North Carolina	Orphans and infants uner 18 years of age. §33-1 (1976)	Same
North Dakota	See Arizona §30.1-01-06(Supp. 1975) §30.1-27-04(Supp. 1975)	See Arizona §30.1-01-06(Supp. 1975) §30.1-29-01(Supp. 1975)
Ohio	A person under the age of 18 and unmarried for whom an appointment is necessary. §2111.02(1976) §2111.45(1976) §2111.06(1976)	A person under the age of 18 for whom an appointment is necessary. §2111.02(1976)
Oklahoma	A person under 18 years of age who has no guardian legally appointed by will or deed and no father or mother competent to have the custody and care of his education, for whom an appointment appears necessary or convenient. 58 § 761(Supp. 1977), 15§13 (1972), 58§803(1965), 30§8(1976)	A person under 18 years of age who has no legally appointed guardian by will or deed and for whom an appointment appears necessary or convenient. 15§13(1972), 58§761(Supp. 1977), 30§8(1976).
Oregon	Any unmarried person under 18 years of age. §126.003(1977)	See Arizona §126.003(1977) §126.157(1977)
Pennsylvania	An individual under the age of 18 years. 20§102(1975) 20§5111(1975)	An individual under the age of 18. 20§102(1975), but see 20§5105(1975)
Rhode Island	An individual under 18 years of age. §15-12-1 (Supp. 1977) §33-15-3(1969)	An individual under 18 years old who has an estate within the state. §33-15-3(1969) §15-12-1(Supp. 1977)
South Carolina	A person under the age of 18 and unmarried. §21-21-20(Supp. 1978)	Same
South Dakota	A person under the age of 18 for whom an appointment appears necessary or convenient. §26-1-1(1976) §30-27-6(1976)	Same
Tennessee	A person under the age of 18 for whom an appointment appears necessary. §34-201 (1977)	Same
Texas	A person under 18 years of age who has not been married or whose disabilities of minority have not been removed and whose rights will be protected by an appointment. Prob. Code 5 §3 (Supp. 1978) Prob. Code 5 §114(1956)	Same
Utah	See Arizona §75-1-201 (Pamphlet 1977) §75-5-204 (Pamphlet 1977)	See Arizona §75-1-201 (Pamphlet 1977) §75-5-401 (Pamphlet 1977)

State	Guardianship	Conservatorship
Vermont	A male under 18 or a female under 18 and unmarried who has no living parent or whose parent is unsuitable or incompetent. 14§2645(1974) 33§632(Supp. 1978) 14§3002(1974)	A person under 18 who is the owner of real or personal property. 14§2645(1974) 33§632(Supp. 1978)
Virginia	A person under the age of 18 with no surviving parents. §31-2(1973) §31-4(1973)	Same
Washington	A person under the age of 18. §11.88.010(Supp. 1977) §11.92.010(Supp. 1977)	Same
West Virginia	A person under the age of 18 who has no guardian appointed by his or her parents. §44-10-3(1966) §44-10-7(Supp. 1978)	Same
Wisconsin	An unmarried person who has not attained the age of 18 who needs a guardian. §880.01 (Pamphlet 1977) §880.03 (Pamphlet 1977) §880.04 (Pamphlet 1977)	A person who has not attained the age of 18 and who has an estate. §880.01(Pamphlet 1977) §880.03(Pamphlet 1977)
Wyoming	A person under the age of 19 and unmarried with no other legally appointed guardian. §14-8-101 (1977) §3-2-101 (1977) §3-2-107 (1977)	A person under the age of 19 with an estate. §14-8-101 (1977) § 3-2-101 (1977)
American Samoa	A person under the age of 18 for whom an appointment appears necessary or convenient. 16§701(Supp. 1977)	Same
Guam	A person under the age of 18 and unmarried for whom an appointment appears necessary or convenient. §1440 (1970) §1590 (1970)	A person under the age of 18 for whom an appointment appears necessary or convenient. §1440(1970) §1590(1970)
Puerto Rico	An unemancipated person under 21 years of age. 31§971(1971) 31§662(1971)	Same
Virgin Islands	A male under the age of 21 or a female under the age of 21 and unmarried for whom an appointment appears necessary or convenient. 15§1242(1964) 15§821 (1964) 15§884 (1964)	Same

[1] All states also provide in their juvenile codes for guardianship of minors who have been neglected or abused.

IX. TYPES OF GUARDIANSHIP and/or CONSERVATORSHIP (MINORS)[1]

State	Power over Person, Property or Both	Testamentary and/or Standby Appointment	Temporary and/or Emergency Appointment
Alabama	No provision for personal guardian only	Testamentary §26-2-23 (1977)	
Alaska	X	Testamentary §13.26.035 (Supp. 1978)	Temporary §13.26.060 (1972) 6 months
Arizona	X	Testamentary §14-5201 (1974)	Temporary §14-5207 (1974) 6 months
Arkansas	X		
California	X	Testamentary Prob. Code §1402, §1403, §1404 (1977)	
Colorado	X	Testamentary §15-14-202 (1973)	Temporary §15-14-207 (1973) 6 months
Connecticut	X	Testamentary §45-51 (1977)	
Delaware	X	Testamentary 12§3904 (1975)	
District of Columbia	X	Testamentary §21-102 (1976)	
Florida	X	Testamentary §744.304 (1976) Standby §744.302 (1976)	
Georgia	X	Testamentary §49-103 (1974)	Temporary §49-112 (1974) §113-1207 (1974) As court directs
Hawaii	X	Testamentary §560:5-207 (Supp. 1977)	Temporary §560:5-207 (Supp. 1977) 90 days
Idaho	X	Testamentary §15-5-201 (Supp. 1978)	Temporary §15-5-207 (Supp. 1978) 6 months
Illinois	X	Testamentary ch. 3§146 (Supp. 1978)	
Indiana	X		Temporary §29-1-18-24 (1972) 60 days

State	Power over Person, Property or Both	Testamentary and/or Standby Appointment	Temporary and/or Emergency Appointment
Iowa	X	Standby §633.560 (1964) §633.591 (1964)	Temporary §633.558 (1964) §633.573 (1964) As court directs
Kansas	X	Testamentary §59-3004 (1976)	
Kentucky	X	Testamentary §387.040 (1969)	
Louisiana	X	Testamentary CC art. 257 (Supp. 1978)	Temporary CCP art. 4070 (1961) As court directs
Maine	X		Special Guardian 4§401 (1964) 18§3510 (1965) 18§3648 (Supp. 1975) As court directs
Maryland	X	Testamentary §13-701	
Massachusetts	No provision for personal guardian only	Testamentary 201§3 (Supp. 1978)	Temporary 201§14 (Supp. 1978) As court directs
Michigan	X	Testamentary §703.10 (1976)	Special Guardian §703.11 (1976) As court directs
Minnesota	X	Testamentary §525.541 (1975)	Special Guardian §525.591 (1975) Until permanent appointment
Mississippi	X	Testamentary §93-13-7 (Supp. 1977)	
Missouri	X	Testamentary §475.045 (1956)	
Montana	X	Testamentary §91A-5-201 (Special Pamphlet 1977)	Temporary §91A-5-207 (Special Pamphlet 1977) 6 months
Nebraska	X	Testamentary §30-2605 (1975)	Temporary §30-2611 (1975) 6 months
Nevada	X	Testamentary §159.061 (1973)	Temporary §159.051 (1973) §159.205 (1975) 6 months
New Hampshire	X	Testamentary §463:1 (1969)	
New Jersey	X	Testamentary §3A:6-18 (1963)	

State	Power over Person, Property or Both	Testamentary and/or Standby Appointment	Temporary and/or Emergency Appointment
New Mexico	X	Testamentary §32A-5-202 (1975)	Temporary §32A-5-207 (1975) 6 months
New York	X	Testamentary SCPA §1711 (1967) DRL §81 (1977)	
North Carolina	X	Testamentary §33-3 (1976)	
North Dakota	X	Testamentary §30.1-27-02 (1976)	Temporary §30.1-27-07 (1976) 6 months
Ohio	X	Testamentary §2111.12 (1976)	
Oklahoma	X	Testamentary 30§6 (1976)	
Oregon	X	Testamentary §126.035 (1977)	Temporary for adoption proceedings §109.335 (1977) Generally §126.070 (1977) 6 months
Pennsylvania	X	Testamentary 20§2519 (1975)	Temporary 20§5513 (1975) As court directs
Rhode Island	X	Testamentary §33-15-6 (1969)	Temporary §33-15-10 (1969) As court directs §33-15-11 (1969)
South Carolina	X	Testamentary §21-21-20 (Supp. 1977)	Temporary §21-19-50 (1976) As court directs
South Dakota	X	Testamentary §30-27-2 (1977)	
Tennessee	X	Testamentary §34-302 (1977)	Special Guardian §34-202 (1977) As court directs
Texas	X	Testamentary Prob. Code 5§117 (1956)	Temporary Prob. Code 5§131 (1956) As court directs
Utah	X	Testamentary §75-5-201 (Pamphlet 1977)	Temporary §75-5-207 (Pamphlet 1977) 6 months
Vermont	No provision for personal guardian only	Testamentary 14§2656 (1974)	

State	Power over Person, Property or Both	Testamentary and/or Standby Appointment	Temporary and/or Emergency Appointment
Virginia	No provision for personal guardian only	Testamentary §31-2 (1973)	
Washington	X	Standby §11.88.125 (Supp. 1977) Testamentary §11.88.080 (1963)	
West Virginia	X	Testamentary §44-10-1 (1966)	Curator. §44-10-6 (1966) As court directs
Wisconsin	X		Temporary §880.15 (Special Pamphlet 1978) As court directs
Wyoming	X	Testamentary §3-2-105 (1977)	
American Samoa	X		
Guam	X	Testamentary §1403 (1970)	Temporary §1442 (1970) As court directs
Puerto Rico	X	Testamentary 31§681 (1971)	
Virgin Islands	X	Testamentary 15§826 (1964)	

[1] Only Maine and Washington Statutorily provide for limited guardians for minors. ME. REV. STAT. tit.18. §3510 (Supp. 1977); WASH. REV. CODE ANN. §11.88.005 (Supp. 1977). All other American jurisdictions provide only for plenary guardianship and conservatorship for minors.

X. PROCEDURES PRECEDING APPOINTMENT (MINORS)

State	Testamentary Designation		Probate Court Appointment			Juvenile Court Appointment		
	appointment effective upon death of parent and acceptance by designee	minor age 14 or over may object to the designee or have designee replaced	notice required to minor age 14 or over	minor may nominate if age 14 or over	minor represented by	court has power to appoint	notice required to minor age 14 or over	minor represented by
Alabama	Must be court approved but has priority over court appointee §26-2-23 (1977)					§12-15-8 (1975)	§12-15-2 (1975)	Guardian *ad litem* §12-15-8 (1975) Lawyer §12-15-63 (1975)
Alaska	§13.26.035 (Supp. 1978)	§13.26.040 (Supp. 1978)	§13.06.120(3) (Supp. 1978) 14 days §13.06.110 (Supp. 1978)	§13.26.055 (1972)	Attorney §13.26.060 (1972)	§47.10.110 (1977)		Guardian *ad litem* §47.10.050 (1975)
Arizona	§14-5202 (1974)	§14-5203 (1974)	§14-5207 (1974) 14 days §14-1401 (1974)	§14-5206 (1974)	Attorney at court's discretion §14-5206 (1974)	§8-241 (Supp. 1977)		Attorney §8-225 (Supp. 1977)
Arkansas			3 Days §57-611	Preference considered §57-608				Guardian *ad litem* at court's discretion §45-428 (1977)
California	Prob. Code §1402, §1403, & §1404 (1977)			Preference considered Prob. Code §1406 (1977)	Attorney Prob. Code §1606.5 (1977)		Welf. & Inst. Code §335 (Supp. 1978) 10 days §337 (Supp. 1978)	Attorney Welf. & Inst. Code §§316, 318.5 (Supp. 1978) Guardian *Ad Litem* Welf. & Inst. Code §326 (Supp. 1978)
Colorado	§15-14-202 (1973)	§15-14-203 (1973)	§15-14-207 (1973) 10 days §15-10-401 (1973)	§15-14-206 (1973)	Attorney §15-14-207 (1973)	§19-3-111 (1973)		Attorney §19-1-106 (Supp. 1977) Guardian *ad litem* §19-3-105 (Supp. 1977)
Connecticut	§45-51 (1977)	§45-51 (1977)	For appointment of guardian of property only §45-47 (1977)	§45-46 (1977)	Guardian *ad litem* §45-54 (1977)	Jurisdiction vested exclusively in probate court §17-59 (1975)		

State	Testamentary Designation		Probate Court Appointment			Juvenile Court Appointment		
	appointment effective upon death of parent and acceptance by designee	minor age 14 or over may object to the designee or have designee replaced	notice required to minor age 14 or over	minor may nominate if age 14 or over	minor represented by	court has power to appoint	notice required to minor age 14 or over	minor represented by
Delaware	Must be court appointed but has priority 12§3902 (1975)	12§3902 (1975)		12§3902 (1975)		10§937 (1974)	10§935 (1974)	
District of Columbia	§21-102 (1976) §21-105 (1976)	§21-108 (1976)		§21-108 (1976)		§16-2320 (Supp. 1978)		Attorney §16-2320 (1973)
Florida	Standby only §744.304 (Supp. 1977)		§731.301 (1976)		Guardian *ad litem* at court's discretion §731.303 (1976)	§39.11 (Supp. 1977)	Only if parent or guardian is unknown §39.06 (1976)	
Georgia				§49-105 (1974)	Guardian *ad litem* §49-111 (1974)	§74-109 (1964) §74-110 (1964)		
Hawaii	Must be court appointed but has priority §560:5-204 (Supp. 1977)		§560:5-207 (Supp. 1977) 14 days §560:1-401 (1976)	§560:5-206 (1976)	Guardian *ad litem* at court's discretion §551-2 (1976) §560:5-207 (Supp. 1977)	§571-46 (1976)		§551-2 (1976)
Idaho	§15-5-201 (Supp. 1978) §15-5-204 (Supp. 1978)	§15-5-203 (Supp. 1978)	14 days §15-1-401 (Supp. 1978)	§15-5-206 (Supp. 1978)	Attorney if court deems necessary §15-5-207 (Supp. 1978)	§16-1610 (Supp. 1978)		Attorney who may be also appointed guardian *ad litem* §16-1618 (Supp. 1978)
Illinois	Must be court appointed ch. 3§146 (Supp. 1978)		3 days ch.110½ §11-10.1 (Supp. 1978)	ch. 3§136 (1976)	Guardian *ad litem* at court's discretion ch. 110½§11-10.1 (Supp. 1978)	ch. 37§705-7 (1972)	3 days ch. 37§704-3 (1972)	Attorney ch. 37§704-5 (Supp. 1978) Guardian *ad litem* ch. 37§704-5 (Supp. 1978)
Indiana	[2]			Preference considered §29-1-18-10 (1972)	Guardian *ad litem* at court's discretion §29-1-18-14 (1972)	§31-5-7-15 (Supp. 1977)		

State	Testamentary Designation		Probate Court Appointment			Juvenile Court Appointment		
	appointment effective upon death of parent and acceptance by designee	minor age 14 or over may object to the designee or have designee replaced	notice required to minor age 14 or over	minor may nominate if age 14 or over	minor represented by	court has power to appoint	notice required to minor age 14 or over	minor represented by
Iowa	Must be court appointed but has priority §633.559 (1964) §633.571 (1964)		If petitioner does not have custody of the ward §633.553, §633.567 (1964)	§633.559 (1964) §633.571 (1964)		§232.33 (1969) §600A.9[3] (Supp. 1978)		Attorney
Kansas	Must be court appointed but has priority §59-3004 (1976)		5 days if court deems necessary §59-3011 §59-3012 (1976)	§59-3014 (1976)	Attorney at court's discretion §59-3-011 (1976)	§38-824 (Supp. 1977)	§38-817 (Supp. 1977)	Attorney §38-817 (Supp. 1977) Guardian *ad litem* §38-821 (Supp. 1977)
Kentucky	Must be court appointed but has priority §387.030 (Supp. 1977)		§387.025 (Supp. 1977) 5 days §395.016 (1969)	§387.050 (Supp. 1977)		208.200 (Supp. 1977)		Attorney §208.060 (1975)
Louisiana	Must be confirmed by court CC art. 248 (Supp. 1978) CCP art. 4062 (1961)	CCP art. 4066 (Supp. 1978)				§13:1580 (Supp. 1978)	If over 12 §13:1575 (Supp. 1978)	Attorney §13:1575 (Supp. 1978)
Maine	Must be court appointed but has priority 18§3552 (Supp. 1977)	18§3552 (Supp. 1977)		18§3552 (Supp. 1977)	Guardian *ad litem* at court's discretion 18§3651 (1965)	22§3793 (1964)		
Maryland	§13-701 (1974)		Rule R. 74a	If over 16, §13-207 (1974), §13-702 (1974) 14 for guardian of person & 16 for guardian of property Rule R. 77a		§3-820 (Supp. 1977) Rule 971a		Attorney §3-821 §3-834 (Supp. 1977) Rule 906a

State	Testamentary Designation		Probate Court Appointment			Juvenile Court Appointment		
	appointment effective upon death of parent and acceptance by designee	minor age 14 or over may object to the designee or have designee replaced	notice required to minor age 14 or over	minor may nominate if age 14 or over	minor represented by	court has power to appoint	notice required to minor age 14 or over	minor represented by
Massachusetts	201§3 (Supp. 1978)			201§2 (1969)	Guardian *ad litem* at court's discretion 201§36 (1969)	119§26 (1975)		Attorney at court's discretion 111§29 (1975)
Michigan	Effective upon probate of the will §703.10 (1976)		14 days §703.2 (1976)	§703.3 (1976)	Guardian *ad litem* at court's discretion §703.12 (1976)	§712A.18 (1976)		Attorney at court's discretion. §712A.17 (1976)
Minnesota	Must be appointed by court but has priority §525.541 (1975)	§525.541 (1975)	Mandatory if minor is over age 14, discretionary if minor is under age 14 §525.55 (1975) 14 days	§525.541 (1975)		§260.111 §260.241 (1971)		Attorney or guardian *ad litem*. §260.155 (1971 & Supp. 1978)
Mississippi	§93-13-9 (Supp. 1977)			§93-13-13 (Supp. 1977)		§3-23-17 (1972)		
Missouri	Must be court appointed but has priority §475.045 (1956)	§475.110 (1956)	§475.070 (Supp. 1978) 14 days §472.100 (Supp. 1978)	§475.045 (1956)	Attorney §475.075 (1956)	§§210.400 210.490 211.501 (1962)		Attorney §210.390 (1962)
Montana	§91A-5-201 (Special Pamphlet 1977)	§91A-5-203 (Special Pamphlet 1977)	§91A-5-207 (Special Pamphlet 1977)	§91A-5-206 (Special Pamphlet 1977)	Attorney at court's discretion §91A-5-207 (Special Pamphlet 1977)	§10-1314 (Supp. 1977)		Guardian *ad litem* at court's discretion. §10-1310 (Supp. 1977)
Nebraska	§30-2605 (1975)	§30-2607 (1975)	§30-2611 (1975)	§30-2610 (1975)	Attorney at court's discretion §30-2611 (1975)	§43-209 (1974)		Attorney §43-205.06 (1974) Guardian *ad litem*, at court's discretion §43-206 (Supp. 1977)

State	Testamentary Designation		Probate Court Appointment			Juvenile Court Appointment		
	appointment effective upon death of parent and acceptance by designee	minor age 14 or over may object to the designee or have designee replaced	notice required to minor age 14 or over	minor may nominate if age 14 or over	minor represented by	court has power to appoint	notice required to minor age 14 or over	minor represented by
Nevada	Must be court appointed but has priority §159.061 §159.063 (1973)		§159.047 (1973) 20 days §159.053 (1973)	§159.061 (1973)		§62.200 (1977)		Attorney at court's dis- cretion. §62.085 (1977)
New Hampshire	Must be court appointed but has priority §463:1 §463:2 (1969)	§463:14 (1969)		Preference considered. §463:11 (1969) §463:11 (1969)	Guardian *ad litem* at court's discretion §170-C:8 (1977) §462:1 (1968)	§169:13a (Supp. 1977)		Attorney §604-A:1-a Guardian *ad litem* at court's discretion. §462:1 (1968)
New Jersey	§3A:6-18 (1963) Court may review §3A:6-23 (1963)	§3A:6-28 (1963)		§3A:6-28 (1963)		§30:4c-12 (1964) Child abuse §9:6-8.54 (Supp. 1978)		Law guardian for child abuse or neglect proceedings §9:6-8.23 (1976)
New Mexico	§32A-5-201 (1975)	§32A-5-203 (1975)	§32A-5-207 (1975)	§32A-5-206 (1975)	Attorney at court's discretion §32A-5-207 (1975)	§13-14-9 (1976)	§13-14-18 (1975)	Attorney §13-14-18 (1975)
New York	Must be approved by court SCPA §1711 (1967) DRL §81 (1977)		SCPA §1705 (1967)	SCPA §1703, §1706 (1967) CPLR §1210 (1977)	Guardian *ad litem* at court's discretion CPLR §1202 (1977)	Jud: Fam Ct Act §634 (Supp. 1978) Soc. Services §384-b (Supp. 1977)		Attorney or Law Guardian at court's dis- cretion. Jud: Fam Ct Act §249 (Supp. 1978) CPLR §1202 (1977)
North Carolina	§33-3 (1976)					§7A-286 (Supp. 1977)	5 days §7A-283 (Supp. 1977)	Guardian *ad litem*. §7A-283 (Supp. 1977)
North Dakota	§30.1-27-01 (1976)	§30.1-27-03 (1976)	§30.1-27-07 (1976)	§30.1-27-06 (1976)	Attorney at court's discretion §30.1-27-07 (1976)	§27-20-30 (1976)	Only if alleged delinquent §27-20-22 (Supp. 1977) 24 hours §27-20-23 (1974)	Attorney §27-20-26 (1974)

State	Testamentary Designation		Probate Court Appointment			Juvenile Court Appointment		
	appointment effective upon death of parent and acceptance by designee	minor age 14 or over may object to the designee or have designee replaced	notice required to minor age 14 or over	minor may nominate if age 14 or over	minor represented by	court has power to appoint	notice required to minor age 14 or over	minor represented by
Ohio	Must be court appointed but has priority §2111.12 (1976)	§2111.46 (1976)	3 days §2111.04 (1976)	§2111.12 (1976)		§2151.32 (1976) §2151.353 (1976)	At court's discretion §2151.28 (1976)	Guardian *ad litem* §2151.281 (1976) Attorney §2151-352 (1976)
Oklahoma	Of the person only 30§6 (1976) 30§7 (1976)	58§768 (1965)		If of sufficient age to form intelligent choice 30§11 (1976) 58§762 (1965)	Guardian *ad litem* at court's discretion 58§891 (1965)	10§1116 (Supp. 1977)	If over 12 10§1104 (Supp. 1977) 48 hours 10§1105 (Supp. 1977)	Attorney 10§1104 (Supp. 1977) 10§1109 (Supp. 1977)
Oregon	Must be court appointed but has priority §126.035 (1977)		§126.070 (1977) 10 days §126.007 (1977)	§126.035 (1977)	Guardian *ad litem* after appointment of a guardian §13.041 (1973) Attorney prior to appointment of a guardian §126.070 (1977)	§419.507 (1977)	If over 12 §419.486 (1977)	Guardian *ad litem* §13.041 (1973) Suitable person §419.494 (1977)
Pennsylvania				20§5113 (1975)	Guardian *ad litem* at court's discretion 20§761 (1975)	42§6351 (Pamphlet 1977)	Only if alleged delinquent 42§6335 (Pamphlet 1977)	Attorney 42§6337 (pamphlet 1977)
Rhode Island	Must be appointed by court §33-15-6 (1969)		14 days §33-15-17 (1969)	§33-15-4 (1969)		§14-1-32 (1970)	§14-1-16 (1970)	
South Carolina	§21-21-20 (Supp. 1977)					§14-21-620 (1976)		
South Dakota	Prob. code §117 (1956) Must be appointed but has priority §30-27-3 (1977)	§30-27-21 (1977)		§30-27-11 (1977) §30-27-20 (1977)		§26-8-35 (1977)		Probation officer §26-8-25 (1977) Attorney §26-8-22.2 (1977)

State	Testamentary Designation		Probate Court Appointment			Juvenile Court Appointment		
	appointment effective upon death of parent and acceptance by designee	minor age 14 or over may object to the designee or have designee replaced	notice required to minor age 14 or over	minor may nominate if age 14 or over	minor represented by	court has power to appoint	notice required to minor age 14 or over	minor represented by
Tennessee	Must qualify before court §34-303 (1977)			§34-203 (1977)		§37-104 §37-105 §37-240 §37-236 (1977)	§37-221 (1977) 3 days §37-223 (1977)	
Texas	Must be appointed by the court Prob. Code 5§117 (1956)	Prob. Code §118 (1956)		Prob. Code 5§118 (1956)		Fam. 2§15.05 (1975)		Fam. 2§11.10 *litem* Fam. §11.10 (Supp. 1978)
Utah	§75-5-202 §75-5-201 (Pamphlet 1977)	§75-5-203 (Pamphlet 1977)	§75-5-207 (Pamphlet 1977)	§75-5-206 (Pamphlet 1977)	Attorney at court's discretion §75-5-207 (Pamphlet 1977)	§78-3a-39 (1977)		Attorney §78-3a-35 (1977)
Vermont	14§2656 (Supp. 1978)	14§2652 (Supp. 1978)		14§265C (Supp. 1978)	Guardian *ad litem* At court's discretion 14§2657 (1974)	33§656 (Supp. 1978)		Guardian *ad litem* at court's dis-cretion 33§653 (Supp. 1978)
Virginia	§31-2 (1973)			§31-5 (1973)		§16.1-279 (Supp. 1978)	If over 12 §16.1-263 (Supp. 1978)	Guardian *ad litem* §16.1-266 (Supp. 1978)
Washington	§11.88.080 (1963)		10 days regard-less of age §11.88.040 (Supp. 1977)		Attorney §11.88.045 (Supp. 1977) Guardian *ad litem* §11.88.090 (Supp. 1977)	§13.04.140 §13.04.110 (1963)		Some suitable person §13.04.070 (1963)
West Virginia	§44-101 §44-10-2 (1966)			§44-10-4 (1966)	Guardian *ad litem* §54-6-10 (1966)	§49-6-5 (1976)		Attorney §49-6-26 (Supp. 1978)

State	Testamentary Designation		Probate Court Appointment			Juvenile Court Appointment		
	appointment effective upon death of parent and acceptance by designee	minor age 14 or over may object to the designee or have designee replaced	notice required to minor age 14 or over	minor may nominate if age 14 or over	minor represented by	court has power to appoint	notice required to minor age 14 or over	minor represented by
Wisconsin	Nomination given due regard §880.09 (Special Pamphlet 1978)	§880.16 (Special Pamphlet 1978)	§880.08 (Special Pamphlet 1978)	§880.09 (Special Pamphlet 1978)	Guardian *ad litem* §879.15E (Supp. 1978), §802.07 (1977)	§48.14 (1957 & Supp. 1978)		Attorney §48.25 (Supp. 1978) Guardian *ad litem* §802.07 (1977)
Wyoming	§3-2-105 (1977)	§3-2-104 (1977)		§3-2-102 (1977)	Guardian *ad litem* at court's discretion §3-2-112 (1977)	§14-8-130 (1977)	§14-8-114 (1977) 2 days §14-8-115 (1977)	Attorney §14-8-123 (1977) Guardian *ad litem* at court's discretion §14-8-117 (1977)
American Samoa								
Guam	§1403 (1970) Must qualify §1484 (1970)	§1406 (1970)		§1406 §1440 (1970)	Guardian *ad litem* at court's discretion §1607 (1970)			
Puerto Rico	31§681 (1971)					34§2506 (1967)		Attorney 34 APP. I R 8.3 Guardian *ad litem*. R 8.4
Virgin Islands	15§826 (1964)			15§822 (1964)	Guardian *ad litem* at court's discretion 15§827 (1964)			

[1] The relevant section was repealed in 1977 and replaced by a section which deals with custody of delinquent children. See, §45-433, §45-501-516 (1977).

[2] Testamentary nomination given due consideration. §29-1-18-10 (Supp. 1977).

[3] Section refers to proceedings for termination of parental rights.

XI. POWERS AND DUTIES IN GUARDIANSHIPS and/or CONSERVATORSHIPS FOR MINORS

State	Same Powers Generally as Parent	Care and Management of Financial Resources	Custody of the Person	Responsibility for Education	Consent to Medical Care	Juvenile Court May Take Child into Custody for Treatment of Serious Injury or Illness	Conservator Must File Bond	Conservators Are Made Fiduciaries by Law	Compensation	Annual Reports or Accounts Must Be Filed
Alabama		§26-4-3 (1977)	§26-2-22 (1977)	§26-4-61 (1977)		§12-15-56 (1977)	§26-3-1 (1977) But testamentary appointee may be relieved §26-3-3, §26-3-4 (1977)	§19-1-2 (1977)		Once every three years §26-5-2 (1977)
Alaska	§13.26.070 (Supp. 1978)	§13.26.070 (Supp. 1978)	§13.26.070 (1972)	§13.26.070 (1972)	§13.26.070 (1972)	§47.10.010 (1977) §47.10.142 (1977)	§13.26.215 (1972)	§13.06.050 (Supp. 1978)	As court orders §13.26.070 (1972)	As court directs §13.26.070 §13.26.255 (1972)
Arizona	§14-5209 (1974)	§14-5209 (1974)	§14-5209 (1974)	§14-5209 (1974)	§14-5209 (1974)	§8-223 (Supp. 1977)	§14-5105 (1974) §14-5411 (1974)	§14-1201 (1974)	As court orders §14-5209 (1974) §14-5413 (1974)	§14-5209 §14-5419 (1974)
Arkansas		§57-624 (1977)	§57-625 (1975)	§80-1502 (1977) §57-624 (1977)		§45-432 (1977)	§57-617 (1977)		Reasonable §57-641 (1977)	§57-624 (1971) §57-642 (1971)
California		Prob. Code §1500 (1977)	Prob.Code §1500 (1977)	Prob. Code §1500 (1977)		Welf. & Inst. Code §305 (Supp. 1978)	Prob. Code §1405 (1977) Testamentary appointee may be relieved Prob. Code §1485 (1977)	Prob. Code §1400 (1977)	Reasonable Prob. Code §1557 (1977)	No less than bi-annually Prob. Code §1553 (Supp. 1978)

State	Same Powers Generally as Parent	Care and Management of Financial Resources	Custody of the Person	Responsibility for Education	Consent to Medical Care	Juvenile Court May Take Child into Custody for Treatment of Serious Injury or Illness	Conservator Must File Bond	Conservators Are Made Fiduciaries by Law	Compen- satioh	Annual Reports or Accounts Must Be Filed
Colorado	§15-14-209 (1973)	§15-14-209 (1973)	§15-14-209 (1973)	§15-14-209 (1973)	§15-14-209 (1973)	§19-1-103 (20)(d) (1973)	§15-14-412 (1973)	§15-1-103 (1973)	As court orders §15-14-209 §15-14-414 (1973)	As court directs §15-14-209 §15-14-419 (1973)
Connecticut	§45-45 (1977)		§45-45 (1977)				§45-48 (1977)		Not to ex- ceed 5% of estate income §45-60 (1977)	
Delaware		12§3921 (1975)	12§3921 (1975)	12§3923 (1975)			12§3907 (1975)	12§4301 (1975)		Not more often than every two years 12§3943 (1975)
District of Columbia		§21-107 (1976)	§21-102 (1976)	§21-143 (1963)			Testamen- tary appointee may be re- lieved §21-115 §21-108 (1966)	§21-1701 (1966)	Reasonable §21-143 (1966)	§21-143 (1963)
Florida		§744.377 (1976)	§744.361 (1976)	§744.361 (1976)		§39.03(1)(c) (Supp. 1977)	§744.351 (1976)		Reasonable §744.108 (1976)	§744.367 (1976)
Georgia	§49-201 (Supp. 1977)	§49-203 (1974)	§49-201 (Supp. 1977)	§49-201 (Supp. 1977)			§49-113 (1974) Testa- mentary appointee may be re- lieved §49-103 (1974)			

State	Same Powers Generally as Parent	Care and Management of Financial Resources	Custody of the Person	Responsibility for Education	Consent to Medical Care	Juvenile Court May Take Child into Custody for Treatment of Serious Injury or Illness	Conservator Must File Bond	Conservators Are Made Fiduciaries by Law	Compensation	Annual Reports or Accounts Must Be Filed
Hawaii	§560:5-209 (1976)	§560:5-209 (1976) §560:5-425 (1976)	§560:5-209 (1976)	§560:5-209 (1976)	§560:5-209 (1976)	§571-31 (1976)	§560:5-410 (1976)	§556-1 (1976)	Reasonable §560:5-209 (Supp. 1978)	
Idaho	§15-5-209 (Supp. 1978)	§15-5-209 (Supp. 1978) §15-5-425 (Supp. 1978)	§15-5-209 (Supp. 1978)	§15-5-209 (Supp. 1978)	§39-4303 (1977) §15-5-209 (Supp. 1973)	§16-1612 (Supp. 1978) §16-1616 (Supp. 1978)	§15-4-411 (Supp. 1978)	§15-1-201 (Supp. 1978)	Reasonable §15-5-414 §15-5-209 (Supp. 1978)	
Illinois		ch.110½ §11-13 (Supp. 1978)	ch.110½ §11-13 (Supp. 1978)	ch.110½ §11-13 (Supp. 1978)		ch.37§703-1 ch.37§703-7 (1974)	ch.3§148 (Supp. 1978)	ch.120 §15-1501 (1974)	ch.3§336 (Supp. 1978)	ch.3§309 (Supp. 1978)
Indiana		§29-1-18-28 (Supp. 1977)	§29-1-18-28 (Supp. 1977)	§29-1-18-28 (Supp. 1977)			§29-1-18-22 (1972)	§29-1-1-3 (Supp. 1977)	Reasonable §29-1-18-45 (1972)	As court directs §29-1-18-28 (1977)
Iowa		§633.641 (1964)				§232.13 §232.15 §232.33 (1969)	§633.634 (1964)	§633.634 §633.3 (1964)		For conservators §633.670 (1964) Generally, as court directs §633.669 (1964)
Kansas		§59-3019 (1976)	§59-3018 (1976)			§38-824 (Supp. 1977)	Testamentary appointee may be relieved §59-3015 (1976)		Reasonable §59-3032 (1976)	Conservators only §59-3029 (1976)

State	Same Powers Generally as Parent	Care and Management of Financial Resources	Custody of the Person	Responsibility for Education	Consent to Medical Care	Juvenile Court May Take Child into Custody for Treatment of Serious Injury or Illness	Conservator Must File Bond	Conservators Are Made Fiduciaries by Law	Compensation	Annual Reports or Accounts Must Be Filed
Kentucky		§387.060 (1969)	§387.060 (1969)	§387.060 (1969)		§208.020(7) §208.060(2)(c) (Supp. 1976)	§387.070 (Supp. 1977)	§395.001 (1969)	5% of the estate income §386.180 (1969)	Conservators only §395.610 (Supp. 1977)
Louisiana		CCP art. 4069 (Supp. 1978) CCP art. 4262 (1961)	CCP art. 4261 (1961)	CCP art. 4261 (1961)	§40:1299.52 (1977)	§13:1578.2 (Supp. 1978) §13:1583 (Supp. 1978)	CCP art. 4131 (1961)		Reasonable (not to exceed 10% of income) CCP art. 4274 (Supp. 1978).	CCP art. 1391(1961) CC art. 356 (1952)
Maine		18§3553 (Supp. 1977)	18§3553 (Supp. 1977)	18§3553 (Supp. 1977)		22§3792 (Supp. 1977)	18§3801 (1965)	18§203 (1965)		Once every three years 18§3901 (1965)
Maryland	2	§3-213 (1974)				§3-822 (Supp. 1977)	At court's discretion §13-208 (1965)	§15-102 (1974)	Expenses only §13-218 (1974)	§3-826 (Supp. 1977)
Massachusetts		201§4 (Supp. 1978)	201§4 (1969)	201§4 (1969)		119§51B 119§51C (1975)	Testamentary appointee may be relieved 205§1 205§5 (1969)		Reasonable 206§16 (1969)	206§1 (1969)
Michigan		§703.8 (1976)	§703.7 (1976)	§703.7 (1976)		§712A.14 §712A.15 §712A.18 (1976)	§704.3 (1976)	§704.11 (1976)	Reasonable §704.33 (1976)	As court directs §704.35 (1976) Annually §704.38 (1976)

State	Same Powers Generally as Parent	Care and Management of Financial Resources	Custody of the Person	Responsibility for Education	Consent to Medical Care	Juvenile Court May Take Child into Custody for Treatment of Serious Injury or Illness	Conservator Must File Bond	Conservators Are Made Fiduciaries by Law	Compen- sation	Annual Reports or Accounts Must Be Filed
Minnesota		§525.56 (1975)	§525.56 (1975)			§260.165 (1971) §260.191 (Supp. 1973)	§525.551 (1975)			§525.58 (1975)
Mississippi		§93-13-1 §93-13-11 (1973)	§93-13-1 §93-13-11 (1973)	§93-13-1 §93-13-11 (1973)			§93-13-17 (Supp. 1977)	§27-7-3(d) (1972)	Minimum $100 a year, as court directs §93-13-67 (Supp. 1977)	§93-13-67 (Supp. 1977)
Missouri		§475.130 (Supp. 1978)	§475.120 (1956)	§475.120 (1956)		§210.125 (Supp. 1978) §211.101 §211.161 §211.181 (1962)	§475.100 (1956)	§472.010 (Supp 1978)	Reasonable §475.265 §211.101 (1956)	§475.190 (Supp. 1977)
Montana	§91A-5-209 (Special Pamphlet 1977)	§91A-5-209 (Special Pamphlet 1977)	§91A-5-209 (Special Pamphlet 1977)	§91A-5-209 (Special Pamphlet 1977)	§91A-5-209 (Special Pamphlet 1977)	§10-1309 §10.1311 §10-1314 (Supp. 1977)	§91A-5-411 (Special Pamphlet 1977)	§91A-1-201 (Special Pamphlet 1977)	Reasonable §91A-5-312 (Special Pamphlet 1977)	
Nebraska	§30-2613 (1975)	§30-2613 §30-2653 (1975)	§30-2613 (1975)	§30-2613 (1975)	§30-2613 (1975)	§43-205.01 §43-210.02 (1974)	§30-2640 (1975)	§30-2209 (1975)	As court directs §30-2613 §30-2643 (1975)	As court directs §30-2613 §30-2648 (1975)
Nevada		§159.083 (1973)	§159.079 (1973)	§159.079 (1973)	§159.079 (1973)	§62.140 §62.170 §62.200 §62.240 (1977)	Testamen- tary appointee may be relieved §159.065 (1973)	§162.020 (1973)	Reasonable §159.183 (1973)	§159.081 §159.177 (1973)

State	Same Powers Generally as Parent	Care and Management of Financial Resources	Custody of the Person	Responsibility for Education	Consent to Medical Care	Juvenile Court May Take Child into Custody for Treatment of Serious Injury or Illness	Conservator Must File Bond	Conservators Are Made Fiduciaries by Law	Compensation	Annual Reports or Accounts Must Be Filed
New Hampshire		§463:18-a (Supp. 1977) §462:4 (1968)	§463:18-a (Supp. 1977) §462:4 (1968)	§463:18-a (Supp. 1977) §462:4 (1968)		§169:17a (Supp. 1977)	§464:18 (Supp. 1978) §462:3 (1968)	§547:11a (1974)		§462:7 (1968)
New Jersey		§3A:6-24 (1963)	§3A:6-24 (1963)			§9:6-8.16 (1976) §9:6-8.29 (Supp. 1978)	Testamentary appointee may be relieved §3A:6-21 (1963)	§3A:2A-1 (Supp. 1978)		
New Mexico	§32A-5-209 (1975)	§32A-5-209 (1975) §32A-5-424 (1975)	§32A-5-209 (1975)	§32A-5-209 (1975)	§32A-5-209 (1976)	§13-14-20 (1976)	§32A-5-411 (Pamphlet 1975)	§32A-1-201 (Pamphlet 1975)	As court directs §32A-5-209 §32A-5-414 (1975)	As court directs §32A-5-209 §32A-5-419 (1975)
New York		SCPA §1723 (Supp. 1977) DRL §82 (1977)	DRL §82 (1977)	DRL §82 (1977)		[3]	May be relieved SCPA §1708 (Supp. 1977)	EPTL §1-2.7(1967) But see SCPA §1901 (Supp. 1977)		SCPA§1719 (1967)
North Carolina		§33-20 (1976)					§33-12 (1976) Testamentary appointee may be relieved §33-2 (1976)	§32-2 (1976)	As court directs §33-43 (1976)	§33-39 (1976)
North Dakota	§30.1-27-09 (1976)	§30.1-27-09 (1976)	§30.1-27-09 (1976)	§30.1-27-09 (1976)	§30.1-27-09 (1976)[4]	§27-20-06 §27-20-13 (1974)	§30.1-29-11 (1976)	§30.1-01-06 (1976)	As court directs §30.1-27-09 (1976) §30.1-29-14 (1976)	As court directs §30.1-27-09 (1976) §30.1-29-19 (1976)

State	Same Powers Generally as Parent	Care and Management of Financial Resources	Custody of the Person	Responsibility for Education	Consent to Medical Care	Juvenile Court May Take Child into Custody for Treatment of Serious Injury or Illness	Conservator Must File Bond	Conservators Are Made Fiduciaries by Law	Compensation	Annual Reports or Accounts Must Be Filed
Ohio		§2111.07 (1976) §2111.14 (1976)	§2111.07 (1976) §2111.13 (1976)	§2111.13 (1976)		§2151.31 (1976) §2151.33 (1976)	Resident guardian of non-resident ward only §2111.38 (1976)	§2109.01 (1976)		
Oklahoma		30§16(1976) 58§805 (1965)	30§15(1976) 58§770(1965) 58§301(1965)	30§15(1976) 58§770 (1965) 58§801 (1965)		10§1104 (Supp. 1977) 10§1120 (Supp. 1977)	58§776 58§778 (1965)		Reasonable 58§884 (Supp. 1977)	As court directs 58§872 (1965)
Oregon	§126.080 (1977)	§126.080 §126.313 (1977)	§126.080 (1977)	§126.080 (1977)	§126.080 (1977)	§419.507 (2) (e), §419.482 (3), §419.569 (1) (b) (1977)	§126.237 (1977)		Reasonable §126.263 (1977) As court directs §126.080 (1977)	§126.283 (1977) As court directs §126.080 (1977)
Pennsylvania		20§5141 (1975)				42§6324 (Pamphlet 1977) 42§6324 (Pamphlet 1977) 42§6339 (Pamphlet 1977)	Testamentary appointee may be relieved 20§5121 20§5122 (1975)	20§102 (1975)		
Rhode Island		§33-15-19 (1969)	§33-15-29 (1969)	§33-15-28 (1969)		§40-11-5 (1977)	§33-17-1 (1969)	§33-18-2 (1969)	Reasonable §33-15-27 (1969)	§33-15-26 (1969)
South Carolina [5]		§21-21-10 (1976)	§21-21-10 (1976)	§21-21-10 (1976)		§14-21-590 §14-21-620(c) (1977)	§21-19-40 (1976)	§21-11-20 (1976)	Same as for executors §21-19-180 (1976)	§21-19-130 §21-19-140 (1976)

State	Same Powers Generally as Parent	Care and Management of Financial Resources	Custody of the Person	Responsibility for Education	Consent to Medical Care	Juvenile Court May Take Child into Custody for Treatment of Serious Injury or Illness	Conservator Must File Bond	Conservators Are Made Fiduciaries by Law	Compensation	Annual Reports or Accounts Must Be Filed
South Dakota	6	§30-29-1 (1977)	§30-28-4 (1977)	§30-28-2 (1977) §30-28-4 (1977)		§26-8-19.1 §26-8-35 §26-8-43 (1977)	§30-27-30 (1977)		Reasonable §30-32-3 (1977)	As court directs §30-32-1 (1977)
Tennessee		§34-401 (1977)				§37-213 (Supp. 1978) §37-228 (1977)	§34-102 (1977) §34-206 (1977) §24-303 (1977)	§35-202 (1977)	Reasonable §34-810 (1977)	§34-306 (1977) §34-803 (1977)
Texas		Prob. Code 5§230 (Supp. 1978)	Prob. Code 5§229 (Supp. 1978)	Prob. Code 5§229 (Supp. 1978)		Fam. Code 2§17.01 (Supp. 1978) 2§17.04 (1975)	Prob. Code 5§193 (Supp. 1978) Testamentary appointee may be relieved Prob. Code 5§195 (1956)		5% of gross income of estate. Prob. Code 5§241 (Supp. 1978)	Prob. Code 5§399 (1956)
Utah	§75-5-209 (Pamphlet 1977)	§75-5-209 (Pamphlet 1977)	§75-5-209 (Pamphlet 1977)	§75-5-209 (Pamphlet 1977)	§75-5-209 (Pamphlet 1977)	§78-3a-23 §78-3a-29 §78-3a-39 §78-3b-6 (1977)	§75-5-105 (Pamphlet 1977) §75-5-411 (Pamphlet 1977)	§75-1-210 (Pamphlet 1977)	Reasonable §75-5-414 (Pamphlet 1977) As court directs §75-5-209 (Pamphlet 1977)	As court directs, §75-5-209 (Pamphlet 1977)
Vermont		14§2653 (1974)	14§2653 (1974)	14§2653 (1974)		33§639 (Supp. 1978)	14§2751 (1974) Testamentary appointee may be relieved 14§2656 (Supp. 1978)		Expenses 14§2847 (1974)	14§2921 (1974)

State	Same Powers Generally as Parent	Care and Management of Financial Resources	Custody of the Person	Responsibility for Education	Consent to Medical Care	Juvenile Court May Take Child into Custody for Treatment of Serious Injury or Illness	Conservator Must File Bond	Conservators Are Made Fiduciaries by Law	Compensation	Annual Reports or Accounts Must Be Filed
Virginia		§31-8 (1973)	§31-8 (1973)	§31-8 (1973)		§16.1-246 §16.1-251 §16.1-275 (Supp. 1978)	Testamentary appointee may be relieved §31-6(1973)	§26-17 (1973)		
Washington		§11.92.040 (Supp. 1977)	§11.92.040 (Supp. 1977)	§11.92.040 (Supp. 1977)	§11.92.040 (Supp. 1977)	§13.34.050 (Supp. 1978) §13.34.140 (Supp. 1978)	§11.88.100 (Supp. 1977)		Reasonable §11.92.180 (Supp. 1977)	§11.92.040 (Supp. 1977)
West Virginia		§44-10-7 (Supp. 1978)	§44-10-7 (Supp. 1978)	§44-10-7 (Supp. 1978)		§49-5-8 §49-6-3 §49-6-4 §49-6A-9 §49-6B-1 (Supp. 1978)	Testamentary appointee may be relieved. §44-10-5 (1966)	§44-5-7 (1966)	Reasonable §44-4-14 (1966)	§44-4-4 (1966)
Wisconsin		§880.19 (Pamphlet 1978)	§880.01 (Pamphlet 1978)			§48.28 §48.35 §48.395 (Supp. 1978)	Court may waive. §880.13 (Pamphlet 1978)	§112.01 (Supp. 1978)	Reasonable §880.24 (Pamphlet 1978)	§880.25 (Pamphlet 1978)
Wyoming		§3-2-108 (1977) §3-2-203 (1977)	§3-2-105 (1977) §3-2-108 (1977)	§3-2-105 (1977) §3-2-108 (1977)		§14-8-106 §14-8-121 (1977)	§3-2-108 (1977) §3-2-111 (1977)	§4-1-102 (1977)	Reasonable §3-2-121 (1977)	As court directs §3-2-207) (1977)
American Samoa							16§704 (1973)		2½% of receipts & disbursements 16§709 (1973)	16§706 (1973)

State	Same Powers Generally as Parent	Care and Management of Financial Resources	Custody of the Person	Responsibility for Education	Consent to Medical Care	Juvenile Court May Take Child into Custody for Treatment of Serious Injury or Illness	Conservator Must File Bond	Conservators Are Made Fiduciaries by Law	Compensation	Annual Reports or Accounts Must Be Filed
Guam		Prob. §1500 §1502 (1970)	Prob. §1500 (1970)	Prob. §1500 (1970)		Prob. §1442 (1970)	Prob. §1480(1970) But testamentary appointee may be relieved Prob. §1485(1970)		Reasonable Prob. §1556(1970)	As court directs Prob.§1553 (1970)
Puerto Rico		31§783 (1967)	31§783 (1967)	31§783 (1967)		34§2010 34§2103 (Supp. 1977)	31§761(1967) Except testamentary appointee may be relieved 31§767 (1976)		As court directs 31§790(1967)	31§801(1967)
Virgin Islands		15§824 (1964)	15§824 (1964)	15§824 (1964)		5§2503 5§2506 (1967)	15§881 (1964)	15§1041 (1964)	Reasonable 15§888 (1964)	15§882 (1964)

1 No definitions of the duties of a guardian are listed in the probate code but §233.2(1974) of the juvenile code lists duties and responsibilities applicable to guardianships under that act.

2 §13-708 (Supp. 1977) lists certain rights, powers and duties of personal guardian but is included in a subchapter entitled "disabled persons" which by definition would not encompass minors.

3 Children covered under this section are those already in the custody of some authorized agency.

4 Minor may also consent personally if parent or guardian unavailable. §14-10-17.1 (Supp. 1977)

5 Provisions apply to natural guardians only.

6 Specific conditions for care and treatment may be inserted in any order of appointment. §30-27-27 (1977)

XII. TERMINATION AND MODIFICATION (MINORS)

State	Terminate Automatically Upon Reaching Majority or Upon Marriage	Minors Age 14 or Over May Petition for Termination	Other Procedures
Alabama	§26-1-1 (1977)	§26-6-3 (1977)	Removal for cause §26-6-2 (1977) Resignation §26-6-1 (1977)
Alaska	§13.26.075 (1972)		Removal or resignation §13.26.085 (Supp. 1978)
Arizona	§14-5211 (1974)	§14-5212 (1974)	Removal or resignation §14-5212 (1974)
Arkansas	§57-643 (1977)	§57-621 (1977)	Removal for cause §62-2203 (1977)
California			Resignation Prob. §1582 (1956) Removal for cause Prob. §1580 (Supp. 1978)
Colorado	§15-14-210 (1973)	§15-14-212 (1973)	Resignation or removal §15-14-212 (1973 & Supp. 1976)
Connecticut	Majority only §45-43 (Supp. 1977) §45-45 (Supp. 1977) §45-47 (Supp. 1977)		Removal §45-43 (Supp. 1977)
Delaware	12§3911 (1975)		Removal or resignation 12§3910 (1975)
District of Columbia	§21-104 (1977)		
Florida			Resignation §744.467 (1976) Removal §744.474 (1976) Periodic Review §744.437 (1976)
Georgia		Upon reaching majority §49-301 (1974)	Removal for cause §49-232 (1974) Resignation §49-313 (1974)
Hawaii	§560:5-210 (1976)	§560:5-430 (1976)	Resignation & Removal §560:5-211 (1976)
Idaho	§15-5-210 (Supp. 1978)	§15-5-212 (Supp. 1978)	Resignation and Removal §15-5-212 (Supp. 1978)
Illinois	Except marriage of female terminates guardianship of person only ch.3§312 (1961)	ch.3§276 (Supp. 1978)	Removal for cause ch.3§276 (Supp. 1978) Resignation ch.3§282 (1961)
Indiana	§29-1-18-47 (Supp. 1977)	§29-1-18-25 (1972)	Removal for cause §29-1-10-6 (1972)

State	Terminate Automatically Upon Reaching Majority or Upon Marriage	Minors Age 14 or Over May Petition for Termination	Other Procedures
Iowa	Majority only §633.675 (1964)	§633.679 (1964)	
Kansas	§59-3028 (1976)	§59-3027 (1976)	
Kentucky			Removal for cause and resignation §387.090 (Supp. 1977)
Louisiana	CCP art. 4206 (1961)[1]	CCP art. 4234 (Supp. 1978)	Resignation general CCP art. 4233 (1961) Resignation at age 70 CC art. 296 (1952) Removal CCP art. 4234 (Supp. 1978)
Maine	18§3509 (Cum Supp. 1978) 18§3553 (Cum Supp. 1978)		Dismissal by judge 18§3509 (1965)
Maryland	§13-214 (1974) Must file petition Rule R. 80	§13-210 §13-221 (1974)	Death, disability resignation of guardian §13-220 (1974)
Massachusetts	Majority only 201§4 (Supp. 1978)	201§33 (1969)	Removal for cause and resignation 205§8 (1969)
Michigan	Majority only §703.8 (1976)		Resignation §704.49 (1976) Removal for cause §704.48 (1976)
Minnesota	Majority only §525.60 (1975)		Court removes when no further need exists §525.60 (1975)
Mississippi	Majority only §93-13-21 (Supp. 1977)		Removal for cause §93-13-23 (Supp. 1977) Resignation §93-13-25(1973)
Missouri	§475.285 (1956)		Removal for cause §475.110(1956) Resignation §475.285(1956)
Montana	§91A-5-210 (Special Pamphlet 1977)	§91A-5-212 (Special Pamphlet 1977)	Resignation or removal §91A-5-212 (Special Pamphlet 1977)
Nebraska	§30-2614 (1975)	§30-2616 (1975)	Resignation or removal §30-2616 (1975)

State	Terminate Automatically Upon Reaching Majority or Upon Marriage	Minors Age 14 or Over May Petition for Termination	Other Procedures
Nevada	Majority only §159.191 (1975)	§159.185 (1975)	Removal for cause §159.185 (1975)
New Hampshire	Majority only §463:18a (Supp. 1977) Marriage of female under age 18 may terminate guardianship §463:17 (1968)	§462:30 (1969)	Removal §462:29 (1969) Resignation §462:31 (1969)
New Jersey	§3A:6-18 (1953) Testamentary appointments only		
New Mexico	§32A-5-210 (1975)	§32A-5-212 (1975)	Resignation or removal §32A-5-212 (1975)
New York	SCPA §1707 (Supp. 1977) DRL §84 (1977)		
North Carolina	Majority only §33-41 (1976)		Removal for cause §33-9 (1976) Resignation §33-11 (1976)
North Dakota	§30.1-27-10 (1976)	§30.1-27-12 (1976)	Resignation or removal §30.1-27-12 (1976)
Ohio	Majority only §2111.46 (1976) §2111.45 (1976)		Removal for cause §2111.46 (1976)
Oklahoma	30§19 30§20 30§22 (1976)	58§876 (1965)	Removal for cause 30§18 (1976) 58§875 (1965) Resignation 58§875 (1965)
Oregon	§126.085 (1977)	§126.095 (1977)	Resignation or removal §126.095 (1977)
Pennsylvania	Majority only 20§2519 (1977)	20§3183 20§5131 (1975)	Removal for cause 20§3182 20§5131 (1975)
Rhode Island	Majority only §33-15-6 (1969)	§33-18-2 (1969)	Resignation or removal §33-15-18 §33-18-4 §33-18-2 (1969)
South Carolina			
South Dakota	One year after majority (Conservatorship not terminated by marriage) §30-27-46 §30-31-2 (1977)		Removal for cause §30-27-41 (1977) Resignation §30-27-42 (1977)
Tennessee	Majority only §34-301 (1977)		Removal for cause §34-419 (1977) Resignation §34-421 (1977)

State	Terminate Automatically Upon Reaching Majority or Upon Marriage	Minors Age 14 or Over May Petition for Termination	Other Procedures
Texas	Prob. Code 5§404 (Supp. 1978)		Resignation Prob. Code 5§221 (1956) Removal for cause Prob. Code 5§222 (Supp. 1978)
Utah	§75-5-210 (Pamphlet 1977)	§75-5-212 (Pamphlet 1977)	Resignation or removal §75-5-210 (Pamphlet 1977)
Vermont	14§3002 (1974)	14§3008 (Supp. 1978)	Resignation or removal 14§3001 (1974) 14§3010 (Supp. 1978)
Virginia	Majority only §31-9 (Supp. 1978)		Removal §31-14 (1973)
Washington	Majority only §11.88.140 (Supp. 1977)		Removal §11.88.120 §11.88.140 (Supp. 1977)
West Virginia	Majority only §44-10-7 (Supp. 1978)		Removal or resig- nation §44-10-7 (Supp. 1978) §44-5-5 (1966)
Wisconsin	Termination of conservator- ship upon marriage within the discretion of the court. Guardianship terminates automatically upon marriage §880.26 (Pamphlet 1978)		Removal for cause §880.16 (Pamphlet 1978) §880.192 (Pamphlet 1978)
Wyoming	Marriage ends guardian- ship of person only. §3-2-108 §3-2-114 (1977)		Removal or resignation §3-2-113 (1977)
American Samoa	Majority only 16§701 (1973)		
Guam	Majority Prob. §1410 (1970) Guardianship of person only terminates upon marriage Prob. §1500 (1970) Prob. §1590 (1970)	Prob. §1440 (1970)	Removal for cause §1580 (1970) Resignation Prob. §1582 (1970)
Puerto Rico	Upon majority or emanci- pation 31§791		Removal for cause 31§742
Virgin Islands			Removal or resig- nation 15§883 (1964)

[1] CC art. 354 *et seq.* (Supp. 1978) provides a mechanism for the continuation of guardianships for mentally retarded children beyond the age of 18.

SUMMARY OF MODEL STATUTE

As is evident from the above analysis, although a number of the more recent state guardianship statutes have addressed some of the concerns outlined in the introduction, none, without additions and changes, constitutes a comprehensive model which legislators can use as a basis for alleviating the inequities, resolving the ambiguities, and eliminating the anachronisms which afflict guardianship today. Hence, portions of current statutes, proposed legislation, and existing models, have been molded, together with some original provisions, into a model Guardianship and Conservatorship Act.

At the conclusion of his essay on "The State as Parent: Social Policy in the Progressive Era," David J. Rothman asks whether our society will be able to recognize and respect the rights of individuals requiring assistance and protection without ignoring their needs.[27] The model statute seeks to answer Mr. Rothman's question affirmatively. Underlying its provisions are the following premises:

- each individual possesses unique skills and talents;
- the level of proficiency in performing the basic human functions varies significantly among all individuals;
- given this variation, restrictions imposed upon individuals to act on their own behalf should be limited to those particular actions which the individual is unable to take and choices which the individual is unable to make; and
- concomitant with any grant of power over an individual must be the establishment of limits on the exercise and duration of that power and mechanisms to assure accountability.

With these guidelines in mind, the Model Guardianship and Conservatorship Act offers a functional rather than a causal definition of who is a partially disabled or disabled person. What is central is that the ability of individuals to receive and evaluate information or communicate decisions is impaired to the extent that they are unable to meet the essential requirements for their physical health or safety or manage their financial resources, and not that they have a developmental disability or mental illness, or have lived to an advanced age. It provides greater flexibility than current statutes in meeting the needs of adults unable to care for themselves and/or their property by encouraging use of such less intrusive means as voluntary services, trusts, and single transaction ratifications by the court, as alternatives to guardianship and/or conservatorship. When a combination of these means would not be sufficient, the statute provides for appointment of a limited personal guardian or personal guardian for the partially disabled or disabled person and/or a limited conservator or conservator for that individual's estate. However, it specifies that only those duties and powers which the individual is unable to perform should be assigned, and prescribes that limited personal guardians, personal guardians, limited conservators and conservators have a responsibility to protect the civil, personal and human rights of the persons whom they are serving. In addition, limited personal guardians, personal guardians, limited conservators or conservators must encourage the persons to participate to the maximum extent of their capabilities in all decisions which affect them, to act on their own behalf on all matters in which they are able to do so, and to develop or regain their ability to meet the essential requirements for their physical health and safety and/or manage their financial resources to the maximum extent possible. A limited personal guardian or personal guardian is prohibited from exercising those rights or performing those legal acts on behalf of a partially disabled or disabled person which are "so private, so personal, and so fundamental that they . . .[should not] be exercised at all unless the individual directly involved is able to provide the essential volition."[28] These include consenting to the commitment, the sterilization or the termination of the parental rights of a partially disabled or disabled person, or prohibiting the marriage or divorce of such persons. Thus, the model statute follows the analogy suggested by the American Association on Mental Deficiency:

> Guardianship can be compared to a

[27] Rothman, *supra* note 3, at 95.

[28] Kindred, *Guardianship and Limitations Upon Capacity*, in PCMR, THE MENTALLY RETARDED CITIZEN AND THE LAW, 62, 76 (1976).

> physical prosthesis which replaces a naturally functioning member. Exercising the member (in this case the legal right of the individual to certain elements of self-determination) should not be undertaken until it is clear that a substitute member, in fact, will give the individual a fuller functional capacity for social living. Like an artificial leg, guardianship necessarily operates in a manner somewhat different than the function it replaces and may even have some advantages which partially compensate for the basic loss of the natural function. In fitting a prosthesis the surgeon makes every effort to conserve the remaining natural functions.
>
> Similarly, in appointing a guardian the responsible body ideally should seek to preserve for the ward the opportunity to exercise those rights which are within his [or her] comprehension and judgment, allowing for the possibility of error to the same degree as is allowed to persons who are not retarded.[29]

Adoption of this act will not substantially alter the results in proceedings involving persons who are severely or profoundly disabled, other than to protect their rights more effectively. However, it is anticipated that the impact will be substantial for those individuals who, although they are suffering from a partial disability, are able to exercise some measure of self-reliance and independence when encouraged and allowed to do so.

With regard to minors, the traditional presumption in favor of the appointment of a personal guardian and conservator is preserved. Provision is still made, however, for use of less restrictive alternatives when they would be sufficient to protect the minor's physical health or safety and financial resources. For example, a limited personal guardian may be appointed when, in a separate proceeding, some or all of the parental rights of the child's parents have been suspended temporarily. A limited conservator may be appointed when a child whose parents are alive and have not had their parental rights terminated or suspended, has some financial resources which require protection. The statute would permit the parents of a minor to designate in a will the individual to serve as limited personal guardian, personal guardian, limited conservator or conservator for their child following their death. But as is currently the practice in many states, there are provisions for objections to a testamentary designation and for appointment by the court of a limited personal guardian, personal guardian, limited conservator or conservator who can better serve the interests of the child.

The model legislation also introduces the full panoply of due process rights and procedures to intervention proceedings. These include an unwaivable right to counsel, to notice of the proceedings, to be present at all hearings, to subpoena witnesses, to present evidence, to cross-examine opposing witnesses, and to have the need for establishment of a guardianship or conservatorship proven beyond a reasonable doubt.[30] Individuals alleged to be partially disabled or disabled persons are entitled to a jury and to closed proceedings. The model act contains, in addition, provisions for monitoring and modifying restrictions imposed on a person's capacity to act on his or her own behalf and for protecting the privacy of individuals subject to intervention proceedings. Included as well are emergency procedures for when there is an imminent danger to the health, life or property of a partially disabled or disabled person or a minor, or apparent malfeasance or misfeasance of a limited guardian, personal guardian, limited conservator or conservator. (Diagrams outlining the procedures established for intervention proceedings are presented in the Appendix .)

Finally, the model act addresses the problems of those individuals requiring guardianship or conservatorship for whom there is no one able and willing to serve in either or both of those capacities. The decision whether or not individuals require guardianship or conservatorship should be based on their abilities and not on whether they have the financial resources to pay for such services or a friend or relative willing to serve.[31] As is indicated on Chart VII, a number of states have adopted various forms of public guardianship utilizing state or local employees to serve as personal guardians or conservators. The model act adopts another alternative. In order to provide a close one-to-one relationship between the individual serving in a guardianship or conservatorship capacity and the person being served and to eliminate the inherent conflicts of interest which occur when the responsibility for public guardianship is placed in the agency providing services to minors and partially disabled or disabled persons, the act proposes that state funds be allocated to compensate private citizens who volunteer to serve as the limited personal guardian, personal guardian, limited conservator or conservator for a partially disabled or disabled person or a minor. While some states with public guardianship agencies may be able to devise other means of providing such relationships and reducing such

[29] AAMD, *supra* note 7, at 14.

[30] The burden of proof in intervention proceedings for minors is generally by a preponderance of the evidence. See §§44(1) and (2) Burden of Proof.

[31] PPMR, *supra* note 7, at 27.

conflicts, the citizen volunteer model proposed in the act offers greater assurance that these objectives can be achieved and avoids creating another costly social service bureaucracy. A volunteer public guardian or conservator may be appointed to serve a minor or a partially disabled or disabled person who has been determined to require guardianship or conservatorship, and for whom:

- there is no qualified private individual or entity able and willing to accept the appointment;
- the cost of private guardianship or conservatorship would substantially impair his or her financial independence, or impede meeting the essential requirements for his or her physical health or safety, managing his or her financial resources or developing his or her abilities;
- public funds provide the primary source of support; or
- the volunteer public guardian and/or conservator is best able to provide the required assistance.

Once appointed, a volunteer public guardian or conservator can be assigned the same duties and powers and has the same responsibilities and limitations as a private limited personal guardian, personal guardian, limited conservator, or conservator. A volunteer may also serve as a representative payee for government assistance payments or as the trustee of a small trust. In addition to the reporting and review procedures applicable to private guardianships and conservatorships, the statute provides for a 5 person, gubernatorally appointed oversight commission to assist in encouraging individuals to volunteer, providing information and training, and monitoring the appointment process. Commission members are to include, *inter alia,* at least one individual who was previously the subject of a guardianship or conservatorship and one member who has served as an advocacy attorney for persons subject to the imposition of a guardianship or conservatorship.

The text of the model act appears in **bold type**. Optional language appears [within brackets/ separated by slashes]. Comments on statutory provisions appear in standard type. The comments explain, elaborate upon or provide a rationale for the text of the act. They are intended to assist the user in interpreting the act and in drafting a statute for a particular state from this model. Following the full Model Guardianship and Conservatorship Act is a short-form version for states which prefer a statute with less procedural and administrative detail.

MODEL STATUTE

A BILL FOR

AN ACT TO REVISE THE METHODS, CRITERIA AND PROCEDURES FOR PROTECTING PARTIALLY DISABLED AND DISABLED PERSONS AND MINORS AND THE PROPERTY OF SUCH PERSONS AND MINORS; TO ESTABLISH THE CRITERIA FOR APPOINTING LIMITED PERSONAL GUARDIANS, PERSONAL GUARDIANS, LIMITED CONSERVATORS AND CONSERVATORS AND THEIR DUTIES AND POWERS; AND TO SPECIFY THE RIGHTS OF INDIVIDUALS SUBJECT TO INTERVENTION PROCEEDINGS AND DISPOSITIONAL ORDERS.

CHAPTER 1 INTRODUCTORY PROVISIONS

Section 1. Title

This act shall be known as the Guardianship and Conservatorship Act.

COMMENT. None necessary.

Section 2. Purpose

Recognizing that every individual has unique needs and differing abilities, the (name of Legislature) declares that it is the purpose of this act to promote the general welfare of all citizens by establishing a system which permits partially disabled and disabled persons and minors to participate as fully as possible in all decisions which affect them, which assists such persons in meeting the essential requirements for their physical health and safety, protecting their rights, managing their financial resources, and developing or regaining their abilities to the maximum extent possible, and which accomplishes these objectives through the use of the least restrictive alternatives. This act shall be liberally construed to accomplish this purpose.

COMMENT. None necessary.

Section 3. Definitions

As used in this act:

(1) "Partially disabled persons" means adults whose ability to receive and evaluate information effectively and/or communicate decisions is impaired to the extent that they lack the capacity to manage at least some of their financial resources and/or meet at least some of the essential requirements for their physical health or safety without court-ordered assistance or appointment of a limited personal guardian or limited conservator.

COMMENT. This act adopts a functional definition of disability. Rather than emphasizing the possible conditions which may cause a partial or total disability,[1] it focuses on the extent to which an impairment of an individual's ability to understand and appreciate the facts necessary to reach an informed decision and to convey that decision impedes that individual from taking those actions necessary to protect his or her physical health or safety and/or manage his or her financial resources. (For related definitions see §§3(3)-(5).) This approach recognizes that the fact that an individual has reached an advanced age, has a developmental disability, is suffering from a mental or debilitating illness, or has injuries of the brain or other portions of the central nervous system is of little relevance to determining whether the legal capacity to act in his or her own behalf should be transferred to another person or a corporate entity. In additon, the definition limits the degree to which value judgments may enter into intervention proceedings.

> Since the purpose of the incapacity standard is to distinguish persons whose decisions . . . must be accepted as final from those whose choices may be validly overridden . . . the standard should focus on the ability to engage in a decisionmaking process rather than on the resulting decision.[2]

Individuals with disabilities should have no less a right to be wrong than those without disabilities.[3] Hence, there is no requirement that deci-

[1] *See e.g.*, ALA. CODE §26-2-1(1975); IND. CODE ANN. §29-1-18-1 (Burns 1975); LA. CIV. CODE ANN. art. 355 and art. 389 (West 1974); N.J. STAT. ANN. §3A:1-1(West 1963); S.C. CODE §21-19-200 (1970).

[2] Note. *Civil Commitment of the Mentally Ill*, 87 HARV. L. REV. 1190, 1216 (1974); *but cf.* National Conference of Commissioners on Uniform State Laws (NCCUSL) Uniform Probate Code (U.P.C.) §5-101 (4th ed. 1975). KAN. STAT. §59-3002 (1965); NEV. REV. STAT. §159.019 (1969); 20 PA. CONS. STAT. ANN. §5501 (Purdon 1974).

[3] The President's Panel on Mental Retardation (PPMR). Report of the Task Force on Law, 25 (1963).

sions regarding person or property be "responsible" ones.[4]

This definition and those contained in sections 3(2), 3(12), 3(15) and 3(26) are based upon the realization that the individuals who may become the subject of an intervention proceeding will have vastly differing abilities and disabilities. Providing a court with the choice of either restricting a person's legal capacity to act totally or doing nothing at all ignores this range of capabilities. Accordingly, the act differentiates between individuals who require some type of assistance in performing certain of the functions necessary to meet particular health care, nutritional, clothing, housing and hygienic needs, or to manage particular property, benefits or income and those who require an individual or entity to make all decisions on their behalf.

Finally, it should be noted that this definition is not intended to refer to individuals who may have difficulty caring for themselves or their property because of an inadequate education.

(2) "Disabled persons" means adults whose ability to receive and evaluate information effectively and/or to communicate decisions is impaired to such an extent that they lack the capacity to manage their financial resources and/or to meet essential requirements for their physical health or safety even with court-ordered assistance or the appointment of a limited personal guardian or limited conservator.

COMMENT. See comment to section 3(1).

(3) "Manage financial resources" means those actions necessary to obtain, administer, and dispose of real and personal property, intangible property, business property, benefits and income.

COMMENT. The term financial resources is intended to encompass any type of asset which the subject of an intervention proceeding owns or is entitled to. Hence, the actions and skills necessary to obtain, administer and dispose of a minor's or a partially disabled or disabled person's financial resources depends upon their size, type and complexity.

(4) "Meet essential requirements for physical health or safety" means those actions necessary to provide the health care, food, shelter, clothing, personal hygiene and other care without which serious physical injury or illness is more likely than not to occur.

COMMENT. This definition adds specificity to the term "caring for self" or "taking care of his [or her] own person" used in many current guardianship statutes.[5] It is intended to discourage disputes over lifestyle by focusing solely on the physical health or safety of the subject of an intervention proceeding. The "more likely than not" standard makes explicit the two-part burden of persuasion operating in intervention proceedings.[6] That is, the petitioner must prove beyond a reasonable doubt both that the respondent is unable to perform certain functions, and that the failure to perform those functions makes it more likely than not that serious illness or injury will occur. (See §44(1) Burden of Proof.)

(5) "Court-ordered assistance" means dispositional alternatives available under section 11 of this act other than the appointment of a limited personal guardian, personal guardian, limited conservator and/or conservator.

COMMENT. In keeping with the principle of favoring the use of the least restrictive or intrusive alternative,[7] this definition distinguishes between those dispositions designed to provide support to individuals without imposing restrictions on their legal capacity to act in their own behalf, and guardianship or conservatorship. Hence, the term court-ordered assistance includes the provision of necessary medical, mental health, counseling, social, advocacy, educational, therapeutic, rehabilitative, homemaking, recreational, financial or other services to the respondent on a noncustodial basis; the establishment of a trust in which the respondent and any individuals legally dependent on the respondent are the beneficiaries; and/or the authorization or ratification of a contract or transaction entered into by a partially disabled or disabled person to meet his or her essential requirements for physical health or safety, to manage his or her financial resources and/or to develop or regain his or her abilities to the maximum extent possible.

(6) "Minor" means an individual [age seventeen (17) or less/under the age of majority].

COMMENT. Consistent with the 26th Amendment to the U.S. Constitution, the age of majority in most states is now 18.[8] However, because some states still retain the common law age of majority of 21 years, or have adopted age 19 as their standard, the definition is phrased in the alternative.

(7) "Respondent" means an individual alleged to be a partially disabled or disabled person.

COMMENT. None necessary.

---

[4] *But see* Council of the American Association on Mental Deficiency (A.A.M.D.). *Guardianship for Mentally Retarded Persons,* POSITION PAPERS OF THE A.A.M.D., 14 (1975); U.P.C. *supra* note 2.

[5] *See e.g.,* ARK. STAT. ANN. §57-601 (Supp. 1977); 18 ME. REV. STAT. tit. 18 §3601 (1975); OKLA. STAT. ANN. tit. 58 §§851 and 852 (West Supp. 1977); VT. STAT. ANN. tit. 14 §2683 (1968); *but see* CAL. PROB. CODE §1460 (West 1976); CONN. GEN. STAT. §45-70a(h) (Supp. 1979).

[6] Monahan and Wexler, *A Definite Maybe: Proof and Probability in Civil Commitment,* 2 LAW & HUMAN BEHAVIOR 37 (1978).

[7] *See* §3(26).

[8] H. BEASER, RUNAWAY YOUTH — FROM WHAT TO WHERE: THE LEGAL STATUS OF RUNAWAY CHILDREN, 44 (1975).

(8) "Subject of an intervention proceeding" means the respondent, a minor for whom a proceeding has been initiated under chapter 3 of this act, or an adult or minor for whom a limited personal guardian, personal guardian, limited conservator or conservator has been appointed.

COMMENT. None necessary.

(9) "Parties" means the individual or entity filing a petition, application, motion, acceptance of a testamentary designation, or objection; the subject of an intervention proceeding; and the limited personal guardian, personal guardian, limited conservator or conservator, if any.

COMMENT. None necessary.

(10) "Court" means the (name of the court(s) or divisions of a court of general jurisdiction which hear(s) intervention proceedings).

COMMENT. Depending on the organization of a state's judicial system, intervention proceedings may be heard in a specialized probate, juvenile, family, surrogate's or chancery court, in divisions of the court of general jurisdiction dedicated to probate or juvenile matters, or in non-specialized divisions of the court of general jurisdiction. This act takes no position on the particular court or courts in which intervention proceedings should be heard. However, the term court is intended to imply a hearing before or determination made by a judge and not a non-judicial hearing officer — e.g., a master or referee — or an administrative official such as the clerk. The restrictions on an individual's right to act on his or her own behalf which can be imposed in conjuction with a intervention proceeding can be as severe, if not more severe in some cases, than those imposed following the conviction of a non-capital criminal offense. As with other types of proceedings subject to the doctrine of *parens patriae*, "there is no place in our system of law for reaching a result of such tremendous consequences without ceremony. . . ."[9] An essential part of that ceremony is that it be presided over by a judge, because of the difficulty of the questions raised, and the need to assure accountability and appropriate legal protection to the respondent.[10]

(11) "Intervention proceeding" means any proceeding under this act.

COMMENT. The term intervention proceeding is intended to include all proceedings following the filing of a petition under sections 6 and 19(1) or an objection to a testamentary designation pursuant to section 20 including review hearings under sections 15 and 24 and the filing of an application pursuant to sections 16, 25, 35 and 40.

(12) "Limited conservator" means an individual or a corporation with general power to serve as trustee, appointed by the court to assist in managing the financial resources of a partially disabled person or of a minor.

COMMENT. This act distinguishes between what has been traditionally known as guardians of property and guardians of the person.[11] This distinction is made to assure that individuals who are determined to be incapable of managing their financial resources are not automatically considered incapable of meeting the essential needs for their physical health or safety. As observed by the President's Committee on Mental Retardation: "There should be a recognition of gradations and separability of competences. . . . [T]he idea that, because a person is incompetent to do one thing he [or she] is therefore incompetent to do all things is absurd, and contrary to the facts of everyday life."[12]

In keeping with this theme, the act draws a further distinction between guardianships and conservatorships for which plenary powers are assigned, i.e. where the subject of the proceeding is wholly incapable of acting on his or her own behalf, and limited guardianships and conservatorships. The role of the limited conservator as stated in the definition and described more fully in sections 18 and 28 is to assist the partially disabled person or the minor whom the limited conservator is serving to manage his or her financial resources. Authority to act in place of the partially disabled person or the minor is restricted to those particular functions and those particular resources which that person or that minor is adjudged to be unable to perform or manage even with assistance. The circumstances in which a limited conservator may be appointed are described in sections 12 and 23(3). The duties, powers and limitations on the authority of limited conservators are prescribed in sections 18 and 28. The individuals who and corporations which may serve as a limited conservator are set forth in sections 13 and 23(4).

(13) "Conservator" means an individual or a corporation with general power to serve as trustee, appointed by the court to manage the financial resources of a disabled person or a minor.

COMMENT. Under this act, a conservator may be appointed only when an individual is adjudged to be wholly unable to manage any of his or her financial resources and it is determined that there is no less restrictive dispositional alternative than

[9] Kent v. United States, 383 U.S. 541, 554 (1966).

[10] *But see* N.C. GEN. STAT. §35-1.8 (Supp. 1977).

[11] *Accord* U.P.C., *supra* note 2, at §§5-301 and 5-401.

[12] PRESIDENT'S COMMITTEE ON MENTAL RETARDATION (PCMR), REPORT TO THE PRESIDENT, MENTAL RETARDATION: CENTURY OF DECISION, 58-59 (1976).

to remove that individual's legal authority to act on his or her own behalf *vis a vis* property and income and reassign that authority to a conservator. (For a discussion of the distinction between conservatorships and guardianships and of the distinction between limited conservators and conservators as well as references to the sections governing the appointment, qualifications, and duties and powers of conservators, see the comment to §3(12).)

(14) "Conservatorship" includes appointment of a limited conservator or a conservator.

COMMENT. None necessary.

(15) "Limited personal guardian" means an individual appointed by the court to assist a partially disabled person or a minor in meeting the essential requirements for the partially disabled person's or the minor's physical health and safety.

COMMENT. As is discussed more fully in the comment to section 3(12), this act draws a distinction between conservatorships which concern the financial resources of a partially disabled or disabled person or a minor, and guardianships which concern the physical well-being of such persons and minors. As is also discussed above, the act is intended to encourage the use of the least restrictive dispositional alternative and promote the self-reliance and habilitation of partially disabled and disabled persons and minors. Accordingly, the primary obligation of a limited personal guardian is to assist the individual to obtain the necessary health care, food, shelter and clothing, enforce his or her legal rights, and develop or regain his or her abilities to the maximum extent possible. Authority to act in place of the partially disabled person or minor is restricted to those particular areas in which that person or minor is adjudged unable to meet the essential needs for his or her physical health or safety even with assistance. Sections 12 and 23(3) set forth the circumstances in which a limited personal guardian may be appointed. Sections 17 and 27 prescribe the duties, powers, and limitations on the authority of limited personal guardians. The individuals qualified to serve as limited personal guardians and the priorities for selection are delineated in sections 13 and 23(4). Related definitions inlcude sections 3(12)-(14) and sections 3(16)-(17).

(16) "Personal guardian" means an individual appointed by the court to act on behalf of a disabled person or a minor to assure that the essential requirements for the physical health and safety of the disabled person or minor are met.

COMMENT. Under this act, a personal guardian may only be appointed if an individual is adjudged to be wholly incapable of meeting the essential requirements for his or her physical health or safety and it is determined that there is no less restrictive alternative than to remove the individual's authority to act on his or her own behalf and reassign that authority to the personal guardian. A discussion of the distinction between guardianships and conservatorships, is included in the comment to section 3(12). In addition, the comment to section 3(15) contains a discussion of the difference between a limited personal guardian and a personal guardian, as well as references to the provisions governing the appointment, selection, duties, powers, and restrictions on the authority of personal guardians.

(17) "Guardianship" includes appointment of a limited personal guardian or a personal guardian.

COMMENT. None necessary.

(18) "Volunteer public guardian," "volunteer public conservator," and "volunteer public trustee," mean an individual appointed pursuant to section 36 of this act to serve as a limited personal guardian or personal guardian, or an individual or corporation appointed to serve as a limited conservator, conservator or trustee for partially disabled or disabled persons and minors who require establishment of a guardianship or conservatorship and who have no one able and willing to serve in that capacity or are unable to afford the cost of a guardianship, conservatorship and/or trust.

COMMENT A more detailed description of the conditions under which volunteer public guardians, volunteer public conservators, and volunteer public trustees may be appointed is contained in section 36. Section 5 establishes a Guardianship/Conservatorship Oversight Commission which is responsible for creating a pool of volunteers from which the court may select, providing training to these individuals, and monitoring the appointment process.

(19) "Guardians *ad litem*" means individuals appointed by the court to assist the subject of an intervention proceeding to determine his or her interests in regard to the intervention proceeding, or to make that determination if the subject of the intervention proceeding is unconscious or otherwise wholly incapable of determining his or her interests in that proceeding even with assistance.

COMMENT. A distinction is made between guardians *ad litem* and counsel. The role of counsel is to serve as a zealous advocate of the legal interests of his or her client but not to determine those interests.[13] The function of the guardian *ad litem* is to assist individuals to determine their

[13] AMERICAN BAR ASSOCIATION (ABA), CODE OF PROFESSIONAL RESPONSIBILITY, EC 7-1, EC 7-7, and EC 7-12 (1976).

interests and, if they are incapable of doing so, of acting in their stead. Just as lawyers are discouraged from representing themselves in important litigation, the attorney who is compelled to represent an individual in an intervention proceeding and serve as that individual's guardian *ad litem* may be caught between conflicting interests and have difficulty maintaining objectivity and avoiding the subrogation of one role to the other. A number of states currently recognize that the rights of individuals in intervention proceedings are better protected when separate individuals serve as counsel and guardian *ad litem*.[14] Sections 33 and 34 address the role of counsel and the appointment and duties of guardians *ad litem* in more detail.

(20) "Counsel" means an attorney admitted to the practice of law in this state.

COMMENT. While a lay advocate can provide valuable assistance in a variety of contexts, given the importance and complexity of the matters at issue in intervention proceedings and their judicial setting, it is essential that individuals subject to intervention proceedings be represented by a licensed attorney. (For discussion of the distinction between counsel and guardians *ad litem*, see §§33 and 34 and the comment to §3(19).)

(21) "Restrictions on the legal capacity of a person to act in his or her own behalf" means powers of a partially disabled or disabled person or a minor which are assigned to a limited personal guardian, personal guardian, limited conservator or conservator.

COMMENT. By assigning to a limited personal guardian, personal guardian, limited conservator, or conservator the power to act on behalf of a partially disabled or disabled person or minor, a court is at the same time removing the authority of the partially disabled or disabled person to act on his or her own. Hence, "restrictions on the legal capacity. . ." appears appropriate as a generic term.

(22) "Disabilities resource officers" means employees of the (name of state/local agency responsible for providing services to partially disabled and disabled persons) who have current knowledge of the full range of public and private medical, mental health, social, advocacy, educational, rehabilitative, counseling, therapeutic, homemaking, recreational and financial services and programs available to assist partially disabled and disabled persons and their families, and who are assigned the duties set forth in this act.

COMMENT. Section 4(1) of this act provides for the appointment of a disabilities resources officer(s) in each local jurisdiction. Section 4(2) as well as sections 6(1)(b)(xiv), 7, 9, and 17(2)(a), 17(6)(b), 18(2)(a), 18(7)(b) and 35(1) outline the duties which disabilities resources officers are to perform.

(23) "Multidisciplinary evaluation team" means a group of at least three (3) individuals which is appointed by the court from a panel of experts approved by the (state name) Guardianship/Conservatorship Oversight Commission, to evaluate an individual subject to an intervention proceeding pursuant to chapter 2 of this act. Each multidisciplinary evaluation team shall include a physician, a social worker with a graduate degree in social work and field training or experience in working with partially disabled or disabled persons, and a psychiatrist, licensed clinical psychologist, and/or other expert with knowledge of the particular disability which an individual is alleged or has been found to have or of the skills required to meet the essential requirements for the individual's physical health or safety or to manage that individual's financial resources.

COMMENT. Prior to the impostition of a guardianship or conservatorship, approximately sixty percent of the states provide for an evaluation of the individual alleged to be a partially disabled or disabled person. As indicated in Chart III, most of these jurisdictions provide that the evaluation be performed by a physican. Those which have adopted the U.P.C. also provide for a "visitor," often a social worker, to interview the proposed guardian and to visit the current and proposed homes of the allegedly disabled or partially disabled person.[15] A few states, recognizing that most physicians may not have the training and experience necessary to evaluate an individual who may be mentally ill or developmentally disabled, provide for examination by a psychiatrist or clinical psychologist. However, as is indicated in section 8, an expert opinion in each of these areas may be essential to the determination whether guardianship or conservatorship is required.[16] Hence, the model statute, following the lead of North Carolina and Texas,[17] provides for a multidisciplinary evaluation team. A physician is included to examine the physical condition of the subject of the intervention proceeding. The social

[14] Indiana, Texas, Washington and Wisconsin provide for appointment of both counsel and a guardian *ad litem* in guardianship proceedings involving adults. Washington has a similar provision applicable to guardianship proceedings involving minors.

[15] U.P.C., *supra* note 2, at §5-303.

[16] INTERNATIONAL LEAGUE OF SOCIETIES FOR THE MENTALLY HANDICAPPED (ILSMH), SYMPOSIUM OF GUARDIANSHIP OF THE MENTALLY RETARDED, 19 (1969).

[17] N.C. GEN. STAT. §35-1.16(b)(1977); TEX. PROB. CODE tit. 5 §130F (Vernon, Supp. 1978); *see also* UNITED STATES SENATE SPECIAL COMMITTEE ON AGING, PROTECTIVE SERVICES FOR THE ELDERLY: A WORKING PAPER, 81-83 (1977).

worker is to focus on the environment, adaptive behavior and social skills of the individual. The additional member or members of the team should have the expertise to evaluate the particular type or types of disability which the subject of the proceeding is alleged to have or the skills needed for that individual in order to be self-reliant. For example, if the petition alleges that the respondent has a mental illness, a psychiatrist or clinical psychologist should be included on the team. If the respondent is alleged to be retarded, a clinical psychologist and/or an expert in retardation should be placed on the team. In some instances, it may be necessary to appoint a certified public accountant or qualified financial counselor to the team to evaluate the skills required to manage the individual's financial resources. Section 5 provides for a Guardianship/Conservatorship Oversight Commission which will, *inter alia,* identify individuals qualified to serve on multidisciplinary evaluation teams and monitor the performances of such teams. Section 8 outlines the form of evaluation reports.

(24) "Evaluation" means an assessment of the ability of an individual to receive and evaluate information effectively and/or communicate decisions and the impact of any impairment of these skills on the individual's capacity to meet the essential requirements for his or her physical health or safety, and/or manage his or her financial resources.

COMMENT. Section 8 governs the appointment of a multidisciplinary evaluation team, prescribes the content and format of the evaluation report, and specifies that evaluations shall be conducted so as to minimize interference with the respondent's activities and intrusion into the respondent's privacy.

(25) '(State name)' Guardianship/Conservatorship Oversight Commission" means the agency established by section 5 of this act.

COMMENT. The Guardianship/Conservatorship Oversight Commission is responsible, *inter alia,* for compiling and maintaining a list of experts qualified to participate in evaluations; encouraging individuals to agree to serve as volunteer public guardians; encouraging individuals and corporations to agree to serve as volunteer public conservator and volunteer public trustees; providing training for individuals serving in a guardianship and/or conservatorship capacity; and assuring that volunteer public guardians, volunteer public conservators, and volunteer public trustees are appointed in a lawful, objective manner. Section 5 prescribes the organization, duties and powers of the Oversight Commission.

(26) "Least restrictive dispositional alternative" means the form of assistance that least interferes with the legal capacity of a respondent to act in his or her own behalf.

COMMENT. The principle of the least restrictive alternative was explained in the report of the President's Committee on Mental Retardation as follows: "When the government. . .[has] a legitimate communal interest to serve by regulating human conduct it should use methods that curtail human freedom to no greater extent than is essential for securing that interest. When you swat a mosquito on a friend's back, you should not use a baseball bat."[18] In the context of this statute, no restriction should be placed on the legal capacity of an individual to act in his or her own behalf unless no combination of voluntary services or other alternatives to guardianship or conservatorship would be sufficient to permit the partially disabled or disabled person to meet the essential requirements for his or her physical health or safety and/or manage his or her financial resources. The dispositional criteria contained in section 12(3) are designed to facilitate implementation of this principle.

(27) "Developmental disability" means a chronic disability of a person which:

(a) Is attributable to a mental or physical impairment or a combination of mental and physical impairments;

(b) Is likely to continue indefinitely;

(c) Results in substantial functional limitations in two or more of the following areas of major life activity:

(i) Self-care;

(ii) Receptive and expressive language;

(iii) Learning;

(iv) Mobility;

(v) Self-direction;

(vi) Capacity for independent living; or

(vii) Economic self-sufficiency; and

(d) Reflects the need for a combination and sequence of special, interdisciplinary or generic care, treatment or other services which are of lifelong or extended duration and individually planned and coordinated.

COMMENT. This definition, like that in the Developmental Disabilities Assistance and Bill of Rights Act, 42 U.S.C. §6001(7) (Cum. Supp. 1978) (Pub. L. 94-103, as amended by Pub. L. 95-1780), adopts a functional rather than a categorical approach. As is noted above in the comment to the definition of partially disabled persons, the effect of the disability is of prime importance for this statute, rather than its cause. The definition differs

---

[18] Chambers, *The Principle of the Least Restrictive Alternative: The Constitutional Issues,* in PCMR, THE MENTALLY RETARDED AND THE LAW, 486 (1976); *accord* Kindred, *Guardianship and Limitations Upon Capacity,* in PCMR, *id.* at 66, *see* ILSMH, *supra* note 16, at 18-19.

from the one contained in the recently amended federal act in that there is no requirement that the disability be "severe" or that it appear before the afflicted person is age 22. Both these changes are intended to expand the group of persons who will have their individual guardianship plan and/or individual conservatorship plan, and the reports of their limited personal guardian, personal guardian, limited conservator and/or conservator sent to the state Developmental Disabilities Advocacy Office.[19] (See §32(3)(b) Notice and Dissemination of Plans and Reports.) The federal definition is to target certain populations for intensive federal support, whereas the purpose of this act's definition is to provide the opportunity for the Developmental Disabilities Advocacy Office to assist all individuals with developmental disabilities who are the subject of an intervention proceeding, regardless of whether they have a "severe" disability or whether their disability is the result of a genetic defect or post-natal trauma or disease.

In considering this act, care should be taken to assure that the definition of developmental disability is consistent with that adopted in the statute establishing the state protection and advocacy system for persons with developmental disabilities.

(28) "Administrator of Veterans' Affairs" means the head of the Veterans' Administration or his or her successor.

COMMENT. This definition is consistent with 38 U.S.C. §§101(1) and 210 (Cum. Supp. 1978).

(29) "Veterans' Administration" means the Veterans' Administration of the United States of America, its predecessors and successors.

COMMENT. The Veterans' Administration was established by 38 U.S.C. §201(1959).

(30) "Identifiable information" means information which is indexed or able to be retrieved by name, identifying code or number, address or other personal characteristic.

COMMENT. This definition applies to information which is collected and files and records which are maintained in conjunction with an intervention proceeding. It includes all such data, files and records which can be located through use of some personal identifiers. Chronological records, such as hospital logs or court dockets, and notations concerning an individual made in a file concerning another person are not covered since the danger of improper disclosures from such records is minimized by the difficulty of finding the entry pertaining to a particular individual. Neither does the definition include published appellate decisions since those are of necessity, public records. The collection, retention, disclosure and disposition of information identifiable to the subject of an intervention proceeding is governed by section 40.

Section 4. Appointment of Disabilities Resources Officers

(1) The head of the [state/local agency responsible for providing services to partially disabled or disabled persons] shall employ individuals on a full-time or part-time basis to serve as disabilities resources officers (in accordance with citation to the state Civil Service Act or the statute, if any, authorizing local jurisdictions to hire personnel). The number of individuals hired to be or assigned the duties of disabilities resources officers shall be commensurate with the volume of intervention proceedings initiated and the number of guardianships and conservatorships for partially disabled or disabled persons established in the [state/county/parish/city], but shall be sufficient to permit easy access for those individuals requiring the services of a disabilities resources officer.

COMMENT. Most individuals who file petitions seeking establishment of a guardianship or conservatorship and many limited personal guardians, personal guardians, limited conservators, and conservators may not be aware of the increasing number of local, state federal, and private programs and services which are designed to assist partially disabled and disabled persons in meeting the essential requirements for their physical health or safety, managing their financial resources, protecting their rights, and developing or regaining their abilities. Since it is anticipated that in many cases such alternatives will obviate the need to seek court-ordered assistance, or permit greater independence for persons whose legal capacity to act on their own behalf has been restricted, the model act provides for a disabilities resources officer who can advise individuals planning to file a petition and limited personal guardians, personal guardians, limited conservators and conservators of the programs and services available. Subsection (1) provides that the number of disabilities resources officers should be directly related to the number of intervention proceedings, guardianships and conservatorships in each jurisdiction. It is not necessary to establish a large new bureaucracy. However, the number of disabilities resources officers should be sufficient to prevent the scheduling of consultations with disabilities resources officers from becoming a serious impediment to providing protection and assistance to partially disabled and disabled persons.

Responsibility for employing or designating disabilities resources officers is assigned to the head of the human resources department since an individual in that agency would be in the best position to develop a familiarity with the programs and ser-

[19] Chapter 9: Advocacy, *infra*.

vices available. Where there is no umbrella agency for health, social and rehabilitative services, the agency providing services to developmentally disabled persons or to the elderly should serve as the lead agency. Coordination with and cooperation by other service-providing public agencies should be required. If disabilities resources officers are state employees, care should be taken to assure access to these officials from each county or local jurisdiction. Finally, although not specified in the statute, disabilities resources officers should have training and experience in working with partially disabled persons, and their families.

(2) (a) Disabilities resources officers shall:

(i) Develop and maintain a current knowledge of the full range of public and private medical, mental health, social, advocacy, educational, rehabilitative, counseling, therapeutic, homemaking, recreational and financial services and programs available to assist partially disabled and disabled persons and their families;

(ii) Discuss personally with individuals seeking to file a petition pursuant to section 6 of this act or an application pursuant to section 35(1), the matters which the individual wishes to allege in the petition or application, the effects of the appointment of a limited personal guardian, personal guardian, limited conservator and/or conservator, and the less restrictive alternatives which are available;

(iii) Discuss with other participants in pre-hearing conferences held pursuant to section 9 of this act, the alternatives to court-ordered assistance or the establishment of a guardianship or conservatorship which can provide the respondent with any necessary assistance and protection;

(iv) Assist partially disabled and disabled persons, and individuals appointed to serve as a limited personal guardian, personal guardian, limited conservator, and/or conservator to develop and update individual guardianship and individual conservatorship plans;

(v) Provide information, guidance, counseling and referral services to individuals regarding the alternatives which are available to assist persons with developmental and/or other disabilities to meet the essential requirements for their physical health or safety, and/or manage their financial resources, and/or protect their rights, and develop or regain their abilities to the maximum extent possible. No referrals shall be made unless the disabilities resources officer has contacted the individual or public or private agency to which the referral is being made, and determined that the individual to be referred meets the eligibility requirements and that the necessary services are available.

COMMENT. This subsection sets out the duties of disabilities resources officers. The range of services and programs with which disabilities resources officers are conversant should include those offered through the private sector as well as those supported by the local, state and federal governments. The duty to explain the impact of guardianship and/or conservatorship to individuals contemplating the filing of a petition and to explore with such individuals the other means available for providing the support, protection and assistance required, is intended to discourage reliance on court-ordered services and restrictions on legal capacity when services offered and accepted on a voluntary basis will adequately safeguard the rights, health, safety and resources of the individual alleged to be a partially disabled or disabled person. The same purpose applies to the consultation required prior to the filing of an application for an emergency appointment pursuant to Section 35(1) and to participation in the pre-hearing conference. Following establishment of a guardianship or conservatorship, sections 17(2)(a) and 18(2)(a) require the limited personal guardian, personal guardian, limited conservator or conservator, the partially disabled or disabled person and the disabilities resources officer to develop jointly an individual guardianship plan or individual conservatorship plan. Sections 17(6)(b) and 18(7)(b) provide for updating the plans annually. Again, the role of the disabilities resources officer is to apprise the partially disabled or disabled person and the individual serving in a guardianship or conservatorship capacity of the programs and services available which may benefit the partially disabled or disabled person and suggest how they can be utilized to maximize that person's independence and the development or regaining of his or her skills in the most normal environment possible. Subparagraph (2)(a)(iv) imposes a duty to respond to inquiries from individuals not contemplating or involved in an intervention proceeding. Such inquiries may come, for example, from the parents of a child with a developmental disability or the spouse of an accident victim regarding the services which are available and the alternatives to guardianship. Finally, the requirement that the disabilities resources officer contact an agency prior to referring an individual to it and that he or she determine in advance the eligibility requirements and availability of services, is included as a means of expediting and simplifying the provision of needed assistance. Hopefully, it will reduce the amount of waiting time and the number of appointments and forms which individuals seeking voluntary services will have to endure.

(b) Disabilities resources officers shall file a motion to dismiss a petition when it appears that the ability of the respondent to receive and evaluate information and/or to communicate decisions is not impaired or when it appears that the

respondent does not require court-ordered assistance or the establishment of a guardianship or conservatorship to meet the essential requirements for his or her physical health or safety and/or to manage his or her financial resources.

COMMENT. This provision authorizes the disabilities resources officer to file a motion with the court to dismiss a petition when the disabilities resources officer believes that the respondent is not a partially disabled or disabled person. (See §7 Motion to Dismiss.) While it is anticipated that this authority will not have to be exercised often, it provides a safeguard against frivolous or malicious petitions. Section 44(1)(a) provides that a motion to dismiss shall be granted unless the petitioner demonstrates at the hearing on the motion that there is probable cause to believe the respondent is a partially disabled or disabled person.

Introduction to Section 5

This section creates an independent agency headed by a part-time commission within the executive branch of the state government. The Oversight Commission is to perform two sets of functions. First, it is charged with identifying physicians, psychiatrists, licensed clinical psychologists, social workers, financial management consultants, and experts on various types of disabilities who are qualified and willing to serve on multidisciplinary evaluation teams; maintaining a list of these individuals for use by the courts hearing intervention proceedings; and deleting names from the list as necessary to assure the accuracy and objectivity of evaluations. (See §8 Evaluation.) Second, the Oversight Commission is responsible for creating a pool of individuals and entities to provide guardianship and/or conservatorship services to partially disabled persons, disabled persons and minors who have no one able and willing to provide such services or whose financial resources are too limited to obtain such services; monitoring the appointment process in order to protect against the development of a system of political patronage in the appointment of volunteers to serve as limited personal guardians, personal guardians, limited conservators, conservators or trustees for such persons; providing training programs for all individuals appointed to serve in a guardianship or conservatorship capacity, and advising the Governor and Legislature on the problems and needs in the area of guardianship and conservatorship. (See §36 Appointment of Volunteer Public Guardians, Volunteer Public Conservators, and Volunteer Public Trustees.)

In view of the competing demands for limited public resources, the Oversight Commission has been designed as an alternative to establishing another state social service agency. Rather than establishing a central diagnostic center or a set of centers, and hiring a staff of professional public guardians, this section, in conjunction with section 36, provides for the organization of a network of private citizens to render the required services on an as-needed basis.[20] It is anticipated that this will not only reduce the cost to the taxpayers, but offer more individualized service than can be provided by a government employee who must deal with a substantial caseload. (See also comment to §5(4)(b) *infra.*)

Section 5. Establishment of the (state name) Guardianship/Conservatorship Oversight Commission

(1) There is hereby established the (state name) Guardianship/Conservatorship Oversight Commission (hereinafter referred to as the Oversight Commission) which shall be independent of all other government agencies, boards, and commissions.

COMMENT. The major commissions and commentators who have recommended establishing a system for providing public guardianship have strongly urged that the agency providing or overseeing public guardianship and conservatorship be wholly separate from the department responsible for providing services to partially disabled and disabled persons.

> A manifest conflict of interest exists when guardianship powers are exercised by a service-providing individual or organization. A service provider makes some decisions in terms of the fiscal and political needs of his or her organization and is not in a position to give . . . sole attention to the best interests of the individual receiving the service.[21]

In some instances, in which the responsibility for public guardianship and the provision of mental health resources have been lodged in the same agency, public guardianship has become a means for circumventing the procedural safeguards which have been erected to protect against unnecessary civil commitments.[22] There is also the danger in such multipurpose agencies that public guardianship would come to be considered a precondition to receiving public services. While some states may be able to develop other means of preventing such inherent conflicts from harming the clients of an agency providing public guardianship

---

[20] *See* ILSMH, *supra* note 16, at 23-24.

[21] Kindred, *supra* note 18, at 85, *accord* ILSMH, *supra* note 16, at 23; GOVERNOR'S COMMISSION FOR REVISION OF THE MENTAL HEALTH CODE OF ILLINOIS REPORT, 135-136 (1976); *see* P.C.M.R., *supra* note 12, at 140; *see also* THE CONSERVATEE RIGHTS COMMITTEE, THE STATUS OF CONSERVATEES IN SANTA CLARA COUNTY [CALIFORNIA] (1979).

[22] *See* Levy, *Protecting the Mentally Retarded: An Empirical Survey and Evaluation of State Guardianship in Minnesota,* 49 MINN. L. REV. 821 (1965); Kindred, *supra* note 18, at 84-87.

among other services, this subsection makes the Oversight Commission a wholly independent agency, and subsection (5) prohibits it from providing services other than those specified in this act.

(2) (a) The Oversight Commission shall consist of five (5) individuals appointed by the Governor, with the advice and consent of the Senate to three (3) year terms, except that of the first appointments, two (2) shall be for two (2) year terms, and one (1) shall be for a one (1) year term. A member of the Oversight Commission shall not serve more than two (2) consecutive terms. Vacancies shall be filled in the same manner as original appointments. Appointments to fill vacancies occurring before the expiration of a term are for the remainder of the unexpired term. Commission members whose term has expired shall continue to serve until a new appointment has been made and confirmed.

COMMENT. The responsibilities and duties of the Oversight Commission require that its members have the status and be subject to the accountability of a Gubernatorial appointee. Accountability is further provided for by the requirement of Senate confirmation and by the relatively short three-year term. The importance of continuity is recognized by staggering the appointments and permitting members to serve two consecutive terms.

(b) At least one (1) member of the Oversight Commission shall have previously been the subject of a guardianship or conservatorship; at least one (1) member shall have served or shall be serving as a limited personal guardian, personal guardian, limited conservator, and/or conservator; and at least one (1) member shall be an attorney who regularly represents individuals with developmental disabilities, individuals who are mentally ill, individuals over age sixty-five (65), or individuals under age eighteen (18).

COMMENT. This provision attempts to assure that the views of those classes of individuals most familiar with and most affected by guardianship and conservatorship will be represented on the Oversight Commission.

(c) The governor shall designate the chairperson of the Oversight Commission.

COMMENT. None necessary.

(3) The Oversight Commission shall meet at least six (6) times per year. Members of the Oversight Commission not employed by the state government shall receive compensation at a rate not to exceed $ ________ for each day, including travel time, they are engaged in the performance of their duties as members of the Oversight Commission. In addition, all members of the Oversight Commission shall be entitled to reimbursement for travel, subsistence and other reasonable expenses incurred by them in carrying out their duties.

COMMENT. The rate of compensation to members of the Oversight Commission should be equal to that paid to non-state employed members of other state supervisory and advisory boards. Some states may wish to accomplish this by inserting a reference to their civil service code rather than specifying a particular rate of compensation.

(4) The Oversight Commission shall:

(a) (i) Establish and annually publish the procedures and criteria for identifying individuals who are qualified to serve on multidisciplinary evaluation teams;

(ii) Prepare and publish, annually, a list of individuals from all sections of the state who are qualified and willing to serve on multidisciplinary evaluation teams; and

(iii) Establish and annually publish the procedures and criteria for removing individuals from the list who prove to be unqualified to serve on multidisciplinary evaluation teams.

COMMENT. The requirement in section 8 that evaluations be performed by a multidisciplinary team including individuals with expertise regarding the type or cause of the disability which a respondent is alleged to have, necessitates that some centralized source of information be developed to assist the courts and the parties in selecting a team with the appropriate qualifications. Accordingly, the list should include each individual's particular area of expertise, and the Oversight Commission should make every effort to identify qualified individuals in all parts of the state, in order to make the examination as convenient as possible for both the respondent and the expert (see §8(3) Evaluation), and to reduce the cost of the evaluation process. The panel of experts may include government employees when their public responsibilities would not limit their willingness to recommend the least restrictive alternative pursuant to sections 11 and 12, and their service on a multidisciplinary evaluation team would be consistent with the terms of their employment. The Oversight Commission is also required to provide a mechanism for receiving and investigating complaints (§5(3)(d) *infra*) about the diligence, competence and objectivity of individuals serving on multidisciplinary evaluation teams and for removing from the eligibility list, those individuals who are no longer willing to accept or have shown themselves to be unqualified for appointment to multidisciplinary evaluation teams. Annual publication of the procedures and criteria for adding and removing individuals from the panel of experts is required in order to ensure that the list is kept current.

Alternative. *Some states may choose to establish a separate Evaluation Oversight Board because of the technical nature of identifying and* *maintaining a panel of experts and the need for a similar panel in conjunction with proceedings to terminate parental rights, to commit an individual involuntarily to a residential facility, and, in criminal cases, to determine competency to stand trial or lack of responsibility by reason of a mental disease or defect. Since the Board would not be performing a general policy-making function, it would not be necessary to include as members, citizen representatives of the various consumer and practitioner interests which may be affected by Board actions. Furthermore, inasmuch as judgments may have to be made regarding the qualifications of individual professionals, it would not be appropriate to include practicing physicians, licensed clinical psychologists, psychiatrists, social workers, etc. on the Board. On the other hand, unless differing perspectives are brought to bear on Board decisions, there is a distinct danger that the list of experts will be limited to those proffering a particular point of view, and that the prestige and legitimacy of the Board will be lessened. Accordingly, the Evaluation Oversight Board should consist of high-ranking state officials including the head of the agency(ies) responsible for providing services to partially disabled and disabled persons, the Director of the state Developmental Disabilities Advocacy Office and the state Attorney General. A provision establishing a separate Evaluation Oversight Board should include the duties and powers set forth in sections 5(3)(a),(d),(g),(h),(i),(j) and 5(5), prohibit the Board or its staff from conducting actual evaluations, and specify a minimum number of meetings.*

(b) (i) Establish programs to encourage individuals to serve as volunteer public guardians, and individuals and corporations to serve as volunteer public conservators or volunteer public trustees;

(ii) Prepare and publish, annually, a list of individuals from all sections of the state who are willing to serve as volunteer public guardians, and a list of individuals and corporations from all sections of the state who are willing to serve as volunteer public conservators or volunteer public trustees;

(iii) Establish procedures to assure that volunteer public guardians, volunteer public conservators, and volunteer public trustees are selected for appointment and compensated in a lawful, equitable and consistent manner;

COMMENT. This provision directs the Oversight Commission to develop a pool of potential volunteer public guardians, volunteer public conservators and volunteer public trustees from which the court may select if the conditions set forth in section 36 arise. The recruitment effort could include advertising in the media and joining with service organizations, professional societies, and labor unions as well as associations with a particular focus on partially disabled and disabled persons to urge their members to volunteer.[23] The names of volunteers may be compiled locally through agreements between the Oversight Commission and local agencies, although a central list should be maintained for those few cases in which a court may need to appoint a volunteer public guardian, volunteer public conservator and/or volunteer public conservator in another part of the state (*e.g.*, when a hospitalized repondent is to be released following appointment of a limited public guardian). Subparagraph (3)(b)(iii) together with paragraph (3)(d) of this section and section 36 provides protection against the appointment as a volunteer public guardian, volunteer public conservator, or volunteer public trustee becoming a patronage reward or a breeding ground for corruption. While misuse of the appointments in intervention proceedings may be rare, such a provision in a system relying on volunteers is a necessary safeguard, given the power which a limited personal guardian, personal guardian, limited conservator, conservator, or trustee may exercise.

Alternative. *As noted earlier, there are a number of other structures which can be employed for providing public guardianship services. For those states considering the establishment of a public guardianship agency within the executive branch of state government, see the entries for Colorado, Georgia, Illinois, Maine, Minnesota, North Carolina and Ohio in Chart VII and the public guardianship sections of Charts I, V, and VI, supra. For those jurisdictions considering implementation of local public guardianship services, see the entries on the above listed Charts for Alabama, Arizona and California. Finally, those interested in a judicially based public guardianship system should refer to the entries for Delaware and Tennessee.*

(c) Establish and maintain training programs which will provide instruction on the general and specific duties of limited personal guardians, personal guardians, limited conservators, and conservators, the nature and extent of the disabilities which most commonly result in establishment of a guardianship or conservatorship, and the programs, services and techniques available for:

(i) Assisting a partially disabled person or minor to meet the essential requirements for his or her physical health and safety, to manage his or her financial resources and to protect his or her rights;

(ii) Assuring that the essential requirements for the physical health and safety of a disabled person are met, that his or her rights are protected, and that his or her financial resources are prudently managed; and

[23] *See* ILSMH, *supra* note 16

(iii) Assisting a partially disabled or disabled person to develop his or her abilities to the maximum extent possible.

COMMENT. Since many individuals who accept an appointment as a limited personal guardian, personal guardian, limited conservator or conservator may not have a full understanding of their role and responsibilities, this paragraph mandates that the Oversight Commission provide training courses.[24] Among the topics covered should be the programs and services available to assist the individual whom they are serving and the cause(s) and effect(s) of the disability which is affecting that individual. Such courses should not be limited only to individuals who participate in the volunteer program. Many individuals serving in a guardianship or conservatorship capacity for a friend or relative or attempting to assist a partially disabled or disabled person — e.g., the parents of a developmentally disabled child — are likely to need and want at least some of the training offered. The Oversight Commission may provide training services through qualified members of its staff or through its contracting power. (See §5(4)(h).)

(d) Conduct investigations and hearings regarding the operation of the volunteer public guardian/conservator/trustee program and the selection, appointment, compensation and removal of volunteer public guardians, volunteer public conservators and volunteer public trustees.

COMMENT. The Oversight Commission is empowered to hold public hearings to help evaluate the programs under its supervision. Such hearings can be an opportunity for those who participate in or are affected by evaluations, or who have agreed to be or are being served by volunteer public guardians, volunteer public conservators, and volunteer public trustees to express their satisfaction and/or concerns. The Oversight Commission is also empowered, as a necessary part of its oversight functions, to investigate through its staff, complaints about multidisciplinary evaluation teams or corruption in the appointment of volunteers. If criminal conduct is involved, the matter should be turned over to the appropriate prosecutorial official.

This investigatory power is not intended to extend to misconduct or neglect by an individual volunteer public guardian, volunteer public conservator, or volunteer public trustee. Providing ombudsperson responsibilities to the Oversight Commission appears unnecessary since the subjects of intervention proceedings will have counsel and, if necessary, an independent guardian *ad litem* (§31 Rights and Privileges; §33 Duties and Compensation of Counsel and §34 Procedures and Criteria for Appointment of Guardians *ad Litem*), since review hearings will be held at least biannually (§15 Periodic Review and §§30(1)(l) and (m) Time Limits), and in view of the advent of advocacy offices and organizations for individuals with developmental disabilities and in some instances for elderly persons and youth as well. However, such added responsibilities should be considered in states which do not have vigorous advocacy programs. The liabilities and penalties applicable under this act to all limited personal guardians, personal guardians, limited conservators and conservators are set forth in section 42.

(e) Testify before the (name of the Legislature) on matters related to the provision of services for and the guardianship and conservatorship of partially disabled persons, disabled persons and minors;

COMMENT. None necessary.

(f) Advise the Governor and other state and local officials and agencies on matters related to the provision of services for and the guardianship and conservatorship of partially disabled persons, disabled persons and minors;

COMMENT. None necessary.

(g) Promulgate such additional policies, rules, regulations and procedures as are necessary for the effective operation of the Oversight Commission;

COMMENT. None necessary.

(h) (i) Employ, in accordance with the (citation to the state civil service code) and the appropriations provided, an Administrative Director and such professional, technical, clerical and administrative personnel as may be necessary to carry out the duties and powers set forth in this act;

(ii) Delegate to the Administrative Director such powers as are necessary to carry out the duties of the Oversight Commission set forth in this act, including the power to redelegate authority when necessary and to designate a member of the staff to serve as Acting Administrative Director during the absence or disability of the Administrative Director or in the event of a vacancy in the office of Administrative Director.

(iii) Contract with individuals, businesses, whether for profit or not-for-profit, and other public and private organizations or agencies to carry out the duties and powers set forth in this act; and

(iv) Assure that all employees and contractors of the Oversight Commission understand their duties and powers.

COMMENT. A part-time commission cannot accomplish all the tasks required to fulfill the duties prescribed above. While it is not intended that the Oversight Commission develop into a large bu-

---

[24] *Id.* at 24.

reaucracy, a multifaceted staff will be necessary. The Administrative Director should have both management skills and knowledge of the issues and problems inherent in organizing and overseeing a public guardianship and evaluation system. Some of the Oversight Commission's responsibilities, e.g., training programs and recruitment of volunteers, may be able to be met or initiated through contracts to businesses and academic and other non-profit institutions. Some states may wish to include a reference to their procurement code and/or regulations in subparagraph (4)(h)(iii).

(i) Prepare and submit a proposed annual budget in accordance with the procedures established by (citation to the relevant statutes, regulations, and executive orders); and

COMMENT. None necessary.

(j) Perform such additional duties and exercise such powers as are prescribed by law or as are necessary for the effective management and operation of the Oversight Commission in accordance with the purpose of this act.

COMMENT. None necessary.

(5) The Oversight Commission shall not have supervisory power over individual intervention proceedings nor shall it manage or operate residential facilities or directly provide services other than those specified in this act for partially disabled or disabled persons or minors.

COMMENT. Under no circumstances should the Oversight Commission be given authority to provide direct services and/or operate residential facilities or institutions. Such authority would greatly heighten the danger that appointment of a volunteer public guardian or volunteer public conservator might be misused as a shortcut to involuntary commitment or as a precondition to the provision of services,[25] thereby decreasing rather than increasing the use of less restrictive alternatives.

(6) (a) The Oversight Commission shall submit an annual report to the Governor, the (name of the Legislature), the Chief Justice of (name of the highest court of civil appeals), the (head of the state agencies responsible for providing services to partially disabled and disabled persons and minors) and the (head of the state Developmental Disabilities Advocacy Office).

(b) The annual report shall describe the actions taken by the Oversight Commission to fulfill its duties, any significant problems which have arisen during the reporting period, and the actions taken or required to resolve those problems.

(c) Copies of the report shall be made available to the public and shall be sent to each court which can hear intervention proceedings.

COMMENT. The annual report is intended as a means of informing the public, the highest level of state government, and the individuals and entities most affected by or involved in the establishment of guardianships and conservatorships and the appointment of volunteer public guardians, volunteer public conservators and volunteer public trustees, of the activities of the Oversight Commission. A description of the problems encountered during reporting period and the actions taken or required to resolve those problems is intended to make this report a substantive document rather than a routine organizational summary or glossy public relations product.

CHAPTER 2 INTERVENTION PROCEEDINGS FOR PARTIALLY DISABLED AND DISABLED PERSONS

Section 6. Initiation of Intervention Proceedings

(1) (a) An intervention proceeding shall be initiated by the filing of a petition with the court by a partially disabled person, an adult interested in the welfare of a partially disabled or disabled person, or by a limited personal guardian, personal guardian, limited conservator or conservator of such a person.

COMMENT. The only means under this act for securing court-ordered assistance for a partially disabled or disabled person (who is not already receiving such assistance) or for appointing a limited personal guardian, personal guardian, limited conservator or conservator is by filing a petition. Once assistance has been ordered or a guardianship or conservatorship established, the procedures prescribed by section 16 may be used to request a change in the types of services provided, the duties and powers assigned to the limited personal guardian, personal guardian, limited conservator and/or conservator, or the individual serving in the guardianship or conservatorship capacity. (See also §6(2) regarding testamentary designations and §35 Emergency and Standby Appointments.) Separate procedures are provided for establishment of guardianships and conservatorships for minors in chapter 3 of this act.

As in most jurisdictions, any adult who is concerned about the well-being of another adult whom he or she has reason to believe is a partially disabled or disabled person may initiate an intervention proceeding. It is anticipated that the formality of the procedures and the required discussion of the petition with the disabilities resources officer will discourage frivolous filings. (§4(2)(a)(ii) Appointment of Disabilities Resources Officers and subparagraph (1)(b)(xiv) of this section; see also §7 Motion to Dismiss.) Individuals may also initiate proceedings in their own behalf. In addition, explicit authority to file a petition is provided

[25] *See* Levy, *supra* note 22; Kindred, *supra* note 18.

to the limited personal guardians, personal guardians, limited conservators and conservators of partially disabled or disabled persons. Such authority would be exercised when, for example, only a conservatorship has been established for a partially disabled or disabled person and the limited conservator or conservator has reason to believe that the ability of the partially disabled or disabled person to meet the essential requirements for his or her physical health or safety is impaired. An explicit expression of the authority to file a petition is necessary because of the specificity with which guardianship and conservatorship duties and powers are prescribed in section 17 and 18. Finally, in order to reduce the likelihood that petitions may be filed maliciously or for purposes of harassment, intervention proceedings may not be initiated by individuals whose only interest in the person alleged to be partially disabled or disabled is that they may be "adversely affected by [the] lack of effective management of. . .[that person's] property or affairs."[26]

(b) The petition shall be verified under oath and shall include:

(i) The name, age, and address of the respondent;

(ii) The nature and degree of the alleged disability;

(iii) The particular type of protection and assistance being sought;

(iv) The duration of the protection and assistance being sought;

(v) The facts and reasons supporting the need for such protection and assistance;

(vi) Any existing restrictions on the legal capacity of the respondent to act in his or her own behalf;

(vii) The approximate value and description of the respondent's financial resources, if the petition alleges that the respondent is unable to manage his or her financial resources;

(viii) The benefits, compensation, pension or insurance payable by the Veterans' Administration to which the respondent is entitled, if known;

(ix) The names and addresses, unless they are unknown and cannot be reasonably ascertained, of the individuals most closely related to the respondent by blood or marriage;

(x) The name and address of the individual or facility, if any, having custody of the respondent;

(xi) The name, address, qualifications and relationship to the respondent of the individual, if any, proposed for appointment as limited person guardian, personal guardian, limited conservator, or conservator;

(xii) The name, address, and interest of the petitioner;

(xiii) The name and address of the petitioner's attorney, if any, or a statement that the petitioner is unable to obtain counsel for financial reasons and requests to be represented by the [district/county/state] attorney; and

(xiv) A statement that the petitioner has discussed with the disabilities resources officer the facts alleged in the petition, the effect of the appointment of a limited personal guardian, personal guardian, limited conservator or conservator, and the alternative services and procedures which are available, and an explanation why the requested type of protection and assistance is the least restrictive dispositional alternative which will meet the needs of the respondent.

COMMENT. The petition is designed to set forth the facts and allegations on which the intervention proceeding is based and the names and addresses of the individuals and entities who are presumed to have sufficient interest in the proceedings to receive notice. (See §32 (1)(a) Notice and Dissemination of Plans and Reports.) In addition, it assures that the petitioner has been informed of the possible consequences of an intervention proceeding and the availability of less restrictive alternatives. Care has been taken to make the petition as complete a resource as possible without placing an undue burden on the petitioner.

Subparagraph (1)(b)(ii) requires petitioners to explain, to the best of their knowledge, the disability they believe the respondent is suffering from and the effects of that disability. Evaluation reports are not required to be filed with the petition in order to avoid imposing this potentially substantial cost on the petitioner and to safeguard the objectivity of evaluation. Subparagraph (1)(b)(iii) requires petitioners to specify whether they are seeking, for example, court-ordered homemaking and vocational training services, or the appointment of a limited conservator to supervise the operation of a business owned by the respondent, or the appointment of a personal guardian with plenary powers. Subparagraph (1)(b)(viii) is included in order to determine whether the Administrator of Veterans' Affairs is entitled to notice pursuant to 38 U.S.C. §3202(b) (1959).[27] While inclusion of this information will be helpful, an indication by the petitioner that he or she does not have access to information regarding the V.A. benefits to which the respondent is entitled should not nullify the petition. Subparagraph (1)(b)(xiii) is to assure that

---

[26] U.P.C., *supra* note 2, at §5-404(a).

[27] *Accord* NCCUSL, UNIFORM VETERANS' GUARDIANSHIP ACT §2 (1942).

petitioners who are indigent receive legal advice concerning the intervention proceeding as early as possible and that the local prosecutor's office receives prompt notice that it will be representing the petitioner pursuant to section 31(1)(c).

The petition is under oath to emphasize the seriousness of initiating an intervention proceeding, and to permit prosecution of petitioners who knowingly file a petition containing false allegations.

(c) No more than five (5) days after the filing of a petition, the court shall set a date for a hearing and shall have notice served on the individuals and entities enumerated in section 32(1)(a) of this act and in the form and manner prescribed in sections 32(1)(a) and 32(2).

COMMENT. Notice of the rights of the parties and of the date, time, place, purpose and possible consequences of the hearing is to be served on all the individuals and entities enumerated in the petition including the Administrator of Veterans' Affairs when applicable. Section 32(2) mandates that, to the maximum extent possible, the contents of the notice and of the petition must be explained to the respondent in the language, mode of communication, and terms which the respondent is most likely to understand.

(2) A parent, spouse, or child of a partially disabled or disabled person who is serving as limited personal guardian, personal guardian, limited conservator or conservator of that person may, by will, designate an individual to serve in that capacity upon the death of the testator. Such a designation shall not become effective if another person is serving as the limited personal guardian, personal guardian, limited conservator or conservator at the time of death, nor until the will has been filed for probate and the court has appointed the designee following a hearing pursuant to section 16 of this act. Pending such appointment, the designee may file for appointment on an emergency basis pursuant to section 35(1) of this act.

COMMENT. This provision permits a parent, spouse or child of a partially disabled or disabled person who is serving in a guardianship or conservatorship capacity for that person to indicate in a will, his or her choice of a successor as limited personal guardian, personal guardian, limited conservator or conservator. However, the process for transferring the guardianship or conservatorship authority is the same as that employed upon the resignation, death or removal of a limited personal guardian, personal guardian, limited conservator or conservator who is not also the parent, spouse or child of the partially disabled or disabled person. This protects against the possibility of having such designations serve as a means for circumventing the safeguards and emphasis on less restrictive alternatives provided in this chapter. Otherwise, two classes of guardianship and conservatorship would be established.[28]

The procedures set forth in section 16 permit the partially disabled or disabled person or another interested individual to request a hearing in order to challenge the necessity or adequacy of the restrictions imposed on the legal capacity of the partially disabled or disabled person to act on his or her own behalf and/or to propose the appointment of an individual and/or corporation other than the designee. (See also §13 regarding the eligibility requirements and priorities which apply to the selection of a limited personal guardian, personal guardian, limited conservator or conservator.) Recognizing, however, that in some situations it is critical that the void caused by the death of a limited personal guardian, personal guardian, limited conservator or conservator is filled without delay, at least on a temporary basis, section 35(1) provides for emergency appointments and section 35(3) authorizes the court to designate eligible individuals or corporations to standby to assume the duties and powers assigned to a limited personal guardian, personal guardian, limited conservator or conservator upon his or her resignation, incapacitation or death.

Section 7. Motion to Dismiss

(1) A motion to dismiss may be filed by any party or the disabilities resources officer when it appears that the ability of the respondent to receive and evaluate information and/or to communicate decisions is not impaired or when it appears that the respondent does not require court-ordered assistance or the appointment of a limited personal guardian, personal guardian, limited conservator, and/or conservator.

COMMENT. This section provides a mechanism for challenging a petition on its face early in the proceedings. Unless the petitioner is able to demonstrate at the hearing on the motion that there is probable cause to believe that the respondent is a partially disabled or disabled person, the case will be dismissed. (§44(1)(c) Burden of Proof.) Pursuant to Section 31, hearings on motions to dismiss are to be on the record and before a judge. That section also provides that each party is entitled to be present, be represented by counsel, subpoena witnesses, present evidence, and cross-examine opposing witnesses.

The motion to dismiss is intended as a protection against the occasional frivolous, malicious or sincere but unwarranted petition, and should be used neither as a discovery technique nor as a means for introducing a probable cause hearing into every intervention proceeding. Although no restriction is imposed on when the motion may be filed, it is anticipated that dismissal will usually be

[28] *E.g.*, U.P.C., *supra* note 2, at §5-310.

sought prior to the respondent's evaluation. The authority afforded the disabilities resources officer to file a motion to dismiss is a logical extension of the screening function assigned to disabilities resources officers by sections 4(2)(a)(ii) and 6(1)(b)(xiv).

(2) No more than five (5) days after the filing of a motion to dismiss, the court shall set a date for a hearing and shall have notice served on the individuals and entities enumerated in section 32(1)(a) in the form and manner prescribed in sections 32(1)(a) and 32(2).

COMMENT. Notice of the rights of the parties and of the date, time, place, purpose and possible consequences of the hearing on the motion to dismiss is to be served on each of the individuals enumerated in the petition including the Administrator of Veterans' Affairs when applicable. (§32(1)(a) Notice and Dissemination of Plans and Reports.) The notice shall be explained to the respondent in the language, mode of communication and terms which the respondent is most likely to understand. (§32(2) Notice and Dissemination of Plans and Reports.)

Section 8. Evaluation

(1) Within the time prescribed in section 30(1)(c) of this act, the court shall appoint a multidisciplinary evaluation team from the list published by the Guardianship/Conservatorship Oversight Commission to evaluate the respondent. Each party may submit names drawn from this list for the court's consideration.

COMMENT. Approximately 60% of American jurisdictions statutorily provide for an evaluation of the subject of an intervention proceeding.[29] Because of the additional dispositional options provided by this act and the emphasis on the choice of the least restrictive alternative, the evaluation report takes on even greater significance than under most existing procedures. Hence, this section makes evaluations mandatory in all intervention proceedings. As in most jurisdictions, the obligation of obtaining and bearing the cost of evaluations is placed on the court, in order to assure the objectivity of the evaluation and to avoid placing an inordinate financial burden on initiation or defense of an intervention proceeding.[30] A list of experts qualified to serve on a multidisciplinary team will be prepared published and maintained by the Guardianship/Conservatorship Oversight Commission pursuant to section 5(3)(a). Each team must consist of a physician to examine the physical condition of the respondent, a social worker to examine the environment, adaptive behavior and social skills of the respondent, and a licensed psychologist, psychiatrist and/or other expert with knowledge of the particular disability which the respondent is alleged to have. As noted in conjunction with section 3(23), a multidisciplinary evaluation team could also include an accountant and/or financial management counselor to assess the skills necessary to manage the respondent's financial resources. In selecting the multidisciplinary evaluation team, the court should accommodate the preferences of the parties to the extent that it is possible to do so without appointing more than one expert with the same specialty. In accordance with section 30(1)(c), the multidisciplinary evaluation team should be appointed within 10 days of the filing of the petition.

(2) The multidisciplinary evaluation team may, with the approval of the court, employ such assistants and incur such expenses as are reasonable and necessary for conducting the evaluation. However, the members of the multidisciplinary team shall be responsible for the contents of and conclusions in the team's report.

COMMENT. This provision permits members of a multidisciplinary evaluation team to employ paraprofessionals or other assistants to perform background or preliminary work which may be necessary to complete the evaluation. The actual examination of the respondent and preparation of the report should be done by the members of the team themselves.

(3) The evaluation shall be conducted so as to minimize interference with the respondent's activities and intrusion into the respondent's privacy.

COMMENT. Except in extraordinary circumstances evaluation of respondents who are living in the community should be at the respondent's home or on an out-patient basis at a time which is mutually convenient to the respondent and the expert performing the evaluation. Especially in investigating the respondent's social skills and adaptive behavior, only that information which is essential should be collected, and great care should be taken so as not to stigmatize or harm the respondent among his or her friends, neighbors and colleagues. (See § 40(1) Records.)

(4) (a) The report of the multidisciplinary evaluation team shall include:

(i) A description of the nature and extent of the respondent's disabilities, if any;

(ii) A description of the respondent's mental, emotional, physical, and educational condition, adaptive behavior and social skills;

(iii) A description of the medical, mental health, social, advocacy, educational, rehabilitative, counseling, therapeutic, homemaking, and financial services, if any, being used by the re-

[29] Chart III, *supra*.

[30] *But see e.g.*, ILL. ANN. STAT. ch. 110 ½, §11a-9 (Smith-Hurd Supp. 1979) which places the burden of evaluating the respondent on the petitioner and N.C. GEN. STAT. §35-1.16 (Cum. Supp. 1977) which makes respondents responsible for the cost of multidisciplinary evaluations.

spondent to meet the essential requirements for his or her physical health or safety, and/or to manage his or her financial resources, and/or to develop his or her abilities;

(iv) An opinion regarding the type(s) and extent of assistance, if any, required by the respondent to manage his or her financial resources, to protect his or her rights, and/or to meet the essential requirements for his or her physical health or safety, and why no less restrictive alternative would be appropriate;

(v) An opinion regarding the probability that the extent of the respondent's disabilities, if any, may significantly lessen, and the type of services or treatment which will facilitate improvement in the respondent's condition, behavior or skills;

(vi) A description of the techniques or tests used;

(vii) The names and addresses of all individuals who examined, interviewed or investigated the respondent in conjunction with the evaluation;

(viii) The signatures of each member of the multidisciplinary evaluation team together with a statement of concurrence or nonconcurrence with the evaluation findings; and

(ix) Any dissenting opinions or other comments by members of the multidisciplinary evaluation team.

COMMENT. The format for the report is designed to assure that the evaluation directly addresses the questions at issue in the proceedings — the respondent's abilities and disabilities, and the means through which the respondent's independence and self-reliance can be maximized — and to provide the court and the parties with the information necessary to assess the conclusions, facts and opinions presented in the report. (See the determinations required to be made at the hearing on the petition pursuant to §10 Hearing.) As a consequence, the statutory requirements are somewhat more detailed than in most current guardianship legislation. Most of the information called for is self-explanatory. Consistent with the other provisions of this act, subparagraph (4)(a)(iv) requires that in suggesting what types of services the respondent may require, if any, the multidisciplinary team must consider less restrictive alternatives and discuss why those alternatives, if any, would not be sufficient to assure that the respondent is able to meet the essential requirements for his or her physical health or safety and/or manage his or her financial resources. Subparagraphs (4)(a)(vii), (viii), and (ix) recognize that there can be differences of opinion among the experts serving on a multidisciplinary evaluation team and that these differences may be of critical importance to the court, the parties, and the jury, if any, in intervention proceedings.

(b) The report of the multidisciplinary evaluation team shall be submitted to the court within the time prescribed in section 30(1)(d) of this act and shall be distributed in the manner and to the individuals specified in section 32(3)(a).

COMMENT. The evaluation report is due 20 days after a petition has been filed. Ordinarily this would provide the multidisciplinary team 10 to 14 days to complete its work. (See §§30(1)(c) and 30(1)(d) Time Limits; see also Figure 1 in the Appendix.) This time period is consistent with the stringent limits proposed throughout this act to assure that intervention proceedings are drawn out no longer than necessary, that respondents who require protection and assistance receive it promptly, and that the cost, anxiety and inconvenience which result from intervention proceedings are limited as much as possible. It is anticipated that 10 to 14 days will be sufficient in most cases. In those instances in which it is not, a continuance may be sought pursuant to section 30(5)(c).

Copies of the report are to be sent to the attorney of record for each party and directly to any party not represented by counsel. The contents of the report must be explained to the respondent in the language, mode of communication and terms which he or she is most likely to understand. (See §32(3) Notice and Dissemination of Plans and Reports.) In some instances, the evaluation report may contain information which could cause severe psychological or physical harm to the respondent or another party to the proceeding. In those cases, the information should be disclosed to the party through a psychiatrist, psychologist or other appropriate professional so as to minimize or nullify any potential negative consequences.

Section 9. Prehearing Conference

Within the time limit prescribed in section 30(1)(g) of this act, the parties shall meet, together with the disabilities resources officer, to discuss whether in light of the report of the multidisciplinary evaluation team, the respondent is in need of protection and assistance and, if so, whether non-custodial treatment or services available on a voluntary basis may obviate the need for requesting the appointment of a limited personal guardian, personal guardian, limited conservator, and/or conservator. If there is agreement that court-ordered assistance or the appointment of a limited personal guardian, personal guardian, limited conservator and/or conservator is not required, the petition shall be withdrawn immediately. If there is agreement that voluntary, noncustodial treatment or services are available which may obviate the need for submission of the matter to the court, the parties may stipulate to a continuance of up to ninety (90) days. If the parties do not agree to withdraw the petition or stipulate to a continuance, the matter shall proceed to hearing.

COMMENT. The prehearing conference provides an opportunity for the parties, together with the disabilities resources officer, to meet and discuss whether in light of the evaluation report, the intervention proceeding needs to continue. Three options are presented: the petition can be withdrawn if the parties agree that the respondent can meet the essential requirements for his or her physical health or safety and/or manage his or her financial resources without court-ordered assistance; the hearing can be postponed for up to 90 days in order to assess whether the respondent will be able to meet those essential needs and/or manage those resources with the aid of voluntary, noncustodial treatment and/or services; or, the hearing on the petition can proceed as scheduled. (See §10 Hearing.) It is anticipated that through the evaluation and the information provided by the disabilities resources officer, many intervention proceedings can be resolved without a hearing and the imposition of restrictions on the respondent's legal capacity to act on his or her own behalf. The prehearing conference is set 30 days after the filing of a petition in order to give the parties sufficient time to study the evaluation reports and information obtained through discovery, and to give the court sufficient time to adjust its calendar should the petition be withdrawn or a continuance sought. (See §30(1)(g) Time Limits.)

Section 10. Hearing

Within the time limit prescribed in section 30(1)(h) of this act, the court shall hold a hearing.

(1) If the petition requests court-ordered assistance not related to the management of the respondent's financial resources and/or the appointment of a limited personal guardian or personal guardian, the following matters shall be determined at the hearing:

(a) The essential requirements for the respondent's physical health and safety;

(b) The skills and knowledge necessary to meet those needs;

(c) The nature and extent of the respondent's disabilities, if any; and

(d) Whether the respondent is a partially disabled or disabled person as defined in sections 3(1) and (2) of this act.

(2) If the petition requests court-ordered assistance related to the management of the respondent's financial resources and/or the appointment of a limited conservator or conservator, the following matters shall be determined at the hearing:

(a) The type and amount of the respondent's financial resources;

(b) The skills and knowledge necessary to manage those financial resources;

(c) The nature and extent of the respondent's disabilities, if any; and

(d) Whether the respondent is a partially disabled or disabled person as defined in sections 3(1) and (2) of this act.

COMMENT. This provision delineates the elements of the central decision in an intervention proceeding. It requires the judge or jury (see §31(1) Rights and Privileges) to consider first the needs which must be met and/or the financial resources which must be managed. Next it requires identification of the skills necessary to meet those needs and/or manage those resources. Third, it directs the judge or jury to determine whether the respondent has the necessary skills by considering the respondent's disabilities, if any. Finally, if an impairment is found, the provision requires the decision-maker to determine whether the respondent is able to meet the essential requirements for his or her physical health or safety, and/or manage his or her financial resources without court-ordered assistance or without the establishment of a guardianship or conservatorship; whether some form of court-ordered assistance and/or some restriction on the respondent's legal capacity to act on his or her own behalf is necessary to assure that these needs are met and/or resources managed; or whether even with court-ordered assistance and a limited personal guardian or limited conservator, the respondent would be unable to protect his or her physical health, safety, and property. This formulation differs from current state codes by requiring explicit consideration of the relationship between the skills needed to protect a respondent's health and financial assets and the respondent's abilities and disabilities. While this may be implied in those state statutes requiring an inquiry into the extent of an individual's disabilities and capacity to care for his or her property and person,[31] spelling out the factors to be considered provides clarity to a difficult and complex decision-making process, and added assurance that individuals requiring protection and assistance will receive it with the least restrictive alternative being utilized.

The hearing is to be held within 35 days of the filing of the petition. Each party is entitled to be represented by counsel, to subpoena witnesses, to present evidence and testimony, to cross-examine opposing witnesses, and to have an impartial decision-maker. (See §31(1)(a) Rights and Privileges.) The subject of the intervention proceeding is also entitled to have a jury make the determinations required by this section and to close the hearing to the general public. (See

[31] *E.g.*, IDAHO. CODE §56-239 (Cum. Supp. 1977); ILL. ANN. STAT. ch. 110½. §11a-11 (Smith-Hurd Supp. 1979); TEX. PROB CODE tit. 5 §130 G (Vernon Cum. Supp. 1978).

§§31(1)(b) and (d) Rights and Privileges.) The petitioner has both the burden of going forward and the burden of persuasion. (See §44(1)(b) Burden of Proof.) If the respondent is a beneficiary of the Veterans' Administration, a certificate of the Administrator of Veterans' Affairs or his or her representative stating that the respondent has been determined to be incompetent by the V.A. and that appointment of a conservator is a condition precedent to the payment of any money due to the respondent by the Veterans' Administration, should constitute *prima facie* evidence of the need for appointment of a limited conservator empowered to accept and disburse the V.A. funds.[32] Such a certificate, however, should not necessarily constitute evidence of the need for imposing greater restrictions on the legal capacity of the respondent to act in his or her own behalf.

Section 11. Dispositional Alternatives

(1) If a respondent is found not to be a partially disabled or disabled person the petition shall be dismissed.

COMMENT. None necessary.

(2) Among the dispositional alternatives available to the court following a finding that the respondent is a partially disabled or disabled person, are:

COMMENT. The list of dispositional alternatives contained in this subsection is arranged in the order of relative restrictiveness beginning with the least restrictive dispositional alternative — ordering services for the respondent — and ending with the most restrictive — appointment of a personal guardian. The alternatives are not intended to be mutually exclusive. A court could, for example, order the provision of educational, vocational and homemaking services in addition to directing the establishment of a trust or appointing a limited conservator. (See §12 Dispositional Hearing and Dispositional Principles and Criteria.) Furthermore, the list of dispositional alternatives should not be considered all-inclusive. It is provided to alert the court and the parties to the range of options which are available.

(a) Directing the provision of necessary medical, mental health, counseling, social, advocacy, educational, therapeutic, rehabilitative, homemaking, recreational, financial or other services to the respondent on a noncustodial basis;

COMMENT. Although it would be preferable for services to be provided on a voluntary basis without court invention, the scarcity of public resources and the reluctance or inability of some individuals to seek out and/or apply for the services which are available require that courts be given authority to order the provision of services to assist partially disabled and disabled persons to meet the essential requirements for their physical health or safety, protect their rights, manage their financial resources, and develop or regain their abilities to the maximum extent possible. Such orders can be directed toward the parents, limited personal guardian, personal guardian, limited conservator, conservator, or primary caretaker of a partially disabled or disabled person, and those state and local agencies which are responsible for providing services to partially disabled and disabled persons, as well as toward partially disabled and disabled persons themselves. When the respondent has sufficient financial resources, he or she may be required to pay for some or all of the costs of such services. However, such payments should not be required when they would substantially impair the financial independence of the partially disabled or disabled person or substantially impede meeting that person's essential requirements for physical health or safety, and/or the protection of his or her rights, and/or the management of his or her financial resources, and/or the provision to that individual of any assistance necessary to develop or regain his or her abilities to the maximum extent possible. While it is true that this provision limits, to some extent, the respondent's right to refuse treatment, such a limitation appears justified in order to make the provision of necessary services an attractive and viable alternative to guardianship and/or conservatorship.

(b) Authorizing or ratifying a contract or transaction entered into by a partially disabled or disabled person to meet his or her essential requirements for physical health or safety, to protect his or her rights, to manage his or her financial resources and/or to develop or regain his or her abilities to the maximum extent possible.

COMMENT. This provision is based upon section 5-409 of the U.P.C.[33] As pointed out in the comment to that section:

> Under present law, a ... [conservatorship] often must be established simply to make possible a valid transfer of land or securities. This section eliminates the necessity of the establishment of long-term arrangements in this situation.[34]

Among the types of arrangements which a court may be called upon to ratify are payment, delivery, deposit or retention of funds or property, sale, mortgage, lease or other transfer of property, management of financial resources, entry into an annuity contract, a contract for life care, a deposit contract, a contract for training and education,

[32] UNIFORM VETERAN'S GUARDIANSHIP ACT, *supra* note 27, at §7.

[33] U.P.C., *supra* note 2.

[34] *Id.* at 250.

and the creation of a trust.[35] The court could also authorize or ratify the appointment of an individual to serve as an agent for a partially disabled or disabled person. The agent's role would be to assist that person to conclude transactions and procure services, but not to act as a substitute decision-maker. Finally, the court may be called upon to ratify a contract following appointment of a limited personal guardian or limited conservator when, for example, a title search reveals the finding of a partial disability and a cautious purchaser or insurer wishes to confirm that the respondent has or had the legal capacity to tranfer title. While under the terms of this act such a confirmation is not necessary (see §14 Dispositional Order and §38 Effect of a Finding of Disability . . .), they may be sought with regard to persons found to be partially disabled or disabled under the guardianship provisions in effect prior to adoption of the model act.

(c) Directing establishment of a trust of which the respondent and any individuals legally dependent on the respondent are the beneficiaries;

COMMENT. The third dispositional alternative available to the court is to order the creation of a trust.

> The trust is a less restrictive and more normal device than general guardianship. . . . [It] is a mechanism to control the use of the donor's own property . . . not an assertion of control over the life of the beneficiary. . . . Trusts are imposed on property by cautious donors and testators in a multitude of situations where some control is desired. It matters little that the beneficiary is a school, a church, a nonretarded individual, or a retarded person. Practically, it is less restrictive in that it provides control only for that property that is made part of the trust corpus, not for all assets of the trust beneficiary.[36]

Care should be taken to assure that the trust does not hinder the partially disabled or disabled person from meeting the essential needs for his or her physical health or safety, protecting his or her rights, managing his or her financial resources, and/or developing his or her abilities to the maximum extent possible.[37] Similarly, care should be taken to select a trustee who has or can develop rapport with the partially disabled or disabled person and who is acquainted with the nature and effects of that person's particular impairment.

(d) Directing the provision of residential services to the respondent in a small, licensed, community-based residential home which houses not more than [six (6)] partially disabled or disabled persons and which provides room and board, personal care, habilitation services, and supervision in a family environment;

COMMENT. In some instances, the need for a guardianship may be obviated if a suitable group home or similar setting can be found for a partially disabled or disabled person. Hence, an alternative is included for provision of residential services. However, this dispositional alternative is not a means for circumventing civil commitment procedures or providing a "voluntary placement" in a closed facility or institution. The section is limited to provision of residential services in a small, non-secure, community-based facility — i.e. a facility which is not characterized by the use of physically restrictive construction, hardware and procedures; which provides for and encourages access by its residents to the surrounding community with minimal supervision, and which maintains community and consumer participation in the planning, operation and evaluation of its programs.[38] By including only smaller homes, the act encourages community-based residences that are most nearly like normal family arrangements. Moreover, the limitation to six residents makes optimum living arrangements such as single bedrooms for some or all residents more possible. Some states, however, may wish to consider raising the maximum size of the family home, for example, to not more than eight residents. The decision as to the upper limit of residents for community homes is best left to a determination based on financial (e.g. appropriate size to ensure economic feasibility of the home), programmatic and political considerations in each state.[39] Finally, consistent with the model act on location for community homes for developmentally disabled persons,[40] this provision requires that these homes be licensed to ensure responsible supervision, density controls, and other requirements designed to protect both residents and neighbors.

(e) Appointing a limited conservator or conservator; and

COMMENT. There is a major qualitative difference between the alternatives listed in paragraphs (e) and (f) and those presented in the earlier paragraphs. Unlike court-ordered assistance,

---

[35] *Id.* at §5-409(a) and (b).

[36] Kindred, *supra* note 18, at 73; *see* PCMR *supra* note 18, at 115-139.

[37] *Cf.* §§17(1) and 18(1); Power, *Cigarettes, Cokes & Candy: Should Mental Patients Beg?*, 2 DISTRICT LAWYER 25 (1978).

[38] *See* 43 Fed. Reg. 36,409 (1978).

[39] Chapter 4: Zoning for Community Homes, *supra*.

[40] *Id.* at §5(1).

imposition of a conservatorship or a guardianship restricts the legal capacity of a partially disabled or disabled person to act in his or her own behalf. Since conservatorship, by definition, restricts legal capacity only with regard to management of financial resources, it is listed as a less restrictive alternative than guardianship. An explanation of the distinction between limited conservator and conservator and a fuller discussion of the distinction between guardianship and conservatorship appears in section 3(12) and the comment thereto. The criteria for establishment of a conservatorship are contained in section 12. The qualifications and priorities for selection of a limited conservator or conservator are in section 13, while section 18 prescribes the duties which may be assigned to a limited conservator or conservator.

(f) Appointing a limited personal guardian or personal guardian.

COMMENT. Because a personal guardian may be assigned the broadest authority over a disabled person, appointment of a personal guardian with plenary powers is considered the most restrictive dispositional alternative. The distinction between a limited personal guardian and a personal guardian is explained in section 3(15) and the comment thereto. The criteria for establishment of a guardianship appear in section 12. The criteria and priorities for selecting an individual to serve as limited personal guardian or personal guardian are in section 13, while section 17 governs the duties and powers which may be assigned to a limited personal guardian or a personal guardian.

Section 12. Dispositional Hearing, and Dispositional Principles and Criteria

(1) Following a finding that the respondent is a partially disabled or disabled person, the parties shall have an opportunity within the time limits prescribed in section 30(1)(i) of this act to present evidence and testimony regarding the types of assistance and protections required, the services available for consideration by the court in making the dispositional decision, and the selection, if any, of a limited personal guardian, personal guardian, limited conservator or conservator.

COMMENT. Section 12 sets forth the procedures and analytic framework for making dispositional decisions. In a departure from current practice, subsection (1) provides for a separate dispositional hearing. A bifurcated intervention proceeding makes possible a more focused inquiry into the facts alleged in the petition and a more complete investigation of the dispositional alternatives available when a respondent has been found to be a partially disabled or disabled person. This distinction between substantive and dispositional issues is particularly important when a jury is determining the matters listed in section 10 since testimony regarding particular services and programs, or who would be best qualified to serve as a personal guardian or conservator is likely to confuse or color the jury's deliberations. The purpose of the dispositional hearing is to assure that all the relevant information and alternatives are presented to the court in order to facilitate selection of the least restrictive dispositional alternative which will assist the partially disabled or disabled person to meet his or her needs.

Each party is entitled to be present at the hearing, be represented by counsel, subpoena witnesses, present evidence and testimony, and cross-examine opposing witnesses. (See §31(1) Rights and Privileges.) In addition, section 31 provides that the subject of the intervention proceeding may close the dispositional hearing to the general public. The dispositional hearing may be held any time within 10 days after a finding that an individual is a partially disabled or disabled person. (See §30(1)(i) Time Limits.) In some cases it may be possible to hold the dispositional hearing immediately following the substantive determination. In others, the parties will need more time to explore the alternatives available.

(2) The court shall impose the least restrictive dispositional alternative which in light of the determinations made at the hearing held pursuant to section 10 of this act and the evidence and testimony presented at the dispositional hearing will be sufficient to assist the respondent to:

(a) Meet the essential requirements for his or her physical health or safety; and/or

(b) Manage his or her financial resources; and

(c) Develop or regain to the maximum extent possible the skills needed to perform the functions listed in a paragraphs (2)(a)-(2)(b) of this section without court-ordered assistance.

Dispositions shall be designed to encourage respondents to participate in all decisions which affect them and to act on their own behalf to the maximum extent possible.

COMMENT. This provision establishes the principles on which dispositions are to be based. It emphasizes use of the least restrictive alternative and makes fostering the respondent's self-reliance and independence one of the primary goals of a disposition. A definition and fuller discussion of the concept of the least restrictive alternative is contained in section 3(26) and the comment thereto.

(3) In selecting among the available dispositional alternatives in accordance with the principles set forth in subsection (2) of this section, the court shall not:

(a) Order that a non-custodial service be provided unless the respondent's need for that particular service has been proven;

(b) Authorize or ratify a contract or transaction unless the need for authorization or ratification has been proven;

(c) Order the provision of residential services or the establishment of a trust or a guardianship or conservatorship unless the need therefor has been proven and it has been proven that no less restrictive dispositional alternative or combination of alternatives will be sufficient to satisfy that need;

(d) Assign a duty or power to a limited personal guardian, personal guardian, limited conservator, or conservator appointed in accordance with paragraph (3)(c) of this section, unless the need therefor has been proven and no less restrictive dispositional alternative or combination of alternatives will be sufficient to satisfy that need;

(e) Appoint a limited personal guardian or personal guardian if the respondent's partial disability or disability affects only the capacity to manage his or her financial resources; and

(f) Appoint a limited conservator or conservator if the respondent's partial disability or disability affects only the capacity to meet the essential requirements of his or her physical health or safety.

COMMENT. This provision is the heart of the model act. It contains a set of criteria to guide implementation and enforcement of the principles set forth in section 12(2). In accordance with the hierarchy of dispositional alternatives established in section 11 and the burden of persuasion prescribed in section 44, it requires that no dispositional alternative be ordered nor a duty or power assigned to a limited personal guardian, personal guardian, limited conservator or conservator, unless the need therefor has been proven beyond a reasonable doubt. (See §44(1) Burden of Proof.) In addition, the provision of residential services or the establishment of a trust, guardianship or conservator, may not be ordered unless there is proof beyond a reasonable doubt that no less restrictive dispositional alternative or combination of alternatives will be sufficient to assist the partially disabled or disabled person to meet the essential requirements for his or her physical health or safety, and/or to manage his or her financial resources. The high burden of proof is not intended to prevent individuals who require guardianship or conservatorship from receiving this protection. However, it is anticipated that along with the conferences required by sections 6(b)(xiv) and 9, it will reduce the number of guardianships and conservatorships sought for reasons of administrative convenience, and encourage the growth and use of less restrictive means of assisting partially disabled and disabled persons.

Section 13. Selection of a Limited Personal Guardian, Personal Guardian, Limited Conservator or Conservator

(1) (a) The court may appoint any non-disabled adult to serve as a limited personal guardian or personal guardian, and any non-disabled adult or any corporation with general power to serve as trustee, to serve as a limited conservator or conservator unless that individual or corporation:

(i) Is performing or is responsible for performing some service other than a guardianship or conservatorship service for the respondent;

(ii) Is providing some form of treatment or is responsible for providing such treatment to the respondent;

(iii) Is a creditor of the respondent; or

(iv) Has interests which may otherwise conflict with those of the respondent.

COMMENT. Section 13 outlines the minimum qualifications for service in a guardianship or conservatorship capacity and the priorities which should guide the court's selection of a limited personal guardian, personal guardian, limited conservator or conservator.

Paragraph (1)(a) permits banks, trust companies and other corporations — either for profit or not-for-profit firms — which under their charters and the applicable provisions of the state corporation, banking, and probate codes are permitted to act as trustees, to serve as limited conservators or conservators. However, such corporations are prohibited from serving in a guardianship capacity. While the personal element is important in assisting a partially disabled or disabled person to manage his or her financial resources, it is critical where decisions over health, safety, residence and services are involved. (See §§17(3) and 17(4) Duties and Powers of Limited Personal Guardians and Personal Guardians.) The statute, however, is not intended to discourage voluntary associations from taking increasing interest in providing guardianship services.[41] Although the associations may not be appointed *per se*, it is hoped that they will encourage their members to serve as volunteer public guardians and/or volunteer public conservators so that partially disabled and disabled persons can have a one-to-one, individualized relationship with their limited personal guardian, personal guardian, limited conservator or conservator.

Subparagraph (1)(a)(i) — (1)(a)(iv) illustrate relationships which would represent at least the potential for a conflict of interest for a limited personal guardian, personal guardian, limited conservator or conservator. As noted in the President's Committee on Mental Retardation volume on *The*

[41] *See* ILSMH, *supra* note 16, at 23.

Mentally Retarded and the Law, "[a] service provider makes some decisions in terms of the fiscal and political needs of his [or her] organization and is not in a position to give . . . sole attention to the best interests of the individual receiving the service."[42] Included among the individuals who and entities which may be precluded from serving in a guardianship or conservatorship capacity for a partially disabled or disabled person are that person's physician, landlord, attorney, broker, or bank, the head of a nursing home, group home, institution or other facility in which that person is residing, and an employee of a service system in which that person is enrolled.[43] While these are the minimum qualifications, in making its selection the court should appoint an individual or entity capable and willing to develop a rapport with the partially disabled or disabled person and perform the assigned duties and wers diligently and good faith.

(b) Nothing in this provision shall exclude the appointment of the respondent's parent, spouse, child or sibling when the conflict of interest is determined to be insubstantial and such an appointment would clearly be the most beneficial to the respondent.

COMMENT. This provision gives the court needed flexibility to balance familial ties and understanding against what may technically constitute a conflict of interest.[44]

(2) (a) The following priorities shall guide the court's selection of a limited personal guardian, personal guardian, limited conservator, or conservator from among those eligible:

(i) The individual or corporation nominated by the respondent;

(ii) The current limited personal guardian, personal guardian, limited conservator, or conservator appointed or recognized by the appropriate court of any other jurisdiction in which the respondent resides;

(iii) The respondent's spouse;

(iv) An adult child of the respondent;

(v) A parent of the respondent;

(vi) An individual or corporation nominated by the will of a deceased parent;

(vii) Any individual with whom the respondent has been living for more than 6 months prior to the filing of the petition;

(viii) A sibling of the respondent;

(ix) A volunteer public guardian and/or a volunteer public conservator.

(b) The court may select the individual or corporation best qualified and most willing to serve from among those of equal priority. In addition, the court may pass over a person or corporation having priority and appoint a person having less priority or no priority when it is demonstrated that the selection will be of greater benefit to the partially disabled person or disabled person. Appointments of volunteer public guardians and volunteer public conservators shall be in accordance with section 36.

COMMENT. Subparagraphs (2)(a)(i)-(ix) are arranged in descending order of priority.[45] In selecting a limited personal guardian, personal guardian, limited conservator or conservator, the court should ordinarily appoint the individual or corporation chosen by the partially disabled or disabled person, unless that individual or corporation is ineligible for appointment or unable or unwilling to serve.[46] A volunteer public guardian and/or volunteer public conservator eligible under section 13(1) may be selected from the list published by the Guardianship/Conservatorship Oversight Commission when the partially disabled or disabled person for whom such an appointment has been found to be necessary, has no individual able or willing to serve as limited personal guardian, personal guardian, limited conservator, or conservator, or is unable to afford the cost of a guardianship or conservatorship. (See §5 Guardianship/Conservatorship Oversight Commission and §36 Appointment of Volunteer Public Guardians, Volunteer Public Conservators, and Volunteer Public Trustees.)

Recognizing that the order of priority established in subparagraphs (2)(a)(iii) through (viii) is somewhat arbitrary, and that there may be some instances in which an individual with a lower priority may have greater interest in and ability to assist the partially disabled or disabled person than another individual whose relationship appears higher on the list, paragraph (2)(b) provides the court with some flexibility. Thus, unlike the dispositional criteria set forth in section 12(3), the list of priorities is intended to serve simply as a guide rather than as a series of mandatory conditions.

Section 14. Dispositional order

(1) The dispositional order shall set forth:

(a) The services, if any, to be provided to the respondent;

(b) The terms or conditions under which a contract or transaction, if any, is authorized or ratified;

---

[42] Kindred, *supra* note 18, at 85; *accord* ILSMH, *supra* note 16, at 23; and AAMD, *supra* note 4, at 20; G.A. Res. 2856, 25 U.N. GAOR 368, UNDOC A/8588 (1971).

[43] ILSMH, *supra* note 16, at 23.

[44] *Accord* Mental Health Law Project (MHLP), *Legal Issues in State Mental Health Care: Proposals for Change – Suggested Statute on Guardianship,* 2 MEN. DIS. L. RPTR. 449, 457 (1978).

[45] *Id.* at 457; *cf.* U.P.C. *supra* note 2, at §§5-311 and 5-410.

[46] MHLP, *supra* note 44, at 457.

(c) The terms of any trust to be established;

(d) The name and addresses of the individual, if any, appointed to serve as the limited personal guardian or personal guardian and/or the individual or corporation, if any, appointed to serve as limited conservator or conservator;

(e) The duties and powers of the limited personal guardian or personal guardian, if any;

(f) The duties and powers of the limited conservator or conservator, if any, and the particular financial resources to which they apply; and

(g) The date of the initial review hearing.

COMMENT. This provision requires that the dispositional order delineate the exact terms of the disposition. The order thus becomes the basic reference document setting forth the responsibilities of those agencies, individuals, and corporations charged with protecting and assisting the partially disabled or disabled person to meet the essential requirements for his or her physical health or safety, to manage his or her financial resources, and to develop or regain his or her abilities to the maximum extent possible, and specifying what if any restrictions have been imposed on the legal capacity of the partially disabled or disabled person to act in his or her own behalf. If a limited personal guardian or limited conservator is appointed, the initial review hearing must be held no more than one year after the dispositional hearing. (§15(1)(a) Periodic Review and §31(1)(1) Time Limits.) If a personal guardian or conservator is appointed, the initial review hearing must be held within 6 months of the dispositional hearing. (§15(1)(b) Periodic Review and §30(1)(m) Time Limits.)

(2) In issuing the dispositional order, the court shall explain on the record the terms and conditions contained therein, the facts and reasons which support the decision not to impose any less restrictive alternative in accordance with the principles and criteria set forth in sections 12(2) and (3) of this act, and if a guardianship or conservatorship has been established, the facts and reasons supporting the selection of the individual and/or corporation appointed. The court shall also explain to the respondent the procedures set forth in section 16 for terminating or modifying the dispositional order.

COMMENT. This provision requires the judge to explain the terms and basis for the dispositional decision. It is anticipated that this not only will help to avoid misunderstandings of the nature and scope of the responsibilities and restrictions imposed and the means for modifying them, but will also help to assure that the dispositional principles and criteria set forth in sections 12(2) and (3) have been followed. The explanation should be in the language, mode of communication, and terms which the respondent is most likely to understand.

(3) Copies of the dispositional order shall be provided to the parties and to each individual, corporation and agency named therein.

COMMENT. None necessary.

Section 15. Periodic Review

(1) (a) The court shall hold a review hearing following the appointment of a limited personal guardian and/or limited conservator within the time limit prescribed in section 30(1)(l) of this act.

(b) The court shall hold a review hearing following the appointment of a personal guardian and/or conservator within the time limits prescribed in section 30(1)(m) of this act.

COMMENT. This section requires periodic hearings to review dispositional orders establishing a guardianship and/or conservatorship for an adult. Currently, 15 states have some form of automatic review procedure. (See Chart VI, *supra*.) The review hearing serves a multifold purpose. It assures that the services, restrictions, duties and powers remain consistent with the abilities of the partially disabled or disabled person as they change over time. Second, it assures that a guardianship or conservatorship continues no longer than necessary. Third, it provides a mechanism for verifying that the partially disabled or disabled person has been receiving the services which were ordered. Fourth, it provides an opportunity for determining whether a limited personal guardian, personal guardian, limited conservator and/or conservator has been performing diligently and in good faith the duties and powers assigned by the court. Finally, it acts as a means for both controlling the costs of guardianship and conservatorship services and guaranteeing that the limited personal guardian, personal guardian, limited conservator and/or conservator receives reasonable compensation and reimbursement for proper out-of-pocket expenses incurred on behalf of the partially disabled or disabled person.

The initial review hearing following appointment of a limited personal guardian or limited conservator must be held within a year of the dispositional hearing. Subsequent review hearings in such cases must be held at intervals of no more than two years until such time as the guardianship and/or conservatorship is terminated. (§30(1)(l) Time Limits.) Following appointment of a personal guardian or conservator with plenary powers, a review hearing must be held within 6 months of the dispositional hearing, and at least yearly thereafter until the guardianship and/or conservatorship is terminated. (§30(1)(m) Time Limits.) More frequent reviews are required when a personal guardian and/or conservator has been appointed because of the greater restrictions on the legal capacity of the disabled person which such guardianships and conservatorships entail, and because it is less likely that a disabled person will be able to apply for modification or termination of the guardianship or conservatorship pursuant to section 16. These stringent review procedures should not be viewed as an encumbrance on a benign proceeding but as a duty which arises whenever an individual has been deprived of his or her rights through state action. Furthermore, the limited personal guardians, personal guardians, limited conservator and/or conservators can use the plan update and review hearing process as an opportunity to share problems, and learn of new techniques and services to assist them and the individual they are serving. The rights and privileges of the parties set forth in section 31 apply to review hearings.

(2) At a review hearing the court shall:

(a) Determine why the appointment should not be terminated or why no less restrictive alternative will be sufficient to assist the respondent to meet the essential requirements for his or her physical health or safety and/or manage his or her financial resources;

COMMENT. Consistent with the dispositional principles and criteria set forth in sections 12 (2) and (3), this provision requires an affirmative showing of why the set of duties, powers, and restrictions then in effect, are still the least restrictive dispositional alternative. Thus, the section avoids placing the burden of proving that the restrictions on legal capacity are no longer necessary on the subject of the intervention proceeding. The review will be based, in large part, on the reports filed by limited personal guardians, personal guardians, limited conservators and conservators pursuant to sections 17(6) and 18(7). Such reports must include a description of any significant changes in the abilities of the partially disabled or disabled person to meet the essential requirements for his or her physical health or safety or to manage his or her financial resources, and a statement of the reasons why no less restrictive dispositional alternative would be sufficient.

(b) Approve the payment of compensation and reimbursement of expenses, if any, from the financial resources of the partially disabled or disabled person to the limited personal guardian, personal guardian, limited conservator and/or conservator; and

COMMENT. Under section 39, a limited personal guardian, personal guardian, limited conservator or conservator is entitled to reasonable compensation and to be reimbursed for reasonable and necessary expenses incurred on behalf of the partially disabled or disabled person. The reports to be submitted under section 17(6) or section 18(7) are to contain a summary of the significant actions taken on behalf of the partially disabled or disabled person, as well as a request for compensation and reimbursement and the basis therefor. All compensation and reimbursements must be approved by the court prior to payment. (§39(5) Compensation and Expenses.)

(c) Resolve any other relevant matters which the parties raise.

COMMENT. The review hearing also presents an opportunity to resolve such matters as questions about the individual guardianship plan and/or individual conservatorship plan, disputes between the limited personal guardian or personal guardian and the limited conservator or conservator if one individual is not serving in both capacities, and differences over the partially disabled or disabled persons's place of residence.

(3) In conjunction with a review hearing, the court may order an evaluation of the partially disabled or disabled person upon a showing that there has been a significant improvement or deterioration of that person's ability to meet the essential requirements for his or her physical health or safety, and/or to manage his or her financial resources.

COMMENT. Both because of the intrusion on the privacy and the expense involved, an evaluation is not required in conjunction with a review hearing except after a *prima facie* showing that there has been a significant change in the abilities of the partially disabled or disabled person. When an evaluation has been ordered, the multidisciplinary evaluation team should be selected, the evaluation conducted, and the report written in the manner prescribed by section 8.

(4) At the conclusion of a review hearing the court may:

(a) Terminate a guardianship or conservatorship;

(b) Modify any or all duties and powers assigned to a limited personal guardian, personal guardian, limited conservator or conservator;

(c) Discharge and/or appoint individuals to serve as the limited personal guardian, personal guardian, limited conservator and/or conservator;

(d) Modify the individual guardianship and/or conservatorship plans; and

(e) Continue the current restrictions upon legal capacity and continue the appointment(s) of the limited personal guardian, personal guardian, limited conservator or conservator until the next review hearing.

All orders shall follow the principles, criteria, and procedures set forth in sections 12 and 13 of this act, and the form and procedures prescribed by section 14.

COMMENT. Following a review hearing, the court has the same options before it as it had following the dispositional hearing plus the option of terminating the guardianship or conservatorship. Thus, the least restrictive dispositional alternative should be imposed which will assist the subject of the intervention proceeding to meet the essential requirements for his or her physical health or safety and/or to manage his or her financial resources.

(5) Failure to hold a review hearing for appoint-

ments made pursuant to this act within the time prescribed in sections 30(1)(l) and 30(1)(m) of this act shall result in the termination of the guardianship and/or conservatorship.

COMMENT. In order to assure that the review procedures are implemented, this subsection provides an incentive to all parties to make certain that the review hearing is held. Section 46(1) contains a parallel provision for guardianships and conservatorships established prior to the effective date of this act.

Section 16. Modification and Termination

(1) In addition to the review procedures prescribed under section 15 of this act, a partially disabled or disabled person, the limited personal guardian, personal guardian, limited conservator, and/ or conservator of that person, or any other interested individual may apply to the court for:

(a) Termination or modification of a dispositional order;

(b) Approval or disapproval of the compensation or reimbursement for expenses requested by the limited personal guardian, personal guardian, limited conservator, or conservator;

(c) Replacement of the limited personal guardian, personal guardian, limited conservator or conservator; or

(d) Resolution of a dispute over the individual guardianship plan or individual conservatorship plan.

COMMENT. Like the guardianship statutes in most states, this section provides a procedure through which a partially disabled or disabled person, or someone interested in the well-being of that person, can seek to change or remove the limited personal guardian, personal guardian, limited conservator or conservator who has been appointed and/or the restrictions which have been imposed on the legal capacity of the partially disabled or disabled person to act in his or her own behalf.

This procedure is not a substitute for the periodic review process through which a court can monitor its transfer of the legal authority of a partially disabled or disabled person to act in his or her own behalf. It simply provides a means for assuring that the powers and the services being provided remain consistent with the abilities and needs of the partially disabled or disabled person in the interval between review hearings. In addition, it allows the limited personal guardian, personal guardian, limited conservator and/or conservator to be compensated and reimbursed in between review hearings. Thus, the court's control of the purse strings is maintained without creating a one to two year delay before compensation and reimbursements can be paid. (See §30(1)(l) and (1)(m) Time Limits, and §39 Compensation and Expenses.)

(2) (a) Except in accordance with section 35, applications may be filed no more than once during any sixty (60) day period and no sooner than thirty (30) days after a review hearing.

COMMENT. In order to encourage the parties and other individuals interested in the welfare of a partially disabled or disabled person to settle all matters in dispute in one proceeding rather than dealing with them *in seriatim*, this provision limits the frequency with which applications may be submitted. Note, however, that these limits do not apply to applications which are submitted in conjunction with emergency or standby appointments pursuant to section 35.

(b) An application from a partially disabled or disabled person may be communicated to the court by any means, including but not limited to oral communication or informal letter. Upon receipt of such an informal request, the court shall contact the partially disabled or disabled person's attorney of record or appoint new counsel to prepare a formal application.

COMMENT. This provision is based on Ill. Ann. Stat. ch. 110½, §11a-20(b). It is intended to make the modification and termination process directly accessible to partially disabled and disabled persons. Such access is of particular significance in cases in which the interests of such persons and those of their limited personal guardian, personal guardian, limited conservator or conservator are in conflict. New counsel should be appointed when the partially disabled or disabled person's attorney of record has become estranged from the case or is no longer available. (See §§31 and 33 on the right to and role of counsel in intervention proceedings.)

(c) Applications not filed by a partially disabled or disabled person and formal applications filed pursuant to paragraph (2)(b) of this section shall set forth:

(i) The name and address of the partially disabled or disabled person;

(ii) The name and address of that person's attorney of record and limited personal guardian, personal guardian, limited conservator, conservator, and/or guardian *ad litem*, if any;

(iii) The name and address of the individuals most closely related to the partially disabled or disabled person unless they are unknown and cannot be reasonably ascertained;

(iv) The name and address of the individual or facility, if any, having custody of the partially disabled or disabled person;

(v) The name, address and interest of the applicant;

(vi) The surety for the bond on the limited conservator or conservator, if any;

(vii) The relief requested; and

(viii) The facts and reasons supporting the request.

COMMENT. Like the initial petition, the form proposed for the application attempts to balance the need to inform other parties of the requests being made and the basis therefor against the burden which additional information places on the applicant. (See §6(1)(b) Initiation of Intervention Proceedings and §§32(1)(a) and 32(2) Notice and Dissemination of Plans and Reports.)

(3) Any party may object to the relief requested in an application within the time limit prescribed by section 30(1)(p) of this act.

COMMENT. Objections should include the name of the objecting party and the facts and reasons underlying the objection. Objections must be filed within 5 days after service of the notice required by subsection (5) of this section.

(4) (a) The court shall hold a hearing on all applications:

(i) Which are filed pursuant to section 35 of this act;

(ii) Which request imposition of additional restrictions upon the legal capacity of a partially disabled or disabled person to act on his or her own behalf; or

(iii) To which an objection has been filed.

(b) All other applications may be granted or denied without a hearing.

COMMENT. It is anticipated that the relief requested in most applications will not raise substantial questions or adversely affect the partially disabled or disabled person. Therefore, this subsection does not require a hearing unless: the partially disabled or disabled person or another party objects to the application within 15 days after notice of the application (see §30(1)(p) Time Limits); additional restrictions are being sought; there is an emergency situation (see §§35(1) and (2) Emergency and Standby Appointments); or a limited personal guardian, personal guardian, limited conservator and/or conservator appointed on a standby basis has assumed the guardianship and/or conservatorship duties and powers assigned by the court. (See §35(3) Emergency and Standby Appointments.) If a hearing is required under subparagraphs (4)(a)(i) and (ii), it must be held within 30 days after the filing of the application. If a hearing is required under subparagraph (4)(a)(iii), it must be held within 30 days of the filing of the objection. (See §30(1)(q) Time Limits.) The parties are entitled to be represented by counsel, be present at the hearing, subpoena witnesses, present evidence and testimony, cross-examine opposing witnesses, have an impartial decisionmaker, and appeal. (See §31 Rights and Privileges.)

Pursuant to section 44(1)(f)(ii), when an application requests additional restrictions on the legal capacity of the partially disabled or disabled person to act on his or her own behalf, the need for such restrictions must be proven beyond a reasonable doubt. When an easing of restrictions or other relief is sought, a clear and convincing evidence standard applies. (See §44(1)(f)(i) Burden of Proof.) Both the burden of going forward and the burden of persuasion are on the applicant with one exception. When an application is filed by a partially disabled or disabled person, the burden of persuasion shifts to the party or parties opposing the application once the partially disabled or disabled person presents evidence and/or testimony constituting a *prima facie* case supporting the requested relief. (See §44(1)(f)(i) Burden of Proof.)

(5) (a) Within five (5) days after the filing of an application for which a hearing is required under subparagraphs (4)(a)(i) and (4)(a)(ii) of this section or the filing of an objection under subsection (3), the court, shall set a date for a hearing and have notice served on the individuals and entities specified in section 32(1)(a) in the form and manner prescribed in sections 32(1)(a) and 32(2).

(b) Within five (5) days after the filing of an application for which a hearing is not required unless an objection is filed, the court shall have notice served on the individuals and entities enumerated in section 32(1)(a) of this act in the form and manner prescribed in sections 32(1)(b) and 32(2).

COMMENT. Notice is to be served on each of the individuals and entities listed in the application, the proposed replacement, if any, for the limited personal guardian, personal guardian, limited conservator or conservator, and on the Administrator of Veterans' Affairs if the financial resources of the partially disabled or disabled person include V.A. benefits. When a hearing is required, section 32(1)(a) requires that the notice must specify the date, time, place, purpose and possible consequences of the hearing and the rights of the parties. When a hearing is not required, in addition to advising the parties of their rights, the notice must specify that an objection to the application may be filed within 10 days and that a hearing will be held if an objection is filed. (§32(1)(b) Notice and Dissemination of Plans and Reports.)

(6) At the request of any party or on its own initiative, the court may order an evaluation of the partially disabled or disabled person identified in the application upon a showing that there has been a significant improvement or deterioration of that person's ability to meet the essential requirements for his or her physical health or safety and/or to manage his or her financial resources. If the application requests imposition of additional restrictions on the legal capacity of the partially disabled or disabled person to act on his or her own behalf, the court shall order an evaluation of that person.

COMMENT. See comment to section 15(3).

(7) After notice to the individuals and entities enumerated in section 32(1)(a) of this act, in the form and manner prescribed in sections 32(1)(c) and 32(2), the court may join the issues raised in separate applications and/or separate objections for determination at a single hearing, unless joinder would be prejudicial to the interests of the partially disabled or disabled person.

COMMENT. This subsection authorizes the court to consolidate applications and objections in order to facilitate expeditious disposition of the issues raised.

(8) The principles, criteria and priorities set forth in sections 12 and 13 of this act, and the form and procedures prescribed by section 14 shall apply to dispositional orders issued following submission of an application under this section.

COMMENT. The least restrictive dispositional alternative should be imposed which will assist the subject of the intervention proceeding to meet the essential requirements for his or her physical health or safety and/or manage his or her financial resources.

Introduction to Section 17

This section sets forth the general duties which inhere to guardianship, the requirements for the individual guardianship plan, the specific duties and powers which may be assigned to limited personal guardians and personal guardians, the limits on the authority of limited personal guardians and personal guardians, and the reports which are to be submitted by limited personal guardians and personal guardians. Section 18 covers the same range of topics with regard to conservatorships.

Section 17. Duties and Powers of Limited Personal Guardians and Personal Guardians

(1) It is the general duty of individuals appointed to serve as a limited personal guardian or personal guardian to carry out diligently and in good faith the specific duties and powers assigned by the court. In carrying out these duties and powers, such individuals shall:

(a) Assure that the personal, civil, and human rights of the partially disabled or disabled persons whom they are serving are protected; and

(b) Encourage those persons to:

(i) Participate to the maximum extent of their abilities in all decisions which affect them;

(ii) Act in their own behalf on all matters in which they are able to do so; and

(iii) Develop or regain, to the maximum extent possible, their capacity to meet the essential requirements for their physical health or safety, and, if impaired, their capacity to manage their financial resources.

COMMENT. This provision makes clear that guardianship should be used not as a means for unnecessarily retaining the subject of the guardianship in a dependent state. Rather it emphasizes that the partially disabled or disabled person should be encouraged to be as self-reliant and independent as possible and, whenever possible, to develop or regain the skills required to obviate the need for the guardianship and/or conservatorship.[47] The traditional analogy between guardianship and the responsibility of parents toward their children[48] may still be applicable, except that the parental role of preparing children to take their place as responsible adults receives greater prominence than it has heretofore. The general duties also impose upon a limited personal guardian or personal guardian the responsibility to make certain that the partially disabled or disabled person whom he or she is serving receives all the benefits and services and is accorded all the rights *vis-a-vis* other individuals, businesses, and agencies to which that person is entitled. This may include assisting the partially disabled or disabled person and that person's limited conservator or conservator, if any, to bring tort, contract, consumer, and antidiscrimination actions.[49]

The subsection requires limited personal guardians and personal guardians to perform their duties diligently and in good faith. Failure to meet this standard may render the limited personal guardian or personal guardian liable to the partially disabled or disabled person for damages. (See §42 Liability of Limited Personal Guardians, Personal Guardians, Limited Conservators and Conservators and Penalties for Violation of this Act.) While stopping short of imposing strict liability, it is anticipated that this provision will adequately protect partially disabled and disabled persons without discouraging individuals from accepting guardianship responsibilities.

(2) (a) Upon appointment, a limited personal guardian or personal guardian shall develop an individual guardianship plan together with the disabilities resources officer and, to the maximum extent possible, the partially disabled or disabled person. The plan shall specify:

(i) The services which are necessary to meet the essential requirements for the partially

---

[47] *See* ILSMH, *supra* note 16, at 20; ILL. ANN. STAT. ch. 110 ½, §11a-3(b) (Smith-Hurd Supp. 1979).

[48] *E.g.*, U.P.C., *supra* note 2, at §5-312.

[49] See Chapter 2: Prohibiting Discrimination, *supra*.

disabled or disabled person's physical health or safety;

(ii) The means for obtaining these services;

(iii) The manner in which the limited personal guardian or personal guardian, the partially disabled or disabled person, and the limited conservator or conservator, if a corporation or another individual has been appointed to serve in that capacity, will exercise and share decision-making authority; and

(iv) Such other items as will assist in fulfilling the needs of the partially disabled or disabled person, the terms of the dispositional order, and the duties of the limited personal guardian or personal guardian.

COMMENT. The individual guardianship plan is intended to fulfill several purposes. First, it encourages a goal-oriented rather than a maintenance approach to assisting the partially disabled or disabled person to meet the essential requirements for his or her physical health or safety, and developing or regaining his or her abilities to the maximum extent possible. Second, it further clarifies for all parties what the limited personal guardian or personal guardian will be doing. Third, it defines how the decision-making authority will be shared in those areas in which restrictions have been imposed on the partially disabled or disabled person's legal capacity to act in his or her own behalf. Disputes over the division of authority which cannot be resolved by the parties may be submitted to the court pursuant to section 16. Finally, the plan provides a reference against which the performance of the limited personal guardian or personal guardian and the delivery of assistance and services can be compared.[50]

The plan is to be developed jointly by the individual serving in a guardianship capacity, the person whom he or she is serving, the limited conservator or conservator for that person, if any, and the disabilities resources officer. (See §4 Appointment of Disabilities Resources Officers.) The participation of the partially disabled or disabled person is consistent with the purposes of this act enumerated in section 2 to permit partially disabled and disabled persons to take as great a role as they are able in making the decisions which affect their lives. (See also the general duties presecribed in subsection (1) of this section.) The disabilities resources officer is included in the planning process to assure that all the appropriate benefits and services which are available are considered.

(b) The individual guardianship plan shall be submitted to the court by the limited personal guardian or personal guardian within the time limit prescribed by section 30(1)(j). The court shall have copies sent to the individuals and entities prescribed by section 32(3)(b).

COMMENT. The initial individual guardianship plan is due 30 days after the dispositional order has been issued. (§30(1)(j) Time Limits.) The partially disabled or disabled person; his or her attorney (see §31 Rights and Privileges); his or her limited conservator, conservator and/or guardian *ad litem,* if any; the individuals most closely related to the partially disabled or disabled person, unless their names and whereabouts are unknown or cannot be reasonably ascertained; the individual having custody of that person; the Administrator of Veterans' Affairs if the financial resources of the partially disabled or disabled person includes V.A. benefits; and the state Developmental Disabilities Advocacy Office if the partially disabled or disabled person has a developmental disability, are all to receive a copy of the plan in order to reduce the likelihood of misunderstandings and further assure accountability. (§32(3)(b) Notice and Dissemination of Plans and Reports.) The individual guardianship plan is to be updated in accordance with the requirements of subsection (6) of this section.

(3) (a) The court may assign to a limited personal guardian any portion of the duties and powers listed in subparagraphs (4)(a)(ii)-(vi) of this section for those particular areas in which the partially disabled person lacks the capacity to meet the essential requirements for his or her physical health or safety.

COMMENT. Consistent with the dispositional principles and criteria in sections 12(2) and (3), this provision directs that a limited personal guardian be authorized to act in place of a partially disabled person in only those areas in which that person lacks the skills necessary to meet the essential requirements for his or her physical health or safety. Hence, if a partially disabled person is able to manage normal housekeeping tasks either alone or with the assistance of homemaking services, but is not able to make judgments regarding the need for medical care, the limited personal guardian should only be assigned the power to provide consent to medical services subject to the limitations imposed by subsection (5) of this section.

> [I]n appointing a guardian the responsible body ideally should seek to preserve for the ward the opportunity to exercise those rights which are within his [or her] comprehension and judgment, allowing for the possibility of error to the same degree as is allowed to persons who are not retarded.[51]

Tailoring the powers and duties to the particular competencies and limitations of a partially disabled

[50] MHLP, *supra* note 44, at 454.

[51] AAMD, *supra* note 4, at 1; ILSMH, *supra* note 16, at 18-19; Kindred, *supra* note 18, at 71.

person may confer a complex additional task on the court in many jurisdictions. However, it is the type of task which judges are regularly called upon to perform in determining the remedy to be ordered in a variety of tort, domestic relations, antitrust and constitutional litigation.

In keeping with the concept of limited guardianship, a limited personal guardian cannot be given custody powers over a partially disabled person. This is not intended to preclude a partially disabled person from living in the home of his or her limited personal guardian. It does signify, however, that the all-inclusive control implied by the concept of custody is inconsistent with limited guardianship.

(b) A limited personal guardian may also be assigned the duty to assist the partially disabled person in those particular areas in which that person's capacity to meet the essential requirements for his or her physical health or safety and/or to protect his or her rights, and/or to obtain necessary services, and/or to fulfill his or her civil duties is impaired, as well as in other ways not specifically prohibited by subsection (5) of this section.

COMMENT. In addition to the powers which may be assigned under the preceding paragraph, the court may direct the limited personal guardian to help the partially disabled person in areas in which that person's legal capacity to act in his or her own behalf has not been restricted. For example, a limited personal guardian may assist the person whom he or she is serving in selecting and obtaining services. In this way, the court can further the self-reliance and independence of the partially disabled person by not transferring legal responsibility for a particular area, while at the same time assuring his or her health and safety through the provision of personal support and encouragement from the limited personal guardian.

(c) If a separate limited conservator or conservator has been appointed for the partially disabled person, the expenditure of funds by the limited personal guardian shall be consistent with the procedures and policies established by the limited conservator or conservator.

COMMENT. Even when a conservatorship has been established, it will be necessary for a limited personal guardian to expend funds on behalf of a partially disabled person during the course of everyday activities. This provision in conjunction with section 17(2) seeks to facilitate the establishment of a smooth-working relationship between the individual with responsibility for the partially disabled person's physical health and safety, and the individual or corporation, if any, with responsibility for that person's financial resources. Disputes over the exercise of this authority which cannot be resolved by the parties may be submitted to the court pursuant to section 16. Nothing in this section is intended to prohibit the appointment of the same individual as limited personal guardian and limited conservator or conservator when both a guardianship and conservatorship is warranted.

(4) (a) The court shall assign to a personal guardian the duty and/or power to:

(i) Take custody of the disabled person;

(ii) Assure that the disabled person has a place of abode in the least restrictive, most normal setting consistent with the essential requirements for his or her physical health or safety;

(iii) Assure the care and comfort of the disabled person;

(iv) Assure that the disabled person receives the services necessary to meet the essential requirements for that person's physical health and safety as well as the services necessary to develop or regain, to the maximum extent possible, his or her capacity to meet those essential requirements;

(v) Provide any required consents or approvals on behalf of the disabled person; and

(vi) Expend sums from the financial resources of the disabled person to cover the reasonable costs incurred on behalf of that person. If a separate limited conservator or conservator has been appointed for the disabled person, the expenditure of funds by the personal guardian shall be consistent with the duties assigned to and procedures and policies established by the limited conservator or conservator.

COMMENT. This paragraph attempts to outline the basic duties and powers which a personal guardian must exercise to assure that the essential requirements are met for the physical health or safety of the disabled person whom they are serving. Each of the specific powers listed must be read in the context of the general duties delineated in section 17(1) and the limitations on authority set forth in section 17(5). Thus, subparagraph (4)(a)(ii) stresses that in selecting a place of residence for a disabled person, preference must be given to community-based residential settings when that person cannot remain in his or her own home. Except pursuant to an order of the court following a formal civil commitment proceeding, disabled persons should not be removed from a setting in which the "patterns of life and conditions of everday living . . . are as close as possible to the regular circumstances and ways of life of society."[52] Similarly, subparagraph (4)(a)(iv) requires the personal guardian to secure services which will help the disabled to become as self-reliant as possible. (For an explanation of subpara-

---

[52] Nirje, *The Normalization Principle,* in CHANGING PATTERNS IN RESIDENTIAL SERVICES FOR THE MENTALLY RETARDED 231 (R. Kugel & A. Shearer eds. 1976); ILSMH, *supra* note 17, at 30; PCMR *supra* note 12, at 60.

graph (vi) see paragraph (3)(c) and subsection (2) of this section.)

(b) A personal guardian may also be assigned the duty to assist the disabled person to fulfill that person's civil duties and in other ways not specifically prohibited by subsection (5) of this section.

COMMENT. See comment to Section 17(3)(b).

(5) A limited personal guardian or a personal guardian shall not have the power:

(a) To place a partially disabled or disabled person in a facility or institution to which an individual without a guardian would have to be commited under (citation to the applicable state civil commitment laws(s)) other than through a formal commitment proceeding in which the partially disabled or disabled person has independent counsel and a separate guardian *ad litem*;

COMMENT. A finding that a person is disabled and requires appointment of a guardian is not tantamount to a determination that he or she requires hospitalization.[53] Yet in many jurisdictions, a person may be committed to a mental health or mental retardation institution by his or her guardian without being afforded the procedural protection to which other citizens are entitled.[54] Since there appears to be no compelling state interest justifying this distinction in the restriction of the fundamental right of liberty, this paragraph requires limited personal guardians and personal guardians to follow the same commitment procedures in placing a partially disabled and disabled person in an institutional and/or secure residential setting as would be required if that person was not subject to guardianship.[55] Changes may have to be made in the pertinent state commitment statute or statutes to conform to this restriction.

(b) To consent on behalf of a partially disabled or disabled person to an abortion, sterilization, psychosurgery, or removal of a bodily organ, except as provided in (citation to the Model Personal and Civil Rights Act) or when necessary to preserve the life or prevent serious impairment of the physical health of that person;

COMMENT. Decisions affecting the control of the reproductive capacity of a partially disabled or disabled person, psychosurgery, and the removal of a bodily organ are so fundamental, so personal, and so subject to conflicts of interest that they cannot be delegated from one individual to another without the creation of checks and safeguards to protect against exploitation and abuse.[56] This provision does not bar such medical procedures altogether. Rather, it limits the instances in which a limited personal guardian or personal guardian may provide consent to when the partially disabled person may die or suffer significant damage to his or her physical health if the particular operation is not performed.[57] Some states may wish to add electro-convulsive therapy (shock treatments) to the list of regulated medical procedures.

(c) To consent on behalf of the partially disabled or disabled person to the withholding of non-heroic life-saving medical procedures;

COMMENT. No right is more basic than the right to survive. Yet it is not unusual for natural and appointed guardians together with doctors to decide to let a partially disabled or disabled person die when they are faced with determining whether to authorize ordinary medical procedures to save the life of that person. Perhaps the most frequently cited case is that of the mentally defective infant who is allowed to die by withholding life-saving medical treatment that would have been provided were it not for the disability.[58] This provision follows the holding in *Maine Medical Center vs. Houle* that a guardian may not withhold non-heroic life-saving medical procedures.[59] It does not address the question of when mechanical life-support systems may be removed.

(d) To consent on behalf of a partially disabled or disabled person to the performance of any experimental biomedical or behavioral medical procedure or participation in any biomedical or behavioral experiment unless:

(i) It is intended to preserve the life or prevent serious impairment of the physical health of the partially disabled or disabled person; or

(ii) It is intended to assist the partially disabled or disabled person to develop or regain his or her abilities and has been approved for that person by the court;

COMMENT. The same principles which require the limitations imposed by the previous two subparagraphs dictate the prohibition of third party consent to the use of partially disabled or disabled persons as the subjects of experiments, when the test is not designed to benefit that person as an individual. The dangerous and dehumanizing abuses documented at the Willowbrook State

[53] U.P.C., *supra* note 2, at comment to §5-304.

[54] *Id.;* Kindred, *supra* note 18, at 82; Levy, *supra* note 22.

[55] PCMR, *supra* note 12, at 67.

[56] Kindred, *supra* note 18, at 76-83; Price & Burt, *Nonconsensual Medical Procedures and the Right to Privacy* in PCMR, *supra* note 18, at 94-110; Wyatt v. Aderholt, 344 F. Supp. 1383 (M.D. Ala. 1974).

[57] *Cf.* Subchapter 1.6: Emergency Medical Care and Treatment for Infants, *supra*.

[58] Kindred, *supra* note 18, at 77.

[59] *Id.* at 78.

School in New York[60] demonstrate what can happen when there are inadequate safeguards.

The proposed federal regulations on research involving institutionalized mentally disabled persons vary the consent procedures according to the predicted level of risk as well as the purpose of the experiment.[61] That approach was not adopted in this statute because of the difficulty of making such predictions accurately and the consequences which errors in those preditions can have.

(e) To prohibit the marriage or divorce of a partially disabled or disabled person;

COMMENT. Another area which is "too personal and individual to be . . . [assigned to] a substitute decision-maker" is that of marriage and divorce.[62] At a minimum, imposition of restrictions on the right to marry must be limited to instances in which it is proven that the partially disabled or disabled person is unable to understand the nature, demands and responsibilities of the marital relationship.[63]

A different problem arises with regard to divorce. In states without no-fault divorce laws, if the personal guardian of a person who has become partially disabled or disabled during the course of a marriage is unable to provide consent to a divorce, the spouse seeking the divorce may have to resort to a contested proceeding, which is likely to result in a far less advantageous settlement for the disabled spouse. Therefore, this provision permits a limited personal guardian or personal guardian to consent to a divorce on behalf of the person whom he or she is serving. However, in keeping with subsection (1), a limited personal guardian or personal guardian should seek the assent of that person before consenting to a divorce. (Compare §40(10) Records.) When the partially disabled or disabled person objects to the divorce, after an explanation of the issues and procedures involved, the limited personal guardian or personal guardian should oppose the divorce petition.

(f) To consent on behalf of a partially disabled or disabled person to the termination of that person's parental rights; and

COMMENT. Many partially disabled persons are able, with support and counseling, to raise and nurture their children.[64] Thus, the same grounds and procedures for termination of parental rights should apply to persons with disabilities as to those who are not partially disabled or disabled.[65]

(g) To assist in managing or to manage the financial resources of the partially disabled or disabled person unless the limited personal guardian or personal guardian has also been appointed as the limited conservator or conservator for that person.

COMMENT. In keeping with the provision distinguishing between the duties and grounds for appointment of a limited conservator or conservator, and those applicable to guardianship, this provision prohibits limited personal guardians or personal guardians from controlling or supervising the financial resources of the individual they are serving unless they have received a dual appointment. Pursuant to section 17(3)(c) and 17(4)(a)(vi), however, a limited personal guardian or personal guardian may expend sums from the financial resources of the person he or she is serving to cover reasonable costs incurred on behalf of that person so long as those expenditures are consistent with the procedures and policies established by the court and the limited conservator or conservator if any. Disputes over such expenditures which cannot be resolved by the parties may be submitted to the court pursuant to section 16.

(6) (a) A limited personal guardian or personal guardian shall submit a report to the court at least ten (10) days before the initial review hearing and:

(i) Annually thereafter;

(ii) When the court orders additional reports to be filed;

(iii) When there is a significant change in the capacity of the partially disabled or disabled person to meet the essential requirements for his or her physical health or safety;

(iv) When the limited personal guardian or personal guardian resigns or is removed; and

(v) When the guardianship is terminated.

COMMENT. This section extends the requirement for a periodic financial accounting to personal guardianship. The report is closely linked to the review hearings required by section 15. It is intended as a means for increasing accountability, by forcing the limited personal guardian or personal guardian to review whether there is a need for continuing the restrictions imposed on a partially disabled or disabled person's legal capacity to act on his or her own behalf, and whether the subject of the guardianship is receiving the services identified or contemplated in the dispositional order. In addition, the report will alert the court and all individuals and

[60] New York State Association for Retarded Children v. Rockefeller, 357 F. Supp. 752 (E.D.N.Y. 1973).

[61] Fed. Reg. 53,950 (1978).

[62] Kindred, *supra* note 18, at 76.

[63] *Cf.* Subchapter 1.1: Marriage, Annulment and Divorce, *supra; see* MHLP, *Legal Issues in State Mental Health Care: Proposals For Change – State Imposed Disabilities.* 2 MENTAL DISABILITY LAW REPORTER 497, 502 (1978); PCMR, *supra* note 12, at 63.

[64] *Id.*; PCMR, *supra* note 12, at 63.

[65] Subchapter 1.2: Relinquishment and Termination of Parental Rights, *supra*.

entities with an interest in the case to major changes and problems, and document the amount of and basis for the compensation and reimbursements to which the limited personal guardian or personal guardian is entitled.

(b) Such reports shall set forth:

(i) The name and address of the partially disabled or disabled person, and of the limited personal guardian or personal guardian;

(ii) Significant changes in the capacity of a partially disabled or disabled person to meet the essential requirements for his or her physical health or safety;

(iii) The services being provided to the partially disabled or disabled person and the relationship of those services to the individual guardianship plan;

(iv) The significant actions taken by the limited personal guardian or personal guardian during the reporting period;

(v) The compensation requested and reasonable and necessary expenses incurred by the limited personal guardian or personal guardian;

(vi) Any significant problems relating to the guardianship which have arisen during the reporting period; and

(vii) The reasons, if any, why the appointment should not be terminated or why no less restrictive alternative will permit the partially disabled or disabled person to meet the essential requirements for his or her physical health or safety.

Attached to the report shall be an accounting of any expenditures made by the limited personal guardian or personal guardian, and an update of the individual guardianship plan developed by the limited personal guardian or personal guardian, the limited conservator or conservator if any, the disabilities resources officer, and to the maximum extent possible, the disabled or partially disabled person.

COMMENT. The report is designed to provide more than *pro forma* administrative information. The topics which should be addressed are listed specifically in order to clarify for the limited personal guardian or personal guardian what is expected of him or her and to assure that the purposes described in the preceding paragraph are achieved. A description is required of the important occurrences during the reporting period including meaningful increases and declines in the partially disabled or disabled person's ability to protect his or her health and safety and the impact of those changes on the need for continuing the guardianship. A discussion of problems which have arisen during the reporting period is included so that the other parties and the court, where possible, will have an opportunity to contribute to a solution. The limited personal guardian or personal guardian must also describe what he or she did to earn the requested compensation. (See §39 Compensation and Expenses.) Finally, nothing in this provision requires that the report be typed or in legal form. Handwritten, informal reports may be submitted so long as they are legible and contain the specified information.

(c) The court shall have copies sent to the individuals and entities specified in section 32(3)(b).

COMMENT. Copies of the report are to be sent to the partially disabled or disabled person; that person s attorney of record (see §31 Rights and Privileges); his or her close relatives; his or her limited conservator, conservator, and/or guardian *ad litem* if any; the Administrator of Veterans' Affairs if the financial assets of that person include V.A. benefits; and the state Developmental Disabilities Advocacy Office if the subject of the guardianship has a developmental disability. Section 32(3)(c) requires that the contents of the report must be explained to the partially disabled person in the language, mode of communication and terms which that person is most likely to understand.

Introduction to Section 18

This section sets forth the general duties which inhere to conservatorship, the requirements for individual conservatorship plans, the specific duties and powers which may be assigned to limited conservators and conservators, the limits on the authority of limited conservators and conservators, and the reporting requirements applicable to conservatorships. Section 17 covers the same range of topics with regard to guardianships.

Section 18. Duties and Powers of Limited Conservators and Conservators

(1) It is the general duty of an individual or corporation appointed to serve as a limited conservator or conservator for a partially disabled or disabled person to carry out diligently and in good faith the specific duties and powers assigned by the court. In carrying out these duties and powers such individuals or corporations shall:

(a) Manage or assist in managing those financial resources placed under their supervision and/or control, as would a prudent person manage his or her own financial resources; and

(b) Encourage the disabled or partially disabled persons whom they are serving to:

(i) Participate, to the maximum extent of their abilities, in all decisions which affect them;

(ii) Act on their own behalf on all matters in which they are able to do so; and

(iii) Develop or regain, to the maximum extent possible, their capacity to manage their financial resources and, if impaired, their capacity to

meet the essential requirements for their physical health or safety.

COMMENT. This provision, together with subsection (3) makes clear that limited conservators and conservators must be concerned not only with fiscal affairs but also with encouraging the partially disabled or disabled person who is being served to be as self-reliant and independent as possible and with utilizing that person's financial resources to develop or regain the skills required to obviate the need for guardianship and/or conservatorship. Included in the general duties is the responsibility to make certain that the partially disabled or disabled person receives all the benefits and services and is accorded all the rights *vis a vis* other individuals, businesses and agencies to which that person is entitled. This may include assisting the partially disabled or disabled person and that person's limited personal guardian or personal guardian to bring tort, contract, consumer, and antidiscrimination actions. (See comment to §17(1) Duties and Powers of Limited Personal Guardian and Personal Guardians.)

The subsection also requires limited personal guardians and personal guardians to perform their duties diligently and in good faith. Failure to meet this standard may render the limited conservator or conservator liable to the partially disabled person or disabled person for damages. (See §42 Liability of Limited Conservators and Conservators and Penalties for Violation of this Act.) While stopping short of imposing strict liability, it is anticipated that this provision will adequately protect partially disabled and disabled persons without discouraging individuals or corporations from accepting conservatorship responsibilities.

(2)(a) Upon appointment, a limited conservator or conservator shall develop an individual conservatorship plan together with the limited personal guardian or personal guardian if any, the disabilities resources officer and, to the maximum extent possible, the partially disabled or disabled person. The plan shall specify:

(i) The services which are necessary to manage the financial resources designated by the court in the dispositional order;

(ii) The means through which those services will be provided;

(iii) The manner in which the limited conservator or conservator, the partially disabled or disabled person and the limited personal guardian or personal guardian, if another individual has been appointed to serve in that capacity, will exercise and share their decision-making authority;

(iv) The policies and procedures governing the expenditure of funds; and

(v) Such other items as will assist in the management of the designated financial resources, and in fulfilling the needs of the partially disabled or disabled person, the terms of the dispositional order, and the duties of the limited conservator or conservator.

COMMENT. The individual conservatorship plan is intended to outline for the partially disabled or disabled person, his or her limited personal guardian or personal guardian, if any, and the court, just how the particular financial resources placed under the limited conservator's or conservator's supervision or control are to be managed. In order to avoid misunderstandings and facilitate the management of these financial resoucres, the plan must contain the policies and procedures governing the expenditure of funds by the partially disabled or disabled person and his or her limited personal guardian or personal guardian, if any. It must also reflect the agreement reached by the limited conservator or conservator, the limited personal guardian or personal guardian, if any, and the individual they are serving, regarding how the decision-making authority will be shared in those areas in which the partially disabled or disabled person's legal capacity has been restricted by the court. Disputes over the agreements, policies and procedures whict cannot be resolved by the parties, may be submitted to the court pursuant to section 16. Like its guardianship counterpart, the individual conservatorship plan will provide a reference against which the performance of the limited conservator or conservator can be compared.

The plan is to be developed jointly by the individual serving in a conservatorship capacity, the person subject to the conservatorship, the limited personal guardian or personal guardian if any, and the disabilities resources officer. The participation of the partially disabled or disabled person in the development of the plan is consistent with the purpose of this act and the general duties imposed by the preceding subsection.[66] The disabilities resources officer is included to assure that all the appropriate benefits and services which are available are considered during the planned process. (See §4 Appointment of Disabilities Resources Officers.)

(b) The individual conservatorship plan shall be submitted to the court within the time limit prescribed in section 30(1)(k), together with a complete inventory of the designated financial resources. The inventory shall include an oath or affirmation that, to the best of the limited conservator's or conservator's knowledge, it is complete and accurate.

COMMENT. Consistent with the U.P.C.[67] and the conservatorship provisions in many states, one of the initial duties of a limited conservator or conservator is to take an inventory of the financial resources placed under his, her or its control. Be-

[66] *See* Power, *supra* note 37.

[67] U.P.C., *supra* note 2, at §5-418.

cause of the inventory requirement, limited conservators and conservators have 60 days following issuance of the dispositional order to submit the initial conservatorship plan rather than the 30 days prescribed for individual guardianship plans. The individual conservatorship plan and inventory is to be updated in accordance with subsection (7) of this section.

(c) The court shall have a copy of the individual conservatorship plan and the inventory sent to the individuals and entities specified in section 32(3)(b).

COMMENT. The partially disabled or disabled person; his or her attorney; his or her limited personal guardian, personal guardian, and/or guardian *ad litem* if any; the individuals most closely related to the partially disabled or disabled person; the individual having custody of that person; the Administrator of Veterans' Affairs if the partially disabled or disabled person receives V.A. benefits; and the state Developmental Disabilities Advocacy Office if the partially disabled or disabled person has a developmental disability are all to receive a copy of the individual conservatorship plan in order to reduce the likelihood of misunderstandings and further assure accountability.

(3) (a) A limited conservator or conservator shall expend or distribute, and/or authorize the expenditure or distribution of, and/or assist in the expenditure or distribution of, the principal of or income from the financial resources placed under his, her, or its supervision and control to assure that:

(i) The essential requirements for the physical health or safety of the partially disabled or disabled person are met;

(ii) The rights of that person are protected;

(iii) The financial resources of that person which are subject to the conservatorship are prudently managed;

(iv) The partially disabled or disabled person has the opportunity to develop or regain the capacity to perform the functions listed in subparagraphs (3)(a)(i)-(iii) of this section, to the maximum extent possible; and

(v) The limited personal guardian or personal guardian for that person, if any, is able to carry out duties and powers assigned by the court.

COMMENT. This provision expands upon subsection (1) of this section and section 5-425(a) (2) of the U.P.C. to clarify the principles which govern the exercise of the duties and powers assigned to a limited conservator or conservator. Preservation of the estate is only one of the cluster of objectives which a limited conservator or conservator must take into account. Prudent management should not be considered synonymous with conserving financial resources for the benefit of the heirs of a partially disabled or disabled person.[68] The primary use of a partially disabled or disabled person's financial resources should be to protect his or her health, safety, rights and assets, and to maximize his or her abilities. These considerations apply to expenditures or distributions made by the limited conservator or conservator directly, as well as to decisions to authorize a limited personal guardian or personal guardian, if any, to expend funds and assistance rendered to a partially disabled or disabled person in disposing of his or her financial resources. Subparagraph (3)(a)(v) places an affirmative duty on a limited conservator or conservator to cooperate with the limited personal guardian or personal guardian, if another individual has been appointed in that capacity, to achieve these goals.

(b) In so doing, the limited conservator or conservator shall consider:

(i) The requests of the partially disabled or disabled person and that person's limited personal guardian or personal guardian, if any;

(ii) The size of the financial resources under the limited conservator's or conservator's supervision or control;

(iii) The probable duration of the conservatorship;

(iv) The likelihood that the partially disabled or disabled person may be able to manage the financial resources in the future; and

(v) The accustomed standard of living of the partially disabled or disabled person and the individuals legally dependent on that person.

In addition, limited conservators shall consider the size of the financial resources of a partially disabled or disabled person which have not been placed under their supervision or control.

COMMENT. This provision is drawn directly from section 5-425(a)(2) of the U.P.C. It sets out the basic factors which must be weighed in making or authorizing any significant expenditure or in advising a partially disabled or disabled person regarding such an expenditure.

(c) Funds under this paragraph may be paid by a limited conservator or conservator to any person, including the partially disabled or disabled person.

COMMENT. None necessary.

(d) Limited conservators or conservators shall establish policies and procedures for the expenditure of the funds under their supervision and control, by the partially disabled or disabled person, or by a limited personal guardian or personal guardian for that person if another individual has been appointed to serve in that capacity.

---

[68] Power, *supra* note 37.

COMMENT. See sections 17(3)(c), 17(4)(b), and 18(2) and the comments thereto.

(4) (a) In addition to the duties and powers prescribed in subsection (3) of this section, the court may assign to a limited conservator any of the duties and powers listed in paragraph (5)(a) which the partially disabled person lacks the capacity to perform, and/or the power to void particular types of contracts, conveyances or dispositions made by the partially disabled person as long as action is taken to void such contracts, conveyances, and dispositions within fifteen (15) days of the transaction. Such assignments shall specify the financial resources and/or the types of transactions to which they apply.

COMMENT. Consistent with the dispositional principles and criteria in sections 12(2) and (3), this provision directs that a limited conservator be authorized to act in place of a partially disabled person only in those areas and for those assets in which that person lacks the necessary management skills. Thus, in the case of a partially disabled person who is able to handle day-to-day financial transactions but whose ability to receive and evaluate information or communicate decisions is impaired to the extent that he or she is unable to operate a business or manage an investment portfolio, the limited conservator's control should extend only to the business or investments, and not to the income paid directly to the partially disabled person.[69] To further encourage the use of dispostional alternatives other than plenary conservatorship, the provision also permits the court to authorize a limited conservator to void specified types of transactions entered into by the partially disabled person — *e.g.,* contracts subject to the statute of frauds — when the terms of the transaction are exploitative or the partially disabled person will be unable to fulfill the contractual obligations. This authority should not be exercised, however, because the limited conservator and the partially disabled person differ over the need for or quality of the item or service being purchased. As noted earlier, partially disabled and disabled persons should be allowed to err to the same degree as individuals without disabilities.[70] In order to protect individuals and business dealing in good faith with partially disabled persons, the limited conservator so empowered cannot exercise this authority more than 15 days after the transaction has been entered into.

(b) The court may assign to a limited conservator the duty of assisting a partially disabled person to perform any of the functions listed in paragraph (5)(a) of this section with regard to any financial resources of the partially disabled person.

COMMENT. In addition to the powers which may be assigned under paragraph (4)(a), the court may direct a limited conservator to help a partially disabled or disabled person to perform certain functions in relation to particular financial resources when that person requires assistance but does not require restrictions of his or her legal capacity. For example, a limited personal guardian could be directed to assist the subject of the conservatorship in budgeting the income which that person is receiving from the assets being managed by the limited conservator.

(c) Upon the death of a partially disabled person, the limited conservator for that person shall have the duties and powers prescribed in paragraph (5)(b) of this section.

COMMENT. See comment to section 18(5)(b).

(5) (a) In addition to the duties and powers prescribed in subsection (3) of this section, the court shall assign to a conservator the duty and/or power to:

(i) Acquire, collect, hold, deposit, retain, operate, develop, repair, improve, insure, subdivide, exchange, partition, alter, lease, convert or dispose of the financial resources of the disabled person;

(ii) Pay, contest, settle or release claims against the disabled person or the financial resources of the disabled person;

(iii) Pay taxes, assessments, compensation and other reasonable expenses incurred in the management of the financial resources of the disabled person;

(iv) Employ persons to perform, advise, or assist in particular aspects of the management of the financial resources of the disabled person;

(v) Borrow money to be repaid from the financial resources of the disabled person;

(vi) Prosecute or defend actions, claims, of proceedings in any jurisdiction for the protection of the financial resources of the disabled person and of the conservator in the performance of the conservator's duties;

(vii) Allocate items of income or expense to either principal or income as provided by law, including the creation of reserves out of income for depreciation, obsolescence, amortization, or depletion of mineral or timber resources;

(viii) Vote a security in person, or by general or limited proxy;

(ix) Establish revocable or irrevocable trusts;

(x) Exercise options;

(xi) Exercise or release powers of appointment or which the disabled person is donor;

(xii) Renounce interests and make gifts;

(xiii) Change beneficiaries under insurance and annuity policies; and

[69] See authorities cited in note 51.

[70] AAMD, *supra* note 4, at 1.

(xiv) Enter into contracts and execute and deliver all instruments which will accomplish or facilitate the exercise of the assigned duties and powers.

COMMENT. The duties and powers enumerated in this paragraph are intended to encompass all those specified in sections 5-408(3), 5-424(b) and (c), and 5-425(b) of the U.P.C. Subject to the limitations imposed in subsection (6), these duties and powers may be exercised during the term of the assignment without returning to the court for specific authorization. It provides the conservator with the authority needed to manage all types of financial resources effectively.[71]

(b) Upon the death of a disabled person:

(i) The conservator shall deliver to the (name of court having probate jurisdiction) any will of the deceased of which the conservator has possession and/or advise that court of the whereabouts of any other will of which the conservator is aware;

(ii) Inform the [executor/personal representative/administrator] or a named beneficiary of the actions taken; and

(iii) Retain those portions of the estate over which the conservator has control for delivery to a duly appointed [executor/personal representative/administrator] of the deceased or the individuals entitled thereto.

The conservator may seek appointment as [executor/personal representative/administrator] of the estate of the disabled person in the manner prescribed in (citation to relevant sections of the probate code).

COMMENT. This provision is derived from section 5-425(e) of the U.P.C. In enacting this subparagraph, care should be taken to assure consistency with the existing statutory provisions on decedent's estates and the probate of wills.

(6) (a) A limited conservator or conservator shall not have the authority to exercise or release powers of appointment of which the partially disabled or disabled person is donee, to renounce interests or to make gifts exceeding twenty (20) percent of any year's income from the financial resources placed under the control of the limited conservator or conservator, nor to change beneficiaries under insurance or annuity policies without the approval of the court. No approval shall be granted unless it is demonstrated at a hearing that the proposed action is consistent with subsections (1) and (3) of this section, and that the partially disabled or disabled person has consented or is unable to provide an informed, voluntary consent. Notice of the hearing shall be served on the individuals and entities enumerated in section 32(1)(a) of this act in the form and manner presecribed in sections 32(1)(a) and 32(2).

COMMENT. The restrictions imposed by this provision are derived from section 5-408(4) of the U.P.C. and are intended to reduce the likelihood that financial resources may be misappropriated to the benefit of a limited conservator and/or to the detriment of the partially disabled or disabled person. Notice of a hearing under this paragraph shall specify the date, time, place, purpose and possible consequences of the hearing, and the rights of the parties. It shall be served on the partially disabled or disabled person; his or her attorney; his or her limited personal guardian, personal guardian and/or guardian *ad litem* if any; and/or the limited conservator or conservator; the close relatives and custodians of the partially disabled or disabled person; the Administrator of Veterans' Affairs if the financial resources of a partially disabled or disabled person include V.A. benefits; and the surety on the bond, if any, for the limited conservator or conservator. The notice must be explained to the partially disabled or disabled person in the language, mode of communication, and terms which that person is most likely to understand. (See §32(3) Notice and Dissemination of Reports and the comment thereto.)

(b) Any sale or encumbrance to a limited conservator or conservator, his or her spouse, agent or attorney, or any corporation or trust in which the limited conservator or conservator has a substantial beneficial interest, or any transaction which is affected by a substantial conflict of interest is voidable unless the transaction is approved by the court after a hearing. Notice of the hearing shall be served on the individuals and entities enumerated in section 32(1)(a) of this act in the form and manner prescribed in sections 32(1)(a) and 32(2).

COMMENT. This provision is based upon section 5-422 of the U.P.C. It provides further protection against exploitation of a partially disabled person by a limited conservator or conservator. (See also §35(2) Emergency and Standby Appointments and §41 Bonds.)

(c) Title to all the designated financial resources of a partially disabled person or disabled person shall remain in that person subject to the possession of the limited conservator or conservator and to the control of the court, unless the court orders otherwise. The provision of title to a limited conservator or conservator is not a transfer or alienation within the meaning of general provisions of any federal or state statute or regulation, insurance policy, pension plan, contract, will, or trust instrument, imposing re-

---

[71] For a discussion of the problems which have occurred with more narrowly drawn statutes, see Fratcher, *Toward Uniform Guardianship Legislation*, 64 MICH. L. REV. 977, 985-987 (1966).

strictions upon or penalties for transfer or alienation by the partially disabled or disabled person of a right or interest. If title is provided to a limited conservator or conservator, the dispositional order shall be evidence of transfer of all designated financial resources. An order terminating a conservatorship or modifying the scope or powers thereof, shall be evidence of transfer of the designated financial resources to the partially disabled or disabled person, the successor of that person, or to a new limited conservator or conservator. A dispositional order providing title to a limited conservator or conservator and an order terminating or modifying a conservatorship shall be [filed/recorded] to record notice of title in accordance with (citation to statute governing the filing/recording of documents to land or other property).

COMMENT. Consistent with the principle that a partially disabled or disabled person should be stripped of the legal capacity to exercise only those powers which it has been proven beyond a reasonable doubt that he or she is incapable of performing (see §§12(2) and (3) Dispositional Hearing and Dispositional Principles and Criteria and §44 Burden of Proof), title to financial resources is not automatically vested in the limited conservator or conservator.[72] Any order providing title to a limited conservator or conservator shall be issued in the form and manner specified in section 14.

(7) (a) A limited conservator or conservator shall submit a report to the court ten (10) days before the initial review hearing and:

(i) At least annually thereafter;

(ii) When the court orders additional reports to be filed;

(iii) When there is a significant change in the capacity of the partially disabled or disabled person to manage his or her financial resources;

(iv) When the limited conservator or conservator resigns or is removed; and

(v) When the conservatorship is terminated.

COMMENT. Most states require limited conservators and conservators to submit periodically an accounting of the financial resources for which they are responsible. This provision requires more than a mere financial statement. Like its counterpart in the section on guardianship (§17(6)(a)) it is intended to serve several purposes. In conjunction with the review hearings held pursuant to section 15, the report will alert individuals and entities with an interest in the case to major changes in the condition of the partially disabled or disabled person and will assure that the restrictions imposed on an individual's legal capacity to act in his or her own behalf continue no longer than necessary. It will also assist in verifying that the subject of the conservatorship is receiving the services identified or contemplated in the dispositional order, and that the limited conservator or conservator is exercising the duties and powers assigned diligently and in good faith. Finally, it provides an opportunity for limited conservators and conservators to notify the court and parties of developing problems, to seek assistance in solving those problems, and to document the compensation and reimbursements to which they are entitled.

(b) Such reports shall set forth:

(i) The name and address of the partially disabled or disabled person and of the limited conservator or conservator;

(ii) Significant changes in the capacity of the partially disabled or disabled person to manage his or her financial resources;

(iii) A complete financial statement of the financial resources under the control and/or supervision of the limited conservator or conservator;

(iv) The services being provided to the partially disabled or disabled person and the relationship of those services to the individual conservatorship plan;

(v) The significant actions taken by the limited conservator or conservator during the reporting period;

(vi) The compensation requested and the reasonable and necessary expenses incurred by the limited conservator or conservator;

(vii) Any significant problems relating to the conservatorship which have arisen during the reporting period; and

(viii) The reasons, if any, why the conservatorship should not be terminated, or why no less restrictive alternative would permit the disabled or partially disabled person to manage his or her financial resources.

Attached to the report shall be an updated individual conservatorship plan developed by the limited conservator or conservator, the limited personal guardian or personal guardian if any, the disabilities resources officer and, to the maximum extent possible, the partially disabled or disabled person.

COMMENT. See preceding comment and the comment to section 17(6)(b).

(c) The court shall have copies of the report sent to the individuals and entities specified in section 32(3)(b).

COMMENT. Copies of the report are to be sent to the partially disabled or disabled person; that person's close relatives, attorney of record, (see §31 Rights and Privileges), custodian and limited personal guardian, personal guardian, and/or guardian *ad litem* if any; the Administrator of Veterans'

[72] U.S. SENATE SPECIAL COMMITTEE ON AGING, *supra* note 17, at 101-102; *but see* U.P.C., *supra* note 2, at §5-420.

Affairs if the financial resources of that person include V.A. benefits; and the state Developmental Disabilities Advocacy Office if the subject of the guardianship has a developmental disability. The contents of the report must be explained to the partially disabled person in the language, mode of communication, and terms which that person is most likely to understand. (See §32(3)(c) Notice and Dissemination of Plans and Reports.)

(d) Following submission of a report or in conjunction with a review hearing held pursuant to section 15 or a proceeding initiated under section 16, a limited conservator or conservator shall submit to a physical check of the financial resources placed under his, her or its control, in the manner specified by the court.

COMMENT. This provision is drawn from section 10(b) of the Uniform Veterans Guardianship Act and section 5-419 of the U.P.C. It permits the physical inspection of securities and investments held by a limited conservator or conservator on behalf of a partially disabled or disabled person to assure that they conform to those listed in the report.

CHAPTER 3 INTERVENTION PROCEEDINGS FOR MINORS

Section 19. Initiation of Intervention Proceedings

(1) (a) An intervention proceeding for a minor may be initiated separately or in conjunction with another proceeding by the filing of a petition with the court by a parent of the minor, the limited personal guardian, personal guardian, limited conservator or conservator for the minor, a person interested in the welfare of the minor, or the Director of (the agency required to provide services to children), if there is reason to believe that:

(i) The parents of the minor are dead;

(ii) Their parental rights have been terminated or suspended by order of the court; or

(iii) The minor has or is entitled to financial resources which may be jeopardized because of the minor's age or which cannot otherwise be protected from damage or dissipation.

Intervention proceedings involving minor siblings may be initiated by a single petition.

COMMENT. It is anticipated that petitions seeking court-ordered assistance or establishment of a guardianship or conservatorship for a minor will ordinarily be filed during the course of a probate, termination of parental rights, neglect and abuse, civil commitment, divorce, adoption, or child custody proceeding. However, in some instances, an intervention proceeding may be initiated independent of another court action — e.g., when a child has inherited directly an operating business, or when a single parent will be unable to care for the child for a substantial time because of illness or injury, incarceration, or a job-connected transfer to a location in which dependants are not allowed.

The paragraph authorizes a number of individuals and officials to initiate an intervention proceeding, including the minor's parent, the head of the child protection agency and/or youth services agency, and an appointed limited personal guardian, personal guardian, limited conservator and/or conservator for a minor. An explicit grant of the authority to file a petition is provided for individuals or entities appointed to serve in a guardianship or conservatorship capacity because of the specificity with which the duties and powers of limited personal guardians, personal guardians, limited conservators and conservators are prescribed by sections 27 and 28.

Guardianship, conservatorship and/or other court-ordered assistance can be sought under this chapter in three circumstances only. The first is when a child has been orphaned. The second is when the court in a separate proceeding has found that the child's parents are unwilling or unable to care for the child and has terminated or temporarily suspended their parental rights.[73] (For a discussion of the circumstances in which termination of parental rights may occur, see the report in this series on personal and civil rights.) The third is when an individual under 18 or the statutory age of majority receives a large sum of money or other financial resources which that individual is unable to manage because of his or her immaturity and/or the complexity of the particular asset(s), and which is not subject to a trust or the provisions of the Uniform Gifts to Minors Act.[74] These grounds, together with section 26 which terminates a guardianship or conservatorship automatically upon the marriage of a minor, combine those found in many state statutes (see Chart VIII) and the U.P.C.[75]

(b) The petition shall be verified under oath and shall specify:

(i) The name, age, and address of the minor;

(ii) The names and addresses of the individuals most closely related to the minor unless they are unknown and cannot be reasonable ascertained;

(iii) The name and address of the individual or facility having custody of the minor;

---

[73] U.P.C., *supra* note 2, at §§5-204 and 5-401(1).

[74] NCCUSL, UNIFORM GIFTS TO MINORS ACT (1956). The Uniform Gifts to Minors Act covers gifts of securities, money, life insurance, or annuity contracts. All gifts under the act are irrevocable. The custodian of the gift has broad discretion over management, investment, and disposition of the particular property given to the minor, as well as over payment to or on behalf of the minor of principal and income derived from that property.

[75] *See* U.P.C., *supra* note 2, at §5-202.

(iv) The nature and description of any existing guardianship and conservatorship and the name and address of the current limited personal guardian, personal guardian, limited conservator and/or conservator;

(v) The particular type of protection and assistance being sought;

(vi) The facts and reasons supporting the need for such protection and assistance;

(vii) The name, address, qualifications and relationship to the minor of the individual, if any, proposed for appointment as the limited personal guardian or personal guardian and/or of the individual or corporation proposed for appointment as the limited conservator or conservator;

(viii) The approximate value and description of the minor's financial resources including any benefits, compensation, pension or insurance payable by the Veterans' Administration to which the minor is entitled;

(ix) The name, address and interest of the petitioner; and

(x) The name and address of the petitioner's attorney, if any, or a statement that the petitioner is unable to obtain counsel for financial reasons and requests to be represented by the [county/state/district] attorney.

COMMENT. With one significant exception, the form of the petition for initiation of an intervention proceeding for a minor is essentially the same as that prescribed by section 6(1)(b) for the initiation of an intervention proceeding for a partially disabled or disabled person. (See the comment to §6(1)(b) for an explanation of the requirements for the petition.) That exception is the absence of the statement indicating that the petitioner has met with the disabilities resources officer, discussed the options available, and concluded that the relief requested is the least restrictive dispositional alternative. (See §6(1)(b)(xiv).) Since it is anticipated that in the vast majority of cases, establishment of a guardianship or conservatorship for a minor will be the least restrictive dispositional alternative, such a meeting and sworn statement appear unnecessary. However, states which have a well-developed system of advocacy, and support services for older minors — e.g., individuals between ages 15 and 17 — may wish to consider requiring an individual who seeks to initiate an intervention proceeding to explore the available options with a youth services official or youth advocate prior to filing the petition.

(c) Within five (5) days of the filing of a petition, the court shall set a date for a hearing, and shall have notice of the petition served on the individuals and entities specified in section 32(1)(a) in the form and manner prescribed by section 32(1)(a) and 32(2).

COMMENT. Notice of the rights of the parties and the date, time, place, purpose and possible consequences of the hearing is to be sent by the court together with a copy of the petition to each of the individuals and entities named in the petition including, when applicable, the Administrator of Veterans' Affairs. The contents of the notice and the petition are to be explained, to the maximum extent possible, to the minor in the language, mode of communication and terms which he or she is most likely to understand. (See §§32 (1)(a) and 32(2) Notice and Dissemination of Plans and Reports.)

(2) (a) A parent of a minor may designate, by will, an individual to serve as limited personal guardian, personal guardian, limited conservator and/or conservator upon the testator's death. A designation of a personal guardian and/or conservator for a minor shall become effective upon the filing of a written acceptance if no objection is filed pursuant to section 20 of this act, and if:

(i) Both parents are dead; or

(ii) One parent is dead and the parental rights of the surviving parent have been suspended or terminated in accordance with (citation to the statute(s) on termination of parental rights).

If both parents are dead, an effective appointment by the parent who died later has priority.

COMMENT. Almost every American jurisdiction permits the parents of a minor to name a person in their will to serve as personal guardian or conservator for their child in the event the child is orphaned before reaching majority. (See chart VIII.) This paragraph provides that such appointments become effective automatically upon the death of the minor's surviving parent and the filing of a written acceptance by the nominee, unless the minor or an individual interested in the minor's well-being files an objection. (See §20 Objections to Testamentary Designations.) Since the parents of a minor are considered his or her natural guardians, the validity of a testamentary designation under this chapter is not conditioned on the testator parent having been formally appointed as limited personal guardian, personal guardian, limited conservator or conservator. (*Contra* §6(2).)

(b) The acceptence shall be filed in the court in which the will is probated within the time limit prescribed in section 30(2)(b). It shall include:

(i) The name, age, and address of the minor;

(ii) The name and address of the individuals most closely related to the minor unless they are unknown and cannot be reasonably ascertained;

(iii) The name and address of the individual or facility having custody of the minor; and

(iv) A statement, verified under oath, that the designee understands and is able and willing to undertake the responsibilities, duties and powers of a limited personal guardian, personal guar-

dian, limited conservator and/or conservator for a minor.

COMMENT. Acceptance of a testamentary designation must be filed within 30 days of the submission for probate of the will containing the designation. (§30(2)(b) Time Limits.) The acceptance must contain the names of those individuals who can be presumed to be the most directly interested in or affected by the establishment of a guardianship and/or conservatorship for a minor, as well as the designee's pledge to perform his or her guardianship and/or conservatorship duties diligently and in good faith. (See §27 Duties and Powers of Limited Personal Guardians and Personal Guardians for Minors, and §28 Duties and Powers of Limited Conservators and Conservators for Minors).

(c) Within five (5) days of the filing of an acceptance, the court shall have notice served on the individuals and entities specified in section 32(1)(d) of this act in the form and manner prescribed by sections 32(1)(d) and 32(2).

COMMENT. A notice describing the meaning of the acceptance, the rights of the parties and the procedures and time limits for filing an objection must be provided to the minor; his or her attorney of record; his or her limited personal guardian, personal guardian, limited conservator, conservator, and/or guardian *ad litem* if any; his or her close relatives; the individual or facility having custody of the minor; the executor, personal representative or administrator of the testator's estate; the Administrator of Veterans' Affairs, if the financial resources of the minor include V.A. benefits; and the surety on the bond for the limited conservator or conservator, if any. Naturally, notice of the acceptance need not be sent to the person who filed it. The minor must be given an explanation of the information contained in the notice in the language, mode of communication and terms which he or she is most likely to understand. (§32(2) Notice and Dissemination of Plans and Reports.)

(d) This state shall recognize a testamentary appointment for which an acceptance has been filed under a will probated in another state which was the testator's domicile.

COMMENT. This provision is based upon section 5-202 of the U.P.C. and is intended to minimize forum shopping and conflicting appointments among states when, for example, the minor is a resident of one state and his or her parent is a domiciliary of another. Together with paragraph (2)(b), it establishes the testator's domicile as the jurisdiction in which objections to a testamentary designation are to be raised.

Section 20. Objections to Testamentary Designations

(1) A minor may file an objection to a testamentary designation of a limited personal guardian, personal guardian, limited conservator and/or conservator before it is accepted or within the time limit prescribed in section 30(2)(e), and propose another individual qualified under section 23(4) of this act for appointment as the limited personal guardian, personal guardian, limited conservator and/or conservator.

COMMENT. This section provides minors with the right to challenge the individual and/or corporation designated in their parents' wills, and to nominate their own choice. Pursuant to section 44(2), the individual or corporation proposed by the minor shall be appointed if the appointment is found to be necessary and the minor's nominee is eligible under section 23(4), unless a preponderance of the evidence proves that the appointment of the minor's nominee would not be beneficial to the minor. Currently, 26 states explicitly permit minors over age 14 to object to a testamentary designation and to select their own personal guardian and/or conservator. (See chart X.) Since the rate of maturation varies significantly among individuals, setting the age at which this right accrues is a somewhat arbitrary decision. However, given the lowering of the age of majority in most states below the common law age of 21, the increasing level of sophistication among youths, and the fundamental nature of the question — who do you want to take care of you — permitting objections by minors under the prevailing statutory level of 14 appears warranted.[76] In view of the lack of consensus regarding the age at which it can be presumed that children are capable of making an informed decision and the fact that while objections under this subsection receive a preference, they still may be overridden, no specific age limit is set. Thus, courts will have to determine on a case-by-case basis the strength, sincerity and soundness of the minor's objection.

(2) An individual interested in the welfare of a minor for whom a testamentary designation has been made, may file an objection to that designation within the time limit prescribed in section 30(2)(e). Such an objection shall specify:

(a) The facts and reasons why:

(i) The individual and/or corporation designated in the will would not be eligible under section 23(4) of this act to serve as limited personal

[76] *See* Wald. *The Rights of Youth.* 4 HUMAN RIGHTS 13 (1974). Keasey and Sales, *An Empirical Investigation of Young Children's Usage of Intentionality in Criminal Situations,* 1 HUMAN BEHAVIOR 45 (1977); TASK FORCE TO DEVELOP STANDARDS AND GOALS FOR JUVENILE JUSTICE AND DELINQUENCY PREVENTION, A COMPARATIVE ANALYSIS OF STANDARDS AND PRACTICES: JURISDICTION — DELINQUENCY, 6-9 (1977); INSTITUTE OF JUDICIAL ADMINISTRATION/AMERICAN BAR ASSOCIATION (IJA/ABA) JOINT COMMISSION, STANDARDS RELATING TO JUVENILE DELINQUENCY AND SANCTIONS, §2.1 (Tentative Draft, 1977), *see also* IJA/ABA. STANDARDS RELATING TO NEGLECT AND ABUSE, §8.4(E) (Tentative Draft, 1977).

guardian, personal guardian, limited conservator and/or conservator of the minor;

(ii) The appointment of that individual and/or corporation would not assure that the duties and powers set forth in section 27 and/or 28 will be exercised diligently and in good faith; or

(iii) The appointment of another individual and/or corporation would be of greater benefit to the minor;

(b) The name, address, qualifications, and relationship to the minor of the individual proposed for appointment as the limited personal guardian or personal guardian for the minor and/or the individual or corporation proposed for appointment as the limited conservator or conservator for the minor; and

(c) The name, address, and interest of the individual filing the objection.

COMMENT. This provision permits an individual other than the minor to challenge a testamentary designation. Such objections may arise, for example, when a child has developed a close relationship with one relative prior to his or her parents' death, but another relative is the testamentary designee. Unlike objections filed by the minor, the individual filing an objection under this subsection must prove by a preponderance of the evidence that another individual and/or corporation should be appointed. (See §44(2) Burden of Proof.) Three grounds for such a substitution are listed. The first is that the nominee does not meet the statutory eligibility requirements. The second is that the nominee will not fulfill the statutory responsibilities of a limited personal guardian, personal guardian, limited conservator or conservator. The third is that another eligible individual would be better able to assist the minor in meeting essential requirements for his or her health or safety and or in developing his or her abilities to the maximum extent possible, or that another individual or corporation would be better able to assist the minor in managing his or her financial resources.

(3) Objections to testamentary appointments shall be filed in the court in which the will is probated. Within five (5) days after the receipt of an objection, the court shall set a date for a hearing and shall have a notice served on the individuals and entities enumerated in section 32(1)(a) of this act in the form and manner prescribed in sections 32(1)(a) and 32(2).

COMMENT. This provision, together with section 19(2)(d), is intended to vest control of an intervention proceeding in a single jurisdiction and discourage forum shopping and conflicting sets of orders concerning the care and custody of a minor. Preference is given to the testator's domicile in proceedings involving testamentary designations since it is the jurisdiction which is best able to determine the validity of the will. The rules of civil procedure should apply when a change of venue is sought because another county within a state or another state may be a more convenient forum.

Following the filing of an objection to a testamentary designation, a notice setting forth the date, time, place, purpose and possible consequences of the hearing and the rights of the parties must be served on the minor; his or her attorney of record; his or her limited personal guardian, personal guardian, limited conservator, conservator, and/or guardian *ad litem* if any; the individuals most closely related to the minor; the individual or facility having custody of the minor; the testamentary designee(s); the individual(s) and/or corporation proposed as a substitute for the testamentary designee(s); the executor of the estate; the Administrator of Veterans' Affairs if the financial resources of the minor includes V.A. benefits; and the individual raising the objection if he or she is not already included in the list of those to be notified. The notice must be explained to the minor in the language, mode of communication and terms which the minor is most likely to be able to understand. (§§32(1)(a) and (2) Notice and Disseminaton of Plans and Reports.)

Section 21. Motion to Dismiss

(1) A motion to dismiss may be filed by any party when it appears that there is no basis for a petition filed pursuant to section 19(1)(a) or an objection filed pursuant to section 20(2).

COMMENT This provision provides a means of quickly challenging a petition when it is clear that the parents of a juvenile are alive, that their parential rights have not been terminated or suspended, or that the juvenile has not received or is not entitled to financial resources that were the basis or part of the basis for the petition. Unless the petitioner is able to demonstrate that there is probable cause to believe that the minor is eligible for appointment of a limited personal guardian, personal guardian, limited conservator and/or conservator, the proceeding will be dismissed. (See §44(2)(a)(i) Burden of Proof.) In addition, it permits a summary challenge to frivolous or malicious objections to a testamentary designation by individuals other than the minor. Pursuant to section 44(2)(a)(i), following the filing of a motion to dismiss, an objection will be dismissed unless the objecting party demonstrates that there is probable cause to believe that the testamentary designee is ineligible or would not exercise the assigned duties and powers diligently and in good faith, or that appointment of another eligible individual or corporation would be of greater benefit to the minor. Since a minor may object to a testamentary objection and propose another nominee as a matter of right, motions to dismiss may not be filed following of an objection pursuant to section 20(1). Hearings on motions to dismiss are to be held within 15 days after the motion has been filed and must be on the record and before a judge. Each party is entitled to be present, be represented by counsel, subpoena witnesses, present evidence, and cross-examine opposing witnesses, have an impartial decision-maker and appeal. (See §30(2)(f) Time Limits and §31 Rights and Privileges.)

(2) No more than five (5) days after the filing of a motion to dismiss the court shall set a date for a hearing and shall have notice served on the individuals and entities enumerated in section 32(1)(a) in the form and manner prescribed in sections 32(1)(a) and 32(2).

COMMENT. Notice of the rights of the parties and of the date, time, place, purpose and possible consequences of the hearing on the motion to dismiss is to be served on each of the individuals enumerated in the petition including the Administrator of Veterans' Affairs when applicable. The notice shall be explained to the minor in the language, mode of communication and terms which he or she is most likely to understand. (§§32(1)(a) and 32(2) Notice and Dissemination of Plans and Reports.)

Section 22. Hearing

Within the time limits prescribed by section 30(2)(g) the court shall hold a hearing to determine:

(1) Whether the parents of a minor are dead, whether they have had their parental rights terminated or suspended by order of the court, or whether the minor has or is entitled to financial resources which may be jeopardized because of the minor's age or which cannot be protected from damage or dissipation other than through court-ordered assistance or the establishment of a conservatorship; and if so,

(2) Whether appointment of a limited personal guardian, personal guardian, limited conservator and/or conservator for the minor is necessary; and if so,

(3) Who should be appointed.

COMMENT. This section sets out, in logical order, the questions to be determined at the hearing following the filing of a petition under section 19(1) or an objection to a testamentary designation pursuant to section 20. While the question of whether the minor is properly before the court is listed first, this provision, unlike sections 10 and 12, permits both the substantive and dispositional questions to be considered at the same hearing. Since it is anticipated that the factual question regarding whether a child is without parents due to circumstance or court order will be rather simple and in many cases uncontested, and since a judge rather than a jury will be the decision-maker, a separate dispositional hearing appears unnecessary. When there is a substantial controversy over whether the financial resources of a minor require special protection, the court can order that the presentation of evidence on the dispositional issues be delayed until after the substantive questions have been determined.

The parties, including the minor, are entitled to be present at the hearing, be represented by counsel, subpoena witnesses, present testimony and evidence, cross-examine opposing witnesses, have the issues determined by an impartial decision-maker, and appeal. Hearings must be before a judge and a verbatim record must be made. (See §31 Rights and Privileges.) The petitioner has the burden of going forward and the burden of supporting the statements and requests in the petition with a preponderance of the evidence. (See §§44(2)(a)(ii) and (b)(i) Burden of Proof.) The burden of proof applicable to hearings following an objection to a testamentary disposition is discussed in sections 44(2)(a)(ii) and (b) and the comments to sections 20(1) and 20(2). The hearing must be held within 30 days of the filing of a petition or an objection to a testamentary designation, or within 15 days of the expiration of the period for accepting a testamentary designation if an acceptance has not been filed. (§30(2)(h) Time Limits.)

Section 23. Dispositional Alternatives

(1) The petition shall be dismissed or the designation declared void if it is not proven that the parents of a minor are dead, that they have had their parental rights of custody terminated or suspended by order of the court, or that the minor has or is entitled to financial resources which may be jeopardized because of the minor's age or which cannot be protected from damage or dissipation other than through court-ordered assistance or the establishment of a conservatorship.

COMMENT None necessary.

(2) If a will is refused probate, any testamentary designation of a personal guardian or conservator contained therein shall be void.

COMMENT. None necessary.

(3) If it is proven that the parents of a minor are dead, that they have had their parental rights terminated or suspended by order of the court, or that the minor has or is entitled to financial resources which may be jeopardized because of the minor's age or cannot be protected from damage or dissipation other than through court-ordered assistance or the establishment of a conservatorship, the court may order any of the dispositional alternatives described in section 11(2) or any combination thereof which will enable the minor to meet the essential requirements for his or her physical health or safety, manage his or her financial resources, and develop his or her abilities to the maximum extent possible.

COMMENT. As noted above, guardianship and/or conservatorship will be the appropriate disposition in most intervention proceedings under this chapter. However, alternatives such as establishment of a trust, ratification of education contracts and/or provision of voluntary services may be sufficient for some juveniles close to majority who are able, in large part, to meet the essential needs for their physical health and/or safety and to manage their financial resources. Because of the traditional presumption of legal incompetence of minors, a set of dispositional principles and criteria has not been included in this chapter. However, the court should impose the least restrictive dispositional alternative which will provide the minor with the protection and assistance which he or she requires. (See §§12(2) and (3) Dispositional . . . Principles and Criteria.)

(4) (a) The court may appoint any individual eligible under section 13(1)(a) of this act to serve as limited personal guardian or personal guardian, and any individual or corporation eligible under that section to serve as limited conservator or conservator for a minor. Nothing in this provision or section 13 shall exclude the appointment of an adult sibling or other close relative of the minor when the conflict of interest is determined to be unsubstantial and such an appointment will clearly be the most beneficial to the minor.

COMMENT. See comment to section 13(1).

(b) The following priorities shall guide the court's selection of a limited personal guardian, personal guardian, limited conservator and/or conservator for a minor from among those eligible:

(i) The individual or corporation proposed by the minor;

(ii) The limited personal guardian, personal guardian, limited conservator and/or conservator appointed or recognized by the appropriate court of the jurisdiction in which the minor resides;

(iii) The individual who or corporation which filed an acceptance of a testamentary designation;

(iv) An individual with whom the minor has been living for more than six (6) months prior to the hearing;

(v) An individual with whom the minor has substantial ties;

(vi) An individual or corporation nominated by any individual or corporation included in subparagraphs (4)(b)(ii)-(v) of this section or listed in the petition or objection; or

(vii) A volunteer public guardian or volunteer public conservator.

The court may select the individual or corporation best qualified and most willing to serve from among those of equal priority. The court may pass over a person or corporation having priority and appoint a person having a lower priority or no priority when it can be demonstrated that the selection will be of greater benefit to the minor. Appointments of volunteer public guardians and volunteer public conservators shall be in accordance with section 36.

COMMENT. Consistent with section 20, the highest preference is given to the individual and/or corporation proposed by the minor. (For an explanation of how this provision is intended to operate, see the comment to section 13(2).)

(5) The dispositional order shall set forth:

(a) The name and address of the individual, if any, appointed to serve as the limited personal guardian or personal guardian for the minor and/or of the individual or corporation, if any, appointed to serve as the limited conservator or conservator of the minor;

(b) The duties and powers of the limited personal guardian or personal guardian for the minor, if any;

(c) The duties and powers of the limited conservator or conservator, for the minor, if any, and the particular financial resources to which they apply;

(d) The terms and conditions of any other dispositional alternatives ordered by the court;

(e) The facts and reasons supporting the selection of the limited personal guardian, personal guardian, limited conservator and/or conservator for the minor; and

(f) The date of the initial review hearing.

The court shall also explain the terms of the order to the minor as well as the procedures set forth in section 25 of this act for terminating or modifying the dispositional order.

COMMENT. This provision requires that the dispositional order delineate the exact terms of the disposition. The order thus becomes the basic reference document setting forth the responsibilities of those agencies, individuals, and corporations charged with protecting and assisting a minor to meet the essential requirements for his or her physical health or safety, to manage his or her financial resources, and to develop his or her abilities to the maximum extent possible.

The provision requires the judge to explain the terms of and basis for the disposition in order to reduce misunderstandings and to assure accountability by providing a basis for review. The judge must also alert the parties to the procedures for requesting a modification or termination of the order. All explanations should be in the language, mode of communication, and terms which the minor is most likely to understand. (§32(2) Notice and Dissemination of Plans and Reports.) Pursuant to sections 24 and 30(2)(j), the initial review hearing must be held no more than 6 months after the dispositional hearing.

Section 24. Periodic Review

(1) Following appointment of a limited personal guardian, personal guardian, limited conservator or conservator for a minor, a review hearing shall be held within the time limit prescribed in section 30(2)(j) to:

(a) Determine whether the duties and powers

set forth in sections 27 and/or 28 are being exercised diligently and in good faith;

(b) Determine whether the dispositional order should be modified or terminated;

(c) Approve payment of compensation and reimbursement of expenses, if any, from the financial resources of the minor; and

(d) Resolve any other relevant matter which the parties raise.

COMMENT. The initial review hearing under this chapter must be held 6 months after a guardianship or conservatorship for a minor has been established and at least annually thereafter. (§30(2)(j) Time Limits.) Such frequent reviews are justified by the rapidly changing needs of minors and by the danger of a child being "'lost' in the foster care system . . . 'in limbo' without a stable placement" after a termination of parental rights because of neglect or abuse.[77]

Like the review hearings prescribed following establishment of a guardianship and/or conservatorship for a partially disabled or disabled person, those required by this section serve a variety of purposes. They provide a means for assuring that a limited personal guardian, personal guardian, limited conservator and/or conservator for a minor is fulfilling his or her responsibilities. They permit the court to ascertain whether the terms of the dispositional order are still adequate and necessary to meet the essential needs for the minor's physical health and safety, to manage his or her financial resources, and to assist the minor to develop his or her abilities to the maximum extent possible. They provide a mechanism for verifying that the minor is receiving the services which were ordered. They act as a means for both controlling the costs of guardianship and/or conservatorship services and guaranteeing that a limited personal guardian, personal guardian, limited conservator and/or conservator receives reasonable compensation and reimbursement for proper out of pocket expenditures on behalf of the minor. Finally, they provide the limited personal guardian, personal guardian, limited conservator or conservator with an opportunity to share problems, learn of new techniques and services, and obtain feedback regarding his or her performance. The review process is closely coordinated with the reports required by sections 27(6) and 28(7).

(2) Following a review hearing the court may issue any of the orders listed in section 15(3) of this act.

COMMENT. Following a review hearing the court may terminate a guardianship or conservatorship; modify any or all of the duties and powers assigned to a limited personal guardian, personal guardian, limited conservator and/or conservator; discharge and/or appoint an individual to serve as the limited personal guardian or personal guardian for the minor and/or individual or corporation to serve as the limited conservator or conservator; modify individual conservatorship plans; direct the provision or cessation of court-ordered assistance; or continue the current disposition until the next review hearing. In issuing an order following a review hearing, the court shall adhere to the priorities, criteria and procedures set forth in sections 23(3)-(5).

(3) Failure to hold a review hearing for appointments made pursuant to this act within the period prescribed in section 30(2)(i) shall result in termination of the guardianship or conservatorship for the minor.

COMMENT. In order to assure that the review procedures are implemented, this section provides an incentive to all parties to make certain that the review hearing is held. Section 15(5) contains a parallel provision for guardianships and conservatorships established for partially disabled and disabled persons.

Section 25. Modification of Dispositional Orders

(1) In addition to the review procedures prescribed under section 24 of this act, the minor, the limited personal guardian, personal guardian, limited conservator, or conservator for the minor, or any other individual interested in the welfare of the minor may apply to the court for the:

(a) Modification or termination of the dispositional order;

(b) Replacement of the limited personal guardian, personal guardian, limited conservator, or conservator;

(c) Approval or disapproval of the compensation or reimbursement for expenses requested by the limited personal guardian, personal guardian, limited conservator, or conservator; or

(d) Resolution of a dispute over the individual conservatorship plan.

COMMENT. Recognizing that questions and problems may arise which cannot wait for resolution until a review hearing, this section provides an avenue for the minor, someone acting in his or her behalf, and the minor's limited personal guardian, personal guardian, limited conservator or conservator to seek immediate relief from the court. Approximately half of the American jurisdictions permit a minor or someone acting in his or her interest to apply to the court for termination of a guardianship or conservatorship. Most statutes on guardianship for minors also provide for removal and replacement of a limited personal

[77] NATIONAL ADVISORY COMMITTEE ON CRIMINAL JUSTICE STANDARDS AND GOALS (NACCJSG), REPORT OF THE TASK FORCE ON JUVENILE JUSTICE AND DELINQUENCY PREVENTION, 476 (1976); TASK FORCE TO DEVELOP STANDARDS AND GOALS FOR JUVENILE JUSTICE AND DELINQUENCY PREVENTION, A COMPARATIVE ANALYSIS OF STANDARDS AND STATE PRACTICES: ABUSE AND NEGLECT, 159-167 (1977).

guardian, personal guardian, limited conservator and/or conservator on the court's own initiative or on motion of the minor or a person interested in his or her welfare. In addition to modification or termination of the dispositional order, and/or removal and replacement of the individual or corporation serving in a guardianship or conservatorship capacity, the provision allows the limited personal guardian, personal guardian, limited conservator and/or conservator to be compensated and reimbursed between review hearings and establishes a means for resolving disputes over the management and control of the minor's financial resources.

(2) (a) Except in accordance with section 35 of this act, applications may not be filed less than thirty (30) days before, or sixty (60) days after a review hearing held under section 24 of this act.

(b) An application from a minor may be communicated to the court by any means, including but not limited to oral communication or informal letter. Upon receipt of such an informal request the court shall contact the minor's attorney of record or appoint new counsel to prepare a formal application.

(c) Applications filed by individuals other than the minor and formal applications filed pursuant to paragraph (2)(b) of this section shall set forth the name, address, and interest of the applicant, if not included below, and the name and address of:

(i) The minor;

(ii) The attorney of record of the minor;

(iii) The limited personal guardian, personal guardian, limited conservator, conservator and guardian *ad litem*, if any, for the minor;

(iv) The individuals most closely related to the minor unless they are unknown or cannot be reasonably ascertained; and

(v) The individual or facility having custody of the minor. In addition, such applications shall specify the relief requested and the facts and reasons supporting the request.

COMMENT. For an explanation of this subsection, see the comments to sections 16 (2)(a)-(c).

(3) Any party may object to the relief requested in an application within the time limit prescribed by section 30(2)(n) of this act.

COMMENT. See comment to section 16(3).

(4) (a) The court shall hold a hearing on all applications:

(i) Which are filed pursuant to section 35 of this act;

(ii) Which request termination of the guardianship or conservatorship of the minor;

(iii) Which request a change in custody of the minor; or

(iv) To which an objection has been filed.

(b) All other applications may be granted or denied without a hearing.

COMMENT. It is anticipated that the relief requested in most applications will not raise substantial questions or adversely affect the minor. Therefore, this section does not require a hearing unless: the minor or another party objects to the application within 15 days after service of the application (see §30(1)(n) Time Limits); the termination of the guardianship or conservatorship is being sought; a shift in the minor's custody is proposed; there is an emergency situation (see §35(1) and (2) Emergency and Standby Appointments); or a limited personal guardian, personal guardian, limited conservator and/or conservator appointed on a standby basis has assumed the guardianship and/or conservatorship duties and powers assigned by the court. (See §35(3) Emergency and Standby Appointments.) In such cases, the hearing must be held within 30 days of the filing of the application or objection. The requirement of a hearing when termination is requested is based on the traditional presumption favoring guardianship for minors. The requirement of a hearing when a shift in the custody of the minor has been requested is to protect against the practice of shuttling children from foster home to foster home which has occurred in many jurisdictions following neglect and abuse proceedings.[78]

At the hearing, the parties are entitled to be represented by counsel, be present at the hearing, subpoena witnesses, present evidence and testimony, cross-examine opposing witnesses, an impartial decisionmaker, and appeal. (See §31 Rights and Privileges.) Pursuant to section 44(2)(b)(iii), both the burden of going forward and the burden of persuasion is on the party filing the application with one exception. When the application is filed by the minor, the burden of persuasion shifts to the party(ies) opposing the minor when he or she presents a *prima facie* case supporting the requested relief.

(5) (a) Within five (5) days after the filing of an application for which a hearing is required under subparagraphs (4)(a)(i)-(4)(a)(iii) of this section, or the filing of an objection, under subsection (3), the court shall have notice served on the individuals and entities specified in section 32(1)(a) in the form and manner prescribed in sections 32(1)(a) and 32(2).

(b) Within five (5) days after the filing of an application for which a hearing is not required unless an objection is filed, the court shall have notice served on the individuals and entities

---

[78] NACCJSG, *supra* note 76, at 496; IJA/ABA, STANDARDS RELATING TO ABUSE AND NEGLECT, *supra* note 76, at §1.6; *see* Katz, *Legal Aspects of Foster Care,* in THE YOUNGEST MINORITY, 72, 85 (S. Katz ed. 1974). For a discussion of the importance to a child of the continuity of a parental type relationship see J. GOLDSTEIN, A. FREUD & A. SOLNIT, BEYOND THE BEST INTERESTS OF THE CHILD (2d ed. 1973).

enumerated in section 32(1)(a) of this act in the form and manner prescribed in sections 32(1)(b) and (2).

COMMENT. See sections 30(2)(k) and (l), and the comments to sections 16(5)(a) and (b).

(6) After notice to the individuals and entities enumerated in section 32(1)(a) of this act in the form and manner prescribed in sections 32(1)(c) and 32(2), the court may join separate applications and separate objections filed under this section, unless joinder would be prejudicial to the interests of the minor.

COMMENT. See comment to section 16(7).

(7) The criteria, priorities and procedures set forth in sections 23(3)-(5) shall apply to dispositional orders issued following submission of an application under this section.

COMMENT. The court may terminate the guardianship or conservatorship for a minor, remove a limited personal guardian, personal guardian, limited conservator or conservator for a minor, or select any of the dispositional alternatives set forth in section 11(2) or any combination thereof, which will enable the minor to meet the essential requirements for his or her physical health or safety, manage his or her financial resources, and develop his or her abilities to the maximum extent possible. If the court is replacing or appointing a new limited personal guardian, personal guardian, limited conservator or conservator, it shall follow the eligibility requirements and priorities set forth in sections 23(4)(a) and (b). The dispositional order following an application for modification or termination shall be presented and explained in the form and manner prescribed by section 23(5).

Section 26. Termination

(1) The guardianship and/or conservatorship of a minor whether established by court order or testamentary designation, shall terminate upon:

(a) The death of the minor;

(b) The minor's [eighteenth birthday/ attaining of majority];

(c) The marriage of the minor;

(d) The adoption of the minor; or

(e) An order of the court.

COMMENT. This section sets forth the four traditional grounds for termination of guardianships and conservatorships for minors. As indicated in Chart XII, the overwhelming majority of American jurisdictions provide for automatic termination when a juvenile attains the age of majority (generally age 18). About 60% of guardianship/ conservatorship statutes provide for termination of a guardianship and conservatorship when a minor marries, presumably because a minor with sufficient maturity to marry no longer requires a guardian or conservator, and because third party control would be detrimental to the formation of a sound marital relationship. In addition, most codes, at least by implication, terminate a guardianship and conservatorship upon the adoption of the minor and following probate of the minor's estate. A court order terminating a guardianship may be issued following a review hearing pursuant section 24, or following submission of an application pursuant to section 25.

(2) Within sixty (60) days of a minor's [eighteenth birthday/attaining of majority], an intervention proceeding may be initiated under section 6 of this act if it appears that upon attaining majority, the minor will be a partially disabled or disabled person.

COMMENT. Since the circumstances in which children who have a developmental disability or are mentally ill require a guardian or conservator are no different from those in which other children may require such protection and assistance, chapter 2 refers only to partially disabled and disabled adults and this chapter makes no distinction between children with disabilities and those without. This subsection provides the procedural link between the two systems of guardianship and conservatorship. Without it, either two types of guardianships and/or conservatorships for minors would have to be created or the parent or guardian of a disabled minor would have to file an application for an emergency appointment on the day the minor attains majority. (See §35 Emergency and Standby Appointments.) The 60 day period should permit the intervention proceeding to be completed without unnecessary haste. (See §30 Time Limits.)

Introduction to Section 27

This section sets forth the general duties which inhere to the guardianship of a minor, the specific duties and powers which may be assigned to limited personal guardians and personal guardians for minors, the limits on the authority of limited personal guardians and personal guardians for minors, and the applicable reporting requirements. It is generally parallel to section 17 on the duties and powers of limited personal guardians and personal guardians for partially disabled and disabled persons. However, unlike section 17, there is no requirement for an individual guardianship plan. Although such plans may be useful when there is a clearly defined problem or objective toward which they can be directed — e.g. in conjunction with an overall dispositional plan when the parental rights of the parents of an abused or neglected child have been temporarily suspended[79] — they appear to be of less value in the case of a non-disabled child for

[79] IJA/ABA STANDARDS RELATING TO ABUSE AND NEGLECT, *supra* note 76, at §6.5.

whom a limited personal guardian or personal guardian has been appointed following the death of his or her parents. While nothing in the act prohibits the court from requiring an individual guardianship plan when it concludes that such a plan is a necessary safeguard for the minor, some states may wish to add a provision to this section making that authority explicit. Section 28 covers the duties and powers of limited conservators and conservators for minors.

Section 27. Duties and Powers of Limited Personal Guardians and Personal Guardians for Minors

(1) It is the general duty of individuals appointed to serve as a limited personal guardian or personal guardian for a minor to carry out diligently and good faith, the specific duties and powers assigned by the court. In carrying out these duties and powers, such individuals shall:

(a) Assure that the personal, civil and human rights of the minors whom they are serving are protected; and

(b) Encourage those minors to:

(i) Participate in all decisions which affect them; and

(ii) Develop their abilities to the maximum extent possible.

COMMENT. This subsection makes clear that accepting the guardianship of a minor carries with it the obligation for developing a personal relationship with the minor, helping to prepare the minor for taking on the responsibilities of adulthood, and ensuring that the minor receives the services, benefits and rights to which he or she is entitled. The provision also sets a high standard of performance for limited personal guardians and personal guardians but does not make them strictly liable for errors in judgment. Making limited guardians liable for any mistake would constitute a significant deterrent to accepting an appointment. It is anticipated that requiring diligence and good faith will provide adequate protection for minors against malfeasance without discouraging eligible individuals from accepting guardianship appointments.

(2) (a) A limited personal guardian for a minor may be assigned any portion of the duties and powers listed in paragraphs (3)(a)-(g):

(i) In those particular areas in which, or for that period of time during which, the parental rights of the parent of the minor have been suspended; or

(ii) In those particular areas in which a minor, whose parents are dead or have had their parental rights terminated or suspended, requires assistance to meet the essential requirements for his or her physical health or safety.

(b) A limited personal guardian may also be assigned the duty to assist the minor in those particular areas in which the minor's ability to meet the essential requirements for his or her physical health or safety, and/or to protect his or her rights, and/or to obtain necessary services, and/or to fulfill his or her civil duties is impaired, as well as in other ways not specifically prohibited by subsections (4) and (5) of this section.

COMMENT. It is anticipated that limited personal guardianships will be used primarily in two circumstances. The first is when some or all parental rights have been temporarily suspended, e.g., when a neglected or abused child is temporarily removed from his or her home or when a single parent is hospitalized or incarcerated for a significant time. The second is when the minor who is subject to guardianship is nearing majority and is capable of meeting at least some of the requirements for his or her physical health or safety. In these instances, only those duties and powers should be assigned to the limited personal guardian which are necessary to protect the minor. Among the powers which may be assigned to a limited personal guardian for a minor is the authority to take custody of the minor. (*Contra* §17(3) Duties and Powers of Limited Personal Guardians and Personal Guardians (for partially disabled and disabled persons).) A limited personal guardian may also be directed to assist a minor in areas in which he or she has not been authorized to act in the minor's behalf. For example, when custody has not been granted to the limited personal guardian for a minor age 17, the limited personal guardian may be directed to assist the minor in locating and maintaining suitable housing.

(3) The specific duties and powers assigned to a personal guardian for a minor by the court shall be those of a parent, whose parental rights have not been terminated or suspended, toward his or her child, including but not limited to:

(a) Taking custody of the minor subject to any limitations and visitation privileges prescribed by the court;

(b) Assuring that the minor receives adequate food, clothing, health and other professional care, shelter, and education;

(c) Assuring through the initiation of court actions and other means, that the minor enjoys the personal and civil rights to which he or she is entitled;

(d) Providing any required consents or approvals on behalf of the minor, including consent to the marriage of the minor or enlistment of the minor in the armed services;

(e) Assisting the minor to the maximum extent possible, to learn and assume the duties and responsibilities of a productive, law-abiding citizen;

(f) Receiving money payable for the support

of the minor if a separate limited conservator or conservator has not been appointed; and

(g) Expending sums from the financial resources of the minor to cover the reasonable costs incurred on behalf of the minor.

COMMENT. This subsection attempts to delineate the primary duties and powers which a personal guardian of a minor must exercise. Each of the specific powers listed should be read in the context of the general duties contained in subsection (1) and the limitations on authority set forth in subsections (4) and (5) of this section.

(4) If a separate limited conservator or conservator has been appointed for the minor, then the receipt and expenditure of funds by the limited personal guardian or personal guardian shall be consistent with the duties assigned to and procedures and policies established by the limited conservator or conservator.

COMMENT. Even when a conservatorship has been established for a minor, it will be necessary for the limited personal guardian or personal guardian to expend funds on behalf of a minor during the course of everyday activities. This provision in conjunction with section 28(2) seeks to facilitate the establishment of a smooth working relationship between the individual with responsibility for the minor's physical health and safety and the individual or corporation with responsibility for that minor's financial resources. Disputes over the exercise of this authority which cannot be resolved by the parties may be submitted to the court pursuant to section 25. Nothing in this section is intended to prohibit the appointment of the same individual as the limited personal guardian or personal guardian and as the limited conservator or conservator for the minor when both guardianship and conservatorship are warranted.

(5) A limited personal guardian or personal guardian shall not be:

(a) Obligated to provide for the minor from his or her own funds;

(b) Liable to third persons by reason of the parental-type relationship for acts of the minor; and

(c) Empowered to take or consent to any of the actions which are prohibited for limited personal guardians for partially disabled and disabled persons by sections 17(5)(a)-(c), (f) and (g).

COMMENT. In accord with section 5-209 of the U.P.C., this provision makes clear that unlike an adoptive parent, a limited personal guardian, or personal guardian for a minor is not required to support that minor from his or her own pocket. The subsection also imposes the same limitations on the authority of limited personal guardians and personal guardians for minors as are imposed on limited personal guardians and personal guardians for partially disabled or disabled adults. These limitations prohibit placing a minor in a facility or institution to which an individual without a guardian has to be committed, other than through a formal commitment proceeding in which the minor has independent counsel and a separate guardian *ad litem;* consenting on behalf of a minor to an abortion, sterilization, psychosurgery, or removal of a bodily organ, except as provided by law or when necessary to preserve the life or prevent serious impairment of the physical health of the minor; consenting on behalf of a minor to the withholding of non-heroic life-saving medical procedures; and consenting on behalf of a minor to termination of the minor's parental rights. In addition, unless a limited personal guardian or personal guardian has also been appointed limited conservator or conservator for a minor, he or she cannot manage the minor's financial resources. An explanation of these restrictions is contained in the comments to sections 17 (5)(a)-(d), (f) and (g). (See also the comment to subsection (4) of this section.) This provision is not intended to limit the right of a minor to consent to certain medical procedures to the extent that such a right is provided under state law.[80] (See §38(2) Effect of . . . a Guardianship or Conservatorship for a Minor.)

Three of the express restrictions on limited personal guardians and personal guardians for partially disabled and disabled adults are not made applicable to guardianship for minors. First, the overall danger to non-institutionalized minors of permitting their appointed limited personal guardian or personal guardian to consent to their participation in a biomedical or behavioral experiment is sufficiently remote that the strict procedural safeguards imposed by section 17(5)(d) appear to be unwarranted. (The participation of institutionalized youngsters in biomedical or behavioral experiments is covered in the report in this series on rights in residential facilities.) Second, since parental or court consent is required in all states for a person below a specified age to marry,[81] guardianships for minors, may include the power to prohibit the marriage of a minor. (See §17(5)(e) Duties and Powers of Limited Personal Guardians and Personal Guardians (for partially disabled or disabled persons).) Finally, no limitation on the power of a limited personal guardian or personal guardian to prohibit a divorce is required since a guardianship, for minors terminates automatically upon the minor's marriage. (See §§26 Termination; *cf.* §17(5)(e) Duties and Powers of Limited Personal Guardians and Personal Guardians (for partially disabled or disabled persons).)

(6) (a) A limited personal guardian or personal

---

[80] BEASER, *supra* note 8, at 205-223.

[81] *Id.* at 302.

guardian for a minor shall submit a report to the court at least ten (10) days before the initial review hearing and:

(i) Annually thereafter;

(ii) When the court orders additional reports to be filed;

(iii) When the limited personal guardian or personal guardian resigns or is removed; or

(iv) When the guardianship is terminated.

COMMENT. This subsection requires limited personal guardians and personal guardians of minors to file annual reports to the court which appointed them or, if they are serving as a result of an uncontested testamentary designation, the court which probated the will. Like the reports required by section 17(6), reports filed under this subsection serve several purposes. They provide the individual serving in a guardianship capacity with a means for obtaining the advice and assistance needed to carry out his or her duties more effectively, for protecting himself or herself by reporting problems to the court before allegations of malfeasance or misfeasance arise, as well as for requesting reasonable compensation and reimbursement for necessary expenses. Together with the review hearings required by section 24, the reports provide the court, the minor, and other persons interested in the minor's welfare with a mechanism for holding the limited personal guardian or personal guardian accountable. Finally, they provide a means for tailoring the guardianship to meet the individual needs and abilities of the minor.

The court may order additional reports when necessary in light of these purposes and the well-being of the minor.

(b) The report shall set forth:

(i) The name and address of the minor, the limited personal guardian or personal guardian, and the individual or facility having custody of the minor;

(ii) Any special services required by and/or being provided to the minor;

(iii) Changes in the residence or custody of the minor within the reporting period and the reason therefor;

(iv) The significant actions taken by the limited personal guardian or personal guardian during the reporting period;

(v) The compensation requested and the reasonable and necessary expenses incurred by the limited personal guardian or personal guardian on behalf of the minor;

(vi) Any significant problems which have arisen during the reporting period;

(vii) Any changes which should be made in the duties and powers of the limited personal guardian or personal guardian; and

(viii) An accounting of the expenditures made and funds received for the minor by the limited personal guardian or personal guardian.

COMMENT. The topics which should be addressed in the report are listed specifically in order to clarify for the limited personal guardian or personal guardian what is expected of him or her and to assure that the purposes described in the comment to the preceding paragraph are achieved. The list of topics is designed to provide more than *pro forma* administrative information. It requires a description of the important occurrences during the reporting period including the significant actions of the limited personal guardian or guardian. Because of the importance to minors of developing and maintaining long-term, close, family type relationships,[82] and the current incidence of multiple foster care placements for children removed from their parent's home,[83] the report must contain an explanation of any change in the child's custody. Nothing in this paragraph is intended to require the report to be typed or in legal form. Informal handwritten reports maybe submitted so long as they cover each of the required topics.

(c) The court shall have copies of the report sent to the individuals and entities enumerated in section 32(3)(b) of this act.

COMMENT. A copy of all reports filed pursuant to this section must be provided to the minor; his or her attorney of record; his or her guardian *ad litem,* limited conservator and/or conservator if any; the individuals most closely related to the minor unless they are unknown and their names and/or whereabouts cannot reasonably be ascertained; the individual or facility having custody of the minor; and the Administrator of Veterans' Affairs if the financial resources of the minor include V.A. benefits. The contents of the report must be explained to the minor in the language, mode of communication, and terms which the minor is most likely to understand. (§§32(3)(b) and (c) Notice and Dissemination of Plans and Reports.)

Introduction to Section 28

This section sets forth the general duties of limited conservators and conservators for minors, the requirements for individual conservatorship plans, the specific duties and powers which may be assigned to limited conservators and conservators for minors, the limits on the authority of limited conservators and conservators for minors, and the reporting requirements applicable to such conservatorships. Section 27 covers the same range of topics with regard to guardianship for minors.

---

[82] GOLDSTEIN, FREUD & SOLNIT, *supra* note 78.

[83] See authorities listed in note 77.

Section 28. Duties and Powers of Limited Conservators and Conservators for Minors.

(1) It shall be the responsibility of the limited conservator or a conservator for a minor to perform diligently and in good faith the specific duties and powers assigned by the court. In carrying out these duties and powers, limited conservators or conservators for minors shall:

(a) Manage the financial resources placed under their supervision and/or control as would a prudent person manage his or her own financial resources; and

(b) Encourage the minors whom they are serving to:

(i) Participate in all decisions which affect them; and

(ii) Develop their abilities to the maximum extent possible.

COMMENT. This subsection makes clear that in addition to prudently managing the particular financial resources subject to conservatorship, limited conservators and conservators for minors must encourage and assist the minor to learn how to manage those financial resources and to develop the capacity to be an independent, responsible adult. The limited conservator or conservator should also make certain that the minor receives all the benefits, services and rights to which he or she is entitled. Limited conservators or conservators are not strictly liable for harm caused to the minor because of their actions or omissions. It is anticipated that the standard of performance set forth in the subsection will provide adequate protection for minors against malfeasance or misfeasance without discouraging eligible individuals and corporations from accepting conservatorship appointments.

(2) A limited conservator or a conservator for a minor shall file an individual conservatorship plan within the time limits prescribed in section 30(2)(i). The plan shall be developed together with the limited personal guardian or personal guardian, if any, and with the minor to the maximum extent possible. It shall cover the matters prescribed for individual conservatorship plans for partially disabled or disabled persons by section 18(2) of this act and shall include a complete inventory of the financial resources under the supervision or control of the limited conservator or conservator. The court shall have copies of the individual conservatorship plan sent to the individuals and entities enumerated in section 32(3)(b) of this act.

COMMENT. In order to reduce the likelihood of misunderstandings and disagreements in an area traditionally subject to dispute, and to protect the minor against waste or misappropriation of his or her financial resources, this subsection requires preparation and submission of an individual conservatorship plan and a complete inventory of the minor's assets. The imposition of an across the board requirement for an individual conservatorship plan appears far less onerous than would a similar requirement in the context of guardianship, since the objectives for financial management are more easily defined than those for guardianship for minors, and an inventory would have to be prepared in any case, (See introduction to §27 Duties and Powers of Limited Personal Guardians and Personal Guardians for Minors.) The requirement for a prompt inventory of the assets subject to conservatorship is consistent with the U.P.C.,[84] the conservatorship statutes of many states, and sound financial management practices.

The individual conservatorship plan is intended to outline for the minor, his or her limited personal guardian or personal guardian if any, and the court, just how the particular financial resources placed under the limited conservator's or conservator's supervision or control are to be managed. In order to avoid misunderstandings and to simplify the management of the minor's financial resources, the plan must contain the policies and procedures governing the expenditure of funds by the minor and his or her limited personal guardian or personal guardian, if any. In order to facilitate the meeting of the essential requirements for the minor's physical health and safety, the development of the minor's abilities, and to foster a smooth working relationship, the plan must also reflect the agreement reached by the limited conservator or conservator, the limited personal guardian or personal guardian, if any, and the individual they are serving, regarding how the decision-making authority will be shared. Disputes over the agreements, policies and procedures to be included in the plan which cannot be resolved by the parties, may be submitted to the court pursuant to section 25.

The initial plan and inventory are due 60 days after the conservatorship has been established and must be updated in accordance with the reporting requirements in section 28(7). A copy of the individual conservatorship plan must be provided to the minor; his or her attorney of record; his or her limited personal guardian, personal guardian and/or guardian *ad litem* if any; the individuals most closely related to the minor unless their names and whereabouts are unknown and cannot reasonably be ascertained; the individual or facility having custody of the minor; and the Administrator of Veterans' Affairs if the financial resources of the minor include V.A. benefits. The contents of the plan must be explained to the minor in the language, mode of communication, and terms which the minor is most likely to under-

---

[84] U.P.C., *supra* note 2, at §5-418.

stand. (§§32(3)(b) and (c) Notice and Dissemination of Plans and Reports.)

(3) (a) A limited conservator or conservator for a minor shall expend or distribute, and/or authorize the expenditure or distribution of, and/or assist in the expenditure or distribution of income or principal from the financial resources of the minor under his, her or its supervision and control to assure that:

(i) The essential requirements for the physical health or safety of the minor and any dependents of the minor are met;

(ii) The rights of the minor are protected;

(iii) The financial resources of the minor are prudently managed;

(iv) The minor has the opportunity to develop his or her abilities to the maximum extent possible; and

(v) The limited personal guardian or personal guardian, if any, for the minor is able to carry out the duties and powers assigned by the court or set forth in the will.

(b) In so doing, a limited conservator or conservator shall consider:

(i) The requests of the minor and the limited personal guardian or personal guardian for the minor;

(ii) The size of the financial resources of the minor;

(iii) The probable duration of the conservatorship; and

(iv) The accustomed standard of living of the minor and individuals legally dependent on the minor.

In addition, limited conservators shall consider the size of the financial resources of a minor which have not been placed under their supervision and control.

COMMENT. An explanation of paragraphs (3)(a) and (3)(b) of this section is contained in the comments to sections 18(3)(a) and (b) regarding conservatorships for partially disabled and disabled persons.

(c) Funds under this subsection may be paid by the limited conservator or conservator to any individual, including the minor.

COMMENT. None necessary.

(d) A limited conservator or conservator shall establish policies and procedures for the expenditure of funds by the minor, or by the limited personal or personal guardian for the minor if a separate individual has been appointed to serve in that capacity, other than those funds payable under section 29.

COMMENT. It is anticipated that the development of clear policies and procedures for the expenditure of assets under the control or supervision of a limited conservator or conservator will facilitate development of a smooth working relationship between the limited conservator, the minor, and the minor's limited personal guardian or personal guardian, if any. Section 29 governs the delivery, receipt and use of funds or personal property under $5,000 in value to which a minor is entitled.

(4) (a) The court may assign to a limited conservator for a minor any of the duties and powers listed in section 18(5)(a) which the minor is unable to perform. Such assignments shall specify the financial resources to which they apply.

(b) The court may assign to a limited conservator the duty of assisting a minor to perform any of the functions listed in section 18(5)(a) of this act with regard to any financial resources of the minor.

(c) If the minor dies during the conservatorship, the limited conservator shall have the duties and powers prescribed in section 18(5)(b) of this act.

COMMENT. This subsection governs the specific areas in which a limited conservator of a minor may be given the authority to act on the minor's behalf, as well as those areas in which a limited conservator may be directed to assist in the minor's efforts to manage his or her own resources. It is anticipated that a limited conservatorship for a minor will usually be established in two situations. The first is when a minor inherits or is given a particular financial resource which he or she is unable to manage. The second is when a minor who is nearing majority and whose parents are dead or have had their parental rights terminated or suspended, is capable of managing at least some of his or her financial resources. In these instances, only those duties and powers should be assigned to a limited conservator which are necessary to protect the minor. (See also the comments to §§18(4)(a)-(b) and 18(5)(a)-(b) Duties and Powers of Limited Conservators and Conservators (for partially disabled and disabled persons).)

(5) The court shall assign to a conservator for a minor the duties and powers of conservators for partially disabled or disabled persons as set forth in section 18(5) of this act.

COMMENT. See comments to sections 18(5)(a)-(b).

(6) The prohibitions which apply to limited conservators and conservators for partially disabled and disabled persons under section 18(6) of this act shall also apply to limited conservators and conservators appointed to serve minors.

COMMENT. There appears to be no basis for imposing greater or fewer restrictions on the powers of a limited conservator or conservator for a minor than for a limited conservator or conservator for a partially disabled or disabled person. An explanation of the limitations imposed by this subsection is contained in the comments to sections 18(6)(a)-(c). Where reference is made to notifying the partially disabled or disabled person, notice should be sent to a minor under this provision.

(7) (a) A limited conservator or conservator for a minor shall submit a report to the court at least ten (10) days before the initial review hearing, and:

(i) Annually thereafter;

(ii) When the court orders additional reports to be filed;

(iii) When the limited conservator or conservator resigns or is removed; and

(iv) When the conservatorship is terminated.

COMMENT. Most states require limited conservators and conservators to submit periodically, an accounting of the financial resources for which they are responsible. This provision requires more than a financial statement. Like the reports filed by limited personal guardians and personal guardians pursuant to section 27(6)(a), reports filed under this section are intended, in conjunction with the review hearings held pursuant to section 24, to alert individuals and entities with an interest in the case to major changes and to assure that: the terms of a conservatorship are consistent with the needs and abilities of the minor; the minor is receiving the services identified or contemplated in the dispositional order; and the limited conservator or conservator is exercising diligently and in good faith, the duties and powers which have been assigned by the court. (See also §18(7)(a) Duties and Powers of Limited Conservators and Conservators (for partially disabled or disabled persons).)

(b) The report shall set forth:

(i) The name and address of the minor and the limited conservator or conservator;

(ii) A complete financial statement of the financial resources under the control and/or supervision of the limited conservator or conservator;

(iii) The services being provided to the minor and the relationship of those services to the individual conservatorship plan;

(iv) The significant actions taken by the limited conservator or conservator during the reporting period;

(v) The compensation requested and the reasonable and necessary expenses incurred by the limited conservator or conservator;

(vi) Any significant problems relating to the conservatorship which have arisen during the reporting period; and

(vii) Any changes which should be made in the duties and powers of the limited conservator or conservator.

Attached to the report shall be an updated individual conservatorship plan developed by the limited conservator or conservator, the limited personal guardian or personal guardian if any, and, to the maximum extent possible, the minor.

COMMENT. See comment to section 27(6)(b).

(c) The court shall have copies of the report sent to the individuals and entities enumerated in section 32(3)(b) of this act.

COMMENT. A copy of the report must be sent by the court to the minor; his or her attorney of record; his or her limited personal guardian, personal guardian, and/or guardian *ad litem* if any; the individuals most closely related to the minor unless they are unknown and their names and whereabouts cannot be reasonably ascertained; the individual or facility having custody of the minor; the Administrator of Veterans' Affairs if the financial resources of the minor include V.A. benefits; and the surety, if any, on the bond for the limited conservator or conservator. The report should be explained to the minor in the language, mode of communication, and terms which he or she is most likely to understand. (§§32(3)(b) and (c) Notice and Dissemination of Plans and Reports.)

(d) Following submission of a report or in conjunction with a review hearing pursuant to section 24 or a proceeding initiated under section 25 of this act, a conservator shall submit to a physical check of the financial resources placed under his, her or its control, in the manner specified by the court.

COMMENT. See comment to section 18(7)(d).

Section 29. Payment or delivery of funds or personal property

(1) Any individual or entity under a duty to pay or deliver money or personal property to a minor may perform this duty, in amounts not exceeding five thousand dollars ($5,000) per annum, by paying or delivering the money or property to:

(a) The minor, if he or she is age sixteen (16) or over or is married;

(b) The individual who has custody of the minor;

(c) The limited personal guardian or personal guardian for the minor; or

(d) A financial institution incident to a deposit in a federally insured savings account in the sole name of the minor. Notice of such a deposit shall be given to the minor.

This section does not apply if the individual or entity making the payment or delivery has actual

knowledge that a conservator has been appointed or that proceedings to establish a conservatorship are pending.

(2) Individuals, other than the minor, receiving money or property for a minor, are obligated to use the money for the purposes set forth in section 28(3)(a) of this act, but may not pay themselves except by way of reimbursement for out-of-pocket expenses for goods and services necessary for the minor's support. Any excess sums shall be preserved for future support of the minor and any balance not so used and any property received for the minor must be turned over to the minor when he or she attains majority.

(3) Individuals or entities who pay or deliver money or personal property in accordance with provisions of this section are not responsible for the proper application thereof.

COMMENT. This section is based upon section 5-103 of the U.P.C. It provides an alternative to establishing a conservatorship when the financial resources of a minor are relatively small and uncomplicated. Subsection (2) offers explicit directions regarding the use of the funds or property paid or delivered under subsection (1). Because the value of funds and property are small, and payment or delivery under this section does not require explicit court authorization, no reporting requirement is imposed. When greater controls are necessary for the limited financial resources of a minor, establishment of a trust or, if necessary, a conservatorship should be considered. (See §23 Dispositional Alternatives.)

CHAPTER 4 GENERAL PROVISIONS

Section 30. Time Limits

(1) The following time limits shall apply to proceedings initiated under chapter 2 of this act:

(a) A date for a hearing shall be set and notice shall be served no more than five (5) days after the filing of a petition pursuant to section 6(1);

(b) Counsel shall be appointed for the subject of an intervention proceeding pursuant to section 31(2) of this act no more than five (5) days after the filing of a petition under sections 6(1), unless the court receives notice that independent counsel has been retained by or on behalf of the subject of the intervention proceeding, or unless an immediate appointment would be in the interests of justice;

(c) A multidisciplinary evaluation team shall be appointed pursuant to section 8(1) of this act no more than ten (10) days after the filing of a petition;

(d) An evaluation report shall be submitted pursuant to section 8(4) of this act no more than twenty (20) days after the filing of a petition;

(e) A date for a hearing shall be set and notice served no more than five (5) days after the filing of a motion to dismiss pursuant to section 7 of this act;

(f) A hearing shall be held on a motion to dismiss no more than fifteen (15) days after the motion has been filed;

(g) A prehearing conference shall be held pursuant to section 9 of this act no more than thirty (30) days after the filing of a petition;

(h) A hearing shall be held pursuant to section 10 of this act no more than thirty-five (35) days after the filing of a petition;

(i) A dispositional hearing shall be held pursuant to section (12)(1) of this act no more than ten (10) days after a finding that an individual is a partially disabled or disabled person;

(j) The initial individual guardianship plan shall be submitted pursuant to section 17(2)(b) of this act no more than thirty (30) days after the limited personal guardian or personal guardian has been appointed. The plan shall be updated in accordance with the reporting requirements set forth in section 17(6);

(k) The initial individual conservatorship plan shall be submitted pursuant to section 18(2)(b) of this act no more than sixty (60) days after the limited conservator or conservator has been appointed. The plan shall be updated in accordance with the reporting requirements set forth in section 18(7);

(l) A review hearing shall be held pursuant to section 15(1)(a) no more than one (1) year after appointment of a limited personal guardian and/or limited conservator, and biennially thereafter;

(m) A review hearing shall be held pursuant to section 15(1)(b) no more than six (6) months after appointment of a personal guardian and/or conservator, and annually thereafter;

(n) Pursuant to section 16(5)(a) of this act, a date for a hearing shall be set and notice shall be served no more than five (5) days after the filing of an objection under section 16(3) or after the filing of an application for which a hearing is required under sections 16(4)(a)(i) and (ii);

(o) Pursuant to section 16(5)(b) of this act, notice shall be served no more than five (5) days after the filing of an application for which a hearing is not required under sections 16(4)(a)(i) and (ii);

(p) An objection to an application shall be filed pursuant to section 16(3) no more than fifteen (15) days after service of the notice required by section 16(5);

(q) If a hearing is required pursuant to sec-

tions 16(4)(a)(i) or (ii) of this act, it shall be held no more than thirty (30) days after the application has been filed. If a hearing is required pursuant to section 16(4)(a)(iii) of this act, it shall be held no more than thirty (30) days after the objection is filed.

(2) The following time limits shall apply to proceedings initiated under chapter 3 of this act:

(a) A date for a hearing shall be set and notice shall be served no more than five (5) days after the filing of a petition pursuant to section 19(1) of this act, the filing of an objection to a testamentary designation pursuant to section 20, or the expiration of the period for filing an acceptance of a testamentary designation if a written acceptance has not been filed;

(b) Acceptance of a testamentary designation shall be filed pursuant to section 19(2)(b) of this act no more than thirty (30) days after the will containing the designation has been filed for probate;

(c) Notice shall be filed no more than five (5) days after the filing of an acceptance of an testamentary appointment pursuant to section 19(2)(b) of this act;

(d) Counsel shall be appointed for the minor pursuant to section 31(2) of this act no more than five (5) days after the filing of a petition pursuant to section 19(1), the filing of an acceptance of a testamentary designation pursuant to section 19(2)(b), or the expiration of the period for filing an acceptance of a testamentary designation;

(e) An objection to a testamentary designation shall be filed pursuant to sections 20(1) or (2) no more than thirty (30) days after service of the notice required by section 19(2)(c);

(f) A date for a hearing shall be set and notice served no more than five (5) days after the filing of a motion to dismiss pursuant to section 21 of this act;

(g) A hearing shall be held on a motion to dismiss no more than fifteen (15) days after the motion has been filed:

(h) A hearing shall be held pursuant to section 22 of this act no more than;

(i) Thirty (30) days after the filing of a petition; or

(ii) Thirty (30) days after the filing of an objection to a testamentary designation; or

(iii) Fifteen (15) days after the expiration of the period for accepting a testamentary designation if a written acceptance has not been filed;

(i) The initial individual conservatorship plan for a minor shall be submitted no more than sixty (60) days after the limited conservator or conservator is appointed by the court or ninety (90) days after the acceptance of a testamentary designation if no objection has been filed. The plan shall be updated in accordance with the reporting requirements set forth in section 28(7);

(j) A review hearing shall be held pursuant to section 24 of this act no more than six (6) months after a limited personal guardian, personal guardian, limited conservator and/or conservator has been appointed for a minor and at least annually thereafter;

(k) Notice shall be served no more than five (5) days after an application has been filed pursuant to section 25 of this act;

(l) Pursuant to section 25(5)(a) of this act, a date for a hearing shall be set and notice shall be served no more than five (5) days after the filing of an objection under section 25(3) or after the filing of an application for which a hearing is required under sections 25(4)(a)(i)-(iii);

(m) Pursuant to section 25(5)(b) of this act, notice shall be served no more than five (5) days after the filing of an application for which a hearing is not required under sections 25(4)(a)(i)-(iii);

n) An objection to an application shall be filed pursuant to section 25(3) of this act no more than fifteen (15) days after service of the notice required by section 25(5);

(o) If a hearing is required pursuant to sections 25(4)(a)(i)-(iii) of this act, it shall be held no more than thirty (30) days after the application has been filed. If a hearing is required pursuant to section 25(4)(a)(iv) of this act, it shall be held no more than thirty (30) days after the objection is filed.

(3) The following time limits shall apply to proceedings for appointment of a guardian *ad litem:*

(a) Notices required by section 34(2) of this act shall be served no more than five (5) days after the filing of an application for appointment of a guardian *ad litem;*

(b) Objections to the appointment of a guardian *ad litem* or to the individual proposed for appointment shall be submitted no more than ten (10) days after service of the notice required by section 34(2);

(c) A date shall be set for the informal, *in camera* hearing required by section 34(4)(b) of this act and notice shall be served no more than five (5) days after an objection has been filed pursuant to section 34(3);

(d) An informal, *in camera* hearing shall be held pursuant to section 34(4)(b) no more than ten (10) days after an objection has been filed.

(4) The following time limits shall apply to proceedings for emergency and standby appointments:

(a) Notices required by section 35(1)(c) of this act shall be served no more than twenty-four (24) hours after the filing of an application pursuant to section 35(1)(a);

(b) A hearing shall be held pursuant to section 35(1)(d) of this act no more than forty-eight (48) hours after an application for an emergency appointment has been filed;

(c) An application shall be filed pursuant to section 35(3) of this act within ten (10) days after an individual appointed on a standby basis assumes guardianship and/or conservatorship duties.

(5) A continuance may be granted:

(a) Upon a stipulation by the parties;

(b) Pursuant to section 9 of this act; or

(c) For good cause shown following a written motion, notice to the individuals and entities enumerated in section 32(1)(a) of this act in the form and manner prescribed in section 32(1)(e) and 32(2), and a hearing if an objection to the continuance is filed or on the court's own initiative.

(6) For purposes of this act all time limits shall be calculated in terms of calendar days. In computing any period of time, the day of the act or event from which the designated period of time begins to run is not to be included. The last day of the period so computed shall be included.

COMMENT. Since an intervention proceeding is likely to be an anxiety producing event for all concerned, both the petitioner and the subject of the intervention proceeding will usually wish to have the issues and uncertainty resolved as quickly as possible. On the other hand, intervention proceedings affect fundamental rights often involve complex and sometimes conflicting diagnoses, and require difficult decisions. Therefore, they ought not to be conducted in a cursory manner.[85] Each party requires time to prepare adequately. Accordingly, this section sets stringent maximum time limits for intervention proceedings. Subsection (1) governs intervention proceedings involving individuals alleged or found to be partially disabled or disabled persons. Subsection (2) governs intervention proceedings for minors. Subsections (3) and (4) cover proceedings for appointment of guardians *ad litem* and proceedings for emergency appointments respectively. The Appendix contains charts which illustrate the sequence and timing of the proceedings.

For intervention proceedings under chapter 2, the maximum time which can elapse between the filing of the petition and the issuance of a dispositional order following a finding that an individual is a partially disabled or disabled person is 45 days, assuming a one day hearing and no continuances. During this period, notice must be provided to all relevant individuals and entities, an evaluation must be conducted, and substantive and dispositional hearings must be held. Because of the absence of an evaluation and of a separate dispositional hearing, an intervention proceeding under chapter 3 can reach the dispositional stage within 30 days after the filing of a petition or of an objection to a testamentary designation, and within 15 days if the period for accepting a testamentary designation expires without an acceptance being filed. It should be noted that these are maximum limits. When possible and necessary, intervention proceedings can be held in less time.

When the parties agree that additional time is required or when good cause is shown that a delay is necessary, subsection (5) authorizes the court to extend the prescribed time limits. This authority should be exercised only when a delay would not be detrimental to the subject of the intervention proceeding.

Subsection (6) sets forth the guidelines for the computation of time periods under this act. In keeping with the effort to avoid unnecessary delay to the maximum extent possible, weekends and legal holidays are not excluded from time period computations. In states in which it would be impossible because of size and caseload to complete intervention proceedings within the prescribed limits, consideration should be given to computing the time limits in terms of court days rather than changing the periods themselves.

In order not to add further length and complexity to the section, no time limits are prescribed for the submission of formal responses to pleadings filed under this act — e.g., petitions, applications, objections, or motions. Some states may wish to set such limits by rule or statute. In others, the existing rules of civil procedure will be sufficient. (See §45 Rules of Evidence/Rules of Procedure.) As with all time limits, the geographic, demographic and administrative realities in each state must be considered in coordinating the response period with the overall time limits set in this section.

Introduction to Section 31

This section sets forth the rights and privileges afforded the parties in all intervention proceedings. Paragraph (1)(a) lists the basic due process guarantees to which all parties are entitled, regardless of whether the proceeding was initiated under chapter 2 or chapter 3 of this act. Paragraphs (1)(b)-1(d) include the rights which are applicable only to the subject of the intervention proceeding or the initiator of the proceeding. Paragraph (1)(e) deals with the difficult question of waiver. Subsections (2) and (3) contain mandatory provisions to protect the integrity of intervention proceedings. Subsection (4) contains three additional safeguards. The first clarifies that all statements made during evaluations and treatment by

---

[85] *See* R. ALLEN, E. FERSTER, & H. WEIHOFEN, MENTAL IMPAIRMENT AND LEGAL INCOMPETENCY, 82 (1968).

individuals alleged or found to be a partially disabled or disabled person are privileged. The second assures that a decision that an individual is a partially disabled or disabled person is not unduly influenced by the effects of psychotropic drugs which have been administered to that individual. The third permits hearings to be held outside the courtroom in order to make it easier for the subject of intervention proceeding to exercise his or her right to be present.

Section 31. Rights and Privileges

(1) (a) In all proceedings under this act the parties shall have a right to:

(i) Prior notice;

(ii) Counsel;

(iii) Be present;

(iv) Compel the attendance of witnesses;

(v) Present evidence and confront and cross-examine witnesses, including the multidisciplinary evaluation team appointed pursuant to section 8 of this act;

(vi) An impartial decision-maker; and

(vii) Appeal adverse orders and judgments in the manner prescribed in (the Rules of Civil Procedure).

COMMENT. In *The Mentally Retarded and the Law,* Michael Kindred observes that "a decision to establish guardianship over an individual involves a very significant restriction of fundamental rights and must be accompanied by procedural safeguards that recognize this fact."[86] In a number of other therapeutic contexts in which liberty and/or other fundamental rights are threatened, the courts have held that basic due process procedures apply.[87] However, as indicated on Charts III, IV and X, the extent to which current guardianship statutes provide for these fundamental safeguards varies considerably from state to state.[88]

Accordingly, this paragraph expressly makes seven key due process guarantees applicable to intervention proceedings. It should be noted that these rights apply to intervention proceedings initiated either under chapter 2 of this act or under chapter 3, and that they apply at all stages of those proceedings, not just at the adjudicatory hearing. (§10 Hearing and §22 Hearing.) Thus, in conjunction with dispositional hearings (§12(1)), review hearings (§§15 and 24), hearings on applications submitted under sections 16 and 25, hearings on emergency appointments (§35(1)), and hearings pursuant to the provisions on records (§40), as well as those held under sections 10 and 22, each party is entitled to a detailed notice of the date, place, time, purpose and possible consequences of the hearing, to be represented by an attorney, to attend the hearing, to subpoena witnesses, to present witnesses and introduce evidence in accordance with the rules of evidence in civil proceedings (§45 Rules of Evidence/Rules of Procedure), to cross-examine opposing witnesses, to have the hearing presided over by an unbiased judge, and to appeal. (See §32 on implementing the right to notice, §33 on the role of counsel, §34 on the appointment and role of guardians *ad litem,* and §44 on the applicable burdens of proof; see also paragraph (1)(c) and subsection (2) of this section.)

(b) In addition to the rights listed in paragraph (1)(a), the subject of an intervention proceeding shall have a right to close proceedings to the public, except that even if a proceeding has been closed to the general public, the court may, for good cause shown, admit persons having a legitimate interest in the proceedings or in the hearing process unless the subject of the intervention proceeding objects.

COMMENT. Following the lead of 14 states and the Uniform Probate Code,[89] this provision adopts a middle position between opening or closing all intervention proceedings to the general public. Closed proceedings enable individuals and families to avoid the embarrassment and possible stigmatization which can result from airing their personal problems publicly. On the other hand:

> Secrecy of judicial action can only breed ignorance and distrust of courts and suspicion concerning the competence and impartiality of judges; free and robust reporting, criticism, and debate can contribute to public under-

---

[86] Kindred, *supra* note 18, at 75; *accord* PCMR, *supra* note 12, at 133; AAMD, *supra* note 4, at 21; ILSMH, *supra* note 16, at 25-26; MHLP, *supra* note 44, at 445; PRESIDENT'S COMMISSION ON MENTAL HEALTH (PCMH), REPORT TO THE PRESIDENT, 43, 70-71 (1978); Horstman, *Protective Services for the Elderly: The Limits of Parens Patriae,* 40 MO. L. REV. 215, 232-259 (1975).

[87] *E.g.,* In re Gault, 387 U.S. 1 (1967); In re Cecelia R., 36 N.Y. 2d 317, 327 N.E. 2d 812 (1975); Dixon v. Attorney General, 325 F. Supp. 966 (M.D. Pa. 1971); Pennsylvania Association for Retarded Children v. Pennsylvania, 334 F. Supp. 1257, 343 F. Supp. 279 (E.D. Pa. 1972); Lessard v. Schmidt, 349 F. Supp. 1078 (E.D. Wis. 1972), *vacated for entry of a definitive decree,* 414 U.S. 473 (1974), *clarified,* 379 F. Supp. 1376 (E.D. Wisc. 1974), *vacated on procedural grounds,* 421 U.S. 957 (1975); Suzuki v. Quisenberry, 411 F. Supp. 1113 (D. Hawaii, 1976); Bartley v. Kremens, 402 F. Supp. 1039 (E.D. Pa. 1975), *rev'd on other grounds,* 431 U.S. 119 (1977); and Saville v. Treadway, 404 F. Supp. 430 (M.D. Tenn., 1974).

[88] See also A. Mitchell, *Involuntary Guardianship for Incompetents: A Strategy for Legal Services Advocates,* 12 CLEARINGHOUSE REV. 451, 453 (1978).

[89] See Chart IV and U.P.C., *supra* note 2, at §5-304.

standing of the rule of law and to comprehension of the functioning of the entire ... justice system, as well as improve the quality of that system by subjecting it to the cleansing effects of exposure and accountability.[90]

Hence, this paragraph gives the subjects of all intervention proceedings the opportunity to close the proceeding to everyone except the judge, necessary court personnel (e.g., the court reporter and clerk), the parties, their families and counsel. It should be noted, in addition, that the rules of procedure in many jurisdictions provide that witnesses may be excluded from the courtroom at the request of the parties in order to reduce the possibility that they may consciously or unconsciously alter their stories to conform to prior testimony. However, even when the general public and press are barred, this paragraph authorizes the court to permit persons with a legitimate interest in the proceedings or in the hearing process, — e.g., researchers and students studying the operation of the court or public interest groups monitoring the court — to attend intervention proceedings[91] unless there is a specific objection by the respondent.

(c) In addition to the rights listed in paragraph (1)(a) of this section, private individuals who initiate an intervention proceeding and who are unable to obtain counsel for financial reasons, shall have a right to have the [district/county/state] attorney appointed to represent them.

COMMENT. As is noted in the introduction of this report, the traditional legal basis for guardianship lies in the doctrine of *parens patriae* — the state serving as the father of its wayward or incapacitated children. Thus, there is a substantial public interest involved in the initiation and disposition of intervention proceedings even though such proceedings are ordinarily initiated by private citizens. In addition, because of the seriousness and complexity of the issues involved in intervention proceedings and the necessary procedural safeguards, it is anticipated that a private citizen filing a petition, pursuant to sections 6 or 19(1) or an objection to a testamentary disposition pursuant to section 20, will ordinarily seek the assistance of counsel. Accordingly, this paragraph provides for public representation of individuals filing a petition or objection when such individuals are unable to obtain the services of an attorney without imposing a substantial hardship on themselves of their families.[92] In this way an adult or minor requiring court-ordered assistance or the establishment of a guardianship or conservatorship will not be denied protection and assistance because the individuals who recognize this need are indigent. A government attorney representing an individual in an intervention proceeding should perform the same advocacy and screening functions as he or she would if the petition or objection had been filed by a public agency.

In enacting this provision, some states may also have to amend the statutes setting forth the responsibilities of [district/county/state] attorneys to provide the necessary authority to represent individuals in intervention proceedings.

(d) An individual alleged or found to be partially disabled or disabled shall also have a right to:

(i) A jury in hearings held pursuant to section 10 of this act; and

(ii) Have an independent expert(s) appointed by the court at the expense of the [state/county/court] to evaluate him or her if that individual is unable for financial reasons to secure independent medical, psychological or other expert opinion regarding the alleged disability and the need for restrictions on the legal capacity to act in his or her own behalf.

COMMENT. This provision extends to respondents in intervention proceedings initiated under chapter 2, the right to have a jury determine whether they are a partially disabled or disabled person, and the right to have independent experts appointed. Currently, 22 states entitle a respondent to a jury upon request as an added safeguard against routinization of intervention proceedings.[93] Because it is anticipated that in most cases the factual issues involved in determining whether a minor is eligible for guardianship will be relatively simple and uncontroversiel, the right to a jury is not extended to intervention proceedings initiated under chapter 3 of the act.

With regard to the appointment of independent experts, it can scarcely be disputed that expert diagnoses, opinions and recommendations, can and often do differ, and that the report of the multidisciplinary evaluation team will often be the primary basis for determining whether an individual is a partially disabled or disabled person, and if so, whether establishment of a guardianship and/or conservatorship is required. Since reports and testimony by mental health and other experts

---

[90] Nebraska Press Asociation v. Stuart, 427 U.S. 539, 587 (1976)(Brennan, J., concurring).

[91] *Accord* NACCJSG, *supra* note 77, at §12.3; IJA/ABA, STANDARDS RELATING TO ADJUDICATION, §§6.1 and 6.2(a) (Tentative Draft, 1977); MODEL ACT FOR FAMILY COURTS, §29(c) (1975).

[92] See ILL. ANN. STAT. ch. 110½, §11a-11(b) (Smith-Hurd Supp. 1979); and ILSMH, *supra* note 16, at 26.

[93] See Chart IV; see also, MHLP, *supra* note 44, at 453; U.S. SENATE SPECIAL COMMITTEE ON AGING, *supra* note 17, at 82; Horstman, *supra* note 86, at 251; and Heap v. Roulet, 152 Cal. Rptr. 425, 590 P.2d 1 (1979).

are a key element of all intervention proceedings initiated under chapter 2 of this act (see §8 Evaluation), the opportunity to have another qualified individual assist counsel in examining the report of the multidisciplinary evaluation team is critical in preparing an adequate defense.[94] Accordingly, this paragraph authorizes the court, upon request, to appoint an expert not serving on the multidisciplinary team to examine an indigent individual alleged or found to be partially disabled or disabled person.[95] Ordinarily, independent experts should be selected from the list prepared pursuant to section 5(4)(b). To the greatest extent possible, the subject of the proceeding should indicate the field of expertise and the names of the individuals whom he or she wishes to have appointed. The court, within reason, should appoint one or more of the experts so designated. A number of states already provide for appointment of independent experts in some circumstances.[96]

(e) No right accorded in this subsection shall be waivable, except for the right to close proceedings to the public set forth in paragraph (1)(b) of this section and the right to a jury set forth in paragraph (1)(d). Any waiver of the right to a closed proceeding or to a jury shall be in writing or on the record, and shall demonstrate that the individual waiving the right has knowingly and voluntarily waived it after being fully informed of the right and of the consequences of relinquishing it.

COMMENT. This provision, in conjunction with subsections (2) and (3), establishes three categories of rights. Those that may be waived — the right to a jury and to close the proceedings; those which may not be waived in advance but which a party may choose not to exercise — e.g., the right to subpoena witnesses or to appeal; and those which cannot be waived or declined — a record of the proceedings and appointed counsel for unrepresented subjects of intervention proceedings. The conceptual and practical difficulty in obtaining a voluntary and knowing waiver from a minor or an individual alleged to be partially disabled or disabled are manifold, and the danger of permitting waiver of essential rights by third parties is great.[97] Accordingly, waiver is limited to those two rights for which a conclusive decision is required at the beginning of a hearing in order to avoid a substantial disruption of the proceedings. In accepting such waivers, the court is required to make certain that the subject of the proceeding understands the right being waived and the impact of the waiver.

The second category includes those rights which parties normally exercise selectively with the aid of counsel. It would be absurd, for example, to require parties either to cross-examine every opposing witness or waive the right to do so. With regard to the right to be present, while it would be equally absurd to require a comatose adult or an infant to attend a hearing, special attention will have to be given to assuring that all subjects of intervention proceedings who are interested in and physically capable of attending are allowed to attend.[98] Unruly respondents should be treated in the same manner as unruly defendants in criminal cases.[99]

As for the third category, both a record and counsel are critical to protecting the individuals involved and the integrity of the intervention process. Without a verbatim transcript of the proceedings, the opportunity for effective review is severely impaired. Without an attorney for the subject of an intervention proceeding, the likelihood of a one-sided hearing and the imposition of unnecessary restrictions on legal capacity is substantially increased. (For a fuller discussion of both these rights, see the comments to subsections (2) and (3).)

(2) An attorney shall be appointed to represent the subject of an intervention proceeding in any matter arising under this act for which independent counsel has not been retained by or on behalf of that individual. Counsel shall be appointed within the time limits prescribed in section 30(1)(b) or 30(2)(d).

COMMENT. This subsection requires that counsel be appointed to represent adults alleged or found to be a partially disabled or disabled person and minors subject to an intervention proceeding. The appointment must be made within 5 days after the initiation of the proceeding unless the court has received notice that independent counsel has been retained or an immediate appointment would be in the interest of justice. Many minors subject to intervention proceedings and individuals alleged or found to be a partially disabled or disabled person will not be able to afford or have access to counsel, or will not know how to retain an attorney. Since representation by a lawyer in a judicial proceeding in which fundamental rights are at stake is essential to protecting both the individuals involved and the society at large from the imposition of unnecessary restrictions on the legal

[94] *Cf.* AMERICAN BAR ASSOCIATION, STANDARDS RELATING TO PROVIDING DEFENSE SERVICES, §1.5 (Approved Draft, 1968).

[95] *Accord* U.S. SENATE SPECIAL COMMITTEE ON AGING, *supra* note 17, at 82.

[96] E.g., ILL. ANN. STAT. ch. 110½, §11a-11(c) (Smith-Hurd Supp. 1979; and WIS. STAT. ANN. §880.33(2)(b) (West 1977).

[97] *See* AAMD, CONSENT HANDBOOK (1977); and Horstman, *supra* note 85, at 235-236.

[98] Mitchell, *supra* note 88, at 454; Horstman, *supra* note 85, at 241-244.

[99] See Illinois v. Allen, 397 U.S. 337 (1970).

capacity to act, a provision authorizing appointment of counsel is required.

> Without adversary counsel to advise the alleged incompetent of available rights, to prepare the best possible case opposing the petition and to advocate his or her interests, the prospective ward is destined to lose. Predictably, the few rights granted are not invoked, and the hearing becomes a meaningless formality.[100]

If the subject of the intervention proceeding is unable for financial reasons to pay reasonable compensation for counsel either in whole or in part, the court is authorized by section 33 to award a private attorney an amount representing reasonable compensation for the services performed from the funds made available to the court for that purpose, or to appoint the agency responsible for providing legal services to individuals who are indigent. While defendants in criminal cases have a right to defend themselves, at least where there are no allegations of incompetence,[101] the analogy between criminal and intervention proceedings breaks down in this instance, since the question of competence is always involved in actions under this act. However, section 33 makes clear that to the maximum extent possible, the subjects of intervention proceedings are to remain responsible for determining their legal interests and are to be encouraged to participate in making significant decisions. Hence, in some cases, counsel may be able to play more of an advisory role by simply assisting the respondent or minor when called upon and calling the judge's attention to matters favorable on which the judge should rule.[102] Moreover, nothing in this subsection is intended to prevent an individual for whom counsel has been appointed from discharging his or her appointed attorney and retaining another, subject to the normal rules regarding delay of the proceedings.

Authorization for appointment of counsel has been enacted in several of the states which provide for counsel in intervention proceedings.[103]

[100] Mitchell, *supra* note 88, at 454-455; *see* Horstman, *supra* note 85, at 244-251; see also sources listed in note 86 and U.S. SENATE SPECIAL COMMITTEE ON AGING, *supra* note 17, at 81-82; with regard to counsel for minors, see IJA/ABA, STANDARDS RELATING TO COUNSEL FOR PRIVATE PARTIES, 72-74 (Tentative Draft 1977); NATIONAL ADVISORY COMMITTEE ON JUVENILE JUSTICE AND DELINQUENCY PREVENTION, REPORT OF THE ADVISORY COMMITTEE TO THE ADMINISTRATOR ON STANDARDS FOR THE ADMINISTRATION OF JUVENILE JUSTICE, §3.132 (1976).

[101] Faretta v. California, 422 U.S. 806 (1975).

[102] *Id.* at 835 n.46.

[103] *E.g.*, N.C. GEN. STAT. §35-1.16(a) (Cum. Supp. 1977); TEX. PROB. CODE tit. 5, §130G (Vernon Cum. Supp. 1978); REV. CODE WASH. §11.88.045 (Supp. 1977); WIS. STAT. ANN. §880.33(2)(a) (West 1977).

The expense to the public which is likely to result from this provision is one of the necessary costs of preserving the integrity of our judicial and service delivery systems.

(3) A stenographic and/or mechanical record shall be made of all proceedings held pursuant to this act.

COMMENT. The availability of a verbatim record of the proceedings is essential for assuring accountability and fairness. As was pointed out by the Supreme Court in a comparable context:

> Failure to make a record, may be . . . to saddle the reviewing process with the burden of attempting to reconstruct a record and to impose upon the . . . [trial] Judge the unseemly duty of testifying under cross-examination as to the events that transpired in the hearings before him [or her].[104]

Accordingly, this subsection mandates that some form of verbatim record be made of all hearings held pursuant to this act. This is not intended to require that a full transcript be prepared in every case. A transcript of the record should be made only on the request of a party or the court. When the subject of an intervention proceeding who is indigent requests a transcript, the cost of its preparation should be at public expense.

(4) (a) Statements of individuals alleged or found to be a partially disabled or disabled person made during the course of evaluations, examinations and treatment pursuant to this act, shall be privileged, confidential, and not admissible without the individual's consent in any civil or criminal proceedings other than proceedings held pursuant to this act.

COMMENT. In order to encourage individuals alleged or found to be a partially disabled or disabled person to be as open, candid, and cooperative as possible during examinations and treatment provided pursuant to this act, the usual doctor/patient privilege is extended to all members of multidisciplinary evaluation teams and to persons providing court-ordered therapeutic services.[105] This extension recognizes that the experts diagnosing and treating particular disabilities may not be physicians. (See §3(23) Definitions and the comment thereto.) It should be noted that this paragraph does not provide for a right against self-incrimination in intervention proceedings,[106] since the danger of coercing individuals to testify against themselves does not appear applicable here. When the subject of an intervention proceed-

[104] In re Gault, *supra* note 87, 387 U.S. at 58; see also, Suzuki v. Quisenberry, *supra* note 87, 411 F. Supp. at 1133; and Lynch v. Baxley, 386 F. Supp. 378, 396 (M.D. Ala. 1974).

[105] MHLP, *supra* note 44, at 452.

[106] *Contra, id.*

ing is also a defendant in a criminal case, the Fifth Amendment's protection would of course apply to any matters related to the criminal charges.

(b) If the subject of an intervention proceeding is under the influence of psychotropic medication during any judicial hearing held pursuant to this act, the court and the jury, if any, shall be advised of this fact and of the purpose of the medication and the effect which it may have on the individual's actions, demeanor and participation at the hearing.

COMMENT. Whenever possible, respondents should not be under the influence of a drug which affects their appearance or ability to participate in a hearing unless administration of that drug is essential to protect the respondent or others from harm at the hearing, and no less drastic means is available.[107] When administration of a psychotropic drug is unavoidable, this provision requires that the trier of fact be made aware of the effects of the drug so that the alteration of the respondent's appearance, demeanor or actions does not improperly influence the decision-making process.

(c) Any hearing held pursuant to this act may be held at such places as the court directs including the place of residence of the subject of the intervention proceeding.

COMMENT. This provision is based on section 11a-10(c) of the recently enacted Illinois guardianship law.[108] It is designed to accomodate respondents for whom it would be physically difficult to attend a hearing at the courthouse. Any request to move the site of the proceeding must be weighed against the cost and inconvenience it may cause to the other parties, witnesses and the court.

Section 32. Notice and Dissemination of Plans and Reports

(1) (a) Notices required by sections 6(1)(c), 7(2), 16(5)(a), 18(6)(a) and (b), 19(1)(c), 20(3), 21(2), 25(5)(a), 34(4)(b) and 35(1)(c) of this act shall be served on:

(i) The parties as defined in section 3(9);

(ii) The attorney of record for each party;

(iii) The individuals most closely related to the subject of the intervention proceeding by blood or marriage unless their names or whereabouts are unknown and cannot be reasonably ascertained;

(iv) The individual or facility, if any, having custody of the subject of the intervention proceeding;

(v) The individual(s), if any, proposed for appointment or testamentarily designated as limited personal guardian or personal guardian;

(vi) The individual(s), if any, appointed or proposed for appointment as guardian *ad litem;*

(vii) The individual(s) and/or corporation(s), if any, proposed for appointment or testamentarily designated as limited conservator conservator;

(viii) The [executor/personal representative/administrator] of the estate of the testator following a testamentary designation;

(ix) The Administrator of Veterans' Affairs if the financial resources of the subject of the intervention proceeding include a benefit, compensation, pension or insurance payable by the Veterans' Administration; and

(x) The surety on the bond for the limited conservator or conservator, if any.

Such notices shall set forth the date, time, place, purpose and possible consequences of the hearing and the rights of the parties.

(b) Notices required by sections 16(5)(b) and 25(5)(b) of this act shall be served on the individuals and entities enumerated in paragraph (1)(a) of this section and shall advise the parties of their rights and of the procedures and time limits for filing an objection. In addition, the notice shall state that if an objection is filed, a hearing will be held to determine whether the application should be granted.

(c) Notices required by section 16(7) or 25(6) shall be served on the individuals and entities enumerated in paragraph (1)(a) of this section and shall describe the applications and/or objections to be joined and advise the parties that joinder will be ordered unless it can be shown to be prejudicial to the interests of the subject of the intervention proceedings.

(d) Notices required by section 19(2)(c) shall be served on the minor, the attorney of record for the minor, the limited personal guardian, personal guardian, limited conservator or conservator for the minor unless that individual filed the acceptance which gave rise to the notice, and the individuals or entities specified in subparagraphs (1)(a)(iii), (iv), (viii), (ix) and (x) of this section, and shall explain:

(i) The meaning of the acceptance;

(ii) The right to file an objection;

(iii) The rights to which the minor and the other recipients of the notice are entitled; and

(iv) The procedure and time limit for, and consequences of filing an objection.

(e) Notices required by section 30(5)(c) of this act shall be served on the individuals enumerated in paragraph 1(a) of this section and shall explain the meaning of the motion, and the

---

[107] *Id.* at 453.

[108] ILL. ANN STAT. ch. 110½, §11a-10(c) (Smith-Hurd Supp. 1979).

procedures for and consequences of, filing an objection.

(f) Notices required by section 34 of this act shall be served on the individuals or entities enumerated in subparagraphs (1)(a)(i)-(iv), (vi) and (viii)-(x), and shall set forth:

(i) The duties of a guardian *ad litem;*

(ii) The possible consequences of the appointment of a guardian *ad litem;*

(iii) The procedures and time limits for objecting to the appointment of a guardian *ad litem* or the individual proposed for appointment; and

(iv) The rights of the parties.

The notice shall further advise the parties that, following submission of an objection, an informal *in camera* hearing will be held to determine whether guardian *ad litem* should be appointed and, if so, who should serve in that capacity.

(g) Notices required by section 35(2)(d) shall be served on the individuals and entities enumerated in paragraph (1)(a) of this section, the individual or corporation whose appointment has been suspended, and the [prosecutor/commonwealth attorney/state attorney/district attorney/county attorney] if the basis for suspension may constitute a criminal offense. Such notices shall set forth:

(i) The actions taken by the court;

(ii) The reasons therefor; and

(iii) The name, address, duties and powers of the emergency appointee.

(h) Notices required by section 40(2) shall be served on the subject of the information retained pursuant to that section, his or her parents if the subject of the information is a minor, his or her attorney of record, and his or her limited personal guardian, personal guardian, and guardian *ad litem*, if any. Such notices shall inform the subject of the information that a record has been retained and explain the purposes for which the record may be used and the procedures for exercising his or her right to:

(i) Inspect the record;

(ii) Challenge its accuracy and retention;

(iii) Include a statement in the file concerning any disputed information; and

(iv) Have the record destroyed, sealed, or have access to it limited pursuant to sections 40(7) and (8) of this act.

(i) Notices required by section 40(5)(b) shall be served on the individual applying for access to information pursuant to that section, the subject of that information, his or her attorney of record, his or her limited personal guardian, personal guardian and/or guardian *ad litem*, if any, and the [executor/personal representative/administrator] of his or her estate, if any. Such notices shall set forth the matters enumerated in paragraph (1)(a) of this section.

COMMENT. In her critique of current guardianship statutes, Annina Mitchell observes that:

> [While most] states, require some form of notification, . . . the notice given is unlikely to be very informative or useful since the notice prescribed by most statutes is a notice of hearing; the recipient thereby learns that a hearing on the petition will be held on a certain date. He or she will not be advised of the serious legal and personal implications of the determination to be made, the importance of countering the allegations, the standard by which competency will be judged, the evidence to be introduced, or rights to legal representation and a jury trial, even if the latter is guaranteed by the statute. In those jurisdictions where a copy of the petition accompanies the notice of hearing the prospective ward will still not be apprised of the facts which underlie the petition, or which actions he or she will be called upon to defend, because the petition itself merely parrots the grounds enunciated in the statute.[109]

This provision together with subsection (2) of this section and sections 6(1)(b), 16(2)(c), 19(1)(b), 25(2)(c), 34(1)(b), 35(1)(b), and 45 are intended to remedy these problems. All notices must explain the purposes, procedures, and significance of the pleading or hearing which they concern, as well as the rights to which the parties are entitled under section 31. In addition, all petitions and applications require a description of the facts and reasons on which they are based.[110] The subsection also seeks to assure that everyone who may have a substantial interest in the intervention proceeding receives notice. When the petitioner or applicant does not know the names or whereabouts of the close relatives of the subject of the intervention proceeding and that information cannot be obtained by checking available records or by asking the subject of the proceedings or others, personal notice is not required, although some states may wish to require that a notice be published in a local general circulation newspaper. As a further precaution, states may wish to add the state offices responsible for providing advocacy services to youth and to the elderly to the list of those being notified and to include a provision similar to section 5-406 of the U.P.C. which permits other individuals interested in an intervention proceeding

---

[109] Mitchell, *supra* note 88, at 453-454; see also Horstman *supra* note 86, at 237-241.

[110] Suzuki v. Quisenberry, *supra* note 87, 411 F. Supp. at 1127.

to request service of any future notices issued in the course of the proceedings.

(2) The court shall have all notices served in the following manner:

(a) Attached to any notice served in conjunction with this act shall be a copy of the petition, application, acceptance, objection or motion which gave rise to the notice.

(b) Personal service shall be made on each of the individuals or entities receiving notice in the manner prescribed by (citation to the relevant Rules of Civil Procedure).

(c) The contents of all documents served shall be explained, to the maximum extent possible, to the subject of the intervention proceeding in the language, mode of communication, and terms which that individual is most likely to understand.

COMMENT. The state has a substantial interest in assuring that intervention proceedings are initiated for individuals who require court-ordered assistance, guardianship and/or conservatorship and conversely, that unnecessary restrictions upon the legal capacity of an individual to act on his or her own behalf are not imposed. Therefore, the burden of providing notice has been assigned to the court. This will also help to assure that notice is provided to the subject of the intervention proceeding in the prescribed manner, that financial considerations do not impose an artificial barrier to initiating or opposing intervention proceedings and that the confidentiality of records of intervention proceedings is maintained. (See §40 Records.)

With two exceptions, personal service may be made by mail, by delivery to the individual's home or office, or in any other manner permitted under the applicable rules of civil procedure. The first exception is for notices resulting from the filing of an application for an emergency appointment pursuant to section 35(1). Since the hearing will be held within 48 hours, the fastest possible means for providing notice should be employed. The second exception is for providing notice to the subject of the intervention proceeding. In order to increase the likelihood that the subject of the intervention proceeding will understand what is being alleged, what the proceedings entail, what the consequences may be, and what his and her rights are, all notices must be explained personally to him or her. While some individuals, either because of youth or disability, will not be able to grasp the meaning of the notice no matter how well it is explained, most will be able to understand if it is explained properly. For example, an explanation of the notice in English to a respondent who understands only Spanish, Vietnamese, or Navajo is of little use whether or not that individual is a partially disabled or disabled person. Attempting to communicate verbally or through a standard printed form to someone who can only understand sign language or read braille, or couching an explanation in technical legal terms is equally futile. Yet, in too many instances, the subjects of intervention proceedings fail to receive proper notice, not because they are unable to understand but because no one has made the effort to present the information in a manner that is comprehensible to them. By specifying that the notice must be in the language, mode of communication and terms which the subject of the intervention proceeding is most likely to understand, this provision mandates that such an effort be made. The individual serving the notice, or someone accompanying that individual, the attorney for the subject of the intervention proceeding, or his or her parent, relative, friend, or teacher can provide the explanation. The form demonstrating receipt of service should include a space for verifying that an explanation was given and indicating by whom it was provided.

Except when an emergency appointment is being sought pursuant to section 35(1), all notices are to be served no more than 5 days after the filing of the pleading which gave rise to the notice requirement. A notice must be served within 24 hours following the filing of an application seeking an emergency appointment. (See §30(4)(a) Time Limits.)

(3) (a) Evaluation reports submitted pursuant to section 8 of this act shall be sent to the attorney of record for each of the parties and directly to any party who is not represented by counsel.

(b) (i) Plans and reports submitted pursuant to sections 17(2), 17(6), 18(2), 18(7), 27(6), 28(2), and 28(7) of this act shall be sent to the partially disabled or disabled person or the minor, the attorney of record for the partially disabled or disabled person or minor, the (name of the state Developmental Disabilities Advocacy Office) if the partially disabled or disabled person has a developmental disability, and the individuals and entities specified in paragraphs (l)(a)(iii), (iv), (vi) and (ix) of this section.

(ii) Plans and/or reports submitted under sections 17(2), 17(6) and 27(6) of this act shall also be sent to the limited conservator or conservator, if any, for the partially disabled or disabled person or for the minor.

(iii) Plans and reports submitted under sections 18(2), 18(7), 28(2) and 28(7) of this act shall also be sent to the limited personal guardian or personal guardian, if any, for the partially disabled or disabled person or for the minor.

(iv) Subparagraphs (3)(b)(ii) and (iii) shall not apply when one individual has been appointed to serve in both a guardianship and conservatorship capacity.

(c) The contents of all plans and reports shall be explained to the subject of the intervention

proceeding in the language, mode of communication and terms which that individual is most likely to understand. Portions of reports which are likely to cause severe psychological or physical harm to the subject of the intervention proceeding may be disclosed to that individual through a psychiatrist, psychologist or other appropriate professional.

COMMENT. The same principles and procedures which govern notices, apply to the distribution of individual guardianship and conservatorship plans and the reports submitted by multidisciplinary evaluation teams, limited personal guardians, personal guardians, limited conservators and conservators. (See subsections (1) and (2) of this section and the comments thereto.) In keeping with the axiom that the subject of the intervention proceeding should participate in all important decisions to the maximun extent possible, the subsection provides for disclosure of all reports and plans to that individual. While some reports may include highly sensitive information which might disturb the minor or the adult alleged or found to be a partially disabled or disabled person, as the court in Suzuki v Quisenberry observed, "on balance, a person's interest in liberty is more important than his [or her] interest in not being perturbed."[111] Thus, provision is made for disclosure of those potentially disturbing portions of the reports through an individual with the professional skills to deal with the questions and emotions involved. Finally, subparagraphs (3)(b)(ii)-(iv) are intended to assure that when the same individual is not serving in both a guardianship and a conservatorship capacity, the limited personal guardian or personal guardian and the limited conservator or conservator each receives a copy of the other's reports.

Section 33. Duties and Compensation of Counsel

(1) The principal duty of an attorney representing the subject of an intervention proceeding is to represent zealously that individual's legitimate interests. At a minimum, this shall include:

(a) Personal interviews with the subject of the intervention proceeding;

(b) Explaining to the subject of the intervention proceeding in the language, mode of communication and terms which that individual is most likely to understand, the nature and possible consequences of the proceeding, the alternatives which are available, and the rights to which the individual is entitled; and

(c) Securing and presenting evidence and testimony and offering arguments to protect the rights of the subject of the intervention proceeding and further that individual's interests.

COMMENT. Reflecting the canons of the ABA Code of Professional Responsibility, this section requires retained or appointed attorneys for the subjects of intervention proceedings to be zealous advocates for the lawful interests of their clients.[112]

> In our government of laws and not of men [and women], each member of our society is entitled to have his [or her] conduct judged and regulated in accordance with the law; to seek any lawful objective through legally permissible means; and to present for adjudication any lawful claims, issue, or defense.[113]

Thus, the attorney for a minor in an intervention proceeding or an adult alleged or found to be a partially disabled or disabled person is to serve as more than a liaison between his or her client, the petitioner and the court. The representational obligations in intervention proceedings are the same as those of attorneys in other civil and criminal proceedings.[114] Paragraphs (1)(a)-(c) set forth the least an attorney should do for his or her client in an intervention proceeding.[115] While these duties appear self-evident, they have often been disregarded in practice.[116] (For an explanation of "language, mode of communication, and terms" see the comment to §32(2) Notice and Dissemination of Plans and Reports.) The same principles apply to representation of the initiator of an intervention proceeding.

(2) To the maximum extent possible, the subject of the intervention proceeding shall remain responsible for determining his or her interests. However, an attorney may seek appointment of a separate guardian *ad litem* when the circumstances set forth in section 34 of this act apply.

COMMENT. Consistent with the position adopted throughout this statute that, to the maximum extent possible, the subjects of intervention proceedings should act in their own behalf and participate in all important decisions which affect them, this subsection provides that an attorney should not automatically assume the role of a guardian *ad litem*. It is for the client to decide what his or her interests are and what course will best promote those interests.

---

[111] *Id.*

[112] ABA, *supra* note 13, at Canon 7; see also Andalman and Chambers, *Effective Counsel for Persons Facing Civil Commitment: A Survey, A Polemic, and a Proposal,* 45 Miss. L. J. 43 (1974).

[113] ABA, *supra* note 13, at EC 7-1.

[114] For suggestions on the approaches which attorneys may take in intervention proceedings see Mitchell, *supra* note 88, at 467-468; Andalman and Chambers, *supra* note 112, at 50-54.

[115] U.S. SENATE SPECIAL COMMITTEE ON AGING, *supra* note 17, at 82-83; ABA, *supra* note 13, at EC 7-8.

[116] Andalman and Chambers, *supra* note 112, at 54-72; Regan, *Protective Services for the Elderly: Commitment, Guardianship and Alternatives,* 13 WM & MARY L. REV. 569, 606 (1972).

> In certain areas of legal representation not affecting the merits of the case or substantially prejudicing the rights of a client, a lawyer is entitled to make decisions on his [or her] own. But otherwise the authority to make decisions is exclusively that of the client and, if made within the framework of the law, such decisions are binding on his [or her] lawyer.[117]

It is anticipated that in most cases, the client will be the one to make those decisions. Attorneys may seek the appointment of a guardian *ad litem* only when they have reason to believe that two sets of conditions have been met. The first is that the subject of an intervention proceeding is unable to determine his or her interests without assistance because of youth or because his or her ability to receive and evaluate information or to communicate decisions is impaired. The second is that a client who is unable to determine his interests without assistance either has no limited personal guardian or personal guardian who can provide such assistance or has interests which conflict with those of the individual serving in a guardianship capacity. (See §34 Procedures and Criteria for Appointment of a Guardian *ad Litem;* Duties and Powers, and §§3(19) and (20) Definitions.) In seeking appointment of a guardian *ad litem* the attorney will have to weigh the possible prejudicial effects on the client's assertion, if any, that court-ordered assistance and/or establishment of a guardianship or conservatorship are unnecessary.[118]

(3) Attorneys appointed to represent the subjects of intervention proceedings shall be entitled to reasonable compensation. If the subject of the intervention proceeding is unable for financial reasons to pay such compensation in whole or in part, the court shall award to a private attorney appointed pursuant to section 31(2), an amount representing reasonable compensation for the services performed, or shall appoint (the name of the Public Defender or Legal Services Office) to represent the individual. In setting the amount of compensation, the court shall take into consideration any sum paid by or on behalf of the subject of the intervention proceeding.

COMMENT. A number of commentators have linked the minimal compensation which appointed attorneys have received in mental health cases with the limited role which counsel has ordinarily played in such proceedings.[119] This subsection makes clear that attorneys are entitled to compensation commensurate with the services they perform. This compensation should be paid by the subject of the intervention proceeding when it would not substantially impair his or her financial independence, or substantially impede: meeting the essential requirements for his or her physical health or safety; managing his or her financial resources; and/or providing him or her with the assistance necessary to develop or regain his or her abilities to the maximum extent possible. When payment of some or all of the compensation would impair the financial independence of the subject of the proceeding or result in any of the impediments listed above, the cost of counsel should be borne by the state. This can be accomplished by compensating private attorneys appointed by the court or appointing the agency normally responsible for providing legal representation to individuals who are indigent. As noted earlier, any additional expense to the public resulting from the appointment of counsel should be viewed as a necessary cost of maintaining the integrity of the judicial intervention process.

Section 34. Procedures and Criteria for Appointment of a Guardian *ad Litem;* Duties and Powers

(1) (a) The subject of an intervention proceeding, the attorney for that individual, the limited personal guardian, personal guardian, limited conservator or conservator if any, for that individual, and/or anyone interested in the welfare of that individual, may file an application to have a guardian *ad litem* appointed.

(b) The application shall set forth:

(i) The name and address of the subject of the intervention proceeding;

(ii) The name, address, and interest of the applicant;

(iii) The name and address of the attorney of record of the subject of the intervention proceeding;

(iv) The name and address of the limited personal guardian, personal guardian, limited conservator or conservator, if any, for the subject of the intervention proceeding;

(v) The names and addresses of the individuals most closely related to the subject of the intervention proceeding by blood or marriage unless they are unknown or cannot reasonably be ascertained;

(vi) The name and address of the individual or facility, if any, having custody of the subject of the intervention proceeding;

(vii) The facts and reasons supporting the request; and

(viii) The name, address and qualifications of the individual proposed for appointment as guardian *ad litem.*

---

[117] ABA, *supra* note 13, at EC 7-7; see also Andalman and Chambers, *supra* note 112 at 48-49.

[118] For the states which permit appointment of separate counsel and guardians *ad litem* see n. 14, *supra*.

[119] Regan, *supra* note 116; Andalman and Chambers, *supra* note 112, at 72.

COMMENT. This subsection permits anyone interested in the well-being of a minor subject to an intervention proceeding or an adult alleged or found to be a partially disabled or disabled person to request the court to appoint a guardian *ad litem* when there is reason to believe the criteria set forth in subsection (5) are met. It is anticipated that such requests will ordinarily be made by either the attorney for the subject of the intervention proceeding or the individual who initiated the proceeding. (See §§3(19) and (20) Definitions and §33(2) Duties and Compensation of Counsel.) The form of the application is similar to that prescribed in this act for other petitions and applications. It is designed to facilitate the provision of notice by including the names of those most likely to have an interest in the application, the specific relief sought, and the basis therefor.

(2) Within five (5) days of the filing of an application for appointment of a guardian *ad litem,* notice shall be served on the individuals and entities enumerated in section 32(1)(f) in the form and manner prescribed in sections 32(1)(f) and 32(2).

COMMENT. The application together with a notice must be served on the parties and their attorneys, the close relatives of the subject of the intervention proceeding, the individual or entity having custody of the subject of the proceeding, the proposed guardian *ad litem,* and when applicable, the Administrator of Veterans' Affairs and the surety on the bond for the limited conservator or conservator. The notice must describe the duties of a guardian *ad litem,* the possible consequences of the appointment of a guardian *ad litem,* the procedures and time limits for filing an objection pursuant to subsection (3), the requirement for an informal hearing following the filing of an objection, and the rights to which the parties are entitled under section 31. The notice and application must be explained to the subject of the intervention proceeding in the language, mode of communication and terms which he or she is most likely to understand. (See §32(2) Notice and Dissemination of Plans and Reports.)

(3) Any party may file an objection to the appointment of a guardian *ad litem* or to the individual proposed for appointment within the time limit prescribed in section 30(3)(b). Objections by the subject of the intervention proceedings may be communicated to the court by any means including but not limited to oral communication or informal letter. Objections by other individuals or entities shall be in writing.

COMMENT. Any party may challenge the need for appointing a guardian *ad litem* or the eligibility of the individual nominated for appointment in the petition. The objection should state the name of the objecting party and the facts and reasons on which it is based. In order to facilitate direct participation by the subject of the intervention proceedings, objections by the subject of the proceeding need not be in writing nor comply with the usual formal requirements for written pleadings. (See §16(2)(b) Modification and Termination of Dispositional Orders and §25(2)(b) Modification of Dispositional Orders.)

(4)(a) The court may grant or deny an application without a hearing unless a timely objection is submitted.

(b) No more than five (5) days after an objection is filed, the court shall set a date for an informal, *in camera* hearing, and have notice served on the individuals and entities enumerated in section 32(1)(f) of this act in the form and manner prescribed in sections 32(1)(a) and 32(2).

COMMENT. If no objection is filed, the court may make its decision on the basis of the information contained in the application. If an objection is filed, the court must hold an informal hearing. The question of what procedural safeguards should be required when there is an objection to a request for appointment of a guardian *ad litem* raises some difficult issues. Even though the criteria for appointment of a guardian *ad litem* are different from those which must be met in order to appoint a limited personal guardian or personal guardian (see §10 Hearing; §12(2) Dispositional Hearing, Dispositional Principles and Criteria; §22 Hearing; and §23 Dispositional Alternatives), in some cases the appointment of a guardian *ad litem* could increase the likelihood that a personal guardianship or conservatorship will be established. On the other hand, given the duties set forth in section 33 for attorneys representing the subjects of intervention proceedings, unwarranted appointments of guardians *ad litem* are unlikely. Moreover, if a formal hearing were required following the filing of a request by the attorney for the subject of the intervention proceeding, it would follow that a separate attorney would have to be appointed to represent the subject of the proceeding at that hearing. If that attorney also had reason to believe that appointment of a guardian *ad litem* was necessary, the cycle would begin anew. Accordingly, this section provides solely for an informal hearing in the judge's chambers where each party can present his or her views and the basis therefor, and the matter can be disposed of without delaying determination of the primary issues. Although the hearing is informal, a record of it must still be made pursuant to section 31(3). Notice of the rights of the parties and the date, time, place, purpose and possible consequences of the hearing should be sent to all the individuals and entities entitled to receive notice of the application, and explained to the subject of the hearing in the language, mode of communication and terms which he or she is most likely to understand. (For an explanation of "language, mode of communications and terms" see the comment to §32(2) Notice and Dissemination of Plans and Reports.)

(5) The court shall appoint guardians *ad litem* to protect the rights and interests of the subjects of intervention proceedings if it is shown that because of youth, because of impaired ability to receive and evaluate information effectively regarding the proceeding, or because of impaired ability to communicate decisions regarding the proceeding, they cannot determine their own interests without assistance, and either:

(a) No limited personal guardian or personal guardian has been appointed;

(b) Their interests and those of their parent, limited personal guardian or personal guardian conflict; or

(c) The appointment is otherwise required in the interests of justice.

COMMENT. This provision sets forth the criteria which the court should use in reviewing both uncontested and contested applications for appointment of a guardian *ad litem*. It requires the judge to make two determinations. The first is that the subject of the proceeding is unable because of age and inexperience or because of a disability to weigh the alternatives and determine without assistance, the course of action which best promotes his or her welfare. The second is that there is no natural or appointed guardian who is able and willing to provide this assistance. The criteria are explicitly limited to a determination of interests in the intervention proceeding. A finding that an individual requires the assistance of a guardian *ad litem* should be wholly irrelevant to the determination under chapter 2 of this act of whether he or she is a partially disabled or disabled person and, if so, whether the establishment of a guardianship or conservatorship is the least restrictive dispositional alternative. Similarly, appointment of a guardian *ad litem* for a minor does not necessitate appointment of a limited personal guardian, personal guardian, limited conservator or conservator for that minor.

(6) The eligibility requirements and priorities set forth in section 13 for selection of a limited personal guardian or personal guardian shall apply to the selection of a guardian *ad litem*.

COMMENT. Any non-disabled adult who is not providing or responsible for providing some service or treatment to the subject of the intervention proceeding, or who is not a creditor of the subject of the proceeding, or whose interests do not otherwise conflict with those of the subject of the proceeding may be appointed as the guardian *ad litem*. Nothing in this provision is intended to prohibit the parent, spouse, child, or sibling of the subject of the intervention proceeding from serving as the guardian *ad litem* when the conflict of interest is determined to be insubstantial and such an appointment would clearly be the most beneficial to that individual. However, contrary to the practice in most states (see Charts III and X), the attorney for the minor or the adult alleged or found to be a partially disabled or disabled person should not be asked to serve as guardian *ad litem* as well. The reason for this change was succinctly stated by the Supreme Court of Vermont in a case involving a minor:

> [A] lawyer attempting to function as both guardian *ad litem* and legal counsel is cast in the quandary of acting as both attorney and client, to the detriment of both capacities and the possible jeopardizing of the infant's [or partially disabled or disabled person's] interests.[120]

(7) Guardians *ad litem* shall assist the individuals for whom they have been appointed to determine their interests in regard to the legal proceedings in which they are involved. If an individual is wholly incapable of determining his or her own interests, the guardian *ad litem* shall make that determination and advise counsel accordingly. In so doing, guardians *ad litem* shall:

(a) Inquire thoroughly into all the circumstances that a prudent individual in the position of the person for whom the guardian *ad litem* has been appointed would consider in determining his or her interests in the proceedings; and

(b) Encourage the individual whom they are serving to participate, to the maximum extent of that individual's ability, in all decisions and to act on his or her own behalf on all matters in which he or she is able.

COMMENT. This subsection sets forth the duties and powers of a guardian *ad litem*. Like sections 17, 18, 27, and 28 on the duties of limited guardians, personal guardians, limited conservators and conservators, this provision stresses that the subject of the intervention proceeding should be encouraged to take as active a role as he or she is able. The extent to which a guardian *ad litem* may act as a substitute decision-maker rather than as an advisor should be indicated by the court in its appointment order.

(8) The court should inform guardians *ad litem*, upon appointment, of their duties and powers.

COMMENT. Since few people are likely to be familiar with the role of a guardian *ad litem* prior to being appointed, the court is required to explain the duties and powers assigned to a guardian *ad litem* at the time of appointment. This explanation should be in non-technical language to the greatest degree possible.

Introduction to Section 35

This section provides procedures for handling crises which may occur during an intervention proceeding. Subsection (1) provides for the

---

[120] In re Dobson, 125 Vt. 165, 168, 212 A. 2d 620, 622 (1965); *cf.* ABA, *supra* note 13, at EC 7-12.

prompt appointment of a limited personal guardian or limited conservator on an emergency basis when the health, safety or financial resources of the subject of an intervention proceeding are in imminent danger of serious harm. Subsection (2) permits the court to suspend and temporarily replace a limited personal guardian, personal guardian, limited conservator or conservator when there is probable cause to believe that as a result of malfeasance or misfeasance, the health, safety or financial resources of the subject of the guardianship are in imminent danger of serious harm. Finally, subsection (3) permits the court to appoint an eligible individual or entity to assume the duties and powers of a limited personal guardian, personal guardian, limited conservator, or conservator in the event the individual serving in that capacity dies, resigns, or becomes incapacitated. Emergency or standby appointments under any of those subsections remain effective until there has been an opportunity for a more thorough review of the situation and a full hearing.

Section 35. Emergency and Standby Appointments

(1) (a) If during the pendency of a proceeding initiated under chapter 2 or 3 of this act, it appears that there is an imminent danger that the physical health or safety of the subject of the intervention proceeding will be seriously impaired or that the financial resources of that individual will be seriously damaged or dissipated unless immediate action is taken, the subject of the intervention proceeding or any adult interested in the welfare of that individual may apply to the court in which the proceeding is pending for the emergency appointment of a limited personal guardian and/or limited conservator.

COMMENT. Although section 30 provides for the expeditious handling of intervention proceedings, there are some instances in which immediate action is required to protect the life, health, or assets of a minor subject to an intervention proceeding or of an adult alleged or found to be a partially disabled or disabled person. Such a situation would arise for example when the parents of a young child die without leaving a will or when the personal guardian or conservator of a disabled person dies or becomes incapacitated and no standby appointment has been made. Accordingly, this subsection permits the subject of an intervention proceeding or any adult interested in his or her well-being to seek appointment of a limited personal guardian or limited conservator on an emergency basis. Applications may be filed simultaneously with a petition under sections 6(1) or 19(1), together with an acceptance of a testamentary designation pursuant to section 19(2), or following the initiation of an intervention proceeding but prior to a hearing on the merits held pursuant to sections 10 or 22. An emergency appointment may also be sought after a determination that court-ordered assistance and/or the establishment of a guardianship and/or conservatorship is required — e.g., by a limited conservator when the health or safety of the partially disabled person are imperiled. States which do not already provide either for appointing a guardian or for authorizing the court to consent to medical treatment in emergency situations when the individual requiring the treatment is incapacitated but no intervention proceeding is pending, may wish to add such a provision to this section. (See Charts II and IX.)

In order to assure that this subsection is not misused as a means of circumventing the safeguards included elsewhere in this act, this paragraph and paragraph (1)(e) require both that the danger be imminent and that the threatened harm would seriously impair the health, safety or financial resources of the subject of the intervention proceeding. Consistent with section 6, when the individual in danger is an adult alleged or found to be a partially disabled or disabled person, the applicant must discuss with the disabilities resources officer whether any less restrictive alternatives are available.

The application must be heard within 48 hours. If a limited personal guardian or limited conservator is appointed, the court may assign only those duties and powers necessary to alleviate the specified danger. Emergency appointments remain in effect for up to 35 days unless a continuance has been granted. (See paragraph 1(f) of this section and §§30 (1)(h) and (q), 30(2)(h) and (o), and 30(5) Time Limits.)

(b) The application shall state:

(i) The name and address of the subject of the intervention proceeding;

(ii) The danger alleged to be imminent;

(iii) The type of appointment and the particular protection and assistance being requested;

(iv) The facts and reasons supporting the allegations and requests;

(v) The name, address and qualifications of the proposed limited personal guardian and/or limited conservator;

(vi) The name, address and interest of the applicant; and

(vii) If the pending proceeding was filed under chapter 2 of this act, a statement that the application has been discussed with the disabilities resources officer and that the restrictions, protections, and assistance requested are the least restrictive alternative which will avert the imminent danger to the respondent and/or the respondent's financial resources.

COMMENT. Because of the time constraints involved in emergency proceedings, the applicant does not have to supply the extensive list of names and addresses which is required for similar applications and petitions filed under this act. However the applicant is required to specify the requested relief and the basis therefor, as well as the names and addresses of the individuals immediately involved. In addition, when the individual who is endangered has been alleged or found to be a partially disabled or disabled person, the applicant must state that he or she has discussed the matter with a disabilities resources officer and that the requested relief is the least restrictive dispositional alternative.

The names and addresses of the other individuals and entities entitled to notice should be readily available in the court case files. (See §40(2)(c) Records.)

(c) Within twenty-four (24) hours of an application for emergency appointment, a notice shall be served on the individuals and entities enumerated in section 32(1)(a) in the form and manner prescribed in sections 32(1)(a) and 32(2) of this act.

COMMENT. Since the hearing on the application must be held within 48 hours, notice must be sent immediately to those individuals and entities with an interest in the case. While the notice will not always arrive in time to permit attendance at the hearing, it is important that all the individuals and entities enumerated in section 32(1)(a) be made aware of the emergency application, since there will be an opportunity to consider the continued need for the limited personal guardian or limited conservator within 35 days. (See paragraph (1)(f).) The notice must describe the rights of the parties and the date, time, place, purpose, and possible consequences of the hearing. A copy of the application must be attached. (§§32(1)(a) and 32(2) Notice and Dissemination of Plans and Reports.) Every effort should be made to assure that the subject of the intervention proceeding, his or her attorney of record, his or her limited personal guardian, personal guardian, limited conservator, and/or conservator if any, and at least some of his or her close relatives within the area served by the court receive notice prior to the hearing.

(d) A hearing shall be held within the time limit prescribed in section 30(4)(b) of this act, at which the rights set forth in section 31 shall apply.

COMMENT. The hearing on the emergency application must be held within 48 hours of the filing the application. At the hearing, the parties are entitled to be present, be represented by counsel, subpoena witness, introduce evidence and testimony, cross-examine opposing witnesses, have an impartial decision-maker and appeal. A record must be made of the hearing. (§31 Rights and Privileges.) The burden of going forward and the burden of persuasion by a preponderance of the evidence is on the applicant. (§44(3) Burden of Proof.)

(e) If the court determines that there is an imminent danger that the physical health or safety of the subject of the intervention proceeding will be seriously impaired and/or that the financial resources of that individual will be seriously damaged or dissipated unless immediate action is taken, and for matters filed under chapter 2 of this act, if the court determines in addition that there is no less restrictive means of protecting the respondent and/or the respondent's financial resources, it shall appoint a limited personal guardian and/or limited conservator in the manner prescribed in section 14 or section 23(5). Individuals and or corporations appointed under this section shall be selected in accordance with the requirements and criteria set forth in section 13 or section 23(4).

COMMENT. This paragraph sets forth the criteria which the court must use in determining whether an emergency appointment should be made, and if so, who should be appointed and what powers and duties should be assigned. (See the comment to paragraph (1)(a) for a discussion of the issues involved.) The court must explain on the record the terms and conditions contained in its order and the facts and reasons on which its decisions are based.

(f) (i) If the application was filed prior to the hearing required by section 10 or section 22, the emergency appointment shall remain in effect until the conclusion of that hearing. However, if at a hearing held pursuant to section 10, the respondent is found to be a partially disabled or disabled person, the court may extend the emergency appointment until the conclusion of the dispositional hearing.

(ii) If the application was filed following a finding that an adult is a partially disabled or disabled person or that a minor meets the criteria set forth in section 23(3), the court shall direct the emergency appointee to file an application in accordance with section 16 or section 25 within five (5) days. The emergency appointment shall remain in effect until the conclusion of the hearing on that application.

COMMENT. Emergency appointments are to remain in effect until there is an opportunity for a more thorough examination of the abilities, disabilities and needs of the subject of the intervention proceeding. For emergency applications filed in conjunction with or shortly after the initiation of an intervention proceeding under Chapter 2, this opportunity would be at the hearing held pursuant to section 10, within 35 days of the filing of the initial petition. (See §30(1)(h) Time Limits.) If the respondent is found to be a partially disabled or disabled person at that hearing, the emergency appointment can be extended up to 10 days in order to provide an opportunity for a full dispositional hearing. (See §12(1) Dispositional Hearing . . ., and §30(1)(i) Time Limits.) For emergency applications filed in conjunction with or shortly after the initiation of an intervention proceeding for a minor, the emergency appointment would remain in effect until the conclusion of the hearing held pursuant to section 22, no more than 30 days after the filing of a petition or an objection to a testamentary designation, or no more than 15 days after the expiration of the period for accepting a testamentary designation when no written acceptance has been filed. (See §19 Initiation of Intervention Proceedings (for minors); §20 Objections to Testamentary Designations, and §30(2)(h) Time Limits.) If the emergency application is filed after a determination that an adult is a partially disabled or disabled person or that a minor is eligible for guardianship or conservatorship, subparagraph (1)(f)(ii) requires that an application to modify or terminate the guardianship and/or conservatorship must be filed within 5 days of the emergency appointment. Pursuant to sections 16(4)(a) and 25(4)(a) a hearing must be held on such an application within 30 days. (See §§30(1)(q) and (2)(o) Time Limits.) The maximum duration of emergency appointments under current statutes varies from 6 days to 6 months. (See Charts II and IX.)

(2) If on the basis of an application filed under sections 16 or 25 of this act, a report or accounting submitted under section 17(6), 18(7), 27(6) or 28(7), or other reliable information, there is probable cause to believe that a limited personal guardian, personal guardian, limited conservator or conservator is not effectively performing his or her duties and powers, and that there is an imminent danger that the physical health or safety of the partially disabled or disabled person or minor will be seriously impaired or that the financial resources of that individual will be seriously damaged or dissipated unless immediate action is taken, the court shall:

(a) Suspend and temporarily replace the limited personal guardian, personal guardian, limited conservator, or conservator, with an individual or corporation meeting the qualifications set forth in section 13 or 23(4);

(b) Reassign the duties and powers of the suspended individual or corporation to the emergency appointee;

(c) Direct the temporary appointee to file an application under section 16 or section 25 of this act within five (5) days, if such an application is not already pending, and submit such reports as may be necessary; and

(d) Provide notice to the individuals and entities enumerated in section 32(1)(a) of this act in the form and manner described in sections 32(1)(g) and 32(2).

COMMENT. This provision empowers the court to act quickly to protect the subject of an intervention proceeding when he or she is in immediate danger of serious harm because of the actions or inaction of his or her limited personal guardian, personal guardian, limited conservator or conservator. Since the court is authorized only to suspend and replace the careless, incompetent or overly avaricious limited personal guardian, personal guardian, limited conservator or conservator, and not to increase the restrictions on the legal capacity of the subject of the intervention proceedings, it may act without first holding a full adversary hearing. (*But cf.* subsection (1) of this section.) However, the court must give notice of its actions, the reasons therefor and the name and address of the emergency appointee to all individuals and entities with an interest in the proceeding, and a hearing must be held pursuant to Section 16(4) or Section 25(4) no more than 35 days after the court has taken action under this provision. (See §30(1)(q) and 30(2)(o) Time Limits.) At that hearing all parties including the suspended individual or corporation will have an opportunity to present their views and supporting evidence, and the court can rescind its suspension, permanently replace the offending individual or corporation, or select any other dispositional alternative consistent with this act. (See §16 Modification and Termination of Dispositional Orders and §25 Modification of Dispositional Orders and the comments thereto for a discussion of the procedures, rights and alternatives which apply.)

(3) On appointing a limited personal guardian, personal guardian, limited conservator or conservator, or following a hearing held pursuant to section 16 or 25 of this act, the court may designate another eligible individual or entity selected in accordance with sections 12 and 13 or section 23 to assume the duties and powers assigned to the limited personal guardian, personal guardian, limited conservator or conservator upon his or her resignation, incapacitation or death. The individual so designated shall submit an application pursuant to section 16 or 25 within the time limits prescribed by section 30(4)(c) of this act. An individual serving on a standby basis may exercise all of the duties and powers assigned to his or her predecessor as a limited personal guardian, personal guardian, limited conservator, or conservator, until the conclusion of the hearing on the application, unless otherwise ordered by the court.

COMMENT. In order to lessen the difficulties of transition caused by the death, incapacitation or resignation of a limited personal guardian, personal guardian, limited conservator or conservator, this subsection permits the court to make an anticipatory appointment of an eligible individual or corporation. This provision might be utilized when the parent of a disabled adult is named as his or her personal guardian and/or conservator in order to reduce the delay and uncertainty following the parent's death. (See also §6(2) Initiation of

Intervention Proceedings.) Upon actually taking over the guardianship and/or conservator duties and powers, a standby appointee is required to file an application pursuant to section 16 or 25, in order to provide an opportunity for the court to confirm that the originally appointed limited personal guardian, personal guardian, limited conservator and/or conservator has died, resigned or become incapacitated, to determine whether there is a continued need for the guardianship or conservatorship, and if so, to ascertain whether the terms and conditions of the guardianship or conservatorship should be modified and whether another eligible individual or corporation should be appointed. The standby appointee is given 10 days to file the application rather than the 5 days accorded emergency appointees, since most of the standby appointments will occur in a non-crisis situation in which all the interested individuals and entities had notice and all parties had an opportunity to object. (For a discussion of the procedures, rights and dispositional alternatives which apply at the hearing on the application, see the comments to §16 Modification and Termination of Dispositional Orders and §25 Modification of Dispositional Orders.)

Section 36. Appointment of Volunteer Public Guardians, Volunteer Public Conservators, and Volunteer Public Trustees

(1) (a) Pursuant to sections 12 and 13 or section 23 of this act, the court may select a volunteer public guardian or volunteer public conservator from the list published by the Guardianship/Conservatorship Oversight Commission for appointment as the limited personal guardian, personal guardian, limited conservator and/or conservator for a minor or a partially disabled or disabled person when:

(i) There is no other private individual or entity qualified under section 13(1) or 23(4) of this act who is able and willing to accept the appointment; or

(ii) The cost of a private limited personal guardian, personal guardian, limited conservator and/or conservator, would substantially impair the financial independence of the minor or the partially disabled or disabled person, or would substantially impede the meeting of the essential requirements for that individual's physical health or safety, and/or the management of his or her financial resources, and/or the provision to that individual of any assistance necessary to develop or regain his or her abilities to the maximum extent possible; or

(iii) The primary source of support for the minor or the partially disabled or disabled person is from public funds; or

(iv) A volunteer public guardian or volunteer public conservator can best assist the minor or the partially disabled or disabled person to meet the essential requirements for his or her physical health or safety, and/or manage his or her financial resources, and/or develop or regain his or her abilities to the maximum extent possible.

(b) A volunteer public guardian may be assigned the same duties and powers and shall be subject to the same requirements and restrictions as a private limited personal guardian or personal guardian. A volunteer public conservator may be assigned the same duties and powers and shall be subject to the same requirements and restrictions as a private limited conservator or conservator.

(c) A volunteer public guardian may be appointed to serve as the limited personal guardian or personal guardian for no more than two (2) individuals. A volunteer public conservator may be appointed to serve as the limited conservator or conservator for not more than four (4) individuals.

COMMENT. This provision, together with section 5, addresses the needs of those partially disabled or disabled persons and minors who require guardianship or conservatorship, but who have no family member or friend qualified and willing to take on this responsibility or who cannot afford such services. Although an increasing number of states are providing public guardianship services in some form (see Chart VII, *supra*), none have yet "successfully developed a mechanism for providing expanded guardianship services without creating serious conflict of interest problems."[121] While some states which have established a state or local public guardianship agency may be able to overcome these problems in the future, this act proposes another alternative. Under this section and section 5, public guardianship and conservatorship services are to be provided by citizen volunteers, identified, trained, and compensated by the state and appointed in individual cases by the court. (See §5 Establishment of the (state name) Guardianship/Conservatorship Oversight Commission and the comments thereto.) By relying on volunteers and limiting the number of individuals for whom a volunteer can serve in a guardianship or conservatorship capacity, it is hoped that partially disabled persons, disabled persons and minors who are alone or who have limited financial resources will be able to receive individualized attention and develop at least some degree of personal rapport with their limited personal guardian, personal guardian, limited conservator or conservator. The Guardianship/Conservatorship Oversight Commission will monitor the appoint-

[121] Kindred, *supra* note 18, at 86; see Levy, *supra* note 22; for other problems which have beset at least one public guardianship agency see THE CONSERVATEE RIGHTS COMMITTEE, *supra* note 21.

ment, compensation and termination of volunteer public guardians and volunteer public conservators to assure honesty and fairness.

Three conditions must be met before a volunteer public guardian may be appointed to serve as a limited personal guardian or personal guardian or a volunteer public conservator may be appointed to serve as a limited conservator or conservator. First, the subject of the intervention proceeding must be eligible to have a limited personal guardian, personal guardian, limited conservator, or conservator appointed. If the intervention proceeding was initiated under chapter 2, the respondent must have been found to be a partially disabled or disabled person for whom a guardianship and/or a conservatorship is the least restrictive dispositional alternative. If these findings are made the question of whether the volunteer public guardian or volunteer public conservator is to have limited or plenary powers is dependent upon the needs and abilities of the subject of the intervention proceeding and must be determined on the basis of the principles and criteria in sections 12(2) and (3). If the intervention proceeding was initiated under chapter 3 of this act, the minor must have been found to meet the criteria set forth in section 23(3) and to require establishment of a guardianship or conservatorship.

Second, the partially disabled or disabled person or the minor must be found to fit one of the categories listed in subparagraphs (1)(a)(i)-(iv). Subparagraphs (1)(a)(i), (iii) and (iv) are self-explanatory. With regard to subparagraph (1)(a)(ii), while the amount required to maintain the financial independence of the subject of the intervention proceeding to protect his or her health, safety and assets, and to enhance his or her abilities, will have to be assessed on a case-by-case basis, a volunteer public guardian and/or volunteer public conservator should be appointed if the cost of the guardianship or conservatorship would cause the subject of the proceeding to become financially dependent upon any governmental agency, private agency or person. Nothing in this section is intended to prohibit an individual or corporation otherwise eligible to serve in a guardianship or conservatorship capacity from declining compensation or reimbursement, or from applying to the Guardianship/Conservatorship Oversight Commission to serve as a volunteer public guardian or volunteer public conservator.

The third condition is that the interests of the volunteer public guardian or volunteer public conservator must not conflict with those of the individual to be served by the guardianship or conservatorship. (See § 13(1) Selection of a Limited Personal Guardian, Personal Guardian, Limited Conservator or Conservator.)

The duties of limited personal guardians, personal guardians, limited conservators and conservators for partially disabled and disabled persons are contained in sections 17 and 18. The comparable provisions concerning guardianships and conservatorships for minors are in sections 27 and 28.

Alternative. *As noted earlier, there are a number of other structures which can be employed for providing public guardianship services. For those states considering the establishment of a public guardianship agency within the executive branch of state government, see the entries for Colorado, Georgia, Illinois, Maine, Minnesota, North Carolina and Ohio in Chart VII and the public guardianship sections of Charts I, V, and VI, supra. For those jurisdictions considering implementation of local public guardianship services, see the entries on the above listed Charts for Alabama, Arizona, and California. Finally, those interested in a judicially based public guardianship system should refer to the entries for Delaware and Tennessee.*

(2) (a) A volunteer public trustee may be selected from the list published by the Guardianship/Conservatorship Oversight Commission for appointment as the trustee or co-trustee of any trust valued at less than $__________, when there are no private individuals or entities qualified, able and willing to serve as trustee, and either:

(i) At least one beneficiary of that trust has been found to be a partially disabled or disabled person, or is a minor for whom a limited personal guardian, personal guardian, limited conservator and/or conservator has been appointed; or

(ii) The creation of the trust would obviate the necessity of appointing or continuing the appointment of a limited conservator or conservator.

(b) No volunteer public trustee may serve as a trustee or co-trustee of a trust when that volunteer public trustee has interests which conflict substantially with those of any beneficiary of the trust. A substantial conflict of interest includes but is not limited to a volunteer public trustee concurrently:

(i) Acting in a guardianship or conservatorship capacity for any beneficiary of the trust;

(ii) Performing or being responsible for performing some service other than a trusteeship service for any beneficiary of the trust;

(iii) Providing some form of treatment or being responsible for providing such treatment to any beneficiary of the trust; or

(iv) Being a creditor of any beneficiary of the trust.

(c) A volunteer public trustee may be assigned the same powers and duties and shall be

subject to the same restrictions as private trustees pursuant to (citation to relevant sections of the state code).

COMMENT. As noted in the comment to section 11(2)(c), a trust is "a less restrictive and more normal device than general guardianship . . . which [can be used as] a mechanism to control the use of . . . property . . . [without asserting] control over the life of the beneficiary."[122] However, the financial resources of many minors and of many partially disabled and disabled persons are not sufficient to pay the costs of trust administration. Thus, unless some form of public trusteeship is available, use of this less restrictive dispositional alternative would be limited to those individuals with substantial assets or with someone willing to serve as trustee at little or no fee. Accordingly, this subsection, together with section 5(4)(b), permits individuals and corporations meeting the criteria and following the procedures established by the Guardianship/Conservatorship Overnight Commission, to offer their services as a volunteer public trustee. A volunteer system is recommended for public trustees for the same reasons as those outlined in the comment to subsection (1) regarding volunteer public guardians and volunteer public conservators. Four conditions must be met in order for a volunteer to be selected to serve as a trustee or co-trustee for a particular trust. First, the value of the trust assets must be below the level at which private management would be economically feasible and attractive to commercial trust companies. Since this level may vary considerably from state to state depending on population, commercial practices, business structure and the state's regulatory pattern, no amount is specified in the act. Second, there must be no individual or corporation qualified under state law to serve as a trustee, who wishes to and can serve as trustee or co-trustee of the particular trust. Third, the trust must be serving at least some individuals who fall within the scope and purpose of this act. (See subparagraphs (2)(a)(i) and (ii).) Finally, the individual or entity selected must be free of relationships which could result in conflicts between the interests of the volunteer public trustee and those of the beneficiaries of the trust. (See §13 Selection of a Limited Personal Guardian, Personal Guardian, Limited Conservator or Conservator.)

Care should be taken in enacting this subsection to assure that it is consistent with existing code provisions on trustees and fiduciaries.

Section 37. Delegation of Powers

(1) For a period not exceeding one (1) month in any twelve (12), a limited personal guardian, personal guardian, limited conservator or conservator, by properly executed power of attorney, may delegate any of the powers which the court has assigned, to another individual or corporation meeting the eligibility criteria set forth in section 13 or 23(4) of this act.

(2) For a period not exceeding one (1) month in any twelve (12) a parent, by properly executed power of attorney, may delegate any of the powers which a court may assign under this act, to another individual or corporation meeting the eligibility criteria set forth in section 23(4) of this act.

COMMENT. This section is based upon section 5-104 of the U.P.C. It is intended to reduce the difficulty of obtaining consent and/or payment of emergency treatment for a minor or a partially disabled or disabled person when the natural guardian or appointed limited personal guardian, personal guardian, limited conservator or conservator for that individual is unavailable. For example, the parents of a minor might delegate to a close friend or relative, their authority to consent to emergency medical care or to provide routine consent signatures required by their child's school, when they are planning a trip overseas. The duration of a delegation under this section is more limited than under the U.P.C. provision. While permitting individuals to take a respite from their responsibilities, it prohibits them from avoiding those responsibilities entirely by perpetually extending a delegation of authority. In enacting this section, reference should be made to the provisions of the state code governing powers of attorney.

Section 38. Effect of a Finding that a Person is Partially Disabled or Disabled and the Effects of a Guardianship or Conservatorship for a Minor

(1) A finding under this act that an individual is a partially disabled or disabled person shall not constitute a finding of legal incompetence. An individual found to be partially disabled or disabled shall retain all legal rights and abilities other than those expressly limited or curtailed in the dispositional order or subsequent order of the court.

COMMENT. This subsection is a necessary complement to those requiring that the least restrictive dispositional alternative be used and that the subjects of intervention proceedings be encouraged to act on their own behalf and to participate in all decisions which affect them to the maximum extent possible. (See e.g., §2 Purpose and §§17(1), 18(1), 27(1) and 28(1) on the duties and powers of limited personal guardians, personal guardians, limited conservators, and conservators.)

> . . . [N]o blanket presumptions should be made. Rather, the application of . . . restrictions to a particular mentally re-

[122] Kindred, *supra* note 18, at 73.

> tarded person should be determined in relation to his [or her] specific qualifications to exercise the specific right or meet the criteria specified by statute; denial should not be an automatic consequence of the appointment of a guardian.[123]

The distinction between court-ordered assistance and guardianship or conservatorship would be wholly unnecessary if a finding that an individual was a partially disabled or disabled person was sufficient in itself to make that individual legally incompetent. Similarly, the concept of a limited guardianship and the differentiation between guardianship and conservatorship would be meaningless if the rights or powers not assigned to a limited personal guardian were presumed lost to the subject of the intervention proceeding, or if individuals for whom only a conservator had been appointed were assumed to be unable to act in their own behalf in other matters.[124] Accordingly, this provision clearly establishes the principle that there can be no implied restrictions on an individual's legal capacity. If the dispositional order or an order issued following a review hearing pursuant to section 15 or following the filing of an application under section 16, does not expressly transfer the authority to perform a particular act or function to a limited personal guardian, personal guardian, limited conservator or conservator, the legal capacity to perform that act or function is retained by the partially disabled or disabled person.

(2) A minor for whom a limited personal guardian, personal guardian, limited conservator and/or conservator has been appointed shall be entitled to the same rights and privileges as a minor whose parents have not died or have not had their parental rights terminated or suspended, or as a minor who does not have or is not entitled to resources which may be jeopardized because of the minor's age.

COMMENT. Despite the traditional presumption of legal incompetence, the rights of minors are being increasingly recognized.[125] This subsection assures that children who, for example, have lost their natural parents, do not lose as well the rights and privileges to which they are entitled under law.

Section 39 Compensation and Expenses

(1) Limited personal guardians, personal guardians, limited conservators and conservators, including volunteer public guardians and volunteer public conservators, are entitled to reasonable compensation for services rendered and to reimbursement for reasonable and necessary expenses incurred in the exercise of their assigned guardianship and/or conservatorship duties and powers.

COMMENT. As is indicated in Charts V and XI *supra*, the overwhelming number of American jurisdictions permit limited personal guardians, personal guardians, limited conservators and conservators to collect reasonable and just compensation. In accordance with this practice, subsection (1) expressly entitles all individuals and corporations serving in a guardianship or conservatorship capacity to be paid for the time and effort they have devoted to assisting the partially disabled or disabled person or the minor whom they have been appointed to serve. In addition, they are entitled to be reimbursed for reasonable out-of-pocket expenses incurred in the course of providing such assistance. Although service in a guardianship or conservatorship capacity under this act is not intended to be a profit-making enterprise (see §13 Selection of a Limited Personal Guardian, Personal Guardian, Limited Conservator or Conservator and §36(1)(c) Appointment of Volunteer Public Guardian, Volunteer Public Conservators, and Volunteer Public Trustees), it is anticipated that the clear right to reasonable compensation will be a significant factor in inducing individuals and corporations to accept guardianship and conservatorship appointments. However, nothing in this provision is intended to prohibit a limited personal guardian, personal guardian, limited conservator or conservator from declining compensation.

By requiring court approval for all compensation (see subsection (5) *infra*) and tying compensation to the exercise of the duties and powers assigned under sections 12 and 14 or section 23, this section provides an effective means for encouraging individuals and corporations serving in a guardianship or conservatorship capacity to meet their obligations. (See §§17 and 27 on the duties and powers of limited personal guardians and personal guardians and §§18 and 28 on the duties and powers of limited conservators and conservators.) Limited personal guardians, personal guardians, limited conservators, or conservators who only partially fulfill their general and specific duties should not receive the same compensation as those who have fully met their obligations. When the failure to meet these responsibilities causes or threatens significant harm to the partially disabled person, disabled person or minor being served, consideration should be given to appointing a replacement pursuant to section 16 or 25. (See also §35(2) Emergency and Standby Appointments.) A description of the important actions taken by a limited personal guardian, personal guardian, limited conservator or conservator and the basis for any compensation or reimbursement requested must be included in the reports required by sec-

---

123 ILSMH, *supra* note 16, at 15.

124 Accord, AAMD, *supra* note 97, at 17; PCMR, *supra* note 12, at 133; and Kindred, *supra* note 18, at 72.

125 See BEASER, *supra* note 8; Wald, *supra* note 76.

tions 17(b), 18(7), 27(6) and 28(7). The supporting facts and reasons must also be included in any requests for compensation and/or reimbursement filed under sections 16 or 25.

(2) Such compensation and reimbursement shall be paid from the financial resources of the partially disabled or disabled person or the minor for whom the guardianship and/or conservatorship has been established unless:

(a) A volunteer public conservator has been appointed in accordance with section 36(1) of this act; and

(b) Either:

(i) Payment of such compensation and reimbursements would substantially impair the financial independence of the minor, partially disabled person, or disabled person, or substantially impede the meeting of the essential requirements for that individual's physical health or safety, the management of his or her financial resources, or the provision to that individual of any assistance necessary to develop or regain his or her abilities to the maximum extent possible; or

(ii) The primary source of support for the minor, partially disabled person, or disabled person is from public funds.

COMMENT. A limited personal guardian, personal guardian, limited conservator or conservator will ordinarily look to the financial resources of the subject of the intervention proceeding for compensation or reimbursement. However, since guardianships and conservatorships are intended to make certain that the individuals being served can meet the essential requirements for their physical health and safety, manage their financial resources, develop or regain their abilities and be as self-reliant and independent as possible, it makes little sense to allow the limited personal guardian's, personal guardian's, limited conservator's or conservator's compensation to thwart or postpone the achievement of these overall purposes.[126] Accordingly, this subsection, together subsection (4) permits payment of compensation and reimbursement by the state when the financial resources of the subject of an intervention proceeding are limited or consist primarily of payments from federal, state and local welfare, social security, or workmen's compensation programs. Compensation and reimbursement from state funds may only be made to individuals and corporations who appear on the list of volunteer public guardians and volunteer public conservators published by the Guardianship/Conservatorship Oversight Commission pursuant to section 5(4)(b). This will help to limit payment of funds to those qualified to serve in a guardianship and/or conservatorship capacity and complying with the restrictions imposed by section 36(1)(c) on the number of appointments which a volunteer public guardian or volunteer public conservator may accept. Nothing in this section or in sections 5 and 36 is intended to prevent an individual or corporation serving in a guardianship or conservatorship capacity from seeking to become a volunteer public guardian or volunteer public conservator pursuant to the procedures established by the Oversight Commission if the financial resources of the individual being served become inadequate to cover the costs of the guardianship or conservatorship.

(3) Volunteer public trustees who have been appointed to serve as a trustee or co-trustee of a trust in accordance with section 36(2) of this act, are entitled to reasonable compensation for services rendered and to reimbursement for reasonable, necessary expenses incurred in the performance of their trusteeship duties. Such compensation and reimbursements shall be paid from the corpus or income of the trust except when:

(a) The corpus of the trust is less than $______________; or

(b) Payment of such compensation and reimbursements would prevent the trust from providing to any beneficiary, funds which will be sufficient, when added to the other income of that beneficiary, to permit him or her to meet the essential requirements for his or her physical health or safety, and/or manage his or her financial resources, and/or develop or regain his or her abilities to the maximum extent possible.

COMMENT. This subsection provides individuals or corporations serving as volunteer public trustees with a right to receive reasonable compensation and reimbursement of reasonable, necessary expenses. The principles and policies governing the source, level and purpose of compensation and reimbursement of volunteer public guardians and volunteer public conservators applies to volunteer public trustees. (See the comments to subsections (1) and (2) of this section.)

(4) The state shall compensate and reimburse:

(a) A volunteer public guardian or volunteer public conservator not entitled under subsection (2) of this section to compensation and reimbursement from the financial resources of the partially disabled or disabled person or the minor; and

(b) A volunteer public trustee not entitled under subsection (3) of this section to compensation and reimbursement from the corpus or income of the trust.

COMMENT. See comments to sections 39(2) and (3)

(5) All compensation and reimbursements shall be approved by the court prior to payment.

[126] MHLP, *supra* note 44, at 448.

COMMENT. As noted in the comment to subsection (2) of this section, the requirement that the court approve the payment of compensation and reimbursements is included as a means for assuring that limited personal guardians, personal guardians, limited conservators, conservators and volunteer public trustees meet their obligations. While it is hoped that an individual's primary motive for accepting a guardianship or conservatorship appointment will not be pecuniary in nature, it is anticipated that court control of the purse strings will provide an effective enforcement tool. Approval for compensation and reimbursement may be given at the periodic review hearings held pursuant to section 15 or 24, and on the basis of an application filed under section 16 or 25.

Introduction to Section 40

Although there has been a marked improvement over the past 25 years, identification as a partially disabled or disabled person and the establishment of a guardianship or conservatorship can still subject individuals to embarrassment, discrimination, and ostracism.[127] While efforts must continue to eradicate this unwarranted stigmatization,[128] measures can be taken to avoid its effects. Thus, this section, in keeping with the growing national concern over privacy and the use of records,[129] closely controls the collection (subsection (1)), retention (subsection (2)), correction (subsection (3)), dissemination (subsections (4)-(6)), and destruction (subsections (7)-(9)) of the records resulting from intervention proceedings. (Subsection (10) defines informed consent for purposes of this section.) Since it is far less likely that minors will be stigmatized because they have lost their parents or because they are entitled to substantial financial resources, the controls on records generated in conjunction with intervention proceedings initiated under chapter 3 are less stringent than those on records regarding individuals alleged or found to be partially disabled or disabled persons. The limits which have been imposed recognize not only the interest in privacy, but also the countervailing need of individuals other than the subject of the information for access to records in particular situations. For example, it is essential for creditors of a partially disabled or disabled person to be able to find out that a conservatorship or trust has been established for that person. In balancing these interests, however, protection of the subject of the record remains of paramount importance.

It should be noted that a distinction has been drawn between the confidentiality of records and the confidentiality of courtroom proceedings. (See §31 Rights and Privileges.) The distinction is based on the impact which "free and robust reporting, criticism, and debate" can have on public understanding of the law and on the accountability of the judicial system,[130] the prohibition announced in the Nebraska Press Association vs. Stuart on imposing a ban on publication of information available to the public,[131] and the knowledge that retention of identifiable information obtained at an open hearing will be subject to the limitation of memory of those relatively few people who attend the hearing or may have been told of the experience. On the other hand, unless access to and retention of records are restricted, the records may remain available long after the court proceedings have been concluded to persons wholly unconnected with the case, and will, whether or not they have been updated, retain an aura of authority which may be out or proportion to their importance at the time of examination.

Finally, in considering this section, reference should be made to the provisions of existing state privacy and freedom of information statutes. Wherever those provisions more effectively promote the principles and policies discussed below, the relevant portions of this section should be omitted.

Section 40. Records

(1) Information identifiable to an individual alleged or found to be a partially disabled or disabled person or identifiable to a minor alleged or found to meet the criteria set forth in section 23(3) shall not be collected in conjunction with any proceeding under this act unless that information is essential to:

(a) Providing necessary services to that individual;

(b) Determining whether that individual is a partially disabled or disabled person or whether a guardianship or conservatorship for a minor should be or has been established;

(c) Issuing, reviewing, modifying or terminat-

[127] PCMH *supra* note 86, at 55; PCMR, *supra* note 12, at 16; MHLP, *supra* note 44, at 445; *Civil Commitment of the Mentally Ill, supra* note 2; Heap v. Roulet, *supra* note 93.

[128] See e.g., Chapter 2: Prohibiting Discrimination, *supra*.

[129] See e.g., PRIVACY PROTECTION STUDY COMMISSION, PERSONAL PRIVACY IN AN INFORMATION SOCIETY: THE REPORT OF THE PRIVACY PROTECTION STUDY COMMISSION (1977); Federal Privacy Act of 1974, 5 U.S.C. §552a (1977); Family Education Privacy Act of 1974, 20 U.S.C. §1231g (1978); Fair Credit Reporting Act, 15 U.S.C. §1681 (1974); LAW ENFORCEMENT ASSISTANCE ADMINISTRATION, COMPENDIUM OF STATE LAWS GOVERNING THE PRIVACY AND SECURITY OF CRIMINAL JUSTICE INFORMATION (1975); IJA/ABA, STANDARDS RELATING TO JUVENILE RECORDS AND INFORMATION SYSTEMS (Tentative Draft, 1977); MHLP, *Legal Issues in State Mental Health Case: Proposals for Change – Therapeutic Confidentiality*, 2 MEN. DIS. L. RPTR. 337 (1977); CONTRACT RESEARCH CORP., MODEL STATE CODE: PRIVACY ACT FOR THE DEVELOPMENTALLY DISABLED (1977).

[130] Nebraska Press Association vs. Stuart, *supra* note 90, 427 U.S. at 587 (Brennan, J., concurring).

[131] *Id.*, at 570.

ing a dispositional order pursuant to sections 12-16 or 23-26 of this act;

(d) Administering, monitoring, and evaluating the court and services provided to partially disabled or disabled persons and minors;

(e) Monitoring and evaluating the guardianship or conservatorship of a partially disabled or disabled person or a minor; or

(f) Conducting a research, evaluation or statistical study authorized under paragraph (5)(d) of this section.

COMMENT. This subsection sets out basic principles to guide the collection of information pertaining to the subjects of intervention proceedings. It is based on the view that limiting the collection of information to that which is essential is one of the most effective means of protecting the privacy of individuals, simplifying the problem of keeping identifiable records confidential and secure, and reducing the cost of data-gathering and record-keeping. While identifiable information is needed to make critical decisions throughout intervention proceedings and to assure accountability, initiation of an intervention proceeding should not be treated as an opportunity to undertake a wholesale investigation of the personal and financial affairs of the individual subject of that proceeding. Accordingly, information may be collected in identifiable form only if that information is directly related and vital to the performance of one of the functions listed in paragraphs (1)(a)-(f).[132] (For a definition of identifiable information see §3(30) Definitions.)

(2) (a) Information identifiable to an individual and collected under subsection (1) of this section shall not be retained in retrievable form unless it is accurate; protected from unauthorized access, disclosure and dissemination; physically secure; and essential to accomplishing one of the purposes specified in subsection (1).

COMMENT. The decision to retain information should be separate from the decision to collect it. Too often information which has proven to be inaccurate, irrelevant, or of only shortlived value, is retained simply because of inertia, and gains importance and credibility by virtue of its existence. Thus, this subsection provides that information should be retained in identifiable form only if it is "essential" to accomplishing at least one of the six purposes for collecting the information, its accuracy and completeness have been verified, its confidentiality has been assured, and it is protected against theft, fire and natural disasters.[133] This will discourage use of identifiable information when non-identifiable data can achieve the same objectives.

The subsection applies to all information retained in conjunction with an intervention proceeding, not just that in court files. It is intended to encompass automated as well as manual record systems, and centralized as well as locally or individually maintained records.

Although not required by this provision, it is urged that identifiable records which are retained not include summary conclusions or labels which describe an individual's social, emotional, medical or behavioral history or predict future development or behavior, unless the use of such professional jargon cannot be avoided, and the underlying factual basis, meaning and implications are explained in terms that are understandable to laymen.

> [Labels] . . . are efficient tools in thinking and communicating, yet because they are such an easy shorthand for complex problems, the temptation to use them too often is great. As a result, the use of labels becomes the lazy alternative to individual descriptions . . . . Therefore, the use of labels should be resisted whenever possible in order to avoid the risk of misuse or misinterpretation, not only by less well-informed persons, but also by other professionals.[134]

(b) Notice that information has been retained shall be served on the individuals and entities enumerated in section 32(1)(h) of this act in the form and manner prescribed in sections 32(1)(h) and 32(2).

COMMENT. In its report to the President, the Privacy Protection Study Commission concluded that providing the subjects of information with access to and the right to challenge the accuracy of their records is the most efficient and cost-effective means of assuring the accuracy of those records and compliance with record-keeping practices.[135] (See subsections (3)-(6).) However, the right to have access to and challenge records is useless if the subject of identifiable information is not aware that a record and those rights exist. Therefore, this provision, together with section 32, requires that the subject of identifiable information; his or her attorney; his or her limited personal guardian, personal guardian, and/or guardian *ad-litem* if any; and, if the subject of the information is a minor, his or her parents must be ad-

[132] *See* PRIVACY PROTECTION STUDY COMMISSION, *supra* note 129, at 388; CONTRACT RESEARCH CORP., *supra* note 129, at §7.

[133] *See* IJA/ABA, *supra* note 129, at §4.2; 28 C.F.R. §20.21(a) (1976); NACJJDP, *supra* note 100, at §0.52.

[134] *See* IJA/ABA, *supra* note 129, at 71. Danto, *Writing Psychiatric Reports for the Court*, 17 INT'L OFFENDER THERAPY & COMP. CRIM. 123 (1973).

[135] PRIVACY PROTECTION STUDY COMMISSION, *supra* note 129, at 31.

vised that a record has been retained. In addition, the notice must explain the purposes for which the record may be used, and the procedures for inspecting the record, challenging its accuracy and retention, inserting a statement concerning any disputed information, and having the record destroyed or sealed, or limiting access to it. This notice must be provided whenever a decision to open an identifiable record is made and should not be made dependent upon a request by the subject of the information.[136] Finally, it must be explained to the subject of the information in the language, mode of communication, and terms which he or she is most likely to understand. (§32(2) Notice and Dissemination of Plans and Reports.)

(c) The court shall maintain a summary file and a case file for identifiable information collected and retained in conjunction with proceedings initiated under chapter 2 of this act. The summary file shall contain only the notice of the filing of a petition, a summary of all formal proceedings, the dispositional orders and any modification or termination thereof. The case file shall contain:

(i) The petition initiating the proceeding;

(ii) The evaluation reports, and other diagnostic, medical, psychological, social history, or financial reports regarding the subject of the petition;

(iii) The individual guardianship and/or conservatorship plan, if any, and all modifications thereof;

(iv) The reports of the limited personal guardian, personal guardian, limited conservator, or conservator, if any;

(v) Any other identifiable information pertaining to the proceedings retained by the court; and

(vi) A log of the individuals and entities which have had access to the case file, and of all requests made pursuant to subsection (3) of this section.

COMMENT. This provision requires the court to maintain two files regarding intervention proceedings — a summary file which is available to the public, and a case file to which access is restricted. The summary file is to contain only the notice of the initiation of an intervention proceeding though not the petition itself, a summary of all hearings — e.g., the minutes maintained by the court clerk on the witnesses who appeared, the findings that were made and the orders which were issued — and most importantly, a copy of the dispositional order and any modifications made thereto. In this way, the personal information and unproven allegations in a petition remain confidential, but individuals and entities dealing with the subject of an intervention proceeding can easily learn what, if any, restrictions have been imposed on the legal capacity of a partially disabled or disabled person to act in his or her own behalf, and the name and address of the limited personal guardian, personal guardian, limited conservator, and/or conservator, if any. If such information were not readily available, normal caution would dictate an assumption that the subject of any intervention proceeding was legally incompetent, thus defeating a major thrust of this act. All other records are to be maintained in the case file together with a log indicating the name, address, and date of access of all individuals who have examined the case file, and the date of any requests under subsection (3) to modify, clarify or delete the records in the case file. This log is intended as a means of assisting the subject of the record to enforce his or her rights under this section.[137] (See §42(4) regarding the penalties for violation of the provisions on records.)

While maintaining a dual set of files may at first seem cumbersome, it is a far simpler and surer means of protecting the privacy of court records than requiring court employees to cull out non-disclosable information each time a request for access is made.

(3) The subject of a record retained under subsection (2) of this section shall be entitled to inspect and copy that record, to request correction, clarification, or deletion of any portion of that record which he or she believes is inaccurate, misleading or improperly retained, and to submit a statement concerning any disputed information for inclusion in that record.

COMMENT. The right of individuals to inspect and challenge the accuracy or retention of identifiable records concerning them is a basic element of most recent sets of recommendations and legislation in the privacy field.[138]

> . . . [B]y opening up record-keeping practices and by giving an individual opportunities to interact easily with a record-keeper, particularly at crucial points in a record-keeping relationship, both individuals and organizations will benefit. The quality of the information in records will be improved while at the same time the individual and the organization will both be protected from errors or other deficiencies that can have untoward consequences for both.[139]

---

136 IJA/ABA, *supra* note 129, at §4.3; *but see* CONTRACT RESEARCH CORP., *supra* note 129, at §6.

137 *See* ILL. ANN. STAT. ch. 91½, §804(d) (Smith-Hurd Supp. 1979).

138 See e.g. the authorities listed in note 129; see also ILL. ANN. STAT. ch. 91½, §§801 *et seq.* (Smith-Hurd Supp. 1979).

139 PRIVACY PROTECTION STUDY COMMISSION, *supra* note 129, at 19.

Accordingly, this subsection entitles the subjects of court and other records resulting from intervention proceedings to examine and make copies of them and to request changes in order to correct errors or resolve significant ambiguities. (See subsection (4) and (5).) Such requests should be handled in the same manner as those made under other state privacy statutes. Review should be available when a request is denied. Statements submitted by the subject of a record should be made part of the record and accessible to the same extent as the portion of the record to which they refer. (See also paragraph (9)(b) of this section.)

(4) Subject to subsections (7)-(9) of this section, identifiable, nonprivileged information collected and retained in conjunction with proceedings initiated under chapter 3 of this act, and the summary file maintained in conjunction with proceedings initiated under chapter 2 of this act, shall be available to the public.

COMMENT. This subsection provides that anyone may examine records of intervention proceeding concerning a minor and the summary file of an intervention proceeding concerning an adult alleged to be a partially disabled or disabled person. The basis for unlimited access to such records is explained in the comments to subsection (1) and (2) of this section. However, nothing in the subsection is intended to exempt these records from subsections (7)-(9) of this section on destroying, sealing or limiting access to records.

(5) (a) Subject to subsections (6), (7) and (9) of this section and consistent with (citation to the pertinent exception, if any, in the state open records law), access to the case file maintained in conjunction with proceedings under chapter 2 of this act shall be limited to:

(i) The subject of the records;

(ii) The attorney and guardian *ad litem*, if any, of that individual;

(iii) The limited personal guardian, personal guardian, limited conservator and/or conservator, if any, and their counsel;

(iv) Other parties to the proceeding and their counsel;

(v) The judge or judges hearing or reviewing the matter;

(vi) A member of the clerical or administrative staff of the court when access is essential for authorized internal administrative purposes;

(vii) Individuals and agencies for the express purpose of conducting a research, evaluative or statistical study authorized under paragraph (5)(d) of this section; and

(viii) An agency or individual, when access is essential to secure evaluation, diagnostic, medical, psychiatric, psychological, counseling, advocacy, therapeutic, rehabilitative, social, educational, financial or other services or benefits for the subject of the record and that individual has provided his or her voluntary, informed, written consent in accordance with subsection (10) of this section.

COMMENT. This provision lists those individuals who are entitled to access to case files maintained under subsection (2) of this section without a special court hearing. Subparagraphs (5)(a)(i)-(v) include the parties to the intervention proceeding, their counsel, the judge or judges presiding over any stage of the proceeding either at the trial court level or on appeal, and the guardian *ad litem*, if any, of the subject of the intervention proceeding. Since the parties and the court are entitled to copies of all the pleadings, plans and reports pursuant to section 32 (see also §45 Rules of Evidence/Rules of Procedures), it would be unnecessary and wasteful to require special procedures before they would be permitted access to the case file.

Similarly, unless there has been a history of sloppy record-keeping practices or misuse of information by court administrative personnel, the cost of enforcing strict controls on access by such personnel far outweighs the protection which stricter procedures afford. On the other hand, there must be some means for assuring that the confidentiality of the information in case files can be maintained. Accordingly, subparagraph (5)(a)(vi) provides that court clerical and administrative personnel may examine the contents of case files only when such access is essential to achieving some management objective or requirement that has been approved by the presiding judge or chief court administrative officer.

The limitations on and procedures for obtaining access under subparagraph (5)(a)(vii) for research, evaluative or statistical purposes are discussed in the comment to paragraph (5)(d) of this section.

The final subparagraph presents the most difficult issues. Opportunities for the subject of a record to obtain needed services should not be impaired by the unavailability of diagnostic or other information. However, once copies of case file records are available to individuals or entities not directly involved in an intervention proceeding or directly subject to court control, the likelihood of unauthorized and unwarranted disclosure increases substantially, especially given the rapid proliferation of information-sharing systems among public and private service providers and among financial and credit institutions.[140] The subparagraph seeks to balance these concerns by permitting access under two conditions. First, it must be demonstrable that provision of the particular service or benefit would be substantially

[140] *Id;* For a more thorough discussion of the various factors to be considered in granting access to records in intervention proceedings, see AAMD, *supra* note 97, at 67-75.

impeded if access to the case file were not provided. Second, the subject of the record must give written consent after having been advised of the information to be disclosed, the individuals or entities requesting disclosure, the purpose of the disclosure and the possible beneficial and harmful consequences thereof. (See paragraph 10(a) of this section.) A limited personal guardian or personal guardian for that individual may provide such consent if empowered to do so, after providing the above explanation to the subject of the record and obtaining his or her agreement to the maximum extent possible. While there are problems in relying on consent as the primary safeguard against unwarranted disclosure of confidential information even when the subject of the record is capable of understanding the conflicting considerations, there appears to be no other practical alternative.[141] (For a discussion of the issues regarding consent, see the comment to subsection (10) of this section.) The requirements imposed by subsection (10) should provide some measure of protection against coercion and abuse.

(b) Access to a case file may be granted to individuals and agencies other than those listed in paragraph (5)(a) of this section only after a hearing at which it is shown that such access will benefit the subject of the records or is otherwise in the interests of justice. Notice of the hearing shall be served on the individuals and entities enumerated in section 32(1)(i) of this act in the form and manner prescribed in sections 32(1)(a) and 32(2).

COMMENT. Few requests for access to a case file by individuals or entities other than those listed in paragraph (5)(a) are anticipated. When they do occur, this provision requires a court hearing at which the applicant, the subject of the record, his or her limited personal guardian or personal guardian, if any, or the executor, personal representative, or administrator of his or her estate can present evidence and arguments regarding the purpose, necessity, and consequences of access. The notice must indicate the date, time, place, purpose and possible consequences of the hearing and the rights to which the parties are entitled under section 31. It must be explained to the subject of the record in the language, mode of communication and terms which he or she is most likely to understand. (See §§32 (1)(a) and (2) Notice and Dissemination of Plans and Reports.)

(c) When the subject of a record requests access to records which the court has reason to believe contain information that is likely to cause severe psychological or physical harm to that individual, such information shall be disclosed through the individual's attorney of record or limited personal guardian or personal guardian, if any, or through a psychiatrist, psychologist or other appropriate professional.

COMMENT. A case file may contain sensitive diagnostic information which could cause severe stress to the subject of the record or impair his or her health or habilitation. In such cases, disclosure should be made through an individual who can deal with the questions and emotions which may arise and reduce or eliminate the possibility of harm.

(d) Access to case files shall not be granted to individuals or agencies for the purpose of conducting a research, evaluative or statistical study unless an application is filed with the court which describes:

(i) The purpose of the study;

(ii) The qualifications of the individuals conducting the study;

(iii) The identifiable information sought and the reasons why the purpose of the study cannot be achieved without using information in identifiable form;

(iv) The methods to be used to assure that the anonymity of the subject of the records is preserved; and

(v) The methods to be used to assure that the information will be physically secure.

Decisions approving or disapproving applications for access shall be in writing and subject to review by the appellate court.

COMMENT. In addition to their critical role in advancing knowledge, research, evaluation, and statistical studies are being increasingly employed in the management and assessment of courts, service agencies and programs and in the monitoring of record-keeping practices. This paragraph provides a means for providing researchers with necessary data while protecting the privacy of the subjects of intervention proceedings. It requires the submission of a detailed application and establishes a preference in favor of using non-identifiable information.[142] The application requirement is not intended as a curb on academic freedom, but simply as a means for assuring that the study has a valid "educational, scientific or public purpose;"[143] that access is granted only to individuals with the training and experience to conduct the proposed study; and that the identifiable information will be used and protected properly.[144] Applications should not be granted if the

[141] Other statutes or proposals which rely strongly on consent include: MHLP, *supra* note 129, at 346; CONTRACT RESEARCH CORP., *supra* note 129, at §7(c); ILL. ANN. STAT. ch. 91½, §805 (Smith-Hurd Supp. 1979).

[142] *See* 28 C.F.R. §22.23 (1976); IJA/ABA, *supra* note 129, at §5.6; *see also* MHLP, *supra* note 129, at 349-350.

[143] IJA/ABA, *supra* note 129, at §5.6(B)(2).

[144] For examples of methods which may be used to preserve the anonymity of the subject of identifiable records, see 28 C.F.R. §§22.23, 22.24, and 22.26 (1976).

purposes of the study can be achieved by using data from which the identities have been stripped. As an added protection, some states may wish to add a provision establishing limited immunity from legal process for identifiable information collected for research, evaluative or statistical purposes.[145]

(e) No disclosure of identifiable information obtained from a case file shall be made to individuals or entities not listed in paragraphs (5)(a) or (5)(b) of this section except by the subject of the record or during a hearing held pursuant to this act.

COMMENT. Without a prohibition on redisclosure of identifiable information, the impact of the provisions limiting access to case files would be substantially weakened. Hence, except for the subject of the record, those permitted to examine a case file cannot reveal the contents of that file other than to the individuals and agencies listed in paragraph (5)(a), granted access pursuant to paragraph (5)(b), or during the course of a hearing.[146] Other redisclosure of information obtained from a case file would subject the individual making that disclosure to liability or prosecution under section 42(4).

(6) Access to records retained under subsection (2) of this section by individuals and agencies providing services to the subject of an intervention proceeding shall be governed by (Citation to the applicable section of the state privacy act) and the applicable legal privileges.

COMMENT. Since the records maintained by service providers on the subjects of intervention proceedings would be indistinguishable from those on other clients or patients, both in terms of their contents and their possible stigmatizing effect, a separate provision governing access to such records is not necessary. Many states have some statutory provisions protecting the confidentiality of records maintained by governmental agencies providing services to partially disabled and disabled persons and minors as well as statutes establishing at least a physician/patient privilege.[147] For those states which do not have privacy statutes covering non-court records retained in conjunction with intervention proceedings, there are a number of reports and actual and model statues which can be drawn upon for assistance.[148]

(7) Following the dismissal or withdrawal of a petition filed under section 6 of this act or the termination of a dispositional order, the subject of any records retained under subsection (2) in conjunction with the petition or order may file an application to have the summary file, the case file and any other identifiable, non-privileged information pertaining to the proceeding destroyed, provided that the individual submits a full release of all claims arising from the proceedings. If such an application and full release is filed, the court shall order the files and records destroyed. If no release is submitted, the court shall order that all files and records containing identifiable, non-privileged information relating to the proceeding be sealed, and that they be disclosed only when and to the extent necessary for the prosecution or defense of a claim made on behalf of the subject of the records.

COMMENT. As indicated above, records can often outlast their usefulness, and may gain unwarranted credibility and importance by reason of their very existence rather than their accuracy and relevance.[149] In order to reduce both the risk of harm from unproven allegations and the likelihood of discrimination against individuals who have overcome or recovered from a disability, this subsection provides for the destruction of all identifiable records relating to a dismissed, withdrawn or terminated intervention proceeding. However, because the records of an intervention proceeding may be needed by the petitioner, a limited personal guardian, personal guardian, limited conservator or conservator, or a service provider to defend against a claim by the individual alleged or found to be a partially disabled or disabled person, a full release is required before a destruction order can be issued. When the subject of the record is unable or unwilling to provide such a release, the court can order that access to the records be limited exclusively to those directly involved in the litigation of a claim made by or on behalf of the subject of the records. Subsection (9) entitles the subjects of records to obtain a copy of their files prior to destruction or sealing so that they will be able to prove that the petition was dismissed or withdrawn or that a dispositional order was terminated.

(8) (a) Following the dismissal of a petition filed under section 19(1) of this act or the termination of a guardianship and/or conser-

---

145 For examples of such a provision, see 42 U.S.C. §3771(a) (1977); NACJJDP, *supra* note 100, at §0.535; NATIONAL ACADEMY OF SCIENCES, PROTECTING INDIVIDUAL PRIVACY IN EVALUATION RESEARCH, 7 (1975); and P. NEJELSKI AND H. PEYSER, A RESEARCHER'S SHIELD STATUTE: GUARDING AGAINST THE COMPULSORY DISCLOSURE OF RESEARCH DATA (1974).

146 For similar provisions see ILL ANN. STAT. ch. 91½, §805(3)(d) (Smith-Hurd Supp. 1979); and 20 U.S.C. §1232g(b)(2)(A) (Supp. 1976).

147 PRIVACY PROTECTION STUDY COMMISSION, *supra* note 129, at Appendix 1: *Privacy Law in the States.*

148 *E.g.*, ILL. ANN. STAT. Ch. 91½, §§801 *et seq.* (Smith-Hurd Supp. 1979); MHLP, *supra* note 129; PRIVACY PROTECTION STUDY COMMISSION, *supra* note 129; and ILLINOIS GOVERNOR'S COMMISSION, *supra* note 21.

149 MHLP, *Legal Issues in State Mental Health Care Proposals for Change - Suggested Statute on Civil Commitment*, 2 MEN. DIS. L. RPTR. 127, 151 (1977); IJA/ABA, *supra* note 129, at 67.

vatorship for a minor, the subject of any records retained under subsection (2) of this section may file an application to have destroyed identifiable non-privileged information contained in those records which is likely to harm or stigmatize that individual.

(b) At any time, the subject of a record retained pursuant to subsection (2) in conjunction with a proceeding initiated under chapter 3 of this act, may file an application to limit access to any identifiable, non-privileged information contained in those records which is likely to harm or stigmatize that individual.

(c) An application to destroy or limit access to information shall set forth:

(i) The name and address of the applicant;

(ii) A description of the identifiable, non-privileged information;

(iii) The name of the court, agency, or individual likely to have custody of the information;

(iv) The relief requested; and

(v) The facts and reasons why destruction of or limited access to the information is necessary.

(d) If the court determines that disclosure of the information specified in the application is likely to cause harm to the subject of that information or expose that individual to social stigma, it shall order that information destroyed or access to it limited to the individuals and entities enumerated in paragraph 5(a) of this section. No order to destroy identifiable information shall be issued unless the subject of the record files a full release of all claims arising from the intervention proceeding.

COMMENT. Subsection (4) of this section, places no general limits on access to records retained in conjunction with intervention proceedings initiated under chapter 3 of this act. However, this subsection provides a measure of protection when it can be proven that the record of the proceeding includes information which could prove harmful to the minor — e.g., the reasons underlying the termination of parental rights. Paragraph (8)(b) permits the court to limit access to such records during the course of the intervention proceeding or following the termination of a dispositional order. Paragraph (8)(a) allows records to be destroyed only after the dismissal of an unsupported petition or after the guardianship or conservatorship has ended, and the subject of the record has filed a release of all claims arising from the proceeding. The application should include all the information needed by the court to make a decision. The order should issue if there is a *prima facie* showing that the information described in the petition is likely to cause harm.

(9) (a) Copies of an order to destroy, seal, or limit access to files and records issued pursuant to subsections (7) and (8) of this section shall be sent to all courts, agencies and individuals which may have those files or records or have copies of or identifiable notations regarding such files and records. Courts, agencies and individuals receiving a copy of such an order shall promptly dispose of all copies of the file or record or the portion or notations thereof which they possess in the manner specified unless the information was obtained for research, evaluation or statistical purposes pursuant to paragraph (5)(d).

COMMENT. This paragraph provides for notifying individuals, courts, and agencies which have had access to a particular record that the record has been destroyed or sealed, or that access to it has been limited. In the event that destruction or sealing of the record has been ordered, they are obligated to destroy any copies or references to it contained in their files. An exception is made for identifiable information obtained for research, evaluation and statistical purposes under subparagraph (5)(a)(vii) and paragraph (5)(d).[150] This exception is to facilitate longitudinal and other long-term research studies as well as after-the-fact assessments — e.g. on the use and effectiveness of guardianship — but it is not intended to exempt persons who have obtained information for research, evaluative or statistical purposes from the civil and criminal sanctions provided by section 42(4) if they redisclose identifiable information in violation of this act.

(b) The subject of a record shall be entitled to receive a copy of that record, upon request, prior to its destruction or sealing.

COMMENT. See comment to subsection (7) of this section.

(10) (a) Subjects of a record shall not be deemed to have provided informed consent unless they have been advised of the information to be disclosed, the individuals and/or entities requesting disclosure, the purpose of the disclosure and the possible beneficial and harmful consequences of that disclosure. This explanation shall be in the language, mode of communication and terms which the subject of the record is most likely to understand.

(b) A limited personal guardian or personal guardian may give the necessary consent if he or she is empowered to do so in the dispositional order; has provided the explanation required in paragraph (10)(a) of this section to the partially disabled or disabled person or the minor; and, has obtained that individual's agreement to the maximum extent possible.

COMMENT. The stringency of the procedures employed to assure that consent to some action is

[150] See e.g., NACJJDP, *supra* note 100, at §0.56.

given knowingly and voluntarily by an individual with the mental capacity to understand the importance of his or her actions, is dependent upon the degree of risk, intrusiveness or irreversibility of the matter to which consent is to be provided.[151] Balancing these considerations in the context of a general rule governing release of identifiable information to service providers is extremely difficult because of the variety of contexts and circumstances involved. On the one hand, the subject of the record must have a basis for deciding whether the benefits from releasing the information outweigh the risks involved. On the other hand, the strictures must not be so severe as to impede the provision of necessary services. Accordingly, this subsection requires that the subject of record be given a full explanation of the reasons and implications of disclosure requested under subparagraph (5)(a)(viii) before consent can be accepted. In addition, in keeping with the requirement throughout this act that individuals subject to a guardianship should be encouraged to take part in making significant decisions which affect their lives, paragraph (10)(b) requires a limited personal guardian or personal guardian who has been assigned the power to consent to release of information, to explain the requested disclosure to the partially disabled or disabled person or the minor and, to the maximum extent possible, to obtain his or her assent to the disclosure. As with other explanations required by this act, the information regarding the proposed disclosure must be presented in the manner which is most likely to be comprehensible to the subject of the record. (See §32(2) Notice and Dissemination of Plans and Reports.)

Section 41. Bonds

(1) The court may require a limited conservator or conservator to have a surety satisfactory to the court furnish a bond conditioned upon the faithful discharge of all duties and powers assigned in the dispositional order or any modification thereof. Unless otherwise directed, the bond shall be in the amount of the aggregate capital value of the property in the limited conservator's or conservator's control plus one (1) year's estimated income minus the value of securities deposited under arrangements requiring an order of the court for their removal and the value of any land which the fiduciary, by express limitation of power, lacks power to sell or convey without court authorization. The court in lieu of sureties on a bond, may accept other security for the performance of the bond, including a pledge of securities or a mortgage of land.

(2) The following requirements and provisions apply to any bond required under this section:

(a) Unless otherwise provided by the terms of the approved bond, sureties are jointly and severally liable with the limited conservator or conservator and with each other;

(b) By executing an approved bond on a limited conservator or conservator, the surety consents to the jurisdiction of the court in any proceeding pertaining to the fiduciary duties of the limited conservator or conservator and naming the surety as a party defendant;

(c) On petition of successor limited conservator or conservator or any interested person, a proceeding may be initiated against a surety for breach of the obligation of the bond of the limited conservator or conservator;

(d) The bond on the limited conservator or conservator is not void after the first recovery but may be proceeded against at subsequent times until the whole penalty is exhausted.

(3) No proceeding may be commenced against the surety on any matter as to which an action or proceeding against the primary obligor is barred by adjudication or limitation.

COMMENT. The guardianship statute in almost every American jurisdiction authorizes the court to order a limited conservator or conservator to file a bond which will indemnify the subject of the intervention proceeding against waste or misappropriation of the financial resources over which the limited conservator or conservator has control. Most of these provisions also permit the court to waive bonding or to require some other form of guarantee. This section is based upon sections 5-411 and 5-412 of the U.P.C. It permits imposition of a bonding requirement, establishes guidelines for determining the amount of the bond, and sets forth the primary obligation of the surety. Section 32(1) provides for notification of sureties of any hearing under this act following their appointment.

Section 42. Liability of a Limited Personal Guardian, Personal Guardian, Limited Conservator or Conservator and Penalties for Violations of the Act

(1) By accepting appointment or designation as a limited personal guardian, personal guardian, limited conservator, or conservator, an individual or corporation submits personally to the jurisdiction of the court in any proceeding relating to the guardianship or conservatorship which may be instituted.

COMMENT. None necessary.

(2) (a) Unless otherwise provided in the contract, limited conservators and conservators shall not be individually liable on a contract properly entered into in their fiduciary capacity in the course of managing the financial resources of a partially disabled or disabled person or a minor, unless they fail to reveal their representative capacity and fail to identify the partially dis-

[151] AAMD, *supra* note 97, at 2.

abled or disabled person or the minor in the contract.

(b) Limited conservators and conservators shall not be individually liable for obligations arising from ownership or control of financial resources of a partially disabled or disabled person or a minor or for torts committed in the course of administration of such financial resources unless they are personally at fault.

(c) Claims based on contracts entered into by limited conservators or conservators in their fiduciary capacity, on obligations arising from ownership or control of the financial resources of partially disabled or disabled persons or minors, or on torts committed in the course of managing the financial resources of partially disabled or disabled persons or minors may be asserted by proceeding against limited conservators and conservators in their fiduciary capacity, whether or not they are individually liable therefor.

(d) Any question of liability between the financial resources of partially disabled or disabled persons or minors and the limited conservators or conservators individually, may be determined in a proceeding for accounting, surcharge, or indemnification, or other appropriate proceeding or action.

COMMENT. This subsection outlines basic principles governing the liability of limited conservators and conservators to third parties and is based upon section 5-429 of the U.P.C. Following the pattern found in many state codes, it applies the general rules applicable to fiduciaries to limited conservators and conservators. Paragraph (2)(c) simplifies the process of asserting a claim by providing that limited conservators and conservators can be named as the defendant whether or not they are personally liable. The issue of whether the claim, if valid, must be paid from the financial resources of the partially disabled or disabled person or minor, or by the limited conservator or conservator can then be determined during the hearing on the merits of the claim or at a separate proceeding not involving the claimant. (See paragraph (2)(d).) Thus, dismissals of valid claims for failing to name either the limited conservator or conservator, or the subject of the conservatorship as a party are avoided.

The same principles should govern the liability of limited personal guardians and personal guardians who are authorized to expend funds from the resources of a partially disabled or disabled person or minors pursuant to section 17(4)(vi) or 27(3)(g). (See also §§18(3)(d) and 28(3)(d) Duties and Powers of Limited Conservators and Conservators.)

(3) Any limited personal guardian, personal guardian, limited conservator or conservator who willfully violates the duties or misuses the powers assigned by the court and thereby causes harm to the subject of the guardianship or conservatorship or to the financial resources of that individual shall be subject to a fine of up to $10,000 for each such violation of duty or misuse of power, in addition to any civil liability to which the limited personal guardian, personal guardian, limited conservator or conservator may be subject.

COMMENT. This provision establishes a criminal penalty when an individual for whom a guardianship or conservatorship has been established, is harmed as a result of his or her limited personal guardian's, personal guardian's, limited conservator's or conservator's willful dereliction of duty or abuse of power. In most circumstances this offense may be charged in addition to other offenses such as embezzlement which may result from the actions of the individual or corporation serving in a guardianship or conservatorship capacity, since the elements for each offense would not be the same.[152] In enacting this provision, consideration should be given to placing it in the state criminal code rather than in the chapter on guardianship and conservatorship, and to adjusting the proposed penalty to be consistent with the existing system of sanctions.

The limited personal guardian, personal guardian, limited conservator or conservator remains civilly liable to the subject of the intervention proceeding even if a criminal penalty is imposed. In a civil proceeding, the subject of the guardianship and/or conservatorship should be entitled to receive both general and special damages, with punitive damages being assessed when the violation involves willful or gross negligence.

(4) (a) Any individual who willfully discloses identifiable information or who willfully gains access to identifiable information in violation of section 40(5) or section 40(8)(b) of this act, or who willfully violates an order issued pursuant to section 40(7) or section 40(8)(a) shall be subject to a fine of up to $5,000 for each violation.

If a public official or public employee is convicted under this paragraph, the conviction shall also be grounds for removal from office.

(b) The subject of identifiable information which has not been destroyed or sealed pursuant to an order issued in accordance with section 40(7) or section 40(8)(a) of this act, may initiate a civil action for damages or to restrain further dissemination. Any individual, court or public or private agency proven to have violated those sections shall be liable for actual damages and exemplary damages for each violation and shall be liable for court costs, expenses and reasonable attorney's fees incurred by the subject of the record.

152 Blockburger v. United States, 284 U.S. 299 (1932).

COMMENT. This subsection establishes criminal penalties, civil liability, and a right to injunctive relief for violations of the provisions restricting the availability of identifiable information collected in conjunction with an intervention proceeding.[153] The subject of the record should be entitled to receive both general and special damages, with punitive damages being assessed when the violation was willful or grossly negligent. The stringent penalties which may be imposed, especially those directed specifically toward public officials, are intended to assure that the privacy safeguards provided by section 40 will not be ignored. As with subsection (2) consideration should be given to placing paragraph (3)(a) in the criminal code and adjusting the maximum fine to be consistent with the sanctions for comparable offenses.

Section 43. Venue of Proceedings Subsequent to a Dispositional Order or an Uncontested Acceptance of a Testamentary Appointment

The court where the subject of an intervention proceeding resides, following issuance of a dispositional order or an uncontested acceptance of a testamentary designation, has concurrent jurisdiction with the court which issued the order or in which the acceptance was filed, over proceedings under sections 16, 25, 34, and 40 of this act and other proceedings related to the imposition of restrictions on the capacity of a person to act in his or her own behalf.

If the court in which a subsequent proceeding is commenced is not the court which issued the dispositional order or in which acceptance of testamentary designation was filed, it shall notify that court, and after consultation with the parties, determine whether to retain or transfer control of the proceedings. Venue shall be in the jurisdiction which is the most convenient for and best serves the interests of the subject of the intervention proceeding. The court which issued the dispositional order or in which acceptance of a testamentary designation is filed shall be sent a copy of any order resulting from a subsequent proceeding in another jurisdiction.

COMMENT. Since a guardianship or conservatorship may extend over a period of years, it is likely that a significant number of partially disabled or diabled persons and minors who have had a limited personal guardian, personal guardian, limited conservator or conservator appointed, may move away from the county or state in which the intervention proceeding was initiated. This section, which is based on section 5-313 of the U.P.C., provides for that contingency. Concurrent jurisdiction is established between the court in which the intervention proceeding was initiated and the court in the county or district to which the subject of the intervention proceeding has moved. Hence, an application for modification of a dispositional order may be filed in either court. If a hearing is required, it must be held in the court which will be the easiest for the subject of the intervention proceeding to attend and which will best serve his or her interests. Since the review hearings required by sections 15 and 24 are automatic — i.e., not initiated by an application or motion — they are not expressly included in this section. However, nothing in this provision is intended to prohibit the subject of an intervention proceeding from seeking a change of venue to the more convenient forum prior to a scheduled review hearing.

In order to assure that the records of the court in which the intervention proceeding was initiated are accurate and up-to-date (see §40(2) Records), it must be sent a copy of any order affecting the proceeding issued by another court.

Section 44. Burden of Proof

(1) (a) A motion to dismiss filed pursuant to section 7 of this act shall be granted unless the petitioner proves there is probable cause to believe that the respondent is a partially disabled or disabled person.

(b) Following a hearing held pursuant to section 10 of this act, a petition shall be dismissed unless the petitioner proves beyond a reasonable doubt that the respondent is a partially disabled or disabled person.

(c) Following a hearing held pursuant to section 12(1) of this act, no guardianship or conservatorship shall be established unless the petitioner proves beyond a reasonable doubt that the guardianship and/or conservatorship is needed and that no less restrictive dispositional alternative or combination of alternatives will be sufficient to satisfy that need. Any other dispositional alternative may be ordered upon clear and convincing proof by any party of the need therefor and in the case of the establishment of a trust or provision of residential services, that no less restrictive or combination of alternatives will be sufficient to satisfy that need.

(d) If a guardianship or conservatorship has been established, no specific restriction shall be imposed on the legal capacity of the partially disabled or disabled person unless the petitioner proves beyond a reasonable doubt that the restriction is needed and that no less restrictive alternative will meet that need.

(e) In conjunction with a review hearing held pursuant to section 15 of this act, no restriction on the legal capacity of the partially disabled or disabled person shall be continued or added unless the limited personal guardian, personal guardian, limited conservator and/or conservator proves beyond a reasonable doubt that the restriction is needed and that no less restrictive alternative will satisfy that need. All other action

[153] *See, e.g.*, MHLP, *supra* note 129, at 350; ILL. ANN. STAT. ch. 91½, §§815-816 (Smith-Hurd Supp. 1979).

may be taken when the basis therefor has been proven by clear and convincing evidence.

(f) (i) An application filed pursuant to section 16 which requests relief other than the imposition of any restriction upon the legal capacity of the partially disabled or disabled person to act on his or her own behalf shall be granted if the need for that relief is proven by clear and convincing evidence. Except when the application is filed by the partially disabled or disabled person, the burden of going forward and the burden of persuasion shall be on the applicant. When the application is filed by the partially disabled or disabled person, the burden of persuasion shall shift to the party or parties opposing the application if that person presents testimony and/or evidence constituting a *prima facie* case supporting the requested relief.

(ii) No application which requests the imposition of any additional restrictions shall be granted unless the applicant proves beyond a reasonable doubt that no less restrictive alternative will be sufficient.

COMMENT. Although often considered merely a procedural technicality, the burden of proof can be a powerful force in implementing a legislative policy such as the use of the least restrictive dispositional alternative. (See §2 Purpose and §12(2) and (3) Dispositional Hearing and Dispositional Principles and Criteria.) For example, by raising the level of proof required before the court orders one of the more restrictive alternatives, a predisposition is built in against utilization of that particular option. Conversely, alternatives which require less stringent proof, are used more frequently since they can be more employed with less difficulty. Perhaps more importantly:

> . . . [T]he choice of the standard for a particular variety of adjudication . . . reflect[s] a very fundamental assessment of the comparative social costs of erroneous factual determinations.[154]

The harsh penalties and opprobrium resulting from a criminal conviction are a major reason why proof beyond a reasonable doubt is constitutionally required in criminal cases.[155] The lesser social costs attached to an erroneous civil judgement permit decisions to be made at a lower level of certainty such as clear and convincing evidence or a mere preponderance of the evidence.[156] Because of the fundamental rights which are at stake and because of the social stigma which is all too often attached to being labeled a partially disabled or disabled person and to being made the subject of a guardianship and/or conservatorship,[157] intervention proceedings initiated under chapter 2 of this act are more analogous to criminal proceedings than civil proceedings.[158]

Accordingly, this subsection requires proof beyond a reasonable doubt before an individual can be declared to be a partially disabled or disabled person and before restrictions can be imposed on the legal capacity of a partially disabled or disabled person to act in his or her own behalf.[159] It does permit, however, the court to order dispositional alternatives which do not restrict the legal capacity of a partially disabled or disabled person and to grant other types of relief — *e.g.*, compensation for or replacement of a limited personal guardian, personal guardian, limited conservator or conservator — on the basis of clear and convincing evidence. For example, if an individual has been found, beyond a reasonable doubt, to be a partially disabled or disabled person, the court may order that a non-custodial service be provided to that individual if the need for that service is proven by clear and convincing evidence. However, a guardianship cannot be established unless there is proof beyond a reasonable doubt that no less restrictive alternative will enable that individual to meet the essential requirements for his or her physical health or safety. (See §12(3) Dispositional Hearings and Dispositional Principles and Criteria.)

It is not anticipated that this higher standard will hinder the establishment of a guardianship or conservatorship for those individuals who are in obvious need of a substitute decision-maker. It simply reinforces the other provisions designed to assure that the right of an individual to act on his or her own behalf is not limited except when no less restrictive alternative will suffice. In states in which a problem has developed with unnecessary placement of partially disabled or disabled persons in residential facilities, consideration should be given to applying the beyond-a-reasonable-doubt standard to residential placements as well.

The one exception to the insistence on a high standard of proof occurs when a motion to dismiss is filed. As noted in the comment to section 7, the motion to dismiss is intended as a mechanism for screening out groundless petitions at an early stage of the proceedings. It is anticipated that the probable cause test — i.e., facts which are sufficient to induce a reasonable belief in the truth of

[154] In re Winship, 397 U.S. 358, 370 (1969) (Harlan, J., concurring).

[155] *Id.* at 371.

[156] *Id.*

[157] See sources listed in note 127 *supra*.

[158] *But see* Addington v. Texas ___ U.S. ___ (1979).

[159] Intervention proceedings under this act can be distinguished from the commitment procedures examined by the Supreme Court in Addington v. Texas because of their greater dependence on factual questions and the availability of other dispositional alternatives to assist partially disabled and disabled persons denied guardianship and/or conservatorship. (See §10 Hearing and §11 Dispositional Alternatives.)

the allegations — will be able to serve this screening function without making the hearing held pursuant to section 10, a wholly duplicative effort.

In addition to setting the required level of proof, this section delineates who has the burden of going forward and the burden of persuasion — i.e., which party is responsible for presenting evidence and testimony which will establish the truth of the allegations before the court. Generally, both these burdens fall on the party who is requesting the court to act — e.g., the petitioner or applicant. If this were not the case, individuals alleged to be a partially disabled or disabled person would have to overcome the presumption that he or she required a personal guardian or conservator, a task which is often extremely difficult and which runs counter to the basic principles of fairness under American law. However, in the dispositional and review stages of intervention proceedings, three exceptions have been made to this general rule, where strict adherence would be detrimental to the partially disabled or disabled persons.

The first exception is the dispositional hearing. Once it has been determined that an individual is a partially disabled or disabled person, he or she should play a major role in identifying the type of services and other dispositional alternatives which should be ordered. Hence, no party is given sole responsibility for proposing a dispositional plan. Each party may propose alternatives from which the court may select, provided the requisite standard of proof has been met.

The second exception applies to the matters which may be raised at review hearings following the presentation by the limited personal guardian, personal guardian, limited conservator or conservator of proof beyond a reasonable doubt that the guardianship or conservatorship should continue and of clear and convincing evidence regarding compensation and reimbursement. Since the issues before the court at the review hearing are identical to those at the dispositional hearing, the same procedures should apply. (See §15 Periodic Review and §24 Periodic Review.)

The third exception is for applications submitted by partially disabled or disabled persons pursuant to section 16 to terminate or lessen the restrictions imposed in a dispositional order. To further enhance the opportunities of the subjects of intervention proceedings to meet their changing needs, paragraph (f)(i) shifts the burden of persuasion to the persons or entities opposing such applications once the partially disabled or disabled person has gone forward with evidence and/or testimony which, absent any evidence to the contrary, is sufficient to support his or her request. In this manner, partially disabled or disabled persons can have a stronger measure of control over who will be their limited personal guardian, personal guardian, limited conservator or conservator, and the restrictions imposed on their legal capacity to act in their own behalf. In order to protect against the misuse of this provision as a means of avoiding the safeguards which apply in other contexts, no shift occurs when the application seeks the imposition of additional restrictions.

In addition, when a motion to dismiss has been filed, the petitioner, rather than the party filing the motion, has both the burden of going forward and the burden of persuasion. This is consistent with the overall responsibility of the petitioner, as the initiator of the intervention proceeding, to support the allegations he or she has made, and with the purpose of the motion to dismiss described above. (See §7 Motion to Dismiss and the comments thereto.)

(2) (a) (i) A motion to dismiss filed pursuant to section 21 shall be granted unless the petitioner or the party objecting to a testamentary designation proves that there is probable cause to believe the allegations in the petition or in the objection are true.

(ii) In all other proceedings under chapter 3 of this act no finding shall be made nor action taken unless it is supported by a preponderance of the evidence.

(b) (i) At hearings held pursuant to section 22 following the filing of a petition under section 19(1) or an objection to a testamentary designation under section 20(2), both the burden of going forward and the burden of persuasion shall be on the petitioner or the individual or entity filing the objection.

(ii) At hearings held pursuant to section 22 following the filing of an objection to a testamentary designation by a minor pursuant to section 20(1), the burden of persuasion shall shift to the party or parties opposing the objection if the minor presents testimony and/or evidence constituting a *prima facie* case supporting the objection.

(iii) Following the filing of an application pursuant to section 25, the burden of going forward and the burden of persuasion shall be on the applicant except when the application is filed by the minor in which case the burden of persuasion shall shift to the party or parties opposing the application if the minor presents testimony and/or evidence constituting a *prima facie* case supporting the requested relief.

COMMENT. Given the traditional presumption in favor of guardianship and conservatorship for minors and the fact that the establishment of a guardianship or conservatorship does not involve the restriction of a minor's legal capacity to act on his or her own behalf (see §38(2) Effect . . . of a Guardianship or Conservatorship for a Minor), the social costs of an erroneous decision that one or more of the circumstances set forth in section 22(1) exist and that establishment of a guardianship or conservatorship for a minor is necessary, are far lower than those of an erroneous decision in an intervention proceeding involving an adult. Hence, the level of certainty required in all proceedings under chapter 3 other than hearings on a motion to dismiss, is a preponderance of the evidence rather than beyond a reasonable doubt.

As in proceedings under chapter 2, the burden of going forward and the burden of persuasion fall on the initiator of any action, except in those instances in which the preferences of the subject of the intervention proceeding are entitled to precedence — e.g., the right of a minor to choose his or her limited personal guardian, personal guardian, limited conservator or conservator. (See §20 and the comments thereto.) Hence, the burden of persuasion shifts to the party or parties opposing an objection to a testamentary designation or an application to modify a dispositional order filed by a minor when the minor presents *prima facie* evidence supporting the objection or application. (See the comment to subsection (1) for an explanation of "*prima facie.*") No specific provision is included for dispositional and review hearings since issues and requests may be raised by any party, although the proponent should have responsibility for presenting the support for any requested finding or court action.

In recognition of the growing concern over the rights of minors and the impact which the establishment of a guardianship or conservatorship may have on their lives, some states may wish to raise the level of proof required in any proceeding under Chapter 3 to clear and convincing evidence. Another alternative would be to strengthen the differential standard now included in the subsection by raising the level of proof and always placing the burden of persuasion on the party opposing the position adopted by the minor. For example, if an application was filed by the limited conservator of a minor and supported by the minor, the limited conservator would have both the burden of going forward and the burden of persuasion by a preponderance of the evidence. If the minor opposed that application, the limited conservator would have to present clear and convincing evidence. If the application was filed by the minor and opposed by the limited conservator and the minor went forward with a *prima facie* case, the application would be granted unless the limited conservator presented clear and convincing evidence in opposition.

(3) In proceedings held pursuant to sections 34(4)(b), 35(1)(d), 40(5)(b) and 40(8), the burden of persuasion shall be by a preponderance of the evidence. Both the burden of going forward and the burden of persuasion shall be on the applicant.

COMMENT. In hearings on the appointment of a guardian *ad litem* pursuant to Section 34, on an emergency appointment pursuant to section 35(1), and on access to or destruction of records, the preponderance of the evidence standard applies. With regard to guardians *ad litem,* this relatively low standard is in keeping with the informal nature of the hearing, and avoids making the decision to appoint a guardian *ad litem* determinative on the issue of whether an individual is a partially disabled or disabled person. With regard to emergency appointments, it is hoped that the requirement that there must be an imminent threat of serious harm will prevent abuse of the lowered standard for establishing a limited guardianship or conservatorship. As for records, there seemed little reason to go beyond the normal civil standard.

Section 45. Rules of Evidence/Rules of Procedure

Except where specified otherwise, the rules of evidence and rules of procedure, including those on discovery, which are applicable in civil matters shall govern all proceedings under this act.

COMMENT. Although in foregoing comments analogies have been drawn between intervention proceedings and criminal cases, the matters under this act remain civil in nature. Hence, the rules of civil procedure and rules of evidence applicable in civil cases apply to intervention proceedings except in those instances in which a specific procedure, right, time limit or standard is set in this act. (See *e.g.*, §30 Time Limits, §31 Rights and Privileges, §32 Notice and Dissemination of Plans and Reports, and §44 Burden of Proof.) Special mention is made of discovery — i.e., pretrial disclosure of evidence by all parties — because of the controversy which has often surrounded its extension outside the usual sphere of civil litigation. There seems little reason not to make discovery applicable to intervention proceedings in order to prevent increased reliance on gamesmanship and surprise as the proceedings become more formalized.

Since in many instances, the exercise of discovery procedures by the subject of an intervention proceeding will be at public expense, some states may wish to add a provision limiting use of depositions and interrogatories to cases in which a witness refuses to talk to opposing counsel or will not be available at the hearing.[160] In addition, consideration should be given to adding a provision which permits disclosure of diagnostic or other information which may harm the subject of an intervention proceeding to be made through a psychiatrist, psychologist, or other professional with the necessary skills and experience to reduce the likelihood and severity of any harmful effects. (See §32(2) Notice and Dissemination of Plans and Reports.)

[160] ARIZ. R. CRIM. P. 15.3 (1973).

Section 46. Effect on Existing Guardianship and Conservatorship

(1) (a) All guardianships and conservatorships established for partially disabled and disabled persons prior to the effective date of this act shall be reviewed by the court pursuant to section 15 within two (2) years. Until the review hearing, all limited personal guardians, personal guardians, limited conservators and conservators appointed prior to the effective date shall retain the powers assigned to them, unless an application has been granted under section 16. Prior to the review hearing, the limited personal guardian, personal guardian, limited conservator and/or conservator shall submit at least one report in the form prescribed in section 17(6) and/or 18(7), and an individual guardianship and/or conservatorship plan developed in the manner set forth in sections 17(2) and/or 18(2).

(b) Failure to hold a review hearing for appointments made prior to the effective date of this act within two (2) years of that date, shall result in the termination of the guardianship and/or conservatorship.

(2) (a) All limited personal guardians, personal guardians, limited conservators and/or conservators for minors appointed by court order or beginning their service following a testamentary appointment prior to the effective date of this act, shall retain the powers assigned to them by the court or by will unless an application has been granted under section 25 of this act, or one of the events specified in section 26 occurs terminating the guardianship or conservatorship. All such limited personal guardians and personal guardians shall submit a report in the form prescribed in section 27 no more than one (1) year after the effective date of this act and annually thereafter for the duration of the guardianship. All such limited conservators and conservators shall submit an individual conservatorship plan in the form specified in section 28(2) and a report in the form specified in section 28(7) no more than one (1) year after the effective date of this act and at least annually thereafter for the duration of the conservatorship.

COMMENT. This section establishes a transition period during which guardianships and conservatorships which exist at the time this act goes into effect can be reviewed and brought into compliance with its requirements. Since a guardianship or conservatorship may continue for a substantial period of time, maintaining two separate systems of intervention procedures would not only be administratively unfeasible, but may also be subject to constitutional challenge in view of the additional rights which this act will provide in many states. On the other hand, it would be detrimental to the well-being of most persons subject to a guardianship or conservatorship to abrogate immediately the duties and powers of their limited personal guardians, personal guardians, limited conservators, and/or conservators upon the effective date of the act. Similarly, it would be totally unrealistic to expect the courts to consider carefully a flood of petitions or review hearings within 30 days of the effective date. Accordingly, the limited personal guardians, personal guardians, limited conservators and conservators of partially disabled and disabled persons are given a two-year grace period in which to develop an individual guardianship and/or conservatorship plan, begin sending reports in the format prescribed by sections 17(6) and 18(7), and schedule a review hearing. The automatic termination of a guardianship or conservatorship if no review hearing has been held before the expiration of the transition period is to provide an incentive to all parties to comply with this provision. Comparable measures are contained in sections 15 and 24 to assure implementation of the review procedures for guardianships and conservatorships established pursuant to this act.

Consistent with the presumption favoring guardianships and conservatorships for minors and the finite termination date of such guardianships and conservatorships, a review hearing is not required for those established prior to the effective date of this act. However, the planning and reporting requirements contained in sections 27 and 28 are made applicable in order to provide minors subject to existing guardianships and conservatorships with a basis for influencing and/or modifying the actions, duties and powers of their limited personal guardians, personal guardians, limited conservators and/or conservators.

Section 47. Authorization of Appropriations

To carry out the purposes of this act, there is authorized to be appropriated $________ for the fiscal year ending ________, $________ for the fiscal year ending ________, and $________ for the fiscal year ending ________.

COMMENT. Although every effort has been made to limit the cost of implementing this act, most states will need to appropriate or transfer some additional funds in order to make the procedures, provisions and safeguards established by this act fully effective. Among the specific provisions which are likely to have some fiscal impact are sections 5, 36, and 39 on public guardianship, sections 5(4)(a) and 8 on evaluation, section 31(2) on provision of counsel, and section 4 on disabilities resources officers. It should be noted, however, that the emphasis on use of the least restrictive alternative and the procedural protections contained in the act will help to eliminate the costs incurred by unnecessary guardianships and conservatorships.

Section 48. Severability of Sections

If any section, subsection, paragraph, sentence, or any other part of this act is adjudged unconstitutional or invalid, such judgment shall not affect, impair or invalidate the remainder of this act, but shall be confined to the section, subsection, paragraph, sentence or other part of this act directly involved in the controversy in which that judgment was rendered.

COMMENT. None necessary.

Section 49. Effective Date

This act takes effect ______

COMMENT. None necessary.

OPTIONAL SHORT FORM MODEL GUARDIANSHIP AND CONSERVATORSHIP ACT

INTRODUCTION

The complete text of the model guardianship and conservatorship bill is detailed and lengthy. Some states may favor a shorter statute. The following short form bill is offered where this is the case.

The short form bill lacks much of the administrative detail of the complete bill but retains the statement of purpose, definitions, organizational structure, procedures, powers and rights of the longer version.

The comment and alternative portions of the complete text should be consulted for a discussion of the issues, options and rationales for choices made. They should also be useful in adapting the short form bill to a particular state's needs.

SHORT FORM MODEL STATUTE

A BILL FOR

AN ACT TO REVISE THE METHODS, CRITERIA AND PROCEDURES FOR PROTECTING PARTIALLY DISABLED AND DISABLED PERSONS AND MINORS AND THE PROPERTY OF SUCH PERSONS AND MINORS; TO ESTABLISH THE CRITERIA FOR APPOINTING LIMITED PERSONAL GUARDIANS, PERSONAL GUARDIANS, LIMITED CONSERVATORS AND CONSERVATORS AND THEIR RESPONSIBILITIES, DUTIES AND POWERS; AND TO SPECIFY THE RIGHTS OF INDIVIDUALS SUBJECT TO INTERVENTION PROCEEDINGS AND DISPOSITIONAL ORDERS.

CHAPTER 1 INTRODUCTORY PROVISIONS

Section 1. Title

This act shall be known as the Guardianship and Conservatorship Act.

Section 2. Purpose

Recognizing that every individual has unique needs and differing abilities, the (name of Legislature) declares that it is the purpose of this act to promote the general welfare of all citizens by establishing a system which permits partially disabled and disabled persons and minors to participate as fully as possible in all decisions which affect them; which assists such persons in meeting the essential requirements for their physical health and safety, protecting their rights, managing their financial resources, and developing or regaining their abilities to the maximum extent possible; and which accomplishes these objectives through the use of the least restrictive alternatives. This act shall be liberally construed to accomplish this purpose.

Section 3. Definitions

As used in this act:

(1) "Partially disabled persons" means adults whose ability to receive and evaluate information effectively and/or communicate decisions is impaired to the extent that they lack the capacity to manage at least some of their financial resources and/or meet at least some of the essential requirements for their physical health or safety without court-ordered assistance or appointment of a limited personal guardian or limited conservator.

(2) "Disabled persons" means adults whose ability to receive and evaluate information effectively and/or to communicate decisions is impaired to such an extent that they lack the capacity to manage their financial resources and/or to meet essential requirements for their physical health or safety even with court-ordered assistance or the appointment of a limited personal guardian or limited conservator.

(3) "Manage financial resources" means those actions necessary to obtain, administer, and dispose of real and personal property, intangible property, business property, benefits and income.

(4) "Meet essential requirements for physical health or safety" means those actions necessary to provide the health care, food, shelter, clothing, personal hygiene and other care without which serious physical injury or illness is more likely than not to occur.

(5) "Court-ordered assistance" means dispositional alternatives available under section 10 of this act other than appointment of a limited personal guardian, personal guardian, limited conservator and/or conservator.

(6) "Minor" means an individual [age seventeen (17) or less/under the age of majority].

(7) "Respondent" means an individual alleged to be a partially disabled or disabled person.

(8) "Subject of an intervention proceeding" means the respondent, a minor for whom a proceeding has been initiated under chapter 3 of this act, or an adult or minor for whom a limited personal guardian, personal guardian, limited conservator or conservator has been appointed.

(9) "Parties" means the individual or entity filing a petition, application, motion, acceptance of a testamentary designation, or objection; the subject of an intervention proceeding; and the limited personal guardian, personal guardian, limited conservator or conservator, if any.

(10) "Intervention proceeding" means any proceeding under this act.

(11) "Limited conservator" means an individual or a corporation with general power to serve as trustee, appointed by the court to assist in managing the financial resources of a partially disabled person or of a minor.

(12) "Conservator" means an individual or a corporation with general power to serve as trustee, appointed by the court to manage the financial resources of a disabled person or a minor.

(13) "Conservatorship" includes appointment of a limited conservator or a conservator.

(14) "Limited personal guardian" means an individual appointed by the court to assist a partially disabled person or a minor in meeting the essential requirements for the partially disabled person's or the minor's physical health and safety.

(15) "Personal guardian" means an individual appointed by the court to act on behalf of a disabled person or a minor to assure that the essential requirements for the health and safety of the disabled person or minor are met.

(16) "Guardianship" includes appointment of a limited personal guardian or a personal guardian.

(17) "Volunteer public guardian;" "volunteer public conservator," and "volunteer public trustee," mean an individual appointed pursuant to section 33 to serve as a limited personal guardian or personal guardian, or an individual or corporation appointed to serve as a limited conservator, conservator or trustee for a partially disabled or disabled person or minor who require establishment of a guardianship or conservatorship and who has no one able and willing to serve in that capacity or is unable to afford the cost of a guardianship, conservatorship and/or trust.

(18) "Guardians *ad litem*" means individuals appointed by the court to assist the subject of an intervention proceeding to determine his or her interests in regard to the intervention proceeding, or to make that determination if the subject of the intervention proceeding is unconscious or otherwise wholly incapable of determining his or her interests in that proceeding even with assistance.

(19) "Counsel" means an attorney admitted to the practice of law in this state.

(20) "Restrictions on the legal capacity of a person to act in his or her own behalf" means powers of a partially disabled or disabled person or a minor which are assigned to a limited personal guardian, personal guardian, limited conservator or conservator.

(21) "Disabilities resource officers" means employees of the (name of state/local agency responsible for providing services to partially disabled and disabled persons) who are assigned the duties set forth in section 4 of this act.

(22) "Multidisciplinary evaluation team" means a group of at least three (3) individuals which is appointed by the court from a panel of experts approved by the (*state name*) Guardianship/Conservatorship Oversight Commission, to evaluate the subject of an intervention proceeding pursuant to chapter 2 of this act.

(23) "Evaluation" means an assessment of the ability of an individual to receive and evaluate information effectively and/or to communicate decisions and the impact of any impairment of these skills on the individual's capacity to meet the essential requirements for his or her physical health or safety, and/or manage his or her financial resources.

(24) "Least restrictive dispositional alternative" means the form of assistance that least interferes with the legal capacity of a respondent to act in his or her own behalf.

(25) "Developmental disability" means a chronic disability of a person which:

(a) Is attributable to a mental or physical impairment or a combination of mental and physical impairments;

(b) Is likely to continue indefinitely;

(c) Results in substantial functional limitations in two (2) or more of the following areas of major life activity:

(i) Self-care;

(ii) Receptive and expressive language;

(iii) Learning;

(iv) Mobility;

(v) Self-direction;

(vi) Capacity for independent living; or

(vii) Economic self-sufficiency; and

(d) Reflects the need for a combination and sequence of special, interdisciplinary or generic care, treatment or other services which are of lifelong or extended duration and individually planned and coordinated.

(26) "Identifiable information" means information which is indexed or able to be retrieved by name, identifying code or number, address or other personal characteristic.

Section 4. Appointment of Disabilities Resources Officers

(1) The head of the (state/local agency responsible for providing services to partially disabled or disabled persons) shall employ individuals on a full-time or part-time basis to serve as disabilities resources officers (in accordance with citation to state Civil Service Act or statute, if any, authorizing local jurisdictions to hire personnel). The number of individuals hired to be or assigned the duties of disabilities resources officers shall be commensurate with the volume of intervention proceedings initiated and the number of guardianships and conservatorships for partially disabled or disabled persons established in the [state/county/parish/city], but shall be sufficient to permit access for those individuals requiring the services of a disabilities resources officer.

(2) (a) Disabilities resources officers shall:

(i) Develop and maintain a current knowledge of the full range of public and private medical, mental health, social, advocacy, educational, rehabilitative, counseling, therapeutic, homemaking, recreational and financial services and programs available to assist partially disabled and disabled persons and their families;

(ii) Discuss personally with individuals seeking to file a petition pursuant to section 6 of this act or an application pursuant to section 32(1), the matters which the individual wishes to allege in the petition or application, the effects of the appointment of a limited personal guardian, personal guardian, limited conservator and/or conservator, and the less restrictive alternatives which are available;

(iii) Assist partially disabled and disabled persons, and individuals appointed to serve as a limited personal guardian, personal guardian, limited conservator, and/or conservator to develop and update individual guardianship and individual conservatorship plans;

(iv) Provide information, guidance, counseling and referral services to individuals regarding the alternatives which are available to assist persons with developmental and/or other disabilities;

(v) Exercise such other duties and powers as are prescribed by this act.

Section 5. Establishment of the (state name) Guardianship/Conservatorship Oversight Commission

(1) There is hereby established the (state name) Guardianship/Conservatorship Oversight Commission (hereinafter referred to as the Oversight Commission) which shall be independent of all other government agencies, boards, and commissions.

(2) (a) The Oversight Commission shall consist of five (5) individuals appointed by the Governor, with the advice and consent of the Senate to three (3) year terms, except that of the first appointments, two (2) shall be for two (2) year terms, and one (1) shall be for a one (1) year term. A member of the Oversight Commission shall not serve more than two (2) consecutive terms. Vacancies shall be filled in the same manner as original appointments. Appointments to fill vacancies occurring before the expiration of a term are for the remainder of the unexpired term. Commission members whose term has expired shall continue to serve until a new appointment has been made and confirmed.

(b) At least one (1) member of the Oversight Commission shall have previously been the subject of a guardianship or conservatorship; at least one (1) member shall have served or shall be serving as a limited personal guardian, personal guardian, limited conservator, and/or conservator; and at least one (1) member shall be an attorney who regularly represents individuals with developmental disabilities, individuals who are mentally ill, individuals over age sixty-five (65), or individuals under age eighteen (18).

(3) The Oversight Commission shall meet at least six (6) times per year. Members of the Oversight Commission not employed by the state government shall receive compensation at a rate not to exceed $________ for each day, including travel time, they are engaged in the performance of their duties as members of the Oversight Commission. In addition, all members of the

Oversight Commission shall be entitled to reimbursement for travel, subsistence and other reasonable expenses incurred by them in carrying out their duties.

(4) The Oversight Commission shall:

(a) (i) Establish and annually publish the procedures and criteria for identifying individuals who are qualified to serve on multidisciplinary evaluation teams;

(ii) Prepare and publish, annually, a list of individuals from all sections of the state who are qualified and willing to serve on multidisciplinary evaluation teams; and

(iii) Establish and annually publish the procedures and criteria for removing individuals from the list who prove to be unqualified to serve on multidisciplinary evaluation teams.

(b) (i) Establish programs to encourage individuals to serve as volunteer public guardians, and individuals and corporations to serve as volunteer conservators or volunteer public trustees;

(ii) Prepare and publish, annually, a list of individuals from all sections of the state who are willing to serve as volunteer public guardians, and a list of individuals and corporations from all sections of the state who are willing to serve as volunteer public conservators and volunteer public trustees;

(iii) Establish procedures to assure that volunteer public guardians, volunteer public conservators, and volunteer public trustees are selected for appointment and compensated in a lawful, equitable and consistent manner;

(c) Establish and maintain training programs which will provide instruction on the general and specific duties of limited personal guardians, personal guardians, limited conservators, and conservators, the nature and extent of the disabilities which most commonly result in establishment of a guardianship or conservatorship, and the programs, services and techniques available to assist and protect partially disabled and disabled persons and minors;

(d) Conduct investigations and hearings regarding the operation of the volunteer public guardian/conservator/trustee program and the selection, appointment, compensation and removal of volunteer public guardians, volunteer public conservators and volunteer public trustees;

(e) Testify before the (name of the Legislature) on matters related to the provision of services for and the guardianship and conservatorship of partially disabled persons, disabled persons and minors;

(f) Advise the Governor and other state and local officials and agencies on matters related to the provision of services for and the guardianship and conservatorship of partially disabled persons, disabled persons, and minors;

(g) Promulgate such additional policies, rules, regulations and procedures as are necessary for the effective operation of the Oversight Commission;

(h) Employ, in accordance with the (citation to the state civil service code) and the appropriations provided, an Administrative Director and such professional, technical, clerical and administrative personnel as may be necessary to carry out the duties and powers set forth in this act;

(i) Contract with individuals, businesses, whether for profit or not-for-profit, and other public and private organizations or agencies to carry out the duties and powers set forth in this act; and

(j) Perform such additional duties and exercise such powers as are prescribed by law or as are necessary for the effective management and operation of the Oversight Commission in accordance with the purpose of this act.

(5) The Oversight Commission shall not have supervisory power over individual intervention proceedings nor shall it manage or operate residential facilities or directly provide services other than those specified in this act for partially disabled or disabled persons or minors.

(6) The Oversight Commission shall submit an annual report to the Governor, the (name of the Legislature), the Chief Justice of (name of the highest court of civil appeals), the (head of the state agencies responsible for providing services to partially disabled and disabled persons and minors) and the (head of the state Developmental Disabilities Advocacy Office). Copies of the report shall be made available to the public and shall be sent to each court which can hear intervention proceedings.

CHAPTER 2 INTERVENTION PROCEEDINGS FOR PARTIALLY DISABLED AND DISABLED PERSONS

Section 6. Initiation of Intervention Proceedings

(1) (a) An intervention proceeding shall be initiated by the filing of a petition with the court by a partially disabled person, an adult interested in the welfare of a partially disabled or disabled person, or by a limited personal guardian, personal guardian, limited conservator or conservator of such a person.

(b) The petition shall be verified under oath and shall specify:

(i) The names and addresses of the individuals and entities entitled to notice pursuant to section 29(1)(a) of this act;

(ii) The nature and degree of the alleged disability;

(iii) The relief requested and the facts and reasons supporting the need for such relief; and

(iv) A statement that the petitioner has discussed with the disabilities resources officer the facts alleged in the petition, the effect of the appointment of a limited personal guardian, personal guardian, limited conservator or conservator, and the alternative services and procedures which are available, and an explanation why the requested protection and assistance is the least restrictive dispositional alternative which will meet the needs of the respondent.

(2) A parent, spouse, or child of a partially disabled or disabled person who is serving as limited personal guardian, personal guardian, limited conservator or conservator may, by will, designate an individual to serve in that capacity upon the death of the testator. Such a designation shall not become effective if another person is serving as the limited personal guardian, personal guardian, limited conservator and/or conservator at the time of death, nor until the will has been filed for probate and the court has appointed the designee following a hearing pursuant to section 15 of this act. Pending such appointment, the designee may file for appointment on an emergency basis pursuant to section 32(1) of this act.

Section 7. Motion to Dismiss

A motion to dismiss may be filed by any party when it appears that the ability of the respondent to receive and evaluate information and/or to communicate decisions is not impaired or when it appears that the respondent does not require court-ordered assistance or the appointment of a limited personal guardian, personal guardian, limited conservator, and/or conservator.

Section 8. Evaluation

(1) The court shall appoint a multidisciplinary evaluation team from the list published by the Guardianship/Conservatorship Oversight Commission to evaluate the respondent. Each party may submit names drawn from this list for the court's consideration.

(2) The evaluation shall be conducted so as to minimize interference with the respondent's activities and intrusion into the respondent's privacy.

(3) The report of the multidisciplinary evaluation team shall include:

(a) A description of the nature and extent of the respondent's disabilities, if any;

(b) A description of the respondent's mental, emotional, physical, and educational condition, adaptive behavior and social skills;

(c) A description of the services, if any, being used by the respondent to meet the essential requirements for his or her physical health or safety, and/or manage his or her financial resources, and/or develop or regain his or her abilities;

(d) An opinion regarding the type(s) and extent of assistance, if any, required by the respondent, and why no less restrictive alternative would be appropriate;

(e) An opinion regarding the probability that the extent of the respondent's disabilities, if any, may significantly lessen, and the type of services or treatment which will facilitate improvement in the respondent's condition, behavior or skills;

(f) A description of the techniques or tests used;

(g) The names and addresses of all individuals who examined, interviewed or investigated the respondent in conjunction with the evaluation;

(h) The signatures of each member of the multidisciplinary evaluation team together with a statement of concurrence or nonconcurrence with the evaluation findings; and

(i) Any dissenting opinions or other comments by members of the multidisciplinary evaluation team.

Section 9. Hearing

Within the time limit prescribed in section 27(1)(b) of this act, the court shall hold a hearing to determine:

(1) The essential requirements for the respondent's physical health and safety and/or the type and amount of the respondent's financial resources;

(2) The skills and knowledge necessary to meet those needs and/or manage those resources;

(3) The nature and extent of the respondent's disabilities, if any; and

(4) Whether the respondent is a partially disabled or disabled person.

Section 10. Dispositional Alternatives

(1) If a respondent is found not to be a partially disabled or disabled person the petition shall be dismissed.

(2) Among the dispositional alternatives available to the court following a finding that the respondent is a partially disabled or disabled person, are:

(a) Directing the provision of necessary services to the respondent on a noncustodial basis;

(b) Authorizing or ratifying a contract or transaction entered into by a partially disabled or disabled person;

(c) Directing establishment of a trust of which the respondent and any individuals legally dependent on the respondent are the beneficiaries;

(d) Directing the provision of residential services to the respondent in a small, licensed, community-based residential home;

(e) Appointing a limited conservator or conservator; and

(f) Appointing a limited personal guardian or personal guardian.

Section 11. Dispositional Hearing, and Dispositional Principles and Criteria

(1) Following a finding that the respondent is a partially disabled or disabled person, the parties shall have an opportunity within the time limits prescribed in section 27(1)(c) of this act, to present evidence and testimony regarding the types of assistance and protection required, the services available for consideration by the court in making the dispositional decision, and the selection, if any, of a limited personal guardian, personal guardian, limited conservator or conservator.

(2) The court shall impose the least restrictive dispositional alternative which will assist the respondent to:

(a) Meet the essential requirements for his or her physical health or safety; and/or

(b) Manage his or her financial resources; and

(c) Develop or regain to the maximum extent possible the skills needed to perform the functions listed in paragraphs (2)(a)-(2)(b) of this section without court-ordered assistance.

Dispositions shall be designed to encourage respondents to participate in all decisions which affect them and to act on their own behalf to the maximum extent possible.

(3) In selecting among the available dispositional alternatives the court shall not:

(a) Order that a non-custodial service be provided unless the respondent's need for that particular service has been proven;

(b) Authorize or ratify a contract or transaction unless the need for authorization or ratification has been proven;

(c) Order the provision of residential services or the establishment of a trust or a guardianship or conservatorship unless the need therefor has been proven and no less restrictive dispositional alternative or combination of alternatives will be sufficient to satisfy that need.

(d) Assign a duty or power to a limited personal guardian, personal guardian, limited conservator, or conservator appointed in accordance with paragraph 3(c) of this section, unless the need therefor has been proven and no less restrictive dispositional alternative or combination of alternatives will be sufficient to satisfy that need.

Section 12. Selection of a Limited Personal Guardian, Personal Guardian, Limited Conservator or Conservator

(1) (a) The court may appoint any non-disabled adult to serve as a limited personal guardian or personal guardian, and any non-disabled adult or any corporation with general power to serve as trustee, to serve as a limited conservator or conservator unless that individual or corporation:

(i) Is providing or is responsible for providing some treatment or service other than a guardianship or conservatorship service for the respondent; or

(ii) Has interests which may otherwise conflict with those of the respondent.

(b) Nothing in this provision shall exclude the appointment of the respondent's parent, spouse, child or sibling when the conflict of interest is determined to be insubstantial and such an appointment would clearly be the most beneficial to the respondent.

(2) (a) The following priorities shall guide the court's selection of a limited personal guardian, personal guardian, limited conservator, or conservator from among those eligible:

(i) The individual or corporation nominated by the respondent;

(ii) The current limited personal guardian, personal guardian, limited conservator, or conservator appointed or recognized by the appropriate court of any other jurisdiction in which the respondent resides;

(iii) The respondent's spouse;

(iv) An adult child of the respondent;

(v) A parent of the respondent;

(vi) An individual or corporation nominated by the will of a deceased parent;

(vii) Any individual with whom the respondent has been living for more than six (6) months prior to the filing of the petition;

(viii) A sibling of the respondent;

(ix) A volunteer public guardian and/or a volunteer public conservator.

(b) The court may select the individual or corporation best qualified and most willing to serve from among those of equal priority. The court may pass over a person or corporation having priority and appoint a person having less priority or no priority when it is demonstrated that the selection will be of greater benefit to the partially disabled person or disabled person. Appointments of volunteer public guardians and volunteer public conservators shall be in accordance with section 33.

Section 13. Dispositional Order

(1) The dispositional order shall set forth:

(a) The specific terms or conditions of the disposition;

(b) The name and address of the individual, if any, appointed to serve as the limited personal guardian or personal guardian and/or the individual or corporation, if any, appointed to serve as limited conservator or conservator;

(c) The duties and powers of the limited personal guardian, personal guardian, limited conservator and/or conservator, if any; and

(d) The date of the initial review hearing.

(2) The court shall explain on the record the terms and conditions contained in the dispositional order, the facts and reasons which support the decision not to impose any less restrictive alternative, and if a guardianship or conservatorship has been established, the facts and reasons supporting the selection of the individual and/or corporation appointed. The court shall also explain to the respondent the procedures set forth in section 15 for terminating or modifying the dispositional order.

Section 14. Periodic Review

(1) (a) The court shall hold a review hearing following the appointment of a limited personal guardian and/or limited conservator within the time limit prescribed in section 27(1)(d) of this act.

(b) The court shall hold a review hearing following the appointment of a personal guardian and/or conservator within the time limits prescribed in section 27(1)(e) of this act.

(2) At a review hearing the court shall:

(a) Determine why the appointment should not be terminated or why no less restrictive alternative will be sufficient to assist the respondent to meet the essential requirements for his or her physical health or safety and/or manage his or her financial resources;

(b) Approve the payment of compensation and reimbursement of expenses, if any, from the financial resources of the partially disabled or to the limited personal guardian, personal guardian, limited conservator and/or conservator; and

(c) Resolve any other relevant matters which the parties raise.

(3) In conjunction with a review hearing, the court may order an evaluation of the partially disabled or disabled person.

(4) At the conclusion of a review hearing the court may:

(a) Terminate a guardianship or conservatorship;

(b) Modify any or all duties and powers assigned to a limited personal guardian, personal guardian, limited conservator or conservator;

(c) Discharge and/or appoint individuals to serve as the limited personal guardian, personal guardian, limited conservator and/or conservator;

(d) Modify the individual guardianship and/or conservatorship plans; and

(e) Continue the current restrictions upon legal capacity and continue the guardianship and/or conservatorship appointments until the next review hearing.

Section 15. Modification and Termination

(1) In addition to the review procedures prescribed under section 14 of this act, a partially disabled or disabled person, the limited personal guardian, personal guardian, limited conservator, and/or conservator of that person, or any other interested individual may apply to the court for:

(a) Termination or modification of the dispositional order;

(b) Approval or disapproval of the compensation or reimbursement for expenses requested by the limited personal guardian, personal guardian, limited conservator or conservator;

(c) Replacement of the limited personal guardian, personal guardian, limited conservator or conservator; or

(d) Resolution of a dispute over the individual guardianship plan or individual conservatorship plan.

(2) (a) An application from a partially disabled or disabled person may be communicated to the court by any means, including but not limited to oral communication or informal letter. Upon receipt of such an informal request, the court shall contact the partially disabled or disabled person's attorney of record or appoint new counsel to prepare a formal application.

(b) Applications not filed by a partially disabled or disabled person and formal applications filed pursuant to paragraph (2)(a) of this section shall set forth:

(i) The names and addresses of the individuals and entities entitled to notice;

(ii) The relief requested; and

(iii) The facts and reasons supporting the request.

(3) Any party may object to the relief requested in an application.

(4) (a) The court shall hold a hearing on all applications:

(i) Which are filed pursuant to section 32 of this act;

(ii) Which request imposition of additional restrictions upon the legal capacity of a partially disabled or disabled person to act on his or her own behalf; or

(iii) To which an objection has been filed.

(b) All other applications may be granted or denied without a hearing.

(5) At the request of any party, the court shall order an evaluation of the partially disabled or disabled person identified in the application.

(6) After notice, the court may join the issues raised in separate applications and/or separate objections for determination at a single hearing, unless joinder would be prejudicial to the interests of the partially disabled or disabled person.

(7) The principles, criteria and priorities set forth in sections 11 and 12 of this act and the form and procedures prescribed in section 13 shall apply to dispositional orders issued following submission of an application under this section.

Section 16. Duties and Powers of Limited Personal Guardians and Personal Guardians

(1) It is the general duty of individuals appointed to serve as a limited personal guardian or personal guardian to carry out diligently and in good faith the specific duties and powers assigned by the court. In carrying out these duties and powers, such individuals shall:

(a) Assure that the personal, civil, and human rights of the partially disabled or disabled persons whom they are serving are protected; and

(b) Encourage those persons to:

(i) Participate to the maximum extent of their abilities in all decisions which affect them;

(ii) Act on their own behalf on all matters in which they are able to do so; and

(iii) Develop or regain, to the maximum extent possible, their capacity to meet the essential requirements for their physical health or safety, and/or to manage their financial resources.

(2) Within thirty (30) days of appointment, a limited personal guardian or personal guardian shall submit to the court an individual guardianship plan developed together with the disabilities resources officers, the limited conservator or conservator if any, and, to the maximum extent possible, the partially disabled or disabled person. The plan shall specify:

(a) The services which are necessary to meet the essential requirements for the partially disabled or disabled person's physical health or safety;

(b) The means for obtaining these services;

(c) The manner in which the limited personal guardian or personal guardian, the partially disabled or disabled person, and the limited conservator or conservator, if a corporation or another individual has been appointed to serve in that capacity, will exercise and share decision-making authority; and

(d) Such other items as will assist in fulfilling the needs of the partially disabled or disabled person, the terms of the dispositional order, and the duties of the limited personal guardian or personal guardian.

(3) (a) The court may assign to a limited personal guardian any portion of the duties and powers listed in subparagraphs (4)(a)(ii)-(vi) of this section for those particular areas in which the partially disabled person lacks the capacity to meet the essential requirements for his or her physical health or safety.

(b) A limited personal guardian may also be assigned the duty to assist the partially disabled person in those particular areas in which that person's capacity to meet the essential requirements for his or her physical health or safety, and/or to protect his or her rights, and/or to obtain necessary services, and/or to fulfill his or her civil duties is impaired, as well as in other ways not specifically prohibited by subsection (5) of this section.

(4) (a) The court shall assign to a personal guardian the duty and/or power to:

(i) Take custody of the disabled person;

(ii) Assure that the disabled person has a place of abode in the least restrictive, most normal setting consistent with the essential requirements for his or her physical health or safety;

(iii) Assure the care and comfort of the disabled person;

(iv) Assure that the disabled person receives the services necessary to meet the essential requirements for that person's physical health and safety and necessary to develop or regain, to the maximum extent possible, the capacity to meet the essential requirements for his or her physical health or safety;

(v) Provide any required consents or approvals on behalf of the disabled person; and

(vi) Expend sums from the financial resources of the disabled person to cover the reasonable costs incurred on behalf of that person subject to the procedures and policies established by the limited conservator or conservator, if any.

(b) A personal guardian may also be assigned the duty to assist the disabled person to fulfill that person's civil duties and in other ways not specifically prohibited by subsection (5) of this section.

(5) A limited personal guardian or a personal guardian shall not have the power:

(a) To place a partially disabled or disabled person in a facility or institution to which an individual without a guardian would have to be committed under (citation to the applicable state civil commitment laws), other than through a formal commitment proceeding in which the partially disabled or disabled person has independent counsel and a separate guardian *ad litem;*

(b) To consent on behalf of a partially disabled or disabled person to an abortion, sterilization, psychosurgery, or removal of a bodily organ, except as provided in (citation to Model Personal and Civil Rights Act) or when necessary to preserve the life or prevent serious impairment of the physical health of that person;

(c) To consent on behalf of the partially disabled or disabled person to the withholding of non-heroic life-saving medical procedures;

(d) To consent on behalf of a partially disabled or disabled person to the performance of any experimental biomedical or behavioral procedure or participation in any biomedical or behavioral experiment unless:

(i) It is intended to preserve the life or prevent serious impairment of the physical health of the partially disabled or disabled person; or

(ii) It is intended to assist the partially disabled or disabled person to develop or regain his or her abilities and has been approved for that person by the court;

(e) To prohibit the marriage or divorce of a partially disabled or disabled person; and

(f) To consent on behalf of a partially disabled or disabled person to the termination of that person's parental rights.

(6) (a) A limited personal guardian or personal guardian shall submit a report to the court at least ten (10) days before the initial review hearing and:

(i) Annually thereafter;

(ii) When the court orders additional reports to be filed;

(iii) When there is a significant change in the capacity of the partially disabled or disabled person to meet the essential requirements for his or her physical health or safety;

(iv) When the limited personal guardian or personal guardian resigns or is removed; and

(v) When the guardianship is terminated.

(b) Such reports shall set forth:

(i) The name and address of the partially disabled or disabled person, and of the limited personal guardian or personal guardian;

(ii) Significant changes in the capacity of a partially disabled or disabled person to meet the essential requirements for his or her physical health or safety;

(iii) The services being provided to the partially disabled or disabled person and the relationship of those services to the individual guardianship plan;

(iv) The significant actions taken by the limited personal guardian or personal guardian during the reporting period;

(v) The compensation requested and reasonable and necessary expenses incurred by the limited personal guardian or personal guardian;

(vi) Any significant problems relating to the guardianship which have arisen during the reporting period; and

(vii) The reasons, if any, why the appointment should not be terminated or why no less restrictive alternative will permit the partially disabled or disabled person to meet the essential requirements for his or her physical health or safety.

Attached to the report shall be an accounting of any expenditures made by the limited personal guardian or personal guardian, and an update of the individual guardianship plan developed by the limited personal guardian or personal guardian, the limited conservator or conservator if any, the disabilities resources officer, and to the maximum extent possible, the partially disabled or disabled person.

Section 17. Duties and Powers of Limited Conservators and Conservators

(1) It is the general duty of an individual or corporation appointed to serve as a limited conservator or conservator for a partially disabled or disabled person to carry out, diligently and in good faith, the specific duties and powers assigned by the court. In carrying out these duties and powers such individuals or corporations shall:

(a) Manage or assist in managing those financial resources placed under their supervision and/or control, as would a prudent person manage his or her own financial resources; and

(b) Encourage the disabled or partially disabled persons whom they are serving to:

(i) Participate, to the maximum extent of their abilities, in all decisions which affect them;

(ii) Act on their own behalf on all matters in which they are able to do so and

(iii) Develop or regain, to the maximum extent possible, their capacity to manage their financial resources and, if impaired, their capacity to meet the essential requirement for their physical health or safety.

(2) (a) Within sixty (60) days of appointment, a limited conservator or conservator shall submit to the court an individual conservatorship plan, developed together with the limited personal guardian or personal guardian if any, and the disabilities resources officer and, to the maximum extent possible, the partially disabled or disabled person. The plan shall specify:

(i) The services which are necessary to manage the financial resources designated by the court in the dispositional order;

(ii) The means through which those services will be provided;

(iii) The manner in which the limited conservator or conservator of the partially disabled or disabled person and the limited personal guardian or personal guardian, if another individual has been appointed to serve in that capacity, will exercise and share their decision-making authority;

(iv) The policies and procedures governing the expenditure of funds; and

(v) Such other items as will assist in the management of the designated financial resources, and in fulfilling the needs of the partially disabled or disabled person, the terms of the dispositional order, and the duties of the limited conservator or conservator.

(b) Attached to the individual conservatorship plan shall be a complete inventory of the designated financial resources. The inventory shall include an oath or affirmation that, to the best of the limited conservator's or conservator's knowledge, it is complete and accurate.

(3) (a) A limited conservator or conservator shall expend or distribute, and/or authorize the expenditure or distribution of, and/or assist in the expenditure or distribution of, the principle of or income from the financial resources placed under his, her, or its supervision and control to assure that:

(i) The essential requirements for the physical health or safety of the partially disabled or disabled person are met;

(ii) The rights of that person are protected;

(iii) The financial resources of that person which are subject to the conservatorship are prudently managed;

(iv) The partially disabled or disabled person has the opportunity to develop or regain the capacity to perform the functions listed in subparagraphs (3)(a)(i)-(iii) of this section; and

(v) The limited personal guardian or personal guardian for that person is able to carry out the duties and powers assigned by the court.

(b) In so doing, the limited conservator or conservator shall consider:

(i) The size of the financial resources under the limited conservator's or conservator's supervision or control;

(ii) The probable duration of the conservatorship;

(iii) The likelihood that the partially disabled or disabled person may be able to manage the financial resources in the future; and

(iv) The accustomed standard of living of the partially disabled or disabled person and the individuals legally dependent on that person.

In addition, limited conservators shall consider the size of the financial resources of a partially disabled or disabled person which have not been placed under their supervision or control.

(c) Funds under this paragraph may be paid by a limited conservator or conservator to any person, including the partially disabled or disabled person.

(4) (a) In addition to the duties and powers prescribed in subsection (3) of this section, the court may assign to a limited conservator any of the duties and powers listed in paragraphs (5)(a) which the partially disabled person lacks the capacity to perform, and/or the power to void particular types of contracts, conveyances or dispositions made by the partially disabled person as long as action is taken to void such contracts, conveyances, and dispositions within fifteen (15) days of the transaction. Such assignments shall specify the financial resources and/or the types of transactions to which they apply.

(b) The court may assign to a limited conservator the duty of assisting a partially disabled person to perform any of the functions listed in paragraph (5)(a) of this section with regard to any financial resource of the partially disabled person.

(c) Upon the death of a partially disabled person, the limited conservator for that person shall have the duties and powers prescribed in paragraph (5)(b) of this section.

(5) (a) In addition to the duties and powers prescribed in subsection (3) of this section, the court shall assign to a conservator the duty and/or power to:

(i) Acquire, collect, hold, deposit, retain, operate, develop, repair, improve, insure, subdivide, exchange, partition, alter, lease, convert or dispose of the financial resources of the disabled person;

(ii) Pay, contest, settle or release claims against the disabled person or the financial resources of the disabled person;

(iii) Pay taxes, assessments, compensation and other reasonable expenses incurred in the management of the financial resources of the disabled person;

(iv) Employ persons to perform, advise, or assist in particular aspects of the management of the financial resources of the disabled person;

(v) Borrow money to be repaid from the financial resources of the disabled person;

(vi) Prosecute or defend actions, claims, or proceedings in any jurisdiction for the protection of the financial resources of the disabled person and of the conservator in the performance of the conservatorship duties;

(vii) Allocate items of income or expense to either principal or income as provided by law, including the creation of reserves out of income for depreciation, obsolescence, amortization, or depletion of mineral or timber resources;

(viii) Vote a security in person, or by general or limited proxy;

(ix) Establish revocable or irrevocable trusts;

(x) Exercise options;

(xi) Exercise or release powers of appointment of which the disabled person is donor;

(xii) Renounce interests and make gifts;

(xiii) Change beneficiaries under insurance and annuity policies; and

(xiv) Enter into contracts and execute and deliver all instruments which will accomplish or facilitate the exercise of the assigned duties and powers.

(b) Upon the death of a disabled person:

(i) The conservator shall deliver to the (name of court having probate jurisdiction) any will of the deceased of which the conservator has possession and/or advise that court of the whereabouts of any other will of which the conservator is aware;

(ii) Inform the [executor/personal representative/administrator] or a named beneficiary of the actions taken; and

(iii) Retain those portions of the estate over which the conservator has control for delivery to a duly appointed [executor/personal representative/administrator] of the deceased or the individuals entitled thereto.

The conservator may seek appointment as [executor/personal representative/administrator] of the estate of the disabled person in the manner prescribed in (citation to relevant sections of the probate code).

(6) (a) A limited conservator or conservator shall not have the authority to exercise or release powers of appointment of which the partially disabled or disabled person is donee, to renounce interests or to make gifts exceeding twenty (20) percent of any year's income from the financial resources placed under the control of the limited conservator or conservator, nor to change beneficiaries under insurance or annuity policies without the approval of the court. No approval shall be granted unless it is demonstrated at a hearing that the proposed action is consistent with subsections (1) and (3) of this section and that the partially disabled or disabled person has consented or is unable to provide an informed, voluntary consent.

(b) Any sale or encumbrance to a limited conservator or conservator, his or her spouse, agent or attorney, or any corporation or trust in which the limited conservator or conservator has a substantial beneficial interest is voidable unless the transaction is approved by the court after a hearing.

(c) Title to all the designated financial resources of a partially disabled person or disabled person shall remain in that person subject to the possession of the limited conservator or conservator and to the control of the court, unless the court orders otherwise. The provision of title to a limited conservator or conservator is not a transfer or alienation within the meaning of general provisions of any federal or state statute or regulation, insurance policy, pension plan, contract, will or trust instrument, imposing restrictions upon or penalities for transfer or alienation by the partially disabled or disabled person of a right or interest. If title is provided to a limited conservator or conservator, the dispositional order shall be evidence of transfer of all designated financial resources. An order terminating a conservatorship or modifying the scope or powers thereof, shall be evidence of transfer of the designated financial resources to the partially disabled or disabled person, the successor of that person, or to a new limited conservator or conservator. A dispositional order and order terminating or modifying a conservatorship shall be [filed/recorded] to give record notice of title in accordance with (citation to state governing the filing/recording of documents to land or other property).

(7) (a) A limited conservator or conservator shall submit a report to the court ten (10) days before the initial review hearing and:

(i) At least annually thereafter;

(ii) When the court orders additional reports to be filed;

(iii) When there is a significant change in the capacity of the partially disabled or disabled person to manage his or her financial resources;

(iv) When the limited conservator or conservator resigns or is removed; and

(v) When the conservatorship is terminated.

(b) Such reports shall set forth:

(i) The name and address of the partially disabled or disabled person and of the limited conservator or conservator;

(ii) Significant changes in the capacity of the partially disabled or disabled person to manage his or her financial resources;

(iii) A complete financial statement of the financial resources under the control and/or supervision of the limited conservator or conservator;

(iv) The services being provided to the partially disabled or disabled person and the relationship of those services to the individual conservatorship plan;

(v) The significant actions taken by the limited conservator or conservator during the reporting period;

(vi) The compensation requested and the reasonable and necessary expenses incurred by the limited conservator or conservator;

(vii) Any significant problems relating to the conservatorship which have arisen during the reporting period; and

(viii) The reasons, if any, why the conservatorship should not be terminated, or why no less restrictive alternative would permit the disabled or partially disabled person to manage his or her financial resources.

Attached to the report shall be an updated individual conservatorship plan developed by the limited conservator or conservator, the limited personal guardian or personal guardian if any, the disabilities resources officer and, to the maximum extent possible, the partially disabled or disabled person.

(c) Following submission of a report or in conjunction with a review hearing held pursuant to section 14 or a proceeding initiated under section 15, a limited conservator or conservator shall submit to a physical check of the financial resources placed under his, her or its control, in the manner specified by the court.

CHAPTER 3 INTERVENTION PROCEEDINGS FOR MINORS

Section 18. Initiation of Intervention Proceedings

(1) (a) An intervention proceeding for a minor may be initiated separately or in conjunction with another proceeding by the filing of a petition with the court by a parent of the minor, the limited personal guardian, personal guardian, limited conservator or conservator for the minor, a person interested in the welfare of the minor, or the director of (the agency required to provide services to children), if there is reason to believe that:

(i) The parents of the minor are dead;

(ii) Their parental rights have been terminated or suspended by order of the court; or

(iii) The minor has or is entitled to financial resources which may be jeopardized because of the minor's age or which cannot otherwise be protected from damage or dissipation.

Intervention proceedings involving minor siblings may be initiated by a single petition.

(b) The petition shall be verified under oath and shall specify:

(i) The names and addresses of the individuals entitled to notice;

(ii) The relief requested; and

(iii) The facts and reasons supporting the requested relief.

(2) (a) A parent of a minor may designate, by will, an individual to serve as limited personal guardian, personal guardian, limited conservator and/or conservator upon the testator's death. A designation of a limited personal guardian, personal guardian, limited conservator and/or conservator for a minor shall become effective upon the filing of a written acceptance if no objection is filed pursuant to section 19 of this act and if:

(i) Both parents are dead; or

(ii) One parent is dead and the parental rights of the surviving parent have been suspended or terminated in accordance with (citation to the statute(s) on termination of parental rights);

If both parents are dead, an effective appointment by the parent who died later has priority.

(b) The acceptance shall be filed in the court in which the will is probated within the time limit prescribed in section 27(2)(a). It shall include the name, age, and address of the minor and the names and addresses of other individuals entitled to notice, as well as a statement, verified under oath, that the designee understands and is able and willing to undertake the responsibilities, duties and powers of a limited personal guardian, personal guardian, limited conservator and/or conservator for a minor.

(c) This state shall recognize a testamentary appointment for which an acceptance has been filed under a will probated in another state which was the testator's domicile.

Section 19. Objections to Testamentary Designations

(1) A minor may file an objection to a testamentary designation before it is accepted or within the time limit prescribed in section 27(2)(b), and propose a person qualified under section 21(4) of this act for appointment as the limited personal guardian, personal guardian, limited conservator and/or conservator.

(2) An individual interested in the welfare of a minor for whom a testamentary designation has been made, may file an objection to that designation within the time limit prescribed in section 27(2)(b). Such an objection shall specify:

(a) The facts and reasons why:

(i) The individual and/or corporation designated in the will would not be eligible under section 21(4) of this act to serve as limited personal guardian, personal guardian, limited conservator and/or conservator of the minor;

(ii) The appointment of that individual and/or corporation would not assure that the duties and powers set forth in section 25 and/or section 26 will be exercised diligently and in good faith; or

(iii) The appointment of another individual and/or corporation would be of greater benefit to the minor;

(b) The name, address, qualifications, and relationship to the minor of the individual proposed for appointment as the limited personal guardian or personal guardian for the minor and/or the individual or corporation proposed for appointment as the limited conservator or conservator for the minor; and

(c) The name, address and interest of the individual filing the objection.

(3) Objections to testamentary appointments shall be filed in the court in which the will is probated.

Section 20. Hearing

Within the time limits prescribed by section 27(2)(c), the court shall hold a hearing to determine:

(1) Whether the parents of a minor are dead, whether they have had their parental rights terminated or suspended by order of the court, or whether the minor has or is entitled to financial resources which may be jeopardized because of the minor's age or which cannot be protected other than through court-ordered assistance or the establishment of a conservatorship; and, if so,

(2) Whether appointment of a limited personal guardian, personal guardian, limited conservator and/or conservator for the minor is necessary; and, if so,

(3) Who should be appointed.

Section 21. Dispositional Alternatives

(1) The petition shall be dismissed if it is not proven that the parents of a minor are dead, that they have had their parental rights of custody terminated or suspended by order of the court, or that the minor has or is entitled to financial resouces which may be jeopardized because of the minor's age or which cannot be protected from damage or dissipation other than through court-ordered assistance or the establishment of a conservatorship.

(2) If a will is refused probate, any testamentary designation of a personal guardian or conservator contained therein shall be void.

(3) If it is proven that the parents of a minor are dead, that they have had their parental rights terminated or suspended by order of the court, or that the minor has or is entitled to financial resources which may be jeopardized because of the minor's age or cannot be protected from damage or dissipation other than through court-ordered assistance or the establishment of a conservatorship, the court may order any of the dispositional alternatives described in section 10(2) or any combination thereof which will enable the minor to meet the essential requirements for his or her physical health or safety, manage his or her financial resources, and develop his or her abilities to the maximum extent possible.

(4) (a) The court may appoint any individual eligible under section 12(1) of this act to serve as limited personal guardian or personal guardian, and any individual or corporation eligible under that section to serve as limited conservator or conservator for a minor. Nothing in this provision or section 12 shall exclude the appointment of an adult sibling or other close relative of the minor when the conflict of interest is determined to be unsubstantial and such an appointment will clearly be the most beneficial to the minor.

(b) The following priorities shall guide the court's selection of a limited personal guardian, personal guardian, limited conservator and/or conservator for a minor from among those eligible:

(i) The individual or corporation proposed by the minor;

(ii) The limited personal guardian, personal guardian, limited conservator and/or conservator appointed or recognized by the appropriate court of the jurisdiction in which the minor resides;

(iii) The individual who/or corporation which filed an acceptance of a testamentary designation;

(iv) An individual with whom the minor has been living for more than six (6) months prior to the hearing;

(v) An individual with whom the minor has substantial ties;

(vi) An individual or corporation nominated by any individual or corporation included in subparagraphs (4)(b)(ii)-(v) of this section;

(vii) A volunteer public guardian or volunteer public conservator.

The court may select the individual or corporation best qualified and most willing to serve from among those of equal priority. The court may pass over a person or corporation having priority and appoint a person having a lower priority or no priority when it can be demonstrated that the selection will be of greater benefit to the minor. Appointments of volunteer public guardians and volunteer public conservators shall be in accordance with section 33.

(5) The dispositional order shall set forth:

(a) The name and address of the individual, if any, appointed to serve as the limited personal guardian or personal guardian for the minor and/or of the individual or corporation, if any, appointed to serve as the limited conservator or conservator of the minor;

(b) The duties and powers of the limited personal guardian, personal guardian, limited conservator and/or conservator for the minor, if any;

(c) The terms and conditions of any other dispositional alternatives ordered by the court;

(d) The facts and reasons supporting the selection of the limited personal guardian, personal guardian, limited conservator and/or conservator for the minor; and

(e) The date of the initial review hearing.

The court shall also explain the terms of the order to the minor as well as the procedures for terminating or modifying the dispositional order.

Section 22. Periodic Review

(1) Following appointment of a limited personal guardian, personal guardian, limited conservator or conservator for a minor, a review hearing shall be held within the time limit prescribed in section 27(2)(d) to:

(a) Determine whether the duties and powers set forth in sections 25 and/or 26 are being exercised diligently and in good faith;

(b) Determine whether the dispositional order should be modified or terminated;

(c) Approve payment of compensation and reimbursement of expense, if any, from the financial resources of the minor; and

(d) Resolve any other relevant matter which the parties raise.

(2) Following a review hearing the court may issue any of the orders listed in section 14(4) of this act.

Section 23. Modification of Dispositional Orders

(1) In addition to the review procedures prescribed under section 22 of this act, the minor, the limited personal guardian, personal guardian, limited conservator, or conservator for the minor, or any other individual interested in the welfare of the minor may apply to the court for:

(a) Modification or termination of the dispositional order;

(b) Replacement of the limited personal guardian, personal guardian, limited conservator, or conservator;

(c) Approval or disapproval of the compensation or reimbursement for expenses requested by the limited personal guardian, personal guardian, limited conservator, or conservator; or

(d) Resolution of a dispute over the individual conservatorship plan.

(2) (a) An application from a minor may be communicated to the court by any means, including but not limited to oral communication or informal letter. Upon receipt of such an informal request the court shall contact the minor's attorney of record or appoint new counsel to prepare a formal application.

(b) Applications filed by individuals other than the minor and formal applications filed pursuant to paragraph (2)(a) of this section shall set forth the name and address of the individuals and entities entitled to notice, the relief requested and the facts and reasons supporting the request.

(3) Any party may object to the relief requested in an application;

(4) (a) The court shall hold a hearing on all applications:

(i) Which are filed pursuant to section 32 of this act;

(ii) Which request termination of the guardianship or conservatorship of the minor;

(iii) Which request a change in custody of the minor; or

(iv) To which an objection has been filed.

(b) All other applications may be granted without a hearing.

(5) After notice, the court may join separate applications and separate objections filed under this section, unless joinder would be prejudicial to the interests of the minor.

(6) The criteria, priorities and procedures set forth in section 21 shall apply to dispositional orders issued following submission of an application under this section.

Section 24. Termination

(1) The guardianship and/or conservatorship of a minor whether established by court order or

testamentary designation, shall terminate upon:

(a) The death of the minor;

(b) The minor's [eighteenth birthday/attaining of majority];

(c) The marriage of the minor;

(d) The adoption of the minor; or

(e) An order of the court.

(2) Within sixty (60) days of a minor's [eighteenth birthday/attaining of majority], an intervention proceeding may be initiated under section 6 of this act if it appears that upon attaining majority, the minor will be a partially disabled or disabled person.

Section 25. Duties and Powers of Limited Personal Guardians and Personal Guardians for Minors

(1) It is the general duty of individuals appointed to serve as a limited personal guardian or personal guardian for a minor to carry out, diligently and in good faith, the specific duties and powers assigned by the court. In carrying out these duties and powers, such individuals shall:

(a) Assure that the personal, civil and human rights of the minors whom they are serving are protected; and

(b) Encourage those minors to:

(i) Participate in all decisions which affect them; and

(ii) Develop their abilities to the maximum extent possible.

(2) (a) A limited personal guardian for a minor may be assigned any portion of the duties and powers listed in paragraphs 3(a)-(g):

(i) In those particular areas in which, or for that period of time during which, the parental rights of the parent of the minor have been suspended; or

(ii) In those particular areas in which a minor, whose parents are dead or have had their parental rights terminated, requires assistance to meet the essential requirements for his or her physical health or safety.

(b) A limited personal guardian may also be assigned the duty to assist the minor in those particular areas in which the minor's ability to meet the essential requirements for his or her physical health or safety, and/or to protect his or her rights, and/or to obtain necessary services, and/or to fulfill his or her civil duties is impaired, as well as in other ways not specifically prohibited by subsections (4) and (5) of this section.

(3) The specific duties and powers assigned to a personal guardian for a minor by the court shall be those of a parent, whose parental rights have not been terminated or suspended, toward his or her child, including but not limited to:

(a) Taking custody of the minor subject to any limitations and visitation privileges prescribed by the court;

(b) Assuring that the minor receives adequate food, clothing, health and other professional care, shelter, and education;

(c) Assuring through the initiation of court actions and other means that the minor enjoys the personal and civil rights to which he or she is entitled;

(d) Providing any required consents or approvals on behalf of the minor, including consent to the marriage of the minor or enlistment of the minor in the armed services;

(e) Assisting the minor to the maximum extent possible, to learn and assume the duties and responsibilities of a productive, law-abiding citizen;

(f) Receiving money payable for the support of the minor if a separate limited conservator or conservator has not been appointed; and

(g) Expending sums from the financial resources of the minor to cover the reasonable costs incurred on behalf of the minor.

(4) If a separate limited conservator or conservator has been appointed for the minor, then the receipt and expenditure of funds by the limited personal guardian or personal guardian shall be consistent with the duties assigned to and procedures and policies established by the limited conservator or conservator.

(5) A limited personal guardian or personal guardian shall not be:

(a) Obligated to provide for the minor from his or her own funds;

(b) Liable to third persons by reason of the parental-type relationship for acts of the minor;

(c) Empowered to take or consent to any of the actions for limited personal guardians for partially disabled and disabled persons by section 16(5)(a)-(c) and (f).

(6) (a) A limited personal guardian or personal guardian for a minor shall submit a report to the court at least ten (10) days before the initial review hearing and:

(i) Annually thereafter;

(ii) When the court orders additional reports to be filed;

(iii) When the limited personal guardian or personal guardian resigns or is removed; or

(iv) When the guardianship is terminated.

(b) The report shall set forth:

(i) The name and address of the minor, the limited personal guardian or personal guardian, and the individual or facility having custody of the minor;

(ii) Any special services required by and/or being provided to the minor;

(iii) Changes in the residence or custody of the minor within the reporting period and the reason therefor;

(iv) The significant actions taken by the limited personal guardian or personal guardian during the reporting period;

(v) The compensation requested and the reasonable and necessary expenses incurred by the limited personal guardian or personal guardian on behalf of the minor;

(vi) Any significant problems which have arisen during the reporting period;

(vii) Any changes which should be made in the duties and powers of the limited personal guardian or personal guardian; and

(viii) An accounting of the expenditures made and funds received for the minor by the limited personal guardian or personal guardian.

Section 26. Duties and Powers of Limited Conservators and Conservators for Minors

(1) It shall be the responsibilty of the limited conservator or a conservator for a minor to perform diligently and in good faith, the specific duties and powers assigned by the court. In carrying out these duties and powers, limited conservators or conservators for minors shall:

(a) Manage the financial resources placed under their supervision and/or control as would a prudent person manage his or her own financial resources; and

(b) Encourage the minors whom they are serving to:

(i) Participate in all decisions which affect them; and

(ii) Develop their abilities to the maximum extent possible.

(2) Within sixty (60) days of appointment, a limited conservator or a conservator for a minor shall submit to the court an individual conservatorship plan developed together with the limited personal guardian or personal guardian if any, and the minor to the maximum extent possible. It shall cover the matters prescribed for individual conservatorship plans for partially disabled or disabled persons by section 17(2) of this act and shall include a complete inventory of the financial resources under the supervision or control of the limited conservator or conservator.

(3) (a) A limited conservator or conservator for a minor shall expend or distribute, and/or authorize the expenditure or distribution of, and/or assist in the expenditure or distribution of income or principal from the financial resources of the minor under his, her or its supervision and control to assure that:

(i) The essential requirements for the physical health or safety of the minor and any dependents of the minor are met;

(ii) The rights of the minor are protected;

(iii) The financial resources of the minor are prudently managed;

(iv) The minor has the opportunity to develop his or her abilities to the maximum extent possible; and

(v) The limited personal guardian or personal guardian, if any, for the minor is able to carry out the duties and powers assigned by the court or set forth in the will.

(b) In so doing, a limited conservator or conservator shall consider:

(i) The requests of the minor and the limited personal guardian or personal guardian for the minor;

(ii) The size of the financial resources of the minor;

(iii) The probable duration of the conservatorship; and

(iv) The accustomed standard of living of the minor and individuals legally dependent on the minor.

In addition, limited conservators shall consider the size of the financial resources of a minor which have not been placed under their supervision and control.

(c) Funds under this subsection may be paid by the limited conservator or conservator to any individual, including the minor.

(d) A limited conservator or conservator shall establish policies and procedures for the expenditure of funds by the minor, or by the limited personal or personal guardian for the minor if a separate individual has been appointed to serve in that capacity.

(4) (a) The court may assign to a limited conservator for a minor any of the duties and powers listed in section 17(5)(a) which the minor is unable to perform. Such assignments shall specify the financial resources to which they apply.

(b) The court may assign to a limited conservator the duty of assisting a minor to perform any of the functions listed in section 17(5)(a) of this act with regard to any financial resources of the minor.

(c) If the minor dies during the conservatorship, the limited conservator shall have the duties and powers prescribed in section 17(5)(b) of this act.

(5) The court shall assign to a conservator for a minor the duties and powers of conservators for partially disabled or disabled persons as set forth in section 17(5) of this act.

(6) The prohibitions which apply to limited

conservators and conservators for partially disabled and disabled persons under section 17(6) of this act shall also apply to limited conservators and conservators appointed to serve minors.

(7) (a) A limited conservator or conservator for a minor shall submit a report to the court at least ten (10) days before the initial review hearing, and:

(i) Annually thereafter;

(ii) When the court orders additional reports to be filed;

(iii) When the limited conservator or conservator resigns or is removed; and

(iv) When the conservatorship is terminated.

(b) The report shall set forth:

(i) The name and address of the minor and the limited conservator or conservator;

(ii) A complete financial statement of the financial resources under the control and/or supervision of the limited conservator or conservator;

(iii) The services being provided to the minor and the relationship of those services to the individual conservatorship plan;

(iv) The significant actions taken by the limited conservator or conservator during the reporting period;

(v) The compensation requested and the reasonable and necessary expenses incurred by the limited conservator or conservator;

(vi) Any significant problems relating to the conservatorship which have arisen during the reporting period; and

(vii) Any changes which should be made in the duties and powers of the limited conservator or conservator.

Attached to the report shall be an updated individual conservatorship plan developed by the limited conservator, conservator, the limited personal guardian or personal guardian if any, and to the maximum extent possible, the minor.

(c) Following submission of a report, or in conjunction with a review hearing pursuant to section 22 or a proceeding initiated under section 23 of this act, a conservator shall submit to a physical check of the financial resources placed under his, her or its control, in the manner specified by the court.

CHAPTER 4 GENERAL PROVISIONS

Section 27. Time Limits

(1) The following time limits shall apply to proceedings initiated under chapter 2 of this act:

(a) An evaluation report shall be submitted pursuant to section 8(3) of this act no more than twenty (20) days after the filing of a petition;

(b) A hearing shall be held pursuant to section 9 of this act no more than thirty-five (35) days after the filing of a petition;

(c) A dispositional hearing shall be held pursuant to section 11(1) of this act no more than ten (10) days after a finding that an individual is a partially disabled or disabled person;

(d) A review hearing shall be held pursuant to section 14(1)(a) no more than one (1) year after appointment of a limited personal guardian and/or limited conservator, and biennially thereafter;

(e) A review hearing shall be held pursuant to section 14(1)(b) no more than six (6) months after appointment of a personal guardian and/or conservator, and annually thereafter;

(f) Hearings required by sections 15(4)(a)(i)-(ii) shall be held no more than thirty (30) days after the application is filed. Hearings required by section 15(4)(a)(iii) shall be held no more than thirty (30) days after the objection is filed.

(2) The following time limits shall apply to proceedings initiated under chapter 3 of this act:

(a) Acceptance of a testamentary designation shall be filed pursuant to section 18(2)(b) no more than thirty (30) days after the will containing the designation has been filed for probate;

(b) An objection to a testamentary designation shall be filed pursuant to section 19 no more than twenty (20) days after service of the notice required by section 29(1)(a);

(c) A hearing shall be held pursuant to section 20 of this act no more than thirty (30) days after the:

(i) Filing of a petition; or

(ii) Filing of an objection to a testamentary designation; or

(iii) No more than fifteen (15) days after expiration of the period for accepting a testamentary designation if a written acceptance has not been filed;

(d) A review hearing shall be held pursuant to section 22 of this act no more than six (6) months after a limited personal guardian, personal guardian, limited conservator and/or conservator has been appointed for a minor and at least annually thereafter;

(e) Hearings required by sections 23(4)(a)(i)-(iii) shall be held no more than thirty (30) days after the application has been filed. Hearings required by section 23(4)(a)(iv) shall be held no more than thirty (30) days after the objection has been filed.

(3) The informal, *in camera* hearing required by section 31(3), shall be held no more than fifteen (15) days after an objection to a request for appointment of a guardian *ad litem* has been filed.

(4) The following time limits shall apply to proceedings for emergency and standby appointments:

(a) A hearing shall be held pursuant to section 32(1)(c) of this act no more than forty-eight (48) hours after an application for an emergency appointment has been filed;

(b) An application shall be filed pursuant to section 32(3) of this act within ten (10) days after an individual appointed on a standby basis assumes guardianship and conservatorship duties.

(5) A continuance may be granted:

(a) Upon a stipulation by the parties; or

(b) For good cause shown following a written motion, notice to the individuals and entities enumerated in section 29(1)(a) of this act in the form and manner prescribed in sections 29(1)(b) and 29(2) and a hearing if an objection to the continuance is filed.

(6) The court shall set the date for a hearing and have notice served within five (5) days of the filing of the pleading requiring the hearing and/or the notice, except following the filing of an application under section 32(1) in which case the time for the hearing shall be set and notice served within twenty-four (24) hours.

Section 28. Rights and Privileges

(1) (a) In all proceedings under this act the parties shall have a right to:

(i) Prior notice;

(ii) Counsel;

(iii) Be present;

(iv) Compel the attendance of witnesses;

(v) Present evidence and confront and cross-examine witnesses, including the multidisciplinary evaluation team appointed pursuant to section 8 of this act;

(vi) An impartial decision-maker; and

(vii) Appeal adverse orders and judgments in the manner prescribed in (the Rules of Civil Procedure).

(b) In addition to the rights listed in paragraph (1)(a), the subject of an intervention proceeding shall have a right to close proceedings to the public, except that even if a proceeding has been closed to the general public, the court may, for good cause shown, admit persons having a legitimate interest in the proceedings or in the hearing process unless the subject of the intervention proceeding objects.

(c) In addition to the rights listed in paragraph (1)(a) of this section, private individuals who initiate an intervention proceeding and who are unable to obtain counsel for financial reasons, shall have a right to have the [district/county/state] attorney appointed to represent them.

(d) An individual alleged or found to be partially disabled or disabled shall also have a right to:

(i) A jury in hearings held pursuant to section 9 of this act; and

(ii) Have an independent expert appointed by the court at the expense of the [state/county/court] to evaluate him or her if that individual is unable for financial reasons to secure independent medical, psychological or other expert opinion regarding the alleged disability and the need for restrictions on the legal capacity to act in his or her own behalf.

(e) No right accorded in this subsection shall be waivable, except for the right to close proceedings to the public set forth in paragraph (1)(b) of this section and the right to a jury set forth in paragraph (1)(d). Any waiver of the right to a closed proceeding or to a jury shall be in writing or on the record, and shall demonstrate that the individual waiving the right has knowingly and voluntarily waived it after being fully informed of the right and of the consequences of relinquishing it.

(2) An attorney shall be appointed to represent the subject of an intervention proceeding in any matter arising under this act for which independent counsel has not been retained by or on behalf of that individual.

(3) A stenographic and/or mechanical record shall be made of all proceedings held pursuant to this act.

(4) (a) Statements of individuals alleged or found to be a partially disabled or disabled person made during the course of evaluations, examinations and treatment pursuant to this act, shall be privileged, confidential, and not admissible without the individual's consent in any civil or criminal proceedings other than proceedings pursuant to this act.

(b) If the subject of an intervention proceeding is under the influence of psychotropic medication during any judicial hearing held pursuant to this act, the court and the jury, if any, shall be advised of this fact and of the purpose of the medication and the effect which it may have on the individual's actions, demeanor and participation at the hearing.

(c) Any hearing held pursuant to this act may be held at such places as the court directs including the place of residence of the subject of the intervention proceeding.

Section 29. Notice and Dissemination of Plans and Reports

(1) (a) The court shall have notice of any pleadings filed, hearings required or actions taken pursuant to sections 6(1), 7, 9, 11, 14, 15, 17(6), 18, 19, 20, 22, 23, 26(6), 27(5), 31, 32(1) and 32(2) of this act served on:

(i) The parties as defined in section 3(9);

(ii) The attorney of record for each party;

(iii) The individuals most closely related to the subject of the intervention proceeding by blood or marriage unless their names or whereabouts are unknown and cannot be reasonably ascertained;

(iv) The individual or facility, if any, having custody of the subject of the intervention proceeding;

(v) The individual(s), if any, appointed or proposed for appointment or testamentarily designated as limited personal guardian or personal guardian;

(vi) The individual(s) and/or corporation(s), if any, appointed or proposed for appointment, or testamentarily designated as limited conservator or conservator;

(vii) The individual(s), if any, appointed or proposed for appointment as guardian *ad litem;*

(viii) The |executor/personal representative/administrator| of the estate of the testator following a testamentary designation;

(ix) The Administrator of Veterans' Affairs if the financial resources of the subject of the intervention proceeding include a benefit, compensation, pension or insurance payable by the Veterans' Administration; and

(x) The surety on the bond for the limited conservator or conservator, if any.

(b) Notices shall set forth the:

(i) Rights of the parties;

(ii) The procedures and time limits for exercising those rights in the particular proceeding to which the notice refers; and

(iii) The date, time, place, purpose and possible consequences of the hearing, if any, to which the notice refers.

(2) (a) Attached to any notice served in conjunction with this act shall be a copy of the pleading which gave rise to the notice.

(b) Personal service shall be made on each of the individuals or entities receiving notice in the manner prescribed by (citation to the relevant Rules of Civil Procedure).

(c) The contents of all documents served shall be explained, to the maximum extent possible, to the subject of the intervention proceeding in the language, mode of communication, and terms which that individual is most likely to understand.

(3) (a) The court shall send evaluation reports submitted pursuant to section 8 of this act to the attorney of record for each of the parties and directly to any party who is not represented by counsel.

(b) The court shall send plans and reports submitted pursuant to sections 16(2), 16(6), 17(2), 17(7), 25(6), 26(2) and 26(7) of this act to:

(i) The partially disabled or disabled person or minor;

(ii) The attorney of record for the partially disabled or disabled person or minor;

(iii) The (name of the state Developmental Disabilities Advocacy Office) if the partially disabled or disabled person has a developmental disability; and

(iv) The individuals and entities specified in subparagraphs (1)(a)(iii)-(vii) and (ix) of this section.

(c) The contents of all plans and reports shall be explained to the subject of the intervention proceeding in the language, mode of communication and terms which that individual is most likely to understand. Portions of reports which are likely to cause severe psychological or physical harm to the subject of the intervention proceeding may be disclosed to that individual through a psychiatrist, psychologist or other appropriate professional.

Section 30. Duties and Compensation of Counsel

(1) The principal duty of an attorney representing the subject of an intervention proceeding is to represent zealously that individual's legitimate interests.

(2) To the maximum extent possible, the subject of the intervention proceeding shall remain responsible for determining his or her interests. However, an attorney may seek appointment of a separate guardian *ad litem* when the circumstances set forth in section 31 of this act apply.

(3) Attorneys appointed to represent the subjects of intervention proceedings shall be entitled to reasonable compensation. If the subject of the intervention proceeding is unable for financial reasons to pay such compensation in whole or in part, the court shall award the appointed attorney an amount representing reasonable compensation for the services performed, or shall appoint the |public defender or legal services office| to represent the individual.

Section 31. Procedures and Criteria for Appointment of a Guardian *ad Litem;* Duties and Powers

(1) (a) The subject of an intervention proceed-

ing, the attorney for that individual, the limited personal guardian, personal guardian, limited conservator or conservator, if any, for that individual, and/or anyone interested in the welfare of that individual, may file an application to have a guardian *ad litem* appointed.

(b) The application shall set forth:

(i) The name and address of the individuals and entities entitled to notice;

(ii) The facts and reasons supporting the request; and

(iii) The name, address and qualifications of the individual proposed for appointment as guardian *ad litem.*

(2) Any party may file an objection to a request for appointment of a guardian *ad litem* or to the individual proposed for the appointment. Objections by the subject of the intervention proceedings may be communicated to the court by any means, including but not limited to oral communication or informal letter. Objections by other individuals or entities shall be in writing.

(3) The court may grant or deny an application without a hearing unless a timely objection is submitted. If an objection is filed, the court shall hold an informal, *in camera* hearing.

(4) The court shall appoint a guardian *ad litem* to protect the rights of the subjects of intervention proceedings if it is shown that because of youth, or because of impaired ability to receive and evaluate information effectively regarding the proceeding, or because of impaired ability to communicate decisions regarding the proceeding, they cannot determine their own interests without assistance, and either:

(a) No limited personal guardian or personal guardian has been appointed;

(b) The interests of the subject of the intervention proceeding and those of his or her parent, limited personal guardian or personal guardian conflict; or

(c) The appointment is otherwise required in the interests of justice.

(5) The eligibility requirements and priorities set forth in section 12 for selection of a limited personal guardian or personal guardian shall apply to the selection of a guardian *ad litem.*

(6) Guardians *ad litem* shall assist the individuals for whom they have been appointed to determine their interests in regard to the legal proceedings in which they are involved. If an individual is wholly incapable of determining his or her own interests, the guardian *ad litem* shall make that determination and advise counsel accordingly. In so doing, guardians *ad litem* shall:

(a) Inquire thoroughly into all the circumstances that a prudent individual in the position of the person for whom the guardian *ad litem* has been appointed would consider in determining his or her interests in the proceedings; and

(b) Encourage the individual whom they are serving to participate, to the maximum extent of that individual's ability, in all decisions and to act on his or her own behalf on all matters in which he or she is able.

(7) The court should inform guardians *ad litem,* upon appointment, of their duties and powers.

Section 32. Emergency and Standby Appointments

(1) (a) If during the pendency of a proceeding initiated under chapter 2 or 3 of this act, it appears that there is an imminent danger that the physical health or safety of the subject of the intervention proceeding will be seriously impaired or that the financial resources of that individual will be seriously damaged or dissipated unless immediate action is taken, the subject of the intervention proceeding, or any adult interested in the welfare of that individual may apply to the court in which the proceeding is pending for the emergency appointment of a limited personal guardian and/or limited conservator.

(b) The application shall state:

(i) The names and addresses of the individuals and entities entitled to notice;

(ii) The relief requested and the facts and reasons supporting that request; and

(iii) If the pending proceeding was filed under chapter 2 of this act, a statement that the application has been discussed with the disabilities resources officer and that the restrictions, protections, and assistance requested are the least restrictive alternative which will avert the imminent danger to the respondent and/or the respondent's financial resources.

(c) A hearing shall be held within the time limit prescribed in section 27(4)(a) of this act, at which the rights set forth in section 28 shall apply.

(d) If the court determines that there is an imminent danger that the physical health or safety of the subject of the intervention proceeding will be seriously impaired and/or that the financial resources of that individual will be seriously damaged or dissipated unless immediate action is taken, and for matters filed under chapter 2 of this act, if the court determines in addition that there is no less restrictive means of protecting the respondent and/or the respondent's financial resources, it shall appoint a limited guardian and/or limited conservator in the manner prescribed in sections 13 or 21. The court shall assign to an emergency appointee only those duties and powers necessary to protect against the imminent danger shown.

(e) (i) If the application was filed prior to the hearing required by section 9 or section 20, the emergency appointment shall remain in effect until the conclusion of that hearing. However, if at a hearing held pursuant to section 9, the respondent is found to be a partially disabled or disabled person, the court may extend the emergency appointment until the conclusion of the dispositional hearing.

(ii) If the application was filed following a finding that an adult is a partially disabled or disabled person or that a minor meets the criteria set forth in section 21(3), the court shall direct the emergency appointee to file an application in accordance with section 15 or section 23 within five (5) days. The emergency appointment shall remain in effect until the conclusion of the hearing on that application.

(2) If at any time the court has probable cause to believe that a limited personal guardian, personal guardian, limited conservator or conservator is not effectively performing his or her duties and powers, and that there is an imminent danger that the physical health or safety of the partially disabled or disabled person or minor will be seriously impaired or that the financial resources of that individual will be seriously damaged or dissipated unless immediate action is taken, it shall:

(a) Suspend and temporarily replace the limited personal guardian, personal guardian, limited conservator, and/or conservator, with an individual or corporation meeting the qualifications set forth in section 12 or 21(4);

(b) Reassign the duties and powers of the suspended individual or corporation to the emergency appointee; and

(c) Direct the temporary appointee to file an application under section 15 or section 23 of this act within five (5) days if such an application is not already pending, and submit such reports as may be necessary.

(3) The court may designate another eligible individual or entity selected in accordance with section 12 or 21(4) to assume the duties and powers assigned to the limited personal guardian, personal guardian, limited conservator or conservator upon his or her resignation, incapacitation or death. The individual so designated shall submit an application pursuant to section 15 or 23 within the time limits prescribed by section 27(4)(b) of this act. An individual serving on a standby basis may exercise all of the duties and powers assigned to his or her predecessor as a limited personal guardian, personal guardian, limited conservator, or conservator, until the conclusion of the hearing on the application, unless otherwise ordered by the court.

Section 33. Appointment of Volunteer Public Guardians, Volunteer Public Conservators, and Volunteer Public Trustees

(1) (a) Pursuant to sections 11 and 12 or section 21 of this act, the court may select a volunteer public guardian or volunteer public conservator from the list published by the Guardianship/Conservatorship Oversight Commission for appointment as the limited personal guardian, personal guardian, limited conservator and/or conservator for a minor or a partially disabled or disabled person when:

(i) There is no other private individual or entity qualified under section 12(1) or 21(4) of this act who is able and willing to accept the appointment; or

(ii) The cost of a private limited personal guardian, personal guardian, limited conservator and/or conservator would substantially impede the minor or the partially disabled or disabled person from meeting the essential requirements for his or her physical health or safety, and/or the management of his or her financial resources, and/or the provision to that individual of any assistance necessary to develop or regain his or her abilities to the maximum extent possible; or

(iii) The primary source of support for the minor, or the partially disabled or disabled person is from public funds; or

(iv) A volunteer public guardian or volunteer public conservator can best assist the minor or the partially disabled or disabled person to meet the essential requirements for his or her physical health or safety, and/or manage his or her financial resources, and/or develop or regain his or her abilities to the maximum extent possible.

(b) A volunteer public guardian may be assigned the same duties and powers and shall be subject to the same requirements and restrictions as a private limited personal guardian or personal guardian. A volunteer public conservator may be assigned the same duties and powers and shall be subject to the same requirements and restrictions as a private limited conservator or conservator.

(c) A volunteer public guardian may be appointed to serve as the limited personal guardian or personal guardian for no more than two (2) individuals. A volunteer public conservator may be appointed to serve as the limited conservator or conservator for no more than four (4) individuals.

(2) (a) A volunteer public trustee may be selected from the list published by the Guardianship/Conservatorship Oversight Commission for appointment as the trustee or co-trustee of any trust valued at less than $ ______ when there are no private individuals or entities qualified, able and willing to serve as trustee, and either:

(i) At least one beneficiary of that trust has been found to be a partially disabled or disabled person, or is a minor for whom a limited personal guardian, personal guardian, limited conservator and/or conservator has been appointed; or

(ii) The creation of the trust would obviate the necessity of appointing or continuing the appointment of a limited conservator or conservator.

(b) No volunteer public trustee may serve as a trustee or co-trustee of a trust when that volunteer public trustee has interests which conflict substantially with those of any beneficiary of the trust.

(c) A volunteer public trustee may be assigned the same powers and duties and shall be subject to the same restrictions as private trustees pursuant to (citation to the relevant section of the state code).

Section 34. Effect of a Finding That a Person Is Partially Disabled or Disabled and the Effect of a Guardianship or Conservatorship for a Minor

(1) A finding under this act that an individual is a partially disabled or disabled person shall not constitute a finding of legal incompetence. An individual found to be partially disabled or disabled shall retain all legal rights and abilities other than those expressly limited or curtailed in the dispositional order or subsequent order of the court.

(2) A minor for whom a limited personal guardian, personal guardian, limited conservator and/or conservator has been appointed shall be entitled to the same rights and privileges as a minor whose parents have not died or have not had their parental rights terminated or suspended, or who does not have or is not entitled to financial resources which may be jeopardized because of the minor's age.

Section 35. Compensation and Expenses

(1) Limited personal guardians, personal guardians, limited conservators and conservators, including volunteer public guardians and volunteer public conservators are entitled to reasonable compensation for services rendered and to reimbursements for reasonable and necessary expenses incurred in the performance of their assigned guardianship and/or conservatorship duties and powers.

(2) Such compensation and reimbursements shall be paid from the financial resources of the partially disabled or disabled person or the minor unless:

(a) A volunteer public guardian and/or volunteer public conservator has been appointed in accordance with section 33(1) of this act; and

(b) Either:

(i) Payment of compensation and reimbursements would substantially impede the minor, partially disabled person, or disabled person from meeting the essential requirements for his or her physical health or safety, managing of his or her financial resources, and/or developing or regaining his or her abilities to the maximum extent possible; or

(ii) The primary source of support for the minor, partially disabled person, or disabled person is from public funds.

(3) Volunteer public trustees who have been appointed to serve as a trustee or co-trustee of a trust in accordance with section 33(2) of this act, are entitled to reasonable compensation for services rendered and to reimbursements for reasonable and necessary expenses incurred in the performance of their trusteeship duties. Such compensation and reimbursements shall be paid from the corpus or income of the trust except when:

(a) The corpus of the trust is less than $____; or

(b) Payment of such compensation and reimbursements would prevent the trust from providing funds to any beneficiary, funds which will be sufficient, when added to the other income of that beneficiary, to permit him or her to meet the essential requirements for his or her physical health or safety, and/or manage his or her financial resources, and/or develop or regain his or her abilities to the maximum extent possible.

(4) The state shall compensate and reimburse:

(a) A volunteer public guardian or volunteer public conservator not entitled under subsection (2) of this section to compensation and reimbursement from the financial resources of the partially disabled or disabled person or the minor; and

(b) A volunteer public trustee not entitled under subsection (3) of this section to compensation and reimbursement from the corpus or income of the trust.

(5) All compensation and reimbursements shall be approved by the court prior to payment.

Section 36. Records

(1) Identifiable information shall not be collected in conjunction with any proceeding under this act unless that information is essential to:

(a) Providing necessary services;

(b) Determining whether an individual is a partially disabled or disabled person or whether a guardianship or conservatorship for a minor should be or has been established;

(c) Issuing, reviewing, modifying or terminating a dispositional order;

(d) Administering, monitoring, and evaluating the court and services provided to partially disabled or disabled persons and minors;

(e) Monitoring and evaluating the guardianship or conservatorship of a partially disabled or disabled person or a minor; or

(f) Conducting an authorized research, evaluation or statistical study.

(2) Information identifiable to an individual alleged or found to be partially disabled or disabled collected under subsection (1) of this section shall not be retained in retrievable form unless it is accurate; protected from unauthorized access, disclosure and dissemination; physically secure; and essential to accomplishing one of the purposes specified in subsection (1).

(3) Identifiable information retained pursuant to subsection (2) in conjunction with a proceeding initiated under chapter 2, shall not constitute a public record. Access to and disclosure of such records shall be strictly controlled.

(4) The subject of any record retained under subsection (2) of this section shall be entitled to inspect and copy that record, request correction, clarification or deletion of any portion of that record which he or she believes is inaccurate, misleading or retained improperly, and to submit for inclusion in that record, a statement concerning any disputed information.

(5) (a) Following the dismissal or withdrawal of a petition filed under section 6 of this act or the termination of a dispositional order, the subject of any record retained under subsection (2) in conjunction with the petition or order may file an application to have any identifiable, non-privileged information pertaining to the proceeding destroyed.

(b) At any time, the subject of a record retained pursuant to subsection (2) in conjunction with a proceeding initiated under chapter 3 of this act, may file an application to limit access to any identifiable, non-privileged information contained in that record which is likely to harm or stigmatize the individual.

(c) Copies of an order to destroy or limit access to a record issued pursuant to paragraphs (5)(a) and (5)(b) of this section shall be sent to all courts, agencies and individuals which may have that record or have copies of or identifiable notations regarding that record. Courts, agencies and individuals receiving a copy of such an order shall promptly dispose of all copies of the record or the portion or notations thereof which they possess in the manner specified unless the information was obtained for authorized research, evaluation or statistical purposes.

(d) The subject of a record shall be entitled to receive a copy of that record, upon request, prior to its destruction or sealing.

Section 37. Bonds

(1) The court may require a limited conservator or conservator to have a surety satisfactory to the court furnish a bond conditioned upon the faithful discharge of all duties and powers assigned in the dispositional order or any modification thereof. The court in lieu of sureties on a bond, may accept other security for the performance of the bond, including a pledge of securities or a mortgage of land.

(2) The following requirements and provisions shall apply to any bond required under this section:

(a) Unless otherwise provided by the terms of the approved bond, sureties are jointly and severally liable with the limited conservator or conservator and with each other;

(b) By executing an approved bond on a limited conservator or conservator, the surety consents to the jurisdiction of the court in any proceeding pertaining to the fiduciary duties of the limited conservator or conservator and the naming of the surety as a party defendent;

(c) On petition of a successor limited conservator or conservator or any interested person, a proceeding may be initiated against a surety for breach of the obligation of the bond of the limited conservator or conservator;

(d) The bond on the limited conservator or conservator is not void after the first recovery but may be proceeded against at subsequent times until the whole penalty is exhausted.

(3) No proceeding may be commenced against the surety on any matter as to which an action or proceeding against the primary obligor is barred by adjudication or limitation.

Section 38. Liability of a Limited Personal Guardian, Personal Guardian, Limited Conservator or Conservator to Third Parties and Penalties for Violations of the Act.

(1) By accepting appointment or designation as a limited personal guardian, personal guardian, limited conservator, or conservator, an individual or corporation submits personally to the jurisdiction of the court in any proceeding relating to the guardianship or conservatorship which may be instituted.

(2) (a) Unless otherwise provided in the contract, limited conservators and conservators shall not be individually liable on a contract properly entered into in their fiduciary capacity in the course of managing the financial resources of a partially disabled or disabled person or a minor, unless they fail to reveal their representative capacity and fail to identify the partially dis-

abled or disabled person or the minor in the contract.

(b) Limited conservators and conservators shall be individually liable for obligations arising from ownership or control of financial resources of a partially disabled or disabled person or a minor or for torts committed in the course of administration of such financial resources only if they are personally at fault.

(c) Claims based on contracts entered into by limited conservators or conservators in their fiduciary capacity, on obligations arising from ownership or control of the financial resources of partially disabled or disabled persons or minors, or on torts committed in the course of managing the financial resources of partially disabled or disabled persons or minors may be asserted by proceeding against the limited conservators and conservators in their fiduciary capacity, whether or not they are individually liable therefor.

(d) Any question of liability between the financial resources of partially disabled or disabled persons or minors and the limited conservators or conservators individually, may be determined in a proceeding for accounting, surcharge, or indemnification, or other appropriate proceeding or action.

(3) Any limited personal guardian, personal guardian, limited conservator or conservator who willfully violates the duties or misuses the powers assigned by court and thereby causes harm to the subject of the guardianship or conservatorship or to the financial resources of that individual shall be subject to a fine of up to $10,000 for each such violation of duty or misuse of power, in addition to any civil liability to which the limited personal guardian, personal guardian, limited conservator or conservator may be subject.

(4) (a) Any individual who willfully discloses identifiable information or who willfully gains access to identifiable information in violation of section 36 of this act or who willfully violates an order issued pursuant to section 36(5)(c) shall be subject to a fine of up to $5,000 for each violation.

(b) The subject of identifiable information which has not been destroyed or sealed in accordance with section 36(5) of this act, may initiate a civil action for damages or to restrain further dissemination.

Section 39. Venue of Proceedings Subsequent to a Dispositional Order or an Uncontested Acceptance of a Testamentary Appointment

The court where the subject of an intervention proceeding resides following issuance of a dispositional order or an uncontested acceptance of a testamentary designation, has concurrent jurisdiction with the court which issued the order or in which the acceptance was filed, over proceedings held pursuant to 15, 23, 31 and 36 of this act and other proceedings related to the imposition of restrictions on the capacity of a person to act in his or her own behalf. Venue shall be in the jurisdiction which is the most convenient for and best serves the interests of the subject of the intervention proceeding. The court which issued the dispositional order or in which acceptance of a testamentary designation is filed shall be sent a copy of any order resulting from a subsequent proceeding in another jurisdiction.

Section 40. Burden of Proof

(1) In proceedings initiated under chapter 2 of this act:

(a) Proof beyond a reasonable doubt shall be required in order for the court to find that an individual is a partially disabled or disabled person and to impose or continue restrictions on that individual's legal capacity to act on his or her own behalf. Any other determination, except as specified in subparagraph (b)(i) of this section, may be made or action taken by the court on the basis of clear and convincing evidence.

(b) Both the burden of going forward and the burden of persuasion shall be on the party filing the petition, application, or motion except:

(i) A motion to dismiss filed pursuant to section 7 of this act shall be granted unless the petitioner proves there is probable cause to believe that the respondent is a partially disabled or disabled person;

(ii) A restriction on the legal capacity of the partially disabled or disabled person shall not be continued or added at a review hearing held pursuant to section 14 of this act unless the limited personal guardian, personal guardian, limited conservator and/or conservator proves beyond a reasonable doubt that the restriction is needed to achieve the purposes set forth in section 11(2) and that no less restrictive alternative will satisfy that need; and

(iii) The burden of persuasion shall shift to the party or parties opposing an application filed pursuant to section 15 by a partially disabled or disabled person, if that person presents testimony and/or evidence constituting a *prima facie* case supporting the requested relief.

(2) In proceedings initiated under chapter 3 of this act:

(a) A preponderance of the evidence shall be required in order for the court to make any determination or take any action.

(b) Both the burden of going forward and the burden of persuasion shall be on the party filing the petition, application, motion, or objection to a testamentary designation, except that the burden of persuasion shall shift to the party or parties opposing an objection to a testamentary designation filed by a minor pursuant to section 19(1) of this act or an application filed by a minor pursuant to section 23 of this act if the minor presents testimony and/or evidence constituting a *prima facie* case supporting the objection or application.

(3) In proceedings held pursuant to sections 31(3) and 32(1)(c), the burden of persuasion shall be by a preponderance of the evidence. Both the burden of going forward and the burden of persuasion shall be on the applicant.

Section 41. Rules of Evidence/Rules of Procedure

Except where specified otherwise, the rules of evidence and rules of procedure, including those on discovery, which are applicable in civil matters shall govern all proceedings under this act.

Section 42. Effect on Existing Guardianships and Conservatorships

(1) All guardianships and conservatorships established for partially disabled and disabled persons prior to the effective date of this act shall be reviewed by the court pursuant to section 14 within two (2) years. Until the review hearing, all limited personal guardians, personal guardians, limited conservators and conservators appointed prior to the effective date shall retain the powers assigned to them, unless an application has been granted under section 15. Prior to the review hearing, the limited personal guardian, personal guardian, limited conservator and/or conservator shall submit at least one (1) report in the form prescribed in sections 16(6) and/or 17(7), and an individual guardianship and/or conservatorship plan developed in the manner set forth in sections 16(2) and/or 17(2).

(2) All limited personal guardians, personal guardians, limited conservators and/or conservators for minors appointed by court order or beginning their service following a testamentary appointment prior to the effective date of this act, shall retain the powers assigned to them by the court or by will unless an application has been granted under section 23 of this act, or one of the events specified in section 24 occurs terminating the guardianship or conservatorship. All such limited personal guardians and personal guardians shall submit a report in the form prescribed in section 25(6) no more than one year after the effective date of this act and annually thereafter for the duration of the guardianship. All such limited conservators and conservators shall submit an individual conservatorship plan in the form specified in section 26(2) and a report in the form specified in section 26(7) no more than one year after the effective date of this act and at least annually thereafter for the duration of the conservatorship.

Section 43. Authorization of Appropriations

To carry out the purposes of this act, there is authorized to be appropriated $__________ for the fiscal year ending__________, $__________ for the fiscal year ending__________, and $__________ for the fiscal year ending__________.

Section 44. Severability of Sections

If any section, subsection, paragraph, sentence, or any other part of this act is adjudged unconstitutional or invalid, such judgment shall not affect, impair or invalidate the remainder of this act, but shall be confined to the section, subsection, paragraph, sentence or other part of this act directly involved in the controversy in which that judgment was rendered.

Section 45. Effective Date

This act takes effect ____________________.

APPENDIX

The following charts illustrate the sequence of events prescribed by the model act for intervention proceedings. Figures 1-4 graphically represent the major actions, decision points, and alternatives required before a guardianship and/or conservatorship may be established and during the review process following imposition of restrictions on an individual's legal capacity to act in his or her own behalf. Figures 5-7 focus on the time limits which apply to intervention proceedings. The time periods indicated on the charts represent the maximum intervals permitted under section 30 of the act. In many instances, the hearing process will be able to be concluded more quickly. In others, a continuance may be required. Citations to particular sections refer to the full Model Guardianship and Conservatorship Act. Section numbers in the optional short-form version differ because of the omission of certain provisions and procedures.

FIGURE 1 depicts the process for determining whether an adult is a partially disabled or disabled person. It begins with the decision by an individual to initiate an intervention proceeding, and includes both the steps leading up to the filing of the petition and the procedures required between the submission of the petition and the dispositional hearing. The chart portrays alternative sets of procedures at two points. The first set applies when a motion to dismiss has been filed. Since the evaluation process will continue while a motion to dismiss is pending, the alternative procedures are connected to the main body of the chart with dotted lines. The second set of a alternatives occurs following the prehearing conference. At the conference the parties can agree either to withdraw the petition, to proceed to the hearing, or to try whether non-custodial services provided on a voluntary basis will be sufficient to assist the respondent to meet the essential requirements for his or her physical health and safety and/or to manage his or her financial resources. When this third alternative is selected, a continuance of up to 90 days may be requested. (See §9 Prehearing Conference).

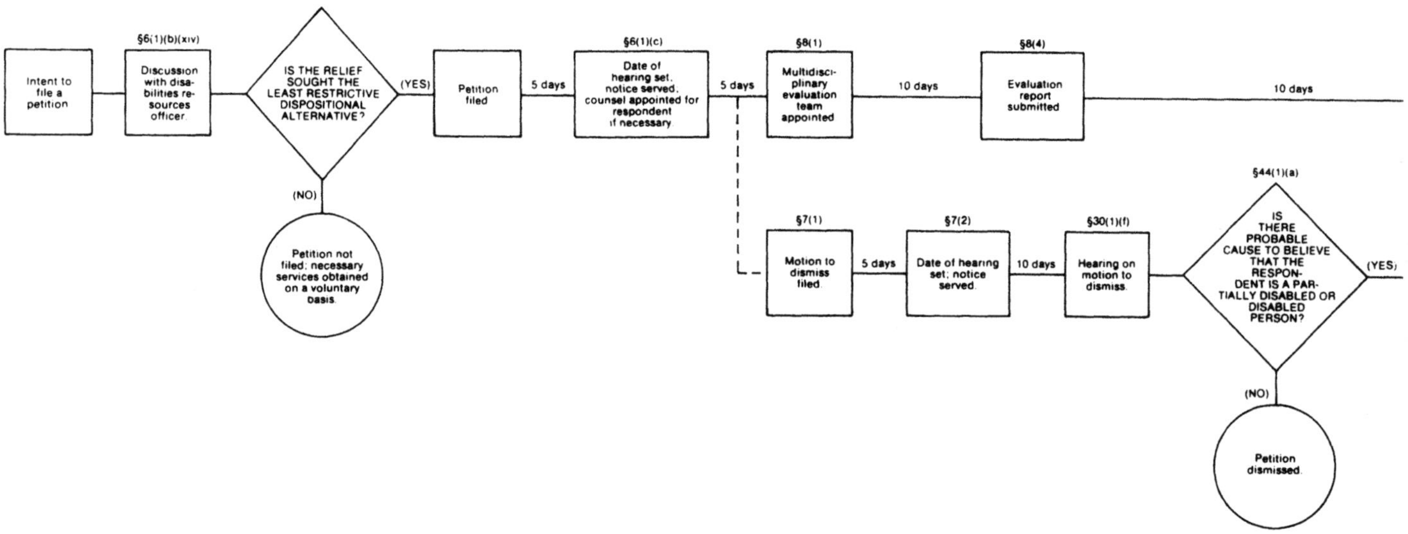

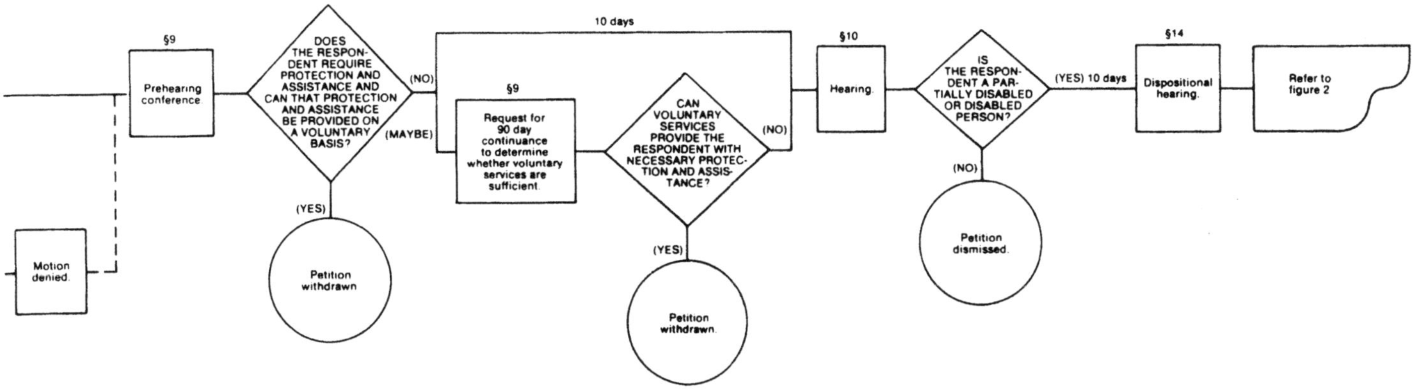

Key to Figure 1

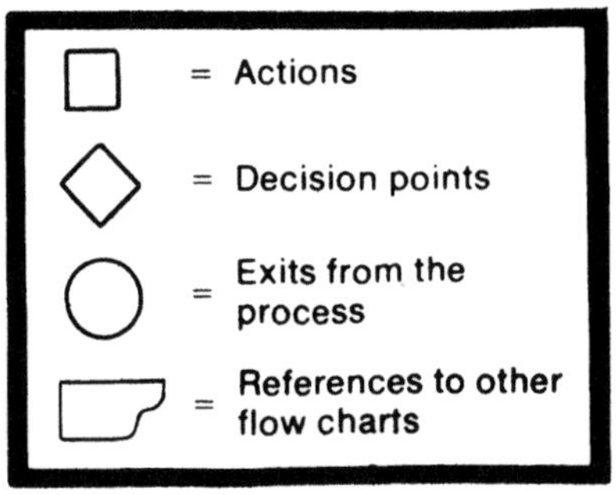

FIGURE 2 portrays the decision-making process at dispositional hearings following a determination that a respondent is a partially disabled or disabled person, as well as the reporting and review procedures which apply following establishment of a guardianship or conservatorship. As discussed in the commentary to sections 11 and 12(2), the alternatives should not be considered to be mutually exclusive. The differences in the intervals between review hearings are explained in the comment to section 15(1). The reporting and review process would continue until the court terminates the guardianship or conservatorship or the partially disabled or disabled person dies.

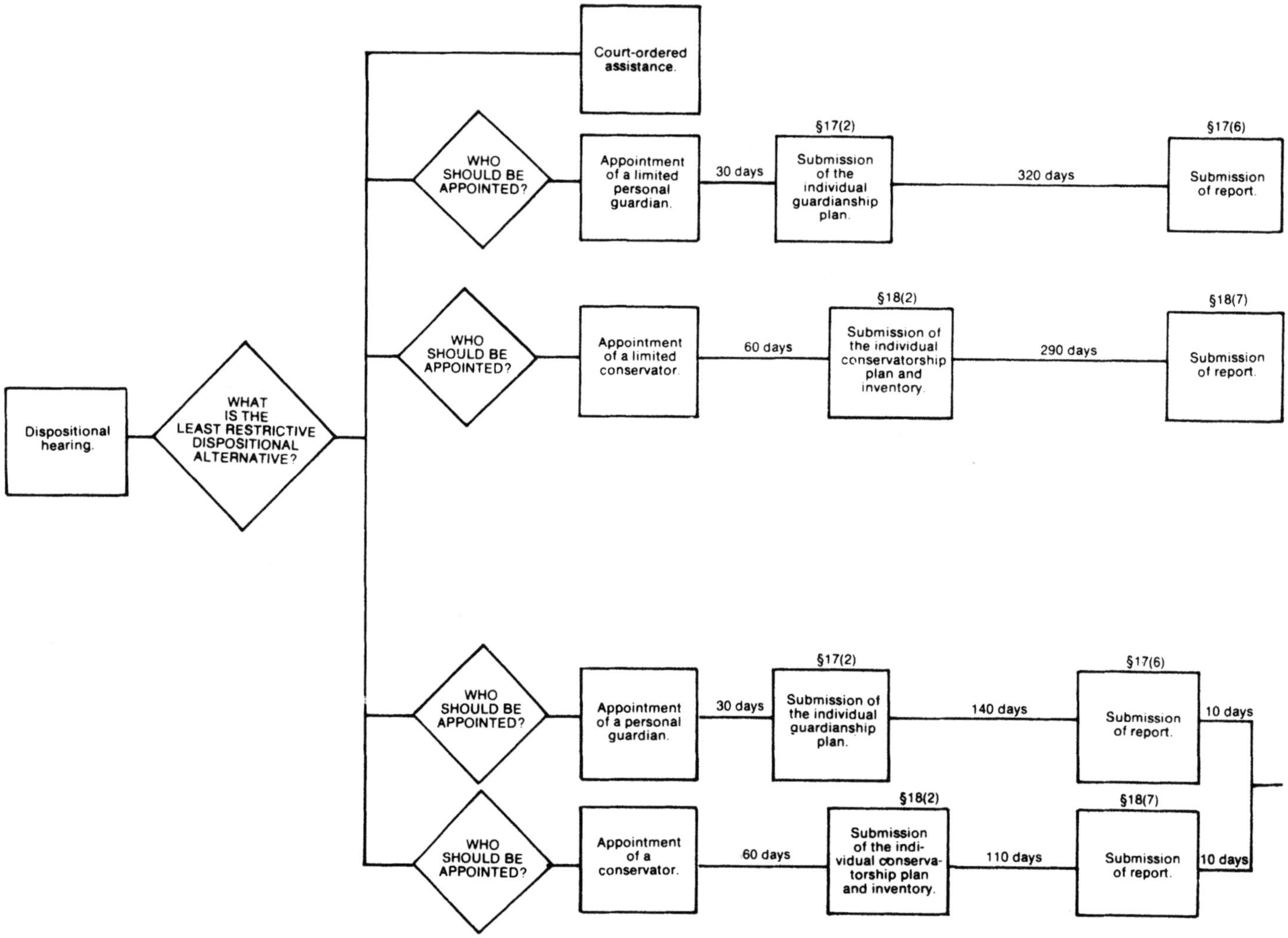

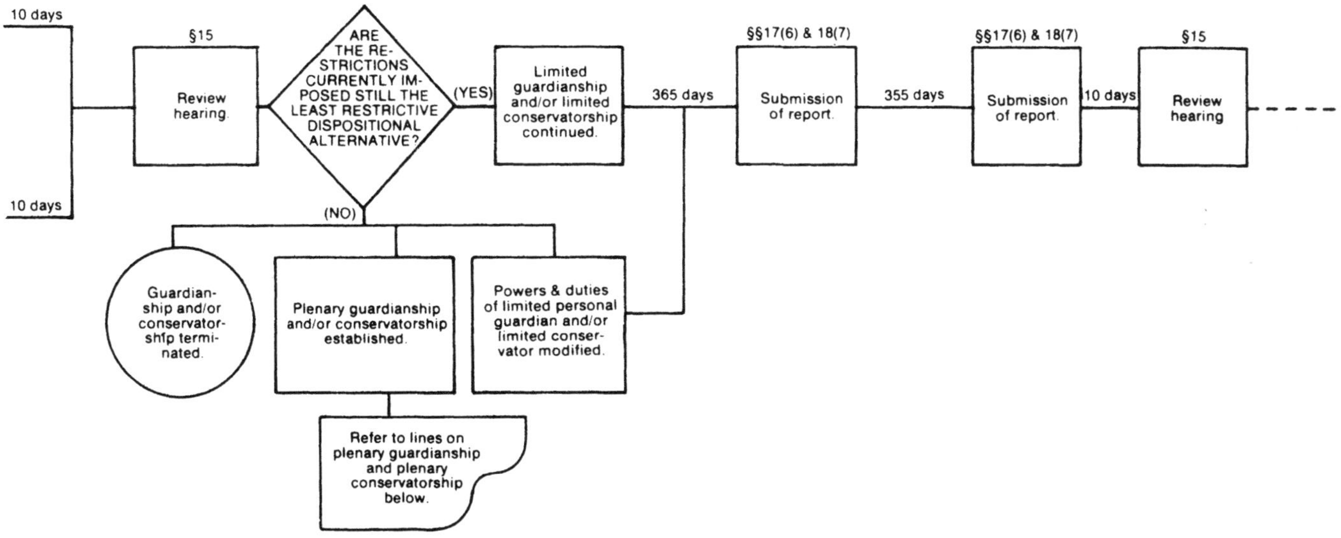

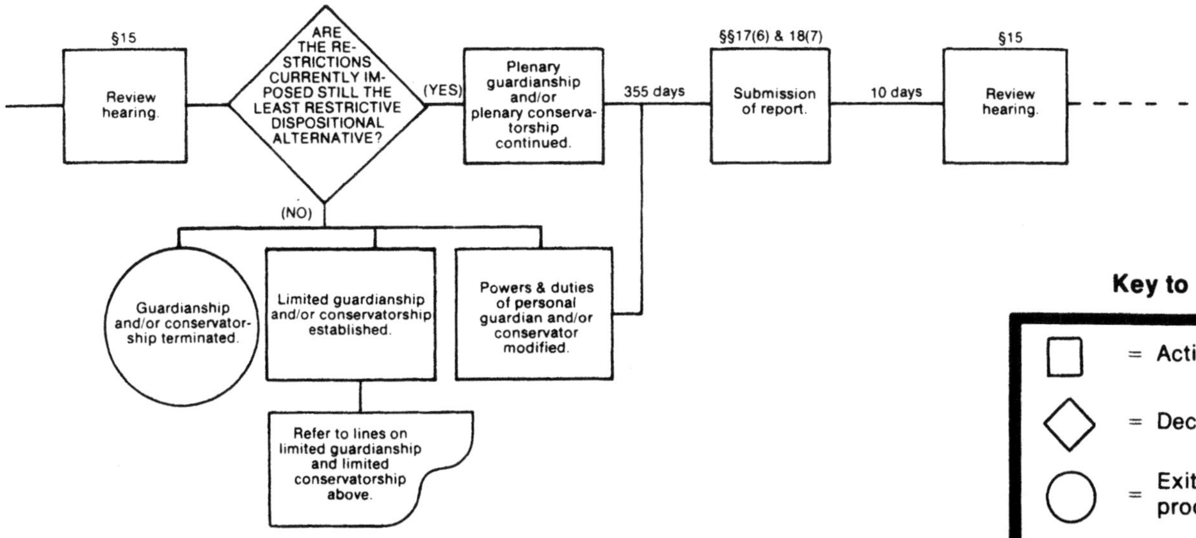

Key to Figure 2

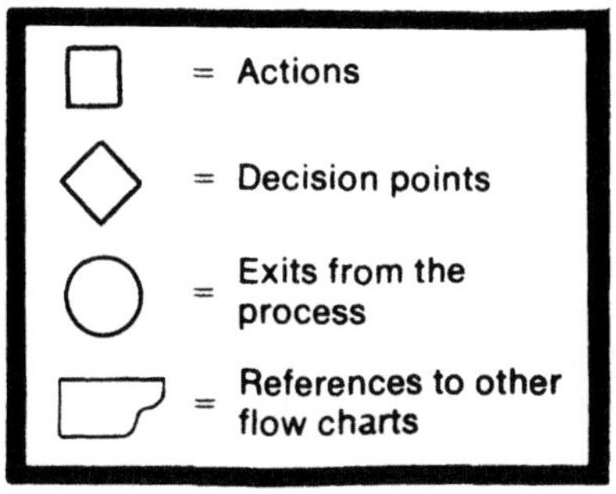

FIGURE 3 illustrates the four paths by which a guardianship or conservatorship may be established for a minor — i.e. by court order following the filing of a petition; by court order following an objection to a testamentary designation; by court order following the expiration of the time period for accepting a testamentary designation; and by an uncontested testamentary designation. The chart assumes that neither an application for an emergency appointment nor a motion to dismiss has been filed. (See §35 Emergency and Standby Appointments and §21 Motion to Dismiss.) As noted above, the time periods indicated on the chart represent the maximum intervals permitted under section 30.

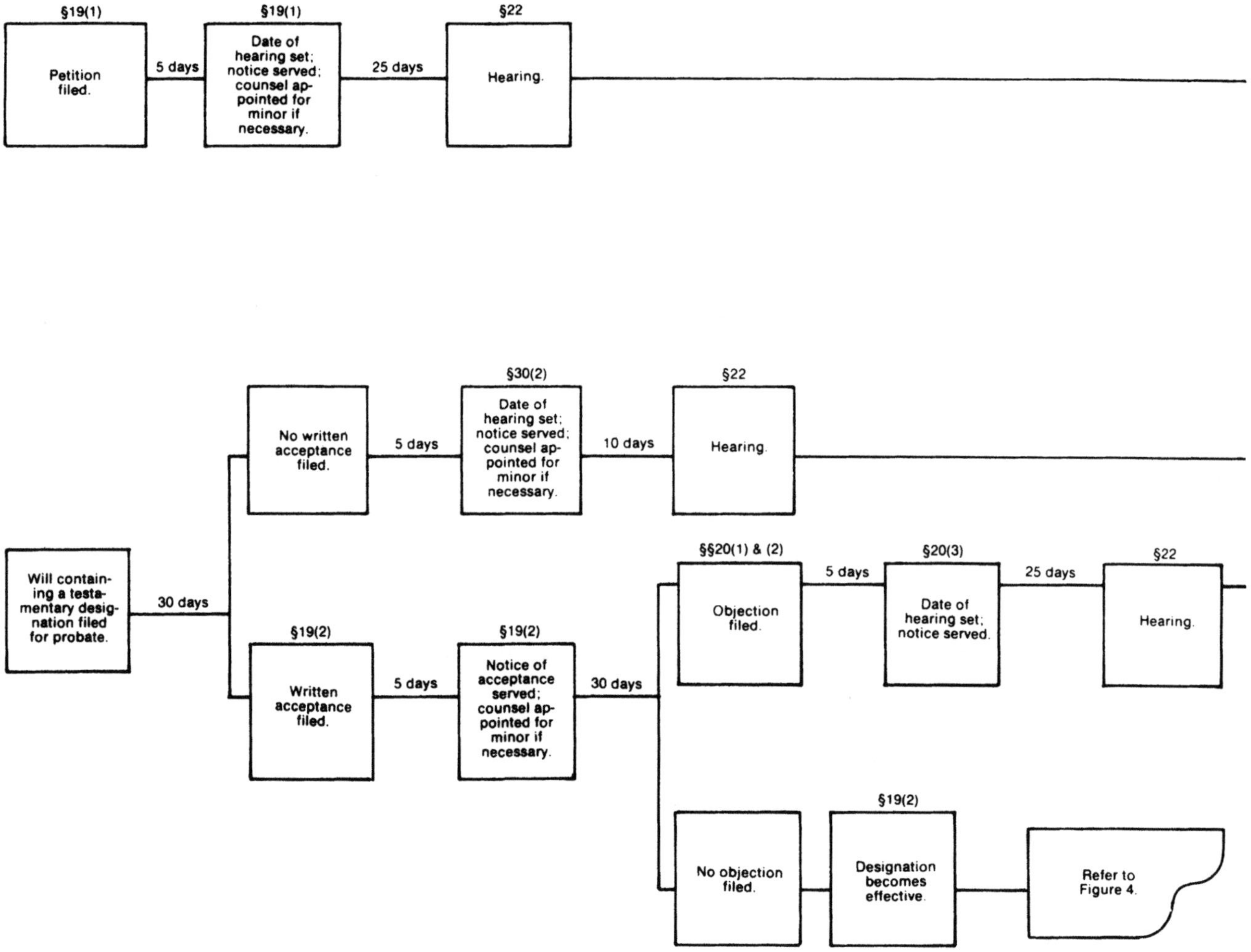

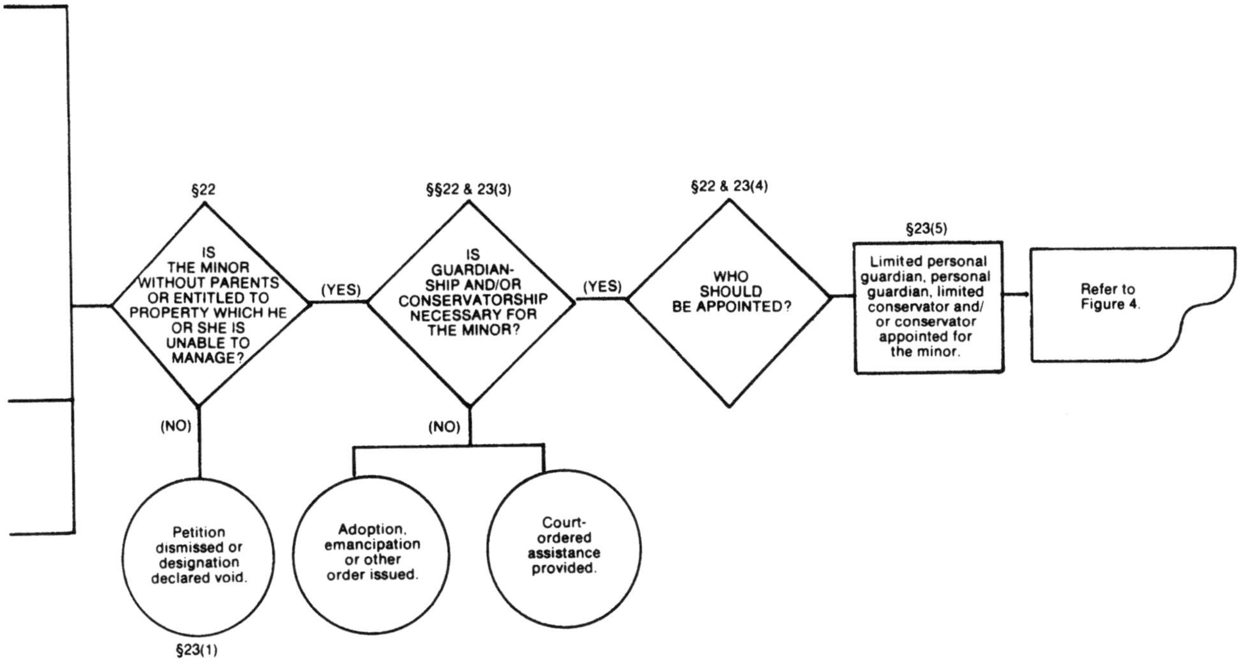

Key to Figure 3

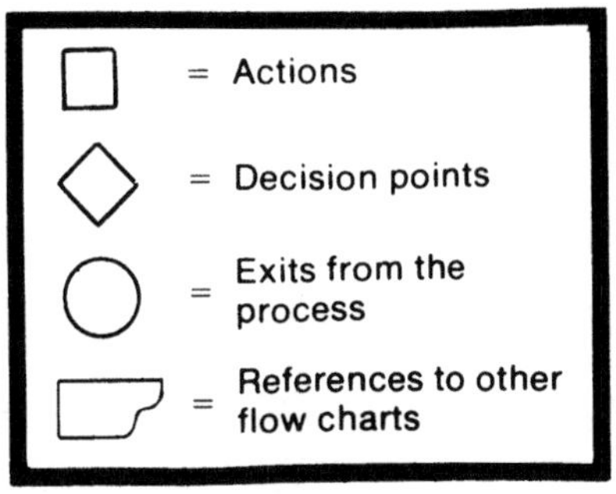

FIGURE 4 traces the review process following establishment of a guardianship or conservatorship for a minor. It is similar to the bottom two lines in Figure 2. The review process would continue until the guardianship or conservatorship for the minor is terminated by court order or by the minor's attaining majority, marriage or death. (See §26 Termination.) The introduction to section 27 explains why individual guardianship plans are not required under chapter 3.

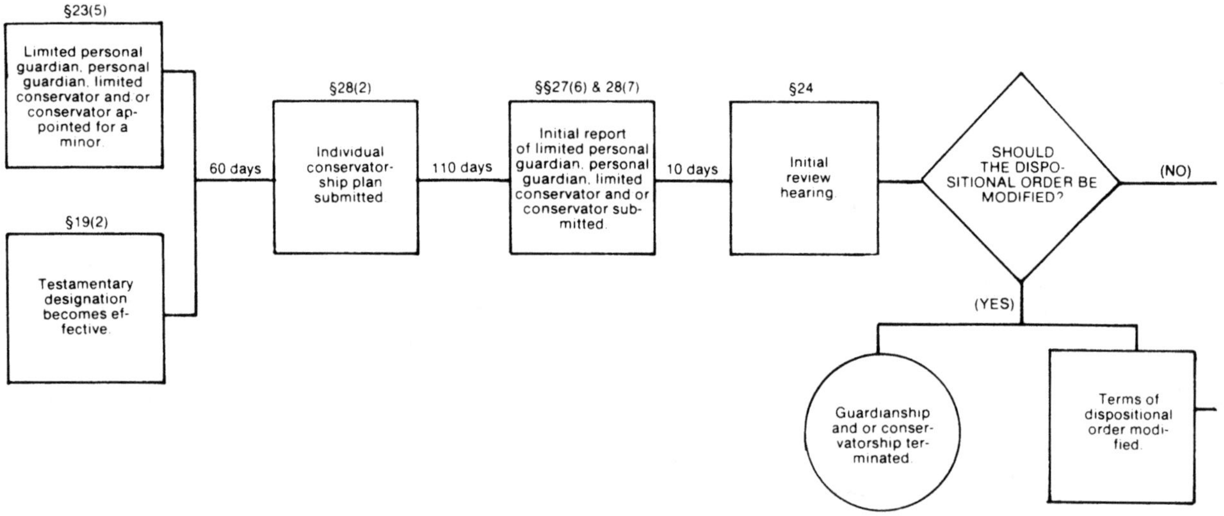

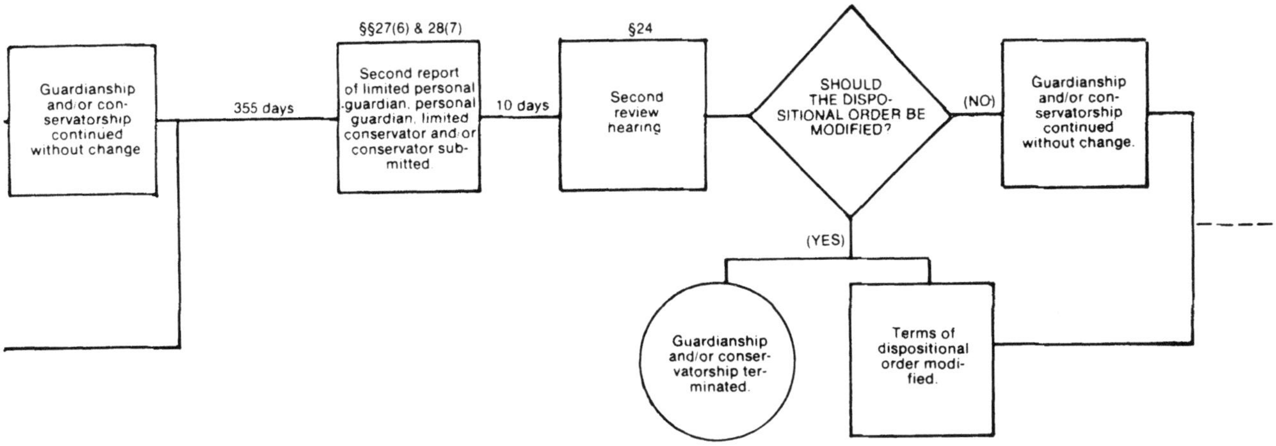

Key to Figure 4

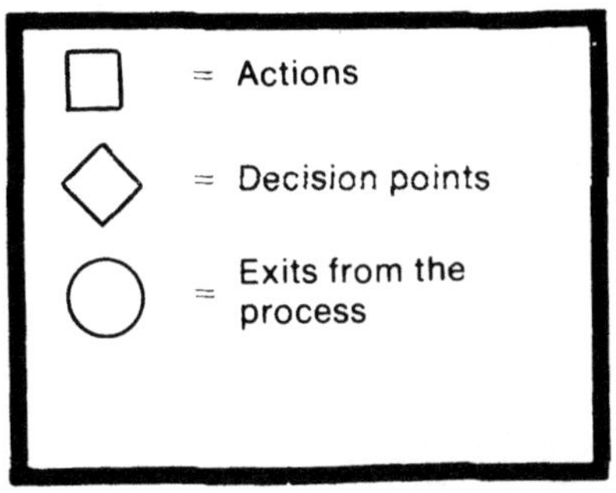

FIGURE 5 sets forth the maximum time which may elapse between the initiation of an intervention proceeding under chapter 2 and each of the major steps in the adjudication process. It assumes that the respondent is found to be a partially disabled or disabled person, that the hearing takes no more than a day, and that no motions to dismiss or for a continuance have been filed.

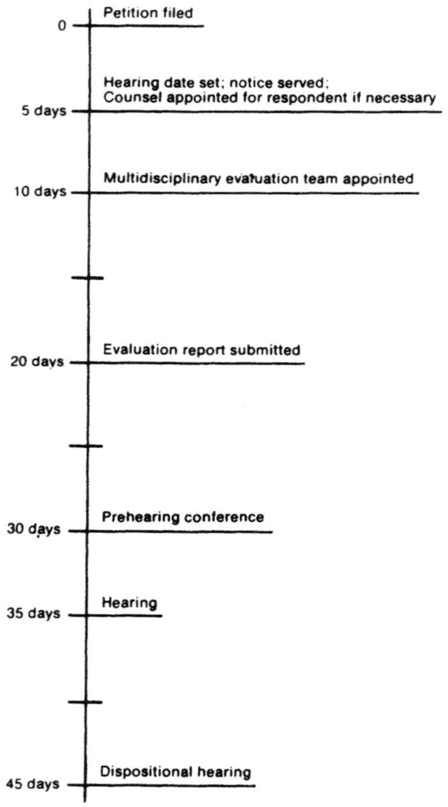

FIGURE 6 sets out the time between reviews following establishment of a guardianship or conservatorship for a partially disabled or disabled person. It assumes that the needs and abilities of that person are found not to have changed significantly at each of the review hearings or following a hearing on an application filed pursuant to section 16 to modify or terminate the guardianship and/or conservatorship.

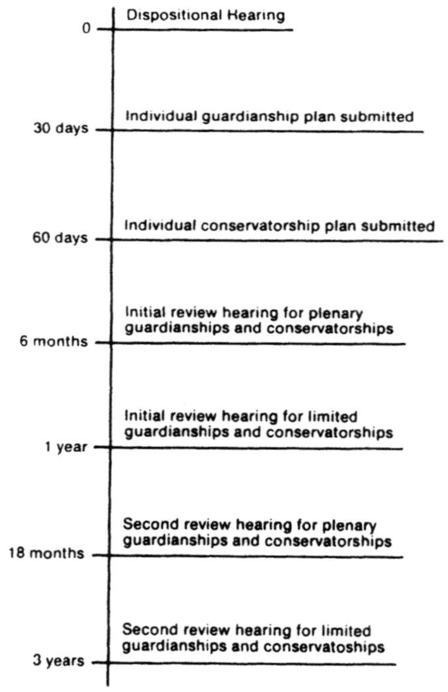

FIGURES 7(A), 7(B), 7(C) & 7(D) delineate the time limits which apply when a guardianship or conservatorship is sought for a minor either through the filing of a petition or by a testamentary designation. No time line is included for the review process established under chapter 3, since it would be identical to those portions of Figure 6 concerning review of plenary guardianships and conservatorships for disabled persons, except that no individual guardianship plan would have to be submitted.

A

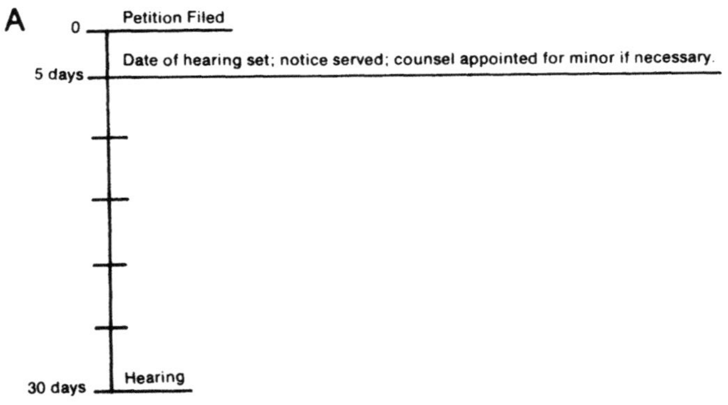

B

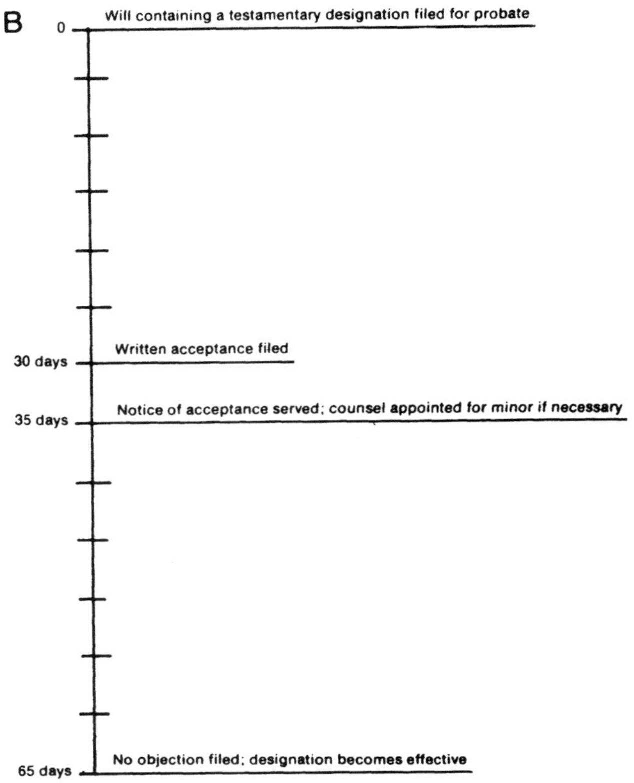

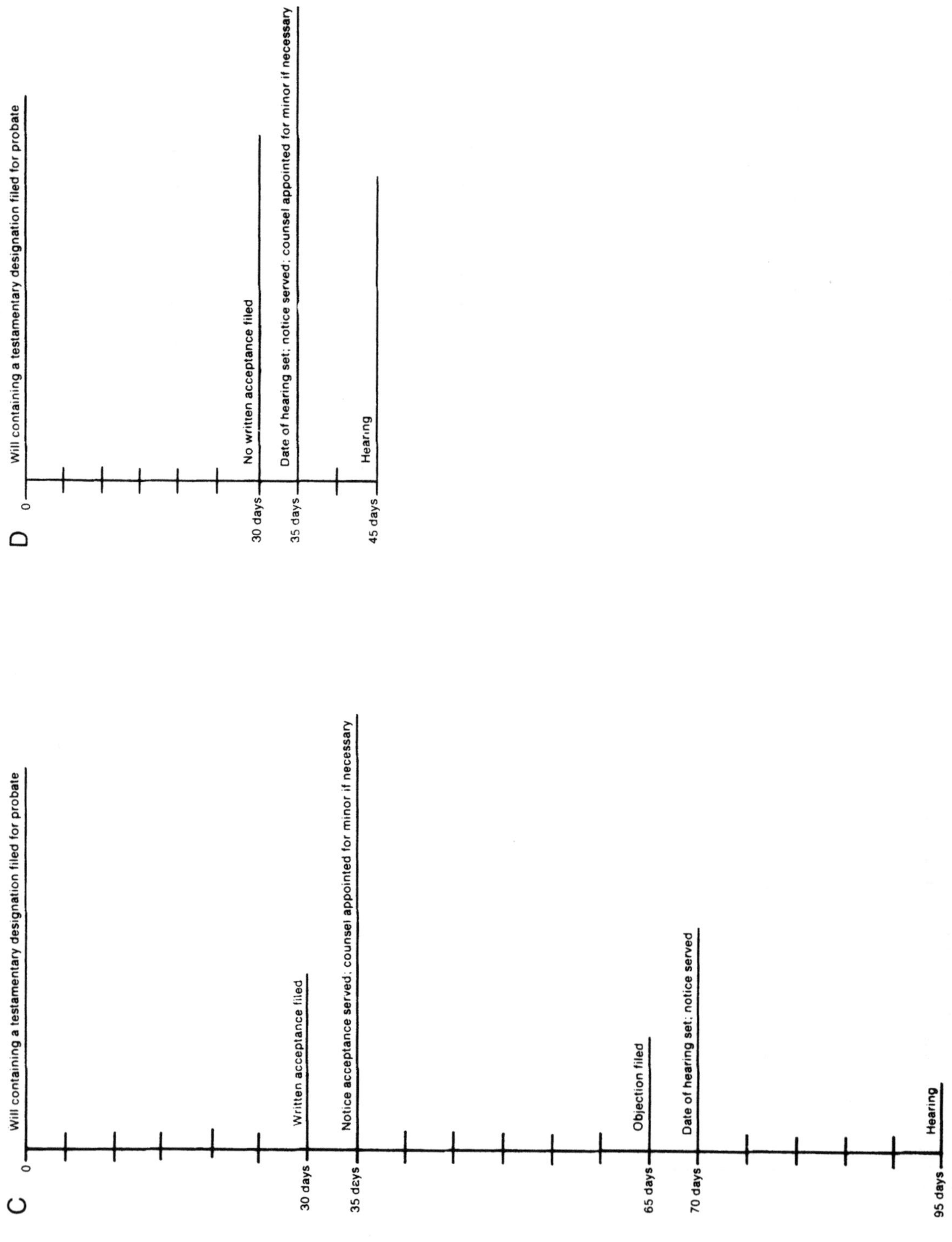
D
0
Will containing a testamentary designation filed for probate
30 days
No written acceptance filed
35 days
Date of hearing set; notice served; counsel appointed for minor if necessary
45 days
Hearing
C
0
Will containing a testamentary designation filed for probate
30 days
Written acceptance filed
35 days
Notice acceptance served; counsel appointed for minor if necessary
65 days
Objection filed
70 days
Date of hearing set; notice served
95 days
Hearing

8
CRIMINAL AND JUVENILE JUSTICE

Table of Contents

INTRODUCTION*

The substantive criminal law states rules of conduct and prescribes sanctions — usually a range of sanctions — for violation of these rules. Enforcement of the criminal law requires police to investigate, apprehend, and sometimes interrogate alleged offenders; prosecutors to determine whether or not a suspect should be charged and the specific offense to prosecute; defense attorneys to provide legal assistance to accused persons; diversion or bail agencies to formulate recommendations concerning release pending trial; judges (sometimes with the assistance of a jury) to determine whether the defendant has committed the offense charged and what sanctions to impose; court investigators to prepare social reports to assist the court in sentencing; correctional officers to supervise offenders; probation and parole officers to supervise offenders placed directly into the community; and parole officials to decide whether institutionalized offenders should be released before the expiration of their sentence. Taken together these officials and processes comprise the criminal justice system.[1] A separate system, commonly referred to as the juvenile justice system, deals with offenses committed by juveniles. All such offenses are subsumed under a broad offender category, such as "delinquent child;" separate juvenile courts and institutions are utilized; and the various bureaucracies involved in enforcement often have separate juvenile divisions.[2]

One goal of the criminal and juvenile justice systems is to treat alleged and adjudicated offenders fairly. The laws underlying these systems, however, are in many respects based upon assumptions about behavior, knowledge, and competence that may be incorrect when applied to a developmentally disabled person.[3] Thus, special accommodations are required in order for these two systems to deal fairly and appropriately with such persons.[4] For example, an individual first encounters the criminal or juvenile justice system through police officers, since they have the responsibility to maintain order, to investigate crimes, and to arrest persons who have committed them.[5] Three special considerations arise when a suspect is a developmentally disabled person.

First, the police officer must seek to ensure that the behavior in question is indeed an offense and not simply disability-related behavior that has created a false appearance of an offense. Speech or mobility impairments often associated with cerebral palsy may be confused by the untutored with drunkenness.[6] Or an individual with epilepsy may have a seizure that causes the person to damage property or to seem violent.[7] It is, of course, appropriate for a police officer to investigate apparently criminal behavior, but it is essential that the officer make reasonable efforts to ascertain whether or not the behavior is explained by non-criminal disability-related characteristics. In that case, no further police intervention is proper,

*Michael Kindred had primary responsibility for drafting this introduction.

[1] *See generally* LAFAVE and SCOTT, HANDBOOK ON CRIMINAL LAW (1972); ISRAEL and LAFAVE, CRIMINAL PROCEDURE IN A NUTSHELL (1975); KADISH and PAULSEN, CRIMINAL LAW AND ITS PROCESSES (3d ed. 1975).

[2] *See generally* FOX, THE LAW OF JUVENILE COURTS IN A NUTSHELL (2d ed. 1977); FOX, CASES AND MATERIALS ON MODERN JUVENILE JUSTICE (1972); MILLER, DAWSON, DIX, and PARNAS, JUVENILE JUSTICE PROCESS (2d ed. 1976).

[3] The term "developmental disability" is a creature of federal and, later, state legislation that was initially created to coordinate funding and provision of services for persons with epilepsy, mental retardation, cerebral palsy, and autism because of a sense that such persons share many common service needs. The term has since come into broad use among professionals and, increasingly, with the lay public.

[4] In a review of major criminal justice reform proposals that have been sponsored by the American Bar Association, the American Law Institute, and the Council of State Governments, Fox notes that little mention is made of the special complications introduced when a defendant is mentally retarded. Fox, *The Criminal Reform Movement,* in PRESIDENT'S COMMITTEE ON MENTAL RETARDATION (PCMR) THE MENTALLY RETARDED CITIZEN AND THE LAW 627, 629-636 (Kindred *et al.* eds. 1976).

[5] On the police function, *see generally* ADAMS, LAW ENFORCEMENT — AN INTRODUCTION TO THE POLICE ROLE IN THE CRIMINAL JUSTICE SYSTEM (2d ed. 1973); AMERICAN BAR ASSOCIATION, STANDARDS RELATING TO THE URBAN POLICE FUNCTION (Approved Draft, 1973); MILLER, THE POLICE FUNCTION (1971); FELKENES and BECKER, LAW ENFORCEMENT: A SELECTED BIBLIOGRAPHY (2d ed. 1977).

[6] *See* CEREBRAL PALSY — ITS INDIVIDUAL AND COMMUNITY PROBLEMS 195 (speech impairment) and 434 (mobility impairments) (2d ed. Cruickshank ed. 1966).

[7] While a susceptibility to seizures is a symptom associated with epilepsy, almost eighty per cent of persons with epilepsy can achieve reasonable control of their seizures through proper medication. BARROW and FABING, EPILEPSY AND THE LAW 2 (2d ed. 1966). On the current status of persons with epilepsy under state laws, see EPILEPSY FOUNDATION OF AMERICA, THE LEGAL RIGHTS OF PERSONS WITH EPILEPSY (1976).

except to offer voluntary assistance or referral to a social service agency if the individual expresses a need for it or is clearly in need of it but unable to express the need.

The second complexity in police interaction with developmentally disabled persons has to do with interrogation. Although police place a high value upon being able to interrogate criminal suspects,[8] a high countervailing value is placed in our society upon the right of an individual to refuse to provide incriminating information against himself or herself. This value is articulated in the Fifth Amendment to the United States Constitution, which states "No person shall . . . be compelled in any criminal case to be a witness against himself. . . ."[9] While this provision could be read literally as only applicable to the courtroom setting, it has long been held applicable to police interrogations as well[10] and has been held to apply to the state criminal process by virtue of the Fourteenth Amendment to the Constitution.[11] Similarly, the Sixth Amendment right to "assistance of counsel" has been held applicable to the states[12] and to apply during interrogation of a suspect in custody.[13] Thus, when the police wish to question a person in custody whom they have probable cause to believe has committed a crime, they are required to advise the person of his or her right to remain silent and to consult counsel and have counsel appointed in case of indigency.[14] After being informed of these rights, the person is free to exercise or waive either of them.[15] The waiver, in order to be valid, must be made voluntarily, knowingly and intelligently.[16] Incriminating statements elicited in violation of these rights and physical evidence secured as a result of them cannot be used against the person in criminal proceedings.[17]

Unfortunately, these rules may be inadequate to protect a developmentally disabled person's constitutional rights. Since many developmentally disabled persons suffer from deficits in intellectual capability,[18] their concepts of attorneys may be rudimentary or non-existent and their ability to evaluate the wisdom of waiving the right to counsel and the privilege against self-incrimination are likely to be limited.[19] Thus, an intelligent waiver of these rights may be precluded. The voluntariness of waivers by developmentally disabled persons should also be subject to question. While any individual can find the atmosphere of a police station interrogation room intimidating,[20] disabled persons may feel even less able to defend themselves and be more intimidated. In fact, it has been recognized that disabled persons may tend to be more susceptible to being led by persons in authority, such as police officers, into giving responses they feel are desired even if such responses are falsely incriminating.[21] Thus it is not surprising that the Supreme Court has often noted the relevance of mental illness and mental retardation to the voluntariness of confessions and waivers of rights.[22]

While the courts have shown a sensitivity to the problems that compromise the validity of confessions and waivers by disabled persons,[23] they have not, as a constitutional matter, held a disability to be a *per se* invalidation of waivers. Rather, once a person is informed of his or her rights, the courts have held that the validity of the waiver, as a constititonal matter, is to be determined in each case

[8] In Miranda vs. Arizona, 384 U.S. 436 (1966), a landmark United States Supreme Court decision prescribing procedures required in police interrogations in order to protect fundamental rights of suspects, the Court noted a substantial body of literature describing police interrogation procedures and instructing police officers in how best to induce a suspect to confess. *Id.* at 448-454.

[9] U.S. Const. amend. V.

[10] Miranda vs. Arizona, 384 U.S. 436 (1966); Bram vs. United States, 168 U.S. 532 (1897).

[11] Miranda vs. Arizona, 384 U.S. 436 (1966); Malloy vs. Hogan, 378 U.S. 1 (1964); Mapp vs. Ohio, 367 U.S. 643 (1961).

[12] Gideon vs. Wainwright, 372 U.S. 335 (1963); Argersinger vs. Hamlin, 407 U.S. 25 (1972).

[13] Escobedo vs. Illinois, 378 U.S. 478 (1964).

[14] Miranda vs. Arizona, 384 U.S. 436, 444 (1966).

[15] *Id.*

[16] *Id.*

[17] *Id.* (with respect to statements); United States v. Cassell, 452 F.2d 533 (7th Cir. 1971) (physical evidence); *see also* Silverthorne Lumber Co. vs. United States, 251 U.S. 385 (1920), and Wong Sun vs. United States, 371 U.S. 471 (1963). While statements made in circumstances that violate the Miranda rules cannot be used as direct evidence against the defendant, they can in some circumstances be used to impeach, or undermine the credibility of, the defendant who testifies on his or her own behalf at trial. Oregon vs. Hass, 420 U.S. 714 (1975); Harris vs. New York, 401 U.S. 222 (1971); *cf.* Mincey vs. Arizona, 437 U.S. 385 (1978).

[18] MANUAL ON TERMINOLOGY AND CLASSIFICATION IN MENTAL RETARDATION 11 (Grossman ed. 1977); *see also* Scheerenberger, *Mental Retardation: Definition, Classification, and Prevalence,* in ROTHSTEIN, MENTAL RETARDATION — READINGS AND RESOURCES 4 (2d ed. 1971).

[19] While the criminal process is sufficiently complex and the risks sufficiently great that it is hard to imagine anyone "intelligently" waiving his or her right to counsel, the developmentally disabled person presents an extreme case of the paradox implicit in the motion of an "intelligent" waiver of counsel.

[20] *See* Miranda vs. Arizona, 384 U.S. 436, 448-456 (1966).

[21] Fikes vs. Alabama, 352 U.S. 191, 196-198 (1957).

[22] Miranda vs. Arizona, 384 U.S. 436, 455-456 (1966); Davis vs. North Carolina, 384 U.S. 737, 742 (1966); Blackburn vs. Alabama, 361 U.S. 199, 207 (1960); Spano vs. New York, 360 U.S. 315, 322 (1959); Fikes vs. Alabama, 352 U.S. 191, 193 (1957); Powell vs. Alabama, 287 U.S. 45, 71 (1932).

[23] *E.g.,* cases cited in note 22; *see also* United States vs. Henderson, 520 F.2d 896, 901 (2d Cir. 1975), *cert. denied* 423 U.S. 998 (1975); Thomas vs. North Carolina, 447 F.2d 1320, 1322 (4th Cir. 1971); Hizel vs. Sigler, 430 F.2d 1398 (8th Cir. 1970); Smallwood v. Warden, 367 F.2d 945, 949 (4th Cir. 1966).

In Blackburn vs. Alabama, 361 U.S. 199, the Court noted that

by reviewing all the circumstances.[24] While this may provide an adequate explication of the Constitution, state legislatures are, of course, free to provide more protective procedures. The difficulty with an after-the-fact individual case determination is that it risks permitting an interrogation in violation of the rights of the disabled persons with this violation to be corrected only through a costly suppression of evidence hearing, while respect for the rights of the individual might have permitted a different investigative tactic and preparation of a valid case.

Third and finally, in order to assure effective police response to developmentally disabled persons, police officers must be aware of social services available to the developmentally disabled. For instance, they are often called upon to assist in difficult situations involving no criminal offense,[25] such as when a person with epilepsy is having a seizure or a mentally retarded individual without adequate housing is wandering the streets. In such cases referral to social service resources is often required. In addition, knowledge of social services is sometimes important where a developmentally disabled person is suspected of a criminal offense. For example, where a developmentally disabled person is apprehended for indecent exposure after urinating in public, the best response may be referral to an agency that can provide instruction on socially appropriate behavior. Where a developmentally disabled person is apprehended for shoplifting from a grocery store because he or she has run out of money to buy food, an appropriate referral to a social agency might provide training in budgeting meager resources that will provide reasonable assurance of an end to the illegal behavior.

Even in those instances, in which the laws underlying the criminal and juvenile justice systems often attempt to address the special needs of developmentally disabled persons, they are often not sufficiently developed to provide the appropriate guidance to the officials involved. For example, following police investigation, possible interrogation, and arrest, the next step in the criminal justice system is the prosecutorial phase. The prosecutor's office exercises considerable discretion in determining whether to press criminal charges against a person, what level of charge to file, and whether to consider possible pre-trial diversion from the criminal process.[26] While the primary factor involved is the prosecutor's view of the accused's guilt, consideration is given to other factors,[27] such as whether the accused is intellectually or emotionally impaired.[28] For example, where a social agency is prepared to provide services to the developmentally disabled person and to take some responsibility for him or her, a prosecutor may feel that referral to that agency is a sufficient discharge of his or her responsibilities. Or recognition that the person already has been sufficiently harshly dealt with in a hostile educational, employment, and social environment may lessen the sense that "justice" requires his or her conviction. The prosecutor may also have less than the usual confidence in his or her ability to secure a conviction in light of a possible finding of incompetence to stand trial,[29] possible defenses of diminished capacity[30] or non-responsibility,[31] and possible attacks on the admissibility of any confession.[32] For these various reasons, the discretionary aspects of the prosecutor's function are of great importance to a developmentally disabled person accused of crime, particularly when that crime is not of a grievous nature.

In recent years a considerable body of literature has developed that is critical of the failure to reg-

---

Blackburn had a history of mental illness (*Id.* at 203) and stated:

> Surely in the present stage of our civilization a most basic sense of justice is affronted by the spectacle of incarcerating a human being on the basis of a statement he made while insane; and this judgment can without difficulty be articulated in terms of the unreliability of the confession, the lack of rational choice of the accused, or simply a strong conviction that our system of law enforcement should not operate so as to take advantage of a person in this fashion. *Id.*, at 207.

[24] *See e.g.,* United States vs. Marchildon, 519 F.2d 337, 343 (8th Cir. 1975); *cf.* Brewer vs. Williams, 430 U.S. 387 (1977).

[25] ADAMS, *supra* note 5, at 145; ABA, *supra* note 5, at 59.

[26] *See generally* MILLER, PROSECUTION: THE DECISION TO CHARGE A SUSPECT WITH A CRIME (1969); ABA, STANDARDS RELATING TO THE PROSECUTION FUNCTION AND THE DEFENSE FUNCTION (Approved Draft 1971); Cox, *Prosecutorial Discretion: An Overview,* 13 AMERICAN CRIMINAL LAW REVIEW 383 (1976); LaFave, *The Prosecutor's Discretion in the United States*, 18 AMERICAN JOURNAL OF COMPARATIVE LAW 532 (1970). NATIONAL INSTITUTE OF LAW ENFORCEMENT AND CRIMINAL JUSTICE, PROSECUTORIAL DISCRETION: THE DECISION TO CHARGE, AN ANNOTATED BIBLIOGRAPHY (1975).

[27] MILLER, *supra* note 26, at 154-292; Lagoy, Senna, and Siegel, *An Empirical Study on Information Usage for Prosecutorial Decision Making in Plea Negotiations,* 13 AMERICAN CRIMINAL LAW REVIEW 435 (1976); Note, *Discretion Exercised by Montana County Attorneys in Criminal Prosecutions,* 28 MONTANA LAW REVIEW 41, 49 (1966); JACOBY, PRE-TRIAL SCREENING IN PERSPECTIVE (1976); INBAU, THOMPSON, HADDAD, ZAGEL, and STARKMAN, CRIMINAL PROCEDURE 487 (1974); ABA, PROSECUTION AND DEFENSE *supra* note 26, at §§3.8 and 3.9.

[28] MILLER, *supra* note 26, at 186 and 214-218; INBAU, *et al., supra* note 27, at 487; Lagoy *et al., supra* note 27, at 446-447 and 450-451; JACOBY, *supra* note 27, at 26; *Prosecutorial Discretion in the Initiation of Criminal Complaints,* 42 SOUTHERN CALIFORNIA LAW REVIEW 519, 529 (1969).

[29] See discussion *infra*, at 8-11.

[30] See discussion *infra*, at 15-16.

[31] See discussion *infra*, at 11-15.

[32] See discussion *supra*, at 3-5.

ularize or regulate the manner in which prosecutorial discretion is exercised.[33] In a few cases of extreme abuse, courts have limited the exercise of that discretion,[34] and some prosecutors have developed guidelines to govern their staffs.[35] Thus, it is of recognized importance to articulate the relevance of a developmental disability to the exercise of discretion, to ensure that prosecutors have the information and training that will permit them to exercise their discretion appropriately, and to ensure sufficient procedural safeguards against the abuse of diversionary agreements.

In some cases, the law has developed specific doctrines that do focus on particular problems of developmentally disabled persons. Unfortunately implementation of these doctrines has revealed numerous inadequacies. For example, once a decision has been made to prosecute an alleged offender, attention turns to the trial process. Historically, disabling conditions have been recognized as pertinent to a person's competence to stand trial and to his or her responsibility under the criminal law.

Competence to stand trial has been held by the United States Supreme Court to be a constitutional prerequisite to adjudication of criminal guilt. The Court has held that due process of law precludes trying a person for a criminal offense where the person's "mental condition is such that he lacks the capacity to understand the nature and object of the proceedings against him, to consult with counsel, and to assist in preparing his defense."[36]

Although this rule appears to recognize the needs of developmentally disabled persons with a mental disability, its implementation historically has worked to the disadvantage of such persons in many cases. State laws typically provide that the issue of incompetence to stand trial could be raised by the prosecutor, the defendant, or the court and that whenever a person was found incompetent to stand trial,[37] the trial court had to commit the defendant to a special institution for the "criminally insane" until his or her competence was restored.[38] This procedure raises several problems.

First, individuals who are totally innocent of the charged offense and whose innocence can readily be proven at trial, are precluded from doing so and could be, or under some statutes would have to be, committed to an institution for life if they were unable to become competent to stand trial.[39] Second, individuals who committed a minor offense ordinarily sanctioned by a fine or short prison sentence likewise could be held in an institution for a long time or permanently.[40] Third,

[33] DAVIS, DISCRETIONARY JUSTICE: A PRELIMINARY INQUIRY (1969). The thrust of Professor Davis' recommendations is toward internal, "administrative law", rather than "legislative", controls on discretion; most of the literature has followed this lead. *See also* Cox, *supra* note 26; Vorenberg, *Narrowing the Discretion of Criminal Justice Officials*, 1976 DUKE LAW JOURNAL 651 (1976); Noll, *Controlling a Prosecutor's Screening Discretion Through Fuller Enforcement*, 29 SYRACUSE LAW REVIEW 697 (1978); *but see* Cardinale and Feldman, *The Federal Courts and the Right to Nondiscriminatory Administration of the Criminal Law: A Critical View*, 29 SYRACUSE LAW REVIEW 659 (1978). For a very thoughtful, albeit general, discussion of the nature and importance of discretion, see Breitel, *Controls in Criminal Law Enforcement*, 27 *University of Chicago Law Review* 427 (1960).

[34] United States vs. Falk, 479 F.2d 616 (7th Cir. *en banc* 1973); Dixon vs. District of Columbia, 394 F.2d 966 (D.C. Cir. 1968); United States vs. Cox, 342 F.2d 167 (5th Cir. 1965), *cert. denied* 381 U.S. 935.

[35] MILLER, *supra* note 26, at 16; JACOBY, *supra*, note 27. See also two Justice Department documents issued under cover of a January 18, 1977, memoranda by then Attorney General Edward Levi to U.S. attorneys and the heads of all Justice Department Offices, to provide guidance in the exercise of prosecutorial discretion, 24 CRIMINAL LAW REPORTER 3001 (1978).

[36] Drope vs. Missouri, 420 U.S. 162, 171 (1975); *see also* Pate vs. Robinson, 383 U.S. 375 (1966); Bishop vs. U.S., 350 U.S. 961 (1956). For federal cases, the court must determine whether the defendant "has sufficient present ability to consult with his lawyer with a reasonable degree of rational understanding — and whether he has a rational as well as factual understanding of the proceeding against him." Dusky vs. United States, 362 U.S. 402 (1960). The Supreme Court also ruled in Drope and Pate that judges must initiate an inquiry on the competence issue when "sufficient indicia" of lack of competence come to their attention. Neither case specifically indicates what constitutes "sufficient indicia", although a statute approved by the Court in Drope required a hearing where the evidence raises a "*bona fide* doubt" concerning competence.

The doctrine regarding competence to stand trial has ancient roots in the criminal justice system. Blackstone stated:

> "[I]f a man . . . commits a capital offense, and before arraignment for it, he becomes mad, he ought not to be arraigned for it; because he is not able to plead to it with the advice and caution that he ought. And if, after he has pleaded, the prisoner becomes mad, he shall not be tried, for how can he make his defense."

4 BLACKSTONE, COMMENTARIES 24 (1665).

Many of the early English and American authorities are cited in Youtsey vs. United States, 97 F.937 (6th Cir. 1899), a case in which a loss of memory due to repeated epileptic seizures was the cause for lack of competence to stand trial.

[37] The term "insane" was often used both in this context and in reference to criminal responsibility, creating considerable confusion between these two very different problems. Comment, *Competency to Stand Trial*, 59 MICHIGAN LAW REVIEW 1078 (1961).

[38] For a thorough, well-documented discussion of this procedure, see WEIHOFEN, MENTAL DISORDER AS A CRIMINAL DEFENSE 428-474 (1954).

[39] BRAKEL and ROCK, THE MENTALLY DISABLED AND THE LAW 412 (2d ed. 1971).

[40] A 1960 study of 755 persons committed to Iona (Michigan) State Hospital as incompetent to stand trial estimated on the basis of past rates that well over one-half of these persons would "expect to spend the rest of their lives at the hospital." Comment, *supra* note 37, at 1088.

While many forms of mental illness can be treated, especially with modern psychotropic drugs, the prospect of long term confinement is probably disproportionately great for severely

the fact that the issue of competence can be raised by the prosecution as well as the defense has led to concern that prosecutors might use the competency issue to secure incarceration where they have a weak case.[41]

In 1972, the United States Supreme Court examined the operation of this process in *Jackson vs. Indiana*,[42] found it to be unconstitutional under both the due process and equal protection clauses of the Fourteenth Amendment, and held that the state "cannot constitutionally commit the petitioner for an indefinite period simply on account of his incompetency to stand trial on the charges against him."[43] The Court held that:

> "[A] person charged by a State with a criminal offense who is committed solely on account of his incapacity to proceed to trial cannot be held more than the reasonable period of time necessary to determine whether there is a substantial probability that he will attain that capacity in the foreseeable future. If it is determined that this is not the case, then the State must either institute the customary civil commitment proceeding that would be required to commit indefinitely any other citizen, or release the defendant. Furthermore, even if it is determined that the defendant probably soon will be able to stand trial, his continued commitment must be justified by progress toward that goal."[44]

While Jackson speaks to several major aspects of the incompetency process, the following are some of the considerable number of other issues it does not address. Who bears the burden of proof in establishing incompetence? What period of time is "reasonable" for a person to be held to determine his competence? How can the determination of competence be made with the least intrusion on the liberty of the still untried and unconvicted defendant? Ought release on bail during this period be allowed? In the absence of competence, what opportunity does a defendant have to demonstrate, through motions to suppress or introduction of exculpatory evidence, the impossibility or extreme unlikelihood of his being convicted.[45] And, how rapid must continued progress toward competence be in order to justify holding for treatment a person whose competence is sought to be restored?[46]

The doctrines limiting criminal responsibility similarly require refinement if they are to provide an appropriate response to a person's developmental disability. Because the criminal law is generally based upon notions of individual responsibility and moral fault,[47] the law recognizes a defense of "insanity", or "not responsible because of mental disease or defect".[48] The defense is available in most jurisdictions to some persons who are mentally retarded as well as to some who are mentally ill,[49] with most states precluding the issue from being raised by anyone but the defendant.[50] It assumes that the defendant has committed the act charged,[51] and asks whether, be-

mentally retarded defendants. HARVARD MEDICAL SCHOOL LABORATORY OF COMMUNITY PSYCHIATRY, COMPETENCY TO STAND TRIAL AND MENTAL ILLNESS, 65 (1973); Gobert, *Competency to Stand Trial: A Pre-Post-Jackson Analysis,* 40 TENNESSEE LAW REVIEW 659, 673 (1973).

[41] The first piece in the modern, critical literature on competency determination procedures was a comment on a case where the dismissal of indictments against three co-defendants made it clear that criminal prosecution was precluded. The fourth co-defendant was found incompetent to stand trial and indefinitely committed. Foote, *A Comment on Pre-Trial Commitment of Criminal Defendants,* 108 UNIVERSITY OF PENNSYLVANIA LAW REVIEW 832 (1960), discussing United States vs. Barnes, 175 F. Supp. 60 (S.D. Cal. 1959).

The negative effects of the doctrine have even led to proposals that it be abolished. *See* Burt and Morris, *A Proposal for the Abolition of the Incompetency Plea,* 40 UNIVERSITY OF CHICAGO LAW REVIEW 66, 75 (1972); Morris, *Special Doctrinal Treatment in Criminal Law,* in P.C.M.R., *supra* note 4, at 682. For a vigorous attack on the Burt-Morris position, see Pizzi, *Competency to Stand Trial in Federal Courts: Conceptual and Constitutional Problems,* 45 UNIVERSITY OF CHICAGO LAW REVIEW 21 (1977).

[42] 406 U.S. 715 (1972).

[43] *Id.*, at 720.

[44] *Id.*, at 738.

[45] Procedures to provide such opportunity have been advocated by commentators (*e.g.* Foote, *supra* note 41, at 845) and by the American Law Institute. See AMERICAN LAW INSTITUTE (ALI), MODEL PENAL CODE §4.06 (3) and §4.06 Alternative (3) and (4). The Supreme Court in Jackson acknowledged, with apparent favor, such suggestions for reform. 406 U.S. 715, 740 (1972).

[46] An additional issue, which in the present state of science concerns primarily defendants where incompetence results from mental illness, is whether a person can be tried while on a drug regimen that reestablishes competence. On this question, see KUNZ, PSYCHOTROPIC MEDICATION AND COMPETENCY TO STAND TRIAL (1974); Winick, *Psychotropic Medication and Competence to Stand Trial,* 1977 AMERICAN BAR FOUNDATION RESEARCH JOURNAL 769 (1977).

[47] The courts have emphasized over the centuries that "free will" is the postulate of responsibility under our jurisprudence. 4 BLACKSTONE'S COMMENTARIES 27. United States vs. Brawner, 471 F.2d 969, 985 (1972). *See also* FLETCHER, RETHINKING CRIMINAL LAW 801-802 (1978); LAFAVE and SCOTT, *supra,* note 1 at 9-10.

[48] The term "insanity defense" has long been used. The Model Penal Code replaced it with "defense of not responsible because of mental disease or defect." The literature on the topic is voluminous.

[49] GOLDSTEIN, THE INSANITY DEFENSE 48 (1967); LAFAVE and SCOTT, *supra* note 1 at 275; WEIHOFEN, *supra* note 38, at 119.

[50] LAFAVE and SCOTT, *supra* note 1, at 305.

[51] Nevertheless, a defendant can plead both "not guilty" and "not guilty by reason of insanity", thus leaving the jury with three possible verdicts, the two foregoing plus "guilty" LAFAVE and SCOTT, *supra* note 1, at 316.

cause of effects of the defendant's mental condition, he or she should be held free of criminal responsibility.

As with the competence to stand trial requirements, this defense recognizes the special limitations and needs of some developmentally and otherwise disabled persons. Yet, the implementation of the defense also often worked to the disadvantage of these persons. For example, most states allow or require an automatic and indefinite commitment upon a finding of non-responsibility.[52] Because of the potentially permanent commitment to a maximum security mental hospital for persons successfully invoking the defense, it is seldom raised by a defendant except in capital cases, where such confinement would be seen as the lesser of two evils.[53]

Various aspects of this process have been attacked in several cases.[54] For example, in 1968 the U.S. Court of Appeals for the District of Columbia, in Bolton vs. Harris,[55] held that a person who has been found "not guilty by reason of insanity" may not be confined beyond a reasonable evaluation period without a further judicial hearing to determine that person's present mental disability and dangerousness.[56] Some courts considering the matter subsequently have followed the lead of Bolton,[57] while others have not.[58] Particularly interesting are two sets of cases — in Wisconsin and New Jersey — in which the respective state supreme courts first rejected the Bolton analysis[59] and then accepted it.[60] Thus, the automatic and indefinite commitment of persons found not guilty by reason of mental disease or defect seems to be giving way to a requirement that a hearing be held on whether the individual is a dangerous person for whom continued confinement is justified.[61]

Several other factors have also led to controversy over the implementation of the defense. For example, should the confinement be subject to some form of periodic review of the defendant's continuing dangerousness and need for confinement?[62] While this approach makes substantial theoretical sense, a significant body of literature has developed that throws great doubt upon the capability of the experts to make reliable predictions of future dangerousness.[63]

There has also been substantial debate over the way in which the defense should be formulated, which in turn may have an effect on the class of persons it covers and kind of expert testimony that is required or permitted on the issue.[64] M'Naghten's case[65] provided an early formulation of what came to be known as the right-and-wrong test, with almost all American jurisdictions adopting some variant of it.[66] In response to criticism that the right-and-wrong test was too narrow, a number of states supplemented it by a test exculpating a person where mental disease created an

[52] See Chart VII, *infra.*

[53] LAFAVE and SCOTT, *supra*, note 1 at 305. *But see* Singer, *Insanity Acquittals in the Seventies: Observations and Empirical Analysis of One Jurisdiction,* 2 MENTAL DISABILITY LAW REPORTER 406 (1978), for a series of counter-intuitive findings, particularly on the frequent use of the insanity defense in minor crimes. The article reviews the cases of persons coming up for review hearings required by State vs. Krol, 344 A.2d 289 (N.J. Sup. Ct. 1975).

[54] *See e.g.,* Lynch vs. Overholser, 369 U.S. 705 (1961); Bolton vs. Harris, 395 F.2d 642 (D.C. Cir. 1968); and cases cited in notes 57 and 58 *infra.*

[55] 395 F.2d 642 (D.C. Cir. 1968).

[56] *Id.*, at 651.

[57] People vs. McNelly, 371 N.Y.S.2d 538 (1975); State vs. Krol, 344 A.2d 289 (N.J. Sup. Ct. 1975); State ex rel. Kovach vs. Schubert, 219 N.W.2d 341 (Wisc. 1974); *app. dismd.* 419 U.S. 1117 (1975) and *cert. den.* 419 U.S. 1130 (1975); People vs. McQuillan, 221 N.W.2d 569 (Mich. 1974); Wilson vs. State, 287 N.E.2d 875 (Ind. 1972).

[58] State vs. Kent, 515 S.W.2d 457 (Mo. 1974); State vs. Kee, 510 S.W.2d 477 (Mo. 1974); State vs. Maik, 287 A.2d 715 (1972); Lindner vs. Peterson, 324 F. Supp. 1261 (W.D. Mo. 1971); Chase vs. Kearns, 278 A.2d 132 (Me. 1971); State ex rel. Schopf vs. Schubert, 173 N.W.2d 673, (Wis. 1970). Pre-Bolton cases are listed in 50 A.L.R.3d 134.

[59] State ex rel. Schopf vs. Schubert, 173 N.W.2d 673 (Wis. 1970); State vs. Maik, 287 A.2d 715 (1972).

[60] State ex rel. Kovack vs. Schubert, 219 N.W.2d 341, *appl. dismd.* 419 U.S. 1117; State vs. Krol, 344 A.2d 289 (N.J. 1975).

[61] In at least one state, the use of the insanity defense seems to be increasing, although it is unclear how this increase is related to the increase in criminal cases in general. A comprehensive study of acquittals "by reason of insanity" in New York state has shown that from 1965 to 1971 the average was eight per year, from 1971 to 1976 the average was forty-seven per year, and from mid-1976 to mid-1978 the average was fifty-six per year. The ratio of its use in non-homicide cases has gone from 47% in 1965-1976 to 65% in 1976-1978. *STEADMAN* and *BURTON*, CHARACTERISTICS AND DISPOSITIONS OF INSANITY ACQUITTALS IN NEW YORK STATE: AN UPDATE [to THE INSANITY DEFENSE IN NEW YORK, A REPORT TO GOVERNOR HUGH L. CASEY, *infra.* note 78] *for 7/1/76 through 6/30/78* (Circular of N.Y. Dept. Mental Hygiene, 1979).

[62] *Bolton vs. Harris,* 395 F.2d 642 (1968).

[63] *STONE, MENTAL HEALTH AND LAW: A SYSTEM IN TRANSITION* 25 (1975); Ennis and Litwack, *Psychiatry and the Presumption of Expertise: Flipping Coins in the Courtroom,* 62 CALIFORNIA LAW REVIEW 693 (1974); Cocozza and Steadman, *The Failure of Psychiatric Predictions of Dangerousness: Clear and Convincing Evidence,* 29 RUTGERS LAW REVIEW 1084 (1976).

[64] The various formulations of the defense and the debates over their effects, merits, and deficiencies are extensively reviewed in BRAKEL and ROCK, *supra* 39, at 376-392.

[65] 8 Eng. Rep. 718 (1843). The test articulated by the court is:

> [I]t must be clearly proved that, at the time of the committing of the act, the party accused was labouring under such a defect of reason, from disease of the mind, as not to know the nature and quality of the act he was doing; or, if he did know it, that he did not know he was doing what was wrong. *Id.* at 722.

[66] WEIHOFEN, *supra*, note 38, at 51, 68-69 (1954). Weihofen notes that "[t]he wording of this . . . test varies so extensively, however, that it is impossible to mention all the ways in which it has been phrased." *Id.* at 69.

"irresistable impulse" to commit the criminal act.[67] New Hampshire rejected the McNaghten formulation at an early date[68] and was later joined by the U.S. Court of Appeals for the District of Columbia, in Durham v. United States,[69] in adopting a very broad test that permits exculpation where "the unlawful act was the product of mental disease or defect."[70] While the McNaghten test had been criticized as too restrictive in concept and too narrow in operation, the New Hampshire — Durham test was criticized as too unspecific in concept and too broad in operation.[71] Shortly after Durham, the American Law Institute's Model Penal Code recommended a formulation that has received broad acceptance and has even been adopted by the U.S. Court of Appeals for the District of Columbia.[72] The Model Penal Code test is:

> A person is not responsible for criminal conduct if at the time of such conduct as a result of mental disease or defect he [or she] lacks substantial capacity either to appreciate the criminality [wrongfulness] of his [or her] conduct or to conform his [or her] conduct to the requirements of the law.[73]

While the McNaghten test's reference to "disease of the mind" would seem to exclude mental retardation, most cases considering the matter seem to have viewed the defense as applicable to a mentally retarded person if the person were "unable to distinguish right from wrong."[74] The Model Penal Code test specifically refers to both mental disease and mental defect, but the Vermont legislature still found it necessary to explicitly indicate that "congenital and traumatic mental conditions" were encompassed by the phrase "mental disease or defect."[75] What is probably most telling is the extent to which this issue has been left unclear is the voluminous litigation, legislation, and literature on the defense. Little attention has been given to the relevance of the various formulations to persons with mental retardation[76] or to the appropriateness of criminal responsibility where mental retardation is a significant causative factor in unlawful behavior.[77]

Concern with the insanity defense has led to development of the concept of "diminished capacity."[78] This concept, is based on the prerequisite for most crimes of a certain intellectual state, such as that the accused acted "intentionally" or "with

---

[67] *See generally* BRAKEL AND ROCK, *supra* note 39, at 380; WEIHOFEN, *supra* note 38; Keedy, *Irresistable Impulse as a Defense in the Criminal Law,* 100 UNIVERSITY OF PENNSYLVANIA LAW REVIEW 956 (1952).

[68] State vs. Jones, 50 N.H. 369, 398 (1871); State vs. Pike, 49 N.H. 399 (1869).

[69] 214 F.2d 862 (D.C. Cir. 1954).

[70] *Id.* at 874. The *Durham* case was the subject of great scholarly interest and discussion, but was adopted only in Maine and the Virgin Islands.

[71] *See e.g.,* GOLDSTEIN, THE INSANITY DEFENSE (1967); Roche, *Criminality and Mental Illness – Two Faces of the Same Coin, 22 University of Chicago Law Review* 320 (1955); Krash, *The Durham Rule and Judicial Administration in the District of Columbia,* 70 YALE LAW JOURNAL 905 (1961). *See also* BRAKEL and ROCK, *supra*, note 39, at 389-391 and works cited therein.

[72] United States vs. Brawner, 471 F.2d 969 (1972).

[73] A.L.I., MODEL PENAL CODE §4.01(1) (Proposed Official Draft, 1962). The brackets in the test indicate optional phrasing for the states.

[74] No jurisdiction by jurisdiction analysis of the case-law on this point exists. However, in Nail vs. State, 328 S.W.2d 836 (Ark. 1959); Reece vs. State, 94 S.E.2d 723 (Ga. 1956), Washington vs. State, 85 N.W.2d 509 (Neb. 1957); State vs. Huff, 102 A.2d 8 (N.J. 1954), a defendant of very low intelligence unsuccessfully claimed that his mental deficiency was sufficient to invoke the "insanity" defense. In each case the court indicated that mental deficiency could bring a defendant within the defense, but only if the defendant was unable to distinguish right from wrong.

[75] BRAKEL and ROCK, *supra* note 39, at 382, n. 50.

[76] A study of court records of thirty-one randomly selected mentally retarded prisoners found that the "insanity defense" had only been raised in one case. Brown and Courtless, *The Mentally Retarded in Penal and Correctional Institutions,* 124 AMERICAN JOURNAL OF PYSCHIATRY 1164 (1968).

[77] *See generally* Person, *The Accused Retardate,* 4 COLUMBIA HUMAN RIGHTS LAW REVIEW 239 (1972); Tupin and Goolishian, *Mental Retardation and Legal Responsibility,* 18 DEPAUL LAW REVIEW 673 (1969); Allen, *The Retarded Offender: Unrecognized in Court and Untreated in Prison,* 32 FEDERAL PROBATION 22 (1968); Burgess, *The Mental Defective and the Law,* 23 INTRAMURAL LAW REVIEW OF NEW YORK UNIVERSITY 115 (1967); Allen, *Toward an Exceptional Offenders Court,* 4 MENTAL RETARDATION 3 (1966).

[78] See e.g., LAFAVE and SCOTT, *supra* note 1, at 325; Arenella, *The Diminished Capacity and Diminished Responsibility Defenses: Two Children of a Doomed Marriage,* 77 COLUMBIA LAW REVIEW 827 (1977); Dix, *Psychological Abnormality as a Factor in Grading Criminal Liability: Diminished Capacity, Diminished Responsibility, and the Like,* 62 JOURNAL OF CRIMINAL LAW, CRIMINOLOGY AND POLITICAL SCIENCE 313 (1971). Comment, *A Punishment Rationale for Diminished Capacity,* 18 U.C.L.A. LAW REVIEW 561 (1971).

Concern with the insanity defense has also led to a number of recommendations for its abolition. *See* Morris, *Psychiatry and the Dangerous Criminal,* 41 SOUTHERN CALIFORNIA LAW REVIEW 514 (1968) and works there cited. *But see* Monahan, *Abolish the Insanity Defense?—Not Yet,* 26 RUTGERS LAW REVIEW 719 (1973).

The United State Justice Department's federal criminal law reform bill, which has not been enacted, originally advocated abolition of insanity as a defense to deferal crimes. S. 1400, 93d Cong. 1st Sess. (1973); S. 1, 94th Cong. 1st Sess. (1975). *See* Reisner and Semmel, *Abolishing the Insanity Defense: A Look at the Proposed Federal Criminal Code Reform Act in Light of the Swedish Experience,* 62 CALIFORNIA LAW REVIEW 753 (1974); Wales, *An Analysis of the Proposal to "Abolish" the Insanity Defense in S. 1: Squeezing a Lemon,* 124 UNIVERSITY OF PENNSYLVANIA LAW REVIEW 687 (1976); Dershowitz, *Abolishing the Insanity Defense: The Most Significant Feature of the Administration's Proposed Criminal Code—An Essay,* 9 CRIMINAL LAW BULLETIN 434 (1973).

The New York Department of Mental Hygiene has also recommended abolition of the insanity defense. NEW YORK STATE DEPARTMENT OF MENTAL HYGIENE, THE INSANITY DEFENSE IN NEW YORK, A REPORT TO GOVERNOR HUGH L. CAREY (1978).

premeditation."[79] Courts have traditionally held that evidence of mental disability is only admissible in connection with the insanity, or non-responsibility defense.[80] Under the "diminished capacity" provision adopted by the Model Penal Code[81] and a number of states,[82] evidence of mental disability may be introduced to prove that the defendant lacked the intellectual state required for the offense charged.[83] It may be that the defense of diminished capacity is of more significance to developmentally disabled persons than the insanity defense, since the lack of intellectual sophistication of a mentally retarded offender may be more relevant to his or her degree of knowledge or forethought about an offense than to the more absolute concepts embodied in the insanity defense.[85] In any case, both the non-responsibility defense and the diminished capacity defense leave for resolution the question of the circumstances in which "civil" incarceration is justified for a person who, without "criminal" guilt, commits what would normally be a criminal act.

Once a developmentally disabled person is tried for and convicted of an offense, a variety of post-conviction issues remain. They are: whether sentencing procedures, and the length and nature of sentences, should be affected by the person's disability; whether the person has any particular rights within the prison setting; whether and under what circumstances a developmentally disabled person should be permitted to transfer, or be required over his or her objection to be transferred, to a special institution; and what effect the disabling condition or special treatment resulting from that condition should have on eligibility for parole.

Presentencing studies and reports have become a common feature of sentencing procedures.[85] The Model Penal Code would require an investigation and report before sentencing in every felony case in which the person convicted is under twenty-two years of age and as a prerequisite to any sentence extending beyond one year.[86] It would provide the trial judge with discretion to order a presentence study in any other case.[87] Where the offender may be developmentally disabled, it is important that the study include certain kinds of information that a judge needs in order to make an intelligent sentencing decision. Two kinds of information should be included that might not otherwise be in a presentencing report. First, it is important to know whether the offender is in fact developmentally disabled and, if so, whether or not the disability may have been causally related to the commission of the offense. For example, it could be significant that the defendant's disability made him or her particularly subject to manipulation by others and thus a follower rather than a leader in the criminal enterprise. Or it could be significant that aggressive behavior of the defendant was the result of frustration at an inability to find satisfactory employment. Second, it is important to know what treatment is available in alternative sentencing placements.

A more difficult issue is whether a presentencing study ought to be required in all, or some sub-category of, cases where the judge has reason to believe that the defendant is a developmentally disabled person. To sentence a developmentally disabled person to a period of imprisonment without any information on the nature and extent of the disability is undesirable. On the other hand, there may be situations where no presentence report is needed even though the defendant is developmentally disabled. If the offense is a relatively minor one in which the penalty is generally a small fine or short-term probation, the cost of a thorough presentence report could exceed its utility. Being subjected to evaluation itself can be a real burden to a defendant, and this burden might not be justified where the offense and sanction are minor. Thus, one might place a limit on the situations in which such a report is required, and require it only where the offense is of a certain degree of seriousness or as a prerequisite to a penalty of imprisonment.

As is implied in the above discussion of presentence reports, a second broad post-conviction issue is whether a developmental disability should, at least in some cases, have an impact on the nature or severity of the sentence to be imposed.

Several attempts to abolish the defense early in this century were held unconstitutional. State vs. Strasburg, 110 P. 1020 (Wash. 1910); Sinclair vs. State, 132 So. 581 (Miss. 1931); *cf.* United States vs. Brawner, 471 F.2d 969, 985 (1972). But see Powell vs. Texas, 392 U.S. 514 (1968), and especially Justice Black's concurring opinion. *Id* at 545. Coupling abolition with adoption of the doctrine of "diminished capacity" would probably meet constitutional objections to crime without fault.

[79] LAFAVE and SCOTT, *supra* note 1, at 191.

[80] See cases collected in annotation at 22 A.L.R. 3d 1228, 1225. See generally, LAFAVE and SCOTT, *supra* note 1, at 325.

[81] A.L.I., MODEL PENAL CODE § 4.02 (Proposed Official Draft, 1962).

[82] See cases collected at 22 A.L.R. 3d 1228, 1238.

[83] Even where the diminished capacity defense is broadly accepted, it is sometimes excluded with respect to particular offenses where it is determined that only a "general intent" and no "specific intent" is required for the offense. For a discussion of this distinction and the policy judgments it may clothe, see opinion of Justice Traynor in People vs. Hood, 462 P.2d 370 (Cal. 1970). See also People vs. Nance, 102 Cal. Rptr. 266 (1972).

[84] *See* United States vs. Brawner, 471 F.2d 969 (D.C. Cir. 1972). *See also Diminished Capacity—Recent Decisions and An Analytical Approach,* 30 VANDERBILT LAW REVIEW 213, 216, 226 (1977).

[85] A.L.I., MODEL PENAL CODE, Comment to § 7.07 (Tentative Draft No. 2 1954).

[86] A.L.I., MODEL PENAL CODE § 7.07 (1) (Proposed Official Draft 1962).

[87] *Id.*, at § 7.07 (2).

This issue arises in several different contexts. At one extreme, where the death penalty is an available sanction for persons convicted of a given crime, ought that penalty to be precluded where the offender is developmentally disabled? The Model Penal Code provides that evidence of impaired capacity is admissible as a mitigating factor.[88] There is, of course, vigorous debate concerning whether the death penalty ought ever to be imposed upon anyone.[89] Putting this broad question aside as beyond the purview of the present study, however, it can be argued that the ultimate sanction should be precluded in the case of a developmentally disabled offender. There are several different lines of argument that support this proposition.[90] First, while the disability may not be sufficiently extreme to negate criminal responsibility, the effect of a developmental disability may be presumed to be sufficiently pervasive to play an indirect or partial role in shaping a person's behavior and to preclude the imposition of the most severe expression of moral judgment, that implied in the imposition of the death penalty. Second, while the disability may not negate the individual's competence to stand trial, one may be concerned that the presence of a disability can trigger subjective discomfort with or prejudice against a defendant or that the effectiveness of the defendant's participation in his or her own defense may be impaired sufficiently to increase the possibility of error in the trial process. Finally, to the extent that the death penalty is seen as an act in which the person executed must participate and accept his or her responsibility, one might be concerned that some developmentally disabled offenders lack the full capacity requisite to achievement of this social satisfaction. The common thread of these arguments is that the death penalty is an extreme penalty which ought to be used, if at all, only in the most extreme and clearest cases.

A lesser issue related to the nature of the sanction to be used relates to the distinction between a prison term and a probationary or suspended sentence. Here again, a kind of informal diminished capacity or sense of lessened individual responsibility can be utilized to argue that the more serious sanction, the prison sentence, ought to be used with greater caution. One may also have concerns about the increased possibility of sexual and other exploitation within the prison context of persons suffering from a disability. Finally, there may be even greater cause for concern about the antitherapeutic nature of the prison environment, which might cause the disability to worsen and thus, in a sense, cause greater damage to the disabled than the able bodied person. While it may be impossible to formulate sufficiently precise standards to govern in detail the use of prison sentences for developmentally disabled persons, lesser steps are possible. The legislature might simply permit the sentencing court to utilize nonprison sanctions where they would otherwise be precluded, *i.e.* where the offense otherwise carries a mandatory prison term. Or, going one step further, the legislature might direct the court always to consider the possibility of a non-prison sentence where the offender is developmentally disabled. Or, finally, the legislature might preclude a prison sentence for developmentally disabled offenders in certain categories of cases, unless the judge makes written findings which justify the sentence.

Assuming a prison sentence is to be ordered, the further question arises as to whether the duration of the sentence should be affected by a person's mental disability. A number of states have adopted,[91] and at least one has subsequently repealed, [92] statutes providing for indefinite (*i.e.* potentially life) sentences where a person is disabled,[93] or is disabled and has committed certain specified offenses.[94] Generally, these statutes provide that the sentence thus imposed (which is sometimes called a "commitment" rather than a "sentence") is to be served in a particular facility, sometimes called the hospital for the criminally insane. One rationale underlying these statutes is that the mentally disabled offender should be provided therapy and, thus, that the time of incarceration ought to be determined by therapeutic rather than penal considerations.[95] An alternative rationale is that the public requires greater protection from the mentally disabled offender than it does from the non-disabled offender. There is no empirical support for the latter position and the former is often belied by the facts.[96]

The opposite side of the coin relates to shorter sentences. The same arguments that favor prohibitions or limitations on the use of prison sen-

[88] *Id.*, at § 4.02 (2).

[89] *See e.g., A Symposium: Capital Punishment in the United States,* 14 CRIMINAL LAW BULLETIN 1 (1978) and works cited there.

[90] *See generally,* Liebman and Shephard, *Guiding Capital Sentencing Discretion Beyond the "Boiler Plate": Mental Disorder as a Mitigating Factor,* 66 GEORGETOWN LAW JOURNAL 757 (1978).

[91] CALIFORNIA WELFARE AND INSTITUTIONS CODE § 6300 (West 1973); OHIO REVISED CODE ANNOTATED § 2947.24 (Page 1975); DISTRICT OF COLUMBIA CODE § 22-3503 (1973).

[92] Ohio House Bill 565, 112th General Assembly, 1977-1978.

[93] Schreiber, *Indeterminate Therapeutic Incarceration of Dangerous Criminals: Perspectives and Problems,* 56 VIRGINIA LAW REVIEW 602 (1970); McGarry and Cotton, *A Study in Civil Commitment: The Massachusetts Sexually Dangerous Persons Act,* 6 HARVARD JOURNAL OF LEGISLATION 263 (1969).

[94] Schreiber, *supra* note 94, at 603, 616.

[95] *Id.*, at 605, 624; Wexler, *Therapeutic Justice* 57 MINNESOTA LAW REVIEW 289, 296 (1972).

[96] Wexler, *supra* note 96, at 323.

tences in some cases support providing for the minimum necessary period of prison incarceration. For example, just as a judgment that a prison setting would be uniquely anti-therapeutic for a developmentally disabled person could justify total avoidance of imprisonment of the person in some cases, such a judgment should also support limiting the duration of imprisonment even though some imprisonment was thought to be required.

The next major post-conviction issue is whether the fact that a prisoner is developmentally disabled gives him or her any special rights within the institutional context. For example, does or should the developmentally disabled prisoner have a "right to treatment or habilitation" during his or her period of incarceration. The United States Supreme Court has so far refused to speak to the right to treatment question even in the context of civil confinement,[97] but there are two sets of lower court cases that speak to the issue to some degree. The first is a line of cases that articulates a right to treatment or habilitation for persons who are civilly confined.[98] The second and most directly relevant is a line of cases that speaks to the rights of prisoners. It has been held that conviction for an offense and subsequent imprisonment do not justify denying normal medical care to the person imprisoned.[99] It has also been held that medical care in this context includes treatment for mental illness.[100] This same rationale could extend to habilitation for developmentally disabled prisoners. The question of whether there is a constitutional right to treatment or habilitation within the correctional setting is relevant to the question of legislative action, but of course not dispositive. Even if there is no such consitutional right, it may be sound legislative policy to articulate such a right. In the first place, to the extent that the developmental disability is contributive to the criminal behavior of the prisoner, the state has an interest in providing such treatment as will lead to a change in the behavior, permit the release of the individual, and make the individual a productive rather than a destructive member of society. Even where the developmental disability is not directly productive of the criminal behavior, the state might reasonably conclude that, if it is already bearing the cost of incarceration, it might as well provide treatment or habilitation that will improve the individual's ability to cope in society upon release.

Related to the question of a right to treatment or habilitation is the question of a right to an inter-institutional transfer. If the state provides some facilities that provide a treatment or habilitation program that is suitable to a particular prisoner, should that individual have a right to be transferred to that facility? A step short of such a right, of course, is simply to provide that the prison shall have the authority to agree to such a placement upon request by the prisoner.

On the other hand, in what circumstances should correctional officials be empowered to transfer a developmentally disabled prisoner to a special institution that is not generally used for prisoners, over the objection of the prisoner. Here, there are several sub-issues. First, does a developmentally disabled prisoner have a right to refuse placement in a therapeutic environment, a right to insist on not being dealt with differentially just because of his developmental disability? Second, if correctional officials are empowered to transfer him or her, what procedural safeguards are available to challenge the propriety of the transfer? Third, what criteria ought to govern the transferability of a prisoner? Fourth, where such a transfer is made over the objection of the developmentally disabled prisoner, does this transfer give rise to a right to treatment or habilitation in the new setting? And finally, to what extent may the length of incarceration be affected by transfer to such a special facility? Thus far, the Supreme Court has addressed only the second issue. In Vitek vs. Jones, the Court held that before a prisoner can be transferred involuntarily to a "mental hospital," there must be:

> A. Written notice to the prisoner that a transfer to a mental hospital is being considered;
>
> B. A hearing, sufficiently after the notice to permit the prisoner to prepare, at which disclosure to the prisoner is made of the evidence being relied upon for the transfer and at which an opportunity to be heard in person and to present documentary evidence is given;
>
> C. An opportunity at the hearing to present testimony of witnesses by the defense and to confront and cross-examine witnesses called by the state, except upon a finding, not arbitrarily made, of good cause for not permitting

---

[97] O'Connor vs. Donaldson, 422 U.S. 563 (1975).

[98] *E.g.*, Wyatt vs. Stickney, 325 F. Supp. 781, 344 F. Supp. 373, 387 (M.D. Ala. 1971), *affd. sub nom* Wyatt vs. Aderholt, 503 F.2d 1305 (5th Cir. 1974); Welsch vs. Likins, 373 F. Supp: 487 D. Minn. 1974); Davis vs. Watkins, 384 F. Supp. 1196 (N.D. Ohio 1974).

[99] Spicer vs. Williamson, 191 N.C. 487, 132 S.E. 291 (1926); Estelle vs. Gamble, 429 U.S. 97 (1976). *See* Klein, *Prisoner's Rights to Physical and Mental Health Care: A Modern Expansion of the Eighth Amendment's Cruel and Unusual Punishment Clause,* 7 FORDHAM URBAN LAW JOURNAL 1 (1979).

[100] Bowring vs. Godwin, 551 F.2d 44, 47 (4th Cir. 1977). The court held that a prisoner is entitled to psychological or psychiatric treatment if a physician or other health care provider, exercising medical certainty concludes (1) that the prisoner's symptoms evidence serious disease or injury; (2) that such disease is curable or may be alleviated; and (3) that the potential for harm to the prisoner by reason of delay or denial of care would be substantial.

such presentation, confrontation, or cross-examination;

D. An independent decisionmaker;

E. A written statement by the factfinder as to the evidence relied on and the reasons for transferring the inmate;

F. Availability of [independent assistance] furnished by the state, if the inmate is financially unable to furnish his own; and

G. Effective and timely notice of all the foregoing rights.[101]

The last post-conviction issue to be considered in this section is what impact a person's mental disability can have upon parole decisions. The United States Supreme Court has recently affirmed that parole decisions are highly discretionary,[102] and, in fact, are often made on a highly individual and subjective basis.[103] Precisely for this reason there is the danger of discrimination on the basis of a developmental disability with the parole board denying a person a parole, where one would otherwise be granted. While in the vast majority of cases, it would be impossible to prove that the individualized and highly informal processes of the parole board were discriminatory in this regard, clear cases could exist and, thus, the concern is real.

Questions, problems and ambiguities similar to those outlined above affect children and adolescents with a developmental disability who come into contact with the juvenile justice system, although until recently they have been less sharply focused. A separate system of courts and services for juveniles in conflict with the law and/or their families was established as a means of ameliorating the harshness of the criminal justice system for "wayward children."

> The goals were to investigate, diagnose, and prescribed treatment, not to adjudicate guilt or fix blame. The individual's background was more important than the facts of a given incident, specific conduct relevant more as symptomatic of a need for the court to bring its helping powers to bear than as prerequisite to exercise of jurisdiction. Lawyers were unnecessary — adversary tactics were out of place, for the mutual aim of all was not to contest or object but to determine the treatment plan best for the child. . . . [D]elinquency was thought of almost as a disease, to be diagnosed by specialists and the patient kindly but firmly dosed.[104]

As a result, questions regarding a juvenile's competence or criminal responsibility were rarely raised.[105] In many cases, services were provided prior to adjudication through diversion on a more or less voluntary basis, or through a formal civil commitment, and the delinquency petition was dismissed.[106] In others, the grounds for the court's jurisdiction shifted. The child was labeled a "person in need of services" or neglected, and services were provided without the need to prove the commission of a delinquent offense.[107] Finally, in some instances in which an adolescent was severely disturbed, the juvenile court would simply transfer jurisdiction over the case to the adult criminal court on the grounds that the youth was unamenable to treatment.[108]

While this system gave juvenile courts great flexibility, it also provided opportunities for unfairness and abuse. Over the past decade and a half, the juvenile justice system has come under increasing scrutiny and criticism for failing to deliver on its promises of individualized attention and fair treatment.[109]

---

[101] Vitek vs. Jones, U.S. (1980).

[102] Morrissey vs. Brewer, 408 U.S. 471, 478, (1971).

[103] *Id.*, at 447.

[104] PRESIDENT'S COMMISSION ON LAW ENFORCEMENT AND THE ADMINISTRATION OF JUSTICE, TASK FORCE REPORT: JUVENILE DELINQUENCY AND YOUTH CRIME, 3 (1967).

[105] Donovan, *The Juvenile Court and the Mentally Disordered Juvenile*, 45 NORTH DAKOTA LAW REVIEW 222, 229-233 (1969); PELC, COMPETENCY TO STAND TRIAL IN JUVENILE COURT PROCEEDINGS (1977) (Unpublished thesis Wellesley College Library).

[106] Donovan, *id.*, at 224-226.

[107] Pelc, *supra* note 105.

[108] Donovan, *supra* note 105, at 240-244; Kent vs. United States, 401 F.2d 408 (D.C. Cir. 1968).

[109] Breed vs. Jones, 421 U.S. 519 (1975); In re Winship, 397 U.S. 358 (1970); In re Gault, 387 U.S. 1 (1967); Kent vs. United States, 383 U.S. 541 (1966). R. KOBETZ and B. BOSARGE, JUVENILE JUSTICE ADMINISTRATION, 218 (1973); AMERICAN LEGION NATIONAL EXECUTIVE COMMITTEE, RESOLUTION No. 43 REMOVAL OF STATUS OFFENSES FROM JUVENILE COURT JURISDICTION (1975); National Council on Crime and Delinquency, *Jurisdiction over Status Offenders Should be removed from the Juvenile Court: A Policy Statement,* 21 CRIME & DELINQUENCY 97 (1975); THE JUVENILE JUSTICE AND DELINQUENCY PREVENTION ACT §3148 and §821, HEARINGS BEFORE THE SUBCOMMITTEE TO INVESTIGATE JUVENILE DELINQUENCY OF THE COMMITTEE ON THE JUDICIARY, UNITED STATES SENATE, 92d Cong., 2d Sess. (1972-1973); E. SCHUR, RADICAL NONINTERVENTION (1973); K. WOODEN, WEEPING IN THE PLAYTIME OF OTHERS: AMERICA'S INCARCERATED CHILDREN (1976); Wald, *Pretrial Detention for Juveniles*, in PURSUING JUSTICE FOR THE CHILD 119 (M. Rosenheim ed. 1976); PCLEAJ, *supra* note 104; WHITE HOUSE CONFERENCE ON CHILDREN, REPORT TO THE PRESIDENT (1970); NATIONAL ADVISORY COMMITTEE ON JUVENILE JUSTICE AND DELINQUENCY PREVENTION (NACJJDP), REPORT OF THE ADVISORY COMMITTEE TO THE ADMINISTRATOR ON STANDARDS FOR THE ADMINISTRATION OF JUVENILE JUSTICE (1976); NACJJDP, REPORT OF THE ADVISORY COMMITTEE TO THE ADMINISTRATOR ON STANDARDS

> Juvenile court history has again demonstrated that unbridled discretion, however benevolently motivated, is frequently a poor substitute for principle and procedure. . . . The absence of substantive standards has not necessarily meant that children receive careful, compassionate, individualized treatment. The absence of procedural rules based upon constitutional principle has not always produced fair, efficient, and effective procedures. Departures from established principles of due process have frequently resulted not in enlightened procedure, but in arbitrariness.[110]

Despite the numerous opportunities for diversion to habilitative services, several recent cases and studies have revealed a disproportionate number of juveniles with a developmental disability in juvenile correctional facilities and an absence of habilitative services.[111]

Accordingly, the courts are being increasingly asked to apply the standards and procedures regarding competency and criminal responsibility to delinquency cases. Predictably, the doctrinal results have been mixed. Some courts have applied the right not to be tried while incompetent and the insanity defense to juvenile delinquency proceedings as a matter of "fundamental fairness."[112] Others have rejected this as antithetical to and unnecessary in light of the purpose of the juvenile court,[113] despite the stigma which is attached to being labeled a delinquent.[114] Yet, regardless of the theoretical position adopted, the children involved have usually been institutionalized.[115] Although alternative, community-based programs have been initiated in some states,[116], in many jurisdictions community habilitative services are not open to juveniles alleged or found to be delinquent,[117] and correctional personnel lack the necessary training and skills.[118] Moreover, even in those jurisdictions where there is at least legal recognition of the impact which a mental disability may have on a juvenile's ability to present a defense and to form the intent required and some available habilitative programs, the procedures,

---

FOR THE ADMINISTRATION OF JUVENILE JUSTICE (1976); NACJJDP, REPORT OF THE ADVISORY COMMITTEE TO THE ADMINISTRATOR ON STANDARDS FOR THE ADMINISTRATION OF JUVENILE JUSTICE (Advance Draft 1977); NATIONAL ADVISORY COMMITTEE ON CRIMINAL JUSTICE STANDARDS AND GOALS, REPORT OF THE TASK FORCE ON JUVENILE JUSTICE AND DELINQUENCY PREVENTION (1976); INSTITUTE OF JUDICIAL ADMINISTRATION AMERICAN BAR ASSOCIATION, JUVENILE JUSTICE STANDARDS PROJECT, STANDARDS FOR JUVENILE JUSTICE: A SUMMARY AND ANALYSIS (1977); [OHIO] ATTORNEY GENERAL'S JUVENILE JUSTICE TASK FORCE, JUSTICE FOR OUR CHILDREN (1976); WISCONSIN COUNCIL ON CRIMINAL JUSTICE SPECIAL STUDY COMMITTEE ON CRIMINAL JUSTICE STANDARDS AND GOALS, JUVENILE JUSTICE STANDARDS AND GOALS (1975); GOVERNOR'S ADULT AND JUVENILE JUSTICE ADVISORY COMMITTEE, STANDARDS AND GOALS FOR THE NEW JERSEY CRIMINAL JUSTICE SYSTEM (Final Draft 1977).

[110] In re Gault, 387 U.S. 1, 18-19 (1967).

[111] Dennis, *Mental Retardation and Corrections: A Research Perspective,* THE MENTALLY RETARDED CITIZENS AND THE LAW, 34 (M. Santamour ed. 1975); Friel, *The Mentally Retarded Offender — Texas CAMIO Research Project, id.,* at 95; Morales vs. Turman, 383 F. Supp. 43 (E.D. Tex. 1974), *rev'd on other grounds* 535 F.2d 864 (5th Cir. 1976), *rev'd and remanded* 430 U.S. 322 (1977), *remanded for further hearings* 562 F.2d 993 (5th Cir. 1977).

[112] For example:

> The dispute over the proper tests of insanity that has so preoccupied the courts in recent times is essentially concerned with the question of fairness. Is it fair to convict of crime, [*sic*] when the defendant, though knowing right from wrong, as the result of mental illness or incapacity, is unable to exercise the restraints upon his [or her] conduct that would enable him [or her] to conform to acceptable standards. It would seem incongruous that this great outpouring of concern should be lavished only upon adults who may be criminals while the children whom we profess to be particular objects of solicitude are bypassed. We conclude that the defense of insanity must be permitted in a juvenile delinquency procedure if those proceedings are to conform to the minimum *Kent* standards of due process and fair treatment. In re Winburn, 32 Wis.2d 152, 145 N.W.2d 178, 184 (1966).

On both competency and the insanity defense, see In re Causey, 363 So.2d 472 (La. S. Ct. 1978).

[113] For example:

> If anti-social acts are committed by children as a result of emotional disturbances, mental retardation or deficiency, or defect of reason due to mental disease, the court is charged by statute with providing the proper treatment, rehabilitative measures or therapy. This cannot be accomplished without first determining that the jurisdictional basis for judicial intervention exists — that is, that the offense was committed. State of New Jersey in the Interest of H.C., 106 N.J. Super. 583, 256 A.2d 322-328 (1969).

[114] In re Gault, 387 U.S. 1, 24-25 (1967).

[115] Donovan, *supra* note 105; Popkin and Lippert, *Is There a Constitutional Right to the Defense in Juvenile Court?,* 10 Journal of Family Law 421, 423-424 (1971).

[116] See THE MENTALLY RETARDED CITIZENS AND THE LAW, *supra* note 111.

[117] *E.g.,* M. PRESCOTT and E. VAN HOUTEN, A STUDY OF THE CHARACTERISTICS AND TREATMENT OF MENTALLY RETARDED ADOLESCENT OFFENDERS IN NEW JERSEY (1979); Donovan, *supra* note 105 at 246-249.

[118] *E.g.,* Morales vs. Turman, 383 F. Supp. 43 (E.D. Tex. 1974), *rev'd on other grounds* 535 F.2d 864 (5th Cir. 1976), *rev'd and remanded* 430 U.S. 322 (1977), *remanded for further hearings* 562 F.2d 993 (5th Cir. 1977); NACJJDP (1977) *supra* note 109.

standards and safeguards to be applied remain uncertain.[119]

With all of these considerations in mind, it is clear that there needs to be a careful rethinking and reformulating of the laws and principles which govern the relationship of the criminal and juvenile justice systems to developmentally disabled persons in order for these systems to deal fairly and appropriately with such individuals. A review of existing state criminal codes reveals the pervasiveness of this need.

[119] Few if any juvenile codes deal with these issues in any detail.

REVIEW OF STATUTES

The following statutory survey is based on an analysis of the codes of the 50 states, the District of Columbia, American Samoa, Guam, Puerto Rico, and the Virgin Islands. The degree to which these codes address the issues, questions and procedures outlined above varies considerably. Accordingly, the review focuses on the two areas which are most frequently and thoroughly covered by legislation — competence of an adult to stand trial (Charts I-IV) and the lack of criminal responsibility of an adult due to a mental disease or defect at the time of the crime (Charts V-VIII).

Competence to Stand Trial

As is noted above, while it is repugnant to our notions of fairness to try persons who are unable to understand the proceedings or assist in their own defense, the inquiry into a defendant's competence and the treatment provided following a finding of incompetence may result in substantial intrusions into that individual's privacy, restraints on his or her liberty, and significant expense to the public. Hence, the procedures regarding competency must assure that persons who are clearly competent are not subjected to unnecessary examinations and hospitalization, without impeding or discouraging the identification and treatment of individuals who are not competent to stand trial or enter a plea. All of the jurisdictions examined except the Virgin Islands have enacted legislation which seeks to achieve the required balance. Charts I-IV examine some of the key issues regarding the initiation of an inquiry into a defendant's competence to stand trial, the determination of competence or incompetence, and the provision of services following a finding of incompetence to stand trial or enter a plea.

Procedures for Raising the Issue of Incompetency (Chart I)

This chart sets forth the provisions governing when the issue of incompetency to stand trial may be raised and who may raise it. Twenty-three jurisdictions statutorily limit the initiation of an inquiry into a defendant's competence until after an indictment or information has been filed. Thus, in these jurisdictions, before an examination may be ordered either on an outpatient or inpatient basis, there must be a finding by a grand jury, magistrate, or prosecutor that there is probable cause to believe that a crime has been committed and the defendant committed it. Ten jurisdictions (Alabama, California, Hawaii, Massachusetts, Oregon, Puerto Rico, Rhode Island, South Carolina, Virginia, and West Virginia), provide that an inquiry into the defendant's competence can be formally initiated only by the court, although there is no prohibition on the prosecutor or defendant bringing the question to the judge's attention informally. Maryland permits both the defendant and the court to initiate the inquiry. The vast majority of American jurisdictions (37), however, permit the prosecutor as well as the defendant and the court to raise the issue.[120]

Procedures for Determining Competency (Chart II)

The criminal code in about three-fourths of the jurisdictions contains a definition of incompetency to guide the decision-making process. Many of these statutory definitions are similar to the standard set forth in Drope v. United States.[121] In addition, almost every jurisdiction provides specifically for an examination by a mental health professional. In about a quarter of these jurisdictions, such an examination is discretionary with the court. In the remainder, it appears to be mandatory.

The statutes in 43 jurisdictions do not require a defendant to be committed to a hospital or other institution in order to be examined. Eight of the jurisdictions which do not require commitment (American Samoa, Indiana, Mississippi, Nebraska, Nevada, Ohio, Rhode Island and Utah) do not even provide for temporary commitment in their criminal codes, although such commitments may be possible under other provisions. The criminal codes of 40 jurisdictions explicitly permit short-term commitment of defendants alleged to be incompetent. The maximum duration of such commitments varies from 15 days in Connecticut, South Carolina and Washington to 120 days (60 days plus a 60 day extension) in Kansas. Eight jurisdictions either provide no statutory limit (Colorado, Florida, Illinois, Louisiana, Missouri) or leave the matter to the discretion of the court (District of Columbia, Guam, Hawaii). In Pennsylvania, the civil commitment procedures apply.

---

[120] Even when there is no explicit statutory authority for the court to initiate an inquiry into the defendant's competence to stand trial, *Pate vs. Robinson*, 383 U.S. 375 (1966) requires courts to undertake such an inquiry when there are "sufficient indicia" of the defendant's lack of competence.

[121] *Supra* note 36.

In 45 jurisdictions, the question of the defendant's competence is determined by the judge. In Alabama and Georgia, the issue must be determined by a jury. Six other jurisdictions (California, Illinois, Kansas, Ohio, Puerto Rico, and Texas) permit either a judge or jury to make the determination, and the South Dakota statute specifies that the judge must determine questions of law and the jury questions of fact. The statutes in 18 jurisdictions specifically permit the introduction of evidence to rebut the results of the examination.

Disposition After a Finding of Incompetence (Chart III)

After a defendant has been found to be incompetent to stand trial, the statutes in all but 3 jurisdictions (Mississippi, New Hampshire, Virgin Islands) provide for furnishing services to the defendant. Twenty-eight of these jurisdictions require that the defendant be committed for treatment; 24 authorize the court to choose between commitment and the provision of services on an outpatient basis. Most jurisdictions (31) rely on special commitment procedures within their criminal codes for the involuntary hospitalization of defendant incompetent to stand trial. Six jurisdictions (Alabama, Minnesota, New Mexico, North Carolina, Virginia, and Washington) rely on their civil commitment procedures. The remaining 12 jurisdictions (Arizona, Florida, Georgia, Indiana, Iowa, Kansas, Massachusetts, South Carolina, Texas, Vermont, West Virginia, Wyoming) provide for a criminal commitment when there is a substantial probability the defendant will become competent within a prescribed period, and for civil commitment when there is not that possibility. Among the jurisdictions which have some provision for a criminal commitment, more than half, (25) set no time limit on the defendant's confinement except the regaining of competency. Ten others (Arkansas, Florida, Indiana, Iowa, Maine, Michigan, South Carolina, Texas, West Virginia, Wisconsin) set a specific time limit ranging from 60 days (South Carolina) to 2 years (Wisconsin), and 10 jurisdictions (American Samoa, California, Connecticut, Illinois, New York, North Dakota, Ohio, Oregon, Pennsylvania, Rhode Island) base their limits on the sentence which could be imposed if the defendant were convicted, usually with some absolute durational limit — e.g., 15 months in Ohio and 30 years in Rhode Island. Of those jurisdictions relying on their civil commitment statutes, only three (Minnesota, North Carolina, Wyoming) provide durational limits other than the regaining of competence.

Procedures Following Commitment Due to a Finding of Incompetence (Chart IV)

Except with regard to the authority to release a defendant, the statutes in most jurisdictions do not include detailed provisions on the procedures which apply after a defendant has been committed following a finding of incompetence. Less than half (22) explicitly provide for regular judicial review of the defendant's condition and whether he or she has become competent. Only 10 jurisdictions have provisions in their criminal codes governing the transfer of the defendant from institutional to community-based settings. Of these jurisdictions, 5 (California, Kansas, Massachusetts, New Jersey, Rhode Island) require a court order to transfer the defendant; and 5 (Colorado, Louisiana, New York, Tennessee and Texas) allow transfers to be carried out administratively. The criminal codes of 6 other jurisdictions (Alabama, Georgia, Indiana, Iowa, Vermont, Wyoming) include cross-references to the applicable sections of their civil commitment laws. On the other hand, only 8 jurisdictions (American Samoa, Kentucky, Mississippi, Nebraska, New Hampshire, Puerto Rico, Virginia, and Virgin Islands) fail to address the question of release authority in their criminal codes or rules of criminal procedure. Seven jurisdictions (Alabama, Guam, North Carolina, Oklahoma, South Dakota, Tennessee and Wyoming) permit the hospital superintendent or other administrative authority to declare defendants to be competent and order their release. In the remainder, the authority to release is ultimately vested in the court. Of course the release authority is tempered by the durational limits discussed in connection with Chart III.

Finally, 16 jurisdictions permit consideration of the matters related to the case which do not require the defendant's assistance while the defendant is incompetent. In this way, issues which may be dispositive of the charges — *e.g.*, the illegality of an arrest or search or a defect in the indictment — can be decided promptly, so that a defendant is not confined or provided services involuntarily on the basis of a criminal charge on which he or she could not be convicted.[122]

Criminal Responsibility

With the exception of Guam, all American jurisdictions have at least some statutory provisions governing either the grounds upon which a defendant may be found not to be criminally responsible because of a mental disability, for what would otherwise be an offense, and/or the procedures which apply when a defense based on those grounds is asserted or results in an acquittal. Generally, statutes provide for the traditional affirmative defense — i.e., one which must be proven by the defendant — of not guilty by reason of insanity which may or may not specifically include mental retardation and other developmental dis-

[122] *Supra* note 45.

abilities.[123] At least a quarter of the jurisdictions also statutorily provide for a defense based on diminished capacity which does not alter the burden of proof. Charts V-VIII provide an overview of the statutory standards defining the grounds for asserting the lack of criminal responsibility, the examination and other procedures which apply following the assertion of this defense, and the dispositional alternatives and procedures available following a verdict of not guilty by reason of insanity.

Definition of Lack of Criminal Responsibility Due to a Mental Disability (Chart V)

Approximately half of the jurisdictions have enacted a standard for the "insanity defense." Twenty-one of these jurisdictions have adopted the standard proposed by the American Law Institute. Four others (Arizona, Georgia, Iowa, Oklahoma) have retained the McNaughten Rule. Mississippi and the Virgin Islands have developed their own standards. The remaining jurisdictions have defined their standard through case law.

In addition, 14 jurisdictions statutorily provide for the defense of diminished responsibility.

Procedures Following Assertion of a Defense of Not Guilty by Reason of Insanity (Chart VI)

After a defendant enters a plea of not quilty by reason of insanity or provides other notice where required of the intent to assert that defense, most jurisdictions (36) make provision for an examination of the defendant. Nine of these jurisdictions permit such examinations on an outpatient basis; 16 provide only for a short-term commitment of the defendant; 7 provide for either outpatient or inpatient examinations; and 4 apply the procedures set forth in their civil commitment statutes. The durational limits on temporary commitments are set statutorily in 20 jurisdictions. They range from 21 days (Texas) to 90 days (Connecticut, Tennessee). The most common time limit is 60 days (Alaska, Idaho, Maine, Maryland, Montana, New Jersey, New York, North Dakota, Vermont). Two states (Missouri and North Carolina) leave the length of the commitment period entirely to the discretion of the court. Finally, 16 jurisdictions provide defendants with the right to an examination independent of that conducted for the court or the prosecution.

Disposition Following Acquittal by Reason of Insanity (Chart VII)

Fifty-three jurisdictions statutorily provide for a disposition following an acquittal by reason of insanity. Sixteen of these jurisdictions make commitment automatic. The remainder either through the criminal code (25) or the civil commitment law (12) or both (1) give the court at least some discretion over whether to institutionalize the defendant. Once a defendant is committed, 5 jurisdictions (American Samoa, Connecticut, Illinois, Kansas, Oregon) limit the duration of the commitment to the maximum sentence which the defendant could have served had he or she been convicted. Four other jurisdictions (Indiana, Montana, Nebraska, North Carolina) limit their commitment under the criminal code to a relatively short period of time (30 to 180 days depending on the state) after which the defendant can only be held in custody if he or she is committed civilly. The remainder provide that the defendant may be held until he or she is cured or no longer dangerous, or may be immediately referred for civil commitment.

Review, Transfer and Release Upon Commitment Following Acquittal By Reason of Insanity (Chart VIII)

The dispositional alternatives for defendants who have been acquitted by reason of insanity are premised on two sometimes conflicting motives. The first is to assist the defendant; the second, to protect society. The statutes governing review, transfer and release procedures appear, to be premised largely, on the second motive. Only 10 jurisdictions (Alaska, California, Connecticut, Illinois, Massachusetts, New Hampshire, Ohio, Rhode Island, Virginia, Washington) provide statutorily for periodic judicial or administrative review of the defendant's condition and need for continued confinement. Only 10 jurisdictions (California, Colorado, Hawaii, Kansas, Massachusetts, New Hampshire, New Jersey, New York, Rhode Island, Wyoming) statutorily provide for transfer to community-based treatment. But 45 jurisdictions set forth special procedures which must be followed before a defendant may be released. Twenty-six of these jurisdictions provide for a court hearing, either at the request of the prosecutor or in every case, to determine whether a defendant can be released. Twenty-two require notice to the prosecutor before a defendant may be released, and 20 require the head of the facility in which the defendant is housed to file a special report on the defendant certifying that he or she is no longer dangerous. In 20 jurisdictions, conditional release is available as an alternative.

[123] As noted above, however, courts have generally interpreted these statutes to include at least mental retardation. See authorities listed in *supra* note 74.

I. PROCEDURES FOR RAISING THE ISSUE OF INCOMPETENCY

State	Information or Indictment Necessary to Raise Incompetency Issue	Anyone May Raise the Issue	Only the Court May Raise the Issue
ALABAMA Code (1975)	tit. 15, §16-21		tit. 15, §16-20 §16-21
ALASKA Stat. (1972)		§12.45.100	
ARIZONA Rev. Stat. Ann. (1973)	R. Crim. P. 11.2	R. Crim. P. 11.2	
ARKANSAS Stat. Ann. (1977)		§41-605	
CALIFORNIA Penal Code (West 1970 & Supp. 1979)	§1368		§1368 §1370.1
COLORADO Rev. Stat. (1973)		§16-8-110	
CONNECTICUT Gen. Stat. Ann. (West Supp. 1978)	§54-40(a)	§54-40(a)	
DELAWARE			
DISTRICT OF COLUMBIA Code Ann. (1973)		§24-301(a)(2)	
FLORIDA Stat. Ann. (West Supp. 1978)		§918.15(2)	
GEORGIA Code Ann. (1978)	§27-1502(a)		
HAWAII Rev. Stat. (1976)			§704-404(1)
IDAHO Code (Supp. 1978)		§18-212	
ILLINOIS Ann. Stat. (Smith-Hurd 1973 & Supp. 1978)		ch. 38, §1005-2-1(b)	
INDIANA Code Ann. (Burns 1975)	§35-5-3.1-1	§35-5-3.1-1	
IOWA Code Ann. (West Special Pamphlet 1978)		§812.3	
KANSAS Stat. Ann. (1974 & Supp. 1978)		§22-3302(1)	

State	Information or Indictment Necessary to Raise Incompetency Issue	Anyone May Raise the Issue	Only the Court May Raise the Issue
KENTUCKY Rev. Stat. (Supp. 1978)		R. Crim. P. 8.06	
LOUISIANA Code Crim. Pro. Ann. (West 1967 & Supp. 1978)		art. 642	
MAINE Rev. Stat. Ann. (Pamphlet 1978)	tit. 15, §101	tit. 15, §101	
MARYLAND Ann. Code (1972 & Supp. 1978)	art. 59 §23		art. 59 §23[1]
MASSACHUSETTS Gen. Laws Ann. (West Supp. 1978)	ch. 123, §15(a)		ch. 123, §15(a)
MICHIGAN Comp. Laws Ann. (1975)	§330.2020	§330.2024	
MINNESOTA Stat. Ann. (West Supp. Pamph. 1978)	[2]	R. Crim. P. 20.01	
MISSISSIPPI			
MISSOURI Ann. Stat. (Vernon Supp. 1978)		§552.020(2)	
MONTANA			
NEBRASKA Rev. Stat. (1975)		§29-1823	
NEVADA Rev. Stat. (1973)	§178.405	§178.405	
NEW HAMPSHIRE Rev. Stat. Ann. (Supp. 1975)	§135:17	§135:17	
NEW JERSEY State Ann. (West 1971 & Supp. 1978)			
NEW MEXICO Stat. Ann. (1978)		§31-9-1	
NEW YORK Crim. Proc. Law (McKinney 1971 & Supp. 1978)		§730.30	

State	Information or Indictment Necessary to Raise Incompetency Issue	Anyone May Raise the Issue	Only the Court May Raise the Issue
NORTH CAROLINA Gen. Stat. (1978)		§15A-1002(a)	
NORTH DAKOTA Cent. Code (1976)		§12.1-04-06	
OHIO Rev. Code Ann. (Page 1977)		§2945.37	
OKLAHOMA Stat. Ann. (West Supp. 1978)		tit. 22, §1171	
OREGON Rev. Stat. (1977)			§161.365
PENNSYLVANIA Stat. Ann. (Purdon Supp. 1977-1978)	tit. 50, §7402	tit. 50, §7402(c)	
RHODE ISLAND Gen. Laws (Cum. Supp. 1978)			§26-4-3(b)
SOUTH CAROLINA Code (1976)	§44-23-410		§44-23-410
SOUTH DAKOTA Compiled Laws Ann. (1967 & Supp. 1978)	§23-38-2	§23-38-2	
TENNESSEE Code Ann. (1977 & Supp. 1978)	§33-708(a)	§33-708(a)	
TEXAS Crim. Proc. Code Ann. (Vernon 1977)	§46.02(2)	§46.02(2)(b)	
UTAH Code Ann. (1978)		§77-48-2	
VERMONT Stat. Ann. (1974)	tit. 13, §4817(b)	tit. 13, §4817(b)	
VIRGINIA Code (1975)			§19.2-169
WASHINGTON Rev. Code Ann. (Supp. 1977)		§10.77.060	
WEST VIRGINIA Code (1976)	§27-6A-1(a)		§27-6A-1(a)

State	Information or Indictment Necessary to Raise Incompetency Issue	Anyone May Raise the Issue	Only the Court May Raise the Issue
WISCONSIN Stat. Ann. (West 1971 & Supp. 1978)	§971.14	§971.14	
WYOMING Stat. (1977)	§7-11-303(a)	§7-11-303(a)	
AMERICAN SAMOA Code (Cum Supp. 1978)		tit. 15, §7803	
GUAM			
PUERTO RICO Laws Ann. (1971)	tit. 34, §240(a)		tit. 34, §240(a)
VIRGIN ISLANDS			

[1] Court and defendant's counsel may raise. MD. ANN. CODE, art. 59, §23 (Supp. 1978).
[2] If the issue is raised prior to a determination of probable cause, the court must find that there is probable cause before ordering an examination of the defendant. MINN. R. CRIM. P. 20.01(2)(1) (1979).

II. PROCEDURES FOR DETERMINING COMPETENCY

State	Standard Defined[1]	Examination by Mental Health Professional Required	Temporary Commitment Not Required for Examination	Temporary Commitment Permitted for Competency Examination[2]	Duration of Temporary Commitment	Determination of Competency Made By Judge	Determination of Competency May Be Made By Jury	Rebuttal Testimony Allowed
ALABAMA Code (1975)	tit.22, §52-30						tit. 15, §16-21	
ALASKA Stat. (1972)	§12.45.100	§12.45.100	§12.45.100	§12.45.100		§12.45.100		
ARIZONA Rev. Stat. (1973)	R.Crim.P. 11.1	R. Crim.P. 11.3	R. Crim. P. 11.3(d)	R.Crim.P. 11.3(d)	30 days R.Crim.P. 11.3(d)	R.Crim.P. 11.3		
ARKANSAS Stat.Ann. (1977)	§41-603	§41-605(2)(a)	§41-605(2)(b)	§41-605(3) §41-605(2)(a), (d)	30 days §41-605(2)(d)	§41-606		§41-610
CALIFORNIA Penal Code (West 1970 & Supp. 1979)	§1367 §1370 §1370.1	§1369		§1370.1	90 days §1370.1	§1368	§1368	§1369 §1370.1
COLORADO Rev. Stat. (1973)	§16-8-102(3)		§16-9-111(1)	§16-8-111(2)		§16-8-111		
CONNECTICUT Gen.Stat.Ann. (West Supp. 1978)	§54-40	§54-40(b)	§54-40(b)	§54-40(b)	15 days §54-40(b)	§54-40		§54-40(b)
DELAWARE Code Ann. (1975)	tit.11,§404					tit.11, §404		
DISTRICT OF COLUMBIA Code Ann. (1973)	§24-301(a)(2)	§24-301(a)(2)		§24-301 (a)(2)	reasonable period §24-301(a)(2)	§24-301(a)(2)		
FLORIDA Stat. Ann. (West Supp. 1978)	§918.15	§925.10	§918.15(3)	§918.15(3) §925.10(1)		§925.10		
GEORGIA Code. Ann. (1978)		§27-1502(b)		§27-1502(a)&(b)	90 days §27-1502(b)		§27-1502(a)	

State	Standard Defined[1]	Examination by Mental Health Professional Required	Temporary Commitment Not Required for Examination	Temporary Commitment Permitted for Competency Examination[2]	Duration of Temporary Commitment	Determination of Competency Made By Judge	Determination of Competency May Be Made By Jury	Rebuttal Testimony Allowed
HAWAII Rev. Stat. (1976)	§704-403	§704-404(2)	§704-404(2)	§704-404(2)	30 days or longer as the court determines §704-404(2)	§704-405		§704-405
IDAHO Code (Supp. 1978)	§18-210	§18-211(1)	§18-211(1)	§18-211(1)	60 days §18-211(1)	§18-212		§18-212
ILLINOIS Ann.Stat. (Smith-Hurd 1973 & Supp. 1978)	ch.38, §1005-2-1	ch. 38, §1005-2-1	ch.38, §1005-2-1(g)	ch.38, §1005-2-2		ch.38, §1005-2-1(d)	ch.38, §1005-2-1(d)	
INDIANA Code Ann. (Burns 1975)	§35-5-3.1-1	§35-5-3.1-1	§35-5-3.1-1			§35-5-3.1-1		§35-5-3.1-1
IOWA Code Ann. (West Special Pamphlet 1978)	§812.3					§812.4		
KANSAS Stat. Ann. (1974 & Supp. 1978)	§22-3301	§22-3302(3)[3]	§22-3302(3)	§22-3302(3)	60 days extended for good cause 60 additional days §22-3302(3)	§22-3302(3)	§22-3302(3)	
KENTUCKY Rev. Stat. (Supp. 1978)[4]		§202.135(3)	§202A.070	§202A.070	60 days §202A.070	§202.135		§202.135(5)
LOUISIANA Code Crim. Pro. Ann. (West 1967 & Supp. 1978)	art. 641	art. 643	art. 644	art. 644		art. 643		
MAINE Rev. Stat. Ann. (1978)		tit. 15, §101	tit. 15, §101	tit. 15, §101	60 days tit. 15, §101	tit. 15, §101		
MARYLAND Ann. Code (1972) & Supp. 1978)	art. 59, §23	art. 59, §23	art. 59, §23	art. 59, §23	30 days art. 59, §23 §26	art. 59, §23		

State	Standard Defined[1]	Examination by Mental Health Professional Required	Temporary Commitment Not Required for Examination	Temporary Commitment Permitted for Competency Examination[2]	Duration of Temporary Commitment	Determination of Competency Made By Judge	Determination of Competency May Be Made By Jury	Rebuttal Testimony Allowed
MASSACHUSETTS Gen. Laws Ann. (West Supp. 1978)	ch. 123, §15	ch. 123, §15[3]	ch. 123, §15(a)	ch. 123, §15(b)	20 days ch. 123, §15(b)	ch. 123, §15		
MICHIGAN Comp. Laws Ann. (1976 & Supp. 1977)	§330.2020	§330.2026		§330.2028	60 days §330.2028	§330.2020		
MINNESOTA Rules of Court (West Supp. Pamph. 1978)	R. Crim. P. 20.01(1)	R. Crim. P. 20.01(2)(3)	R. Crim. P. 20.01(2)(3)	R. Crim. P. 20.01(2)(3)	60 days R. Crim. P. 20.01(2)(3)	R. Crim. P. 20.01		R. Crim. P. 20.01
MISSISSIPPI Code Ann. (1972)		§99-13-11[3]	§99-13-11			§99-13-3		
MISSOURI Ann. Stat. (Vernon Supp. 1978)	§552.020	§552.020(2)	§552.020(2)	§552.020(2)		§552.020		§552.020
MONTANA Rev. Codes Ann. (1969)	§95-504	§95-505(1)	§95-505(1)	§95-505(1)	60 days or longer §95-505(1)	§95-506		
NEBRASKA Rev. Stat. (1975)		§29-1823	§29-1823			§29-1823		
NEVADA Rev. Stat. (1973)		§178.415	§178.415			§178.415		
NEW HAMPSHIRE Rev. Stat. Ann. (1974)		§135:17	§135:17	§135:17	30 days §135:17	§135:17		
NEW JERSEY Stat. Ann. (West 1971 & Supp. 1979)	§2C:4-4	§2C:4-5		§2C:4-5	30 days §2C:4-5	§2C:4-5		
NEW MEXICO Stat. Ann. (1978)		§31-9-2	§31-9-2			§31-9-1		
NEW YORK Crim. Proc. Law (McKinney Supp. 1978)	§730.10	§730.30(1)	§730.20(2)	§730.20(3)	30-60 days §730.20(4)	§730.30(2)		

State	Standard Defined[1]	Examination by Mental Health Professional Required	Temporary Commitment Not Required for Examination	Temporary Commitment Permitted for Competency Examination[2]	Duration of Temporary Commitment	Determination of Competency Made By Judge	Determination of Competency May Be Made By Jury	Rebuttal Testimony Allowed
NORTH CAROLINA Gen. Stat. (1978)	§15A-1001	§15A-1002(b)[3]	§15A-1002(b)(1)	§15A-1002(b)(2)	60 days §15A-1002(b)(2)	§15A-1002		
NORTH DAKOTA Cent. Code (1976)	§12.1-04-04	§12.1-04-06[3]	§12.1-04-06	§12.1-04-06	30-60 days §12.1-04-06	§12.1-04-08		§12.1-04-07
OHIO Rev. Code Ann. (Page 1977)		§2945.37	§2945.37			§2945.37	§2945.37	
OKLAHOMA Stat. Ann. (West Supp. 1978)		tit. 22, §1171		tit. 22, §1171	60 days tit. 22, §1171	tit. 22, §1172		
OREGON Rev. Stat. (1977)	§161.30	§161.365	§161.365(1)	§161.365(2)	30 days §161.365(2)	§161.370(1)		
PENNSYLVANIA Stat. Ann. (Purdon Supp. 1977-1978)	tit. 50, §7402	tit. 50, §7402(c)	tit. 50, §7402(c)(1)	tit. 50, §7402(c)(1)	[5]	tit. 50, §7403(a)		tit. 50, §7402(f)
RHODE ISLAND Gen. Laws (Cum. Supp. 1978)	§26-4-3	§26-4-3(b)	§26-4-3(b)			§26-4-3(d)		§26-4-3(d)
SOUTH CAROLINA Code (1976)	§44-23-410	§44-23-410	§44-23-410(1)	§44-23-410(2)	15 days §44-23-410(2)	§44-23-430		
SOUTH DAKOTA Compiled Laws Ann. (1967 & Supp. 1978)						§23-38-4 (Questions of law)	§23-38-4 (Questions of fact)	
TENNESSEE Code Ann. (1977) & Supp. 1978)		§33-708(a)	§33-708(a)	§33-708(a)	30 days §33-708(a)	§33-708		
TEXAS Crim. Pro. Code Ann. (Vernon 1977)	§46.02	§46.02(3)	§46.02(3)(b)	§46.02(3)(b)	21 days §46.02(3)(b)	§46.02(4)	§46.02(4)	§46.02(3)(h)

State	Standard Defined[1]	Examination by Mental Health Professional Required	Temporary Commitment Not Required for Examination	Temporary Commitment Permitted for Competency Examination[2]	Duration of Temporary Commitment	Determination of Competency Made By Judge	Determination of Competency May Be Made By Jury	Rebuttal Testimony Allowed
UTAH (1978)		§77-48-4	§77-48-4			§77-48-4		§77-48-4
VERMONT Stat. Ann. (1974)		tit. 13, §4817(b)		tit. 13, §4815	60 days tit. 13, §4815	tit. 13, §4817(b)		
VIRGINIA Code (1975)	§19.2-169	§19.2-169	§19.2-169	§19.2-169	45 days §19.2-169	§19.2-169		
WASHINGTON Rev. Code Ann. (Supp. 1977)		§10.77.060	§10.77.060	§10.77.060	15 days §10.77.060	§10.77.060		
WEST VIRGINIA Code (1976)	§27-6A-2	§27-6A-1	§27-6A-1(a)	§27-6A-1(b)	20 days §27-6A-1(b)	§27-6A-2(b)		§27-6A-2(a)
WISCONSIN Stat. Ann. (West 1971)	§971.13	§971.14(2)	§971.14(2)	§971.14(2)	60 days §971.14(2)	§971.14(4)		§971.14
WYOMING Stat. (1977)	§7-11-302	§7-11-303(b)	§7-11-303(b)	§7-11-303(b)	30 days §7-11-303(b)	§7-11-303(b)		§7-11-303(f)
AMERICAN SAMOA Code (Cum. Supp. 1978)		tit. 15, §7802[6]	tit. 15, §7802			tit. 15, §7802		tit. 15, §7802
GUAM Penal Code (1970)	§1368	§1368	§1368	§1368	reasonable period §1368	§1368		
PUERTO RICO Laws Ann. (1977)		tit. 34, R.240	tit. 34, R.240			tit. 34, R.240	tit. 34, R.240	
VIRGIN ISLANDS								

[1] While *all* states provide that persons may not be tried while incompetent only some include a specific statutory definition of incompetence. In those states where provided, the standard is generally as follows: As a result of mental disease or defect the defendant lacks the capacity to understand the nature of the charges against him and is unable to assist in his own defense.

[2] Five states provide for a hearing prior to a temporary commitment. GA CODE ANN. §27-1502(a) (1978); Ill. ANN. STAT. ch. 38, §1005-2-1 (Smith-Hurd Supp. 1978); MD. ANN. CODE art. 59, §23 (Supp. 1978); MINN. R. CRIM. P. 20.01(2) (West Pamphlet 1978); WIS. STAT. ANN. §971.14(1) (West 1971).

[3] At court's discretion.

[4] Criminal rules refer to civil determination of insanity.

[5] *See* civil statute.

[6] "A psychiatrist or other person medically or otherwise qualified to give an opinion on the defendant's mental condition." A.S. Code, tit. 15, §7804 (Cum. Supp. 1978).

III. DISPOSITION AFTER A FINDING OF INCOMPETENCY

State	Commitment Automatic or Discretionary	Type of Commitment		Maximum Duration of Commitment		
		Criminal Commitment	Civil Commitment	Until Competent	Maximum Length of Sentence	Other
ALABAMA Code (1975)	automatic tit. 15, §16-21		tit. 22, §52-31	tit. 15, §16-21		
ALASKA Stat. (1972)	discretionary §12.45.110	§12.45.110		§12.45.110		
ARIZONA Rev. Stat. (1973)	discretionary R. Crim. P. 11.5(b)(3)	R. Crim. P. 11.5(b)(3)	R. Crim. P. 11.5(b)(2)(1)[1]	R. Crim. P. 11(b)(3)		
ARKANSAS Stat. Ann. (1977)	discretionary §41.607(1)	§41-607(1)				1 year §41-607(2)
CALIFORNIA Penal Code (West 1970 & Supp. 1979)	automatic §1370.1	§1370.1			Lesser of 3 years or maximum sentence §1370.1	
COLORADO Rev. Stat. (1973)	automatic §16-8-112(2)	§16-8-112(2)		§16-8-112(2)		
CONNECTICUT Gen. Stat. Ann. (West Supp. 1978)	automatic ·§54-40(c)	§54-40(c)			Lesser of maximum sentence or 18 months §54-40(c)	
DELAWARE Code Ann. (1975)	discretionary tit. 11, §404	tit. 11, §404		tit. 11, §404		
DISTRICT OF COLUMBIA Code Ann. (1973)	automatic §24-301(a)(2)	§24-301(a)(2)		§24-301(b)		
FLORIDA Stat. Ann. (West Supp. 1978)	Court must find that defendant meets criteria for involuntary admission §925.10		§925.10(1)(a)(1)(b)	§925.10[1]		180 days §925.10(1)(a) (1)(b)[2]
GEORGIA Code Ann. (1978)	discretionary §27-1502(d)	§27-1502(d)	§27-1502(c)[1]			§27-1502(d)[3]
HAWAII Rev. Stat. (1976)	discretionary §704-406	§704-406		§704-406		
IDAHO Code (Supp. 1978)	automatic §18-212	§18-212		§18-212(2)		
ILLINOIS Ann. Stat. (Smith-Hurd 1973 & Supp. 1978)	discretionary ch. 38, §1005-2-2	ch. 38, §1005-2-2(a)			ch. 38, §1005-2-2(c)	

State	Commitment Automatic or Discretionary	Type of Commitment		Maximum Duration of Commitment		
		Criminal Commitment	Civil Commitment	Until Competent	Maximum Length of Sentence	Other
INDIANA Code Ann. (Burns 1975)	automatic §35-5-3.1-1	§35-5-3.1-1	§35-5-3.1-3[1]			90 days §35-5-3.1-3[1]
IOWA Code Ann. (West Special Pamphlet 1978)	discretionary §812.4	§812.5[1]	§812.5[1]			§812.5[4]
KANSAS Stat. Ann. (1974 & Cum. Supp. 1978)	automatic §22-3303(1)	§22-3303	§22-3303[1]	§22-3303		
KENTUCKY Rev. Stat. (Supp. 1978)	automatic R. Crim. P. 8.06	R. Crim. P. 8.06		R. Crim. P. 8.06		
LOUISIANA Code Crim. Pro. Ann. (West 1967 & Supp. 1978)	automatic art. 648	art. 648		art. 648		
MAINE Rev. Stat. Ann. (Supp. Pamphlet 1978)	discretionary tit. 15, §§101(1)&(2)	tit. 15, §101(1)				1 year tit. 15, §101(1)
MARYLAND Ann. Code (1972 & Supp. 1978)	discretionary art. 59, §24(a)			art. 59, §24(a)		
MASSACHUSETTS Gen. Laws Ann. (West Supp. 1978)	discretionary ch. 123, §16	ch. 123, §16		ch. 123, §17(a)		
MICHIGAN Comp. Laws Ann. (West 1975)	automatic §330.2031	§330.2031				18 months §330.2031
MINNESOTA Stat. Ann. (West Supp. Pamph. 1978)	discretionary R. Crim. P. 20.01(4)(2)		R. Crim. P. 20.01(4)(2)(a)(b)[5]			Until not dangerous R. Crim. P. 20.01(5)
MISSISSIPPI						
MISSOURI Ann. Stat. (Vernon Supp. 1978)	automatic §552.020(7)	§552.020(7)		§552-020(7)		
MONTANA Rev. Codes Ann. (1969 & Supp. 1975)	automatic §95-506(2)	§95-506(2)		§95-506(2)		

State	Commitment Automatic or Discretionary	Type of Commitment		Maximum Duration of Commitment		
		Criminal Commitment	Civil Commitment	Until Competent	Maximum Length of Sentence	Other
NEBRASKA Rev. Stat. (1975)	automatic §29-1823	§29-1823		§29-1823		
NEVADA Rev. Stat. (1973)	automatic §178.425	§178.425		§178.450		
NEW HAMPSHIRE						
NEW JERSEY Stat. Ann. (West Supp. 1979)	discretionary §2:4-6	§2:4-6(b)		§2C:4-6		
NEW MEXICO Stat. Ann. (1978)	discretionary §31-9-1		§31-9-1[6]	§31-9-1		
NEW YORK Crim. Proc. Law (McKinney 1971 & Supp. 1978)	automatic §730.40(1)	§730.40(1)				2/3 reasonable length of sentence §730:50(3)
NORTH CAROLINA Gen. Stat. (1977)	automatic §15A-1003		§15A-1003		Lesser of maximum sentence or 5 years for a misdemeanor or 10 years for a felony §15A-1008	
NORTH DAKOTA Cent. Code (1976)	automatic §12.1-04-08	§12.1-04-08			Lesser of maximum sentence or 3 years §12.1-04-08	
OHIO Rev. Code Ann. (Page Supp. 1978)	automatic §2945.38	§2945.38				Lesser of 15 months or 2/3 of minimum sentence if charged with a felony or 2/3 of max. sentence if charged with a misdemeanor. §2945.38
OKLAHOMA Stat. Ann. (West Supp. 1978)	automatic tit. 22, §1172	tit. 22, §1174		tit. 22, §1173		
OREGON Rev. Stat. (1977)	discretionary §161.370(3)	§161.370(2)			Lesser of maximum sentence or 5 years §161.370(3)	
PENNSYLVANIA Stat. Ann. (Purdon Supp. 1977-1978)	discretionary tit. 50, §7403(d)	tit. 50, §7403			tit. 50, §7403(f) Lesser of maximum length of sentence or 10 years[7]	
RHODE ISLAND Gen. Laws (Cum. Supp. 1978)	automatic §26-4-3(e)	§26-4-3(e)			2/3 maximum sentence up to 30 years §26-4-3(f)	

State	Commitment Automatic or Discretionary	Type of Commitment		Maximum Duration of Commitment		
		Criminal Commitment	Civil Commitment	Until Competent	Maximum Length of Sentence	Other
SOUTH CAROLINA Code (1976)	automatic §44-23-430(2)&(3)[8]	§44-23-430(3)	§44-23-430(2)[1]			60 days §44-23-430(3)
SOUTH DAKOTA Comp. Laws Ann. (Supp. 1978)	automatic §23-38-6	§23-38-6		§23-38-6		
TENNESSEE Code Ann. (1977 & Supp. 1978)	discretionary §33-708[9]			§33-712		
TEXAS Crim. Pro. Code Ann. (Vernon 1977)	discretionary §46-02(5)(d)	§46-02(5)	§46.02(6)[1]			18 months §46.02(5)(a)
UTAH Code Ann. (1978)	automatic §77-48-5	§77-48-5		§77-48-5		
VERMONT Stat. Ann. (1974)	discretionary tit. 13, §4822	tit. 13, §4822	tit. 13, §4822[1]	tit. 13, §4822		
VIRGINIA Code (1975)	automatic §19.2-169		§19.2-169	§19.2-169		
WASHINGTON Rev. Code Ann. (Supp. 1977)	discretionary §10.77.090		§10.77.090	§10.77.090		
WEST VIRGINIA Code (1976)	discretionary §27-6A-2(b)	§27-6A-2(b)	§27-6A-2(c)[1]			6-9 months §27-6A-2(b)
WISCONSIN Stat. Ann. (West. 1971 & Supp. 1978)	automatic §971-14(5)	§971.14(5)				24 months §971.14(5)
WYOMING Stat. (1977)	automatic §7-11-303(g)	§7-11-303(g)	§7-11-303(g)[1]			§7-11-303(g)[10]
AMERICAN SAMOA Code (Cum. Supp. 1978)	discretionary tit. 5, §7809[11]	tit. 5, §7805(1)			Lesser of 18 months or the maximum sentence tit. 15, §7805(1)	

State	Commitment Automatic or Discretionary	Type of Commitment		Maximum Duration of Commitment		
		Criminal Commitment	Civil Commitment	Until Competent	Maximum Length of Sentence	Other
GUAM Penal Code (1970)	automatic §1370	§1370		§1370		
PUERTO RICO Laws Ann. (1971)	discretionary tit. 34, R. 240(b)			tit. 34, R.240(b)		
VIRGIN ISLANDS						

1 Defendant is civilly committed if he or she meets the civil standard and there is not a substantial probability that he or she will become competent.
2 Six months if the defendant is ordered to receive outpatient treatment rather than being committed.
3 Initial 90 day period but court may grant a 9 month extension.
4 If after 6 months there is not a substantial probability that the defendant will become competent, he or she is subject to civil commitment proceedings.
5 With court supervision.
6 Court order required for release.
7 No time limit when the defendant is charged with first or second degree murder.
8 Only if likely to become competent.
9 Civil standard applies.
10 Time reasonably necessary to determine if the individual is likely to regain competency.
11 Automatic unless parties and court agree to outpatient treatment.

IV. PROCEDURES FOLLOWING COMMITMENT DUE TO A FINDING OF INCOMPETENCE

State	Periodic Review	Authority to Transfer to Community Treatment			Release Authority		Consideration of Merits Of Case After Finding of Incompetence Permitted
		Administrative	Court Ordered	Civil Statute Applies	Administrative	Court	
ALABAMA Code (1975)				tit. 22, §52-31	tit. 15, §16-21		
ALASKA Stat. (1972)	§12.45.110					§12.45.110	
ARIZONA Rev. Stat. (1973)	R. Crim. P. 11.6					R. Crim. P. 11.5	
ARKANSAS Stat. Ann. (1977)	§41-607(2)					§41-607	
CALIFORNIA Penal Code (West 1970 & Supp. 1979)			§1370.3			§1372	
COLORADO Rev. Stat. (1973)	§16-8-113(1)	§16-8-112				§16-8-113	
CONNECTICUT Gen. Stat. Ann. (West Supp. 1978)	§54-40(e)					§54-40(e)	
DELAWARE Code Ann. (1975)						tit. 11, §404	tit. 11, §404
DISTRICT OF COLUMBIA Code Ann. (1973)						§24-301(b)	
FLORIDA Stat. Ann. (West Supp. 1978)	§925.10					§925.10	
GEORGIA Code Ann. (1978)				after 90 days §27-1502(c)&(d)		§27-1502	
HAWAII Rev. Stat. (1976)						§704-406	§704-407
IDAHO Code (Supp. 1978)	§18-212(2)					§18-212	§18-212
ILLINOIS Ann. Stat. (Smith Hurd 1973 & Supp. 1978)	ch. 38, §1005-2-2(b)					ch. 38, §1005-2-2(b)	

State	Periodic Review	Authority to Transfer to Community Treatment			Release Authority		Consideration of Merits Of Case After Finding of Incompetence Permitted
		Administrative	Court Ordered	Civil Statute Applies	Administrative	Court	
INDIANA Code Ann. (Burns 1975)	§35-5-3.1-3			after 90 days §35-5-3.1-3	§35-5-3.1-3	§35-5-3.1-3	
IOWA Code Ann. (West Special Pamphlet 1978)				after 6 months §812.5	§812.5[1]	§812.5	
KANSAS Stat. Ann. (1974 & Supp. 1978)			§22-3303			§22-3303	
KENTUCKY							
LOUISIANA Rev. Stat. Ann. (West 1967 & Supp. 1978)	§15:211(a)	§15:211(b)				§15:211(a)	
MAINE Rev. Stat. Ann. (Pamphlet 1978)	tit. 15, §101(1) §104					tit. 15, §104	
MARYLAND Ann. Code (1972 & Supp. 1978)	art. 59, §24(b)					art. 59, §24(a)	art. 59, §24(a)
MASSACHUSETTS Gen. Laws Ann. (West Supp. 1978)	ch. 123, §17		ch. 123, §16(e)			ch. 123, §16	ch. 123, §17(b)
MICHIGAN Comp. Laws Ann. (West 1975)	§330.2038					§330.2040(6)	
MINNESOTA Stat. Ann. (West Supp. Pamph. 1978)	R. Crim. P. 20.10(5)					R. Crim. P. 20.01(5)	R. Crim. P. 20.10(7)
MISSISSIPPI							
MISSOURI Ann. Stat. (Vernon Supp. 1978)						§552-020(7)	
MONTANA Rev. Codes Ann. 1969 & Supp. 1975)						§95-506(2)	§§95-506(3)&(4)

State	Periodic Review	Authority to Transfer to Community Treatment			Release Authority		Consideration of Merits Of Case After Finding of Incompetence Permitted
		Administrative	Court Ordered	Civil Statute Applies	Administrative	Court	
NEBRASKA							
NEVADA Rev. Stat. (1973)						§178.450	
NEW HAMPSHIRE							
NEW JERSEY Stat. Ann. (West Supp. 1979)			§2C:4-6			2C:4-6[2]	
NEW MEXICO Stat. Ann. (1978)						§31-9-1	
NEW YORK Crim. Proc. Law (McKinney 1971 & Supp. 1978)	§730.50	§730.60(3)				§730.50	
NORTH CAROLINA[3] Gen. Stat. (1977)					§122-84		
NORTH DAKOTA Cent. Code (1976)						§12.1-04-08	§12.1-04-09
OHIO Rev. Code Ann. (Page 1977)						§2945.38[4]	
OKLAHOMA Stat. Ann. (West Supp. 1978)					tit. 22, §1173		
OREGON Rev. Stat. (1977)						§161.370(2)	§161.370(4)
PENNSYLVANIA Stat. Ann. (Purdon Supp. 1977 & 1978)						tit. 50, §§7403(d)&(f)	tit. 50, §7403(b)
RHODE ISLAND Gen. Laws (Cum. Supp. 1978)	§26-4-4(f)		§26-4-7			§26-4-9	
SOUTH CAROLINA Code (1976)						§44-23-460	§44-23-440

State	Periodic Review	Authority to Transfer to Community Treatment			Release Authority		Consideration of Merits Of Case After Finding of Incompetence Permitted
		Administrative	Court Ordered	Civil Statute Applies	Administrative	Court	
SOUTH DAKOTA Comp. Laws Ann. (Supp. 19778)					§23-38-9		
TENNESSEE Code Ann. (1977 & Supp. 1978)	§33-708(c)	§33-711			§33-712		
TEXAS Crim. Proc. Code Ann. (Vernon 1977)	§46.02(5)(c)	§46.02(8)(a)				§46.02(5)(d)	§46.02
UTAH Code Ann. (1978)						§77-48-5	
VERMONT Stat. Ann. (1974)			tit. 13, §4822(b)			tit. 13, §4822(b)[5]	
VIRGINIA Code (1975)[6]							§19.2-169
WASHINGTON Rev. Code Ann. (Supp. 1977)						§10.77.150	
WEST VIRGINIA Code (1976)	§27-6A-5					§27-6A-4	§27-6A-6
WISCONSIN Stat. Ann. (West 1971 & Supp. 1978)	§971.14(5)					§971.14(5)	§971.14
WYOMING Stat. (1977)	§7-11-303(g)			§7-11-303(g)	§7-11-303(g)		§7-11-303(j)
AMERICAN SAMOA							
GUAM Penal Code (1970)					§1372		
PUERTO RICO							
VIRGIN ISLANDS							

[1] Civil statute applies after six months.
[2] May be discharged by administrative determination alone after maximum length of sentence has passed.
[3] Civil statute applies. *See* §15A-1003.
[4] Court reviews after one year; if not competent or likely to regain competence then must be civilly committed.
[5] When the offense charged involves a personal injury, the court may require a hearing before discharge.
[6] Civil statutes apply. *See* §19.2-169.

V. DEFINITION OF LACK OF CRIMINAL RESPONSIBILITY DUE TO A MENTAL DISABILITY

State	Not Guilty by Reason of Insanity		Diminished Capacity
	A.L.I.[1]	McNaughton[2]	
ALABAMA Code (Special Pamphlet, 1978)	tit. 13A, §3-1		
ALASKA Stat. (1972)	§12.45.083		§12.45.085
ARIZONA Rev. Stat. (1978)		§13-502	
ARKANSAS Stat. Ann. (1977)	§41-601		§41-602
CALIFORNIA Penal Code (West 1970 & Supp. 1979)			§1027
COLORADO Rev. Stat. (1973)	§16-8-101		
CONNECTICUT Gen. Stat. Ann. (West 1972 & Supp. 1978)	§53a-13		
DELAWARE Code (1975)	tit. 11, §401(a)		tit, 11, §407
DISTRICT OF COLUMBIA			
FLORIDA			
GEORGIA Code Ann. (1978)		§26-702	
HAWAII Rev. Stat. (1976)	§704-400		§704-401
IDAHO Code (Supp. 1978)	§18-207		§18-208
ILLINOIS Ann. Stat. (Smith-Hurd 1973 & Supp. 1978)	ch. 38, §6-2(a)		
INDIANA Code Ann. (Burns Supp. 1978)	§35-41-3-6		
IOWA Code Ann. (West Special Pamphlet 1978)		§701.4	
KANSAS			
KENTUCKY			
LOUISIANA			
MAINE Rev. Stat. (Pamphlet 1978)	tit. 17-A, §57(1)		tit. 17-A, §58(1-A)
MARYLAND Ann. Code (1972 & Supp. 1978)	art. 59, §25(a)[3]		

State	Not Guilty by Reason of Insanity		Diminished Capacity
	A.L.I.[1]	McNaughton[2]	
MASSACHUSETTS			
MICHIGAN			
MINNESOTA			
MISSISSIPPI		[4]	
MISSOURI Ann. Stat. (Vernon Supp. 1978)	§552.030(1)		§552.030(3)
MONTANA Rev. Code Ann. (1969 & Supp. 1975)	§95-501		
NEBRASKA			
NEVADA			
NEW HAMPSHIRE			
NEW JERSEY Stat. Ann. (West 1971 & Supp. 1978)			§2C:4-2
NEW MEXICO Stat. Ann. (1978)			R. Crim. P. 35(e)
NEW YORK			
NORTH CAROLINA Gen. Stat. (1977)			§15A-959(b)
NORTH DAKOTA Cent. Code (1976)	§12.1-04-03		
OHIO			
OKLAHOMA Stat. Ann. (West 1958 & Supp. 1978)		tit. 21, §152	
OREGON Rev. Stat. (1977)	§161.295		
PENNSYLVANIA			
RHODE ISLAND			
SOUTH CAROLINA			
SOUTH DAKOTA			
TENNESSEE			
TEXAS Crim. Proc. Code Ann. (Vernon 1977)	§8.01		
UTAH Code Ann. (1978)	§76-2-305		

State	Not Guilty by Reason of Insanity		Diminished Capacity
	A.L.I.[1]	McNaughton[2]	
VERMONT Stat. Ann. (1974)	tit. 13, §4801		
VIRGINIA Code (Supp. 1978)			§19.2-168
WASHINGTON Rev. Code Ann. (Supp. 1977)			10.77.060
WEST VIRGINIA			
WISCONSIN Stat. Ann. (West 1971)	§971.15		
WYOMING Stat. (1977)	§7-11-304(a)		§7-11-305(e)
AMERICAN SAMOA			
GUAM			
PUERTO RICO			
VIRGIN ISLANDS		5	

[1] The A.L.I. Rule states that: "A person is not responsible for criminal conduct if at the time of such conduct as a result of mental disease or defect he [or she] lacks substantial capacity either to appreciate the criminality (wrongfulness) of his [or her] conduct or to conform his [or her] conduct to the requirements of law. MODEL PEN. CODE §4.01 (Proposed Official Draft, 1962).

[2] The McNaughton Rule states that at the time the offense is alleged to have been committed, the defendent was "labouring under such a defect of reason . . ., as not to know the nature and quality of the act . . .; or if he [or she] did know it, that he [or she] did not know he [or she] was doing what was wrong. 10 Cl. & F. 200, 8 Eng. Rep. 718 (1843).

[3] Excludes mental retardation.

[4] Defined as "insane or feebleminded to such an extent as not to be responsible for his or her acts." MISSISSIPPI CODE ANN. §99-13-3 (1972).

[5] "No capacity for judgment, discretion and control of his [or her] conduct." VIRGIN ISLANDS CODE ANN., tit.5,§3637.

VI. PROCEDURES FOLLOWING ASSERTION OF A DEFENSE OF NOT GUILTY BY REASON OF INSANITY

State	Provision for Examination Without Commitment	Temporary Commitment for Observation Permitted	Maximum Duration of Temporary Commitment	Right to Alternative Examination
ALABAMA				
ALASKA Stat. (1972)	§12.45.087	§12.45.087	60 days §12.45.087	§12.45.087
ARIZONA Rev. Stat. (1978)	§13-3993			
ARKANSAS Stat. Ann. (1977)		§41-605	30 days §41-605(2)(d)	§41-610
CALIFORNIA Penal Code (West 1970 & Supp. 1979)	§1027			§1027
COLORADO				
CONNECTICUT Gen. Stat. Ann. (West 1972 & Supp. 1978)		§53a-47(a)(1)	90 days §53a-47(a)(1)	§53a-47(2)
DELAWARE Code (1975)				tit. 11, §402[1]
DISTRICT OF COLUMBIA				
FLORIDA Stat. Ann. (West Supp. 1978)				§918.017
GEORGIA				
HAWAII Rev. Stat. (1976)	§704-404	§704-404	30 days §704-404(2)	§704-409
IDAHO Code (Supp. 1978)	§18-211	§18-211	60 days §18-211	§18-211
ILLINOIS Ann. Stat. (Smith-Hurd 1977)	ch. 38, §115-6			
INDIANA Code Ann. (Burns Supp. 1976)		§35-5-3.2-1		§35-5-3.2-1
IOWA Code Ann. (West 1979)	§813.2, Rule 10			

State	Provision for Examination Without Commitment	Temporary Commitment for Observation Permitted	Maximum Duration of Temporary Commitment	Right to Alternative Examination
KANSAS Stat. Ann. (1974 & Supp. 1978)	§22-3219	§22-3219[2] §22-3302		§22-3219(2)
KENTUCKY				
LOUISIANA				
MAINE Rev. Stat. (Pamphlet 1978)	tit. 15, §101	tit. 15, §101	60 days tit. 15, §101	
MARYLAND Ann. Code (1972 & Supp. 1978)		art 59, §25 §26	60 days art. 59, §26	
MASSACHUSETTS Gen. Laws Ann. (West Supp. 1978		ch. 123, §16(a)	40 days ch. 123, §16(a)	
MICHIGAN Comp. Laws Ann. (1975)	§768.20(a)	§768.20(a)		§768.20(a)
MINNESOTA Stat. Ann. (West Pamphlet 1978		R. Crim. P. 20.02(2)	60 days R. Crim. P. 20.02(2)	R. Crim. P. 20.02(2)
MISSISSIPPI		[3]		
MISSOURI Ann. Stat. (Vernon Supp. 1978)		§552.030(4)	§552.030[2]	§552.030(4)
MONTANA Rev. Codes Ann. (1969 & Supp. 1975)		§95-505(1)	60 days[4] §95-505(1)	§95-507
NEBRASKA Rev. Stat. (Cum Supp. 1978)	§29-2203			
NEVADA				
NEW HAMPSHIRE				
NEW JERSEY Stat. Ann. (West Supp. 1978)		§2C:4-2	60 days §2C:4-2	
NEW MEXICO				
NEW YORK Crim. Proc. Law McKinney 1971 & Supp. 1978)		§330.20(1)-(2)	60 days §330.20	

State	Provision for Examination Without Commitment	Temporary Commitment for Observation Permitted	Maximum Duration of Temporary Commitment	Right to Alternative Examination
NORTH CAROLINA Gen. Stat. (1977)		§15A.1322 §122-58.3 §122-58.18	Pending hearing §122-58.7	
NORTH DAKOTA Cent. Code (1976)		§12.1-04-06	30-60 days §12.1-04-06	
OHIO Rev. Code Ann. (Page 1977)		§2945.39[3]		
OKLAHOMA				
OREGON Rev. Stat. (1977)	§161.315	§161.315	30 days §161.315	
PENNSYLVANIA Stat. Ann. (Purdon 1964 & Cum. Supp. 1978)		tit. 50, §7404[3]		
RHODE ISLAND		[3]		
SOUTH CAROLINA				
SOUTH DAKOTA Compiled Laws (Pamphlet 1978)	§23A-10-4			
TENNESSEE				
TEXAS Crim. Proc. Code Ann. (Vernon 1977)		§46.031(3)(b)	21 days §46.03(3)(b)	§46.03(3)(f)
UTAH Code Ann. (1978)				§77-24-7
VERMONT Stat. Ann. (1974)		tit. 13, §4815	60 days tit. 13, §4815	
VIRGINIA				
WASHINGTON Rev. Code Ann. (Supp. 1977)	§10.77.060			§10.77.070
WEST VIRGINIA Code (Cum. Supp. 1979)		§27-6A-1(b)	20 days + 20 day extension §27-6A-1(b) §27-6A-3(a)	
WISCONSIN Stat. Ann. (West 1971)	§971.16			§971.16(3)

State	Provision for Examination Without Commitment	Temporary Commitment for Observation Permitted	Maximum Duration of Temporary Commitment	Right to Alternative Examination
WYOMING Stat. (1977)		§7-11-303(b)	30 days §7-11-303(b)	
AMERICAN SAMOA Code (Cum. Supp. 1978)	tit. 15, §7804			
GUAM				
PUERTO RICO				
VIRGIN ISLANDS				

[1] Rules for examination entirely at court's discretion.
[2] At court's discretion.
[3] Procedures governed by civil commitment statute.
[4] May be extended by order of the court.

VII. DISPOSITION FOLLOWING ACQUITTAL BY REASON OF INSANITY

State	Commitment			Maximum Duration	
	Automatic	Judicial Discretion	Civil Statute Applies[1]	Until Cured or Not Dangerous	Other
ALABAMA Code (1975)			tit. 22, §52-33		
ALASKA Stat. (1972)		§12.45.090		§12.45.090	
ARIZONA Rev. Stat. (1978)		§13-3991		§13-3991	
ARKANSAS Stat. Ann. (1977)		§41-612(1)(a)		§41-613	
CALIFORNIA Penal Code (West 1970 & Supp. 1979)		§1026		§1026a	
COLORADO Rev. Stat. (1973)	§16-8-105(4)			§16-8-120	
CONNECTICUT Gen. Stat. Ann. (West 1972 & Supp. 1978)		§53a-47			maximum sentence §53a-47[2]
DELAWARE Code (1975)	tit. 11, §403(a)			tit. 11, §403(b)	
DISTRICT OF COLUMBIA Code Ann. (1973)	§24-301(d)(1) & (e)			§24-301(e)[3]	
FLORIDA Stat. Ann. (West Supp. 1978)			§918.017(2)	§918.017(2)	
GEORGIA Code Ann. (1978)			§27-1503(a)	§27-1503(a)	
HAWAII Rev. Stat. (1976)		§704-411		§704-412[4]	
IDAHO Code (Supp. 1978)	§18-214(1)			§18-214	

State	Commitment			Maximum Duration	
	Automatic	Judicial Discretion	Civil Statute Applies[1]	Until Cured or Not Dangerous	Other
ILLINOIS Ann. Stat. (Smith-Hurd 1973) & Supp. 1978)		ch. 38, §1005-2-4[5]			maximum sentence ch. 38, §1005-2-4
INDIANA Code Ann. (Burns 1975)			§§35-5-3.2-1(b) & (c)		180 days unless civilly committed §16-14-9.1-9
IOWA Code Ann. (West 1979)		813.2 Rule 21		813.2 Rule 21	
KANSAS Stat. Ann. (1974 & Supp. 1978)	§22-3428				maximum sentence §22-3430
KENTUCKY Rev. Stat. (Supp. 1978)	R. Crim. P. 9.90			R. Crim. P. 9.90	
LOUISIANA Code of Crim. Pro. (1967 and Cum. Supp. 1978)		art. 654[6]		art. 657	
MAINE Rev. Stat. (Pamphlet 1978)	tit. 15, §103			tit. 15, §104	
MARYLAND Ann. Code (1972 & Supp. 1978)		art. 59, §27			art. 59, §27[7]
MASSACHUSETTS Gen. Laws Ann. (West Supp. 1978)		ch. 123, §16(b)		ch. 123, §16	
MICHIGAN Comp. Laws Ann. 1968 & Supp. 1977)	§330.2050			§330.2050	
MINNESOTA Stat. Ann. (West Pamphlet 1978)			R. Crim. P. 20.02(8)[8]	R. Crim. P. 20.02(8)(4)	

State	Commitment			Maximum Duration	
	Automatic	Judicial Discretion	Civil Statute Applies[1]	Until Cured or Not Dangerous	Other
MISSISSIPPI Code Ann. (1973)		Jury's discretion §99-13-7	for "feeble-minded" §99-13-9	§99-13-7	
MISSOURI Ann. Stat. (Vernon Supp. 1978)	§552.040(1)			§552.040(1)	
MONTANA Rev. Codes Ann. (1969 & Supp. 1975)	§95-508(1)				50 days §95-508
NEBRASKA Rev. Stat. (Cum. Supp. 1978)		§29-2203			30 days[9] §29-2203
NEVADA Rev. Stat. (1977)	§178.425			§§178.450 *et seq*	
NEW HAMPSHIRE Rev. Stat. Ann. (Supp. 1977)		§651:9[10]		§651:9-a	
NEW JERSEY Stat. Ann. (West 1971 & Supp. 1978)		§2C:4-8		§2C:4-8	
NEW MEXICO					
NEW YORK Crim. Proc. Law (McKinney Supp. 1978)		§330.20(3)		§330.20(3)	
NORTH CAROLINA Gen. Stat. (1977)			§122-84.1		§122-84.1[11]
NORTH DAKOTA Cent. Code (1976)		§29-22-36			
OHIO Rev. Code Ann. (Page 1977)	§2935.39			§2945.39	

State	Commitment			Maximum Duration	
	Automatic	Judicial Discretion	Civil Statute Applies[1]	Until Cured or Not Dangerous	Other
OKLAHOMA Stat. Ann. (West Supp. 1978)			tit. 22, §1167		
OREGON Rev. Stat. (1977)		§161.325 §161.329 §161.336			maximum sentence §161.341
PENNSYLVANIA Stat. Ann. (Purdon 1964 & Cum. Supp. 1978)		tit. 19, §1351		tit. 19, §1351	
RHODE ISLAND Gen. Laws (Cum. Supp. 1978)	§26-4-4(b)			§26-4-4	
SOUTH CAROLINA Code (1976)			§44-23-610		
SOUTH DAKOTA Codified Laws Ann. (Pamphlet 1978)			§23A-26-13		
TENNESSEE Code Ann. (1977 & Supp. 1978)			§33-709	§33-709	
TEXAS Crim. Proc. Code Ann. (Vernon 1977)			§46.03(4)(a)		
UTAH Code Ann. (1978)	§77-24-15			§77-24-16	
VERMONT Stat. Ann. (1974)		tit. 13, §4820		tit. 13, §4822	
VIRGINIA Code (1975 & Supp. 1978)		§19.2-181		§19.2-181	

State	Commitment			Maximum Duration	
	Automatic	Judicial Discretion	Civil Statute Applies[1]	Until Cured or Not Dangerous	Other
WASHINGTON Rev. Code Ann. (Supp. 1977)		§10.77.110		§10.77.120	
WEST VIRGINIA Code (1976 & Cum. Supp. 1979)			§27-6A-3(b)	§27-6A-3(b)	
WISCONSIN Stat. Ann. (West 1971 & Supp. 1978)	§971.17(1)			§971.17(2)	
WYOMING Stat. (1977)		§7-11-306		§7-11-306	
AMERICAN SAMOA Code (Cum. Supp. 1978)		tit. 15, §7805			maximum sentence tit. 15, §7805
GUAM					
PUERTO RICO Laws Ann. (1971)		tit. 34, R. 241		tit. 34, R. 241	
VIRGIN ISLANDS Code Ann. (Cum. Supp. 1978)	tit. 5, §3637			tit. 5, §3637	

[1] In states where the civil commitment statute applies, the civil provisions for periodic review, transfers and release are generally also applicable. Where exceptions exist they are cited on Chart VIII.
[2] Twenty-five years for Class A felony.
[3] Hearing after fifty days to determine whether the defendant is entitled to release from custody. §24-301(d).
[4] Minimum of 90 days.
[5] Also at discretion of jury.
[6] Mandatory commitment in capital cases.
[7] Civil commitment standard applies.
[8] Subject to judicial supervision.
[9] During the 30 days, the county board of mental health must determine whether the defendant should be civilly committed.
[10] Once committed, the civil commitment statute applies.
[11] Civil commitment procedures apply after ninety days.

VIII. REVIEW, TRANSFER AND RELEASE UPON COMMITMENT FOLLOWING ACQUITTAL BY REASON OF INSANITY

State	Periodic Review	Transfer Authority		Release Procedures			Conditional Release
		Administrative	Court	Court Hearing	Notice to Prosecutor	Superintendent Report	
ALABAMA							
ALASKA Stat. (1972)	§12.45.110						
ARIZONA Rev. Stat. (1978)				§13-3991 (implied)		§13-3991	
ARKANSAS Stat. Ann. (1977)				§41-613	§41-613	§41-613	§41-613 §41-614
CALIFORNIA Penal Code (West 1970-Supp. 1979)	§1026		§1026	§1026	§1026	§1026[1]	§1026
COLORADO Rev. Stat. (1973)		§16-8-105(4)		§16-8-115	§16-8-116	§16-8-115	
CONNECTICUT Gen. Stat. Ann. (West 1972 & Supp. 1978)	§53a-47			§53a-47	§53a-47	§53a-47	§53a-47
DELAWARE Code (1975)				tit. 11, §403(b)		tit. 11, §403(b)	
DISTRICT OF COLUMBIA Code Ann. (1973)				§§24-301(e) & (k)	§§24-301(e) & (k)	§§24-301(e) & (k)	§24-301(k)
FLORIDA Stat. Ann. (West Supp. 1978)							
GEORGIA Code Ann. (1978)				§27-1503 (a)		§27-1503(a)	
HAWAII Rev. Stat. (1976)			§704-412	§704-411	§704-411		§704-411(b)

State	Periodic Review	Transfer Authority		Release Procedures			Conditional Release
		Administrative	Court	Court Hearing	Notice to Prosecutor	Superintendent Report	
IDAHO Code (Supp. 1978)				§18-214			§18-214(4)
ILLINOIS Ann. Stat. (Smith-Hurd 1973 & Supp. 1978)	ch. 38, §1005-2-4(b)				ch. 38, §1005-2-4(b)	ch. 38, §1005-2-4(b)	ch. 38, §1005-2-4
INDIANA Code Ann. (Burns 1975)							
IOWA							
KANSAS Stat. Ann. (1974 & Supp. 1978)		§22-3428		§22-3428			§22-3428
KENTUCKY							
LOUISIANA							
MAINE Rev. Stat. (Pamphlet 1978)				tit. 15, §103			tit. 15, §104(3)
MARYLAND Ann. Code (1972 & Supp. 1978)					art. 59, §27	art. 59, §27	art. 59, §28
MASSACHUSETTS Gen. Law Ann. (West Supp. 1978)			ch. 123, §16(e)		ch. 123, §16		
MICHIGAN Comp. Laws Ann. (1975 & Supp. 1977)				§330.2050			
MINNESOTA Stat. Ann. (West Pamphlet 1973)						R. Crim. P. 20.02(8)	R. Crim. P. 20.02(8)
MISSISSIPPI							

State	Periodic Review	Transfer Authority		Release Procedures			Conditional Release
		Administrative	Court	Court Hearing	Notice to Prosecutor	Superintendent Report	
MISSOURI Ann. Stat. (Vernon Supp. 1978)				§552.040(1)			§552.040(4)
MONTANA Rev. Codes Ann. (1969 & Supp. 1975)				§95-508(1)			§95-508
NEBRASKA Rev. Stat. (Cum. Supp. 1978)						§29-2203[2]	
NEVADA Rev. Stat. (1977)				§178.425			
NEW HAMPSHIRE Rev. Stat. Ann. (Supp. 1977)	§651:11(a)		§651:9-a §651:11-a	§651:9-a §651:11-a			
NEW JERSEY Stat. Ann. (West 1971 & Supp. 1978)		§2C:4-9		§2C:4-9	§2C:4-9	2C:4-9	
NEW MEXICO							
NEW YORK Crim. Proc. Law (McKinney Supp. 1978)		§330.20(6)		§330.20(3)	§330.20(3)		§330.20(3)
NORTH CAROLINA Gen. Stat. (1977)				§122-84.1	§122-84.1		
NORTH DAKOTA Cent. Code (1976)					§29-22-36	§12.1-04-10	
OHIO Rev. Code Ann. (Page 1977)	§2945.39			§2945.39		§2945.39	§2945.38
OKLAHOMA Stat. Ann. (West Supp. 1978)						tit. 22, §1167	
OREGON Rev. Stat. (1977)					§161.336		§161.341

State	Periodic Review	Transfer Authority		Release Procedures			Conditional Release
		Administrative	Court	Court Hearing	Notice to Prosecutor	Superintendent Report	
PENNSYLVANIA Stat. Ann. (Purdon 1964 & Supp. 1977-1978)					tit. 19, §1351		tit. 19, §1381
RHODE ISLAND Gen. Laws (Cum. Supp. 1978)	§26-4-4(f)		§26-4-7 §26-4-8	§26-4-4		§26-4-4	
SOUTH CAROLINA						§44-23-610	
SOUTH DAKOTA							
TENNESSEE Code Ann. (1977 & Supp. 1978)					§33-709		
TEXAS Crim. Proc. Code Ann. (Vernon 1977)						§46.03(4)(a)	
UTAH Code Ann. (1978)				§77-24-15			
VERMONT Stat. Ann. (1974)				tit. 13, §4820	tit. 13, §4820(4)		
VIRGINIA Code (1975 & Supp. 1978)	§19.2-181			§19.2-181	§19.2-181		
WASHINGTON Rev. Code Ann. (Supp. 1977)	§10.77.140			§10.77.150			
WEST VIRGINIA Code (1976)						§27-6A-3(b)	
WISCONSIN Stat. Ann (West 1971 & Supp. 1978)				§971.17			
WYOMING Stat. (1977)		§7-11-306(e)			§7-11-306	§7-11-306(d)	§7-11-306

State	Periodic Review	Transfer Authority		Release Procedures			Conditional Release
		Administrative	Court	Court Hearing	Notice to Prosecutor	Superintendent Report	
AMERICAN SAMOA Code (Cum. Supp. 1978)				tit. 15, §7805			
GUAM							
PUERTO RICO Laws Ann. (1971)					tit. 34, R. 241		
VIRGIN ISLANDS Code Ann. (Cum. Supp. 1978)				tit. 5, §3637			tit. 5, §3637(c)

[1] Patient can also apply for release.
[2] Decision by county board of mental health on whether the defendant should be committed civilly must be made within 30 days.

SUMMARY OF MODEL STATUTES

As indicated above, despite the fact that a significant percentage of the persons who come into contact with the criminal and juvenile justice systems have a developmental disability,[124] few current statutes address their unique problems. While procedures and programs designed to protect and assist developmentally disabled individuals could exist within the criminal and juvenile justice processes without special statutory provisions, the absence of such provisions is indicative of a general lack of awareness of the problems presented by suspects and offenders with a developmental disability.

> [W]hile the last ten years has seen significant reform in the substance and procedures of the criminal law, very few of those reforms have been directed at the problem of the retarded. Judicial decisions protecting pre-trial rights of defendants are of little practical use to an unrecognized retarded defendant, for these protections do not "operate automatically; rather [they] must be affirmatively raised, and the system works in such a way as to virtually insure that the issue will *not* be raised. . . ." Substantive revisions in penal codes may expand the "insanity" defense to include "mental defects," but corollary changes in procedures or correctional treatment are rarely, if ever, provided.[125]

In light of this lack of statutory recognition and the need in this area for clearly articulated policy, two model statutes are presented. They are designed to cover the matters and issues which particularly affect persons with a developmental disability throughout the criminal and juvenile justice processes, and should be inserted in, not substituted for, the enacting jurisdiction's criminal and juvenile codes. The objective of these models is to:

— enhance the awareness of judges and other criminal justice and juvenile justice professionals of the causes, indications, and effects of developmental disabilities, and their skill in dealing humanely, equitably, and effectively with victims, witnesses, suspects, and offenders who have such a disability;

— assure the availability of necessary evaluative and habilitative services for suspects and offenders with a developmental disability; and

— provide protections designed to accommodate to the greatest extent possible, the special needs of developmentally disabled persons within the context of the rights, procedures, and penalties applicable to all individuals alleged or found to have violated the criminal law.

Two model statutes are required because of the philosophical, administrative, and semantic distinctions which have been drawn between the handling of adults accused or convicted of a crime, and juveniles alleged or adjudicated to be delinquent.

The Model Developmentally Disabled Offenders Act covers those areas which the criminal justice system and the juvenile justice system have in common — police investigation procedures and the need for training regarding developmental disabilities[126] — and those aspects which apply only to adults. More specifically, it focuses on the key decision points in the criminal justice process — arrest, charging, diversion, trial or entry of a guilty plea, sentencing, and parole — to make certain that developmentally disabled suspects or offenders are not treated more harshly or have options foreclosed solely because of their disability. In each instance following arrest and the filing of the initial charge, provision is made to furnish the decision-maker with the necessary information regarding the defendant's or offender's abilities and disabilities.[127] Whenever a disposition is required, the use of the least restrictive alternative which is consistent with the interests of the public and the individual is called for.[128] In addition, procedures are established to assure that the array of habilitative services are provided which will assist the defendant or offender, depending on the context, to

---

[124] BROWN and COURTLESS, THE MENTALLY RETARDED OFFENDER (1971); SANTAMOUR and WEST, THE MENTALLY RETARDED OFFENDER AND CORRECTIONS (1977).

[125] Plotkin, *Stranded in the Jungle: The Mentally Retarded Person in the Criminal Justice System,* 1 DISTRICT LAWYER 37-38 (1977); Fox, *The Criminal Reform Movement,* PRESIDENT'S COMMITTEE ON MENTAL RETARDATION 636 (1976); Haggerty, Kane and Udall, *An Essay on the Legal Rights of the Mentally Retarded,* 6 FAMILY LAW QUARTERLY 59 (1972).

[126] See sections 4-6, 14, and 16.

[127] See sections 7(2), 8(4), 10(5), 10(10), 11(2), 12(2) and 15.

[128] See sections 3(15), 6, 8(4), 8(5), 8(6), 10(6), 10(7), 10(11), and 11(3).

become competent to stand trial or to function in the community as self-reliant, law-abiding citizen to the extent and within the shortest time possible.[129] These procedures include provisions for evaluation, the development and periodic updating of an individual Program Plan, and judicial review and enforcement.[130] In each instance, the responsibility for and authority to provide or purchase services is made clear.[131]

While the section on incompetence to stand trial relies on the standard for competence set forth in the enacting jurisdiction's criminal code,[132] the section on criminal responsibility makes the insanity defense, in any of its current forms, inapplicable to defendants with a developmental disability unless they also have a mental illness.[133] It substitutes the concept of diminished capacity and establishes dispositional authority to cover those few cases in which the judge or jury find the prosecutor has proven beyond a reasonable doubt that the defendant committed all the physical acts required as elements of the criminal offense charged or a lesser included offense, but that the defendant lacked the prescribed state of mind. This approach focuses the proceedings on questions which are the most relevant to the impairments resulting from developmental disabilities. It also eliminates the procedural complexities which have resulted from use of the insanity defense, and should reduce the almost total reliance on expert testimony when the issue of criminal responsibility is raised.

In addition, the model act clarifies that the rights accorded defendants in criminal cases apply in all judicial proceedings held pursuant to its provisions.[134] It also establishes a mechanism for identifying individuals qualified to conduct examinations and evaluations of persons alleged or found to have a developmental disability, and for maintaining the quality and integrity of such examinations and evaluations.[135] The procedures for selecting examiners and evaluators are delineated as well as the contents, distribution and use of evaluation and examination reports.[136] Finally, and perhaps most importantly, the model act establishes a program and requirements for training all individuals with significant official responsibilities in the criminal and juvenile justice systems regarding:

- — the nature, effects, and causes of developmental disabilities;
- — the techniques for recognizing a person who has a developmental disability;
- — the most common problems which occur when a person with a developmental disability comes in contact with the criminal and/or juvenile justice systems as a victim, witness, suspect, defendant, or offender;
- — the availability of diagnostic and habilitative services and the procedures for obtaining such services; and
- — the rights to which persons with a developmental disability are entitled and the procedures for protecting those rights.[137]

The provisions of the Model Developmentally Disabled Juvenile Offender Act are parallel to those of the Model Developmentally Disabled Offender Act, but take into account the differences in the procedures, dispositional philosophy, rights, and terminology generally applicable in delinquency proceedings. As discussed in the introduction and the comments to the applicable sections, there has been considerable uncertainty and confusion regarding the extent to which the concepts of competency to stand trial and intent to commit an offense apply to delinquency proceedings. The model act attempts to provide some guidance. Following the precepts set forth in the line of Supreme Court cases beginning with *Kent* vs. *United States*,[138] it takes the position that it is fundamentally unfair to adjudicate juveniles who, because of a developmental disability, are unable to understand the nature of the juvenile court proceedings and assist in their defense[138] and to disregard the element of intent in determining whether a juvenile violated the criminal law.[140] Accordingly, it permits the issues of competency and diminished capacity to be raised in delinquency proceedings, but, to the extent consistent with equity and the need for accountability, does so in a manner in keeping with the traditional dispositional flexibility of the juvenile court.[141]

Unlike the Model Developmentally Disabled Offender Act, the model act on developmentally disabled juveniles does not contain provisions on training and on police investigation and interrogation. The sections of the adult act on those topics cover both juveniles and adults.[142] Since there is

[129] See sections 8(6), 10(7) and 12(2).

[130] See sections 8(6)-(8); 10(8)-(11); and 12(2)(f) and (3).

[131] See sections 8(5)(d)(ii) and (6); 10(6)(c)(ii) and (7); and 12(1) and (2).

[132] Section 8(1).

[133] Section 10(2).

[134] Section 13.

[135] Section 14.

[136] Section 15.

[137] Section 16.

[138] 383 U.S. 541 (1966); In re Gault, 387 U.S. 1 (1967); In re Winship, 397 U.S. 538 (1970); McKeiver vs. Pennsylvania 403 U.S. 528 (1971); Breed vs. Jones, 421 U.S. 519 (1975).

[139] Section 7.

[140] Section 9.

[141] See sections 7(5)-(8) and 9(6)-(11).

[142] See sections 4 and 16.

need for greater understanding of developmental disabilities and the proper means of dealing with persons who have such disabilities by members of both the criminal justice and the juvenile justice systems, and since the agencies and procedures involved in the investigation of crimes committed by juveniles and the apprehension of suspected juvenile offenders are the same, in most cases, as those employed when adults are involved, there was no need to include training and investigation sections in both acts. However, jurisdictions which enact a statute based only on the Model Developmentally Disabled Juvenile Offenders Act should make certain that these two critical areas are covered.

The text of the model acts appears in **bold type.** Optional language appears [within brackets/ separated by slashes]. Comments on statutory provisions appear in standard type. The comments explain, elaborate upon or provide a rationale for the text of the acts. They are intended to assist the user in interpreting the acts and in drafting a statute for a particular state from these models.

MODEL STATUTE*

A BILL FOR

AN ACT TO ESTABLISH STANDARDS AND PROCEDURES FOR HANDLING PERSONS WITH A DEVELOPMENTAL DISABILITY WHO COME IN CONTACT WITH THE CRIMINAL JUSTICE OR JUVENILE JUSTICE SYSTEM; TO ESTABLISH STANDARDS AND PROCEDURES FOR DETERMINING THE COMPETENCE TO STAND TRIAL AND THE CRIMINAL RESPONSIBILITY OF ADULTS WITH A DEVELOPMENTAL DISABILITY; TO ESTABLISH STANDARDS AND PROCEDURES FOR THE DISPOSITION OF ADULTS WITH A DEVELOPMENTAL DISABILITY WHO HAVE BEEN FOUND INCOMPETENT TO STAND TRIAL OR UNABLE TO FORM THE REQUISITE INTENT TO COMMIT A CRIME; TO CLARIFY AND PROTECT THE RIGHTS OF ADULTS WITH A DEVELOPMENTAL DISABILITY IN DETENTION AND CORRECTIONAL FACILITIES; AND TO PROVIDE TRAINING FOR INDIVIDUALS WITH SIGNIFICANT OFFICIAL RESPONSIBILITIES IN THE CRIMINAL JUSTICE SYSTEM REGARDING DEVELOPMENTAL DISABILITIES AND THE RIGHTS OF PERSONS WITH SUCH A DISABILITY.

Section 1. Title

This act shall be known as the "Developmentally Disabled Offender Act."

COMMENT. None necessary.

Section 2. Purpose

The (name of the Legislature) declares that it is the purpose of this act to assure that all members of the criminal justice system deal fairly with persons who have a developmental disability and recognize fully the rights of such persons; that the determinations of the competence to stand trial and of the criminal responsibility of adults with a developmental disability are fair, accurate, and expeditious; that adults with a developmental disability who are found to be incompetent to stand trial or unable to form the requisite intent to commit a crime receive the appropriate habilitative services in the least restrictive manner which is consistent with their needs and public safety; that adults who have a developmental disability and who are in detention or correctional facilities are protected from abuse and exploitation; that adult offenders with a developmental disability are offered the services to which they are entitled; and that all individuals having significant official responsibilities in the criminal justice and juvenile justice systems receive training regarding the nature and recognition of developmental disabilities and the handling and rights of persons with such a disability.

This act shall be liberally construed in order to achieve this purpose.

COMMENT. None necessary.

Section 3. Definitions

As used in this act:

(1) "Community-based facility" means a small, open, residential facility which relies primarily on services available in the surrounding community rather than on in-house services and which maintains community and consumer participation in the planning, operation, and evaluation of its programs.

COMMENT. This definition is drawn from the federal Juvenile Justice and Delinquency Prevention Act.[1] It is intended to include such facilities as group homes, foster homes, and non-secure halfway houses which house six or less persons who are not on the staff,[2] and which are not characterized by physically restrictive construction or location, or by procedures designed to prevent residents from enjoying access to the surrounding community and its resources.[3]

(2) "Correctional agency" means a unit of state or local government responsible for the supervision and/or custody of adults sentenced and/or placed on probation after being found guilty of a criminal offense.

COMMENT. This term is intended to include state and local departments of corrections, state and local probation offices, the state parole au-

* Richard Van Duizend had primary responsibility for drafting this model statute aided by a prior draft prepared by Michael Kindred.

[1] 42 UNITED STATES CODE §5603(1) (Supp. 1979).

[2] *See* Chapter 4: Zoning for Community Homes, *supra*.

[3] *See* INSTITUTE OF JUDICIAL ADMINISTRATION/AMERICAN BAR ASSOCIATION JOINT COMMISSION ON JUVENILE JUSTICE STANDARDS (IJA/ABA) STANDARDS RELATING TO INTERIM STATUS: THE RELEASE, CONTROL, AND DETENTION OF ACCUSED JUVENILE OFFENDERS BETWEEN ARREST AND DISPOSITION, §§ 2.10 and 2.11 (Tentative Draft, 1977).

thority, and sheriff offices and police departments which are responsible for operating jails and other correctional facilities. In enacting this statute, some states may wish to omit this term and insert the name of the correctional agencies responsible for carrying out the duties set forth in section 12. The phrase "sentenced and/or placed on probation" reflects differences in the status of probation among the states. In some jurisdictions a convicted offender may receive a sentence of probation;[4] in others, the sentence must be suspended in order to place an offender on probation.[5] The model statute is compatible with either practice.

(3) "Correctional facility" means a state or local place of confinement for adults found guilty of having committed a criminal offense.

COMMENT. This term includes prisons, jails and camps as well as community-based facilities used primarily to house convicted criminal offenders.

(4) "Court" means the court having jurisdiction over the criminal offense charged.

COMMENT. This term is meant to include both courts of general jurisdiction and courts of limited jurisdiction which are authorized to hear criminal cases and convict and sentence persons found guilty of a felony and/or misdemeanor.

(5) "Criminal offense" means a violation by an adult of a statute or ordinance punishable by a period of incarceration.

COMMENT. This term is intended to include felonies, misdemeanors and traffic offenses for which a term of incarceration can be imposed. Incarceration is used as the limiting criterion in this definition since it would be incongruous to invoke the provisions of this act and potentially interfere with liberty and privacy[6] when the offense is sufficiently minor that a person could not be sentenced to a period of incarceration upon being convicted. Enacting states should conform the terminology in the definition to that contained in their criminal code.

(6) "Defendant" means a person charged with committing a criminal offense.

COMMENT. None necessary.

(7) "Delinquent offense" means a criminal offense committed by a juvenile.

COMMENT. As noted in the introduction, the investigation of crimes committed by juveniles is generally handled by the same law enforcement agencies that investigate crimes committed by adults. Hence, there is no need to separate the provisions on police investigation and interrogation relating to juveniles from those covering adults.[7] It is necessary, therefore, to include a definition of delinquent offense in this act as well as in the Model Developmentally Disabled Juvenile Offender Act.[8]

(8) "Designated diversion unit" means an organization, program, or governmental agency authorized to offer defendants, alternatives to the formal criminal justice process prior to an adjudication of guilt.

COMMENT. Since diversion authority and procedures vary considerably from jurisdiction to jurisdiction,[9] designated diversion unit is included as a generic term encompassing all of the organizational variations. An enacting state may wish to omit this term and substitute the name of the state or local organizational unit(s) which provides diversion services in Section 7.

(9) "Developmental disability" means a chronic disability of a person which:

(a) is attributable to a mental or physical impairment or a combination of mental and physical impairments;

(b) is likely to continue indefinitely;

(c) results in substantial functional limitations in two or more of the following areas of major life activity:

(i) self-care;

(ii) receptive and expressive language;

(iii) learning;

(iv) mobility;

(v) self-direction;

(vi) capacity for independent living; or

(vii) economic self-sufficiency; and

(d) reflects the need for a combination and sequence of special, interdisciplinary or generic care, treatment or other services which are of lifelong or extended duration and individually planned and coordinated.

COMMENT. This definition, like that in the Developmental Disabilities Assistance and Bill of Rights Act,[10] adopts a functional rather than a categorical approach. Thus, it places greater significance on the effect of a disability than on its particular cause. The definition differs from the one contained in the recently amended federal act in that there is no requirement that the disability be "severe" or that it appear before the afflicted person is age 22. The federal definition is to target certain populations for intensive federal support, whereas the purpose of this act's definition is to assure that persons who have a developmental disability and who come into contact with the criminal justice or juvenile justice systems are

[4] *E.g.*, ILLINOIS ANNOTATED STATUTES, ch. 38, *§1005-6.1 (Smith-Hurd Cum. Supp. 1979).*

[5] *E.g.*, DISTRICT OF COLUMBIA CODE ANN. §16-710 (1973).

[6] See sections 8(5) and 10(6).

[7] See section 4.

[8] For a discussion of the intent of and basis for this definition, see section 3(2) of the Developmentally Disabled Juvenile Offenders Act, *infra*.

[9] NATIONAL ASSOCIATION OF PRETRIAL SERVICES AGENCIES (NAPSA), PERFORMANCE STANDARDS AND GOALS FOR PRETRIAL RELEASE AND DIVERSION: DIVERSION, 131 (Approved Draft 1978).

[10] 42 UNITED STATES CODE §6001(7) (Cum. Supp. 1978) (Pub. L. 94-103 as amended by Pub. L. 95-1780).

treated fairly and equitably. Hence, the federal definition has been broadened in order to increase the number of persons accorded the safeguards set forth in this act.

Despite the problems of imprecision and overbreadth inherent in categorical definitions, some states may wish to use a definition which enumerates the causes of the covered disabilities — e.g., autism, cerebral palsy, epilepsy, and mental retardation — in order to facilitate training and the transformation of the requirements of section 4 into operational guidelines. In any case, care should be taken to assure that the definition adopted is consistent with that used in the provisions governing admission to services and establishing the state protection and advocacy agency.

(10) "Developmental disability agency" means the unit of state and/or local government responsible for providing or coordinating services to persons with a developmental disability.

COMMENT. The administrative organization of services for developmentally disabled individuals varies among the states. The critical requirement for the effective operation of this act is that the responsibility for discharging the duties set forth below be clearly placed. Enacting states may wish to delete this term and substitute the name of the responsible department or agency in sections 8, 10 and 12. Where no one agency is responsible for services to persons with a developmental disability, a lead agency should be designated.

(11) "Evaluation" means an assessment by a multidisciplinary evaluation team of a person's mental, emotional, physical, and educational condition, adaptive behavior, and social skills, and of the habilitative services, if any, required by that person.

COMMENT. See comment to section 3 (12).

(12) "Examination" means a diagnosis by a person with expertise regarding developmental disabilities to determine whether an individual has a developmental disability and if so, the effect of the disability if any, on that individual's responsibility for the alleged criminal offense.

COMMENT. This statute makes a distinction between evaluations and examinations. The key difference is in the scope of the assessment. An evaluation is intended to be a comprehensive investigation of the abilities and disabilities of an individual and the types of habilitative services, if any, which will assist that person to function in the community as a self-reliant, law-abiding citizen or to become competent to assist in his or her own defense. Accordingly, evaluations are conducted by a multidisciplinary team of experts.[11] An examination, on the other hand, is a determination by a single expert whether or not an individual has a developmental disability, and whether the alleged offense may be innocent disability-related behavior.

Thus, examination reports do not include a recommendation regarding what, if any, habilitative services may be appropriate. Section 15 sets forth the procedures and requirements, applicable to evaluations and examinations under this act.

(13) "Habilitative services" means medical, mental health, social, vocational, educational, physical therapy, counseling, financial management or other services which will assist a person who has a developmental disability, to the extent and within the shortest time possible, to become competent to stand trial, or to function in the community as a self-reliant, law-abiding citizen.

COMMENT. Persons who have a developmental disability, no less than persons who do not, are unique individuals. Their skills, talents, abilities, disabilities and problems vary. No one type of service or one set of services can help all defendants who because of a developmental disability are incompetent to stand trial, to become competent. Neither can any single service assist all offenders who have a developmental disability to function as responsible members of the community. Hence, the term habilitative services is intended to encompass a broad range of services which, through the procedures set forth in this act, can be tailored to the abilities and needs of an individual defendant or offender. While leaving the types of services which may be offered subject only to the limitations on psychosurgery, electroconvulsive therapy and certain types of aversive conditioning set forth in section 13, the definition brings the objective of the services into sharp focus — i.e., depending on the context, to assist the individual to meet the standards of competency set forth in section 8, or to function in the community as a self-reliant, law-abiding citizen as quickly as possible. This is intended not only to facilitate the preparation of Program Plans, but also to help to reduce the danger of the criminal justice system becoming the primary means for providing services to persons with a developmental disability. Commission of a criminal offense should not have to be the key which unlocks the door to needed services.

(14) "Innocent disability-related behavior" means uncontrollable conduct by a person which is the direct result of a developmental disability, including but not limited to slurred speech, awkward movement, or involuntary movements associated with a seizure.

COMMENT. Behavior that is a result of a developmental disability can and has been mistaken for conduct which is punishable under the criminal or juvenile codes. The most frequently cited exam-

[11] ACCREDITATION COUNCIL FOR SERVICES FOR MENTALLY RETARDED AND OTHER DEVELOPMENTALLY DISABLED PERSONS (ACMRDD), STANDARDS FOR SERVICES FOR DEVELOPMENTALLY DISABLED INDIVIDUALS §1.2 (1977).

ples of such mistakes involve persons who are mentally retarded, or have cerebral palsy, or are in the midst of or still groggy after an epileptic seizure, and who are arrested under the misapprehension that they are under the influence of alcohol or illicit drugs.[12] This definition, together with sections 4, 6, and 16, are intended to heighten the awareness of this problem and provide means for dealing with it.

(15) "Least restrictive alternative" means that habilitative services and the conditions under which they are administered are no more harsh, hazardous or intrusive and involve no more restrictions on a person's physical freedom, social interaction, or individual choice than are essential.

COMMENT. The principle of the least restrictive alternative was explained in the report of the President's Committee on Mental Retardation as follows:

> "When the government. . .(has) a legitimate communal interest to serve by regulating human conduct it should use methods that curtail human freedom to no greater extent than is essential for securing that interest. When you swat a mosquito on a friend's back, you should not use a baseball bat."[13]

In the context of this statute, no condition or restraint should be imposed on a person's liberty following a finding of incompetency pursuant to section 8, or acquittal solely because of diminished capacity pursuant to section 10, unless: the services to be provided fit the defendant's individual needs; those services meet the definition of habilitative services set forth in section 3(13); and there is no less intrusive way in which those services can be delivered safely and effectively. Specific dispositional criteria are set forth in sections 8(5) and 10(6) in order to assist courts in implementing this principle.

(16) "Multidisciplinary evaluation team" means a group of at least three (3) individuals which is appointed by the court from a panel of experts approved by the Evaluation Oversight Board, to evaluate an individual pursuant to this act. Each multidisciplinary evaluation team shall include a physician, a social worker with a graduate degree in social work and field training or experience in working with persons who have a developmental disability, and another expert with knowledge of the particular disability which an individual is alleged or has been found to have.

COMMENT. As is indicated in charts II and VI, most jurisdictions provide for some type of examination after a question has been raised regarding a defendant's competence to stand trial or enter a plea, or his or her lack of criminal responsibility. However, since most of these statutes were drafted with mental illness in mind, they specify or at least suggest that the examination must be conducted by a mental health professional. Because many persons who have a developmental disability suffer from a combination of physical, mental and adaptive impairments, such a unidimensional approach is inadequate. Hence, the model statute provides for a multidisciplinary team of experts to conduct evaluations. A physician is included to examine the physical condition of the defendant offender. The social worker is to focus on the environment, adaptive behavior and social skills of the individual. The additional member or members of the team should have the expertise to evaluate the particular type or types of disability which the individual is alleged or has been found to have. For example, if the person is alleged to be mentally retarded, a clinical psychologist and/or an expert in the education of retarded persons should be placed on the team. Section 14 establishes an Evaluation Oversight Board which will identify individuals qualified to serve on multidisciplinary evaluation teams and monitor the performance of such teams. Section 15 outlines the procedures applicable to evaluations and the form of evaluation reports.

(17) "Offender" means an adult convicted of committing a criminal offense;

COMMENT. See section 3(5) and the comment thereto.

(18) "Pretrial diversion agreement" means a written agreement between a defendant and a designated diversion unit containing the terms and conditions for diversion from the criminal justice process.

COMMENT. See sections 3(8) and 7.

(19) "Program Plan" means a comprehensive individualized plan for the provision of habilitative services which sets forth specific goals, the time and habilitative services required to meet those goals, the manner and location in and frequency with which those services will be provided, and how the cost for those services will be allocated, and which does not contain standardized provisions applicable to general categories of persons without regard to their individual abilities and needs.

COMMENT. The Program Plan is modelled after the Individual Education Plans (IEP) required by the federal Education for All Handicapped Chil-

[12] Norley, *Brick Savvy and Retardation*, in THE MENTALLY RETARDED CITIZENS AND THE CRIMINAL JUSTICE SYSTEM, 125, 128 (M. Santamour, Ed. 1975); Norley, *The Least Restrictive Alternative and the Police Investigatory Process*, in THE PRESIDENT'S COMMITTEE ON MENTAL RETARDATION (PCMR), THE MENTALLY RETARDED CITIZEN AND THE LAW, 525-526 (1976); *See* MISSOURI ASSOCIATION FOR RETARDED CITIZENS, INC., THE MENTALLY RETARDED OFFENDER IN MISSOURI (1976).

[13] Chambers, *The Principle of the Least Restrictive Alternative: The Constitutional Issues*, in PCMR, *supra* note 12, at 486.

dren.[14] It is intended to fulfill several purposes. First, it specifies the objectives which are being sought. Second, it clarifies for the benefit of all parties, the types of services to be offered, whether they will be offered in an institutional or community setting, and in the context of a plan prepared in conjunction with a pretrial diversion agreement, whether the cost of the services will be borne by the public, the defendant, or both. Third, it provides a reference against which the progress of the defendant, or offender and the delivery of services can be compared.[15] In keeping with the definition of habilitative services in section 3(13), this definition encourages use of a goal-oriented rather than a maintenance approach. In addition, it makes clear that the array of services offered is to be tailored to the individual's needs. Use of boilerplate language is expressly prohibited.

(20) "Prosecutor" means the official representing the government in criminal and delinquency proceedings.

COMMENT. This term is intended to include members of the state/county/district/city attorney's office or attorney general's office who appear on behalf of the government in criminal proceedings, or other government officials who perform that function — e.g., some jurisdictions permit law enforcement officers to represent the state in minor or preliminary criminal proceedings.[16]

(21) "Restraint on liberty" means custody of and any bond or other condition of release imposed on a defendant or juvenile, and any conditions or commitment contained in a dispositional order issued in accordance with section 11.

COMMENT. None necessary.

Section 4. Police Investigation and Interrogation

(1) Law enforcement officers shall make reasonable efforts to determine whether a person who is in apparent need of assistance, or who is a victim of or witness to a crime, or who is suspected of committing a criminal or delinquent offense, and is being interrogated or taken into custody, has a developmental disability.

COMMENT. In our society, law enforcement agencies are expected to perform a variety of functions in addition to enforcing the criminal law.[17] For example, police officers are called upon to intervene in domestic disputes; assist accident victims, crime victims and other endangered persons; provide instruction on how to avoid becoming a victim of a crime; and conduct investigations of crimes committed by juveniles and adults. In the course of any of these duties, a law enforcement officer may encounter an individual with a developmental disability and be uncertain about how to deal with him or her appropriately. This is especially true when an officer comes upon an individual in the midst of an epileptic seizure, particularly a psychomotor seizure resulting in bizarre or erratic behavior,[18] or stops a persons whose speech and movements are somewhat impaired by a developmental disability.[19]

In response to these problems, this provision together with section 16 on training is intended to make law enforcement officers more aware of the charactertistics and effects of developmental disabilities, and more sensitive to the needs of persons who have a developmental disability, as well as to assure that the precautions and procedures set forth below are applied.[20] Inasmuch as the principles which apply to investigation and interrogation of juveniles with a developmental disability are identical to those which apply to developmentally disabled adults, and since most law enforcement agencies investigate offenses committed by juveniles as well as those committed by adults, this provision covers both and no comparable section has been included in the Model Developmentally Disabled Juvenile Offender Act.

(2) When there is reason to believe that a person suspected of committing a criminal or delinquent offense has a developmental disability, the investigating officer shall verify prior to taking that person into custody that there is probable cause to believe the person's conduct constitutes a criminal or delinquent offense rather than innocent disability-related behavior.

COMMENT. As noted previously, the effects of some developmental disabilities can be mistaken in certain circumstances for conduct constituting a criminal offense. While these errors usually occur in connection with minor offenses, they are a major source of irritation to persons with a developmental disability and a significant deterrent to their use of community resources, as well as a waste of law enforcement resources. Thus, this provision requires that when a police officer is informed or has other reason to believe that a suspect has a developmental disability, the officer must make a reasonable effort to assure that the individual's conduct was not merely uncontrollable actions directly resulting from the developmental disability.[21] It is intended to prevent, for

[14] 20 UNITED STATES CODE §§1401 *et seq.* (1978).

[15] 20 UNITED STATES CODE §1401 (1978); 42 U.S.C. §6012(b) (Supp. 1979).

[16] ABA, STANDARDS RELATING TO THE PROSECUTION FUNCTION, §2.1 (Approved Draft 1971).

[17] ABA, STANDARDS RELATING TO THE URBAN POLICE FUNCTION, §§1.1(b) and 2.2 (Approved Draft 1973); PRESIDENT'S COMMISSION ON LAW ENFORCEMENT AND THE ADMINISTRATION OF JUSTICE (PLEAJ) TASK FORCE REPORT: THE POLICE, 13-17 (1967).

[18] *E.g.*, a case reported of a man who was jailed after dropping his pants and running from a restaurant due to a psychomotor seizure which caused him to feel that he had to urinate. The Kansas City Times, September 15, 1976 at 2B; The Jefferson City (Mo.) Post-Tribune, September 17, 1976.

[19] See authority cited *supra* note 12.

[20] A number of states already have a similar provision. See EPILEPSY FOUNDATION OF AMERICA, THE LEGAL RIGHTS OF PERSONS WITH EPILEPSY, 3-5 (1976).

[21] See section 3(14).

example, a person who has slurred speech and awkward movement due to cerebral palsy, from being erroneously charged with and taken into custody[22] for drunken driving after being stopped for speeding, when that person informs the law enforcement officer of his or her disability and there is no other evidence of intoxication. Like subsection (1) of this section, this provision covers both juveniles and adults.

(3) If subsequent to issuance of a summons or arrest warrant or the making of a warrantless arrest but prior to the filing of formal charges, the investigating officer concludes that a person suspected of committing a criminal or delinquent offense has a developmental disability and that the person's conduct was innocent disability-related behavior rather than a criminal or delinquent offense, the officer shall terminate the investigation and immediately initiate the process for cancelling the summons or arrest warrant or terminating all restraints on the person's liberty imposed in connection with the arrest.

COMMENT. This provision addresses the situation in which a law enforcement officer, pursuant to subsection (1) of this section or otherwise, discovers before the criminal justice or juvenile process has been fully set in motion but after an individual has been arrested, or after an arrest warrant has been issued but not served, or after a summons has been issued whether or not it has been served, that what appeared to be a crime was instead innocent disability related behavior as defined in section 3(14). An example is the case cited earlier involving a man who was arrested after he partially disrobed in public in the course of a psychometer epileptic seizure.[23] Upon making this discovery, the officer must take whatever steps are necessary to halt initiation of the criminal justice or juvenile justice process, to terminate any bond or other conditions of release imposed pursuant to the erroneous charge and if the individual is in custody, to have him or her released unless there are other grounds for detention. This may include, depending on state procedure and practice, requesting the prosecutor to have the warrant or summons quashed or making that request directly to the issuing magistrate. In addition, the record of the arrest should be expunged to the extent permissible under state law. Again, this provision covers both juvenile and adults since most law enforcement officers are called upon to investigate crimes committed by offenders of all ages.

(4) A person suspected of a criminal or delinquent offense who is in custody shall not be interrogated in the absence of counsel when there is reason to believe that he or she has a developmental disability.

COMMENT. A cardinal principle of American jurisprudence is that persons accused of committing a crime have the right to remain silent.[24] This principle applies on the streets and in the police station as well as in the courtrooms.[25]

> [T]he constitutional function underlying the privilege [to remain silent] is the respect a government — state or federal — must accord to the dignity and integrity of its citizens. . .
>
> [T]o respect the inviolability of the human personality, our accusatory system of criminal justice demands that the government seeking to punish an individual produce the evidence against him by its own independent labors, rather than by the cruel, simple expedient of compelling it from his own mouth. . . . In sum, the privilege is fulfilled only when the person is guaranteed the right "to remain silent unless he chooses to speak in the unfettered exercise of his own will."[26]

Other court decisions have established that a waiver of a consitutional right is not valid unless it was made knowingly and voluntarily.[27] Thus, in Miranda vs. Arizona, the Court held that evidence obtained as a result of interrogation of persons who are in custody, cannot be used against those persons unless prior to the interrogation they were advised of their right to remain silent, their right to the presence of an attorney, and their right to have an attorney appointed if they are indigent, and told that anything they say might be used against them in court.[28]

While the impact of the Miranda warning on the actions of defendants in general remains subject to question,[29] the inadequacy of this procedure for assuring a valid waiver by many persons with developmental disabilities is clear.

> For most retarded offenders, recitation by a law officer of the "Miranda Card" explaining their rights becomes an empty ceremony.
>
> The retarded, who usually want approval and thus seek to oblige, often "confess" to alleged crimes without understanding the consequences. It is rare

[22] The term "taking into custody" is used in place of "arrest" because it is more compatible with the language of the juvenile codes of many states. The terms are used interchangeably in the adult context.

[23] *Supra* note 18.

[24] U.S. CONST. amend. V; Miranda vs. Arizona, 384 U.S. 436 (1966).

[25] Escobedo vs. Illinois, 378 U.S. 478 (1964).

[26] Miranda vs. Arizona, 384 U.S. 436, 460 (1966).

[27] *E.g.*, Johnson vs. Zerbst, 304 U.S. 458 (1938).

[28] 384 U.S. 436, (1966); see also In re Gault, 387 U.S. 1 (1967).

[29] *E.g.*, Medalie, Zeitz and Alexander, *Custodial Interrogation in Our Nation's Capital: The Attempt to Implement Miranda*, 66 MICH. LAW REVIEW 1347, 1375 (1979); Leiken, *Police Interrogation in Colorado: The Implementation of Miranda*, 47 DENVER LAW JOURNAL 1, 15-16, 33 (1970).

when they understand their right to legal representation.[30]

Accordingly, this provision prohibits interrogation of an individual who has or appears to have a developmental disability, unless an attorney representing the person is present. In considering the effect of this prohibition, it should be noted that the standard triggering its application is an objective one. That is, if from the suspect's statements, actions, and identification a reasonable person would conclude that he or she may have a developmental disability as defined in section 3(9) of this act, then the prohibition applies. This standard and definition protects both the suspect — by not requiring proof of the investigating officer's unstated conclusions should the introduction of a statement be challenged in court; and the police and prosecutor — by not having evidence subject to challenge by, for example, a defendant who happens to have epilepsy but who did not have a seizure at the time of arrest or while in custody, who was not carrying an identification card or bracelet, and who did not mention this condition during the interrogation.

On the other hand, this provision would prohibit a person who is not mentally retarded but who because of cerebral palsy has substantial functional limitations in two or more areas of major life activity, from agreeing, in the absence of counsel, to answer a police officer's questions when he or she is in custody. The inconvenience and cost of securing counsel for the relatively few criminal suspects who fit that description and would otherwise forego their right to counsel appears to be far outweighed by the protection provided to both developmentally disabled individuals and the integrity of the criminal justice and juvenile justice processes. Moreover, the use of a flat rule avoids time-consuming court proceedings reviewing determinations by police officers that a defendant or juvenile with mental retardation made a knowing, voluntary waiver. An alternative may be to permit interrogation if a knowing and voluntary waiver has been accepted by a magistrate although in most cases, the cost and delay involved in arranging and holding even an informal hearing before a magistrate would exceed that required by allowing the defendant to retain counsel or if he or she is indigent, to arrange for appointment of an attorney. Thus, the provision seeks to strike a practical balance between the current inadequate protection afforded developmentally disabled persons and an outright ban on interrogation of a person who has any developmental disability, however mild.

Since law enforcement officers generally interrogate both adults and juveniles and the danger of uninformed, involuntary confessions by developmentally disabled youngsters is as great if not greater than it is for adults with a developmentally disability, this provision,[31] like the others in this sections covers juveniles suspected of committing a delinquent offense as well as adult alleged offenders.

(5) A consent to a search by a person who there is reason to believe has a developmental disability shall not be valid unless:

- **(a) The purpose and nature of the search and the right to refuse consent has been explained to that person; and**
- **(b) The person has consulted with:**
  - **(i) A friend, relative, or other trusted advisor, if the person is not in custody; or**
  - **(ii) Counsel if he or she is in custody.**

COMMENT. The Fourth Amendment to the U.S. Constitution protects individuals from unreasonable searches and seizures. While there is a strong legal preference in favor of obtaining a search warrant before the police conduct a search, most searches are performed under one of the exceptions which have been established over time.[32] One of these exceptions is when a person, in effect, waives his rights by consenting to the search. Under the current interpretation of the Constitution, in order for such consent to be valid, it must be shown to have been voluntarily given, and not the result of duress or coercion, express or implied. In determining whether consent was voluntary, courts are to look at the totality of the circumstances including the conduct of the law enforcement officers at the time, whether the individual was in custody, his or her age, schooling, intelligence, education, and sophistication, and whether the individual knew that he or she could refuse to consent.[33]

Consent searches are often problematical for law enforcement personnel.

> The issue of a voluntary waiver is always open to uncertainty and thus litigation. . . . Of all forms of searches, the consent type . . . entails the highest legal risk of being held improper under the Fourth Amendment . . . Often *consent searches* [*sic*] substitute for more

[30] Krause, *The Retarded Citizen and The Criminal Justice System*, in THE MENTALLY RETARDED CITIZENS AND THE CRIMINAL JUSTICE SYSTEM, *supra* note 12, at 23, 24; PRESIDENT'S PANEL ON MENTAL RETARDATION (PPMR), REPORT OF THE TASK FORCE ON LAW, 33 (1963); *see* Argrisani, *Police Assistance in Mental Retardation Cases* in NEW ENGLAND SEMINAR ON RETARDED YOUTH AND THE LAW ENFORCEMENT PROCESS, THE NAIVE OFFENDER: FORMAT AND ESSAYS, 41, 49 (1971); AMERICAN ASSOCIATION ON MENTAL DEFICIENCY (AAMD), CONSENT HANDBOOK (1977); T. GRISSO, COMPETENCE OF JUVENILES TO WAIVE RIGHTS (1979).

[31] GRISSO, *id.*; In re Gault, 387 U.S. 1 (1967).

[32] For a discussion of these exceptions, *see e.g.*, J. LASOTA and G. BROMLEY, MODEL RULES FOR LAW ENFORCEMENT: WARRANTLESS SEARCHES OF PERSONS AND PLACES (1974); Y. KAMISAR, W. LAFAVE, J. ISRAEL, MODERN CRIMINAL PROCEDURES: CASES, COMMENTS, QUESTIONS, 266-372 (4th Ed. 1974).

[33] *Schneckcloth vs. Bustamonte*, 412 U.S. 218, 248 (1973).

thorough preparation and investigation.[34]

They also present problems for persons with a developmental disability, particularly those who are mentally retarded. Because of the strong desire to placate authority figures evidenced by many persons with mental retardation[35] and their lack of sophistication in legal matters, they are particularly vulnerable to consenting to intrusions of their privacy and that of others, which would not be otherwise lawful.

Accordingly, this provision adapts the principles suggested in the American Association on Mental Deficiency handbook on consent, to the search and seizure context.[36] Like section 4(4), it establishes a broad rule covering all developmentally disabled persons, not just those who have mental retardation or autism. The added inconvenience and delay which this may engender, is more than offset by the greater protections afforded to both developmentally disabled individuals and the integrity of the investigatory process, and the greater ease of implementing a policy which applies across-the-board.

Specifically, the subsection provides that if there is reason to believe that an individual has a developmental disability,[37] the officer must explain to the person in the language, mode of communication, and terms which that person is most likely to understand[38] what the search will entail, that he or she is not required to allow the officer(s) to conduct the search, and that anything found may be taken and used in court as evidence. This is similar to the requirements proposed in the American Law Institute Model Code of Pre-Arrangement Procedure.[39] It applies whether it is the person from whom consent is sought or a member of his or her household who is suspected of a crime. The provision also requires that the person talk over the request to search with someone whom he or she trusts to gain added perspective and understanding. If consent is requested while the individual is in custody, consultation must be with counsel because of the added level of coercion and risk involved in custodial situations.[40] The rights to counsel and to appointment of counsel established in Section 13, cover custodial consent search situations involving developmental disability.

(6) Any evidence obtained as a result of an interrogation or search conducted in violation of subsection (4) or (5) of this section shall not be admissible in court against a defendant except during cross-examination to impeach his or her testimony.

COMMENT. This provision embodies the so-called exclusionary rule. It prohibits introduction during the prosecutor's case-in-chief of statements and evidence obtained as a result of a violation of the protections set forth in the two preceeding paragraphs.[41] While questions remain regarding its wisdom and effectiveness, it remains the primary sanction against lawful investigatory techniques.[42]

Section 5. Pretrial Release

Persons alleged to have committed a criminal offense shall not be disqualified from being released on recognizance nor shall the amount, if any, set for bail be increased because they have a developmental disability.

COMMENT. This provision is intended to assure that defendants with a developmental disability are not discriminated against when the conditions for pre-trial release are determined. Persons who have a developmental disability should be no less (and no more) entitled to release or bail than anyone else accused of committing a crime. If for example, there is concern that because a person appears to be mentally retarded he or she may have trouble remembering when and where to appear for future court dates, the answer is to condition release on the agreement of another person or an organization to assist the defendant to appear, rather than setting bail so high that he or she will remain in custody.[43] This becomes all the more important when it is realized that:

> [I]ndividuals accused of violations of law — whether they be defendants in adult criminal proceedings or respondents in proceedings technically denominated delinquency — have, curiously enough, fared far worse in matters of detention and treatment than those already adjudged guilty or "involved." Those convicted of offenses are normally assigned to institutions that are reasonably spacious, where there are ongoing programs of education and recreation, and where, at least on a relative basis, a reasonable number of counselors and other professionals are

[34] LASOTA and BROMLEY, *supra* note 32, at 49.

[35] *See* authorities listed *supra* in note 30.

[36] AAMD, *supra* note 30.

[37] For a discussion of this standard, see the comment to section 4(4).

[38] See section 13(5).

[39] AMERICAN LAW INSTITUTE (ALI) MODEL CODE OF PREARRAIGNMENT PROCEDURE, §240.2 (1975); *accord* LASOTA and BROMLEY, *supra* note 32, at 51 and 53; IJA/ABA, *POLICE, supra* note 22 at 67-68.

[40] ALI, *id.*; IJA/ABA, POLICE, *supra* note 22, at 67.

[41] Harris vs. New York, 401 U.S. 222 (1971).

[42] ABA, POLICE, *supra* note 17, at §5.3; *see, e.g.*, Mapp vs. Ohio 367 U.S. 643 (1961); Miranda vs. Arizona 304 U.S. 436 (1966); Harris vs. New York, 401 U.S. 222 (1971); Wong Sun vs. United States, 371 U.S. 471 (1963).

[43] *See* NATIONAL ADVISORY COMMISSION ON CRIMINAL JUSTICE STANDARDS AND GOALS, CORRECTIONS, §4.4 (1973); ABA, STANDARDS RELATING TO PRETRIAL RELEASE (Approved Draft 1968); NATIONAL ASSOCIATION OF PRETRIAL SERVICE AGENCIES, PERFORMANCE STANDARDS AND GOALS FOR PRETRIAL RELEASE AND DIVERSION: PRETRIAL RELEASE (1978); IJA/ABA, INTERIM STATUS, *supra* note 3.

> present to assist with problems and to help on the road to rehabilitation. On the other hand, those merely accused of wrongdoing are typically detained in structures that are little more than cages, with scarcely any programs, professional help, or space for anything but sleeping, eating, and an occasional walk in a narrow prison yard.[44]

Section 6. Dismissal of Charges

If the prosecutor concludes subsequent to the filing of formal charges but prior to entry of a judgment that a defendant has a developmental disability and that the alleged conduct constitutes innocent disability-related behavior rather than a criminal offense, or that the interests of justice will not be served by proceeding further the prosecutor shall move to dismiss the charges and to have all restraints on the defendant's liberty terminated.

COMMENT. The existence and legitimacy of prosecutorial discretion is now widely accepted.

> There is no obligation or duty upon a district attorney to prosecute all complaints that may be filed with him [or her]. While it is his [or her] duty to prosecute criminals, it is obvious that a great portion of the power of the state has been placed in his [or her] hands . . . to use in the furtherance of justice . . . .[45]

This provision specifically authorizes a prosecutor to investigate whether an apparent offense is, in reality, uncontrollable conduct by a person which is the direct result of a developmental disability, and to seek dismissal of the charges if it is.[46] Furthermore, it suggests that in evaluating a case involving a person who has a developmental disability, the prosecutor consider whether the use of the resources of the criminal justice system and the imposition of criminal sanctions are the appropriate remedy. Especially where the offense is a minor one or where the defendant is already receiving habilitative services or is willing to receive them on a voluntary basis, outright dismissal or at least diversion appear to be more appropriate then prosecution.[47]

While the term "move to dismiss" is used in the model provision, enacting jurisdictions should substitute the terminology and procedures employed in their criminal codes or codes of criminal procedure — e.g., *nolle prosequi.*

Although the provision does not mandate that the motion to dismiss be granted, it is anticipated that the courts will continue to defer to the prosecutors choice except when it is evident that discretion has been abused.[48]

INTRODUCTION TO SECTION 7

Diversion is now recognized as a necessary and proper part of the criminal justice process.[49] While diversion has been seen as an appropriate means of handling alleged offenders who have a developmental disability,[50] all too often, the paucity of available community-based resources has resulted in a choice for between institutionalization through civil commitment and institutionalization through criminal prosecution.[51] In addition, serious questions have been raised regarding the use of coercion to induce a defendant to accept supervision without trial.[52] Accordingly, subsection (1) of this section states that defendants shall not be disqualified from consideration for diversion solely because they may have a developmental disability. Subsection (2) provides for examinations of defendants who there is reason to believe have a developmental disability. Subsection (3) establishes procedural safeguards to assure that the defendant's decision to accept diversion is made knowingly and understandingly. Finally, subsection (4) includes alternative means for supporting the establishment of community-based habilitative services for divertees with a developmental disability.

These provisions are drafted to fit into an existing statute or rule, rather than as a comprehensive diversion authorization statute. In states without a rule or statute governing diversion generally, provisions should be added which, at a minimum, formally authorize pretrial diversion, set eligibility criteria or establish a mechanism for doing so, provide for personnel to staff the program, limit the length of time for which a defendant may be subject to a diversion agreement, and create procedures for termination proceedings in the event the pretrial diversion agreement has been violated.[53]

---

[44] In the Matter of Savoy, Juvenile Case No. J-4808-70, at 30-31 (D.C. Super. Ct. January 11, 1973) as quoted in IJA/ABA, INTERIM STATUS, *supra* note 3, at 3 n.5.

[45] State ex rel. Kurkierewicz vs. Cannon, 42 Wis. 2d 368, 378, 166 N.W. 2d 255 (1969); NATIONAL DISTRICT ATTORNEYS ASSOCIATION (NDAA), NATIONAL PROSECUTION STANDARDS, 126-128 (1977); THE PRESIDENT'S COMMISSION OF LAW ENFORCEMENT AND THE ADMINISTRATION OF JUSTICE (PCLEAJ), THE CHALLENGE OF CRIME IN A FREE SOCIETY, 133 (1967).

[46] See section 3(14) and the comment thereto and sections 4(1)-4(3).

[47] Consider, for example, the case described *supra*, note 28.

[48] *See* NDAA, *supra* note 45, at 126.

[49] *See, e.g.,* (PCLEAJ), *supra* note 45, at 133-134; NATIONAL ADVISORY COMMISSION, *supra* note 43 at 72-97. For a definition of diversion, see the comment to section 3(8).

[50] PCLEAJ, *id.*

[51] *See generally* Leeke, *Mentally Retarded Adult Offenders in the South Carolina Criminal Justice System,* in THE MENTALLY RETARDED CITIZENS AND THE CRIMINAL JUSTICE SYSTEM, *supra* note 12, at 75; NATIONAL INSTITUTE OF MENTAL HEALTH (NIMH), DIVERSION FROM THE CRIMINAL JUSTICE SYSTEM (1973).

[52] *E.g.,* NAPSA, *supra* note 9, at 19.

[53] For a discussion of authorization techniques and examples of existing diversion rules and statutes, see ABA, AUTHORIZING TECHNIQUES FOR PRETRIAL INTERVENTION PROGRAMS: A SURVIVAL KIT (1977).

Section 7. Informal Disposition

(1) Adults alleged to have committed a criminal offense shall not be disqualified from being counseled and released or from being offered an opportunity to enter into a pretrial diversion agreement because they have a developmental disability.

COMMENT. This provision together with subsection (4) of this section is intended to assure that true diversion opportunities are available for developmentally disabled defendants. The opening of existing habilitative services to persons alleged to have committed a crime as well as the establishment of necessary services available to all persons with a developmental disability would benefit both the individual defendant and society as a whole by reducing: the need for correctional agencies to operate a parallel system of services; the delay in delivering a program of services designed to assist the individual to function in the community as a self-reliant law-abiding citizen; and the abuse which persons with a developmental disability often suffer in detention and correctional facilities.[54]

(2) (a) In determining whether to negotiate a pretrial diversion agreement, the [designated diversion unit/prosecutor] may request a defendant to undergo an examination, if there is a reason to believe that the defendant has a developmental disability. Such a request shall not be made unless:

(i) The defendant is represented by counsel;

(ii) The purpose, nature, location, duration, and possible beneficial and harmful consequences of the examination have been explained to the defendant and his or her attorney; and

(iii) The defendant will be able to consult with his or her attorney before deciding whether or not to consent to the examination.

(b) An examination shall be conducted at [state/county/court] expense on the request of a defendant who there is reason to believe has a developmental disability and who is unable to obtain such an examination for financial reasons.

(c) An examination provided under this section shall be performed by an expert selected jointly by the [designated diversion unit/prosecutor] and the defendant from the list published by the Evaluation Oversight Board.

COMMENT. This subsection authorizes examination of a defendant at government expense when a question arises as to whether he or she has a developmental disability.[54] The examination can be provided only on a voluntary basis since diversion is premised on an agreement between the [designated diversion unit/prosecutor] and the defendant and rather than imposed by court order.[55] An examination is called for rather than a full evaluation in order to encourage diversion by minimizing the cost and delay of verifying whether an individual does indeed have a developmental disability. If it is determined that the defendant has a developmental disability and is otherwise eligible for diversion, consent to an evaluation can be included as part of the pretrial diversion agreement.[56]

In keeping with the voluntary nature of the examination, paragraph (2) (a) prohibits attempts to obtain consent to an examination from a defendant who is not represented by an attorney and who has not been able to discuss the question of consent with counsel. The ramifications of an examination on the determination of the defendants' competency and criminal responsibility should the case go to trial, and the importance of the advice of counsel in determining the advisability of diversion make representation critical throughout the negotiation process.[57] To facilitate diversion and the implementation of the protections offered by this subsection, section 13(1) establishes a right to appointed counsel for individuals who there is reason to believe have a developmental disability in all matters arising under this act including diversion.

Paragraph (2) (a) also requires a full explanation of what the examination will entail, the dangers if any involved, and how the results may be used in order to assure that the defendant has the information necessary to provide valid consent.[58] In accordance with section 13(5), the explanation must be in the language, mode of communication, and terms which the defendant is most likely to understand.

Paragraph (2) (b) permits indigent defendant to initiate the examination process. Defendants who are not indigent may of course obtain an examination at their own expense and present the findings to the appropriate official. Examinations sought by the [designated diversion unit/prosecutor] are always at government expense. While theoretically, this provision could be abused by defendants who seek an examination in the hopes of being mistakenly diverted even though they clearly do not have a developmental disability or by diversion officials who fail to request consent for an examination in the hope that the defendant will get one on his or her own, it is not anticipated that either of these abuses will often occur.

Subsection (2) (c) further emphasizes the volun-

[54] See comment to section 12(1)(b).

[55] NAPSA, *supra* note 9, at §1.3.

[56] See subsection (3) of this section regarding the contents of pretrial diversion agreements and section 3(11) for a definition of evaluation. See also section 15-Evaluations and Examinations.

[57] NAPSA, *supra* note 9, at 39-41, 56-57; see also subsection (3) of this section and sections 13(1) and 13(2).

[58] AAMD, *supra* note 30, at 8-10.

tariness of diversion by providing that the expert who is to conduct the examination must be acceptable to both the [designated diversion unit/prosecutor] and the defendant. Pursuant to sections 14 and 15(1), the expert must be selected from the list of qualified individuals published by the Evaluations Oversight Board. The format for and distribution or examination reports are set forth in section 15(4).

(3) (a) In addition to the requirements prescribed by [citation to the relevant statute, court rules, or regulations], no diversion agreement with a defendant who has a developmental disability shall be entered into unless:

(i) It is in writing;

(ii) The defendant is represented by an attorney;

(iii) The agreement has been reviewed by the defendant's attorney; and

(iv) The terms of the agreement, the consequences of signing it, and the consequences of violating it have been explained to the defendant.

(b) A pretrial diversion agreement with a defendant who has a developmental disability shall specify in addition to the items prescribed in [citation to relevant statute, court rules or regulations]:

(i) The agreed-upon goals;

(ii) Whether an evaluation is necessary, and if so, whether the defendant consents to be evaluated;

(iii) The habilitative services which the defendant agrees to accept in order to meet those goals if an evaluation has already been performed; and

(iv) The location and manner in which the evaluation or habilitative services will be provided, and the allocation of the cost of those services.

(c) In order to be enforceable, any modification to a pretrial diversion agreement shall be in writing, reviewed by the defendant's attorney, and explained in the manner specified in subparagraph (3) (a) (iv) of this section.

(d) Evaluations and habilitative services provided through a pretrial diversion agreement shall use the least restrictive alternatives consistent with the defendant's needs. No pretrial diversion agreement shall include a term or condition or shall be based on an informal understanding regarding the placement of the defendant in a residential facility which is secure or not community-based, or in any facility whatsoever to which individuals are committed under (citation to the Model Right to the Admission to Services Act/applicable state civil commitment law(s)). Any such term, condition, or understanding shall be invalid and unenforceable.

COMMENT. This subsection provides a series of safeguards intended to assure that the diversion agreement is entered into knowingly and voluntarily; that the defendant understands his or her obligations under the agreement and any modification thereof; and that a pretrial diversion agreement is not used as a means of circumventing the formal civil commitment procedures.

Paragraph (3) (a) provides procedural protections similar to those required for consent to an examination. Since by agreeing to diversion, a defendant waives the right to a speedy trial and subjects him or herself to at least some degree of governmental supervision, it is essential that the choice to accept a pretrial diversion agreement be made voluntarily and with a full understanding of its possible beneficial and harmful consequences. Besides:

> Legal considerations aside, common sense dictates that a defendant's participation in pretrial diversion be voluntary. Since one of diversion's primary goals is to minimize arrest-provoking behavior on the part of program participants, failure on the part of the defendant to be interested in changing that behavior would obviously hinder progressive change and jeopardize successful completion of the pretrial diversion process.[59]

Thus, in addition to an assessment of his or her chances at trial, the defendant should be informed, in the language, mode of communication and terms which he or she is most likely to understand, about the nature and requirements of the program, its duration, the restrictions if any on liberty and privacy, the confidentiality of statements made during the program, what if any right will be waived by agreeing to diversion, whether the record of arrest will be expunged, whether there may be a stigma attached to participating in the program, what actions may result in a termination proceeding, and what consequences the termination of diversion may have on a subsequent trial.[60] Once this information is revealed, some defendants may prefer unsupervised probation where that is the likely sentence to a more demanding diversion program.

Paragraph (3) (b) requires a pretrial diversion agreement for defendants with a developmental disability to contain the equivalent to a Program Plan.[61] Specifying these items in the agreement makes clear the obligation of all the parties involved. Since an evaluation will ordinarily be needed to assist in the determination of the habilitative services to be offered during the diversion period, the pretrial diversion agreement must

---

[59] NAPSA, *supra* note 9, at 38.

[60] *Id.*, at 57.

[61] For an outline and discussion of the content of Program Plans, see section 3(19) and the comment thereto.

specify whether an evaluation will be required and the place and manner in which it will be performed. Only when there is already a current and complete evaluation of the defendant, should a provision for an evaluation be omitted from the agreement.

Paragraph (3) (c) assures that the defendant's assent to a modification of a pretrial diversion agreement — *e.g.*, to certain habilitative services following an evaluation — is as voluntary and informed as his or her assent to the original agreement. Thus, no oral modification of a pretrial diversion agreement or any written modification which has not been reviewed and explained in the specified manner can serve as a basis for termination of diversion, reinstitution of the criminal prosecution, or any other punitive measure against the defendant.

Paragraph (3) (d) applies the principle of using the least restrictive alternative to diversion,[62] and assures that in signing a pretrial diversion agreement, a defendant is not exchanging an indefinite institutionalization by way of a voluntary civil commitment, for a relatively finite sentence. The humane purposes of diversion should not be used to avoid the due process safeguards applicable when confinement is at issue.[63] Procedures similar to those employed in accepting a guilty plea could be applied to acceptance of a pretrial diversion agreement providing for or based on a voluntary civil commitment.[64] Such an alternative, however, would destroy the attractiveness of diversion as an informal procedure requiring relatively little prosecutorial[65] and court time, while doing little to limit the possible period of confinement. An outright ban on commitment except through the formal civil commitment procedures[66] provides surer protection for both the integrity of the diversion and civil commitment processes and the well-being of the defendant, without infringing upon the protection of public safety.

(4) (Alternative 1)

The state developmental disability agency shall establish community-based programs and/or contract for services throughout the state to assist individuals diverted under this section and section 6 of the Developmentally Disabled Juvenile Offenders Act to function in the community as self-reliant, law-abiding citizens within the shortest time and to the extent possible.

(Alternative 2)

(a) (i) Within one (1) year of the effective date of this act, each county with a population of [50,000] or more shall provide and/or purchase community-based programs and services to assist persons diverted pursuant to this section and section 6 of the Developmentally Disabled Juvenile Offenders Act to function in the community as self-reliant, law-abiding citizens within the shortest time and extent possible.

(ii) Within two (2) years of the effective date of this act, counties with a population of less than [50,000] shall provide and/or purchase community-based programs and services, either singly or pursuant to an intercounty agreement, to assist persons diverted pursuant to this section and section 6 of the Developmentally Disabled Juvenile Offenders Act to function in the community as self-reliant, law-abiding citizens within the shortest time and to the extent possible.

(b) Within six (6) months of the effective date of this act, the head of the state developmental disability agency shall promulgate guidelines governing the scope, content, and staffing of such programs and services and encouraging the use of the least restrictive alternatives.

(c) The state shall reimburse seventy-five percent (75%) of the cost of programs and services provided or purchased under paragraph (4) (a) of this section which comply with the guidelines issued under paragraph (4) (b).

COMMENT. The proposed diversion procedures will be useless unless there are community-based services available to assist persons with a developmental disability who have been accused of a crime. Even where community-based services exist, they are often closed to persons with the taint of the criminal justice system upon them, leaving a choice between doing nothing and institutionalization.[67] This subsection contains two alternative methods for establishing a statewide network of community-based services. One calls for the state agency generally responsible for providing services to persons with a developmental disability to establish the programs directly. The other provides for reimbursement by the state of a percentage of the cost of locally established programs. Both permit the purchase of services from public agencies and impose no restrictions on the population served, so long as it includes divertees. Whether either of these alternatives or another model is employed depends on the characteristics, governmental structure, and traditions of the enacting jurisdiction. For purposes of this act, the form adopted is unimportant, so long as it is effective in providing the necessary alternatives.

---

[62] NAPSA, *supra* note 9, at 75; see the definition of least restrictive alternative in §3(15).

[63] See NIMH, *supra* note 51.

[64] (PCLEAJ), TASK FORCE REPORT: COURTS, 9 (1967).

[65] *See e.g.*, Lottman, *Jumping on the Bandwagon*, in THE MENTALLY RETARDED CITIZENS AND THE CRIMINAL JUSTICE SYSTEM, *supra* note 12, at 257, 258.

[66] *Cf.* Chapter 6: Right to Services, *supra*.

[67] See M. PRESCOTT AND E. VAN HOUTEN, A STUDY OF THE CHARACTERISTICS AND TREATMENT OF MENTALLY RETARDED ADOLESCENT OFFENDERS IN NEW JERSEY (1979).

INTRODUCTION TO SECTION 8

Under both the common law and the United States Constitution:

> It is "fundamental to an adversary system of justice that an individual whose "mental condition is such that he lacks the capacity to understand the nature and object of the proceedings against him, to consult with counsel, and to assist in preparing his defense" may not be subjected to a trial. The conviction of an accused who is mentally incompetent violates due process.[68]

In addition to protecting individual defendants, the prohibition against trying persons who are incompetent also helps to protect the integrity of the criminal justice system.[69] However, a finding that an individual is incompetent to stand trial may result in his or her confinement for a lengthy period of time and impairment of both the defendant's and prosecution's ability to present an effective case.[70] Hence, any statute addressing this topic must provide sufficient latitude to assure that individuals who are incompetent are promptly identified and properly assisted through use of the least restrictive alternative, while providing sufficient safeguards to prevent competent defendants from being denied their right to a speedy trial, having their liberty unnecessarily restrained, and their privacy violated.

This section seeks to provide the necessary balance. It sets forth the procedures and standards applicable to, and the dispositions available following an inquiry into the competency of a defendant to stand trial. The provision covers only those defendants who have a developmental disability. Nothing in the section, however, would prevent states wishing to have a single provision governing competency to stand trial from applying it to persons alleged or found to be mentally ill. Subsection (1) establishes a presumption of competency and the standard for finding a defendant incompetent to stand trial. Subsections (2) and (4) set forth the procedures for initiating an inquiry into the competence of a defendant to stand trial, and subsection (3) assures that defendants are not retained in custody solely because of such an inquiry. The key provision within this section is subsection (5) which outlines the range or possible findings and the dispositional choices applicable to each. Subsection (6) governs the development of a Program Plan following a finding that a defendant is incompetent to stand trial and that there is a substantial likelihood that he or she will become competent within a year of the competency hearing. Subsections (7) and (8) establish reporting and review procedures. Subsection (9) permits legal challenges to the charges despite a finding that a defendant is incompetent to stand trial, and subsection (10) provides for the termination of a dispositional order issued under this section when the underlying charges are dismissed. Finally, subsection (11) requires that the time spent in custody because of an allegation or finding of incompetency be credited against the sentence of a defendant who is subsequently found to be competent and convicted, and subsection (12) provides a means for applying the procedures and protections established by this act to persons with a developmental disability who are found to be incompetent to stand trial prior to its enactment and who as a result of that finding are in custody.

Section 8. Incompetence to Stand Trial

(1) Defendants who have a developmental disability shall be presumed to be competent to stand trial, whether or not they require, are receiving or have received psychotropic drugs, other medication, or treatment related to the disability. A defendant who has a developmental disability shall be found to be incompetent to stand trial only if it is proven by a preponderance of the evidence that he or she [**meets the standard set forth in (citation to statutory section defining incompetence to stand trial)/lacks the present ability to understand the nature and purpose of the proceedings and assist in preparing his or her own defense.**]

COMMENT. This provision, together with subsection (5) of this section, makes clear that a defendant who has a developmental disability should be considered competent to stand trial unless proven otherwise. A developmental disability should not be automatically equated with incompetence.[71] Neither should a defendant be considered incompetent solely because he or she is being treated with psychotropic drugs or is receiving another form of treatment.

> In some jurisdictions, the defendant's need of tranquilizing or sedating drugs is conclusive evidence of trial competency. . . .[This] reaction . . . is based on the false assumption that psychotropic

[68] United States vs. Masthers, 539 F. 2d 721, 725, (1976); Drope v. Missouri, 420 U.S. 162 (1975); Note, *Competence to Plead Guilty: A New Standard;* 1974 DUKE LAW JOURNAL 149, 151-152 (1974); *but see* Burt and Morris, *A Proposal for the Abolition of the Incompetency Plea* 40 UNIVERSITY OF CHICAGO LAW REVIEW 66 (1972).

[69] An adversary trial presupposes an accused who is capable of defending himself against the accusations of the state. But the trial of a person, totally out of contact with reality, would be arbitrary and irrational, and would tend to undermine public confidence in the entire system. Bennett, *Competency to Stand Trial: A Call for Reform, 59 JOURNAL OF CRIMINAL LAW, CRIMINOLOGY AND POLICE SCIENCE 569, 570 (1968); see Comment Incompetency to Stand Trial,* 81 HARVARD LAW REVIEW 454, 458 (1967).

[70] Bennett, *id.*, at 570-573.

[71] A. ROBEY, CRITERIA FOR COMPETENCY TO STAND TRIAL: A CHECKLIST FOR PSYCHIATRISTS (1965); Brakel, *Presumption, Bias, and Incompetency in the Criminal Process,* 1974 WISCONSIN LAW REVIEW 1105, 1129-1130 (1974); see also Bennett, *supra* note 69, at 574; Ennis, *CLU Client Crazy but Competent,* 17 CIVIL LIBERTIES IN NEW YORK 2 (1969).

drugs produce a kind of "chemical sanity" — a pejorative term used mistakenly to characterize a state of mind unacceptable for participation in a trial. Yet many persons in responsible positions throughout society function effectively despite their continued need for similar medication, and defendants may also.[72]

The subsection proposes no major changes in the legal standard for competence to stand trial. Hence, the existing state standard should be inserted to avoid confusion over minor differences in wording. For states in which the criminal code does not include a specific standard, the traditional ability-to-assist counsel-and-understand-the-proceedings standard is included in the form of an alternative. The provision thus recognizes that competence to stand trial is a legal, not a medical or psychological determination. Hence, in order for an individual to be found incompetent to stand trial, it is necessary to prove that he or she is not aware of the court surroundings, the principal participants in the proceedings, the nature of the charges and the possible consequences of conviction, and unable to communicate with his or her attorney regarding the facts of the case, the names of witnesses, and the evidence presented.[73] In addition, some states may wish to include for both defendants alleged to have a developmental disability and those alleged to be mentally ill, "the inability to make a reasoned choice among the alternatives presented."[74] Not only is this volitional factor critical in determining the validity of a plea of guilty,[75] but also in many trial decisions as well.[76]

(2) The prosecutor or a defendant who there is reason to believe has a developmental disability may move for an evaluation to determine whether the defendant is incompetent to stand trial. In cases in which an indictment or information is required in order to proceed to trial, the motion shall be made following the filing of the indictment or information but prior to sentencing. In cases in which an indictment or information is not required, the motion shall be made following the filing of a complaint but prior to sentencing.

COMMENT. This provision governs who may initiate an inquiry into the defendant's competency and when the issue may be raised. It allows both the prosecutor and the defendant (presumably through counsel) to request an evaluation. Subsection (4) permits the court to hold a hearing on its own initiative in order to determine whether an evaluation should be conducted.[77] While questions have been raised regarding the misuse of incompetency procedures by prosecutors as a means for disposing of weak cases and keeping troublesome defendants off the street,[78] the state has an interest in preserving the integrity of the criminal process and in avoiding post-conviction challenges. In addition, the prosecutor may, at times, have information regarding the defendant's competence which is "unnoticed by or unavailable to the defense."[79] The rights and procedures provided in this section and sections 13 and 15 of this act are designed, *inter alia,* to minimize the chances for abuse.

One of those safeguards is the requirement that there must, at least in felony cases, be a determination that there is probable cause to believe that a crime has been committed and that the defendant committed it, before the issue of incompetency can be formally raised. Since defendant may be hospitalized for up to a year under subsections (5) and (8),[80] fundamental fairness demands that there must be some assurance that a trial will be held before a defendant's liberty is so restricted.[81] Since a probable cause of determination is often not required for misdemeanors and other minor offenses — offenses which come under the purview of this act purusant to section 3(5) — the provision permits the issue of incompetency to be raised following the filing of a complaint if an indictment or information is not needed to proceed to trial. However, two factors help to offset the reduced protection afforded in such cases. The first is the greater likelihood of diversion when nonserious offenses are involved. The second is the requirement in section 8(5)(d)(ii) that the case be dismissed when a defendant who has a developmental disability is found to be incompetent to stand trial and the maximum period of incarceration which could be imposed upon conviction of the most serious offense with which that defendant is charged is one year or less.

(3) Except as otherwise provided by this section, a defendant who there is reason to believe has a developmental disability and who is awaiting determination of his or her competence to stand trial shall be entitled to release on bail or recognizance to the same extent and on the

[72] Mental Health Law Project (MHLP), *Issues in State Mental Health Care: Proposals for Change – Incompetence to Stand Trial on Criminal Charges,* 2 MENTAL DISABILITY LAW REPORTER 617, 626-627 (1978).

[73] *See* Bennett, *supra* note 69, at 573-577; ROBEY *supra* note 71, at 619; Wieter vs. Settle, 193 F. Supp. 318, 321-322 (W.D. Mo. 1961); Brakel, *supra* note 71.

[74] Schoeller vs. Dunbar, 423 F. 2d 1183, 1194 (9th Cir., 1970), (Hufstedler J. dissenting) *cert. denied* 400 U.S. 834 (1970); Seiling vs. Eyman, 478 F. 2d 211, 215 (9th Cir. 1973).

[75] Seiling vs. Eyman, *id.*; United States vs. Masthers 539 F. 2d 721, 726 (D.C. Cir. 1976).

[76] Note, *supra* note 68, at 168.

[77] Pate vs. Robinson 383 U.S. 375 (1966) held that trial judges have a duty to raise the question themselves when it appears the defendant is not competent.

[78] MHLP, *supra* note 72, at 620; Ennis, *supra* note 71 at 3; A. STONE, MENTAL HEALTH AND LAW: A SYSTEM IN TRANSITION, 207 (1975).

[79] MHLP, *id.*

[80] Section 8(5)(e).

[81] Ennis, *supra* note 71, at 2.

same terms and conditions as if the issue of incompetency had not been raised.

COMMENT. The Eighth Amendment to the Constitution provides that "excessive bail shall not be set." In discussing this provision, the Supreme Court has stated that:

> The right to release before trial is conditioned upon the accused's giving adequate assurance that he [or she] will stand trial and submit to sentence if found guilty . . . . Bail set at a figure higher than an amount reasonably calculated to fulfill this purpose is 'excessive' under the Eighth Amendment.[82]

Despite the limited grounds upon which a defendant may be legitimately detained prior to trial, the allegation or a determination of incompetency has often disqualified defendants for release on bail or on their own recognizance.[83]

This provision, together with the criteria set forth in paragraph (5)(e) of this section seeks to assure that the liberty of defendants who have a developmental disability and are alleged or found to be incompetent to stand trial, is not unnecessarily restricted.[84] It provides no special privilege for such defendants, but states simply that their release or detention should be determined on the same basis as defendants whose competency has not been placed in question.[85]

(4) (a) Following the filing of a motion pursuant to subsection (2) of this section, or on its own initiative, the court shall hold a hearing to determine whether there is a reasonable doubt that the defendant is competent to stand trial. The burdens of going forward and of persuasion shall be on the moving party except that when the hearing is initiated by the court, both burdens shall be on the prosecutor.

(b) If the court determines that a reasonable doubt has not been raised regarding the defendant's competence to stand trial, the case shall proceed as provided by law.

(c) If the court determines that a reasonable doubt has been raised, it shall:

(i) Dismiss the charges and remove any restraints on the defendant's liberty imposed as a result of those charges if neither the interests of justice nor those of the defendant will be served by proceeding further; or

(ii) Order the defendant to undergo an evaluation and appoint a multidisciplinary evaluation team to evaluate the defendant.

COMMENT. The mere assertion that a defendant is incompetent to stand trial is not sufficient to trigger the constitutionally mandated inquiry into his or her competency. In order to prevent unnecessary delay in the proceedings and unwarranted intrusion into the defendant's privacy, there must be at least reasonable cause to believe that the defendant may not be competent before an evaluation is ordered and a competency hearing held.[86] Hence, this provision requires the party initiating the inquiry to introduce testimony and evidence which can support a finding that there is a reasonable doubt regarding the defendant's competence to stand trial. The opposing party may, of course, seek to demonstrate the defendant's competence through cross-examination and through the introduction of evidence and testimony. When the issue is raised by the court itself rather than by either of the parties, the prosecutor as the representative of the state, has the responsibility for sustaining the burden of proof. As indicated above, proof that the defendant has a developmentally disability is not sufficient. It must be demonstrated that because of that disability, there is a reasonable doubt that the defendant can meet the standard set forth in subsection (1) of this section.

In view of the fact that most defendants whose competency is called into question are found to be competent, a number of commentators have endorsed the use of quick screening tests to sift out those whose are clearly competent.[87] The use of an appropriate and reliable screening device attuned to persons who may have a developmental disability is not prohibited by this subsection. If such a test is developed, however, care must be taken to assure that it is administered properly, reflects the legal standard of competency to stand trial rather than a clinical standard of disability,[88] is free from racial or cultural bias, is employed as a screening mechanism and not as a substitiute for an evaluation, and finally, that, no matter how reliable, it does not take on the aura of infallibility.

Upon determining that a reasonable doubt of the defendant's competency exists, the court has two options. It may order an evaluation of the defendant and appoint a multidisplinary evaluation team,[89], or, if the alleged offense is so minor or the defendant's impairments so substantial that the intrusion, delay and expense of the competency inquiry is not justified, it may dismiss the criminal proceedings. While special advantage should not accrue to defendants who happen to have a developmental disability, neither should criminal

[82] Stack vs. Boyle, 342 U.S. 1, 4-5 (1951).

[83] Burt and Morris, *supra* note 68, at 88; STONE, *supra* note 78, at 208-209; MHLP, *supra* note 72, at 620-621.

[84] Also section 5 of this act.

[85] MHLP, *supra* note 72, at 621 and 637; STONE *supra* note 78, at 209.

[86] Pate vs. Robinson, 383 U.S. 375 (1966); Drope vs. Missouri 420 U.S. 162 (1975), *e.g.*, Massie vs. Blankenship, 469 F. Supp. 686 (E.D. Va. 1979).

[87] MHLP, *supra* note 72, at 621; ROBEY, *supra* note 71; *but see* Brakel, *supra* note 71.

[88] Brakel, *supra* note 71.

[89] For a definition of multidisciplinary evaluation team, see section 3(16); see also sections 3(11) and 15.

proceedings be continued simply as a means of providing needed services.[90]

If an evaluation is ordered, the report of the multidisciplinary evaluation team must be submitted within 10 days. The report must contain among other information, the opinion of the multidisciplinary disability, and if so, on the extent to which he or she meets the standard set forth in subsection (1). The report must also include the names and addresses of all the individuals who examined the defendant, a description of the techniques or tests used during the evaluation, the signatures of the members of the team and an indication of whether each member concurs or disagrees with the conclusions in the report. If the opinion of the team is that the defendant may not meet the requisite standard, the report must contain, in addition, an opinion regarding the likelihood that the defendant will be able to understand the proceedings and assist counsel within one year if appropriate habilitative services are provided, and a description of the type(s) of habilitative services required.[91]

(5) (a) The court shall promptly convene a competency hearing following submission of the evaluation report. At the hearing, the burdens of going forward and of persuasion shall be on the party who initiated the inquiry into the defendant's competency to stand trial except that when the inquiry was initiated by the court, both burdens shall be on the prosecutor.

COMMENT. While the Supreme Court has not prescribed the exact procedures which must be followed in determining the competency of a defendant to stand trial, a court hearing is clearly required to protect the interests of both the defendant and the prosecution.[92] In order to minimize delay, the hearing should be held as soon as possible after the report and prepare their positions. In few cases should more than 10 days elapse between the submission of the evaluation report and the hearing. The allocation of the burdens of going forward and of persuasion remain the same as in the initial hearing held pursuant to subsection (4). Paragraphs (5)(b)-(e) set forth the possible findings and dispositions which may result from the hearing.

(b) If on the basis of the testimony and evidence presented at the competency hearing, the judge or the jury, if any, find(s) that there is a preponderance of the evidence that the defendant is found competent to stand trial and is receiving psychotropic drugs or other medication or services, the court may order the continued administration of such drugs, medication or services under appropriate supervision in order to maintain the defendant's competence, unless it is demonstrated that the continuation of such drugs, medication or treatment would pose a serious risk to the health of the defendant. If such a risk is shown, the court may order the discontinuance of the drug, medication or treatment and reassess the defendant's competence to stand trial.

COMMENT. This provision sets the applicable standard of proof at a preponderance of the evidence. The purpose of a standard of proof is to "instruct the factfinder concerning the degree of confidence our society thinks he [or she] should have in the correctness of factual conclusions".[93] Inasmuch as the question of competency is usually raised by the defense and is intended to protect the right to a fair trial, and in light of the review procedures, time limits and other protections against abuse provided by this section, requiring proof beyond a preponderance of the evidence appears unnecessary.[94]

In addition, this paragraph delineates the procedures and considerations when a defendant is receiving psychotropic drugs, or other medication or services. As noted in connection with subsection (1), the fact that a defendant is being treated with psychotropic drugs does not require that he or she must be considered incompetent.

> Treating defendants in criminal actions with drugs is no different from treating anyone else; only the special circumstances are different. . .By applying some general rules in regard to the use of the various types of drugs, with an awareness of their special actions, one should be able to make a valid, common-sense judgment of court competence.[95]

Thus, this provision would permit the trial to proceed even if the defendant requires psychotropic medication to remain competent, unless the continued administration of those drugs threatens the defendant's physical health. However, it is not intended to imply that psychotropic drugs should be administered to every developmentally disabled defendant. The court must remain alert for instances in which use of tranquilizing or other medications impair a defendant's competence or affect the fairness of the trial.[96] Moreover, section 13 requires that when the defendant is being given

[90] Chapter 6: Right to Services, *supra*.

[91] See sections 15(4)(a) and (4)(c).

[92] Pate vs. Robinson 383 U.S. 375 (1966).

[93] In re Winship, 397 U.S. 358, 370 (1970).

[94] MHLP, *supra* note 72, at 624; GOVERNOR'S COMMISSION FOR REVISION OF THE MENTAL HEALTH CODE OF ILLINOIS, REPORT 177 (1976); *e.g.*, State vs. Buie 254 S.E. 2d 26 (N.C. Sup. Ct. 1979).

[95] Hollister, *Psychotropic Drugs and Court Competence*, LAW, PSYCHIATRY, AND THE MENTALLY DISORDERED OFFENDER (I. Irvine and T. Brelje, eds.) as quoted in MHLP, *supra* note 72, at 626; Stone, *supra* note 78, at 213-215.

[96] *E.g.*, State vs. Murphy, 56 Wash. 2d 761, 355 P. 2d 323 (1960); Burt and Morris, *supra* note 68, at 85-86.

psychotropic medication, he or she is entitled to have the judge and the jury, if any, informed of the purpose of the medication and the effect which it may have on his or her actions, demeanor and participation in the hearing in order to minimze the possible prejudice which may result from some types of drug-induced demeanor.[97]

(c) If on the basis of the testimony and evidence presented at the competency hearing, the judge or the jury, if any, find(s) there is a preponderance of the evidence that the defendant is incompetent and that there is not a substantial probability that he or she will become competent to stand trial within one (1) year if provided with appropriate habilitative services, the court shall dismiss the charges against the defendant and either:

(i) Terminate any restraints on the defendant's liberty imposed because of the charges; or

(ii) If the prosecutor indicates the intent to initiate proceedings pursuant to [citation to the Model Right to and Admission to Services Act/ state civil commitment law(s)] and if the defendant is in custody and there is probable cause to believe that the defendant meets the criteria for placement in a residential facility as defined in that [Act/law(s)], the court may exonerate the defendant's bond, if any, and order that he or she be held for up to forty-eight (48) hours in the least restrictive facility consistent with the defendant's needs, pending initiation of the civil proceeding.

COMMENT. In Jackson vs. Indiana, the Supreme Court held that:

> . . . .[A] person charged by a State with a criminal offense who is committed solely on account of his incapacity to proceed to trial cannot be held more than the reasonable period of time necessary to determine whether there is a substantial probability that he will attain that capacity in the foreseeable future. If it is determined that this is not the case, then the State must either institute the customary civil commitment proceeding that would be required to commit indefinitely any other citizen, or release the defendant.[98]

Accordingly, this provision requires the court to dismiss the charges against a defendant who is incompetent to stand trial and who is unlikely to become competent within one year. Upon dismissing the charges the court must either have the defendant released or order that he or she be retained in custody for up to two days pending initiation of a proceeding to provide services on an involuntary basis. There is little consensus on what constitutes "a reasonable period of time," although the Court indicated that three and a half years was probably more than enough.[99] Some commentators have suggested 6 months as an outside limit, at least for mentally ill defendants.[100] The one year maximum proposed in this act appears to be a reasonable compromise. Empirical studies of competency procedures, however, should be continued to determine whether, in fact, a shorter period more accurately reflects the time required for most defendants who are incompetent to stand trial because of a developmental disability to become competent, or for determining that there is not substantial probability that such defendants will become competent to stand trial within a reasonable time.

The requirement that the charges be dismissed is a key part of this provision, especially for defendants subsequently placed in a residential facility pursuant to the state procedures for civil commitment. As Burt and Morris point out:

> Hospital authorities are exceedingly reluctant to discharge, or even to authorize brief community leaves for, mental patients confined while criminal charges are outstanding. In many states, their release is specifically forbidden and, in practice, hospital authorities must take the initiative of negotiating with prosecuting officials to dismiss charges. Even where release is not formally prescribed, the practical obstacles are formidable; only a zealous state hospital administrator will have much success in freeing a patient confined with criminal charges outstanding.[101]

A number of states have sought to address this problem by enacting special commitment procedures within the context of their criminal codes. Some include durational limits based on a specific time period or the sentence which could have been imposed.[102] While these approaches may address the problem, basing the length of a commitment on the possible sentence flies in the face of the presumption of innocence. As a way of avoiding this shortcoming, some commentators have suggested adjudicating the criminal charges against such defendants, either through a criminal trial to which extra procedural protections have been added to compensate for the defendant's disability,[103] or through a quasi-criminal commitment process.[104] However, both the adequacy of

---

[97] See section 13(3); MHLP, *supra* note 72, at 626.

[98] Jackson vs. Indiana, 406 U.S. 715, 738 (1971).

[99] *Id.* at 738-739; see chart III *supra*.

[100] Burt and Morris, *supra* note 68, at 92; MHLP, *supra* note 72, at 62; Stone, *supra* note 78, at 212.

[101] Burt and Morris, *id.*, at 77-78; Bennett, *supra* note 69, at 571.

[102] See Chart III, *supra*.

[103] Burt and Morris, *supra* note 68, at 93-95; MHLP *supra* note 72, at 629-631.

[104] MHLP, *id.* at 631-632.

the proposed extra procedural protections and the fairness of any adjudication of guilt or innocence based on a standard of beyond a reasonable doubt are highly questionable. Thus, the civil process for providing services on an involuntary basis remains one of the dispositional options when a defendant who has a developmental disability is found incompetent to stand trial and unlikely to become competent within one year if appropriate habilitative services are provided.

It should be clearly understood, however, that this proposal does not contemplate nor intend that long-term hospitalization should be the automatic outcome for defendants unlikely to become competent to stand trial. A criminal charge in and of itself should not be proof that an individual meets the traditional "dangerous to self or others" standard contained in many civil commitment statutes. Neither should such a charge be presumed to be conclusive evidence that community-based habilitative services are inadequate both to assist the individual and protect the public.[105] Enacting jurisdictions are strongly urged to consider simultaneously the Model Right to Services/Admission to Services Act contained in this series which presents a practical approach to remedying the documented problems and abuses of civil commitment.[106]

(d) If on the basis of the testimony and evidence presented at the competency hearing, the judge or the jury, if any, find that there is a preponderance of the evidence that the defendant is incompetent but that there is a substantial probability that the defendant will become competent to stand trial or enter a plea within one (1) year of the competency hearing if appropriate habilitative services are provided, the court shall either:

(i) Dismiss the charges and terminate any restraints on the defendant's liberty imposed because of the charges if the most serious offense with which the defendant is charged carries a maximum sentence of incarceration of less than one (1) year, or if it concludes that neither the interests of justice nor those of the defendant will be served by continuing the proceeding; or

(ii) Order a continuance in the proceedings of up to six (6) months, the development of a Program Plan in the manner set forth in subsection (6) of this section, and the provision and/or purchase of the necessary habilitative services by the state developmental disability agency using the least restrictive alternatives.

(e) An order issued pursuant to subparagraph (5)(d)(ii) of this section shall specify whether the Program Plan may include placement of the defendant in a residential facility which is secure or not community-based, or to which nonoffenders are committed pursuant to (citation to Model Right to and Admission to Services Act/state civil commitment law(s)). No such placement shall be permitted unless the judge or jury find that the prosecutor has presented clear and convincing proof that it is the least restrictive alternative which will:

(i) Assure the defendant's presence at future hearings;

(ii) Provide the defendant with the habilitative services which are most likely to assist him or her to become competent to stand trial; and

(iii) Protect the defendant from suffering serious bodily harm.

COMMENT. Subparagraphs (5)(d)(i) and (ii) set forth the procedures and dispositional options available when a defendant with a developmental disability is found to be incompetent to stand trial but to have a substantial possibility of becoming competent within one year if the appropriate habilitative services are provided.[107] As under paragraphs 8(5)(b) and (c), the standard of proof is "by a preponderance of the evidence."

Again, two dispositional options are presented. First, the court may dismiss the charges and order the defendant to be released, any bond exonerated and any conditions of release terminated. This would be the appropriate course of action when the defendant is charged with a misdemeanor or other minor offense for which the delay, public expense and intrusion that would result from the effort to assist the defendant to become competent is simply not warranted.

If a more serious charge is involved, the court may instead order that the state developmental disability agency[108] develop a Program Plan with the defendant and his or her attorney, and provide habilitative services for the defendant. The particular medical, mental health, social, educational, physical therapy, counseling, or other services required to assist the defendant to become competent to stand trial,[109] are to be specified in the order of the court, on the basis of the evaluation report and the evidence and testimony adduced at the hearing. The mannner in which the services are to be provided is to be spelled out in the Program Plan.[110] The agency may either provide those services directly or purchase them from other public agencies or private organizations. The court

[105] *N.B.* Once it has been determined that a defendant is unable to become competent to stand trial within a reasonable time, the purpose of providing services is to assist that person to be able to live safely in the community.

[106] See also section 8(5)(e).

[107] For a discussion of the 1 year time limitation, see the comment to section 8(5)(c).

[108] See definition in section 3(10).

[109] See definition of habilitative services in section 3(13) *supra* and the comment thereto.

[110] See definition of program plan in section 3(19); see also section 8(6).

also may delay the proceedings for up to 6 months. Prior to the end of this initial continuance a review hearing must be held pursuant to subsection (8) of this section. In this way inadequate or inappropriate services can be discovered before the expiration of the one year maximum period of delay, and unnecessary delay and restrictions can be avoided.[111] Twenty-two states currently provide for some form of judicial review following a finding that a defendant is incompetent to stand trial.[112]

Paragraph (5)(e) sets forth special criteria governing the circumstances in which defendants may be placed in secure, closed or remote facilities, or in facilities in which individuals who have been civilly committed are placed. Such criteria are necessary because of the presumption in many jurisdictions that institutionalization of defendants who are incompetent to stand trial is required.[113]

> The real tragedy of the incompetency process is that is has, in many places, lost its distinct purpose of protecting defendants and has become merely another element in the array of techniques used by the State to effect the same result: involuntary confinement of worrisome individuals in grossly substandard facilities . . .
>
> Consistent with this ruling [Jackson vs. Indiana], new less ambiguous legislation is required based on these premises: *that institutional commitment of the incompetent accused should be a last resort, and that competency should always be evaluated with regard to the minimal real needs of the given trial.*[114] [Emphasis in the original.]

Accordingly, before a defendant who is incompetent to stand trial may be placed in a facility meeting the description outlined above,[115] the prosecutor must provide clear and convincing proof[116] that an alternative which is less intrusive or imposes fewer restraints on the defendant's liberty[117] would not provide adequate protection against the defendant fleeing the jurisdiction and against serious impairment of the defendant's physical health and safety, and would not provide those particular habilitative services which hold the greatest chance of assisting the defendant to become competent to stand trial within on year. Assuring that an incompetent defendant will be present for future hearings corresponds to the basis for detaining any criminal defendant prior to trial.[118] Protecting the defendant from suffering serious physical harm is the primary purpose of civil commitment authority, and protects the trial process as well. Provision of the requisite habilitative services is the duty assumed by the state when trial is delayed and a defendant is required to accept services. At the least, due process requires that the nature and duration of commitment bear some reasonable relation to the purpose for which the individual is committed.[119] Absent from the list of criteria is that the defendant presents a substantial risk of serious bodily harm to others. While this may be a constitutionally permissible basis for commitment,[120] it is an unreliable one in the pretrial context.

> One might reasonably assume that the medical and psychiatric experts who typically make predictions about "dangerousness" must have some demonstrated skills in this regard. . . . The studies which are available indicate quite clearly and consistently that psychiatrists and other clinicians uniformly over-predict dangerousness, i.e., the great majority of the persons predicted to be dangerous turn out to be "false positives" (persons predicted as likely to be dangerous but who is fact will not display such behavior.)
>
> Ironically, it is a statistical truism and most predictable that predictions about events which have very low rates of occurrence in a given population (i.e., low baserates) — e.g., violent and other dangerous behavior by various groups — will tend to have a very high rate of "false positive" errors. This issue has been discussed in the literature for at least the past twenty years. Yet, it would seem accurate to say that a large majority of the physicians and psychiartists who testify on issues of dangerousness very probably are not familiar with this literature nor with its obvious implications. . . .
>
> The clear policy implication of these research findings and the high rate of "false positives" is this: for every person who may accurately be detained on a preventive basis as dangerous (viz., a true positive), we would have to also confine anywhere from five to even 99

[111] See MHLP, *supra* note 72, at 625; *accord* Burt and Morris, *supra* note 68, at 94; see also sections 8(7)(a) and (b).

[112] Chart IV *supra*.

[113] Chart III *supra*.

[114] Stone, *supra* note 78, at 205-206; Bennett, *supra* note 69, at 570-571.

[115] See section 3(1) for a definition of community-based facility.

[116] Addington vs. Texas, 441 U.S. 418 (1979).

[117] See section 3(15) for the definition of least restrictive alternative.

[118] See the comment to section 5 on pretrial release.

[119] Jackson vs. Indiana, 406 U.S. 715, 738 (1971).

[120] *Id.*, at 737 (*dicta*); Addington vs. Texas, 441 U.S. 418 (1979) (*dicta*).

or more non-dangerous (also positive) individuals. . . .[121]

States which choose to adopt preventive detention as a basis for placement under this paragraph, should do so in a manner which will require clear and convincing proof both that the defendant presents a substantial risk of serious bodily harm, and that no less restrictive alternative can adequately prevent the infliction of that harm.[122]

(6) (a) Following issuance of an order under subparagraph (5)(d)(ii) of this section, the state developmental disabilities agency shall develop, in consultation with the defendant and the defendant's attorney, a Program Plan for the defendant. The objective of the Program Plan shall be to assist the defendant to become competent to stand trial within the shortest time possible, utilizing the least restrictive alternatives. During the development of the plan, the purpose, nature, and possible beneficial and harmful consequences of the various alternatives shall be explained to the defendant to the maximum extent possible.

(b) The Program Plan shall be submitted for approval to the court no more than ten (10) days after issuance of the order. A copy shall be provided to the prosecutor, the defendant, the defendant's attorney, and the (name of the state Developmental Disabilities Advocacy Office). It shall be approved by the court without a hearing within ten (10) days of its submission unless the court on its own initiative sets a hearing or the prosecutor or defendant move for a hearing. In such a motion, the moving party shall specify the reasons why the plan is alleged to be inadequate or inappropriate and the proposed modifications.

(c) Following the hearing, the court shall approve the plan or modify it so as to further the objectives set forth in paragraph (6)(a) of this section.

COMMENT. This subsection governs the development, dissemination and approval of a Program Plan following a finding that a defendant is incompetent to stand trial but is likely to become competent within a year if appropriate habilitative services are provided. The purposes and origin of Program Plans are discussed in the comment to section 3(19). The responsibility for developing the Program Plan and providing services is assigned to the state developmental disabilities agency.[123] Since a defendant found to be incompetent to stand trial is presumed to be innocent, and since, therefore, the purpose of the habilitative services is to assist the defendant to become competent rather than to assist him or her to function in the community as a law-abiding citizen, it would be inappropriate to charge a corrections agency with this duty. In order to assure that the plan addresses the defendant's needs and falls within the bounds of the court's order, and that the goals and time limits stated in the plan are realistic, the provision calls for participation of the defendant and his or her attorney in development of the plan. Moreover, since the efficacy of treatment is usually enhanced when participation is voluntary, it is anticipated that inclusion of the defendant in the plan development process will further assist habilitation.[124] To facilitate the defendant's participation in this process, the subsection requires a full explanation of various options in the language, mode of communication and terms which the defendant is most likely to understand.[125] To further clarify the process, paragraph (6)(a) sets forth the objective of the Program Plan and habilitative services — i.e. to assist the defendant to become competent to stand trial as quickly as possible through the least restrictive alternative. While the defendant may require a number of other services to assist him or her to live in the community as a self-reliant, law-abiding citizen, these goals are relevant only at the diversion and dispositional stages of the criminal process[126] and not to the question of competency to stand trial.[127] The Program Plan should of course, take into account any habilitative services in which the defendant is already participating.

In order that the provision of services may begin promptly, strict time limits are imposed by paragraph (6)(b) on the development and approval of the plan. In keeping with the right of the parties to object to the plan, and with the plan's function as a means for monitoring the provision of services, a copy is to be sent to the prosecutor, the defendant, and his or her attorney, and to the state Developmental Disability Advocacy Office. The Advocacy Office is included in order to assist defense counsel to assess the plan and to provide a means of identifying systemic problems and deficiencies in the provision of habilitative services.[128]

The Program Plan can be approved as written by the court without a hearing if neither party files an objection to it. A hearing must be held, however, if it is requested by a party or if the judge has questions about or wishes to modify the plan.

---

121 Shah, *Social Interactions of Law and Mental Health in Handling Social Deviance*, 23 CATHOLIC UNIVERSITY LAW REVIEW 674, 705-706 (1974).

122 See section 10(6)(d); of D.C. Code §23-1322 (1973).

123 See section 3(10) for the definition of developmental disabilities agency.

124 See PCMR, REPORT TO THE PRESIDENT: MENTAL RETARDATION, CENTURY OF DECISION, 92-93 (1976); ACMRDD *supra* note 11, at xix, 11 (1978).

125 PCMR, *id.*, at 93; see section 13(5).

126 But see section 7 on informal dispositions, section 10(7) on criminal responsibility, section 11 on sentencing and section 12 on corrections.

127 *See* Bennett, *supra* note 69, at 571.

128 See Chapter 9: Advocacy, *infra*.

(7) (a) The individual in the state developmental disabilities agency responsible for implementing the defendant's Program Plan shall file a written report with the court and send copies to the prosecutor, the defendant, the defendant's attorney, and the (name of the state Developmental Disabilities Advocacy Office):

(i) At least once every ninety (90) days following the competency hearing;

(ii) Whenever the individual believes the defendant is competent to stand trial;

(iii) Whenever the individual believes that there is not a substantial probability that the defendant will become competent to stand trial within one (1) year of the competency hearing; and

(iv) Fourteen (14) days before the expiration date of the continuance ordered pursuant to subparagraph (5)(d)(iii) of this section.

(b) The report shall contain:

(i) A description of the services being provided;

(ii) A list of the habilitative services prescribed in the Program Plan which have not been provided and the reasons therefor;

(iii) A description of the progress which has been made in achieving the goals set forth in the Program Plan;

(iv) An opinion on whether there is a substantial probability that the defendant will become competent to stand trial within one (1) year of the competency hearing;

(v) The facts and reasons supporting that opinion; and

(vi) A description of the modifications, if any, of the Program Plan which may increase the likelihood that the defendant will become competent within the prescribed period.

COMMENT. This subsection requires the periodic submission of reports outlining the habilitative services which have been provided to the defendant, the effect which those services have had, and what if any changes may be necessary in the type or intensity of the services or the manner and location in which they are provided. This reporting requirement is closely linked to the Program Plan developed under subsection (6) of this section, and the review hearings required by subsection (8). The reports have a twofold purpose. First, they serve to alert the court and counsel to significant changes in the defendant's condition and requirements. Second, they increase the accountability of the state developmental disabilities agency by documenting whether the defendant has received the requisite services and, if not, the reasons why those services have not been provided in accordance with the order of the court.[129] To further assure accountability, the provision suggests that there should be an individual personally responsible for arranging for the services identified in the order of the court and the Program Plan, rather than having this duty spread among several persons or offices with all reports being submitted *pro forma*, by the head of the agency or his designee.

Reports must be submitted at least once each quarter during the delay in the criminal proceedings, two weeks before the end of the continuance ordered by the court, and whenever there is a significant change in defendant's condition. Thus, if a six month delay is ordered and there is nothing to indicate that the defendant has become competent to stand trial or that he or she will not become competent within one year of the competency hearing, a report would be due 3 months and approximately 5½ months after the competency hearing.

(8) (a) The court shall hold a hearing:

(i) Following submission of a report pursuant to paragraph (7)(a) of this section which indicates that the defendant is competent to stand trial or that there is not a substantial probability that the defendant will become competent within one (1) year of the competency hearing;

(ii) At the request of the prosecutor or defendant following submission of a report pursuant to paragraph (7)(a);

(iii) At the end of a continuance ordered pursuant to subparagraph (5)(d)(ii) of this section; and

(iv) Within thirty (30) days after receiving a notice pursuant to subsection (12) of this section.

(b) At the hearing, the court shall redetermine whether the defendant is competent to stand trial, and if not, whether there is a substantial probability that the defendant will become competent within one (1) year of the initial competency hearing if provided with appropriate habilitative services. Upon making the determination, the court shall proceed in the manner specified in the applicable paragraph of subsection (5) of this section.

COMMENT. As indicated earlier, a requirement for judicial review is needed to assure for the benefit of both the prosecution and the defendant that the criminal proceeding is delayed no longer than necessary and that the defendant is receiving the habilitative services which are most likely to assist him or her to become competent to stand trial within 1 year of the finding of incompetence.[130] Hence, this provision calls for a review hearing at the end of any continuance granted following a finding of incompetency, and following submission of report by the individual responsible for implementing the defendant's program plan — automatically when the report indicates either that

[129] *See* Bennett, *supra* note 69, at 582.

[130] See sources listed *supra* notes 111 and 112.

the defendant is now competent to stand trial or that there is no longer a substantial likelihood that the defendant will become competent to stand trial within one year of being found incompetent, or at the prosecutor's or defendant's request when the report contains a different conclusion.[131] Thus, defendants who, under this act, are found incompetent to stand trial and likely to become competent within a year, but who in fact do not become competent within that period will have at least two review hearings: one following the expiration of the original continuance, and another at the end of the one year maximum delay period.

At the hearing, the court must redetermine on the basis of the reports, evidence, and testimony presented, whether the defendant is still incompetent to stand trial, and if not whether there is a substantial likelihood that he or she will become competent before the end of the maximum delay period. If the defendant is found to be competent to stand trial, then the criminal proceeding should go forward without delay. If the defendant is found to be incompetent to stand trial and there is not a substantial likelihood that he or she will become competent within one year of the competency hearing, or a year has passed since the competency hearing, then the criminal charges must be dismissed and the restraints on the defendant's liberty terminated, unless the circumstances set forth in section 8(5)(c)(ii) exist. If the maximum delay period has not expired and it is determined that the defendant is still incompetent but that there is a substantial likelihood that he or she will become competent to stand trial within a year of the competency hearing, then the court must determine whether it is still appropriate to provide services to the defendant to assist him or her to become competent. If it is, the court must review its original dispositional order and the Program Plan to determine whether changes are required. If it is neither in the interests of justice nor in those of the defendant to continue the criminal proceeding further, the charges are to be dismissed and all restraints on the defendant's liberty, terminated. In reviewing the dispositional order and Program Plan, the court must ascertain whether changes in the habilitative services being provided will be more likely to assist the defendant to become competent to stand trial within the shortest period of time possible and whether the services can be provided through less restrictive alternatives.[132]

(9) A defendant who has a developmental disability shall have the right at any time after an initial finding of incompetence:

(i) To file motions for discovery and for suppression of confessions and other evidence and to receive a prompt hearing on such motions;

(ii) To raise any legal objection to the charges that is susceptible of fair determination without his or her personal participation; and

(iii) To request a review by the court of whether there is probable cause to believe that the changes contained in the indictment, information, or complaint can be proven at trial.

A motion may be reheard upon a showing that evidence is available which was not available due to the defendant's incompetence when the motion was first decided.

COMMENT. In 1949, four soldiers were convicted by a military court-martial. In 1959, the Supreme Court overturned the conviction because the military court lacked jurisdiction over the case, and indictments against the defendants were filed in federal District Court. Subsequently, the indictments were dismissed against three of the defendants on the grounds that their right to a speedy trial has been denied. However, because the fourth defendant was incompetent to stand trial, the case against him was continued and he was returned to the federal instituion for the criminally insane until he was sufficiently competent to participate in the dismissal of the indictment.[133]

This provision is intended to prevent a reoccurrence of such anomalies by permitting defense counsel to raise and have decided any issue "susceptible of fair determination prior to trial and without the personal participation of the defendant."[134] Thus, each dispositive questions as the constitutionality of the statute on which the charges are based, the sufficiency of the indictment, and the expiration of the statute of limitations, as well as some motions to suppress improperly obtained evidence can be decided while the defendant is incompetent.[135] This approach has been adopted by at least 16 states[136] and was endorsed by the Supreme Court in Jackson vs. Indiana.[137] The provision permitting a rehearing of an issue decided against the defendant when relevant evidence becomes available which was not known at the time of the original hearing on the motion because of the incompetence of the defendant to stand trial, is intended to encourage use of this means of early disposition and to avoid

[131] In addition, this paragraph together with section 8(12) requires a hearing upon a receipt of notice that a person with a developmental disability is residing in a facility pursuant to an order of the court following a finding prior to the effective date of this act, that he or she is incompetent to stand trial.

[132] See sections 8(5)(d) and (e), and 8(6)(a).

[133] United States vs. Barnes, 175 F. Supp. 60 (S.D. Cal. 1959); for a discussion of this case see Foote, *A Comment on Pretrial Commitment of Criminal Defendants,* 108 UNIVERSITY OF PENNSYLVANIA LAW REVIEW 832 (1960).

[134] ALI, MODEL PENAL CODE §4.06(3) (Proposed Official Draft 1962); see ILLINOIS GOVERNOR'S COMMISSION *supra* note 94, at 177.

[135] MHLP, *supra* note 72 at 640.

[136] Chart IV *supra*.

[137] 406 U.S. 715, 740-741 (1971).

further forcing counsel to choose between seeking dismissal of the charges and obtaining a continuance and treatment for a client who is incompetent to stand trial.[138]

(b) If at any time the court determines that there is not probable cause to believe that the charges contained in the indictment, information, or complaint can be proven against the defendant, it shall dismiss the indictment, information or complaint.

COMMENT. In accord with the above paragraph and subsections (2) and (10) of this section, this provision clearly establishes the principle that the incompetence of a defendant to stand trial should not preclude reexamination of the charges and dismissal of any charge which cannot be supported. It is intended as a further means of preventing misuse of the procedures in this section as a way of detaining individuals by filing minor or weakly supported charges.[139]

(10) No order issued under this section authorizing the provision of services or retention in custody of a defendant other than an order issued pursuant to subsection (5)(c)(ii), shall remain in effect after the indictment, information, or complaint is dismissed. The court shall notify the defendant, the defendant's attorney, the prosecutor, the [sheriff/marshall/or other appropriate official], and the individual overseeing the provision of services to the defendant whenever an indictment, information, or complaint against a defendant is dismissed.

COMMENT. This subsection together with subsections (5), (9) and (11) are designed to reduce the likelihood that defendants who are incompetent to stand trial will have their liberty restrained pursuant to the criminal charges for longer than if they had not been found incompetent. It provides that upon dismissal of the charges, an order for an evaluation issued under subsection (3) and, with one exception, any dispositional order issued pursuant to subsections (5) or (8) automatically terminates. Once the charges are dismissed, the court no longer has jurisdiction to maintain control over the defendant unless an action is initiated under the civil provisions on services to persons with a developmental disability. Accordingly, the one exception is when the court under section 8(5)(c)(ii) orders that a defendant who is in custody, who has been found to be incompetent and unlikely to become competent within a year of the competency hearing, and who meets the criteria for residential placement, be detained for up to 48 hours pending the filing of a petition under the Model Right to and Admission to Services Act or state civil commitment laws.

[138] ILLINOIS GOVERNOR'S COMMISSION, *supra* note 94, at 177; Foote, *supra* note 133 at 846.

[139] *See* MHLP, *supra* note 72, at 618; Ennis, *supra* note 71, at 3; STONE *supra* note 78, at 207; Bennett, *supra* note 69, at 573.

(11) A defendant convicted of a crime and sentenced to a period of incarceration shall have his or her sentence reduced by the total number or days he or she was confined for examination, evaluation, or services under this section.

COMMENT. While it is not unusual for a court to moderate the sentence of a defendant who was confined prior to conviction for treatment to assist in restoring competency, the degree to which credit against a sentence is given for such confinement varies considerably from case to case, and judge to judge.[140] Mandatory credit for pretrial confinement in general, and credit for confinement due to an evaluation or services provided as a result of an inquiry into or finding of incompetency to stand trial in particular, has been strongly urged by a number of groups and commissions which have examined the issue, in order to promote equity, fairness and consistency.[141] A number of states already provide for such credit.[142]

(12) The head of any facility housing a person who:

(a) There is reason to believe may have a developmental disability;

(b) Has been found incompetent to stand trial or enter a plea prior to the effective date of this act; and

(c) As a result of that finding has been placed in custody; shall notify the court that committed the person no more than thirty (30) days after the effective date of this act. The notice shall contain the information specified in subparagraphs (7)(b)(i), (iii) and (iv) of this section. Within thirty (30 days after receiving the notice, the court shall conduct a hearing and proceed in accordance with subsections (8)-(11) of this section.

COMMENT. This is a transition section, applying the rights and procedures accorded by this act to defendants with a developmental disability who as a result of having been found incompetent to stand trial, are in custody pursuant to the laws and procedures in effect prior to enactment of this statute. The provision requires the head of the facility in which the defendant is housed to advise the committing court of the defendant's where-

[140] *See* Steadman and Braff, *Effects of Incompetency Determinations on Subsequent Criminal Processing: Implications for Due Process*, 23 CATHOLIC UNIVERSITY LAW REVIEW 754 (1974); S. BRAKEL and R. ROCK, THE MENTALLY DISABLED AND THE LAW, 417 (Rev. ed. 1971).

[141] ABA, STANDARDS RELATING TO THE ADMINISTRATION OF CRIMINAL JUSTICE: SENTENCING ALTERNATIVES AND PROCEDURES, §18-4.7 (2d Ed. Approved Draft 1979); NATIONAL ADVISORY COMMISION, *supra* note 43, at §5.8; ILLINOIS GOVERNOR'S COMMISSION, *supra* note 94, at 185; MHLP, *supra* note 72, at 643; *see also* ALI, MODEL PENAL CODE *supra* note 134, at §7.09.

[142] *E.g.*, ILLINOIS ANNOTATED STATUTES ch. 38, §1005-2-2 (Smith-Hurd 1973) and TEXAS CRIMINAL PROCEDURE CODE ANNOTATED §46.02(9) (Vernon 1977) (mandatory credit); DELAWARE CODE ANNOTATED tit. 11, §404(1975) (credit within discretion of the court).

about, the services being provided, whether in his or her opinion the defendant is competent to stand trial, and if not, whether there is a substantial likelihood that the defendant will become competent within a year of the original competency hearing, and of the facts and reasons supporting that opinion. The court must then conduct a review hearing and based on the evidence and testimony adduced during that hearing, order a disposition in accordance with subsection (5) of this section. The subsection does not apply to individuals who, following the finding of incompetence to stand trial, have been placed in a residential facility pursuant to the state civil commitment laws, and who may be released on the same basis and pursuant to the same procedures as persons who were not subject to criminal prosecution prior to placement.

Section 9. Pleas of Guilty or No Contest

(1) The court shall not accept a plea of guilty or of no contest from a defendant who there is reason to believe has a developmental disability, unless that defendant is represented by counsel, and the court, at a minimum:

(a) Explains to the defendant on the record:

(i) The charge(s) to which the plea is offered;

(ii) That upon acceptance of the plea the defendant will be considered to have committed the criminal offense(s) specified;

(iii) The nature and range of possible sentences for the criminal offense(s) to which the plea is offered, including any special conditions regarding sentence, parole, or commutation imposed by statute;

(iv) The rights which the defendant will waive by entering the plea, including the right to plead not guilty; and

(v) The terms of the plea agreement if any;

(b) Diligently determines through questioning the defendant personally that:

(i) The defendant understands the matters listed in subparagraphs (a) (i)-(v); and

(ii) The plea is not a product of force or threats, or of promises other than those included in the plea agreement if any; and

(c) Requires the prosecutor to present affidavits, witnesses, and/or evidence demonstrating that there is a factual basis for the plea.

COMMENT. While, almost all state criminal codes contain provisions on competency to stand trial, the procedures to be applied when a defendant with a developmental disability wishes to plead guilty are virtually unaddressed. Yet in most jurisdictions, the majority of criminal cases are resolved by a plea of guilty. Often these pleas are the result of an agreement between the prosecution and the defense through which a charge is dropped or reduced in seriousness, or through which a sentencing recommendation is made in exchange for the defendant's willingness to forego a trial. While there has been much controversy over the propriety and necessity of plea-bargaining,[143] it remains a significant feature of the American criminal justice process, and therefore, must be covered in any consideration of the problems faced by defendants with developmental disabilities.

The Supreme Court has long recognized that by pleading guilty, defendants waive the right to remain silent, to trial by jury, and to confront their accusers.[144]

> [A] plea of guilty is more than an admission of conduct; it is a conviction. Ignorance, incomprehension, coercion, terror, inducements, subtle or blatant threats might be a perfect coverup of unconstitutionality.[145]

Accordingly, when a defendant pleads guilty, the judge must assure on the record that the defendant has "a full understanding of what the plea connotes and of its consequences, and that he or she is entering the plea voluntarily.[146] The Court has also held that a defendant may plead no contest — *i.e.* waive trial without admitting guilt — if the "knowing and voluntary" criteria are met and there is a "strong evidence of actual guilt."[147]

While it could be argued that because the risk of coercion is too great, courts should not be authorized to accept any guilty plea from a defendant who there is reason to believe has a developmental disability, or, more narrowly, guilty pleas of such defendants resulting from a plea agreement,[148] such a policy would place defendants with a developmental disability, at a severe disadvantage in many American jurisdictions. Hence, this subsection prescribes the four sets of procedures and requirements designed to assure, to the maximum extent possible, that pleas of guilty by defendants with a developmental disability are knowing and voluntary. They are similar to but

[143] *See, e.g.,* (pro) ABA, STANDARDS RELATING TO PLEAS OF GUILTY (Approved Draft 1968); NATIONAL DISTRICT ATTORNEYS ASSOCIATION (NDAA) *supra* note 45, at 214-277; Santobello vs. New York, 404 U.S. 275 (1971); (con) NATIONAL ADVISORY COMMISSION ON CRIMINAL JUSTICE STANDARDS AND GOALS, COURTS 42-49 (1973); Alschuler, *The Defense Attorney's Role in Plea Bargaining,* 84 YALE LAW JOURNAL 1179 (1975); Alschuler, *The Prosecutor's Role in Plea Bargaining,* 36 UNIVERSITY OF CHICAGO LAW REVIEW 50 (1968); Casper, AMERICAN CRIMINAL JUSTICE: THE DEFENDANT'S PERSPECTIVE (1972).

[144] Kercheval vs. United States, 274 U.S. 220, 223 (1927); McCarthy vs. United States, 394 U.S. 459 (1969).

[145] Boykin vs. Alabama, 395 U.S. 238, 242-243 (1969).

[146] *Id.,* at 244.

[147] North Carolina vs. Alford, 400 U.S. 25 (1970).

[148] *Cf.* section 4(4) and 4(5); NATIONAL ADVISORY COMMITTEE *supra* note 28, at §13.1; NACJJDP, REPORT OF THE ADVISORY COMMITTEE TO THE ADMINISTRATOR ON STANDARDS FOR THE ADMINISTRATION OF JUVENILE JUSTICE, §3.175 (1976).

more stringent than those contained in most provisions on acceptance of pleas of guilty and no contest.

The first safeguard contained in the provision is that the defendant be represented by counsel. The complexity of the issues involved, especially when the plea is based on an agreement with the prosecutor, and the consequences of the waiver demand that the defendant have an attorney to zealously protect his or her interests.[149]

The second set of requirements is contained in paragraph (1)(a), and is intended to assure that the defendant has the information necessary to understand what he or she is agreeing to, giving up, and risking by pleading guilty.[150] The required explanations must be in the language, mode of communication and terms which the defendant is most likely to understand,[151] and are intended as a supplement to any other admonitions required by law.[152] The provision requires the terms of the plea agreement to be revealed in open court as a matter of good practice, to protect both the public and the defendant and to facilitate the enforceability of the agreement.[153]

The third safeguard is the colloquy required in paragraph (1)(b) between the judge and the defendant to determine whether the defendant understands the explanations which have been provided, and is changing his or her plea voluntarily. When a defendant with a developmental disability is involved, this colloquy must be more thorough than the usual plea exchange in order to assure that the defendant is able to and does understand the nature and consequences of his actions and has not been coerced or misled.

> . . . [I]t is apparent that the standard Rule 11 colloquy may prove an inadequate measure of the validity of a plea proffered by a defendant of questionable mental competence. As the Supreme Court observed, "[t]he nature of the inquiry . . . must necessarily vary from case to case. . . . In all such inquiries, '[m]atters of reality and not mere ritual should be controlling."[154]

The Ninth Circuit Court of Appeals has suggested that a higher standard of competence should be required for defendants pleading guilty than for defendants going to trial.[155] The more thorough explanations and questioning required by paragraphs (a) and (b) appear to provide the same protection without the ambiguities and complications introduced by a having dual standards of competence.[156]

Finally, in order to assure the accuracy of the plea paragraph (1)(c) requires the prosecutor to provide a factual basis for the conviction. This is similar to the requirement imposed by the Supreme Court for pleas of no contest[157] and has been adopted for both guilty and *nolo contendere* pleas by a number of jurisdictions as well as several standards.[158] Judging from the *Alford* standard of "strong evidence of actual guilt", a court must find evidence which shows more than probable cause for believing the defendant is actually guilty but less than belief of guilt beyond a reasonable doubt.[159] How the court determines whether there is a factual basis may vary from case to case.

(2) Upon the filing of a timely motion for withdrawal, the court shall permit a defendant who there is reason to believe has a developmental disability to withdraw a plea of guilty or of no contest:

(a) If it is shown that:

(i) The defendant did not understand the matters explained in subparagraphs (1)(a)(i)-(v) of this section;

(ii) The defendant was denied effective assistance of counsel; or

(iii) The plea was involuntary;

(b) If the court decides to impose a sentence less favorable to the defendant than that contemplated in the plea agreement; or

(c) To correct or prevent manifest injustice.

COMMENT. This subsection establishes a liberal rule for withdrawal of pleas of guilty or no contest by defendants with a developmental disability.[160] The criteria are based on those recommended in the ABA standards on pleas of guilty.[161] It is anticipated, however, that there will be few withdrawals, except under paragraph (b), if the procedures prescribed by subsection (1) are followed.

Motions to withdraw a plea of guilty or no con-

[149] See section 13(1) and (2).

[150] See AAMD, *supra* note 30.

[151] See section 13(5).

[152] *E.g.*, ALI, MODEL PRE-ARRANGEMENT CODE, *supra* note 39, at §§350.4 and 350.5; FEDERAL RULES OF CRIMINAL PROCEDURE 11(e) (1975); ARIZONA RULES OF CRIMINAL PROCEDURE 17.4 (1973).

[153] See sources cited in note 152 and ABA, *supra* note 143, at §1.5; Santobello vs. New York, 404 U.S. 257 (1971).

[154] United States vs. Masthers, 539 F. 2d 721, 728-729 (D.C. Cir. 1976); see AAMD, *supra* note 30.

[155] Seiling vs. Eyman, 478 F. 2d 211 (9th Cir. 1973); *see also* United States vs. Masthers, *id.*; but see Malinauskus vs. United States 531 F.2d 766 (5th Cir. 1974).

[156] Note, *supra* note 68, at 170-171.

[157] North Carolina vs. Alford 400 U.S. 25 (1970).

[158] *E.g.*, FEDERAL RULES OF CRIMINAL PROCEDURE 11(f) (1975); ARIZONA RULES OF CRIMINAL PROCEDURE 17.3 (1973); ALI, MODEL PRE-ARRANGEMENT CODE, *supra* note 39, at §350.4(3); ABA, GUILTY PLEAS *supra* note 143, at §1.6.

[159] *But see* ALI, *id.*

[160] *Accord* United States vs. Masthers, 539 F. 2d 721 (D.C. Cir. 1976).

[161] *Supra* note 143, at §2.1.

test may be made either before or after judgment and sentence.

> It does not follow, however, that the time of the defendant's motion is totally irrelevant. The fact that it comes after sentence and judgment or a considerable time thereafter may have a bearing upon whether the motion is timely, considering the nature of the allegations in the motion. For example, if the allegation is that the prosecuting attorney did not seek the sentence concessions promised in a plea agreement following discussions with defendant's attorney, it is reasonable to expect the motion to be made promptly upon learning of the prosecutor's inaction in this regard.[162]

Such motions should be granted when the defendant demonstrates that the plea was not entered knowingly or voluntarily, that by action of the court, prosecutor, or his or her own attorney, the defendant was denied effective assistance of counsel,[163] or that the interests of justice required it.[164] In addition, the court must give the defendant an opportunity to withdraw a plea when the court rejects the proposed disposition that formed the basis for the plea agreement and intends to impose a more restrictive alternative.[165]

INTRODUCTION TO SECTION 10

The question of the circumstances under which an individual may be found not guilty of having committed a crime because of a "mental disease or defect"[166] is one of the most hotly debated criminal law issues.[167] As noted in the introduction to this report, there are two legal theories within American jurisprudence which defendants may employ when they contend that because of a mental disease or defect they should not be held responsible for criminal conduct. The first is the so-called insanity defense which in its most modern form holds that:

> A person is not responsible for criminal conduct if at the time of such conduct as a result of mental disease or defect he [or she] lacks substantial capacity either to appreciate the criminality [wrongfulness] of his [or her] conduct or to conform his [or her] conduct to the requirements of law.[168]

Some form of this defense is in effect in every American jurisdiction.

The second theory is variously known as diminished capacity, diminished responsibility, or subjective liability. It is premised on the requirement for most crimes that particular physical conduct be coupled with a particular state of mind — *e.g.*, for first degree murder, "malice" and "premeditation" is required. Under this theory, the defendant is permitted to introduce evidence to prove that he or she did not have the requisite state of mind.[169] About half of the American jurisdictions permit the introduction of such evidence either by statute or pursuant to common law.[170]

Proponents and critics of both theories have, in the words of one commentator, expended "[r]ivers of ink, mountains of printer's lead, [and] forests of paper. . .on this issue. . . .[171] Almost all the attention, however, has been focused on defendants claiming to be mentally ill at the time the offense is alleged to have been committed. Other than subsuming mental retardation within the scope of the "insanity defense," little consideration has been given to the impact and appropriateness of the current rules for defendants with a developmental disability.[172]

As a result, this section is perhaps the most innovative and controversial in the model act. It seeks to address both the general problems of existing mental disease and defect defense provisions, and the specific concerns of defendants with developmental disabilities. Subsections (1) and (2) adopt the diminished capacity theory for defendants who have a developmental disability. While the insanity defense in one of its incarnations may be needed to take account of persons such as Daniel M'Naughten who as a result of a mental illness intentionally engage in criminal conduct while suffering from a delusion,[173] the question of whether the defendant had or did not have the state of mind required as an element of

---

162 *Id.* at 55.

163 *Id.* at 56; *see* Haggerty, Kane, and Udall, *An Essay on the Legal Rights of the Mentally Retarded*, 6 FAMILY LAW QUARTERLY 59 (1972).

164 FEDERAL RULES OF CRIMINAL PROCEDURE, 32(d) (1975).

165 ABA, GUILTY PLEAS, *supra* note 143, at §2.1; ARIZONA RULES OF CRIMINAL PROCEDURE, 17.4 (1975).

166 ALI, *supra* note 134, at §4.01.

167 *E.g.*, Goldstein and Katz, *Abolish the Insanity Defense – Why Not?* 72 YALE LAW JOURNAL 853 (1963); Brady, *Abolish the Insanity Defense - No!*, 8 HOUSTON LAW REVIEW 629 (1971); Monahan, *Abolish the Insanity Defense? – Not Yet*, 26 RUTGERS LAW REVIEW 719 (1973); Allen, *The Brawner Rule – New Lyrics for an Old Tune*, 1973; WASHINGTON UNIVERSITY LAW REVIEW 67 (1973); for a discussion of the development of the rules, policies, and practices governing criminal responsibility and the summary of the various arguments, see BRAKEL and ROCK, *supra* note 140, at 376-408.

168 ALI, MODEL PENAL CODE *supra* note 134, at §4.01(1).

169 *Id.*, at §4.02(1).

170 Comment, *Diminished Capacity – Recent Cases and an Analytical Approach*, 130 VANDERBILT LAW REVIEW 213, 217 (1977); Chart V, *supra*.

171 Morris, *Psychiatry and the Dangerous Criminal* 41 SOUTHERN CALIFORNIA LAW REVIEW 514, 516 (1968); see the introduction to this report and BRAKEL and ROCK, *supra* note 140, at 376-408.

172 *But see* Note, *The Mentally Retarded Offender in Omaha-Douglas County*, 8 CREIGHTON LAW REVIEW 622, 644-649 (1975) and the sources cited in footnotes 75 and 76 of the introduction to this report, *supra*.

173 M'Naughten's Case 8 Eng Rep. 718 (1843); *see* Arenella, *The Diminished Capacity and Diminished Responsibility Defenses: Two Children of a Doomed Marriage;* 77 COLUMBIA LAW REVIEW 827, 834 (1977).

the offense is far more germane, for example, to persons with mental retardation or epilepsy who are accused of committing a crime.[174] In addition, since the diminished capacity defense capacity falls within the usual pattern for the presentation of evidence, it eliminates the often criticized use of special burdens of proof[175] and bifurcated trials,[176] and over-dependence on expert testimony.[177]

Subsection (3) assures that defendants are not held in custody solely as a result of asserting a defense of diminished capacity due to a developmental disability. Subsection (4) lists the range of possible verdicts and the dispositional provisions applicable to each. Subsections (5) and (6) establish special dispositional provisions for defendants who are acquitted solely by reason of diminished capacity. The lack of such a dispositional mechanism has been cited as a major factor in the reluctance of states to permit the diminished capacity defense.[178] Those subsections, together with subsection (7), which limits the length of the dispositional period, and subsections (8)-(11), which establish review procedures, address many of the shortcomings of current dispositional practices following a verdict of not guilty by reason of insanity that are outlined in the introduction to this report.[179] To help to assure that defendants are offered the habilitative services they require, subsection (12) expressly makes the state developmental disabilities agency a party to all dispositional proceedings held under this section. Finally, subsection (13) provides a means for providing the protections and procedures established by this section to persons with a developmental disability who were found not guilty by reason of insanity and, as a result, placed in custody prior to its enactment.

[174] *See* PCMR, *supra* note 124, at 63; Note, *supra* note 172, at 646-647.

[175] *See e.g.*, Comment, *supra* note 170; at 231-236; Note, *Due Process and the Insanity Defense: The Supreme Court's Retreat from Winship and Mullaney,* 54 INDIANA LAW JOURNAL 95 (1978); Comment, *Oregon's Partial Responsibility Defense: Disposition of the Defendant and Burden of Proof,* 13 WILLAMETTE LAW JOURNAL 347 (1977); BRAKEL and ROCK, *supra* note 140, at 399-401.

[176] *E.g.*, BRAKEL and ROCK, *id.*, at 398-399; comment, *Mens Rea and Insanity,* 28 MAINE LAW REVIEW 500 (1976); State vs. Shaw, 106 Ariz. 103, 471 P. 2d 715 (1970).

[177] *E.g.*, Bazelon, *Psychiatrists and the Adversary Process,* 230 SCIENTIFIC AMERICAN 18 (1974); STONE, *supra* note 78, at 233-228; NEW YORK DEPARTMENT OF MENTAL HYGIENE, A REPORT TO GOVERNOR HUGH L. CAREY ON THE INSANITY DEFENSE IN NEW YORK, 108-113 (1978).

[178] Arenella, *supra* note 173, at 832 fn.25; Note, *Insanity – Guilty But Mentally Ill-Diminished Capacity: An Aggregate Approach to Madness,* 12 JOHN MARSHALL JOURNAL OF PRACTICE AND PROCEDURE 351, 377-378 (1976); Comment, *supra* note 170, at 237; Comment, *supra* note 176, at 522-523; Comment, *supra* note 175, at 355-356; *e.g.*, Bethea vs. United States, 365 A.2d 64 (D.C.C.A. 1976).

[179] *E.g.*, BRAKEL and ROCK, *supra* note 140, at 404-407; Note, *supra* note 178, at 366-367; MHLP, *Legal Issues in State Mental Health Cases: Proposals for Change – Insanity Defense,* 2 MENTAL DISABILITY LAW REPORTER 651, 655 (1978); Bolton vs. Harris, 395 F.2d 642 (D.C. Cir. 1968); Allen vs. Radack, 426 F. Supp. 1052 (D.S.D. 1977); Eckerhart vs. Hensley, 475 F. Supp. 908 (W.D. Mo. 1979); State vs. Fields, 77 N.J. 282, 390 A.2d 574 (1978).

As noted above the section is drafted to apply solely to defendants with a developmental disablity. Those states seeking to extend or clarify the application of the diminished capacity defense to persons who were voluntarily intoxicated at the time of the crime, and/or in conjunction with or as a substitute for the insanity defense to persons who were mentally ill at the time of the offense,[180] can easily expand its coverage by inserting the appropriate term after references to a developmental disability and habilitative services.

Section 10. Criminal Responsibility

(1) In order for a defendant to be found guilty of committing a criminal offense, the prosecutor shall prove beyond a reasonable doubt every element of that offense.

COMMENT. This provision echoes the constitutional mandate protecting persons accused of a crime from conviction "except upon proof beyond a reasonable doubt of every fact necessary to constitute the crime."[181]

> There is always in litigation a margin of error, representing error in factfinding, which both parties must take into account. Where one party has at stake an interest of transcending value — as a criminal defendent his [or her] liberty — this margin of error is reduced as to him [or her] by the process of placing on the other party the burden of. . .persuading the factfinder at the conclusion of the trial of his [or her] guilt beyond a reasonable doubt. Due process commands that no man [or woman] shall lose his [or her] liberty unless the Government has borne the burden of. . .convincing the factfinder of his [or her] guilt.[182]

Also basic to American jurisprudence is that in order for physical conduct to constitute a crime it must be accompanied by a particular state of mind.[183] For example, the Model Penal Code provides that:

> . . .[A] person is not guilty of an offense unless he [or she] acted purposely, knowingly, recklessly or negligently as the law may require, with respect to each material element of the offense.[184]

[180] *See e.g.*, A.L.I., MODEL, PENAL CODE, *supra* note 134, at §§4.01 and 4.02; Morris, *Special Doctrinal Treatment in Criminal Law,* in PCMR, *supra* note 12, at 682, 687 (1976); NEW YORK DEPARTMENT OF MENTAL HYGIENE, *supra* note 177, at 9-10; *but see* Arenella, *supra* note 173.

[181] In re Winship, 397 U.S. 358, 364 (1969).

[182] Speiser vs. Randall 357 U.S. 513, 525-526 (1958).

[183] *E.g.*, W. LAFAVE and A. SCOTT, JR., CRIMINAL LAW, 191-192 (1972).

[184] ALI, MODEL PENAL CODE, *supra* note 134, at §2.02(1).

The Model Penal Code contains definitions of each of these culpable states of mind.[185] This format has also been adopted in a number of state criminal codes.[186]

Hence, this subsection makes clear that although the defendant's state of mind may be implied from his or her conduct, it is a point which is subject to challenge by the defense and which the prosecution bears the burden of proving beyond a reasonable doubt.[187]

(2) (a) Defendants who have a developmental disability are entitled to assert a defense of diminished capacity by introducing testimony and evidence demonstrating that due to their disability, they lacked the state of mind required as an element of the criminal offense(s) with which they are charged.

(b) Defendants who have a developmental disability but who do not also have a mental illness are not entitled to assert as a defense that they [are not guilty by reason of insanity/lack substantial capacity either to appreciate the criminality of their conduct or to conform their conduct to the requirements of the law/are guilty but mentally ill] pursuant to (citation to the relevant sections of the state criminal code).

COMMENT. This subsection replaces the insanity defense with the defense of diminished capacity for defendants who have a developmental disability.[188] It does not affect, however, the availability of the insanity defense to persons who are mentally ill or who have both a mental illness and a developmental disability. This approach will be advantageous to both developmentally disabled defendants and the criminal justice system as a whole. First, it focuses on the legal issues most likely to be affected by the impairments related to developmental disabilities. The insanity defense has been constructed to deal with individuals who because of delusions resulting from a mental illness — *e.g.*, a voice telling them to kill — are unable to appreciate that the particular type of criminal conduct is wrong or who are unable to stop themselves. They may fully intend to commit the acts involved and understand their consequences, but because of their illness they lack the required moral blameworthiness. The diminished capacity defense on the other hand, centers on a defendant's knowledge and understanding of the consequences of his or her actions — *e.g.*, a person who strikes another during the course of an epileptic seizure or a person with mental retardation who sets fire to a building without comprehending the result and the risks involved but who knows that destroying another person's property is wrong.

Second, it avoids many of the procedural complexities which accompany the insanity defense. Most states require defendants pleading not guilty by reason of insanity to bear the burden of proving this defense, usually by a preponderance of the evidence.[189] In light of this, some states have attempted to delay the presentation of evidence regarding the defendant's mental disabilities until after a finding that he or she committed an offense.[190] Generally this bifurcation of the trial process has proven unsuccessful,[191] and, where strictly enforced, unconstitutional.[192] Since by asserting a defense of diminished capacity a defendant is simply challenging an essential point in the prosecution's case, these extra layers on an already complex trial process are unnecessary.

Third, since evidence of the defendant's state of mind is routinely proven and challenged in court without reliance on expert testimony, the dependence on mental disability experts inherent in the insanity defense and the confusion often wrought in applying psychiatric and psychological theory to legal axioms,[193] are substantially reduced through the use of the diminished capacity approach. The need for special notice of the defense and compulsory examinations of the defendant is also eliminated.[194]

Fourth, rather than the all or nothing imperative of the insanity defense, diminished capacity recognizes that the degree to which a developmental disability affects the ability to form the state of mind required to commit a criminal offense varies

[185] *Id.*, at §§2.02(2)(a)-(d).

[186] *See e.g.*, MAINE REVISED STATUTES ANNOTATED, tit. 17-A, §10 (Supp. 1976); NEW YORK PENAL LAW §15.05 (McKinney 1975). For other terms see Morrissette vs. United States, 342 U.S. 237, 251-252 (1952).

[187] Morrissette vs. United States, *id.*, at 275; State vs. Shaw 106 Ariz. 103, 471 P. 2d. 715 (1970).

[188] Santamour and West, *The Retarded Offender: Problems, Programs, and Prescriptions*, in THE MENTALLY RETARDED CITIZENS AND THE CRIMINAL JUSTICE SYSTEM, *supra* note 12, at 1, 14.

[189] Comment, *supra* note 176, at 505; Note, *supra* note 175; Leland vs. Oregon, 343 U.S. 790 (1952); Patterson vs. New York, 432 U.S. 197 (1977).

[190] See *e.g.*, CALIFORNIA PENAL CODE §1026 (1970); MAINE REVISED STATUTES ANNOTATED tit. 17-A, §59(1) (Supp. 1976); WISCONSIN CRIMINAL PROCEDURE CODE §971.175 (1971).

[191] Comment, *supra* note 176.

[192] *E.g.*, State vs. Shaw, 106 Ariz. 103, 471 P. 2d 715 (1970); Hughes vs. Matthews, 440 F. Supp. 1272 (E.D. Wis. 1977).

[193] See sources cited *supra* note 177.

[194] See BRAKEL and ROCK , *supra* note 140, at 396-398. Enacting states which nevertheless choose to require an examination of the defendent and which provide the prosecutor with access to the results should provide in addition that: the examining experts be selected from the list published by the Evaluation Oversight Board; the defendent and the prosecution have input into the expert selection process; the examination report contains an opinion on whether the defendant has a developmental disability as defined in this act and if so, an opinion regarding the capacity of the defendant at the time the offense is alleged to have been committed to form the state of mind required for each criminal offense charged and for any lesser included offenses, and the facts and reasons supporting the opinions, as well as the information required by section 15(4)(1) *infra;* and the procedures and protections offered by section 15 apply including those in section 15(6)(b) relating to the non-admissibility of statements and evidence obtained as a result of the examination on the issue of whether the defendant committed the physical acts required to commit the crime charged.

from individual to individual, and therefore, that the degree of culpability should vary.

Finally, it is anticipated that providing separate defenses for developmental disabilities and mental illnesses will help to increase awareness of the distinctions between these conditions and facilitate the development of necessary habilitative services.

There have been previous attempts to eliminate the use of the insanity defense. In the first third of this century, three states enacted statutes abolishing the defense. These provisions, however, were found to violate the constitutionally protected right to a trial by jury,[195] due to process of law,[196] or both.[197] However, since this provision substitutes a related defense addressing the same concern, and since it does not prohibit or attempt to alter the constituency of the jury, it does not suffer from the constitutional infirmities which affected the statutes struck down in the Strasburg, Sinclair, and Lange cases.

(3) Except as provided in section 15(3)(a) of this act, a defendant who asserts a defense of diminished capacity due to a developmental disability is entitled to release on bail or recognizance prior to or following trial to the same extent and on the same terms as if the defense had not been raised.

COMMENT. As indicated in the comment to Section 8(3) of this act, conditions of release or bail prior to trial should be based solely on the likelihood that a defendant may flee the jurisdiction of the court.[198] Since the presumption of innocence is no longer operative following conviction, other factors such as the likelihood that the defendant may commit further crimes may be considered in determining whether and under what conditions a defendant may be released pending the sentencing proceeding and/or appeal.[199] This subsection does not alter the applicable standards. It merely assures that defendants asserting a defense based on diminished capacity receive the same opportunities for release as those who rely on a different defense and are not penalized by having their liberty unnecessarily restricted.

(4) (a) If following the assertion of a defense of diminished capacity due to a developmental disability the judge or jury finds that the prosecutor has not proven beyond a reasonable doubt that the defendant committed all the physical acts required as elements of the criminal offense(s) charged or for a lesser included offense, the defendant shall be acquitted and all restraints on his or her liberty imposed as a result of that charge/those charges shall be terminated.

(b) If following an assertion of a defense of diminished responsibility due to a developmental disability, the judge or jury finds that the prosecutor has proven beyond a reasonable doubt that the defendant committed all the physical acts and had the state of mind required for the criminal offense(s) charged or for a lesser included offense, the defendant shall be found guilty and the court shall proceed in the manner set forth in section 11 of this act if the defendant has a developmental disability and as otherwise prescribed by law if the defendant does not have a developmental disablity.

(c) If following an assertion of a defense of diminished capacity due to a developmental disability the judge or jury finds that the prosecutor has proven beyond a reasonable doubt that the defendant committed all the physical acts required as elements of the criminal offense charged or of a lesser included offense, but that the defendant lacked the state of mind prescribed as an element of that offense or of lesser included offense, it shall render a verdict of acquittal solely by reason of diminished capacity and proceed in the manner set forth below.

COMMENT. This subsection sets forth the possible determinations which can be made following assertion of a defense of diminished capacity due to a developmental disability, and the consequences which flow from each determination. Paragraph (4)(a) provides that if the prosecutor fails to prove beyond a reasonable doubt that a crime was committed or that the defendant committed it, the defendant must be found innocent. The fact that the accused interposed a defense of diminished capacity is of absolutely no consequence. Accordingly, a defendant with a developmental disability who has been acquitted should be free of any restraints on his or her liberty just like any other acquitted defendant.

Paragraph (4)(b) states that if the prosecutor proves that a defendant engaged in the prohibited acts with the requisite state of mind, the defendant must be found guilty whether or not he or she has raised a defense of diminished capacity. The fact that the defendant may have a developmental disability is then relevant only to the determination of the sentence to be imposed. Special sentencing

---

[195] State vs. Strasburg, 60 Wash. 106, 110 P. 1020 (1910). The Washington statute struck down in Strasburg eliminated the insanity defense but authorized the judge to determine whether accused persons were insane and to commit them to a mental institution. BRAKEL and ROCK, *id.*, at 378.

[196] Sinclair vs. State, 161 Miss. 142, 132 So. 581 (1936). The Mississippi statute abolished insanity as a defense for murder, but authorized transfer to a mental institution following conviction. BRAKEL and ROCK, *id.*

[197] State vs. Lange, 168 La. 958, 123 So. 639 (1929). The Louisiana provision required defendants raising the insanity defense to be tried by a "lunacy commission." If the commission found the defendant to be insane it could have him or her committed. If the defendant was found to be sane, he or she was referred for trial. BRAKEL and ROCK, *id.*, at 378-379.

[199] See note 82 *supra*, and the accompanying text.

[199] *See e.g.*, ARIZONA RULES OF CRIMINAL PROCEDURE 7.2 (1973); DISTRICT OF COLUMBIA CODE §23-1325 (1973); ABA, STANDARDS RELATING TO CRIMINAL APPEAL §2.5 (2d Ed., Approved Draft 1978).

procedures and considerations are set forth in Section 11 for developmentally disabled offenders.

Subsection (4)(c) presents an alternative not currently available, but analogous to findings of not guilty by reason of insanity. It applies in what is anticipated to be the relatively few cases in which the prosecution proves beyond a reasonable doubt that the defendant engaged in criminal conduct but the judge or jury find that due to a developmental disability, the defendant did not have the particular state of mind required for conviction. As noted in the introduction to this section, the absence of specific authority to require the provision of services to persons acquitted solely by reason of diminished capacity has been a major factor in the reluctance of courts to permit the assertion of this defense.[200] While civil commitment could be relied upon in such cases,[201] that approach has been rejected for a number of reasons. First, there is great public concern over the premature release or the inadequate supervision of persons acquitted solely on the grounds of insanity or diminished capacity.[202] As a result, when such persons are handled through the civil commitment process additional restrictions have been imposed on the liberty of and the opportunities and services for all individuals who have been admitted to civil mental health facilities whether or not they have been involved with the criminal justice system. These restriction, in turn, impede the drive toward use of the least restrictive alternative, normalization, and increased public awareness that mental disability and criminality are not synonymous.[203] Second, given the vagueness of and lack of alternatives available through many current civil commitment statutes, reliance on current laws may result in unnecessary or unnecessarily lengthy confinement without adequate habilitative services.[204] Hence, special dispositional procedures are proposed below, which address these concerns.[205] They seek to protect both the interests of the public and those of the defendant while leaving efforts to reform current civil commitment procedures unimpeded.

The term lesser included offense used in this provision means an offense for which "some of the elements of the crime charged themselves constitute a lesser crime."[206] For example, homicide without the premeditation and malice required for murder constitutes the lesser included offense of manslaughter. Similarly, a person who unlawfully enters a building without the intent to commit a crime therein, is not guilty of burglary but of the lesser included crime of breaking and entering or criminal trespass.

(5) When a defendant who has a developmental disability is acquitted solely by reason of diminished capacity, the court shall appoint a multidisciplinary evaluation team to evaluate the defendant or to supplement a prior examination or evaluation conducted pursuant to this act.

COMMENT. The question before the court in determining the disposition of a defendant acquitted solely by reason of diminished capacity is the defendant's condition and needs at the time of disposition, not at the time the offense was committed. The provision of habilitative services on an involuntary basis, particularly in residential facilities, is no more constitutionally valid for offenders who have a developmental disability then it is for offenders who are mentally ill unless there are due process safeguards including a judicial determination that such services are required by the needs of the defendant and/or public safety.[207] An evaluation pursuant to section 15 of this act is the first step in determining whether the defendant is able to function in the community as a self-reliant law-abiding citizen, and if not, what habilitative services should be provided.

The multidisciplinary evaluation team should be appointed from the nominees selected by the prosecutor and defendant from the list published by the Evaluation Oversight Board.[208] Unlike the current practice in many states,[209] there is a presumption in favor of conducting the evaluation on an outpatient basis. The evaluation report is to include an opinion on whether the defendant has a developmental disability and if so, on the habilitative services which would assist the defendant, to the extent possible and within the shortest time possible, to function in the community as a self-reliant law-abiding citizen; descriptions of the defendant's mental, emotional, physical, and educational condition, adaptive behavior and social skills, and the basis for these opinions and conclusions; a list of the individuals who examined, interviewed or investigated the defendant; and the signatures of the concurring and dissenting members of the multidisciplinary evaluation.[210] In addition, the report may contain an

200 See sources cited *supra* note 178.

210 See Comment, *supra* note 170, at 227.

202 See Charts VII and VIII *supra*; BRAKEL and ROCK, *supra* note 140, at 422.

203 NEW YORK DEPARTMENT OF MENTAL HYGIENE, *supra* note 177, at 9 and 113-117.

204 See Chapter 6: Right to Services, *supra; see also* MHLP, *Legal Issues in State Mental Health CARE: Proposals for Change – Civil Commitment,* 2 MENTAL DISABILITY LAW REPORTER 77, 83–89 (1977); BRAKEL and ROCK, *supra* note 140 at 39–49; Morris, *supra* note 180, at 686.

205 These procedures are substantially similar to, but distinct from those contained in the Model Right to Developmental Disabilities Services Act, in Chapter 6, *supra; see* Comment, *supra* note 175, at 355–356.

206 Sansone vs. United States 380 U.S. 343 (1965).

207 *See e.g.,* Bolton vs. Harris 395 F.2d 642 (D.C. Cir. 1968); Allen vs. Radack 426 F. Supp. 1052 (D.S.D. 1977); State vs. Krol, 68 N.J. 236, 344 A.2d 289 (1975).

208 See sections 14(3)(b) and 15(1).

209 BRAKEL and ROCK, *supra* note 140, at 404.

210 Sections 15(4)(a) and (d).

opinion on whether the defendant would gain greater benefit from the proposed services if they were provided in a secure residential setting instead of in the community. The report must be submitted to the court within 10 days after the team was appointed with copies sent to the defendant, his or her attorney, the state Developmental Disabilities Advocacy Office and, following the dispositional hearing, to the individual(s) having direct responsibility for development of a Program Plan for the defendant.[211]

(6) (a) Following submission of the evaluation report, the court shall hold a dispositional hearing.

(b) If at the hearing the prosecutor fails to prove by a preponderance of the evidence that the provision of habilitative services on an involuntary basis is required by the needs of the defendant and/or of public safety, the court shall terminate the proceeding and all restraints on the defendant's liberty imposed in conjunction with the criminal proceeding.

(c) If at the hearing the prosecutor proves by a preponderance of the evidence that the provision of habilitative services on an involuntary basis is required by the needs of the defendant and/or of public safety, the court shall order:

(i) The development of a Program Plan in the manner set forth in subsection (7) of this section;

(ii) The provision and/or purchase by the state developmental disabilities agency of the necessary habilitative services using the least restrictive alternatives; and

(iii) The participation of the defendant in those necessary habilitative services.

(d) An order issued pursuant to paragraph (6)(c) of this section shall specify whether the Program Plan may include placement of the defendant in a residential facility which is secure or not community-based, or to which nonoffenders are committed pursuant to (citation to the Model Right to end Admissions to Services Act/state civil commitment law(s)). No such placement shall be permitted unless the defendant has been acquitted solely by reason of diminished capacity of homicide, rape, armed robbery, kidnapping, arson or another criminal offense involving serious bodily harm or the substantial risk of such harm; the defendant's role in the commission of that offense was not merely as a conspirator, abettor or accessory; and the judge, or the jury, if any, finds that the prosecutor has presented clear and convincing proof that:

(i) The defendant currently presents a substantial risk of serious bodily harm to others;

(ii) The defendant will be provided at the facility with the habilitative services, identified through the evaluation report and the testimony and evidence presented at the hearing, which will assist him or her to function in the community as a self-reliant, law-abiding citizen to the extent and within the shortest time possible; and

(iii) There is no less restrictive setting in which the defendant can be provided those services without seriously endangering public safety.

COMMENT. This subsection outlines the dispositional procedures and alternatives available following the acquittal of a defendant with a developmental disability solely by reason of diminished capacity. Paragraph (6)(a) requires a dispositional hearing. While such a hearing is a normal part of criminal procedure following conviction, the statutes in 14 American jurisdictions still provide for automatic commitment following an acquittal by reason of insanity.[212] Some of these provisions have been overturned on constitutional grounds[213] but many remain in effect. As the decision is Bolton vs. Harris makes clear, the equal protection clause demands that judicial procedures "substantially similar to those in civil commitment proceedings" be provided to defendants acquitted because of a mental disease or defect.[214] Moreover, since the purpose of the dispositional proceeding is to determine whether the defendant is able to function in the community as a self-reliant, law-abiding citizen, and if not, what habilitative services are necessary to assist the defendant in becoming such a citizen, the individual abilities and disabilities of the defendant must be determined in order to develop the appropriate Program Plan.[215] Thus, on both legal and habilitative grounds, sound policy dictates the holding of a dispositional hearing.

At the hearing the court must decide whether the needs of the defendant and of public safety require that the defendant be ordered to participate in habilitative services provided or purchased by the state developmental disabilities agency — i.e. that such services are necessary to assist the defendant to function in the community as a self-reliant law-abiding citizen. Under current procedures, the hearing following a not guilty by reason of insanity verdict is to determine whether the defendant is mentally ill at the time of the hearing and if not, whether he or she is dangerous to himself or others.[216] Since, by definition, a developmental disability is likely to continue indefinite-

[211] Section 15(4)(d).

[212] See Chart VII *supra*.

[213] See authorities listed *supra* note 207.

[214] 395 F.2d 642, 651 (D.C. Cir. 1968); see Baxtrom vs. Herold, 383 U.S. 107 (1966).

[215] See Davis vs. Watkins, 384 F. Supp. 1196 (N.D. Ohio 1974); Eckerhart vs. Hensley 475 F. Supp. 908 (W.D. Mo. 1979).

[216] See authorities listed *supra*, note 207.

ly,[217] and such a disability must be proven at trial in order for a defendant to be acquitted under this section, it is unnecessary to require proof at the dispositional hearing that the defendant still has a developmental disability. However, as indicated above, proof of the nature and extent of the defendant's abilities and disabilities, and the implications of those abilities and disabilities both for the defendant personally and for the public in light of the criminal conduct in which the defendant engaged, remains essential.

If the court determines that habilitative services are not required — *e.g.*, for a defendant with epilepsy who was acquitted solely because of diminished capacity of an assault committed during the course of a seizure and who is now voluntarily taking anticonvulsive medication — the proceedings are to be dismissed, all conditions on release or bail requirements lifted, any bond exonerated, and the defendant if he or she is in custody must be released unless there is another basis for detention. If the court concludes that habilitative services are required, it must then determine what services are necessary, direct the responsible agency to provide or purchase those services and to develop a Program Plan in conjunction with the defendant, and finally, order the defendant to participate.

The prosecutor bears the burdens of going forward and of persuasion, since if habilitative services are required, they will be imposed by the state, for state purposes and utilizing state funds. The level of proof required to find that services are necessary is set at a preponderance of the evidence. However, a higher degree of certainty is required before the defendant may be placed in a secure or non-community based residential facility[218] or a facility in which persons are placed pursuant to the Model Right to Services/Admission to Services Act or the state's civil commitment provision. While there is a strong argument that clear and convincing proof is necessary before any service can be provided on an involuntary basis,[219] the preponderance standard was selected in light of the relatively mild impact on the defendant of an erroneous decision,[220] the quasi-criminal context of the dispositional proceeding, the need to strongly differentiate non-residential and institutional services, and to encourage use of the least restrictive alternative.

The criteria set forth in paragraph (6)(d) are to assure to the greatest extent possible that only those acquittees who actually pose a substantial threat to the public are placed in a secure, remote or institution-like residential facility, and that if such placement is warranted, a facility offering the appropriate habilitative services is selected. The threshold condition which must be met before such a placement may be considered is that the defendant was found beyond a reasonable doubt to have engaged in the conduct required for conviction of an offense involving serious bodily harm or the substantial risk thereof.

> The nonviolent nature of other offenses cannot justify a substantial determination that a person is likely to cause physical injury to others.[221]

A representative list is included in the provision. The terms used should reflect those contained in the criminal code of the enacting jurisdiction. Next, on the basis of the evaluation report and the evidence and testimony presented at the hearing, there must be clear and convincing proof that the defendant remains a threat to community safety. In some cases, the unique circumstances of the offense — *e.g.*, an intra-family crime — make reoccurence unlikely. While there are inherent problems with predictions of dangerousness,[222] they are somewhat lessened in this context. Since there has been proof beyond a reasonable doubt that the defendant inflicted or threatened serious bodily harm to each person. Besides, unless the defendant is found to pose a present danger, there is no basis for restricting his or her liberty, since the judicial process has had an opportunity to function and there has been no conviction.[223]

The third factor is that the habilitative services required by the defendant will be available at the facility in which the defendant will be placed.

> . . .[D]ue process requires that. . .patients. . . who have been involuntarily committed to a state mental hospital, be provided with such treatment as will give each of them a reasonable opportunity to be cured or to improve his mental condition. . . . The fact that plaintiffs have been deemed dangerous does not deprive them of a constitutional right to treatment. . . . On the contrary, if a mental patient is involuntarily confined because he [or she] is dangerous, due process requires a specific focus to the treatment which is his [or her] right. Treatment efforts must be directed to that aspect of his [or her] behavior which caused him [or her] to be classified as dangerous so that he [or she] has a reasonable opportunity to be eventually discharged from maximum security confinement. To hold otherwise would certainly "transform the hospital

[217] Section 3(9).

[218] For a definition of these terms see section 3(1) and the comment thereto.

[219] Chapter 6: Right to Services, *supra;* see Addington vs. Texas 441 U.S. 418 (1979).

[220] *See* In re Winship, 397 U.S. 358, 364 (1969).

[221] MHLP, *supra* note 204, at 656.

[222] See note 121 and accompanying text.

[223] *Cf.* section 8 on incompetence to stand trial.

> into a penitentiary where one could be held indefinitely for no convicted offense." Ragsdale v. Overholser, 108 U.S. App. D.C. 308, 315, 281 F.2d 943, 950 (1950).[224]

The final factor is that no less restrictive alternative would adequately protect the public.[225]

It is anticipated that these criteria, coupled with the limits on the length of the dispositional period, the provisions for review and for development of an individualized Program Plan, will help to remedy the current incongruous situation in which persons acquitted by reason of a mental illness or defect may be dealt with more harshly than if they had been convicted.[226]

(7) (a) Following issuance of an order under paragraph 6(c) of this section, the state developmental disabilities agency shall develop a Program Plan in consultation with the defendant and his or her attorney. The objective of the Program Plan shall be to assist the defendant, to the extent and within the shortest time possible, to function in the community as a self-reliant law-abiding citizen. The purpose, nature and possible beneficial and harmful consequences of the various alternatives shall be explained to the defendant to the maximum extent possible, and to the defendant's attorney.

(b) The plan shall cover the lesser of:

(i) The minimum parole eligibility period for the most serious offense of which the defendant was acquitted solely by reason of diminished capacity; or

(ii) Two (2) years.

(c) The Program Plan shall be submitted for approval to the court no more than ten (10) days after issuance of the order. A copy shall be provided to the prosecutor, the defendant, the defendant's attorney, and the (name of state Developmental Disabilities Advocacy Office). It shall be approved by the court without a hearing within ten (10) days of its submission unless the court on its own initiative sets a hearing or the prosecutor or defendant move for a hearing. In such a motion, the moving party shall specify the reasons why the plan is alleged to be inadequate or inappropriate and the proposed modifications.

(d) Following the hearing, the court shall approve the plan or modify it so as to further the objectives set forth in paragraph (7)(a) of this section.

COMMENT. This subsection governs the development, purpose, coverage and approval of Program Plans following a determination that habilitative services are required for a defendant who has been acquitted solely by reason of diminished capacity due to a developmental disability. With two exceptions, it corresponds with section 8(6) on the development and approval of Program Plans following a finding of incompetency to stand trial. The first difference is that the objective of the plan is focused on the defendant's ability to live safely and lawfully in the community. The second is that the length of time under this provision for which the defendant may be required to participate in habilitative services may be greater than the one year maximum possible following a finding of incompetency to stand trial.

Since subsection (8) of this section limits the length of the disposition under this section to the maximum sentence which could have been imposed for the most serious offense of which the defendant was acquitted solely by reason of diminished capacity,[227] since the maximum sentence for many offenses is several years, and since as a result of natural changes and the impact of the habilitative services the defendant's condition and needs can be expected to change more rapidly, this provision together with subsection (9) and (10) of this section call for periodic reviews and updates of the Program Plan to assure that the habilitative services prescribed continue to match the defendant's condition and needs and to reflect the least restrictive alternative for achieving the purpose of the disposition. Because of the wide range of maximum sentences, it is necessary to provide some flexibility in the review system. Thus, paragraph (7)(c) provides that the initial Program Plan should cover a period of the lesser of two years or the period between disposition and the date on which the defendant would first be eligible for parole if he or she were sentenced to the maximum term of imprisonment possible for the most serious offense for which he or she was found to have committed the requisite physical acts. For example, in states in which offenders become eligible for parole after serving one-third of their maximum sentence, the initial Program Plan under this section for a defendant who could have been sentenced to a maximum of five years in prison would cover a period of 20 months. If the maximum possible sentence had been 20 years, the initial Program Plan would cover two years. Since the parole eligibility date is a benchmark triggering a significant review process for persons convicted of a crime, it appears logical and equitable to have to review process for dispositions under this section correspond, especially for those cases in which the possible maximum sen-

[224] Eckerhart vs. Hensley, 475 F. Supp. 908, 914-915 (W.D. Mo. 1979).

[225] See section 3(15) for a definition and discussion of least restrictive alternative.

[226] Morris, *supra* note 180, at 686; Monahan, *supra* note 167, at 738-739; Caulfield, *Ohio Commitments of the Mentally Ill Offender*, 4 CAPITAL LAW REVIEW 1, 22 (1974).

[227] See *id.*; In re Moye, 149 Cal. Rptr. 491, 584 P. 2d 1097 (1978); see Chart VII *supra*.

tence is three years or less.[228] The two year limit on the duration of a Program Plan is to assure that the Plan does not become dated and to facilitate oversight of the provision of habilitative services under subsection (10). While it can be argued that even two years is too long, the mechanism provided in subsection (11) which permits judicial review under certain circumstances at the request of the defendant, the Plan monitor, or the head of a facility in which the defendant has been placed, should provide sufficient protection without the substantial expense of mandatory yearly or even more frequent reviews.

As is the case with dispositions following a finding of incompetency to stand trial, the state developmental disabilities agency rather than the state correctional agency is responsible for developing the Program Plan and for providing or purchasing the necessary habilitative services, since the defendant was not actually convicted of a crime. However, in light of the finding that the defendant did engage in the physical conduct necessary for conviction, the provisions on corrections contained in section 12 of this act, and the arguments for differentiating civil and criminal commitments outlined in the comment to subsection (4) of this section, placing the responsibility with the correctional system would not be inconsistent with the purposes of the model act, so long as all the procedural protections continue to apply and adequate funds are allocated to the correctional agency to provide or purchase the necessary habilitative services.

A further explanation of the operation and intent of this subsection is contained in the comment to section 8(6).

(8) No disposition ordered pursuant to subsections (4) or (11) of this section shall exceed the maximum period of incarceration to which the defendant could have been sentenced for the most serious criminal offense of which the defendant was acquitted solely by reason of diminished capacity, minus the number of days he or she was confined prior to issuance of the dispositional order as a result of the charges or any examination or evaluation under this act.

COMMENT. This subsection limits the duration of any habilitative services or restraints on a defendant's liberty imposed under this section. In so doing, it follows the principle that defendants should not be penalized for asserting the defense of diminished capacity by having their liberty restrained for a longer time than if they had been convicted of the crime with which they were charged.[229] Accordingly, the authority to impose, provide, or continue both non-residential and residential habilitative services as a result of an acquittal by reason of diminished capacity ceases upon passage of the maximum period of confinement which could have been imposed upon conviction. The statutes in four states and American Samoa currently limit disposition in this manner following a verdict of not guilty by reason of insanity.[230]

In addition, credit for time spent in confinement as a result of the charges prior to disposition must be provided. Such credit against the sentence is provided or at least considered in the sentencing provisions of many states and has been recommended by a number of commentators.[231]

(9) Six (6) months after development of the initial Program Plan and every six (6) months thereafter until termination of the dispositional order or expiration of the dispositional period pursuant to subsection (8) of this section, or at such intervals as the court requires, the official of the state developmental disabilities agency who is monitoring the provision of habilitative services to the defendant shall file a report with the court. The report shall list the services specified in the most recent Program Plan, describe the defendant's response to those services, and indicate which, if any, of those services have not been offered to the offender and the reasons therefor. A copy of the report shall be provided to the defendant, the (name of the state Developmental Disabilities Advocacy Office), and the state developmental disabilities agency. At the defendant's request, the report shall also be sent to his or her attorney or a designated representative. The information contained in the report shall be explained to the defendant to the maximum extent possible.

COMMENT. To assure the continuing accountability of the service providers during the dispositional period, this provision requires submission of periodic accountings of the habilitative services provided, how they compare to those required by the Program Plan, the reason for any discrepancies, and whether habilitative services are having the effect anticipated. These reports will provide feedback to the courts on the impact of their orders and shortcomings of the service delivery system, warning to the director of the state developmental disabilities agency of problem areas and

[228] In keeping with the premise of this act that the procedures for persons with a developmental disability should be comparable to those governing persons without such a disability, enacting jurisdictions which have abolished parole should omit subparagraph (7)(c)(i). See D. FOGEL, WE ARE THE LIVING PROOF: THE JUSTICE MODEL FOR CORRECTIONS (1975).

[229] Morris, *supra* note 180, at 686; Monahan, *supra* note 167, at 538-539.

[230] CONNECTICUT GENERAL STATUTE ANNOTATED §53a-47 (West 1972 & Supp. 1978); ILLINOIS ANNOTATED STATUTES ch. 38, §1005-2-4 (Smith -Hurd 1973 & Supp. 1978); KANSAS STATUTES ANNOTATED §22-3430 (1974 & Supp. 1978); OREGON REVISED STATUTES §161.341 (1977); AMERICAN SAMOA CODE, tit. 15, §7805 (Cum. Supp. 1978); *accord* In re Moye, 149 Cal. Rptr. 491, 584 P. 2d 1047 (1978).

[231] *E.g.*, ABA, *supra* note 141; NATIONAL ADVISORY COMMISSION, *supra* note 43, at §5.8; ALI, MODEL PENAL CODE, *supra* note 134, at §7.09.

potential liability, and a basis for motions for modification or termination pursuant to subsection (11). The term designated representative is intended to include, *inter alia,* employees of the state Developmental Disability Advocacy Office, a family member, a friend or a guardian.

(10) (a) The court shall hold a review hearing:

(i) Not less than two (2) weeks before expiration of the initial Program Plan under subsection (7) of this section;

(ii) Every two (2) years thereafter until termination of the proceedings or expiration of the maximum dispositional period pursuant to subsection (8);

(iii) Following submission of a motion pursuant to subsection 11; and

(iv) Within ninety (90) days after receipt of a notice pursuant to subsection (13).

(b) At the hearing the court shall review:

(i) Whether the services prescribed in the dispositional order have been provided to the defendant and if so, whether they have been provided in the manner prescribed in the defendant's Program Plan;

(ii) Whether provision of habilitative services to the defendant on an involuntary basis continues to be required by the needs of the defendant and of public safety; and if so,

(iii) Whether the dispositional order and Program Plan should be modified to assure that the necessary habilitative services are provided and that they are provided by means of the least restrictive alternative.

(c) In conjunction with the hearing the court may appoint a multidisciplinary evaluation team to update the evaluation performed pursuant to subsection (5) of this section. Such an evaluation shall be ordered upon a showing that there has been a significant change in the defendant's mental, emotional, physical, or educational condition, adaptive behavior or social skills.

(d) (i) Except as specified in subparagraph (10)(d)(ii), the prosecutor must prove by a preponderance of the evidence that provision of services on an involuntary basis continues to be required by the needs of the defendant and of public safety, and that a particular habilitative service requested by the defendant is not so required. The prosecutor must present clear and convincing proof of the requirements for commitment set forth in paragraph (6)(d) of this section in order for a commitment to continue.

(ii) When the hearing is held pursuant to a motion filed by the defendant under subsection (11) of this section, the defendant shall have the burden of going forward with a *prima facie* case supporting the requested relief. If such support is presented, the court shall grant the relief requested by the defendant unless the prosecutor meets the burden of pursuasion set forth in paragraph (10)(d)(i).

(e) (i) If the court concludes that the provision of habilitative services on an involuntary basis is not required, it shall terminate the proceeding and all restraints on the defendant's liberty imposed in conjunction with the proceedings.

(ii) If the court concludes that provision of habilitative services on an involuntary basis is required, it shall determine what changes, if any, are necessary in the habilitative services being provided to the defendant and the manner and location in which those services are provided. The court shall then issue a revised dispositional order directing development of an updated Program Plan and the provision of services in accordance with that Plan. The updated Program Plan shall be developed in the manner prescribed in subsection (7) of this section.

(iii) If the court concludes that services have not been provided in accordance with the dispositional order or the Program Plan, it may order that contempt proceedings be initiated against the state developmental disabilities agency. If such services cannot be provided, it may terminate the proceeding and all restraints on the defendant's liberty imposed in conjunction with the proceeding.

COMMENT. The state's authority to maintain restrictions on the liberty of an individual continues only so long as the individual meets the legal criteria for imposing those restrictions.[232] Furthermore, as indicated by the Supreme Court in Jackson vs. Indiana, the nature of the restraints must "bear some reasonable relation to the purpose" for which they were imposed.[233] Accordingly, this provision establishes a procedure for periodic judicial review. Ten states statutorily prescribe some form of periodic review for defendants committed following a verdict of not guilty by reason of insanity.[234] However, since the principles and purposes underlying periodic review hearings which are outlined below apply with equal strength to the provision of services on an involuntary basis in any setting, judicial review is required for dispositions which rely on community services as well as those involving placement in secure facilities.

The review hearings are intended to serve a multifold purpose. First, they are designed to determine whether at the time of the review hearing the defendant can function in the community as a self-reliant, law-abiding citizen without habilitative

---

232 *See* O'Connor vs. Donaldson, 422 U.S. 563 (1975); State vs. Fields, 77 N.J. 282, 390 A.2d 574 (1978).

233 406 U.S. 715, 738 (1971).

234 *See* Chart VIII *supra;* see also United States *ex rel.* Schuster vs. Herold, 410 F.2d 1071 (2d Cir. 1969); State vs. Fields, 77 N.J. 282, 390 A.2d 574 (1978).

services imposed on a nonvoluntary basis. If so, the proceedings and all restraints on the defendant's liberty arising therefrom must be terminated.

Second, when the court determines that involuntary habilitative services are still required, it must then determine what changes in the type, location, and intensity of the habilitative services being offered may be necessary in order to assist the defendant to function as a self-reliant, law-abiding citizen within the shortest time possible and with no more restriction of physical freedom, social interaction and individual choice than is essential.[235]

Third, the hearing is designed to provide added accountability to the dispositional process by assuring that the habilitative services ordered by the court have been provided and that they have been presented in the manner which was approved. The failure to follow the dispositional order and Program Plan is neither in the public interest nor those of defendants, and flaunts both the authority of the court and the purpose of this act. Thus, pursuant to subparagraph (10)(e)(iii) of this section, it may result in a contempt citation against the individuals responsible and termination of all restraints on the defendant's liberty.

In order to assist the court in making first and second determinations, paragraph (10)(c) permits the court to order a new evaluation of the defendant. However, in order to reduce both the intrusion upon the privacy of the defendant and the expense to the state inherent in an evaluation, the provision requires a special showing of need before an evaluation may be ordered. The selection and appointment of the multidisciplinary evaluation team and the content of the evaluation report are governed by section 15.

Under paragraph (10)(a) review hearings must be held in three instances. In keeping with the above purposes, the first instance is in conjunction with the expiration of the defendant's Program Plan.[236] Second, to prevent: the defendant from having his or her liberty restrained unnecessarily; the provision of unneeded services; the public from being endangered gratuitously; and the circumvention or disregarding of the dispositional order and Program Plan, an interim review hearing must be held upon request of the defendant or the individual overseeing the provision of and the response to services. Finally, to facilitate the extension of the rights and procedures established by this act to individuals with a developmental disability committed following a not guilty by reason of insanity or diminished capacity verdict prior to the act's effective date, a review hearing must be held upon receipt of the notice required by subsection (13).

Except when the review hearing is held at the defendant's request, the prosecutor retains both the burden of going forward and the burden of persuasion.[237] To do otherwise would impermissibly force the defendant to prove his or her fitness to be free of restraints.[238] In order to limit frivolous requests for hearings, however, defendants requesting a review hearing persuant to subsection (11) are required to present evidence which, unless rebutted, supports the need for the relief requested. Following the presentation of such evidence, the burden of persuasion reverts to the prosecutor. The levels of proof required are the same as those prescribed by subsection (6) for the initial dispositional hearing.

(11) (a) A defendant receiving habilitative services pursuant to this section, the official of the state developmental disabilities agency who is monitoring the provision of habilitative services to the defendant, or the head of the facility in which a defendant has been placed pursuant to paragraph (7)(d) may file a motion for termination of the services or placement or modification of the Program Plan on the grounds that:

(i) The defendant no longer requires such habilitative services to function in the community as a self-reliant, law-abiding citizen; or

(ii) The habilitative services being provided are no longer consistent with the defendant's needs and other habilitative services are required; or

(iii) The manner in which the habilitative services are being provided or the facility to which the defendant has been committed is not the least restrictive alternative consistent with his or her needs and public safety; or

(iv) The habilitative services specified in the most recent Program Plan are not being provided.

(b) A defendant may submit a motion to the court pursuant to this subsection by any means including but not limited to oral communication or informal letter. Except in extraordinary circumstances, motions may be filed by the defendant no more than once during any sixty (60) day period and no sooner than thirty (30) days after a review hearing.

COMMENT. As indicated in the comment to the preceding subsection, this provision is included to provide added flexibility and accountability to the dispositional process as well as added protection to both the defendant and the public. It is intended to replace, for defendants falling within the scope of this statute, the existing provisions on transfer, conditional release and full release following an acquittal by reason of insanity.[239] As in disposi-

[235] See section 3(15) and 10(b), (c) and (d).

[236] See section 10(7).

[237] United States *ex. rel.* Schuster vs. Herold, 410 F.2d 1071 (2d Cir. 1969); State vs. Fields, 77 N.J. 282, 390 A. 2d 574 (1978).

[238] *See* Baxtrom vs. Herold, 383 U.S. 107 (1966).

[239] See Chart VIII *supra*.

tions following a finding of incompetency to stand trial, the subsection requires that individual employees of the state developmental disabilities agency be made responsible for overseeing the provision of habilitative services to a defendant with a developmental disability who was acquitted solely by reason of diminished capacity. Making an identifiable individual responsible rather than having this duty spread among several persons and offices or vested *pro forma* in the head of the agency, is intended to enhance both responsiveness to the defendant's needs, and accountability when problems occur.[240]

Paragraph (11)(b) is intended to make the review process more accessible to individual defendants. However, to encourage that all relevant issues are raised in a single motion and to reduce repetitive and frivolous filings, it limits the frequency with which a defendant may request a review hearing. The exception for "extraordinary circumstances" is intended to cover instances in which, for example, a defendant is denied necessary services or has been transferred in violation of the Program Plan.

(12) The state developmental disabilities agency shall be a party in any hearing held pursuant to subsections (7) and (10) of this section.

COMMENT. Because of the responsibilities and potential liability imposed by this section on the state agency responsible for providing services to persons with a developmental disability,[241] the agency is made a party to the dispositional proceedings with the same rights as other parties to notice, to be represented, to present testimony and evidence, to cross examine witnesses, and to appeal adverse orders.[242] This not only permits the agency to be heard, but also makes clear that the agency is subject to the orders of the court, thus enhancing the practicality, clarity and the enforceability of dispositional orders.[243]

(13) The head of any facility housing a person who has a developmental disability and who was found not criminally responsible and placed in custody prior to the effective date of this act shall, within thirty (30) days after the effective date of this act, notify the court that committed the person. Within ninety (90) days after receiving the notice, the court shall conduct a review hearing in accordance with subsection (10) of this section.

COMMENT. Like section 8(12), this is a transition provision designed to facilitate the extension of the rights and procedures established by this section to persons with a developmental disability who as a result of that disability were found not guilty by reason of insanity or a "mental deficit" prior to the effective date of this act. It imposes a duty on the director of a facility housing NGI acquittees to review the files and identify which of the persons who were criminally committed, have a developmental disability and which are mentally ill. The director must then notify the court that committed the developmentally disabled defendant of the defendant's presence in the facility. Upon receiving the notice, the court must hold a review hearing and determine the appropriate disposition in light of the criteria and procedures established by this act.[244] While this provision may impose an initial burden on facilities without adequate records,[245] and while there will undoubtedly be cases involving defendants who were both developmentally disabled and mentally ill at the time of commitment, it appears to be the only effective means of assuring that developmentally disabled defendants who have been committed under the prior code provision are not denied the opportunities and rights provided by this act, and that the public is not forced to pay the high cost of institutionalization for individuals able to function in the community as self-reliant, law abiding citizens.

INTRODUCTION TO SECTION 11

Sentencing is said to have four objectives: rehabilitation of the offender, protection of the public, deterrance of potential offenders, and retribution or punishment.[246] The priority afforded each of these objectives and the means for achieving them is the subject of a vigorous national debate.[247] Although there is much needed to be done to increase the effectiveness, fairness and equity of current sentencing practices and procedures, a wholesale revision of the laws governing sentencing is far beyond the scope of this model statute. Thus, this section has the more limited goal of assuring that the sentencing court has the information necessary to determine the appropriate sentence for an offender with a developmental disability in light of his or her special needs and vulnerabilities, and that the duration and nature of the sentence imposed is no more restrictive than is necessary to achieve the objectives of sentencing, particularly that of assisting the defendant to function in the community as a self-reliant, law-

240 See section 8(7).

241 See sections 10(4)-(10).

242 Section 13.

243 *Cf.* NACJJDP, *supra* note 148, §3.112.

244 See section 10(10).

245 Morris, *supra* note 180, at 684-685.

246 JUSTICE IN SENTENCING: PAPERS AND PROCEEDINGS OF THE SENTENCING INSTITUTE FOR THE FIRST AND SECOND JUDICIAL CIRCUITS, 4 (L. Orland and H. Tyler eds. 1974); Pugsley, *Retributivism: A Just Basis for Criminal Sentences,* 7 HOFSTRA LAW REVIEW 379, 381 (1979); PCLEAJ, TASK FORCE REPORT: THE COURTS 14 (1967); NATIONAL ADVISORY COMMISSION, *supra* note 43, at 3.

247 *See e.g., Symposium on Sentencing,* 7 HOFSTRA LAW REVIEW 1-138, 243-456 (1978-1979); FOGEL, *supra* note 228; DETERMINATE SENTENCING: REFORM OR REGRESSION? (1978); N. MORRIS AND G. HAWKINS, THE HONEST POLITICIAN'S GUIDE TO CRIME CONTROL (1970); NATIONAL ADVISORY COMMISSION, *supra* note 213; PCLEAJ, *supra* note 246.

abiding citizen. It is intended to fit within the enacting jurisdiction's existing sentencing framework.

Section 11. Sentencing

(1) (a) When there is reason to believe that a defendant has a developmental disability, a presentence report shall be prepared following conviction for any criminal offense for which an individual may be sentenced to imprisonment for one (1) year or more. The court may require preparation of a presentence report following the conviction of such a defendant for any other criminal offense.

(b) In addition to the matters prescribed in (citation to a statute or rule, if any, governing contents of presentence reports), a presentence report prepared pursuant to paragraph (1)(a) of this section shall set forth:

(i) Whether the defendant has a developmental disability; and if so

(ii) The nature and effect of that disability;

(iii) The type of habilitative services, if any, which will assist the defendant, to the extent and within the shortest time possible, to function in the community as a self-reliant, law-abiding citizen;

(iv) The availability of the necessary habilitative services in the community;

(v) The availability of the necessary habilitative services within the correctional system; and

(vi) The amount of time, if any, for which the defendant was confined prior to the sentencing hearing for examination, evaluation, or services under this act.

COMMENT. Most jurisdictions provide for the preparation of presentence reports, either on a mandatory or discretionary basis, for at least some offenders, and expanded use of such reports has been urged repeatedly over the past 20 years.[248] As one group noted:

> ...[S]erious consequences turn on the correct resolution of the sentencing decision, and it is sheer folly to attempt such a decision without more information than is typically provided by the guilt determining process.[249]

This subsection requires preparation of a presentence report in any case in which a sentence of incarceration for a year or more may be imposed upon an offender who there is reason to believe has a developmental disability, and permits preparation of such reports in less serious cases. This policy is premised on the fact that offenders with a developmental disability, contrary to the notion of inherent criminality which prevailed in the early part of this century,[250] do respond positively to appropriate habilitative services presented under the proper conditions and circumstances, and can become productive, self-reliant, law-abiding citizens.[251] Thus, it is in the interest of both the offender and the public, that the sentence be designed to facilitate habilitation.[252] The problem for the court then is to match the offender with the appropriate services.[253] The distinction based on the length of the potential sentence is included in recognition that some cases may be so minor as not to warrant the intrusion and expense of a presentence investigation and the provision of habilitative services at public expense.[254]

The information required to be added to the standard presentence report by paragraph (1)(b) is intended to apprize the judge of the defendant's needs and the community and correctional services available which can address those needs. It is anticipated that in most cases, an evaluation will have to be ordered pursuant to subsection (2) to assist in obtaining this information.[255] In order to clarify the time which must be credited against the defendant's sentence pursuant to section 8(11), the report must also advise the court of the total period, if any, the defendant was confined pursuant to section 8 and/or 15.

(2) In conjunction with the preparation of the presentence report, the court may order the defendant to undergo an evaluation and appoint a multidisciplinary evaluation team.

COMMENT. As noted in the preceding comment, it is anticipated that an evaluation will be required to assist in sentencing a convicted defendant with a developmental disability. Even when a defendant has been evaluated in conjunction with an inquiry into his or her competency to stand trial,[256] an update will be ncessary to aid in determining the habilitative services, if any, which are needed to assist the defendant to function in the community as a self-reliant, law-abiding citizen.

The multidisciplinary evalution team should be

---

248 *E.g.*, ALI, MODEL PENAL CODE, *supra* note 134, at §7.07; PCLEAJ, *supra* note 246, at 18; ABA *supra* note 141, at §§4.1-4.5; NATIONAL ADVISORY COMMISSION, *supra* note 43, §5.14; Santamour and West *supra* note 188 at 11; Rowan, *Corrections* in PCMR, *supra* note 12, at 649, 674.

249 ABA, STANDARDS RELATING TO SENTENCING ALTERNATIVES AND PROCEDURES, 205 (Approved Draft 1968); *accord* ABA, SENTENCING, *supra* note 141, at 112.

250 Fernald, *The Imbecile with Criminal Instincts*, 64 AMERICAN JOURNAL OF INSANITY 65 (1909).

251 *See e.g.*, McGovern, *The Incarcerated Retarded Offender: A Conceptual Analysis*, in LAW, PSYCHIATRY, AND THE MENTALLY DISORDERED OFFENDER (T. Brelje and W. Craine, eds. 1974); THE MENTALLY RETARDED CITIZENS AND THE CRIMINAL JUSTICE SYSTEM, *supra* note 12; M. SANTAMOUR and B. WEST, THE MENTALLY RETARDED OFFENDER AND CORRECTIONS, 9-11 (1977).

252 Allen, *The Retarded Offender: Unrecognized in Court and Untreated in Prison*, FEDERAL PROBATION 22, 27 (1978).

253 Martinson, *New Findings, New Views: A Note of Caution Regarding Sentencing Reform*, 7 HOFSTRA LAW REVIEW 243, 254 (1979).

254 *Cf.* ALI, MODEL PENAL CODE, *supra* note 134, at §7.07(1); ABA, SENTENCING, *supra* note 141, at §18-5.1(b).

255 For the procedures applicable to evaluations and the contents of evaluation reports, see section 15.

256 See section 8(4)(c)(ii).

selected in the manner specified in section 15 from the list of experts published by the Evaluation Oversight Board.[257] The evaluation report should contain descriptions of the nature and extent of the defendant's disabilities, if any, including his or her mental, emotional, physical and educational condition, adaptive behavior and social skills, and the techniques, tests and other examination or evaluation reports used to develop the diagnoses; an opinion regarding the habilitative services which would assist the defendant, to the extent and within the shortest time possible, to function in the community as a self-reliant, law-abiding citizen; the names and addresses of individuals who examined, interviewed or investigated the defendant, and the concurring or dissenting signatures of the members of the multidisciplinary disciplinary team.[258] The report must be submitted to the court within ten days. Copies should be provided to the prosecutor, the defendant, the defendant's attorney, the state Developmental Disabilities Advocacy Office and, following sentencing, to the individual responsible for the supervision and habilitation of the defendant.[259]

(3) In sentencing a defendant who is developmentally disabled, the court shall impose the least restrictive alternative consistent with the needs of the defendant and of public safety.

COMMENT. The sentencing judge has at least some, and often considerable discretion over the sentence to be imposed. Too often in the past, this discretion has worked against developmentally disabled offenders, particularly those with mental retardation. Because of a lack of community-based programs,[260] the failure to recognize that a defendant is mentally retarded,[261] the reluctance of both correctional and developmental disabilities agencies to establish means for dealing with aggressive mentally retarded individuals,[262] and lack of information and lingering suspicion,[263] mentally retarded offenders are grossly overrepresented in the American prison population.[264] Although the conditions in many prisons and jails are hardly conducive to health and safety, much less rehabilitation, for non-disabled offenders, they pose even a greater threat to offenders with a developmental disability.

> Every study that has been made of the prison system has indicated that while prisons may be dehabilitating, dehumanizing, and exasperating for most of the offenders they are infinitely more so for the retarded who cannot profit even by the meager programs of education and rehabilitation that are there for the so-called 'normal resident'. A retarded person is likely to be exploited and more bitterly dehumanized than his [or her] intellectually normal [colleagues]. . . .[265]

In recent years, a number of jurisdictions across the country have begun to develop alternative correctional programs for developmentally disabled offenders.[266] In order to encourage the growth of such programs, protect the safety of developmentally disabled offenders, and facilitate their habilitation, this subsection applies the principle of employing the least restrictive alternative to sentencing decisions. The provision does not rule out imprisonment or any other sanction, nor make inapplicable any of the sentencing objectives listed above.[267] It simply legitimizes consideration of developmental disability in determining the degree and duration of the restraints on liberty to be imposed, and establishes a presumption in favor of selecting a sentence or conditions of probation which are no more harsh, hazardous or intrusive and involve no more restrictions on the defendant's physical freedom or social interaction than are absolutely essential. Thus, it provides a guide

---

[257] See section 14(3)(b).

[258] See section 15(4)(a) and (e).

[259] See section 15(4)(e).

[260] See Santamour and West, *supra* note 188, at 9 and 12; SANTAMOUR and WEST, *supra* note 251, at 44-45; Rowan, *supra* note 248, at 673.

[261] Haggerty, Kane and Udall, *supra* note 163; SANTAMOUR and WEST, *supra* note 251, at 5-6; Allen, *supra* note 252, at 25; McGovern, *supra* note 251; Plotkin, *Stranded in the Jungle: The Mentally Retarded Person in the Criminal Justice System,* 1 DISTRICT LAWYER 35 (1977).

[262] SANTAMOUR and WEST, *supra* note 251, at 18; Rowan, *supra* note 248, at 661; Bicklen, *Advocacy for a Nonexceptional Approach,* in THE MENTALLY RETARDED CITIZENS AND THE CRIMINAL JUSTICE SYSTEM, *supra,* note 12 at 185, 187.

[263] See SANTAMOUR and WEST, *supra* note 251, at 10; Rowan, *supra* note 248, at 652 and 666.

[264] B. BROWN and T. COURTLESS, THE MENTALLY RETARDED OFFENDER (1971); SANTAMOUR and WEST, *supra* note 251, at 3-5, 7-8, 16-17; Rowan, *supra* note 248, at 658-659; 662-664; McGovern, *supra* note 251.

[265] Richard Allen as quoted in Krause, *supra* note 30, at 26; *see also* Haggerty, Kane and Udall *supra* note 163, at 68; Bicklen, *supra* note 262, at 187-188; SANTAMOUR and WEST, *supra* note 251, at 9; M. Harris and F. Dunbaugh, *Premise for a Sensible Sentencing Debate: Giving Up Imprisonment,* 7 HOFSTRA LAW REVIEW 417, 423 (1979).

[266] SANTAMOUR and WEST, *supra* note 251, at 18-20; THE MENTALLY RETARDED CITIZENS AND THE CRIMINAL JUSTICE SYSTEM, *supra* note 12; Note, *supra* note 172.

[267] The danger of overreliance on rehabilitation as a correctional objective is illustrated by the following excerpt from the Reaction Comment by Vincent Ziccardi, PCMR, *supra* note 12, at 687, 688:

> For example, I recently walked into a courtroom where one of our attorneys was representing a client at sentencing. The judge was saying to the attorney, "I want to help your client. I think he can be rehabilitated and he should learn a trade. I am going to send him to the state penitentiary for 5 to 15 years."
>
> The defense attorney jumped up and said, "Judge, this man cannot be rehabilitated. What he needs is punishment. Send him away for 11 ½ to 23 months. He can never learn a trade."
>
> This scene would be humorous if it did not occur so frequently. Unfortunately, "helping" a defendant, in the eyes of the court, often results in a person's being sentenced to a very long period of incarceration and seldom results in the provision of constructive services to him.

to the parties, the court, correctional personnel and the public, a basis for use of the court's inherent power to order services,[268] and a measure for appellate review.

No provision is made for a special "defective offender" category or indeterminate sentencing procedures for defendants with a developmental disability.[269] The concept of indeterminate treatment-oriented correctional programs which enjoyed significant support in both adult and juvenile corrections over the past few decades is being increasingly rejected by legislatures, courts, and commentators around the country[270] because of the failure of special institutions to provide the treatment and the negative impact of indeterminate sentences on offenders.

> The indeterminate sentence is used as an instrument of inmate control, as it gives the institution complete power to release an inmate at any time or to keep him [or her] incarcerated for life. This type of sentence is fraught with potential for abuse, and there is a real danger that it can be used to punish political beliefs and views unpopular with the staff. Even when the sentence does not turn out to be for life, the result is a longer confinement than would have occurred had the inmate gone to a regular prison. The indeterminate sentence is just another device to hide society's dehumanizing treatment of criminals, particularly those who are mentally disabled.
>
> Clever inmates quickly discover what is expected of them, go through the motions, play the game, and fool the therapist. Thus, the indeterminate sentence structure encourages cunning offenders and is more favorable to them than to the less intelligent. . . .
>
> Treatment and habilitation are either token or nonexistent, and educational opportunities are poor. Also, there is not enough vocational training available to meet the needs of the inmates. And when there is not enough vocational training to go around, the retarded are the first among those excluded.
>
> As negative as the results are for the nonretarded segment of the population, they are that much more damaging for the retarded. Their frustrations are greater, and they are continually faced with their own inability to cope with structures and attitudes within the institution.[271]

Developmentally disabled offenders should neither be excused nor receive extra punishment for crimes which they have committed. As stated previously, to the greatest extent possible they should be handled in the same manner as nondisabled offenders except when their added vulnerability to abuse and need for specialized habilitative services require otherwise. Indeterminate sentencing and special correctional categories and facilities have not realized these objectives. This provision and the protection and procedures in section 12 provide an alternative approach which, with the requisite support, holds out the promise of doing so.

Section 12. Corrections

(1) (a) An offender who has a developmental disability shall be entitled to receive habilitative services as defined in this act. The state correctional agency has the duty to provide or purchase the habilitative services specified in the Program Plans prepared for all offenders with a developmental disability whether they are in custody, on parole, or on probation.

COMMENT. The record of correctional agencies regarding the provision of services tailored to the needs of developmentally disabled offenders is not a good one.

> In most states, appropriate institutional, probation, and other treatment facilities are not available for differential treatment of the retarded, and they therefore receive little or none of the special care needed if they are to be . . . habilitated. This shortage of programs and services is a result of several factors; disagreement as to whether retarded offenders are most appropriately handled as part of an integrated prison system or in specialized facilities; a lack of [personnel] resources in the areas of special education, psychiatry, and psychology; a lack of alternative resources in the community; and a lack of coordination among and within the agencies dealing with mentally retarded offenders.

[268] *See* N. WEINSTEIN, THE INHERENT POWER OF THE COURT, 1 (1978).

[269] *E.g.*, MARYLAND ANNOTATED CODE art. 31B, §5 (1971); *see* BRAKEL and ROCK, *supra* note 140, at 341-359.

[270] *E.g.*, MARYLAND ANNOTATED CODE art. 31B, §5 (Cum. Supp. 1979); California Penal Code §1170 *et reg.* (West Supp. 1979); Grindle vs. Miller 400 A.2d 787 (N.H. Sup. Ct. 1979); Nebraska vs. Shaw, 277 N.W. 2d. 106 (Neb. Sup. Ct. 1979); Rowan, *supra* note 248; Bicklin, *supra* note 262; STONE, *supra* note 78, at 179-198; D. WEXLER, CRIMINAL COMMITMENTS AND DANGEROUS MENTAL PATIENTS: LEGAL ISSUES OF CONFINEMENT, TREATMENT AND RELEASE (1976); DETERMINATE SENTENCING, *supra* note 247; Rector and Van Duizend, *New Directions for Juvenile Justice*, 39 OHIO STATE LAW JOURNAL 347; NACJJDP, REPORT TO THE ADMINISTRATOR ON STANDARDS FOR THE ADMINISTRATION OF JUVENILE JUSTICE, §§3.181-3.182 (1976). W. GAYLIN, I. GLASSER, S. MARCUS, and D. ROTHMAN, DOING GOOD: THE LIMITS OF BENEVOLENCE (1978).

[271] Rowan, *supra* note 248, at 668-669.

> Undifferentiated handling has resulted not only in neglect of retarded offenders, but in positive damage to them. In these situations they vegetate in institutions, only occasionally participating in any programs at all. They are assigned to menial maintenance tasks having no vocational training potential and are required to conform to standard rules and share work assignments with more intelligent inmates. They tend to react by withdrawing from competition completely, thus making it more difficult to prepare them for life in the community.[272]

The most recent decision in the continuing litigation over Alabama's prison system,[273] suggests that despite the efforts in a few jurisdictions,[274] the situation remains bleak in many areas.

Accordingly, this subsection establishes a statutory right to treatment — i.e., it requires the state correctional agency to provide or procure the medical, mental health, social, vocational, educational, physical therapy, counseling, financial management or other services which will assist a person who has a developmental disability, to the extent and within the shortest time possible, to function in the community as a self-reliant, law-abiding citizen.[275] Many of the services required by developmentally disabled offenders do not differ generically from those which should be offered to non-disabled offenders, although the structure and manner of presentation may vary. Others relate to problems specifically related to the individual offender's disabilities.[276] Whether or not a constitutional right to treatment for offenders is found to lie within the Eighth and/or Fourteenth Amendments to the U.S. Constitution, sound correctional policy as well as fiscal concern over the cost of maintaining offenders in correctional facilities and reconvicting repeating offenders argue strongly for the provision of habilitative services. As is noted in Newman vs. Alabama:

> Quite apart from humanitarian reasons... the education of prisoners, and the chance of their rehabilitation, directly involve the safety of society. There is clearly no guarantee that a prisoner — adequately educated — will be a better person when he gets out because he has enough education and training to obtain and hold a job. But if he gets no training or education at all, it is absolutely certain that he — and another 98 percent of the prison population like him — will commit other crimes when they get out.[277]

Traditionally, developmentally disabled offenders have been unwanted by both the correctional and services systems.[278] The duty to assure that the appropriate habilitative services are offered is assigned by this subsection to the correctional agency rather than the developmental disabilities agency, in order to improve accountability by not dividing responsibility for the supervision, protection, and habilitation of the offender among two or more bureaucracies and to assure that incarcerated offenders with a developmental disability receive the same opportunities for release as other offenders.

> Since up to one-third of the population of correctional facilities may be at least psychometrically retarded, those whose professional interest is in mental retardation have a special responsibility for the corrections system. In discharging this responsibility they must not be content merely to construct laws that will relieve commissioners of corrections of the problems raised by mentally retarded offenders. They must address themselves to changing the corrections system so that it can provide more adequate services to mentally retarded offenders.... Grouping persons together may intensify the maladaptive consequences of retardation while the delinquent patterns remain untreated. . . . [M]entally retarded citizens have a right to be treated as nearly as possible like other citizens, and this right includes bearing responsibility for one's own behavior. Thus, there should be an insistence that present correctional facilities improve their habilitative procedures, incorporating nonsegregated programming for retarded offenders.[279]

---

272 *Id.* at 673; see also the other sources listed in note 260 and Haggerty, Kane and Udall, *supra* note 163 at 68; McGovern, *supra* note 251; *see* MHLP, *Legal Issues in State Mental Health Care: Proposals for Change – Mental Health Services for Prisoners,* 2 MENTAL DISABILITY LAW REPORTER 665, 666-667 (1978).

273 Newman vs. Alabama, 466 F. Supp. 628, 631 (M.D. Ala. 1979).

272 *See e.g.*, THE MENTALLY RETARDED CITIZENS AND THE CRIMINAL JUSTICE SYSTEM, *supra* note 12; SANTAMOUR and WEST, *supra* note 251, at 18-23.

275 PCMR, *supra* note 124, at 63; SANTAMOUR and WEST, *supra* note 251, at 30; NATIONAL ADVISORY COMMISSION, *supra* note 43, at §2.9; Rowan, *supra* note 248, at 673; Haywood, *Reaction Comment* PCMR, *supra* note 12, at 679-680; *cf.* UNITED STATES DEPARTMENT OF JUSTICE (DOJ), FEDERAL STANDARDS FOR CORRECTIONS, 30 (Draft 1978); Rouse vs. Cameron 373 F.2d, 451 (D.C. Cir. 1966); *cf.*, ABA COMMISSION ON THE MENTALLY DISABLED, *Rights Of Disabled Persons In Residential Facilities,* 3 MENTAL DISABILITY LAW REPORTER, 348, 357-359, (1979).

276 SANTAMOUR and WEST, *supra* note 251, at 35-44.

277 466 F. Supp. 628, 638.

278 SANTAMOUR and WEST, *supra* note 251 at 18; Wald, *Advocacy For the Mentally Retarded Offender,* in THE MENTALLY RETARDED CITIZENS AND THE CRIMINAL JUSTICE SYSTEM, *supra* note 12, at 247, 256.

279 Haywood, *supra,* note 274, at 679; *accord* SANTAMOUR and WEST, *supra* note 251; NEW YORK DEPARTMENT OF MENTAL HYGIENE, *supra* note 177; Rowan, *supra* note 248; *but see* MHLP, *supra* note 272.

However, nothing in the subsection is intended to prevent the correctional agency from contracting with state or local developmental disabilities agencies to provide the required services, so long as accountability remains in the correctional system. Responsibility is vested at the state rather than the local level to assure the habilitative services can be offered statewide in both rural and urban areas.

(b) A person who has a developmental disability shall be entitled to protection against physical abuse while housed in a correctional facility. The head of a correctional facility which houses one (1) or more persons who have a developmental disability shall establish programs and procedures to protect that/those persons(s) from abuse and exploitation by other residents of the facility to the maximum extent possible.

COMMENT. This provision establishes a right to protection from abuse and exploitation for developmentally disabled persons housed in jails, prisons and other facilities. It applies regardless of whether that person is a suspect, defendant or convicted offender. While developmentally disabled persons have been "among the last to receive specialized services. . ." in correctional facilities, they have been "among the first to be brutalized and victimized by both the prison administration and fellow inmates."[280] A number of cases have held that unless reasonably adequate measures have been taken to assure the personal safety of prisoners, their incarceration constitutes cruel and unusual punishment in violation of the Eighth Amendment.[281] Accordingly, this provision requires the individual in charge of each correctional facility[282] to develop means of protecting developmentally disabled residents of the facility from physical abuse, sexual abuse, and harassment. This includes the deprivation of necessary medication to control seizures or using access to such medication as an enforcement mechanism.

This provision is not intended to endorse isolation of developmentally disabled persons either within a facility or in a special segregated facility. Although segregation may be administratively easier, it runs directly counter to the principles of normalization and use of the least restrictive alternative which form the basis of most recommendations regarding habilitation of developmentally disabled offenders, and has usually resulted in confinement without the requisite services.[283] "The 'special education' error of segregating individuals merely on the basis of diagnostic tests" should not be repeated.[284]

(c) An offender who has a developmental disability shall be entitled to the assistance of an advocate while housed in a correctional facility. The state correctional agency shall provide or purchase advocacy services in each secure or non-community-based correctional facility housing one or more offenders with a developmental disability. Such services shall include direct personal contact on at least a monthly basis between each offender with a developmental disability and an advocate with the training and the qualifications set forth by the (name of the state Developmental Disabilities Advocacy Office). In providing advocacy services under this section, advocates shall be entitled to support, advise, espouse the rights of, or directly intercede on behalf of offenders with a developmental disability in any matter affecting the habilitative services provided to, the protection of, or the restraints imposed on such offenders, except for:

(i) An appeal of the offender's conviction or sentence;

(ii) Post-conviction collateral attack on the offender's conviction; or

(iii) Proceedings arising from criminal charges other than those which resulted in the offender's incarceration.

COMMENT. Following passage of the Developmentally Disabled Assistance and Bill of Rights Act,[285] an extensive network of protection and advocacy agencies was established to assist persons with a developmental disability.[286] For the most part, these agencies have focused on developmentally disabled individuals in the community or in mental health and mental retardation facilities, with little outreach to persons with developmental disabilities incarcerated in correctional facilities.[287] Yet because of their isolation and particular vulnerability, and the lack of habilitative services,[288] incarcerated developmentally disabled offenders are in as great or even greater need of access to trained advocates than persons with a developmental disability in the community.

---

[280] Katz, *The Criminal Reform Movement*, in PCMR, *supra* note 180, at 626, 634; see also Allen, *supra* note 252 at 27; SANTAMOUR and WEST, *supra* note 251 at 9; *e.g.*, *Pugh vs. Locke* 406 F. Supp. 318, 325 (M.D. Ala. 1976), *affd. sub nom.* Newman vs. Alabama, 559 F. 2d 283 (5th Cir. 1977), *cert. denied* 438 U.S. 915 (1978).

[281] *E.g.*, Newman vs. Alabama, *id.*; Holt v. Sarver 300 F. Supp. 825, 830-831 (E.D. Ark. 1969); *cf.* ABA COMMISSION, Residential Facilities, *supra* note 275, at 354-356.

[282] See section 3(3) for a definition of the term "correctional facility."

[283] *See* Rowan, *supra* note 248, at 673; Schwartz, *Reaction Comment*, PCMR, *supra* note 12, at 675-677; Haywood, *supra* note 274; SANTAMOUR and WEST, *supra* note 251, at 10-12.

[284] Haywood, *supra* note 274, at 680; Burgdorf, *The Legal Rights of Mentally Retarded Offenders: Some Underlying Issues*, in THE MENTALLY RETARDED CITIZENS AND THE CRIMINAL JUSTICE SYSTEM, *supra* note 12, at 233, 240-241.

[285] 42 U.S.C. §§6001 *et seq.* (Supp. 1979).

[286] 42 U.S.C. §6012 (Supp. 1979); Chapter 9: Advocacy, *infra.*

[287] Chapter 9: Advocacy, *infra;* D. SKOLER and A. COHEN, ADVOCACY SYSTEMS FOR THE DEVELOPMENTALLY DISABLED: LEGISLATIVE UNDERPINNINGS (1979).

[288] See authorities cited in notes 265 and 272 *supra*.

The need for such advocates should not be confused with the need for and right to an attorney.[289] Most attorneys do not have the knowledge of developmental disabilities[290] nor the easy frequent access to the defendant contemplated in this provision. Moreover, the advocate's duty under this paragraph is to assure that developmentally disabled offenders receive the habilitative services and protection to which they are entitled under paragraphs (1) (a) and (1) (b) of this section, while the attorneys role centers on challenges to the offenders conviction and sentence.[291]

Unlike the preceding subsection, the scope of this provision includes only offenders, since habilitative services are not generally offered to pretrial detainees unless they have been found to be incompetent to stand trial, and incarcerated defendants usually have relatively frequent contact with counsel. Some jurisdictions, particularly those in which the conditions in jails and other pretrial detention facilities have come under the scrutiny of the courts, may wish to expand coverage of the advocacy services to include defendants alleged or found to have a developmental disability.

The state correctional agency may either provide the advocacy service directly — e.g. through an established ombudsmen program[292] — or contract with the state developmental disabilities advocacy office.[293] While no offender to advocate ratio is set in the provision, there should be a sufficient number available to permit at least monthly visits to each developmentally disabled offender who does not have the freedom of movement personally to seek out an advocate in the community, and to perform the necessary investigation and followup after each visit. An advocacy service with resources sufficient to permit only reaction to complaints or requests for visits by a developmentally disabled offender is inadequate, since prisoners with a developmental disability are far less likely to utilize a grievance mechanism than other inmates. Advocates for developmentally disabled offenders should have at least the same training and qualifications, and the same duties and responsibilities as those employed by the state developmentally disabled advocacy office,[294] except to the extent that additional duties are required by this section.[295]

(2) (a) Within thirty (30) days after a term of probation or sentence of over ninety (90) days has been imposed on an offender who has a developmental disability, the correctional agency responsible for custody or supervision of the offender shall develop a Program Plan and submit it to the sentencing court.

(b) The Program Plan shall be developed in consultation with the offender, the offender's attorney, and if the offender resides in a correctional facility, a developmental disabilities advocate for that facility. The objective of the plan shall be to assist the offender, to the extent and within the shortest time possible, to function in the community as a self-reliant, law-abiding citizen. The purpose, nature and possible beneficial and harmful consequences of the various alternatives shall be explained to the defendant to the maximum extent possible, and to the defendant's attorney.

(c) The plan shall cover the lesser of:

(i) The minimum parole eligibility period, if any, applicable to the defendant;

(ii) The maximum sentence or term of probation; or

(iii) Two years.

(d) The Program Plan shall be submitted for approval to the court no more than thirty (30) days after issuance of the order. A copy shall be provided to the prosecutor, the defendant, and the defendant's attorney, the chief developmental disabilities advocate of the correctional facility in which the offender resides, or if the offender is on probation, to the (name of state Developmental Disabilities Advocacy Office).

(e) (i) The Program Plan shall be approved by the court without a hearing within ten (10) days of its submission unless the court on its own initiative sets a hearing or the prosecutor or defendant move for a hearing. In such a motion, the moving party shall specify the reasons why the plan is alleged to be inadequate or inappropriate and the proposed modifications.

(ii) Following a hearing held pursuant to subparagraph (2)(e)(i), the court shall approve the plan as written, or modify it so as to further the objectives set forth in paragraph (2)(b) of this section, and order its implementation.

---

289 Johnson vs. Avery 393 U.S. 483 (1969); Douglas vs. California 372 U.S. 353 (1963).

290 Haggerty, Kane and Udall, *supra* note 163.

291 For a further discussion of the possible role of developmental disabilities advocates in prisons, see MISSOURI ASSOCIATION FOR RETARDED CITIZENS, *supra* note 12, at 56-57; and Golten, *Reaction Comment*, PCMR, *supra* note 12, at 645-648. One possible area of overlap between the developmental disability advocate and an attorney is where the deprivation of services and conditions of confinement are so poor that they constitute cruel and unusual punishment and form the basis of a *habeas corpus* petition challenging further confinement.

292 *See e.g.*, M. KANNERSEHN, A REPORT ON THE NEW YORK DIVISION FOR YOUTH OMBUDSMAN PROJECT (1974); ABA, STANDARDS RELATING TO THE LEGAL STATUS OF PRISONERS, §§8.3-8.5 (Tentative Draft 1977); NATIONAL ADVISORY COMMISSION, *supra* note 43, at 459; Tibbles, *Ombudsmen for American Prisons*, 48 N. DAKOTA LAW REVIEW 383 (1972); L. ORLAND, PRISONS HOUSES OF DARKNESS, 155, 207-215 (1975); FOGEL, *supra* note 228, at 230-236.

293 *See* Model Developmental Disabilities Advocacy Act, §5 and 5(11) in Chapter 9: Advocacy, *infra;* see also 42 U.S.C. §6012 (Supp. 1979).

294 Model Developmental Disabilities Advocacy Act, in Chapter 9: Advocacy, *infra*, at §§9(6) and 12.

295 See §§12(2)(b), (d), and (f); and §§12(3)(b) and (c).

(f) An updated Program Plan shall be developed at the end of the period specified in paragraph (2)(c)(i) or (iii) of this section and:

(i) At least every two (2) years thereafter until the end of the offender's sentence or term of probation; and

(ii) When the offender is released on parole or is incarcerated following revocation of probation or parole; and

(iii) Upon request of the offender, or with the offender's consent, the advocate, when the services being provided no longer meet the objective of the Program Plan set forth in paragraph (2)(b) or this section.

Updated plans shall be developed jointly by the correctional agency and the offender and shall be approved by the court in the manner specified in paragraph (2)(e). Offenders who reside in a correctional facility shall be entitled to assistance of a developmental disabilities advocate for that facility. Offenders on probation or parole shall be entitled to the assistance of a representative of their own choosing in updating the plan. In conjunction with a hearing on the revised Program Plan, the court may appoint a multidisciplinary evaluation team to update the evaluation performed pursuant to section 11 of this act, upon a showing that there has been a significant change in the offender's mental, emotional, physical, or educational condition, adaptive behavior or social skills.

COMMENT. In order to achieve the habilitative objectives of corrections and fulfill the rights to treatment and protection established in paragraphs (1) (a) and (b) of this section, this provision requires development of a Program Plan for all developmentally disabled offenders sentenced to confinement or placed under supervision in the community. While Program Plans have not been an integral part of most correctional programs for adults, they are a critical means for assuring the provision of a program of habilitative services tailored to the individual offender's abilities and needs.[296] The Plan should be based on the information contained in the presentence report and adduced at the sentencing hearing, as well as on any information obtained through the reception/assessment process at the correctional facility. It is not due until 30 days after the inception of the sentence or term of probation, in order to permit its development during the course of the classification procedure employed by many correctional systems. While development of a Program Plan is not required for minor offenses when the period of confinement and/or supervision is brief, nothing in this subsection is intended to prohibit the development and use of such a plan in those cases and the referral of the offender to community services which are available on a voluntary basis.[297]

The procedures for developing, reviewing, and updating the Program Plan for an offender with a developmental disability are similar to those set forth in Section 10 (7) for defendants acquitted solely by reason of diminished capacity and are explained more fully in the comments to that section. For defendants sentenced to a period of incarceration in either a community-based or non-community based facility, both the attorney who represented the defendant at trial and sentencing and the facility's developmental disabilities advocate[298] should be involved in the plan development process in order to assure that all the relevant information is considered and interpreted correctly, that the defendant's wishes and concerns are addressed, and that the delivery of services is properly monitored.

Since the offender's needs are likely to change during the course of the sentence or term of probation, provision is made for periodic updating of the plan at least biennially. If a change is needed in the manner or type of services provided under the Program Plan before the scheduled review date, the offender or, if the offender has given his or her consent, the developmental disabilities advocate may request that the plan be modified. A modification in the Program Plan only affects the habilitative services to be offered, their intensity, and particularly when the offender is on probation or parole, the manner in which they are delivered. It is not a means for obtaining or revoking full release from confinement. The procedures for updating a Program Plan are identical to those used in initially developing it, except that there is no requirement that the attorney who represented the offender at trial be involved since, in most cases, he or she will not still be in frequent contact with the offender. For offenders in correctional facilities, the facility developmental disabilities advocate must take part in the Plan updating process. Since offenders under community supervision have access to a greater spectrum of advocacy services, paragraph (2)(f) provides that they are entitled to the assistance of a representative of their own choosing. This representative could be an attorney, a member of the staff of the state developmental disabilities advocacy office, a person experienced in working with persons with a developmental disability, a family member, or a friend.

As in dispositions following acquittals solely by reason of diminished capacity, judicial oversight of the development and updating of Program Plans is provided to enhance the accountability of

[296] *See SANTAMOUR and WEST, supra* note 251, at 30; Rowan, *supra* note 248, at 624.

[297] Chapter 6: Right to Services, *supra*.

[298] See section 12(1)(c).

the service delivery system and the enforcement of the right to habilitative services. However, because habilitation is only one of the objectives of sentencing, the finite length of sentences, and the availability of other sentence adjustment measures such as parole, the oversight is limited to approval and monitoring of the Program Plan and periodic judicial review hearings are not required.

(3) (a) Six (6) months after development of the initial Program Plan and every six (6) months thereafter until the end of the offender's sentence or period of probation, the head of the correctional facility, probation office or parole office responsible for the offender's custody or supervision, shall file a report with the sentencing court. The report shall list the services specified in the most recent Program Plan, and indicate which, if any, of those services have not been offered to the offender and the reasons therefor.

(b) A copy of the report shall be provided to:

(i) The offender;

(ii) The chief developmental disabilities advocate for the correctional facility, if any, in which the offender is housed;

(iii) The (name of the state Developmental Disabilities Advocacy Office) if the offender is on probation or parole;

(iv) The (head of the state correctional agency); and

(v) At the offender's request, to his or her designated representative.

The information contained in the report shall be explained to the offender to the maximum extent possible.

(c) Within thirty (30) days after submission of a report under paragraph (3) (a) of this section, the offender, or an advocate, attorney, or representative on the offender's behalf and with his or her consent, may petition the court for an order directing the state correctional agency to provide or purchase the omitted services immediately. The court shall issue the order following a hearing at which all interested parties are represented, unless it has been established by clear and convincing evidence that the service is not necessary.

(d) An offender shall not be penalized in any way by the correctional agency responsible for his or her custody or supervision, or by any employee or official thereof as a result of filing a petition under this subsection.

COMMENT. Given the dearth of habilitative services in the past in many correctional programs,[299] the lack of attention and inadequate funding which many correctional systems have faced,[300] the general absence of court oversight following sentencing, and the importance to the developmentally disabled offender, the correctional system and the public of offering adequate habilitative services,[301] this subsection provides a mechanism for monitoring the provision of habilitative services. It requires the individual administratively responsible for the offender's custody or supervision to file a report biannually indicating which, if any, of the habilitative services specified in the offender's Program Plan have not been offered and the reasons why. These reports, like those required by section 10(9) will provide feedback to the courts on the availability of habilitative services in the community and at the various correctional facilities, and an indication to correctional officials of problem areas and resource needs, as well as a basis for enforcing the offender's rights under paragraph (1)(a).[302] A copy of the report must be sent and explained to the offender in the language, terms and mode of communication which he or she is most likely to understand. Copies are also to be sent to the sentencing court, the state corrections chief, the relevant advocacy service, and if the offender requests, to an attorney or other individual whom he or she has designated as a representative.

The final paragraph seeks to assure that offenders will not be penalized for exercizing their rights. Given the inherently coercive atmosphere of correctional facilities and the broad discretion of probation and parole officials, such assurance is necessary if offenders are to come forward and enforce their rights. Both the explanation of the Program Plan required by paragraph (2)(b) of this section, and the explanation of the report should advise the offender of the procedures for filing a petition and of the protection from such forms of reprisal as transfer to less desirable units or denials of parole or good time because of filing the petition.

(4) (a) No incarcerated offender who has a developmental disability shall be required or coerced to participate in any habilitative services. Failure or refusal to participate in a habilitative service shall not constitute the basis for penalizing such an offender.

(b) The informed written consent of an offender who has a developmental disability shall be obtained before the offender is provided with a program or service not included in his or her Program Plan unless the program or service is essential to protect the offender's physical health and the offender is unable to consent. In order for the consent to be informed, the purpose, nature, duration, and possible beneficial

[299] See authorities listed *supra* note 272.

[300] SANTAMOUR and WEST, *supra* note 251, at 10.

[301] *Id.*, at 30.

[302] *See* WEINSTEIN, *supra* note 268; Holt vs. Sarver 309 F. Supp. 362, 385 (E.D. Ark. 1970); *aff'd* 442 F. 2d 304 (8th Cir. 1971).

and harmful consequences of the program or service must be explained to the offender and the offender must be given an opportunity to consult with a developmental disabilities advocate if the offender is incarcerated or a designated representative if he or she is not.

COMMENT. A necessary correlate to the right to habilitative services provided in subsection (1)(a) is the right of prisoners to decline services and simply "do their time." A number of commentators have questioned the wisdom and effectiveness of coerced treatment,[303] and depending on the services involved, serious constitutional questions may be raised by use or threat of punishment to induce acceptance of or participation in allegedly rehabilitative services.[304]

> . . .[T]he recognition of a right to refuse treatment in a non-emergency situation is practical. . . . [I]nvoluntary treatment is much less effective then the same treatment voluntarily received . . . . [P]sychiatric experts [have] stated that it is more likely that a patient will consent to desirable treatment when consulted before action is taken, and when he [or she] has some real control over his [or her] fate, than when he [or she] feels totally at the mercy of the hospital doctors.[305]

Accordingly, paragraph (4)(a) provides a right to refuse habilitative services without reprisal. It is anticipated that in light of the offender's involvement in developing the Program Plan, this right will be exercized rarely. Nevertheless, it provides a necessary safeguard, together with the consent provision in paragraph (4)(b) and the provisions on transfer in subsection (5), against the unconsented to use of psychotropic drugs, psychosurgery, electro-convulsive therapy, and aversive conditioning employing pain-inducing or health-threatening stimuli.

Paragraph (4) (b) is designed to prevent circumvention of the program planning process. That process would be of little value if programs and services can be added or altered solely at the behest of the correctional authorities. In addition, the provision helps to assure that offenders with a developmental disability will not be subjected to experimental, intrusive, or health-threatening programs without their informed, voluntary consent. The use of prisoners as subjects for experimental biomedical or behavioral procedures has been widespread and its abuses increasingly recognized.[306] Explicitly barring nontherapeutic experimentation entirely is beyond the scope of this report.[307] By requiring, however that the the proposed service and its ramifications must be explained to the offender, that the offender must have an opportunity to discuss the matter with an advocate or designated representative,[308] and that the offender's consent to the procedure must be free of coercion, the paragraph provides substantial protection against exploitation. To protect offenders further, some jurisdictions may wish to add a provision for court review before certain types of services — *e.g.,* psychosurgery, electroconvulsive therapy, or aversive conditioning utilizing pain-producing or health-threatening stimuli — or experimental biomedical or behavioral procedures can be provided.[309]

Consent for medical services is not required in emergency situations such as when the offender is unconscious and the services are required to preserve his or her life.

(5) (a) An offender who has a developmental disability and who is incarcerated or on probation or parole, shall not be admitted into a facility to which nonoffenders are committed under (citation to Model Right to and Admission to Services Act/applicable state civil commitment law(s)) other than through a formal commitment proceeding in which the offender is represented by counsel or pursuant to paragraph (5)(b) of this section.

COMMENT. The transfer of offenders to mental health or mental retardation facilities has been a continuing subject of litigation and controversy, particularly when a transfer is sought by the correctional authorities.[310] In part this is due to the reluctance of mental health and mental retardation facilities and agencies to deal with criminal offenders, the reluctance of correctional agencies to deal with offenders who may be mentally ill or developmentally disabled, and the more restrictive conditions which are imposed on mentally ill or mentally retarded offenders as a result.[311] How-

---

[303] FOGEL, *supra* note 228, at 202; N. MORRIS, THE FUTURE OF IMPRISONMENT (1974); Martinson, *supra* note 253; NACJJDP, *supra* note 28, at §4.410.

[304] NACJJDP, *id.*; *see* Rennie vs. Klein, 462 F. Supp. 1131 (D.N.J. 1978); Note, *The Right Against Treatment: Behavior Modification and the Involuntarily Committed*, 23 CATHOLIC UNIVERSITY LAW REVIEW 774 (1974).

[305] Rennie vs. Klein, *id.* at 1144 and 1145 (1978); NATIONAL ADVISORY COMMITTEE, *supra* note 43, at 45; See MHLP, *supra* note 272; for a survey of state statutory provisions on the right to refuse treatment, see 2 MENTAL DISABILITY LAW REPORTER 240 (1977).

[306] SUBCOMMITTEE TO INVESTIGATE JUVENILE DELINQUENCY OF THE COMMITTEE ON THE JUDICIARY, UNITED STATES SENATE, 94TH CONG., 1st SESS., DRUGS IN INSTITUTIONS-Vol. III, 339-353; 356-365; 439-447 (1977); DOJ, *supra* note 275, at 33-34; ABA, LEGAL STATUS OF PRISONERS, *supra* note 292, at §5.8.

[307] *But see* ABA, *id.*; DOJ, *id.*, at 33.

[308] For an explanation of the term "designated representative," see the comment to section 12(2).

[309] See AAMD, *supra* note 30, at 25-27; DOJ, *supra* note 275, at 34.

[310] See *e.g.,* Vitek vs. Jones, U.S. (1980); Sites vs. McKenzie, 423 F. Supp. 1190 (N.D. W.Va. 1976); Cruz vs. Ward, 424 F. Supp. 1277 (S.D.N.Y. 1976); New York ex rel. Overton vs. Director, Central New York Psychiatric Center, 418 N.Y.S. 2d 254 (App. Div. 1979).

[311] See authorities cited *supra* note 262 and NEW YORK DEPARTMENT OF MENTAL HYGIENE, *supra* note 177, at 115; Ziccardi, *supra* note 267, at 690.

ever, the Supreme Court has made clear that individuals are not stripped of all their rights upon conviction for a crime. The protection offered by the due process and equal protection clauses of the United States Constitution still apply.[312] This includes the protection against involuntary commitment to a mental health or mental retardation facility without full notice, a hearing before an impartial decision-maker, the opportunity to call and cross-examine witnesses, the opportunity to be heard in person and present evidence, a written statement of the basis for commitment, and assistance of a qualified advocate.[313]

Accordingly, this subsection makes explicit that developmentally disabled offenders, whether they are imprisoned or subject to community supervision in lieu of or following imprisonment, may not be placed in a facility to which non-offenders are civilly committed unless they meet the statutory criteria for commitment and the formal civil commitment procedures are invoked and followed. These procedures must include at a minimum, the constitutionally mandated due process safeguards listed in the preceding paragraph.[314] In addition, in light of the variation in current laws governing commitment, and the abuses to which transfer and transferred offenders have been subject, the provision assures that offenders for whom involuntary commitment is sought will have the assistance of an attorney.[315]

It is anticipated that as a result of the services and programs to be provided by the state correctional agency under this act and the emphasis which the act places on use of the least restrictive alternatives, the civil commitment of offenders will occur only rarely. In addition, relying on a judicial hearing in such cases is likely, in reality, to yield less of an administrative and financial burden and a far simpler statutory structure than following the administrative model outlined in Vitek vs. Jones. First, it avoids requiring the establishment or expansion of a hearing officer and administrative appeal system within the correctional agency. Second, it resolves a question left open by Vitek by making clear that the admission procedures for incarcerated offenders also apply to offenders on probation or parole.

To assure that the necessary services are provided and to protect further against abuse of the commitment authority, enacting jurisdictions are urged to consider as well, the Model Right to Services/Admission to Services Act included in this series.

(b) The request of an offender with a developmental disability who is incarcerated, on probation or on parole for admittance to a facility to which nonoffenders are committed pursuant to (citation to Model Right to and Admission to Services Act/state civil commitment statute(s)) shall not be granted unless the offender is represented by counsel and establishes by a preponderance of the evidence at a hearing before the (court authorized to grant commitment petitions) that:

(i) Services are required in accordance with his or her Program Plan which cannot be provided by the state correctional agency;

(ii) The required services will be provided in the facility to which admission is sought;

(iii) There is no less restrictive alternative for obtaining the required services; and

(iv) Admission to the facility will not endanger the safety of the public.

COMMENT. In order to prevent circumvention of the requirements of paragraph (5)(a), and in light of the inherently coercive atmosphere of correctional facilities for persons with a developmental disability,[316] this provision requires a special judicial proceeding prior to the voluntary admission of a developmentally disabled offender into a mental health or mental retardation facility to which non-offenders are civilly committed. It is anticipated that this procedure will forestall attempts to impose probation and parole conditions requiring an offender to seek a voluntary commitment, and bring to light instances in which a correctional agency is failing to meet its obligations under subsection (1) of this section.

When voluntary commitment is appropriately sought, the information necessary to sustain the low burden of persuasion should be readily obtainable by counsel from the developmental disability advocate, and officials from the appropriate correctional and mental health or mental retardation agencies.

Four criteria are set forth to guide voluntary admission decisions. The criterion concerning public safety recognizes that a criminal offense has been committed, and is intended to assure that this provision is not misused by some offenders as a means of facilitating an escape.

(c) Time spent by offenders, who have a developmental disability, in a facility to which they were admitted under paragraph (5)(a) or (5)(b) of this section shall be credited toward their sentence. While in such a facility, the offender shall remain eligible for good time credits.

[312] Wolff vs. McDonnell, 418 U.S. 539, 556 (1974); Gagnon vs. Scarpelli, 411 U.S. 778 (1973); Morrissey vs. Brewer, 408 U.S. 471 (1972).

[313] Vitek vs. Jones, U.S. (1980).

[314] *Id.*

[315] Chapter 6: Right to Services, *supra*. It should be noted that the Vitek decision is ambiguous on the right to counsel in transfer proceedings. Four members of the majority appear to favor the appointment of counsel, but the holding of the case requires only that an inmate receive "qualified and independent assistance." Vitek vs. Jones, U.S. (1980).

[316] See authorities listed *supra* note 280.

COMMENT. An offender should not be penalized for receiving necessary services nor have to choose between treatment and release.

> The policy of denying statutory good time has the effect of requiring prisoners to choose between their statutory good time, on the one hand, and their constitutional right to receive medical treatment, including relief from working, on the other. Requiring that choice by the disabled necessarily has a chilling effect upon procurement of the constitutional rights to medical treatment.[317]

Accordingly, this provision together with paragraph (5)(d) and subsection (6) of this section assures that any period spent in a mental health or mental retardation facility pursuant to either an involuntary commitment or a voluntary admission is included in determining the date on which an offender is eligible for parole or has served the maximum sentence imposed.

(d) If on the expiration date of his or her sentence, an offender who has a developmental disability resides in a facility to which he or she was admitted pursuant to paragraph (5)(a) or (5)(b) of this subsection, the offender shall be immediately released unless a formal commitment proceeding has been initiated in accordance with (citation to the Model Right to and Admission to Services Act/state civil commitment law(s)).

COMMENT. Once offenders have served their sentence or term of probation, their liberty may not continue to be restrained unless there is another basis justifying custody or the imposition of restraints, and the procedures applicable to nonoffenders have been invoked.[318] This provision reflects the holding in the Baxtrom case that the procedures necessary to retain an offender in a treatment facility following expiration of his or her sentence must be identical to those necessary to civilly commit non-offenders.

(6) The fact that an offender has a developmental disability or that he or she is being provided habilitative services in a correctional facility or in another facility pursuant to subsection (5) of this section, shall not affect his or her eligibility for parole. The (name of the state parole authority) shall hold parole release hearings in facilities which house one (1) or more offenders admitted pursuant to subsection (5) of this section.

COMMENT. A number of studies have shown that incarcerated offenders with a developmental disability tend to serve longer sentences than non-disabled offenders for the same crimes and are less likely to be released on parole.[319] While the rights and services provided by this section should assist developmentally disabled offenders in overcoming the difficulty in preparing release plans and in adjusting to prison rules which have hindered parole in the past,[320] this provision is included to assure that as in the case of pretrial release and diversion, offenders are not discriminated against solely because they have a developmental disability.[321] Parole for disabled offenders should be determined on the same basis as it is for offenders without such a disability. In addition, developmentally disabled offenders who have been admitted to special facilities pursuant to subsection (5) of this section should not be denied the opportunity for parole release or have such release contingent on their return to a correctional program or facility, when the services if any which they require are available in the community and they are otherwise eligible for parole.[322]

Pursuant to subsection (1)(a) of this section, special intensive parole programs and habilitative services must be provided to paroled offenders with a developmental disability who require them.[323]

Section 13. Rights and Privileges

(1) In all judicial hearings provided by this act, the parties shall have a right to:

(a) Receive prior notice;

(b) Be represented by counsel;

(c) Be present;

(d) Compel the attendance of witnesses;

(e) Present evidence and confront and cross-examine witnesses, including experts appointed to conduct examinations or evaluations;

(f) Have an impartial decision-maker; and

(g) Appeal adverse orders and judgments, except that the prosecutor shall not be entitled to appeal dispositional orders issued pursuant to section 10 nor sentences or terms of probation imposed pursuant to section 11 unless otherwise provided by law.

---

[317] Sawyer vs. Sigler, 320 F. Supp. 690, 699 (D. Neb. 1970), *aff'd* 445 F.2d 818 (8th Cir. 1971); *accord* MHLP, *supra* note 72; *but see* Trivento vs. Commissioner of Corrections, 2 MENTAL DISABILITY LAW REPORTER 370 (1978); Bush vs. Ciccone, 325 F. Supp. 669 (W.D. Mo. 1971).

[318] Baxtrom vs. Herold, 383 U.S. 107 (1966).

[319] SANTAMOUR and WEST, *supra* note 251, at 9 and 25; MHLP, *supra* note 272, at 668.

[320] SANTAMOUR and WEST, *id.*

[321] See sections 5 and 7(1).

[322] *See* MHLP, *supra* note 272, at 668-670 and 674.

[323] In keeping with the premise of this act that the procedures for persons with a developmental disability should be comparable to those governing persons without such a disability, enacting jurisdictions which have abolished parole should omit this subsection and all other references to parole in this section. *See* FOGEL, *supra* note 228.

COMMENT. This subsection applies the basic due process rights afforded criminal defendants[324] to all proceedings under this act. There appears little basis for denying these protections in hearings and review hearings pursuant to section 8 of this act regarding the competency of defendants to stand trial; dispositional and review hearings pursuant to section 10 following an acquittal solely by reason of diminished capacity due to a developmental disability; sentencing hearings pursuant to section 11; and hearings regarding Program Plans or the failure to provide services in accordance with section 12, since fundamental interests of the defendant or offender and the integrity and fairness of the judicial and correctional process are at stake in each of these circumstances. In addition, the provision assures that these same basic rights are accorded to the state developmental disabilities agency in dispositional and review hearings held pursuant to sections 10 (7) and (10),[325] and the state correctional agency in hearings held pursuant to sections 12, as well as to the defendant or offender and, with one exception, the prosecutor.

The exception concerns the authority of the prosecutor to appeal the sentence or disposition. Traditionally, prosecutorial appeal of sentences has been prohibited so as not to deter defendants from exercizing their right to appeal nor violate the constitutional protection against double jeopardy.[326] However, in light of the clarification of the double jeopardy clause and the concern about the disparity of sentences, there has been increasing interest in permitting prosecutorial appeals of sentences at least to a limited extent.[327] This provision is intended to provide authority for prosecutorial appeal of dispositions involving defendants and offenders with a developmental disability to the same extent as that permitted under current state law for other criminal cases.

(2) In addition to the rights listed in subsection (1) of this section, a person who there is reason to believe has a developmental disability and who is alleged or found to have committed a criminal offense shall be entitled to have:

(a) An attorney appointed to represent him or her in conjunction with any judicial proceeding and any matter arising from that allegation, finding or the disposition thereof, if he or she is not represented by counsel. An attorney appointed pursuant to this paragraph shall be compensated by the [state/county/court] if the person is unable to retain counsel for financial reasons.

(b) An independent expert appointed by the court at the expense of the [state/county/court] to examine or evaluate him or her if that person requests such an appointment and is unable for financial reasons to secure independent medical, psychological or other expert opinions; and

(c) A jury determine, upon request, whether he or she:

(i) Is competent to stand trial pursuant to section 8(5) of this act; and

(ii) Meets the criteria for placement in a secure or not community based residential facility pursuant to section 10(6).

COMMENT. This subsection clarifies the scope of the rights of persons, with a developmental disability who are alleged or found to have committed a crime, to have counsel and experts appointed to assist them and to have a jury rather than a judge make the initial determination of their competence to stand trial and the determination of the need for secure placement following an acquittal solely by reason of diminished capacity. Indigent defendants are entitled to have counsel appointed to represent them at "every stage of a criminal proceeding where substantial rights of a criminal accused may be affected."[328] This provision, together with sections 4(4), 7(2) and (3), 8(6), 10(7), and 12(2) makes clear that for defendants and offenders who there is at least a reason to believe have a developmental disability, such critical stages include negotiations concerning pretrial diversion; all hearings including review hearings concerning competency to stand trial; the development of Program Plans following a finding of incompetence to stand trial or an acquittal solely by reason of diminished capacity; the development of the initial Program Plan following sentencing; and all hearings to review Program Plans or the provision of services as well as police interrogation, arraignment, trial, sentencing and appeal.[329] At each of these points the rights to liberty, a fair trial, and adequate treatment are at issue. Moreover, the need for the "guiding hand of counsel"[330] in cop[ing] with problems of law, mak-

---

[324] See In re Oliver, 333 U.S. 257 (1948), and Mullane vs. Central Hanover Trust Co., 339 U.S. 306 (1950) regarding notice; Gideon vs. Wainwright 372 U.S. 335 (1963), Argersinger vs. Hamlin, 407 U.S. 25 (1972), and Mempa vs. Rhay, 389 U.S. 128 (1967) regarding counsel; Lewis vs. United States, 146 U.S. 370 (1892), and Illinois vs. Allen, 397 U.S. 337 (1970) regarding presence; Pointer vs. Texas, 380 U.S. 400 (1965) regarding confrontation and cross-examination; Tumey vs. Ohio, 273 U.S. 510 (1927), and Sheppard vs. Maxwell, 389 U.S. 333 (1966), regarding impartial decision-makers.

[325] See also section 10(12).

[326] ABA, STANDARDS RELATING TO THE ADMINISTRATION OF CRIMINAL JUSTICE: APPELLATE REVIEW OF SENTENCES, 2 (2d ed., Approved Draft 1978).

[327] *Id.* at §20-1.1; NDAA, *supra* note 45, at §18.2; see United States vs. Di Francesco 79-567, 26 CRIMINAL LAW REPORTER 4064 (October 10, 1979).

[328] Mempa vs. Rhay, 389 U.S. 128, 138 (1967); *see e.g.*, Miranda vs. Arizona, 384 U.S. 436 (1966); Gideon vs. Wainwright, 372 U.S. 335 (1963); Douglas vs. California, 372 U.S. 353 (1963).

[329] See the comments to the sections listed in the text and the authorities cited in the preceding note.

[330] Powell vs. Alabama, 287 U.S. 45, 69 (1932).

[ing] skilled inquiry into the facts, insist[ing] upon regularity of the proceedings," and preparing and presenting supportive evidence and arguments[331] is even more critical for most persons with a developmental disability than it is for non-disabled defendants and offenders.[332]

With regard to the appointment of independent experts, it can scarcely be disputed that expert diagnoses, opinions and recommendations, can and often do differ, and that the report of an examiner or a multidisciplinary evaluation team will often be the primary basis for determining whether a defendant or offender has a developmental disability and the effect of the disability, if any, on his or her competency to stand trial or on the habilitative services which he or she may need.[333] Thus, the opportunity to have another qualified individual assist counsel in examining the report of the multidisciplinary evaluation team is critical in preparing an adequate defense.[334] Accordingly, this paragraph authorizes the court, upon request, to appoint an expert not serving as an examiner or a member of a multidisciplinary team to examine or evaluate an indigent defendant or offender. Independent experts should be selected from the list prepared by the Evaluation Oversight Board pursuant to section 14(3)(b). To the greatest extent possible, the defendant or offender should indicate the field of expertise and the names of the individuals whom he or she wishes to have appointed. The court, within reason, should appoint one or more of the experts so designated. A number of states already provide for appointment of independent experts in some circumstances.[335]

The right to a jury determination of incompetence to stand trial and eligibility for secure placement is included as an added protection against routinization of incompetency and placement determinations. Currently, the statutes of nine American jurisdictions permit jury participation in competency determinations.[336]

(3) If a defendant or offender alleged or found to have a developmental disability is under the influence of psychotropic medication during any judicial hearing, and court and the jury, if any, shall be advised of this fact and of the purpose of the medication and the effect which it may have on the person's actions, demeanor and participation at the hearing.

COMMENT. Whenever possible, defendants should not be under the influence of a drug which affects their appearance or ability to participate in a hearing unless administration of that drug is essential to protect the respondent or others from harm at the hearing, and no less drastic means is available.[337] When administration of a psychotropic drug is unavoidable, this provision requires that the trier of fact be made aware of the effects of the drug so that the alteration of the respondent's appearance, demeanor or actions does not improperly influence the decision-making process.[338]

(4) A stenographic and/or mechanical record shall be made of all proceedings held pursuant to this act.

COMMENT. The availability of a verbatim record of the proceedings is essential for assuring accountability and fairness. As was pointed out by the Supreme Court:

> Failure to make a record, may be . . . to saddle the reviewing process with the burden of attempting to reconstruct a record and to impose upon the . . . [trial] Judge the unseemly duty of testifying under cross-examination as to the events that transpired in the hearings before him [or her].[339]

Accordingly, this subsection mandates that some form of verbatim record be made of all hearings held pursuant to this act not just of the trial itself. This is not intended to require that a full transcript be prepared in every case. A transcript of the record should be made only on the request of a party or the court. When the subject of an intervention proceeding who is indigent requests a transcript, the cost of its preparation should be at public expense.[340]

(5) All explanations required by this act to be given to a person alleged or found to have a developmental disability shall be in the language, mode of communication, and terms which that person is most likely to understand.

COMMENT. At decision points throughout this act, an explanation must be given to the defendant or offender of the nature, and purpose of the proceeding, services, or other matter at issue; his or her rights; the possible beneficial or harmful consequences; and the alternatives available.[341] While some defendants or offenders may have difficulty grasping the meaning of what is occurring and the choices before them no matter how well the ex-

[331] In re Gault, 387 U.S. 1, 36 (1967).

[332] *See* PCMR *supra* note 124, at 63, 133; THE MENTALLY RETARDED CITIZENS AND THE CRIMINAL JUSTICE SYSTEM, *supra* note 12, at 13.

[333] See sections 7(2), 8(4)(c), 10(5), 10(10)(c), 11(2), 12(2)(f), and 15(4).

[334] ABA, AMERICAN BAR ASSOCIATION STANDARDS RELATING TO THE ADMINISTRATION OF CRIMINAL JUSTICE — PROVIDING DEFENSE SERVICES, §5-1.4 (2d ed. Approved Draft 1979).

[335] *Id.* at 7; Chart VI *supra*.

[336] See Chart II *supra;* see also Chapter 6: Right to Services, *supra*.

[337] See comments to sections 8(1) and 8(5)(d).

[338] Psychotropic drugs, also known as antipsychotic drugs include prolixin, thorazine, mellaril, haldol, trilafon, and lithium among others.

[339] In re Gault, 387 U.S. 1, 58 (1967).

[340] *See* Griffin vs. Illinois, 351 U.S. 12 (1955).

[341] *See* sections 4(5), 7(2), 7(3), 8(6), 9(1), 10(7), 10(9), 12(2), 12(3) and 15(5).

planation is presented, most will be able to understand if the information is explained properly. For example, an explanation in English to a respondent who understands only Spanish, Vietnamese, or Navajo is of little use whether or not that individual has a developmental disability. Attempting to communicate verbally or through a standard printed form to someone who can only understand sign language or read braille, or couching an explanation in technical legal terms is equally futile. Yet, in too many instances, the defendants fail to receive proper notice or make competent choices or waivers not because they are unable to understand but because no one has made the effort to present the information in a manner tha is comprehensible to them. By specifying that the explanation must be in the language, mode of communication and terms which the defendant is most likely to understand, this provision mandates that such an effort be made.

(6) No pretrial diversion agreement, Program Plan, or dispositional order entered into, developed, or issued pursuant to this act shall provide for or approve psychosurgery, electro-convulsive therapy, or aversive conditioning utilizing pain-producing or health-threatening stimuli.

COMMENT. Given the public concern about crime, such otherwise dubious techniques as psychosurgery, electrical or chemical stimulation of the brain, or drastic forms of aversive conditioning have been tolerated in the hope that they will lead to a simple solution to the problem of recidivism.[342] However, in view of the questionable results of these techniques, their obvious side effects, and their high potential for misuse, particularly in correctional settings, permitting their inclusion as part of a Program Plan for a defendant or offender with a developmental disability is untenable.

> While the use of these . . . techniques in prison is not yet widespread, their potential proliferation is a matter for real concern. Given the prisons with their second-rate techniques and highly vulnerable population, conservatism must be the fashion and caution an important principle. In order to halt the new versions of old failures, a moratorium on the use of the new techniques is desirable.[343]

Therefore, this provision bars inclusion of the most intrusive forms of treatment as a routine condition of diversion or probation or as part of a Program Plan for defendants or offenders with a developmental disability.[344] Given the inherent atmosphere of coercion even in diversion programs, the assent of the defendant or offender during the program planning process is insufficient protection. While the possibility for consent to the prohibited forms of treatment exists under section 12 (4)(b), it is anticipated that the special procedural protections required will be adequate to prevent abuse.

Section 14. The (state name) Evaluation Oversight Board

(1) There is hereby established the (state name) Evaluation Oversight Board (hereinafter referred to as the Board) which shall be independent of all other state boards and commissions.

(2) The Board shall consist of the Director of the (name of the state Developmental Disabilities Advocacy Office) who shall serve as chairperson, the (Director/Administrator/Secretary) of the state developmental disability agency and the Attorney General.

(3) The Board shall:

(a) Establish and annually publish the procedures and criteria for identifying individuals who are qualified to perform examinations and evaluations of persons alleged or found to have a developmental disability;

(b) Prepare and publish, annually, a list of individuals from all sections of the state who are qualified to perform examinations and evaluations;

(c) Establish and annually publish the procedures and criteria for removing individuals from the list of those qualified to perform examinations and evaluations;

(d) Conduct investigations to assure the objectivity and thoroughness of examinations and evaluations;

(e) (i) Employ, in accordance with the (citation to state civil service code), an Administrative Director and such professional, technical, clerical, and administrative personnel as may be necessary to carry out the responsibilities, duties and powers set forth in this act;

(ii) Delegate to the Administrative Director such powers as are necessary to carry out the duties of the Board set forth in this act, including the power to redelegate authority when necessary;

(iii) Contract with individuals, businesses, whether for profit or not for profit, and other public or private organizations or agencies to carry out the responsibilities, duties and powers set forth in this act; and

(iv) Assure that all employees and contractors of the Board understand their duties and powers;

---

[342] Gaylin and Blatte, *Behavior Modification in Prisons*, 13 AMERICAN CRIMINAL LAW REVIEW 11 (1975).

[343] Id., at 34-35; see also Chorover, *The Pacification of the Brain*, 8 PSYCHOLOGY TODAY 59 (1974); 43 Fed. Reg. 53242 (1978); ABA LEGAL STATUS OF PRISONERS, *supra* note 292, at §5.8; DOJ, *supra* note 275, at 33.

[344] See sections 7(2), 8(6), 10(7), 11(3), and 12(2); see also section 12(4)(a) regarding the right to refuse services.

(f) Prepare and submit a proposed annual budget in accordance with the procedures established by (citation to the relevant statutes, regulations, and executive orders);

(g) (i) Prepare an annual report describing the actions taken by the Board to fulfill its duties, any significant problems which have arisen during the reporting period, and the actions taken or required to resolve those problems;

(ii) Submit the report to the Governor, the (name of the Legislature), and the Chief Justice of (name of the highest state court); and

(iii) Send copies of the report to each court which can hear proceedings for which an examination or evaluation may be required and make copies of the report available to the public;

(h) Promulgate such policies, rules, regulations and procedures, in addition to those required above, as are necessary for the effective operation of the Board; and

(i) Perform such additional duties and exercise such powers are prescribed by law or as are necessary for the effective management and operation of the Board in accordance with the purpose of this act.

(4) The Board or employees thereof shall not supervise or conduct individual examinations and evaluations.

(5) The Board shall meet at least three (3) times per year. All members of the Board shall be entitled to reimbursement for travel, subsistence, and other reasonable expenses incurred by them in carrying out their duties.

COMMENT. This section creates an independent agency within the executive branch of the state government, charged with identifying experts on various types and aspects of disabilities who are qualified and willing to serve as examiners or members of multidisciplinary evaluation teams; maintaining a list of these individuals for use by the courts; and deleting names from the list as necessary to assure the accuracy and objectivity of examinations and evaluations. In view of the competing demands for limited public resources, the Oversight Board has been designed as an alternative to creating another state social service agency. Rather than establishing a central diagnostic center or a set of centers, and hiring a staff of examiners and evaluators, this section provides for the organization of a network of professionals to render the required services on an as-needed basis. It is anticipated that this approach will not only reduce the cost to the taxpayers, but offer more individualized service than can be provided by a government employee who must deal with a substantial caseload. While the Board established by this section is tailored to facilitating examinations and evaluations in the context of criminal justice and juvenile justice proceedings, it can play the same identification and monitoring roles with regard to other types of proceedings involving examinations or evaluations such as those to establish guardianships, commit an individual involuntarily to a residential facility, and in some cases, to terminate parental rights.

Policy direction and supervision is provided by the heads of the state executive agencies most directly affected by examinations and evaluations. Since the Board would not be performing a general policy-making function, it is not necessary to include citizen representatives of the various consumer and practitioner interests which may be affected by Board actions. Furthermore, inasmuch as judgments may have to be made regarding the qualifications of individual professionals, it would not be appropriate to include practicing physicians, clinical psychologists, psychiatrists, social workers, etc. on the Board. On the other hand, unless differing perspectives of the particular agencies involved are brought to bear on Board decisions, there is a distinct danger that the list of experts will be limited to those proffering a particular point of view, and that the prestige and legitimacy of the Board will be lessened. It should be noted that this is not the only feasible model, nor even the sole model offered in this series of reports.[345] The specific structure and identity of the body performing the requisite identification, clearinghouse and monitoring functions is unimportant so long as its independence, competence and integrity is assured.

In keeping with the definition of a multidisciplinary evaluation team in section 3(16) and the requirements for examinations and evaluations in section 15, the Board must identify physicians, social workers, and other professionals such as clinical psychologists, special education teachers who have had training and experience in diagnosing and assessing developmental disabilities and working with individuals who have such a disability. Many states, reflecting the emphasis in current laws on mental illness, provide for an examination by a psychiatrist or other mental health professional.[346] However, most psychiatrists have had little contact with the multifaceted effects of developmental disabilities. The special expertise required by the act necessitates that some centralized source of information be developed to assist the courts and the parties in selecting persons with the appropriate qualifications.

Accordingly, the list should include each individual's particular area of expertise, and the Oversight Board should make every effort to identify qualified individuals in all parts of the state, in order to make the examination or evaluation as convenient as possible for both the respondent

---

[345] See *e.g.*, Chapter 6: Right to Services, *supra;* Chapter 7: Guardianship and Conservatorship, *supra*.

[346] See *e.g.*, Chart II.

and the expert and to reduce the cost of the examination or evaluation process. The panel of experts may include government employees when their public responsibilities would not affect their willingness to recommend the least restricitive alternative and their service on a multidisciplinary evaluation team would be consistent with the terms of their employment. The Oversight Board is also required to provide a mechanism for receiving and investigating complaints about the diligence, competence and objectivity of individuals serving as examiners or members of a multidisciplinary evaluation team and for removing from the eligibility list, those individuals who are no longer willing to accept or have shown themselves to be unqualified for appointment to multidisciplinary evaluation teams. Annual publication of the procedures and criteria for adding and removing individuals from the panel of experts is required in order to ensure that the list is kept current.

Obviously a Board of public officials who have a host of other responsibilities cannot accomplish all the tasks required to fulfill the duties prescribed above without assistance. While a large bureaucracy is neither needed nor intended, a small staff will be necessary. The Administrative Director should have both management skills and knowledge of the issues and problems inherent in organizing and overseeing a referral system.

The annual report required by paragraph (3) (g) is intended as a means of informing the public, the highest level of state government, and the individuals and entities most affected by or involved in the establishment of guardianships and conservatorships and the appointment of volunteer public guardians, volunteer public conservators and volunteer public trustees, of the activities of the Oversight Commission. A description of the problems encountered during the reporting period and the actions taken or required to resolve those problems is intended to make this report a substantive document rather than a routine organizational summary or glossy public relations product.

Section 15. Examinations and Evaluations

(1) (a) All members of multidisciplinary evaluation teams appointed pursuant to Sections 8, 10, 11 and 12 of this act, and experts designated to examine a defendant pursuant to section 7 of this act shall be selected from the list published by the Evaluation Oversight Board.

(b) The prosecutor and the person to be evaluated shall be entitled to nominate experts from the list for the court's consideration prior to the appointment of a multidisciplinary evaluation team.

COMMENT. For the reasons discussed in the comments to sections 3(11), (12) and (16), this statute distinguishes between examinations and evaluations, and requires evaluations to be performed by a multidisciplinary team. In order to assist in locating experts in various disciplines qualified to conduct examinations and evaluations of persons alleged or found to have a developmental disability, section 14 establishes an Evaluation Oversight Board which will maintain a list of the qualified experts in all parts of the state. To assure further that only individuals with the necessary training and experience will be asked to render the complex opinions and recommendations required, this provision mandates that only persons approved by the Evaluation Oversight Board be appointed to serve as examiners or members of a multidisciplinary evaluation team. However, in recognition of the significance of the examination or evaluation to each of the parties and to the criminal justice process as a whole, the uncertainty of the art,[347] and the likelihood that the attitudes of experts toward the parties will differ,[348] paragraph (b) permits each party to nominate experts from the list for consideration by the court in appointing a multidisciplinary evaluation team. The court should accommodate the preferences of the parties to the extent that it is possible to do so without appointing more than one expert with the same specialty. Given the relative informality of the pretrial diversion process and the limited scope of examinations under section 3(12), the formal safeguards established for selection of a multidisciplinary evaluation team are less critical for the designation of examiners, though soliciting and considering the defendant's choice of an examiner should be a matter of good practice.

(2) Each expert appointed or designated to examine or evaluate a person pursuant to this act, shall receive at the time of appointment or designation a written description of his or her duties and of the matters to be addressed in the examination or evaluation report.

COMMENT. One of the problems which have plagued determinations of competency and criminal responsibility in the past is the failure of examining experts to understand the legal concepts involved and their role in the proceedings, and the corresponding lack of understanding by judges and attorneys of mental health terminology. Too often, psychosis has been inappropriately equated with incompetence to stand trial, and the need for treatment considered, without more, to be a justification for institutionalization.[349] As one means of remedying the problem,[350] this subsection requires that each examiner or member of a multidisciplinary evaluation team receive a written explanation of the matters to be covered in the examination or evaluation report under subsection

[347] See Addington vs. Texas 441 U.S. 418 (1979).

[348] *Cf.* Bazelon, *supra* note 177; NEW YORK DEPARTMENT OF MENTAL HYGIENE, *supra* note 177, at 112.

[349] *See e.g.*, Bennett, *supra* note 69, at 573-574; ROBEY, *supra* note 71 at 617; STONE, *supra* note 78, at 202 and 224; Ennis, *supra* note 71.

[350] See also section 16 on training.

(4) of this section. For evaluations pursuant to section 8, this should include a definition of the applicable legal standards for competence to stand trial with examples of the ways in which defendants ordinarily assist counsel and the decisions which defendants are ordinarily requested to make.

(3) (a) All examinations and evaluations shall be conducted on an outpatient basis unless the person to be examined or evaluated fails or refuses to keep at least two (2) appointments for an examination or evaluation ordered under this act. Any confinement ordered under this paragraph shall not exceed five (5) days.

(b) If the person to be examined or evaluated is in custody, the court shall order, or pursuant to section 7 the (head of the designated diversion unit/prosecutor) shall request in writing, that the (sheriff/marshall/other appropriate official) permit the person to be examined or evaluated in the place of detention or transport the person to and from an appropriate facility for examination or evaluation.

COMMENT. Until recently, mental disability examinations of criminal defendants has often meant hospitalization for 30 days or longer; yet in the vast majority of cases, inpatient examinations are not necessary to obtain sufficient information regarding a defendant's abilities and disabilities.[351] Thus, in keeping with the principle which runs throughout this act of using the least restrictive alternative, and the specific provisions establishing a preference for pretrial release,[352] this section makes clear that examinations and evaluations of defendants and offenders should be conducted on an outpatient basis. Currently, the statutes in 43 American jurisdictions explicitly or implicitly permit outpatient examinations regarding competency to stand trial.[353] A lesser number permit outpatient examinations in other portions of criminal proceedings.[354] Inpatient examinations are permitted only when the presumption that the defendant can and will appear has been overcome by his or her failure to do so on two occasions. While a single failure to appear may be attributable to a simple mix up or misunderstanding, after two such failures, the interest in avoiding unnecessary delay of the proceedings becomes paramount.

For those defendants and offenders who are in custody at the time of an examination, paragraph (3) (b) clarifies the procedures and authority for providing transportation to and from the appropriate facility for an examination or evaluation.

[351] *See e.g.*, Bendt, Balcanoff, and Tragellis, *Incompetency to Stand Trial: Is Psychiatry Necessary?*, 130 AMERICAN JOURNAL OF PSYCHIATRY 11 (1973); MHLP, *supra*, note 721, at 619.

[352] Sections 5, 8(3), and 10(3).

[353] Chart II, *supra*.

[354] Chart VI, *supra*.

(4) (a) All examination and evaluation reports prepared under this act shall include:

(i) The names and addresses of all individuals who examined, interviewed or investigated the person alleged or found to have a developmental disability in conjunction with the examination or evaluation;

(ii) A description of the techniques or tests used; and

(iii) When a multidisciplinary evaluation team has been appointed, the signatures of each member of the team together with a statement of concurrence or nonconcurrence with the evaluation findings.

Examination and evaluation reports shall be submitted to the court no more than ten (10) days after the designation of the examiner or appointment of the multidisciplinary evaluation team.

(b) Reports for examinations conducted pursuant to section 7(2) of this act shall include in addition:

(i) An opinion on whether the defendant has a developmental disability as defined in this act; and if so,

(ii) A description of the nature and effect of that disability; and

(iii) An opinion on whether a full evaluation should be conducted.

Copies of the report shall be provided to the (designated diversion unit/prosecutor), the defendant, and the defendant's attorney.

(c) Reports for evaluations conducted pursuant to section 8(4)(c) of this act shall include in addition:

(i) An opinion on whether the defendant has a developmental disability as defined in this act; and if so,

(ii) A description of the defendant's mental, emotional, physical, and educational condition, adaptive behavior and social skills;

(iii) An opinion regarding the extent to which the defendant has the present ability to (state standard/understand the nature and purpose of the proceedings and assist in preparing his or her own defense); and if the defendant lacks that ability,

(iv) A description of the habilitative services which will help him or her develop that ability within one year; and

(v) An opinion on the likelihood of the defendant developing that ability within one (1) year.

A copy of the report shall be provided to the prosecutor, the defendant, the defendant's attorney, and the (name of the state Developmental Disabilities Advocacy Office).

(d) Reports for evaluations conducted pursuant to sections 10(5) and 10(10)(c) of this act shall include in addition:

(i) A description of the examination and evaluation reports regarding the defendant prepared by other individuals which were utilized by the multidisciplinary evaluation team;

(ii) An opinion regarding whether the defendant has a developmental disability as defined in this act, and if so;

(iii) A description of the defendant's mental, emotional, physical, and educational condition, adaptive behavior and social skills;

(iv) An opinion regarding the habilitative services which would assist the defendant to the extent and within the shortest time possible, to function in the community as a self-reliant, law-abiding citizen; and

(v) At the specific request of the court, an opinion regarding whether the defendant is more likely to benefit from the proposed habilitative services if they are provided in a community based or in a non-community based setting.

A copy of the evaluation report shall be provided to the prosecutor, the defendant, the defendant's attorney, the (name of the state Developmental Disabilities Advocacy Office) and, following disposition, to the individual(s) having direct responsibility for development of a Program Plan for the defendant.

(e) Reports for evaluations conducted pursuant to sections 11(2) and 12(2)(f) of this act, shall contain the information required by paragraphs (4)(a) and (d) of this section. A copy of any evaluation report prepared pursuant to section 11(2) shall be provided to the prosecutor, the defendant, the defendant's attorney, the (name of the state Developmental Disabilities Advocacy Office), and following sentencing, to the individual(s) having direct responsibility for the custody, supervision, habilitation and release of the defendant. A copy of evaluation reports prepared pursuant to section 12(2)(f) shall be provided the offender, the offender's attorney, the person having direct responsibility for the custody, supervision, habilitation and release of the offender, and if the offender is incarcerated, the chief developmental disabilities advocate for the correctional facility in which he or she is housed.

COMMENT. This provision outlines the contents of examination and evaluation reports, and specifies when they are to be submitted and to whom they are to be sent. Paragraph (a) contains the requirements applicable to all examination and evaluation reports and sets a 10 day time limit for preparation and submission. Paragraphs (b)-(e) set forth the requirements which vary according to the purpose of the examination or evaluation and the point in the criminal proceeding at which it is ordered.

The 10 day time limit is consistent with the stringent time constraints imposed generally in criminal proceedings. It is anticipated that this period will be sufficient in most cases to complete a thorough evaluation. In those instances in which it is not, a continuance may be sought in accordance with the procedures and standards established for criminal proceedings.

The format of the report is designed to assure that the examination or evaluation directly addresses the questions at issue in the proceedings and to provide the court and the parties with the information necessary to assess the conclusions, facts and opinions presented. As a consequence, the statutory requirements are somewhat more detailed than in most current legislation. Most of the information called for is self-explanatory. Subparagraphs (4)(a)(i)-(iii) recognize that there can be differences of opinion among the experts serving on a multidisciplinary evaluation team and that these differences may be of critical importance to the court, the parties, and the jury, if any.

Subparagraph (4)(d)(v) acknowledges that the court will often have to decide whether to place a defendant in a residential facility. However, it requires evaluators to provide an opinion on the matters which they are best able to answer — the efficacy of services in various settings — rather than calling for judgments regarding public safety or public policy which are properly those of the court.

The distribution of examination and evaluation reports is strictly limited to those persons directly involved in the proceedings — the parties, their attorneys and/or advocates, and individuals responsible for providing habilitative services or supervising the defendant or offender. This is in keeping with the privacy protections afforded by subsection (6) of this section. The state Developmental Disabilities Advocacy Office is included under paragraphs (4)(d) and (e) as part of that office's outreach and coordination responsibility[355] and to assure that the court is made aware of all the alternatives for the provision of habilitative services.

(5) Examination and evaluation reports shall be explained to the person examined or evaluated to the maximum extent possible. Portions of such reports which are likely to cause severe psychological harm to that person may be disclosed and explained to him or her through a psychiatrist, psychologist or other appropriate professional.

COMMENT. In keeping with the axiom that defendants and offenders alleged or found to have a developmental disability should participate in all

[355] See Chapter 9: Advocacy, *infra.*

important decisions to the maximum extent possible, subsection (4) provides for disclosure of all examination and evaluation reports to the person examined or evaluated and this subsection requires that the contents of those reports be explained in the language, terms and mode of communication which he or she is most likely to understand.[356] While some reports may include highly sensitive information which might be disturbing to the defendant or offender, as the court in Suzuki v. Quisenberry observed, "on balance, a person's interest in liberty is more important than his [or her] interest in not being perturbed."[357] Thus, provision is made for disclosure of those potentially disturbing portions of the reports through an individual with the professional skills to deal with the questions and emotions involved so as to minimize any potential negative consequences.

(6) (a) Examination or evaluation reports shall not constitute public records.

(b) Statements made by a person alleged or found to have a developmental disability and other information developed in the course of a screening test, examination or evaluation provided pursuant to this act and physical evidence obtained as a result of such statements or information shall not be admissible in evidence against that person on the issue of guilt in any criminal proceeding. In addition, such statements, information or evidence shall not be admissible in any civil proceeding without the person's informed consent.

COMMENT. This provision is intended to encourage individuals to be as candid and cooperative as possible during examinations and evaluations by safeguarding the privacy and constitutional protection against self-incrimination of the person being examined. Examination and evaluation reports may contain highly sensitive personal information concerning an individual's physical and mental condition and living situation. There appears to be little benefit in permitting public access to such information and serious costs in terms of the invasion of the privacy of the defendant or offender and his or her family, and the increased reluctance of individuals to provide information to the examiner or multidisciplinary evaluation team.[358]

It should be noted that a distinction is drawn between the confidentiality of records and the confidentiality of courtroom proceedings. The distinction is based on the impact which "free and robust reporting, criticism, and debate" can have on public understanding of the law and on the accountability of the judicial system,[359] the prohibition announced in the Nebraska Press Association vs. Stuart on imposing a ban on publication of information available to the public,[360] and the knowledge that retention of identifiable information obtained at an open hearing will be subject to the limitation of memory of those relatively few people who attend the hearing or may have been told of the experience. On the other hand, unless access to and retention of records are restricted, the records may remain available long after the court proceedings have been concluded to persons wholly unconnected with the case, and will, whether or not they have been updated, retain an aura of authority which may be out of proportion to their importance at the time of examination.

Paragraph (b) extends the privilege against self-incrimination guaranteed by the Fifth Amendment to the U.S. Constitution to the context of examination and evaluations, and the traditional doctor/patient privilege to examiners and all members of multidisciplinary evaluation teams, whether or not they are physicians. Without this protection, defendants, particulary in examinations and evaluations conducted for purposes of determining eligibility for diversion and competency to stand trial, would be required to choose between a thorough examination or evaluation and possibly waiving his or her Constitutional right to remain silent. The disruption of the evaluation or examination which would occur if Miranda warnings and the presence of counsel were required, and the resulting diminution of the reliability of examination and evaluation reports, more than justify the broad immunity accorded the statements of individuals being examined or evaluated under this act.

(7) The cost of any examination or evaluation conducted by an expert appointed pursuant to this act shall be borne by the (name of state/county developmental disabilities agency/name of court(s)).

COMMENT. Inasmuch as the determination of competency of a defendant to stand trial protects the interests of the state as well as those of the accused, in light of the obligation of the state to provide necessary habilitative services pursuant to sections 10 and 12, to persons acquitted solely by reason of diminished capacity and developmentally disabled offenders, and in order not to discriminate against persons on the basis of wealth in making pretrial diversion decisions, this provision makes examinations and evaluations a public expense. Whether this cost is paid by a local or state agency, by the court, or shared among various entities depends on the organization and tradition of the enacting jurisdiction. Nothing in this subsection is intended to prohibit defendants and offenders from being examined or

[356] Section 13(5).

[357] 411 F. Supp. 1113, 1127 (D. Hawaii 1976).

[358] ABA, *supra* note 141, at 120-121.

[359] Nebraska Press Association vs. Stuart, 427 U.S. 539, 587 (Brennan, J. concurring).

[360] *Id.*, at 570.

evaluated at their own expense by experts of their own choice.[361]

Section 16. Training

(1) The state developmental disabilities agency shall plan and prepare a developmental disabilities training program. It shall provide appropriate training without charge to those individuals required to receive it under subsections (3) and (4) of this section and upon request, to judges, criminal defense attorneys, attorneys who defend juveniles alleged to be delinquent and other persons having official responsibilities within the criminal and/or juvenile justice systems. This training responsibility may be carried out by the developmental disabilities agency through qualified employees or through contractual arrangements.

(2) At a minimum, the training program should include instruction regarding:

(a) The nature, effects, and causes of developmental disabilities;

(b) The techniques for recognizing a person who has a developmental disability;

(c) The most common problems which occur when a person with a developmental disability comes in contact with the criminal and/or juvenile justice systems as a victim, witness, suspect, defendant, or offender;

(d) The availability of diagnostic and habilitative services and the procedures for obtaining such services; and

(e) The rights to which persons with a developmental disability are entitled and the procedures for protecting those rights.

(3) The following officials shall ensure that each employee having responsibilities within the criminal and/or juvenile justice systems receives at least three (3) hours of instruction provided pursuant to subsection (1) of this section within one (1) year after the effective date of this act, or within six (6) months after first employment if hired more than six (6) months after the effective date of this act:

(a) Chiefs of police;

(b) Sheriffs;

(c) (Title of the head of the state highway patrol);

(d) (Title of the head of the state bureau of investigation);

(e) The Attorney General;

(f) Prosecutors;

(g) (Title of the head(s) of the state/local public defender office(s);

(h) (Title of the heads(s) of the state/local adult probation agency(ies));

(i) (Title of the head(s) of the state/local juvenile probation agency(ies));

(j) (Title of the head(s) of the state agency(ies) responsible for adult and juvenile corrections).

(4) Each official listed in subsection (4) of this section shall also ensure that each employee receives, in addition, at least three (3) hours of inservice training provided pursuant to subsection (1) every two (2) years after receiving initial instruction through the developmental disabilities training program.

COMMENT. Nearly every conference and commentator addressing the problems faced by persons with developmental disabilities in the criminal justice system has pointed out the failure of criminal justice professionals to recognize persons with developmental disabilities, and the need to provide training for law enforcement officers, judges, prosecutors, defense attorneys and correctional personnel regarding the causes and characteristics of developmental disabilities and the needs, methods for handling, and services available to defendants and offenders with such disabilities.[362] This section seeks to address that need. It sets a minimum level of initial and inservice training for all law enforcement personnel; prosecutors; public defenders; and corrections, probation, and parole personnel, and authorizes the provision of such training to other criminal justice personnel, judges, and the private defense bar. In addition, since such training is critical for juvenile court judges and other juvenile justice system personnel whether or not legislation similar to the Model Developmentally Disabled Juvenile Offender Act[363] is enacted, they are included in the training requirements of this section as well. Enacting jurisdictions should insert the terms which fit the structure of their criminal and juvenile justice systems and exclude clerical and maintenance personnel from the training requirements unless they are likely to come into contact with developmentally disabled citizens during the course of their duties.

Responsibility for developing and presenting training programs is assigned to the state agency responsible for providing services to persons with a developmental disability. Authorization is provided for hiring contractors to meet this responsibility. It is intended that the training required by

[361] *Cf.* section 13(2) and comment thereto.

[362] See *e.g.*, PCMR, *supra* note 124, at 63 and 133; SANTAMOUR and WEST, *supra* note 188, at 11; MISSOURI ASSOCIATION FOR RETARDED CITIZENS, *supra* note 12, at 39-43; Allen, *supra*, note 252; Haggerty, Udall and Kane, *supra* note 163; Norley, *Retarded Youth and the Law Enforcement Process* in NEW ENGLAND SEMINAR ON RETARDED YOUTH AND THE LAW ENFORCEMENT PROCESS, *supra* note 30, at 12; Rowan, *supra*, note 248, at 674; SANTAMOUR and WEST, *supra* note 251, at 24.

[363] Since such training is critical for juvenile justice personnel whether or not legislation similar to the Model Developmentally Disabled Juvenile Offender Act, *infra*, is enacted, it is included in this section.

this act be integrated into existing training mechanisms whenever possible such as police academies, judicial conference sessions, and continuing legal education programs. Funds to support development of the training program may be obtainable from federal revenue sharing and discretionary grants as well as from state appropriations.

Section 17. Authorization of Appropriations

To carry out the purposes of this act, there is authorized to be appropriated $____ for the fiscal year ending ________, $ ____ for the fiscal year ending ________, $____ for the fiscal year ending ________, and $____ for the fiscal year ending ________.

COMMENT. Although every effort has been made to limit the cost of implementing this act, most states will need to appropriate or transfer some additional funds to make the sections on training, evaluation, advocacy, and the provision of habilitative services fully effective. It should be noted, however, that the provisions encouraging early disposition of cases,[364] and discouraging confinement in secure or non-community based facilities[365] should help to offset some of these costs. Moreover, in light of the number of persons with a developmental disability currently imprisoned and the direct cost of crime to the public, if the provision of the habilitative services called for in this act is successful in reducing the rate of recidivism even to a small extent; then the required investment will have reaped a significant reward.

Section 18. Severability of Sections

If any section, subsection, paragraph, sentence, or any other part of this act is adjudged unconstitutional or invalid, such judgment shall not affect, impair or invalidate the remainder of this act, but shall be confined to the section, subsection, paragraph, sentence or other part of this act directly involved in the controversy in which that judgment was rendered.

COMMENT. None necessary.

Section 19. Effective Date

This act takes effect ________________ .

COMMENT. None necessary.

[364] *E.g.*, sections 4(2), 4(3), 6, 7, 8(4)(c), 8(5)(c), 8(5)(d), 8(9), 8(10), and 10(6).

[365] *E.g.*, sections 5, 8(3), 8(5)(d), 8(5)(e), 10(6)(d) and 11(3).

MODEL STATUTE *

A BILL FOR

AN ACT TO ESTABLISH STANDARDS AND PROCEDURES FOR HANDLING JUVENILES WITH A DEVELOPMENTAL DISABILITY WHO COME IN CONTACT WITH THE JUVENILE JUSTICE SYSTEM; TO ESTABLISH STANDARDS AND PROCEDURES FOR DETERMINING THE COMPETENCE OF JUVENILES TO ASSIST COUNSEL AND THE ABILITY OF JUVENILES WITH A DEVELOPMENTAL DISABILITY TO FORM THE INTENT NEEDED TO COMMIT A DELINQUENT OFFENSE; TO ESTABLISH STANDARDS AND PROCEDURES FOR THE DISPOSITION OF JUVENILES WITH A DEVELOPMENTAL DISABILITY ALLEGED OR FOUND TO HAVE COMMITTED A DELINQUENT OFFENSE; AND TO CLARIFY AND PROTECT THE RIGHTS OF JUVENILES WITH A DEVELOPMENTAL DISABILITY IN DETENTION AND CORRECTIONAL FACILITIES.

Section 1. Title

This act shall be known as the "Developmentally Disabled Juvenile Offender Act."

COMMENT. None necessary.

Section 2. Purpose

The (name of the Legislature) declares that it is the purpose of this act to assure: that all members of the juvenile justice system deal fairly with juveniles who have a developmental disability and fully recognize their rights; that determinations of the competence to assist counsel and the responsibility for delinquent offenses of juveniles with a developmental disability are fair, accurate, and expeditious; that juveniles with a developmental disability who are found to be incompetent to assist counsel or unable to form the requisite intent to commit a delinquent offense receive the appropriate habilitative services in the least restrictive manner which is consistent with their needs and public safety; that juveniles who have a developmental disability and who are in detention or correctional facilities are protected from abuse and exploitation; and that juvenile offenders with a developmental disability are offered the habilitative services to which they are entitled.

This act shall be liberally construed in order to achieve this purpose.

COMMENT. As indicated in the Introduction, state codes provide for a separate justice system for handling juveniles alleged or found to have committed a crime, to be in conflict with their families or to be neglected or abused. Provision is generally made for specialized juvenile or family courts and juvenile correctional agencies. Some states also provide for a separate system of youth services.[366] Although, parallel in most instances to the criminal justice system established for adults, the philosophical basis of the juvenile justice system, its focus and terminology, is quite different. As originally conceived, the juvenile justice system:

> [W]as one in which a fatherly judge touched the heart and conscience of the erring youth by talking over his [or her] problems, by paternal advice and admonition, and in which, in extreme situations, benevolent and wise institutions of the State provided guidance and help "to save him [or her] from a downward career."[367]

During the past 15 years, many of the problems of the *parens patriae* doctrine and a discretionary system governed only by beneficent motives have been revealed and judicial and legislative action has been taken to assure that juveniles receive both the "protections accorded to adults. . . [and] the solicitous care and regenerative treatment postulated for children."[368] In examining juvenile justice proceedings, the Supreme Court has sought to strike balance in determining what rights and procedures are to be applied, so as not to "diminish flexibility and informality to the extent

* Richard Van Duizend had primary responsibility for drafting this model statute.

[366] *See* NATIONAL TASK FORCE TO DEVELOP STANDARDS AND GOALS FOR JUVENILE JUSTICE AND DELINQUENCY PREVENTION, A COMPARATIVE ANALYSIS OF STANDARDS AND STATE PRACTICES: COURT STRUCTURE, JUDICIAL AND NON-JUDICIAL PERSONNEL, AND JUVENILE RECORDS (1977); NACJJDP, *supra* note 28, at §§0.111 and 0.121; NATIONAL ADVISORY COMMITTEE, *supra* note 28, at 7; PRESCOTT and VAN HOUTEN, *supra* note 67.

[367] In re Gault, 387 U.S. 1, 26 (1967); Mack, *The Juvenile Court*, 23 HARVARD LAW REVIEW 104, 120 (1909).

[368] *E.g.*, Kent vs. United States, 383 U.S. 541, 566 (1966); In re Gault, *id.*; In re Winship, 397 U.S. 358 (1970); Breed vs. Jones, 421 U.S. 519 (1975); for a discussion of current criticisms, changes, and trends in juvenile justice, see Rector and Van Duizend, *supra* note 270; GAYLIN *et al.*, *supra* note 270.

that those qualities relate uniquely to the goals of the juvenile court system."[369]

In recognition of the distinctness of the juvenile justice system and the evidence that juveniles with a developmental disability who come into contact with that system have often been faced with the same lack of recognition, understanding, and services as developmentally disabled adults alleged or found to have committed a crime,[370] this model statute seeks to translate the concepts, procedures and protections contained in the Model Developmentally Disabled Offenders Act into the context of juvenile delinquency proceedings. As is discussed in more detail in the comments to subsequent sections, many of the procedures and services required by the model act lie well within the traditional scope of juvenile justice practices and theory.[371] Others may break new ground in some jurisdictions.[372] However, the purpose section makes clear that the overall intent is to enhance fairness, equity and effectiveness of the juvenile justice system for the benefit of both youths who have a developmental disability and the public at large.

Section 3. Definitions

As used in this act:

(1) "Community-based facility" means a small, open, residential facility which relies primarily on services available in the surrounding community rather than on in-house services and which maintains community and consumer participation in the planning, operation, and evaluation of its programs;

COMMENT. See comment to section 3(1) of the Model Developmentally Disabled Offenders Act.

(2) "Delinquent offense" means a violation of a statute or ordinance by a juvenile, which if committed by an adult would be punishable by a period of incarceration;

COMMENT. With two possible exceptions, this term is intended to coincide with the jurisdiction of the enacting state's juvenile court over delinquency. Hence, the maximum chronological age at which a person is considered a juvenile under this act should be the same as the maximum age set by the state's juvenile codes.[373] The first possible exception is that this definition limits the term delinquent offenses to offenses which would be a crime if committed by an adult.[374] Thus, conduct that subjects only juveniles to court jurisdiction—*e.g.*, truancy, running away, and "incorrigibility"—is not covered,[375] even though it is subsumed under the rubric of delinquency, in some states. This is because neither the concept of diminished responsibility nor the insanity defense are applicable to status offenses. Furthermore, although compentency to assist counsel would be relevant in such cases, the major premise underlying court jurisdiction over status offenses is to assist young people and fam-
problems and abuses to which the jurisdiction over non-
ilies in need of services. Given this premise and the criminal misbehavior has been subject,[376] it would be preferrable to provide any necessary habilitative services in accordance with the Model Right to Developmental Disabilities Services Act.[377]

The second possible exception is that the definition covers only those offenses for which an adult could be incarcerated, thus excluding those offenses punishable only by a fine. As noted in the comment to section 3(5) of the Model Developmentally Disabled Offenders Act, the intrusion and expense involved in determining competency and providing habilitative services on an involuntary basis following a determination that a juvenile is incompetent to assist counsel or is not responsible solely by reason of diminished capacity, is not warranted for such minor law violations. Again, if services are required they should be provided in accordance with the Model Right to Services/Admission to Services Act.

(3) "Developmental disability" means a chronic disability of a juvenile which:

(a) is attributable to a mental or physical impairment or a combination of mental and physical impairments;

(b) is likely to continue indefinitely;

(c) results in substantial functional limitations in two or more of the following areas of major life activity:

(i) self-care;

(ii) receptive and expressive language;

(iii) learning;

(iv) mobility;

(v) self-direction;

(vi) capacity for independent living; or

(vii) economic self-sufficiency; and

---

[369] Breed vs. Jones, *id.*, at 535.

[370] Norley, *Retarded Youth and the Law Enforcement Process, supra* note 12; PRESCOTT and VAN HOUTEN, *supra* note 67; Morales vs. Turman, 383 F. Supp. 53 (E.D. Tex. 1974), *rev'd on other grounds* 535 F.2d 864 (5th Cir. 1976), *rev'd and remanded* 430 U.S. 322 (1977), *remanded for further hearings* 562 F.2d 993 (5th Cir. 1977); Friel, *The Mentally Retarded Offender – Texas CAMIO Research Project*, in THE MENTALLY RETARDED CITIZENS AND THE CRIMINAL JUSTICE SYSTEM, *supra* note 12, at 95-124.

[371] See sections 6, 10, and 11.

[372] *E.g.*, sections 7 and 9.

[373] For a discussion of the age limits set for juvenile/family court jurisdiction over delinquency, see NATIONAL TASK FORCE TO DEVELOP STANDARDS AND GOALS FOR JUVENILE JUSTICE AND DELINQUENCY PREVENTION, A COMPARATIVE ANALYSIS OF STANDARDS AND STATE PRACTICES: JURISDICTION DELINQUENCY, 9-19 (1977).

[374] See section 3(5) of the Model Developmentally Disabled Offenders Act.

[375] NATIONAL TASK FORCE, *supra* note 373, at 2-5.

[376] NATIONAL TASK FORCE TO DEVELOP STANDARDS AND GOALS FOR JUVENILE JUSTICE AND DELINQUENCY PREVENTION, A COMPARATIVE ANALYSIS OF STANDARDS AND STATE PRACTICES JURISDICTION — STATUS OFFENSES, 4-16 (1977).

[377] Chapter 6: Right to Services, *supra*.

(d) reflects the need for a combination and sequence of special, interdisciplinary or generic care, treatment or other services which are of lifelong or extended duration and individually planned and coordinated.

COMMENT. See comment to section 3(3) of the Model Developmentally Disabled Offenders Act.

(4) "Developmental disability agency" means the unit of state and/or local government responsible for providing or coordinating services to juveniles with a developmental disability;

COMMENT. See comment to section 3(10) of the Model Developmentally Disabled Offenders Act.

(5) "Evaluation" means an assessment, by a multidisciplinary evaluation team appointed in accordance with section 13, of a juvenile's mental, emotional, physical, and educational condition, adaptive behavior, and social skills, and of the habilitative services, if any, required by the juvenile;

COMMENT. See the comment to section 3(12) of the Model Developmentally Disabled Offenders Act.

(6) "Examination" means a diagnosis by a person, with expertise regarding developmental disabilities, selected in accordance with section 13, to determine whether a juvenile has a developmental disability and if so, the effect of the disability if any, on the juvenile's responsibility for the alleged delinquent offense;

COMMENT. See the comment to section 3(12) of the Model Developmentally Disabled Offenders Act.

(7) "Habilitative services" means medical, mental health, social, vocational, educational, physical therapy, counseling, financial management or other services which will assist a juvenile, who has a developmental disability, to the extent and within the shortest time possible, to become competent to assist counsel, or to function in the community as a self-reliant, law-abiding citizen;

COMMENT. See the comment to section 3(13) of the Model Developmentally Disabled Offenders Act.

(8) "Innocent disability-related behavior" means uncontrollable conduct by a juvenile which is the direct result of a developmental disability, including but not limited to slurred speech, awkward movement, or the involuntary movements associated with a seizure;

COMMENT. See comment to section 3(14) of the Model Developmentally Disabled Offenders Act.

(9) "Intake unit" means the unit of state or local government responsible for reviewing complaints submitted pursuant to the jurisdiction of the juvenile court;

COMMENT. Generally, incoming cases involving juveniles are screened by an "intake officer" who makes a decision or recommendation on whether a petition should be filed against the juvenile, or whether the juvenile should be counseled and released or diverted, and if a petition is to be filed, whether, at least initially, the juvenile should be held in custody or released. In most jurisdictions, intake officers and their support staff are court employees. In others they are part of an executive branch agency.[378] This definition is intended to encompass any governmental unit authorized to perform the intake screening function. Enacting jurisdictions may wish to omit this term and substitute the name of their intake unit.

(10) "Juvenile correctional agency" means a unit of state or local government responsible for the supervision and/or custody of juvenile offenders;

COMMENT. This term is intended to include state and local departments of corrections, state and local departments of juvenile corrections, state and local youth services departments, state and local probation and/or juvenile probation offices, state juvenile parole authorities, sheriff offices and police departments which are responsible for operating jails or other juvenile correctional facilities, and any other agency which operates a facility in which juveniles found to be delinquent are held or which is responsible for supervising such juveniles in the community. Where the juvenile or family court operates a juvenile correctional facility or employs probation officers, the duties and limitations set forth for juvenile correctional agencies pursuant to section 11 also apply to the juvenile court. Enacting jurisdictions may wish to omit this term and substitute the name of the relevant agency(ies) or court in section 11.

(11) "Juvenile correctional facility" means a state or local place of confinement for juvenile offenders operated by a juvenile correctional agency;

COMMENT. This term is intended to include training schools and detention centers as well as community-based facilities housing juveniles alleged or found to be delinquent.[379]

(12) "Juvenile court" means the court with jurisdiction over matters relating to juvenile delinquency;

COMMENT. See section 3(2) and the comment thereto.

(13) "Juvenile offender" means a juvenile adjudicated delinquent;

---

[378] NACJJDP, *supra* note 148, at 60-61.

[379] *See* 42 UNITED STATES CODE §5603(12) (1977).

COMMENT. See section 3(2) and the comment thereto.

(14) "Least restrictive alternative" means that habilitative services and the conditions under which they are administered are no more harsh, hazardous or intrusive and involve no more restrictions on a juveniles's physical freedom, social interaction, or individual choice than are essential;

COMMENT. See comment to section 3(15) of the Model Developmentally Disabled Offenders Act.

(15) "Multidisciplinary evaluation team" means a group of at least three (3) individuals which is appointed by the juvenile court from a panel of experts approved by the Evaluation Oversight Board, to evaluate a juvenile pursuant to this act. Each multidisciplinary evaluation team shall include a physician, a social worker with a graduate degree in social work and field training or experience in working with persons who have a developmental disability, and another expert with knowledge of the particular disability which a juvenile is alleged or has been found to have;

COMMENT. See comment to section 3(16) of the Model Developmentally Disabled Offenders Act.

(16) "Preadjudication diversion agreement" means a written agreement between a juvenile and an intake unit containing the terms and conditions for diversion from the juvenile justice process;

COMMENT. See sections 3(9) and 6 of this act and the comment to section 3(8) of the Model Developmentally Disabled Offenders Act.

(17) "Program Plan" means a comprehensive individualized plan for the provision of habilitative services which sets forth specific goals, the time and habilitative services required to meet those goals, and the manner and location in and frequency with which those services will be provided, and which does not contain standardized provisions applicable to general categories of persons without regard to their individual abilities and needs;

COMMENT. See comment to section 3(18) of the Model Developmentally Disabled Offenders Act.

(18) "Prosecutor" means the official representing the government in delinquency proceedings;

COMMENT. See comment to section 3(20) of the Model Developmentally Disabled Offenders Act.

(19) "Restraint on liberty" means custody of and any [bond or other] condition of release imposed on a juvenile, and any conditions or placement contained in a dispositional order issued in accordance with section 9.

COMMENT. The reference to bail bonds is bracketed, since the right to release on bail does not apply to delinquency proceedings in many jurisdictions.

Section 4. Pretrial Release

Juveniles alleged to have committed a delinquent offense shall not be disqualified from being released on recognizance [nor shall the amount, if any, set for bail be increased] because they have a developmental disability.

COMMENT. See comment to section 3(19) of this act and section 5 of the Model Developmentally Disabled Offenders Act.

Section 5. Dismissal of Charges

If the prosecutor concludes subsequent to the filing of formal charges but prior to entry of a judgment that a juvenile has a developmental disability and that the alleged conduct constitutes innocent disability-related behavior rather than a delinquent offense, or that the interests of justice will not be served by proceeding further the prosecutor shall move to dismiss the charges and to have all restraints on the juvenile's liberty terminated.

COMMENT. See comment to section 6 of the Model Developmentally Disabled Offenders Act.

Section 6. Informal Disposition

(1) Juveniles alleged to have committed a delinquent offense shall not be disqualified from being counseled and released by the intake unit, or from being offered an opportunity to enter a preadjudication diversion agreement because they have a developmental disability.

(2) (a) In determining whether to negotiate a preadjudication diversion agreement or to recommend that a petition be filed, the intake unit may request a juvenile alleged to have committed a delinquent offense to undergo an examination by a qualified expert if there is reason to believe the juvenile has a developmental disability. Such a request shall not be made unless:

(i) The juvenile is represented by counsel;

(ii) The purpose, nature, location, duration, and possible beneficial and harmful consequences of the examination have been explained to the juvenile, and described to his or her attorney, and his or her parent or guardian; and

(iii) Consent is obtained from the juvenile's parent or guardian and from the juvenile following consultation with his or her attorney.

(b) An examination by an expert selected from the list published by the Evaluation Oversight Board shall be conducted at [state/county/court] expense on the request of a juvenile who there is reason to believe has a developmental disability and who is unable to

obtain such an examination for financial reasons.

(3) (a) In addition to the requirements prescribed by [citation to statutes, court rules, and/or regulations governing diversion agreements for juveniles alleged to have committed a delinquent offense], a preadjudication diversion agreement between the intake unit and a juvenile who has a developmental disability shall not be entered into unless:

(i) It is in writing;

(ii) The juvenile is represented by an attorney;

(iii) The agreement has been reviewed by the juvenile's attorney; and

(iv) The terms of the agreement, the consequences of signing it, and the consequences of violating it have been explained to the juvenile and his or her parents or guardian.

(b) A preadjudication diversion agreement with a juvenile who has a developmental disability shall specify in addition to the items prescribed in [citation to relevant statute, court rules or regulations]:

(i) The agreed-upon goals;

(ii) Whether an evaluation is necessary, and if so, whether the juvenile and his or her parents or guardian consent to the evaluation;

(iii) If an evaluation has already been performed, the habilitative services which the juvenile and his or her parents or guardian agree to accept in order to meet those goals; and

(iv) The location and manner in which the evaluation or habilitative services will be provided.

(c) In order to be enforceable, any modification to a preadjudication diversion agreement shall be in writing, reviewed by the juvenile's attorney, and explained in a manner specified in subparagraph (3)(a)(iv) of this section.

(d) Evaluations and habilitative services provided through a preadjudication diversion agreement shall use the least restrictive alternatives consistent with the juvenile's needs. No preadjudication diversion agreement shall include a term or condition or shall be based on an informal understanding regarding the placement of the juvenile in a residential facility which is secure or not community-based, or in any facility whatsoever to which individuals are committed under [citation to the Model Right to and Admission to Services Act/applicable state civil commitment law(s)]. Any such term, condition, or understanding shall be invalid and unenforceable.

COMMENT. Diversion is an established part of the juvenile justice process in most American jurisdictions. Recent statistics indicate that:

> The human aspect of a criminal case exerts a powerful pull when the suspect is young. If the harshness of the criminal process supports use of discretion with adults, the argument is even more persuasive when applied to young offenders. Furthermore, underlying the special handling of the young offender is the belief that he [or she] can be changed. For him [or her], therefore, the criminal process, because of its ineffectiveness as an agent of change, seems particularly inappropriate.
>
> Similarly, the overcriminalizing or hidden welfare aspects of the criminal law support discretion even more strongly in the juvenile than in the adult context. Non-crime-prevention provisions are far more prevalent in juvenile court law than in the most antiquated of criminal codes.
>
> Discretionary judgment also provides a necessary steam valve in the juvenile justice system. Just as the administration of criminal justice is strained by growing criminal populations, so too the increasing volume of juvenile offenders builds up pressure within police departments, juvenile courts, and all the other institutions with responsibility to control delinquency. Neither police ranks nor the number of judges and auxiliary staff of juvenile courts has expanded at a rate commensurate with the increase in recorded delinquency. Informal determinations as to which cases demand official action are inevitable, and in the process many juveniles are diverted to nonjudicial sources of control and service.[380]

Over 40% of the juveniles coming into contact with the police as a result of allegations of delinquent conduct or non-criminal misbehavior are counseled and released or referred to a social service providing agency.[381] In addition, a significant number of those juveniles who are referred to the juvenile court by police are diverted prior to adjudication.[382] But as is the case for adults, diversion of juveniles with a developmental disability all too often results in a choice between institutionalization in a mental retardation facility and the risk or continuation of institutionalization in a juvenile detention or correctional facility.[383] Like

[380] PCLEAJ, TASK FORCE REPORT: JUVENILE DELINQUENCY AND YOUTH CRIME, 10-11 (1967).

[381] W. WEBSTER, UNIFORM CRIME REPORTS: CRIME IN THE UNITED STATES — 1977, 219 (1978); see also NATIONAL ADVISORY COMMITTEE, *supra* note 28, at 209.

[382] IJA/ABA, STANDARDS RELATING TO THE JUVENILE PROBATION FUNCTION: INTAKE AND PREDISPOSITION INVESTIGATIVE SERVICES 34-35 (Tentative Draft 1977); NACJJDP, *supra* note 148, at §3.142.

[383] PRESCOTT and VAN HOUTEN, *supra* note 67.

section 7 of the Model Developmentally Disabled Offenders Act, this section seeks to assure that juveniles with a developmental disability are not prosecuted solely because of their disability or as a means to provide necessary services.[384]

The procedures and protections offered by subsections (2) and (3) are identical to those required by sections 7(2) and (3) of the Model Developmentally Disabled Offenders Act, except that provision is made for participation of the juvenile's parent(s) or guardian.[385] These procedures are intended to assure that there is knowing and voluntary consent to a preadjudication diversion agreement.

> The non-judicial disposition may result in substantial intervention in the juvenile's life without any judicial determination that such intervention is warranted, and if the juvenile and his or her parents do not voluntarily consent to the non-judicial disposition, the intervention it produces is in essence coercive.[386]

Besides, without parental cooperation, the achievement of the goals of the preadjudication agreement is far less likely. If because of age and/or disability it is concluded that a juvenile is unable to provide voluntary informed consent, a substantial question arises regarding his or her competence to assist counsel and ability to have formed the requisite intent to commit a delinquent offense. In such instances, therefore, the more appropriate course of action would appear to be to dismiss the allegations and to have the necessary services provided pursuant to the Model Right to and Admission to Services Act or its equivalent.

Even the best diversion procedures will be of little use unless there are community-based habilitative services which are open to juveniles with a developmental disability who have come in contact with the juvenile jurisdiction.[387] Hence, if a provision comparable to section 7(4) of the Model Developmentally Disabled Offenders Act is not in force at the time this act is considered, one should be included in this section.

Introduction to Section 7

The competency of a juvenile alleged to have committed a delinquent offense is seldom formally raised.[388] This is attributable to a number of factors. The first is the relatively informal nature of juvenile court proceedings. For example, when an alleged delinquent is thought to be mentally ill or developmentally disabled, the delinquency petition is often withdrawn, and services are provided through diversion or by refiling the case under the juvenile court's jurisdiction over neglect, "children in need of services," or civil commitment.[389] The second factor is the emphasis in juvenile court proceedings on a child's treatment needs. As a result, a mental disability is sometimes viewed to be a matter relevant solely to the disposition, and the question of its impact on the adjudication stage of the proceedings is not considered. A third factor is the authority of the juvenile court to transfer cases to the adult criminal court when a juvenile is considered unamenable to treatment within the juvenile justice system.[390] However, as the Supreme Court of Louisiana recently held:

> Since the right not to be tried while incompetent is a due process-fundamental fairness right, see Pate [vs. Robinson 383 U.S. 375 (1966] and Drope [vs. Missouri 420 U.S. 162 (1975)] . . . it should . . . be applicable to juvenile proceedings, unless some essential end of the juvenile justice system will be thwarted by its application. Indeed, the right — rooted as it is in traditional notions of fair play, viewed from the early days of the common law as central to a defendant's ability to present his defense, and akin to the right not to be tried *in absentia* — is not only historically "fundamental", but also functionally "essential", and therefore cannot be outweighed under the *Gault/McKeiver/Winship* test by other desirable aspects of the juvenile system.[391]

In recognition of the applicability of this fundamental right, this section translates the procedures and requirements set forth in section 8 of the Model Developmentally Disabled Offenders Act to fit the problems and terminology of delinquency proceedings.

Specifically, the section sets forth the procedures and standards applicable to and the dispositions available following an inquiry into the competency of a juvenile to assist counsel. It covers only those juveniles who have a developmental disability. Like the section pertaining to adults nothing in this provision would prevent states from applying it to persons alleged or found to be mentally ill. Subsection (1) establishes a presumption of competency and the standard for finding a juvenile incompetent to assist counsel. Subsec-

---

[384] See comment to sections 7(1) and (4) of the Model Developmentally Disabled Offenders Act.

[385] See IJA/ABA, PROBATION, *supra* note 382, at 33-53.

[386] *Id.*, at 47.

[387] PRESCOTT and VAN HOUTEN, *supra* note 67.

[388] Donovan, *The Juvenile Court and the Mentally Disordered Juvenile*, 45 NORTH DAKOTA LAW REVIEW 222, 229-232 (1969); PELC. COMPETENCY TO STAND TRIAL IN JUVENILE COURT PROCEEDINGS (1977) (Unpublished thesis Wellesley College library).

[389] *Id.*

[390] Donovan, *id.*, at 240-244; Kent vs. United States, 401 F. 2d 408 (D.C. Cir., 1968).

[391] In re Causey, 363 So. 2d 472, 476 (La.S.Ct. 1978); also In re S.W.T., 277 N.W.2d 507 Minn. S.Ct 1979).

tions (2) and (4) set forth the procedures for initiating an inquiry into the competence of a juvenile to assist counsel, and subsection (3) assures that juveniles are not retained in custody solely because of such an inquiry. Subsection (5) outlines the range of possible findings and the dispositional choices applicable to each. Subsection (6) governs the development of a Program Plan following a finding that a juvenile is incompetent to assist counsel when there is a substantial likelihood that he or she will become competent within a year of the competency hearing. Subsections (7) and (8) establish reporting and review procedures. Subsection (9) permits legal challenges to the allegations despite a finding that a juvenile is incompetent to assist counsel. Finally, subsection (10) provides for the termination of a dispositional order issued under this section when the underlying charges are dismissed.

Two provisions contained in section 8 of the Model Developmentally Disabled Offenders Act have been omitted from this section. Because most juvenile codes provide for indeterminate dispositional terms rather than finite sentences, the section does not contain a provision requiring reduction of the length of time spent confined for examination, evaluation or habilitative services from the length of any sentence of incarceration imposed.[392] Section 8 of this act, however, requires that such periods of confinement be considered in determining the disposition if a juvenile is found to be competent and adjudicated delinquent.

Also omitted is the transition provision contained in Section 8 (12) of the adult offenders act. Since competency to assist counsel determinations have not been common in delinquency proceedings, a subsection extending the rights and procedures provided by this act to those found to be incompetent and institutionalized prior to its enactment, is unnecessary.

Section 7. Incompetence to Assist Counsel

(1) Juveniles with a developmental disability who are alleged to have committed a delinquent offense shall be presumed to be competent to assist counsel, whether or not they require or are receiving or have received psychotropic drugs, other medication, or treatment related to that disability. A juvenile who has a developmental disability and who is alleged to have committed a delinquent offense shall be found to be incompetent to assist counsel only if it is proven by a preponderance of the evidence that he or she [meets the standard set forth in (citation to the statutory section defining incompetence of a juvenile to assist counsel)/lacks the present ability to understand the nature and purpose of the proceedings and assist in preparing his or her own defense.].

COMMENT. Like section 8(1) of the Model Developmentally Disabled Offenders Act, this provision establishes that a juvenile should be considered competent unless proven otherwise, even if he or she has a developmental disability. For purposes of this act, competence is phrased in terms of assisting counsel rather than "standing trial" for both idiomatic and substantive reasons. Traditionally, juvenile justice procedures avoid terms identified with the criminal justice process so as to reduce the stigma of being found delinquent. Thus, adjudication hearing or a similar term is usually substituted for "trial" making the phrase "competent to stand trial" inapposite. More significantly, "assisting counsel" places the emphasis on the more objective portion of the usual test for competency in criminal proceedings — present ability to consult with counsel and to understand the proceedings.[393] While many non-disabled youngsters as well as those with a developmental disability may not grasp the full import of delinquency proceedings to the same extent as an adult, most juveniles are able, though not always willing, to provide their attorney with a version of the facts, respond to opposing testimony, and cooperate in preparing a defense. It should be noted, however, the ability to assist counsel cannot be the sole test, since the juvenile must still be able to assist in determining such questions as whether to oppose the petition, and to assist in developing an appropriate disposition if he or she adjudicated delinquent.[394] Both these responsibilities require the ability to understand the nature and consequences of the proceeding.[395]

(2) Following the filing of the petition. A juvenile who there is reason to believe has a developmental disability or the prosecutor may move for an evaluation to determine whether the juvenile is incompetent to assist counsel.

COMMENT. Since indictments are not required for prosecution of a juvenile alleged to have committed a delinquent offense, the issue of incompetency may be raised any time after the filing of the petition or an equivalent charging document initiating juvenile court proceedings.[396]

(3) Except as otherwise provided by this sec-

[392] See section 8(11) of the Model Developmentally Disabled Offenders Act and the comment thereto.

[393] Dusky vs. United States 362 U.S. 402 (1960); Drope vs. Missouri, 420 U.S. 162 (1975).

[394] IJA/ABA, STANDARDS RELATING TO COUNSEL FOR PRIVATE PARTIES, §§3.1(6); 7.1 and 9.3 (Tentative Draft 1977); NATIONAL ADVISORY COMMITTEE, *supra* note 28, at §16.2; NACJJDP, *supra* note 148, at §3.134.

[395] For a discussion of other aspects of the operation and meaning of this subsection, see the comment to section 8(1) of the Model Developmentally Disabled Offenders Act.

[396] See also the comment to section 8(2) of the Model Developmentally Disabled Offenders Act.

tion, a juvenile who there is reason to believe has a developmental disability and who is awaiting determination of his or her competence to assist counsel is entitled to release on recognizance [or bail] to the same extent and on the same terms and conditions as if the competency issue had not been raised.

COMMENT. The fact that a juvenile's competence to assist counsel may be in question should have no relevance to whether he or she is detained or released prior to adjudication. While the grounds for detaining juveniles are often somewhat broader than those for pretrial detention of adults,[397] there is little basis for believing that the jurisdiction of the court, public safety, or the juvenile will be better protected through the detention of a juvenile alleged to have a developmental disability and to be incompetent to assist counsel. As noted in the comment to section 8(3) of the Model Developmentally Disabled Offenders Act, there are usually less restrictive alternatives than detention for assuring that a juvenile will appear for further hearings, and preadjudication predictions of dangerousness are fraught with error. Moreover, given the lack of services available in most instances during preadjudication detention, the poor conditions in many juvenile detention facilities, the adverse impact of incarceration on juveniles in general and its cost, protective custody in delinquency cases involving developmentally disabled juveniles should be employed sparingly.[398]

The reference to bail in this and subsequent subsections has been placed in brackets since juveniles in many states do not have a right to release on bail.[399] This statute takes no position on whether release conditioned on the posting of a bail bond or some other financial restriction is appropriate for juveniles in delinquency proceedings.[400] The intent is simply to assure that the release options available to juveniles with a developmental disability are the same as those available to non-disabled juveniles alleged to have committed a delinquent offense.

(4) (a) Following the filing of a motion pursuant to subsection (2) of this section or on its own initiative, the juvenile court shall hold a hearing to determine whether there is a reasonable doubt that the juvenile is competent to assist counsel. The burdens of going forward and of persuasion shall be on the moving party, except that when the hearing is initiated by the juvenile court, both burdens shall be on the prosecutor.

(b) If the juvenile court determines that a reasonable doubt has not been raised regarding the competence of the juvenile to assist counsel, the case shall proceed as provided by law.

(c) If the juvenile court determines that a reasonable doubt has been raised, it shall:

(i) Dismiss the petition and remove any restraints on the juvenile's liberty if neither the interests of justice not those of the juvenile will be served by proceeding further; or

(ii) Order the juvenile to undergo an evaluation and appoint a multidisciplinary evaluation team.

COMMENT. See comment to section 8(4) of the Model Developmentally Disabled Offenders Act.

(5) (a) The court shall promptly convene a competency hearing following submission of the evaluation report. At the hearing, the burdens of going forward and of persuasion shall be on the party who initiated the inquiry into the juvenile's competency to assist counsel except that when the inquiry was initiated by the court, both burdens shall be on the prosecutor.

(b) If on the basis of the testimony and evidence presented at the competency hearing the judge [or the jury if any] find there is a preponderance of the evidence that the juvenile is competent to assist counsel, the case shall proceed without delay. If the juvenile is found competent to assist counsel and is receiving psychotropic drugs or other medication or services, the juvenile court may order the continued administration of such drugs, medication or services under appropriate supervision in order to maintain the juvenile's competence, unless it is demonstrated that continuation of such drugs or other medication or treatment would pose a serious risk to the juvenile's health. If such a risk is shown, the court may order the discontinuance of the drug, medication or treatment and reassess the juvenile's competence to assist counsel.

(c) If on the basis of the testimony and evidence presented at the competency hearing, the judge [or the jury, if any] find there is a preponderance of the evidence that the juvenile is incompetent and that there is not a substantial probability that he or she will become competent to assist counsel within one (1) year if provided with appropriate habilitative services, the court shall dismiss the petition and either:

(i) Terminate any restraints on the juvenile's liberty imposed because of the petition; or

(ii) If the prosecutor indicates the in-

---

[397] NATIONAL TASK FORCE TO DEVELOP STANDARDS AND GOALS FOR JUVENILE JUSTICE AND DELINQUENCY PREVENTION, A COMPARATIVE ANALYSIS OF STANDARDS AND STATE PRACTICES: PRE-ADJUDICATION AND ADJUDICATION PROCESSES, 16-21 (1977).

[398] NACJJDP, *supra* note 148, at comment to §3.151; Martarella vs. Kelley, 349 F. Supp. 565 (S.D.N.Y. 1972); Donovan, *supra* note 388, at 228-229.

[399] NATIONAL TASK FORCE, *supra* note 397, at 32-40.

[400] *But see* NACJJDP, *supra* note 148, at §3.151; IJA/ABA, INTERIM STATUS, *supra* note 3, at §4.7; NATIONAL ADVISORY COMMITTEE, *supra* note 28, at §12.12.

tent to initiate proceedings pursuant to [citation to the Model Right to the Admission to Services Act/state civil commitment law(s)], and if the juvenile is in custody and there is probable cause to believe that he or she meets criteria for placement in a residential facility as defined in that [Act/laws], the court may [exonerate the juvenile's bond if any and] order that the juvenile be held for up to forty-eight (48) hours in the least restrictive facility consistent with his or her needs pending initiation of the civil commitment proceeding.

(d) If on the basis of the testimony and evidence presented at the competency hearing, the judge [or the jury if any] find that there is a preponderance of the evidence the juvenile is incompetent but that there is a substantial probability that the juvenile will become competent to assist counsel within one (1) year of the competency hearing if appropriate habilitative services are provided, the juvenile court shall either:

(i) Dismiss the petition and terminate any restraints on the juvenile's liberty imposed because of the petition if the juvenile is alleged to have committed an offense for which the maximum term of incarceration that could be imposed upon an adult is no more than one (1) year, or if neither the interests of justice nor those of the juvenile will be served by continuing the proceedings; or

(ii) Order a continuance in the proceedings of up to six (6) months, the development of a Program Plan in the manner set forth in subsection (6) of this section, and the provision and/or purchase of necessary habilitative services by the developmental disabilities agency using the least restrictive alternatives.

(e) An order issued pursuant to subparagraph (5)(d)(ii) of this section shall specify whether the Program Plan may include placement of the juvenile in a residential facility which is secure or not community-based, or in which adults or juveniles are placed pursuant to [citation to Model Right to and Admission to Services Act/state civil commitment law(s)]. No such placement shall be permitted unless the judge [or jury if any] finds that the prosecutor has presented clear and convincing proof that it is the least restrictive alternative which will:

(i) Assure the juvenile's presence at future hearings;

(ii) Provide the juvenile with the habilitative services which are most likely to assist him or her to become competent to assist counsel; and

(iii) Protect the juvenile from suffering serious bodily harm.

COMMENT. This subsection sets forth the possible findings following a hearing to determine whether a juvenile is competent to assist counsel and the dispositional alternatives which flow from those findings. The findings and disposition are the same as those outlined in sections 8(5)(a) - (e) of the Model Developmentally Disabled Offenders Act and explained in the comments thereto. However, in view of the shorter time limits often applicable in juvenile proceedings, some enacting jurisdictions may wish to reduce the one (1) year maximum time for which a case may be delayed under this section to six (6) months. The references to a jury are placed in brackets since not all states extend the right to a jury to juveniles in delinquency proceedings.[401] If juveniles are permitted to request a jury trial in the enacting jurisdiction, juries should be permitted to make the determinations required by this provision.

The subsection rejects the alternative contained in recently proposed juvenile justice standards of appointing a guardian *ad litem* for a juvenile and going forward with the delinquency proceeding when defense counsel finds that the juvenile "cannot understand the nature and consequences of the proceedings."[402] While this may be the appropriate approach in neglect, custody, and adoption proceedings involving very young children, it does not take into account the role of the juvenile in assisting counsel to prepare the defense to allegations of delinquency which is a *raison d'etre* of the right not to be tried while incompetent.[403] Also rejected, as is discussed more fully in the comment to sections 8(5)(d) and (e) of the Model Developmentally Disabled Offenders Act, is the suggestion that civil commitment is always the appropriate response when it is determined that a juvenile is incompetent to assist counsel.[404] The preference for delivering necessary habilitative services on a non-custodial basis is as applicable, if not more applicable, when a juvenile with a developmental disability is involved, as when such services are required by an adult.[405]

(6) (a) Following issuance of an order under subparagraph (5)(d)(ii) of this section, the state developmental disabilities agency shall develop a Program Plan in consultation with the juvenile, the juvenile's parent or guardian, and the juvenile's attorney. The objective of the Program Plan shall be to assist the juvenile to become competent to assist counsel within the shortest time possible using the least restrictive alternatives. During the development of the Plan, the

---

[401] NATIONAL TASK FORCE, *supra* note 397, at 69-88; McKeiver vs. Pennsylvania, 403 U.S. 528 (1971).

[402] NACJJDP, *supra* note 148, at §3.134; NATIONAL ADVISORY COMMITTEE, *supra* note 28 at §16.3; IJA/ABA, COUNSEL, *supra* note 394, at 81-82.

[403] See in re Causey, 363 So. 2d 472, 474-475 (La.S.Ct. 1978); and the introduction to this section.

[404] NATIONAL ADVISORY COMMITTEE, *supra* note 28, §14.18.

[405] See section 3(14); *cf.* 42 UNITED STATES CODE §5602 (Supp. 1979); IJA/ABA, STANDARDS RELATING TO DISPOSITIONS, §2.1 (Tentative DRAFT 1977); NACJJDP, *supra* note 148, at §3.182.

purpose, nature, and possible beneficial and harmful consequences of the various alternatives available shall be explained to the juvenile to the maximum extent possible, and to his or her parents or guardian and attorney.

(b) The Program Plan shall be submitted for approval to the juvenile court no more than ten (10) days after issuance of the order. A copy shall be provided to the prosecutor, the juvenile, the juvenile's parent or guardian, the juvenile's attorney, and the (name of the state Developmental Disabilities Advocacy Office). It shall be approved by the court within ten (10) days of submission unless the court on its own initiative sets a hearing or the prosecutor, the juvenile, or the juvenile's parents or guardian move for a hearing. In such a motion, the moving party shall specify the reasons why the plan is alleged to be inadequate or inappropriate and the proposed modifications.

(c) Following the hearing, the juvenile court shall approve the plan or modify it so as to further the objectives set forth in paragraph (6)(a) of this section.

COMMENT. As in section 8(6) of the Model Developmentally Disabled Offenders Act, responsibility for developing the Program Plan and providing and/or purchasing necessary habilitative services is assigned to the state developmental disabilities agency rather than the agency responsible for juvenile corrections or youth services in order to avoid duplication services and infringement on the presumption of innocence. Like section 4 on informal dispositions, this provision includes the juvenile's parent(s) or guardian in the plan development and approval process since their custodial rights may be affected and their cooperation is essential if not custodial alternatives are employed. For a further explanation of the operation and intent of this section, see the comment to section 8(6) of the Model Developmentally Disabled Offenders Act.

(7) (a) The individual in the state developmental disabilities agency responsible for monitoring implementation of the juvenile's Program Plan shall file a written report with the court and send copies to the prosecutor, the juvenile, the juvenile's parents or guardian, and the juvenile's attorney:

(i) At least once every ninety (90) days following the competency hearing;

(ii) Whenever the individual believes the juvenile is competent to assist counsel;

(iii) Whenever the individual believes that there is not a substantial probability that the juvenile will become competent to assist counsel within one (1) year of the competency hearing; and

(iv) Fourteen (14) days before the expiration date of the continuance ordered pursuant to subparagraph (5)(d)(ii) of this section.

(b) The report shall contain:

(i) A description of the services being provided;

(ii) A list of the habilitative services prescribed in the Program Plan which have not been provided and the reasons therefor;

(iii) A description of the progress which has been made in achieving the goals set forth in the Program Plan;

(iv) An opinion on whether there is a substantial probability that the juvenile will become competent to assist counsel within one (1) year of the competency hearing;

(v) The facts and reasons supporting that opinion; and

(vi) A description of the modifications, if any, of the Program Plan which may increase the likelihood that the juvenile will become competent within the prescribed period.

COMMENT. See the comment to section 8(7) of the Model Developmentally Disabled Offenders Act.

(8) (a) The juvenile court shall hold a hearing:

(i) Following submission of a report pursuant to paragraph (7)(d) of this section indicating that the juvenile is competent to assist counsel or that there is not a substantial probability that the juvenile will become competent within one (1) year of the competency hearing;

(ii) At the request of the prosecutor or the juvenile following submission of a report pursuant to paragraph (7)(d); and

(iii) At the end of a continuance ordered pursuant to subparagraph (5)(d)(ii) of this section.

(b) At the hearing, the juvenile court shall redetermine whether the juvenile is competent to assist counsel, and if not, whether there is a substantial probability that the juvenile will become competent within one (1) year of the competency hearing if provided with the appropriate habilitative services. Upon making the determination, the juvenile court shall proceed in the manner specified in the applicable paragraph of subsection (5) of this section.

COMMENT. As noted above, some jurisdictions may wish to set the period for gaining competency in delinquency proceedings at six months rather than one year. For a further discussion of the operation and intent of this provision see the comment to section 8(8) of the Model Developmentally Disabled Offenders Act.

(9) (a) A juvenile who has a developmental disability shall have the right at any time after an initial finding of incompetence:

(i) To file motions for discovery and for

suppression of confessions and other evidence and to receive a prompt hearing on such motions;

(ii) To raise any legal objection that is susceptible of fair determination without his or her personal participation; [and]

[(iii) To request a review by the juvenile court of whether there is probable cause to believe that the allegations contained in the petition can be proven.]

A motion may be reheard upon a showing that evidence is available which was not available due to the juvenile's incompetence when the motion was first decided.

[(b) If at any time the juvenile court determines that there is not probable cause to believe that the allegations in the petition can be proven, it shall dismiss the petition.]

COMMENT. The references to determination of probable cause are bracketed since not all jurisdictions permit such determinations during delinquency proceedings, even when the juvenile is being detained.[406] If probable cause determinations are permitted in the enacting jurisdiction, the bracketed material should be retained; otherwise, it should be omitted. For an explanation of the operation and intent of this provision, see the comment to section 8(9) of the Model Developmentally Disabled Offenders Act.

(10) No order issued under this section authorizing the provision of services or the retention in custody of a juvenile other than an order issued pursuant to subsection (5)(4)(ii) shall remain in effect after the petition is dismissed. The juvenile court shall notify the juvenile, his or her parents or guardian, his or her attorney, the prosecutor, and the individual overseeing the provision of services to the juvenile when the petition is dismissed.

COMMENT. See comment to section 8(10) of the Model Developmentally Disabled Offenders Act.

Section 8. Uncontested Adjudications

(1) The juvenile court shall not accept an admission to the allegations in the petition from a juvenile who there is reason to believe has a developmental disability unless that juvenile is represented by counsel and the juvenile court at a minimum:

(a) Explains to the juvenile on the record:

(i) The allegations which are being admitted;

(ii) That upon acceptance of the admission, the juvenile will be considered delinquent;

(iii) The nature and range of possible dispositions including any special conditions imposed by statute;

(iv) The rights which the juvenile will waive by entering the admission including the right to contest the petition; and

(v) The terms of the plea agreement if any;

(b) Diligently determines through questioning the juvenile personally that:

(i) The juvenile understands the matters listed in subparagraphs (a)(i) - (iv) [(v)] of this section; and

(ii) The admission is not a product of force or threats, or of promises [other than those included in the plea agreement if any]; and

(c) Requires the prosecutor to present affidavits, witnesses, and/or evidence demonstrating that there is a factual basis for the admission.

(2) The juvenile court shall permit a juvenile who there is reason to believe has a developmental disability to withdraw an admission to the allegations in the petition:

(a) If it shown that:

(i) The juvenile did not understand the matters explained in subparagraphs (1)(a)(i) - (iv) [(v)] of this section;

(ii) The juvenile was denied effective assistance of counsel; or

(iii) The admission was involuntary;

[(b) If the juvenile court decides to impose a disposition less favorable to the juvenile than that contemplated in the plea agreement;] or

(c) To correct or prevent manifest injustice.

COMMENT. As in criminal cases involving adults, a significant number of delinquency cases are resolved by the uncontested adjudications.[407] Although by admitting the allegations in the petition a juvenile gives up the same rights as adult defendants who plead guilty, the applicable procedures have generally been less defined.[408] In order to assure that juveniles who have a developmental disability do not unknowingly or involuntarily waive these rights and submit themselves to the penalties and stigma which arise from being adjudged a juvenile offender, this section establishes procedures for accepting admissions in delinquency proceedings similar to those required for adult defendants under Boykin vs. Alabama[409] and North Carolina vs. Alford.[410] In fact, such pro-

---

406 NATIONAL TASK FORCE, *supra* note 397, at 54-58; *but see* NACJJDP at §§3.155 and 3.165; IJA/ABA, STANDARDS RELATING TO PRETRIAL COURT PROCEDURES, §4.1 (Tentative Draft, 1977).

407 IJA/ABA, STANDARDS RELATING TO ADJUDICATIONS, 35-36 (Tentative Draft 1977).

408 NATIONAL TASK FORCE, *supra* note 397, at 64-68; D. BESHAROV, JUVENILE JUSTICE ADVOCACY 33 (1974).

409 395 U.S. 238 (1969).

410 400 U.S. 25 (1970).

cedures have been required for all juveniles by a number of the state appellate courts which have addressed the issue [411] and recommended by several recent juvenile justice standards-setting groups.[412]

The references to plea negotiations have been placed in brackets. While plea-bargaining in delinquency proceedings does not appear to be as widespread as in adult criminal cases, it does occur and is subject to the same criticisms and rationales.[413] This provision is intended neither to endorse nor reject plea-negotiations in delinquency cases. In jurisdictions in which this practice is recognized, the bracketed language will assure that when a juvenile with a developmental disability is involved, the terms of the agreement will be clear to all parties and that the admission may be withdrawn if the juvenile court judge declines to accept the proposed disposition. In jurisdictions in which plea negotiations is not common practice, the language should be omitted.

A detailed explanation of the operation of this section is contained in the comments to section 9 of the Model Developmentally Disabled Offenders Act.

Introduction to Section 9

As noted in a recent law review article:

> It is not entirely clear in juvenile law today whether the government must establish that the child possessed the requisite *mens rea* [intent] to commit the offense, or simply whether [he or] she performed an act which was proscribed by the criminal law regardless of [his or] her intent.[414]

Under the traditional governing principles of the juvenile court, the intent was nearly irrelevant since the question was not whether a specific wrong had been committed, but how the child could best be helped.[415] However under Gault and Winship, "civil labels and good intentions do not themselves obviate the need for criminal due process safeguards in juvenile court,"[416] and the prosecution is required to prove every necessary fact — presumably including intent — beyond a reasonable doubt.[417]

Similarly, although the defenses of infancy and insanity have close historical roots,[418] challenges to a juvenile's responsibility for a delinquent offense on the grounds of a mental illness or developmental offense on the grounds of a mental illness or developmental disability have been made infrequently and with mixed success.[419] This is attributable to the general absence of statutory dispositional authority for the juvenile court following an acquittal by reason of insanity, and the reliance on diversion, alternative forms of jurisdictional authority such as neglect or civil commitment, and transfer to the adult criminal justice system, as well as to the conflict in philosophical perspectives between the two defenses.[420]

This section, in subsection (1), takes the position that proof is required of the state of mind required by law for the criminal offense underlying an allegation of delinquency.[421] Juvenile courts, since their inception, have recognized that accidental actions cannot justify a finding of delinquency.[422] Moreover:

> At a time when there is an ever increasing emphasis upon the rights of juveniles in juvenile court proceedings, *mens rea* does not seem out of place in the protective atmosphere of the juvenile courts.[423]

Subsection (2) permits juveniles with a developmental disability to challenge the prosecution's evidence of intent by asserting a defense of diminished capacity. However, like section 10 of the Model Developmentally Disabled Offenders Act, it prohibits those juveniles from asserting any form of the insanity defense unless they are also mentally ill. There seems little reason to transfer the problems and criticisms which have arisen under the various formulations of the insanity defense to jurisdictions in which it has not already been adopted, when, at least for juveniles with a developmental disability, the diminished capacity theory can provide comparable protection with

---

411 See In re Appeal No. 544, 25 Md. App. 26, 332 A.2d 680 (1965); G.M.K. vs. State, 312 S.2d 538 (Fla. App. 1975); State ex rel. Juvenile Dept. of Coos County vs. Welch, 501 P.2d 991 (Ore. App. 1972); In re Mary B, 20 Cal. App. 3d 816, 98 Cal. Rptr. 178 (1971).

412 IJA/ABA, ADJUDICATIONS, *supra* note 407, at §§3.1-3.7; NACJJDP, *supra* note 148, at §§3.175-3.177; NATIONAL ADVISORY COMMITTEE, *supra* note 28, at §§13.1-13.3.

413 See BESHAROV, *supra* note 408, at 311; (pro) IJA/ABA, *id.*, at 35-40; (con) NACJJDP, *supra* note 148, at §3.175; NATIONAL ADVISORY COMMITTEE, *supra* note 28, at §13.1; see also the sources listed *supra* note 143.

414 McCarthy, *Responsibility in Juvenile Delinquency,* 10 MICHIGAN JOURNAL OF LAW REFORM 131, 192-193 (1977).

415 Mack, *supra* note 367.

416 In re Gault, 387 U.S. 1, 36 (1967); In re Winship, 397 U.S. 358, 365-366 (1969).

417 Winship, *id.*, at 364, 368.

418 Fox, *Responsibility in Juvenile Court,* 11 WILLIAM AND MARY LAW REVIEW 636 (1970).

419 Donovan, *supra* note 388; State of New Jersey in the Interest of H.C., 106 N.J. Super. 583, 256 A.2d 322 (1969) (insanity defense does not apply); In re Winburn, 32 Wis. 2d 152, 145 N.W. 2d 178 (1966) (insanity defense does apply); In re M.G.S. 72 Cal. Rptr. 808 (1968) (insanity defense does apply); In re Causey, 363 So. 2d 472 (La.S.Ct. 1978) (insanity defense does apply).

420 Donovan, *supra* note 388, at 233.

421 See section 9(1).

422 Popkin and Lippert, *Is There a Constitutional Right to the Insanity Defense in Juvenile Court?,* 10 JOURNAL OF FAMILY LAW 421, 426 (1971).

423 Westbrook, *Mens Rea in the Juvenile Court,* 5 JOURNAL OF FAMILY LAW 121, 132 (1965).

less procedural and theoretical complexity.[424] In jurisdictions in which the insanity defense does apply to delinquency cases, this statute will limit its application to juveniles who were mentally ill at the time the offense allegedly occurred.

Subsection (3) assures that juveniles are not held in custody solely as a result of asserting a defense of diminished capacity due to a developmental disability. Subsections (4)-(6) list the range of possible verdicts and the dispositional provisions applicable to each, including explicit authority for the juvenile court to develop a disposition for juveniles whose petitions have been dismissed solely by reason of diminished capacity. Providing such authority will correct the problem faced by the New Jersey courts in In re H.C.,[425] and the corresponding need to view a juvenile's disability solely as another factor to be taken into account following adjudication rather than as a matter which may negate a finding of delinquency or lessen the seriousness of the alleged offense.[426]

Subsection 7 provides for the development of a Program Plan for juveniles with a developmental disability whose petitions have been dismissed solely because of diminished capacity. Subsection (8) limits the length of the dispositional period, and subsections (9)-(11), establish review procedures, thus addressing many of the shortcomings of the dispositional practices following a verdict of not guilty by reason of insanity in a criminal proceeding. Finally, to help to assure that juveniles are offered the habilitative services they require, subsection (11) expressly makes the state developmental disabilities agency a party to all dispositional proceedings held under this section.

The intent in establishing these procedures is not to denigrate or diminish the use of diversion of juveniles with a developmental disability who are alleged to be delinquent. When the provision of habilitative services on a voluntary basis is appropriate, this alternative should be used. However, there will continue to be instances in which voluntary services are not appropriate and in which the juvenile's responsibility for the alleged delinquent offense will be in question. It is critical that juvenile courts have the authority to handle such cases in a manner that will protect both the interests of the juvenile and those of the public.

While, as noted above, the section is drafted to apply solely to juveniles with a developmental disability, those states seeking to extend or clarify the application of the diminished capacity defense to youths who were voluntarily intoxicated at the time of the offense, and/or in conjunction with or as a substitute for the insanity defense to juveniles who were mentally ill at the time of the offense, can easily expand its coverage by inserting the appropriate term after references to a developmental disability and habilitative services.

Section 9. Diminished Capacity in Delinquency Proceedings

(1) In order for a juvenile to be found delinquent, the prosecutor shall prove beyond a reasonable doubt every element of the offense which the juvenile is alleged to have committed.

COMMENT. See the introduction to this section and the comment to section 10(1) of the Model Developmentally Disabled Offenders Act.[427]

(2) (a) Juveniles who have a developmental disability and who are alleged to be delinquent are entitled to assert a defense of diminished capacity by introducing testimony and evidence that due to their disability, they lacked the state of mind required as an element of the delinquent offense(s) alleged in the petition.

(b) Juveniles who have a developmental disability but who do not also have a mental illness are not entitled to assert as a defense that they [are not delinquent by reason of insanity/lack substantial capacity either to appreciate the criminality of their conduct or to conform their conduct to the requirements of the law/are delinquent but mentally ill] pursuant to (citation to the relevant sections of the state juvenile code).

COMMENT. See the introduction to this section and the comment to section 10(2) of the Model Developmentally Disabled Offenders Act.

(3) Except as provided in section 13(3)(a) of this act, a juvenile who asserts the defense of diminished capacity due to a developmental disability is entitled to release on recognizance [or bail] to the same extent and on the same terms and conditions as if the defense had not been raised.

COMMENT. As indicated in the comment to section 5(3) of this act, although the grounds for detaining juveniles alleged to be delinquent are often broader than those for the pretrial detention of adults accused of a crime, there is little basis for believing that solely because a juvenile has a developmental disability the jurisdiction of the court, public safety, or the juvenile will be better protected if he or she is housed in a detention facility. Thus, this provision requires that juveniles who assert a defense of diminished capacity be eligible for release on the same grounds and conditions as juveniles who offer another defense to the petition. The reference to release on bail is bracketed since juveniles in many states do not

[424] For an explanation of the diminished capacity and insanity defenses, and of the greater applicability of the concept of diminished capacity to persons with a developmental disability, see the introduction to section 10 of the Model Developmentally Disabled Offenders Act and the comment to section 10(2) of that act.

[425] 106 N.J. Super. 583, 256 A.2d 322 (1969).

[426] *See* Popkin and Lippert, *supra* note 422, at 440.

[427] See also In re Winship, 397 U.S. 358 (1969).

have a right to release on bail.[428] The intent of this provision is simply to assure that juveniles with a developmental disability have the same release options available to them as do non-disabled juveniles alleged to have committed a delinquent offense.

(4) (a) If following the assertion of a defense of diminished capacity due to a developmental disability the judge [or jury] finds that the prosecutor has not proven beyond a reasonable doubt that the juvenile committed all the physical acts required as elements of the delinquent offense(s) or a lesser included offense alleged in the petition, the petition shall be dismissed and all restraints on the juvenile's liberty imposed as a result of that petition shall be terminated.

(b) If following an assertion of a defense of diminished capacity due to a developmental disability the judge [or jury] finds that the prosecutor has proven beyond a reasonable doubt that the juvenile committed all the physical acts and had the state of mind required for the delinquent offense(s) alleged in the petition or for a lesser included offense, the juvenile shall be adjudged delinquent and the juvenile court shall proceed in the manner set forth in section 10 of this act if the juvenile has a developmental disability and as otherwise prescribed by law if the juvenile does not have a developmental disability.

(c) If following an assertion of a defense of diminished capacity due to a developmental disability the judge [or jury] finds that the prosecutor has proven beyond a reasonable doubt that the juvenile committed all the physical acts required as elements of the delinquent offense alleged in the petition or a lesser included offense, but that the juvenile lacked the prescribed state of mind, the petition shall be dismissed solely by reason of the juvenile's diminished capacity and the juvenile court shall proceed in the manner set forth below.

COMMENT. This subsection sets forth the possible determinations following assertion of a diminished capacity defense in a delinquency proceeding. In accordance with the decisions of the Supreme Court in *Gault* and *Winship*,[429] paragraph (4)(a) requires that the allegations in the petition be proven beyond a reasonable doubt before a juvenile can be adjudicated delinquent. Thus, a youth cannot be found to be delinquent simply because he or she may require habilitative services. Of course nothing in the paragraph is intended to prohibit the provision of services on a voluntary basis or otherwise in accordance with the procedures and limitations set forth in the Model Right to and Admission to Services Act.[430]

[428] See authority cited *supra* note 399.

[429] In re Gault, 387 U.S. 1, 36 (1967); In re Winship, 358, 365-366 (1969).

[430] *Supra* note 66.

Paragraph (4)(b) states that if the prosecutor proves that a juvenile engaged in the prohibited acts with the requisite state of mind, the juvenile must be found to be delinquent whether or not he or she has raised a defense of diminished capacity. The fact that the youth may have a developmental disability is then relevant only to the determination of the disposition to be imposed. Special dispositional procedures and considerations are set forth in section 8 for developmentally disabled juvenile offenders.

Subsection (4)(c) presents an alternative not currently available, but analogous to findings of not guilty by reason of insanity in adult criminal cases. It applies in what is anticipated to be the relatively few cases in which the prosecution proves beyond a reasonable doubt that the juvenile engaged in criminal conduct but that due to a developmental disability, he or she did not have the particular state of mind required. A fuller explanation of the intent and operation of this provision is contained in the comment to section 10(4)(c) of the Model Developmentally Disabled Offenders Act.

References to jury determinations are bracketed since juveniles are not entitled to a jury in delinquency proceedings under the laws of many states.[431] The authority to request a jury under this provision should be made comparable to that enjoyed by juveniles in delinquency proceedings under the enacting jurisdiction's juvenile code.

(5) When a petition is dismissed pursuant to paragraph (4)(c) of this section, the juvenile court shall appoint a multidisciplinary evaluation team to evaluate the juvenile or to supplement an examination or evaluation previously conducted pursuant to this act.

COMMENT. See sections 3(5), 3(6), and 3(15), and 13 of this act and the comment to section 10(5) of the Model Developmentally Disabled Offenders Act.

(6) (a) Following submission of the evaluation report, the juvenile court shall hold a dispositional hearing.

(b) If at the hearing the prosecutor fails to prove by a preponderance of the evidence that the provision of habilitative services on an involuntary basis is required by the needs of the juvenile and/or of public safety, the juvenile court shall terminate the proceeding and all restraints on the juvenile's liberty imposed in conjunction with the delinquency proceeding.

(c) If at the hearing the prosecutor proves by a preponderance of the evidence that the provision of habilitative services on an involuntary basis is required by the needs of the juvenile and/or of public safety, the court shall order:

[431] *Supra* note 401.

(i) The development of a Program Plan in the manner set forth in subsection (7) of this section;

(ii) The provision and/or purchase of the necessary habilitative services using the least restrictive alternatives by the state developmental disabilities agency; and

(iii) The participation of the juvenile in those necessary habilitative services.

(d) An order issued pursuant to paragraph (6)(c) of this section shall specify whether the Program Plan may include placement of the juvenile in a residential facility which is secure or non-community based or in which adults or juveniles are placed pursuant to (citation to the Model Right to and Admission to Services Act/ state civil commitment law(s)). No such placement shall be permitted unless the petition has been dismissed solely by reason of the juvenile's diminished capacity and the delinquent conduct the juvenile was shown to have performed, would if committed by an adult, constitute homicide, rape, armed robbery, kidnapping, aggravated assault, arson or another criminal offense involving serious bodily harm or the substantial risk of such harm; the juvenile's role was not merely as a conspirator, abettor or accessory, and the judge [or jury] finds that the prosecutor has presented clear and convincing proof that:

(i) The juvenile currently presents a substantial risk of serious bodily harm to others or to him or herself;

(ii) The juvenile will be provided at the facility with the habilitative services identified through the evaluation report and the testimony and evidence presented at the hearing which will assist him or her to function in the community as a self-reliant, law-abiding citizen to the extent and within the shortest time possible; and

(iii) There is no less restrictive setting in which the juvenile can be provided with those services without seriously endangering public safety.

COMMENT. This subsection sets forth the procedures for determining the disposition following the dismissal of a delinquency petition solely because of the diminished capacity of the juvenile. It is parallel to section 10(6) of the Model Developmentally Disabled Offenders Act. Although the dispositional portion of delinquency proceedings remains informal in many jurisdictions, there is a clear trend toward the provision of more procedural protections to assure equity and the appropriateness of the disposition.[432] In addition, as is discussed in the comment to section 10(6) of the Model Developmentally Disabled Offenders Act and the introduction to the report on The Right to and Admission to Services,[433] a similar trend is evident in proceedings concerning the provision of services on an involuntary basis, both for juveniles and adults.

> No longer is there validity to the assertion that ". . . a child, unlike an adult, has a right not to liberty but to custody," . . . as a postulate for involuntary psychiatric institutionalization, even of a juvenile delinquent. "[W]hatever may be their precise impact, neither the Fourteenth Amendment nor the Bill of Rights is for adults alone."[434]

Careful scrutiny is all the more warranted in light of the traditional reluctance of public and private agencies, providing services to juveniles with a developmental disability, to deal with those who have become involved in criminal-type activity;[435] the abuses to which young people have been subjected in institutions like the Willowbrook State School in New York;[436] and the necessity of matching the habilitative services to be provided to the abilities and disabilities of the youngster involved, if those services are to be effective. Moreover, unlike the situation examined by the Supreme Court in Parham vs. J. L.,[437] the provision of habilitative services under this section is on an involuntary rather than a voluntary basis.[438]

Accordingly, the subsection requires a dispositional hearing at which the court must decide whether the needs of the juvenile and of public safety require that the juvenile be ordered to participate in habilitative services provided or purchased by the state developmental disabilities agency — i.e. that such services are necessary to assist the juvenile to function in the community as a self-reliant law-abiding citizen. A further explanation of the operation and intent of the provision is contained in the comment to section 10(6) of the Model Developmentally Disabled Offenders Act.

(7) (a) Following issuance of an order under paragraph 6(c) of this section, the state developmental disabilities agency shall develop a Program Plan in consultation with the juvenile,

---

[432] NATIONAL TASK FORCE TO DEVELOP STANDARDS AND GOALS FOR JUVENILE JUSTICE AND DELINQUENCY PREVENTION, A COMPARATIVE ANALYSIS OF STANDARDS AND STATE PRACTICES: JUVENILE DISPOSITIONS AND CORRECTIONS 49 (1977); see also IJA/ ABA STANDARDS RELATING TO DISPOSITION (Tentative Draft 1977); IJA/ABA STANDARDS RELATING TO DISPOSITIONAL PROCEDURES (Tentative Draft 1977); NACJJDP, *supra* note 148, at §§3.181-3.1813 (1976); NATIONAL ADVISORY COMMITTEE, *supra* note 28, at §§14.1-14.15.

[433] *Supra* note 66.

[434] In re Ralph M., 417 N.Y.S. 2d 608 (N.Y. County Fam. Ct. 1979); *see* Addington vs. Texas 441 U.S. 418 (1979).

[435] PRESCOTT and VAN HOUTEN, *supra* note 67; Donovan, *supra* note 388 at 246-249.

[436] New York Association for Retarded Children vs. Rockefeller 357 F. Supp. 752 (E.D.N.Y. 1973).

[437] 442 U.S. 584 (1979).

[438] See In re Ralph M., 417 N.Y.S. 2d 608 (N.Y. County Fam. Ct. 1979).

his or her parents or guardian, and his or her attorney. The objective of the Program Plan shall be to assist the juvenile, to the extent and within the shortest time possible, to function in the community as a self-reliant law-abiding citizen. The purpose, nature and possible beneficial and harmful consequences of the various alternatives shall be explained to the juvenile to the maximum extent possible, and to his or parents or guardian, and attorney.

(b) The Program Plan shall cover a period of one year unless a lesser time is specified by the juvenile court in its dispositional order.

(c) The Program Plan shall be submitted for approval to the juvenile court no more than ten (10) days after issuance of the order. A copy shall be provided to the prosecutor, the juvenile, the juvenile's parents or guardian, the juvenile's attorney, and the (name of state Developmental Disabilities Advocacy Office). It shall be approved by the juvenile court without a hearing within ten (10) days of its submission unless the juvenile court on its own initiative sets a hearing or the prosecutor, the juvenile, or the juvenile's parents or guardian move for a hearing. In such a motion, the moving party shall specify the reasons why the plan is alleged to be inadequate or inappropriate and the proposed modifications.

(d) Following the hearing, the court shall approve the plan or modify it so as to further the objectives set forth in paragraph (7)(a) of this section.

COMMENT. In the context of both delinquency and disability proceedings, the development of an individualized Program Plan is viewed as one of the critical means for encouraging the provision of appropriate services through use of the least restrictive alternative.[439]

This provision differs from section 10(7) of the Model Developmentally Disabled Offenders Act in two ways. First, consistent with sections 6 and 7(6) of this act, it provides for participation of the juvenile's parents or guardian in the Program Plan development and approval process, since their interests are likely to be affected. Second, paragraph (7)(b) provides time limit provisions which are more consistent with the indeterminate nature of dispositions in delinquency proceedings in most jurisdictions and the more rapid rate of change applicable to juveniles.[440]

(8) The juvenile court shall not retain jurisdiction over the juvenile nor shall habilitative services be provided or restraints imposed on the juvenile's liberty pursuant to the section after the juvenile has attained [majority/age 21].

COMMENT. This provision simply makes dispositions under the act coterminous with the dispositional authority of the juvenile court.[441] If the enacting jurisdiction has adopted or adopts a determinate dispositional system for delinquency cases, a measure similar to section 10(8) of the Model Developmentally Disabled Offenders Act should be substituted.

(9) Six (6) months after development of the initial Program Plan and every six (6) months thereafter until termination of the dispositional order or expiration of the dispositional period pursuant to subsection (8) of this section, or at such intervals as the juvenile court requires, the official of the state developmental disabilities agency who is monitoring the provision of habilitative services to the juvenile shall file a report with the juvenile court. The report shall list the services specified in the most recent Program Plan, describe the juvenile's response to those services, and indicate which, if any, of those services have not been offered to the juvenile and the reasons therefor. A copy of the report shall be provided to the juvenile, his or her parents or guardian, the (name of the state Developmental Disabilities Advocacy Office), and the state developmental disabilities agency. At the juvenile's request, the report shall also be sent to his or her attorney or a designated representative. The information contained in the report shall be explained to the juvenile to the maximum extent possible.

COMMENT. See comment to section 10(9) of the Model Developmentally Disabled Offenders Act.

(10) (a) The juvenile court shall hold a review hearing:

(i) Not less than two (2) weeks before expiration of the initial Program Plan under subsection (7) of this section;

(ii) At least annually thereafter until termination of the proceedings or expiration of the maxmium dispositional period pursuant to the subsection (8); and

(iii) Following submission of a motion pursuant to subsection (11).

---

[439] (delinquency) Morales vs. Turman, 383 F. Supp. 53, 88, 92-93 (E.D. Tex. 1974), *rev'd on other grounds* 535 F. 2d 864 (5th Cir. 1976), *rev'd and remanded* 430 U.S. 322 (1977), *remanded for further hearings* 562 F. 2d 993 (5th Cir. 1977); Morgan vs. Sproat, 431 F. Supp. 1130 (S.D. Miss. 1977); NACJJDP, *supra* note 28, at §4.214; NATIONAL ADVISORY COMMITTEE, *supra* note 28 at §§23.3 and 24.6 (1976); (developmental disability) ACMRDD, *supra* note 11 at §1.3; Eckerhart vs. Hensley, 475 F. Supp. 908 (W.D. Mo. 1979); New York State Association for Retarded Citizens vs. Carey, 393 F. Supp. 715 (E.D.N.Y. 1975); 20 UNITED STATES CODE §1401 (1978) (Education for All Handicapped Children Act); 42 UNITED STATES CODE §6012(b) (Supp. 1979) (Developmental Disabilities Assistance and Bill of Rights Act).

[440] As of 1976, 41 jurisdictions permit a disposition ordered in a delinquency proceeding to continue until the juvenile is age 21; three set a limit of age 20; three others set the limit at 18; and one uses age 17. NATIONAL TASK FORCE, *supra* note 432 at 28. It should be noted however, that the limits set forth in paragraph (3)(b) are also compatible with a determinate dispositional system.

[441] NATIONAL TASK FORCE, *id.*

(b) At the hearing the court shall review:

(i) Whether the prescribed services have been provided to the juvenile and if so, whether they have been provided in the manner set forth in the juvenile's Program Plan;

(ii) Whether provision of habilitative services to the juvenile on an involuntary basis continues to be required by the needs of the juvenile and of public safety; and if so,

(iii) Whether the dispositional order and Program Plan should be modified to assure that the necessary habilitative services are provided and that they are provided by means of the least restrictive alternative.

(c) In conjunction with that hearing, the juvenile court may appoint a multidisciplinary evaluation team to update the evaluation performed pursuant to subsection (5) of this section. Such an evaluation shall be ordered upon a showing that there has been a significant change in the juvenile's mental, emotional, physical or educational condition, adaptive behavior or social skills.

(d) (i) Except as specified in subparagraph (10)(d)(ii), the prosecutor must prove by a preponderance of the evidence that provision of services on an involuntary basis continues to be required by the needs of the juvenile and of public safety, and that a particular habilitative service requested by the juvenile is not so required. The prosecutor must present clear and convincing proof of the requirements set forth in paragraph (6)(d) of this section in order for a placement in a residential facility which is secure or non-community based or in which adults or juveniles are placed pursuant to (citation to Model Right to Services/Admission to Services Act/state civil commitment law(s)) to continue.

(ii) When the hearing is held pursuant to a motion filed by a juvenile or his or her parent or guardian pursuant to subsection (11) of this section, the moving party shall have the burden of going forward with a *prima facie* case supporting the requested relief. If such support is presented, the juvenile court shall grant the relief requested unless the prosecutor meets the burden of persuasion set forth in subparagraph (10)(d)(i).

(e) (i) If the juvenile court concludes that the provision of habilitative services on an involuntary basis is not required, it shall terminate the proceeding and all restraints on the juvenile's liberty imposed in connection with the proceeding.

(ii) If the juvenile court concludes that provision of habilitative services on an involuntary basis is required, it shall determine what changes, if any, are necessary in the habilitative services being provided to the juvenile and in the manner and location in which those services are being provided. The juvenile court shall then issue a revised dispositional order directing development of an updated Program Plan and the provision of services in accordance with that plan. The updated Program Plan shall be developed in the manner prescribed in paragraph (7) of this section.

(iii) If the juvenile court concludes that services have not been provided in accordance with the Program Plan, it may order that contempt proceedings be initiated against the state developmental disabilities agency and if such services cannot be provided, it may terminate the dispositional order and all restraints on the juvenile's liberty imposed in connection with that proceeding.

COMMENT. This subsection establishes a periodic review of the provision of habilitative services to and the continued need for such services for, a juvenile who was found not responsible for delinquent conduct solely by reason of diminished capacity. It exactly parallels section 10(10) of the Model Developmentally Disabled Offenders Act except for increasing the frequency of review hearings and omitting the reference to the transition provision contained in section 10(13)[442].

A number of states provide for annual or biannual review of dispositions in delinquency, neglect, and abuse cases.[443] Recent standards setting groups have generally recommended more frequent reviews.[444] The one year maximum, coupled with the provision in this subsection and subsection (11) for an earlier review appears to be a reasonable compromise.

An explanation of the intent and operation of other aspects of this subsection is contained in the comment to section 10(10) of the Model Developmentally Disabled Offenders Act.

(11) (a) A juvenile receiving habilitative services pursuant to this section, his or her parents or guardian, the official of the state developmental disabilities agency who is monitoring the provision of habilitative services to the juvenile, or the head of the facility in which a juvenile has been placed pursuant to paragraph (7)(d) of this section, may file a motion for termination of the services or placement or for modification of the Program Plan on the grounds that:

(i) The juvenile no longer requires such habilitative services to function in the community as a self-reliant, law-abiding citizen; or

---

[442] For an explanation of this omission, see the introduction to this section.

[443] *Id.* and NATIONAL TASK FORCE TO DEVELOP STANDARDS AND GOALS FOR JUVENILE JUSTICE AND DELINQUENCE PREVENTION, COMPARATIVE ANALYSIS OF STANDARDS AND STATE PRACTICES: ABUSE AND NEGLECT, 159-160 (1977).

[444] *E.g.*, NACJJDP, *supra* note 28, at §4.214 and *supra* note 148, at §3.1812; NATIONAL ADVISORY COMMITTEE, *supra* note 28, at §14.30.

(ii) The habilitative services being provided are no longer consistent with the juvenile's needs and other habilitative services are required; or

(iii) The manner in which the services are being provided or the facility in which the juvenile has been placed is not the least restrictive alternative consistent with his or her needs and public safety; or

(iv) The habilitative services specified in the most recent Program Plan are not being provided.

(b) A juvenile may submit a motion to the juvenile court pursuant to this subsection by any means including but not limited to oral communication or informal letter. Except in extraordinary circumstances, motions may be filed by the juvenile no more than once during any sixty (60) day period and no sooner than thirty (30) days after a review hearing.

COMMENT. See comment to section 10(11) of the Model Developmentally Disabled Offenders Act.

(12) The state developmental disabilities agency shall be a party in any hearing held pursuant to subsections (7) and (10) of this section.

COMMENT. See comment to section 10(12) of the Model Developmentally Disabled Offenders Act.

Section 10. Dispositions Following an Adjudication of Delinquency

(1) (a) A predisposition report shall be prepared for juveniles adjudicated delinquent who there is reason to believe have a developmental disability.

(b) In addition to the matters prescribed in (citation to the statute or rule, if any, governing the content of predisposition reports), a predisposition report prepared pursuant to paragraph (1)(a) of this section shall set forth:

(i) Whether the juvenile has a developmental disability; and if so

(ii) The nature and effect of that disability;

(iii) The type of habilitative services, if any, which will assist the juvenile to the extent and within the shortest time possible to function in the community as a self-reliant, law-abiding citizen;

(iv) The availability of the necessary habilitative services in the community;

(v) The availability of the necessary habilitative services within the juvenile correctional system; and

(vi) The amount of time, if any, for which the juvenile was confined prior to the dispositional hearing for examination, evaluation, or services under this act.

(2) In conjunction with the preparation of the disposition report, the juvenile court may order the juvenile to undergo an evaluation and appoint a multidisciplinary evaluation team.

(3) (a) In determining the disposition to be imposed on a juvenile who has a developmental disability and who has been adjudicated delinquent, the juvenile court shall consider the amount of time, if any, for which the juvenile was confined for examination, evaluation, or services under this act. The court shall select the least restrictive alternative consistent with the needs of the juvenile and of public safety.

(b) The dispositional order shall not provide for placement of a juvenile offender who has a developmental disability in a facility in which juveniles or adults are placed pursuant to (citation to Model Right to and Admission to Services Act/state civil commitment law(s)) unless the juvenile was adjudicated for committing a delinquent offense which if committed by an adult would constitute homicide, rape, armed robbery, kidnapping, aggravated assault, arson, or another criminal offense involving serious bodily harm; the juvenile's role in the commission of that offense was not merely as a conspirator, abettor or accessory; and there is clear and convincing proof that:

(i) The juvenile currently presents a substantial risk of serious bodily harm to others or to him or herself;

(ii) The juvenile will be provided at the facility with the habilitative services identified through the evaluation report and the testimony and evidence presented at the hearing which will assist him or her to function in the community as a self-reliant, law-abiding citizen to the extent and within the shortest time possible; and

(iii) There is no less restrictive setting in which the juvenile can be provided those services without seriously endangering public safety.

A juvenile offender placed under this paragraph shall be entitled to the rights, procedures, and habilitative services set forth in section 9 of this act. The agency operating the facility shall be subject to the duties and limitations set forth in that section for the juvenile correctional agency. No placement under this paragraph may extend beyond the date on which the juvenile attains [majority/age 21].

COMMENT. While as noted earlier, a number of states have enacted and several standards setting groups have recommended measures to promote equity and acknowledge punishment as one of the purposes of dispositions:[445]

---

445 See in addition to the sources listed *supra* note 432, Rector and Van Duizend, *supra* note 270, at 360-361; IJA/ABA, STANDARDS RELATING TO JUVENILE DELINQUENCY AND SANCTIONS (Tentative Draft 1977).

> Most juvenile codes state that a disposition is to be made "in the best interests of the child and for the protection of the community."
>
> * * *
>
> However the statutes have been framed, it is clear that the benevolent purpose or rehabilitative ideal has dominated juvenile corrections in this century. . . . [T]he structure of dispositions, the allocation of discretion, the roles assigned the key participants, and the nature and duration of dispositions take their cue from rehabilitative objectives. . . . [F]or example, . . . a premium is placed on expertise. . . . an identity of interest is said to exist between the State and the child (this has enormous implications for the amount of adversariness deemed appropriate), and. . .dispositions are age-based rather than offense-based. . . . The mildest and the most aggravated offenses are subject to the same dispositions and the juvenile is subject to State authority until he or she attains majority. (A few states do impose durational limits but [with one exception] those limits are not offense-related).[446]

This provision is consistent with either approach. Like section 11 of the Model Developmentally Disabled Offenders Act, it seeks to assure that the juvenile court has the information regarding a juvenile with a developmental disability that is necessary for determining what habilitative services, if any, should be provided and whether or not the juvenile should be placed in a residential facility, as well as applying the principle of using the least restrictive alternative to dispositional decisions.

Although not always required by statute, predisposition reports of some type are prepared in most jurisdictions.[447] The information required to be added to the standard predisposition report by paragraph (1)(b) is intended to apprise the juvenile court judge of the youth's needs and the community and correctional services available which can address those needs. It is anticipated that in most cases, an evaluation will have to be ordered pursuant to subsection (2) to assist in obtaining this information. Even when a juvenile has been evaluated in conjunction with an inquiry into his or her competency to assist counsel, an update will be necessary to aid in determining the habilitative services, if any, which are needed to assist him or her to function in the community as a self-reliant, law-abiding citizen.[448] In order to clarify the time which the court must consider pursuant to paragraph (3)(a), the report must advise the court of the total period, if any, during which the juvenile was confined pursuant to section 7 and/or 13.

Use of the least restrictive alternative has been urged for delinquency dispositions by many recent sets of recommendations for improvement of the juvenile justice system.[449] There is reason to believe that juveniles with developmental disabilities, for the same reasons as their adult counterparts, are overrepresented in secure correctional facilities and face greater risks of harm and the absence of necessary services than juveniles without such disabilities.[450] Like section 10(3) of the Model Developmentally Disabled Offenders Act, this provision does not rule out incarceration or any other sanction. It simply legitimizes consideration of developmental disabilities in determining the degree and duration of the restraints on liberty to be imposed, and establishes a presumption in favor of selecting a placement or conditions of probation which are no more harsh, hazardous or intrusive and involve no more restrictions on the juvenile's physical freedom or social interaction than are absolutely essential. Thus, it provides a guide to the parties, the court, correctional personnel and the public, a basis for use of the court's inherent power to order services[451] and a measure for appellate review.

The provision also sets forth criteria for determining whether to place a juvenile with a developmental disability who has been adjudicated delinquent in a facility to which juveniles or adults are committed pursuant to the state's civil commitment statutes. Many state juvenile codes permit such placements as part of a delinquency disposition.[452] As the United States Supreme Court recently recognized, however, placement in a mental hospital involves deprivation of liberty and stigma beyond that inherent in an adjudication of delinquency and incarceration, and therefore, additional due process protections must be applied.[453] The traditional therapeutic emphasis of delin-

---

[446] NATIONAL TASK FORCE, *supra* note 432, at 3-5.

[447] *Id.* at 44.

[448] See Section 14 of the Model Developmentally Disabled Offenders Act. For the requirements regarding the contents and distribution of the evaluation report, see section 13(4) of this act.

[449] NACJJDP, *supra* note 148, at §3.182; NATIONAL ADVISORY COMMITTEE, *supra*, note 28 at §14.15; IJA/ABA, DISPOSITIONS, *supra* note 432, at §2.1; NATIONAL ADVISORY COMMISSION, CORRECTIONS, *supra* note 43, at 575.

[450] See Dennis, Mental Retardation and Corrections: A Research Perspective, THE MENTALLY RETARDED CITIZENS AND THE LAW, *supra* note 12, at 34, 38, 48; Wald, *supra* note 278, at 248-249; Friel, *supra* note 370, at 107, 109-119; Morales vs. Turman, 383 F. Supp. 43 (E.D. Tex. 1974), *rev'd on other grounds* 535 F.2d 864 (5th Cir. 1976), *rev'd and remanded* 430 U.S. 322 (1977), *remanded for further hearings* 562 F.2d 993 (5th Cir. 1977).

[451] WEINSTEIN, *supra* note 268.

[452] See *e.g.*, State of New Jersey in the Interest of H.C., 106 N.J. Super. 583, 256 A.2d 322 (1969); MISSISSIPPI YOUTH COURT ACT, §73 (1978).

[453] Vitek vs. Jones, U.S. (1980).

quency dispositions does not exempt them from type of procedures required in Vitek v. Jones. "The rehabilitative goals of the system are admirable, but they do not change the drastic nature of the action taken."[454] As noted earlier, delinquency proceedings are adversary in nature and placements cannot be considered, voluntary admissions like those reviewed in Parham v. J. L.[455] Moreover, juveniles who are committed to mental retardation facilities are likely to have fewer opportunities for community placement than those retained in the juvenile corrections system and, unlike other juvenile offenders, may be subject to automatic procedures for renewing the commitment when they attain the age of majority.[456]

Accordingly, the criteria set forth in paragraph (3)(b) parallel those for dispositions following dismissal of a petition solely because of diminished capacity under section 9(6)(d).[457] In addition, it makes applicable to juveniles placed under this provision the rights and procedures established for developmentally disabled juvenile offenders in juvenile correctional programs and facilities, and clarifies that the duration of such placement cannot exceed the length of the juvenile court's maximum dispositional authority in delinquency cases.[458] The provision of habilitative services to developmentally disabled juvenile offenders not committed under this section is covered in section 9 of this act.

Section 11. Corrections

(1) (a) A juvenile offender who has a developmental disability shall be entitled to receive habilitative services as defined in this act. The state juvenile correctional agency has the duty to provide or purchase the habilitative services specified in the Program Plans prepared for all juvenile offenders with a developmental disability whether they are in custody or under supervision in the community in lieu of or following custody.

(b) A juvenile who has a developmental disability shall be entitled to protection against physical abuse while housed in a juvenile correctional facility. The head of a juvenile correctional facility which houses one (1) or more juveniles who have a developmental disability shall establish programs and procedures to protect the juvenile(s) from abuse and exploitation by other residents of the facility to the maximum extent possible.

(c) A juvenile offender who has a developmental disability shall be entitled to the assistance of an advocate while housed in a juvenile correctional facility. The state juvenile correctional agency shall provide or purchase advocacy services in each secure or non-community based juvenile correctional facility housing one or more juvenile offenders with a developmental disability. Such services shall include direct personal contact on at least a monthly basis between the juvenile offender with a developmental disability and an advocate with the training and the qualifications set forth by the (name of the state Developmental Disabilities Advocacy Office). In providing advocacy services under this section, advocates shall be entitled to support, advise, espouse the rights of, or directly intercede on behalf of juvenile offenders with a developmental disability in any matter affecting the habilitative services provided to, the protection of, or the restraints imposed on such offenders, except for:

(i) An appeal of the adjudication of delinquency or the disposition;

(ii) Post-conviction collateral attack on the adjudication of delinquency; or

(iii) Proceedings arising from allegations of delinquency other than those which resulted in the juvenile offender's incarceration.

COMMENT. A series of court decisions reveal that despite the rehabilitative ideal of the juvenile justice system, the conditions procedures, and available services in many juvenile detention and correctional facilities fail to meet even the minimum requirements necessary to avoid constitutional censure.[459] These cases and a series of reports and recommendations have called for the provision of the staff and services necessary to provide individualized treatment for all juveniles in detention and correctional facilities; procedural safeguards to prevent abuses and protect the rights of juvenile offenders, and improvement of the physical conditions under which the juveniles are housed.[460]

This subsection is intended to provide a statutory guarantee that juvenile offenders with a developmental disability receive the habilitative services and protection which they need to become productive, law-abiding citizens. Studies in-

[454] Fain vs. Duff, 488 F.2d 218, 225 (5th Cir. 1973).

[455] 442 U.S. 584 (1979).

[456] See PRESCOTT and VAN HOUTEN, *supra* note 67.

[457] See also section 10(6)(d) of the Model Developmentally Disabled Offenders Act and the comment thereto.

[458] See authorities cited in note 440 *supra*.

[459] See *e.g.*, Nelson vs. Heyne, 355 F. Supp. 451 (N.D. Indiana 1972), *aff'd* 491 F.2d 352 (7th Cir. 1974) *cert. denied* 417 U.S. 976 (1974); Martarella vs. Kelley, 349 F. Supp. 565 (S.D.N.Y. 1972); Morales vs. Turman, 383 F. Supp 53 (E.D. Tex. 1974) *rev'd on other grounds, 535 F.2d 864 (5th Cir. 1976), rev'd and remanded* 430 U.S. 322 (1977) *remanded for further hearings* 562 F.2d 993 (5th Cir. 1977); Pena v. New York State Division for Youth, 419 F. Supp 203 (S.D.N.Y. 1976); Morgan v. Sproat, 432 F. Supp. 1130 (S.D. Miss. 1977).

[460] In addition to the sources cited in note 459, *supra*, see Rector and Van Duizend, *supra* note 270; NACJJDP, *supra*, note 28, at §§4.21-4.82; IJA/ABA, STANDARDS RELATING TO CORRECTIONS ADMINISTRATION (Tentative Draft 1977); NATIONAL ADVISORY COMMITTEE, *supra*, note 28, at §§23.1-24.16.

dicate that developmentally disabled juvenile offenders like their adult counterparts are more likely to lack the services designed to address their needs and to have more difficulty adjusting to the facility environment than other juvenile offenders.[461] While the establishment of a statutory right to habilitative services, protection from abuse, and advocacy assistance will hopefully help individual juvenile offenders, it should reap significant benefits for the public as well:

> When a State assumes the place of a juvenile's parents, it assumes as well the parental duties, and its treatment of its juveniles should, so far as can be reasonably required, be what proper parental care would provide. Without a program of individual treatment the result may be that the juveniles will not be rehabilitated, but warehoused, and that at the termination of detention they will likely be incapable of taking their proper places in free society; their interests and those of the state thereby being defeated.[462]

The phrase under supervision in the community in lieu of or following custody is intended to encompass what in the context of adult corrections would be known as probation and parole. While probation is an accepted juvenile justice term, parole is used in only a few jurisdictions. Enacting jurisdictions which refer to parole in delinquency may wish to substitute the wording used in section 11 of the Model Developmentally Disabled Offenders Act in this subsection and below. A further explanation of the operation and intent of this entire subsection is contained in the comments to sections 11(1)(a)-(c) of that act.

(2) (a) Within thirty (30) days after issuance of the dispositional order, the juvenile correctional agency responsible for custody or supervision of a juvenile offender who has a developmental disability shall develop a Program Plan and submit it to the juvenile court.

(b) The Program Plan shall be developed in consultation with the juvenile, the juvenile's parent or guardian, and the juvenile's attorney, and, if the juvenile resides in a correctional facility, a developmental disabilities advocate for that facility. The objective of the plan shall be to assist the juvenile, to the extent and within the shortest time possible, to function in the community as a self-reliant law-abiding citizen.

The purpose, nature, and possible beneficial and harmful consequences of the various alternatives available shall be explained to the juvenile to the maximum extent possible and to the juvenile's parents or guardian, and attorney.

(c) The plan shall cover the lesser of:

(i) The maximum dispositional period; or

(ii) One (1) year.

(d) The Program Plan shall be submitted for approval to the juvenile court no more than thirty (30) days after issuance of the dispositional order. A copy shall be provided to the prosecutor, the juvenile, the juvenile's parents or guardian, the juvenile's attorney, and the chief developmental disabilities advocate of the correctional facility in which the juvenile resides, or if the offender is on probation, to the (name of state Developmental Disabilities Advocacy Office).

(e) (i) The Program Plan shall be approved by the juvenile court without a hearing within ten (10) days of its submission unless the court on its own initiative sets a hearing or the prosecutor, juvenile, or the juvenile's parents or guardian, move for a hearing. In such a motion, the moving party shall specify the reasons why the plan is alleged to be inadequate or inappropriate and the proposed modifications.

(ii) Following a hearing held pursuant to subparagraph (2)(c)(i), the juvenile court shall approve the plan as written, or modify it so as to further the objectives set forth in paragraph (2)(b) of this section, and order its implementation.

(f) An updated Program Plan shall be developed at the end of the period specified in paragraph (2)(c)(ii); and

(i) At least annually thereafter until the end of the dispositional period; and

(ii) When the juvenile is placed on conditional release or is incarcerated following revocation of probation or another form of conditional release;

(iii) Upon request of the juvenile offender, or with the juvenile's consent, the advocate, when the services being provided no longer meet the objective of the Program Plan set forth in paragraph (2)(b) of this section.

Updated plans shall be developed jointly by the correctional agency, the juvenile, and the juvenile's parents or guardian, and shall be approved by the juvenile court in the manner specified in paragraph (2)(e). Juveniles who reside in a juvenile correctional facility shall be entitled to assistance of a developmental disabilities advocate for that facility. Juvenile offenders, under community supervision in lieu of or following custody shall be entitled to the assistance of a representative of their own choosing in updating the plan. In conjunction with a hearing on the

---

461 See PRESCOTT and VAN HOUTEN, *supra*, note 67 and the authorities cited in note 450.

462 Nelson vs. Heyne, 491 F.2d 353, 360 (7th Cir. 1976); See also Pugh vs. Locke, 406 F. Supp. 318, 325 (M.D. Ala. 1976), *aff'd sub. nom.* Newman vs. Alabama, 559 F.2d 283 (5th Cir. 1977), *cert. denied* 438 U.S. 915 (1978).

revised Program Plan, the juvenile court may appoint a multidisciplinary evaluation team to update the evaluation performed pursuant to section 10 of this act, upon a showing that there has been a significant change in the juvenile offender's mental, emotional, physical, or educational condition, adaptive behavior or social skills.

COMMENT. As indicated earlier, development of a Program Plan is one of the critical means for encouraging the provisions of appropriate services in delinquency and disability proceedings.[463] Like the Program Plans developed following dismissal of a petition pursuant to section 9 of this act, plans developed under this provision are to be reviewed at least annually to determine what, if any, changes in the array of services will more quickly and effectively assist the juvenile to be able to function in the community as a self-reliant, law-abiding citizen. The involvement of the juvenile court in approving initial and updated Program Plans is in keeping with both its traditional role in individualizing the dispositional decisions[464] and the need to enhance accountability in the dispositional process.[465]

Unlike the parallel provision in the Model Developmentally Disabled Offenders Act, this subsection requires development of a Program Plan in any delinquency disposition involving a developmentally disabled juvenile offender and provides a role for persons other than the correctional agency, offender, and his or her advocates. Since in most jurisdictions, there are few limits on the length of the disposition other than the offender's age,[466] a distinction between major and minor offenses is irrelevant.[467] Because the willingness and ability of a juvenile's parents or guardian to assist or at least cooperate with the dispositional program may often be crucial to the determination to place a juvenile offender on probation or at least in a local community-based facility as well as to the success of the services, it is appropriate, if not essential, to include the parents or guardian in developing and determining the adequacy of the Program Plan.[468]

(3) (a) Six (6) months after development of the initial Program Plan and every six (6) months thereafter until the expiration of the dispositional period, the head of the facility or office responsible for the juvenile's custody or supervision, shall file a report with the juvenile court. The report shall list the services specified in the most recent Program Plan, and indicate which, if any, of those services have not been offered to the juvenile and the reasons therefor.

(b) A copy of the report shall be provided to:

(i) the juvenile;

(ii) the juvenile's parent or guardian;

(iii) the (name of the state Developmental Disabilities Advocacy Office) if the juvenile is under community supervision in lieu of or following custody;

(iv) the (head of the state juvenile correctional agency); and

(v) At the juvenile's request, to his or her designated representative.

(c) Within thirty (30) days after submission of a report under paragraph (3)(a) of this section, the juvenile or an advocate, attorney, or representative on the juvenile's behalf and with his or her consent, may petition the juvenile court for an order directing the state juvenile correctional agency to provide or purchase the omitted service immediately. The court shall issue the order following a hearing at which all interested parties are represented, unless it is established by clear and convincing evidence that the service is not necessary.

(d) A juvenile offender shall not be penalized in any way by the juvenile correctional agency responsible for his or her custody or supervision, by any employee or official thereof, or by the juvenile court for filing a petition under this subsection.

COMMENT. See comment to section 12(3) of the Model Developmentally Disabled Offenders Act.

(4) (a) No juvenile offender who is in custody and who has a developmental disability shall be required or coerced to participate in any habilitative services. Failure or refusal to participate in a habilitative service shall not constitute a basis for penalizing such a juvenile offender.

(b) The informed written consent of a juvenile offender who has a developmental disability and his or her parents or guardian shall be obtained before the juvenile is provided with a program or service not included in his or her Program Plan unless the program or service is essential to protect the juvenile's physical health and the juvenile and his parents or guardian are unable to provide a timely consent. In order for the consent to be informed, the purpose, nature, duration, and possible beneficial and harmful consequences of the program or service must be explained to the juvenile and his or her parents or guardian and the juvenile must be given an opportunity to consult with a developmental dis-

[463] See authorities cited *supra*, note 439.

[464] See *e.g.*, L. ARTHUR and W. GAUGER, DISPOSITION HEARINGS: HEARTBEAT OF THE JUVENILE COURT (1974).

[465] See NACJJDP, *supra* note 148, at §3.189 and the comment thereto.

[466] NATIONAL TASK FORCE, *supra* note 440.

[467] *Cf.* Model Developmentally Disabled Offenders Act, §12(2).

[468] *Cf.* IJA/ABA, STANDARDS RELATING TO PRETRIAL COURT PROCEEDINGS, 113 (Tentative Draft 1977); NACJJDP, *supra*, note 148, at §3.134; NATIONAL ASSESSMENT OF JUVENILE CORRECTIONS, BROUGHT TO JUSTICE? JUVENILES THE COURTS, AND THE LAW, 148 (R. Sarri and Y. Hasenfeld, eds. 1975).

abilities advocate if the juvenile is in custody or with a designated representative if he or she is not.

COMMENT. This subsection, with one important difference, tracks section 12(4) of the Model Developmentally Disabled Offenders Act. Its intent and the need for a right to refuse treatment are explained in the comment to the adult provision.[469] The difference is the requirement for the informed consent of both the juvenile offender and his or her parents or guardian. This dual consent requirement is included both as added protection against the use of particularly intrusive or dangerous forms of treatment,[470] and in recognition of the "parents' traditional interests in and responsibility for the upbringing of their child."[471]

(5) (a) Following the dispositional hearing, a juvenile offender who has a developmental disability shall not be admitted into a facility to which adults or juveniles are committed under (citation to Model Right to and Admission to Services Act/applicable state civil commitment law(s)) other than through a formal commitment proceeding in which the juvenile is afforded the rights set forth in sections 10(1)-(5) of this act.

(b) If on the expiration date of a dispositional order issued in conjunction with an adjudication of delinquency, a juvenile offender who has a developmental disability resides in a facility as a result of an admission pursuant to subsection (5)(a), he or she shall be immediately released unless a formal involuntary commitment proceeding has been initiated in accordance with (citation to Model Right to and Admission to Services Act/state civil commitment law(s)).

COMMENT. In a recent decision the Supreme Court recognized that: "involuntary commitment to a mental hospital is not within the range of conditions of confinement to which a prison sentence subjects an individual," and therefore, that "a transfer to a mental hospital for involuntary. . . treatment. . .constitute[s] the kind of deprivation of liberty that requires procedural protection.[472] The transfer of juvenile offenders with a developmental disability from a juvenile correctional facility to an institution for the developmentally disabled does not appear to be as frequent a practice as it is for adults.[473] This is probably due to the greater opportunities for commitment prior to and at disposition in delinquency cases. However, with the increasing formality of juvenile court proceedings and the increasing pressure on juvenile correctional agencies to provide the array of services necessary to realize the ideals on which the juvenile justice system was established, such transfers may be attempted more frequently in the future. Thus it is necessary to address in this statute, the procedures and standards which should be applied when a transfer is sought.

As noted in the comment to section 10 of this act, there appears to be little basis for distinguishing between adult and juvenile offenders on this issue. Since a transfer of a juvenile offender from a juvenile correctional facility is part of an adversary delinquency proceeding, it is more analogous to the circumstances covered in Vitek vs. Jones rather than to the voluntary admission of an orphaned or neglect child under Parham vs. J. R.[474] Accordingly, like section 12(5) of the Model Developmental Disabled Offenders Act, this provision prohibits transfer of a juvenile offender unless the substantive standards for the involuntary admission of a juvenile who has not been adjudicated delinquent have been met at a judicial hearing and the juvenile has been afforded the due process rights provided in all other proceedings under this act.[475]

As in comparable provisions under this act, the length of a commitment is limited to the maximum dispositional period set forth in the court's original dispositional order so as to assure that transfer is not misused as a means to lengthen the period of confinement beyond that which could be imposed as a result of the adjudication of delinquency.[476]

Section 12. Rights and Privileges

(1) In all judicial hearings provided by this act, the parties shall have a right to:

(a) Receive prior notice;

(b) Be represented by counsel;

(c) Be present;

(d) Compel the attendance of witnesses;

(e) Present evidence and confront and cross-examine witnesses, including experts appointed to conduct examinations or evaluations;

(f) Have an impartial decision-maker; and

(g) Appeal adverse orders and judgments, expect that the prosecutor shall not be entitled to appeal dispositional orders issued pursuant to section 10 or 11 unless otherwise provided by law.

COMMENT. This provision assures that the basic due process rights accorded to juveniles in

[469] See also IJA/ABA, CORRECTIONS ADMINISTRATION, *supra* note 460, at 86-94; NACJJDP, *supra* note 28, at §4.410.

[470] IJA/ABA, *id.*

[471] Parham vs. J. L. 442 U.S. 584 (1979).

[472] Vitek vs. Jones, U.S. (1980).

[473] PRESCOTT and VAN HOUTEN, *supra*, note 67 at 19; Friel, *supra* note 370, at 111-112.

[474] 442 U.S. 584 (1979).

[475] For a further discussion of the basis and intent of this provision, see the comment to section 12(5) of the Model Developmentally Disabled Offenders Act. See also the Model Right to Services Admission to Services Act, *supra*, note 66.

[476] See Sections 9(8) and 10(3) of this act; *see also* Baxtrom vs. Herold, 383 U.S. 107 (1966).

delinquency proceedings by In re Gault[477] plus the right to appeal, apply to all proceedings under this act including competency hearings, dispositional hearings, review hearings, and transfer hearings.[478] These rights apply to all parties including the state developmental disabilities agency in dispositional and review hearings, the state juvenile correctional agency for hearings under section 11 and the juvenile's parents or guardian in hearings challenging a Program Plan pursuant to sections 7(6), 9(7), and 11(2),[479] as well as the juvenile and, with one exception, the prosecutor.[480] There appears to be little justification for denying these rights in such proceedings since fundamental interests of the juvenile, and in some instances the juvenile's parents or guardian as well as the integrity and fairness of the judicial and correctional process are at stake.

(2) In addition to the rights listed in subsection (1) of this section, a juvenile who there is reason to believe has a developmental disability and who is alleged or found to have committed a delinquent offense shall be entitled to have:

(a) An attorney appointed to represent him or her in conjunction with any judicial proceeding and any matter arising from that allegation, finding or the disposition thereof, if he or she is not represented by counsel. An attorney appointed pursuant to this paragraph shall be compensated by the [state/county/court] if the juvenile is unable to retain counsel for financial reasons.

(b) An independent expert appointed by the court at the expense of the [state/county/court] to examine or evaluate him or her if the juvenile requests such an appointment and is unable for financial reasons to secure independent medical, psychological or other expert opinions; [and]

[(c) A jury determine, upon request, whether he or she

(i) Is competent to stand trial pursuant to section 7(5) of this act; and

(ii) Meets the criteria set forth in section 9(6) for placement in a residential facility which is secure or non-community based or in which adults or juveniles are placed pursuant to (citation to the Model Right to and Admission to Services Act/state civil commitment law(s)).]

COMMENT. A general discussion of the scope and basis of this provision is contained in the comment to section 13(2) of the Model Developmentally Disabled Offenders Act.[481] Enacting jurisdictions which provide for appointed counsel for parents in delinquency proceedings should expand the scope of paragraph (2)(a) to include parents when their interests are directly at issue.[482] Paragraph (2)(c) is placed in brackets since not all jurisdictions extend the right to a trial by jury to juveniles in delinquency proceedings.[483] Where that right is extended, juries should be permitted upon the request of the juvenile to determine whether he or she is competent to assist counsel or meets the criteria for commitment following dismissal of the delinquency petition solely by reason of diminished capacity.

(3) If a juvenile with a developmental disability who is alleged or found to be delinquent is under the influence of psychotropic medication during any judicial hearing, the juvenile court [and the jury, if any,] shall be advised of this fact and of the purpose of the medication and the effect which it may have on the juvenile's actions, demeanor and participation at the hearing.

COMMENT. See the comment to section 13(3) of the Model Developmentally Disabled Offenders Act.

(4) A stenographic and/or mechanical record shall be made of all proceedings held pursuant to this act.

COMMENT. See the comment to section 13(4) of the Model Developmentally Disabled Offenders Act.

(5) All explanations required by this act to be given to a juvenile alleged or found to have a developmental disability shall be in the language, mode of communication, and terms which that juvenile is most likely to understand.

COMMENT. See the comment to section 13(5) of the Model Developmentally Disabled Offenders Act.

(6) No pre-adjudication diversion agreement, Program Plan, or dispositional order entered into, developed, or issued pursuant to this act shall provide for or approve psychosurgery, electroconvulsive therapy, or aversive conditioning utilizing pain-producing or health-threatening stimuli.

COMMENT. See comment to section 13(6) of the Model Developmentally Disabled Offenders Act.

Section 13 Examinations and Evaluations

(1) (a) All members of multidisciplinary evalu-

---

[477] 387 U.S. 1 (1967).

[478] See sections 7(4) and (8): 9(6) and (10): 10(3) and 11(5). For a discussion of the importance of the right to appeal in delinquency, see In re Gault, *id.*, at.

[479] *See NACJJDP*, *supra* note 148 at §3.133; see also IJA/ABA, DISPOSITIONAL PROCEDURE, note 432, at §3.1.

[480] For a discussion of the limitation on prosecutorial appeals, see the comment to section 13(1) of the Model Developmentally Disabled Offenders Act.

[481] On the need for appointed counsel in delinquency and disability proceedings, *see* In re GAULT, 387 U.S. 1, 41 (1967); NACJJDP, *supra* note 148, at §3.132; IJA-ABA, COUNSEL, *supra* note 394, at §§2.3-2.4; Vitek vs. Jones, U.S. (1980) (opinion of White J. for the court) *but see* Vitek vs. Jones, *id.*, at (Powell, J. concurring).

[482] See sections 7(6), 9(7), and 11(2).

[483] See authorities cited *supra* note 401.

ation teams appointed pursuant to Sections 7, 9, 10, and 11 of this act, and experts designated to examine a defendant pursuant to section 6 of this act shall be selected from the list published by the Evaluation Oversight Board.

(b) The prosecutor and the juvenile to be evaluated shall be entitled to nominate experts from the list for the court's consideration prior to the appointment of a multidisciplinary evaluation team.

COMMENT. The intent and operation of this provision are discussed in the comment to section 15(1) of the Model Developmentally Disabled Offenders Act. Enacting jurisdictions which have not adopted section 14 of that Act establishing and prescribing the duties of the Evaluation Oversight Board should insert that provision into this statute, or provide an alternative system which will assure the independence, competence, and integrity of the evaluations and examinations.[484]

(2) Each expert appointed or designated to examine or evaluate a person pursuant to this act, shall receive at the time of his or her appointment or designation, a written description of his or her duties and of the matters to be addressed in the examination or evaluation report.

COMMENT. See comment to section 15(2) of the Model Developmentally Disabled Offenders Act.

(3) (a) All examinations and evaluations shall be conducted on an outpatient basis unless the juvenile to be examined or evaluated fails or refuses to keep at least two (2) appointments for an examination or evaluation ordered under this act. Any confinement ordered under this paragraph shall not exceed five (5) days.

(b) If the juvenile to be examined or evaluated is in custody, the court shall order, or pursuant to section 4, the head of the intake unit shall request in writing, that the (sheriff/marshall/other appropriate official) permit the juvenile to be examined or evaluated in the place of detention or to transport the juvenile to and from an appropriate facility for examination or evaluation.

COMMENT. See the comment to section 15(3) of the Model Developmentally Disabled Offenders Act.

(4) (a) All examination and evaluation reports prepared under this act shall include:

(i) The names and addresses of all individuals who examined, interviewed or investigated the juvenile alleged or found to have a developmental disability in conjunction with the examination or evaluation;

(ii) A description of the techniques or tests used; and

(iii) When a multidisciplinary evaluation team has been appointed, the signatures of each member of the team together with a statement of concurrence or nonconcurrence with the evaluation findings.

Examination and evaluation reports shall be submitted to the juvenile court no more than ten (10) days after the designation of the examiner or appointment of the multidisciplinary evaluation team.

(b) Reports for examinations conducted pursuant to section 6(2) of this act shall include in addition:

(i) An opinion on whether the juvenile has a developmental disability as defined in this act; and if so,

(ii) A description of the nature and effect of that disability; and

(iii) An opinion on whether a full evaluation should be conducted.

Copies of the report shall be provided to the intake unit, the juvenile, the juvenile's parents or guardian, and the juvenile's attorney.

(c) Reports for evaluations conducted pursuant to section 7(4)(c) of this act shall include in addition:

(i) An opinion on whether the juvenile has a developmental disability as defined in this act; and if so,

(ii) A description of the defendant's mental, emotional, physical, and educational condition, adaptive behavior and social skills;

(iii) An opinion regarding the extent to which the juvenile has the present ability to [state standard/understand the nature and purpose of the proceedings and assist in preparing his or her own defense]; and if the juvenile lacks that ability,

(iv) A description of the habilitative services which will help him or her develop that ability within one year; and

(v) An opinion on the likelihood of the juvenile developing that ability within one (1) year.

A copy of the report shall be provided to the prosecutor, the juvenile, the juvenile's parents or guardian, and the juvenile's attorney.

(d) Reports for evaluations conducted pursuant to sections 9(5) and 9(10)(c) of this act shall include in addition:

(i) A description of the examination and evaluation reports regarding the juvenile prepared by other individuals which were utilized by the multidisciplinary evaluation team;

(ii) An opinion regarding whether the

---

[484] See section 14 of the Model Developmentally Disabled Offfenders Act and the comment thereto.

juvenile has a developmental disability as defined in this act, and if so,

(iii) A description of the juvenile's mental, emotional, physical, and educational condition, adaptive behavior and social skills;

(iv) An opinion regarding the habilitative services which would assist the juvenile to the extent and within the shortest time possible, to function in the community as a self-reliant, law-abiding citizen; and

(v) At the specific request of the court, an opinion regarding whether the juvenile is more likely to benefit from the proposed habilitative services if they are provided in a community based or in a non-community based setting.

A copy of the evaluation report shall be provided to the prosecutor, the juvenile, the juvenile's parents or guardian, the juvenile's attorney, the (name of the state Developmental Disabilities Advocacy Office) and, following disposition, to the individual(s) having direct responsibility for development of a Program Plan for the juvenile.

(e) Reports for evaluations conducted pursuant to section 10(2) and 11(2)(f) of this act, shall contain the information required by paragraphs (4)(a) and (d) of this section. A copy of any evaluation report prepared pursuant to section 10(2) shall be provided to the prosecutor, the juvenile, the juvenile's parents or guardian, the juvenile's attorney, the (name of the state Developmental Disabilities Advocacy Office), and following disposition, to the individual(s) having direct responsibility for the custody, supervision, habilitation and release of the juvenile. A copy of evaluation reports prepared pursuant to section 11(2)(f) shall be provided the juvenile offender, his or her parents or guardian, and the juvenile's attorney, the person having direct responsibility for the custody, supervision, habilitation and release of the juvenile offender, and if the juvenile is in custody, the chief developmental disabilities advocate for the juvenile correctional facility in which he or she is housed.

COMMENT. See the comment to section 15(4) of the Model Developmentally Disabled Offenders Act.[485]

(5) Examination and evaluation reports shall be explained to the juvenile examined or evaluated to the maximum extent possible. Assistance shall also be provided to the juvenile's parents to facilitate their understanding of such reports. Portions of examination and evaluation reports which are likely to cause severe psychological harm to the juvenile may be disclosed and explained to him or her through a psychiatrist, psychologist or other appropriate professional.

COMMENT. This subsection parallels section 15(5) of the Model Developmentally Disabled Offenders Act except that it provides for assistance to the juvenile's parents to aid them in understanding and interpreting the results of evaluation or examination. The comment to the adult provision contains a description of the intent and operation of this subsection.

(6) (a) Examination or evaluation reports shall not constitute public records.

(b) Statements made by a juvenile alleged or found to have a developmental disability and other information developed in the course of a screening test, examination or evaluation provided pursuant to this act and physical evidence obtained as a result of such statements or information shall not be admissible in evidence against the juvenile on the issue of guilt in any delinquency proceeding. In addition, such statements, information or evidence shall not be admissible in any civil proceeding without the juvenile's informed consent.

COMMENT. Most states currently limit access to the records resulting from juvenile court proceedings.[486] Hence, paragraph (6)(a) should not impose any additional burden on juvenile courts and juvenile correctional agencies. The purpose and implication of paragraph (6)(b) are discussed in the comment to section 15(6) of the Model Developmentally Disabled Offenders Act. It should be noted that the constitutional protection against self-incrimination was extended to juveniles in delinquency proceedings by In re Gault.[487]

(7) The cost of any examination or evaluation conducted by an expert appointed pursuant to this act shall be borne by the (name of state/county developmental disabilities agency/juvenile court).

COMMENT. See the comment to section 16 (7) of the Model Developmentally Disabled Offenders Act.

Section 14. Authorization of Appropriations

To carry out the purposes of this act, there is authorized to be appropriated $ for the fiscal year ending , $ for the fiscal year ending , $ for the fiscal year ending , and $ for the fiscal year ending

COMMENT. Although every effort has been made to limit the cost of implementing this act, most states will need to appropriate or transfer some additional funds to make the sections on

[485] Enacting jurisdictions which reduce the maximum period for delay because of a finding of incompetence to assist counsel, should modify paragraph (4)(c) accordingly.

[486] NATIONAL TASK FORCE, COURT STRUCTURE, *supra* note 366, at 51; *see also* NACJJDP, *supra* note 28 §§0.53 and 0.532; IJA-ABA, STANDARDS RELATING TO JUVENILE RECORDS AND INFORMATION SYSTEMS (Tentative Draft 1977).

[487] 387 U.S. 1, 47 (1967).

evaluation, advocacy, and the provision of habilitative services fully effective. It should be noted, however, that the provisions encouraging early disposition of cases,[488] and discouraging confinement in secure or non-community based facilities[489] should help to offset some of these costs. Moreover, as is noted in the comment to section 17 of the Model Developmentally Disabled Offenders Act, in light of the number of persons with a developmental disability, both juveniles and adults, currently incarcerated and the direct cost of crime to the public, if the provision of the habilitative services called for in this act is successful in reducing the rate of recidivism even to a small extent, then the required investment will have reaped a significant reward.

Section 15. Severability of Sections

If any section, subsection, paragraph, sentence, or any other part of this act is adjudged unconstitutional or invalid, such judgment shall not affect, impair or invalidate the remainder of this act, but shall be confined to the section, subsection, paragraph, sentence or other part of this act directly involved in the controversy in which that judgment was rendered.

COMMENT. None necessary.

Section 16. Effective Date

This act takes effect

COMMENT. None necessary.

[488] *E.g.*, sections 6, 7(4)(c), 7(5)(c), 7(5)(d), 7(9), 7(10), and 9(6).

[489] *E.g.*, sections 7(3), 7(5)(d), 7(5)(e), 9(6)(f), and 10(3).

9
ADVOCACY

TABLE OF CONTENTS

A REVIEW OF STATE ADVOCACY LEGISLATION

INTRODUCTION

In 1975 the United States Congress passed legislation, known as the Developmentally Disabled Assistance and Bill of Rights Act[1] (hereinafter DD Act), which requires that the states and territories [hereinafter states] have in effect by October 1, 1977, a protection and advocacy system in order to qualify for receipt of federal funds for developmental disabilities programs. Such systems must be able to (1) "protect and advocate the rights of persons with developmental disabilities," (2) "pursue legal, administrative, and other appropriate remedies" and (3) "be independent of any state agency which provides treatment, services, or habilitation to persons with developmental disabilities."[2] This is the first time that Congress has conditioned receipt of federal funds on the establishment of a system which must advocate for and protect individuals' rights.

No definition of "advocacy" per se is provided in this law. Guidelines, developed by the U.S. Department of Health, Education and Welfare (hereinafter HEW) to assist states in developing protection and advocacy systems under section 113 of the law, however, define advocacy as:

> "speaking for, pleading for, supporting, advising, espousing the rights of, or interceding on behalf of persons with developmental disabilities before any public or private individuals, agencies, organizations or institutions serving such people."[3]

Neither the DD Act nor the Guidelines specify a method for establishing an advocacy system. Rather, they give states flexibility to utilize different approaches.[4] Statutory creation is one alternative. A private not-for-profit corporation is another. A quasi-public corporation is yet another.

The choice among the alternative approaches will be made based upon a broad range of factors, including state constitutional limitations on the number of state departments, political climate of the state, support of competing interest groups, and support given by the state's governor to the establishment of an advocacy agency. Unfortunately, there is no empirical data indicating which approach is most successful and under what conditions.

Review of Statutes

The HEW Guidelines suggest that an advocacy system may be created by statute but that legislation is not the only alternative.[5] If a state does wish to create its system statutorily then it should be helpful for it to consider those advocacy statutes that have been enacted in other states. This section will review the advocacy legislation currently in existence which should be useful to all other states since elements of the statutorily created systems discussed here may be used in planning other types of systems as well.

A review of the statutes of all 50 states revealed that twelve states have advocacy systems created through legislation. An analysis of these twelve statutes follows. States which have created advocacy systems through means other than legislation are not included.[6] In addition, three types of statutes were excluded even though they create entities which may provide services in the nature of protection and advocacy. First, states which have statutorily created mechanisms such as a "human rights committee" are not included because such committees do not have independent authority to pursue legal remedies and because they may be too closely affiliated with service providers to qualify under § 113.[7] Second, state legislation creating developmental disabilities state planning councils has been excluded for similar reasons. Third, legislation which created legal aid and public defender services has been excluded because these agencies generally are not statutorily required to provide assistance to developmentally disabled persons. Rather, developmentally disabled persons are treated like any other clients and are provided with legal assistance if they meet the criteria, such as indigency, which are applicable to all clients. Addition-

---

[1] Pub. L. No. 94-103, 42 U.S.C. §§6001-6081 (Supp. V 1975).

[2] Pub. L. No. 94-103, §113, 42 U.S.C. §6012 (Supp. V 1975).

[3] Interim Guidelines, "Development and Implementation of a System for Protection of Individual Rights and Advocacy for Persons with Developmental Disabilities," Part A §5(b), November 1, 1976 (hereinafter HEW Guidelines). Although guidelines are not legally binding, they do indicate the criteria which HEW personnel will consider when determining whether a system qualifies under §113.

[4] *See* HEW Guidelines, Part A, §1 and Part B, §§4 and 5.

[5] HEW Guidelines, B §5(a).

[6] Minnesota and Massachusetts have advocacy systems which were established through non-legislative mechanisms. See S. Herr, *Advocacy Under the Developmental Disabilities Act*, Developmental Disabilities Office, Office of Human Development, U.S. Dept. Of Health, Education and Welfare (1977).

[7] *See, e.g.,* OHIO REV. CODE ANN. §5119.801, .802, .81 & .82 (Page) (1976), and FLA. STAT. ANN. §20.19. These committees, however, may be an important component of a protection and advocacy system because of their investigative powers.

ally, where public defender services do provide assistance to mentally ill or developmentally disabled persons, it generally is limited to representation in commitment or release hearings. Consequently, although these agencies provide valuable services to developmentally disabled clients, they are limited by their resources and their legislative mandates.[8]

There are other limits to this review that must be kept in mind. First, most of the advocacy statutes were passed before the creation of the Developmental Disabilities Act and thus they often were designed to address other state needs; for example, the persons covered in these enactments may not include developmentally disabled persons.[9] This legislation is appropriate for review here, however, because such statutes otherwise meet many of the requirements of § 113 and easily could be amended or expanded to include developmentally disabled persons. Second, three advocacy statutes provide that persons be indigent before they can qualify for advocacy services.[10] Section 113 is silent on this consideration but the language of the law taken as a whole and related legislative history suggest that Congress meant for coverage to be broad. A provision in state legislation which limits protection and advocacy services to indigent persons may, therefore, disqualify a state system from receiving developmental disabilities funds. Where these discrepancies exist in respect to § 113 requirements, they have been noted in the subsequent commentary and chart.

Review of State Statutes

The remainder of this section will present an analysis of the elements of each state advocacy statute. Both brief commentary and a chart are presented and are organized under three main headings: 1) persons covered as clients; 2) the scope of services included in the statute; and 3) the administrative structure of the advocacy system. These categories roughly mirror the requirements of § 113 as previously noted.

Persons Covered

Public Law 94-103, §102(7) defines a developmental disability as that which

"(A) (i) is attributable to mental retardation, cerebral palsy, epilepsy, or autism;

(ii) is attributable to any other condition of a person found to be closely related to mental retardation because such condition results in similar impairment of general intellectual functioning or adaptive behavior to that of mentally retarded persons or requires treatment and services similar to those required for such persons or;

(iii) is attributable to dyslexia resulting from a disability described in clause (i) or (ii) . . .;

(B) originates before such person attains age 18;

(C) has continued or can be expected to continue indefinitely; and

(D) constitutes a substantial handicap to such person's ability to function normally in society."

Three of the five states which provide advocacy services to "developmentally disabled" persons (California, Indiana, Ohio) specifically include persons with mental retardation, cerebral palsy and epilepsy. Two of these three states (California, Indiana) also specifically include autism. One of the three states (California) also includes persons "with other neurological handicapping conditions" and another (Ohio) includes persons with neurological conditions closely related to mental retardation or requiring treatment similar to that required for mental retardation. The other two states (Connecticut, North Carolina) do not define the term "developmentally disabled" in the advocacy statute.

There are two other broad categories of persons covered in some of these statutes — mentally disabled and mentally ill — neither of which, without further clarification, would be likely to qualify a state under §113 because persons with epilepsy or cerebral palsy might not be covered. For statutes which have ambiguous definitions of the persons covered, but which clarify the ambiguity elsewhere in the statute, the clients are listed in brackets in the chart. There is also a chart entry for the three states (Massachusetts, New Jersey, Rhode Island) that make client indigency a prerequisite for receipt of advocacy services. The remaining state statutes have no such provision, although Louisiana provides that non-indigent clients must pay for services rendered.

Seven states (Iowa, Louisiana, Maryland, Massachusetts, New Jersey, North Carolina, Rhode Island) provide services only to persons who at the time are admitted to a facility which provides treatment services or who are in the process of being admitted. Advocacy services in these states may be available to persons who, although technically listed as institutional admittees, actually are living in the community on leave or out-patient care status (conditional release, for instance). Such persons, because of their technical status as admittees, may meet the state's advocacy statute criteria for receipt of services. One state statute (New York) specifically notes that persons on conditional release may receive services from that state's advocacy system.

Services Provided

Advocacy systems reviewed here provide three

---

[8] Although the legislation creating these agencies will not be discussed here, it should be considered for integration into a protection and advocacy system.

[9] It is not entirely clear in many cases that developmentally disabled persons would not be included in the class of persons covered; however, reading particular statutes as a whold strongly suggests that they would not be included.

[10] An indigency test also is common in legislation which establishes legal aid and public defender services.

basic kinds of services to the clients whom they represent: "direct legal services," "referral services," and "investigative services." "Direct legal services" for purposes of this analysis means services which normally would be provided by an attorney (e.g., conducting litigation and representation at administrative hearings). "Referral services" enable the advocacy system to refer the client to a lawyer for help with legal problems when representation for such matters is not authorized by the statute (e.g., a rent or employment problem). "Investigative services" refers to an explicit statutory authorization to conduct an investigation either into the facts of a complaint or into the practices, procedures or conditions generally prevailing at a facility, in the community or in the state system. In addition, there is a broad range of other kinds of services (e.g., educational, information dissemination) which each of the advocacy systems may be authorized to undertake. The chart entry on this topic makes note of these other services.

In those states which provide for "direct legal services" there often is additional text in the statute regarding specific activities which may be undertaken. Examples include authorization to sue and be sued and to take action to secure any legal or equitable remedy. Where there is statutory language, the chart reflects this by indicating the range of legal services which each statute has authorized. Only four of the statutes (California, Connecticut, Indiana and North Carolina) clearly were drafted in response to P.L. 94-103 and consequently the other statutes may not authorize the advocate to undertake the full range of legal activity required under the federal law.

Administration

Administrative matters addressed in the statutes reviewed include who appoints persons to positions in the advocacy system; what particular qualifications, if any, those persons must have; who supervises such persons; whether such supervision is exercised on a statewide or regional basis; and what funding provisions are noted in the statute.

Two broad approaches are utilized in appointing the advocate. First, some statutes provide for appointment of a commission or advisory-type board which then hires the staff director. The commission or board continues to have a role to play in the actual provision or oversight of advocacy services, although in one of the four states which uses this approach (California) the board has a very active role and actually reviews information and makes findings which have a direct effect on clients. In two other states (Indiana, Louisiana) the commission primarily is charged with responsibility for setting up a system, hiring staff, and setting policy which the staff then carries out. In the remaining state (Connecticut) the board's primary function is to hire a director who then sets up the office. Second, other statutes give the judicial or executive branch the appointment authority for hiring the director of the advocacy system.

The section of the chart which deals with staff qualifications is largely self-explanatory. The chart covers both the qualifications required for persons appointed to the board or commission and the qualifications required for the director and staff hired for the advocacy system. Specific qualifications, where noted in the statute, are included in the chart entries.

Supervision and reporting requirements for the advocate or the system are addressed in the next chart section. The entry enumerates who, or what body, makes policy decisions to guide the advocacy staff. Three of the four states which have established boards (California, Indiana and Louisiana) require that it set policies, while the remaining states provide that individuals such as the director of the advocacy system have this authority.

Another aspect of supervision is the geographical scope of the authority which makes policy and determines whether policy is being followed. The two basic approaches used here are regionalizing (dispersing) and centralizing (concentrating) such authority. The choice which a state makes between these two basic approaches can have significant impact on the effectiveness of the system in providing services. One significant factor in making this choice is population patterns within the state. The chart reflects each state's decision although statutory citations are not noted because most statutes must be read as a whole in order to reach a conclusion on this subject.

The final elements noted in the supervision/reporting section of the chart are whether the statute requires that a report be submitted for advocacy system activities, to whom the report must be submitted, and whether provision is made for a review of the system by an organization not affiliated with the advocacy system.

The last chart entry addresses the question of whether or not there are provisions in the statute which indicate means and sources for funding. One statute (New York) is silent on this issue while others provide a variety of alternatives. These include: (1) legislative appropriation, (2) monies from state department budgets, (3) federal monies through the developmental disabilities council, (4) federal and private monies through grants and, (5) gifts.

STATE STATUTE CHARTS

STATE Statute and (Subject)	I. PERSONS COVERED: DISABILITY	I. PERSONS COVERED: ADMISSION REQUIREMENT	II. SERVICES PROVIDED: SERVICES AUTHORIZED	II. SERVICES PROVIDED: RANGE OF LEGAL SERVICES
CALIFORNIA:				
Cal. Welf. & Inst. Code §38150 *et. seq.*, (1976) (Developmentally Disabled; Area Boards)	developmental disabilities: MR, CP, epilepsy, autism and "other neurological handicapping conditions" §38003	need not be a patient or admitted — "area boards shall protect and advocate the rights of all persons in the area" §38170	conduct public hearings after investigating practices §38170; §38172; encourage growth of citizen advocacy programs §38171.5	must seek authorization from state council on Developmental Disabilities in order "to initiate legal action" §38172
CONNECTICUT:				
Conn. Public Act 77-589 (1977) (An Act Concerning a Protection and Advocacy System for the Handicapped)	handicapped persons: persons who have a physical, mental, emotional or other disability or dysfunction which constitutes a significant handicap to such person's ability to function normally in society, including developmentally disabled persons §2	need not be a patient or admitted §4(4)	investigative, educational, informational, referral services, coordinational and cooperative services with other agencies, client needs, research, and legal §4	represent, appear, intervene in, or bring an action on behalf of any handicapped person or class in any proceeding before any court, agency, board or commission in the state §4(7)
INDIANA:				
Indiana Code §16-13-1 Ch. 19 (1977) (Protection & Advocacy for the Developmentally Disabled)	developmental disabilities: attributable to MR, CP, epilepsy or to any other condition found to be closely related to MR because such condition results in similar impairment §2	need not be a patient or admitted — must assure that services "are accessible to such [developmentally disabled] persons throughout the state" §4	information and referral services §5 (7); promote, protect and advocate the rights of developmentally disabled persons §4 speak for, plead for, support, advise, espouse the rights of, intercede on behalf of — before public or private agencies, organizations, institutions or individuals serving the developmentally disabled §2	"Provide legal . . . services throughout the state to individuals or organizations on matters related to the protection of the legal and human rights" of DD persons §5 (2) to sue and be sued §5 (5)
IOWA:				
Iowa Code Ann. §229.19 (1975) (Mental Health Advocate)	"mentally ill or impaired" §229.19 mentally retarded persons specifically excluded §229.1	must be involuntarily hospitalized patient §229.19	review periodic reports required under statute to be undertaken by medical personnel regarding continued need for treatment §229.19	not clearly delineated — only requires that if the advocate is not an attorney he must advise the court when the services of an attorney are required §229.19
LOUISIANA:				
La. Rev. Stats. Title 28 §64 (1977)	mentally ill; unclear whether it applies to m.r. or d.d.	patients who are admitted pursuant to state law who request assistance (A)	Direct legal representation (C)(3); contract with others for legal services (C)(7)	Programs to provide legal counsel and representation for mentally disabled to ensure that rights are protected (C)(1); such legal proceedings as are necessary to fulfill duties (C)(3)

	I. PERSONS COVERED		II. SERVICES PROVIDED	
STATE Statute and (Subject)	**DISABILITY**	**ADMISSION REQUIREMENT**	**SERVICES AUTHORIZED**	**RANGE OF LEGAL SERVICES**
MARYLAND:				
Md. Ann. Code Art. 59 §53 *et seq.* (1971) (Mental Health Advocate)	mentally disordered or allegedly mentally disordered §54	involuntarily committed patients, voluntary and informal patients §55	study and review the admission and retention of patients; inform patients of their rights; provide court with all relevant information regarding patient's case §54	other than informing patients of their rights, no provisions
MASSACHUSETTS:				
Mass. Gen. Laws Ann. Ch. 221, §34E (1973) (Mentally Ill Indigents — Legal Assistance)	mentally ill and mentally retarded persons; client indigency is a condition of representation §34E	must be a patient or resident in a mental health or mental retardation facility §34E	study and review the admission and retention of patients; inform patients of their rights; provide court with all relevant information regarding patient's case §54	other than informing patients of their rights, no provisions
NEW JERSEY:				
N.J. Stat. Ann. §52-27E-21 to 27 (1974) (Dept. of Public Advocate; Div. of Mental Health Advocate)	mentally ill — "mental hospital admittee" §52:27E-23 (does not appear to cover mentally retarded persons) client indigency is a condition of representation §52:27E-26	must be a patient in a mental hospital or similar facility or subject to admission §52:27E-23	"such legal representation and medical consultation as the director [of the Div. M.H. Advocate] deems appropriate" §52:27E-24	may represent persons in any proceeding concerning the admittee's admission to, retention in, or release from confinement" §52:27E-24 may participate in class action suits §52:27E-25
NEW YORK:				
N.Y. Mental Hygiene Law §§29.09 & 88 (1972) (Mental Health Information Service)	mentally ill or alleged mentally ill, mentally disabled or alleged mentally disabled §29.09(b)	patients conditionally released have a "right to avail themselves of the facilities" of MHIS §29.15(c) (1976)	investigative services: case research and legal research for court system; study and review the admission and retention of patients; inform patients of their rights	scope of legal services to be provided are to be determined by the chief judges of each of the judicial departments §29.09(b)(4); initiate and take any legal action necessary to safeguard abuse and mistreatment §29.09(b)(5)
NORTH CAROLINA:				
N.C. Gen. Stats. §143B-399 *et seq.* (1977) (Advocacy Council for the Handicapped)	mentally ill and developmentally disabled persons §13B-399	must be receiving treatment, services or habilitation from any state, local or area program §143B-399	investigative and legal services; review and recommend changes in laws, rules, regulation, programs and policies; and contract with public or private agencies to fulfill duties §143B-399	pursue legal, administrative or other appropriate remedies to insure protection of rights §143B-399(2)
OHIO:				
Ohio Rev. Code Ann. §5123.94 (1976) (Legal Rights Service)	developmentally disabled persons 5123.94 (A)	"all persons detained, hospitalized, discharged or institutionalized, and all persons whose detention, hospitalization, discharge or institutionalization is sought" §5123.94 (A)	receive and act upon complaints, inform persons of their rights, and assure that all persons are adequately represented by counsel §5123.94 (A)	actions in mandamus and other legal and equitable remedies §5123.94 (G)

	I. PERSONS COVERED		II. SERVICES PROVIDED	
STATE Statute and (Subject)	**DISABILITY**	**ADMISSION REQUIREMENT**	**SERVICES AUTHORIZED**	**RANGE OF LEGAL SERVICES**
RHODE ISLAND:				
R.I. Gen. Laws §40.1-5-13 to 29 (1974) (Mental Health Advocate)	mentally disabled: (having) a mental disorder in which the capacity of a person to exercise self-control or judgment in the conduct of his affairs and social relations or to care for his own personal needs, is significantly impaired (Does not appear to cover mentally retarded persons) §40.1-5-2(13); client indigency is a condition of representation §40.1-5-17	involuntarily committed persons and persons voluntarily admitted §40.1-5-17	Apprise persons of rights, review procedures at facilities, investigate complaints, direct legal services relating to commitment, lawyer referral, initiate programs of public education and legislative advocacy §40.1-5-22	"take all possible action including . . . legislative advocacy and formal legal action, to secure and ensure the legal, civil and special rights" of clients §40.1-5-22(7)

III. ADMINISTRATION

STATE Statute and (Subject)	APPOINTMENT	PERSONNEL QUALIFICATIONS	SUPERVISION/ REPORTING	APPROPRIATION/ FUNDING
CALIFORNIA:				
Cal. Welf. & Inst. Code §38150 *et. seq.* (1976) (Developmentally Disabled; Area Boards)	*area boards:* Governor and county governments appoint §38155; *staff:* selected by the executive secretary, who is appointed by the area board §38187; no state employee may be hired §38185	*area boards:* drawn from the following groups: (a) persons with DD, their parents or guardians — 50% (b) general public — 50% §38156; *staff:* area board determines §38187	no supervision requirement, but must submit summary of activities to State DD Council annually and State legislative analyst conducts a cost effectiveness study every 3 years §§38180-81 supervision is regionalized	area boards present annual funding proposals to State DD Council, which allots funds including federal monies available under P.L. 94-103 §38186
CONNECTICUT:				
Public Act 77-589 (1977) (An Act Concerning a Protection and Advocacy System for the Handicapped)	*board:* 15 members appointed by the governor; service providers excluded §5; *staff:* director appointed by governor; director hires additional staff §3	*board:* 10 members must be handicapped persons or guardians or parents of handicapped persons, at least four of whom represent developmentally disabled persons; remaining five members must be knowledgeable in problems of handicapped persons §5; *staff:* director must be knowledgeable in problems of the handicapped, DD, or advocacy §3	Director of office adopts rules and regulations to carry out act's purposes, with board approval §3 Annual report submitted jointly by director and board to governor and joint committee on human services of general assembly §7; supervision is centralized	$75,000 appropriated for 1977-78 fiscal year §8; may "receive and spend" gifts; bequests; state, private, or federal grants §4(2)
INDIANA:				
Indiana Code §16-13-1 Ch. 19 (1977) (Protection & Advocacy for the Developmentally Disabled)	*commission* of 13 members appointed by the Governor, and non-voting advisory board members appointed by Speaker of the House and Pres. of the Senate (one ea.); service providers are excluded §3(a); *staff:* appointed by the commission §5(2)	*commission:* one member representing each of the major DD's — M.R., C.P., epilepsy, and autism — who either has such a DD or is a parent, guardian, or family member; remaining members must be experienced or expert in DD matters §3(a); *staff:* attorneys may be appointed §5(2)	commission may "promulgate rules and regulations . . . it deems appropriate for the establishment and operation of local protection and advocacy service units" and the service as a whole §5(4) legislature will review need for service in 1980. Service will automatically terminate 6/30/81 without action to continue §7; supervision is centralized	"apply for, solicit, and accept contributions or grants of money, property or services . . . from any source" §5(6)
IOWA:				
Iowa Code Ann. §229.19 (1975) (Mental Health Advocate)	county district court appoints advocate §19	advocate must have "demonstrated by prior activities an informed concern for the welfare and rehabilitation" of clients — need not be an attorney §19	no provisions	county district court periodically prescribes compensation based on advocate's work; funds are drawn from county mental health and institution funding §19
LOUISIANA:				
La. Rev. Stats. Title 28 §64 (1977)	*board of trustees:* 12 members; automatic appointment as designated in statute (A); *staff:* board appoints a director who hires staff (B)(1) & (C)(4)	*board of trustees:* deans of law and medical schools; Pres. of Mental Health Assoc. and member of La. Medical Soc. and state bar assoc. (A); *staff:* must be members of La. Bar (C) & (C)(4)	*board of trustees* review and approve annual report to be submitted to the legislature, governor, and chief justice of the supreme court (B)(5); review and evaluate operation of service (B)(3); supervision is centralized.	Director with board approval is to take action necessary to secure private, federal and other public funds (C)(6)

III. ADMINISTRATION

STATE Statute and (Subject)	APPOINTMENT	PERSONNEL QUALIFICATIONS	SUPERVISION/ REPORTNG	APPROPRIATION/ FUNDING
MARYLAND: Md. Ann. Code Art. 59 §53 *et. seq.* (1971) (Mental Health Information & Review Service)	for each appellate judicial circuit, the head of the service is to be appointed by the senior judge of the circuit §53	staff qualifications established in agreement with the Commissioner of Mental Hygiene §53	the senior judge may remove the head of the service in his judicial district; rules and regulations for the service are established by the Court of Appeals and Commissioner of Mental Hygiene §53	"such funds from the general funds of the state as necessary from time to time for carrying on the functions contained" in the statute §53
MASSACHUSETTS: Mass. Gen. Laws Ann. ch. 221 §34E (1973) (Mentally Ill Indigents — Legal Assistance)	*legal advisors committee* of 14 members appointed by State Supreme Court Justices; *attorneys:* the legal advisors committee is to establish a list of attorneys from each m.h. district and court appoints attorneys from this list to represent patients seeking legal assistance §34E	*committee:* must be attorneys, one from each mental health district in the State; *attorneys:* an attorney from the m.h. district of the patient or from an adjoining district is appointed to represent §34E	no provisions — this is essentially a lawyer referral service; supervision is the same as in any other attorney-client relationship	the committee establishes a fee schedule and the court awards fees based on the schedule; "shall be eligible for federal funds and may accept gifts, grants or contributions from any source" §34E
NEW JERSEY: N.J. Stat. Ann. §52-27E-21 to 27 (1974) (Dept. of Public Advocate: Div. of Mental Health Advocate)	Public Advocate appointed by Gov. with advice and consent of Senate §52:27E-3; Public Advocate appoints Dir. of the Div. of M.H. Adv. §52:27E-4(1); *staff* appointed by dir. with approval of Pub. Adv. — may temporarily retain experts §52:27E-22	Public Advocate must be an attorney §52:27E-3; director need not be an attorney but is to have appropriate experience (and probably is an attorney) §52:27E-4(1)	Public Advocate adopts rules and regulations to run the Dept. §52:27E-4(c) & (d); Pub. Adv. makes such reports as the Governor and Legislature request §52:27E-4(g); supervision is centralized	Public Advocate may solicit and accept federal and private foundation grants §52:27E-4(m); Div. of Administration (within Dept. of Pub. Adv.) proposes budgets §52:27E-8
NEW YORK:* N.Y. Mental Hygiene Law §§29.09 & 88 (1972) (Mental Health Information Service)	for each branch, staff appointments are to be made by the presiding justice of the appellate div. of the judicial dept. affected §29.09(a)	staff qualifications established in agreement with the Commissioner of M.H./M.R. §29.09(a)	activities supervised by the affiliated court system §29.09(a) & (b); supervision is regionalized	no provisions

*for additional information see: Andalman and Chambers, "Effective Counsel for Persons Facing Civil Commitment: a Survey, a Polemic and a Proposal," 45 Miss. L.J. 43, 64-74 (1974)

III. ADMINISTRATION

STATE Statute and (Subject)	APPOINTMENT	PERSONNEL QUALIFICATIONS	SUPERVISION/ REPORTING	APPROPRIATION/ FUNDING
NORTH CAROLINA:				
N.C. Gen. Stats. §143B-399 *et seq.* (1977) (Advocacy Council for the Handicapped)	*council:* 11 members; 7 members appointed by Gov; 2 members appointed by Speaker of House; 2 members appointed by Pres. of Senate; service providers excluded §143B-400 (no advocacy staff provided for)	*council:* 3 members from legal, citizen or advisory groups; 2 members to be parents of developmentally disabled persons; 1 developmentally disabled member; 1 member who has been a volunteer advocate for the developmentally disabled; and 4 members who have demonstrated an interest in the mentally ill and developmentally disabled §143B-400	Annual report submitted by Council to Governor and the Secretary of Administration §143B-399(6)	no provisions
OHIO:				
Ohio Rev. Code Ann. §5123.94 (1974) (Legal Rights Service)	Chief Justice of the Ohio Supreme Court appoints an administrator §5123.94(D); requires that Ohio Legal Rights Serv. be independent of Dept. of M.H./M.R. and Atty. Gen. §5123.94(E)	administrator of Service must be a member of the Ohio Bar §5123.94(D)	report must be submitted annually to Gov., Chief Justice, Pres. of Senate, and Speaker of House, and the Dir. of the Dept. of M.H./M.R. §5123.94(H); adm. of service promulgates rules §5123.94(F); supervision is centralized	administrator recommends budget to the General Assembly §5123.94(F); may receive charitable gifts, grants and bequests §5123.94(I)
RHODE ISLAND:				
R.I. Gen. Laws §40.1-5-13 to 29 (1974) (Mental Health Advocate)	Governor, with advice and consent of Senate, appoints mental health advocate from a list submitted by four attorneys, one psychologist, one psychiatrist, one social worker, one mental health assoc. designee, and the dir. of the Dept. of M.H./M.R. and Hospitals §40.1-5-14; mental health adv. may appoint subordinate staff §40.1-5-15	mental health advocate must be a member of the R.I. Bar §40.1-5-14	Annual report submitted to Governor, Dir. of Dept. of M.H./M.R. and Hospitals, and the General Assembly §40.1-5-21; supervision is centralized	General Assembly annually shall appropriate sums to cover office expenses §40.1-5-16

SUMMARY OF MODEL ACT

Statutory Authorization

Statutory authorization has major advantages. It provides a means for giving the advocacy agency the required authority to pursue legal, administrative and other remedies. In addition, it provides a means for giving Advocacy Office personnel the right to access to information, materials, records, and clients critical to the discovery, consultation and confrontation necessary for effective advocacy. Statutory authorization can give this authority in a clear and undisputed manner while no other mechanism for establishing the advocacy agency could grant the broad right to access.

Finally, the General Assembly makes a commitment to the advocacy agency by creating it. The statute, in turn, signifies a commitment by the state to providing protection and advocacy to those developmentally disabled persons who need them. This commitment can aid the advocacy agency in two ways. First, if the advocacy agency is to be effective, other funds must be obtained to supplement the minimal federal appropriations that will be awarded to carry out the program mandated by §113. This commitment may assist in securing state appropriations, thereby giving the agency a more stable financial base. Effective long-range planning of activity requires a minimal certainty of funding availability from year to year.

Second, if the advocacy agency is to be successful, its functions must be accepted as legitimate by state agencies, service providers, and developmentally disabled persons. The establishment of the agency by the state's General Assembly lends a stamp of approval to the agency it would not otherwise have, for the General Assembly has declared that the provision of advocacy to developmentally disabled persons is a part of the official policy of the state. This official sanction may supplement the agency's actual public authority and should have a positive effect on its acceptance by others.

In addition to the advantages which may be had from statutory creation, the public employee status of the Office personnel gives them the additional advantage of greater familiarity with the operation of state government on an informal, practical level as well as a formal, organizational level. Being a part of state government will subject them to and give them a greater opportunity to learn about public budgeting, financial administration, public personnel administration, and most importantly, the politics and practice of public administration. This knowledge should enhance their ability to obtain information and to promote the concerns of developmentally disabled persons by allowing staff to better identify problem areas within public service agencies, by enhancing their bargaining position by greater familiarity with potential adversaries, and by enabling them to take more informal, non-legal approaches to solving problems. Even in an atmosphere of mutual trust and program interest, inaccessibility is always a problem for an agency outside the government. Public status may help to negate this problem.

Independence of the Agency

The advocacy agency is required by §113 to be independent of service providers. Although a private, non-profit organization will provide the greatest independence, a public advocacy agency also can be adequately insulated from service providers. The act proposes a public advocacy office having maximum independence as a major feature.

An independent public advocacy office avoids its inappropriate placement within another office of state government. For example, the advocacy agency should not be placed in an unrelated state agency (e.g., state department of administration, a state planning office or the auditor's office) since protection and advocacy have nothing to do with these functions. This in turn may cause the director of the umbrella agency to de-emphasize the Advocacy Office's functions, needs, and fiscal priorities each year in budgeting for and administering the total agency. This is particularly likely since the duties of the umbrella agency and the interests of its director would pre-exist the introduction of the Advocacy Office into it. Further intervention by the state on behalf of individuals for the purpose of asserting their rights vis-a-vis other state agencies and private agencies providing services or goods may not be seen as being equal in importance to traditional state functions.

Since many state agencies provide services to developmentally disabled persons (e.g., health, mental retardation, mental health, social welfare or social services), they are inappropriate for containing the Advocacy Office. The state developmental disabilities planning council has also been held to be a service provider. The state office of the attorney general also would be inappropriate since the attorney general represents the state agencies that provide services and would have to defend the validity or constitutionality of any statutes, rules or regulations that were challenged by the Advocacy Office. Similarly, the Office of the Governor is not an appropriate organizational placement for the Advocacy Office since the Governor is likely to be a defendant in any suit which might be brought against the state and state service providers.

Although not required, the advocacy agency should be independent of partisan political considerations as well. Thus, placement in the governor's control, even though not directly within his office, is inappropriate and not recommended. Such placement could allow the appointment of Office personnel to be based upon patronage rather than merit. In addition, the governor has political interests which may conflict and administrative responsibility for state service agencies which do conflict with the purpose and functions of any advocacy agency. The public advocacy agency's independence will guard against such partisan politics and conflicts of interests.

Apart from the reasons stated above, making the advocacy agency independent of other state agencies has additional advantages. The visibility of the advocacy agency will be substantially increased, thereby improving public awareness of the problems of developmentally disabled persons, increasing the ability of persons needing protection and advocacy to more easily locate this help, and enhancing the ability of the advocacy agency to mobilize public and interest group support on behalf of its protection and advocacy efforts. Independence also will prevent the advocacy agency from being tucked away out of sight in a potentially hostile environment.

The Office may, of course, cooperate with other state agencies to achieve their desired goals. For example, the Office can provide education and information to the state developmental disabilities planning council. The council in turn can be a valuable ally in pursuit of support and resources for the Office. The Office also may cooperate with the attorney general by providing information and technical assistance where possible when it does not create a conflict of interest which may compromise the interests of Office clients.

Persons Covered by the Act

An independent advocacy agency need not be created solely to represent developmentally disabled persons. Although this act has been written to cover only these individuals, other persons (e.g., mentally ill, elderly and physically handicapped individuals or consumers) need the same type of advocacy services. Therefore, states considering adoption of this act should also consider expanding its coverage to include other classes of persons, if legislative appropriations permit.

Combining advocacy functions for different types of clients into one agency is legitimate and may provide better administration of all advocacy. There is a danger, however, that developmentally disabled persons may be discriminated against whether consciously or unconsciously vis-a-vis other groups such as consumers or mentally ill persons if they are all served by a combined protection and advocacy office. There must be a mechanism to insure that funds appropriated for advocacy for developmentally disabled persons are actually used for that purpose.

Alternative Approaches

It is recognized that there are problems with the approach espoused in this introduction. The general assemblies of many states will be reluctant to create a new state agency at a time when government continues to grow. Where the new state agency consideration is a major one, states should explore the use of a special public not-for-profit corporation to carry out the advocacy program. Because it would be statutorily created, the special corporation could have most of the advantages of a statutorily created public agency. It would be a firmly established entity with clear authority, high visibility, a right to access, maximum independence, and legislative commitment. In addition, it could be integrated into the state's government for fiscal purposes without creating another agency of government.

The disadvantages to this approach are that a legislature may not be any more willing to create a public corporation than a private agency, the employees would not have the familiarity with state government, and a commission or board would have to run the corporation. With a public agency, a state can choose whether it wants its agency controlled by a board or by the agency director. In the latter case, the director could be aided by an advisory board.

At least two states, Illinois and Pennsylvania, have adopted as a goal of their developmental disability plans the establishment of a special public corporation to operate the states' protection and advocacy systems. Illinois has specific statutory language covering these special corporations.[1] Since a bill establishing a special public corporation would look very similar to the proposed act which follows, this act could be easily adapted to this purpose.[2]

---

[1] ILL. ANN. STAT. ch. 32, §§198-199.4 (Smith-Hurd 1970).

[2] Other examples which may be helpful include 42 U.S.C. §§996-29986(/) (Supp. V 1975); ILL. ANN. STAT. ch. 32, §§415-1 to 415-15 (Smith-Hurd Supp. 1977); ILL. ANN. STAT. ch. 32 §§595-624, 1001-1028 (Smith-Hurd 1970 and Supp. 1977).

It is also recognized that many states will have utilized other non-statutory approaches (e.g., an executive order) to create an advocacy agency by the October 1, 1977, deadline. This does not, however, preclude the subsequent adoption of a bill establishing an independent public advocacy agency. It is urged that states consider adoption of this approach. The special public corporation is offered as a good alternative.

Statutory creation is not essential, however, for those states which have made a decision to utilize a private non-profit organization. They will be looking to the state's not-for-profit corporation act or some other mechanism. For those states it is hoped that the statute which follows will generate thoughts and raise issues which will assist in the establishment or operation of the advocacy agency.

The text of the act appears in **bold type.** Optional language appears [within brackets/ separated by slashes]. Comments on statutory provisions appear in standard type. Alternatives to the text appear in *italics.* The comments explain, elaborate upon or provide a rationale for the text of the act. They are intended to assist the user in interpreting the act, in drafting an act for a particular state from this model act, and in operating an advocacy agency. The alternatives suggest other approaches which might be considered.

MODEL STATUTE*

A BILL FOR

AN ACT TO ESTABLISH A DEVELOPMENTAL DISABILITIES ADVOCACY OFFICE TO PROVIDE PROTECTION, LEGAL AND NON-LEGAL ADVOCACY, AND PROCEDURES FOR THE ADMINISTRATIVE RESOLUTION OF CONFLICT, TO PROMOTE AND PROTECT THE LEGAL AND HUMAN RIGHTS OF DEVELOPMENTALLY DISABLED PERSONS, AND TO PROVIDE PENALTIES FOR VIOLATION OF THE PROTECTION PROVISIONS OF THIS ACT.

Section 1. Title

This act shall be known as the "Developmental Disabilities Advocacy Act."

COMMENT. None necessary.

Section 2. Statement of Purpose

(1) The general assembly declares that it is the purpose of this act to improve the quality of life of all developmentally disabled persons by providing investigation, protection, representation, and other advocacy to developmentally disabled persons to promote, protect and advocate their interests and to advocate for delivery of services.

COMMENT. The ultimate goal of this act is to improve the quality of life of all developmentally disabled persons within the state, whether they live within the community or an institution. Protection and advocacy are means to achieve this goal. This act establishes a right to protection and advocacy for developmentally disabled persons that will promote, protect, and advocate their interests. Developmentally disabled persons are frequently not able to advocate for their human needs and their legal rights whether they arise before, during or after institutionalization, or during community residence. Thus, advocacy is critically needed. This act provides for establishing an agency which would assist developmentally disabled persons in acquiring needed services and in securing their rights when they are threatened or violated, thereby helping achieve the stated goal of the act.

(2) The general assembly intends by adoption of this act to secure for the state of ________, its local units of government, and its citizens, maximum benefits from the Developmentally Disabled Assistance and Bill of Rights Act, 42 U.S.C. §§6001-6081 (Supp. V 1975) and this act should be liberally construed toward that end.

COMMENT. None necessary.

Section 3. Definitions

As used in this act, unless the context requires otherwise, the terms have the meaning ascribed to them in this section.

(1) "Protection" means preventive or supportive assistance given with the purpose of preventing or alleviating neglect, abuse or exploitation of children or adults who are substantially impaired in their capacity to protect their own self interest. It includes investigation and legal representation in cases of reported or suspected neglect, abuse or exploitation, or referral to service agencies which have professional or legal authority to intervene on behalf of clients in such cases.

COMMENT. Protective assistance is intended to be limited to cases where basic needs are being neglected, a person is being subjected to physical or mental abuse, a person is being personally or financially exploited, or other similar cases. The Office may investigate and seek remedies on its own or refer the matter to other agencies which have the authority to intervene and employ sanctions to prevent or stop neglect, abuse or exploitation. It is intended that the Office be aggressive in discovering such clients and offering assistance where appropriate. It is not intended that the Office control client actions or substitute its judgment for that of clients capable of deciding for themselves.

(2) "Advocacy" means speaking for, pleading for, supporting, advising, espousing the rights of, or interceding on behalf of, persons with developmental disabilities before public or private agencies, organizations, institutions, or individuals serving developmentally disabled persons.

COMMENT. This broad definition is intended to encompass the full range of advocacy provided through or in cooperation with the Advocacy Office. This advocacy includes lay or citizen advocacy, i.e., a one-to-one relationship between a capable volunteer and a person who is devel-

* D. Matthew Powell had primary responsibility for drafting this model statute.

opmentally disabled in which the volunteer advocates and defends the rights and interests of the other and provides practical or emotional reinforcement for him or her; investigative advocacy, i.e., fact finding and informed non-legal conflict resolution on behalf of developmentally disabled persons; legal advocacy, i.e., representing developmentally disabled persons in the litigation or legal negotiation process concerning rights, grievances, or appeals, including class actions; and class advocacy through representation before bodies that enact legislation or rules or before other forums. Advocacy does not include such functions as coordination of service delivery or case management. A good case worker should, however, be a good advocate as well.

Advocacy is normally sought to protect the rights and interests held by a developmentally disabled person and to expand those rights to the maximum extent. This should be the focus of the Advocacy Office. There may be situations where advocating for limited guardianship or conservatorship is appropriate, for without the assistance of a surrogate acting on behalf of some developmentally disabled persons certain of their rights may go unexercised. Thus, advocacy may include advocating for or against guardianship or conservatorship. It is expected that advocating for such a restriction would be the exception to the rule.

(3) "Developmental Disability" means a disability of a person which:

(a)(1) is attributable to autism, cerebral palsy, epilepsy, or mental retardation;

(2) is attributable to any other condition found to be closely related to autism, cerebral palsy, epilepsy or mental retardation or requires treatment and services similar to those required for such persons; or

(3) is attributable to dyslexia resulting from a disability described in clause (1) or (2) of this paragraph; and

(b) has continued or can be expected to continue indefinitely.*

COMMENT. This act is intended to meet the requirements of an advocacy agency specified by the Developmental Disabilities Act. Therefore, the definition of a developmental disability contained within the Developmental Disabilities Act has been adapted. However, changes have been made in the federal definition to broaden its coverage. Paragraph (2) has been broadened to include conditions closely related to autism, cerebral palsy and epilespy as well as mental retardation. In addition, the requirement that the disability originate before the person reaches age eighteen has been deleted. This was done since adults become epileptic, cerebral palsied, or mentally retarded from a number of sources, including trauma and disease. They may be no less in need of advocacy and protection. Finally, the requirement that the disability constitute a substantial handicap to such person's ability to function normally in society has been deleted. The reactions and attitudes of other persons may be the cause of inability to function normally in society rather than the disability. Therefore, limiting protection and advocacy to those who are substantially handicapped by their developmental disability is too restrictive.

The definition adopts a categorical approach rather than a functional approach. Thus, the disability must be caused by one or more of the conditions described in paragraph (a). In addition, the condition contained in paragraph (b) must also be met for the disability to qualify as a developmental disability.

Alternative. *Some states may choose to adopt the existing federal definition or the revised definition developed by the Task Force on Definitions – HEW and recently submitted to the Secretary of HEW.*[3]

The existing definition states

The term "developmental disability" means a disability of a person which–

(A) (i) is attributable to mental retardation, cerebral palsy, epilepsy, or autism;

(ii) is attributable to any other condition of a person found to be closely related to mental retardation because such condition results in similar impairment of general intellectual functioning or adaptive behavior to that of mentally retarded persons or requires treatment and services similar to those required for such persons; or

(iii) is attributable to dyslexia resulting from a disability described in clause (i) or (ii) of this subparagraph;

(B) originates before such person attains age eighteen;

(C) has continued or can be expected to continue indefinitely; and

(D) constitutes a substantial handicap to such person's ability to function normally in society.

* This provision and commentary were prepared prior to the amendment of the Developmental Disabilities Act.

[3] *Final Report of the National Task Force on the Definition of Developmental Disabilities* (1977). (Copies of this report may be obtained through a State's Developmental Disabilities Council or a regional office of the Developmental Disabilities Office, Department of Health, Education and Welfare.)

The Task Force's recommended appropriate basis of the definition states [4]

> *For purposes of the Developmental Disabilities Act, a developmental disability is a severe, chronic disability of a person which:*
>
> *1) is attributable to a mental or physical impairment or a combination of mental and physical impairments;*
>
> *2) is manifest before age 22;*
>
> *3) is likely to continue indefinitely;*
>
> *4) results in substantial functional limitations in three or more of the following areas of major life activity:*
>
> *a) self-care,*
>
> *b) receptive and expressive language,*
>
> *c) learning,*
>
> *d) mobility,*
>
> *e) self-direction,*
>
> *f) capacity for independent living, or*
>
> *g) economic self-sufficiency; and*
>
> *5) reflects the need for a combination and sequence of special, interdisciplinary or generic care, treatment or other services which are*
>
> *a) of lifelong or extended duration and*
>
> *b) individually planned and coordinated.*

Another approach is to reference the federal definition and let the act's definition change with the changes in the federal law. This approach will be unconstitutional in some states since their state law must be enacted by their state legislature. These states will be limited to the definition adopted at the time of the passage of the statute in the absence of a subsequent amendment.

Finally, as noted in the introduction to this act, some states may wish to expand the coverage of the act to include non-developmentally disabled persons. In this event care should be taken to (1) insure that the state's definition of persons covered by the act is consistent with the federal definition so as not to lose eligibility for federal funds under §113, (2) insure adequate funding for the expansion, and (3) protect against discrimination against developmentally disabled persons vis-a-vis the other persons served.

[4] Id. at p. 9. Note that this is a recommended basis of a majority of the Task Force. The Report also includes a minority proposal to substitute the following for Part 1 of the revised definition:

> "Is attributable to mental retardation, cerebral palsy, epilepsy, or autism; or is attributable to any other condition of a person similar to mental retardation, cerebral palsy, epilepsy, or autism *because* such condition results in similar impairment of general intellectual functioning and adaptive behavior and requires treatment and services similar to those required for such persons." (Emphasis in original.)

This constitutes a kind of hybrid definition which includes both catagorical and functional elements.

(4) "Developmentally Disabled Person" means a person with a developmental disability.

COMMENT. None necessary.

(5) "Office" means the Developmental Disabilities Advocacy Office.

COMMENT. None necessary.

(6) "Commission" means the Advocacy Commission of the Office.

COMMENT. None necessary.

(7) "Chief Justice" means the Chief Justice of the Supreme Court of the State.

COMMENT. The intent is to specify the Chief Justice of the highest court in the state. Since the supreme court is not the highest court in all states, the reference should be changed accordingly.

(8) "Director" means the Director of the Office.

COMMENT. None necessary.

(9) "Advocacy Attorney" means any attorney either employed directly by or working under a contract or written agreement with the Office. For purposes of this act advocacy attorney shall also include attorneys who have been assigned, in writing, through the Office to represent a developmentally disabled client.

COMMENT. This broad definition includes all attorneys who are providing legal advocacy through the Office regardless of whether they are employed by the Office, are working under contract or written agreement, or have been assigned in writing through the Office to individual clients. This approach gives all attorneys representing developmentally disabled persons through the Office access to information, materials, and records pertaining to their clients and access to their clients by section 10 of this act subject to the privacy and confidentiality limits in section 11. It also makes clear that all of these attorneys have a duty to represent their clients' rights as required by section 12 of the act and subjects all of these attorneys to the penalty provisions of section 16 for violation of a client's privacy and confidentiality.

(10) "Office Personnel" means the director, advocacy attorneys, and all other personnel working for the Office who are directly employed by, under contract with, or under agreement with the Office whose responsibilities include protection or advocacy.

COMMENT. This term includes the Director and all persons who are carrying out protection and advocacy through the Office, including all advocacy attorneys. They are given the rights and duties specified in sections 10, 11 and 12.

(11) "Person" means an individual, corporation, partnership, association, unincorporated organization, or a government or any subdivision, agency, or instrumentality thereof.

COMMENT. This term is defined to confer

broad jurisdiction on the Office to investigate and pursue possible abuses or denials of rights of developmentally disabled persons by any individual, corporation, or any governmental agency, including service providers.

(12) "Client" means any developmentally disabled person or class or group of persons who are receiving protection or advocacy through the Office.

COMMENT. The Advocacy Office is intended to represent developmentally disabled persons, not their parents or guardians. It is recognized that in some instances parents or guardians may be acting on behalf of their developmentally disabled children or wards and the Office will be working with parents and guardians in many instances along with the client. However, the Office is to represent the interests of the client. Developmentally disabled persons receiving protection or advocacy assistance from the Office are its clients.

(13) "Rights" means all human rights and all legal rights including, but not limited to, rights, benefits, and privileges guaranteed by law, the constitution of this state, and the constitution of the United States.

COMMENT. Rights as defined are not limited to specific rights as such and must be construed in the widest way to include any human, legal or other right which other citizens enjoy. The rights of due process and equal protection of the laws will be the basis for many more specific rights such as the right to treatment or services. Therefore, rights are defined in broad terms rather than attempting to list each and every one of them. These rights are not static and change with changes in statutory, administrative and case law. Therefore, this definition provides flexibility in a changing area within the law.

Section 4. Creation of the Advocacy Office

The Developmental Disabilities Advocacy Office is created as an agency of state government. The Office shall be independent of all other governmental agencies and (not withstanding section__ of the code) shall also be independent of the Office of the Attorney General.

COMMENT. See introduction. One additional comment is necessary. The office of the state attorney general cannot provide legal representation for the Advocacy Office, since it may have to represent the state in defending the validity or constitutionality of any challenge to a state statute, rule or regulation affecting developmentally disabled persons. Therefore, any state which has a requirement that the state attorney general's office be the legal counsel for every state agency unless otherwise exempted by statute, must exempt the Advocacy Office from that requirement in order to create an independent agency eligible for funds under the Developmental Disabilities Act.

Alternative. *(1) Some states may wish to combine the Advocacy Office with an existing state agency. For example, if an agency's primary function is to provide advocacy to other types of clients, the Office could be situated within it. This approach would have the advantage of providing an existing structure that already offers similar services. It is not recommended, however, that the Office be situated in a non-related agency (e.g., Agriculture), or in a related agency that provides services to the developmentally disabled (e.g., Mental Health and Mental Retardation). The former situation would most likely stifle the visibility and overall effectiveness of the Office while the latter would violate the provisions of §113 and create severe conflicts of interest. (See Introduction.)*

(2) Other states may have a private, non-profit corporation which has a proven record of providing advocacy. This corporation may be supplemented or adapted to provide advocacy to developmentally disabled persons. This approach has the advantage of an existing structure, reduced start-up costs, current visibility and maximum independence. State and federal funds also could be channeled to a non-profit corporation. This may be a more desirable approach when the political environment is hostile to the creation of an effectively functioning state advocacy agency for developmentally disabled persons.

(3) A special public corporation may provide a middle ground between these two approaches. (See Introduction.)

Section 5. Office Authority

The Office shall have authority to:

(1) Assess the protection and advocacy needs of developmentally disabled persons within the state, set priorities among those needs consistent with subsection 7.2 of this act based upon available resources, devise strategies for meeting the needs, and implement these strategies.

COMMENT. The Office can best achieve its goals and purposes by utilizing a well designed and well focused approach. Priorities must be chosen and resources carefully utilized to avoid the early dissipation of resources over too wide a range of tasks. The Office should expressly adopt a set of operational guidelines for determining the activities and cases in which to become involved. Therefore, the Office is first authorized to assess the protection and advocacy needs of developmentally disabled persons and set priorities among those needs. Subsection 7.2 of the act specifies the priorities for selection of cases for legal advocacy. Priorities within the other functions of the Office are to be set based upon available resources.

(2) Publicize the availability of the Office's assistance, create an awareness of the problems, needs, and rights of developmentally disabled persons within the state, and engage in outreach

efforts to discover developmentally disabled persons in need of protection and advocacy.

COMMENT. The Office's success will depend in part upon its ability to publicize the availability of assistance offered and conduct an effective outreach program. This will assist developmentally disabled persons in two ways. First, one of the major obstacles to full expression by developmentally disabled persons is a general public ignorance of their problems, needs, and rights and a paternalistic or hostile attitude toward them. Public education and awareness can assist in alleviating this condition. Second, developmentally disabled persons are frequently not able to acquire the protection or advocacy needed. It is incumbent upon the Office to seek out these individuals, rather than waiting for such persons to come to the Office.

(3) Coordinate the efforts of the Office with other agencies, groups, and persons serving the protection and advocacy needs of developmentally disabled persons.

COMMENT. Many other agencies or groups of persons are likely to be serving the protection and advocacy needs of developmentally disabled persons within the state. The Office should coordinate all advocacy efforts to the maximum extent possible and incorporate them in an overall advocacy system. This will protect against duplication of efforts while ensuring that protection and advocacy are delivered most effectively. This coordination is subject, of course, to the voluntary participation of other advocacy agencies.

(4) Provide an organizational base and support necessary to establish and encourage programs through which persons may volunteer their services as citizen advocates for individual developmentally disabled persons. In addition, the Office shall take such steps as may be needed to coordinate with and cooperate with existing citizen advocacy programs. Such citizen advocacy programs shall, to the extent practical and desirable, be operated in conjunction or in coordination with the legal advocacy performed by the Office.

COMMENT. Citizen advocacy is an important part of the total advocacy system. The Office is in a unique position to organize and support such efforts, and thus, is empowered to cooperate with and coordinate with existing citizen programs and to operate these programs in conjunction with its advocacy functions.

(5) Receive complaints of violations of rights or denial of services and investigate and make recommendations on such complaints in accor-ˡance with section 10 of this act.

(6) Initiate investigations and make recommendations in accordance with section 10 of this act.

(7) Organize and administer programs to provide legal counsel and representation to developmentally disabled persons as individuals and classes to enforce their rights. These legal services may include negotiations, representation in individual and class suits, amicus curiae briefs, and all other legal services.

(8) Provide direct advice, legal counsel and representation to developmentally disabled persons as individuals and classes to enforce their rights. In the course of such representation, the Office may negotiate or initiate litigation in any appropriate court or administrative hearing tribunal, against any person.

COMMENT. The Office has authority to pursue both legal and non-legal protection and advocacy. It combines two approaches in pursuing remedies by containing both investigation and legal advocacy. The Office has authority to receive complaints, investigate and make recommendations as well as to initiate its own investigations through the investigative authority, and to provide legal counsel and representation, including the initiation of litigation in the appropriate court or administrative hearing against any person. Person is broadly defined in this act. It includes any service provider or state agency.

The combination of investigation and legal advocacy in one agency as opposed to separate administrative structures, is designed to provide for more efficient administration by coordinating these efforts. The Office can investigate specific complaints, initiate examination of areas which the Office believes may involve abuses of or denial of rights of developmentally disabled persons and make recommendations in an attempt to resolve disputes without resort to litigation. The investigator can examine disputes from a broader perspective and consider many more factors than may be the case with a legal advocate advocating for the rights of an individual client. Those complaints or disputes which cannot be resolved can then be referred to the advocacy attorneys for potential litigation.

In this approach, litigation serves as a last resort. This procedure allows for a reduction of expenses because attorneys usually need not be used to investigate the factual merits of claims and are only used for negotiation periodically, since investigators will also engage in this activity. It recognizes that many complaints are not appropriate for redress through formal legal advocacy. It should be noted that the act does not establish a new administrative forum before which developmentally disabled persons may assert their rights. The Office does not regulate state agencies or service providers; it asserts the rights of developmentally disabled persons vis-a-vis them.

(9) Comment upon, draft, or advocate for or

against legislation affecting developmentally disabled citizens.

COMMENT. Legislative advocacy is an important type of class advocacy. It is intended that the Office be a focal point for legislative advocacy after input from various consumer groups and interested persons through the mechanism of the Commission. Thus, the Office is given the authority to comment upon, draft and advocate for or against legislation.

(10) Comment upon or recommend to any state agency procedures or regulations for the purpose of safeguarding the rights of developmentally disabled persons. A state agency shall notify the Director within 60 days of the receipt of such recommendations or comments of the action taken thereon and the reason therefor.

COMMENT. Another important type of class advocacy is rule-making. The method of administration of a statute is frequently as important as the terms of the statute itself. Therefore, it is important that the Office participate in rulemaking activity on behalf of developmentally disabled persons to insure that their rights are safeguarded in the administration of the state statutes. Included within this authority is the authority to both recommend regulations and procedures to state agencies, and to require that the Office be informed of the action taken by a state agency and the reason thereof. This provision is not intended to give the Office the authority to enforce its proposed regulations and procedures, but merely to require that the receiving state agency give the suggestions serious consideration and respond to the Office with a reason for its actions.

(11) Prepare manuals or other materials and conduct training seminars for staff members, outside attorneys, professionals, lay advocates, developmentally disabled persons, and others as needed.

COMMENT. Education is a basic function of the advocacy system. These education and training efforts should extend beyond the Office personnel to outside attorneys, other professionals who are serving developmentally disabled persons, lay advocates, the public, and to developmentally disabled persons.

(12) Provide counsel and representation to developmentally disabled persons with legal problems not directly related to a developmental disability or refer them to appropriate legal services.

COMMENT. Many developmentally disabled persons who are in need of legal services for problems which do not arise out of their developmental disabilities do not have access to adequate legal counsel because they are confined to institutions or facilities, are unable to afford such services, or do not know how to access them. The Office is given the authority to provide counsel and representation to these persons or to refer them to other attorneys or to legal service agencies.

Introduction to Section 6
Organization of the Advocacy Office

The act establishes an Office organization consisting of an Advocacy Commission, a Director and Office personnel. Within this general structure two options for the division of authority to carry out the functions of the Office are presented.

The first option places primary responsibility for the functions of the Office with the Director. The Advocacy Commission advises the Director, provides input from various sources and maintains accountability over the Director by its informal, internal evaluation and by contracting for outside evaluation of the Office.

This option recognizes the importance of the Director and the personnel which s/he selects. The ultimate success of the Office will depend largely on the ability and initiative of the Director and Office personnel to organize and carry out an effective advocacy program. Thus, the Office's authority and responsibility are placed in the hands of the Director. In exercising his/her authority the Director will be receiving direct input and guidance from the advisory Commission to insure that the concerns of developmentally disabled persons are heard and included in the advocacy program.

By placing the authority and responsibility with the Director, this approach takes advantage of the day-to-day experience in receiving complaints, identifying problems and pursuing solutions which the Director and Office personnel will develop. This experience, coupled with input from the Commission, should give them the necessary knowledge of the problems and needs of developmentally disabled persons. The Director's knowledge of how state government and service providers operate, developed through experience, will give her/him the needed perception of which solutions are most feasible.

This option avoids what can be a major pitfall for the advocacy program, the inability of a diverse group to agree upon a focus for the program. Developmentally disabled persons, their parents, and their guardians are a diverse group of people, just as any other large group. They have different interests and needs, which can, and not infrequently do, conflict with one another. What is good for one subgroup may truly not be good for another. One may gain only at the expense of another.

These differences may make it difficult for a Commission (composed of representatives of these subgroups) to establish priorities for the Office and decide on an overall advocacy program. Requiring the Director to seek Commission approval of all major activities can greatly hinder an otherwise assertive and highly motivated staff. A strong Director can lead to a strong advocacy program. This approach has been used with success in New Jersey (N.J. Stat. Ann. §52:27E-21 to 27 (1974)) and Ohio (Ohio Rev. Code Ann. §5123.94 (1974)). See also Connecticut Public Act. No. 77-589.

The option may have a disadvantage, however, in that it places even greater reliance on the Director than would be the case with a governing Commission. Where it is believed that the Director is not doing a good job, the Commission has no formal authority over the Director short of seeking removal. A governing Commission would have the authority to set policies and priorities and to redirect the administration of the Office if it believes this is necessary. It is unlikely, however, that the Commission could truly push a Director whom it felt was not assertive enough. It is more likely that a Commission could be more successful in curbing a Director who is considered too aggressive than in stimulating a Director to greater activity.

The second option places primary responsibility for directing the functions of the Office with a governing Commission. It would set policies and priorities for the Office, approve plans, rules, and regulations and conduct evaluations of the Office's performance. The Director would administer the Office and develop plans, draft rules, hire personnel and handle all contracting, budgeting, and reporting.

This approach places great value on the power of developmentally disabled persons, their parents and family members, guardians, and the general public represented on the Commission to decide the direction of the Office. Supporters of this option may argue that Commission members are better able to set over-all policy, since they are likely to have a broader collective background and be more intimately involved with the field of developmental disabilities. Therefore, they should be given governing authority, not just advisory authority. In conjunction with this, the option also reflects a value preference in favor of citizen governing boards rather than individual administrators. This value preference seems to be based upon a mistrust of bureaucrats. This approach has been adopted in different forms in California (Cal. Welf. & Inst. Code §38150 et. seq. (1976)) and Indiana (Ind. Code §16-13-1 Ch. 19 (1977)). This option can have the disadvantage mentioned above of conflicting and competing interests possibly resulting in a lack of direction and inertia. In addition, there is no guarantee that members of the Commission will favor a strong advocacy program. Advocacy has many facets and is not well understood. Some may view advocacy as better case management and oppose active advocacy programs. Others may have a narrow view of advocacy limited only to citizen advocacy or to legal advocacy. A truly effective advocacy program will include a number of different types of advocacy.

Since there is no evaluation data available for the success of either of the two options discussed here, this act does not propose one option over the other. Rather both options are presented within brackets in the text of the model statute below, the first or governing Director, advisory Commission option preceeding the slash, and the second or governing Commission option following the slash, e.g., The Commission shall: [advise on/establish] general policies and priorities. . . .

Section 6. Advocacy Commission

(1) There is established an advocacy Commission (hereinafter referred to as Commission) consisting of members appointed by the Chief Justice.

COMMENT. An Advocacy Commission is established for the Office for two major reasons. First, it provides a better means of insuring input from consumers and other persons interested in the area of developmental disabilities. Second, it provides a means for maintaining the accountability of the Director during his or her term of office.

The Chief Justice was selected as the appointing authority because s/he is least involved in the state's politics. The Governor and the leadership of the General Assembly maintain positions which are more likely to subject the selection process to partisan political considerations. A Chief Justice, on the other hand, has a greater likelihood of making his or her selection on the basis of merit without regard to the political impact of the choices.

Alternative. *Members of the Commission could also be appointed by the Governor or the leadership of the General Assembly or a combination thereof. This approach would give the Governor, the General Assembly, or both, input into the governing of the Office. This, in turn, could create greater interest and support on their part for the Office.*

(2) The Commission shall consist of [fourteen (14)/seven (7)] members. At least [six (6)/three (3)] members of the commission must have cerebral palsy, epilepsy, mental retardation or have been autistic. The remaining members of the Commission shall be cerebral palsied, epileptic, mentally retarded, or formerly autistic, be an immediate family member or a guardian of a person who is autistic, cerebral palsied, epileptic or mentally retarded or have demonstrated an interest in the developmental disability field. No employee or member of the governing board of any agency or organization from which developmentally disabled persons receive treatment, services or habilitation, or member of the general assembly shall be a member of the Commission. The Chief Justice shall strive to ensure representation of minority groups.

COMMENT. A governing Commission would have significant authority with respect to policy making and accountability. Therefore, its size must be kept manageable in order to insure that it has the capability to carry out these functions. At the same time, however, it must be large enough to represent the various groups interested in having input into the advocacy system. Therefore, a seven member Commission was selected as a compromise between these two considerations where the Commission is intended to have governing authority.

If the advisory Commission is selected, the number can be expanded without making it unmanageable. Therefore, it is proposed that the number be doubled to fourteen, if the advisory Commission option is selected, since this will increase the breadth of available input.

The Commission membership is designed to reflect a strong consumer input. These persons would be invaluable in the assessment of needs and the establishment of priorities by the Office because they are from the same population as the Office's clientele. Therefore, a significant portion of the Commission is required to have one or more of the conditions included within the definition of developmental disability. The remaining members may be appointed from among this group, or from among other persons with a demonstrated interest in developmental disabilities. These persons may include parents or other immediate family members, persons who have been active volunteers with developmentally disabled persons, lawyers, doctors, psychologists, therapists and other professionals as long as they are not an employee or member of the governing board of any agency or organization from which developmentally disabled persons receive treatment, services or habilitation. The Chief Justice is prohibited from selecting any state employee or any individual connected with a service provider since this may result in a conflict of interest for the Commission member. In addition, he is required to seek minority representation.

The Chief Justice should seek recommendations for Commission membership from all consumer organizations representing developmentally disabled persons within the state as well as other persons. These recommendations should be made with full disclosure of the recommended individual's prior or current relationships to state agencies, service providers, or consumer organizations so that the Chief Justice will be aware of any possible conflicts of interest. The burden should be upon the recommender and not the Chief Justice to list the individual's qualifications for appointment. These recommendations should be of great value in selecting members of the Commission.

(3) A member of the Commission shall serve a term of three (3) years, except that of the members who are cerebral palsied, epileptic, mentally retarded or who have been autistic [two (2)/one (1)] shall serve an initial term of one (1) year and [two (2)/one (1)] shall serve an initial term of two (2) years, and of the remaining members of the commission [two (2)/one (1)] shall serve an initial term of one (1) year and [two (2)/one (1)] shall serve an initial term of two (2) years. A member of the Commission shall not serve more than two (2) consecutive terms.

COMMENT. The members of the Commission are to serve staggered three year terms. Three years is believed to be long enough to make a significant contribution but not long enough to allow the Commission member's interest and contributions to wane. Three year terms and the prohibition against serving two consecutive terms will guard against the possibility of certain individuals becoming controlling members of the Commission.

(4) Vacancies on the Commission shall be filled in the same manner as original appointments. The individual appointed must be from the same of class of persons from which the original position was filled. Appointments to fill vacancies occuring before the expiration of a term are for the remainder of the unexpired term.

COMMENT. Any vacancies occurring must be filled by individuals coming from the same class as the individual previously appointed. This insures continued consumer representation on the Commission.

(5) The Commission shall annually elect from its membership, by a majority vote of a quorum of the membership, a chairperson and such other officers as it determines to be appropriate.

COMMENT. None necessary.

(6) The Commission shall organize itself and adopt procedures as it deems appropriate. The Commission shall meet at least six (6) times per year. A majority of the Commission shall constitute a quorum. Meetings may also be held upon petition to the chairperson by [six/three] or more members of the Commission.

COMMENT. None necessary.

(7) A member of the Commission is not entitled to compensation, but shall receive reimbursement for actual expenses incurred in the performance of his or her duties.

COMMENT. It is intended that the member of the Commission receive expenses actually incurred in the performance of their duties but not per diem compensation, since advocacy systems will have severely limited funds. A state should give per diem if other commissions or boards performing similar duties within the state receive such compensation.

Section 7. Duties of the Advocacy Commission

The Commission shall:

(1) Appoint a Director of the Office in accordance with section 8 of the act.

COMMENT. The Commission is given authority to appoint the Director of the Office. The composition of the Commission should insure adequate input from developmentally disabled and other interested persons. Selection by the Commission will hopefully be made based upon a consensus of the interests represented on the Commission and foster a cooperative relationship between the Commission and the Director.

Alternative. *The Director could also be appointed by the Chief Justice, the Governor or some other person. (See subsection 8.1 infra.)*

(2) Review and advise on [and approve] general policies and priorities for the operation of the Office on an annual basis.

COMMENT. See subsection 9.2.

(3) Review and advise on [and approve] plans for protection and advocacy developed by the Director on an annual basis.

COMMENT. The Commission is given the authority to provide input on all plans developed annually by the Director for protection and advocacy. A governing Commission would also be given approval power. These annual reviews are intended to keep the Commission and Director aware of the goals and purposes of the Office and to periodically evaluate and revise their programs in light of changed circumstances. There is always a danger of a public agency becoming preoccupied with a bureaucratic routine and losing sight of the purposes for which it was created. This procedure may help guard against such an occurrence.

(4) Review and advise on [and approve] rules and regulations drafted by the Director.

COMMENT. The Commission must advise on all rules and regulations drafted by the Director. These are not necessarily reviewed or altered on an annual basis, but must be coordinated with policies and priorities and the Director's protection and advocacy plans which are developed annually. A governing Commission would also be given approval power.

(5) Review and advise the Director on the operations of the Office. However, the Commission shall not have supervisory powers over the selection or conduct of particular cases or personnel of the Office.

COMMENT. As a part of its accountability function, the Commission is directed to review and advise the Director, on all operations of the Office. It is not intended, however, that the Commission have any supervisory power over particular cases being handled by the Director or Office personnel. The Director is to be free to make judgments on individual cases within established policies and priorities. This procedure insures that a governing Commission will not become overwhelmed with the day to day operations of the Office and lose sight of their real responsibilities. It also recognizes the importance of the individual attorney-client relationship. (See text of Alternative under subsection 8.2.)

(6) Review the annual report on the operations of the Office prepared by the Director.

COMMENT. None necessary.

(7) Establish advisory subcommittees, the membership of which may be supplemented by other interested persons. These persons may include employees of the state or a subdivision thereof or employees or members of the governing board of an agency or organization from which developmentally disabled persons receive services where the Commission deems it appropriate to have input from such persons.

COMMENT. The membership of the Commission cannot include any service providers or state employees. These individuals, however, may at times be helpful in advising the Commission on matters within the Commission's authority. For example, the Commission may wish to have the input of service providers before making general recommendations to all service providers within the state with respect to procedures or revised regulations. Faculty of the state's colleges or universities may also provide helpful information and manpower which could be utilized through the mechanism of a subcommittee.

(8) Provide for an annual independent evaluation by an outside person. This report shall be transmitted in entirety to the general assembly, the governor, the chief justice, and be available to each of the consumer organizations within the state that represent developmentally disabled persons.

COMMENT. The interim guidelines governing § 113 state that the system should have a mechanism for both internal and external evaluation. This subsection provides independent outside evaluation of the Office on an annual basis. The report is required to be transmitted in entirety to the General Assembly, the Governor, and the Chief Justice, and be made available to each of the consumer organizations representing developmentally disabled persons within the state. These organizations are the most directly interested in the functioning of the Office and should help promote a discussion of the Office's operation among developmentally disabled persons throughout the state. Note that this evaluative responsibility is given to both governing and advisory commissions in order to insure that an independent evaluation of the Office is conducted.

Section 8. Director

(1) The Commission shall appoint a Director of the Office and set his/her salary. The Commission shall not reduce the Director's salary during his or her term of office.

COMMENT. See subsection 7.1. The Commission should notify consumer organizations so they have an opportunity to suggest potential applicants. The burden is upon the applicant to present his/her qualifications and to reveal any relationship which may create conflicts of interest. Since attorneys are trained to represent all clients, the fact an attorney previously represented a service provider, state agency, professional, or consumer organization does not render him or her unqualified to serve. This information may help the Commission identify cases, however, where it will want to probe into the attorney's attitudes toward developmentally disabled persons, their rights, service providers and other relevant issues.

Alternative. *The Director might also be appointed by a person outside the Office structure such as the Chief Justice, the Governor, the Speaker of the House, or President Protem of the State Senate. The members of a governing Commission may tend to select a Director who will "rubber stamp" their decisions and show no personal initiative. A Director who was independently appointed could balance the governing Commission members' ideas about the operation of the Office. In addition, appointment by the Governor or leadership of the General Assembly could give them some input into and commitment to the Office.*

(2) The Director shall be an attorney at law licensed in this state, or eligible to be licensed, and shall be qualified by experience to perform the duties of the position. An attorney who is not licensed in the state shall become licensed within one (1) year of appointment as Director. The Director shall devote full time to his or her duties and shall not engage in the part-time practice of law or in any other employment.

COMMENT. Since the Office includes a large legal component and the legal process is the ultimate remedy available under the Office's authority, it is necessary that the Director understand and evaluate the Office's activities from a legal standpoint. An attorney can better recognize the legal issues involved and thereby integrate the legal and non-legal protection and advocacy provided by the Office. This is particularly significant since the purpose of the Office is to advocate for legal and human rights of developmentally disabled persons.

The Director is required to be licensed to practice law in the state or to be eligible to be licensed. This would not prohibit an attorney who is not currently licensed within the state from accepting the job of Director of the Office as long as a state license was obtained within the year. It is not wise to limit the availability of potential candidates for the directorship of those currently licensed within the state since the number of attorneys with education or experience in the area is limited.

Because of the large responsibility placed upon the Director, and the limited resources that always hamper the hiring of sufficient supportive staff, it is critical that the Director devote his/her full energies to his/her duties. Thus s/he is precluded from engaging in the part-time practice of law or any other employment. This section should not limit the Director, however, from teaching a course on developmental disabilities law at the local law school since education and training is part of his or her job responsibilities.

Alternative. *The Director may not be required to be an attorney since some may feel an attorney would overemphasize the legal advocacy of the Office. Removing the restriction would also expand the number of available candidates for the position. This may be appropriate where the Office is large enough to hire a sufficient full-time attorney to handle all legal matters. The Director could spend full-time on administration and coordination.*

However, it should be noted that a non-attorney Director could not exercise supervisory authority over attorneys employed directly by the Office or through contract or written agreement while in the exercise of their legal activity. Attorneys are required by the American Bar Association's Code of Professional Responsibility, Cannon 5, to exercise individual professional judgment. Ethical Consideration 5-24 says

> *To assist a lawyer in preserving his professional independence, a number of courses are available to him. For example, a lawyer should not practice with or in the form of a professional legal corporation, even though the corporate form is permitted by law, if any director, officer, or stockholder of it is a non-lawyer. Although a lawyer may be employed by a business corporation with non-lawyers serving as directors or officers, and they necessarily have the right to make decisions of business policy, a lawyer must decline to accept direction of his professional judgment from any layman. Various types of legal aid offices are administered by boards of directors composed of lawyers and laymen. A lawyer should not accept employment from such an organization unless the board sets only broad policies and there is no interference in the relationship of the lawyer and the individual client he serves. Where a lawyer is employed by an organization, a written agreement that defines the relationship between him and the organization and provides for his independence is desirable since it may serve to prevent misunderstanding as to their respective roles. Although other innovations in the means of supplying legal counsel may develop, the responsibility of the lawyer to maintain his professional independence remains constant, and the legal profession must insure that changing circumstances do*

not result in loss of the professional independence of the lawyer.

(4) The Director shall be appointed for a term of five (5) years, subject to removal by the Chief Justice or his judicial designee only for good cause shown after notice and hearing.

COMMENT. The Director is appointed for a specified term of years to help to insulate him or her from disagreements over the conduct of individual cases and from any undue pressures which may otherwise threaten his or her removal. It should also give the Director greater freedom to administer the Office as s/he sees fit consistent with established policies and priorities. The Chief Justice, a neutral person, or his/her judicial designee is authorized to remove the Director during his or her term only for incompetent performance. In that event, the Director is to be given a hearing before the Chief Justice or a judicial designee with full due process rights attached. No appeal from the judicial decision is contemplated unless some mechanism for appeal is provided by some other state statute.

Alternative. *The Director might serve at the pleasure of the appointing authority rather than serve a specified term subject to removal for cause. This would give greater flexibility to remove a Director whose performance was not satisfactory.*

Section 9. Duties of the Director

The Director shall:

(1) Perform the duties and exercise the powers of the Office.

COMMENT. The Director is responsible for administering all duties and powers of the Office specified in section 5 of the act.

(2) [Adopt/Draft] general policies and priorities for the operation of the Office and review these annually.

In setting priorities with respect to legal representation of clients, the Director and Commission shall give priority in the following order to:

(a) cases where the legal problem is clearly attributable to a developmental disability which is severe and where the action may serve as a precedent or have a system-wide effect;

(b) cases where the legal problem is clearly attributable to a developmental disability which is severe;

(c) cases where the legal problem is clearly attributable to a developmental disability and where the action may serve as a precedent or have a system wide effect;

(d) cases where the legal problem is clearly attributable to a developmental disability.

COMMENT. Every public agency must establish its policy and set priorities for the operation of the agency. This is particularly important to the Office due to its limited resources and the possibility of spreading these resources too thinly. Although the Office is given a wide variety of duties to carry out, it is not intended to perform all of these duties at any one time. Rather, the Office will specify which of its duties will be carried out during each fiscal year within the limited resources available to it. The priorities among the duties may change from year to year based upon the needs of developmentally disabled persons within the state and an assessment of which of these needs should be met first.

The priorities for selection of cases to be handled by the legal advocacy section of the Office are specified within the priorities of the Office. First priority is given to cases clearly attributable to a developmental disability which is severe and where the case may set a precedent or have a system-wide effect. Severity may be determined by examining the effect which the disability has on such things as self-care, mobility, learning ability, and capacity for independent living and economic self-sufficiency. When resources are limited the effect that the individual case will have upon the class of developmentally disabled persons is also an important consideration. This priority, however, should not be interpreted to require class actions. The intent is to choose individual cases that could affect the class. Although, by definition, class actions affect the class, they are often too expensive and time consuming to be cost-effective for the Office running on a limited budget. Second priority is given to cases clearly attributable to a developmental disability which is severe. It is intended that limited resources of the Office be concentrated on those whose disabilities and resulting problems are most severe. Third priority is given to cases clearly attributable to a developmental disability when the case may have value as precedence. Fourth priority is given to cases clearly attributable to a developmental disability. Each priority recognizes the Office's primary responsibility to serve individuals whose problems arise because of their disability in advance of giving any legal assistance to a person whose problems are not disability related.

(3) [Adopt/Draft] plans for protection and advocacy on an annual basis. In developing a plan for protection and advocacy throughout the state, the Director shall consider the availability of existing protection and advocacy resources and the extent to which these could be adapted to support and supplement any staff of the Office. The system proposed by the Director shall encompass all parts of the state and may be designed in part to meet the special needs of particular areas or groups within the state.

COMMENT. The Director is required to assess, annually, the availability of existing protection and advocacy resources and prepare a plan for protection and advocacy during the coming year.

(4) Establish such offices as are required to efficiently and effectively carry out the duties and powers of the Office.

COMMENT. The Director is given the authority to establish a centralized or regionalized system to meet priorities and to serve special groups or areas within the jurisdiction of the Office. This subsection recognizes that the Director can determine the organizational structure which will best meet the geographic and demographic needs of the state. He may also determine if separate offices are needed to serve those developmentally disabled persons residing within an institution. The best decision will vary between states and thus cannot be made a priori by this act.

It is intended, however, that regional offices be under the control of the Director and not function as separate entities. Employees working out of regional offices would be delegated duties and act as the Director's agents in carrying out the purposes of this act pursuant to subsection 7 of this section. This will insure that the policies and priorities are implemented uniformly throughout the state.

Alternative. *A state may wish to mandate by statute a system of regional offices and specify a greater degree of independence from the central office. Such an organizational approach might better fit the state's size, demography, or existing structure, or better reflect local needs.*

(5) [Adopt/Draft] rules and regulations as appropriate.

COMMENT. The Director is responsible for drafting all rules and regulations for the conduct of the Office consistent with established priorities and the plans for protection and advocacy.

(6) Employ attorneys and other professional and non-professional staff as may be necessary to carry out the functions of the Office, without regard to the state's merit system. Staff may be employed on a part-time basis with the limitation that attorneys be employed for no less than half-time.

COMMENT. The Director is given wide discretion in hiring attorneys and other professional and non-professional staff to carry out the functions of the Office. This hiring is exempted from the state's merit system for two reasons. First, the professional positions will be unique and thus require the creation of new positions within the merit system prior to hiring personnel. This process will create substantial delay in hiring the original staff. Second, this exemption insures that the Director will not be limited in his selection of Office personnel by the merit system's pre-classification of available candidates. Thus, he will be able to hire those persons with the special training, experience, and attitudes desirable for the Office who may not have fit the merit system's more general position description.

It is not envisioned that the Director will hire attorneys for investigation. These personnel are intended to be nonlegal since they can rely upon the advocacy attorneys for legal advice. All attorneys are required to be employed by the Office at least half time, since attorneys employed less than that are not likely to devote sufficient time to be effective. Individual attorneys or other persons needed for particular cases or projects can be hired through contracting pursuant to subsections 8 through 11 of this section.

(7) Designate Office personnel to act as agents to carry out the functions of the Office.

COMMENT. The Director is given the authority to delegate his/her duties to such Office personnel as s/he deems appropriate to carry out the functions of the Office.

(8) Contract or enter into other agreements with other persons who provide nonlegal advocacy to developmentally disabled persons, when the Director determines that such agreements serve the purposes of enhancing the coordination of legal and nonlegal protection and advocacy, providing more economical or efficacious training to Office personnel, or otherwise improving the ability of the Office and Office personnel to provide protection and advocacy.

COMMENT. The Director is given broad authority to contract or enter into agreements as appropriate to provide nonlegal protection and advocacy where they are currently available or may be developed through joint efforts. This nonlegal protection and advocacy includes citizen or lay advocacy and consumer coalition class advocacy. They can supplement the Office's direct assistance, often at minimal cost. For example, lay advocates can work with Office clients on a one-to-one volunteer basis while consumer coalitions can exert pressure as interest groups in conjunction with the Office's legislative and rule making efforts. The Office should always be encouraged to utilize such existing resources to supplement the resources of the Office.

(9) Contract or enter into other agreements to use the legal services of local legal-aid societies or public defender offices or comparable organizations providing legal services, and devise a system of individual case-by-case appointment of private attorneys approved by the Office and assigned to individual clients, as appropriate. In utilizing these alternative approaches the Director shall provide a method for screening such attorneys to ensure that they are qualified to serve clients of the Office, shall provide legal resources and training to such attorneys as appropriate to carry out the purposes of this Act, and shall provide a mechanism for periodic evalua-

tion of the performance of such attorneys. The Director also shall arrange for such professional and non-professional assistance to attorneys as deemed necessary to perform their duties under this Act.

COMMENT. Many parts of the state may not have sufficient work loads to utilize full- or half-time attorneys. Thus, other mechanisms must be provided for serving clients in these areas. The Director is given authority to contract with or enter into agreements with other attorneys or legal offices where this is the most practical means of providing legal advocacy to developmentally disabled persons. The Director is required to screen such attorneys, to provide them with legal resources and professional and non-professional assistance, and to periodically evaluate the performance of these attorneys in order to insure that clients being served by non-office personnel are receiving the same quality of legal advocacy as would be available through the provision of full-time attorneys.

(10) Contract or enter into agreements with law schools and universities in the state for the purpose of establishing and operating a program whereby the students are assigned to the Office, where appropriate. Such agreements may, to the extent feasible and in the best interests of the Office, provide for advocacy attorneys to teach courses or parts of courses related to developmental disabilities and the law or provide for clinical education courses in law school to take cases attributable to a person's developmental disability or provide other legal services to developmentally disabled persons.

COMMENT. Law schools and universities within the state may be a helpful source of additional manpower in carrying out the functions of the Office. Advocacy attorneys employed by the Office may also assist in the training of future lawyers within the state in the area of developmental disabilities (See the comment to subsection 8(2)). The Director is authorized to contract with or enter into agreements with the law schools and universities toward this end.

(11) Contract or enter into other agreements as they are needed to carry out the objectives of the Office.

COMMENT. None necessary.

(12) Institute legal proceedings as may be necessary to enforce and give effect to the other duties or powers of the Office.

COMMENT. The Director is given specific authorization to institute legal proceedings as s/he deems necessary to enforce and give effect to the powers and duties of the Office.

(13) Take any other action as may be appropriate and reasonable to carry out the purposes of this act.

COMMENT. None necessary.

(14) Insure that all employees of the Office and any other persons providing protection or advocacy through the Office are informed of their responsibilities to protect the confidential information of clients of the Office in accordance with section 11 of this act.

COMMENT. None necessary.

(15) Prepare an annual budget for the Office for submission to the General Assembly.

COMMENT. None necessary.

(16) Prepare an annual report on the operation of the Office. This report shall be submitted to the Commission, General Assembly, Governor, and Chief Justice.

COMMENT. None necessary.

(17) Solicit and accept contributions of money, property, or services of whatever nature made by gift, devise, bequest, grant or other means made to assist the Office in carrying out the provisions of this act, if the Director determines acceptance of any such contribution, and the performance of any conditions of a contribution are consistent with the provisions of this act and are in the best interest of the Office.

COMMENT. It is recognized that nearly every state will have a difficult time implementing the authority of the advocacy agency with the level of funding available to it. Therefore, the Director is given specific authority to solicit and accept contributions to supplement the funding of the Office as long as the acceptance of such contributions is not inconsistent with the best interests of the Office. The Director has a right to reject any contribution which may comprise the authority of the Office, be inconsistent with this act, or be donated by an individual or group which may cast a shadow over the Office.

Section 10. Investigations

(1) When the Office receives a complaint that the rights of a developmentally disabled person have been, may have been, or may be violated, an investigation may be conducted unless the complaint is beyond the scope of the Office's authority or not within the Office's priorities established pursuant to subsection 7.2. The complainant shall be informed whether an investigation will be conducted, and if not, the reason therefor and whether any other appropriate mechanisms for remedy exist. A complaint may be referred to another person as appropriate.

COMMENT. The Office is authorized to investigate any complaint unless it is considered to be trivial, beyond the Office's authority, or not within the Office's priorities. The complainant must be informed whether an investigation will be conducted and if not, the reason for the decision. The complainant is required to be advised of other remedies which may exist so that s/he will be aware of the options available and be able to make a choice on how s/he wishes to proceed. Where other agencies or persons may be of assistance, the Office is authorized to refer complaints to them.

(2) The Office may conduct investigations upon its own initiative if there is reason to believe that the rights of a developmentally disabled person have been, may have been, or may be violated.

COMMENT. The power to initiate investigations on its own assures that problems can be investigated without a complaint being filed by a developmentally disabled or other person. It is recognized that such outside action is not always forthcoming. When investigations are initiated by the Office, it is important that the privacy and confidentiality of developmentally disabled persons are protected. (See section 11 of the act.)

(3) In the course of an investigation, for the purpose of investigating actual or possible neglect, abuse, exploitation, or violation of rights of developmentally disabled persons, Office personnel may enter at any reasonable time, in and upon the premises of any state agency, commission, board or office, or any subdivision of the state, speak privately with any individual therein and inspect and copy any information, materials, or records relevant to an investigation or case within reasonable limits and in a reasonable manner, subject to section 11 of this act and notwithstanding section(s)____ of the code.

COMMENT. Adequate fact finding is essential to effective protection and advocacy. This requires a broad right to access. Subject to section 11 provisions or confidentiality, this section gives Office personnel the right to enter the premises of any public agency at reasonable times to talk to individuals therein about an investigation and to inspect and copy any information, materials, and records held by the public agency so long as they are relevant to the investigation or case.

The access granted to Office personnel is likely to provide an exception to the state's public records law. This law should be checked for consistency and specifically referenced for purposes of clarity.

Note that this broad right of access is limited to public agencies only. The primary intent of this section is to allow Office personnel, during self-initiated investigations, access to developmentally disabled residents residing in state and county institutions, and to relevant records of such institutions. Residents of these institutions are likely to be most severely disabled and least able to articulate their needs, lodge a complaint against abuse or denial of rights, or contact outside assistance. Thus, access to these residents and institutional records is necessary if the Office is to effectively initiate investigations and advocate for this group of developmentally disabled persons.

This need for access also applies to private institutions. However, granting such broad access to private agencies raises constitutional questions of violation of their right to privacy. Therefore, the applicability of this section is restricted to public agencies. However, States should consider putting an access clause in all contracts with private service providers.

(4) Any Office personnel conducting an investigation or handling a case shall have ready access to view and copy all records pertaining to a client held by any person, subject to section 11. Such personnel shall have the opportunity to consult with clients whenever necessary for the performance of their duties. Persons shall not refuse access to client records or opportunity for consultation as long as it is requested at a reasonable time and in a reasonable manner. A hospital, institution, or mental health or developmental disability facility shall provide adequate space and privacy for the purpose of consultation with clients and examination of client records.

COMMENT. This section applies to all agencies, public and private. Office personnel are given access to all clients and to all records of clients when acting on their behalf. Thus, these agencies cannot refuse Office personnel, who are acting on behalf of a client, access to view and copy all client records if requested at a reasonable time and in a reasonable manner. This access is subject to the provisions of section 11 on confidentiality.

Office personnel must, of course, have the opportunity to consult with clients. Thus, these persons have an affirmative duty to allow this opportunity. Institutions or facilities are required to provide adequate space and privacy since these may not otherwise be available. The communications between clients and advocacy attorneys are privileged and should not be required to take place in the presence of other persons.

This access is considerably narrower than that granted within public agencies. The access is granted only where Office personnel are acting on behalf of a specific client with respect to a specific investigation, case or complaint. Access is limited to clients and their records only. Office personnel may be required to show proof they are acting on behalf of a client prior to access. If shown, access must be granted.

(5) During the course of an investigation, the client shall be periodically informed of the status of the investigation, subject to section 12 of this act.

COMMENT. Since investigation may occur over a long period of time, clients may be frustrated by a lack of information as to the status of their particular case before an agency. This section puts a burden on the Office to periodically inform any client of the status of the investigation currently being conducted. These persons need to know they are being served.

(6) Within fifteen (15) days of completion of an investigation, Office personnel shall inform the client of the findings and advise him or her of recommended courses of action. The client shall approve any further course of action subject to section 12 of this act. Nothing in this section shall be interpreted to limit the powers of the Director and the Office as enumerated in section 5 and subsection 10.2 of this act.

COMMENT. A prompt transmittal of the results of the investigation to the client is required. The Office also is required to advise the client of any recommended courses of action and consult with him/her on which course s/he wishes to pursue. The client has final approval of any further course of action, subject to section 12 of this act. This provision does not affect or limit the powers of the Office to conduct investigations and pursue remedies in its own name as provided for in this section.

(7) If the Office finds that

1) a matter should be considered further;

2) a statute or regulation should be altered;

3) a person should give reasons for their actions;

4) an action of a person should be ceased or modified; or

5) any other action should be taken;

The Director shall report any recommendations to the persons investigated. Such persons investigated shall notify the Director within 30 days of the receipt of such recommendation, of the action taken thereon and the reason therefor.

COMMENT. The Director is required to report any recommendations resulting from an investigation to those state agencies, service providers, or other persons who were investigated. They in turn are required to report to the Director any actions taken and the reasons therefor within thirty days. This procedure should promote an informal resolution of the conflict, since it will force the persons to seriously consider the recommendations and their subsequent actions.

(8) The Director may make findings and recommendations public subject to the provisions of section 11 of this act. S/he shall include in any such public statement the substance of any reply made by the person investigated.

COMMENT. Public disclosure may provide one of the remedies which the Director may wish to take with respect to any findings following an investigation. The Director is therefore given specific authority to make public his or her findings and recommendations subject to the provisions of section 11 of this act. In making such public statements, s/he is required to include the substance of any reply made by the investigated person so that the public disclosure will be fair and complete.

(9) Notwithstanding any of the procedures noted in this section, if the Director determines that other action is required, s/he may pursue such action, subject to the approval of the client or pursuant to section 12, including but not limited to referring a matter to the advocacy attorneys, any federal, state, or local agency, or other persons, as appropriate. No approval is required where the Director and Office initiate an action pursuant to an investigation authorized within section 5 and subsection 10.2 of this act.

COMMENT. In addition to making recommendations to the state agency, service providers, or other persons investigated, or making the findings and recommendations public, the Director may decide that further action is required. Therefore, s/he is given specific authority to pursue the matter through the advocacy attorneys or through some other means which may be available. In addition, this subsection specifies that the Director may pursue these actions immediately and bypass investigative procedures if s/he determines this to be the most appropriate approach to the problem. Action taken pursuant to this section must have approval of the client or pursuant to section 12 unless the Office is initiating an investigation/action in its own name.

Section 11. Confidentiality of Information, Materials, and Records

(1) Office personnel inspecting information, materials, and records pursuant to an investigation, may copy all such information, materials, and records when personally identifying data can be removed for the purpose of such copying. Where personally identifying data cannot be removed, then the Office personnel may copy such information, materials, and records only after consent is received pursuant to subsection 4 of this section.

COMMENT. The purpose of this subsection is to allow Office personnel to copy viewed information, materials, and records so that the investigation can continue back at the Office or some other suitable work area. The ability to view and transport copies of these items will facilitate the work of the Office and is necessary to build an appropriate case file prior to negotiation or litigation.

It is recognized, however, that the broad investigative authority necessary for carrying out the purpose of this act must be tempered by the right to confidentiality and privacy for developmentally disabled persons. This section attempts to balance these two rights. It gives the Office personnel the right to inspect all information, materials or records relevant to an investigation or case. These persons are not, however, allowed to copy such materials unless personally identifying data can be removed prior to copying. When this cannot be done, Office personnel must receive knowing consent pursuant to subsection 4 of this section prior to copying.

(2) Office personnel may utilize viewed or copied information, materials and records which do not contain personally identifiable data in the course of negotiations, administrative hearings, suits, or other legal or non-legal actions.

COMMENT. This subsection allows Office personnel within the scope of their authority to make public through legal action, the media, or otherwise all information, materials and records which do not contain personally identifiable data. There is no need to balance the right to confidentiality and privacy where individuals cannot be identified.

(3) Office personnel may utilize viewed or copied information, materials, and records which contain personally identifiable data, in the course of negotiations, administrative hearings, suits, or other legal actions if consent is obtained pursuant to subsection 4 of this section.

COMMENT. This subsection allows Office personnel to utilize personally identifiable information, materials and records as evidence in legal actions where they obtain consent. The information, materials, or records may be relevant to legal actions not brought on behalf of the identifiable persons and need to be used in such cases. The need is balanced against the person's right to confidentiality and privacy and thus consent is required. (See also subsection 11(5).)

(4) Consent to copy or utilize information, materials, and records containing personally identifiable data must be given knowingly, voluntarily, and in writing. For the purposes of subsection 1 and 3 of this section the following persons may give written consent:

(a) a developmentally disabled person who is the subject of the information, materials, or records and who is 18 years or older unless the person lacks the capacity to understand the purpose, significance and result of giving his or her consent;

(b) the parent or guardian of a developmentally disabled person who is the subject of the information, materials or records and who is under 18 years old;

(c) the guardian of a developmentally disabled person who is the subject of the information, materials or records and who is 18 years old or older, provided the guardian has such powers;

(d) the guardian ad litem of a developmentally disabled person who is the subject of the information whether appointed in conjunction with section 12 or not, provided the guardian has such powers.

COMMENT. This subsection recognizes that only the developmentally disabled person may consent to release of information, materials, or records identifying him or her, if s/he has reached the age of majority and has sufficient understanding of the act of consent. The parent has the power to consent for a minor while a guardian may have the power to consent for those adults who have been adjudicated incompetent. Where there has been no guardian appointed for an adult developmentally disabled person who lacks actual capacity to consent or where consent cannot be obtained from a parent or guardian, it may be given by a guardian ad litem having the power to give such consent whether the appointment has been sought by the Office in accordance with section 12 or not. Where consent is necessary, it must be in writing to insure that a record is made.

(5) Office personnel may not disclose any personally identifiable information, materials, and records, unless authorized by the subject of such information, materials, and records provided s/he is legally competent to give consent.

COMMENT. This subsection recognizes that the Office personnel usually have no legitimate need to make public personally identifiable information, materials, and records outside the context of legal action. The developmentally disabled person is recognized as the only person who can give consent to release of his or her identifiable records in other forums such as the media. Where s/he refuses or is incompetent, no such disclosure can be made by the Office. This does not prohibit a parent or guardian from personally making such a disclosure directly.

Section 12. Duties of Advocacy Office Personnel

(1) Office personnel, including advocacy attorneys employed by the Office or working under contract or through an agreement with the Office or representing a client obtained by referral through the Office, shall be active advocates for the rights of the client.

(2) When informing any client of the status of an investigation or of the findings of an investigation, or when advising any client on recommended courses of action, Office personnel shall employ all reasonable methods calculated to maximize the client's understanding of the investigation, its results, and the options available for resolution. Where the client is a minor or lacks the legal capacity to make an informed judgment in the matter, the parent, guardian or other responsible person also shall be so informed.

(3) If a client is legally unable to make an informed judgment in the matter, Office personnel shall nevertheless involve the client in determining a course of action to the maximum extent of

the client's capability of understanding the matter in question. Office personnel shall also consult with the parent or guardian of a client to determine his or her recommended course of action.

Where the courses of action selected by the client and the parent or guardian of a client who is a minor or who is legally unable to make an informed judgment in the matter are in conflict and the conflict cannot be resolved so as to protect the rights of the client, an advocacy attorney may seek appointment of a guardian ad litem in accordance with subsection 5 of this section to decide what course of action shall be taken on behalf of the client.

(4) All legal actions brought by an advocacy attorney must be brought in the name of a client or by a parent, guardian, guardian ad litem or next friend on behalf of those clients who are minors or who are legally unable to make an informed judgment in the matter.

(5) For purposes of this act, the attorney-client relationship may be established between an advocacy attorney and a developmentally disabled person without regard to the person's competence or the existence of a guardianship or conservatorship, for the purpose of bringing an action for appointment of a guardian ad litem to act on behalf of the developmentally disabled person:

(a) in instances in which a client lacks actual capacity to consent pursuant to subsection 11(4) of this act and consent cannot be obtained from a parent or guardian;

(b) in a proceeding to establish a guardianship or conservatorship on behalf of a developmentally disabled person;

(c) in a proceeding to determine the legal competency of a developmentally disabled person to perform an act;

(d) in a proceeding to commit a developmentally disabled person;

(e) in a proceeding to renew a commitment, guardianship or conservatorship;

(f) in an action brought against a guardian or conservator for his or her dismissal or a reduction of the authority of the guardian or conservator;

(g) in an action for termination of commitment;

(h) in any other proceeding or action involving any other legally imposed restraint on personal freedom or affecting the rights of a developmentally disabled person.

An advocacy attorney may represent a developmentally disabled person through the guardian ad litem, whether as plaintiff or defendant in subsequent proceedings.

COMMENT. Office personnel should be made specifically aware that their duty is to represent the client of the Office. At times in the past, legal procedures have been utilized by parents, guardians, or other individuals to serve their needs with respect to developmentally disabled persons. Office personnel are specifically directed to refrain from this practice and to advocate for the rights of the developmentally disabled person.

Thus, it follows that Office personnel are required to communicate to the client the status and findings of an investigation and of available courses of action so that the client can indicate a choice from among them. When the client's capability for understanding and making an informed judgment is limited, Office personnel shall make every reasonable effort to explain matters to the client in a manner in which the client can best understand and indicate a preference in the matter.

Clients of age who do not lack legal capacity are represented directly by the Office. Where the client is a minor or otherwise lacks the capacity to make an informed judgment in the matter, Office personnel are required to consult with the parent or guardian prior to taking a course of action in a particular case. Where the selected courses of action of the client and the parent or guardian are in conflict, Office personnel shall attempt to resolve the conflict without jeopardizing the rights of the client.

All legal actions must be brought in the name of a competent client or by a parent or guardian or other representative for those clients who lack legal capacity because of minority or lack of understanding. Where there is no legal representative available to act or who will act on behalf of an incompetent client, an advocacy attorney must seek appointment of a guardian ad litem prior to undertaking legal action on behalf of a client.

Advocacy attorneys are given specific authority to seek appointment of a guardian ad litem in certain cases to protect the interests of a developmentally disabled person regardless of the person's competency or the existence of a guardian or conservatorship. The advocacy attorney then may represent the client through the guardian ad litem as a plaintiff or defendant. For example, these cases could include the advocacy attorney defending against a guardianship sought by a parent, or advocating for the appointment of a guardian with limited authority and no conflict of interest.

This procedure is intended to cover any situation where a client lacks capacity and has no one to act or who will act on his or her behalf or where the interests of the client are at odds with the parent or guardian exercising control over the client. In the latter case the parent or guardian has an initial opportunity to resist the appointment of a guardian ad litem. If the guardian ad litem is appointed s/he will then have the limited power to act on behalf of the client in the specific matter in question within the power granted in the order of appointment.

Section 13. Private Representation of Clients

Nothing in this act shall be construed to prohibit a developmentally disabled person from being represented by privately retained counsel. If an advocacy attorney has been assigned and the client secures his own counsel, the Office shall discharge the advocacy attorney.

COMMENT. This section makes it clear that anyone who wishes to retain his/her own counsel shall not be required to utilize the assistance of the Office. This assistance is intended to supplement that currently available to developmentally disabled persons.

Section 14. Other Remedies Available

The remedies provided by this act shall be in addition to any other remedies which may be available to any person, and the failure to pursue or exhaust the remedies or engage in the procedures provided by this act shall not preclude the invocation of any other remedy.

COMMENT. The procedures contained within the act are intended to supplement currently existing remedies. It is not intended that any developmentally disabled person be required to pursue any of these procedures, however, prior to or in lieu of any other remedy or procedure which may be available to that person.

Section 15. Sanctions Prohibited

Any person participating in filing a complaint or providing information pursuant to this act or participating in a judicial proceeding resulting therefrom shall be presumed to be acting in good faith and unless the presumption is rebutted shall be immune from any liability, civil or criminal, and shall be immune from any penalty, sanction, or restriction that otherwise might be incurred or imposed by the person investigated.

COMMENT. Persons are to be encouraged to report abuses of or violations of rights of developmentally disabled persons without fear of recrimination. This section is intended to insure that persons reporting such abuses or participating in the resolution of complaints suffer no adverse consequences as a result of their actions.

Section 16. Penalties

(1) No person shall obstruct any Office personnel in the performance of duties as described in this act, including but not limited to access to Office clients or potential clients, or to their records, whether financial, medical, or otherwise, or to other information, materials, or records, or otherwise violate the provisions of this act. The Director may institute legal proceedings necessary to enforce the penalties provision of this section.

COMMENT. No right is of value unless there is a remedy available when it has been violated. This subsection, when combined with the penalty provision in subsection (3), creates such a remedy for persons whose right to access granted in subsections 10(3) and 10(4), is being interfered with by service providers, state agencies, or other persons. It also creates a remedy against state agencies, service providers and others who refuse to comply with other requirements of the act such as responding to the Director within 30 days of receipt of recommendations as required by subsection 10(9) of this act. The Director, rather than the developmentally disabled person, is given the authority to institute legal proceedings because 1) there may not be a readily identifiable person affected by the violation such as in an Office initiated investigation; and 2) the Director and his staff are in a better position to know when a violation has occurred and what effect it may have on the advocacy effort involved.

(2) No Office personnel shall knowingly violate any provision of section 11 of this act. The Attorney General may at the request of an Office client, or any other aggrieved individual, institute legal proceedings necessary to enforce the penalties provision of this section.

COMMENT. This subsection, when combined with the penalty provision in subsection (3), provides a remedy for developmentally disabled persons whose rights to confidentiality and privacy are violated by Office personnel through their overzealous efforts on behalf of clients. Naturally the Director cannot institute legal proceedings to enforce this subsection because of the obvious conflict of interest. Therefore, the attorney general is given the authority upon the request of the Office client or some other person who has been aggrieved. This procedure will insure that the client or other aggrieved person will not be discouraged from pursuing a remedy because of a need to retain private counsel.

(3) Any person found in violation of subsections (1) and (2) of this section shall pay a civil penalty of not less than one hundred dollars ($100.00) nor more than one thousand dollars ($1,000.00) for each offense.

COMMENT. It is intended that any person convicted of a violation provided for in subsection (1) or (2) of this section shall pay a minimum fine for the offense and that greater fines be imposed for subsequent offenses or blatant offenses.

Section 17. Appropriations

The general assembly is authorized to appropriate _____ dollars for the immediate fiscal year to cover the initial expenses incurred in implementing this act.

COMMENT. None necessary.

Section 18. Severability of Sections

If any section, subsection, paragraph, sentence, or any other part of this act is adjudged unconstitutional or invalid, such judgment shall not affect, impair or invalidate the remainder of this act, but shall be confined to the section, subsection, paragraph, sentence or any other part of this act directly involved in the controversy in which said judgment shall have been rendered.

COMMENT. None necessary.

Section 19. Emergency Clause

Because an emergency exists, this act takes effect on __________.

COMMENT. None necessary.

OPTIONAL SHORT FORM MODEL STATUTE FOR ESTABLISHING A DEVELOPMENTAL DISABILITIES ADVOCACY AGENCY

INTRODUCTION

The complete text of the advocacy bill is lengthy and detailed. Some states may favor a shorter statute or one which contains less administrative detail. The following short form bill is offered where this is the case.

The short form bill lacks most of the administrative detail of the complete bill but retains the statement of purpose, definitions, organizational structure, powers and confidentiality limitations of the longer version. Adoption of the short form will provide the basic authority for the advocacy office. The longer version may be used in determining the administrative practices of an established office.

The comment and alternative portions of the complete text should be consulted for a discussion of issues, options and rationales for choices made. They should also be useful in adapting the short form bill to a particular state's needs.

SHORT FORM MODEL STATUTE

A BILL FOR

AN ACT TO ESTABLISH A DEVELOPMENTAL DISABILITIES ADVOCACY OFFICE TO PROVIDE PROTECTION, LEGAL AND NON-LEGAL ADVOCACY, AND PROCEDURES FOR THE ADMINISTRATIVE RESOLUTION OF CONFLICT TO PROMOTE AND PROTECT THE LEGAL AND HUMAN RIGHTS OF DEVELOPMENTALLY DISABLED PERSONS, AND TO PROVIDE PENALTIES FOR VIOLATION OF THE PROTECTION PROVISIONS OF THIS ACT.

Section 1. Title

This act shall be known as the "Developmental Disabilities Advocacy Act."

Section 2. Statement of Purpose

(1) The General Assembly declares that it is the purpose of this act to improve the quality of life of all developmentally disabled persons by providing investigation, protection, representation, and other advocacy to developmentally disabled persons to promote, protect and advocate their interests and to advocate for delivery of services.

(2) The General Assembly intends by adoption of this act to secure for the state of __________, its local units of government, and its citizens, maximum benefits from the Developmentally Disabled Assistance and Bill of Rights Act, 42 U.S.C. §§6001-6081 (Supp.V 1975) and this act should be liberally construed toward that end.

Section 3. Definitions

As used in this act, unless the context requires otherwise, the terms have the meaning ascribed to them in this section.

(1) "Protection" means preventive or supportive assistance given with the purpose of preventing or alleviating neglect, abuse or exploitation of children or adults who are substantially impaired in their capacity to protect their own self interest. It includes investigation and legal representation in cases of reported or suspected neglect, abuse or exploitation, or referral to service agencies which have professional or legal authority to intervene on behalf of clients in such cases.

(2) "Advocacy" means speaking for, pleading for, supporting, advising, espousing the rights of, or interceding on behalf of, persons with developmental disabilities before public or private agencies, organizations, institutions, or individuals serving developmentally disabled persons.

(3) "Developmental Disability" means a disability of a person which:

(a) (1) is attributable to autism, cerebral palsy, epilepsy or mental retardation;

(2) is attributable to any other condition found to be closely related to autism, cerebral palsy, epilepsy or mental retardation or requires treatment and services similar to those required for such persons; or

(3) is attributal to dyslexia resulting from a disability described in clause (1) or (2) of this paragraph; and

(b) has continued or can be expected to continue indefinitely.

(4) "Developmentally Disabled Person" means a person with a developmental disability.

(5) "Office" means the Developmental Disabilities Advocacy Office.

(6) "Commission" means the Advocacy Commission of the Office.

(7) "Chief Justice" means the Chief Justice of the Supreme Court of the State.

(8) "Director" means the Director of the Office.

(9) "Advocacy Attorney" means any attorney either employed directly by or working under a contract or written agreement with the Office. For purposes of this act advocacy attorney shall also include attorneys who have been assigned, in writing, through the Office to represent a client.

(10) "Office Personnel" means the Director, advocacy attorneys, and all other personnel working for the Office who are directly employed by, under contract with, or under agreement with the Office whose responsibilities include protection or advocacy.

(11) "Person" means an individual, corporation, partnership, association, unincorporated organization, or a government or any subdivision, agency, or instrumentality thereof.

(12) "Client" means any developmentally disabled person or class or group of such persons who are receiving protection or advocacy through the Office.

(13) "Rights" means all human rights and all legal rights including but not limited to rights, benefits, and privileges guaranteed by law, the constitution of this state, and the constitution of the United States.

Section 4. Creation of the Advocacy Office

The Developmental Disabilities Advocacy Office is created as an agency of state government. The Office shall be independent of all other governmental agencies and (not withstanding section ____of the code) shall also be independent of the Office of the Attorney General.

Section 5. Office Authority

The Office shall have authority to:

(1) assess protection and advocacy needs within the state and set priorities among those needs;

(2) publicize the availability of the Office's assistance and engage in outreach efforts;

(3) coordinate the efforts of the Office with persons serving the protection and advocacy needs of developmentally disabled persons;

(4) establish citizen advocacy programs and coordinate and cooperate with existing citizen advocacy programs;

(5) receive complaints of violations of rights or denial of services, conduct investigations and make recommendations in accordance with section 10 of this act;

(6) organize and administer programs to provide legal counsel and representation to developmentally disabled persons as individuals and classes to enforce their rights;

(7) provide direct advice, legal counsel and representation to developmentally disabled persons as individuals and classes to enforce their rights. In the case of such representation, the Office may negotiate or initiate litigation in any appropriate court or administrative hearing tribunal, against any person;

(8) comment upon, draft, and advocate for or against legislation, rules and regulations affecting developmentally disabled persons.

Section 6. Advocacy Commission

(1) There is established an Advocacy Commission (hereinafter referred to as Commission) consisting of members appointed by the Chief Justice.

(2) The Commission shall consist of [fourteen (14)/seven (7)] members. At least [six (6)/three (3)] members of the commission must have cerebral palsy, epilepsy, mental retardation or have been autistic. The remaining members of the Commission shall be cerebral palsied, epileptic, mentally retarded, or formerly autistic, be an immediate family member or a guardian of a person who is autistic, cerebral palsied, epileptic or mentally retarded or have demonstrated an interest in the developmental disability field. No employee or member of the governing board of any agency or organization from which developmentally disabled persons receive treatment, services or habilitation, or member of the General Assembly shall be a member of the Commission. The Chief Justice shall strive to ensure representation of minority groups.

(3) A member of the Commission shall serve a term of three (3) years, except that of the members who are cerebral palsied, epileptic, mentally retarded or who have been autistic [two (2)/one (1)] shall serve an initial term of one (1) year and [two (2)/one (1)] shall serve an initial term of two (2) years, and of the remaining members of the Commission [two (2)/one (1)] shall serve an initial term of one (1) year and [two (2)/one (1)] shall serve an initial term of two (2) years. A member of the Commission shall not serve more than two (2) consecutive terms.

(4) Vacancies on the Commission shall be filled in the same manner as original appointments. The individual appointed must be from the same class of persons from which the original position was filled. Appointments to fill vacancies occurring before the expiration of a term are for the remainder of the unexpired term.

(5) The Commission shall organize itself and adopt procedures as it deems appropriate. The Commission shall meet at least six (6) times per year. A majority of the Commission shall constitute a quorum.

(6) A member of the Commission is not entitled to compensation, but shall receive reimburse-

ment for actual expenses incurred in the performance of his or her duties.

Section 7. Duties of the Advocacy Commission

The Commission shall:

(1) appoint a Director of the Office in accordance with section 8 of the act;

(2) review and advise on [and approve] annually the Office's general policies and priorities and plans for protection and advocacy developed by the Director;

(3) review and advise on [and approve] rules and regulations drafted by the Director;

(4) review, and advise the Director on the operations of the Office. However, the Commission shall not have supervisory powers over the selection or conduct of particular cases or personnel of the Office;

(5) review the annual report on the operations of the Office;

(6) provide for an annual independent evaluation by an outside person.

Section 8. Director

(1) The Commission shall appoint a Director of the Office and set his/her salary. The Commission shall not reduce the Director's salary during his or her term of office.

(2) The Director shall be an attorney at law licensed in this state, or eligible to be licensed. An attorney who is not licensed in the state shall become licensed within one (1) year of appointment as Director.

(3) The Director shall be appointed for a term of five (5) years subject to removal by the Chief Justice or his judicial designee only for good cause shown after notice and hearing.

Section 9. Duties of the Director

The Director shall:

(1) perform the duties and exercise the powers of the Office;

(2) [adopt/draft] general policies and priorities for the operation of the Office and review these annually;

(3) [adopt/draft] plans for protection and advocacy on an annual basis;

(4) [adopt/draft] rules and regulations as appropriate;

(5) employ Office personnel as may be necessary to carry out the functions of the Office, without regard to the state's merit system;

(6) designate Office personnel to act as agents to carry out the functions of the Office;

(7) take any legal or nonlegal action and enter in any contract or agreement which may be appropriate and reasonable to carry out the purposes of this act;

(8) prepare an annual budget and an annual report;

(9) solicit and accept contributions of money, property, or services of whatever nature made by gift, devise, bequest, grant or other means as appropriate to assist the Office in carrying out the provisions of this act.

Section 10. Investigations

(1) The Office may conduct investigations upon its own initiative or on the basis of a complaint if there is reason to believe that the rights of a developmentally disabled person have been, may have been, or may be violated.

(2) In the course of an investigation, for the purpose of investigating actual or possible neglect, abuse, exploitation, or violation of rights of developmentally disabled persons, Office personnel may enter at any reasonable time, in and upon the premises of any state agency, commission, board or office, or any subdivision of the state, speak privately with any individual therein and inspect and copy any information, materials, or records relevant to an investigation or case within reasonable limits and in a reasonable manner, subject to section 11 of this act and notwithstanding section(s)____ of the code.

(3) Any Office personnel conducting an investigation or handling a case shall have ready access to view and copy all records pertaining to a client held by any person subject to section 11. Such personnel shall have the opportunity to consult with clients whenever necessary for the performance of their duties. Persons shall not refuse access to client records or opportunity for consultation as long as it is requested at a reasonable time and in a reasonable manner. A hospital, institution, or mental health or developmental disability facility shall provide adequate space and privacy for the purpose of consultation with clients and examination of client records.

(4) The Director shall report any recommendations to the persons investigated. Such persons investigated shall notify the Director within 30 days of the receipt of such recommendation, of the action taken thereon and the reason therefor.

(5) Notwithstanding any of the procedures noted in this section, if the Director determines that other action is required, s/he may pursue such action, including but not limited to referring a matter to the advocacy attorneys, any federal, state, or local agency, or other persons, as appropriate.

Section 11. Confidentiality of Information, Materials, and Records

(1) Office personnel inspecting information, materials, and records pursuant to an investigation, may copy all such information, materials,

and records when personally identifying data can be removed for the purposes of such copying. Where personally identifying data cannot be removed, then the Office personnel may copy such information, materials, and records only after consent is received pursuant to subsection 4 of this section.

(2) Office personnel may utilize viewed or copied information, materials and records which do not contain personally identifiable data, in the course of negotiations, administrative hearings, suits, or other legal or nonlegal actions.

(3) Office personnel may utilize viewed or copied information, materials, and records which contain personally identifiable data, in the course of negotiations, administrative hearings, suits or other legal actions if consent is obtained pursuant to subsection 4 of this section.

(4) Consent to copy or utilize information, materials, and records containing personally identifiable data must be given knowingly, voluntarily, and in writing. For the purposes of subsections 1 and 3 of this section the following persons may give written consent:

(a) a developmentally disabled person who is the subject of the information, materials, or records and who is 18 years or older unless the person lacks the capacity to understand the purpose, significance and result of giving his or her consent;

(b) the parent or guardian of a developmentally disabled person who is the subject of the information, materials, or records and who is under 18 years old;

(c) the guardian of a developmentally disabled person who is the subject of the information, materials, or records and who is 18 years old or older, provided the guardian has such powers;

(d) the guardian ad litem of a developmentally disabled person who is the subject of the information whether appointed in conjunction with section 12 or not, provided the guardian has such powers.

(5) Office personnel may not disclose any personally identifiable information, materials, and records, unless authorized by subsection 3 of this section or by the subject of such information, materials, and records provided s/he is legally competent to give consent.

Section 12. Duties of Advocacy Office Personnel

(1) Office personnel, including advocacy attorneys employed by the Office or working under contract or through an agreement with the Office or representing a client obtained by referral through the Office, shall be active advocates for the rights of the client.

(2) If a client is legally unable to make an informed judgment in the matter, Office personnel shall nevertheless involve the client in determining a course of action to the maximum extent of the client's capability of understanding the matter in question. Office personnel shall also consult with the parent or guardian of a client to determine his or her recommended course of action.

Where the courses of action selected by the client and the parent or guardian of a client who is a minor or who is legally unable to make an informed judgment in the matter are in conflict and the conflict cannot be resolved so as to protect the rights of the client, an advocacy attorney may seek appointment of a guardian ad litem in accordance with subsection 5 of this section to decide what course of action shall be taken on behalf of the client.

(3) All legal actions brought by an advocacy attorney must be brought in the name of a client or by a parent, guardian, guardian ad litem or next friend on behalf of those clients who are minors or who are legally unable to make an informed judgment in the matter.

(4) For purposes of this act, the attorney-client relationship may be established between an advocacy attorney and a developmentally disabled person without regard to the person's competence or the existence of a guardianship or conservatorship, for the purpose of bringing an action for appointment of a guardian ad litem to act on behalf of the developmentally disabled person:

(a) In instances in which a client lacks actual capacity to consent pursuant to subsection 11(4) of this act and consent cannot be obtained from a parent or guardian;

(b) In a proceeding to establish a guardianship or conservatorship on behalf of a developmentally disabled person;

(c) In a proceeding to determine the legal competency of a developmentally disabled person to perform an act;

(d) In a proceeding to commit a developmentally disabled person;

(e) In a proceeding to renew a commitment, guardianship or conservatorship;

(f) In an action brought against a guardian or conservator for his or her dismissal or a reduction of the authority of the guardian or conservator;

(g) In an action for termination of commitment;

(h) In any other proceeding or action involving any other legally imposed restraint on personal freedom or affecting the rights of a developmentally disabled person.

An advocacy attorney may represent a developmentally disabled person through the guardian ad litem, whether as plaintiff or defendant in subsequent proceedings.

Section 13. Other Remedies Available

The remedies provided by this act shall be in addition to any other remedies which may be available to any person, and the failure to pursue or exhaust the remedies or engage in the procedures provided by this act shall not preclude the invocation of any other remedy.

Section 14. Sanctions Prohibited

Any person participating in filing a complaint or providing information pursuant to this act or participating in a judicial proceeding resulting therefrom shall be presumed to be acting in good faith and unless the presumption is rebutted shall be immune from any liability, civil or criminal, and shall be presumed to be acting in good faith and unless the presumption is rebutted shall be immune from any penalty, sanction, or restriction that otherwise might be incurred or imposed by the person investigated.

APPENDIX
RIGHTS IN RESIDENTIAL FACILITIES

Rights of Disabled Persons in Residential Facilities

State	I. GENERAL PROVISIONS: Persons Covered	I. GENERAL PROVISIONS: Retention of Rights after Institutionalization	II. RECORDS: Complete and Confidential Records	II. RECORDS: Access to Records
ALABAMA Code (1975)	Mentally or emotionally ill		§22-50-62	
ALASKA Stat. (1975)	Mentally ill/ mentally retarded	§47.30.150	§47.30.260	
ARIZONA Rev. Stat. Ann. (Supp. 1978-79)	Mentally retarded	§36-551.01	§36-568.01	
ARKANSAS Stat. Ann. (1971 & Supp. 1977)	Mentally retarded			
CALIFORNIA Welf. & Inst. Code (West Supp. 1979)	Mentally ill/ mentally retarded	§5325.1	§5328	
COLORADO Rev. Stat. (Supp 1978)	Developmentally disabled	§27-10.5-110	§27-10.5-120	
CONNECTICUT Gen. Stat. (1979)	Mentally retarded	§19-575a		
DELAWARE Code Ann. (Supp. 1978)	Mentally retarded		tit. 16, §5161	
DISTRICT OF COLUMBIA Code Ann. (1973)	Mentally retarded			
FLORIDA Stat. Ann. (West Supp. 1978)	Mentally retarded	§393.13	§393.13	§393.13
GEORGIA Code Ann. (1979)	Mentally retarded	§88-2503.1	§88-2503.12	§§88-2503.12, 88-2503.13
HAWAII	Mentally retarded			
IDAHO Code (1978)	Mentally retarded/ mentally deficient	§56-243(a) (6)		
ILLINOIS Ann. Stat. (Smith-Hurd Supp. 1979)	Mentally retarded	Ch. 91½, §§2-100, 2-101	ch. 91½, §803	ch. 91½, §804
INDIANA Code Ann. (Burns 1974 & Supp. 1976)	Mentally ill includes mentally retarded epileptic	§16-14-1.5-5	§16-14-1.5-7	

State	Persons Covered	Retention of Rights after Institutionalization	Complete and Confidential Records	Access to Records
IOWA Code Ann. (West 1969)	Mentally retarded			
KANSAS Stat. Ann. (1976)	Mentally retarded/ mentally ill	§§59-2930, 59-2933	§59-2931	
KENTUCKY Rev. Stat. Ann. (Baldwin Supp. 1978)	Mentally retarded	§20213.050		
LOUISIANA Rev. Stat. Ann. (West Supp. 1979)	Mentally retarded	§390(A)		
MAINE Rev. Stat. Ann. (1978)	Mentally retarded		tit. 34, §2143	tit. 34, §2143
MARYLAND Ann. Code (Supp. 1978)	Mentally retarded	art. 59A, §35	art. 59A, §17	
MASSACHUSETTS Gen. Laws Ann. (West Supp. 1978)	Mentally retarded	ch. 123, §§23, 25	ch. 123, §36	ch. 123, §36
MICHIGAN Comp. Laws. Ann. (1975)	Mentally retarded/ mentally ill	§330.1702	§§330.1746, 330.1748	
MINNESOTA Stat. Ann. (West 1971)	Mentally retarded/ mentally ill	§253A.18		
MISSISSIPPI	No rights for MR			
MISSOURI Ann. Stat. (Vernon Supp. 1978)	Developmentally disabled		§202.900[1]	
MONTANA Rev. Codes Ann. (Supp. 1977)	Developmentally disabled	§38-1217	§38-1223	
NEBRASKA Rev. Stat. (1976)	Mentally retarded	§83-390		
NEVADA Rev. Stat. (1975)	Mentally retarded	§435.340	§433.474[2]	§433.504
NEW HAMPSHIRE Rev. Stat. Ann. (1977)	Developmentally disabled	§171-A:14		
NEW JERSEY Stat. Ann. (West Supp. 1978)	Developmentally disabled	§30:6D-4	§30:6D-4	

State	Persons Covered	Retention of Rights after Institutionalization	Complete and Confidential Records	Access to Records
NEW MEXICO Stat. Ann. (1978)	Developmentally disabled		§43-1-19	§43-1-19
NEW YORK Mental Hyg. Law (McKinney 1978)	Mentally disabled	§33.01	§33.13	
NORTH CAROLINA Gen. Stat. (Supp. 1977)	Mentally ill/ mentally retarded	§122-55.2		
NORTH DAKOTA Cent. Code (1978)	Mentally retarded		§25-16-07	
OHIO Rev. Code Ann. (Page Supp. 1978)	Mentally retarded	§5123.83	§5123.89	
OKLAHOMA Stat. Ann. (West 1979)	Mentally retarded			
OREGON Rev. Stat. (1977)	Mentally deficient		§179.505	
PENNSYLVANIA Stat. Ann. (Purdon 1969)	Mentally retarded		tit. 50, §4602	
RHODE ISLAND	Mentally retarded			
SOUTH CAROLINA Code (1976)	Mentally retarded/ mentally ill	§44-23-1040	§44-23-1090	
SOUTH DAKOTA Compiled Laws Ann. (1976)	Mentally retarded	§27B-1-5	§§27B-8-27, 27B-8-28	§27B-8-30
TENNESSEE Code Ann. (1977 & Supp. 1978)	Mentally ill/ mentally retarded	§§33-306, 33-1601	§33-306	
TEXAS Rev. Civ. Stat. Ann. (Vernon Supp. 1978-79)	Mentally retarded	art. 5547-300, §5	art. 5547-300, §57	
UTAH	No DD or MR rights as of 1978 Supp.[3]			
VERMONT Stat. Ann. (1968)	Mentally ill/ mentally retarded	tit. 18, §7705	tit. 18, §7103	
VIRGINIA Code (1976)	Mentally retarded/ mentally ill	§37.1-84.1	§37.1-84.1	§37.1-84.1

State	Persons Covered	Retention of Rights after Institutionalization	Complete and Confidential Records	Access to Records
WASHINGTON Rev. Code Ann. (1962)	Mentally retarded/ developmentally disabled (not included in MI rights)[4]	§72.33.900		
WEST VIRGINIA Code (Supp. 1979)	Mentally ill/ mentally retarded	§27-5-9	§27-5-9	
WISCONSIN Stat. Ann. (West Supp. 1979-80)	Developmentally disabled/ mentally ill	§51.59	§§51.30, 51.61	§51.61
WYOMING	No MR or DD rights as of 1977 Interim Supp. DD not included in MI rights			
AMERICAN SAMOA	No MR or DD rights			
CANAL ZONE	No MR or DD rights			
GUAM	No MR or DD rights			
PUERTO RICO	No MR or DD rights			
VIRGIN ISLANDS	No MR or DD rights			

[1] Administrative department to establish rules and regulations.
[2] Administrative department to establish rules and regulations.
[3] See Utah Code Ann. §§64-7-47 through 64-7-50 (1978) for MI provisions.
[4] See Wash. Rev. Code Ann. §71.05.040 (Supp. 1977); §§71.05.360 through 71.05.420 (1976).

III. PHYSICAL AND PSYCHOLOGICAL ENVIRONMENT

State	Humane Care or Environment	Privacy and Dignity	Protection from Harm	Personal Belongings	Individual Storage Space	Restraints and/or Isolation	Corporal Punishment	Social Interaction	Adequate Diet	Religious Worship
ALABAMA										
ALASKA Stat. (1975)	§47.30.130					§47.30.140				
ARIZONA Rev. Stat. Ann. (Supp. 1978 & 1979)	§36-551.01 (Q)		§36-551.01 (B)	§36-551.01 (Q)						
ARKANSAS										
CALIFORNIA Welf. & Inst. Code (West Supp. 1979)	§5325.1(b)	§5325.1(b)	§5325.1(c)	§5325(a)	§5325(b)	§5325(c)		§5325.1(g)		§5325.1(e)
COLORADO Rev. Stat. (Supp. 1978)	§27-10.5-115			§27-10.5-121		§27-10.5-115	§27-10.5-115	§27-10.5-117		§27-10.5-116
CONNECTICUT Gen. Stat. (1979)		§19-575a	§19-575a	§19-575a		§19-575a	§19-575a			
DELAWARE Code Ann. (Supp. 1978)		tit. 16, §5161	tit. 16, §5506	tit. 16, §5161						
DISTRICT OF COLUMBIA										
FLORIDA Stat. Ann. (West Supp. 1978)	§393.13(2)	§393.13		§393.13		§393.13		§393.13		§393.13
GEORGIA Code Ann. (1979)		§88-2503.4		§88-2503.8		§88-2503.5				
HAWAII										
IDAHO Code (1976 & Supp. 1978)	§56-238			§56-243(a) (3)	§56-243(a) (3)	§56-238				§56-243(a) (8)
ILLINOIS	ch. 91½, §2-102			Ch. 91½, §2-104	ch. 91½, §2-104	ch. 91½, §§2-108, 2-109				ch. 91½, §2-102
INDIANA Code Ann. (Burns 1974 & Supp. 1976)	§16-14-1.5-2			§16-14-1.5-3(a)	§16-14-1.5-3 (a)	§16-14-1.5-6				§16-14-1.5-3 (h)
IOWA										
KANSAS Stat. Ann. (1976)	§59-2927			§59-2929		§59-2928		§59-2929[1]		
KENTUCKY Rev. Stat. Ann. (Baldwin Supp. 1978)				§202B.060(4)		§202B.060(10)				
LOUISIANA	§390(B)		§449	§391(C)(3)	§391(C)(3)					§391(C)(6)
MAINE Rev. Stat. Ann. (1978)	tit. 34, §2143	tit. 34, §2143		tit. 34, §2143		tit. 34, §2143	tit. 34, §2143	tit. 34, §2143	tit. 34, §2143	tit. 34, §2143

State	Humane Care or Environment	Privacy and Dignity	Protection from Harm	Personal Belongings	Individual Storage Space	Restraints and/or Isolation	Corporal Punishment	Social Interaction	Adequate Diet	Religious Worship
MARYLAND Ann. Code (Supp. 1978)	Legis. intent art. 59A, §2			art. 59A, §34						art. 59A, §37
MASSACHUSETTS Gen. Laws Ann. (West Supp. 1978)					ch. 123, §23	ch. 123, §23				
MICHIGAN Comp. Laws Ann. (1975)	§330.1708			§§330.1728, 330.1730	§330.1728	§§330.1740, 330.1742				
MINNESOTA Stat. Ann. (West (1971)						§253A.17(1)				§253A.17(6)
MISSISSIPPI										
MISSOURI Ann. Stat. (Vernon 1972 & Supp. 1978)			§202.900[2]			§202.843				
MONTANA Rev. Codes Ann. (Supp. 1977)	§§38-1221(1), (6)	§38-1221(1)	§38-1225	§38-1221(6)[3]		§§38-1227, 38-1221(9)	§38-1221	§38-1221	§38-1221(11)	§38-1221(10)
NEBRASKA										
NEVADA Rev. Stat. (1975)				§433.484	§433.484	§433.484				
NEW HAMPSHIRE Rev. Stat. Ann. (1971)	§171-A:13	§171-A:14		§171-A:14	§171-A:14	§171-A:9				
NEW JERSEY Stat. Ann. (West Supp. 1978)	§30:6D-10					§30:6D-5	§30:6D-5	§30:6D-4	§30:6D-5	§30:6D-4
NEW MEXICO Stat. Ann. (1978)	§43-1-6		§43-1-6			§43-1-6			§43-1-6	§43-1-6
NEW YORK Mental Hyg. Law (McKinney 1978)	§33.03	§33.03		§33.07		§33.04				
NORTH CAROLINA Gen. Stat. (Supp. 1977)	§122-55.1 Legis. intent	§122-55.1 Legis. intent §122-55.13 (minors)		§§122-55.2, 122-55.14 (minors)	§§122-55.2, 122-55.14 (minors)	§122-55.3 (1974)	§122-55.4 (1974)	§§122-55.2 122-55.14 (minors)		§§122-55.2, 122-55.14
NORTH DAKOTA										
OHIO										
OKLAHOMA Stat. Ann. (West 1979)	tit. 43A, §91					tit. 43A, §92	tit. 43A, §91		tit. 43A, §91	
OREGON										
PENNSYLVANIA Stat. Ann. (Purdon 1969)						tit. 50, §4422				tit. 50, §4423(2)

State	Humane Care or Environment	Privacy and Dignity	Protection from Harm	Personal Belongings	Individual Storage Space	Restraints and/or Isolation	Corporal Punishment	Social Interaction	Adequate Diet	Religious Worship
RHODE ISLAND										
SOUTH CAROLINA Code (1976)				§44-23-1030	§44-23-1030	§44-23-1020				
SOUTH DAKOTA Compiled Laws Ann. (1976)		§27B-8-1	§27B-8-18	§27B-8-2	§27B-8-2	§27B-8-5				§27B-8-8
TENNESSEE Code Ann. (1977)	§33-306		§33-1601			§§33-306, 33-307				
TEXAS Rev. Civ. Stat. Ann. (Vernon Supp. 1978-79)	art. 5547-300, §25		art. 5547-300, §§6, 20	art. 5547-300 §25		art. 5547-300, §21				
UTAH										
VERMONT Stat. Ann. (1968)	tit. 18, §7703					tit. 18, §7704				tit. 18, §7710
VIRGINIA Code (1976)		§37.1-84.1		§37.1-84.1[4]		§37.1-84.1			§37.1-84.1	§37.1-84.1
WASHINGTON										
WEST VIRGINIA Code (Supp. 1979)	§27-5-9	§27-5-9								
WISCONSIN Stat. Ann. (West Supp. 1979-80)	§51.61	§51.61		§51.61	§51.61	§51.61				§51.61
WYOMING										
AMERICAN SAMOA										
CANAL ZONE										
GUAM										
PUERTO RICO										
VIRGIN ISLANDS										

1 Conjugal visits.
2 Administrative department to establish rules.
3 Suitable clothing also.
4 Administrative department to establish rules.

IV. HABILITATION/STANDARDS OF CARE

State	Right to Treatment	Suitable Habilitative Services	Individual Treatment Plan	Least Restrictive Alternatives	Medical Care	Periodic Review of Exams	Daily Exercise Outdoor Recreation
ALABAMA							
ALASKA Stat. (1975)	§47.30.130				§47.30.130		
ARIZONA Rev. Stat. Ann. (Supp. 1978-79)	§36-551.01		§36-551.01(J)	§36-551.01(C)		§36-551.01(K)	
ARKANSAS							
CALIFORNIA Welf. & Inst. Code (West Supp. 1979)	§5325.1(a)		§5325.1(a)	§5325.1(a)	§5325.1(d)		§5325.1(h)
COLORADO Rev. Stat. (Supp. 1978)	§27-10.5-112		§27-10.5-113	§27-10.5-112	§27-10.5-114	§27-10.5-109	
CONNECTICUT Gen. Stat. (1979)	§19-575a	§19-575a					
DELAWARE Code Ann. (Supp. 1978)	tit. 16, §5161			tit. 16, §5504		tit. 16, §5161	
DISTRICT OF COLUMBIA							
FLORIDA Stat. Ann. (West Supp. 1978)	§393.13	§393.13	§393.13	§393.13	§393.13	§393.13	§393.13
GEORGIA Code Ann. (1979)	§88-2503.4	§88-2503.4	§88-2503.4	§88-2503.4	§88-2503.4		
HAWAII							
IDAHO Code (1976 & Supp. 1978)	§56-238		§56-244		§56-238		
ILLINOIS							
INDIANA Code Ann. (Burns 1974 & Supp. 1976)	§16-14-1.5-2	§16-14-1.5-3(i)			§16-14-1.5-2		
IOWA							
KANSAS Stat. Ann. (1976)	§59-2927					§59-2917a	
KENTUCKY Rev. Stat. Ann. (Baldwin Supp. 1978)			§202B.060	§202B.060			
LOUISIANA							
MAINE Rev. Stat. Ann. (1978)	tit. 34, §§2143, 2147	tit. 34, §2147	tit 34, §2655	tit. 34, §2147	tit. 34, §2143	tit. 34, §2143	tit. 34, §2143

State	Right to Treatment	Suitable Habilitative Services	Individual Treatment Plan	Least Restrictive Alternatives	Medical Care	Periodic Review of Exams	Daily Exercise Outdoor Recreation
MARYLAND Ann. Code (Supp. 1978)			art. 59A, §8A			art. 59A, §12	
MASSACHUSETTS Gen. Laws Ann. (West Supp. 1978)						ch. 123, §4	
MICHIGAN Comp. Laws Ann. (1975)		§330.1708	§330.1712	§330.1744		§330.1710	
MINNESOTA Stat. Ann. (West 1971)	§253A.17		§253A.17(9) (Supp. 1978)			§§253A.17(7), (9)	
MISSISSIPPI							
MISSOURI Ann. Stat. (Vernon Supp. 1978)		§202.900[1]			§202.900		
MONTANA Rev. Codes Ann. (Supp. 1977)	§38-1221(1)	§38-1222(1)	§38-1222(4)	§38-1222	§§38-1221(8), 38-1222(1)	§38-1222(7)	§38-1221(12)
NEBRASKA							
NEVADA Rev. Stat. (1975)	§433.484	§433.484	§433.494	§433.494	§433.484	§433.494	
NEW HAMPSHIRE Rev. Stat. Ann. (1977)	§171-A:13	§171-A:13	§171-A:12	§171-A:12	§§171-A:13, 171-A:11	§171-A:11	
NEW JERSEY Stat. Ann. (West Supp. 1978)		§30:6D-9	§30:6D-10		§30:6D-5	§30:6D-12	
NEW MEXICO Stat. Ann. (1978)	§43-1-7	§43-1-8	§43-1-7	§43-1-8	§43-1-6	§43-1-6	§43-1-6
NEW YORK Mental Hyg. Law (McKinney 1978)	§33.03	§33.11	§33.03			§33.03	
NORTH CAROLINA Gen. Stat. (Supp. 1977)	§§122-55.6 122-55.14 (minors)		§§122-55.6, 122-55.5		§§122-55.5(1974), 122-55.6		§§122-55.2, 122-55.14 (minors)
NORTH DAKOTA							
OHIO Rev. Code Ann. (Page Supp. 1978)	§5123.85	§5123.85	§5123.85		§5123.85	§5123.85	
OKLAHOMA Stat. Ann. (West 1979)	tit. 43A, §91				tit. 43A, §91	tit. 43A, §411	
OREGON							
PENNSYLVANIA							

State	Right to Treatment	Suitable Habilitative Services	Individual Treatment Plan	Least Restrictive Alternatives	Medical Care	Periodic Review of Exams	Daily Exercise Outdoor Recreation
RHODE ISLAND Gen. Laws (Supp. 1978)						§23-43.1-11	
SOUTH CAROLINA							
SOUTH DAKOTA Compiled Laws Ann. (1976)	§27B-8-11	§27B-8-11	§27B-8-10	§27B-8-14	§27B-8-18	§§27B-8-10, 27B-8-15	§§27B-8-13, 27B-8-2
TENNESSEE Code Ann. (1977)	§§33-306, 33-1601	§§33-1601, 33-306		§33-1601	§§33-1601, 33-306		
TEXAS Rev. Civ. Stat. Ann. (Vernon Supp. 1978-79)	art. 5547-300, §11	art. 5547-300, §§8, 11	art. 5547-300, §§16, 18	art. 5547-300, §§15, 7	art. 5547-300, §24	art. 5547-300, §17	
UTAH							
VERMONT Stat. Ann. (1968)	tit. 18, §7703				tit. 18, §7703	tit. 18, §7802 (Supp. 1978)	
VIRGINIA Code (1976)				§37.1-84.1		§37.1-84.2	
WASHINGTON							
WEST VIRGINIA Code (Supp. 1979)	§27-5-9		§27-5-9			§27-5-9	
WISCONSIN Stat. Ann. (West Supp. 1979-80)	§51.61	§51.61		§51.61	§51.61		
WYOMING							
AMERICAN SAMOA							
CANAL ZONE							
GUAM							
PUERTO RICO							
VIRGIN ISLANDS							

[1] Administrative department to establish rules and regulations.

State	V. COMMUNICATION				VI. EMPLOYMENT		
	Visitors	Writing	Telephone	Access to Courts, Atty., etc.	Restrictions	Right to Refuse Work Assignment	Compensation
ALABAMA							
ALASKA Stat. (1975)	§47.30.150	§47.30.150		§47.30.150	§47.30.150		
ARIZONA Rev. Stat. Ann. (Supp. 1978-79)	§36-551.01(Q)	§36-551.01(Q)	§36.551-01(Q)	§36-551.01(R)			§36-551.01(I)
ARKANSAS							
CALIFORNIA Welf. & Inst. Code (West Supp. 1979)	§5325(c)	§5325(e)	§5325(d)		§5325		
COLORADO Rev. Stat. (Supp. 1978)	§27-10.5-117	§27-10.5-117	§27-10.5-117		§27-10.5-111	§27-10.5-118	§27-10.5-118
CONNECTICUT							
DELAWARE Code Ann. (Supp. 1978)	tit. 16, §5161	tit. 16, §5161	tit. 16, §5161				tit. 16, §5161[1]
DISTRICT OF COLUMBIA							
FLORIDA Stat. Ann. (West Supp. 1978)	§393.13	§393.13	§393.13		§393.13		§393.13
GEORGIA Code Ann. (1979)	§88-2503.7[2]	§88-2503.7	§88-2503.7	§§88-2503.14, 88-2503.15			§88-2503.10 (outside the facility only)
HAWAII Rev. Stat. (1976)							§333-37[3]
IDAHO Code (1976 & Supp. 1978)	§56-243(a) (2)	§56-243(a) (1)		§56-243(a) (5)	§56-243		
ILLINOIS Ann. Stat. (Smith-Hurd Supp. 1979)	ch. 91½, §2-103	ch. 91½, §2-103	ch. 91½, §2-103	ch. 91½, §2-103			ch. 91½, §2-106
INDIANA Code Ann. (Burns 1974 & Supp. 1976)	§16-14-1.5-3(b)	§16-14-1.5-3(c)		§§16-14-1.5-3(d), (e)	§16-14-1.5-4		§16-13-12.8-6
IOWA Code Ann. (West 1969 & Supp. 1979)		§222.8		§222.8			
KANSAS Stat. Ann. (1976)	§59-2929	§59-2929	§59-2929	§59-2929	§59-2929	§59-2929	§59-2929
KENTUCKY Rev. Stat. Ann. (Baldwin Supp. 1978)	§202B.060(5)						§202B.060(6)
LOUISIANA	§391(C)(2)	§391(C)(1)	§391(C)(1)	§391(D)			

State	Visitors	Writing	Telephone	Access to Courts, Atty., etc.	Restrictions	Right to Refuse Work Assignment	Compensation
MAINE Rev. Stat. Ann. (1978)	tit. 34, §2143	tit. 34, §2143	tit. 34, §2143				tit. 34, §2143
MARYLAND Ann. Code (Supp. 1978)	art. 59A, §34	art. 59A, §34	art. 59A, §34	art. 59A, §34	art. 59A, §34		
MASSACHUSETTS Gen. Laws Ann. (West Supp. 1978)	ch. 123, §23	ch. 123, §23	ch. 123, §23	ch. 123, §23	ch. 123, §23		
MICHIGAN Comp. Laws Ann. (1975)	§330.1726	§330.1726	§330.1726	§§330.1715, 330.1726	§330.1726	§330.1736	§330.1736
MINNESOTA Stat. Ann. (West 1971)	§253A.17(6)	§§253A.17(3), 253A.17(4)		§253A.17(2)	§253A.17(5)		
MISSISSIPPI							
MISSOURI							
MONTANA Rev. Codes Ann. (Supp. 1977)	§38-1221(3)	§38-1221(2)	§38-1221(3)		§38-1221	§38-1229	§38-1229
NEBRASKA Rev. Stat. (1976)	§83-390[4]	§83-390	§83-390	§83-390	§83-390		
NEVADA Rev. Stat. (1975)	§433.484	§433.484	§433.484		§§433.484, 433.534	§433.524	§433.524
NEW HAMPSHIRE Rev. Stat. Ann. (1977)	§171-A:14	§171-A:14	§171-A:14		§171-A:14		§171-A:14
NEW JERSEY Stat. Ann. (West Supp. 1978)	§30:6D-4	§30:6D-4	§30:6D-4				
NEW MEXICO Stat. Ann. (1978)	§43-1-6	§43-1-6	§43-1-6	§43-1-4	§43-1-6		
NEW YORK Mental Hyg. Law (McKinney 1978)	§33.05	§33.05	§33.05	§33.05	§33.05		§33.09
NORTH CAROLINA Gen. Stat. (Supp. 1977)	§§122-55.2, 122-55.14 (minors)	§§122-55.2, 122-55.14 (minors)	§§122-55.2, 122-55.14 (minors)	§§122-55.2, 122-55.14 (minors)	§122-55.2		
NORTH DAKOTA Cent. Code (1978)	§25-04-06						
OHIO Rev. Code Ann. (Page Supp. 1978)	§5123.84	§5123.84	§5123.84	§5123.84		§5123.87	§5123.87
OKLAHOMA Stat. Ann. (West 1979)	tit. 43A, §93	tit. 43A, §93	tit. 43A, §93	tit. 43A, §93	tit. 43A, §93		tit. 43A, §191

State	Visitors	Writing	Telephone	Access to Courts, Atty., etc.	Restrictions	Right to Refuse Work Assignment	Compensation
OREGON							
PENNSYLVANIA Stat. Ann. (Purdon 1969)	tit. 50, §§4423(1), (5)	tit. 50, §§4423(1), (5)		tit. 50, §4423(1)			
RHODE ISLAND Gen. Laws (Supp. 1978)	§23-43.1-13	§23-43.1-12					
SOUTH CAROLINA Code (1976)	§44-23-1030	§44-23-1030	§44-23-1030	§44-23-1030	§44-23-1030	§44-23-1060	
SOUTH DAKOTA Codified Laws Ann. (1976)	§§27B-8-2, 27B-8-6	§§27B-8-2, 27B-8-6	§§27B-8-2, 27B-8-6	§27B-8-7	§§27B-8-2, 27B-8-6	§27B-8-23	§27B-8-23
TENNESSEE Code Ann. (1977)	§33-306	§33-306		§33-306	§33-306		
TEXAS Rev. Civ. Stat. Ann. (Vernon Supp. 1978-79)	art. 5547-300, §25[5]	art. 5547-300, §25	art. 5547-300, §25				
UTAH							
VERMONT Stat. Ann. (1968)	tit. 18, §7705	tit. 18, §7705	tit. 18, §7705	tit. 18, §§7710, 7705	tit. 18, §7707		
VIRGINIA Code (1976)		§37.1-84.1	§37.1-84.1[6]	§37.1-84.1		§37.1-84.1	
WASHINGTON							
WEST VIRGINIA							
WISCONSIN Stat. Ann. (West Supp. 1979-80)	§51.61	§51.61	§51.61		§51.61	§51.61	§51.61
WYOMING							
AMERICAN SAMOA							
CANAL ZONE							
GUAM							
PUERTO RICO							
VIRGIN ISLANDS							

[1] Not compensated for therapeutic work.
[2] Administrative department to establish rules and regulations.
[3] Determined by director.
[4] Administrative department to establish rules and regulations.
[5] Right to communicate.
[6] Administrative department to establish rules and regulations.

VII. ENFORCEMENT

State	Informed of Rights	Resident Grievance Procedure	Administrative Action	Civil Actions	Criminal Penalties
ALABAMA					
ALASKA					
ARIZONA Rev. Stat. Ann. (Supp. 1978-79)			§36-551.01(p)		§36-569
ARKANSAS Stat. Ann. (1971)			§59-601		§59-602
CALIFORNIA Welf. Inst. Code (West Supp. 1979)	§5325		§5326.9		
COLORADO Rev. Stat. (Supp. 1978)	§27-10.5-111	§27-10.5-122		§27-10.5-134	
CONNECTICUT					
DELAWARE Code Ann. (Supp. 1978)	tit. 16, §5161			tit. 16, §5162	
DISTRICT OF COLUMBIA					
FLORIDA Stat. Ann. (West Supp. 1978)	§393.13[1]	§393.13		§393.13	
GEORGIA Code Ann. (1979)			§88-2503.22	§88-2503.14	
HAWAII					
IDAHO Code (1976 & Supp. 1978)	§56-243(d)				
ILLINOIS Ann. Stat. (Smith-Hurd Supp. 1979)	ch. 91½, §2-200		ch. 91½, §5-103	ch. 91½, §708	
INDIANA Code Ann. (Burns 1974)	§16-14-1.5-3				§16-14-1-2
IOWA					
KANSAS Stat. Ann. (1976)	§59-2929				§59-2929
KENTUCKY Rev. Stat. Ann. (Baldwin Supp. 1978)	§202A.170				§202B.990
LOUISIANA	§§390(B)(9), 391(E)	§447.1	§447.1		§449
MAINE Rev. Stat. Ann. (1978)	tit. 34, §2145	tit. 34, §2146	tit. 34, §2144	tit. 34, §2144	tit. 34, §2144
MARYLAND Ann. Code (Supp. 1978)					art. 59A, §36
MASSACHUSETTS					

State	Informed of Rights	Resident Grievance Procedure	Administrative Action	Civil Actions	Criminal Penalties
MICHIGAN Comp. Laws Ann. (1975)	§330.1706		§330.1752		
MINNESOTA					
MISSISSIPPI					
MISSOURI					
MONTANA Rev. Codes Ann. (Supp. 1977)			§§38-1225, 38-1232		
NEBRASKA					
NEVADA Rev. Stat. (1975)	§433.484		§433.534		§433.554
NEW HAMPSHIRE Rev. Stat. Ann. (1977)	§171-A:15	§171-A:17[2]			
NEW JERSEY Stat. Ann. (West Supp. 1978)	§30:6D-8				
NEW MEXICO Stat. Ann. (1978)				§43-1-23	
NEW YORK Mental Hyg. Law (McKinney 1978)					
NORTH CAROLINA Gen. Stat. (1974)				§122-55.7	
NORTH DAKOTA					
OHIO Rev. Code Ann. (Page Supp. 1978)			§§5123.94, 5123.98	§5123.94	
OKLAHOMA Stat. Ann. (West 1979)			tit. 43A, §92		
OREGON					
PENNSYLVANIA Stat. Ann. (Purdon 1969)					tit. 50, §4605
RHODE ISLAND Gen. Laws (Supp. 1978)			§23-43.1-27		§23-43.1-26
SOUTH CAROLINA Code (1976)					§§44-23-1070, 44-23-1090
SOUTH DAKOTA Compiled Laws Ann. (1976)	§27B-8-3				§27B-1-8
TENNESSEE Code Ann. (1977)			§33-308		
TEXAS Rev. Civ. Stat. Ann. (Vernon Supp 1978-79)	art. 5547-300, §23	art. 5547-300, §§22, 56	art. 5547-300, §§56, 58	art. 5547-300, §64	art. 5547-300, §63

State	Informed of Rights	Resident Grievance Procedure	Administrative Action	Civil Actions	Criminal Penalties
UTAH					
VERMONT Stat. Ann. (1968)	tit. 18, §7701				tit. 18, §7104
VIRGINIA Code (1976)			§37.1-84.1		§37.1-150
WASHINGTON					
WEST VIRGINIA					
WISCONSIN Stat. Ann. (West Supp. 1979-80)	§51.61	§51.61	§51.61[3]	§51.61	
WYOMING					
AMERICAN SAMOA					
CANAL ZONE					
GUAM					
PUERTO RICO					
VIRGIN ISLANDS					

[1] Must be informed in clear language if able to understand.
[2] Human rights committee.
[3] Administrative department to establish rules and regulations.

INDEX

GPSR Compliance

The European Union's (EU) General Product Safety Regulation (GPSR) is a set of rules that requires consumer products to be safe and our obligations to ensure this.

If you have any concerns about our products, you can contact us on ProductSafety@springernature.com

In case Publisher is established outside the EU, the EU authorized representative is:

Springer Nature Customer Service Center GmbH
Europaplatz 3
69115 Heidelberg, Germany

Batch number: 10762468

Printed by Printforce, the Netherlands